AF697463

Prepare for Class "Read the Boo...

Feature	Description	Benefit	Page
	Every chapter begins with....		
Chapter Opening Article & Project	Each chapter begins with a current article and ends with a related project.	The Article describes a real situation. The Project lets you apply what you learned to solve a related problem.	401, 502
	Every section begins with....		
Learning Objectives 1	Each section begins with a list of objectives. Objectives also appear in the text where the objective is covered.	These focus your studying by emphasizing what's most important and where to find it.	423
	Most sections contain...		
PREPARING FOR THIS SECTION	Most sections begin with a list of key concepts to review, with page numbers.	Ever forget what you've learned? This feature highlights previously learned material to be used in this section. Review it, and you'll always be prepared to move forward.	423
Now Work the 'Are You Prepared?' Problems	Special problems that support the Preparing for This Section feature.	Not sure you need the Preparing for This Section review? Work the 'Are You Prepared? problems. If you get one wrong, you'll know exactly what you need to review and where to review it!	423, 432
"Now Work" PROBLEMS	These follow most examples, and direct you to a related exercise.	We learn best by doing. You'll solidify your understanding of examples if you try a similar problem right away, to be sure you understand what you've just read.	431
WARNING	Warnings are provided in the text.	These point out common mistakes and help you to avoid them.	424
Explorations and Seeing the Concept	These features often suggest optional graphing utility activities.	You will obtain a deeper and more intuitive understanding of theorems and definitions.	424, 427
In Words	These provide alternative descriptions of select definitions and theorems.	Does math ever look foreign to you? This feature translates math into plain English.	425
Calculus Icon	These appear next to information essential for the study of calculus.	Pay attention– if you spend extra time now, you'll do better later!	429
Step-by-Step, Annotated EXAMPLES	Examples contain detailed, intermediate steps. Many include additional annotations.	Work the examples on your own, uncovering the solution line-by-line as you go. Each line will verify your work. For additional help, consult the blue annotations, which provide the reason why the statement to the left is valid.	417

Practice "Work the Problems"

Feature	Description	Benefit	Page
"Assess Your Understanding" contains a variety of problems at the end of each section.			
'Are You Prepared?' Problems	These assess your retention of the prerequisite material you'll need. Answers are given at the end of the section exercises. This feature is related to the Preparing for This Section feature.	Do you always remember what you've learned? Working these problems is the best way to find out. If you get one wrong, you'll know exactly what you need to review and where to review it!	423, 432
Concepts and Vocabulary	These short-answer questions, mainly Fill-in-the-Blank and True/False items, assess your understanding of key definitions and concepts in the current section.	Learning math is more than memorization – it's about discovering connections. These problems help you understand the 'big ideas' before diving into skill building.	433
Skill Building	Correlated to section examples, these problems provide straightforward practice.	It's important to dig in and develop your skills. These problems provide you with ample practice to do so.	433–435
Applications and Extensions	These problems allow you to apply your skills to real-world problems. These problems also allow you to extend concepts leamed in the section.	You will see that the material learned within the section has many uses in everyday life.	435–437
Discussion and Writing	"Discussion and Writing" problems are colored red. These support class discussion, verbalization of mathematical ideas, and writing and research projects.	To verbalize an idea, or to describe it clearly in writing, shows real understanding. These problems nurture that understanding. Many are challenging but you'll get out what you put in.	437
"Now Work" Problems	Many examples refer you to a related homework problem. These related problems are marked by a pencil and yellow numbers.	If you get stuck while working problems, look for the closest Now Work problem and refer back to the related example to see if it helps.	429
Graphing Calculator	These **optional problems** require the use of a graphing utility, and are marked by a special icon and green numbers.	Your instructor will usually provide guidance on whether or not to do these problems. If so, these problems help to verify and visualize your analytical results.	463

Review "Study for Quizzes and Tests"

Feature	Description	Benefit	Page
Chapter Reviews at the end of each chapter contain...			
"Things to Know"	A detailed list of important theorems, formulas, and definitions from the chapter.	Review these and you'll know the most important material in the chapter!	498-499
"You Should be Able To..."	Contains a complete list of objectives by section, examples that illustrate the object, and practice exercises that test your understanding of the objective.	Do the recommended exercises and you'll have mastery over the key material. If you get something wrong, review the suggested page numbers and try again.	499-500
Review Exercises	These provide comprehensive review and practice of key skills, matched to the Learning Objectives for each section.	Practice makes perfect.These problems combine exercises from all sections, giving you a comprehensive review in one place.	500-503
CHAPTER TEST	About 15–20 problems that can be taken as a Chapter Test. Be sure to take the Chapter Test under test conditions—no notes!	Be prepared. Take the sample practice test. This will get you ready for your instructor's test. If you get a problem wrong, watch the Chapter Test Prep video.	504
CUMULATIVE REVIEW	These problem sets appear at the end of each chapter, beginning with Chapter 2. They combine problems from previous chapters, providing an ongoing cumulative review.	These are really important. They will ensure that you are not forgetting anything as you go. These will go a long way toward keeping you constantly primed for quizzes and tests.	504-505
CHAPTER PROJECTS	The Chapter Project applies what you've learned in the chapter. Additional projects are available on the Instructor's Resource Center (IRC).	The Project gives you an opportunity to apply what you've learned in the chapter to solve a problem related to the opening article. If your instructor allows, these make excellent opportunities to work in a group, which is often the best way of learning math.	505

College Algebra
Enhanced with Graphing Utilities

Michael Sullivan
Michael Sullivan, III

Second Custom Edition
Miami Dade College – Wolfson
MAC 1105

Taken from:
College Algebra: Enhanced with Graphing Utilities, Fifth Edition
by Michael Sullivan and Michael Sullivan, III

Learning Solutions

New York Boston San Francisco
London Toronto Sydney Tokyo Singapore Madrid
Mexico City Munich Paris Cape Town Hong Kong Montreal

Cover Art: Illustration by Kristen Kiley

Taken from:

College Algebra: Enhanced with Graphing Utilities, Fifth Edition
by Michael Sullivan and Michael Sullivan, III

Published by Prentice Hall
Upper Saddle River, New Jersey 07458

This special edition published in cooperation with Pearson Learning Solutions.

Pearson Learning Solutions, 501 Boylston Street, Suite 900, Boston, MA 02116
A Pearson Education Company
www.pearsoned.com

Printed in the United States of America

8 9 10 V092 15 14 13 12

000200010270593883

CG

PEARSON

ISBN 10: 0-558-82856-6
ISBN 13: 978-0-558-82856-1

In Memory of Mary . . .

Wife and Mother

Contents

Three Distinct Series

Students have different goals, learning styles, and levels of preparation. Instructors have different teaching philosophies, styles, and techniques. Rather than write one series to fit all, the Sullivans have written three distinct series. All share the same goal—to develop a high level of mathematical understanding and an appreciation for the way mathematics can describe the world around us. The manner of reaching that goal, however, differs from series to series.

Contemporary Series, Eighth Edition

The Contemporary Series is the most traditional in approach, yet modern in its treatment of precalculus mathematics. Graphing utility coverage is optional and can be included or excluded at the discretion of the instructor: *College Algebra, College Algebra Essentials, Algebra & Trigonometry, Trigonometry, Precalculus.*

Enhanced with Graphing Utilities Series, Fifth Edition

This series provides a more thorough integration of graphing utilities into topics, allowing students to explore mathematical concepts and foreshadow ideas usually studied in later courses. Using technology, the approach to solving certain problems differs from the Contemporary Series, while the emphasis on understanding concepts and building strong skills does not: *College Algebra, Algebra & Trigonometry, Trigonometry, Precalculus.*

Concepts through Functions Series, First Edition

This series differs from the others, utilizing a functions approach that serves as the organizing principle tying concepts together. Functions are introduced early in various formats. This approach supports the Rule of Four, which states that functions are represented symbolically, numerically, graphically, and verbally. Each chapter introduces a new type of function and then develops all concepts pertaining to that particular function. The solutions of equations and inequalities, instead of being developed as stand-alone topics, are developed in the context of the underlying functions. Graphing utility coverage is optional and can be included or excluded at the discretion of the instructor: *College Algebra; Precalculus, with a Unit Circle Approach to Trigonometry; Precalculus, with a Right Triangle Approach to Trigonometry.*

To the Student

As you begin, you may feel anxious about the number of theorems, definitions, procedures, and equations. You may wonder if you can learn it all in time. Don't worry, your concerns are normal. This textbook was written with you in mind. If you attend class, work hard, and read and study this book, you will build the knowledge and skills you need to be successful. Here's how you can use the book to your benefit.

Read Carefully

When you get busy, it's easy to skip reading and go right to the problems. Don't... the book has a large number of examples and clear explanations to help you break down the mathematics into easy-to-understand steps. Reading will provide you with a clearer understanding, beyond simple memorization. Read before class (not after) so you can ask questions about anything you didn't understand. You'll be amazed at how much more you'll get out of class if you do this.

Use the Features

We use many different methods in the classroom to communicate. Those methods, when incorporated into the book, are called "features." The features serve many purposes, from providing timely review of material you learned before (just when you need it), to providing organized review sessions to help you prepare for quizzes and tests. Take advantage of the features and you will master the material.

To make this easier, we've provided a brief guide to getting the most from this book. Refer to the 'Prepare for Class,' 'Practice,' and 'Review' pages on the inside front cover of this book. Spend fifteen minutes reviewing the guide and familiarizing yourself with the features by flipping to the page numbers provided. Then, as you read, use them. This is the best way to make the most of your textbook.

Please do not hesitate to contact us, through Pearson Education, with any questions, suggestions, or comments that would improve this text. We look forward to hearing from you, and good luck with all of your studies.

Best Wishes!

Michael Sullivan

Michael Sullivan, III

Preface to the Instructor

As professors at both an urban university and a community college, Michael Sullivan and Michael Sullivan, III, are aware of the varied needs of College Algebra students, ranging from those who have little mathematical background and a fear of mathematics courses, to those having a strong mathematical education and a high level of motivation. For some of your students, this will be their last course in mathematics, whereas others will further their mathematical education. This text is written for both groups.

As a teacher, and as an author of precalculus, engineering calculus, finite mathematics, and business calculus texts, Michael Sullivan understands what students must know if they are to be focused and successful in upper-level math courses. However, as a father of four, he also understands the realities of college life. As an author of a developmental mathematics series, Michael's co-author and son, Michael Sullivan, III, understands the trepidations and skills students bring to the College Algebra course. Michael, III also believes in the value of technology as a tool for learning that enhances understanding without sacrificing math skills. Together, both authors have taken great pains to ensure that the text contains solid, student-friendly examples and problems, as well as a clear and seamless writing style.

A tremendous benefit of authoring a successful series is the broad-based feedback we receive from teachers and students. We are sincerely grateful for their support. Virtually every change in this edition is the result of their thoughtful comments and suggestions. We are sincerely grateful for this support and hope that we have been able to take these ideas and, building upon a successful fourth edition, make this series an even better tool for learning and teaching. We continue to encourage you to share with us your experiences teaching from this text.

Features in the Fifth Edition

Rather than provide a list of new features here, that information can be found on the endpapers in the front of this book. This places the new features in their proper context, as building blocks of an overall learning system that has been carefully crafted over the years to help students get the most out of the time they put into studying. Please take the time to review the features listed in the front of this book and to discuss them with your students at the beginning of your course. Our experience has been that when students utilize these features, they are more successful in the course.

New to the Fifth Edition

- **Showcase Examples** are used to present examples in a guided, step-by-step format. Students can immediately see how each of the steps in a problem are employed. The "How To" examples have a two-column format in which the left column describes the step in solving the problem and the right column displays the algebra complete with annotations.
- **Model It** examples and exercises are clearly marked with a icon. These examples and exercises are meant to develop the student's ability to build models from both verbal descriptions and data. Many of the problems involving data require the students to first determine the appropriate model (linear, quadratic, and so on) to fit to the data and justify their choice.
- **Exercise Sets** at the end of each section have been classified according to purpose. Where appropriate, more problems to challenge the better student have been added. In addition, **Mixed Practice** exercises have been added where appropriate so that students may synthesize skills from a variety of sections.
- **Applied Problems** have been updated and many new problems involving sourced information as well as data sets have been added to bring relevance and timeliness to these exercises.
- **Chapter 4 Linear and Quadratic Functions** New to this edition, this chapter gives more emphasis to linear and quadratic functions. The discussions allow for a fuller discussion of linear and quadratic functions, their properties, models involving the functions, and building models from verbal descriptions and data. A full section on quadratic inequalities is also provided with both graphical and algebraic analysis of the inequality.
- **New Content**
  - Section R.3, Geometry Essentials, now contains a discussion of congruent and similar triangles.
  - Examples and exercises that involve a more in-depth analysis of graphing and exponential and logarithmic functions are provided.

Organizational Changes in the Fifth Edition

- **Chapter 1** introduces graphing and the graphing calculator along with solving equations in one variable.
- **Chapter 2** expands on the discussion of graphs in two variables and introduces lines and circles. Plus, Section 2.4 presents a complete discussion of variation.
- **Chapter 4** consolidates the discussion on linear and quadratic functions.
- **Chapter 8** discusses systems of equations and inequalities. This chapter includes both linear and nonlinear systems.

Using the Fifth Edition Effectively with Your Syllabus

To meet the varied needs of diverse syllabi, this book contains more content than is likely to be covered in a *College Algebra* course. As the chart illustrates, this book has been organized with flexibility of use in mind. Within a given chapter, certain sections are optional (see the detail following the flowchart) and can be omitted without loss of continuity.

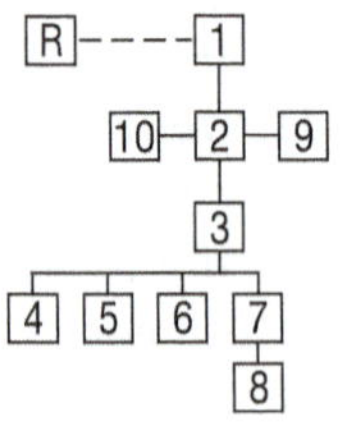

Chapter R Review
This chapter consists of review material. It may be used as the first part of the course or later as a just-in-time review when the content is required. Specific references to this chapter occur throughout the book to assist in the review process.

Chapter 1 Equations and Inequalities
Primarily a review of Intermediate Algebra topics, with the exception of the introduction to the graphing utility, this material is prerequisite for later topics. The coverage of complex numbers and quadratic equations with a negative discriminant is optional and may be postponed or skipped entirely without loss of continuity.

Chapter 2 Graphs
This chapter lays the foundation for functions. Section 2.4 is optional.

Chapter 3 Functions and Their Graphs
Perhaps the most important chapter. Section 3.6 is optional.

Chapter 4 Linear and Quadratic Functions
Topic selection depends on your syllabus. Sections 4.2 and 4.4 may be omitted without a loss of continuity.

Chapter 5 Polynomial and Rational Functions
Topic selection depends on your syllabus.

Chapter 6 Exponential and Logarithmic Functions
Sections 6.1–6.6 follow in sequence. Sections 6.7, 6.8, and 6.9 are optional.

Chapter 7 Analytic Geometry
Sections 7.1–7.4 follow in sequence.

Chapter 8 Systems of Equations and Inequalities
Sections 8.2–8.7 may be covered in any order, but each requires Section 8.1. Section 8.8 requires Section 8.7.

Chapter 9 Sequences; Induction; The Binomial Theorem
There are three independent parts: Sections 9.1–9.3; Section 9.4; and Section 9.5.

Chapter 10 Counting and Probability
The sections follow in sequence.

Acknowledgments

Textbooks are written by authors but evolve from an idea to final form through the efforts of many people. It was Don Dellen who first suggested this book and series. Don is remembered for his extensive contributions to publishing and mathematics.

Thanks are due the following people for their assistance and encouragement in the preparation of this edition:

- From Pearson Education: Adam Jaworski for his substantial contributions, ideas, and enthusiasm; Kate Valentine, who remains a huge fan and supporter; Dawn Murrin, for her unmatched talent at getting the details right; Bob Walters for his superb organizational skills in directing production; Chris Hoag for her continued support and genuine interest; Greg Tobin for his leadership and commitment to excellence; and the Pearson Math and Science Sales team, for their continued confidence and personal support of our books.
- Accuracy checkers: Kevin Bodden and Randy Gallaher, for creating the Solutions Manuals and accuracy checking answers; LaurelTech's editors, Teri Lovelace, Harold Whipple, Jill McClain-Wardynski and Peggy Irish who read the entire manuscript and checked all the answers.
- Reviewers: Debra Kopcso, *Louisiana State University;* Lynn Marecek, *Santa Ana College;* Kathleen Miranda, *SUNY at Old Westbury;* Karla Neal, *Louisiana State University;* Leticia Oropesa, *University of Miami;* Laura Pyzdrowski, *West Virginia University;* Mike Rosenthal, *Florida International University;* Phoebe Rouse, *Louisiana State University;* Brenda Santistevan, *Salt Lake Community College;* Larissa Williamson, *University of Florida.*

Finally, we offer our grateful thanks to the dedicated users and reviewers of our books, whose collective insights form the backbone of each textbook revision.

Our list of indebtedness just grows and grows. And, if we've forgotten anyone, please accept our apology. Thank you all.

James Africh, College of DuPage
Steve Agronsky, Cal Poly State University
Grant Alexander, Joliet Junior College
Dave Anderson, South Suburban College
Joby Milo Anthony, University of Central Florida
James E. Arnold, University of Wisconsin-Milwaukee
Adel Arshaghi, Center for Educational Merit
Carolyn Autray, University of West Georgia
Agnes Azzolino, Middlesex County College
Wilson P Banks, Illinois State University
Sudeshna Basu, Howard University
Dale R. Bedgood, East Texas State University
Beth Beno, South Suburban College
Carolyn Bernath, Tallahassee Community College
William H. Beyer, University of Akron
Annette Blackwelder, Florida State University
Richelle Blair, Lakeland Community College
Kevin Bodden, Lewis and Clark College
Larry Bouldin, Roane State Community College
Bob Bradshaw, Ohlone College
Trudy Bratten, Grossmont College
Tim Bremer, Broome Community College
Tim Britt, Jackson State Community College
Joanne Brunner, Joliet Junior College
Warren Burch, Brevard Community College
Mary Butler, Lincoln Public Schools
Melanie Butler, West Virginia University
Jim Butterbach, Joliet Junior College
William J. Cable, University of Wisconsin-Stevens Point
Lois Calamia, Brookdale Community College
Jim Campbell, Lincoln Public Schools
Roger Carlsen, Moraine Valley Community College
Elena Catoiu, Joliet Junior College
Mathews Chakkanakuzhi, Palomar College
John Collado, South Suburban College
Nelson Collins, Joliet Junior College
Jim Cooper, Joliet Junior College
Denise Corbett, East Carolina University
Carlos C. Corona, San Antonio College
Theodore C. Coskey, South Seattle Community College
Donna Costello, Plano Senior High School
Paul Crittenden, University of Nebraska at Lincoln
John Davenport, East Texas State University
Faye Dang, Joliet Junior College
Antonio David, Del Mar College
Duane E. Deal, Ball State University
Timothy Deis, University of Wisconsin-Platteville
Vivian Dennis, Eastfield College
Deborah Dillon, R. L. Turner High School
Guesna Dohrman, Tallahassee Community College
Cheryl Doolittle, Iowa State University
Karen R. Dougan, University of Florida
Louise Dyson, Clark College
Paul D. East, Lexington Community College
Don Edmondson, University of Texas-Austin
Erica Egizio, Joliet Junior College
Laura Egner, Joliet Junior College
Jason Eltrevoog, Joliet Junior College
Christopher Ennis, University of Minnesota
Kathy Eppler, Salt Lake Community College
Ralph Esparza, Jr., Richland College
Garret J. Etgen, University of Houston
Scott Fallstrom, Shoreline Community College
Pete Falzone, Pensacola Junior College
W.A. Ferguson, University of Illinois-Urbana/Champaign
Iris B. Fetta, Clemson University
Mason Flake, student at Edison Community College
Timothy W. Flood, Pittsburg State University
Merle Friel, Humboldt State University
Richard A. Fritz, Moraine Valley Community College
Dewey Furness, Ricke College
Randy Gallaher, Lewis and Clark College
Tina Garn, University of Arizona
Dawit Getachew, Chicago State University
Wayne Gibson, Rancho Santiago College
Robert Gill, University of Minnesota Duluth
Sudhir Kumar Goel, Valdosta State University
Joan Goliday, Sante Fe Community College
Frederic Gooding, Goucher College
Donald Goral, Northern Virginia Community College
Sue Graupner, Lincoln Public Schools
Jennifer L. Grimsley, University of Charleston
Ken Gurganus, University of North Carolina
James E. Hall, University of Wisconsin-Madison
Judy Hall, West Virginia University
Edward R. Hancock, DeVry Institute of Technology
Julia Hassett, DeVry Institute-Dupage
Christopher Hay-Jahans, University of South Dakota
Michah Heibel, Lincoln Public Schools
LaRae Helliwell, San Jose City College
Celeste Hernandez, Richland College
Brother Herron, Brother Rice High School
Robert Hoburg, Western Connecticut State University
Lynda Hollingsworth, Northwest Missouri State University
Charla Holzbog, Denison High School
Lee Hruby, Naperville North High School
Miles Hubbard, St. Cloud State University
Kim Hughes, California State College-San Bernardino
Ron Jamison, Brigham Young University
Richard A. Jensen, Manatee Community College
Sandra G. Johnson, St. Cloud State University
Tuesday Johnson, New Mexico State University
Moana H. Karsteter, Tallahassee Community College
Donna Katula, Joliet Junior College
Arthur Kaufman, College of Staten Island
Thomas Kearns, North Kentucky University
Jack Keating, Massasoit Community College
Shelia Kellenbarger, Lincoln Public Schools
Rachael Kenney, North Carolina State University
Debra Kopcso, Louisiana State University
Lynne Kowski, Raritan Valley Community College
Keith Kuchar, Manatee Community College
Tor Kwembe, Chicago State University
Linda J. Kyle, Tarrant Country Jr. College
H.E. Lacey, Texas A & M University
Harriet Lamm, Coastal Bend College
James Lapp, Fort Lewis College
Matt Larson, Lincoln Public Schools
Christopher Lattin, Oakton Community College
Julia Ledet, Lousiana State University
Adele LeGere, Oakton Community College
Kevin Leith, University of Houston
JoAnn Lewin, Edison College
Jeff Lewis, Johnson County Community College
Janice C. Lyon, Tallahassee Community College
Jean McArthur, Joliet Junior College
Virginia McCarthy, Iowa State University
Karla McCavit, Albion College
Michael McClendon, University of Central Oklahoma
Tom McCollow, DeVry Institute of Technology
Marilyn McCollum, North Carolina State University
Jill McGowan, Howard University
Dave McGuire, Joliet Junior College
Angela McNulty, Joliet Junior College
Laurence Maher, North Texas State University
Jay A. Malmstrom, Oklahoma City Community College
Rebecca Mann, Apollo High School
Lynn Marecek, Santa Ana College
Sherry Martina, Naperville North High School
Alec Matheson, Lamar University
Nancy Matthews, University of Oklahoma
James Maxwell, Oklahoma State University-Stillwater
Marsha May, Midwestern State University
Judy Meckley, Joliet Junior College
David Meel, Bowling Green State University
Carolyn Meitler, Concordia University
Samia Metwali, Erie Community College
Rich Meyers, Joliet Junior College
Eldon Miller, University of Mississippi
James Miller, West Virginia University
Michael Miller, Iowa State University
Kathleen Miranda, SUNY at Old Westbury
Thomas Monaghan, Naperville North High School
Maria Montoya, Our Lady of the Lake University
Craig Morse, Naperville North High School
Samad Mortabit, Metropolitan State University
Pat Mower, Washburn University
A. Muhundan, Manatee Community College
Jane Murphy, Middlesex Community College
Richard Nadel, Florida International University
Gabriel Nagy, Kansas State University
Bill Naegele, South Suburban College
Karla Neal, Lousiana State University
Lawrence E. Newman, Holyoke Community College
Dwight Newsome, Pasco-Hernando Community College
James Nymann, University of Texas-El Paso
Mark Omodt, Anoka-Ramsey Community College
Seth F. Oppenheimer, Mississippi State University
Leticia Oropesa, University of Miami
Linda Padilla, Joliet Junior College
E. James Peake, Iowa State University
Kelly Pearson, Murray State University
Philip Pina, Florida Atlantic University
Michael Prophet, University of Northern Iowa
Laura Pyzdrowski, West Virginia University
Neal C. Raber, University of Akron

Thomas Radin, San Joaquin Delta College
Ken A. Rager, Metropolitan State College
Kenneth D. Reeves, San Antonio College
Elsi Reinhardt, Truckee Meadows Community College
Jane Ringwald, Iowa State University
Stephen Rodi, Austin Community College
William Rogge, Lincoln Northeast High School
Howard L. Rolf, Baylor University
Mike Rosenthal, Florida International University
Phoebe Rouse, Lousiana State University
Edward Rozema, University of Tennessee at Chattanooga
Dennis C. Runde, Manatee Community College
Alan Saleski, Loyola University of Chicago
Susan Sandmeyer, Jamestown Community College
Brenda Santistevan, Salt Lake Community College
Linda Schmidt, Greenville Technical College
A.K. Shamma, University of West Florida
Martin Sherry, Lower Columbia College
Tatrana Shubin, San Jose State University
Anita Sikes, Delgado Community College
Timothy Sipka, Alma College
Lori Smellegar, Manatee Community College
Gayle Smith, Loyola Blakefield
Vicki Smith, Nicholls State University
John Spellman, Southwest Texas State University
Karen Spike, University of North Carolina, Wilmington
Rajalakshmi Sriram, Okaloosa-Walton Community College
Becky Stamper, Western Kentucky University
Judy Staver, Florida Community College-South
Neil Stephens, Hinsdale South High School
Sonya Stephens, Florida A&M Univeristy
Patrick Stevens, Joliet Junior College
Matthew TenHuisen, University of North Carolina, Wilmington
Christopher Terry, Augusta State University
Tommy Thompson, Brookhaven College
Martha K. Tietze, Shawnee Mission Northwest High School
Richard J. Tondra, Iowa State University
Suzanne Topp, Salt Lake Community College
Marvel Townsend, University of Florida
Jim Trudnowski, Carroll College
Robert Tuskey, Joliet Junior College
Richard G. Vinson, University of South Alabama
Mary Voxman, University of Idaho
Jennifer Walsh, Daytona Beach Community College
Donna Wandke, Naperville North High School
Kathryn Wetzel, Amarillo College
Darlene Whitkenack, Northern Illinois University
Suzanne Williams, Central Piedmont Community College
Larissa Williamson, University of Florida
Christine Wilson, West Virginia University
Brad Wind, Florida International University
Mary Wolyniak, Broome Community College
Canton Woods, Auburn University
Tamara S. Worner, Wayne State College
Terri Wright, New Hampshire Community Technical College, Manchester
George Zazi, Chicago State University
Steve Zuro, Joliet Junior College

Michael Sullivan
Chicago State University

Michael Sullivan, III
Joliet Junior College

STUDENT RESOURCES

Available to students are the following supplements:

- ***Student Solutions Manual (ISBN 10: 0136001963; ISBN 13: 9780136001966)***
  Fully worked solutions to odd-numbered exercises.
- ***Pearson Tutor Center (ISBN 10: 0130646040; ISBN 13: 9780130646040)***
  Tutors provide one-on-one tutoring for any problem with an answer at the back of the book. Students access the Tutor Center via toll-free phone, fax, or email. Available only to college students in the U.S. and Canada.
- ***Algebra Review (ISBN 10: 0131480065; ISBN 13: 9780131480063)***
  Four chapters of Intermediate Algebra review. Perfect for a slower-paced course or for individual review.
- ***CD Lecture Series (ISBN 10: 0136002013; ISBN 13: 9780136002017)***
  A comprehensive set of CD-ROMs, tied to the textbook, containing short video clips of an instructor working key book examples. The CD Lecture Series can be packaged with the book or is available for purchase standalone.
- ***Chapter Test Prep Video CD***
  Packaged at the back of the text, this video CD provides students with step-by-step solutions for each of the exercises in the book's chapter tests.

MathXL®

MathXL® is a powerful online homework, tutorial, and assessment system that accompanies Pearson Education's textbooks in mathematics or statistics. With MathXL, instructors can create, edit, and assign online homework and tests using algorithmically generated exercises correlated at the objective level to the textbook. They can also create and assign their own online exercises and import TestGen tests for added flexibility. All student work is tracked in MathXL's online gradebook. Students can take chapter tests in MathXL and receive personalized study plans based on their test results. The study plan diagnoses weaknesses and links students directly to tutorial exercises for the objectives they need to study and retest. Students can also access supplemental animations and video clips directly from selected exercises. MathXL is available to qualified adopters. For more information, visit our website at **www.mathxl.com,** or contact your sales representative.

MyMathLab®

MyMathLab® is a series of text-specific, easily customizable online courses for Pearson Education's textbooks in mathematics and statistics. Powered by CourseCompass™ (our online teaching and learning environment) and MathXL® (our online homework, tutorial, and assessment system), MyMathLab gives you the tools you need to deliver all or a portion of your course online, whether your students are in a lab setting or working from home. MyMathLab provides a rich and flexible set of course materials, featuring free-response exercises that are algorithmically generated for unlimited practice and mastery. Students can also use online tools, such as video lectures, animations, and a multimedia textbook, to independently improve their understanding and performance. Instructors can use MyMathLab's homework and test managers to select and assign online exercises correlated directly to the textbook, and they can also create and assign their own online exercises and import TestGen tests for added flexibility. MyMathLab's online gradebook—designed specifically for mathematics and statistics—automatically tracks students' homework and test results and gives the instructor control over how to calculate final grades. Instructors can also add offline (paper-and-pencil) grades to the gradebook. MyMathLab is available to qualified adopters. For more information, visit our website at **www.mymathlab.com** or contact your sales representative.

MathXL® Tutorials on CD

This interactive tutorial CD-ROM provides algorithmically generated practice exercises that are correlated at the objective level to the exercises in the textbook. Every practice exercise is accompanied by an example and a guided solution designed to involve students in the solution process. Selected exercises may also include a video clip to help students visualize concepts. The software provides helpful feedback for incorrect answers and can generate printed summaries of students' progress.

InterAct Math Tutorial Website: www.interactmath.com

Get practice and tutorial help online! This interactive tutorial website provides algorithmically generated practice exercises that correlate directly to the exercises in the textbook. Students can retry an exercise as many times as they like with new values each time for unlimited practice and mastery. Every exercise is accompanied by an interactive guided solution that provides helpful feedback for incorrect answers, and students can also view a worked-out sample problem that steps them through an exercise similar to the one they're working on.

List of Applications

Calculus

Carpentry. *See also* **Construction**

Chemistry, 111

Combinatorics

Communications

Computers and computing

Construction

Crime

Cryptography

Decorating

Demographics

Design

Direction

Distance

Photo Credits

Chapter R	Page 1, Gary Connor/PhotoEdit, Inc.; Page 31, AP Wide World Photos; Page 38, problem #51, Whole Spirit Press; problem # 52, Red River Press, Inc.; Page 84, Corbis/Bettmann.
Chapter 1	Pages 85 and 166, Superstock, Inc.; Page 112, PhotoEdit, Inc.; Page 123, David Young-Wolf/PhotoEdit, Inc.; Page 148, Alamy Images; Page 159, Nancy R. Cohen/Getty Images, Inc. Page 164, Kent Wood/Photo Researchers, Inc.-PhotoDisc.
Chapter 2	Pages 169 and 208, Alamy Images; Page 171, Peter Arnold, Inc.; Page 177, CORBIS-NY; Kathy McLaughlin/The Image Works; Page 199, AP Wide World Photos.
Chapter 3	Pages 211 and 280, CORBIS-NY; Page 225, NASA/Jet Propulsion Laboratory; Page 267, Tony Freeman/PhotoEdit, Inc.
Chapter 4	Pages 281 and 328, MonicaGraff/The Image Works; Page 317, Erin Garvey/Photolibrary.com.
Chapter 5	Pages 329 and 400, Dr. Jeremy Burgess/Photo Researchers, Inc.; Page 368, Corbis Digital Stock.
Chapter 6	Pages 403 and 505, Alamy Images; Page 460, The Granger Collection; Page 468, NASA/GESC/Tom Stack & Associates, Inc.; Page 469, Tony Freeman/PhotoEdit, Inc.; Page 474, Jim Pickerell/The Image Works; Page 489, ©Theo Allofs/ Corbis/Bettmann.
Chapter 7	Pages 507 and 545, Lynette Cook/Photo Researchers, Inc.; Page 526, Scott McDonald/Hedrich Blessing.
Chapter 8	Pages 547 and 643, Steven Rubin/The Image Works; Page 601, Corbis/Bettmann.
Chapter 9	Pages 645 and 689, NASA/GFSC/Tom Stack & Associates, Inc.; Page 671, The Granger Collection; Page 684, Corbis/Bettmann.
Chapter 10	Pages 691 and 719, Photofest; Page 712, The Granger Collection.

College Algebra

Review

R

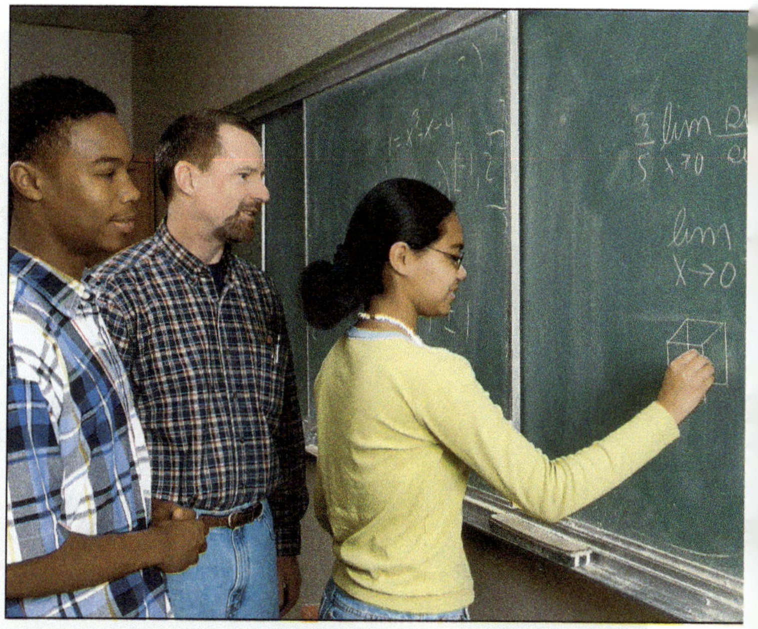

A Look Ahead

Chapter R, as the title states, contains review material. Your instructor may choose to cover all or part of it as a regular chapter at the beginning of your course or later as a just-in-time review when the content is required. Regardless, when information in this chapter is needed, a specific reference to this chapter will be made either at the beginning of the section or within the section, so you can review.

Outline

R.1 Real Numbers

PREPARING FOR THIS BOOK *Before getting started, read "To the Student" on page xxvii at the front of this book.*

OBJECTIVES
1. Work with Sets (p. 2)
2. Classify Numbers (p. 4)
3. Evaluate Numerical Expressions (p. 8)
4. Work with Properties of Real Numbers (p. 9)

1 Work with Sets

A **set** is a well-defined collection of distinct objects. The objects of a set are called its **elements.** By **well-defined,** we mean that there is a rule that enables us to determine whether a given object is an element of the set. If a set has no elements, it is called the **empty set,** or **null set,** and is denoted by the symbol $\varnothing$.

For example, the set of *digits* consists of the collection of numbers 0, 1, 2, 3, 4, 5, 6, 7, 8, and 9. If we use the symbol D to denote the set of digits, then we can write

$$D = \{0, 1, 2, 3, 4, 5, 6, 7, 8, 9\}$$

In this notation, the braces { } are used to enclose the objects, or **elements,** in the set. This method of denoting a set is called the **roster method.** A second way to denote a set is to use **set-builder notation,** where the set D of digits is written as

$$D = \{ \quad x \quad | \quad x \text{ is a digit}\}$$

Read as "D is the set of all x such that x is a digit."

EXAMPLE 1 **Using Set-builder Notation and the Roster Method**

(a) $E = \{x | x \text{ is an even digit}\} = \{0, 2, 4, 6, 8\}$
(b) $O = \{x | x \text{ is an odd digit}\} = \{1, 3, 5, 7, 9\}$

Because the elements of a set are distinct, we never repeat elements. For example, we would never write $\{1, 2, 3, 2\}$; the correct listing is $\{1, 2, 3\}$. Because a set is a collection, the order in which the elements are listed is immaterial. $\{1, 2, 3\}$, $\{1, 3, 2\}$, $\{2, 1, 3\}$, and so on, all represent the same set.

If every element of a set A is also an element of a set B, then we say that A is a **subset** of B and write $A \subseteq B$. If two sets A and B have the same elements, then we say that A **equals** B and write $A = B$.

For example, $\{1, 2, 3\} \subseteq \{1, 2, 3, 4, 5\}$ and $\{1, 2, 3\} = \{2, 3, 1\}$.

DEFINITION If A and B are sets, the **intersection** of A with B, denoted $A \cap B$, is the set consisting of elements that belong to both A and B. The **union** of A with B, denoted $A \cup B$, is the set consisting of elements that belong to either A or B, or both.

EXAMPLE 2 **Finding the Intersection and Union of Sets**

Let $A = \{1, 3, 5, 8\}$, $B = \{3, 5, 7\}$, and $C = \{2, 4, 6, 8\}$. Find:

(a) $A \cap B$ (b) $A \cup B$ (c) $B \cap (A \cup C)$

Solution

(a) $A \cap B = \{1, 3, 5, 8\} \cap \{3, 5, 7\} = \{3, 5\}$

(b) $A \cup B = \{1, 3, 5, 8\} \cup \{3, 5, 7\} = \{1, 3, 5, 7, 8\}$

(c) $B \cap (A \cup C) = \{3, 5, 7\} \cap [\{1, 3, 5, 8\} \cup \{2, 4, 6, 8\}]$
$= \{3, 5, 7\} \cap \{1, 2, 3, 4, 5, 6, 8\} = \{3, 5\}$

Now Work PROBLEM 13

Usually, in working with sets, we designate a **universal set** U, the set consisting of all the elements that we wish to consider. Once a universal set has been designated, we can consider elements of the universal set not found in a given set.

DEFINITION

If A is a set, the **complement** of A, denoted $\overline{A}$, is the set consisting of all the elements in the universal set that are not in A.*

EXAMPLE 3 Finding the Complement of a Set

If the universal set is $U = \{1, 2, 3, 4, 5, 6, 7, 8, 9\}$ and if $A = \{1, 3, 5, 7, 9\}$, then $\overline{A} = \{2, 4, 6, 8\}$.

It follows that $A \cup \overline{A} = U$ and $A \cap \overline{A} = \varnothing$. Do you see why?

Now Work PROBLEM 17

It is often helpful to draw pictures of sets. Such pictures, called **Venn diagrams,** represent sets as circles enclosed in a rectangle, which represents the universal set. Such diagrams often help us to visualize various relationships among sets. See Figure 1.

Figure 1

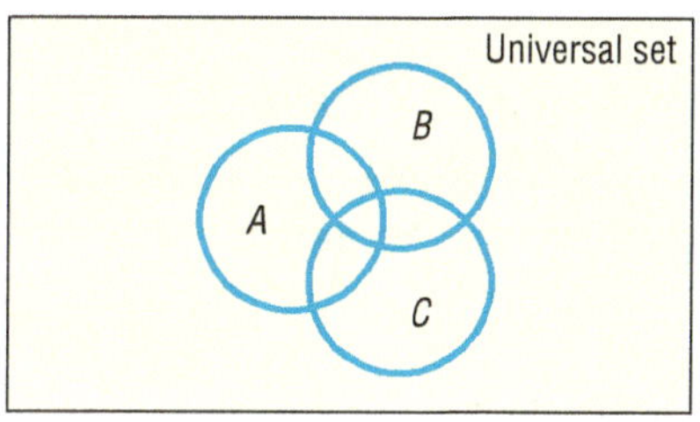

If we know that $A \subseteq B$, we might use the Venn diagram in Figure 2(a). If we know that A and B have no elements in common, that is, if $A \cap B = \varnothing$, we might use the Venn diagram in Figure 2(b). The sets A and B in Figure 2(b) are said to be **disjoint.**

Figure 2

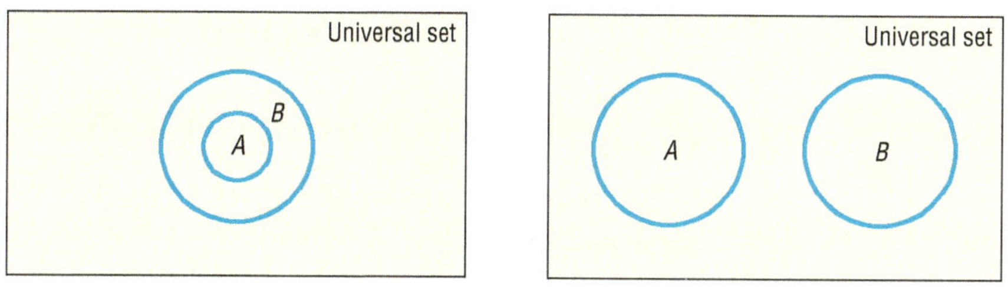

(a) $A \subseteq B$ subset

(b) $A \cap B = \varnothing$ disjoint sets

Figures 3(a), 3(b), and 3(c) use Venn diagrams to illustrate the definitions of intersection, union, and complement, respectively.

Figure 3

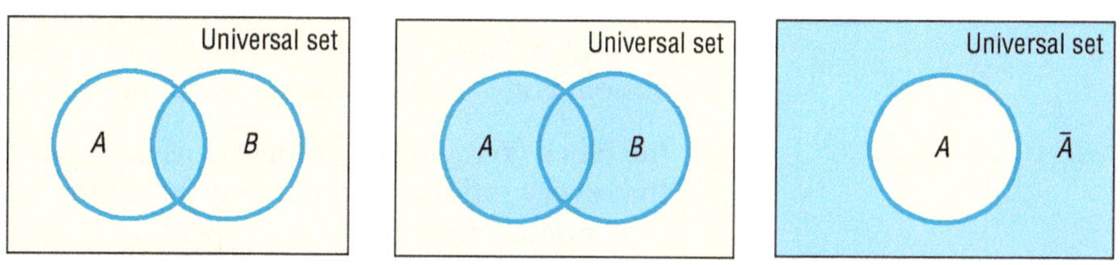

(a) $A \cap B$ intersection

(b) $A \cup B$ union

(c) $\bar{A}$ complement

*Some books use the notation A' or A^c for the complement of A.

2 Classify Numbers

It is helpful to classify the various kinds of numbers that we deal with as sets. The **counting numbers,** or **natural numbers,** are the numbers in the set $\{1, 2, 3, 4 \ldots\}$. (The three dots, called an **ellipsis,** indicate that the pattern continues indefinitely.) As their name implies, these numbers are often used to count things. For example, there are 26 letters in our alphabet; there are 100 cents in a dollar. The **whole numbers** are the numbers in the set $\{0, 1, 2, 3, \ldots\}$, that is, the counting numbers together with 0.

DEFINITION The **integers** are the set of numbers $\{\ldots, -3, -2, -1, 0, 1, 2, 3, \ldots\}$.

These numbers are useful in many situations. For example, if your checking account has \$10 in it and you write a check for \$15, you can represent the current balance as $-\$5$.

Notice that the set of counting numbers is a subset of the set of whole numbers. Each time we expand a number system, such as from the whole numbers to the integers, we do so in order to be able to handle new, and usually more complicated, problems. The integers allow us to solve problems requiring both positive and negative counting numbers, such as profit/loss, height above/below sea level, temperature above/below 0°F, and so on.

But integers alone are not sufficient for *all* problems. For example, they do not answer the question "What part of a dollar is 38 cents?" To answer such a question, we enlarge our number system to include *rational numbers*. For example, $\frac{38}{100}$ answers the question "What part of a dollar is 38 cents?"

DEFINITION A **rational number** is a number that can be expressed as a quotient $\frac{a}{b}$ of two integers. The integer a is called the **numerator,** and the integer b, which cannot be 0, is called the **denominator.** The rational numbers are the numbers in the set $\{x \mid x = \frac{a}{b}$, where a, b are integers and $b \neq 0\}$.

Examples of rational numbers are $\frac{3}{4}, \frac{5}{2}, \frac{0}{4}, -\frac{2}{3}$, and $\frac{100}{3}$. Since $\frac{a}{1} = a$ for any integer a, it follows that the set of integers is a subset of the set of rational numbers.

Rational numbers may be represented as **decimals.** For example, the rational numbers $\frac{3}{4}, \frac{5}{2}, -\frac{2}{3}$, and $\frac{7}{66}$ may be represented as decimals by carrying out the indicated division:

$$\frac{3}{4} = 0.75 \qquad \frac{5}{2} = 2.5 \qquad -\frac{2}{3} = -0.666\ldots = -0.\overline{6} \qquad \frac{7}{66} = 0.1060606\ldots = 0.1\overline{06}$$

Notice that the decimal representations of $\frac{3}{4}$ and $\frac{5}{2}$ terminate, or end. The decimal representations of $-\frac{2}{3}$ and $\frac{7}{66}$ do not terminate, but they do exhibit a pattern of repetition. For $-\frac{2}{3}$, the 6 repeats indefinitely, as indicated by the bar over the 6; for $\frac{7}{66}$, the block 06 repeats indefinitely, as indicated by the bar over the 06. It can be shown that every rational number may be represented by a decimal that either terminates or is nonterminating with a repeating block of digits, and vice versa.

On the other hand, some decimals do not fit into either of these categories. Such decimals represent **irrational numbers.** Every irrational number may be represented by a decimal that neither repeats nor terminates. In other words, irrational numbers cannot be written in the form $\frac{a}{b}$, where a, b are integers and $b \neq 0$.

Irrational numbers occur naturally. For example, consider the isosceles right triangle whose legs are each of length 1. See Figure 4. The length of the hypotenuse is $\sqrt{2}$, an irrational number.

Also, the number that equals the ratio of the circumference C to the diameter d of any circle, denoted by the symbol π (the Greek letter pi), is an irrational number. See Figure 5.

Figure 4

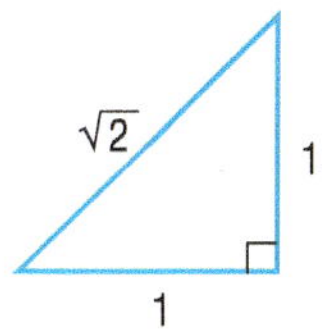

Figure 5 $\pi = \frac{C}{d}$

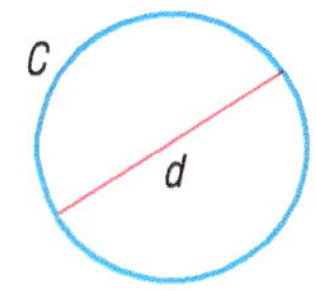

DEFINITION The set of real numbers is the union of the set of rational numbers with the set of irrational numbers.

Figure 6 shows the relationship of various types of numbers.*

Figure 6

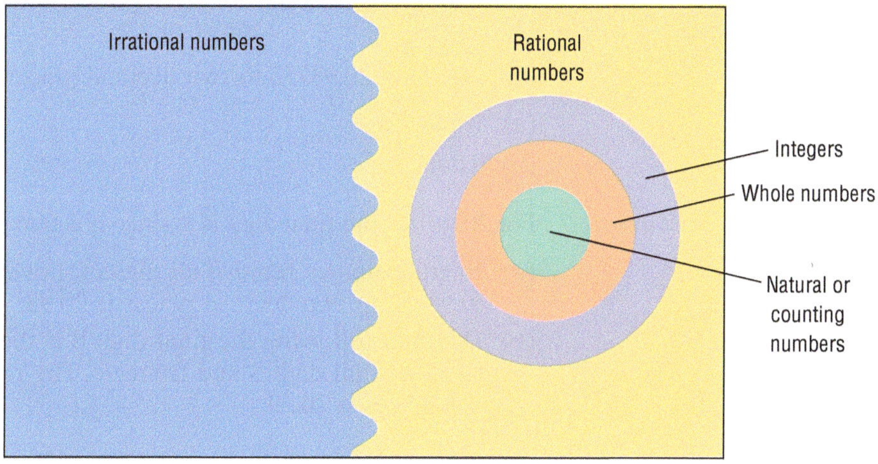

EXAMPLE 4 Classifying the Numbers in a Set

List the numbers in the set

$$\{-3, \frac{4}{3}, 0.12, \sqrt{2}, \pi, 10, 2.151515\ldots = 2.\overline{15} \text{ (where the block 15 repeats)}\}$$

that are

(a) Natural numbers (b) Integers (c) Rational numbers
(d) Irrational numbers (e) Real numbers

Solution

(a) 10 is the only natural number.

(b) -3 and 10 are integers.

(c) $-3, 10, \frac{4}{3}, 0.12$, and $2.151515\ldots = 2.\overline{15}$ are rational numbers.

(d) $\sqrt{2}$ and π are irrational numbers.

(e) All the numbers listed are real numbers.

Now Work PROBLEM 23

*The set of real numbers is a subset of the set of complex numbers. We discuss complex numbers in Chapter 1, Section 4.

Approximations

Every decimal may be represented by a real number (either rational or irrational), and every real number may be represented by a decimal.

In practice, the decimal representation of an irrational number is given as an approximation. For example, using the symbol $\approx$ (read as "approximately equal to"), we can write

$$\sqrt{2} \approx 1.4142 \qquad \pi \approx 3.1416$$

In approximating decimals, we either *round off* or *truncate* to a given number of decimal places.* The number of places establishes the location of the *final digit* in the decimal approximation.

Truncation: Drop all the digits that follow the specified final digit in the decimal.

Rounding: Identify the specified final digit in the decimal. If the next digit is 5 or more, add 1 to the final digit; if the next digit is 4 or less, leave the final digit as it is. Then truncate following the final digit.

EXAMPLE 5 Approximating a Decimal to Two Places

Approximate 20.98752 to two decimal places by

(a) Truncating

(b) Rounding

Solution For 20.98752, the final digit is 8, since it is two decimal places from the decimal point.

(a) To truncate, we remove all digits following the final digit 8. The truncation of 20.98752 to two decimal places is 20.98.

(b) The digit following the final digit 8 is the digit 7. Since 7 is 5 or more, we add 1 to the final digit 8 and truncate. The rounded form of 20.98752 to two decimal places is 20.99.

EXAMPLE 6 Approximating a Decimal to Two and Four Places

Number	Rounded to Two Decimal Places	Rounded to Four Decimal Places	Truncated to Two Decimal Places	Truncated to Four Decimal Places
(a) 3.14159	3.14	3.1416	3.14	3.1415
(b) 0.056128	0.06	0.0561	0.05	0.0561
(c) 893.46125	893.46	893.4613	893.46	893.4612

Now Work PROBLEM 27

Significant Digits

There are two types of numbers—*exact* and *approximate*. **Exact numbers** are numbers whose value is known with 100% certainty and accuracy. For example, there are 12 donuts in a dozen donuts, or there are 50 states in the United States. **Approximate numbers** are numbers whose value is not known with 100% certainty, or whose measurement is inexact. When values are determined from measurement they are typically approximate numbers because the exact measurement is limited by the accuracy of the measuring device and the skill of the individual obtaining the measurement. The **number of significant digits** in a number represents the level of accuracy of the measurement.

*Sometimes we say "correct to a given number of decimal places" instead of "truncate."

The following rules are used to determine the number of significant digits in approximate numbers.

The Number of Significant Digits

- Leading zeros are not significant. For example, 0.0034 has two significant digits.
- Embedded zeros are significant. For example, 208 has three significant digits.
- Trailing zeros are significant only if the decimal point is specified. For example, 2800 has two significant digits. However, if we specify the measurement is accurate to the ones digit, then 2800 has four significant digits.

When performing computations with approximate numbers, it is important not to report the result with more accuracy than the measurements used in the computation.

When performing computations using significant digits, proceed with the computation as you normally would, then round the final answer to the number of significant digits as the least accurately known number. For example, suppose we want to find the area of a rectangle whose width is 1.94 inches (three significant digits) and whose length is 2.7 inches (two significant digits). Because the length has two significant digits, we report the area to two significant digits. The area, (1.94 inches)(2.7 inches) = 5.238 square inches, can only be written to two significant digits and is reported as 5.2 square inches.

Calculators

Calculators are finite machines. As a result, they are incapable of displaying decimals that contain a large number of digits. For example, some calculators are capable of displaying only eight digits. When a number requires more than eight digits, the calculator either truncates or rounds. To see how your calculator handles decimals, divide 2 by 3. How many digits do you see? Is the last digit a 6 or a 7? If it is a 6, your calculator truncates; if it is a 7, your calculator rounds.

There are different kinds of calculators. An **arithmetic** calculator can only add, subtract, multiply, and divide numbers; therefore, this type is not adequate for this course. **Scientific** calculators have all the capabilities of arithmetic calculators and also contain **function keys** labeled ln, log, sin, cos, tan, x^y, inv, and so on. As you proceed through this text, you will discover how to use many of the function keys. **Graphing** calculators have all the capabilities of scientific calculators and contain a screen on which graphs can be displayed.

Figure 7

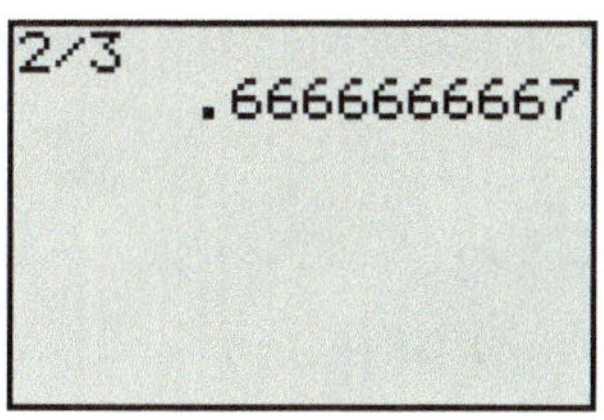

Figure 7 shows $\frac{2}{3}$ on a T1-84 Plus graphing calculator. How many digits are displayed? Does a T1-84 Plus round or truncate? What does your calculator do?

Operations

In algebra, we use letters such as x, y, a, b, and c to represent numbers. The symbols used in algebra for the operations of addition, subtraction, multiplication, and division are $+$, $-$, $\cdot$, and $/$. The words used to describe the results of these operations are **sum, difference, product,** and **quotient.** Table 1 summarizes these ideas.

Table 1

Operation	Symbol	Words
Addition	$a + b$	Sum: a plus b
Subtraction	$a - b$	Difference: a minus b
Multiplication	$a \cdot b$, $(a) \cdot b$, $a \cdot (b)$, $(a) \cdot (b)$, ab, $(a)b$, $a(b)$, $(a)(b)$	Product: a times b
Division	a/b or $\frac{a}{b}$	Quotient: a divided by b

In algebra, we generally avoid using the multiplication sign $\times$ and the division sign $\div$ so familiar in arithmetic. Notice also that when two expressions are placed next to each other without an operation symbol, as in ab, or in parentheses, as in $(a)(b)$, it is understood that the expressions, called **factors,** are to be multiplied.

We also prefer not to use mixed numbers in algebra. When mixed numbers are used, addition is understood; for example, $2\frac{3}{4}$ means $2 + \frac{3}{4}$. In algebra, use of a mixed number may be confusing because the absence of an operation symbol between two terms is generally taken to mean multiplication. The expression $2\frac{3}{4}$ is therefore written instead as 2.75 or as $\frac{11}{4}$.

The symbol $=$, called an **equal sign** and read as "equals" or "is," is used to express the idea that the number or expression on the left of the equal sign is equivalent to the number or expression on the right.

EXAMPLE 7 Writing Statements Using Symbols

(a) The sum of 2 and 7 equals 9. In symbols, this statement is written as $2 + 7 = 9$.
(b) The product of 3 and 5 is 15. In symbols, this statement is written as $3 \cdot 5 = 15$.

Now Work PROBLEM 39

3 Evaluate Numerical Expressions

Consider the expression $2 + 3 \cdot 6$. It is not clear whether we should add 2 and 3 to get 5, and then multiply by 6 to get 30; or first multiply 3 and 6 to get 18, and then add 2 to get 20. To avoid this ambiguity, we have the following agreement.

In Words
Multiply first, then add.

We agree that whenever the two operations of addition and multiplication separate three numbers, the multiplication operation always will be performed first, followed by the addition operation.

For $2 + 3 \cdot 6$, we have

$$2 + 3 \cdot 6 = 2 + 18 = 20$$

EXAMPLE 8 Finding the Value of an Expression

Evaluate each expression.

(a) $3 + 4 \cdot 5$ (b) $8 \cdot 2 + 1$ (c) $2 + 2 \cdot 2$

Solution

(a) $3 + 4 \cdot 5 = 3 + 20 = 23$ (Multiply first)

(b) $8 \cdot 2 + 1 = 16 + 1 = 17$ (Multiply first)

(c) $2 + 2 \cdot 2 = 2 + 4 = 6$

Figure 8

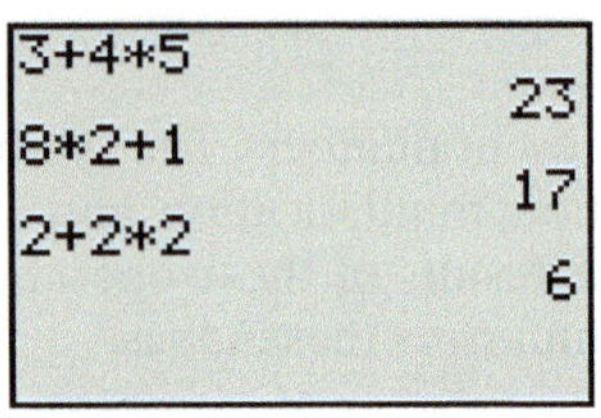

Figure 8 shows the solutions to Example 8 using a graphing calculator. Notice that the calculator follows the order of operations agreed to.

Now Work PROBLEM 51

To first add 3 and 4 and then multiply the result by 5, we use parentheses and write $(3 + 4) \cdot 5$. Whenever parentheses appear in an expression, it means "perform the operations within the parentheses first!"

EXAMPLE 9 Finding the Value of an Expression

(a) $(5 + 3) \cdot 4 = 8 \cdot 4 = 32$
(b) $(4 + 5) \cdot (8 - 2) = 9 \cdot 6 = 54$

When we divide two expressions, as in

$$\frac{2+3}{4+8}$$

it is understood that the division bar acts like parentheses; that is,

$$\frac{2+3}{4+8} = \frac{(2+3)}{(4+8)}$$

The following list gives the rules for the order of operations.

Rules for the Order of Operations

1. Begin with the innermost parentheses and work outward. Remember that in dividing two expressions the numerator and denominator are treated as if they were enclosed in parentheses.
2. Perform multiplications and divisions, working from left to right.
3. Perform additions and subtractions, working from left to right.

EXAMPLE 10 Finding the Value of an Expression

Evaluate each expression.

(a) $8 \cdot 2 + 3$ (b) $5 \cdot (3+4) + 2$

(c) $\dfrac{2+5}{2+4 \cdot 7}$ (d) $2 + [4 + 2 \cdot (10+6)]$

Solution

(a) $8 \cdot 2 + 3 = 16 + 3 = 19$
↑ Multiply first

(b) $5 \cdot (3+4) + 2 = 5 \cdot 7 + 2 = 35 + 2 = 37$
↑ Parentheses first ↑ Multiply before adding

(c) $\dfrac{2+5}{2+4 \cdot 7} = \dfrac{2+5}{2+28} = \dfrac{7}{30}$

(d) $2 + [4 + 2 \cdot (10+6)] = 2 + [4 + 2 \cdot (16)] = 2 + [4 + 32] = 2 + [36] = 38$

Figure 9

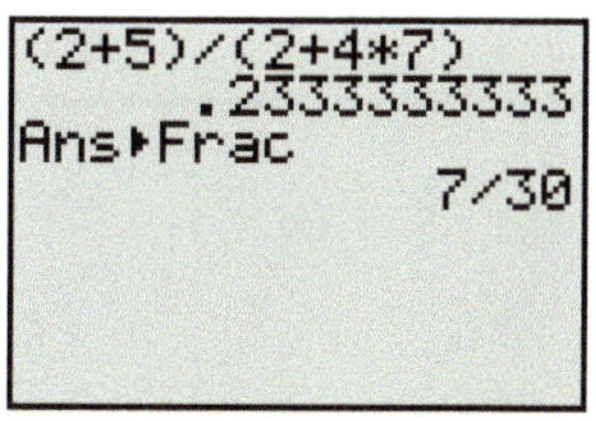

Be careful if you use a calculator. For Example 10(c), you need to use parentheses. See Figure 9.* If you don't, the calculator will compute the expression

$$2 + \frac{5}{2} + 4 \cdot 7 = 2 + 2.5 + 28 = 32.5$$

giving a wrong answer.

Now Work PROBLEMS 57 AND 65

4 Work with Properties of Real Numbers

We have used the equal sign to mean that one expression is equivalent to another. Four important properties of equality are listed next. In this list, a, b, and c represent real numbers.

1. The **reflexive property** states that a number always equals itself; that is, $a = a$.
2. The **symmetric property** states that if $a = b$ then $b = a$.
3. The **transitive property** states that if $a = b$ and $b = c$ then $a = c$.
4. The **principle of substitution** states that if $a = b$ then we may substitute b for a in any expression containing a.

*Notice that we converted the decimal to its fraction form. Consult your manual to see how your calculator does this.

Now, let's consider some other properties of real numbers. We begin with an example.

EXAMPLE 11

Commutative Properties

(a) $3 + 5 = 8$
$5 + 3 = 8$
$3 + 5 = 5 + 3$

(b) $2 \cdot 3 = 6$
$3 \cdot 2 = 6$
$2 \cdot 3 = 3 \cdot 2$

This example illustrates the **commutative property** of real numbers, which states that the order in which addition or multiplication takes place will not affect the final result.

Commutative Properties

$$a + b = b + a \quad \textbf{(1a)}$$
$$a \cdot b = b \cdot a \quad \textbf{(1b)}$$

Here, and in the properties listed next and on pages 11–13, a, b, and c represent real numbers.

EXAMPLE 12

Associative Properties

(a) $2 + (3 + 4) = 2 + 7 = 9$
$(2 + 3) + 4 = 5 + 4 = 9$
$2 + (3 + 4) = (2 + 3) + 4$

(b) $2 \cdot (3 \cdot 4) = 2 \cdot 12 = 24$
$(2 \cdot 3) \cdot 4 = 6 \cdot 4 = 24$
$2 \cdot (3 \cdot 4) = (2 \cdot 3) \cdot 4$

The way we add or multiply three real numbers will not affect the final result. Expressions such as $2 + 3 + 4$ and $3 \cdot 4 \cdot 5$ present no ambiguity, even though addition and multiplication are performed on one pair of numbers at a time. This property is called the **associative property.**

Associative Properties

$$a + (b + c) = (a + b) + c = a + b + c \quad \textbf{(2a)}$$
$$a \cdot (b \cdot c) = (a \cdot b) \cdot c = a \cdot b \cdot c \quad \textbf{(2b)}$$

The next property is perhaps the most important.

Distributive Property

$$a \cdot (b + c) = a \cdot b + a \cdot c \quad \textbf{(3a)}$$
$$(a + b) \cdot c = a \cdot c + b \cdot c \quad \textbf{(3b)}$$

The **distributive property** may be used in two different ways.

EXAMPLE 13

Distributive Property

(a) $2 \cdot (x + 3) = 2 \cdot x + 2 \cdot 3 = 2x + 6$ Use to remove parentheses.

(b) $3x + 5x = (3 + 5)x = 8x$ Use to combine two expressions.

(c) $(x + 2)(x + 3) = x(x + 3) + 2(x + 3) = (x^2 + 3x) + (2x + 6)$
$= x^2 + (3x + 2x) + 6 = x^2 + 5x + 6$

Now Work PROBLEM 87

The real numbers 0 and 1 have unique properties.

EXAMPLE 14 **Identity Properties**

(a) $4 + 0 = 0 + 4 = 4$ (b) $3 \cdot 1 = 1 \cdot 3 = 3$

The properties of 0 and 1 illustrated in Example 14 are called the **identity properties.**

Identity Properties

$$0 + a = a + 0 = a \quad \textbf{(4a)}$$
$$a \cdot 1 = 1 \cdot a = a \quad \textbf{(4b)}$$

We call 0 the **additive identity** and 1 the **multiplicative identity.**

For each real number a, there is a real number $-a$, called the **additive inverse** of a, having the following property:

Additive Inverse Property

$$a + (-a) = -a + a = 0 \quad \textbf{(5a)}$$

EXAMPLE 15 **Finding an Additive Inverse**

(a) The additive inverse of 6 is -6, because $6 + (-6) = 0$.
(b) The additive inverse of -8 is $-(-8) = 8$, because $-8 + 8 = 0$.

The additive inverse of a, that is, $-a$, is often called the *negative* of a or the *opposite* of a. The use of such terms can be dangerous, because they suggest that the additive inverse is a negative number, which may not be the case. For example, the additive inverse of -3, or $-(-3)$, equals 3, a positive number.

For each *nonzero* real number a, there is a real number $\frac{1}{a}$, called the **multiplicative inverse** of a, having the following property:

Multiplicative Inverse Property

$$a \cdot \frac{1}{a} = \frac{1}{a} \cdot a = 1 \qquad \text{if } a \neq 0 \quad \textbf{(5b)}$$

The multiplicative inverse $\frac{1}{a}$ of a nonzero real number a is also referred to as the **reciprocal** of a.

EXAMPLE 16 **Finding a Reciprocal**

(a) The reciprocal of 6 is $\frac{1}{6}$, because $6 \cdot \frac{1}{6} = 1$.

(b) The reciprocal of -3 is $\frac{1}{-3}$, because $-3 \cdot \frac{1}{-3} = 1$.

(c) The reciprocal of $\frac{2}{3}$ is $\frac{3}{2}$, because $\frac{2}{3} \cdot \frac{3}{2} = 1$.

With these properties for adding and multiplying real numbers, we can now define the operations of subtraction and division as follows:

DEFINITION

The **difference** $a - b$, also read "a less b" or "a minus b," is defined as

$$a - b = a + (-b) \quad (6)$$

To subtract b from a, add the opposite of b to a.

DEFINITION

If b is a nonzero real number, the **quotient** $\frac{a}{b}$, also read as "a divided by b" or "the ratio of a to b," is defined as

$$\frac{a}{b} = a \cdot \frac{1}{b} \quad \text{if } b \neq 0 \quad (7)$$

EXAMPLE 17 Working with Differences and Quotients

(a) $8 - 5 = 8 + (-5) = 3$ (b) $4 - 9 = 4 + (-9) = -5$ (c) $\frac{5}{8} = 5 \cdot \frac{1}{8}$

For any number a, the product of a times 0 is always 0; that is,

In Words
The result of multiplying by zero is zero.

Multiplication by Zero

$$a \cdot 0 = 0 \quad (8)$$

For a nonzero number a,

Division Properties

$$\frac{0}{a} = 0 \qquad \frac{a}{a} = 1 \quad \text{if } a \neq 0 \quad (9)$$

WARNING Division by 0 is *not defined.* One reason is to avoid the following difficulty: $\frac{2}{0} = x$ means to find x such that $0 \cdot x = 2$. But $0 \cdot x$ equals 0 for all x, so there is *no* unique number x such that $\frac{2}{0} = x$.

Rules of Signs

$$a(-b) = -(ab) \qquad (-a)b = -(ab) \qquad (-a)(-b) = ab$$

$$-(-a) = a \qquad \frac{a}{-b} = \frac{-a}{b} = -\frac{a}{b} \qquad \frac{-a}{-b} = \frac{a}{b} \quad (10)$$

EXAMPLE 18 Applying the Rules of Signs

(a) $2(-3) = -(2 \cdot 3) = -6$

(b) $(-3)(-5) = 3 \cdot 5 = 15$

(c) $\frac{3}{-2} = \frac{-3}{2} = -\frac{3}{2}$

(d) $\frac{-4}{-9} = \frac{4}{9}$

(e) $\frac{x}{-2} = \frac{1}{-2} \cdot x = -\frac{1}{2}x$

Cancellation Properties

$$ac = bc \text{ implies } a = b \quad \text{if } c \neq 0$$

$$\frac{ac}{bc} = \frac{a}{b} \quad \text{if } b \neq 0, c \neq 0 \qquad \textbf{(11)}$$

EXAMPLE 19 Using the Cancellation Properties

(a) If $2x = 6$, then

$$\begin{aligned} 2x &= 6 \\ 2x &= 2 \cdot 3 && \text{Factor 6.} \\ x &= 3 && \text{Cancel the 2's.} \end{aligned}$$

(b) $\dfrac{18}{12} = \dfrac{3 \cdot \not{6}}{2 \cdot \not{6}} = \dfrac{3}{2}$

↑ Cancel the 6's.

NOTE We follow the common practice of using slash marks to indicate cancellations. ■

In Words
If a product equals 0, then one or both of the factors is 0.

Zero-Product Property

If $ab = 0$, then $a = 0$, or $b = 0$, or both. **(12)**

EXAMPLE 20 Using the Zero-Product Property

If $2x = 0$, then either $2 = 0$ or $x = 0$. Since $2 \neq 0$, it follows that $x = 0$.

Arithmetic of Quotients

$$\frac{a}{b} + \frac{c}{d} = \frac{ad}{bd} + \frac{bc}{bd} = \frac{ad + bc}{bd} \quad \text{if } b \neq 0, d \neq 0 \qquad \textbf{(13)}$$

$$\frac{a}{b} \cdot \frac{c}{d} = \frac{ac}{bd} \quad \text{if } b \neq 0, d \neq 0 \qquad \textbf{(14)}$$

$$\frac{\frac{a}{b}}{\frac{c}{d}} = \frac{a}{b} \cdot \frac{d}{c} = \frac{ad}{bc} \quad \text{if } b \neq 0, c \neq 0, d \neq 0 \qquad \textbf{(15)}$$

EXAMPLE 21 Adding, Subtracting, Multiplying, and Dividing Quotients

(a) $\dfrac{2}{3} + \dfrac{5}{2} = \dfrac{2 \cdot 2}{3 \cdot 2} + \dfrac{3 \cdot 5}{3 \cdot 2} = \dfrac{2 \cdot 2 + 3 \cdot 5}{3 \cdot 2} = \dfrac{4 + 15}{6} = \dfrac{19}{6}$

↑ By equation (13)

(b) $\dfrac{3}{5} - \dfrac{2}{3} = \dfrac{3}{5} + \left(-\dfrac{2}{3}\right) = \dfrac{3}{5} + \dfrac{-2}{3}$

↑ By equation (6) ↑ By equation (10)

$= \dfrac{3 \cdot 3 + 5 \cdot (-2)}{5 \cdot 3} = \dfrac{9 + (-10)}{15} = \dfrac{-1}{15} = -\dfrac{1}{15}$

↑ By equation (13)

NOTE Slanting the cancellation marks in different directions for different factors, as shown here, is a good practice to follow, since it will help in checking for errors. ■

(c) $\frac{8}{3} \cdot \frac{15}{4} = \frac{8 \cdot 15}{3 \cdot 4} = \frac{2 \cdot \cancel{4} \cdot \cancel{3} \cdot 5}{\cancel{3} \cdot \cancel{4} \cdot 1} = \frac{2 \cdot 5}{1} = 10$

↑ By equation (14) ↑ By equation (11)

(d) $\dfrac{\frac{3}{5}}{\frac{7}{9}} = \frac{3}{5} \cdot \frac{9}{7} = \frac{3 \cdot 9}{5 \cdot 7} = \frac{27}{35}$

↑ By equation (15) ↑ By equation (14)

NOTE In writing quotients, we shall follow the usual convention and write the quotient in **lowest terms.** That is, we write it so that any common factors of the numerator and the denominator have been removed using the cancellation properties, equation (11). As examples,

$$\frac{90}{24} = \frac{15 \cdot \cancel{6}}{4 \cdot \cancel{6}} = \frac{15}{4}$$

$$\frac{24x^2}{18x} = \frac{4 \cdot \cancel{6} \cdot x \cdot \cancel{x}}{3 \cdot \cancel{6} \cdot \cancel{x}} = \frac{4x}{3} \qquad x \neq 0$$

■

Now Work PROBLEMS 67, 71, AND 81

Sometimes it is easier to add two fractions using *least common multiples* (LCM). The LCM of two numbers is the smallest number that each has as a common multiple.

EXAMPLE 22 Finding the Least Common Multiple of Two Numbers

Find the least common multiple of 15 and 12.

Solution To find the LCM of 15 and 12, we look at multiples of 15 and 12.

15, 30, 45, 60, 75, 90, 105, 120, ...

12, 24, 36, 48, 60, 72, 84, 96, 108, 120, ...

The *common* multiples are in blue. The *least* common multiple is 60.

EXAMPLE 23 Using the Least Common Multiple to Add Two Fractions

Find: $\frac{8}{15} + \frac{5}{12}$

Solution We use the LCM of the denominators of the fractions and rewrite each fraction using the LCM as a common denominator. The LCM of the denominators (12 and 15) is 60. Rewrite each fraction using 60 as the denominator.

$$\begin{aligned} \frac{8}{15} + \frac{5}{12} &= \frac{8}{15} \cdot \frac{4}{4} + \frac{5}{12} \cdot \frac{5}{5} \\ &= \frac{32}{60} + \frac{25}{60} \\ &= \frac{32 + 25}{60} \\ &= \frac{57}{60} \\ &= \frac{19 \cdot 3}{20 \cdot 3} = \frac{19}{20} \end{aligned}$$

Now Work PROBLEM 75

Historical Feature

The real number system has a history that stretches back at least to the ancient Babylonians (1800 BC). It is remarkable how much the ancient Babylonian attitudes resemble our own. As we stated in the text, the fundamental difficulty with irrational numbers is that they cannot be written as quotients of integers or, equivalently, as repeating or terminating decimals. The Babylonians wrote their numbers in a system based on 60 in the same way that we write ours based on 10. They would carry as many places for π as the accuracy of the problem demanded, just as we now use

$$\pi \approx 3\frac{1}{7} \quad \text{or} \quad \pi \approx 3.1416 \quad \text{or} \quad \pi \approx 3.14159$$
$$\text{or} \quad \pi \approx 3.14159265358979$$

depending on how accurate we need to be.

Things were very different for the Greeks, whose number system allowed only rational numbers. When it was discovered that $\sqrt{2}$ was not a rational number, this was regarded as a fundamental flaw in the number concept. So serious was the matter that the Pythagorean Brotherhood (an early mathematical society) is said to have drowned one of its members for revealing this terrible secret. Greek mathematicians then turned away from the number concept, expressing facts about whole numbers in terms of line segments.

In astronomy, however, Babylonian methods, including the Babylonian number system, continued to be used. Simon Stevin (1548–1620), probably using the Babylonian system as a model, invented the decimal system, complete with rules of calculation, in 1585. [Others, for example, al-Kashi of Samarkand (d. 1429), had made some progress in the same direction.] The decimal system so effectively conceals the difficulties that the need for more logical precision began to be felt only in the early 1800s. Around 1880, Georg Cantor (1845–1918) and Richard Dedekind (1831–1916) gave precise definitions of real numbers. Cantor's definition, although more abstract and precise, has its roots in the decimal (and hence Babylonian) numerical system.

Sets and set theory were a spin-off of the research that went into clarifying the foundations of the real number system. Set theory has developed into a large discipline of its own, and many mathematicians regard it as the foundation upon which modern mathematics is built. Cantor's discoveries that infinite sets can also be counted and that there are different sizes of infinite sets are among the most astounding results of modern mathematics.

Historical Problems

The Babylonian number system was based on 60. Thus 2,30 means $2 + \frac{30}{60} = 2.5$, and 4,25,14 means

$$4 + \frac{25}{60} + \frac{14}{60^2} = 4 + \frac{1514}{3600} = 4.42055555\ldots$$

1. What are the following numbers in Babylonian notation?

   (a) $1\frac{1}{3}$ **(b)** $2\frac{5}{6}$

2. What are the following Babylonian numbers when written as fractions and as decimals?

   (a) 2,20 **(b)** 4,52,30 **(c)** 3,8,29,44

R.1 Assess Your Understanding

Concepts and Vocabulary

1. The numbers in the set $\{x \mid x = \frac{a}{b}$, where a, b are integers and $b \neq 0\}$, are called __________ numbers.

2. The value of the expression $4 + 5 \cdot 6 - 3$ is __________.

3. The fact that $2x + 3x = (2 + 3)x$ is a consequence of the __________ Property.

4. "The product of 5 and $x + 3$ equals 6" may be written as __________.

5. ***True or False*** Rational numbers have decimals that either terminate or are nonterminating with a repeating block of digits.

6. ***True or False*** The Zero-Product Property states that the product of any number and zero equals zero.

7. ***True or False*** The least common multiple of 12 and 18 is 6.

8. ***True or False*** No real number is both rational and irrational.

Skill Building

In Problems 9–20, use U = *universal set* = $\{0, 1, 2, 3, 4, 5, 6, 7, 8, 9\}$, $A = \{1, 3, 4, 5, 9\}$, $B = \{2, 4, 6, 7, 8\}$, *and* $C = \{1, 3, 4, 6\}$ *to find each set.*

9. $A \cup B$ **10.** $A \cup C$ **11.** $A \cap B$ **12.** $A \cap C$

13. $(A \cup B) \cap C$ **14.** $(A \cap B) \cup C$ **15.** $\overline{A}$ **16.** $\overline{C}$

17. $\overline{A \cap B}$ **18.** $\overline{B \cup C}$ **19.** $\overline{A} \cup \overline{B}$ **20.** $\overline{B} \cap \overline{C}$

In Problems 21–26, list the numbers in each set that are (a) Natural numbers, (b) Integers, (c) Rational numbers, (d) Irrational numbers, (e) Real numbers.

21. $A = \left\{-6, \frac{1}{2}, -1.333\ldots = -1.\overline{3}\text{ (the 3's repeat)}, \pi, 2, 5\right\}$

22. $B = \left\{-\frac{5}{3}, 2.060606\ldots = 2.\overline{06}\text{ (the block 06 repeats)}, 1.25, 0, 1, \sqrt{5}\right\}$

23. $C = \left\{0, 1, \frac{1}{2}, \frac{1}{3}, \frac{1}{4}\right\}$

24. $D = \{-1, -1.1, -1.2, -1.3\}$

25. $E = \left\{\sqrt{2}, \pi, \sqrt{2} + 1, \pi + \frac{1}{2}\right\}$

26. $F = \left\{-\sqrt{2}, \pi + \sqrt{2}, \frac{1}{2} + 10.3\right\}$

In Problems 27–38, approximate each number (a) rounded and (b) truncated to three decimal places.

27. 18.9526 **28.** 25.86134 **29.** 28.65319 **30.** 99.05249 **31.** 0.06291 **32.** 0.05388

33. 9.9985 **34.** 1.0006 **35.** $\frac{3}{7}$ **36.** $\frac{5}{9}$ **37.** $\frac{521}{15}$ **38.** $\frac{81}{5}$

In Problems 39–48, write each statement using symbols.

39. The sum of 3 and 2 equals 5.

40. The product of 5 and 2 equals 10.

41. The sum of x and 2 is the product of 3 and 4.

42. The sum of 3 and y is the sum of 2 and 2.

43. The product of 3 and y is the sum of 1 and 2.

44. The product of 2 and x is the product of 4 and 6.

45. The difference x less 2 equals 6.

46. The difference 2 less y equals 6.

47. The quotient x divided by 2 is 6.

48. The quotient 2 divided by x is 6.

In Problems 49–86, evaluate each expression.

49. $9 - 4 + 2$ **50.** $6 - 4 + 3$ **51.** $-6 + 4 \cdot 3$ **52.** $8 - 4 \cdot 2$

53. $4 + 5 - 8$ **54.** $8 - 3 - 4$ **55.** $4 + \frac{1}{3}$ **56.** $2 - \frac{1}{2}$

57. $6 - [3 \cdot 5 + 2 \cdot (3 - 2)]$ **58.** $2 \cdot [8 - 3(4 + 2)] - 3$ **59.** $2 \cdot (3 - 5) + 8 \cdot 2 - 1$ **60.** $1 - (4 \cdot 3 - 2 + 2)$

61. $10 - [6 - 2 \cdot 2 + (8 - 3)] \cdot 2$ **62.** $2 - 5 \cdot 4 - [6 \cdot (3 - 4)]$

63. $(5 - 3)\frac{1}{2}$ **64.** $(5 + 4)\frac{1}{3}$ **65.** $\frac{4 + 8}{5 - 3}$ **66.** $\frac{2 - 4}{5 - 3}$

67. $\frac{3}{5} \cdot \frac{10}{21}$ **68.** $\frac{5}{9} \cdot \frac{3}{10}$ **69.** $\frac{6}{25} \cdot \frac{10}{27}$ **70.** $\frac{21}{25} \cdot \frac{100}{3}$

71. $\frac{3}{4} + \frac{2}{5}$ **72.** $\frac{4}{3} + \frac{1}{2}$ **73.** $\frac{5}{6} + \frac{9}{5}$ **74.** $\frac{8}{9} + \frac{15}{2}$

75. $\frac{5}{18} + \frac{1}{12}$ **76.** $\frac{2}{15} + \frac{8}{9}$ **77.** $\frac{1}{30} - \frac{7}{18}$ **78.** $\frac{3}{14} - \frac{2}{21}$

79. $\frac{3}{20} - \frac{2}{15}$ **80.** $\frac{6}{35} - \frac{3}{14}$ **81.** $\dfrac{\frac{5}{18}}{\frac{11}{27}}$ **82.** $\dfrac{\frac{5}{21}}{\frac{2}{35}}$

83. $\frac{1}{2} \cdot \frac{3}{5} + \frac{7}{10}$ **84.** $\frac{2}{3} + \frac{4}{5} \cdot \frac{1}{6}$ **85.** $2 \cdot \frac{3}{4} + \frac{3}{8}$ **86.** $3 \cdot \frac{5}{6} - \frac{1}{2}$

In Problems 87–98, use the Distributive Property to remove the parentheses.

87. $6(x + 4)$ **88.** $4(2x - 1)$ **89.** $x(x - 4)$ **90.** $4x(x + 3)$

91. $2\left(\frac{3}{4}x - \frac{1}{2}\right)$ **92.** $3\left(\frac{2}{3}x + \frac{1}{6}\right)$ **93.** $(x + 2)(x + 4)$ **94.** $(x + 5)(x + 1)$

95. $(x - 2)(x + 1)$ **96.** $(x - 4)(x + 1)$ **97.** $(x - 8)(x - 2)$ **98.** $(x - 4)(x - 2)$

Discussion and Writing

99. Explain to a friend how the Distributive Property is used to justify the fact that $2x + 3x = 5x$.

100. Explain to a friend why $2 + 3 \cdot 4 = 14$, whereas $(2 + 3) \cdot 4 = 20$.

101. Explain why $2(3 \cdot 4)$ is not equal to $(2 \cdot 3) \cdot (2 \cdot 4)$.

102. Explain why $\frac{4 + 3}{2 + 5}$ is not equal to $\frac{4}{2} + \frac{3}{5}$.

103. Is subtraction commutative? Support your conclusion with an example.

104. Is subtraction associative? Support your conclusion with an example.

105. Is division commutative? Support your conclusion with an example.

106. Is division associative? Support your conclusion with an example.

107. If $2 = x$, why does $x = 2$?

108. If $x = 5$, why does $x^2 + x = 30$?

109. Are there any real numbers that are both rational and irrational? Are there any real numbers that are neither? Explain your reasoning.

110. Explain why the sum of a rational number and an irrational number must be irrational.

111. A rational number is defined as the quotient of two integers. When written as a decimal, the decimal will either repeat or terminate. By looking at the denominator of the rational number, there is a way to tell in advance whether its decimal representation will repeat or terminate. Make a list of rational numbers and their decimals. See if you can discover the pattern. Confirm your conclusion by consulting books on number theory at the library. Write a brief essay on your findings.

112. The current time is 12 noon CST. What time (CST) will it be 12,997 hours from now?

113. Both $\frac{a}{0}$ $(a \neq 0)$ and $\frac{0}{0}$ are undefined, but for different reasons. Write a paragraph or two explaining the different reasons.

R.2 Algebra Essentials

OBJECTIVES

1 Graph Inequalities (p. 18)
2 Find Distance on the Real Number Line (p. 19)
3 Evaluate Algebraic Expressions (p. 20)
4 Determine the Domain of a Variable (p. 21)
5 Use the Laws of Exponents (p. 21)
6 Evaluate Square Roots (p. 23)
7 Use a Graphing Calculator to Evaluate Exponents (p. 24)
8 Use Scientific Notation (p. 24)

The Real Number Line

The real numbers can be represented by points on a line called the **real number line.** There is a one-to-one correspondence between real numbers and points on a line. That is, every real number corresponds to a point on the line, and each point on the line has a unique real number associated with it.

Pick a point on the line somewhere in the center, and label it O. This point, called the **origin,** corresponds to the real number 0. See Figure 10. The point 1 unit to the right of O corresponds to the number 1. The distance between 0 and 1 determines the **scale** of the number line. For example, the point associated with the number 2 is twice as far from O as 1. Notice that an arrowhead on the right end of the line indicates the direction in which the numbers increase. Points to the left of the origin correspond to the real numbers -1, -2, and so on. Figure 10 also shows the points associated with the rational numbers $-\frac{1}{2}$ and $\frac{1}{2}$ and with the irrational numbers $\sqrt{2}$ and π.

Figure 10
Real number line

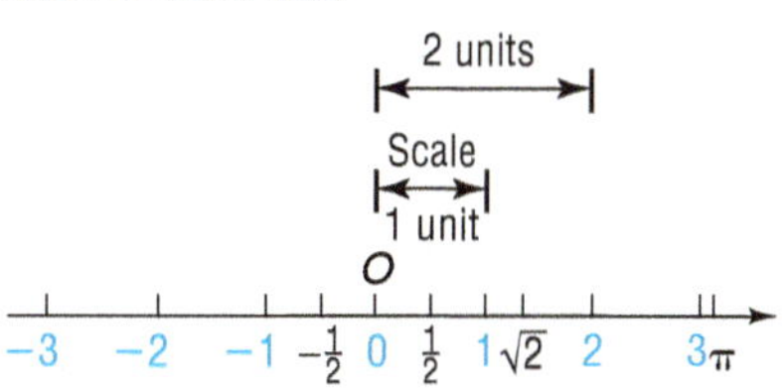

DEFINITION

The real number associated with a point P is called the **coordinate** of P, and the line whose points have been assigned coordinates is called the **real number line.**

Now Work PROBLEM 11

Figure 11

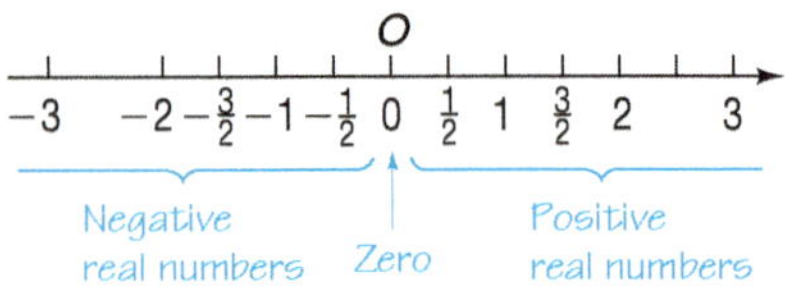

The real number line consists of three classes of real numbers, as shown in Figure 11.

1. The **negative real numbers** are the coordinates of points to the left of the origin O.
2. The real number **zero** is the coordinate of the origin O.
3. The **positive real numbers** are the coordinates of points to the right of the origin O.

Negative and positive numbers have the following multiplication properties:

Multiplication Properties of Positive and Negative Numbers

1. The product of two positive numbers is a positive number.
2. The product of two negative numbers is a positive number.
3. The product of a positive number and a negative number is a negative number.

1 Graph Inequalities

Figure 12

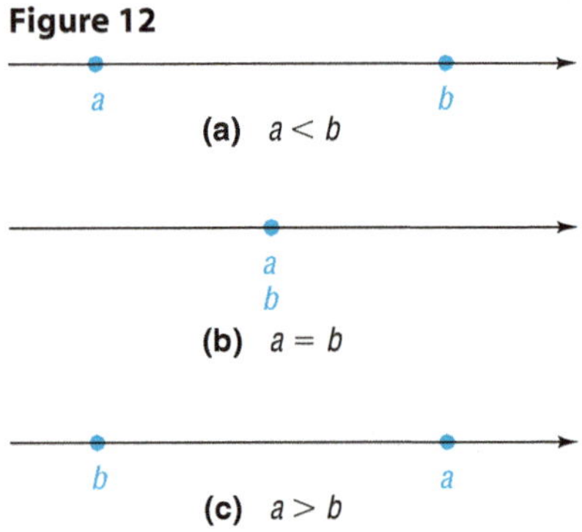

An important property of the real number line follows from the fact that, given two numbers (points) a and b, either a is to the left of b, or a is at the same location as b, or a is to the right of b. See Figure 12.

If a is to the left of b, we say that "a is less than b" and write $a < b$. If a is to the right of b, we say that "a is greater than b" and write $a > b$. If a is at the same location as b, then $a = b$. If a is either less than or equal to b, we write $a \leq b$. Similarly, $a \geq b$ means that a is either greater than or equal to b. Collectively, the symbols $<, >, \leq$, and $\geq$ are called **inequality symbols.**

Note that $a < b$ and $b > a$ mean the same thing. It does not matter whether we write $2 < 3$ or $3 > 2$.

Furthermore, if $a < b$ or if $b > a$, then the difference $b - a$ is positive. Do you see why?

EXAMPLE 1 **Using Inequality Symbols**

(a) $3 < 7$ (b) $-8 > -16$ (c) $-6 < 0$
(d) $-8 < -4$ (e) $4 > -1$ (f) $8 > 0$

In Example 1(a), we conclude that $3 < 7$ either because 3 is to the left of 7 on the real number line or because the difference, $7 - 3 = 4$, is a positive real number.

Similarly, we conclude in Example 1(b) that $-8 > -16$ either because -8 lies to the right of -16 on the real number line or because the difference, $-8 - (-16) = -8 + 16 = 8$, is a positive real number.

Look again at Example 1. Note that the inequality symbol always points in the direction of the smaller number.

An **inequality** is a statement in which two expressions are related by an inequality symbol. The expressions are referred to as the **sides** of the inequality. Statements of the form $a < b$ or $b > a$ are called **strict inequalities,** whereas statements of the form $a \leq b$ or $b \geq a$ are called **nonstrict inequalities.**

Based on the discussion so far, we conclude that

$$a > 0 \quad \text{is equivalent to } a \text{ is positive}$$
$$a < 0 \quad \text{is equivalent to } a \text{ is negative}$$

We sometimes read $a > 0$ by saying that "a is positive." If $a \geq 0$, then either $a > 0$ or $a = 0$, and we may read this as "a is nonnegative."

Now Work PROBLEMS 15 AND 25

We shall find it useful in later work to graph inequalities on the real number line.

EXAMPLE 2 **Graphing Inequalities**

(a) On the real number line, graph all numbers x for which $x > 4$.
(b) On the real number line, graph all numbers x for which $x \leq 5$.

Solution

(a) See Figure 13. Notice that we use a left parenthesis to indicate that the number 4 is *not* part of the graph.

Figure 13

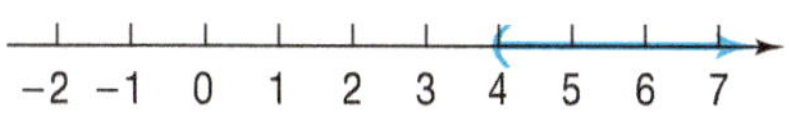

(b) See Figure 14. Notice that we use a right bracket to indicate that the number 5 *is* part of the graph.

Figure 14

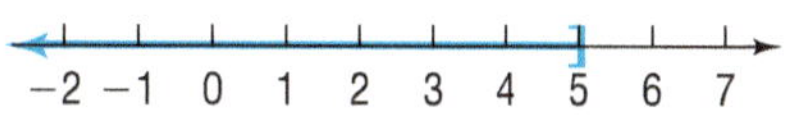

Now Work PROBLEM 31

2 Find Distance on the Real Number Line

Figure 15

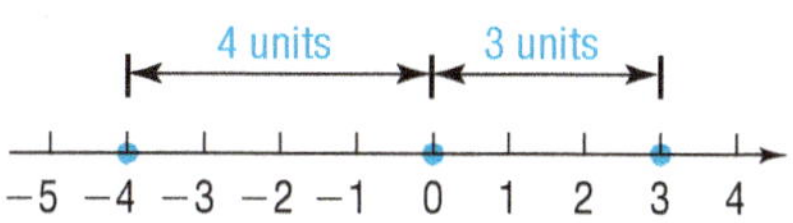

The *absolute value* of a number a is the distance from 0 to a on the number line. For example, -4 is 4 units from 0, and 3 is 3 units from 0. See Figure 15. Thus, the absolute value of -4 is 4, and the absolute value of 3 is 3.

A more formal definition of absolute value is given next.

DEFINITION

The **absolute value** of a real number a, denoted by the symbol $|a|$, is defined by the rules

$$|a| = a \quad \text{if } a \geq 0 \qquad \text{and} \qquad |a| = -a \quad \text{if } a < 0$$

For example, since $-4 < 0$, the second rule must be used to get $|-4| = -(-4) = 4$.

EXAMPLE 3 **Computing Absolute Value**

(a) $|8| = 8$ (b) $|0| = 0$ (c) $|-15| = -(-15) = 15$

Look again at Figure 15. The distance from -4 to 3 is 7 units. This distance is the difference $3 - (-4)$, obtained by subtracting the smaller coordinate from the larger. However, since $|3 - (-4)| = |7| = 7$ and $|-4 - 3| = |-7| = 7$, we can use absolute value to calculate the distance between two points without being concerned about which is smaller.

DEFINITION

If P and Q are two points on a real number line with coordinates a and b respectively, the **distance between P and Q,** denoted by $d(P, Q)$, is

$$d(P, Q) = |b - a|$$

Since $|b - a| = |a - b|$, it follows that $d(P, Q) = d(Q, P)$.

EXAMPLE 4 Finding Distance on a Number Line

Let P, Q, and R be points on a real number line with coordinates -5, 7, and -3, respectively. Find the distance

(a) between P and Q (b) between Q and R

Solution See Figure 16.

Figure 16

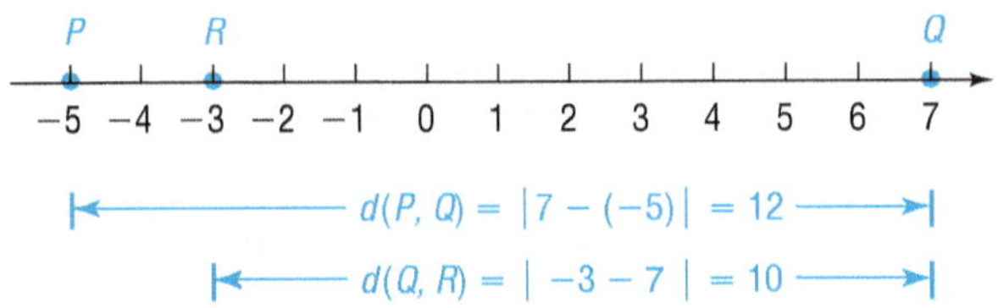

(a) $d(P, Q) = |7 - (-5)| = |12| = 12$
(b) $d(Q, R) = |-3 - 7| = |-10| = 10$

Now Work PROBLEM 37

3 Evaluate Algebraic Expressions

Remember, in algebra we use letters such as x, y, a, b, and c to represent numbers. If the letter used is to represent *any* number from a given set of numbers, it is called a **variable.** A **constant** is either a fixed number, such as 5 or $\sqrt{3}$, or a letter that represents a fixed (possibly unspecified) number.

Constants and variables are combined using the operations of addition, subtraction, multiplication, and division to form *algebraic expressions*. Examples of algebraic expressions include

$$x + 3 \qquad \frac{3}{1 - t} \qquad 7x - 2y$$

To evaluate an algebraic expression, substitute for each variable its numerical value.

EXAMPLE 5 Evaluating an Algebraic Expression

Evaluate each expression if $x = 3$ and $y = -1$.

(a) $x + 3y$ (b) $5xy$ (c) $\dfrac{3y}{2 - 2x}$ (d) $|-4x + y|$

Solution (a) Substitute 3 for x and -1 for y in the expression $x + 3y$.

$$x + 3y \underset{\substack{\uparrow \\ x = 3,\, y = -1}}{=} 3 + 3(-1) = 3 + (-3) = 0$$

(b) If $x = 3$ and $y = -1$, then

$$5xy = 5(3)(-1) = -15$$

(c) If $x = 3$ and $y = -1$, then

$$\frac{3y}{2 - 2x} = \frac{3(-1)}{2 - 2(3)} = \frac{-3}{2 - 6} = \frac{-3}{-4} = \frac{3}{4}$$

(d) If $x = 3$ and $y = -1$, then

$$|-4x + y| = |-4(3) + (-1)| = |-12 + (-1)| = |-13| = 13$$

A graphing calculator can be used to evaluate an algebraic expression. Figure 17 shows the results of Example 5 using a T1-84 Plus.

Figure 17

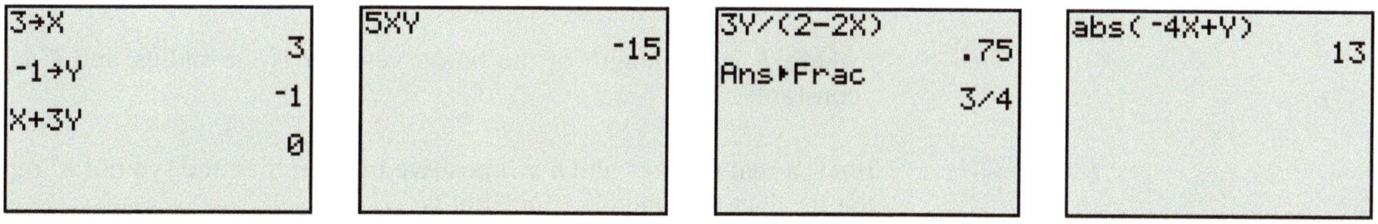

(a) (b) (c) (d)

Now Work PROBLEMS 39 AND 47

4 Determine the Domain of a Variable

In working with expressions or formulas involving variables, the variables may be allowed to take on values from only a certain set of numbers. For example, in the formula for the area A of a circle of radius r, $A = \pi r^2$, the variable r is restricted to the positive real numbers since a radius cannot be negative or zero. In the expression $\frac{1}{x}$, the variable x cannot take on the value 0, since division by 0 is not defined.

DEFINITION The set of values that a variable may assume is called the **domain of the variable.**

EXAMPLE 6 Finding the Domain of a Variable

The domain of the variable x in the expression

$$\frac{5}{x - 2}$$

is $\{x|x \neq 2\}$, since, if $x = 2$, the denominator becomes 0, which is not defined.

EXAMPLE 7 Circumference of a Circle

In the formula for the circumference C of a circle of radius r,

$$C = 2\pi r$$

the domain of the variable r, representing the radius of the circle, is the set of positive real numbers. The domain of the variable C, representing the circumference of the circle, is also the set of positive real numbers.

In describing the domain of a variable, we may use either set notation or words, whichever is more convenient.

Now Work PROBLEM 57

5 Use the Laws of Exponents

Integer exponents provide a shorthand device for representing repeated multiplications of a real number. For example,

$$3^4 = 3 \cdot 3 \cdot 3 \cdot 3 = 81$$

Many formulas have exponents. For example,

- The formula for the horsepower rating H of an engine is

$$H = \frac{D^2 N}{2.5}$$

where D is the diameter of a cylinder and N is the number of cylinders.

- A formula for the resistance R of blood flowing in a blood vessel is

$$R = C\frac{L}{r^4}$$

where L is the length of the blood vessel, r is the radius, and C is a positive constant.

DEFINITION

If a is a real number and n is a positive integer, then the symbol $\boldsymbol{a^n}$ represents the product of n factors of a. That is,

$$a^n = \underbrace{a \cdot a \cdot \ldots \cdot a}_{n \text{ factors}} \tag{1}$$

Here it is understood that $a^1 = a$.

Then $a^2 = a \cdot a$, $a^3 = a \cdot a \cdot a$, and so on. In the expression a^n, a is called the **base** and n is called the **exponent,** or **power.** We read a^n as "a raised to the power n" or as "a to the nth power." We usually read a^2 as "a squared" and a^3 as "a cubed."

In working with exponents, the operation of *raising to a power* is performed before any other operation. As examples,

$$4 \cdot 3^2 = 4 \cdot 9 = 36 \qquad 2^2 + 3^2 = 4 + 9 = 13$$
$$-2^4 = -16 \qquad 5 \cdot 3^2 + 2 \cdot 4 = 5 \cdot 9 + 2 \cdot 4 = 45 + 8 = 53$$

Parentheses are used to indicate operations to be performed first. For example,

$$(-2)^4 = (-2)(-2)(-2)(-2) = 16 \qquad (2 + 3)^2 = 5^2 = 25$$

NOTE Be careful with negatives and exponents.

$$-2^4 = -1 \cdot 2^4 = -16$$

whereas

$$(-2)^4 = (-2)(-2)(-2)(-2) = 16$$

DEFINITION

If $a \neq 0$, we define

$$a^0 = 1 \quad \text{if } a \neq 0$$

DEFINITION

If $a \neq 0$ and if n is a positive integer, then we define

$$a^{-n} = \frac{1}{a^n} \quad \text{if } a \neq 0$$

Whenever you encounter a negative exponent, think "reciprocal."

EXAMPLE 8 **Evaluating Expressions Containing Negative Exponents**

(a) $2^{-3} = \frac{1}{2^3} = \frac{1}{8}$ (b) $x^{-4} = \frac{1}{x^4}$ (c) $\left(\frac{1}{5}\right)^{-2} = \frac{1}{\left(\frac{1}{5}\right)^2} = \frac{1}{\frac{1}{25}} = 25$

Now Work PROBLEMS 75 AND 95

The following properties, called the **Laws of Exponents,** can be proved using the preceding definitions. In the list, a and b are real numbers, and m and n are integers.

THEOREM

Laws of Exponents

$$a^m a^n = a^{m+n} \qquad (a^m)^n = a^{mn} \qquad (ab)^n = a^n b^n$$

$$\frac{a^m}{a^n} = a^{m-n} = \frac{1}{a^{n-m}} \quad \text{if } a \neq 0 \qquad \left(\frac{a}{b}\right)^n = \frac{a^n}{b^n} \quad \text{if } b \neq 0$$

EXAMPLE 9 Using the Laws of Exponents

(a) $x^{-3} \cdot x^5 = x^{-3+5} = x^2 \quad x \neq 0$

(b) $(x^{-3})^2 = x^{-3 \cdot 2} = x^{-6} = \dfrac{1}{x^6} \quad x \neq 0$

(c) $(2x)^3 = 2^3 \cdot x^3 = 8x^3$

(d) $\left(\dfrac{2}{3}\right)^4 = \dfrac{2^4}{3^4} = \dfrac{16}{81}$

(e) $\dfrac{x^{-2}}{x^{-5}} = x^{-2-(-5)} = x^3 \quad x \neq 0$

Now Work PROBLEM 77

EXAMPLE 10 Using the Laws of Exponents

Write each expression so that all exponents are positive.

(a) $\dfrac{x^5 y^{-2}}{x^3 y} \quad x \neq 0, \quad y \neq 0$ (b) $\left(\dfrac{x^{-3}}{3y^{-1}}\right)^{-2} \quad x \neq 0, \quad y \neq 0$

Solution

(a) $\dfrac{x^5 y^{-2}}{x^3 y} = \dfrac{x^5}{x^3} \cdot \dfrac{y^{-2}}{y} = x^{5-3} \cdot y^{-2-1} = x^2 y^{-3} = x^2 \cdot \dfrac{1}{y^3} = \dfrac{x^2}{y^3}$

(b) $\left(\dfrac{x^{-3}}{3y^{-1}}\right)^{-2} = \dfrac{(x^{-3})^{-2}}{(3y^{-1})^{-2}} = \dfrac{x^6}{3^{-2}(y^{-1})^{-2}} = \dfrac{x^6}{\frac{1}{9}y^2} = \dfrac{9x^6}{y^2}$

Now Work PROBLEM 87

6 Evaluate Square Roots

A real number is squared when it is raised to the power 2. The inverse of squaring is finding a **square root.** For example, since $6^2 = 36$ and $(-6)^2 = 36$, the numbers 6 and -6 are square roots of 36.

The symbol $\sqrt{}$, called a **radical sign,** is used to denote the **principal,** or nonnegative, square root. For example, $\sqrt{36} = 6$. We read $\sqrt{36}$ as "give me the nonnegative number whose square is 36."

DEFINITION

If a is a nonnegative real number, the nonnegative number b for which $b^2 = a$ is the **principal square root** of a, and is denoted by $b = \sqrt{a}$.

In Words
The symbol $\sqrt{a}$ means "give me the nonnegative number whose square is a."

The following comments are noteworthy:

1. Negative numbers do not have square roots (in the real number system), because the square of any real number is *nonnegative.* For example, $\sqrt{-4}$ is not a real number, because there is no real number whose square is -4.
2. The principal square root of 0 is 0, since $0^2 = 0$. That is, $\sqrt{0} = 0$.
3. The principal square root of a positive number is positive.
4. If $c \geq 0$, then $(\sqrt{c})^2 = c$. For example, $(\sqrt{2})^2 = 2$ and $(\sqrt{3})^2 = 3$.

EXAMPLE 11 Evaluating Square Roots

(a) $\sqrt{64} = 8$ (b) $\sqrt{\dfrac{1}{16}} = \dfrac{1}{4}$ (c) $(\sqrt{1.4})^2 = 1.4$

Examples 11(a) and (b) are examples of square roots of perfect squares, since $64 = 8^2$ and $\frac{1}{16} = \left(\frac{1}{4}\right)^2$.

Consider the expression $\sqrt{a^2}$. Since $a^2 \geq 0$, the principal square root of a^2 is defined whether $a > 0$ or $a < 0$. However, since the principal square root is nonnegative, we need an absolute value to ensure the nonnegative result. That is,

$$\sqrt{a^2} = |a| \quad a \text{ any real number} \tag{2}$$

EXAMPLE 12 **Using Equation (2)**

(a) $\sqrt{(2.3)^2} = |2.3| = 2.3$

(b) $\sqrt{(-2.3)^2} = |-2.3| = 2.3$

(c) $\sqrt{x^2} = |x|$

Now Work PROBLEM 83

7 Use a Graphing Calculator to Evaluate Exponents

Your calculator has either a caret key, $\boxed{\wedge}$ or an $\boxed{x^y}$ key, which is used for computations involving exponents.

EXAMPLE 13 **Exponents on a Graphing Calculator**

Evaluate: $(2.3)^5$

Solution Figure 18 shows the result using a TI-84 graphing calculator.

Figure 18

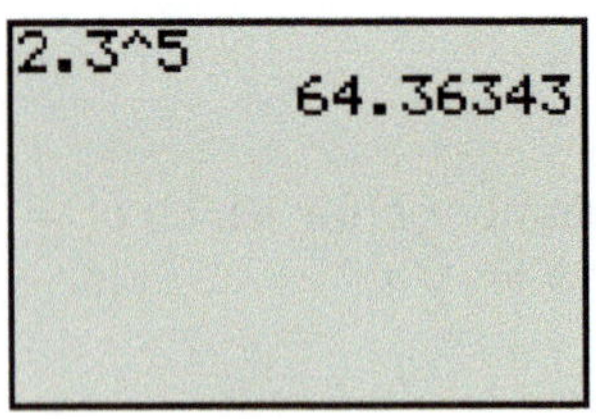

Now Work PROBLEM 113

8 Use Scientific Notation

Measurements of physical quantities can range from very small to very large. For example, the mass of a proton is approximately 0.00000000000000000000000000167 kilogram and the mass of Earth is about 5,980,000,000,000,000,000,000,000 kilograms. These numbers obviously are tedious to write down and difficult to read, so we use exponents to rewrite each.

DEFINITION When a number has been written as the product of a number x, where $1 \leq x < 10$, times a power of 10, it is said to be written in **scientific notation.**

In scientific notation,

$$\text{Mass of a proton} = 1.67 \times 10^{-27} \text{ kilogram}$$
$$\text{Mass of Earth} = 5.98 \times 10^{24} \text{ kilograms}$$

Converting a Decimal to Scientific Notation

To change a positive number into scientific notation:

1. Count the number N of places that the decimal point must be moved to arrive at a number x, where $1 \le x < 10$.
2. If the original number is greater than or equal to 1, the scientific notation is $x \times 10^N$. If the original number is between 0 and 1, the scientific notation is $x \times 10^{-N}$.

EXAMPLE 14 Using Scientific Notation

Write each number in scientific notation.

(a) 9582 (b) 1.245 (c) 0.285 (d) 0.000561

Solution (a) The decimal point in 9582 follows the 2. We count left from the decimal point

9 5 8 2 .
3 2 1

stopping after three moves, because 9.582 is a number between 1 and 10. Since 9582 is greater than 1, we write

$$9582 = 9.582 \times 10^3$$

(b) The decimal point in 1.245 is between the 1 and 2. Since the number is already between 1 and 10, the scientific notation for it is $1.245 \times 10^0 = 1.245$.

(c) The decimal point in 0.285 is between the 0 and the 2. We count

0 . 2 8 5
1

stopping after one move, because 2.85 is a number between 1 and 10. Since 0.285 is between 0 and 1, we write

$$0.285 = 2.85 \times 10^{-1}$$

(d) The decimal point in 0.000561 is moved as follows:

0 . 0 0 0 5 6 1
1 2 3 4

As a result,

$$0.000561 = 5.61 \times 10^{-4}$$

Now Work PROBLEM 119

EXAMPLE 15 Changing from Scientific Notation to Decimals

Write each number as a decimal.

(a) 2.1×10^4 (b) 3.26×10^{-5} (c) 1×10^{-2}

Solution (a) $2.1 \times 10^4 = 2\ .\ 1\ 0\ 0\ 0 \times 10^4 = 21{,}000$
1 2 3 4

(b) $3.26 \times 10^{-5} = 0\ 0\ 0\ 0\ 0\ 3\ .\ 2\ 6 \times 10^{-5} = 0.0000326$
5 4 3 2 1

(c) $1 \times 10^{-2} = 0\ 0\ 1\ .\ 0 \times 10^{-2} = 0.01$
2 1

Now Work PROBLEM 127

EXAMPLE 16 **Using Scientific Notation**

(a) The diameter of the smallest living cell is only about 0.00001 centimeter (cm).* Express this number in scientific notation.

(b) The surface area of Earth is about 1.97×10^8 square miles.† Express the surface area as a whole number.

Solution (a) $0.00001 \text{ cm} = 1 \times 10^{-5} \text{ cm}$ because the decimal point is moved five places and the number is less than 1.

(b) 1.97×10^8 square miles $=$ 197,000,000 square miles.

Now Work PROBLEM 153

COMMENT On a calculator, a number such as 3.615×10^{12} is usually displayed as 3.615E12.

Powers of Ten, Philip and Phylis Morrison.

†*1998 Information Please Almanac.*

Historical Feature

The word *algebra* is derived from the Arabic word *al-jabr*. This word is a part of the title of a ninth century work, "Hisâb al-jabr w'al-muqâbalah," written by Mohammed ibn Músâ al-Khowârizmî. The word *al-jabr* means "a restoration," a reference to the fact that, if a number is added to one side of an equation, then it must also be added to the other side in order to "restore" the equality. The title of the work, freely translated, is "The Science of Reduction and Cancellation." Of course, today, algebra has come to mean a great deal more.

R.2 Assess Your Understanding

Concepts and Vocabulary

1. A(n) __________ is a letter used in algebra to represent any number from a given set of numbers.
2. On the real number line, the real number zero is the coordinate of the __________.
3. An inequality of the form $a > b$ is called a(n) __________ inequality.
4. In the expression 2^4, the number 2 is called the __________ and 4 is called the __________.
5. In scientific notation, $1234.5678 =$ __________.
6. ***True or False*** The product of two negative real numbers is always greater than zero.
7. ***True or False*** The distance between two distinct points on the real number line is always greater than zero.
8. ***True or False*** The absolute value of a real number is always greater than zero.
9. ***True or False*** When a number is expressed in scientific notation, it is expressed as the product of a number x, $0 \le x < 1$, and a power of 10.
10. ***True or False*** To multiply two expressions having the same base, retain the base and multiply the exponents.

Skill Building

11. On the real number line, label the points with coordinates $0, 1, -1, \frac{5}{2}, -2.5, \frac{3}{4}$, and 0.25.

12. On the real number line, label the points with coordinates $0, -2, 2, -1.5, \frac{3}{2}, \frac{1}{3}$, and $\frac{2}{3}$.

In Problems 13–22, replace the question mark by <, >, or =, whichever is correct.

13. $\frac{1}{2}$? 0
14. 5 ? 6
15. −1 ? −2
16. $-3 ? -\frac{5}{2}$
17. π ? 3.14
18. $\sqrt{2}$? 1.41
19. $\frac{1}{2}$? 0.5
20. $\frac{1}{3}$? 0.33
21. $\frac{2}{3}$? 0.67
22. $\frac{1}{4}$? 0.25

In Problems 23–28, write each statement as an inequality.

23. x is positive **24.** z is negative **25.** x is less than 2

26. y is greater than -5 **27.** x is less than or equal to 1 **28.** x is greater than or equal to 2

In Problems 29–32, graph the numbers x on the real number line.

29. $x \geq -2$ **30.** $x < 4$ **31.** $x > -1$ **32.** $x \leq 7$

In Problems 33–38, use the given real number line to compute each distance.

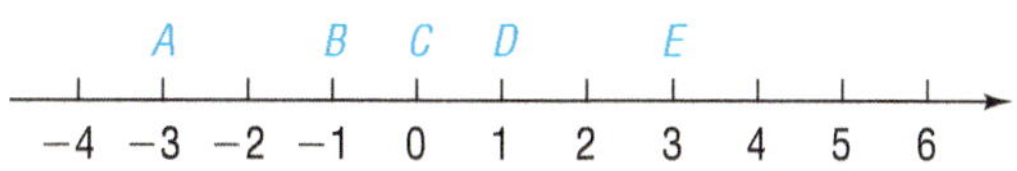

33. $d(C, D)$ **34.** $d(C, A)$ **35.** $d(D, E)$ **36.** $d(C, E)$ **37.** $d(A, E)$ **38.** $d(D, B)$

In Problems 39–46, evaluate each expression if $x = -2$ and $y = 3$.

39. $x + 2y$ **40.** $3x + y$ **41.** $5xy + 2$ **42.** $-2x + xy$

43. $\dfrac{2x}{x - y}$ **44.** $\dfrac{x + y}{x - y}$ **45.** $\dfrac{3x + 2y}{2 + y}$ **46.** $\dfrac{2x - 3}{y}$

In Problems 47–56, find the value of each expression if $x = 3$ and $y = -2$.

47. $|x + y|$ **48.** $|x - y|$ **49.** $|x| + |y|$ **50.** $|x| - |y|$ **51.** $\dfrac{|x|}{x}$

52. $\dfrac{|y|}{y}$ **53.** $|4x - 5y|$ **54.** $|3x + 2y|$ **55.** $||4x| - |5y||$ **56.** $3|x| + 2|y|$

In Problems 57–64, determine which of the value(s) (a) through (d), if any, must be excluded from the domain of the variable in each expression:

(a) $x = 3$ (b) $x = 1$ (c) $x = 0$ (d) $x = -1$

57. $\dfrac{x^2 - 1}{x}$ **58.** $\dfrac{x^2 + 1}{x}$ **59.** $\dfrac{x}{x^2 - 9}$ **60.** $\dfrac{x}{x^2 + 9}$

61. $\dfrac{x^2}{x^2 + 1}$ **62.** $\dfrac{x^3}{x^2 - 1}$ **63.** $\dfrac{x^2 + 5x - 10}{x^3 - x}$ **64.** $\dfrac{-9x^2 - x + 1}{x^3 + x}$

In Problems 65–68, determine the domain of the variable x in each expression.

65. $\dfrac{4}{x - 5}$ **66.** $\dfrac{-6}{x + 4}$ **67.** $\dfrac{x}{x + 4}$ **68.** $\dfrac{x - 2}{x - 6}$

In Problems 69–72, use the formula $C = \frac{5}{9}(F - 32)$ for converting degrees Fahrenheit into degrees Celsius to find the Celsius measure of each Fahrenheit temperature.

69. $F = 32°$ **70.** $F = 212°$ **71.** $F = 77°$ **72.** $F = -4°$

In Problems 73–84, simplify each expression.

73. $(-4)^2$ **74.** -4^2 **75.** 4^{-2} **76.** -4^{-2} **77.** $3^{-6} \cdot 3^4$ **78.** $4^{-2} \cdot 4^3$

79. $(3^{-2})^{-1}$ **80.** $(2^{-1})^{-3}$ **81.** $\sqrt{25}$ **82.** $\sqrt{36}$ **83.** $\sqrt{(-4)^2}$ **84.** $\sqrt{(-3)^2}$

In Problems 85–94, simplify each expression. Express the answer so that all exponents are positive. Whenever an exponent is 0 or negative, we assume that the base is not 0.

85. $(8x^3)^2$ **86.** $(-4x^2)^{-1}$ **87.** $(x^2y^{-1})^2$ **88.** $(x^{-1}y)^3$ **89.** $\dfrac{x^2y^3}{xy^4}$

90. $\dfrac{x^{-2}y}{xy^2}$ **91.** $\dfrac{(-2)^3x^4(yz)^2}{3^2xy^3z}$ **92.** $\dfrac{4x^{-2}(yz)^{-1}}{2^3x^4y}$ **93.** $\left(\dfrac{3x^{-1}}{4y^{-1}}\right)^{-2}$ **94.** $\left(\dfrac{5x^{-2}}{6y^{-2}}\right)^{-3}$

In Problems 95–106, find the value of each expression if $x = 2$ and $y = -1$.

95. $2xy^{-1}$ **96.** $-3x^{-1}y$ **97.** $x^2 + y^2$ **98.** x^2y^2

99. $(xy)^2$ **100.** $(x + y)^2$ **101.** $\sqrt{x^2}$ **102.** $(\sqrt{x})^2$

103. $\sqrt{x^2 + y^2}$ **104.** $\sqrt{x^2} + \sqrt{y^2}$ **105.** x^y **106.** y^x

107. Find the value of the expression $2x^3 - 3x^2 + 5x - 4$ if $x = 2$. What is the value if $x = 1$?

108. Find the value of the expression $4x^3 + 3x^2 - x + 2$ if $x = 1$. What is the value if $x = 2$?

109. What is the value of $\dfrac{(666)^4}{(222)^4}$?

110. What is the value of $(0.1)^3(20)^3$?

In Problems 111–118, use a graphing calculator to evaluate each expression. Round your answer to three decimal places.

111. $(8.2)^6$ **112.** $(3.7)^5$ **113.** $(6.1)^{-3}$ **114.** $(2.2)^{-5}$

115. $(-2.8)^6$ **116.** $-(2.8)^6$ **117.** $(-8.11)^{-4}$ **118.** $-(8.11)^{-4}$

In Problems 119–126, write each number in scientific notation.

119. 454.2 **120.** 32.14 **121.** 0.013 **122.** 0.00421

123. 32,155 **124.** 21,210 **125.** 0.000423 **126.** 0.0514

In Problems 127–134, write each number as a decimal.

127. 6.15×10^4 **128.** 9.7×10^3 **129.** 1.214×10^{-3} **130.** 9.88×10^{-4}

131. 1.1×10^8 **132.** 4.112×10^2 **133.** 8.1×10^{-2} **134.** 6.453×10^{-1}

Applications and Extensions

In Problems 135–144, express each statement as an equation involving the indicated variables.

135. Area of a Rectangle The area A of a rectangle is the product of its length l and its width w.

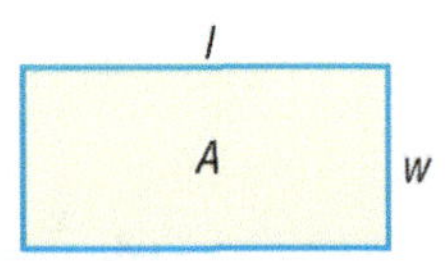

136. Perimeter of a Rectangle The perimeter P of a rectangle is twice the sum of its length l and its width w.

137. Circumference of a Circle The circumference C of a circle is the product of π and its diameter d.

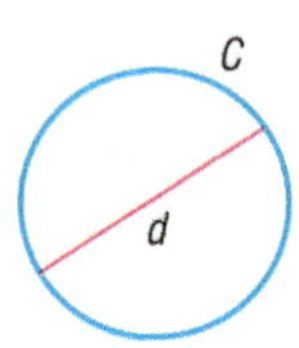

138. Area of a Triangle The area A of a triangle is one-half the product of its base b and its height h.

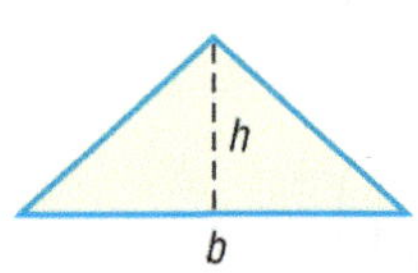

139. Area of an Equilateral Triangle The area A of an equilateral triangle is $\dfrac{\sqrt{3}}{4}$ times the square of the length x of one side.

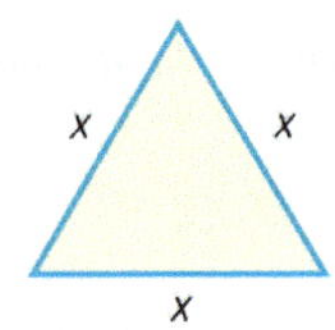

140. Perimeter of an Equilateral Triangle The perimeter P of an equilateral triangle is 3 times the length x of one side.

141. Volume of a Sphere The volume V of a sphere is $\dfrac{4}{3}$ times π times the cube of the radius r.

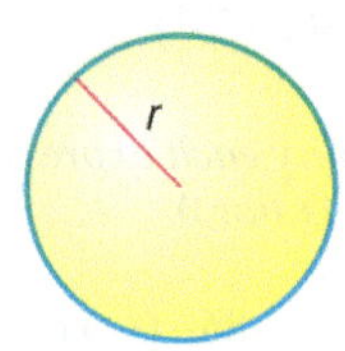

142. Surface Area of a Sphere The surface area S of a sphere is 4 times π times the square of the radius r.

143. Volume of a Cube The volume V of a cube is the cube of the length x of a side.

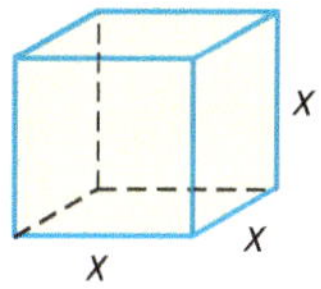

144. Surface Area of a Cube The surface area S of a cube is 6 times the square of the length x of a side.

145. Manufacturing Cost The weekly production cost C of manufacturing x watches is given by the formula $C = 4000 + 2x$, where the variable C is in dollars.

(a) What is the cost of producing 1000 watches?

(b) What is the cost of producing 2000 watches?

146. Balancing a Checkbook At the beginning of the month, Mike had a balance of \$210 in his checking account. During the next month, he deposited \$80, wrote a check for \$120, made another deposit of \$25, wrote two checks: one for \$60 and the other for \$32. He was also assessed a monthly service charge of \$5. What was his balance at the end of the month?

In Problems 147 and 148, write an inequality using an absolute value to describe each statement.

147. x is at least 6 units from 4.

148. x is more than 5 units from 2.

149. U.S. Voltage In the United States, normal household voltage is 110 volts. It is acceptable for the actual voltage x to differ from normal by at most 5 volts. A formula that describes this is

$$|x - 110| \leq 5$$

(a) Show that a voltage of 108 volts is acceptable.

(b) Show that a voltage of 104 volts is not acceptable.

150. Foreign Voltage In other countries, normal household voltage is 220 volts. It is acceptable for the actual voltage x to differ from normal by at most 8 volts. A formula that describes this is

$$|x - 220| \leq 8$$

(a) Show that a voltage of 214 volts is acceptable.

(b) Show that a voltage of 209 volts is not acceptable.

151. Making Precision Ball Bearings The FireBall Company manufactures ball bearings for precision equipment. One of its products is a ball bearing with a stated radius of 3 centimeters (cm). Only ball bearings with a radius within 0.01 cm of this stated radius are acceptable. If x is the radius of a ball bearing, a formula describing this situation is

$$|x - 3| \leq 0.01$$

(a) Is a ball bearing of radius $x = 2.999$ acceptable?

(b) Is a ball bearing of radius $x = 2.89$ acceptable?

152. Body Temperature Normal human body temperature is 98.6°F. A temperature x that differs from normal by at least 1.5°F is considered unhealthy. A formula that describes this is

$$|x - 98.6| \geq 1.5$$

(a) Show that a temperature of 97°F is unhealthy.

(b) Show that a temperature of 100°F is not unhealthy.

153. Distance from Earth to Its Moon The distance from Earth to the Moon is about 4×10^8 meters.* Express this distance as a whole number.

154. Height of Mt. Everest The height of Mt. Everest is 8872 meters.* Express this height in scientific notation.

155. Wavelength of Visible Light The wavelength of visible light is about 5×10^{-7} meter.* Express this wavelength as a decimal.

156. Diameter of an Atom The diameter of an atom is about 1×10^{-10} meter.* Express this diameter as a decimal.

157. Diameter of Copper Wire The smallest commercial copper wire is about 0.0005 inch in diameter.† Express this diameter using scientific notation.

158. Smallest Motor The smallest motor ever made is less than 0.05 centimeter wide.† Express this width using scientific notation.

159. Astronomy One light-year is defined by astronomers to be the distance that a beam of light will travel in 1 year (365 days). If the speed of light is 186,000 miles per second, how many miles are in a light-year? Express your answer in scientific notation.

160. Astronomy See Problem 159. How long does it take a beam of light to reach Earth from the Sun when the Sun is 93,000,000 miles from Earth? Express your answer in seconds, using scientific notation.

161. Does $\frac{1}{3}$ equal 0.333? If not, which is larger? By how much?

162. Does $\frac{2}{3}$ equal 0.666? If not, which is larger? By how much?

Discussion and Writing

163. Is there a positive real number "closest" to 0?

164. **Number game** I'm thinking of a number! It lies between 1 and 10; its square is rational and lies between 1 and 10. The number is larger than π. Correct to two decimal places (that is, truncated to two decimal places) name the number. Now think of your own number, describe it, and challenge a fellow student to name it.

165. Write a brief paragraph that illustrates the similarities and differences between "less than" ($<$) and "less than or equal to" ($\leq$).

166. Give a reason why the statement $5 < 8$ is true.

**Powers of Ten,* Philip and Phylis Morrison.
†*1998 Information Please Almanac.*

R.3 Geometry Essentials

OBJECTIVES
1 Use the Pythagorean Theorem and Its Converse (p. 30)
2 Know Geometry Formulas (p. 31)
3 Understand Congruent Triangles and Similar Triangles (p. 32)

In this section we review some topics studied in geometry that we shall need for our study of algebra.

1 Use the Pythagorean Theorem and Its Converse

Figure 19

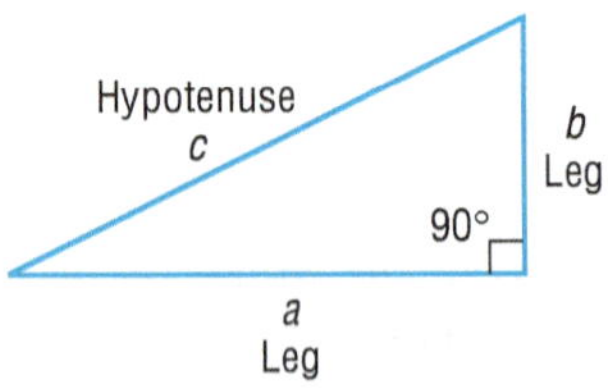

The *Pythagorean Theorem* is a statement about *right triangles*. A **right triangle** is one that contains a **right angle,** that is, an angle of 90°. The side of the triangle opposite the 90° angle is called the **hypotenuse;** the remaining two sides are called **legs.** In Figure 19 we have used c to represent the length of the hypotenuse and a and b to represent the lengths of the legs. Notice the use of the symbol ⌐ to show the 90° angle. We now state the Pythagorean Theorem.

PYTHAGOREAN THEOREM

In a right triangle, the square of the length of the hypotenuse is equal to the sum of the squares of the lengths of the legs. That is, in the right triangle shown in Figure 19,

$$c^2 = a^2 + b^2 \qquad \textbf{(1)}$$

A proof of the Pythagorean Theorem is given at the end of this section.

EXAMPLE 1 Finding the Hypotenuse of a Right Triangle

In a right triangle, one leg has length 4 and the other has length 3. What is the length of the hypotenuse?

Solution Since the triangle is a right triangle, we use the Pythagorean Theorem with $a = 4$ and $b = 3$ to find the length c of the hypotenuse. From equation (1), we have

$$c^2 = a^2 + b^2$$
$$c^2 = 4^2 + 3^2 = 16 + 9 = 25$$
$$c = \sqrt{25} = 5$$

Now Work PROBLEM 13

The converse of the Pythagorean Theorem is also true.

CONVERSE OF THE PYTHAGOREAN THEOREM

In a triangle, if the square of the length of one side equals the sum of the squares of the lengths of the other two sides, the triangle is a right triangle. The 90° angle is opposite the longest side.

A proof is given at the end of this section.

EXAMPLE 2 Verifying That a Triangle Is a Right Triangle

Show that a triangle whose sides are of lengths 5, 12, and 13 is a right triangle. Identify the hypotenuse.

Solution We square the lengths of the sides.

$$5^2 = 25, \qquad 12^2 = 144, \qquad 13^2 = 169$$

Figure 20

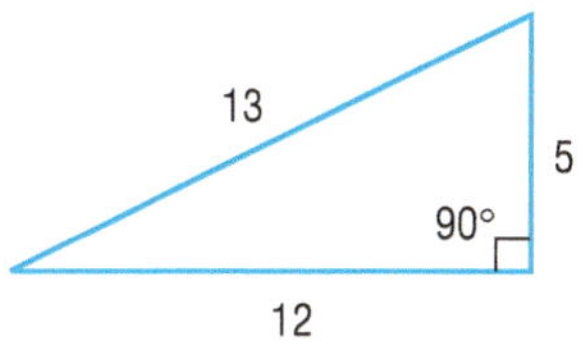

Notice that the sum of the first two squares (25 and 144) equals the third square (169). Hence, the triangle is a right triangle. The longest side, 13, is the hypotenuse. See Figure 20.

Now Work PROBLEM 21

EXAMPLE 3 Applying the Pythagorean Theorem

Excluding antenna, the tallest inhabited building in the world is Taipei 101 in Taipei, Taiwan. See Figure 21. If the indoor observation deck is 1437 feet above ground level, how far can a person standing on the observation deck see (with the aid of a telescope)? Use 3960 miles for the radius of Earth.

Source: Council on Tall Buildings and Urban Habitat, 2006.

Figure 21

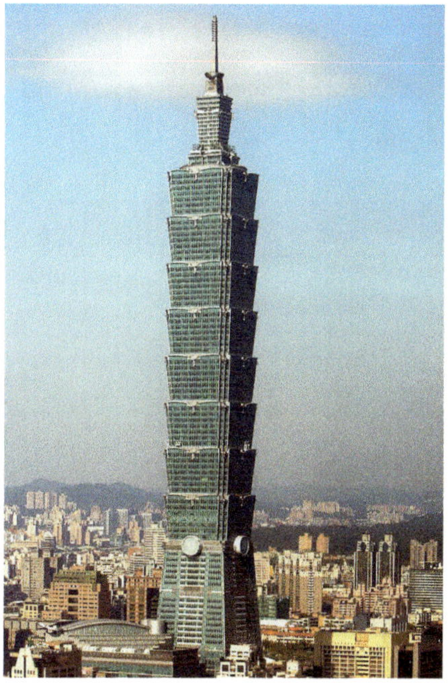

Solution From the center of Earth, draw two radii: one through Taipei 101 and the other to the farthest point a person can see from the observation deck. See Figure 22. Apply the Pythagorean Theorem to the right triangle.

Since 1 mile = 5280 feet, then 1437 feet $= \dfrac{1437}{5280}$ mile. So we have

$$d^2 + (3960)^2 = \left(3960 + \frac{1437}{5280}\right)^2$$

$$d^2 = \left(3960 + \frac{1437}{5280}\right)^2 - (3960)^2 \approx 2155.57$$

$$d \approx 46.43$$

A person can see about 46 miles from the observation tower.

Figure 22

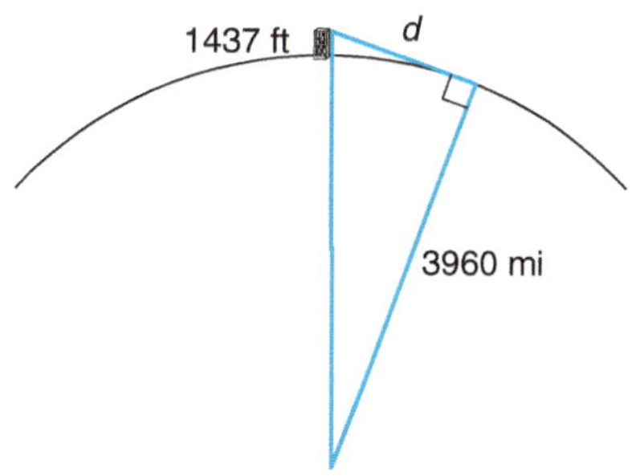

Now Work PROBLEM 53

2 Know Geometry Formulas

Certain formulas from geometry are useful in solving algebra problems. We list some of these next.

For a rectangle of length l and width w,

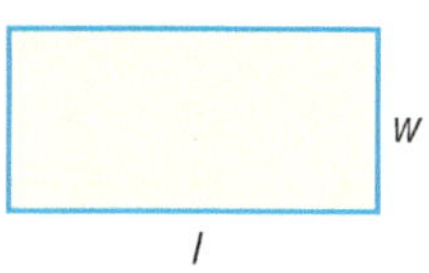

Area = lw Perimeter = $2l + 2w$

For a triangle with base b and altitude h,

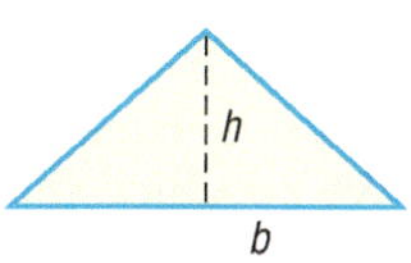

Area $= \dfrac{1}{2}bh$

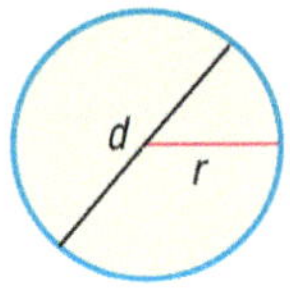

For a circle of radius r (diameter $d = 2r$),

$$\text{Area} = \pi r^2 \qquad \text{Circumference} = 2\pi r = \pi d$$

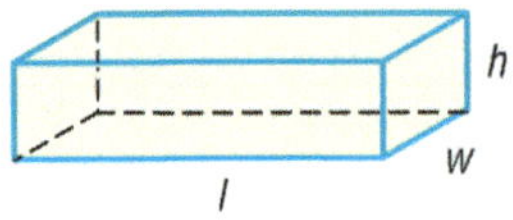

For a closed rectangular box of length l, width w, and height h,

$$\text{Volume} = lwh \qquad \text{Surface area} = 2lh + 2wh + 2lw$$

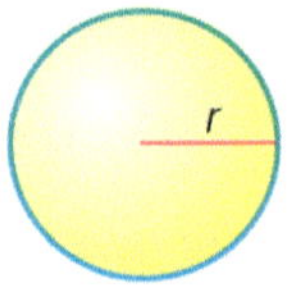

For a sphere of radius r,

$$\text{Volume} = \frac{4}{3}\pi r^3 \qquad \text{Surface area} = 4\pi r^2$$

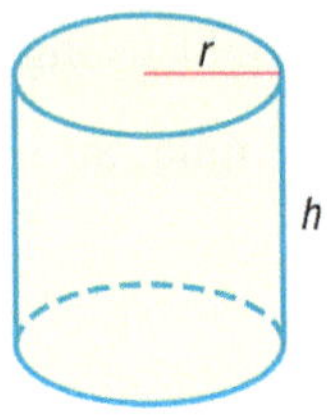

For a right circular cylinder of height h and radius r,

$$\text{Volume} = \pi r^2 h \qquad \text{Surface area} = 2\pi r^2 + 2\pi rh$$

Now Work PROBLEM 29

EXAMPLE 4 Using Geometry Formulas

A Christmas tree ornament is in the shape of a semicircle on top of a triangle. How many square centimeters (cm) of copper is required to make the ornament if the height of the triangle is 6 cm and the base is 4 cm?

Solution See Figure 23. The amount of copper required equals the shaded area. This area is the sum of the area of the triangle and the semicircle. The triangle has height $h = 6$ and base $b = 4$. The semicircle has diameter $d = 4$, so its radius is $r = 2$.

Figure 23

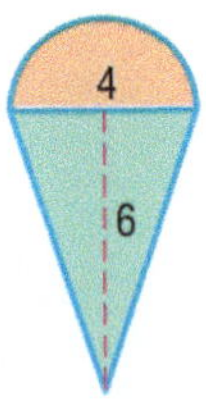

$$\begin{aligned} \text{Area} &= \text{Area of triangle} + \text{Area of semicircle} \\ &= \frac{1}{2}bh + \frac{1}{2}\pi r^2 = \frac{1}{2}(4)(6) + \frac{1}{2}\pi \cdot 2^2 \qquad b = 4;\ h = 6;\ r = 2 \\ &= 12 + 2\pi \approx 18.28\ \text{cm}^2 \end{aligned}$$

About 18.28 cm^2 of copper is required.

Now Work PROBLEM 47

3 Understand Congruent Triangles and Similar Triangles

Throughout the text we will make reference to triangles. We begin with a discussion of *congruent* triangles. According to dictionary.com, the word **congruent** means coinciding exactly when superimposed. For example, two angles are congruent if they have the same measure and two line segments are congruent if they have the same length.

In Words
Two triangles are congruent if they have the same size and shape.

DEFINITION

Two triangles are **congruent** if each of the corresponding angles is the same measure and each of the corresponding sides is the same length.

In Figure 24, corresponding angles are equal and the lengths of the corresponding sides are equal: $a = d$, $b = e$, and $c = f$. We conclude that these triangles are congruent.

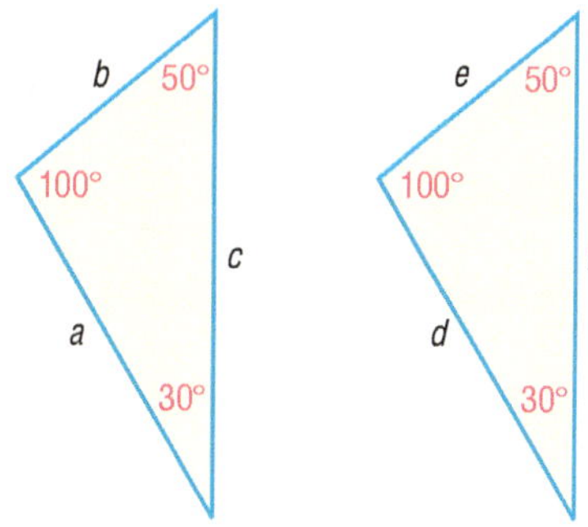

Figure 24
Congruent triangles

It is not necessary to verify that all three angles and all three sides are the same measure to determine whether two triangles are congruent.

Determining Congruent Triangles

1. **Angle–Side–Angle Case** Two triangles are congruent if two of the angles are equal and the lengths of the corresponding sides between the two angles are equal.
   For example, in Figure 25(a), the two triangles are congruent because two angles and the included side are equal.
2. **Side–Side–Side Case** Two triangles are congruent if the lengths of the corresponding sides of the triangles are equal.
   For example, in Figure 25(b), the two triangles are congruent because the three corresponding sides are all equal.
3. **Side–Angle–Side Case** Two triangles are congruent if the lengths of two corresponding sides are equal and the angles between the two sides are the same.
   For example, in Figure 25(c), the two triangles are congruent because two sides and the included angle are equal.

Figure 25

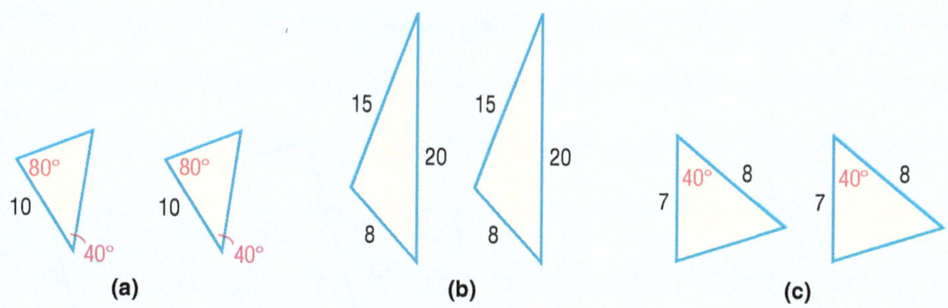

We contrast congruent triangles with *similar* triangles.

DEFINITION

Two triangles are **similar** if the corresponding angles are equal and the lengths of the corresponding sides are proportional.

In Words
Two triangles are similar if they have the same shape, but (possibly) different sizes.

For example, the triangles in Figure 26 on page 34 are similar because the corresponding angles are equal. In addition, the lengths of the corresponding sides are proportional because each side in the triangle on the right is twice as long as each corresponding side in the triangle on the left. That is, the ratio of the corresponding sides is a constant: $\frac{d}{a} = \frac{e}{b} = \frac{f}{c} = 2$.

Figure 26

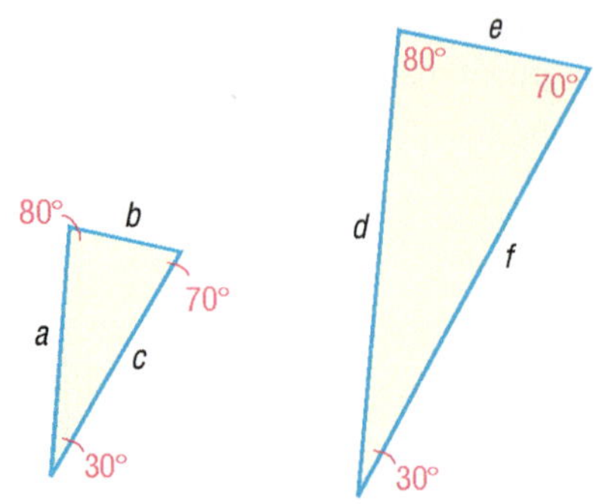

It is not necessary to verify that all three angles are equal and all three sides are proportional to determine whether two triangles are similar.

Determining Similar Triangles

1. **Angle–Angle Case** Two triangles are similar if two of the corresponding angles are equal.
   For example, in Figure 27(a), the two triangles are similar because two angles are equal.
2. **Side–Side–Side Case** Two triangles are similar if the lengths of all three sides of each triangle are proportional.
   For example, in Figure 27(b), the two triangles are similar because
   $$\frac{10}{30} = \frac{5}{15} = \frac{6}{18} = \frac{1}{3}.$$
3. **Side–Angle–Side Case** Two triangles are similar if two corresponding sides are proportional and the angles between the two sides are equal.
   For example, in Figure 27(c), the two triangles are similar because $\frac{4}{6} = \frac{12}{18} = \frac{2}{3}$ and the angles between the sides are equal.

Figure 27

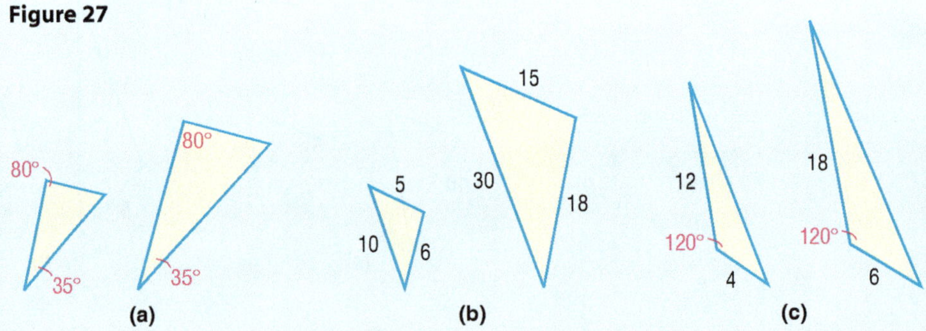

EXAMPLE 5 Using Similar Triangles

Given that the triangles in Figure 28 are similar, find the missing length x and the angles A, B, and C.

Figure 28

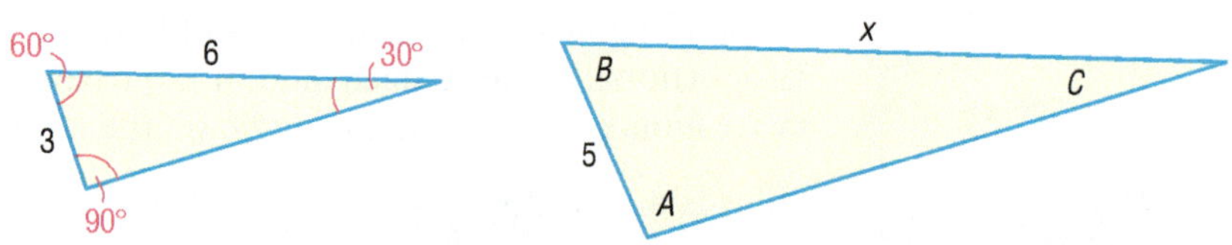

Solution Because the triangles are similar, corresponding angles are equal. So $A = 90°$, $B = 60°$, and $C = 30°$. Also, the corresponding sides are proportional. That is, $\frac{3}{5} = \frac{6}{x}$. We solve this equation for x.

$$\frac{3}{5} = \frac{6}{x}$$

$$5x \cdot \frac{3}{5} = 5x \cdot \frac{6}{x} \quad \text{Multiply both sides by } 5x.$$

$$3x = 30 \quad \text{Simplify.}$$

$$x = 10 \quad \text{Divide both sides by 3.}$$

The missing length is 10 units.

Now Work PROBLEM 41

Proof of the Pythagorean Theorem We begin with a square, each side of length $a + b$. In this square, we can form four right triangles, each having legs equal in length to a and b. See Figure 29. All these triangles are congruent (two sides and their included angle are equal). As a result, the hypotenuse of each is the same, say c, and the pink shading in Figure 29 indicates a square with an area equal to c^2.

Figure 29

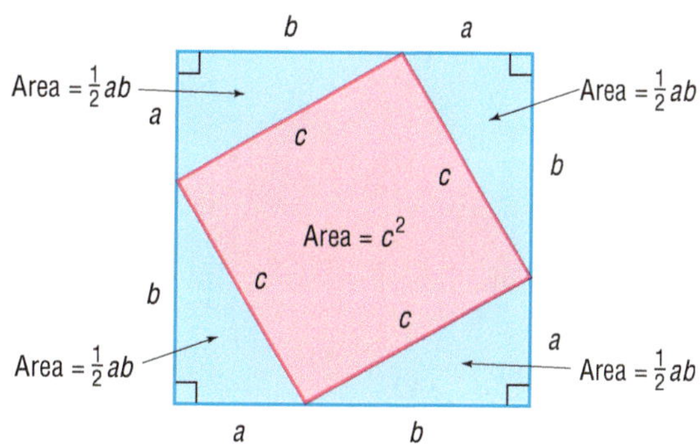

The area of the original square with sides $a + b$ equals the sum of the areas of the four triangles (each of area $\frac{1}{2}ab$) plus the area of the square with side c. That is,

$$(a + b)^2 = \frac{1}{2}ab + \frac{1}{2}ab + \frac{1}{2}ab + \frac{1}{2}ab + c^2$$

$$a^2 + 2ab + b^2 = 2ab + c^2$$

$$a^2 + b^2 = c^2$$

The proof is complete. ■

Figure 30

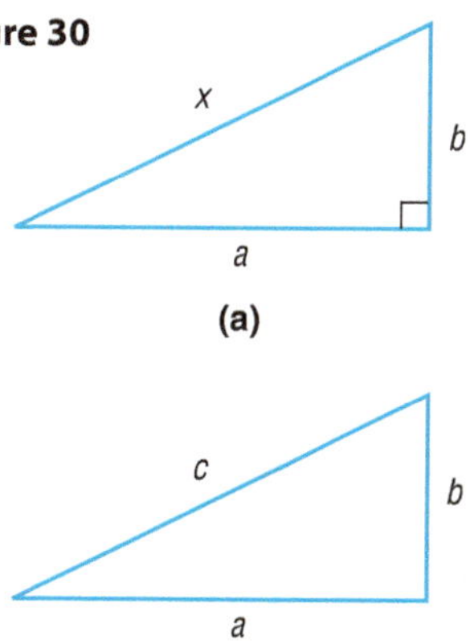

(b) $c^2 = a^2 + b^2$

Proof of the Converse of the Pythagorean Theorem We begin with two triangles: one a right triangle with legs a and b and the other a triangle with sides a, b, and c for which $c^2 = a^2 + b^2$. See Figure 30. By the Pythagorean Theorem, the length x of the third side of the first triangle is

$$x^2 = a^2 + b^2$$

But $c^2 = a^2 + b^2$. Hence,

$$x^2 = c^2$$

$$x = c$$

The two triangles have sides with the same length and are therefore congruent. This means corresponding angles are equal, so the angle opposite side c of the second triangle equals 90°.

The proof is complete. ■

R.3 Assess Your Understanding

Concepts and Vocabulary

1. A(n) ________ triangle is one that contains an angle of 90 degrees. The longest side is called the ________.

2. For a triangle with base b and altitude h, a formula for the area A is ________.

3. The formula for the circumference C of a circle of radius r is ________.

4. Two triangles are ________ if corresponding angles are equal and the lengths of the corresponding sides are proportional.

5. ***True or False*** In a right triangle, the square of the length of the longest side equals the sum of the squares of the lengths of the other two sides.

6. ***True or False*** The triangle with sides of length 6, 8, and 10 is a right triangle.

7. ***True or False*** The volume of a sphere of radius r is $\frac{4}{3}\pi r^2$.

8. ***True or False*** The triangles shown are congruent.

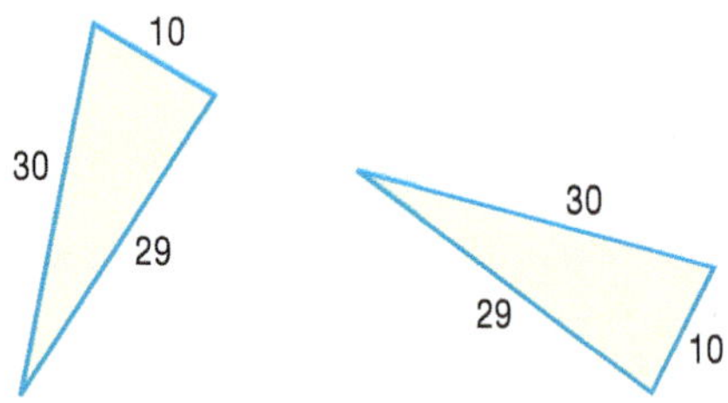

9. ***True or False*** The triangles shown are similar.

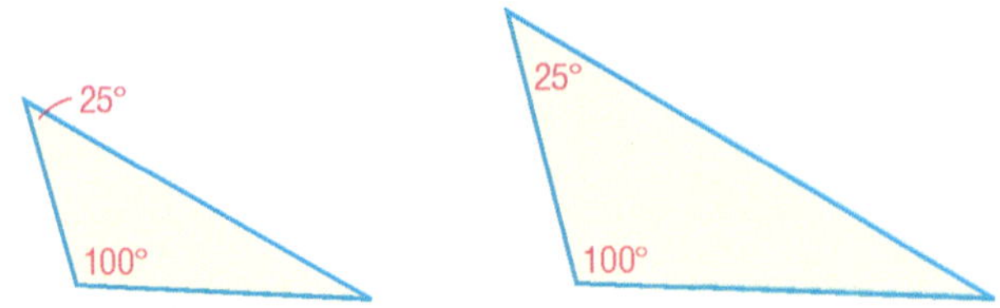

10. ***True or False*** The triangles shown are similar.

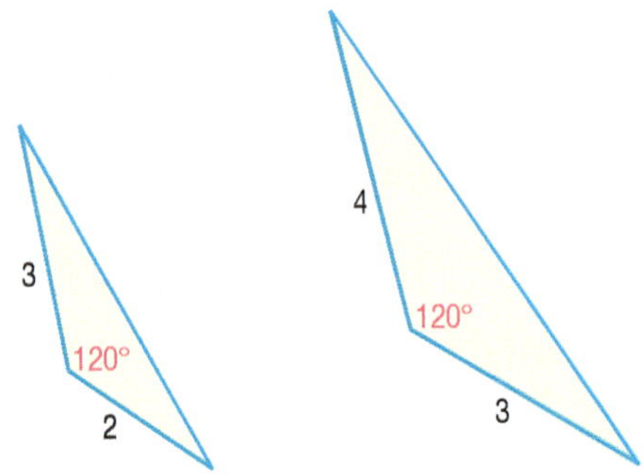

Skill Building

In Problems 11–16, the lengths of the legs of a right triangle are given. Find the length of the hypotenuse.

11. $a = 5, \quad b = 12$ **12.** $a = 6, \quad b = 8$ **13.** $a = 10, \quad b = 24$

14. $a = 4, \quad b = 3$ **15.** $a = 7, \quad b = 24$ **16.** $a = 14, \quad b = 48$

In Problems 17–24, the lengths of the sides of a triangle are given. Determine which are right triangles. For those that are, identify the hypotenuse.

17. 3, 4, 5 **18.** 6, 8, 10 **19.** 4, 5, 6 **20.** 2, 2, 3

21. 7, 24, 25 **22.** 10, 24, 26 **23.** 6, 4, 3 **24.** 5, 4, 7

25. Find the area A of a rectangle with length 4 inches and width 2 inches.

26. Find the area A of a rectangle with length 9 centimeters and width 4 centimeters.

27. Find the area A of a triangle with height 4 inches and base 2 inches.

28. Find the area A of a triangle with height 9 centimeters and base 4 centimeters.

29. Find the area A and circumference C of a circle of radius 5 meters.

30. Find the area A and circumference C of a circle of radius 2 feet.

31. Find the volume V and surface area S of a rectangular box with length 8 feet, width 4 feet, and height 7 feet.

32. Find the volume V and surface area S of a rectangular box with length 9 inches, width 4 inches, and height 8 inches.

33. Find the volume V and surface area S of a sphere of radius 4 centimeters.

34. Find the volume V and surface area S of a sphere of radius 3 feet.

35. Find the volume V and surface area S of a right circular cylinder with radius 9 inches and height 8 inches.

36. Find the volume V and surface area S of a right circular cylinder with radius 8 inches and height 9 inches.

In Problems 37–40, find the area of the shaded region.

37.

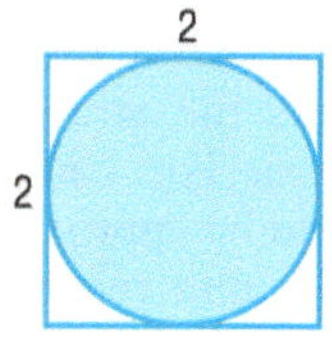

38.

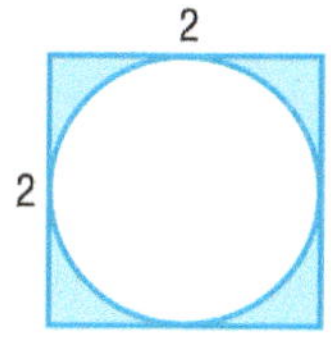

39.

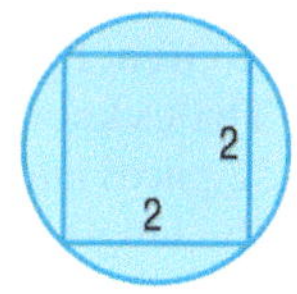

40.

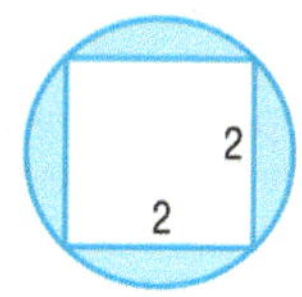

In Problems 41–44, each pair of triangles is similar. Find the missing length x and the missing angles A, B, and C.

41.

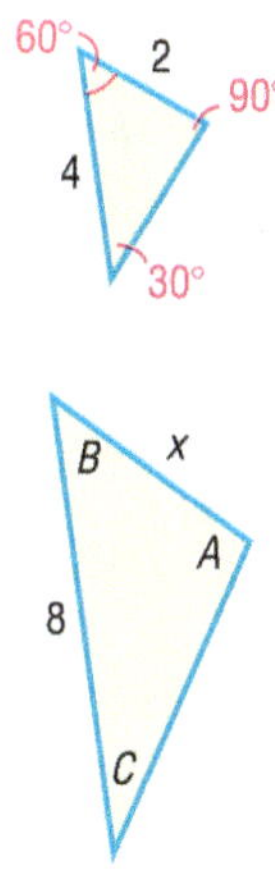

42.

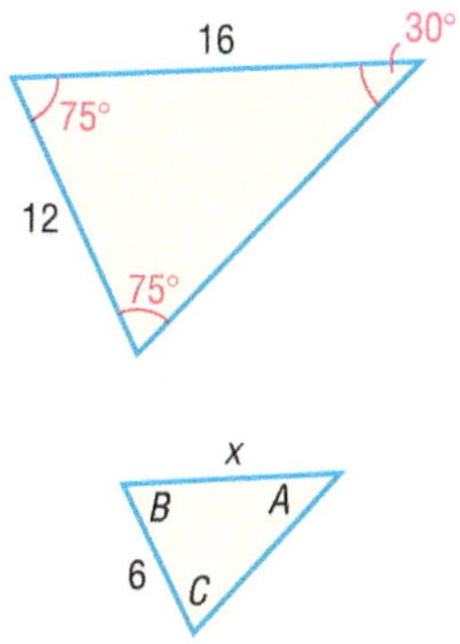

43.

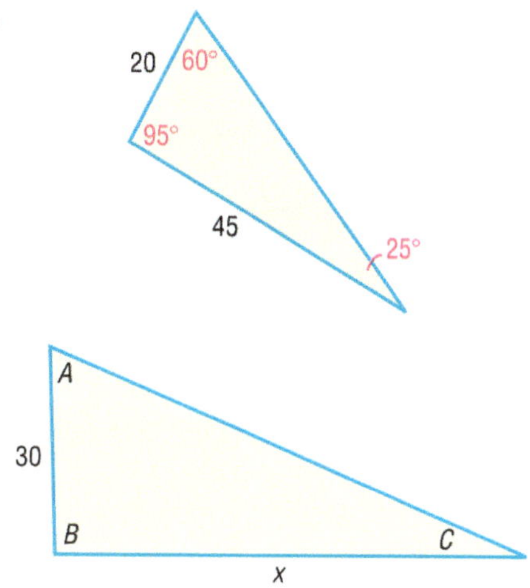

44.

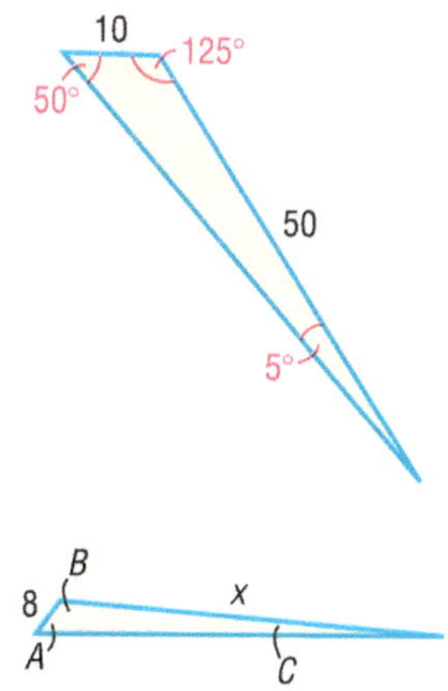

Applications and Extensions

45. How many feet does a wheel with a diameter of 16 inches travel after four revolutions?

46. How many revolutions will a circular disk with a diameter of 4 feet have completed after it has rolled 20 feet?

47. In the figure shown, $ABCD$ is a square, with each side of length 6 feet. The width of the border (shaded portion) between the outer square $EFGH$ and $ABCD$ is 2 feet. Find the area of the border.

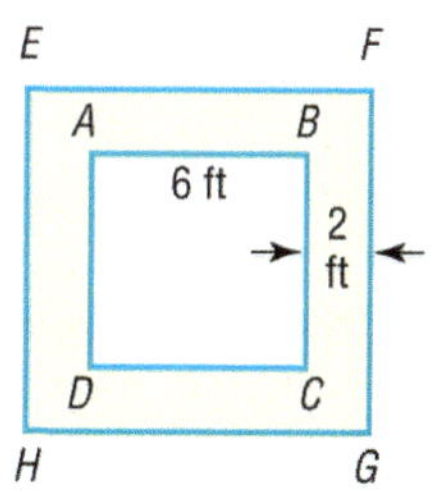

48. Refer to the figure. Square $ABCD$ has an area of 100 square feet; square $BEFG$ has an area of 16 square feet. What is the area of the triangle CGF?

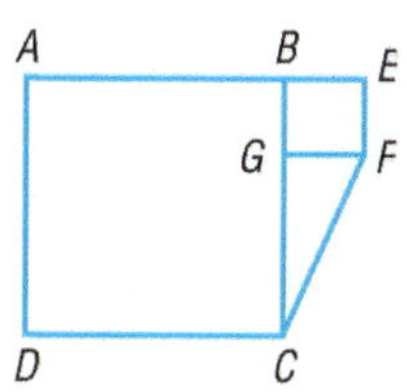

49. Architecture A **Norman window** consists of a rectangle surmounted by a semicircle. Find the area of the Norman window shown in the illustration. How much wood frame is needed to enclose the window?

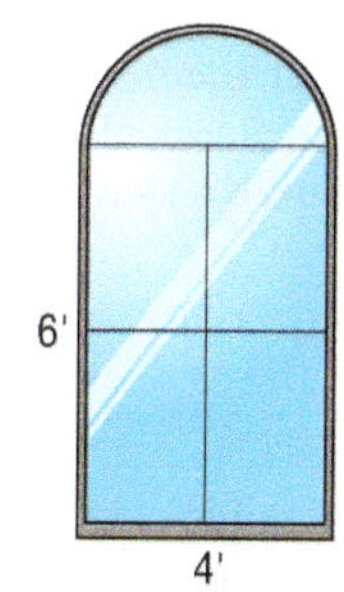

50. Construction A circular swimming pool, 20 feet in diameter, is enclosed by a wooden deck that is 3 feet wide. What is the area of the deck? How much fence is required to enclose the deck?

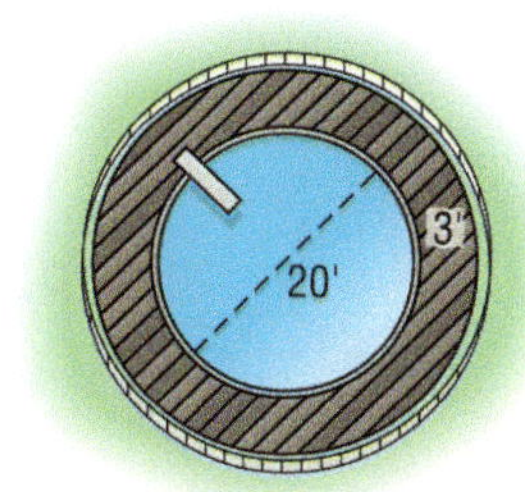

51. How Tall Is the Great Pyramid? The ancient Greek philosopher Thales of Miletus is reported on one occasion to have visited Egypt and calculated the height of the Great Pyramid of Cheops by means of shadow reckoning. Thales knew that each side of the base of the pyramid was 252 paces and that his own height was 2 paces. He measured the length of the pyramid's shadow to be 114 paces and determined the length of his shadow to be 3 paces. See the illustration. Using similar triangles, determine the height of the Great Pyramid in terms of the number of paces.

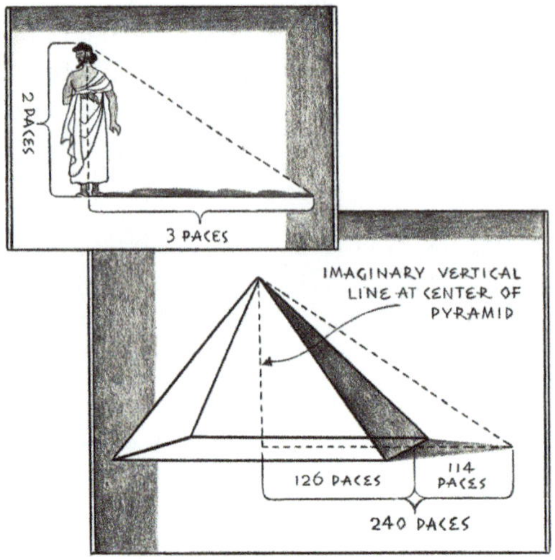

Source: *www.anselm.edu/homepage/dbanach/ thales.htm.* This site references another source: Selections, from Julia E. Diggins, *String, Straightedge, and Shadow,* Viking Press, New York, 1965, Illustrations by Corydon Bell.

52. The Bermuda Triangle Karen is doing research on the Bermuda Triangle which she defines roughly by Hamilton, Bermuda; San Juan, Puerto Rico; and Fort Lauderdale, Florida. On her atlas Karen measures the straight-line distances from Hamilton to Fort Lauderdale, Fort Lauderdale to San Juan, and San Juan to Hamilton to be approximately 57 millimeters (mm), 58 mm, and 53.5 mm respectively. If the actual distance from Fort Lauderdale to San Juan is 1046 miles, approximate the actual distances from San Juan to Hamilton and from Hamilton to Fort Lauderdale.

Source: *www.worldatlas.com*

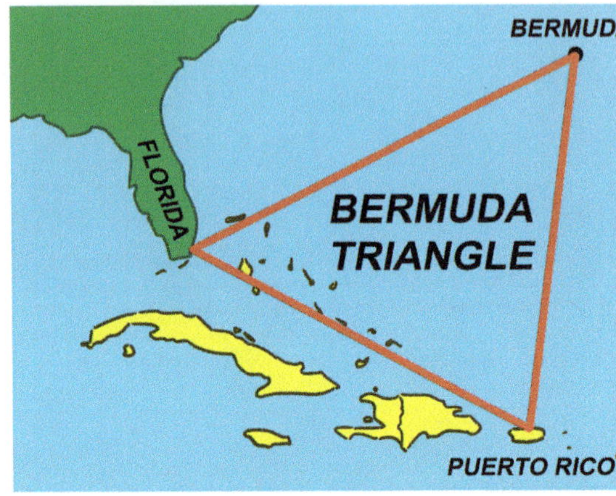

Source: *www.en.wikipedia.org/wiki/Bermuda_Triangle.*

In Problems 53–55, use the facts that the radius of Earth is 3960 miles and 1 mile = 5280 feet.

53. How Far Can You See? The conning tower of the U.S.S. *Silversides,* a World War II submarine now permanently stationed in Muskegon, Michigan, is approximately 20 feet above sea level. How far can you see from the conning tower?

54. How Far Can You See? A person who is 6 feet tall is standing on the beach in Fort Lauderdale, Florida, and looks out onto the Atlantic Ocean. Suddenly, a ship appears on the horizon. How far is the ship from shore?

55. How Far Can You See? The deck of a destroyer is 100 feet above sea level. How far can a person see from the deck? How far can a person see from the bridge, which is 150 feet above sea level?

56. Suppose that m and n are positive integers with $m > n$. If $a = m^2 - n^2$, $b = 2mn$, and $c = m^2 + n^2$, show that a, b, and c are the lengths of the sides of a right triangle. (This formula can be used to find the sides of a right triangle that are integers, such as 3, 4, 5; 5, 12, 13; and so on. Such triplets of integers are called **Pythagorean triples.**)

Discussion and Writing

57. You have 1000 feet of flexible pool siding and wish to construct a swimming pool. Experiment with rectangular-shaped pools with perimeters of 1000 feet. How do their areas vary? What is the shape of the rectangle with the largest area? Now compute the area enclosed by a circular pool with a perimeter (circumference) of 1000 feet. What would be your choice of shape for the pool? If rectangular, what is your preference for dimensions? Justify your choice. If your only consideration is to have a pool that encloses the most area, what shape should you use?

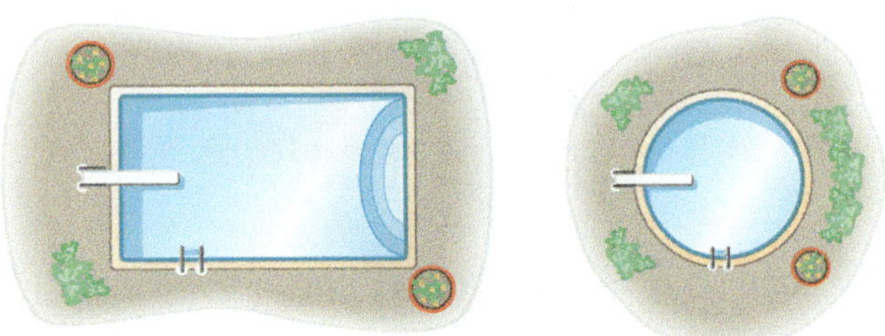

58. The Gibb's Hill Lighthouse, Southampton, Bermuda, in operation since 1846, stands 117 feet high on a hill 245 feet high, so its beam of light is 362 feet above sea level. A brochure states that the light itself can be seen on the horizon about 26 miles distant. Verify the correctness of this information. The brochure further states that ships 40 miles away can see the light and planes flying at 10,000 feet can see it 120 miles away. Verify the accuracy of these statements. What assumption did the brochure make about the height of the ship?

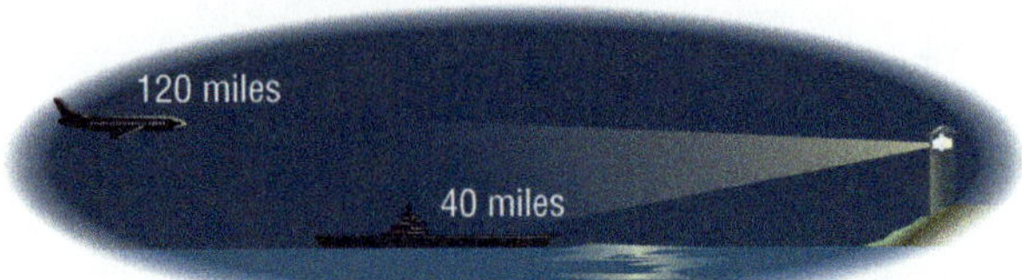

R.4 Polynomials

OBJECTIVES

We have described algebra as a generalization of arithmetic in which letters are used to represent real numbers. From now on, we shall use the letters at the end of the alphabet, such as x, y, and z, to represent variables and the letters at the beginning of the alphabet, such as a, b, and c, to represent constants. In the expressions $3x + 5$ and $ax + b$, it is understood that x is a variable and that a and b are constants, even though the constants a and b are unspecified. As you will find out, the context usually makes the intended meaning clear.

1 Recognize Monomials

DEFINITION

A **monomial** in one variable is the product of a constant and a variable raised to a nonnegative integer power. A monomial is of the form

$$ax^k$$

where a is a constant, x is a variable, and $k \geq 0$ is an integer. The constant a is called the **coefficient** of the monomial. If $a \neq 0$, then k is called the **degree** of the monomial.

NOTE The nonnegative integers are the integers 0, 1, 2, 3, ...

EXAMPLE 1 Examples of Monomials

Monomial	Coefficient	Degree
(a) $6x^2$	6	2
(b) $-\sqrt{2}x^3$	$-\sqrt{2}$	3
(c) 3	3	0 Since $3 = 3 \cdot 1 = 3x^0$, $x \neq 0$
(d) $-5x$	-5	1 Since $-5x = -5x^1$
(e) x^4	1	4 Since $x^4 = 1 \cdot x^4$

Now let's look at some expressions that are not monomials.

EXAMPLE 2 Examples of Nonmonomial Expressions

(a) $3x^{1/2}$ is not a monomial, since the exponent of the variable x is $\frac{1}{2}$ and $\frac{1}{2}$ is not a nonnegative integer.

(b) $4x^{-3}$ is not a monomial, since the exponent of the variable x is -3 and -3 is not a nonnegative integer.

Now Work PROBLEM 7

2 Recognize Polynomials

Two monomials with the same variable raised to the same power are called **like terms.** For example, $2x^4$ and $-5x^4$ are like terms. In contrast, the monomials $2x^3$ and $2x^5$ are not like terms.

We can add or subtract like terms using the Distributive Property. For example,

$$2x^2 + 5x^2 = (2 + 5)x^2 = 7x^2 \quad \text{and} \quad 8x^3 - 5x^3 = (8 - 5)x^3 = 3x^3$$

The sum or difference of two monomials having different degrees is called a **binomial.** The sum or difference of three monomials with three different degrees is called a **trinomial.** For example,

$x^2 - 2$ is a binomial.

$x^3 - 3x + 5$ is a trinomial.

$2x^2 + 5x^2 + 2 = 7x^2 + 2$ is a binomial.

DEFINITION

A **polynomial** in one variable is an algebraic expression of the form

$$a_nx^n + a_{n-1}x^{n-1} + \cdots + a_1x + a_0 \qquad \textbf{(1)}$$

where $a_n, a_{n-1}, \ldots, a_1, a_0$ are constants,* called the **coefficients** of the polynomial, $n \geq 0$ is an integer, and x is a variable. If $a_n \neq 0$, it is called the **leading coefficient,** and n is called the **degree** of the polynomial.

In Words
A polynomial is a sum of monomials.

The monomials that make up a polynomial are called its **terms.** If all the coefficients are 0, the polynomial is called the **zero polynomial,** which has no degree.

Polynomials are usually written in **standard form,** beginning with the nonzero term of highest degree and continuing with terms in descending order according to degree. If a power of x is missing, it is because its coefficient is zero.

EXAMPLE 3 Examples of Polynomials

Polynomial	Coefficients	Degree
$-8x^3 + 4x^2 - 6x + 2$	$-8, 4, -6, 2$	3
$3x^2 - 5 = 3x^2 + 0 \cdot x + (-5)$	$3, 0, -5$	2
$8 - 2x + x^2 = 1 \cdot x^2 + (-2)x + 8$	$1, -2, 8$	2
$5x + \sqrt{2} = 5x^1 + \sqrt{2}$	$5, \sqrt{2}$	1
$3 = 3 \cdot 1 = 3 \cdot x^0$	3	0
0	0	No degree

Although we have been using x to represent the variable, letters such as y or z are also commonly used.

$3x^4 - x^2 + 2$ is a polynomial (in x) of degree 4.

$9y^3 - 2y^2 + y - 3$ is a polynomial (in y) of degree 3.

$z^5 + \pi$ is a polynomial (in z) of degree 5.

Algebraic expressions such as

$$\frac{1}{x} \quad \text{and} \quad \frac{x^2 + 1}{x + 5}$$

*The notation a_n is read as "a sub n." The number n is called a **subscript** and should not be confused with an exponent. We use subscripts to distinguish one constant from another when a large or undetermined number of constants is required.

are not polynomials. The first is not a polynomial because $\frac{1}{x} = x^{-1}$ has an exponent that is not a nonnegative integer. Although the second expression is the quotient of two polynomials, the polynomial in the denominator has degree greater than 0, so the expression cannot be a polynomial.

Now Work PROBLEM 17

3 Add and Subtract Polynomials

Polynomials are added and subtracted by combining like terms.

EXAMPLE 4 Adding Polynomials

Find the sum of the polynomials:

$$8x^3 - 2x^2 + 6x - 2 \quad \text{and} \quad 3x^4 - 2x^3 + x^2 + x$$

Solution We shall find the sum in two ways.

Horizontal Addition: The idea here is to group the like terms and then combine them.

$$\begin{aligned}&(8x^3 - 2x^2 + 6x - 2) + (3x^4 - 2x^3 + x^2 + x)\\ &\quad = 3x^4 + (8x^3 - 2x^3) + (-2x^2 + x^2) + (6x + x) - 2\\ &\quad = 3x^4 + 6x^3 - x^2 + 7x - 2\end{aligned}$$

Vertical Addition: The idea here is to vertically line up the like terms in each polynomial and then add the coefficients.

$$\begin{array}{rrrrrr} & x^4 & x^3 & x^2 & x^1 & x^0\\ & & 8x^3 & -\,2x^2 & +\,6x & -\,2\\ + & 3x^4 & -\,2x^3 & +\,x^2 & +\,x & \\ \hline & 3x^4 & +\,6x^3 & -\,x^2 & +\,7x & -\,2\end{array}$$

We can subtract two polynomials horizontally or vertically as well.

EXAMPLE 5 Subtracting Polynomials

Find the difference: $(3x^4 - 4x^3 + 6x^2 - 1) - (2x^4 - 8x^2 - 6x + 5)$

Solution ***Horizontal Subtraction:***

$$\begin{aligned}&(3x^4 - 4x^3 + 6x^2 - 1) - (2x^4 - 8x^2 - 6x + 5)\\ &\quad = 3x^4 - 4x^3 + 6x^2 - 1 + \underbrace{(-2x^4 + 8x^2 + 6x - 5)}_{\text{Be sure to change the sign of each term in the second polynomial.}}\\ &\quad \underset{\text{Group like terms.}}{=} (3x^4 - 2x^4) + (-4x^3) + (6x^2 + 8x^2) + 6x + (-1 - 5)\\ &\quad = x^4 - 4x^3 + 14x^2 + 6x - 6\end{aligned}$$

COMMENT Vertical subtraction will be used when we divide polynomials. ■

Vertical Subtraction: We line up like terms, change the sign of each coefficient of the second polynomial, and add vertically.

$$\begin{array}{rl} & 3x^4 - 4x^3 + 6x^2 \qquad - 1 \\ - & [2x^4 \qquad - 8x^2 - 6x + 5] \end{array} = \begin{array}{r} 3x^4 - 4x^3 + \ \ 6x^2 \qquad\quad - 1 \\ + \ -2x^4 \qquad\quad + \ \ 8x^2 + 6x - 5 \\ \hline x^4 - 4x^3 + 14x^2 + 6x - 6 \end{array}$$

The choice of which of these methods to use for adding and subtracting polynomials is left to you. To save space, we shall most often use the horizontal format.

Now Work PROBLEM 29

4 Multiply Polynomials

Two monomials may be multiplied using the Laws of Exponents and the Commutative and Associative Properties. For example,

$$(2x^3)\cdot(5x^4) = (2\cdot 5)\cdot(x^3\cdot x^4) = 10x^{3+4} = 10x^7$$

Products of polynomials are found by repeated use of the Distributive Property and the Laws of Exponents. Again, you have a choice of horizontal or vertical format.

EXAMPLE 6 **Multiplying Polynomials**

Find the product: $(2x + 5)(x^2 - x + 2)$

Solution ***Horizontal Multiplication:***

$$(2x + 5)(x^2 - x + 2) = 2x(x^2 - x + 2) + 5(x^2 - x + 2)$$

↑ Distributive Property

$$= (2x\cdot x^2 - 2x\cdot x + 2x\cdot 2) + (5\cdot x^2 - 5\cdot x + 5\cdot 2)$$

↑ Distributive Property

$$= (2x^3 - 2x^2 + 4x) + (5x^2 - 5x + 10)$$

↑ Law of Exponents

$$= 2x^3 + 3x^2 - x + 10$$

↑ Combine like terms.

Vertical Multiplication: The idea here is very much like multiplying a two-digit number by a three-digit number.

$$\begin{array}{rrl} & x^2 - \ \ x + \ \ 2 & \\ & 2x + \ \ 5 & \\ \hline & 5x^2 - 5x + 10 & \text{This line is } 5(x^2 - x + 2). \\ (+) & 2x^3 - 2x^2 + 4x \qquad & \text{This line is } 2x(x^2 - x + 2). \\ \hline & 2x^3 + 3x^2 - \ \ x + 10 & \text{Sum of the above two lines.} \end{array}$$

Now Work PROBLEM 45

5 Know Formulas for Special Products

Certain products, which we call **special products,** occur frequently in algebra. We can calculate them easily using the **FOIL** (*F*irst, *O*uter, *I*nner, *L*ast) method of multiplying two binomials.

$$
\begin{aligned}
(ax + b)(cx + d) &= ax(cx + d) + b(cx + d) \\
&= \underbrace{ax \cdot cx}_{\text{First}} + \underbrace{ax \cdot d}_{\text{Outer}} + \underbrace{b \cdot cx}_{\text{Inner}} + \underbrace{b \cdot d}_{\text{Last}} \\
&= acx^2 + adx + bcx + bd \\
&= acx^2 + (ad + bc)x + bd
\end{aligned}
$$

EXAMPLE 7 Using FOIL

(a) $(x - 3)(x + 3) = \overset{F}{x^2} + \overset{O}{3x} - \overset{I}{3x} - \overset{L}{9} = x^2 - 9$

(b) $(x + 2)^2 = (x + 2)(x + 2) = x^2 + 2x + 2x + 4 = x^2 + 4x + 4$

(c) $(x - 3)^2 = (x - 3)(x - 3) = x^2 - 3x - 3x + 9 = x^2 - 6x + 9$

(d) $(x + 3)(x + 1) = x^2 + x + 3x + 3 = x^2 + 4x + 3$

(e) $(2x + 1)(3x + 4) = 6x^2 + 8x + 3x + 4 = 6x^2 + 11x + 4$

Now Work PROBLEMS 47 AND 55

Some products have been given special names because of their form. The following special products are based on Examples 7(a), (b), and (c).

Difference of Two Squares

$$(x - a)(x + a) = x^2 - a^2 \quad \textbf{(2)}$$

Squares of Binomials, or Perfect Squares

$$(x + a)^2 = x^2 + 2ax + a^2 \quad \textbf{(3a)}$$

$$(x - a)^2 = x^2 - 2ax + a^2 \quad \textbf{(3b)}$$

EXAMPLE 8 Using Special Product Formulas

(a) $(x - 5)(x + 5) = x^2 - 5^2 = x^2 - 25$ — Difference of two squares

(b) $(x + 7)^2 = x^2 + 2 \cdot 7 \cdot x + 7^2 = x^2 + 14x + 49$ — Square of a binomial

(c) $(2x + 1)^2 = (2x)^2 + 2 \cdot 1 \cdot 2x + 1^2 = 4x^2 + 4x + 1$ — Notice that we used $2x$ in place of x in formula (3a).

(d) $(3x - 4)^2 = (3x)^2 - 2 \cdot 4 \cdot 3x + 4^2 = 9x^2 - 24x + 16$ — Replace x by $3x$ in formula (3b).

Now Work PROBLEMS 65, 67, AND 69

Let's look at some more examples that lead to general formulas.

EXAMPLE 9 **Cubing a Binomial**

(a) $(x + 2)^3 = (x + 2)(x + 2)^2 = (x + 2)(x^2 + 4x + 4)$ Formula (3a)

$= (x^3 + 4x^2 + 4x) + (2x^2 + 8x + 8)$

$= x^3 + 6x^2 + 12x + 8$

(b) $(x - 1)^3 = (x - 1)(x - 1)^2 = (x - 1)(x^2 - 2x + 1)$ Formula (3b)

$= (x^3 - 2x^2 + x) - (x^2 - 2x + 1)$

$= x^3 - 3x^2 + 3x - 1$

Cubes of Binomials, or Perfect Cubes

$$(x + a)^3 = x^3 + 3ax^2 + 3a^2x + a^3 \quad \textbf{(4a)}$$

$$(x - a)^3 = x^3 - 3ax^2 + 3a^2x - a^3 \quad \textbf{(4b)}$$

Now Work PROBLEM 85

EXAMPLE 10 **Forming the Difference of Two Cubes**

$$\begin{aligned}(x - 1)(x^2 + x + 1) &= x(x^2 + x + 1) - 1(x^2 + x + 1)\\ &= x^3 + x^2 + x - x^2 - x - 1\\ &= x^3 - 1\end{aligned}$$

EXAMPLE 11 **Forming the Sum of Two Cubes**

$$\begin{aligned}(x + 2)(x^2 - 2x + 4) &= x(x^2 - 2x + 4) + 2(x^2 - 2x + 4)\\ &= x^3 - 2x^2 + 4x + 2x^2 - 4x + 8\\ &= x^3 + 8\end{aligned}$$

Examples 10 and 11 lead to two more special products.

Difference of Two Cubes

$$(x - a)(x^2 + ax + a^2) = x^3 - a^3 \quad \textbf{(5)}$$

Sum of Two Cubes

$$(x + a)(x^2 - ax + a^2) = x^3 + a^3 \quad \textbf{(6)}$$

6 Divide Polynomials Using Long Division

The procedure for dividing two polynomials is similar to the procedure for dividing two integers.

EXAMPLE 12 Dividing Two Integers

Divide 842 by 15.

Solution

$$\begin{array}{rrl} & 56 & \leftarrow \text{Quotient} \\ \text{Divisor} \rightarrow & 15\overline{)842} & \leftarrow \text{Dividend} \\ & \underline{75} & \leftarrow 5 \cdot 15 \text{ (subtract)} \\ & 92 & \\ & \underline{90} & \leftarrow 6 \cdot 15 \text{ (subtract)} \\ & 2 & \leftarrow \text{Remainder} \end{array}$$

So, $\dfrac{842}{15} = 56 + \dfrac{2}{15}$.

In the long division process detailed in Example 12, the number 15 is called the **divisor,** the number 842 is called the **dividend,** the number 56 is called the **quotient,** and the number 2 is called the **remainder.**

To check the answer obtained in a division problem, multiply the quotient by the divisor and add the remainder. The answer should be the dividend.

$$\text{(Quotient)(Divisor)} + \text{Remainder} = \text{Dividend}$$

For example, we can check the results obtained in Example 12 as follows:

$$(56)(15) + 2 = 840 + 2 = 842$$

To divide two polynomials, we first must write each polynomial in standard form. The process then follows a pattern similar to that of Example 12. The next example illustrates the procedure.

EXAMPLE 13 Dividing Two Polynomials

Find the quotient and the remainder when

$$3x^3 + 4x^2 + x + 7 \quad \text{is divided by} \quad x^2 + 1$$

Solution Each polynomial is in standard form. The dividend is $3x^3 + 4x^2 + x + 7$, and the divisor is $x^2 + 1$.

STEP 1: Divide the leading term of the dividend, $3x^3$, by the leading term of the divisor, x^2. Enter the result, $3x$, over the term $3x^3$, as follows:

$$\begin{array}{r} 3x \phantom{{}+4x^2+x+7} \\ x^2 + 1\overline{)3x^3 + 4x^2 + x + 7} \end{array}$$

NOTE Remember, a polynomial is in standard form when each term is written according to descending degree. ■

STEP 2: Multiply $3x$ by $x^2 + 1$ and enter the result below the dividend.

$$\begin{array}{rl} 3x \phantom{{}+4x^2+x+7} & \\ x^2 + 1\overline{)3x^3 + 4x^2 + x + 7} & \\ \underline{3x^3 \phantom{{}+4x^2} + 3x \phantom{{}+7}} & \leftarrow 3x \cdot (x^2 + 1) = 3x^3 + 3x \end{array}$$

↑ Notice that we align the $3x$ term under the x to make the next step easier.

STEP 3: Subtract and bring down the remaining terms.

$$\begin{array}{rl} 3x \phantom{{}+4x^2+x+7} & \\ x^2 + 1\overline{)3x^3 + 4x^2 + x + 7} & \\ \underline{3x^3 \phantom{{}+4x^2} + 3x \phantom{{}+7}} & \leftarrow \text{Subtract (change the signs and add).} \\ 4x^2 - 2x + 7 & \leftarrow \text{Bring down the } 4x^2 \text{ and the } 7. \end{array}$$

STEP 4: Repeat Steps 1–3 using $4x^2 - 2x + 7$ as the dividend.

$$\begin{array}{r}
3x \quad + 4 \\
x^2 + 1\overline{)3x^3 + 4x^2 + \quad x + 7} \\
\underline{3x^3 \qquad\quad + 3x} \\
4x^2 - 2x + 7 \\
\underline{4x^2 \qquad\quad + 4} \\
-2x + 3
\end{array}$$

Divide $4x^2$ by x^2 to get 4.
Multiply $x^2 + 1$ by 4; subtract.

Since x^2 does not divide $-2x$ evenly (that is, the result is not a monomial), the process ends. The quotient is $3x + 4$, and the remainder is $-2x + 3$.

Check: (Quotient)(Divisor) + Remainder

$$= (3x + 4)(x^2 + 1) + (-2x + 3)$$
$$= 3x^3 + 3x + 4x^2 + 4 + (-2x + 3)$$
$$= 3x^3 + 4x^2 + x + 7 = \text{Dividend}$$

Then

$$\frac{3x^3 + 4x^2 + x + 7}{x^2 + 1} = 3x + 4 + \frac{-2x + 3}{x^2 + 1}$$

The next example combines the steps involved in long division.

EXAMPLE 14 Dividing Two Polynomials

Find the quotient and the remainder when

$$x^4 - 3x^3 + 2x - 5 \quad \text{is divided by} \quad x^2 - x + 1$$

Solution In setting up this division problem, it is necessary to leave a space for the missing x^2 term in the dividend.

$$\begin{array}{r}
x^2 - 2x \quad - 3 \\
x^2 - x + 1\overline{)x^4 - 3x^3 \qquad\qquad + 2x - 5} \\
\underline{x^4 - \;\, x^3 + \;\, x^2 \qquad\qquad} \\
-2x^3 - \;\, x^2 + 2x - 5 \\
\underline{-2x^3 + 2x^2 - 2x \quad} \\
-3x^2 + 4x - 5 \\
\underline{-3x^2 + 3x - 3} \\
x - 2
\end{array}$$

Quotient; Divisor; Dividend; Subtract; Subtract; Subtract; Remainder

Check: (Quotient)(Divisor) + Remainder

$$= (x^2 - 2x - 3)(x^2 - x + 1) + x - 2$$
$$= x^4 - x^3 + x^2 - 2x^3 + 2x^2 - 2x - 3x^2 + 3x - 3 + x - 2$$
$$= x^4 - 3x^3 + 2x - 5 = \text{Dividend}$$

As a result,

$$\frac{x^4 - 3x^3 + 2x - 5}{x^2 - x + 1} = x^2 - 2x - 3 + \frac{x - 2}{x^2 - x + 1}$$

The process of dividing two polynomials leads to the following result:

THEOREM

Let Q be a polynomial of positive degree and let P be a polynomial whose degree is greater than or equal to the degree of Q. The remainder after dividing P by Q is either the zero polynomial or a polynomial whose degree is less than the degree of the divisor Q.

Now Work PROBLEM 93

7 Work with Polynomials in Two Variables

A **monomial in two variables** x and y has the form ax^ny^m, where a is a constant, x and y are variables, and n and m are nonnegative integers. The **degree** of a monomial is the sum of the powers of the variables.

For example,

$$2xy^3, \quad x^2y^2, \quad \text{and} \quad x^3y$$

are monomials, each of which has degree 4.

A **polynomial in two variables** x and y is the sum of one or more monomials in two variables. The **degree of a polynomial** in two variables is the highest degree of all the monomials with nonzero coefficients.

EXAMPLE 15 Examples of Polynomials in Two Variables

$3x^2 + 2x^3y + 5$	$\pi x^3 - y^2$	$x^4 + 4x^3y - xy^3 + y^4$
Two variables, degree is 4.	Two variables, degree is 3.	Two variables, degree is 4.

Multiplying polynomials in two variables is handled in the same way as polynomials in one variable.

EXAMPLE 16 Using a Special Product Formula

To multiply $(2x - y)^2$, use the Square of a Binomial formula (3b) with $2x$ instead of x and y instead of a.

$$\begin{aligned}(2x - y)^2 &= (2x)^2 - 2 \cdot y \cdot 2x + y^2 \\ &= 4x^2 - 4xy + y^2\end{aligned}$$

Now Work PROBLEM 79

R.4 Assess Your Understanding

Concepts and Vocabulary

1. The polynomial $3x^4 - 2x^3 + 13x^2 - 5$ is of degree _____. The leading coefficient is _____.
2. $(x^2 - 4)(x^2 + 4) =$ _____.
3. $(x - 2)(x^2 + 2x + 4) =$ _____.
4. *True or False* $4x^{-2}$ is a monomial of degree -2.
5. *True or False* The degree of the product of two nonzero polynomials equals the sum of their degrees.
6. *True or False* $(x + a)(x^2 + ax + a) = x^3 + a^3$.

Skill Building

In Problems 7–16, tell whether the expression is a monomial. If it is, name the variable(s) and the coefficient and give the degree of the monomial. If it is not a monomial, state why not.

7. $2x^3$
8. $-4x^2$
9. $\frac{8}{x}$
10. $-2x^{-3}$
11. $-2xy^2$
12. $5x^2y^3$
13. $\frac{8x}{y}$
14. $-\frac{2x^2}{y^3}$
15. $x^2 + y^2$
16. $3x^2 + 4$

In Problems 17–26, tell whether the expression is a polynomial. If it is, give its degree. If it is not, state why not.

17. $3x^2 - 5$
18. $1 - 4x$
19. 5
20. $-\pi$
21. $3x^2 - \frac{5}{x}$
22. $\frac{3}{x} + 2$
23. $2y^3 - \sqrt{2}$
24. $10z^2 + z$
25. $\frac{x^2 + 5}{x^3 - 1}$
26. $\frac{3x^3 + 2x - 1}{x^2 + x + 1}$

In Problems 27–46, add, subtract, or multiply, as indicated. Express your answer as a single polynomial in standard form.

27. $(x^2 + 4x + 5) + (3x - 3)$
28. $(x^3 + 3x^2 + 2) + (x^2 - 4x + 4)$
29. $(x^3 - 2x^2 + 5x + 10) - (2x^2 - 4x + 3)$
30. $(x^2 - 3x - 4) - (x^3 - 3x^2 + x + 5)$
31. $(6x^5 + x^3 + x) + (5x^4 - x^3 + 3x^2)$
32. $(10x^5 - 8x^2) + (3x^3 - 2x^2 + 6)$
33. $(x^2 - 3x + 1) + 2(3x^2 + x - 4)$
34. $-2(x^2 + x + 1) + (-5x^2 - x + 2)$
35. $6(x^3 + x^2 - 3) - 4(2x^3 - 3x^2)$
36. $8(4x^3 - 3x^2 - 1) - 6(4x^3 + 8x - 2)$
37. $(x^2 - x + 2) + (2x^2 - 3x + 5) - (x^2 + 1)$
38. $(x^2 + 1) - (4x^2 + 5) + (x^2 + x - 2)$
39. $9(y^2 - 3y + 4) - 6(1 - y^2)$
40. $8(1 - y^3) + 4(1 + y + y^2 + y^3)$
41. $x(x^2 + x - 4)$
42. $4x^2(x^3 - x + 2)$
43. $-2x^2(4x^3 + 5)$
44. $5x^3(3x - 4)$
45. $(x + 1)(x^2 + 2x - 4)$
46. $(2x - 3)(x^2 + x + 1)$

In Problems 47–64, multiply the polynomials using the FOIL method. Express your answer as a single polynomial in standard form.

47. $(x + 2)(x + 4)$
48. $(x + 3)(x + 5)$
49. $(2x + 5)(x + 2)$
50. $(3x + 1)(2x + 1)$
51. $(x - 4)(x + 2)$
52. $(x + 4)(x - 2)$
53. $(x - 3)(x - 2)$
54. $(x - 5)(x - 1)$
55. $(2x + 3)(x - 2)$
56. $(2x - 4)(3x + 1)$
57. $(-2x + 3)(x - 4)$
58. $(-3x - 1)(x + 1)$
59. $(-x - 2)(-2x - 4)$
60. $(-2x - 3)(3 - x)$
61. $(x - 2y)(x + y)$
62. $(2x + 3y)(x - y)$
63. $(-2x - 3y)(3x + 2y)$
64. $(x - 3y)(-2x + y)$

In Problems 65–88, multiply the polynomials using the special product formulas. Express your answer as a single polynomial in standard form.

65. $(x - 7)(x + 7)$
66. $(x - 1)(x + 1)$
67. $(2x + 3)(2x - 3)$
68. $(3x + 2)(3x - 2)$
69. $(x + 4)^2$
70. $(x + 5)^2$
71. $(x - 4)^2$
72. $(x - 5)^2$
73. $(3x + 4)(3x - 4)$
74. $(5x - 3)(5x + 3)$
75. $(2x - 3)^2$
76. $(3x - 4)^2$

77. $(x + y)(x - y)$

78. $(x + 3y)(x - 3y)$

79. $(3x + y)(3x - y)$

80. $(3x + 4y)(3x - 4y)$

81. $(x + y)^2$

82. $(x - y)^2$

83. $(x - 2y)^2$

84. $(2x + 3y)^2$

85. $(x - 2)^3$

86. $(x + 1)^3$

87. $(2x + 1)^3$

88. $(3x - 2)^3$

In Problems 89–104, find the quotient and the remainder. Check your work by verifying that

$$(Quotient)(Divisor) + Remainder = Dividend$$

89. $4x^3 - 3x^2 + x + 1$ divided by $x + 2$

90. $3x^3 - x^2 + x - 2$ divided by $x + 2$

91. $4x^3 - 3x^2 + x + 1$ divided by x^2

92. $3x^3 - x^2 + x - 2$ divided by x^2

93. $5x^4 - 3x^2 + x + 1$ divided by $x^2 + 2$

94. $5x^4 - x^2 + x - 2$ divided by $x^2 + 2$

95. $4x^5 - 3x^2 + x + 1$ divided by $2x^3 - 1$

96. $3x^5 - x^2 + x - 2$ divided by $3x^3 - 1$

97. $2x^4 - 3x^3 + x + 1$ divided by $2x^2 + x + 1$

98. $3x^4 - x^3 + x - 2$ divided by $3x^2 + x + 1$

99. $-4x^3 + x^2 - 4$ divided by $x - 1$

100. $-3x^4 - 2x - 1$ divided by $x - 1$

101. $1 - x^2 + x^4$ divided by $x^2 + x + 1$

102. $1 - x^2 + x^4$ divided by $x^2 - x + 1$

103. $x^3 - a^3$ divided by $x - a$

104. $x^5 - a^5$ divided by $x - a$

Discussion and Writing

105. Explain why the degree of the product of two nonzero polynomials equals the sum of their degrees.

106. Explain why the degree of the sum of two polynomials of different degrees equals the larger of their degrees.

107. Give a careful statement about the degree of the sum of two polynomials of the same degree.

108. Do you prefer adding two polynomials using the horizontal method or the vertical method? Write a brief position paper defending your choice.

109. Do you prefer to memorize the rule for the square of a binomial $(x + a)^2$ or to use FOIL to obtain the product? Write a brief position paper defending your choice.

R.5 Factoring Polynomials

OBJECTIVES

1 Factor the Difference of Two Squares and the Sum and Difference of Two Cubes (p. 50)
2 Factor Perfect Squares (p. 51)
3 Factor a Second-Degree Polynomial: $x^2 + Bx + C$ (p. 52)
4 Factor by Grouping (p. 53)
5 Factor a Second-Degree Polynomial: $Ax^2 + Bx + C, A \neq 1$ (p. 54)

Consider the following product:

$$(2x + 3)(x - 4) = 2x^2 - 5x - 12$$

The two polynomials on the left side are called **factors** of the polynomial on the right side. Expressing a given polynomial as a product of other polynomials, that is, finding the factors of a polynomial, is called **factoring.**

We shall restrict our discussion here to factoring polynomials in one variable into products of polynomials in one variable, where all coefficients are integers. We call this **factoring over the integers.**

Any polynomial can be written as the product of 1 times itself or as -1 times its additive inverse. If a polynomial cannot be written as the product of two other polynomials (excluding 1 and -1), then the polynomial is said to be **prime.** When a polynomial has been written as a product consisting only of prime factors, it is said to be **factored completely.** Examples of prime polynomials (over the integers) are

$$2, \quad 3, \quad 5, \quad x, \quad x+1, \quad x-1, \quad 3x+4 \quad x^2+4$$

COMMENT Over the real numbers, $3x + 4$ factors into $3\left(x + \frac{4}{3}\right)$. It is the noninteger $\frac{4}{3}$ that causes $3x + 4$ to be prime over the integers. ■

The first factor to look for in a factoring problem is a common monomial factor present in each term of the polynomial. If one is present, use the Distributive Property to factor it out.

EXAMPLE 1 Identifying Common Monomial Factors

Polynomial	Common Monomial Factor	Remaining Factor	Factored Form
$2x + 4$	2	$x + 2$	$2x + 4 = 2(x + 2)$
$3x - 6$	3	$x - 2$	$3x - 6 = 3(x - 2)$
$2x^2 - 4x + 8$	2	$x^2 - 2x + 4$	$2x^2 - 4x + 8 = 2(x^2 - 2x + 4)$
$8x - 12$	4	$2x - 3$	$8x - 12 = 4(2x - 3)$
$x^2 + x$	x	$x + 1$	$x^2 + x = x(x + 1)$
$x^3 - 3x^2$	x^2	$x - 3$	$x^3 - 3x^2 = x^2(x - 3)$
$6x^2 + 9x$	$3x$	$2x + 3$	$6x^2 + 9x = 3x(2x + 3)$

Notice that, once all common monomial factors have been removed from a polynomial, the remaining factor is either a prime polynomial of degree 1 or a polynomial of degree 2 or higher. (Do you see why?)

Now Work PROBLEM 5

1 Factor the Difference of Two Squares and the Sum and Difference of Two Cubes

When you factor a polynomial, first check for common monomial factors. Then see whether you can use one of the special formulas discussed in the previous section.

Difference of Two Squares	$x^2 - a^2 = (x - a)(x + a)$
Perfect Squares	$x^2 + 2ax + a^2 = (x + a)^2$
	$x^2 - 2ax + a^2 = (x - a)^2$
Sum of Two Cubes	$x^3 + a^3 = (x + a)(x^2 - ax + a^2)$
Difference of Two Cubes	$x^3 - a^3 = (x - a)(x^2 + ax + a^2)$

EXAMPLE 2 Factoring the Difference of Two Squares

Factor completely: $x^2 - 4$

Solution We notice that $x^2 - 4$ is the difference of two squares, x^2 and 2^2.

$$x^2 - 4 = (x - 2)(x + 2)$$

■

EXAMPLE 3 **Factoring the Difference of Two Cubes**

Factor completely: $x^3 - 1$

Solution Because $x^3 - 1$ is the difference of two cubes, x^3 and 1^3, we find that

$$x^3 - 1 = (x - 1)(x^2 + x + 1)$$

EXAMPLE 4 **Factoring the Sum of Two Cubes**

Factor completely: $x^3 + 8$

Solution Because $x^3 + 8$ is the sum of two cubes, x^3 and 2^3, we have

$$x^3 + 8 = (x + 2)(x^2 - 2x + 4)$$

EXAMPLE 5 **Factoring the Difference of Two Squares**

Factor completely: $x^4 - 16$

Solution Because $x^4 - 16$ is the difference of two squares, $x^4 = (x^2)^2$ and $16 = 4^2$, we have

$$x^4 - 16 = (x^2 - 4)(x^2 + 4)$$

But $x^2 - 4$ is also the difference of two squares. Thus,

$$x^4 - 16 = (x^2 - 4)(x^2 + 4) = (x - 2)(x + 2)(x^2 + 4)$$

Now Work PROBLEMS 15 AND 33

2 Factor Perfect Squares

When the first term and third term of a trinomial are both positive and are perfect squares, such as x^2, $9x^2$, 1, and 4, check to see whether the trinomial is a perfect square.

EXAMPLE 6 **Factoring Perfect Squares**

Factor completely: $x^2 + 6x + 9$

Solution The first term, x^2, and the third term, $9 = 3^2$, are perfect squares. Because the middle term $6x$ is twice the product of x and 3, we have a perfect square.

$$x^2 + 6x + 9 = (x + 3)^2$$

EXAMPLE 7 **Factoring Perfect Squares**

Factor completely: $9x^2 - 6x + 1$

Solution The first term, $9x^2 = (3x)^2$, and the third term, $1 = 1^2$, are perfect squares. Because the middle term, $-6x$, is -2 times the product of $3x$ and 1, we have a perfect square.

$$9x^2 - 6x + 1 = (3x - 1)^2$$

EXAMPLE 8 **Factoring Perfect Squares**

Factor completely: $25x^2 + 30x + 9$

Solution The first term, $25x^2 = (5x)^2$, and the third term, $9 = 3^2$, are perfect squares. Because the middle term, $30x$, is twice the product of $5x$ and 3, we have a perfect square.

$$25x^2 + 30x + 9 = (5x + 3)^2$$

Now Work PROBLEM 25

If a trinomial is not a perfect square, it may be possible to factor it using the technique discussed next.

3 Factor a Second-Degree Polynomial: $x^2 + Bx + C$

The idea behind factoring a second-degree polynomial like $x^2 + Bx + C$ is to see whether it can be made equal to the product of two, possibly equal, first-degree polynomials.

For example, we know that

$$(x + 3)(x + 4) = x^2 + 7x + 12$$

The factors of $x^2 + 7x + 12$ are $x + 3$ and $x + 4$. Notice the following:

$$x^2 + 7x + 12 = (x + 3)(x + 4)$$

12 is the product of 3 and 4

7 is the sum of 3 and 4

In general, if $x^2 + Bx + C = (x + a)(x + b) = x^2 + (a + b)x + ab$, then $ab = C$ and $a + b = B$.

To factor a second-degree polynomial $x^2 + Bx + C$, find integers whose product is C and whose sum is B. That is, if there are numbers a, b, where $ab = C$ and $a + b = B$, then

$$x^2 + Bx + C = (x + a)(x + b)$$

EXAMPLE 9 Factoring Trinomials

Factor completely: $x^2 + 7x + 10$

Solution First, determine all integers whose product is 10 and then compute their sums.

Integers whose product is 10	1, 10	−1, −10	2, 5	−2, −5
Sum	11	−11	7	−7

The integers 2 and 5 have a product of 10 and add up to 7, the coefficient of the middle term. As a result, we have

$$x^2 + 7x + 10 = (x + 2)(x + 5)$$

EXAMPLE 10 Factoring Trinomials

Factor completely: $x^2 - 6x + 8$

Solution First, determine all integers whose product is 8 and then compute each sum.

Integers whose product is 8	1, 8	−1, −8	2, 4	−2, −4
Sum	9	−9	6	−6

Since −6 is the coefficient of the middle term,

$$x^2 - 6x + 8 = (x - 2)(x - 4)$$

EXAMPLE 11 **Factoring Trinomials**

Factor completely: $x^2 - x - 12$

Solution First, determine all integers whose product is -12 and then compute each sum.

Integers whose product is -12	1, −12	−1, 12	2, −6	−2, 6	3, −4	−3, 4
Sum	−11	11	−4	4	−1	1

Since -1 is the coefficient of the middle term,

$$x^2 - x - 12 = (x + 3)(x - 4)$$

EXAMPLE 12 **Factoring Trinomials**

Factor completely: $x^2 + 4x - 12$

Solution The integers -2 and 6 have a product of -12 and have the sum 4. Thus,

$$x^2 + 4x - 12 = (x - 2)(x + 6)$$

To avoid errors in factoring, always check your answer by multiplying it out to see if the result equals the original expression.

When none of the possibilities works, the polynomial is prime.

EXAMPLE 13 **Identifying Prime Polynomials**

Show that $x^2 + 9$ is prime.

Solution First, list the integers whose product is 9 and then compute their sums.

Integers whose product is 9	1, 9	−1, −9	3, 3	−3, −3
Sum	10	−10	6	−6

Since the coefficient of the middle term in $x^2 + 9 = x^2 + 0x + 9$ is 0 and none of the sums equals 0, we conclude that $x^2 + 9$ is prime.

Example 13 demonstrates a more general result:

THEOREM Any polynomial of the form $x^2 + a^2$, a real, is prime.

Now Work PROBLEM 39

4 Factor by Grouping

Sometimes a common factor does not occur in every term of the polynomial, but in each of several groups of terms that together make up the polynomial. When this happens, the common factor can be factored out of each group by means of the Distributive Property. This technique is called **factoring by grouping.**

EXAMPLE 14 **Factoring by Grouping**

Factor completely by grouping: $(x^2 + 2)x + (x^2 + 2) \cdot 3$

Solution Notice the common factor $x^2 + 2$. By applying the Distributive Property, we have

$$(x^2 + 2)x + (x^2 + 2) \cdot 3 = (x^2 + 2)(x + 3)$$

Since $x^2 + 2$ and $x + 3$ are prime, the factorization is complete.

The next example shows a factoring problem that occurs in calculus.

EXAMPLE 15 **Factoring by Grouping**

Factor completely by grouping: $3(x-1)^2(x+2)^4 + 4(x-1)^3(x+2)^3$

Solution Here, $(x-1)^2(x+2)^3$ is a common factor of $3(x-1)^2(x+2)^4$ and of $4(x-1)^3(x+2)^3$. As a result,

$$\begin{aligned} 3(x-1)^2(x+2)^4 + 4(x-1)^3(x+2)^3 &= (x-1)^2(x+2)^3[3(x+2) + 4(x-1)] \\ &= (x-1)^2(x+2)^3[3x + 6 + 4x - 4] \\ &= (x-1)^2(x+2)^3(7x+2) \end{aligned}$$

EXAMPLE 16 **Factoring by Grouping**

Factor completely by grouping: $x^3 - 4x^2 + 2x - 8$

Solution To see if factoring by grouping will work, group the first two terms and the last two terms. Then look for a common factor in each group. In this example, we can factor x^2 from $x^3 - 4x^2$ and 2 from $2x - 8$. The remaining factor in each case is the same, $x - 4$. This means that factoring by grouping will work, as follows:

$$\begin{aligned} x^3 - 4x^2 + 2x - 8 &= (x^3 - 4x^2) + (2x - 8) \\ &= x^2(x-4) + 2(x-4) \\ &= (x-4)(x^2+2) \end{aligned}$$

Since $x^2 + 2$ and $x - 4$ are prime, the factorization is complete.

Now Work PROBLEMS 51 AND 121

5 Factor a Second-degree Polynomial: $Ax^2 + Bx + C, A \neq 1$

To factor a second-degree polynomial $Ax^2 + Bx + C$, when $A \neq 1$ and A, B, and C have no common factors, follow these steps:

Steps for Factoring $Ax^2 + Bx + C$, when $A \neq 1$ and A, B, and C Have No Common Factors

STEP 1: Find the value of AC.
STEP 2: Find integers whose product is AC that add up to B. That is, find a and b so that $ab = AC$ and $a + b = B$.
STEP 3: Write $Ax^2 + Bx + C = Ax^2 + ax + bx + C$.
STEP 4: Factor this last expression by grouping.

EXAMPLE 17 **Factoring Trinomials**

Factor completely: $2x^2 + 5x + 3$

Solution Comparing $2x^2 + 5x + 3$ to $Ax^2 + Bx + C$, we find that $A = 2$, $B = 5$, and $C = 3$.

STEP 1: The value of AC is $2 \cdot 3 = 6$.
STEP 2: Determine the integers whose product is $AC = 6$ and compute their sums.

Integers whose product is 6	1, 6	−1, −6	2, 3	−2, −3
Sum	7	−7	5	−5

STEP 3: The integers whose product is 6 that add up to $B = 5$ are 2 and 3.

$$2x^2 + 5x + 3 = 2x^2 + 2x + 3x + 3$$

STEP 4: Factor by grouping.

$$\begin{aligned} 2x^2 + 2x + 3x + 3 &= (2x^2 + 2x) + (3x + 3) \\ &= 2x(x + 1) + 3(x + 1) \\ &= (x + 1)(2x + 3) \end{aligned}$$

As a result,

$$2x^2 + 5x + 3 = (x + 1)(2x + 3)$$

EXAMPLE 18 Factoring Trinomials

Factor completely: $2x^2 - x - 6$

Solution Comparing $2x^2 - x - 6$ to $Ax^2 + Bx + C$, we find that $A = 2$, $B = -1$, and $C = -6$.

STEP 1: The value of AC is $2 \cdot (-6) = -12$.

STEP 2: Determine the integers whose product is $AC = -12$ and compute their sums.

Integers whose product is −12	1, −12	−1, 12	2, −6	−2, 6	3, −4	−3, 4
Sum	−11	11	−4	4	−1	1

STEP 3: The integers whose product is −12 that add up to $B = -1$ are −4 and 3.

$$2x^2 - x - 6 = 2x^2 - 4x + 3x - 6$$

STEP 4: Factor by grouping.

$$\begin{aligned} 2x^2 - 4x + 3x - 6 &= (2x^2 - 4x) + (3x - 6) \\ &= 2x(x - 2) + 3(x - 2) \\ &= (x - 2)(2x + 3) \end{aligned}$$

As a result,

$$2x^2 - x - 6 = (x - 2)(2x + 3)$$

Now Work PROBLEM 57

SUMMARY

Type of Polynomial	Method	Example
Any polynomial	Look for common monomial factors. (Always do this first!)	$6x^2 + 9x = 3x(2x + 3)$
Binomials of degree 2 or higher	Check for a special product:	
	Difference of two squares, $x^2 - a^2$	$x^2 - 16 = (x - 4)(x + 4)$
	Difference of two cubes, $x^3 - a^3$	$x^3 - 64 = (x - 4)(x^2 + 4x + 16)$
	Sum of two cubes, $x^3 + a^3$	$x^3 + 27 = (x + 3)(x^2 - 3x + 9)$
Trinomials of degree 2	Check for a perfect square, $(x \pm a)^2$	$x^2 + 8x + 16 = (x + 4)^2$
		$x^2 - 10x + 25 = (x - 5)^2$
	Factoring $x^2 + Bx + C$ (p. 52)	$x^2 - x - 2 = (x - 2)(x + 1)$
	Factoring $Ax^2 + Bx + C$ (p. 54)	$6x^2 + x - 1 = (2x + 1)(3x - 1)$
Four or more terms	Grouping	$2x^3 - 3x^2 + 4x - 6 = (2x - 3)(x^2 + 2)$

R.5 Assess Your Understanding

Concepts and Vocabulary

1. If factored completely, $3x^3 - 12x =$ _____.
2. If a polynomial cannot be written as the product of two other polynomials (excluding 1 and -1), then the polynomial is said to be _____.
3. ***True or False*** The polynomial $x^2 + 4$ is prime.
4. ***True or False*** $3x^3 - 2x^2 - 6x + 4 = (3x - 2)(x^2 + 2)$.

Skill Building

In Problems 5–14, factor each polynomial by removing the common monomial factor.

5. $3x + 6$
6. $7x - 14$
7. $ax^2 + a$
8. $ax - a$
9. $x^3 + x^2 + x$
10. $x^3 - x^2 + x$
11. $2x^2 - 2x$
12. $3x^2 - 3x$
13. $3x^2y - 6xy^2 + 12xy$
14. $60x^2y - 48xy^2 + 72x^3y$

In Problems 15–22, factor the difference of two squares.

15. $x^2 - 1$
16. $x^2 - 4$
17. $4x^2 - 1$
18. $9x^2 - 1$
19. $x^2 - 16$
20. $x^2 - 25$
21. $25x^2 - 4$
22. $36x^2 - 9$

In Problems 23–32, factor the perfect squares.

23. $x^2 + 2x + 1$
24. $x^2 - 4x + 4$
25. $x^2 + 4x + 4$
26. $x^2 - 2x + 1$
27. $x^2 - 10x + 25$
28. $x^2 + 10x + 25$
29. $4x^2 + 4x + 1$
30. $9x^2 + 6x + 1$
31. $16x^2 + 8x + 1$
32. $25x^2 + 10x + 1$

In Problems 33–38, factor the sum or difference of two cubes.

33. $x^3 - 27$
34. $x^3 + 125$
35. $x^3 + 27$
36. $27 - 8x^3$
37. $8x^3 + 27$
38. $64 - 27x^3$

In Problems 39–50, factor each polynomial.

39. $x^2 + 5x + 6$
40. $x^2 + 6x + 8$
41. $x^2 + 7x + 6$
42. $x^2 + 9x + 8$
43. $x^2 + 7x + 10$
44. $x^2 + 11x + 10$
45. $x^2 - 10x + 16$
46. $x^2 - 17x + 16$
47. $x^2 - 7x - 8$
48. $x^2 - 2x - 8$
49. $x^2 + 7x - 8$
50. $x^2 + 2x - 8$

In Problems 51–56, factor by grouping.

51. $2x^2 + 4x + 3x + 6$
52. $3x^2 - 3x + 2x - 2$
53. $2x^2 - 4x + x - 2$
54. $3x^2 + 6x - x - 2$
55. $6x^2 + 9x + 4x + 6$
56. $9x^2 - 6x + 3x - 2$

In Problems 57–68, factor each polynomial.

57. $3x^2 + 4x + 1$
58. $2x^2 + 3x + 1$
59. $2z^2 + 5z + 3$
60. $6z^2 + 5z + 1$
61. $3x^2 + 2x - 8$
62. $3x^2 + 10x + 8$
63. $3x^2 - 2x - 8$
64. $3x^2 - 10x + 8$
65. $3x^2 + 14x + 8$
66. $3x^2 - 14x + 8$
67. $3x^2 + 10x - 8$
68. $3x^2 - 10x - 8$

Mixed Practice

In Problems 69–116, factor completely each polynomial. If the polynomial cannot be factored, say it is prime.

69. $x^2 - 36$
70. $x^2 - 9$
71. $2 - 8x^2$
72. $3 - 27x^2$
73. $x^2 + 11x + 10$
74. $x^2 + 5x + 4$
75. $x^2 - 10x + 21$
76. $x^2 - 6x + 8$
77. $4x^2 - 8x + 32$
78. $3x^2 - 12x + 15$
79. $x^2 + 4x + 16$
80. $x^2 + 12x + 36$
81. $15 + 2x - x^2$
82. $14 + 6x - x^2$
83. $3x^2 - 12x - 36$
84. $x^3 + 8x^2 - 20x$
85. $y^4 + 11y^3 + 30y^2$
86. $3y^3 - 18y^2 - 48y$
87. $4x^2 + 12x + 9$
88. $9x^2 - 12x + 4$
89. $6x^2 + 8x + 2$
90. $8x^2 + 6x - 2$
91. $x^4 - 81$
92. $x^4 - 1$
93. $x^6 - 2x^3 + 1$
94. $x^6 + 2x^3 + 1$
95. $x^7 - x^5$
96. $x^8 - x^5$
97. $16x^2 + 24x + 9$
98. $9x^2 - 24x + 16$
99. $5 + 16x - 16x^2$
100. $5 + 11x - 16x^2$

101. $4y^2 - 16y + 15$ **102.** $9y^2 + 9y - 4$ **103.** $1 - 8x^2 - 9x^4$ **104.** $4 - 14x^2 - 8x^4$

105. $x(x + 3) - 6(x + 3)$ **106.** $5(3x - 7) + x(3x - 7)$ **107.** $(x + 2)^2 - 5(x + 2)$ **108.** $(x - 1)^2 - 2(x - 1)$

109. $(3x - 2)^3 - 27$ **110.** $(5x + 1)^3 - 1$ **111.** $3(x^2 + 10x + 25) - 4(x + 5)$

112. $7(x^2 - 6x + 9) + 5(x - 3)$ **113.** $x^3 + 2x^2 - x - 2$ **114.** $x^3 - 3x^2 - x + 3$

115. $x^4 - x^3 + x - 1$ **116.** $x^4 + x^3 + x + 1$

Applications and Extensions

In Problems 117–126, expressions that occur in calculus are given. Factor completely each expression.

117. $2(3x + 4)^2 + (2x + 3)\cdot 2(3x + 4)\cdot 3$ **118.** $5(2x + 1)^2 + (5x - 6)\cdot 2(2x + 1)\cdot 2$

119. $2x(2x + 5) + x^2\cdot 2$ **120.** $3x^2(8x - 3) + x^3\cdot 8$

121. $2(x + 3)(x - 2)^3 + (x + 3)^2\cdot 3(x - 2)^2$ **122.** $4(x + 5)^3(x - 1)^2 + (x + 5)^4\cdot 2(x - 1)$

123. $(4x - 3)^2 + x\cdot 2(4x - 3)\cdot 4$ **124.** $3x^2(3x + 4)^2 + x^3\cdot 2(3x + 4)\cdot 3$

125. $2(3x - 5)\cdot 3(2x + 1)^3 + (3x - 5)^2\cdot 3(2x + 1)^2\cdot 2$ **126.** $3(4x + 5)^2\cdot 4(5x + 1)^2 + (4x + 5)^3\cdot 2(5x + 1)\cdot 5$

127. Show that $x^2 + 4$ is prime. **128.** Show that $x^2 + x + 1$ is prime.

Discussion and Writing

129. Make up a polynomial that factors into a perfect square.

130. Explain to a fellow student what you look for first when presented with a factoring problem. What do you do next?

R.6 Synthetic Division

OBJECTIVE 1 Divide Polynomials Using Synthetic Division (p. 57)

1 Divide Polynomials Using Synthetic Division

To find the quotient as well as the remainder when a polynomial of degree 1 or higher is divided by $x - c$, a shortened version of long division, called **synthetic division,** makes the task simpler.

To see how synthetic division works, we will use long division to divide the polynomial $2x^3 - x^2 + 3$ by $x - 3$.

$$\begin{array}{r|l} & 2x^2 + 5x + 15 \quad \leftarrow \text{Quotient} \\ x - 3 & 2x^3 - x^2 \qquad + 3 \\ & \underline{2x^3 - 6x^2} \\ & 5x^2 \\ & \underline{5x^2 - 15x} \\ & 15x + 3 \\ & \underline{15x - 45} \\ & 48 \quad \leftarrow \text{Remainder} \end{array}$$

CHECK: (Divisor) $\cdot$ (Quotient) + Remainder

$$= (x - 3)(2x^2 + 5x + 15) + 48$$
$$= 2x^3 + 5x^2 + 15x - 6x^2 - 15x - 45 + 48$$
$$= 2x^3 - x^2 + 3$$

The process of synthetic division arises from rewriting the long division in a more compact form, using simpler notation. For example, in the long division on page 57, the terms in blue are not really necessary because they are identical to the terms directly above them. With these terms removed, we have

$$\begin{array}{r|rrrr}
 & 2x^2 & +\,5x & +\,15 & \\ \hline
x-3 & 2x^3 & -\,x^2 & & +\,3 \\
 & & -\,6x^2 & & \\ \cline{2-3}
 & & 5x^2 & & \\
 & & & -\,15x & \\ \cline{3-4}
 & & & 15x & \\
 & & & & -\,45 \\ \cline{4-5}
 & & & & 48
\end{array}$$

Most of the x's that appear in this process can also be removed, provided that we are careful about positioning each coefficient. In this regard, we will need to use 0 as the coefficient of x in the dividend, because that power of x is missing. Now we have

$$\begin{array}{r|rrrr}
 & 2x^2 & +\,5x & +\,15 & \\ \hline
x-3 & 2 & -\,1 & 0 & 3 \\
 & & -\,6 & & \\ \cline{2-3}
 & & 5 & & \\
 & & & -\,15 & \\ \cline{3-4}
 & & & 15 & \\
 & & & & -\,45 \\ \cline{4-5}
 & & & & 48
\end{array}$$

We can make this display more compact by moving the lines up until the numbers in blue align horizontally.

$$\begin{array}{r|rrrrl}
 & 2x^2 & +\,5x & +\,15 & & \text{Row 1} \\ \hline
x-3 & 2 & -\,1 & 0 & 3 & \text{Row 2} \\
 & & -\,6 & -\,15 & -\,45 & \text{Row 3} \\ \cline{2-5}
 & \bigcirc & 5 & 15 & 48 & \text{Row 4}
\end{array}$$

Because the leading coefficient of the divisor is always 1, we know that the leading coefficient of the dividend will also be the leading coefficient of the quotient. So we place the leading coefficient of the quotient, 2, in the circled position. Now, the first three numbers in row 4 are precisely the coefficients of the quotient, and the last number in row 4 is the remainder. Thus, row 1 is not really needed, so we can compress the process to three rows, where the bottom row contains both the coefficients of the quotient and the remainder.

$$\begin{array}{r|rrrrl}
x-3 & 2 & -\,1 & 0 & 3 & \text{Row 1} \\
 & & -\,6 & -\,15 & -\,45 & \text{Row 2 (subtract)} \\ \cline{2-5}
 & 2 & 5 & 15 & 48 & \text{Row 3}
\end{array}$$

Recall that the entries in row 3 are obtained by subtracting the entries in row 2 from those in row 1. Rather than subtracting the entries in row 2, we can change the sign of each entry and add. With this modification, our display will look like this:

$$\begin{array}{r|rrrrl}
x-3 & 2 & -\,1 & 0 & 3 & \text{Row 1} \\
 & & 6 & 15 & 45 & \text{Row 2 (add)} \\ \cline{2-5}
 & 2 & 5 & 15 & 48 & \text{Row 3}
\end{array}$$

Notice that the entries in row 2 are three times the prior entries in row 3. Our last modification to the display replaces the $x - 3$ by 3. The entries in row 3 give the quotient and the remainder, as shown next.

$$\begin{array}{r|rrrr} 3 & 2 & -1 & 0 & 3 \\ & & 6 & 15 & 45 \\ \hline & 2 & 5 & 15 & 48 \end{array}$$

Row 1
Row 2 (add)
Row 3

Quotient: $2x^2 + 5x + 15$ Remainder: 48

Let's go through an example step by step.

EXAMPLE 1 Using Synthetic Division to Find the Quotient and Remainder

Use synthetic division to find the quotient and remainder when

$$x^3 - 4x^2 - 5 \quad \text{is divided by} \quad x - 3$$

Solution

STEP 1: Write the dividend in descending powers of x. Then copy the coefficients, remembering to insert a 0 for any missing powers of x.

$$1 \quad -4 \quad 0 \quad -5 \qquad \text{Row 1}$$

STEP 2: Insert the usual division symbol. In synthetic division, the divisor is of the form $x - c$, and c is the number placed to the left of the division symbol. Here, since the divisor is $x - 3$, we insert 3 to the left of the division symbol.

$$\begin{array}{r|rrrr} 3 & 1 & -4 & 0 & -5 \end{array} \qquad \text{Row 1}$$

STEP 3: Bring the 1 down two rows, and enter it in row 3.

$$\begin{array}{r|rrrrl} 3 & 1 & -4 & 0 & -5 & \text{Row 1} \\ & \downarrow & & & & \text{Row 2} \\ \hline & 1 & & & & \text{Row 3} \end{array}$$

STEP 4: Multiply the latest entry in row 3 by 3, and place the result in row 2, one column over to the right.

$$\begin{array}{r|rrrrl} 3 & 1 & -4 & 0 & -5 & \text{Row 1} \\ & & 3 & & & \text{Row 2} \\ \hline & 1 & & & & \text{Row 3} \end{array}$$

STEP 5: Add the entry in row 2 to the entry above it in row 1, and enter the sum in row 3.

$$\begin{array}{r|rrrrl} 3 & 1 & -4 & 0 & -5 & \text{Row 1} \\ & & 3 & & & \text{Row 2} \\ \hline & 1 & -1 & & & \text{Row 3} \end{array}$$

STEP 6: Repeat Steps 4 and 5 until no more entries are available in row 1.

$$\begin{array}{r|rrrrl} 3 & 1 & -4 & 0 & -5 & \text{Row 1} \\ & & 3 & -3 & -9 & \text{Row 2} \\ \hline & 1 & -1 & -3 & -14 & \text{Row 3} \end{array}$$

STEP 7: The final entry in row 3, the -14, is the remainder; the other entries in row 3, the 1, -1, and -3, are the coefficients (in descending order) of a polynomial whose degree is 1 less than that of the dividend. This is the quotient. Thus,

$$\text{Quotient} = x^2 - x - 3 \qquad \text{Remainder} = -14$$

Check: (Divisor)(Quotient) + Remainder

$$\begin{aligned} &= (x - 3)(x^2 - x - 3) + (-14) \\ &= (x^3 - x^2 - 3x - 3x^2 + 3x + 9) + (-14) \\ &= x^3 - 4x^2 - 5 = \text{Dividend} \end{aligned}$$

Let's do an example in which all seven steps are combined.

EXAMPLE 2 Using Synthetic Division to Verify a Factor

Use synthetic division to show that $x + 3$ is a factor of

$$2x^5 + 5x^4 - 2x^3 + 2x^2 - 2x + 3$$

Solution If the remainder is 0, then the divisor is a factor of the dividend, and the quotient is the other factor. The divisor is $x + 3 = x - (-3)$, so we place -3 to the left of the division symbol. Then the row 3 entries will be multiplied by -3, entered in row 2, and added to row 1.

$$\begin{array}{r|rrrrrrl} -3 & 2 & 5 & -2 & 2 & -2 & 3 & \text{Row 1} \\ & & -6 & 3 & -3 & 3 & -3 & \text{Row 2} \\ \hline & 2 & -1 & 1 & -1 & 1 & 0 & \text{Row 3} \end{array}$$

Because the remainder is 0, we have

$$\text{Dividend} = (\text{Divisor})(\text{Quotient}) + \text{Remainder}$$

$$2x^5 + 5x^4 - 2x^3 + 2x^2 - 2x + 3 = (x + 3)(2x^4 - x^3 + x^2 - x + 1)$$

As we see, $x + 3$ is a factor of $2x^5 + 5x^4 - 2x^3 + 2x^2 - 2x + 3$.

As Example 2 illustrates, the remainder after division gives information about whether the divisor is, or is not, a factor. We shall have more to say about this in Chapter 5.

Now Work PROBLEMS 7 AND 17

R.6 Assess Your Understanding

Concepts and Vocabulary

1. To check division, the expression that is being divided, the dividend, should equal the product of the ______ and the ______ plus the ______.
2. To divide $2x^3 - 5x + 1$ by $x + 3$ using synthetic division, the first step is to write ______$\overline{)______}$.
3. ***True or False*** In using synthetic division, the divisor is always a polynomial of degree 1, whose leading coefficient is 1.
4. ***True or False*** $\begin{array}{r|rrrr} -2 & 5 & 3 & 2 & 1 \\ & & -10 & 14 & -32 \\ \hline & 5 & -7 & 16 & -31 \end{array}$ means $\dfrac{5x^3 + 3x^2 + 2x + 1}{x + 2} = 5x^2 - 7x + 16 + \dfrac{-31}{x + 2}$.

Skill Building

In Problems 5–16, use synthetic division to find the quotient and remainder when:

5. $x^3 - x^2 + 2x + 4$ is divided by $x - 2$
6. $x^3 + 2x^2 - 3x + 1$ is divided by $x + 1$
7. $3x^3 + 2x^2 - x + 3$ is divided by $x - 3$
8. $-4x^3 + 2x^2 - x + 1$ is divided by $x + 2$
9. $x^5 - 4x^3 + x$ is divided by $x + 3$
10. $x^4 + x^2 + 2$ is divided by $x - 2$
11. $4x^6 - 3x^4 + x^2 + 5$ is divided by $x - 1$
12. $x^5 + 5x^3 - 10$ is divided by $x + 1$
13. $0.1x^3 + 0.2x$ is divided by $x + 1.1$
14. $0.1x^2 - 0.2$ is divided by $x + 2.1$
15. $x^5 - 1$ is divided by $x - 1$
16. $x^5 + 1$ is divided by $x + 1$

In Problems 17–26, use synthetic division to determine whether $x - c$ is a factor of the given polynomial.

17. $4x^3 - 3x^2 - 8x + 4$; $x - 2$
18. $-4x^3 + 5x^2 + 8$; $x + 3$
19. $3x^4 - 6x^3 - 5x + 10$; $x - 2$
20. $4x^4 - 15x^2 - 4$; $x - 2$
21. $3x^6 + 82x^3 + 27$; $x + 3$
22. $2x^6 - 18x^4 + x^2 - 9$; $x + 3$

23. $4x^6 - 64x^4 + x^2 - 15; \quad x + 4$

24. $x^6 - 16x^4 + x^2 - 16; \quad x + 4$

25. $2x^4 - x^3 + 2x - 1; \quad x - \frac{1}{2}$

26. $3x^4 + x^3 - 3x + 1; \quad x + \frac{1}{3}$

Applications and Extensions

27. Find the sum of a, b, c, and d if

$$\frac{x^3 - 2x^2 + 3x + 5}{x + 2} = ax^2 + bx + c + \frac{d}{x + 2}$$

Discussion and Writing

28. When dividing a polynomial by $x - c$, do you prefer to use long division or synthetic division? Does the value of c make a difference to you in choosing? Give reasons.

R.7 Rational Expressions

OBJECTIVES

1 Reduce a Rational Expression to Lowest Terms

If we form the quotient of two polynomials, the result is called a **rational expression.** Some examples of rational expressions are

(a) $\frac{x^3 + 1}{x}$ (b) $\frac{3x^2 + x - 2}{x^2 + 5}$ (c) $\frac{x}{x^2 - 1}$ (d) $\frac{xy^2}{(x - y)^2}$

Expressions (a), (b), and (c) are rational expressions in one variable, x, whereas (d) is a rational expression in two variables, x and y.

Rational expressions are described in the same manner as rational numbers. In expression (a), the polynomial $x^3 + 1$ is called the **numerator,** and x is called the **denominator.** When the numerator and denominator of a rational expression contain no common factors (except 1 and -1), we say that the rational expression is **reduced to lowest terms,** or **simplified.**

The polynomial in the denominator of a rational expression cannot be equal to 0 because division by 0 is not defined. For example, for the expression $\frac{x^3 + 1}{x}$, x cannot take on the value 0. The domain of the variable x is $\{x | x \neq 0\}$.

A rational expression is reduced to lowest terms by factoring completely the numerator and the denominator and canceling any common factors by using the Cancellation Property:

$$\frac{a\not{c}}{b\not{c}} = \frac{a}{b} \quad \text{if } b \neq 0, c \neq 0 \tag{1}$$

EXAMPLE 1 **Reducing a Rational Expression to Lowest Terms**

Reduce to lowest terms: $\frac{x^2 + 4x + 4}{x^2 + 3x + 2}$

Solution We begin by factoring the numerator and the denominator.

$$x^2 + 4x + 4 = (x + 2)(x + 2)$$
$$x^2 + 3x + 2 = (x + 2)(x + 1)$$

WARNING Apply the Cancellation Property only to rational expressions written in factored form. Be sure to cancel only common factors, not common terms!

Since a common factor, $x + 2$, appears, the original expression is not in lowest terms. To reduce it to lowest terms, we use the Cancellation Property:

$$\frac{x^2 + 4x + 4}{x^2 + 3x + 2} = \frac{\cancel{(x + 2)}(x + 2)}{\cancel{(x + 2)}(x + 1)} = \frac{x + 2}{x + 1} \qquad x \neq -2, -1$$

EXAMPLE 2 **Reducing Rational Expressions to Lowest Terms**

Reduce each rational expression to lowest terms.

(a) $\dfrac{x^3 - 8}{x^3 - 2x^2}$ (b) $\dfrac{8 - 2x}{x^2 - x - 12}$

Solution (a) $\dfrac{x^3 - 8}{x^3 - 2x^2} = \dfrac{\cancel{(x - 2)}(x^2 + 2x + 4)}{x^2\cancel{(x - 2)}} = \dfrac{x^2 + 2x + 4}{x^2} \qquad x \neq 0, 2$

(b) $\dfrac{8 - 2x}{x^2 - x - 12} = \dfrac{2(4 - x)}{(x - 4)(x + 3)} = \dfrac{2(-1)\cancel{(x - 4)}}{\cancel{(x - 4)}(x + 3)} = \dfrac{-2}{x + 3} \qquad x \neq -3, 4$

Now Work PROBLEM 5

2 Multiply and Divide Rational Expressions

The rules for multiplying and dividing rational expressions are the same as the rules for multiplying and dividing rational numbers. If $\frac{a}{b}$ and $\frac{c}{d}$, $b \neq 0$, $d \neq 0$, are two rational expressions, then

$$\frac{a}{b} \cdot \frac{c}{d} = \frac{ac}{bd} \qquad \text{if } b \neq 0, d \neq 0 \qquad \textbf{(2)}$$

$$\frac{\frac{a}{b}}{\frac{c}{d}} = \frac{a}{b} \cdot \frac{d}{c} = \frac{ad}{bc} \qquad \text{if } b \neq 0, c \neq 0, d \neq 0 \qquad \textbf{(3)}$$

In using equations (2) and (3) with rational expressions, be sure first to factor each polynomial completely so that common factors can be canceled. Leave your answer in factored form.

EXAMPLE 3 **Multiplying and Dividing Rational Expressions**

Perform the indicated operation and simplify the result. Leave your answer in factored form.

(a) $\dfrac{x^2 - 2x + 1}{x^3 + x} \cdot \dfrac{4x^2 + 4}{x^2 + x - 2}$

(b) $\dfrac{\dfrac{x + 3}{x^2 - 4}}{\dfrac{x^2 - x - 12}{x^3 - 8}}$

Solution (a) $\dfrac{x^2-2x+1}{x^3+x}\cdot\dfrac{4x^2+4}{x^2+x-2} = \dfrac{(x-1)^2}{x(x^2+1)}\cdot\dfrac{4(x^2+1)}{(x+2)(x-1)}$

$$= \frac{(x-1)^2(4)\cancel{(x^2+1)}}{x\cancel{(x^2+1)}(x+2)\cancel{(x-1)}}$$

$$= \frac{4(x-1)}{x(x+2)} \qquad x \neq -2, 0, 1$$

(b) $\dfrac{\dfrac{x+3}{x^2-4}}{\dfrac{x^2-x-12}{x^3-8}} = \dfrac{x+3}{x^2-4}\cdot\dfrac{x^3-8}{x^2-x-12}$

$$= \frac{x+3}{(x-2)(x+2)}\cdot\frac{(x-2)(x^2+2x+4)}{(x-4)(x+3)}$$

$$= \frac{\cancel{(x+3)}\cancel{(x-2)}(x^2+2x+4)}{\cancel{(x-2)}(x+2)(x-4)\cancel{(x+3)}}$$

$$= \frac{x^2+2x+4}{(x+2)(x-4)} \qquad x \neq -3, -2, 2, 4$$

Now Work PROBLEMS 17 AND 25

3 Add and Subtract Rational Expressions

The rules for adding and subtracting rational expressions are the same as the rules for adding and subtracting rational numbers. So, if the denominators of two rational expressions to be added (or subtracted) are equal, we add (or subtract) the numerators and keep the common denominator.

> **In Words**
> To add (or subtract) two rational expressions with the same denominator, keep the common denominator and add (or subtract) the numerators.

If $\dfrac{a}{b}$ and $\dfrac{c}{b}$ are two rational expressions, then

$$\frac{a}{b}+\frac{c}{b}=\frac{a+c}{b} \qquad \frac{a}{b}-\frac{c}{b}=\frac{a-c}{b} \qquad \text{if } b \neq 0 \qquad \textbf{(4)}$$

EXAMPLE 4 Adding and Subtracting Rational Expressions with Equal Denominators

Perform the indicated operation and simplify the result. Leave your answer in factored form.

(a) $\dfrac{2x^2-4}{2x+5}+\dfrac{x+3}{2x+5} \quad x \neq -\dfrac{5}{2}$ (b) $\dfrac{x}{x-3}-\dfrac{3x+2}{x-3} \quad x \neq 3$

Solution (a) $\dfrac{2x^2-4}{2x+5}+\dfrac{x+3}{2x+5} = \dfrac{(2x^2-4)+(x+3)}{2x+5}$

$$= \frac{2x^2+x-1}{2x+5} = \frac{(2x-1)(x+1)}{2x+5}$$

(b) $\dfrac{x}{x-3}-\dfrac{3x+2}{x-3} = \dfrac{x-(3x+2)}{x-3} = \dfrac{x-3x-2}{x-3}$

$$= \frac{-2x-2}{x-3} = \frac{-2(x+1)}{x-3}$$

EXAMPLE 5 Adding Rational Expressions Whose Denominators Are Additive Inverses of Each Other

Perform the indicated operation and simplify the result. Leave your answer in factored form.

$$\frac{2x}{x-3} + \frac{5}{3-x} \qquad x \neq 3$$

Solution Notice that the denominators of the two rational expressions are different. However, the denominator of the second expression is the additive inverse of the denominator of the first. That is,

$$3 - x = -x + 3 = -1 \cdot (x - 3) = -(x - 3)$$

Then

$$\frac{2x}{x-3} + \frac{5}{3-x} \underset{3-x=-(x-3)}{=} \frac{2x}{x-3} + \frac{5}{-(x-3)} \underset{\frac{a}{-b}=\frac{-a}{b}}{=} \frac{2x}{x-3} + \frac{-5}{x-3}$$

$$= \frac{2x + (-5)}{x-3} = \frac{2x-5}{x-3}$$

Now Work PROBLEMS 37 AND 43

If the denominators of two rational expressions to be added or subtracted are not equal, we can use the general formulas for adding and subtracting quotients.

$$\frac{a}{b} + \frac{c}{d} = \frac{a \cdot d}{b \cdot d} + \frac{b \cdot c}{b \cdot d} = \frac{ad + bc}{bd} \qquad \text{if } b \neq 0, d \neq 0 \qquad \textbf{(5a)}$$

$$\frac{a}{b} - \frac{c}{d} = \frac{a \cdot d}{b \cdot d} - \frac{b \cdot c}{b \cdot d} = \frac{ad - bc}{bd} \qquad \text{if } b \neq 0, d \neq 0 \qquad \textbf{(5b)}$$

EXAMPLE 6 Adding and Subtracting Rational Expressions with Unequal Denominators

Perform the indicated operation and simplify the result. Leave your answer in factored form.

(a) $\frac{x-3}{x+4} + \frac{x}{x-2} \qquad x \neq -4, 2$

(b) $\frac{x^2}{x^2-4} - \frac{1}{x} \qquad x \neq -2, 0, 2$

Solution (a)

$$\frac{x-3}{x+4} + \frac{x}{x-2} \underset{(5a)}{=} \frac{x-3}{x+4} \cdot \frac{x-2}{x-2} + \frac{x+4}{x+4} \cdot \frac{x}{x-2}$$

$$= \frac{(x-3)(x-2) + (x+4)(x)}{(x+4)(x-2)}$$

$$= \frac{x^2 - 5x + 6 + x^2 + 4x}{(x+4)(x-2)} = \frac{2x^2 - x + 6}{(x+4)(x-2)}$$

(b) $$\frac{x^2}{x^2-4} - \frac{1}{x} \underset{\substack{\uparrow \\ (5b)}}{=} \frac{x^2}{x^2-4}\cdot\frac{x}{x} - \frac{x^2-4}{x^2-4}\cdot\frac{1}{x} = \frac{x^2(x) - (x^2-4)(1)}{(x^2-4)(x)}$$

$$= \frac{x^3 - x^2 + 4}{x(x-2)(x+2)}$$

Now Work PROBLEM 47

4 Use the Least Common Multiple Method

If the denominators of two rational expressions to be added (or subtracted) have common factors, we usually do not use the general rules given by equations (5a) and (5b). Just as with fractions, we apply the **least common multiple (LCM) method.** The LCM method uses the polynomial of least degree that has each denominator polynomial as a factor.

The LCM Method for Adding or Subtracting Rational Expressions

The Least Common Multiple (LCM) Method requires four steps:

STEP 1: Factor completely the polynomial in the denominator of each rational expression.
STEP 2: The LCM of the denominator is the product of each of these factors raised to a power equal to the greatest number of times that the factor occurs in the polynomials.
STEP 3: Write each rational expression using the LCM as the common denominator.
STEP 4: Add or subtract the rational expressions using equation (4).

Let's work an example that only requires Steps 1 and 2.

EXAMPLE 7 Finding the Least Common Multiple

Find the least common multiple of the following pair of polynomials:

$x(x-1)^2(x+1)$ and $4(x-1)(x+1)^3$

Solution **STEP 1:** The polynomials are already factored completely as

$$x(x-1)^2(x+1) \quad \text{and} \quad 4(x-1)(x+1)^3$$

STEP 2: Start by writing the factors of the left-hand polynomial. (Or you could start with the one on the right.)

$$x(x-1)^2(x+1)$$

Now look at the right-hand polynomial. Its first factor, 4, does not appear in the polynomial above, so we insert it.

$$4x(x-1)^2(x+1)$$

The next factor, $x-1$, is already in our polynomial, so no change is necessary. The final factor is $(x+1)^3$. Since our polynomial has $x+1$ to the first power only, we replace $x+1$ in the polynomial by $(x+1)^3$. The LCM is

$$4x(x-1)^2(x+1)^3$$

Notice that the LCM is, in fact, the polynomial of least degree that contains $x(x-1)^2(x+1)$ and $4(x-1)(x+1)^3$ as factors.

Now Work PROBLEM 53

EXAMPLE 8 Using the Least Common Multiple to Add Rational Expressions

Perform the indicated operation and simplify the result. Leave your answer in factored form.

$$\frac{x}{x^2+3x+2} + \frac{2x-3}{x^2-1} \qquad x \neq -2, -1, 1$$

Solution **STEP 1:** Factor completely the polynomials in the denominators.

$$x^2+3x+2 = (x+2)(x+1)$$
$$x^2-1 = (x-1)(x+1)$$

STEP 2: The LCM is $(x+2)(x+1)(x-1)$. Do you see why?

STEP 3: Write each rational expression using the LCM as the denominator.

$$\frac{x}{x^2+3x+2} = \frac{x}{(x+2)(x+1)} = \frac{x}{(x+2)(x+1)} \cdot \frac{x-1}{x-1} = \frac{x(x-1)}{(x+2)(x+1)(x-1)}$$

Multiply numerator and denominator by $x - 1$ to get the LCM in the denominator.

$$\frac{2x-3}{x^2-1} = \frac{2x-3}{(x-1)(x+1)} = \frac{2x-3}{(x-1)(x+1)} \cdot \frac{x+2}{x+2} = \frac{(2x-3)(x+2)}{(x-1)(x+1)(x+2)}$$

Multiply numerator and denominator by $x + 2$ to get the LCM in the denominator

STEP 4: Now we can add by using equation (4).

$$\frac{x}{x^2+3x+2} + \frac{2x-3}{x^2-1} = \frac{x(x-1)}{(x+2)(x+1)(x-1)} + \frac{(2x-3)(x+2)}{(x+2)(x+1)(x-1)}$$
$$= \frac{(x^2-x)+(2x^2+x-6)}{(x+2)(x+1)(x-1)}$$
$$= \frac{3x^2-6}{(x+2)(x+1)(x-1)} = \frac{3(x^2-2)}{(x+2)(x+1)(x-1)}$$

EXAMPLE 9 Using the Least Common Multiple to Subtract Rational Expressions

Perform the indicated operations and simplify the result. Leave your answer in factored form.

$$\frac{3}{x^2+x} - \frac{x+4}{x^2+2x+1} \qquad x \neq -1, 0$$

Solution **STEP 1:** Factor completely the polynomials in the denominators.

$$x^2+x = x(x+1)$$
$$x^2+2x+1 = (x+1)^2$$

STEP 2: The LCM is $x(x+1)^2$.

STEP 3: Write each rational expression using the LCM as the denominator.

$$\frac{3}{x^2+x} = \frac{3}{x(x+1)} = \frac{3}{x(x+1)} \cdot \frac{x+1}{x+1} = \frac{3(x+1)}{x(x+1)^2}$$

$$\frac{x+4}{x^2+2x+1} = \frac{x+4}{(x+1)^2} = \frac{x+4}{(x+1)^2} \cdot \frac{x}{x} = \frac{x(x+4)}{x(x+1)^2}$$

STEP 4: Subtract, using equation (4).

$$\frac{3}{x^2+x} - \frac{x+4}{x^2+2x+1} = \frac{3(x+1)}{x(x+1)^2} - \frac{x(x+4)}{x(x+1)^2}$$
$$= \frac{3(x+1) - x(x+4)}{x(x+1)^2}$$
$$= \frac{3x+3-x^2-4x}{x(x+1)^2}$$
$$= \frac{-x^2-x+3}{x(x+1)^2}$$

Now Work PROBLEM 63

5 Simplify Complex Rational Expressions

When sums and/or differences of rational expressions appear as the numerator and/or denominator of a quotient, the quotient is called a **complex rational expression.*** For example,

$$\frac{1+\frac{1}{x}}{1-\frac{1}{x}} \quad \text{and} \quad \frac{\frac{x^2}{x^2-4}-3}{\frac{x-3}{x+2}-1}$$

are complex rational expressions. To **simplify** a complex rational expression means to write it as a rational expression reduced to lowest terms. This can be accomplished in either of two ways.

Simplifying a Complex Rational Expression

METHOD 1: Treat the numerator and denominator of the complex rational expression separately, performing whatever operations are indicated and simplifying the results. Follow this by simplifying the resulting rational expression.

METHOD 2: Find the LCM of the denominators of all rational expressions that appear in the complex rational expression. Multiply the numerator and denominator of the complex rational expression by the LCM and simplify the result.

We will use both methods in the next example. By carefully studying each method, you can discover situations in which one method may be easier to use than the other.

EXAMPLE 10 Simplifying a Complex Rational Expression

Simplify: $\dfrac{\frac{1}{2}+\frac{3}{x}}{\frac{x+3}{4}} \qquad x \neq -3, 0$

*Some texts use the term **complex fraction.**

Solution *Method 1:* First, we perform the indicated operation in the numerator, and then we divide.

$$\frac{\frac{1}{2}+\frac{3}{x}}{\frac{x+3}{4}} = \frac{\frac{1\cdot x + 2\cdot 3}{2\cdot x}}{\frac{x+3}{4}} = \frac{\frac{x+6}{2x}}{\frac{x+3}{4}} = \frac{x+6}{2x}\cdot\frac{4}{x+3}$$

Rule for adding quotients; Rule for dividing quotients

$$= \frac{(x+6)\cdot 4}{2\cdot x\cdot (x+3)} = \frac{\cancel{2}\cdot 2\cdot (x+6)}{\cancel{2}\cdot x\cdot (x+3)} = \frac{2(x+6)}{x(x+3)}$$

Rule for multiplying quotients

Method 2: The rational expressions that appear in the complex rational expression are

$$\frac{1}{2} \qquad \frac{3}{x} \qquad \frac{x+3}{4}$$

The LCM of their denominators is $4x$. We multiply the numerator and denominator of the complex rational expression by $4x$ and then simplify.

$$\frac{\frac{1}{2}+\frac{3}{x}}{\frac{x+3}{4}} = \frac{4x\cdot\left(\frac{1}{2}+\frac{3}{x}\right)}{4x\cdot\left(\frac{x+3}{4}\right)} = \frac{4x\cdot\frac{1}{2} + 4x\cdot\frac{3}{x}}{\frac{4x\cdot(x+3)}{4}}$$

Multiply the numerator and denominator by 4x. Use the Distributive Property in the numerator.

$$= \frac{\cancel{2}\cdot 2x\cdot\frac{1}{\cancel{2}} + 4\cancel{x}\cdot\frac{3}{\cancel{x}}}{\frac{\cancel{4}x\cdot(x+3)}{\cancel{4}}} = \frac{2x+12}{x(x+3)} = \frac{2(x+6)}{x(x+3)}$$

Simplify; Factor

EXAMPLE 11 Simplifying a Complex Rational Expression

Simplify: $\dfrac{\frac{x^2}{x-4}+2}{\frac{2x-2}{x}-1} \qquad x \neq 0, 2, 4$

Solution We will use Method 1.

$$\frac{\frac{x^2}{x-4}+2}{\frac{2x-2}{x}-1} = \frac{\frac{x^2}{x-4}+\frac{2(x-4)}{x-4}}{\frac{2x-2}{x}-\frac{x}{x}} = \frac{\frac{x^2+2x-8}{x-4}}{\frac{2x-2-x}{x}}$$

$$= \frac{\frac{(x+4)(x-2)}{x-4}}{\frac{x-2}{x}} = \frac{(x+4)\cancel{(x-2)}}{x-4}\cdot\frac{x}{\cancel{x-2}} = \frac{(x+4)\cdot x}{x-4}$$

Now Work PROBLEM 73

Application

EXAMPLE 12 **Solving an Application in Electricity**

An electrical circuit contains two resistors connected in parallel, as shown in Figure 31. If the resistance of each is R_1 and R_2 ohms, respectively, their combined resistance R is given by the formula

Figure 31

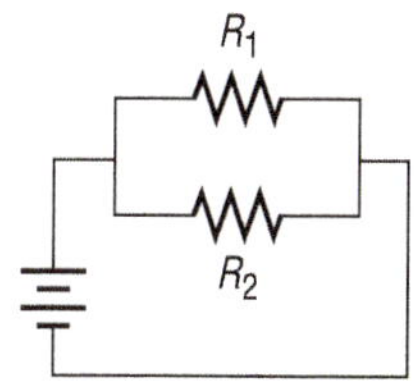

$$R = \frac{1}{\frac{1}{R_1} + \frac{1}{R_2}}$$

Express R as a rational expression; that is, simplify the right-hand side of this formula. Evaluate the rational expression if $R_1 = 6$ ohms and $R_2 = 10$ ohms.

Solution We will use Method 2. If we consider 1 as the fraction $\frac{1}{1}$, then the rational expressions in the complex rational expression are

$$\frac{1}{1} \quad \frac{1}{R_1} \quad \frac{1}{R_2}$$

The LCM of the denominators is R_1R_2. We multiply the numerator and denominator of the complex rational expression by R_1R_2 and simplify.

$$R = \frac{1}{\frac{1}{R_1} + \frac{1}{R_2}} = \frac{1 \cdot R_1R_2}{\left(\frac{1}{R_1} + \frac{1}{R_2}\right) \cdot R_1R_2} = \frac{R_1R_2}{\frac{1}{R_1} \cdot R_1R_2 + \frac{1}{R_2} \cdot R_1R_2} = \frac{R_1R_2}{R_2 + R_1}$$

Thus,

$$R = \frac{R_1R_2}{R_2 + R_1}$$

If $R_1 = 6$ and $R_2 = 10$, then

$$R = \frac{6 \cdot 10}{10 + 6} = \frac{60}{16} = \frac{15}{4} \text{ ohms}$$

R.7 Assess Your Understanding

Concepts and Vocabulary

1. When the numerator and denominator of a rational expression contain no common factors (except 1 and −1), the rational expression is in _____ _____.
2. LCM is an abbreviation for _____ _____ _____.
3. ***True or False*** The rational expression $\frac{2x^3 - 4x}{x - 2}$ is reduced to lowest terms.
4. ***True or False*** The LCM of $2x^3 + 6x^2$ and $6x^4 + 4x^3$ is $4x^3(x + 1)$.

Skill Building

In Problems 5–16, reduce each rational expression to lowest terms.

5. $\frac{3x + 9}{x^2 - 9}$
6. $\frac{4x^2 + 8x}{12x + 24}$
7. $\frac{x^2 - 2x}{3x - 6}$
8. $\frac{15x^2 + 24x}{3x^2}$
9. $\frac{24x^2}{12x^2 - 6x}$
10. $\frac{x^2 + 4x + 4}{x^2 - 4}$
11. $\frac{y^2 - 25}{2y^2 - 8y - 10}$
12. $\frac{3y^2 - y - 2}{3y^2 + 5y + 2}$
13. $\frac{x^2 + 4x - 5}{x^2 - 2x + 1}$
14. $\frac{x - x^2}{x^2 + x - 2}$
15. $\frac{x^2 + 5x - 14}{2 - x}$
16. $\frac{2x^2 + 5x - 3}{1 - 2x}$

In Problems 17–34, perform the indicated operation and simplify the result. Leave your answer in factored form.

17. $\frac{3x+6}{5x^2} \cdot \frac{x}{x^2-4}$

18. $\frac{3}{2x} \cdot \frac{x^2}{6x+10}$

19. $\frac{4x^2}{x^2-16} \cdot \frac{x^3-64}{2x}$

20. $\frac{12}{x^2+x} \cdot \frac{x^3+1}{4x-2}$

21. $\frac{4x-8}{-3x} \cdot \frac{12}{12-6x}$

22. $\frac{6x-27}{5x} \cdot \frac{2}{4x-18}$

23. $\frac{x^2-3x-10}{x^2+2x-35} \cdot \frac{x^2+4x-21}{x^2+9x+14}$

24. $\frac{x^2+x-6}{x^2+4x-5} \cdot \frac{x^2-25}{x^2+2x-15}$

25. $\frac{\frac{6x}{x^2-4}}{\frac{3x-9}{2x+4}}$

26. $\frac{\frac{12x}{5x+20}}{\frac{4x^2}{x^2-16}}$

27. $\frac{\frac{8x}{x^2-1}}{\frac{10x}{x+1}}$

28. $\frac{\frac{x-2}{4x}}{\frac{x^2-4x+4}{12x}}$

29. $\frac{\frac{4-x}{4+x}}{\frac{4x}{x^2-16}}$

30. $\frac{\frac{3+x}{3-x}}{\frac{x^2-9}{9x^3}}$

31. $\frac{\frac{x^2+7x+12}{x^2-7x+12}}{\frac{x^2+x-12}{x^2-x-12}}$

32. $\frac{\frac{x^2+7x+6}{x^2+x-6}}{\frac{x^2+5x-6}{x^2+5x+6}}$

33. $\frac{\frac{2x^2-x-28}{3x^2-x-2}}{\frac{4x^2+16x+7}{3x^2+11x+6}}$

34. $\frac{\frac{9x^2+3x-2}{12x^2+5x-2}}{\frac{9x^2-6x+1}{8x^2-10x-3}}$

In Problems 35–52, perform the indicated operation and simplify the result. Leave your answer in factored form.

35. $\frac{x}{2}+\frac{5}{2}$

36. $\frac{3}{x}-\frac{6}{x}$

37. $\frac{x^2}{2x-3}-\frac{4}{2x-3}$

38. $\frac{3x^2}{2x-1}-\frac{9}{2x-1}$

39. $\frac{x+1}{x-3}+\frac{2x-3}{x-3}$

40. $\frac{2x-5}{3x+2}+\frac{x+4}{3x+2}$

41. $\frac{3x+5}{2x-1}-\frac{2x-4}{2x-1}$

42. $\frac{5x-4}{3x+4}-\frac{x+1}{3x+4}$

43. $\frac{4}{x-2}+\frac{x}{2-x}$

44. $\frac{6}{x-1}-\frac{x}{1-x}$

45. $\frac{4}{x-1}-\frac{2}{x+2}$

46. $\frac{2}{x+5}-\frac{5}{x-5}$

47. $\frac{x}{x+1}+\frac{2x-3}{x-1}$

48. $\frac{3x}{x-4}+\frac{2x}{x+3}$

49. $\frac{x-3}{x+2}-\frac{x+4}{x-2}$

50. $\frac{2x-3}{x-1}-\frac{2x+1}{x+1}$

51. $\frac{x}{x^2-4}+\frac{1}{x}$

52. $\frac{x-1}{x^3}+\frac{x}{x^2+1}$

In Problems 53–60, find the LCM of the given polynomials.

53. $x^2-4,\ \ x^2-x-2$

54. $x^2-x-12,\ \ x^2-8x+16$

55. $x^3-x,\ \ x^2-x$

56. $3x^2-27,\ \ 2x^2-x-15$

57. $4x^3-4x^2+x,\ \ 2x^3-x^2,\ \ x^3$

58. $x-3,\ \ x^2+3x,\ \ x^3-9x$

59. $x^3-x,\ \ x^3-2x^2+x,\ \ x^3-1$

60. $x^2+4x+4,\ \ x^3+2x^2,\ \ (x+2)^3$

In Problems 61–72, perform the indicated operations and simplify the result. Leave your answer in factored form.

61. $\frac{x}{x^2-7x+6}-\frac{x}{x^2-2x-24}$

62. $\frac{x}{x-3}-\frac{x+1}{x^2+5x-24}$

63. $\frac{4x}{x^2-4}-\frac{2}{x^2+x-6}$

64. $\frac{3x}{x-1}-\frac{x-4}{x^2-2x+1}$

65. $\frac{3}{(x-1)^2(x+1)}+\frac{2}{(x-1)(x+1)^2}$

66. $\frac{2}{(x+2)^2(x-1)}-\frac{6}{(x+2)(x-1)^2}$

67. $\frac{x+4}{x^2-x-2}-\frac{2x+3}{x^2+2x-8}$

68. $\frac{2x-3}{x^2+8x+7}-\frac{x-2}{(x+1)^2}$

69. $\frac{1}{x}-\frac{2}{x^2+x}+\frac{3}{x^3-x^2}$

70. $\frac{x}{(x-1)^2}+\frac{2}{x}-\frac{x+1}{x^3-x^2}$

71. $\frac{1}{h}\left(\frac{1}{x+h}-\frac{1}{x}\right)$

72. $\frac{1}{h}\left[\frac{1}{(x+h)^2}-\frac{1}{x^2}\right]$

In Problems 73–84, perform the indicated operations and simplify the result. Leave your answer in factored form.

73. $\dfrac{1+\dfrac{1}{x}}{1-\dfrac{1}{x}}$ **74.** $\dfrac{4+\dfrac{1}{x^2}}{3-\dfrac{1}{x^2}}$ **75.** $\dfrac{2-\dfrac{x+1}{x}}{3+\dfrac{x-1}{x+1}}$ **76.** $\dfrac{1-\dfrac{x}{x+1}}{2-\dfrac{x-1}{x}}$

77. $\dfrac{\dfrac{x+4}{x-2}-\dfrac{x-3}{x+1}}{x+1}$ **78.** $\dfrac{\dfrac{x-2}{x+1}-\dfrac{x}{x-2}}{x+3}$ **79.** $\dfrac{\dfrac{x-2}{x+2}+\dfrac{x-1}{x+1}}{\dfrac{x}{x+1}-\dfrac{2x-3}{x}}$ **80.** $\dfrac{\dfrac{2x+5}{x}-\dfrac{x}{x-3}}{\dfrac{x^2}{x-3}-\dfrac{(x+1)^2}{x+3}}$

81. $1-\dfrac{1}{1-\dfrac{1}{x}}$ **82.** $1-\dfrac{1}{1-\dfrac{1}{1-x}}$ **83.** $\dfrac{2(x-1)^{-1}+3}{3(x-1)^{-1}+2}$ **84.** $\dfrac{4(x+2)^{-1}-3}{3(x+2)^{-1}-1}$

In Problems 85–92, expressions that occur in calculus are given. Reduce each expression to lowest terms.

85. $\dfrac{(2x+3)\cdot 3-(3x-5)\cdot 2}{(3x-5)^2}$ **86.** $\dfrac{(4x+1)\cdot 5-(5x-2)\cdot 4}{(5x-2)^2}$ **87.** $\dfrac{x\cdot 2x-(x^2+1)\cdot 1}{(x^2+1)^2}$

88. $\dfrac{x\cdot 2x-(x^2-4)\cdot 1}{(x^2-4)^2}$ **89.** $\dfrac{(3x+1)\cdot 2x-x^2\cdot 3}{(3x+1)^2}$ **90.** $\dfrac{(2x-5)\cdot 3x^2-x^3\cdot 2}{(2x-5)^2}$

91. $\dfrac{(x^2+1)\cdot 3-(3x+4)\cdot 2x}{(x^2+1)^2}$ **92.** $\dfrac{(x^2+9)\cdot 2-(2x-5)\cdot 2x}{(x^2+9)^2}$

Applications and Extensions

93. The Lensmaker's Equation The focal length f of a lens with index of refraction n is

$$\frac{1}{f}=(n-1)\left[\frac{1}{R_1}+\frac{1}{R_2}\right]$$

where R_1 and R_2 are the radii of curvature of the front and back surfaces of the lens. Express f as a rational expression. Evaluate the rational expression for $n = 1.5$, $R_1 = 0.1$ meter, and $R_2 = 0.2$ meter.

94. Electrical Circuits An electrical circuit contains three resistors connected in parallel. If the resistance of each is R_1, R_2, and R_3 ohms, respectively, their combined resistance R is given by the formula

$$\frac{1}{R}=\frac{1}{R_1}+\frac{1}{R_2}+\frac{1}{R_3}$$

Express R as a rational expression. Evaluate R for $R_1 = 5$ ohms, $R_2 = 4$ ohms, and $R_3 = 10$ ohms.

Discussion and Writing

95. The following expressions are called **continued fractions:**

$$1+\frac{1}{x} \qquad 1+\cfrac{1}{1+\cfrac{1}{x}} \qquad 1+\cfrac{1}{1+\cfrac{1}{1+\cfrac{1}{x}}} \qquad 1+\cfrac{1}{1+\cfrac{1}{1+\cfrac{1}{1+\cfrac{1}{x}}}} \quad \dots$$

Each simplifies to an expression of the form

$$\frac{ax+b}{bx+c}$$

Trace the successive values of a, b, and c as you "continue" the fraction. Can you discover the patterns that these values follow? Go to the library and research Fibonacci numbers. Write a report on your findings.

96. Explain to a fellow student when you would use the LCM method to add two rational expressions. Give two examples of adding two rational expressions, one in which you use the LCM and the other in which you do not.

97. Which of the two methods given in the text for simplifying complex rational expressions do you prefer? Write a brief paragraph stating the reasons for your choice.

R.8 *n*th Roots; Rational Exponents

PREPARING FOR THIS SECTION *Before getting started, review the following:*

- Exponents, Square Roots (Section R.2, pp. 21–24)

Now Work the 'Are You Prepared?' problems on page 77.

OBJECTIVES
1. Work with *n*th Roots (p. 72)
2. Simplify Radicals (p. 73)
3. Rationalize Denominators (p. 74)
4. Simplify Expressions with Rational Exponents (p. 75)

1 Work with *n*th Roots

DEFINITION

The **principal *n*th root of a real number *a*,** $n \geq 2$ an integer, symbolized by $\sqrt[n]{a}$, is defined as follows:

$$\sqrt[n]{a} = b \quad \text{means} \quad a = b^n$$

where $a \geq 0$ and $b \geq 0$ if n is even and a, b are any real numbers if n is odd.

In Words

The symbol $\sqrt[n]{a}$ means "give me the number, when raised to the nth power, that results in a."

Notice that if a is negative and n is even then $\sqrt[n]{a}$ is not defined. When it is defined, the principal nth root of a number is unique.

The symbol $\sqrt[n]{a}$ for the principal nth root of a is called a **radical;** the integer n is called the **index,** and a is called the **radicand.** If the index of a radical is 2, we call $\sqrt[2]{a}$ the **square root** of a and omit the index 2 by writing $\sqrt{a}$. If the index is 3, we call $\sqrt[3]{a}$ the **cube root** of a.

EXAMPLE 1 **Simplifying Principal *n*th Roots**

(a) $\sqrt[3]{8} = \sqrt[3]{2^3} = 2$ The number whose cube is 8 is 2.

(b) $\sqrt[3]{-64} = \sqrt[3]{(-4)^3} = -4$

(c) $\sqrt[4]{\frac{1}{16}} = \sqrt[4]{\left(\frac{1}{2}\right)^4} = \frac{1}{2}$

(d) $\sqrt[6]{(-2)^6} = |-2| = 2$

These are examples of **perfect roots,** since each simplifies to a rational number. Notice the absolute value in Example 1(d). If n is even, the principal nth root must be nonnegative.

In general, if $n \geq 2$ is an integer and a is a real number, we have

$$\sqrt[n]{a^n} = a \qquad \text{if } n \geq 3 \text{ is odd} \qquad \textbf{(1a)}$$

$$\sqrt[n]{a^n} = |a| \qquad \text{if } n \geq 2 \text{ is even} \qquad \textbf{(1b)}$$

Now Work PROBLEM 7

Radicals provide a way of representing many irrational real numbers. For example, there is no rational number whose square is 2. Using radicals, we can say that $\sqrt{2}$ is the positive number whose square is 2.

EXAMPLE 2 **Using a Calculator to Approximate Roots**

Use a calculator to approximate $\sqrt[5]{16}$.

Solution Figure 32 shows the result using a TI-84 Plus graphing calculator.

Figure 32

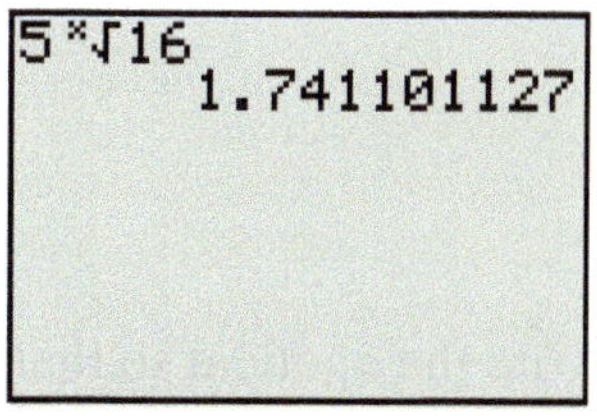

Now Work PROBLEM 101

2 Simplify Radicals

Let $n \geq 2$ and $m \geq 2$ denote positive integers, and let a and b represent real numbers. Assuming that all radicals are defined, we have the following properties:

Properties of Radicals

$$\sqrt[n]{ab} = \sqrt[n]{a}\sqrt[n]{b} \qquad \textbf{(2a)}$$

$$\sqrt[n]{\frac{a}{b}} = \frac{\sqrt[n]{a}}{\sqrt[n]{b}} \qquad \textbf{(2b)}$$

$$\sqrt[n]{a^m} = (\sqrt[n]{a})^m \qquad \textbf{(2c)}$$

When used in reference to radicals, the direction to "simplify" will mean to remove from the radicals any perfect roots that occur as factors. Let's look at some examples of how the preceding rules are applied to simplify radicals.

EXAMPLE 3 **Simplifying Radicals**

(a) $\sqrt{32} = \sqrt{16 \cdot 2} = \sqrt{16} \cdot \sqrt{2} = 4\sqrt{2}$

↑ Factor out 16, a perfect square. ↑ (2a)

(b) $\sqrt[3]{16} = \sqrt[3]{8 \cdot 2} = \sqrt[3]{8} \cdot \sqrt[3]{2} = \sqrt[3]{2^3} \cdot \sqrt[3]{2} = 2\sqrt[3]{2}$

↑ Factor out 8, a perfect cube. ↑ (2a)

(c) $\sqrt[3]{-16x^4} = \sqrt[3]{-8 \cdot 2 \cdot x^3 \cdot x} = \sqrt[3]{(-8x^3)(2x)}$

↑ Factor perfect cubes inside radical. ↑ Group perfect cubes.

$$= \sqrt[3]{(-2x)^3 \cdot 2x} = \sqrt[3]{(-2x)^3} \cdot \sqrt[3]{2x} = -2x\sqrt[3]{2x}$$

↑ (2a)

(d) $\sqrt[4]{\frac{16x^5}{81}} = \sqrt[4]{\frac{2^4x^4x}{3^4}} = \sqrt[4]{\left(\frac{2x}{3}\right)^4 \cdot x} = \sqrt[4]{\left(\frac{2x}{3}\right)^4} \cdot \sqrt[4]{x} = \left|\frac{2x}{3}\right|\sqrt[4]{x}$

Now Work PROBLEMS 11 AND 17

Two or more radicals can be combined, provided that they have the same index and the same radicand. Such radicals are called **like radicals.**

EXAMPLE 4 **Combining Like Radicals**

(a) $-8\sqrt{12} + \sqrt{3} = -8\sqrt{4 \cdot 3} + \sqrt{3}$

$$= -8 \cdot \sqrt{4}\sqrt{3} + \sqrt{3}$$

$$= -16\sqrt{3} + \sqrt{3} = -15\sqrt{3}$$

$$\text{(b) } \sqrt[3]{8x^4} + \sqrt[3]{-x} + 4\sqrt[3]{27x} = \sqrt[3]{2^3x^3x} + \sqrt[3]{-1\cdot x} + 4\sqrt[3]{3^3x}$$
$$= \sqrt[3]{(2x)^3}\cdot\sqrt[3]{x} + \sqrt[3]{-1}\cdot\sqrt[3]{x} + 4\sqrt[3]{3^3}\cdot\sqrt[3]{x}$$
$$= 2x\sqrt[3]{x} - 1\cdot\sqrt[3]{x} + 12\sqrt[3]{x}$$
$$= (2x+11)\sqrt[3]{x}$$

Now Work PROBLEM 33

3 Rationalize Denominators

When radicals occur in quotients, it is customary to rewrite the quotient so that the new denominator contains no radicals. This process is referred to as **rationalizing the denominator.**

The idea is to multiply by an appropriate expression so that the new denominator contains no radicals. For example:

If a Denominator Contains the Factor	Multiply by	To Obtain a Denominator Free of Radicals
$\sqrt{3}$	$\sqrt{3}$	$(\sqrt{3})^2 = 3$
$\sqrt{3}+1$	$\sqrt{3}-1$	$(\sqrt{3})^2 - 1^3 = 3 - 1 = 2$
$\sqrt{2}-3$	$\sqrt{2}+3$	$(\sqrt{2})^2 - 3^2 = 2 - 9 = -7$
$\sqrt{5}-\sqrt{3}$	$\sqrt{5}+\sqrt{3}$	$(\sqrt{5})^2 - (\sqrt{3})^2 = 5 - 3 = 2$
$\sqrt[3]{4}$	$\sqrt[3]{2}$	$\sqrt[3]{4}\cdot\sqrt[3]{2} = \sqrt[3]{8} = 2$

In rationalizing the denominator of a quotient, be sure to multiply both the numerator and the denominator by the expression.

EXAMPLE 5 Rationalizing Denominators

Rationalize the denominator of each expression:

(a) $\dfrac{1}{\sqrt{3}}$ (b) $\dfrac{5}{4\sqrt{2}}$ (c) $\dfrac{\sqrt{2}}{\sqrt{3}-3\sqrt{2}}$

Solution (a) The denominator contains the factor $\sqrt{3}$, so we multiply the numerator and denominator by $\sqrt{3}$ to obtain

$$\frac{1}{\sqrt{3}} = \frac{1}{\sqrt{3}}\cdot\frac{\sqrt{3}}{\sqrt{3}} = \frac{\sqrt{3}}{(\sqrt{3})^2} = \frac{\sqrt{3}}{3}$$

(b) The denominator contains the factor $\sqrt{2}$, so we multiply the numerator and denominator by $\sqrt{2}$ to obtain

$$\frac{5}{4\sqrt{2}} = \frac{5}{4\sqrt{2}}\cdot\frac{\sqrt{2}}{\sqrt{2}} = \frac{5\sqrt{2}}{4(\sqrt{2})^2} = \frac{5\sqrt{2}}{4\cdot 2} = \frac{5\sqrt{2}}{8}$$

(c) The denominator contains the factor $\sqrt{3}-3\sqrt{2}$, so we multiply the numerator and denominator by $\sqrt{3}+3\sqrt{2}$ to obtain

$$\frac{\sqrt{2}}{\sqrt{3}-3\sqrt{2}} = \frac{\sqrt{2}}{\sqrt{3}-3\sqrt{2}}\cdot\frac{\sqrt{3}+3\sqrt{2}}{\sqrt{3}+3\sqrt{2}} = \frac{\sqrt{2}(\sqrt{3}+3\sqrt{2})}{(\sqrt{3})^2-(3\sqrt{2})^2}$$
$$= \frac{\sqrt{2}\sqrt{3}+3(\sqrt{2})^2}{3-18} = \frac{\sqrt{6}+6}{-15} = -\frac{6+\sqrt{6}}{15}$$

Now Work PROBLEM 47

4 Simplify Expressions with Rational Exponents

Radicals are used to define rational exponents.

DEFINITION

If a is a real number and $n \geq 2$ is an integer, then

$$a^{1/n} = \sqrt[n]{a} \qquad \textbf{(3)}$$

provided that $\sqrt[n]{a}$ exists.

Note that if n is even and $a < 0$ then $\sqrt[n]{a}$ and $a^{1/n}$ do not exist.

EXAMPLE 6 **Writing Expressions Containing Fractional Exponents as Radicals**

(a) $4^{1/2} = \sqrt{4} = 2$

(b) $8^{1/2} = \sqrt{8} = 2\sqrt{2}$

(c) $(-27)^{1/3} = \sqrt[3]{-27} = -3$

(d) $16^{1/3} = \sqrt[3]{16} = 2\sqrt[3]{2}$

DEFINITION

If a is a real number and m and n are integers containing no common factors, with $n \geq 2$, then

$$a^{m/n} = \sqrt[n]{a^m} = (\sqrt[n]{a})^m \qquad \textbf{(4)}$$

provided that $\sqrt[n]{a}$ exists.

We have two comments about equation (4):

1. The exponent $\dfrac{m}{n}$ must be in lowest terms and n must be positive.

2. In simplifying the rational expression $a^{m/n}$, either $\sqrt[n]{a^m}$ or $(\sqrt[n]{a})^m$ may be used, the choice depending on which is easier to simplify. Generally, taking the root first, as in $(\sqrt[n]{a})^m$, is easier.

EXAMPLE 7 **Using Equation (4)**

(a) $4^{3/2} = (\sqrt{4})^3 = 2^3 = 8$

(b) $(-8)^{4/3} = (\sqrt[3]{-8})^4 = (-2)^4 = 16$

(c) $(32)^{-2/5} = (\sqrt[5]{32})^{-2} = 2^{-2} = \dfrac{1}{4}$

(d) $25^{6/4} = 25^{3/2} = (\sqrt{25})^3 = 5^3 = 125$

Now Work PROBLEM 55

It can be shown that the Laws of Exponents hold for rational exponents. The next example illustrates using the Law of Exponents to simplify.

EXAMPLE 8 **Simplifying Expressions Containing Rational Exponents**

Simplify each expression. Express your answer so that only positive exponents occur. Assume that the variables are positive.

(a) $(x^{2/3}y)(x^{-2}y)^{1/2}$

(b) $\left(\dfrac{2x^{1/3}}{y^{2/3}}\right)^{-3}$

(c) $\left(\dfrac{9x^2y^{1/3}}{x^{1/3}y}\right)^{1/2}$

Solution (a) $(x^{2/3}y)(x^{-2}y)^{1/2} = (x^{2/3}y)[(x^{-2})^{1/2}y^{1/2}]$ $\quad (ab)^n = a^n b^n$

$$= x^{2/3}yx^{-1}y^{1/2} \quad (a^m)^n = a^{mn}$$
$$= (x^{2/3}\cdot x^{-1})(y\cdot y^{1/2})$$
$$= x^{-1/3}y^{3/2} \quad a^m\cdot a^n = a^{m+n}$$
$$= \frac{y^{3/2}}{x^{1/3}}$$

(b) $$\left(\frac{2x^{1/3}}{y^{2/3}}\right)^{-3} = \left(\frac{y^{2/3}}{2x^{1/3}}\right)^3 = \frac{(y^{2/3})^3}{(2x^{1/3})^3} = \frac{y^2}{2^3(x^{1/3})^3} = \frac{y^2}{8x}$$

(c) $$\left(\frac{9x^2y^{1/3}}{x^{1/3}y}\right)^{1/2} = \left(\frac{9x^{2-(1/3)}}{y^{1-(1/3)}}\right)^{1/2} = \left(\frac{9x^{5/3}}{y^{2/3}}\right)^{1/2} = \frac{9^{1/2}(x^{5/3})^{1/2}}{(y^{2/3})^{1/2}} = \frac{3x^{5/6}}{y^{1/3}}$$

Now Work PROBLEM 71

The next two examples illustrate some algebra that you will need to know for certain calculus problems.

EXAMPLE 9 Writing an Expression as a Single Quotient

Write the following expression as a single quotient in which only positive exponents appear.

$$(x^2+1)^{1/2} + x\cdot\frac{1}{2}(x^2+1)^{-1/2}\cdot 2x$$

Solution

$$(x^2+1)^{1/2} + x\cdot\frac{1}{2}(x^2+1)^{-1/2}\cdot 2x = (x^2+1)^{1/2} + \frac{x^2}{(x^2+1)^{1/2}}$$
$$= \frac{(x^2+1)^{1/2}(x^2+1)^{1/2} + x^2}{(x^2+1)^{1/2}}$$
$$= \frac{(x^2+1) + x^2}{(x^2+1)^{1/2}}$$
$$= \frac{2x^2+1}{(x^2+1)^{1/2}}$$

Now Work PROBLEM 77

EXAMPLE 10 Factoring an Expression Containing Rational Exponents

Factor: $\frac{4}{3}x^{1/3}(2x+1) + 2x^{4/3}$

Solution We begin by writing $2x^{4/3}$ as a fraction with 3 as denominator.

$$\frac{4}{3}x^{1/3}(2x+1) + 2x^{4/3} = \frac{4x^{1/3}(2x+1)}{3} + \frac{6x^{4/3}}{3} = \frac{4x^{1/3}(2x+1) + 6x^{4/3}}{3}$$

Add the two fractions

$$= \frac{2x^{1/3}[2(2x+1) + 3x]}{3} = \frac{2x^{1/3}(7x+2)}{3}$$

2 and $x^{1/3}$ are common factors; Simplify

Now Work PROBLEM 89

Historical Feature

The radical sign, $\sqrt{\ }$, was first used in print by Christoff Rudolff in 1525. It is thought to be the manuscript form of the letter *r* (for the Latin word *radix* = *root*), although this is not quite conclusively confirmed. It took a long time for $\sqrt{\ }$ to become the standard symbol for a square root and much longer to standardize $\sqrt[3]{\ }$, $\sqrt[4]{\ }$, $\sqrt[5]{\ }$, and so on. The indexes of the root were placed in every conceivable position, with

$$\sqrt{\overset{3}{8}},\quad \sqrt{③8},\quad \text{and}\quad \underset{3}{\sqrt{8}}$$

all being variants for $\sqrt[3]{8}$. The notation $\sqrt{\ }\sqrt{16}$ was popular for $\sqrt[4]{16}$. By the 1700s, the index had settled where we now put it.

The bar on top of the present radical symbol, as follows,

$$\sqrt{a^2 + 2ab + b^2}$$

is the last survivor of the **vinculum,** a bar placed atop an expression to indicate what we would now indicate with parentheses. For example,

$$a\overline{b + c} = a(b + c)$$

R.8 Assess Your Understanding

'Are You Prepared?' *Answers are given at the end of these exercises. If you get a wrong answer, read the pages in red.*

1. $(-3)^2 =$ ____; $-3^2 =$ ____ (pp. 21–23)

2. $\sqrt{16} =$ ____; $\sqrt{(-4)^2} =$ ____ (pp. 23–24)

Concepts and Vocabulary

3. In the symbol $\sqrt[n]{a}$, the integer n is called the ____.

4. ***True or False*** $\sqrt[5]{-32} = -2$

5. We call $\sqrt[3]{a}$ the ____ ____ of a.

6. ***True or False*** $\sqrt[4]{(-3)^4} = -3$

Skill Building

In Problems 7–42, simplify each expression. Assume that all variables are positive when they appear.

7. $\sqrt[3]{27}$

8. $\sqrt[4]{16}$

9. $\sqrt[3]{-8}$

10. $\sqrt[3]{-1}$

11. $\sqrt{8}$

12. $\sqrt[3]{54}$

13. $\sqrt[3]{-8x^4}$

14. $\sqrt[4]{48x^5}$

15. $\sqrt[4]{x^{12}y^8}$

16. $\sqrt[5]{x^{10}y^5}$

17. $\sqrt[4]{\dfrac{x^9y^7}{xy^3}}$

18. $\sqrt[3]{\dfrac{3xy^2}{81x^4y^2}}$

19. $\sqrt{36x}$

20. $\sqrt{9x^5}$

21. $\sqrt{3x^2}\sqrt{12x}$

22. $\sqrt{5x}\sqrt{20x^3}$

23. $(\sqrt{5}\sqrt[3]{9})^2$

24. $(\sqrt[3]{3}\sqrt{10})^4$

25. $(3\sqrt{6})(2\sqrt{2})$

26. $(5\sqrt{8})(-3\sqrt{3})$

27. $3\sqrt{2} + 4\sqrt{2}$

28. $6\sqrt{5} - 4\sqrt{5}$

29. $-\sqrt{18} + 2\sqrt{8}$

30. $2\sqrt{12} - 3\sqrt{27}$

31. $(\sqrt{3} + 3)(\sqrt{3} - 1)$

32. $(\sqrt{5} - 2)(\sqrt{5} + 3)$

33. $5\sqrt[3]{2} - 2\sqrt[3]{54}$

34. $9\sqrt[3]{24} - \sqrt[3]{81}$

35. $(\sqrt{x} - 1)^2$

36. $(\sqrt{x} + \sqrt{5})^2$

37. $\sqrt[3]{16x^4} - \sqrt[3]{2x}$

38. $\sqrt[4]{32x} + \sqrt[4]{2x^5}$

39. $\sqrt{8x^3} - 3\sqrt{50x}$

40. $3x\sqrt{9y} + 4\sqrt{25y}$

41. $\sqrt[3]{16x^4y} - 3x\sqrt[3]{2xy} + 5\sqrt[3]{-2xy^4}$

42. $8xy - \sqrt{25x^2y^2} + \sqrt[3]{8x^3y^3}$

In Problems 43–54, rationalize the denominator of each expression. Assume that all variables are positive when they appear.

43. $\dfrac{1}{\sqrt{2}}$

44. $\dfrac{2}{\sqrt{3}}$

45. $\dfrac{-\sqrt{3}}{\sqrt{5}}$

46. $\dfrac{-\sqrt{3}}{\sqrt{8}}$

47. $\dfrac{\sqrt{3}}{5 - \sqrt{2}}$

48. $\dfrac{\sqrt{2}}{\sqrt{7} + 2}$

49. $\dfrac{2 - \sqrt{5}}{2 + 3\sqrt{5}}$

50. $\dfrac{\sqrt{3} - 1}{2\sqrt{3} + 3}$

51. $\dfrac{5}{\sqrt[3]{2}}$ 52. $\dfrac{-2}{\sqrt[3]{9}}$ 53. $\dfrac{\sqrt{x+h}-\sqrt{x}}{\sqrt{x+h}+\sqrt{x}}$ 54. $\dfrac{\sqrt{x+h}+\sqrt{x-h}}{\sqrt{x+h}-\sqrt{x-h}}$

In Problems 55–66, simplify each expression.

55. $8^{2/3}$ 56. $4^{3/2}$ 57. $(-27)^{1/3}$ 58. $16^{3/4}$ 59. $16^{3/2}$ 60. $25^{3/2}$

61. $9^{-3/2}$ 62. $16^{-3/2}$ 63. $\left(\dfrac{9}{8}\right)^{3/2}$ 64. $\left(\dfrac{27}{8}\right)^{2/3}$ 65. $\left(\dfrac{8}{9}\right)^{-3/2}$ 66. $\left(\dfrac{8}{27}\right)^{-2/3}$

In Problems 67–74, simplify each expression. Express your answer so that only positive exponents occur. Assume that the variables are positive.

67. $x^{3/4}x^{1/3}x^{-1/2}$ 68. $x^{2/3}x^{1/2}x^{-1/4}$ 69. $(x^3y^6)^{1/3}$ 70. $(x^4y^8)^{3/4}$

71. $\dfrac{(x^2y)^{1/3}(xy^2)^{2/3}}{x^{2/3}y^{2/3}}$ 72. $\dfrac{(xy)^{1/4}(x^2y^2)^{1/2}}{(x^2y)^{3/4}}$ 73. $\dfrac{(16x^2y^{-1/3})^{3/4}}{(xy^2)^{1/4}}$ 74. $\dfrac{(4x^{-1}y^{1/3})^{3/2}}{(xy)^{3/2}}$

Applications and Extensions

In Problems 75–88, expressions that occur in calculus are given. Write each expression as a single quotient in which only positive exponents and/or radicals appear.

75. $\dfrac{x}{(1+x)^{1/2}} + 2(1+x)^{1/2} \quad x > -1$

76. $\dfrac{1+x}{2x^{1/2}} + x^{1/2} \quad x > 0$

77. $2x(x^2+1)^{1/2} + x^2\cdot\dfrac{1}{2}(x^2+1)^{-1/2}\cdot 2x$

78. $(x+1)^{1/3} + x\cdot\dfrac{1}{3}(x+1)^{-2/3} \quad x \neq -1$

79. $\sqrt{4x+3}\cdot\dfrac{1}{2\sqrt{x-5}} + \sqrt{x-5}\cdot\dfrac{1}{5\sqrt{4x+3}} \quad x > 5$

80. $\dfrac{\sqrt[3]{8x+1}}{3\sqrt[3]{(x-2)^2}} + \dfrac{\sqrt[3]{x-2}}{24\sqrt[3]{(8x+1)^2}} \quad x \neq 2, x \neq -\dfrac{1}{8}$

81. $\dfrac{\sqrt{1+x} - x\cdot\dfrac{1}{2\sqrt{1+x}}}{1+x} \quad x > -1$

82. $\dfrac{\sqrt{x^2+1} - x\cdot\dfrac{2x}{2\sqrt{x^2+1}}}{x^2+1}$

83. $\dfrac{(x+4)^{1/2} - 2x(x+4)^{-1/2}}{x+4} \quad x > -4$

84. $\dfrac{(9-x^2)^{1/2} + x^2(9-x^2)^{-1/2}}{9-x^2} \quad -3 < x < 3$

85. $\dfrac{\dfrac{x^2}{(x^2-1)^{1/2}} - (x^2-1)^{1/2}}{x^2} \quad x < -1 \text{ or } x > 1$

86. $\dfrac{(x^2+4)^{1/2} - x^2(x^2+4)^{-1/2}}{x^2+4}$

87. $\dfrac{\dfrac{1+x^2}{2\sqrt{x}} - 2x\sqrt{x}}{(1+x^2)^2} \quad x > 0$

88. $\dfrac{2x(1-x^2)^{1/3} + \dfrac{2}{3}x^3(1-x^2)^{-2/3}}{(1-x^2)^{2/3}} \quad x \neq -1, x \neq 1$

In Problems 89–98, expressions that occur in calculus are given. Factor each expression. Express your answer so that only positive exponents occur.

89. $(x+1)^{3/2} + x\cdot\dfrac{3}{2}(x+1)^{1/2} \quad x \geq -1$

90. $(x^2+4)^{4/3} + x\cdot\dfrac{4}{3}(x^2+4)^{1/3}\cdot 2x$

91. $6x^{1/2}(x^2 + x) - 8x^{3/2} - 8x^{1/2} \quad x \geq 0$

92. $6x^{1/2}(2x + 3) + x^{3/2} \cdot 8 \quad x \geq 0$

93. $3(x^2 + 4)^{4/3} + x \cdot 4(x^2 + 4)^{1/3} \cdot 2x$

94. $2x(3x + 4)^{4/3} + x^2 \cdot 4(3x + 4)^{1/3}$

95. $4(3x + 5)^{1/3}(2x + 3)^{3/2} + 3(3x + 5)^{4/3}(2x + 3)^{1/2} \quad x \geq -\frac{3}{2}$

96. $6(6x + 1)^{1/3}(4x - 3)^{3/2} + 6(6x + 1)^{4/3}(4x - 3)^{1/2} \quad x \geq \frac{3}{4}$

97. $3x^{-1/2} + \frac{3}{2}x^{1/2} \quad x > 0$

98. $8x^{1/3} - 4x^{-2/3} \quad x \neq 0$

In Problems 99–106, use a calculator to approximate each radical. Round your answer to two decimal places.

99. $\sqrt{2}$

100. $\sqrt{7}$

101. $\sqrt[3]{4}$

102. $\sqrt[3]{-5}$

103. $\dfrac{2 + \sqrt{3}}{3 - \sqrt{5}}$

104. $\dfrac{\sqrt{5} - 2}{\sqrt{2} + 4}$

105. $\dfrac{3\sqrt[3]{5} - \sqrt{2}}{\sqrt{3}}$

106. $\dfrac{2\sqrt{3} - \sqrt[3]{4}}{\sqrt{2}}$

Applications and Extensions

107. Calculating the Amount of Gasoline in a Tank A Shell station stores its gasoline in underground tanks that are right circular cylinders lying on their sides. See the illustration. The volume V of gasoline in the tank (in gallons) is given by the formula

$$V = 40h^2\sqrt{\frac{96}{h} - 0.608}$$

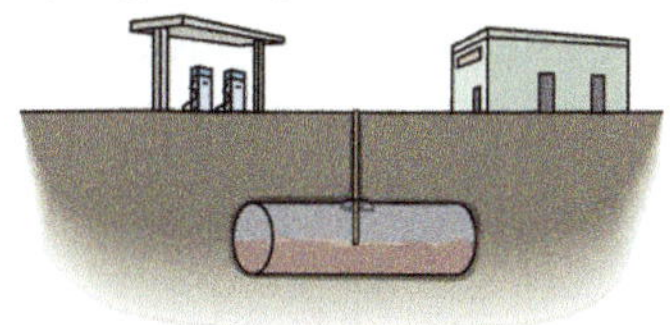

where h is the height of the gasoline (in inches) as measured on a depth stick.

(a) If $h = 12$ inches, how many gallons of gasoline are in the tank?

(b) If $h = 1$ inch, how many gallons of gasoline are in the tank?

108. Inclined Planes The final velocity v of an object in feet per second (ft/sec) after it slides down a frictionless inclined plane of height h feet is

$$v = \sqrt{64h + v_0^2}$$

where v_0 is the initial velocity (in ft/sec) of the object.

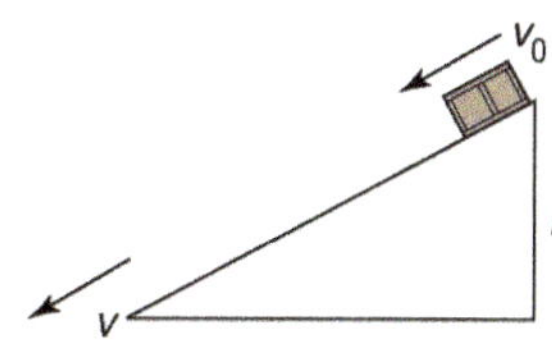

(a) What is the final velocity v of an object that slides down a frictionless inclined plane of height 4 feet? Assume that the initial velocity is 0.

(b) What is the final velocity v of an object that slides down a frictionless inclined plane of height 16 feet? Assume that the initial velocity is 0.

(c) What is the final velocity v of an object that slides down a frictionless inclined plane of height 2 feet with an initial velocity of 4 ft/sec?

Problems 109–112 require the following information.

Period of a Pendulum *The period T, in seconds, of a pendulum of length l, in feet, may be approximated using the formula*

$$T = 2\pi\sqrt{\frac{l}{32}}$$

In Problems 109–112, express your answer both as a square root and as a decimal.

109. Find the period T of a pendulum whose length is 64 feet.

110. Find the period T of a pendulum whose length is 16 feet.

111. Find the period T of a pendulum whose length is 8 inches.

112. Find the period T of a pendulum whose length is 4 inches.

Discussion and Writing

113. Give an example to show that $\sqrt{a^2}$ is not equal to a. Use it to explain why $\sqrt{a^2} = |a|$.

'Are you Prepared?' Answers

1. 9; −9 **2.** 4; 4

CHAPTER REVIEW

Things to Know

Set (pp. 2–3) Well-defined collection of distinct objects, called elements

Empty set or null set	$\varnothing$	Set that has no elements
Equality	$A = B$	A and B have the same elements.
Subset	$A \subseteq B$	Each element of A is also an element of B.
Intersection	$A \cap B$	Set consisting of elements that belong to both A and B
Union	$A \cup B$	Set consisting of elements that belong to either A or B, or both
Universal set	U	Set consisting of all the elements that we wish to consider
Complement	$\overline{A}$	Set consisting of elements of the universal set that are not in A

Classify numbers (pp. 4–5)

Counting numbers	$1, 2, 3 \ldots$
Whole numbers	$0, 1, 2, 3 \ldots$
Integers	$\ldots, -2, -1, 0, 1, 2, \ldots$
Rational numbers	Quotients of two integers (denominator not equal to 0); terminating or repeating decimals
Irrational numbers	Nonrepeating nonterminating decimals
Real numbers	Rational or irrational numbers

Work with properties of real numbers (pp. 9–13)

Commutative properties	$a + b = b + a, \quad a \cdot b = b \cdot a$
Associative properties	$a + (b + c) = (a + b) + c, \quad a \cdot (b \cdot c) = (a \cdot b) \cdot c$
Distributive property	$a \cdot (b + c) = a \cdot b + a \cdot c$
Identity properties	$a + 0 = a \quad a \cdot 1 = a$
Inverse properties	$a + (-a) = 0 \quad a \cdot \dfrac{1}{a} = 1$ where $a \neq 0$
Zero-Product Property	If $ab = 0$, then $a = 0$ or $b = 0$ or both.

Absolute value (p. 19) $|a| = a$ if $a \geq 0 \quad |a| = -a$ if $a < 0$

Exponents (p. 22)

$a^n = \underbrace{a \cdot a \cdot \ldots \cdot a}_{n \text{ factors}}$ $\quad n$ a positive integer $\qquad a^0 = 1 \quad a \neq 0 \qquad a^{-n} = \dfrac{1}{a^n} \quad a \neq 0, \quad n$ a positive integer

Laws of exponents (p. 22)

$$a^m \cdot a^n = a^{m+n} \qquad (a^m)^n = a^{mn} \qquad (a \cdot b)^n = a^n \cdot b^n \qquad \frac{a^m}{a^n} = a^{m-n} = \frac{1}{a^{n-m}} \qquad \left(\frac{a}{b}\right)^n = \frac{a^n}{b^n}$$

Pythagorean Theorem (p. 30)

In a right triangle, the square of the length of the hypotenuse is equal to the sum of the squares of the lengths of the legs.

Converse of the Pythagorean Theorem (p. 30)

In a triangle, if the square of the length of one side equals the sum of the squares of the lengths of the other two sides, then the triangle is a right triangle.

Polynomial (p. 40)

Algebraic expression of the form $a_n x^n + a_{n-1}x^{n-1} + \cdots + a_1 x + a_0$, $\quad n$ a nonnegative integer

Special products/factoring formulas (pp. 43–44)

$(x - a)(x + a) = x^2 - a^2$

$(x + a)^2 = x^2 + 2ax + a^2 \qquad (x - a)^2 = x^2 - 2ax + a^2$

$(x + a)(x + b) = x^2 + (a + b)x + ab$

$(ax + b)(cx + d) = acx^2 + (ad + bc)x + bd$

$(x + a)^3 = x^3 + 3ax^2 + 3a^2x + a^3 \qquad (x - a)^3 = x^3 - 3ax^2 + 3a^2x - a^3$

$(x - a)(x^2 + ax + a^2) = x^3 - a^3 \qquad (x + a)(x^2 - ax + a^2) = x^3 + a^3$

Radicals; Rational exponents (pp. 72 and 75)

$\sqrt[n]{a} = b$ means $a = b^n$ where $a \geq 0$ and $b \geq 0$ if $n \geq 2$ is even and a, b are any real numbers if $n \geq 3$ is odd.

$a^{m/n} = \sqrt[n]{a^m} = (\sqrt[n]{a})^m \quad m, n \geq 2 \quad m/n$ in lowest terms

Objectives

Section		You should be able to ...	Example(s)	Review Exercises
R.1	1	Work with sets (p. 2)	1–3	1–8
	2	Classify numbers (p. 4)	4	9, 10
	3	Evaluate numerical expressions (p. 8)	8–10	11–14, 109
	4	Work with properties of real numbers (p. 9)	11–23	17, 18
R.2	1	Graph inequalities (p. 18)	2	19, 20
	2	Find distance on the real number line (p. 19)	3, 4	21, 22
	3	Evaluate algebraic expressions (p. 20)	5	23–27, 108
	4	Determine the domain of a variable (p. 21)	6, 7	29, 30
	5	Use the Laws of Exponents (p. 21)	8–10	83, 84
	6	Evaluate square roots (p. 23)	11, 12	15, 27
	7	Use a calculator to evaluate exponents (p. 24)	13	33
	8	Use scientific notation (p. 24)	14–16	36, 105
R.3	1	Use the Pythagorean Theorem and its converse (p. 30)	1–3	106, 107, 112
	2	Know geometry formulas (p. 31)	4	110, 111
	3	Understand congruent triangles and similar triangles (p. 32)	5	31, 32
R.4	1	Recognize monomials (p. 39)	1, 2	34
	2	Recognize polynomials (p. 40)	3	35
	3	Add and subtract polynomials (p. 41)	4, 5	37, 38
	4	Multiply polynomials (p. 42)	6	39–44
	5	Know formulas for special products (p. 43)	7–11	41, 42
	6	Divide polynomials using long division (p. 44)	12–14	45–50
	7	Work with polynomials in two variables (p. 47)	15, 16	39–40
R.5	1	Factor the difference of two squares and the sum and difference of two cubes (p. 50)	2–5	57, 58, 61, 62
	2	Factor perfect squares (p. 51)	6–8	65, 66
	3	Factor a second-degree polynomial: $x^2 + Bx + C$ (p. 52)	9–13	51, 52, 64
	4	Factor by grouping (p. 53)	14–16	59, 60
	5	Factor a second-degree polynomial: $Ax^2 + Bx + C, A \neq 1$ (p. 54)	17, 18	53–56, 63
R.6	1	Divide polynomials using synthetic division (p. 57)	1, 2	45, 46, 49, 50
R.7	1	Reduce a rational expression to lowest terms (p. 61)	1, 2	67, 68
	2	Multiply and divide rational expressions (p. 62)	3	69, 70, 75
	3	Add and subtract rational expressions (p. 63)	4–6	71–74
	4	Use the least common multiple method (p. 65)	7–9	71–74
	5	Simplify complex rational expressions (p. 67)	10–12	76
R.8	1	Work with nth roots (p. 72)	1, 2	28, 87–90
	2	Simplify radicals (p. 73)	3, 4	77–82, 87–90, 101, 102
	3	Rationalize denominators (p. 74)	5	91–96
	4	Simplify expressions with rational exponents (p. 75)	6–10	85, 86, 97–100, 103, 104

Review Exercises

In Problems 1–8, use U = *universal set* = $\{1, 2, 3, 4, 5, 6, 7, 8, 9\}$, $A = \{1, 3, 5, 7\}$, $B = \{3, 5, 6, 7, 8\}$, *and* $C = \{2, 3, 7, 8, 9\}$ *to find each set.*

1. $A \cup B$ **2.** $B \cup C$ **3.** $A \cap C$ **4.** $A \cap B$

5. $\overline{A}$ **6.** $\overline{B}$ **7.** $\overline{B \cap C}$ **8.** $\overline{A \cup B}$

In Problems 9 and 10, list the numbers in the set that are (a) Natural numbers, (b) Integers, (c) Rational numbers, (d) Irrational numbers, (e) Real numbers.

9. $A = \left\{-10, 0.65, 1.343434\ldots = 1.3\overline{4}, \sqrt{7}, \frac{1}{9}\right\}$

10. $B = \left\{0, -5, \frac{1}{3}, 0.59, 1.333\ldots = 1.\overline{3}, 2\sqrt{2}, \frac{\pi}{2}\right\}$

In Problems 11–16, evaluate each expression.

11. $-6 + 4 \cdot (8 - 3)$

12. $\frac{5}{18} + \frac{1}{12}$

13. $\frac{4}{3} \cdot \frac{9}{16}$

14. $\dfrac{\frac{5}{18}}{\frac{11}{27}}$

15. $\sqrt{(-3)^2}$

16. $(-3)^5$

In Problems 17 and 18, use the Distributive Property to remove the parentheses.

17. $4(x - 3)$

18. $(x - 2)(3x + 1)$

In Problems 19 and 20, graph the numbers x on the real number line.

19. $x > 3$

20. $x \le 5$

In Problems 21 and 22, let P, Q, and R be points on a real number line with coordinates −2, 3, and 9, respectively.

21. Find the distance between P and Q.

22. Find the distance between Q and R.

In Problems 23–28, evaluate each expression if $x = -5$ and $y = 7$.

23. $\frac{4x}{x + y}$

24. $|2x - 3y|$

25. $5x^{-1}y^2$

26. $2x^2y^{-2}$

27. $\sqrt{x^2}$

28. $\sqrt[3]{x^3}$

In Problems 29 and 30, determine the domain of the variable.

29. $\frac{3}{x - 6}$

30. $\frac{x + 1}{x + 5}$

In Problems 31 and 32, the triangles are similar. Find the missing length x and the missing angles A, B, and C.

31.

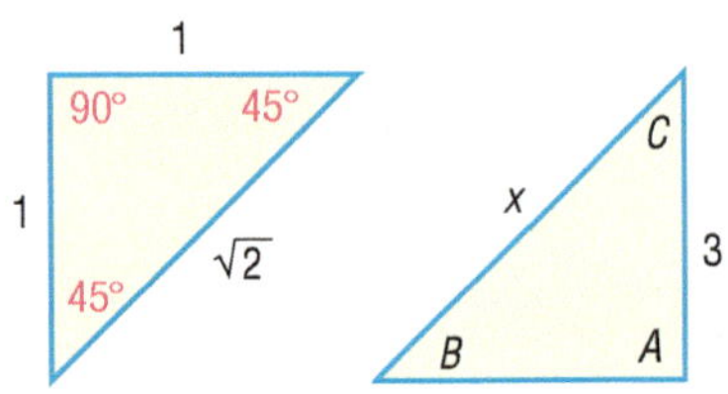

32.

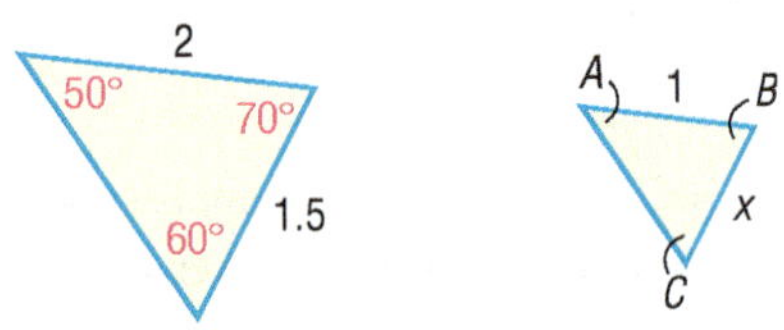

33. Use a calculator to evaluate $(1.5)^4$.

34. Identify the coefficient and degree of the monomial $-5x^3$.

35. Identify the coefficients and degree of the polynomial $3x^5 + 4x^4 - 2x^3 + 5x - 12$.

36. Write 3.275×10^5 as an integer.

In Problems 37–44, perform the indicated operation.

37. $(2x^4 - 8x^3 + 5x - 1) + (6x^3 + x^2 + 4)$

38. $(x^3 + 8x^2 - 3x + 4) - (4x^3 - 7x^2 - 2x + 3)$

39. $(2x + y)(3x - 5y)$

40. $(2x - 5y)(3x + 2y)$

41. $(4x + 1)(4x - 1)$

42. $(5x + 2)^2$

43. $(x + 1)(x + 2)(x - 3)$

44. $(x + 1)(x + 3)(x - 5)$

In Problems 45–50, find the quotient and remainder. Check your work by verifying that

$$(\textit{Quotient})(\textit{Divisor}) + \textit{Remainder} = \textit{Dividend}$$

45. $3x^3 - x^2 + x + 4$ divided by $x - 3$

46. $2x^3 - 3x^2 + x + 1$ divided by $x - 2$

47. $-3x^4 + x^2 + 2$ divided by $x^2 + 1$

48. $-4x^3 + x^2 - 2$ divided by $x^2 - 1$

49. $x^5 + 1$ divided by $x + 1$

50. $x^5 - 1$ divided by $x - 1$

In Problems 51–66, factor each polynomial completely (over the integers). If the polynomial cannot be factored, say it is prime.

51. $x^2 + 5x - 14$

52. $x^2 - 9x + 14$

53. $6x^2 - 5x - 6$

54. $6x^2 + x - 2$

55. $3x^2 - 15x - 42$

56. $2x^3 + 18x^2 + 28x$

57. $8x^3 + 1$

58. $27x^3 - 8$

59. $2x^3 + 3x^2 - 2x - 3$

60. $2x^3 + 3x^2 + 2x + 3$

61. $25x^2 - 4$

62. $16x^2 - 1$

63. $9x^2 + 1$

64. $x^2 - x + 1$

65. $x^2 + 8x + 16$

66. $4x^2 + 12x + 9$

In Problems 67–76, perform the indicated operation and simplify. Leave your answer in factored form.

67. $\dfrac{2x^2 + 11x + 14}{x^2 - 4}$

68. $\dfrac{x^2 - 5x - 14}{4 - x^2}$

69. $\dfrac{9x^2 - 1}{x^2 - 9} \cdot \dfrac{3x - 9}{9x^2 + 6x + 1}$

70. $\dfrac{x^2 - 25}{x^3 - 4x^2 - 5x} \cdot \dfrac{x^2 + x}{1 - x^2}$

71. $\dfrac{x + 1}{x - 1} - \dfrac{x - 1}{x + 1}$

72. $\dfrac{x}{x + 1} - \dfrac{2x}{x + 2}$

73. $\dfrac{3x + 4}{x^2 - 4} - \dfrac{2x - 3}{x^2 + 4x + 4}$

74. $\dfrac{x^2}{2x^2 + 5x - 3} + \dfrac{x^2}{2x^2 - 5x + 2}$

75. $\dfrac{\dfrac{x^2 - 1}{x^2 - 5x + 6}}{\dfrac{x + 1}{x - 2}}$

76. $\dfrac{\dfrac{x + 4}{3}}{\dfrac{1}{4} + \dfrac{2}{x}}$

In Problems 77–86, simplify each expression. All exponents must be positive. Assume that $x > 0$ and $y > 0$, when they appear.

77. $\sqrt{32}$

78. $\sqrt{75}$

79. $\sqrt[3]{-16}$

80. $\sqrt[5]{64}$

81. $5\sqrt{8} - 2\sqrt{32}$

82. $4\sqrt{12} + 5\sqrt{27}$

83. $\dfrac{(x^2y)^{-4}}{(xy)^{-3}}$

84. $\left(\dfrac{x^2y^2}{x^{-1}}\right)^2$

85. $(25x^{-4/3}y^{-2/3})^{3/2}$

86. $(27x^{-3/2}y^{5/2})^{2/3}$

In Problems 87–90, simplify each expression.

87. $\sqrt{\dfrac{9x^2}{25y^4}} \quad x \geq 0, y > 0$

88. $\dfrac{\sqrt{2x}\sqrt{5xy^3}}{\sqrt{10y}} \quad x \geq 0, y > 0$

89. $\sqrt[3]{27x^4y^{12}}$

90. $\dfrac{\sqrt[4]{243x^5y}}{\sqrt[4]{3xy^9}} \quad x > 0, y > 0$

In Problems 91–96, rationalize the denominator of each expression.

91. $\dfrac{4}{\sqrt{5}}$

92. $\dfrac{-2}{\sqrt{3}}$

93. $\dfrac{2}{1 - \sqrt{2}}$

94. $\dfrac{-4}{1 + \sqrt{3}}$

95. $\dfrac{1 + \sqrt{5}}{1 - \sqrt{5}}$

96. $\dfrac{4\sqrt{3} + 2}{2\sqrt{3} + 1}$

In Problems 97–102, write each expression as a single quotient in which only positive exponents and/or radicals appear.

97. $(2 + x^2)^{1/2} + x \cdot \dfrac{1}{2}(2 + x^2)^{-1/2} \cdot 2x$

98. $(x^2 + 4)^{2/3} + x \cdot \dfrac{2}{3}(x^2 + 4)^{-1/3} \cdot 2x$

99. $\dfrac{(x + 4)^{1/2} \cdot 2x - x^2 \cdot \dfrac{1}{2}(x + 4)^{-1/2}}{x + 4} \quad x > -4$

100. $\dfrac{(x^2 + 4)^{1/2} \cdot 2x - x^2 \cdot \dfrac{1}{2}(x^2 + 4)^{-1/2} \cdot 2x}{x^2 + 4}$

101. $\dfrac{\dfrac{x^2}{\sqrt{x^2 - 1}} - \sqrt{x^2 - 1}}{x^2}$

102. $\dfrac{\dfrac{4 + x^2}{2\sqrt{x}} - 2x\sqrt{x}}{(4 + x^2)^2}$

In Problems 103 and 104, factor each expression.

103. $3(x^2 + 4)^{4/3} + x \cdot 4(x^2 + 4)^{1/3} \cdot 2x$

104. $2x(3x + 5)^{5/3} + x^2 \cdot 4(3x + 5)^{2/3}$

---

105. U.S. Population According to the U.S. Census Bureau, the U.S. population in 2000 was 281,421,906. Write the U.S. population in scientific notation.

106. Find the length of the hypotenuse of a right triangle whose legs are of lengths 5 and 8.

107. The lengths of the sides of a triangle are 12, 16, and 20. Is this a right triangle?

108. Manufacturing Cost The weekly production cost C of manufacturing x calculators is given by the formula

$$C = 3000 + 6x - \frac{x^2}{1000}$$

(a) What is the cost of producing 1000 calculators?
(b) What is the cost of producing 3000 calculators?

109. **Quarterly Corporate Earnings** In the first quarter of its fiscal year, a company posted earnings of \$1.20 per share. During the second and third quarters, it posted losses of \$0.75 per share and \$0.30 per share, respectively. In the fourth quarter, it earned a modest \$0.20 per share. What were the annual earnings per share of this company?

110. **Design** A window consists of a rectangle surmounted by a triangle. Find the area of the window shown in the illustration. How much wood frame is needed to enclose the window?

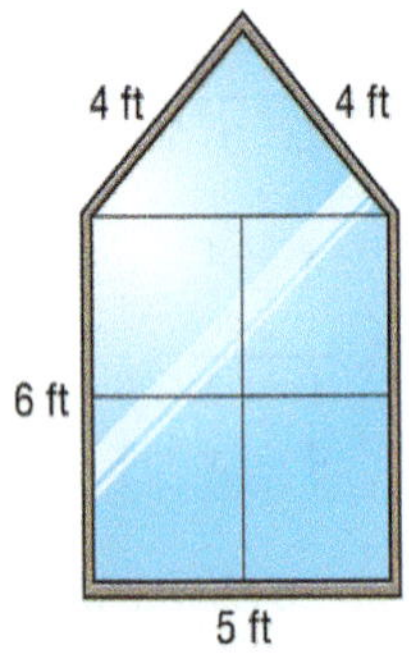

111. **Construction** A statue with a circular base of radius 3 feet is enclosed by a circular pond as shown in the illustration. What is the surface area of the pond? How much fence is required to enclose the pond?

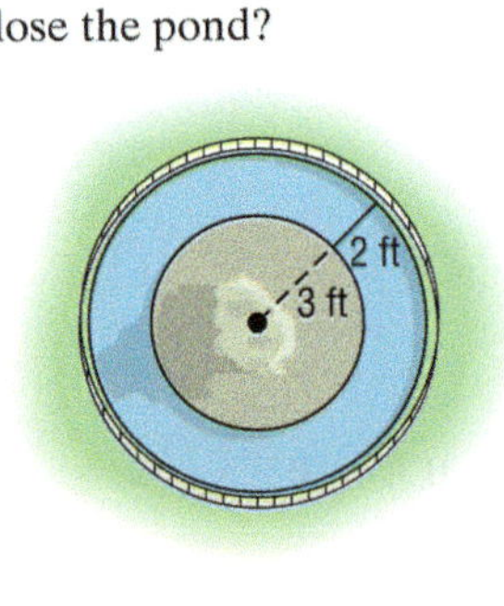

112. **How Far Can a Pilot See?** On a recent flight to San Francisco, the pilot announced that we were 139 miles from the city, flying at an altitude of 35,000 feet. The pilot claimed that he could see the Golden Gate Bridge and beyond. Was he telling the truth? How far could he see?

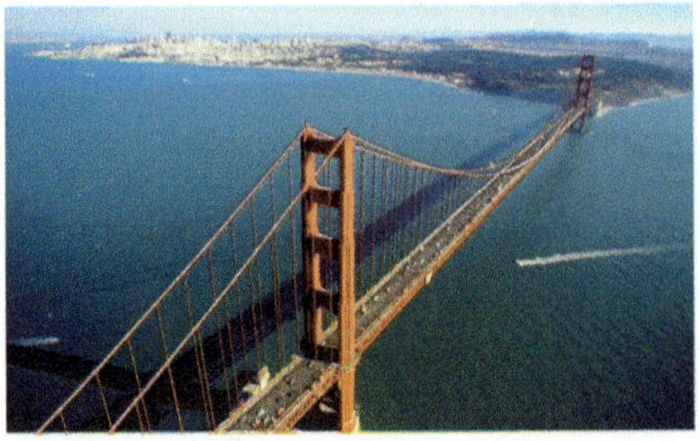

113. Use the material in this chapter to create a problem that uses each of the following words:
(a) Simplify (b) Factor (c) Reduce

CHAPTER TEST

1. List the numbers in the set that are (a) Natural numbers, (b) Integers, (c) Rational numbers, (d) Irrational numbers, (e) Real numbers.

$$\left\{0, 1.2, \sqrt{2}, 7, \frac{1}{2}, \pi\right\}$$

2. Evaluate each expression if $x = -3$ and $y = 4$.
(a) $3x^{-1}y^2$ (b) $|2x - 3y|$
(c) $\sqrt{x^2 + y^2}$ (d) $5x^3 - 3x^2 + 2x - 6$

3. The triangles below are similar. Find the missing length x and the missing angles A, B, and C.

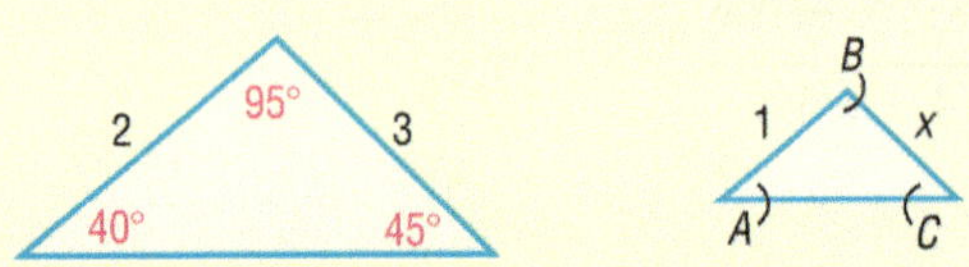

4. Perform the indicated operation.
(a) $(-2x^3 + 4x^2 - 6x + 10) - (6x^3 - 7x^2 + 8x - 1)$
(b) $(2x - 3)(-5x + 2)$

5. Factor each polynomial.
(a) $x^2 - 6x + 8$
(b) $4x^2 - 25$
(c) $6x^2 - 19x - 7$

6. Find the quotient and remainder if $x^3 - 3x^2 + 8x - 10$ is divided by $x - 2$.

7. Simplify each expression.
(a) $\sqrt{27}$ (b) $\sqrt[3]{-8}$ (c) $\dfrac{(xy^2)^{-3}}{(x^2y)^{-2}}$ (d) $(16x^{4/3}y^{-2/3})^{3/2}$

8. Rationalize the denominator of $\dfrac{3}{5 + \sqrt{3}}$.

9. Graph the numbers x on the real number line: $x > -3$.

10. The Zero-Product Property states that if $a \cdot b = 0$ then either ____ or ____, or both.

11. A rectangular swimming pool, 20 feet long and 10 feet wide, is enclosed by a wooden deck that is 3 feet wide. What is the area of the deck? How much fence is required to enclose the deck?

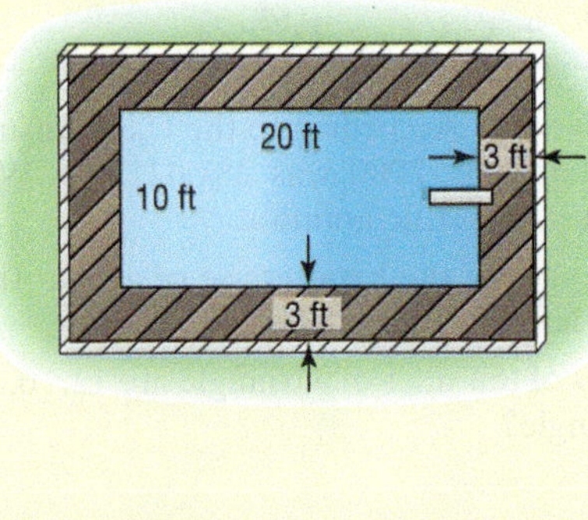

Graphs, Equations, and Inequalities

1

Interest Rates Fall on 15- and 30-Year Loans

From the Associated Press May 12, 2006 Rates on 30-year fixed-rate mortgages averaged 6.58% this week, down from 6.59% last week, mortgage company Freddie Mac said.

Rates on 15-year, fixed-rate mortgages, a popular choice for refinancing a home mortgage, fell to 6.17%, down from 6.22% last week. One-year adjustable-rate mortgages (ARM) fell to 5.62%, down from 5.67%. Rates on 5-year hybrid adjustable-rate mortgages averaged 6.22%, up slightly from 6.21% last week.

The mortgage rates do not include add-on fees known as points. The 1-year ARM carried a nationwide average fee of 0.7 point, and the three other mortgage categories had an average fee of 0.5 point.

Source: Adapted from *Martin Crutsinger, "Rates on 30-year Mortgages Edge Down Slightly." Associated Press Financial Wire, May 11, 2006. ©2006 The Associated Press.*

—See Chapter Project I—

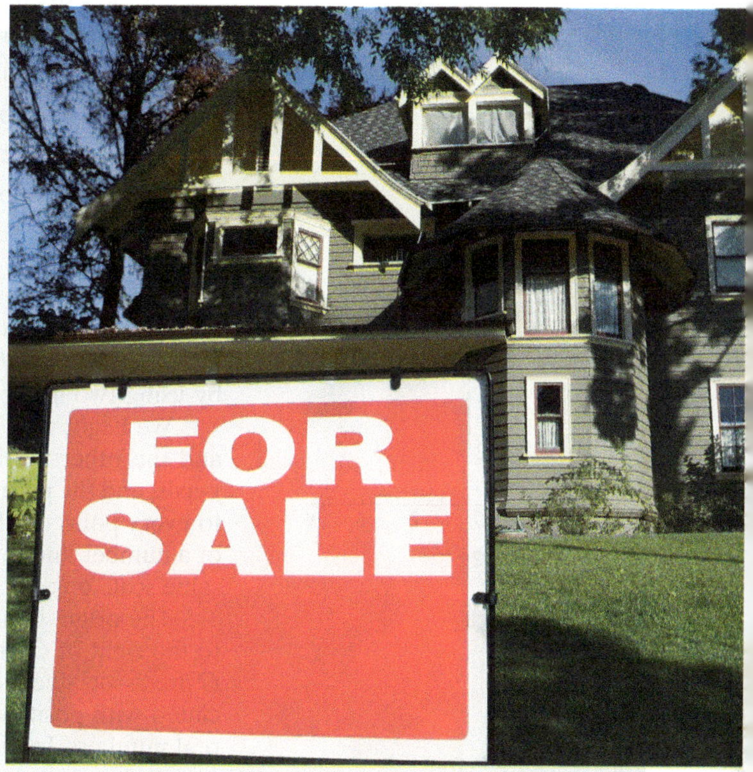

A Look Back

In Chapter R, we reviewed skills from Intermediate Algebra.

A Look Ahead

In this chapter, we use the rectangular coordinate system to graph equations. Then, we algebraically solve equations and use the rectangular coordinate system to visualize the solution. The idea of using a system of rectangular coordinates dates back to ancient times, when such a system was used for surveying and city planning. Apollonius of Perga, in 200 BC, used a form of rectangular coordinates in his work on conics, although this use does not stand out as clearly as it does in modern treatments. Sporadic use of rectangular coordinates continued until the 1600s. By that time, algebra had developed sufficiently so that René Descartes (1596–1650) and Pierre de Fermat (1601–1665) could take the crucial step, which was the use of rectangular coordinates to translate geometry problems into algebra problems, and vice versa. This step allowed both geometers and algebraists to gain new insights into their subjects, which previously had been regarded as separate, but now were seen to be connected in many important ways.

With the advent of modern technology, in particular, graphing utilities, now not only are we able to visualize the dual roles of algebra and geometry, but we are also able to solve many problems that required advanced methods before this technology.

Outline

1.1 Rectangular Coordinates; Graphing Utilities; Introduction to Graphing Equations

PREPARING FOR THIS SECTION *Before getting started, review the following:*

- Algebra Essentials (Section R.2, pp. 17–26)
- Geometry Essentials (Section R.3, pp. 30–35)

Now Work the 'Are You Prepared?' problems on page 98.

OBJECTIVES
1. Use the Distance Formula (p. 88)
2. Use the Midpoint Formula (p. 91)
3. Graph Equations by Hand by Plotting Points (p. 91)
4. Graph Equations Using a Graphing Utility (p. 94)
5. Use a Graphing Utility to Create Tables (p. 96)
6. Find Intercepts from a Graph (p. 96)
7. Use a Graphing Utility to Approximate Intercepts (p. 97)

Rectangular Coordinates

Figure 1

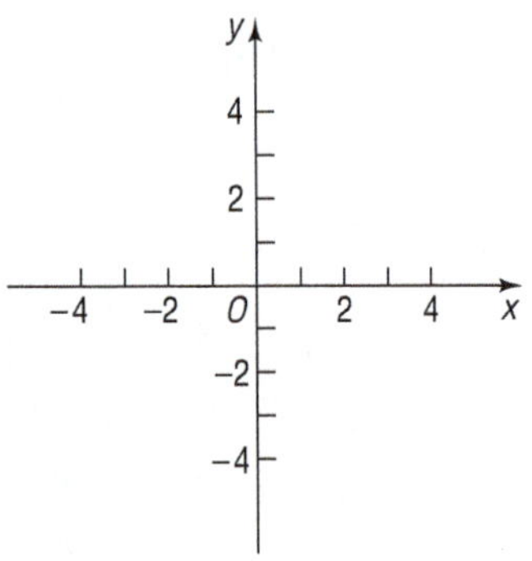

We locate a point on the real number line by assigning it a single real number, called the *coordinate of the point.* For work in a two-dimensional plane, we locate points by using two numbers.

We begin with two real number lines located in the same plane: one horizontal and the other vertical. We call the horizontal line the ***x*-axis,** the vertical line the ***y*-axis,** and the point of intersection the **origin *O*.** See Figure 1. We assign coordinates to every point on these number lines using a convenient scale. Recall that the scale of a number line is the distance between 0 and 1. In mathematics, we usually use the same scale on each axis; in applications, a different scale is often used on each axis.

The origin O has a value of 0 on both the x-axis and y-axis. Points on the x-axis to the right of O are associated with positive real numbers, and those to the left of O are associated with negative real numbers. Points on the y-axis above O are associated with positive real numbers, and those below O are associated with negative real numbers. In Figure 1, the x-axis and y-axis are labeled as x and y, respectively, and we have used an arrow at the end of each axis to denote the positive direction.

The coordinate system described here is called a **rectangular** or **Cartesian*** **coordinate system.** The plane formed by the x-axis and y-axis is sometimes called the ***xy*-plane,** and the x-axis and y-axis are referred to as the **coordinate axes.**

Any point P in the xy-plane can then be located by using an **ordered pair** (x, y) of real numbers. Let x denote the signed distance of P from the y-axis (*signed* means that, if P is to the right of the y-axis, then $x > 0$, and if P is to the left of the y-axis, then $x < 0$); and let y denote the signed distance of P from the x-axis. The ordered pair (x, y), also called the **coordinates** of P, then gives us enough information to locate the point P in the plane.

Figure 2

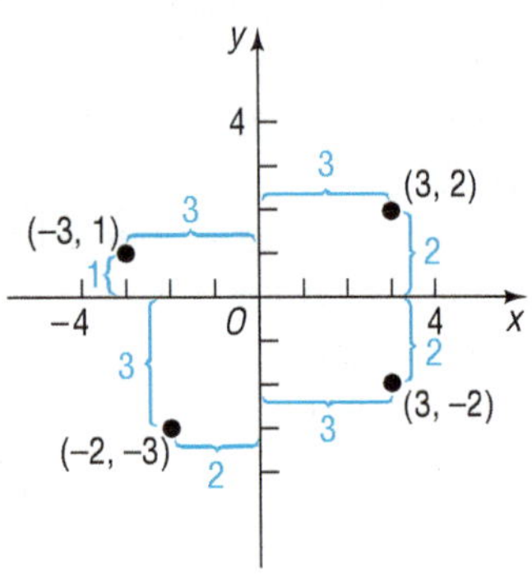

For example, to locate the point whose coordinates are $(-3, 1)$, go 3 units along the x-axis to the left of O and then go straight up 1 unit. We **plot** this point by placing a dot at this location. See Figure 2, in which the points with coordinates $(-3, 1)$, $(-2, -3)$, $(3, -2)$, and $(3, 2)$ are plotted.

The origin has coordinates $(0, 0)$. Any point on the x-axis has coordinates of the form $(x, 0)$, and any point on the y-axis has coordinates of the form $(0, y)$.

If (x, y) are the coordinates of a point P, then x is called the ***x*-coordinate,** or **abscissa,** of P and y is the ***y*-coordinate,** or **ordinate,** of P. We identify the point P by its coordinates (x, y) by writing $P = (x, y)$. Usually, we will simply say "the point (x, y)" rather than "the point whose coordinates are (x, y)."

The coordinate axes divide the xy-plane into four sections called **quadrants,** as shown in Figure 3. In quadrant I, both the x-coordinate and the y-coordinate of all points are positive; in quadrant II, x is negative and y is positive; in quadrant III,

*Named after René Descartes (1596–1650), a French mathematician, philosopher, and theologian.

Figure 3

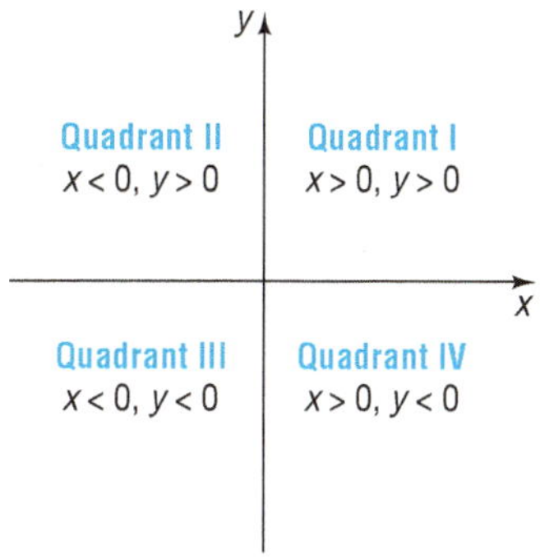

both x and y are negative; and in quadrant IV, x is positive and y is negative. Points on the coordinate axes belong to no quadrant.

Now Work PROBLEM 11

Graphing Utilities

All graphing utilities, that is, all graphing calculators and all computer software graphing packages, graph equations by plotting points on a screen. The screen itself actually consists of small rectangles, called **pixels.** The more pixels the screen has, the better the resolution. Most graphing calculators have 48 pixels per square inch; most computer screens have 32 to 108 pixels per square inch. When a point to be plotted lies inside a pixel, the pixel is turned on (lights up). The graph of an equation is a collection of lighted pixels. Figure 4 shows how the graph of $y = 2x$ looks on a TI-84 Plus graphing calculator.

Figure 4
$Y = 2X$

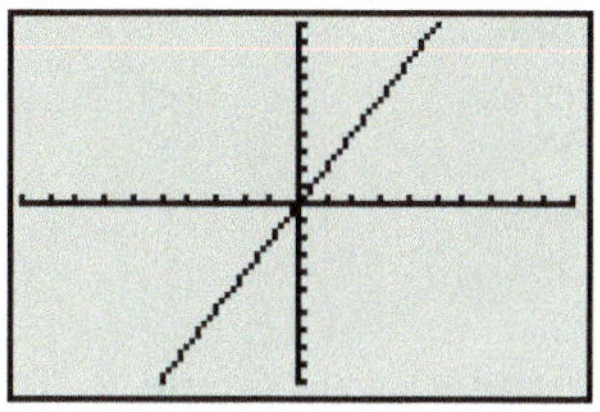

The screen of a graphing utility will display the coordinate axes of a rectangular coordinate system. However, you must set the scale on each axis. You must also include the smallest and largest values of x and y that you want included in the graph. This is called **setting the viewing rectangle** or **viewing window.** Figure 5 illustrates a typical viewing window.

Figure 5

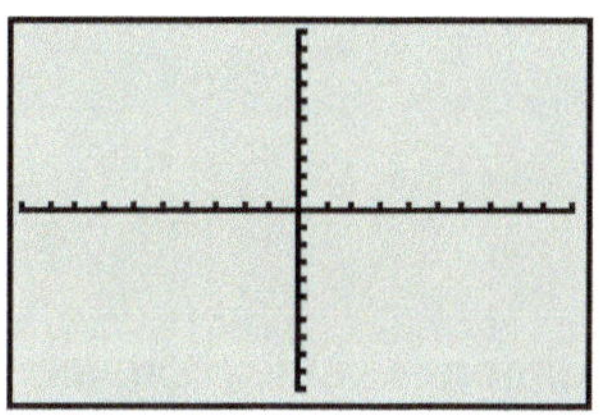

To select the viewing window, we must give values to the following expressions:

*X*min: the smallest value of x shown on the viewing window
*X*max: the largest value of x shown on the viewing window
*X*scl: the number of units per tick mark on the x-axis
*Y*min: the smallest value of y shown on the viewing window
*Y*max: the largest value of y shown on the viewing window
*Y*scl: the number of units per tick mark on the y-axis

Figure 6 illustrates these settings and their relation to the Cartesian coordinate system.

Figure 6

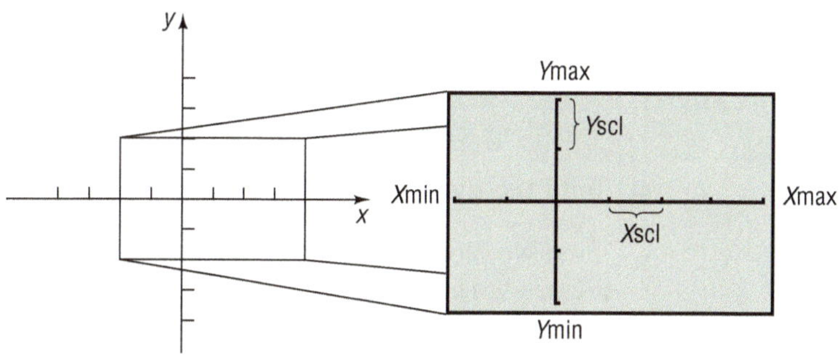

If the scale used on each axis is known, we can determine the minimum and maximum values of x and y shown on the screen by counting the tick marks. Look again at Figure 5. For a scale of 1 on each axis, the minimum and maximum values of x are -10 and 10, respectively; the minimum and maximum values of y are also -10 and 10. If the scale is 2 on each axis, then the minimum and maximum values of x are -20 and 20, respectively; and the minimum and maximum values of y are -20 and 20, respectively.

Conversely, if we know the minimum and maximum values of x and y, we can determine the scales being used by counting the tick marks displayed. We shall follow the practice of showing the minimum and maximum values of x and y in our illustrations so that you will know how the window was set. See Figure 7.

Figure 7

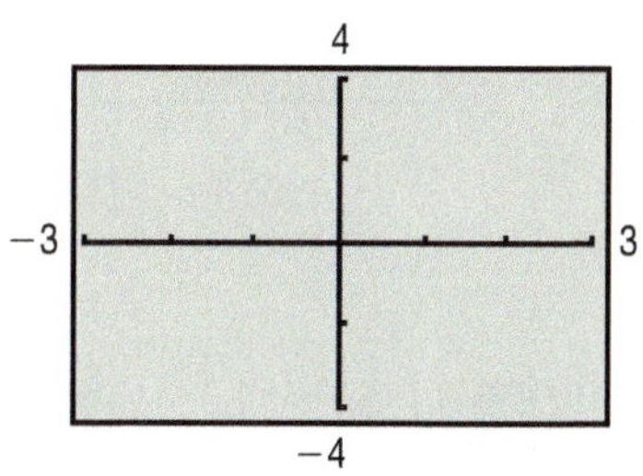

means

$X\text{min} = -3$	$Y\text{min} = -4$
$X\text{max} = 3$	$Y\text{max} = 4$
$X\text{scl} = 1$	$Y\text{scl} = 2$

EXAMPLE 1 **Finding the Coordinates of a Point Shown on a Graphing Utility Screen**

Find the coordinates of the point shown in Figure 8. Assume the coordinates are integers.

Figure 8

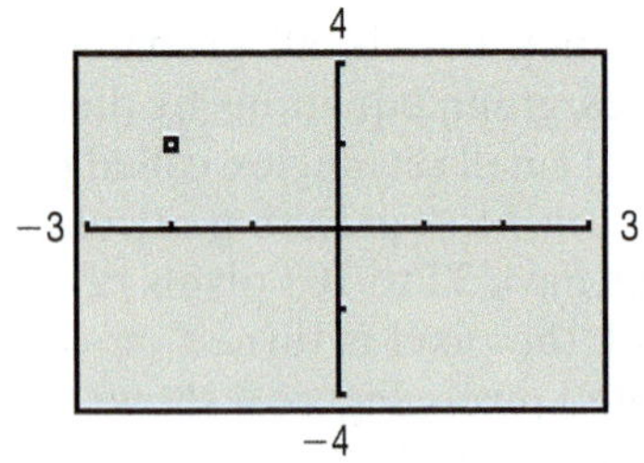

Solution First we note that the viewing window used in Figure 8 is

$$X\text{min} = -3 \qquad Y\text{min} = -4$$
$$X\text{max} = 3 \qquad Y\text{max} = 4$$
$$X\text{scl} = 1 \qquad Y\text{scl} = 2$$

The point shown is 2 tick units to the left on the horizontal axis (scale = 1) and 1 tick up on the vertical scale (scale = 2). The coordinates of the point shown are $(-2, 2)$.

Now Work PROBLEMS 15 AND 25

1 Use the Distance Formula

If the same units of measurement, such as inches, centimeters, and so on, are used for both the x-axis and y-axis, then all distances in the xy-plane can be measured using this unit of measurement.

EXAMPLE 2 **Finding the Distance between Two Points**

Find the distance d between the points $(1, 3)$ and $(5, 6)$.

Solution First we plot the points $(1, 3)$ and $(5, 6)$ and connect them with a straight line. See Figure 9(a). We are looking for length d. We begin by drawing a horizontal line from $(1, 3)$ to $(5, 3)$ and a vertical line from $(5, 3)$ to $(5, 6)$, forming a right triangle, as in Figure 9(b). One leg of the triangle is of length 4 (since $|5 - 1| = 4$) and the other is of length 3 (since $|6 - 3| = 3$). By the Pythagorean Theorem, the square of the distance d that we seek is

$$d^2 = 4^2 + 3^2 = 16 + 9 = 25$$
$$d = \sqrt{25} = 5$$

Figure 9

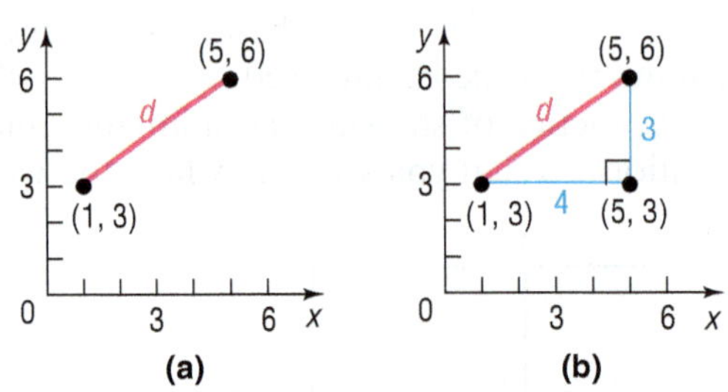

The **distance formula** provides a straightforward method for computing the distance between two points.

THEOREM

Distance Formula

The distance between two points $P_1 = (x_1, y_1)$ and $P_2 = (x_2, y_2)$, denoted by $d(P_1, P_2)$, is

$$d(P_1, P_2) = \sqrt{(x_2 - x_1)^2 + (y_2 - y_1)^2} \quad (1)$$

In Words

To compute the distance between two points, find the difference of the x-coordinates, square it, and add this to the square of the difference of the y-coordinates. The square root of this sum is the distance.

Figure 10 illustrates the theorem.

Figure 10

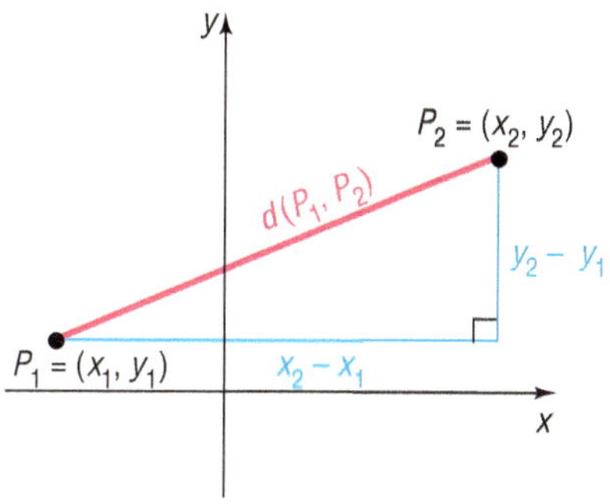

Proof of the Distance Formula Let (x_1, y_1) denote the coordinates of point P_1, and let (x_2, y_2) denote the coordinates of point P_2. Assume that the line joining P_1 and P_2 is neither horizontal nor vertical. Refer to Figure 11(a). The coordinates of P_3 are (x_2, y_1). The horizontal distance from P_1 to P_3 is the absolute value of the difference of the x-coordinates, $|x_2 - x_1|$. The vertical distance from P_3 to P_2 is the absolute value of the difference of the y-coordinates, $|y_2 - y_1|$. See Figure 11(b). The distance $d(P_1, P_2)$ that we seek is the length of the hypotenuse of the right triangle, so, by the Pythagorean Theorem, it follows that

$$\begin{aligned}[d(P_1, P_2)]^2 &= |x_2 - x_1|^2 + |y_2 - y_1|^2 \\ &= (x_2 - x_1)^2 + (y_2 - y_1)^2 \\ d(P_1, P_2) &= \sqrt{(x_2 - x_1)^2 + (y_2 - y_1)^2}\end{aligned}$$

Figure 11

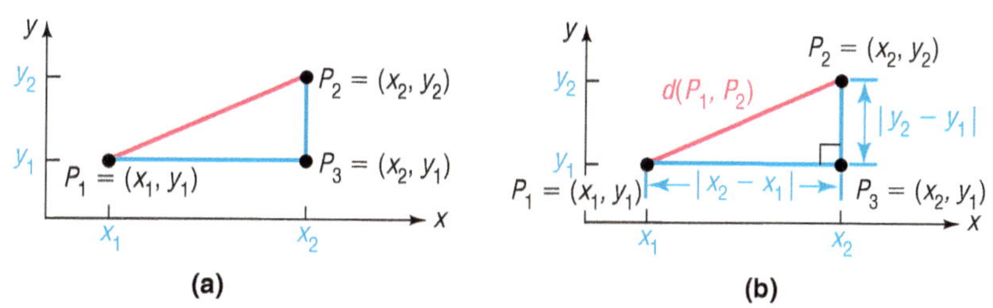

Now, if the line joining P_1 and P_2 is horizontal, then the y-coordinate of P_1 equals the y-coordinate of P_2; that is, $y_1 = y_2$. Refer to Figure 12(a). In this case, the distance formula (1) still works, because, for $y_1 = y_2$, it reduces to

$$d(P_1, P_2) = \sqrt{(x_2 - x_1)^2 + 0^2} = \sqrt{(x_2 - x_1)^2} = |x_2 - x_1|$$

Figure 12

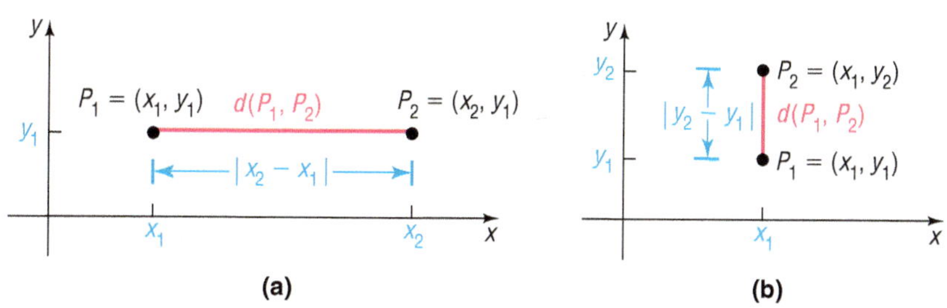

A similar argument holds if the line joining P_1 and P_2 is vertical. See Figure 12(b). The distance formula is valid in all cases. ■

EXAMPLE 3 **Finding the Length of a Line Segment**

Find the length of the line segment shown in Figure 13 on page 90.

Solution The length of the line segment is the distance between the points $P_1 = (x_1, y_1) = (-4, 5)$ and $P_2 = (x_2, y_2) = (3, 2)$. Using the distance formula (1) with $x_1 = -4$, $y_1 = 5$, $x_2 = 3$, and $y_2 = 2$, the length d is

Figure 13

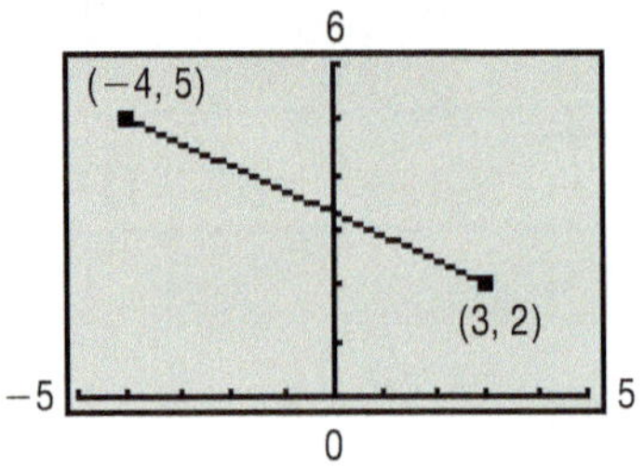

$$d = \sqrt{(x_2 - x_1)^2 + (y_2 - y_1)^2} = \sqrt{[3 - (-4)]^2 + (2 - 5)^2}$$
$$= \sqrt{7^2 + (-3)^2} = \sqrt{49 + 9} = \sqrt{58} \approx 7.62$$

Now Work PROBLEM 31

The distance between two points $P_1 = (x_1, y_1)$ and $P_2 = (x_2, y_2)$ is never a negative number. Furthermore, the distance between two points is 0 only when the points are identical, that is, when $x_1 = x_2$ and $y_1 = y_2$. Also, because $(x_2 - x_1)^2 = (x_1 - x_2)^2$ and $(y_2 - y_1)^2 = (y_1 - y_2)^2$, it makes no difference whether the distance is computed from P_1 to P_2 or from P_2 to P_1; that is, $d(P_1, P_2) = d(P_2, P_1)$.

The introduction to this chapter mentioned that rectangular coordinates enable us to translate geometry problems into algebra problems, and vice versa. The next example shows how algebra (the distance formula) can be used to solve geometry problems.

EXAMPLE 4 Using Algebra to Solve Geometry Problems

Consider the three points $A = (-2, 1)$, $B = (2, 3)$, and $C = (3, 1)$.

(a) Plot each point and form the triangle ABC.
(b) Find the length of each side of the triangle.
(c) Verify that the triangle is a right triangle.
(d) Find the area of the triangle.

Solution (a) Points A, B, C and triangle ABC are plotted in Figure 14.

Figure 14

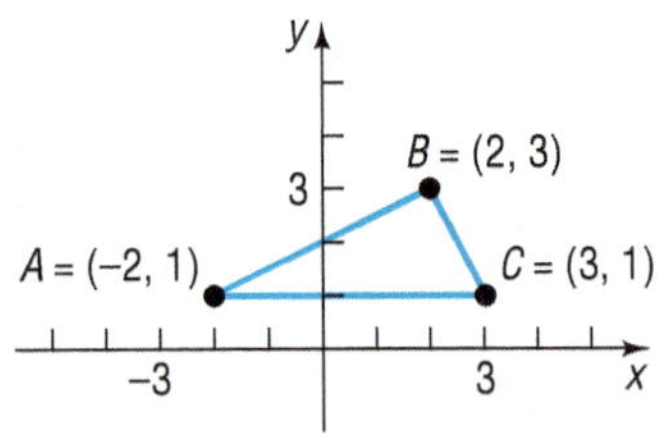

(b) $d(A, B) = \sqrt{[2 - (-2)]^2 + (3 - 1)^2} = \sqrt{16 + 4} = \sqrt{20} = 2\sqrt{5}$

$d(B, C) = \sqrt{(3 - 2)^2 + (1 - 3)^2} = \sqrt{1 + 4} = \sqrt{5}$

$d(A, C) = \sqrt{[3 - (-2)]^2 + (1 - 1)^2} = \sqrt{25 + 0} = 5$

(c) To show that the triangle is a right triangle, we need to show that the sum of the squares of the lengths of two of the sides equals the square of the length of the third side. (Why is this sufficient?) Looking at Figure 14, it seems reasonable to conjecture that the right angle is at vertex B. We shall check to see whether

$$[d(A, B)]^2 + [d(B, C)]^2 = [d(A, C)]^2$$

We find that

$$[d(A, B)]^2 + [d(B, C)]^2 = \left(2\sqrt{5}\right)^2 + \left(\sqrt{5}\right)^2$$
$$= 20 + 5 = 25 = [d(A, C)]^2$$

so it follows from the converse of the Pythagorean Theorem that triangle ABC is a right triangle.

(d) Because the right angle is at vertex B, the sides AB and BC form the base and height of the triangle. Its area is

$$\text{Area} = \frac{1}{2}(\text{Base})(\text{Height}) = \frac{1}{2}\left(2\sqrt{5}\right)\left(\sqrt{5}\right) = 5 \text{ square units}$$

Now Work PROBLEM 49

2 Use the Midpoint Formula

Figure 15

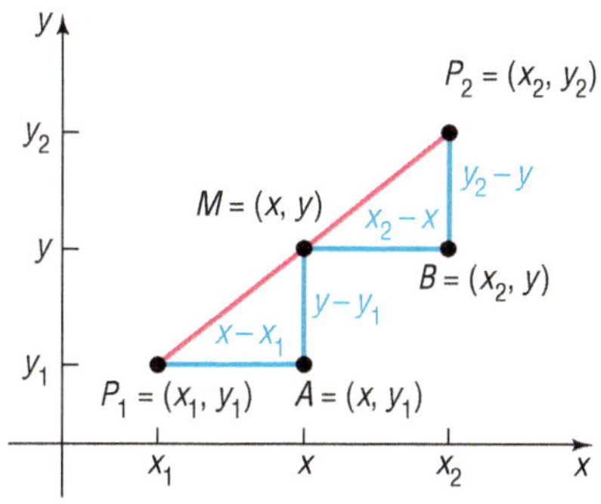

We now derive a formula for the coordinates of the **midpoint of a line segment.** Let $P_1 = (x_1, y_1)$ and $P_2 = (x_2, y_2)$ be the endpoints of a line segment, and let $M = (x, y)$ be the point on the line segment that is the same distance from P_1 as it is from P_2. See Figure 15. The triangles P_1AM and MBP_2 are congruent. [Do you see why? Angle AP_1M = angle BMP_2,* angle P_1MA = angle MP_2B, and $d(P_1, M) = d(M, P_2)$ is given. So, we have angle–side–angle.] Hence, corresponding sides are equal in length. That is,

$$x - x_1 = x_2 - x \quad \text{and} \quad y - y_1 = y_2 - y$$
$$2x = x_1 + x_2 \qquad\qquad 2y = y_1 + y_2$$
$$x = \frac{x_1 + x_2}{2} \qquad\qquad y = \frac{y_1 + y_2}{2}$$

THEOREM

Midpoint Formula

The midpoint $M = (x, y)$ of the line segment from $P_1 = (x_1, y_1)$ to $P_2 = (x_2, y_2)$ is

$$M = (x, y) = \left(\frac{x_1 + x_2}{2}, \frac{y_1 + y_2}{2}\right) \tag{2}$$

In Words

To find the midpoint of a line segment, average the x-coordinates and average the y-coordinates of the endpoints.

EXAMPLE 5 Finding the Midpoint of a Line Segment

Find the midpoint of a line segment from $P_1 = (-5, 5)$ to $P_2 = (3, 1)$. Plot the points P_1 and P_2 and their midpoint.

Solution

We apply the midpoint formula (2) using $x_1 = -5$, $y_1 = 5$, $x_2 = 3$, and $y_2 = 1$. Then the coordinates (x, y) of the midpoint M are

$$x = \frac{x_1 + x_2}{2} = \frac{-5 + 3}{2} = -1 \quad \text{and} \quad y = \frac{y_1 + y_2}{2} = \frac{5 + 1}{2} = 3$$

That is, $M = (-1, 3)$. See Figure 16.

Figure 16

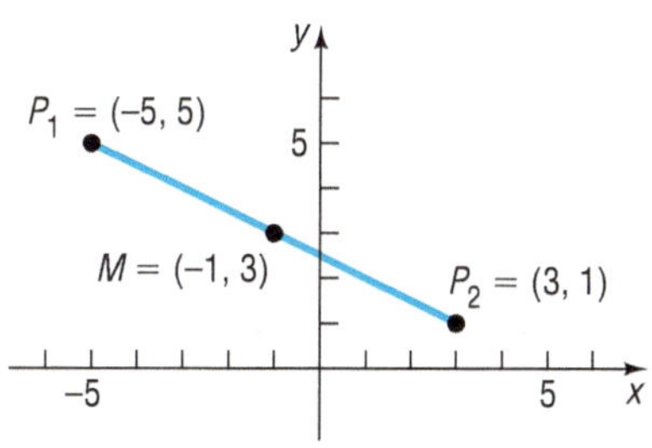

Now Work PROBLEM 55

3 Graph Equations by Hand by Plotting Points

An **equation in two variables,** say x and y, is a statement in which two expressions involving x and y are equal. The expressions are called the **sides** of the equation. Since an equation is a statement, it may be true or false, depending on the value of the variables. Any values of x and y that result in a true statement are said to **satisfy** the equation.

For example, the following are all equations in two variables x and y:

$$x^2 + y^2 = 5 \qquad 2x - y = 6 \qquad y = 2x + 5 \qquad x^2 = y$$

The first of these, $x^2 + y^2 = 5$, is satisfied for $x = 1, y = 2$, since $1^2 + 2^2 = 1 + 4 = 5$. Other choices of x and y also satisfy this equation. It is not satisfied for $x = 2$ and $y = 3$, since $2^2 + 3^2 = 4 + 9 = 13 \neq 5$.

*A postulate from geometry states that the transversal $\overline{P_1P_2}$ forms congruent corresponding angles with the parallel line segments $\overline{P_1A}$ and $\overline{MB}$.

The **graph of an equation in two variables** x and y consists of the set of points in the xy-plane whose coordinates (x, y) satisfy the equation.

Graphs play an important role in helping us to visualize the relationships that exist between two variables or quantities. Figure 17 shows the relation between the level of risk in a stock portfolio and the average annual rate of return. From the graph, we can see that, when 30% of a portfolio of stocks is invested in foreign companies, risk is minimized.

Figure 17
Source: T. Rowe Price

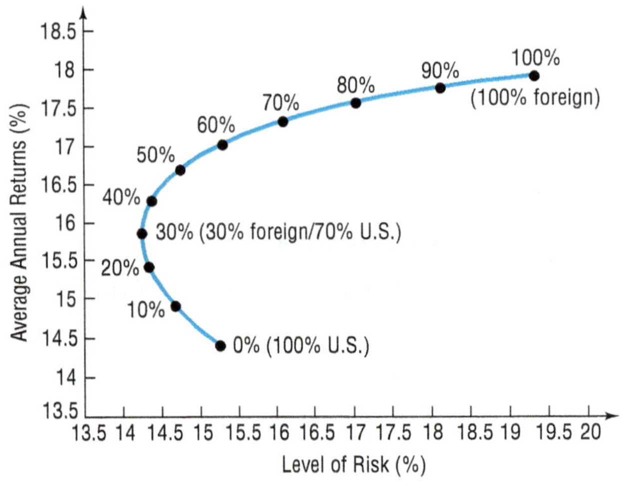

EXAMPLE 6 Determining Whether a Point Is on the Graph of an Equation

Determine if the following points are on the graph of the equation $2x - y = 6$.

(a) $(2, 3)$ (b) $(2, -2)$

Solution

(a) For the point $(2, 3)$, we check to see if $x = 2$, $y = 3$ satisfies the equation $2x - y = 6$.

$$2x - y = 2(2) - 3 = 4 - 3 = 1 \neq 6$$

The equation is not satisfied, so the point $(2, 3)$ is not on the graph.

(b) For the point $(2, -2)$, we have

$$2x - y = 2(2) - (-2) = 4 + 2 = 6$$

The equation is satisfied, so the point $(2, -2)$ is on the graph.

Now Work PROBLEM 65

EXAMPLE 7 How to Graph an Equation by Hand by Plotting Points

Graph the equation: $y = -2x + 3$

Step-by-Step Solution

STEP 1 Find all points (x, y) that satisfy the equation. To determine these points, choose values of x and use the equation to find the corresponding values for y. See Table 1.

Table 1

x	$y = -2x + 3$	(x, y)
-2	$-2(-2) + 3 = 7$	$(-2, 7)$
-1	$-2(-1) + 3 = 5$	$(-1, 5)$
0	$-2(0) + 3 = 3$	$(0, 3)$
1	$-2(1) + 3 = 1$	$(1, 1)$
2	$-2(2) + 3 = -1$	$(2, -1)$

STEP 2 Plot the points listed in the table as shown in Figure 18(a). Now connect the points to obtain the graph of the equation (a *line*), as shown in Figure 18(b).

Figure 18

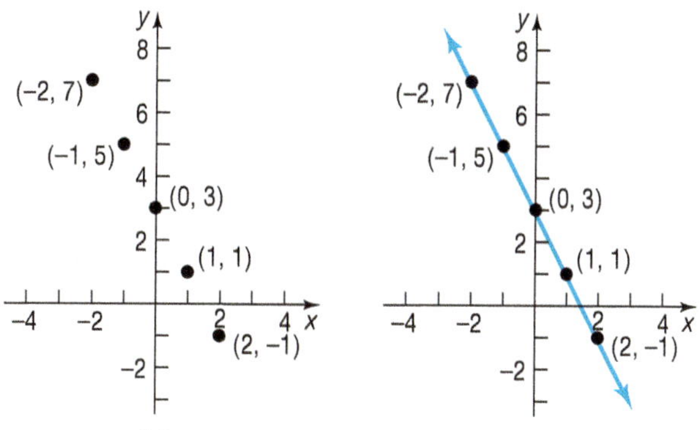

EXAMPLE 8 Graphing an Equation by Hand by Plotting Points

Graph the equation: $y = x^2$

Solution Table 2 provides several points on the graph. In Figure 19 we plot these points and connect them with a smooth curve to obtain the graph (a *parabola*).

Table 2

x	$(y = x^2)$	(x, y)
−4	16	(−4, 16)
−3	9	(−3, 9)
−2	4	(−2, 4)
−1	1	(−1, 1)
0	0	(0, 0)
1	1	(1, 1)
2	4	(2, 4)
3	9	(3, 9)
4	16	(4, 16)

Figure 19

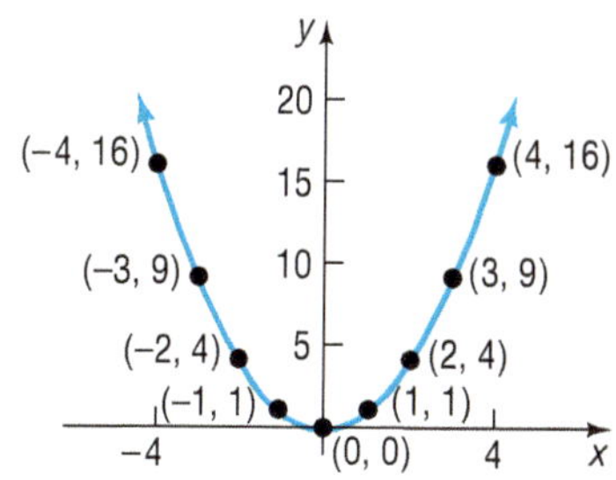

The graphs of the equations shown in Figures 18(b) and 19 do not show all the points that are on the graph. For example, in Figure 18(b), the point (20, −37) is a part of the graph of $y = -2x + 3$, but it is not shown. Since the graph of $y = -2x + 3$ could be extended out as far as we please, we use arrows to indicate that the pattern shown continues. It is important when illustrating a graph to present enough of the graph so that any viewer of the illustration will "see" the rest of it as an obvious continuation of what is actually there. This is referred to as a **complete graph.**

One way to obtain a complete graph of an equation is to plot a sufficient number of points on the graph until a pattern becomes evident. Then these points are connected with a smooth curve following the suggested pattern. But how many points are sufficient? Sometimes knowledge about the equation tells us. For example, we will learn in Section 2.2 that, if an equation is of the form $y = mx + b$, then its graph is a line. In this case, two points would suffice to obtain the graph.

One purpose of this book is to investigate the properties of equations in order to decide whether a graph is complete. Sometimes we shall graph equations by plotting a sufficient number of points on the graph until a pattern becomes evident and then connect these points with a smooth curve, following the suggested pattern. (Shortly, we shall investigate various techniques that will enable us to graph an equation without plotting so many points.) Other times we shall graph equations using a graphing utility.

4 Graph Equations Using a Graphing Utility

From Examples 7 and 8, we see that a graph can be obtained by plotting points in a rectangular coordinate system and connecting them. Graphing utilities perform these same steps when graphing an equation. For example, the TI-84 Plus determines 95 evenly spaced input values,* uses the equation to determine the output values, plots these points on the screen, and finally (if in the connected mode) draws a line between consecutive points.

To graph an equation in two variables x and y using a graphing utility requires that the equation be written in the form $y = \{\text{expression in } x\}$. If the original equation is not in this form, replace it by equivalent equations until the form $y = \{\text{expression in } x\}$ is obtained. In general, there are four ways to obtain equivalent equations.

Procedures That Result in Equivalent Equations

1. Interchange the two sides of the equation:

$$\text{Replace} \quad 3x + 5 = y \quad \text{by} \quad y = 3x + 5$$

2. Simplify the sides of the equation by combining like terms, eliminating parentheses, and so on:

$$\begin{aligned} \text{Replace} \quad 2y + 2 + 6 &= 2x + 5(x + 1) \\ \text{by} \quad 2y + 8 &= 7x + 5 \end{aligned}$$

3. Add or subtract the same expression on both sides of the equation:

$$\begin{aligned} \text{Replace} \quad y + 3x - 5 &= 4 \\ \text{by} \quad y + 3x - 5 + 5 &= 4 + 5 \end{aligned}$$

4. Multiply or divide both sides of the equation by the same nonzero expression:

$$\begin{aligned} \text{Replace} \quad 3y &= 6 - 2x \\ \text{by} \quad \frac{1}{3} \cdot 3y &= \frac{1}{3}(6 - 2x) \end{aligned}$$

EXAMPLE 9 **Expressing an Equation in the Form $y = \{\text{expression in } x\}$**

Solve for y: $2y + 3x - 5 = 4$

Solution We replace the original equation by a succession of equivalent equations.

$$\begin{aligned} 2y + 3x - 5 &= 4 \\ 2y + 3x - 5 + 5 &= 4 + 5 && \text{Add 5 to both sides.} \\ 2y + 3x &= 9 && \text{Simplify.} \\ 2y + 3x - 3x &= 9 - 3x && \text{Subtract } 3x \text{ from both sides.} \\ 2y &= 9 - 3x && \text{Simplify.} \\ \frac{2y}{2} &= \frac{9 - 3x}{2} && \text{Divide both sides by 2.} \\ y &= \frac{9 - 3x}{2} && \text{Simplify.} \end{aligned}$$

Now we are ready to graph equations using a graphing utility.

*These input values depend on the values of Xmin and Xmax. For example, if $X\text{min} = -10$ and $X\text{max} = 10$, then the first input value will be -10 and the next input value will be $-10 + (10 - (-10))/94 = -9.7872$, and so on.

EXAMPLE 10

How to Graph an Equation Using a Graphing Utility

Use a graphing utility to graph the equation: $6x^2 + 3y = 36$

Step-by-Step Solution

STEP 1 Solve the equation for y in terms of x.

$$6x^2 + 3y = 36$$
$$3y = -6x^2 + 36 \quad \text{Subtract } 6x^2 \text{ from both sides of the equation.}$$
$$y = -2x^2 + 12 \quad \text{Divide both sides of the equation by 3 and simplify.}$$

STEP 2 Enter the equation to be graphed into your graphing utility. Figure 20 shows the equation to be graphed entered on a TI-84 Plus.

Figure 20

STEP 3 Choose an initial viewing window. Without any knowledge about the behavior of the graph, it is common to choose the *standard viewing window* as the initial viewing window. The standard viewing window is

Xmin = −10 Ymin = −10
Xmax = 10 Ymax = 10
Xscl = 1 Yscl = 1

See Figure 21.

Figure 21

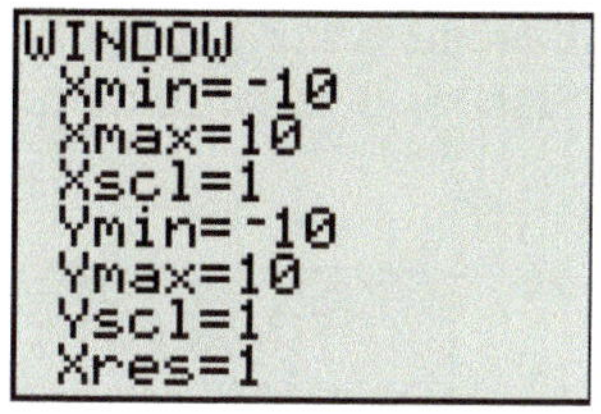

STEP 4 Graph the equation. See Figure 22.

Figure 22

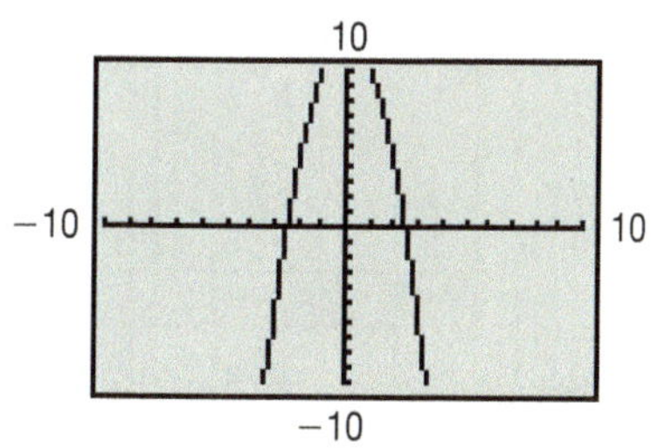

STEP 5 Adjust the viewing window until a complete graph is obtained.

The graph of $y = -2x^2 + 12$ is not complete. The value of Ymax must be increased so that the top portion of the graph is visible. After increasing the value of Ymax to 12, we obtain the graph in Figure 23. The graph is now complete.

Figure 23

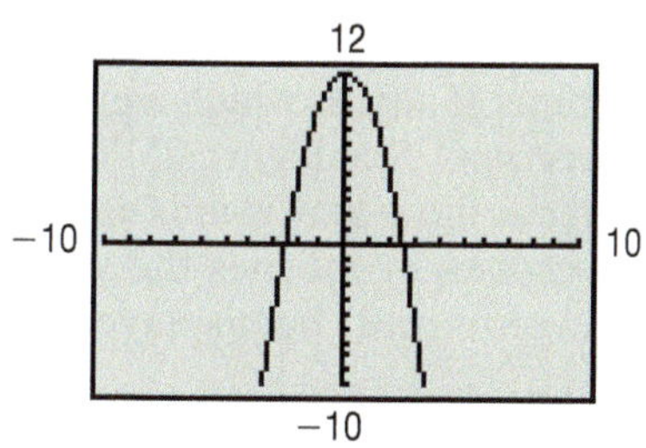

NOTE Some graphing utilities have a ZOOM-STANDARD feature that automatically sets the viewing window to the standard viewing window. In addition, some graphing utilities have a ZOOM-FIT feature that determines the appropriate Ymin and Ymax for a given Xmin and Xmax. Consult your owner's manual for the appropriate keystrokes. ■

Now Work PROBLEM 83

5 Use a Graphing Utility to Create Tables

In addition to graphing equations, graphing utilities can also be used to create a table of values that satisfy the equation. This feature is especially useful in determining an appropriate viewing window when graphing an equation.

EXAMPLE 11 How to Create a Table Using a Graphing Utility

Create a table that displays the points on the graph of $6x^2 + 3y = 36$ for $x = -3, -2, -1, 0, 1, 2,$ and 3.

Step-by-Step Solution

STEP 1 Solve the equation for y in terms of x.

We solved the equation for y in terms of x in Example 10 and obtained $y = -2x^2 + 12$.

STEP 2 Enter the expression in x following the $Y =$ prompt of the graphing utility.

See Figure 20 on page 95.

STEP 3 Set up the table. Graphing utilities typically have two modes for creating tables. In the AUTO mode, the user determines a starting point for the table (TblStart) and ΔTbl (pronounced "delta table"). The ΔTbl feature determines the increment for x in the table. The ASK mode requires the user to enter values of x, and then the utility determines the corresponding value of y.

We will create a table using AUTO mode. The table we wish to create starts at -3, so TblStart $= -3$. The increment in x is 1, so ΔTbl $= 1$. See Figure 24.

Figure 24

STEP 4 Create the table. See Table 3.

Table 3

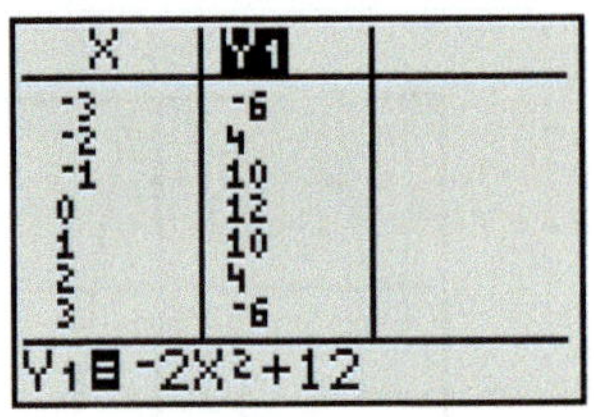

X	Y1
-3	-6
-2	4
-1	10
0	12
1	10
2	4
3	-6

Y1 = -2X²+12

The user can scroll within the table if it is created in AUTO mode.

In looking at Table 3, we notice that $y = 12$ when $x = 0$. This information could have been used to help to create the initial viewing window by letting us know that Ymax needs to be at least 12 in order to get a complete graph.

6 Find Intercepts from a Graph

The points, if any, at which a graph crosses or touches the coordinate axes are called the **intercepts.** See Figure 25. The x-coordinate of a point at which the graph crosses or touches the x-axis is an ***x*-intercept,** and the y-coordinate of a point at which the graph crosses or touches the y-axis is a ***y*-intercept.** For a graph to be complete, all its intercepts must be displayed.

Figure 25

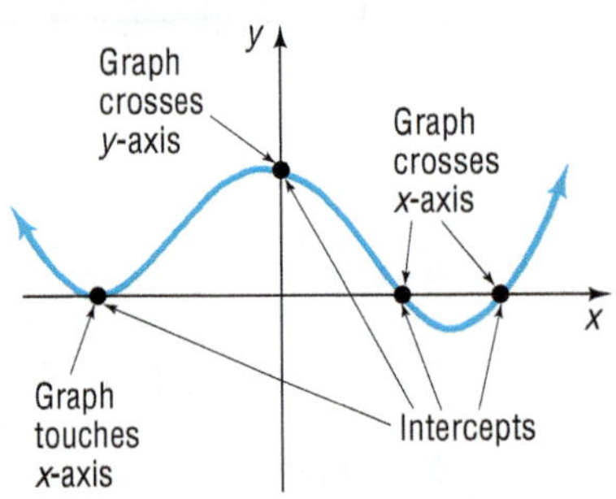

EXAMPLE 12 Finding Intercepts from a Graph

Find the intercepts of the graph in Figure 26. What are its x-intercepts? What are its y-intercepts?

Figure 26

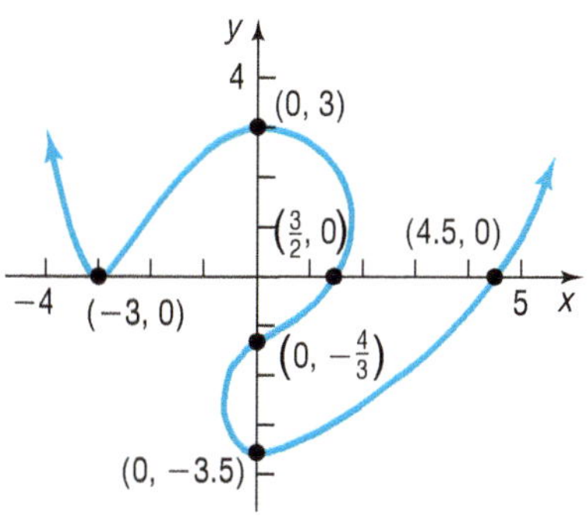

Solution The intercepts of the graph are the points

$$(-3, 0), \quad (0, 3), \quad \left(\frac{3}{2}, 0\right), \quad \left(0, -\frac{4}{3}\right), \quad (0, -3.5), \quad (4.5, 0)$$

The x-intercepts are $-3, \frac{3}{2}$, and 4.5; the y-intercepts are $-3.5, -\frac{4}{3}$, and 3.

In Example 12, you should notice the following usage: If we do not specify the type of intercept (x- versus y-), then we report the intercept as an ordered pair. However, if we specify the type of intercept, then we only report the coordinate of the intercept. For x-intercepts, we report the x-coordinate of the intercept; for y-intercepts, we report the y-coordinate of the intercept.

Now Work PROBLEM 71

7 Use a Graphing Utility to Approximate Intercepts

We can use a graphing utility to approximate the intercepts of the graph of an equation.

EXAMPLE 13 Approximating Intercepts Using a Graphing Utility

Use a graphing utility to approximate the intercepts of the equation $y = x^3 - 16$.

Solution Figure 27(a) shows the graph of $y = x^3 - 16$.

The eVALUEate feature of a TI-84 Plus graphing calculator accepts as input a value of x and determines the value of y. If we let $x = 0$, we find that the y-intercept is -16. See Figure 27(b).

The ZERO feature of a TI-84 Plus is used to find the x-intercept(s). See Figure 27(c). Rounded to two decimal places, the x-intercept is 2.52.

Figure 27

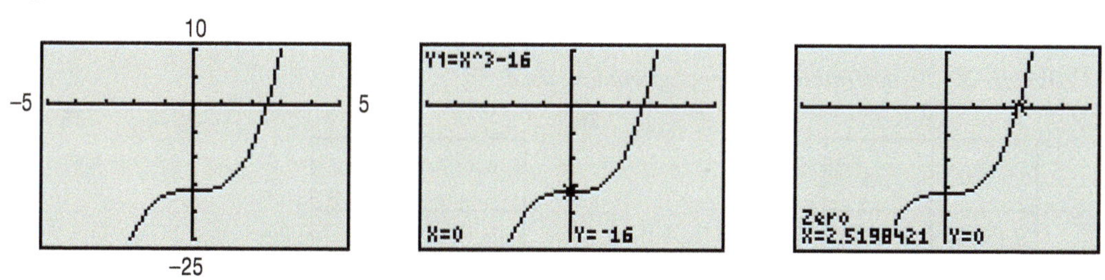

(a) (b) (c)

Now Work PROBLEM 93

To find the intercepts algebraically requires the ability to solve equations, the subject of the following four sections.

1.1 Assess Your Understanding

'Are You Prepared?' *Answers are given at the end of these exercises. If you get a wrong answer, read the pages listed in red.*

1. If -3 and 5 are the coordinates of two points on the real number line, the distance between these points is ______. (p. 20)

2. If 3 and 4 are the legs of a right triangle, the hypotenuse is ______. (p. 30)

3. Use the converse of the Pythagorean Theorem to show that a triangle whose sides are of lengths 11, 60, and 61 is a right triangle. (pp. 30–31)

4. The ______-______-______ case states that two triangles are congruent if two of the angles are congruent and the lengths of the corresponding sides between the two angles are equal. (p. 33)

Concepts and Vocabulary

5. If (x, y) are the coordinates of a point P in the xy-plane, then x is called the ______ of P and y is the ______ of P.

6. The coordinate axes divide the xy-plane into four sections called ______.

7. If three distinct points P, Q, and R all lie on a line and if $d(P, Q) = d(Q, R)$, then Q is called the ______ of the line segment from P to R.

8. ***True or False*** The distance between two points is sometimes a negative number.

9. ***True or False*** The point $(-1, 4)$ lies in quadrant IV of the Cartesian plane.

10. ***True or False*** The midpoint of a line segment is found by averaging the x-coordinates and averaging the y-coordinates of the endpoints.

Skill Building

In Problems 11 and 12, plot each point in the xy-plane. Tell in which quadrant or on what coordinate axis each point lies.

11. (a) $A = (-3, 2)$ (b) $B = (6, 0)$ (c) $C = (-2, -2)$ (d) $D = (6, 5)$ (e) $E = (0, -3)$ (f) $F = (6, -3)$

12. (a) $A = (1, 4)$ (b) $B = (-3, -4)$ (c) $C = (-3, 4)$ (d) $D = (4, 1)$ (e) $E = (0, 1)$ (f) $F = (-3, 0)$

13. Plot the points $(2, 0)$, $(2, -3)$, $(2, 4)$, $(2, 1)$, and $(2, -1)$. Describe the set of all points of the form $(2, y)$, where y is a real number.

14. Plot the points $(0, 3)$, $(1, 3)$, $(-2, 3)$, $(5, 3)$, and $(-4, 3)$. Describe the set of all points of the form $(x, 3)$, where x is a real number.

In Problems 15–18, determine the coordinates of the points shown. Tell in which quadrant each point lies. Assume the coordinates are integers.

15.

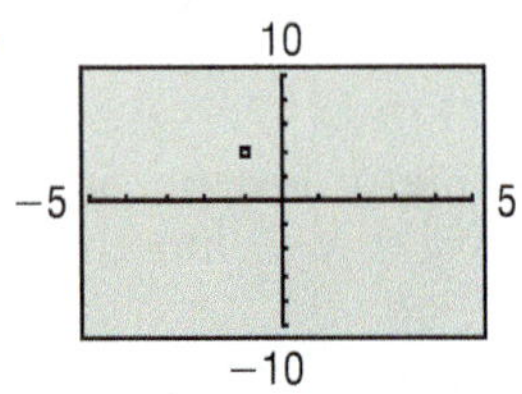

16.

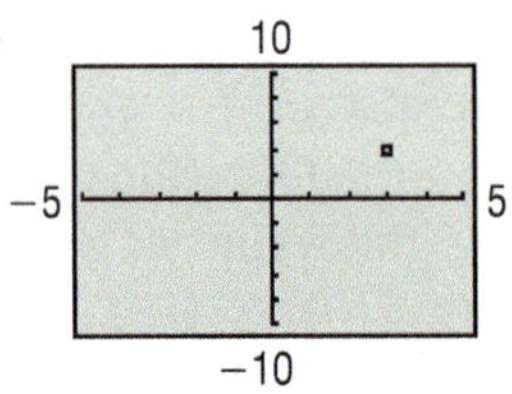

17.

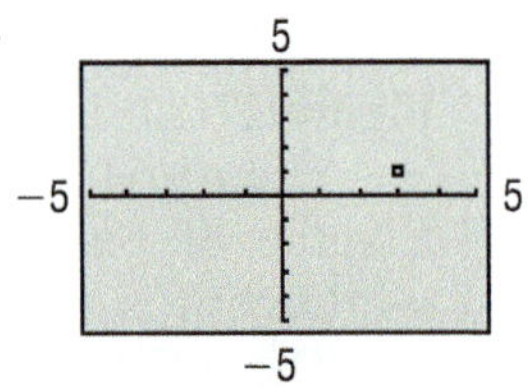

18.

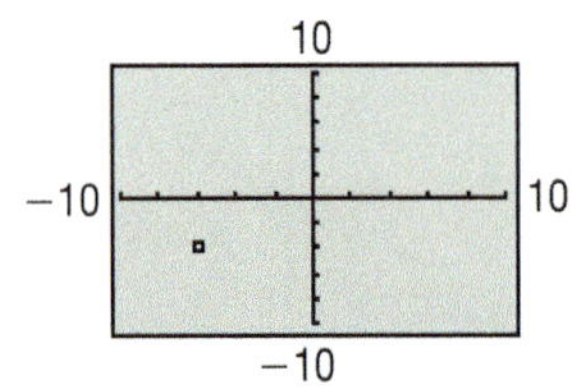

In Problems 19–24, select a setting so that each given point will lie within the viewing window.

19. $(-10, 5)$, $(3, -2)$, $(4, -1)$

20. $(5, 0)$, $(6, 8)$, $(-2, -3)$

21. $(40, 20)$, $(-20, -80)$, $(10, 40)$

22. $(-80, 60)$, $(20, -30)$, $(-20, -40)$

23. $(0, 0)$, $(100, 5)$, $(5, 150)$

24. $(0, -1)$, $(100, 50)$, $(-10, 30)$

In Problems 25–30, determine the viewing window used.

25.

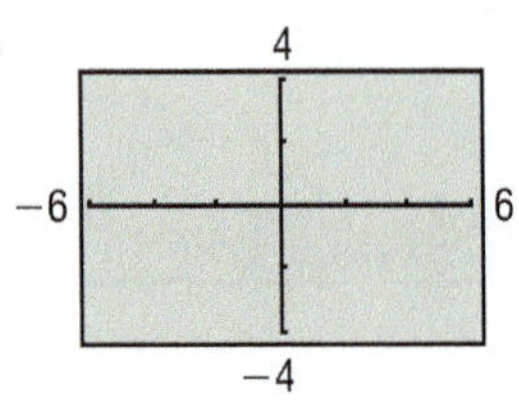

26.

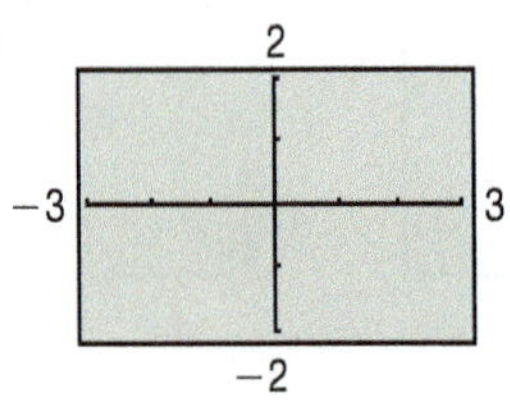

27.

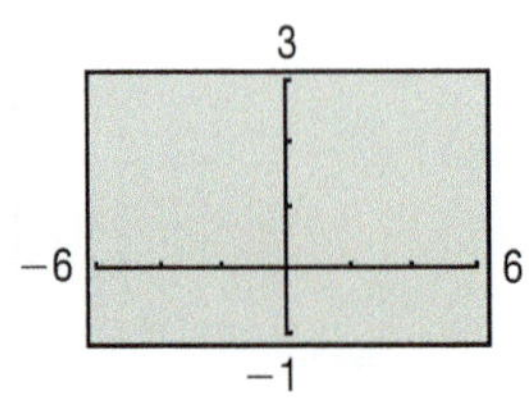

28.

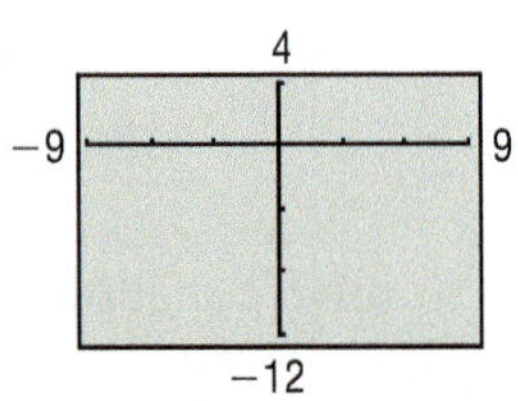

29.

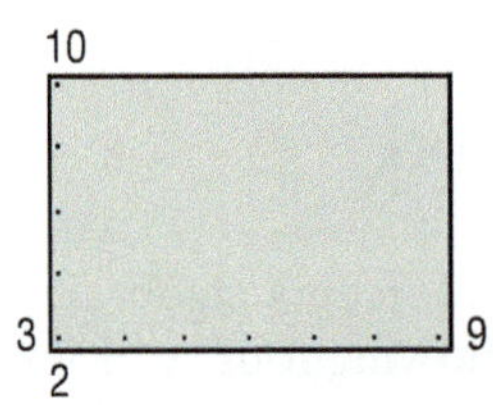

30.

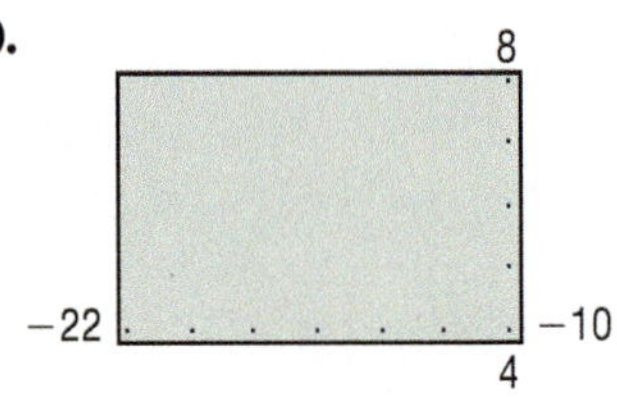

In Problems 31–44, find the distance $d(P_1, P_2)$ between the points P_1 and P_2.

31.

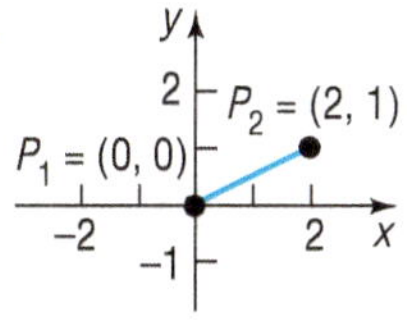

32.

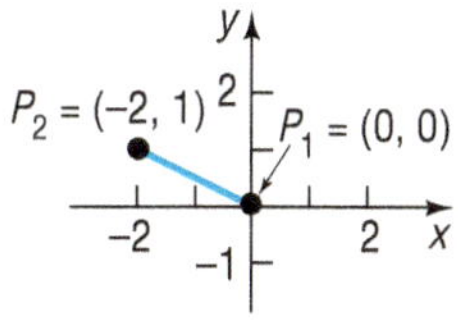

33.

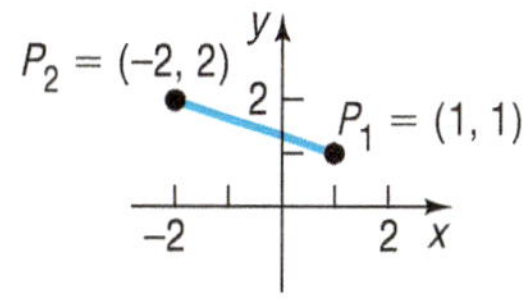

34.

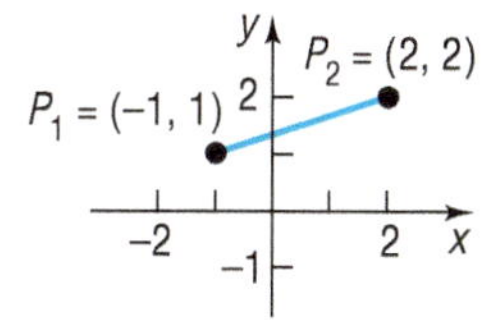

35. $P_1 = (3, -4);\quad P_2 = (5, 4)$

36. $P_1 = (-1, 0);\quad P_2 = (2, 4)$

37. $P_1 = (-5, -3);\quad P_2 = (11, 9)$

38. $P_1 = (2, -3);\quad P_2 = (10, 3)$

39. $P_1 = (4, -3);\quad P_2 = (6, 4)$

40. $P_1 = (-4, -3);\quad P_2 = (6, 2)$

41. $P_1 = (-0.2, 0.3);\quad P_2 = (1.3, 2.1)$

42. $P_1 = (1.2, 2.3);\quad P_2 = (-0.5, 1.6)$

43. $P_1 = (a, b);\quad P_2 = (0, 0)$

44. $P_1 = (a, a);\quad P_2 = (0, 0)$

In Problems 45–48, find the length of the line segment. Assume that the endpoints of each line segment have integer coordinates.

45.

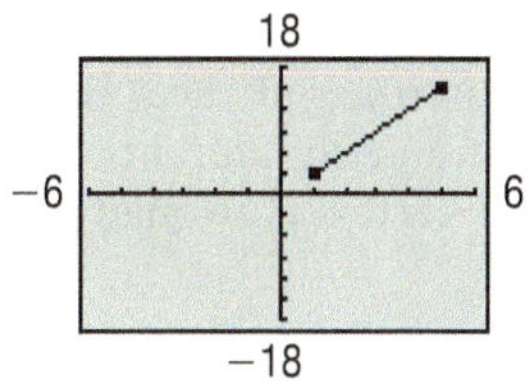

46.

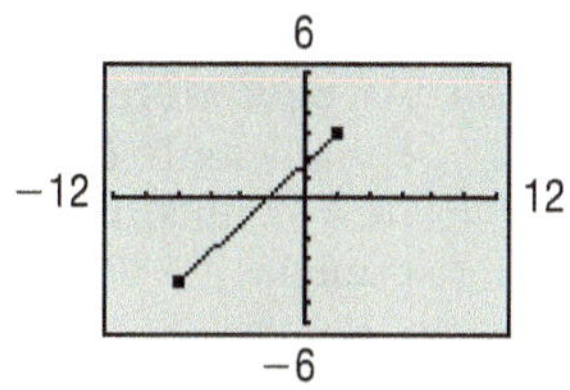

47.

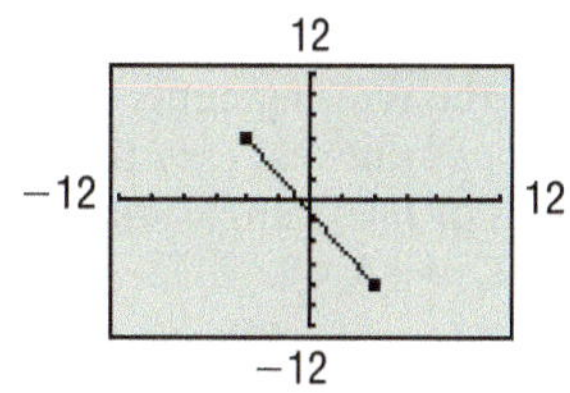

48.

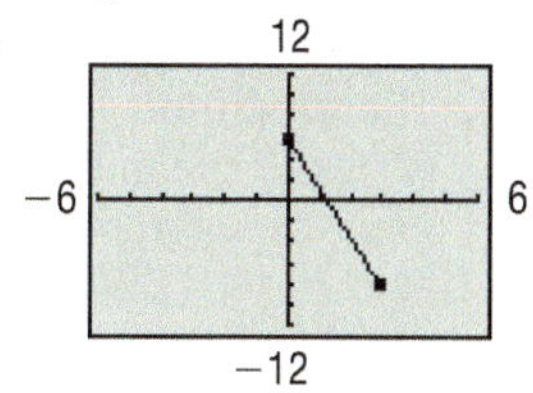

In Problems 49–54, plot each point and form the triangle ABC. Verify that the triangle is a right triangle. Find its area.

49. $A = (-2, 5);\quad B = (1, 3);\quad C = (-1, 0)$

50. $A = (-2, 5);\quad B = (12, 3);\quad C = (10, -11)$

51. $A = (-5, 3);\quad B = (6, 0);\quad C = (5, 5)$

52. $A = (-6, 3);\quad B = (3, -5);\quad C = (-1, 5)$

53. $A = (4, -3);\quad B = (0, -3);\quad C = (4, 2)$

54. $A = (4, -3);\quad B = (4, 1);\quad C = (2, 1)$

In Problems 55–64, find the midpoint of the line segment joining the points P_1 and P_2.

55. $P_1 = (3, -4);\quad P_2 = (5, 4)$

56. $P_1 = (-2, 0);\quad P_2 = (2, 4)$

57. $P_1 = (-5, -3);\quad P_2 = (11, 9)$

58. $P_1 = (2, -3);\quad P_2 = (10, 3)$

59. $P_1 = (4, -3);\quad P_2 = (6, 1)$

60. $P_1 = (-4, -3);\quad P_2 = (2, 2)$

61. $P_1 = (-0.2, 0.3);\quad P_2 = (1.3, 2.1)$

62. $P_1 = (1.2, 2.3);\quad P_2 = (-0.5, 1.6)$

63. $P_1 = (a, b);\quad P_2 = (0, 0)$

64. $P_1 = (a, a);\quad P_2 = (0, 0)$

In Problems 65–70, tell whether the given points are on the graph of the equation.

65. Equation: $y = x^4 - \sqrt{x}$
Points: $(0, 0); (1, 1); (-1, 0)$

66. Equation: $y = x^3 - 2\sqrt{x}$
Points: $(0, 0); (1, 1); (1, -1)$

67. Equation: $y^2 = x^2 + 9$
Points: $(0, 3); (3, 0); (-3, 0)$

68. Equation: $y^3 = x + 1$
Points: $(1, 2); (0, 1); (-1, 0)$

69. Equation: $x^2 + y^2 = 4$
Points: $(0, 2); (-2, 2); \left(\sqrt{2}, \sqrt{2}\right)$

70. Equation: $x^2 + 4y^2 = 4$
Points: $(0, 1); (2, 0); \left(2, \frac{1}{2}\right)$

In Problems 71–78, the graph of an equation is given. List the intercepts of the graph.

71.

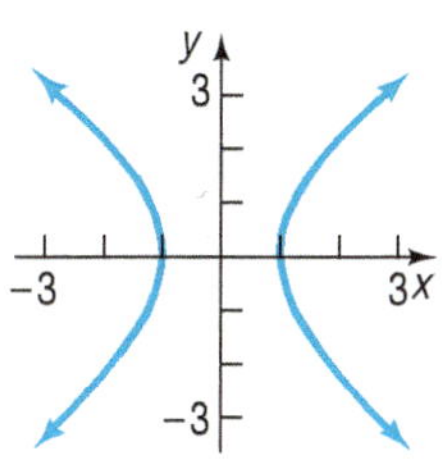

72.

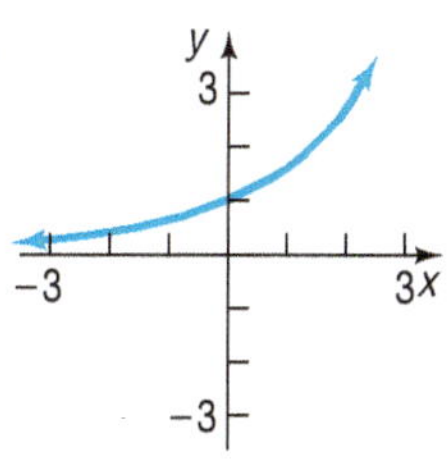

73.

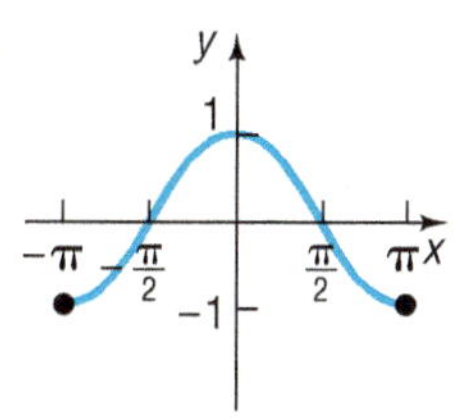

74.

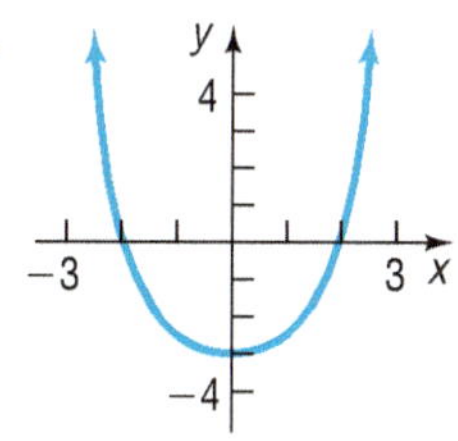

75.

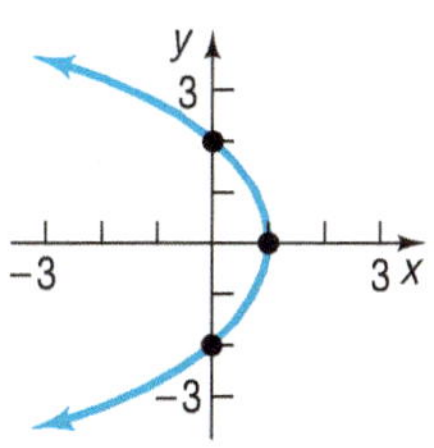

76.

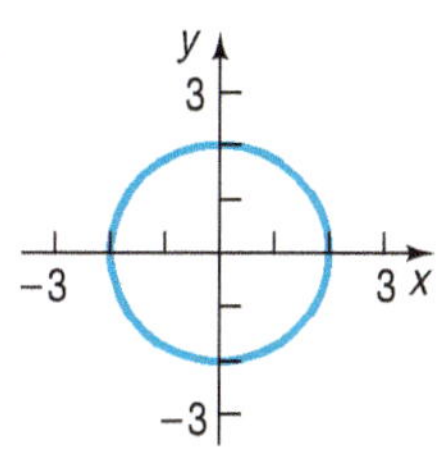

77.

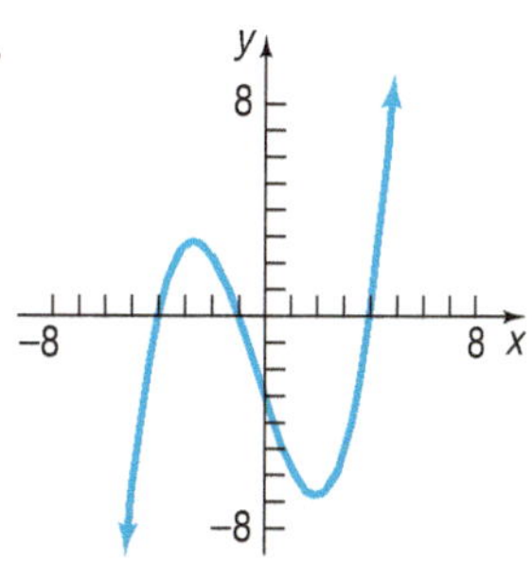

78.

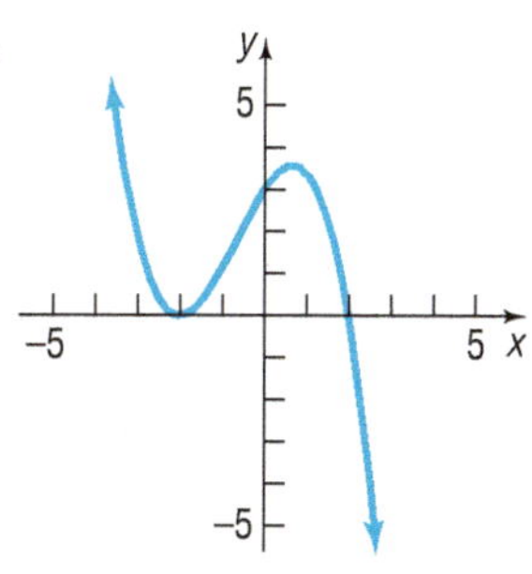

In Problems 79–90, graph each equation by hand by plotting points. Verify your results using a graphing utility.

79. $y = x + 2$ **80.** $y = x - 6$ **81.** $y = 2x + 8$ **82.** $y = 3x - 9$

83. $y = x^2 - 1$ **84.** $y = x^2 - 9$ **85.** $y = -x^2 + 4$ **86.** $y = -x^2 + 1$

87. $2x + 3y = 6$ **88.** $5x + 2y = 10$ **89.** $9x^2 + 4y = 36$ **90.** $4x^2 + y = 4$

In Problems 91–98, graph each equation using a graphing utility. Use a graphing utility to approximate the intercepts rounded to two decimal places. Use the TABLE feature to help to establish the viewing window.

91. $y = 2x - 13$ **92.** $y = -3x + 14$ **93.** $y = 2x^2 - 15$ **94.** $y = -3x^2 + 19$

95. $3x - 2y = 43$ **96.** $4x + 5y = 82$ **97.** $5x^2 + 3y = 37$ **98.** $2x^2 - 3y = 35$

Applications and Extensions

99. The **medians** of a triangle are the line segments from each vertex to the midpoint of the opposite side (see the figure). Find the lengths of the medians of the triangle with vertices at $A = (0, 0)$, $B = (6, 0)$, and $C = (4, 4)$.

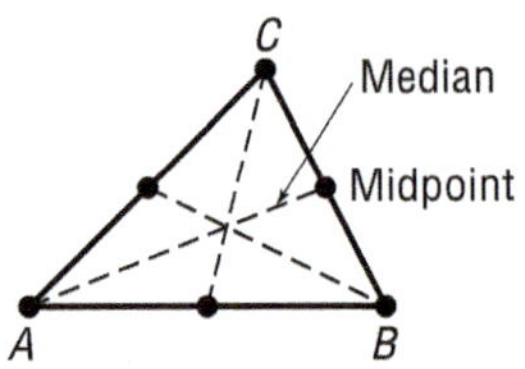

100. An **equilateral triangle** is one in which all three sides are of equal length. If two vertices of an equilateral triangle are $(0, 4)$ and $(0, 0)$, find the third vertex. How many of these triangles are possible?

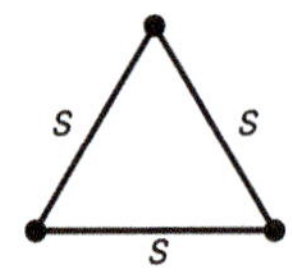

In Problems 101–104, find the length of each side of the triangle determined by the three points P_1, P_2, and P_3. State whether the triangle is an isosceles triangle, a right triangle, neither of these, or both. (An ***isosceles triangle*** *is one in which at least two of the sides are of equal length.)*

101. $P_1 = (2, 1)$; $P_2 = (-4, 1)$; $P_3 = (-4, -3)$

102. $P_1 = (-1, 4)$; $P_2 = (6, 2)$; $P_3 = (4, -5)$

103. $P_1 = (-2, -1)$; $P_2 = (0, 7)$; $P_3 = (3, 2)$

104. $P_1 = (7, 2)$; $P_2 = (-4, 0)$; $P_3 = (4, 6)$

105. **Baseball** A major league baseball "diamond" is actually a square, 90 feet on a side (see the figure). What is the distance directly from home plate to second base (the diagonal of the square)?

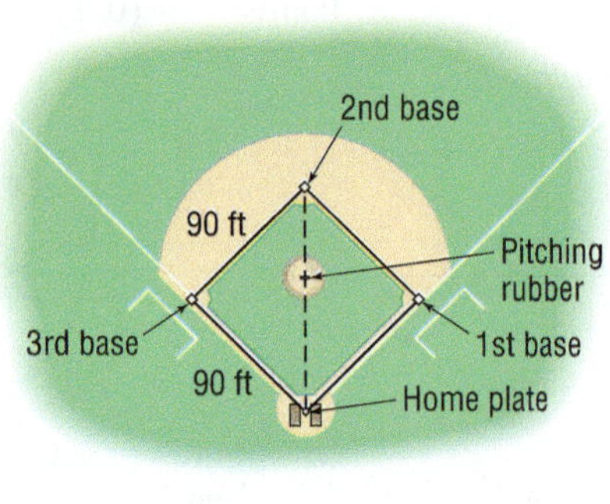

106. **Little League Baseball** The layout of a Little League playing field is a square, 60 feet on a side. How far is it directly from home plate to second base (the diagonal of the square)?

Source: *Little League Baseball, Official Regulations and Playing Rules,* 2007.

107. **Baseball** Refer to Problem 105. Overlay a rectangular coordinate system on a major league baseball diamond so that the origin is at home plate, the positive x-axis lies in the direction from home plate to first base, and the positive y-axis lies in the direction from home plate to third base.

(a) What are the coordinates of first base, second base, and third base? Use feet as the unit of measurement.

(b) If the right fielder is located at $(310, 15)$, how far is it from there to second base?

(c) If the center fielder is located at $(300, 300)$, how far is it from there to third base?

108. **Little League Baseball** Refer to Problem 106. Overlay a rectangular coordinate system on a Little League baseball diamond so that the origin is at home plate, the positive x-axis lies in the direction from home plate to first base, and the positive y-axis lies in the direction from home plate to third base.

(a) What are the coordinates of first base, second base, and third base? Use feet as the unit of measurement.

(b) If the right fielder is located at $(180, 20)$, how far is it from there to second base?

(c) If the center fielder is located at $(220, 220)$, how far is it from there to third base?

109. **Distance between Moving Objects** A Dodge Neon and a Mack truck leave an intersection at the same time. The Neon heads east at an average speed of 30 miles per hour, while the truck heads south at an average speed of 40 miles per hour. Find an expression for their distance apart d (in miles) at the end of t hours.

110. Distance of a Moving Object from a Fixed Point A hot-air balloon, headed due east at an average speed of 15 miles per hour at a constant altitude of 100 feet, passes over an intersection (see the figure). Find an expression for its distance d (measured in feet) from the intersection t seconds later.

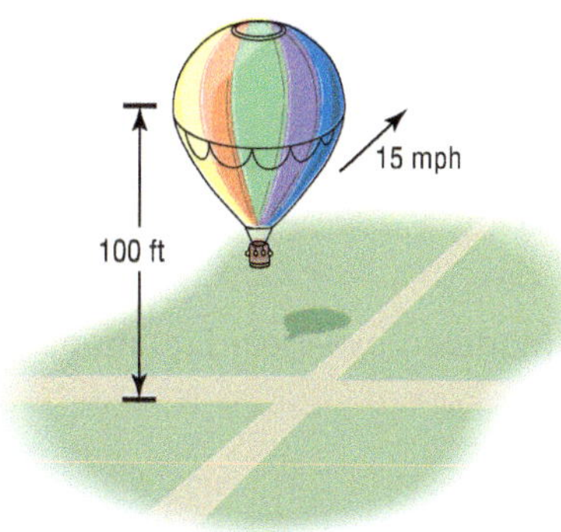

111. Drafting Error When a draftsman draws three lines that are to intersect at one point, the lines may not intersect as intended and subsequently will form an **error triangle.** If this error triangle is long and thin, one estimate for the location of the desired point is the midpoint of the shortest side. The figure shows one such error triangle.

(a) Find an estimate for the desired intersection point.

(b) Find the length of the median for the midpoint found in part (a). See Problem 99.

Source: *www.uwgb.edu/dutchs/STRUCTGE/s100.htm*

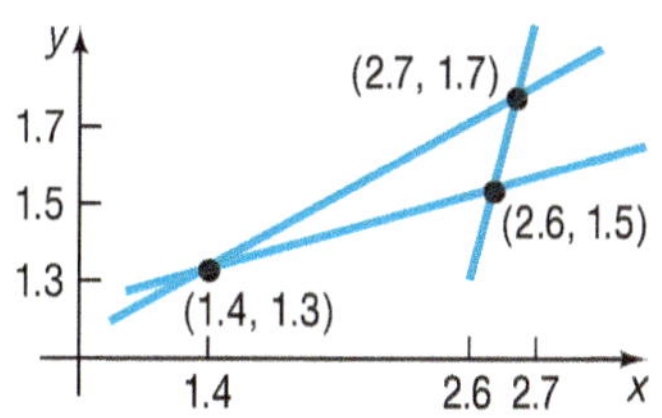

112. Net Sales The figures illustrates how net sales of Wal-Mart Stores, Inc., have grown from 2002 through 2006. Use the midpoint formula to estimate the net sales of Wal-Mart Stores, Inc., in 2004. How does your result compare to the reported value of \$256 billion?

Source: Wal-Mart Stores, Inc., 2006 Annual Report

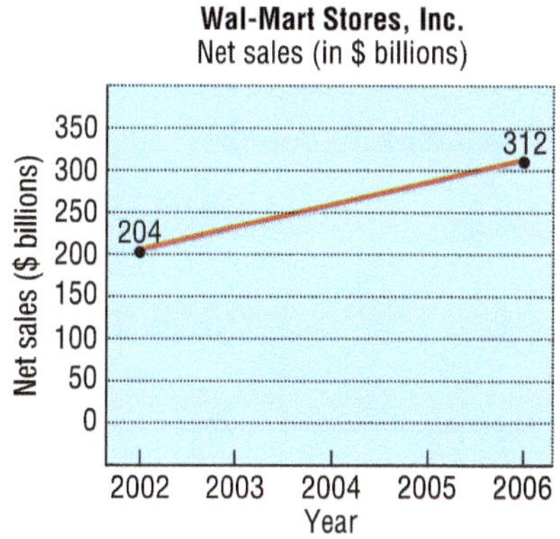

113. Poverty Threshold Poverty thresholds are determined by the U.S. Census Bureau. A poverty threshold represents the minimum annual household income for a family not to be considered poor. In 1995, the poverty threshold for a family of four with two children under the age of 18 years was \$15,455. In 2005, the poverty threshold for a family of four with two children under the age of 18 years was \$19,806. Assuming poverty thresholds increase in a straight-line fashion, use the midpoint formula to estimate the poverty threshold of a family of four with two children under the age of 18 in 2000. How does your result compare to the actual poverty threshold in 2000 of \$17,463?

Source: U.S. Census Bureau

114. Horizontal and Vertical Shifts Suppose that $A = (2, 5)$ are the coordinates of a point in the xy-plane.

(a) Find the coordinates of the point if A is shifted 3 units to the right and 2 units down.

(b) Find the coordinates of the point if A is shifted 2 units to the left and 8 units up.

115. Completing a Line Segment Plot the points $A = (-1, 8)$ and $M = (2, 3)$ in the xy-plane. If M is the midpoint of a line segment AB, find the coordinates of B.

Discussion and Writing

In Problem 116, you may use a graphing utility, but it is not required.

116. (a) Graph $y = \sqrt{x^2}$, $y = x$, $y = |x|$, and $y = (\sqrt{x})^2$, noting which graphs are the same.

(b) Explain why the graphs of $y = \sqrt{x^2}$ and $y = |x|$ are the same.

(c) Explain why the graphs of $y = x$ and $y = (\sqrt{x})^2$ are not the same.

(d) Explain why the graphs of $y = \sqrt{x^2}$ and $y = x$ are not the same.

117. Make up an equation with the points $(2, 0)$, $(4, 0)$, and $(0, 1)$. Compare your equation with a friend's equation. Comment on any similarities.

118. Draw a graph that contains the points $(-2, -1)$, $(0, 1)$, $(1, 3)$, and $(3, 5)$. Compare your graph with those of other students. Are most of the graphs almost straight lines? How many are "curved"? Discuss the various ways that these points might be connected.

119. Explain what is meant by a complete graph.

120. Write a paragraph that describes a Cartesian plane. Then write a second paragraph that describes how to plot points in the Cartesian plane. Your paragraphs should include the terms "coordinate axes," "ordered pair," "coordinates," "plot," "x-coordinate," and "y-coordinate."

'Are You Prepared?' Answers

1. 8 **2.** 5 **3.** $11^2 + 60^2 = 61^2$ **4.** angle–side–angle

1.2 Solving Equations Using a Graphing Utility; Linear and Rational Equations

PREPARING FOR THIS SECTION *Before getting started, review the following:*

- Domain of a Variable (Section R.2, p. 21)
- Rational Expressions (Section R.7, pp. 61–69)

Now Work the 'Are You Prepared?' problems on page 110.

OBJECTIVES
1. Solve Equations Using a Graphing Utility (p. 103)
2. Solve Linear Equations (p. 105)
3. Solve Rational Equations (p. 106)
4. Solve Problems That Can Be Modeled by Linear Equations (p. 108)

An **equation in one variable** is a statement in which two expressions, at least one containing the variable, are equal. The expressions are called the **sides** of the equation. Since an equation is a statement, it may be true or false, depending on the value of the variable. Unless otherwise restricted, the admissible values of the variable are those in the domain of the variable. Those admissible values of the variable, if any, that result in a true statement are called **solutions,** or **roots,** of the equation. To **solve an equation** means to find all the solutions of the equation.

For example, the following are all equations in one variable, x:

$$x + 5 = 9 \qquad x^2 + 5x = 2x - 2 \qquad \frac{x^2 - 4}{x + 1} = 0 \qquad x^2 + 9 = 5$$

The first of these statements, $x + 5 = 9$, is true when $x = 4$ and false for any other choice of x. We say that 4 is a solution of the equation $x + 5 = 9$ or that 4 **satisfies** the equation $x + 5 = 9$, because, when we substitute 4 for x, a true statement results.

Sometimes an equation will have more than one solution. For example, the equation

$$\frac{x^2 - 4}{x + 1} = 0$$

has $x = -2$ and $x = 2$ as solutions.

Usually, we will write the solution of an equation in set notation. This set is called the **solution set** of the equation. For example, the solution set of the equation $x^2 - 9 = 0$ is $\{-3, 3\}$.

Unless indicated otherwise, we will limit ourselves to real solutions, that is, solutions that are real numbers. Some equations have no real solution. For example, $x^2 + 9 = 5$ has no real solution, because there is no real number whose square, when added to 9, equals 5.

An equation that is satisfied for every choice of the variable for which both sides are defined is called an **identity.** For example, the equation

$$3x + 5 = x + 3 + 2x + 2$$

is an identity, because this statement is true for any real number x.

In this text, we present two methods for solving equations: algebraic and graphical. We shall see as we proceed through this book that some equations can be solved using algebraic techniques that result in *exact* solutions. For other equations, however, there are no algebraic techniques that lead to an exact solution. For such equations, a graphing utility can often be used to investigate possible solutions. When a graphing utility is used to solve an equation, usually *approximate* solutions are obtained.

One goal of this text is to determine when equations can be solved algebraically. If an algebraic method for solving an equation exists, we shall use it to obtain an exact solution. A graphing utility can then be used to support the algebraic result.

However, if an equation must be solved for which no algebraic techniques are available, a graphing utility will be used to obtain approximate solutions.

1 Solve Equations Using a Graphing Utility

When a graphing utility is used to solve an equation, usually *approximate* solutions are obtained. Unless otherwise stated, we shall follow the practice of giving approximate solutions as decimals *rounded to two decimal places.*

The ZERO (or ROOT) feature of a graphing utility can be used to find the solutions of an equation when one side of the equation is 0. In using this feature to solve equations, we make use of the fact that when the graph of an equation crosses or touches the x-axis then $y = 0$. For this reason, any value of x for which $y = 0$ will be a solution to the equation. That is, solving an equation for x when one side of the equation is 0 is equivalent to finding where the graph of the corresponding equation crosses or touches the x-axis.

EXAMPLE 1 Using ZERO (or ROOT) to Approximate Solutions of an Equation

Find the solution(s) of the equation $x^3 - x + 1 = 0$. Round answers to two decimal places.

Solution The solutions of the equation $x^3 - x + 1 = 0$ are the same as the x-intercepts of the graph of $Y_1 = x^3 - x + 1$. We begin by graphing the equation. Figure 28 shows the graph.

From the graph there appears to be one x-intercept (solution to the equation) between -2 and -1. Using the ZERO (or ROOT) feature of our graphing utility, we determine that the x-intercept, and thus the solution to the equation, is $x = -1.32$ rounded to two decimal places. See Figure 29.

Figure 28

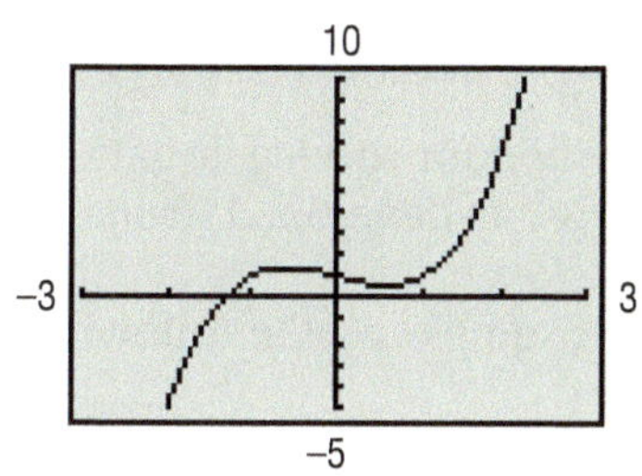

Figure 29

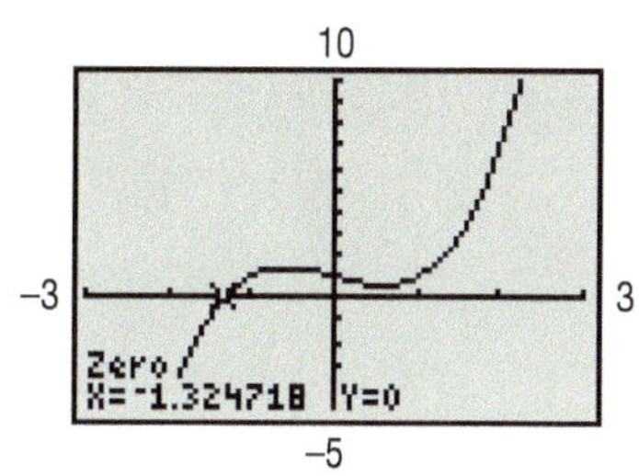

Now Work PROBLEM 17

A second method for solving equations using a graphing utility involves the INTERSECT feature of the graphing utility. This feature is used most effectively when one side of the equation is not 0.

EXAMPLE 2 Using INTERSECT to Approximate Solutions of an Equation

Find the solution(s) to the equation $4x^4 - 3 = 2x + 1$. Round answers to two decimal places.

Solution We begin by graphing each side of the equation as follows: graph $Y_1 = 4x^4 - 3$ and $Y_2 = 2x + 1$. See Figure 30 on page 104.

At the point of intersection of the graphs, the value of the y-coordinate is the same. Thus, the x-coordinate of the point of intersection represents the solution to the equation. Do you see why? The INTERSECT feature on a graphing utility determines the point of intersection of the graphs. Using this feature, we find that the graphs intersect at $(-0.87, -0.73)$ and $(1.12, 3.23)$ rounded to two decimal places. See Figure 31(a) and (b). The solutions of the equation are $x = -0.87$ and $x = 1.12$ rounded to two decimal places.

Figure 30

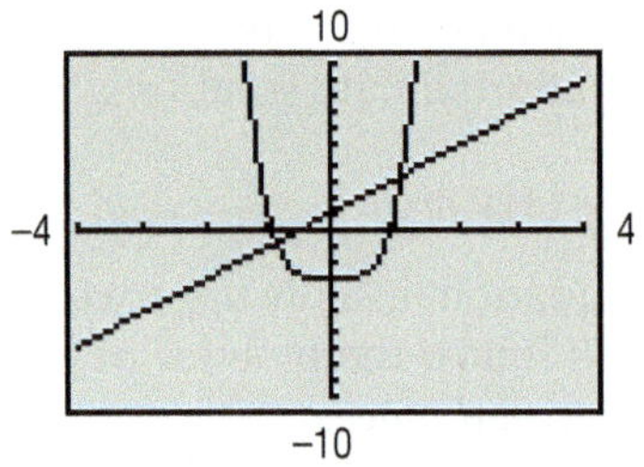

Figure 31

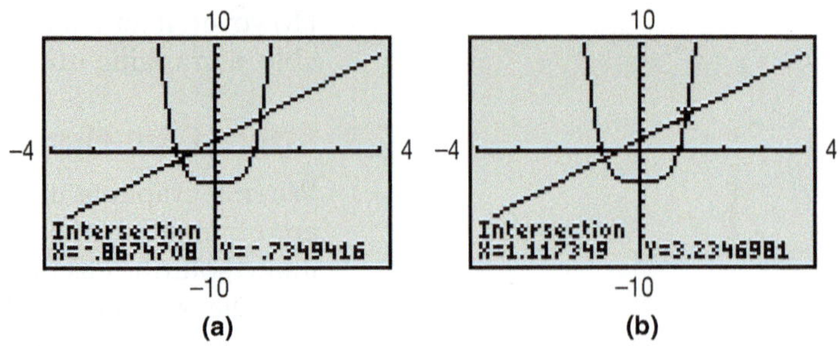

Now Work PROBLEM 19

SUMMARY

Steps for Approximating Solutions of Equations Using ZERO (or ROOT)

STEP 1: Write the equation in the form {expression in x} = 0.
STEP 2: Graph $Y_1 =$ {expression in x}.
STEP 3: Use ZERO (or ROOT) to determine each x-intercept of the graph.

Steps for Approximating Solutions of Equations Using INTERSECT

STEP 1: Graph $Y_1 =$ {expression in x on the left side of the equation}
$Y_2 =$ {expression in x on the right side of the equation}
STEP 2: Use INTERSECT to determine the x-coordinate of each point of intersection.

Solving Equations Algebraically

One method for solving equations algebraically requires that a series of *equivalent equations* be developed from the original equation until an obvious solution results.

For example, all the following equations are equivalent.

$$2x + 3 = 13$$
$$2x = 10$$
$$x = 5$$

The question, though, is "How do I obtain an equivalent equation?" In general, there are five ways to do so.

Procedures That Result in Equivalent Equations

1. Interchange the two sides of the equation:

$$\text{Replace} \quad 3 = x \quad \text{by} \quad x = 3$$

2. Simplify the sides of the equation by combining like terms, eliminating parentheses, and so on:

$$\text{Replace} \quad x + 2 + 6 = 2x + 3(x + 1)$$
$$\text{by} \quad x + 8 = 5x + 3$$

3. Add or subtract the same expression on both sides of the equation:

$$\text{Replace} \quad 3x - 5 = 4$$
$$\text{by} \quad (3x - 5) + 5 = 4 + 5$$

4. Multiply or divide both sides of the equation by the same nonzero expression:

$$\text{Replace} \quad \frac{3x}{x-1} = \frac{6}{x-1} \qquad x \neq 1$$

$$\text{by} \quad \frac{3x}{x-1} \cdot (x-1) = \frac{6}{x-1} \cdot (x-1)$$

5. If one side of the equation is 0 and the other side can be factored, then we may use the Zero-Product Property and set each factor equal to 0:

$$\text{Replace} \quad x(x-3) = 0$$

$$\text{by} \quad x = 0 \quad \text{or} \quad x - 3 = 0$$

WARNING Squaring both sides of an equation does not necessarily lead to an equivalent equation. ■

Whenever it is possible to solve an equation in your head, do so. For example:

The solution of $2x = 8$ is $x = 4$.

The solution of $3x - 15 = 0$ is $x = 5$.

Now Work PROBLEM 9

We now introduce a specific type of equation that can be solved algebraically to obtain exact solutions.

2 Solve Linear Equations

Linear equations are equations such as

$$3x + 12 = 0 \qquad \frac{3}{4}x - \frac{1}{5} = 0 \qquad 0.62x - 0.3 = 0$$

A **linear equation in one variable** is equivalent to an equation of the form

$$ax + b = 0$$

where a and b are real numbers and $a \neq 0$.

Sometimes a linear equation is called a **first-degree equation,** because the left side is a polynomial in x of degree 1.

EXAMPLE 3 Solving a Linear Equation

Solve the equation: $3(x-2) = 5(x-1)$

Algebraic Solution

$3(x-2) = 5(x-1)$

$3x - 6 = 5x - 5$ — Use the Distributive Property.

$3x - 6 - 5x = 5x - 5 - 5x$ — Subtract 5x from each side.

$-2x - 6 = -5$ — Simplify.

$-2x - 6 + 6 = -5 + 6$ — Add 6 to each side.

$-2x = 1$ — Simplify.

$\frac{-2x}{-2} = \frac{1}{-2}$ — Divide each side by −2.

$x = -\frac{1}{2}$ — Simplify.

Graphing Solution

Graph $Y_1 = 3(x-2)$ and $Y_2 = 5(x-1)$. See Figure 32. Using INTERSECT, we find the point of intersection to be $(-0.5, -7.5)$. The solution of the equation is $x = -0.5$.

Figure 32

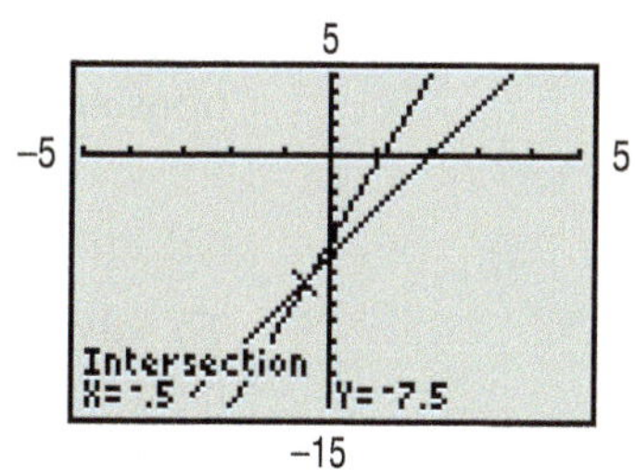

Check: Let $x = -\frac{1}{2}$ in the expression in x on the left side of the equation and simplify. Let $x = -\frac{1}{2}$ in the expression in x on the right side of the equation and simplify. If the two expressions are equal, the solution checks.

$$3(x-2) = 3\left(-\frac{1}{2} - 2\right) = 3\left(-\frac{5}{2}\right) = -\frac{15}{2}$$
$$5(x-1) = 5\left(-\frac{1}{2} - 1\right) = 5\left(-\frac{3}{2}\right) = -\frac{15}{2}$$

Since the two expressions are equal, the solution $x = -\frac{1}{2}$ checks.

The solution set is $\left\{-\frac{1}{2}\right\}$.

Now Work PROBLEM 35

The next example illustrates the solution of an equation that does not appear to be linear, but leads to a linear equation upon simplification.

EXAMPLE 4 Solving an Equation That Leads to a Linear Equation

Solve the equation: $(2x-1)(x-1) = (x-5)(2x-5)$

Algebraic Solution

$(2x-1)(x-1) = (x-5)(2x-5)$	
$2x^2 - 3x + 1 = 2x^2 - 15x + 25$	Multiply and combine like terms.
$2x^2 - 3x + 1 - 2x^2 = 2x^2 - 15x + 25 - 2x^2$	Subtract $2x^2$ from each side.
$-3x + 1 = -15x + 25$	Simplify.
$-3x + 1 - 1 = -15x + 25 - 1$	Subtract 1 from each side.
$-3x = -15x + 24$	Simplify.
$-3x + 15x = -15x + 24 + 15x$	Add 15x to each side.
$12x = 24$	Simplify.
$\frac{12x}{12} = \frac{24}{12}$	Divide each side by 12.
$x = 2$	Simplify.

Graphing Solution

Graph $Y_1 = (2x-1)(x-1)$ and $Y_2 = (x-5)(2x-5)$. See Figure 33. Using INTERSECT, we find the point of intersection to be (2, 3). The solution of the equation is $x = 2$.

Figure 33

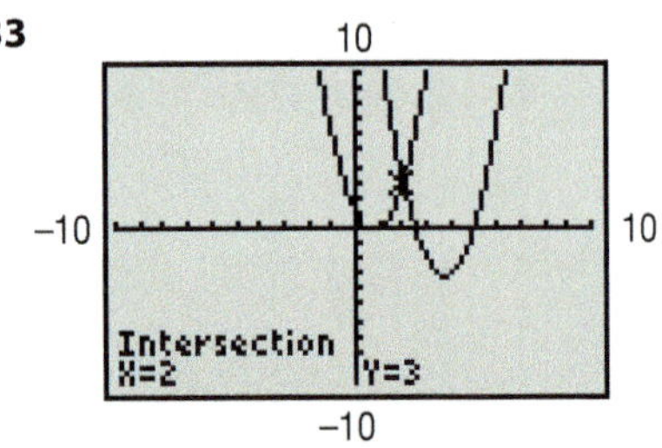

Check: $(2x-1)(x-1) = (2\cdot 2 - 1)(2-1) = (3)(1) = 3$

$(x-5)(2x-5) = (2-5)(2\cdot 2 - 5) = (-3)(-1) = 3$

Since the two expressions are equal, the solution checks. The solution set is $\{2\}$.

Now Work PROBLEM 55

3 Solve Rational Equations

We now introduce another type of equation, the *rational equation*. A **rational equation** is an equation that contains a rational expression. Examples of rational equations are

$$\frac{3}{x+1} = \frac{2}{x-1} + 7 \quad \text{and} \quad \frac{x-5}{x-4} = \frac{3}{x+2}$$

To solve a rational equation, multiply both sides of the equation by the least common multiple of the denominators of the rational expressions that make up the rational equation. We shall only show the algebraic solution to rational equations. We leave it to you to verify the results using a graphing utility.

EXAMPLE 5 Solving a Rational Equation

Solve the equation: $\dfrac{3}{x-2} = \dfrac{1}{x-1} + \dfrac{7}{(x-1)(x-2)}$

Solution First, we note that the domain of the variable is $\{x \mid x \neq 1, x \neq 2\}$. We clear the equation of rational expressions by multiplying both sides by the least common multiple of the denominators of the three rational expressions, $(x-1)(x-2)$.

$$\frac{3}{x-2} = \frac{1}{x-1} + \frac{7}{(x-1)(x-2)}$$

$$(x-1)\cancel{(x-2)}\frac{3}{\cancel{x-2}} = (x-1)(x-2)\left[\frac{1}{x-1} + \frac{7}{(x-1)(x-2)}\right]$$ Multiply both sides by $(x-1)(x-2)$. Cancel on the left.

$$3x - 3 = \cancel{(x-1)}(x-2)\frac{1}{\cancel{x-1}} + \cancel{(x-1)(x-2)}\frac{7}{\cancel{(x-1)(x-2)}}$$ Use the Distributive Property on each side; cancel on the right.

$$3x - 3 = (x-2) + 7$$

$$3x - 3 = x + 5$$ Combine like terms.

$$2x = 8$$ Add 3 to each side; subtract x from each side.

$$x = 4$$ Divide each side by 2.

Check: $\dfrac{3}{x-2} = \dfrac{3}{4-2} = \dfrac{3}{2}$

$$\frac{1}{x-1} + \frac{7}{(x-1)(x-2)} = \frac{1}{4-1} + \frac{7}{(4-1)(4-2)} = \frac{1}{3} + \frac{7}{3 \cdot 2} = \frac{2}{6} + \frac{7}{6} = \frac{9}{6} = \frac{3}{2}$$

Since the two expressions are equal, the solution $x = 4$ checks. The solution set is $\{4\}$.

Now Work PROBLEM 71

Sometimes the process of creating equivalent equations leads to apparent solutions that are not solutions of the original equation. These are called **extraneous solutions.**

EXAMPLE 6 A Rational Equation with No Solution

Solve the equation: $\dfrac{3x}{x-1} + 2 = \dfrac{3}{x-1}$

Solution First, we note that the domain of the variable is $\{x \mid x \neq 1\}$. Since the two rational expressions in the equation have the same denominator, $x - 1$, we can simplify by multiplying both sides by $x - 1$. The resulting equation is equivalent to the original equation, since we are multiplying by $x - 1$, which is not 0 (remember, $x \neq 1$).

$$\frac{3x}{x-1} + 2 = \frac{3}{x-1}$$

$$\left(\frac{3x}{x-1} + 2\right)\cdot(x-1) = \frac{3}{\cancel{x-1}}\cdot\cancel{(x-1)}$$ Multiply both sides by $x - 1$; cancel on the right.

$$\frac{3x}{\cancel{x-1}}\cdot\cancel{(x-1)} + 2\cdot(x-1) = 3$$ Use the Distributive Property on the left side; cancel on the left.

$$3x + 2x - 2 = 3$$ Simplify.

$$5x - 2 = 3$$ Combine like terms.

$$5x = 5$$ Add 2 to each side.

$$x = 1$$ Divide both sides by 5.

The solution appears to be 1. But recall that $x = 1$ is not in the domain of the variable, so $x = 1$ is an extraneous solution. The equation has no solution. The solution set is $\varnothing$.

Now Work PROBLEM 61

4 Solve Problems That Can Be Modeled by Linear Equations

Although each situation has unique features, we can provide an outline of the steps to follow in setting up applied problems.

NOTE The icon is a *Model It!* icon. It indicates that the discussion or problem involves modeling.

NOTE It is a good practice to choose a variable that reminds you of the unknown. For example, use t for time.

Steps for Setting Up Applied Problems

STEP 1: Read the problem carefully, perhaps two or three times. Pay particular attention to the question being asked in order to identify what you are looking for. If you can, determine realistic possibilities for the answer.

STEP 2: Assign a letter (variable) to represent what you are looking for, and, if necessary, express any remaining unknown quantities in terms of this variable.

STEP 3: Make a list of all the known facts, and translate them into mathematical expressions. These may take the form of an equation (or, later, an inequality) involving the variable. The equation (or inequality) is called the **model.** If possible, draw an appropriately labeled diagram to assist you. Sometimes a table or chart helps.

STEP 4: Solve the equation for the variable, and then answer the question using a complete sentence.

STEP 5: Check the answer with the facts in the problem. If it agrees, congratulations! If it does not agree, try again.

Let's look at two examples.

EXAMPLE 7 How to Solve an Applied Problem: Investments

A total of \$18,000 is invested, some in stocks and some in bonds. If the amount invested in bonds is half that invested in stocks, how much is invested in each category?

Step-by-Step Solution

STEP 1 Determine what you are looking for.

We are being asked to find the amount of two investments. These amounts must total \$18,000, because a total of \$18,000 is invested.

STEP 2 Assign a variable to represent what you are looking for. If necessary, express any remaining unknown quantities in terms of this variable.

Let s represent the amount invested in stocks. Now, if we invested \$10,000 in stocks, how much would be left to invest in bonds? In general, if s represents the amount in stocks, then $18{,}000 - s$ is the amount invested in bonds.

STEP 3 Translate the English into mathematical statements. It may be helpful to draw a figure that represents the situation. Also, you may want to set up a table to organize your thoughts. Use the information to build your model.

Let's set up a table.

Amount in Stocks	Amount in Bonds	Reason
s	$18{,}000 - s$	Total invested is \$18,000

We also know that

Total amount invested in bonds is one-half that in stocks

$$18{,}000 - s = \frac{1}{2}(s) \quad \text{The Model}$$

STEP 4 Solve the equation and answer the original question.

$$18{,}000 - s = \frac{1}{2}s$$

$$18{,}000 = s + \frac{1}{2}s \quad \text{Add } s \text{ to both sides.}$$

$$18{,}000 = \frac{3}{2}s \quad \text{Simplify.}$$

$$\left(\frac{2}{3}\right)18{,}000 = \left(\frac{2}{3}\right)\left(\frac{3}{2}s\right) \quad \text{Multiply both sides by } \frac{2}{3}.$$

$$12{,}000 = s \quad \text{Simplify.}$$

So \$12,000 is invested in stocks and \$18,000 − \$12,000 = \$6000 is invested in bonds.

Step 5 Check your answer with the facts presented in the problem. If it agrees, congratulations. If it does not agree, try again.

The total invested is \$12,000 + \$6000 = \$18,000, and the amount invested in bonds, \$6000, is half the amount invested in stocks, \$12,000.

Now Work PROBLEM 95

EXAMPLE 8 Determining an Hourly Wage

Shannon grossed \$609 one week by working 52 hours. Her employer pays time-and-a-half for all hours worked in excess of 40 hours. With this information, can you determine Shannon's regular hourly wage?

Solution

Step 1: We are looking for an hourly wage. Our answer will be in dollars per hour.

Step 2: Let w represent the regular hourly wage; w is measured in dollars per hour.

Step 3: We set up a table:

	Hours Worked	Hourly Wage	Salary
Regular	40	w	$40w$
Overtime	12	$1.5w$	$12(1.5w) = 18w$

The sum of regular salary plus overtime salary will equal \$609. From the table, $40w + 18w = 609$. This is our model.

Step 4:

$$40w + 18w = 609$$
$$58w = 609$$
$$w = 10.50$$

Shannon's regular hourly wage is \$10.50 per hour.

Step 5: Forty hours yields a salary of $40(10.50) = \$420$, and 12 hours of overtime yields a salary of $12(1.5)(10.50) = \$189$, for a total of \$609.

Now Work PROBLEM 99

Historical Feature

Solving equations is among the oldest of mathematical activities, and efforts to systematize this activity determined much of the shape of modern mathematics.

Consider the following problem and its solution using only words: Solve the problem of how many apples Jim has, given that

"Bob's five apples and Jim's apples together make twelve apples" by thinking,

"Jim's apples are all twelve apples less Bob's five apples" and then concluding,

"Jim has seven apples."

The mental steps translated into algebra are

$$5 + x = 12$$
$$x = 12 - 5$$
$$= 7$$

The solution of this problem using only words is the earliest form of algebra. Such problems were solved exactly this way in Babylonia in 1800 BC. We know almost nothing of mathematical work before this date, although most authorities believe the sophistication of the earliest known texts indicates that a long period of previous development must have occurred. The method of writing out equations in words persisted for thousands of years, and although it now seems extremely cumbersome, it was used very effectively by many generations of mathematicians. The Arabs developed a good deal of the theory of cubic equations while writing out all the equations in words. About AD 1500, the tendency to abbreviate words into written equations began to lead in the direction of modern notation; for example, the Latin word *et* (meaning *and*) developed into the plus sign, +. Although the occasional use of letters to represent variables dates back to AD 1200, the practice did not become common until about AD 1600. Development thereafter was rapid, and by 1635 algebraic notation did not differ essentially from what we use now.

1.2 Assess Your Understanding

'Are You Prepared?' *Answers are given at the end of these exercises. If you get a wrong answer, read the pages listed in red.*

1. Is $x = 4$ in the domain of $\frac{3}{x-4}$? (p. 21)

2. Find the least common multiple of the denominators of $\frac{1}{x^2-1}$ and $\frac{3}{x^2+4x+3}$. (pp. 65–66)

Concepts and Vocabulary

3. ***True or False*** Multiplying both sides of an equation by any number results in an equivalent equation.

4. An equation that is satisfied for every value of the variable for which both sides are defined is called a(n) _____.

5. An equation of the form $ax + b = 0$ is called a(n) _____ equation or a(n) _____ equation.

6. ***True or False*** The solution of the equation $3x - 8 = 0$ is $\frac{3}{8}$.

7. ***True or False*** Some equations have no solution.

8. ***True or False*** When solving equations using a graphing utility, the solutions are always exact.

Skill Building

In Problems 9–16, mentally solve each equation.

9. $7x = 21$

10. $6x = -24$

11. $3x + 15 = 0$

12. $6x + 18 = 0$

13. $2x - 3 = 0$

14. $3x + 4 = 0$

15. $\frac{1}{3}x = \frac{5}{12}$

16. $\frac{2}{3}x = \frac{9}{2}$

In Problems 17–28, use a graphing utility to approximate the real solutions, if any, of each equation rounded to two decimal places. All solutions lie between −10 and 10.

17. $x^3 - 4x + 2 = 0$

18. $x^3 - 8x + 1 = 0$

19. $-2x^4 + 5 = 3x - 2$

20. $-x^4 + 1 = 2x^2 - 3$

21. $x^4 - 2x^3 + 3x - 1 = 0$

22. $3x^4 - x^3 + 4x^2 - 5 = 0$

23. $-x^3 - \frac{5}{3}x^2 + \frac{7}{2}x + 2 = 0$

24. $-x^4 + 3x^3 + \frac{7}{3}x^2 - \frac{15}{2}x + 2 = 0$

25. $-\frac{2}{3}x^4 - 2x^3 + \frac{5}{2}x = -\frac{2}{3}x^2 + \frac{1}{2}$

26. $\frac{1}{4}x^3 - 5x = \frac{1}{5}x^2 - 4$

27. $x^4 - 5x^2 + 2x + 11 = 0$

28. $-3x^4 + 8x^2 - 2x - 9 = 0$

In Problems 29–76, solve each equation algebraically. Verify your results using a graphing utility.

29. $3x + 4 = x$

30. $2x + 9 = 5x$

31. $2t - 6 = 3 - t$

32. $5y + 6 = -18 - y$

33. $6 - x = 2x + 9$

34. $3 - 2x = 2 - x$

35. $2(3 + 2x) = 3(x - 4)$

36. $3(2 - x) = 2x - 1$

37. $8x - (3x + 2) = 3x - 10$

38. $7 - (2x - 1) = 10$

39. $2(3x - 5) + 6(x - 3) = -3(4 - 5x) + 5x - 6$

40. $5(x - 2) - 2(3x + 1) = 4(1 - 2x) + x$

41. $\frac{3}{2}x + 2 = \frac{1}{2} - \frac{1}{2}x$

42. $\frac{1}{3}x = 2 - \frac{2}{3}x$

43. $\frac{1}{2}x - 5 = \frac{3}{4}x$

44. $1 - \frac{1}{2}x = 6$

45. $\frac{2}{3}p = \frac{1}{2}p - \frac{1}{3}$

46. $\frac{1}{2} - \frac{1}{3}p = \frac{4}{3}$

47. $0.9t = 0.4 + 0.1t$

48. $0.9t = 1 + t$

49. $\frac{x+1}{3} + \frac{x+2}{7} = 2$

50. $\frac{2x+1}{3} + 16 = 3x$

51. $\frac{2}{y} + \frac{4}{y} = 3$

52. $\frac{4}{y} - 5 = \frac{5}{2y}$

53. $\frac{1}{2} + \frac{2}{x} = \frac{3}{4}$

54. $\frac{3}{x} - \frac{1}{3} = \frac{1}{6}$

55. $(x + 7)(x - 1) = (x + 1)^2$

56. $(x + 2)(x - 3) = (x + 3)^2$

57. $x(2x - 3) = (2x + 1)(x - 4)$

58. $x(1 + 2x) = (2x - 1)(x - 2)$

59. $z(z^2 + 1) = 3 + z^3$

60. $w(4 - w^2) = 8 - w^3$

61. $\frac{x}{x - 2} + 3 = \frac{2}{x - 2}$

62. $\frac{2x}{x + 3} = \frac{-6}{x + 3} - 2$

63. $\frac{2x}{x^2 - 4} = \frac{4}{x^2 - 4} - \frac{3}{x + 2}$

64. $\frac{x}{x^2 - 9} + \frac{4}{x + 3} = \frac{3}{x^2 - 9}$

65. $\frac{x}{x + 2} = \frac{3}{2}$

66. $\frac{3x}{x - 1} = 2$

67. $\frac{5}{2x - 3} = \frac{3}{x + 5}$

68. $\frac{-4}{x + 4} = \frac{-3}{x + 6}$

69. $\frac{6t + 7}{4t - 1} = \frac{3t + 8}{2t - 4}$

70. $\frac{8w + 5}{10w - 7} = \frac{4w - 3}{5w + 7}$

71. $\frac{4}{x - 2} = \frac{-3}{x + 5} + \frac{7}{(x + 5)(x - 2)}$

72. $\frac{-4}{2x + 3} + \frac{1}{x - 1} = \frac{1}{(2x + 3)(x - 1)}$

73. $\frac{2}{y + 3} + \frac{3}{y - 4} = \frac{5}{y + 6}$

74. $\frac{5}{5z - 11} + \frac{4}{2z - 3} = \frac{-3}{5 - z}$

75. $\frac{x}{x^2 - 1} - \frac{x + 3}{x^2 - x} = \frac{-3}{x^2 + x}$

76. $\frac{x + 1}{x^2 + 2x} - \frac{x + 4}{x^2 + x} = \frac{-3}{x^2 + 3x + 2}$

Applications and Extensions

77. If $(a, 2)$ is a point on the graph of $y = 5x + 4$, what is a?

78. If $(2, b)$ is a point on the graph of $y = x^2 + 3x$, what is b?

79. If (a, b) is a point on the graph of $2x + 3y = 6$, write an equation that relates a to b.

80. If $(2, 0)$ and $(0, 5)$ are points on the graph of $y = mx + b$, what are m and b?

In Problems 81–86, solve each equation. The letters a, b, and c are constants.

81. $ax - b = c, \quad a \neq 0$

82. $1 - ax = b, \quad a \neq 0$

83. $\frac{x}{a} + \frac{x}{b} = c, \quad a \neq 0, b \neq 0, a \neq -b$

84. $\frac{a}{x} + \frac{b}{x} = c, \quad c \neq 0, \quad a + b \neq 0$

85. $\frac{1}{x - a} + \frac{1}{x + a} = \frac{2}{x - 1}$

86. $\frac{b + c}{x + a} = \frac{b - c}{x - a}, \quad c \neq 0, a \neq 0$

87. Find the number a for which $x = 4$ is a solution of the equation

$$x + 2a = 16 + ax - 6a$$

88. Find the number b for which $x = 2$ is a solution of the equation

$$x + 2b = x - 4 + 2bx$$

Problems 89–94 list some formulas that occur in applications. Solve each formula for the indicated variable.

89. **Electricity** $\frac{1}{R} = \frac{1}{R_1} + \frac{1}{R_2}$ for R

90. **Finance** $A = P(1 + rt)$ for r

91. **Mechanics** $F = \frac{mv^2}{R}$ for R

92. **Chemistry** $PV = nRT$ for T

93. **Mathematics** $S = \frac{a}{1 - r}$ for r

94. **Mechanics** $v = -gt + v_0$ for t

95. **Finance** A total of \$20,000 is to be invested, some in bonds and some in certificates of deposit (CDs). If the amount invested in bonds is to exceed that in CDs by \$3000, how much will be invested in each type of investment?*

96. **Finance** A total of \$10,000 is to be divided between Sean and George, with George to receive \$3000 less than Sean. How much will each receive?

97. **Finance** An inheritance of \$900,000 is to be divided among Scott, Alice, and Tricia in the following manner: Alice is to receive $\frac{3}{4}$ of what Scott gets, while Tricia gets $\frac{1}{2}$ of what Scott gets. How much does each receive?

*The icon is a Model It! icon. It indicates that the problem involves modeling.

98. Sharing the Cost of a Pizza Judy and Tom agree to share the cost of an \$18 pizza based on how much each ate. If Tom ate $\frac{2}{3}$ the amount that Judy ate, how much should each pay?

[**Hint:** Some pizza may be left.]

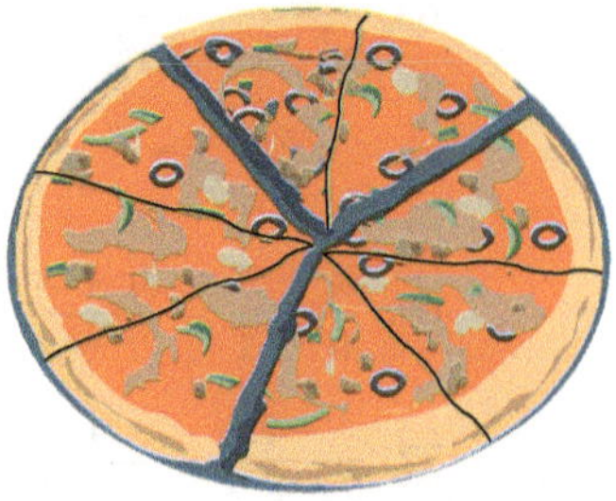

99. Computing Hourly Wages Sandra, who is paid time-and-a-half for hours worked in excess of 40 hours, had gross weekly wages of \$598 for 48 hours worked. What is her regular hourly rate?

100. Computing Hourly Wages Leigh is paid time-and-a-half for hours worked in excess of 40 hours and double-time for hours worked on Sunday. If Leigh had gross weekly wages of \$808.50 for working 50 hours, 4 of which were on Sunday, what is her regular hourly rate?

101. Computing Grades Going into the final exam, which will count as two tests, Brooke has test scores of 80, 83, 71, 61, and 95. What score does Brooke need on the final in order to have an average score of 80?

102. Computing Grades Going into the final exam, which will count as two-thirds of the final grade, Mike has test scores of 86, 80, 84, and 90. What score does Mike need on the final in order to earn a B, which requires an average score of 80? What does he need to earn an A, which requires an average of 90?

103. Business: Discount Pricing A builder of tract homes reduced the price of a model by 15%. If the new price is \$170,000, what was its original price? How much can be saved by purchasing the model?

104. Business: Discount Pricing At a year-end clearance, a car dealer reduces the list price of last year's models by 12%. If a certain four-door model has a discounted price of \$28,160, what was its list price? How much can be saved by purchasing last year's model?

105. Business: Marking up the Price of Books A college book store marks up the price that it pays the publisher for a book by 35%. If the selling price of a book is \$108, how much did the bookstore pay for this book?

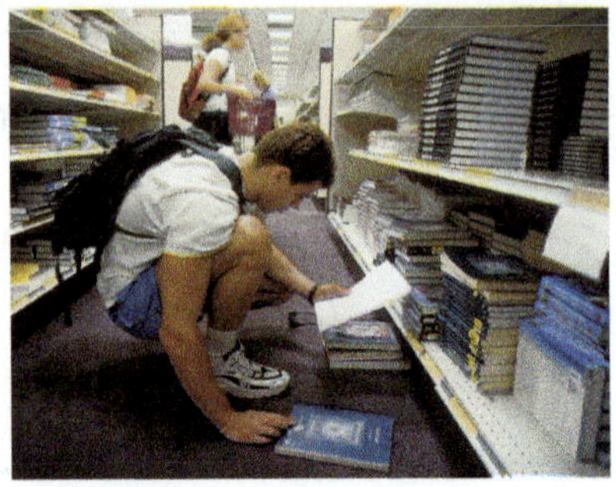

106. Personal Finance: Cost of a Car The suggested list price of a new car is \$34,000. The dealer's cost is 85% of list. How much will you pay if the dealer is willing to accept \$100 over cost for the car?

107. Business: Theater Attendance The manager of the Coral Theater wants to know whether the majority of its patrons is adults or children. During a week in July, 5200 tickets were sold and the receipts totaled \$32,200. The adult admission is \$8.50, and the children's admission is \$6.00. How many adult patrons were there?

108. Business: Discount Pricing A wool suit, discounted by 30% for a clearance sale, has a price tag of \$399. What was the suit's original price?

109. Geometry The perimeter of a rectangle is 60 feet. Find its length and width if the length is 8 feet longer than the width.

110. Geometry The perimeter of a rectangle is 42 meters. Find its length and width if the length is twice the width.

111. MySpace.com In November 2006, there were 4.55 billion hits on *MySpace.com* that originated at colleges and universities. This accounted for 11.8% of all hits to *MySpace.com*. Altogether, how many hits were there to *MySpace.com* in November 2006?

Source: comScore Networks

Discussion and Writing

112. One step in the following list contains an error. Identify it and explain what is wrong.

$$
\begin{aligned}
x &= 2 && (1)\\
3x - 2x &= 2 && (2)\\
3x &= 2x + 2 && (3)\\
x^2 + 3x &= x^2 + 2x + 2 && (4)\\
x^2 + 3x - 10 &= x^2 + 2x - 8 && (5)\\
(x-2)(x+5) &= (x-2)(x+4) && (6)\\
x + 5 &= x + 4 && (7)\\
5 &= 4 && (8)
\end{aligned}
$$

113. The equation

$$\frac{5}{x+3} + 3 = \frac{8+x}{x+3}$$

has no solution, yet when we go through the process of solving it we obtain $x = -3$. Write a paragraph to explain what causes this to happen.

114. Make up an equation that has no solution and give it to a fellow student to solve. Ask the fellow student to write a critique of your equation.

115. Explain the difference between the directions "solve," "evaluate," and "simplify." Write an example using each direction with the expression $3(x + 2) - x$.

'Are You Prepared?' Answers

1. No **2.** $(x + 1)(x - 1)(x + 3)$

1.3 Quadratic Equations

PREPARING FOR THIS SECTION *Before getting started, review the following:*

- Zero-Product Property (Section R.1, p. 13)
- Square Roots (Section R.2, pp. 23–24)
- Factoring (Section R.5, pp. 49–55)

Now Work the 'Are You Prepared?' problems on page 121.

OBJECTIVES
1 Solve Quadratic Equations by Factoring (p. 113)
2 Solve Quadratic Equations Using the Square Root Method (p. 114)
3 Solve Quadratic Equations by Completing the Square (p. 115)
4 Solve Quadratic Equations Using the Quadratic Formula (p. 117)
5 Solve Problems That Can Be Modeled by Quadratic Equations (p. 120)

Quadratic equations are equations such as

$$2x^2 + x + 8 = 0 \qquad 3x^2 - 5x = 0 \qquad x^2 - 9 = 0$$

A general definition is given next.

A **quadratic equation** is an equation equivalent to one of the form

$$ax^2 + bx + c = 0 \qquad \textbf{(1)}$$

where a, b, and c are real numbers and $a \neq 0$.

A quadratic equation written in the form $ax^2 + bx + c = 0$ is said to be in **standard form.**

Sometimes, a quadratic equation is called a **second-degree equation,** because the left side is a polynomial of degree 2. We shall discuss three algebraic ways of solving quadratic equations: by factoring, by completing the square, and by using the quadratic formula.

1 Solve Quadratic Equations by Factoring

When a quadratic equation is written in standard form, $ax^2 + bx + c = 0$, it may be possible to factor the expression on the left side as the product of two first-degree polynomials. Then, by setting each factor equal to 0 and solving the resulting linear equations, we obtain the *exact* solutions of the quadratic equation. This approach leads us to a basic premise in mathematics. Whenever a problem is encountered, use techniques that reduce the problem to one you already know how to solve. In this instance, we are reducing quadratic equations to linear equations using the technique of factoring.

Let's look at an example.

EXAMPLE 1 Solving a Quadratic Equation by Factoring and by Graphing

Solve the equation: $x^2 = 12 - x$

Algebraic Solution

We put the equation in standard form by adding $x - 12$ to each side:

$$x^2 = 12 - x$$
$$x^2 + x - 12 = 0$$

Graphing Solution

Graph $Y_1 = x^2$ and $Y_2 = 12 - x$. See Figure 34(a) on page 114. From the graph it appears that there are two points of intersection: one near -4, the other near 3. Using INTERSECT, the points of intersection are $(-4, 16)$ and $(3, 9)$, so the solutions of the equation are $x = -4$ and $x = 3$. See Figure 34(b) and (c). The solution set is $\{-4, 3\}$.

The left side of the equation may now be factored as

$$(x + 4)(x - 3) = 0$$

so that

$$x + 4 = 0 \quad \text{or} \quad x - 3 = 0$$
$$x = -4 \qquad\qquad x = 3$$

The solution set is $\{-4, 3\}$.

Figure 34

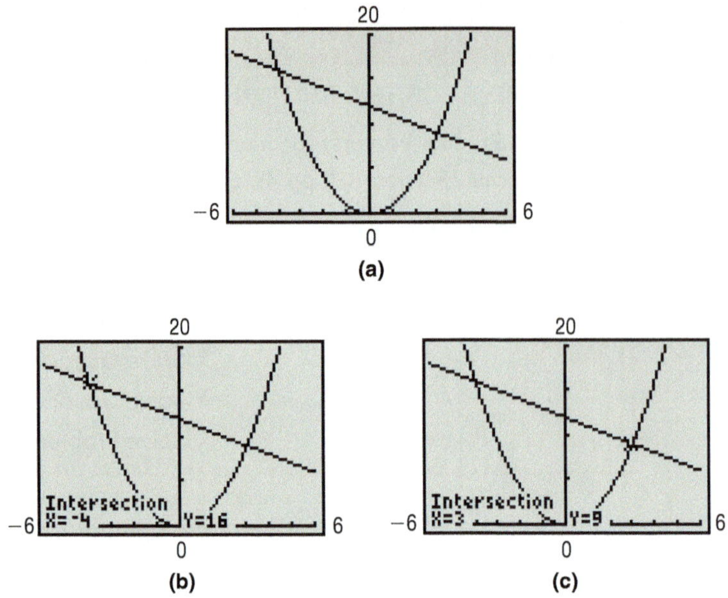

Now Work PROBLEM 11

When the left side factors into two linear equations with the same solution, the quadratic equation is said to have a **repeated solution.** We also call this solution a **root of multiplicity 2,** or a **double root.**

EXAMPLE 2 Solving a Quadratic Equation by Factoring and by Graphing

Solve the equation: $x^2 - 6x + 9 = 0$

Algebraic Solution

This equation is already in standard form, and the left side can be factored:

$$x^2 - 6x + 9 = 0$$
$$(x - 3)(x - 3) = 0$$

so

$$x = 3 \quad \text{or} \quad x = 3$$

This equation has only the repeated solution 3. The solution set is $\{3\}$.

Graphing Solution

Figure 35 shows the graph of the equation

$$Y_1 = x^2 - 6x + 9$$

From the graph it appears that there is one x-intercept, 3. Using ZERO (or ROOT), we find that the only x-intercept is 3. The equation has only the repeated solution 3. The solution set is $\{3\}$.

Figure 35

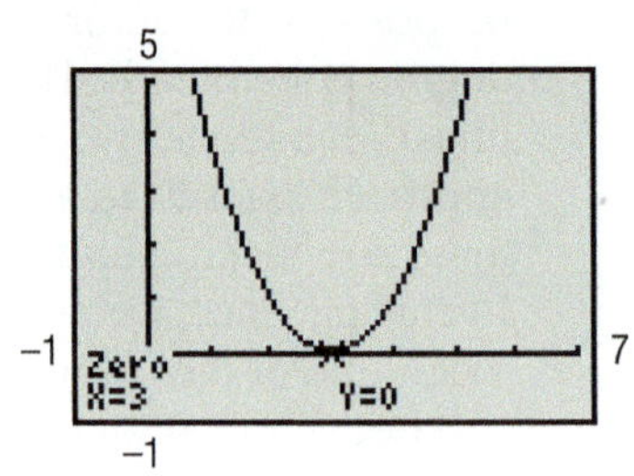

Now Work PROBLEM 23

2 Solve Quadratic Equations Using the Square Root Method

Suppose that we wish to solve the quadratic equation

$$x^2 = p \qquad (2)$$

where $p \geq 0$ is a nonnegative number. We proceed as in the earlier examples:

$$x^2 - p = 0 \quad \text{Put in standard form.}$$
$$(x - \sqrt{p})(x + \sqrt{p}) = 0 \quad \text{Factor (over the real numbers).}$$
$$x = \sqrt{p} \quad \text{or} \quad x = -\sqrt{p} \quad \text{Solve.}$$

We have the following result:

If $x^2 = p$ and $p \geq 0$, then $x = \sqrt{p}$ or $x = -\sqrt{p}$. **(3)**

When statement (3) is used, it is called the **Square Root Method.** In statement (3), note that if $p > 0$ the equation $x^2 = p$ has two solutions, $x = \sqrt{p}$ and $x = -\sqrt{p}$. We usually abbreviate these solutions as $x = \pm\sqrt{p}$, read as "x equals plus or minus the square root of p." For example, the two solutions of the equation

$$x^2 = 4$$

are

$$x = \pm\sqrt{4}$$

and, since $\sqrt{4} = 2$, we have

$$x = \pm 2$$

The solution set is $\{-2, 2\}$.

EXAMPLE 3 Solving Quadratic Equations by Using the Square Root Method

Solve each equation: (a) $x^2 = 5$ (b) $(x - 2)^2 = 16$

Solution (a)

$$x^2 = 5$$
$$x = \pm\sqrt{5} \quad \text{Use the Square Root Method.}$$
$$x = \sqrt{5} \quad \text{or} \quad x = -\sqrt{5}$$

The solution set is $\{-\sqrt{5}, \sqrt{5}\}$.

(b) $(x - 2)^2 = 16$

$$x - 2 = \pm\sqrt{16} \quad \text{Use the Square Root Method.}$$
$$x - 2 = \sqrt{16} \quad \text{or} \quad x - 2 = -\sqrt{16}$$
$$x - 2 = 4 \qquad x - 2 = -4$$
$$x = 6 \qquad x = -2$$

The solution set is $\{-2, 6\}$.

Check: Verify the solutions using a graphing utility. Are the solutions provided by the utility exact?

Now Work PROBLEM 33

3 Solve Quadratic Equations by Completing the Square

We now introduce the method of **completing the square.** The idea behind this method is to "adjust" the left side of a quadratic equation, $ax^2 + bx + c = 0$, so that it becomes a perfect square, that is, the square of a first-degree polynomial. For example, $x^2 + 6x + 9$ and $x^2 - 4x + 4$ are perfect squares because

$$x^2 + 6x + 9 = (x + 3)^2 \quad \text{and} \quad x^2 - 4x + 4 = (x - 2)^2$$

How do we "adjust" the left side? We do it by adding the appropriate number to create a perfect square. For example, to make $x^2 + 6x$ a perfect square, we add 9.

Let's look at several examples of completing the square when the coefficient of x^2 is 1.

Start	Add	Result
$x^2 + 4x$	4	$x^2 + 4x + 4 = (x + 2)^2$
$x^2 + 12x$	36	$x^2 + 12x + 36 = (x + 6)^2$
$x^2 - 6x$	9	$x^2 - 6x + 9 = (x - 3)^2$
$x^2 + x$	$\frac{1}{4}$	$x^2 + x + \frac{1}{4} = \left(x + \frac{1}{2}\right)^2$

Do you see the pattern? Provided that the coefficient of x^2 is 1, we complete the square by adding the square of one-half of the coefficient of x.

Start	Add	Result
$x^2 + mx$	$\left(\frac{m}{2}\right)^2$	$x^2 + mx + \left(\frac{m}{2}\right)^2 = \left(x + \frac{m}{2}\right)^2$

EXAMPLE 4 Completing the Square

Determine the number that must be added to each expression to complete the square. Then factor the expression.

Start	Add	Result	Factored Form
$y^2 + 8y$	$\left(\frac{1}{2}\cdot 8\right)^2 = 16$	$y^2 + 8y + 16$	$(y + 4)^2$
$x^2 + 12x$	$\left(\frac{1}{2}\cdot 12\right)^2 = 36$	$x^2 + 12x + 36$	$(x + 6)^2$
$a^2 - 20a$	$\left(\frac{1}{2}\cdot(-20)\right)^2 = 100$	$a^2 - 20a + 100$	$(a - 10)^2$
$p^2 - 5p$	$\left(\frac{1}{2}\cdot(-5)\right)^2 = \frac{25}{4}$	$p^2 - 5p + \frac{25}{4}$	$\left(p - \frac{5}{2}\right)^2$

Figure 36

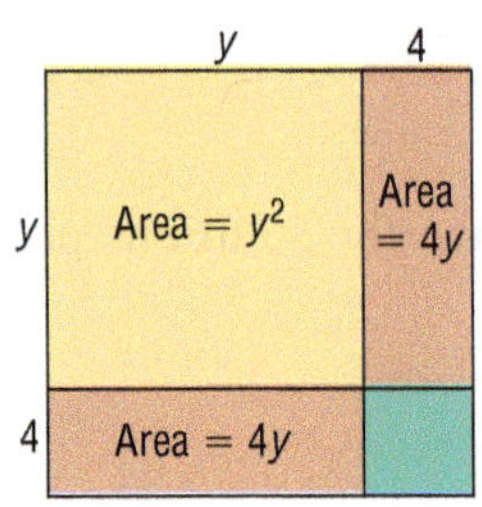

Notice that the factored form of a perfect square is either

$$x^2 + bx + \left(\frac{b}{2}\right)^2 = \left(x + \frac{b}{2}\right)^2 \quad \text{or} \quad x^2 - bx + \left(\frac{b}{2}\right)^2 = \left(x - \frac{b}{2}\right)^2$$

Are you wondering why we call making an expression a perfect square "completing the square"? Look at the square in Figure 36. Its area is $(y + 4)^2$. The yellow area is y^2 and each orange area is $4y$ (for a total area of $8y$). The sum of these areas is $y^2 + 8y$. To complete the square, we need to add the area of the green region: $4 \cdot 4 = 16$. As a result, $y^2 + 8y + 16 = (y + 4)^2$.

Now Work PROBLEM 37

The next example illustrates how the procedure of completing the square can be used to solve a quadratic equation.

EXAMPLE 5 Solving a Quadratic Equation by Completing the Square

Solve by completing the square: $x^2 + 5x + 4 = 0$

Solution We always begin this procedure by rearranging the equation so that the constant is on the right side.

$$x^2 + 5x + 4 = 0$$
$$x^2 + 5x = -4$$

Since the coefficient of x^2 is 1, we can complete the square on the left side by adding $\left(\frac{1}{2}\cdot 5\right)^2 = \frac{25}{4}$. Of course, in an equation, whatever we add to the left side must also be added to the right side. We add $\frac{25}{4}$ to *both* sides.

$$x^2 + 5x + \frac{25}{4} = -4 + \frac{25}{4} \quad \text{Add } \frac{25}{4} \text{ to both sides.}$$
$$\left(x + \frac{5}{2}\right)^2 = \frac{9}{4} \quad \text{Factor; simplify.}$$
$$x + \frac{5}{2} = \pm\sqrt{\frac{9}{4}} \quad \text{Use the Square Root Method.}$$
$$x + \frac{5}{2} = \pm\frac{3}{2}$$
$$x = -\frac{5}{2} \pm \frac{3}{2}$$
$$x = -\frac{5}{2} + \frac{3}{2} = -1 \quad \text{or} \quad x = -\frac{5}{2} - \frac{3}{2} = -4$$

NOTE If the coefficient of the square term is not 1, divide through by the coefficient of the square term before attempting to complete the square. For example, to solve $2x^2 - 8x = 5$ by completing the square, divide both sides of the equation by 2 and obtain $x^2 - 4x = \frac{5}{2}$. ■

The solution set is $\{-4, -1\}$.

Check: Verify the solution using a graphing utility.

The solution of the equation in Example 5 can also be obtained by factoring. Rework Example 5 using factoring.

Now Work PROBLEM 41

4 Solve Quadratic Equations Using the Quadratic Formula

We can use the method of completing the square to obtain a general formula for solving the quadratic equation

$$ax^2 + bx + c = 0, \qquad a > 0$$

NOTE There is no loss in generality to assume that $a > 0$, since if $a < 0$ we can multiply both sides by -1 to obtain an equivalent equation with a positive leading coefficient. ■

As in Example 5, we begin by rearranging the terms as

$$ax^2 + bx = -c$$

Since $a > 0$, we can divide both sides by a to get

$$x^2 + \frac{b}{a}x = -\frac{c}{a}$$

Now the coefficient of x^2 is 1. To complete the square on the left side, add the square of $\frac{1}{2}$ of the coefficient of x; that is, add

$$\left(\frac{1}{2}\cdot\frac{b}{a}\right)^2 = \frac{b^2}{4a^2}$$

to each side. Then

$$x^2 + \frac{b}{a}x + \frac{b^2}{4a^2} = \frac{b^2}{4a^2} - \frac{c}{a}$$
$$\left(x + \frac{b}{2a}\right)^2 = \frac{b^2 - 4ac}{4a^2} \qquad \frac{b^2}{4a^2} - \frac{c}{a} = \frac{b^2}{4a^2} - \frac{4ac}{4a^2} = \frac{b^2 - 4ac}{4a^2} \tag{4}$$

Provided that $b^2 - 4ac \geq 0$, we now can use the Square Root Method to get

$$x + \frac{b}{2a} = \pm\sqrt{\frac{b^2 - 4ac}{4a^2}}$$

$$x + \frac{b}{2a} = \frac{\pm\sqrt{b^2 - 4ac}}{2a}$$

The square root of a quotient equals the quotient of the square roots. Also, $\sqrt{4a^2} = 2a$ since $a > 0$.

$$x = -\frac{b}{2a} \pm \frac{\sqrt{b^2 - 4ac}}{2a}$$

Add $-\frac{b}{2a}$ to both sides.

$$= \frac{-b \pm \sqrt{b^2 - 4ac}}{2a}$$

Combine the quotients on the right.

What if $b^2 - 4ac$ is negative? Then equation (4) states that the left expression (a real number squared) equals the right expression (a negative number). Since this occurrence is impossible for real numbers, we conclude that if $b^2 - 4ac < 0$ the quadratic equation has no *real* solution.*

We now state the *quadratic formula*.

THEOREM

Quadratic Formula

Consider the quadratic equation

$$ax^2 + bx + c = 0 \qquad a \neq 0$$

If $b^2 - 4ac < 0$, this equation has no real solution.

If $b^2 - 4ac \geq 0$, the real solution(s) of this equation is (are) given by the **quadratic formula.**

$$x = \frac{-b \pm \sqrt{b^2 - 4ac}}{2a}$$

The quantity $\mathbf{b^2 - 4ac}$ is called the **discriminant** of the quadratic equation, because its value tells us whether the equation has real solutions. In fact, it also tells us how many solutions to expect.

Discriminant of a Quadratic Equation

For a quadratic equation $ax^2 + bx + c = 0$:

1. If $b^2 - 4ac > 0$, there are two unequal real solutions.
2. If $b^2 - 4ac = 0$, there is a repeated real solution, a root of multiplicity 2.
3. If $b^2 - 4ac < 0$, there is no real solution.

When asked to find the real solutions, if any, of a quadratic equation, always evaluate the discriminant first to see how many real solutions there are.

*We consider quadratic equations where $b^2 - 4ac$ is negative in the next section.

EXAMPLE 6 Solving a Quadratic Equation by Using the Quadratic Formula and by Graphing

Find the real solutions, if any, of the equation $3x^2 - 5x + 1 = 0$.

Algebraic Solution

The equation is in standard form, so we compare it to $ax^2 + bx + c = 0$ to find a, b, and c.

$$3x^2 - 5x + 1 = 0$$
$$ax^2 + bx + c = 0 \quad a = 3, b = -5, c = 1$$

With $a = 3, b = -5,$ and $c = 1$, we evaluate the discriminant $b^2 - 4ac$.

$$b^2 - 4ac = (-5)^2 - 4(3)(1) = 25 - 12 = 13$$

Since $b^2 - 4ac > 0$, there are two real solutions.

We use the quadratic formula with $a = 3, b = -5, c = 1$, and $b^2 - 4ac = 13$.

$$x = \frac{-b \pm \sqrt{b^2 - 4ac}}{2a} = \frac{-(-5) \pm \sqrt{13}}{2(3)} = \frac{5 \pm \sqrt{13}}{6}$$

The solution set is $\left\{\frac{5 - \sqrt{13}}{6}, \frac{5 + \sqrt{13}}{6}\right\}$. These solutions are exact.

Graphing Solution

Figure 37 shows the graph of the equation

$$Y_1 = 3x^2 - 5x + 1$$

There are two x-intercepts: one between 0 and 1, the other between 1 and 2. Using ZERO (or ROOT), we find the solutions to the equation are 0.23 and 1.43, rounded to two decimal places. These solutions are approximate.

Figure 37

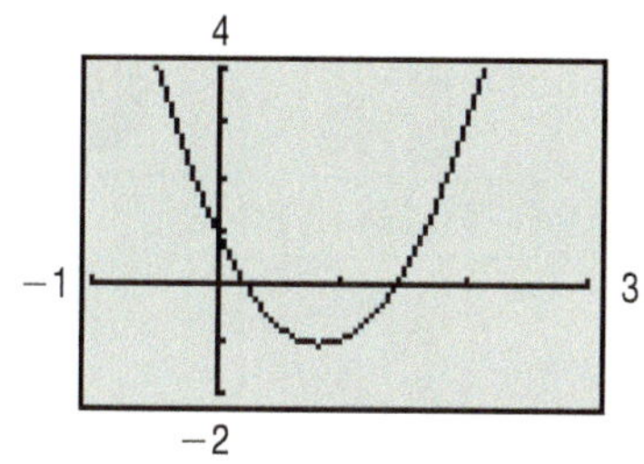

Now Work PROBLEMS 51 AND 61

EXAMPLE 7 Solving a Quadratic Equation by Using the Quadratic Formula and by Graphing

Find the real solutions, if any, of the equation $3x^2 + 2 = 4x$.

Algebraic Solution

The equation, as given, is not in standard form.

$$3x^2 + 2 = 4x$$
$$3x^2 - 4x + 2 = 0 \quad \text{Subtract } 4x \text{ from both sides to put the equation in standard form.}$$
$$ax^2 + bx + c = 0 \quad \text{Compare to standard form.}$$

With $a = 3, b = -4$, and $c = 2$, we find that

$$b^2 - 4ac = (-4)^2 - 4(3)(2) = 16 - 24$$
$$= -8$$

Since $b^2 - 4ac < 0$, the equation has no real solution.

Graphing Solution

We use the standard form of the equation and graph

$$Y_1 = 3x^2 - 4x + 2$$

See Figure 38. We see that there are no x-intercepts, so the equation has no real solution, as expected based on the value of the discriminant.

Figure 38

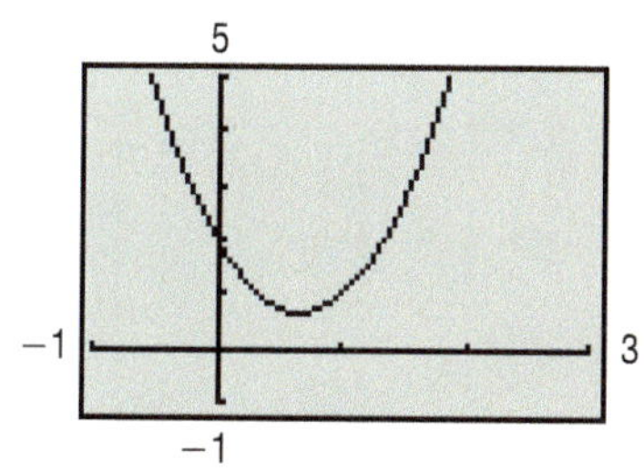

Now Work PROBLEM 53

SUMMARY Procedure for Solving a Quadratic Equation Algebraically

To solve a quadratic equation algebraically, first put it in standard form:

$$ax^2 + bx + c = 0$$

Then:

STEP 1: Identify a, b, and c.

STEP 2: Evaluate the discriminant, $b^2 - 4ac$.

STEP 3: (a) If the discriminant is negative, the equation has no real solution.
(b) If the discriminant is zero, the equation has one real solution, a repeated root.
(c) If the discriminant is positive, the equation has two distinct real solutions.

If you can easily spot factors, use the factoring method to solve the equation. Otherwise, use the quadratic formula or the method of completing the square.

5 Solve Problems That Can Be Modeled by Quadratic Equations

Many applied problems require the solution of a quadratic equation. Let's look at one that you will probably see again in a slightly different form if you study calculus.

EXAMPLE 8 Constructing a Box

From each corner of a square piece of sheet metal, remove a square of side 9 centimeters. Turn up the edges to form an open box. If the box is to hold 144 cubic centimeters (cm^3), what should be the dimensions of the piece of sheet metal?

Solution We use Figure 39 as a guide. We have labeled by x the length of a side of the square piece of sheet metal. The box will be of height 9 centimeters, and its square base will have $x - 18$ as the length of a side. The volume (Length $\times$ Width $\times$ Height) of the box is therefore

$$(x - 18)(x - 18) \cdot 9 = 9(x - 18)^2 \quad \text{The Model}$$

Figure 39

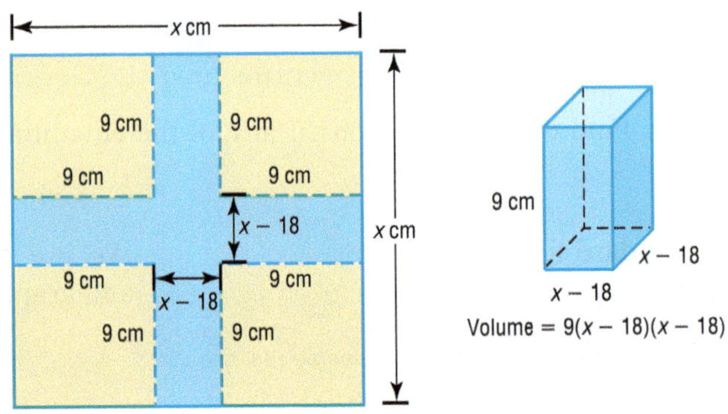

Since the volume of the box is to be 144 cm^3, we have

$$9(x - 18)^2 = 144$$
$$(x - 18)^2 = 16 \quad \text{Divide each side by 9.}$$
$$x - 18 = \pm 4 \quad \text{Use the Square Root Method.}$$
$$x = 18 \pm 4$$
$$x = 22 \quad \text{or} \quad x = 14$$

We discard the solution $x = 14$ (do you see why?) and conclude that the sheet metal should be 22 centimeters by 22 centimeters.

Check: If we begin with a piece of sheet metal 22 centimeters by 22 centimeters, cut out a 9 centimeter square from each corner, and fold up the edges, we get a box whose dimensions are 9 by 4 by 4, with volume $9 \times 4 \times 4 = 144$ cm^3, as required.

Now Work PROBLEM 97

Historical Feature

Problems using quadratic equations are found in the oldest known mathematical literature. Babylonians and Egyptians were solving such problems before 1800 BC. Euclid solved quadratic equations geometrically in his *Data* (300 BC), and the Hindus and Arabs gave rules for solving any quadratic equation with real roots. Because negative numbers were not freely used before AD 1500, there were several different types of quadratic equations, each with its own rule. Thomas Harriot (1560–1621) introduced the method of factoring to obtain solutions, and François Viète (1540–1603) introduced a method that is essentially completing the square.

Until modern times it was usual to neglect the negative roots (if there were any), and equations involving square roots of negative quantities were regarded as unsolvable until the 1500s.

Historical Problems

1. *One solution of al-Khwăiizmi* We solve $x^2 + 12x = 85$ by drawing the square shown. The area of the four white rectangles and the yellow square is $x^2 + 12x$. We then set this expression equal to 85 to get the equation $x^2 + 12x = 85$. If we add the four blue squares, we will have a larger square of known area. Complete the solution.

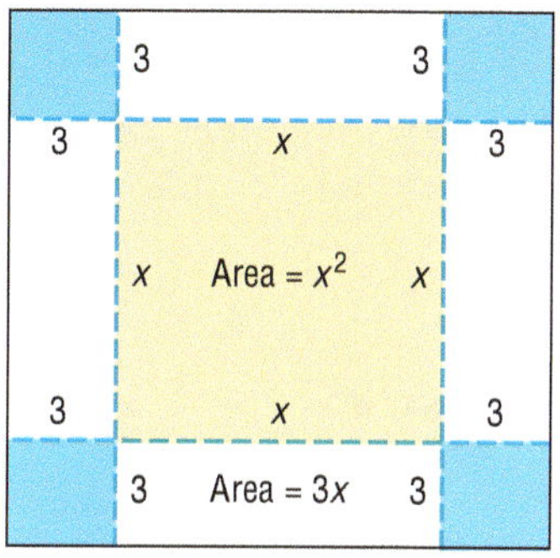

2. *Viète's method* We solve $x^2 + 12x - 85 = 0$ by letting $x = u + z$. Then

$$(u + z)^2 + 12(u + z) - 85 = 0$$
$$u^2 + (2z + 12)u + (z^2 + 12z - 85) = 0$$

Now select z so that $2z + 12 = 0$ and finish the solution.

3. *Another method to get the quadratic formula* Look at equation (4) on page 117. Rewrite the right side as $\left(\frac{\sqrt{b^2 - 4ac}}{2a}\right)^2$ and then subtract it from each side. The right side is now 0 and the left side is a difference of two squares. If you factor this difference of two squares, you will easily be able to get the quadratic formula and, moreover, the quadratic expression is factored, which is sometimes useful.

1.3 Assess Your Understanding

'Are You Prepared?' *Answers are given at the end of these exercises. If you get a wrong answer, read the pages listed in red.*

1. Factor: $x^2 - 5x - 6$ (pp. 52–53)

2. Factor: $2x^2 - x - 3$ (pp. 54–55)

3. The solution set of the equation $(x - 3)(3x + 5) = 0$ is _____. (p. 13)

4. Simplify: $\sqrt{8^2 - 4 \cdot 2 \cdot 3}$ (pp. 23–24)

Concepts and Vocabulary

5. When a quadratic equation has a repeated solution, it is called a(n) _____ root or a root of _____ _____.

6. The quantity $b^2 - 4ac$ is called the _____ of a quadratic equation. If it is _____, the equation has no real solution.

7. Quadratic equations are sometimes called _____ -degree equations.

8. ***True or False*** To complete the square in the expression $x^2 + 8x$, add 16.

9. ***True or False*** Quadratic equations always have two real solutions.

10. ***True or False*** A quadratic equation that is in the form $ax^2 + bx + c = 0$ is said to be in standard form.

Skill Building

In Problems 11–30, solve each equation by factoring. Verify your solution using a graphing utility.

11. $x^2 - 9x = 0$

12. $x^2 + 4x = 0$

13. $x^2 - 25 = 0$

14. $x^2 - 9 = 0$

15. $z^2 + z - 6 = 0$

16. $v^2 + 7v + 6 = 0$

17. $2x^2 - 5x - 3 = 0$

18. $3x^2 + 5x + 2 = 0$

19. $3t^2 - 48 = 0$

20. $2y^2 - 50 = 0$

21. $x(x + 8) + 12 = 0$

22. $x(x - 4) = 12$

23. $4x^2 + 9 = 12x$

24. $25x^2 + 16 = 40x$

25. $6(p^2 - 1) = 5p$

26. $2(2u^2 - 4u) + 3 = 0$

27. $6x + 7 = \frac{20}{x}$

28. $x + \frac{12}{x} = 7$

29. $\frac{4(x - 2)}{x - 3} + \frac{3}{x} = \frac{-3}{x(x - 3)}$

30. $\frac{5}{x + 4} = 4 + \frac{3}{x - 2}$

In Problems 31–36, solve each equation by the Square Root Method. Verify your solution using a graphing utility.

31. $x^2 = 50$
32. $x^2 = 20$
33. $(x - 1)^2 = 4$
34. $(x + 2)^2 = 1$
35. $(2x + 3)^2 - 9 = 0$
36. $(3x - 2)^2 - 4 = 0$

In Problems 37–40, what number should be added to complete the square of each expression?

37. $x^2 + 8x$
38. $x^2 - 4x$
39. $y^2 - \frac{1}{2}y$
40. $z^2 + \frac{1}{3}z$

In Problems 41–46, solve each equation by completing the square. Verify your solution using a graphing utility.

41. $x^2 + 4x = 8$
42. $x^2 - 6x = 9$
43. $x^2 - \frac{1}{2}x - \frac{3}{16} = 0$
44. $x^2 + \frac{2}{3}x - \frac{1}{3} = 0$
45. $3x^2 + x - \frac{1}{2} = 0$
46. $2x^2 - 3x - 1 = 0$

In Problems 47–66, find the real solutions, if any, of each equation. Use the quadratic formula. Verify your solution using a graphing utility.

47. $x^2 - 4x + 2 = 0$
48. $x^2 + 4x + 2 = 0$
49. $x^2 - 4x - 1 = 0$
50. $x^2 + 6x + 1 = 0$
51. $2x^2 - 5x + 3 = 0$
52. $2x^2 + 5x + 3 = 0$
53. $4y^2 - y + 2 = 0$
54. $4t^2 + t + 1 = 0$
55. $4x^2 = 1 - 2x$
56. $2x^2 = 1 - 2x$
57. $2x(x + 2) = 3$
58. $3x(x + 2) = 1$
59. $9t^2 - 6t + 1 = 0$
60. $4u^2 - 12u + 9 = 0$
61. $\frac{3}{4}x^2 - \frac{1}{4}x = \frac{1}{2}$
62. $\frac{2}{3}x^2 - x = 3$
63. $4 - \frac{1}{x} - \frac{2}{x^2} = 0$
64. $4 + \frac{1}{x} - \frac{1}{x^2} = 0$
65. $\frac{3x}{x - 2} + \frac{1}{x} = 4$
66. $\frac{2x}{x - 3} + \frac{1}{x} = 4$

In Problems 67–72, determine whether each quadratic equation has two unequal real solutions, a repeated real solution, or no real solution, without solving the equation. **[Hint:** *Use the discriminant.*]

67. $2x^2 - 6x + 7 = 0$
68. $x^2 + 4x + 7 = 0$
69. $9x^2 - 30x + 25 = 0$
70. $25x^2 - 20x + 4 = 0$
71. $3x^2 + 5x - 8 = 0$
72. $2x^2 - 3x - 7 = 0$

Mixed Practice

In Problems 73–88, find the exact real solutions, if any, of each equation. Use any method. Verify your solution using a graphing utility.

73. $x^2 - 8 = 0$
74. $x^2 - 12 = 0$
75. $16x^2 - 8x + 1 = 0$
76. $4x^2 - 20x + 25 = 0$
77. $10x^2 - 19x - 15 = 0$
78. $6x^2 + 7x - 20 = 0$
79. $2 + z = 6z^2$
80. $2 = y + 6y^2$
81. $3(x - 4) + 5 = 0$
82. $2(x + 4) - 3 = 0$
83. $x^2 + x = 4$
84. $x^2 + x = 1$
85. $5x(x - 1) = -7x^2 + 2$
86. $10x(x + 2) = -3x + 5$
87. $\frac{3x}{x + 2} - \frac{5}{x + 1} = \frac{-5}{x^2 + 3x + 2}$
88. $\frac{2x}{x - 3} - \frac{3}{x + 1} = \frac{2x + 18}{x^2 - 2x - 3}$

Applications and Extensions

89. Dimensions of a Window The area of the opening of a rectangular window is to be 143 square feet. If the length is to be 2 feet more than the width, what are the dimensions?

90. Dimensions of a Window The area of a rectangular window is to be 306 square centimeters. If the length exceeds the width by 1 centimeter, what are the dimensions?

91. Geometry Find the dimensions of a rectangle whose perimeter is 26 meters and whose area is 40 square meters.

92. Watering a Field An adjustable water sprinkler that sprays water in a circular pattern is placed at the center of a square field whose area is 1250 square feet (see the figure). What is the shortest radius setting that can be used if the field is to be completely enclosed within the circle?

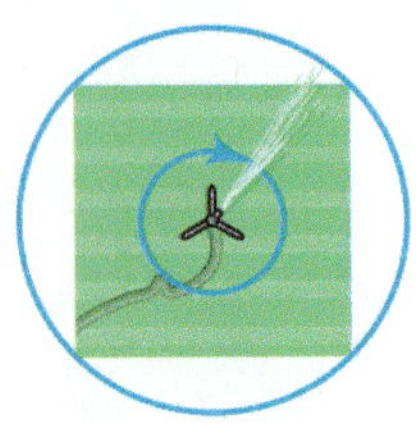

93. Pythagorean Theorem How many right triangles might have a hypotenuse that measures $2x + 3$ meters and legs that measure $2x - 5$ meters and $x + 7$ meters? What are the dimensions of the triangle(s)?

94. Pythagorean Theorem How many right triangles might have a hypotenuse that measures $4x + 5$ inches and legs that measure $3x + 13$ inches and x inches? What are the dimensions of the triangle(s)?

95. Comparing TVs The screen size of a television is determined by the length of the diagonal of the rectangular screen. See the figure.

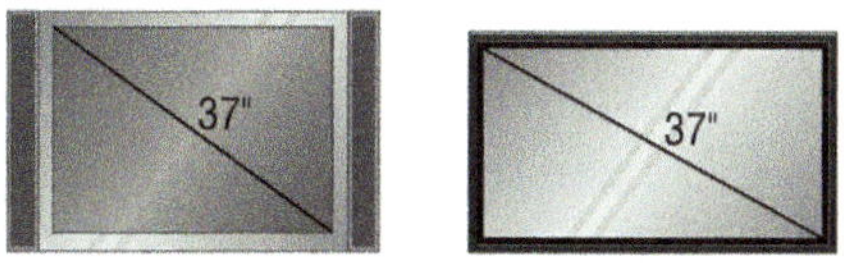

Traditional 4:3 LCD 16:9

Traditional TVs come in a 4:3 format, meaning that the ratio of the length to the width of the rectangle is 4 to 3. What is the area of a 37-inch traditional TV screen? What is the area of a 37-inch LCD TV whose screen is in a 16:9 format? Which screen is larger?

[**Hint:** If x is the length of a 4:3 format screen, then $\frac{3}{4}x$ is the width.]

96. Comparing TVs Refer to Problem 95. Find the screen area of a traditional 50-inch TV and compare it with a 50-inch plasma TV whose screen is in a 16:9 format. Which screen is larger?

97. Constructing a Box An open box is to be constructed from a square piece of sheet metal by removing a square of side 1 foot from each corner and turning up the edges. If the box is to hold 4 cubic feet, what should be the dimensions of the sheet metal?

98. Constructing a Box Rework Problem 97 if the piece of sheet metal is a rectangle whose length is twice its width.

99. Physics A ball is thrown vertically upward from the top of a building 96 feet tall with an initial velocity of 80 feet per second. The distance s (in feet) of the ball from the ground after t seconds is $s = 96 + 80t - 16t^2$.

(a) After how many seconds does the ball strike the ground?
(b) After how many seconds will the ball pass the top of the building on its way down?

100. Physics An object is propelled vertically upward with an initial velocity of 20 meters per second. The distance s (in meters) of the object from the ground after t seconds is $s = -4.9t^2 + 20t$.

(a) When will the object be 15 meters above the ground?
(b) When will it strike the ground?
(c) Will the object reach a height of 100 meters?

101. Reducing the Size of a Candy Bar A jumbo chocolate bar with a rectangular shape measures 12 centimeters (cm) in length, 7 cm in width, and 3 cm in thickness. Due to escalating costs of cocoa, management decides to reduce the volume of the bar by 10%. To accomplish this reduction, management decides that the new bar should have the same 3-cm thickness, but the length and width of each should be reduced an equal number of centimeters. What should be the dimensions of the new candy bar?

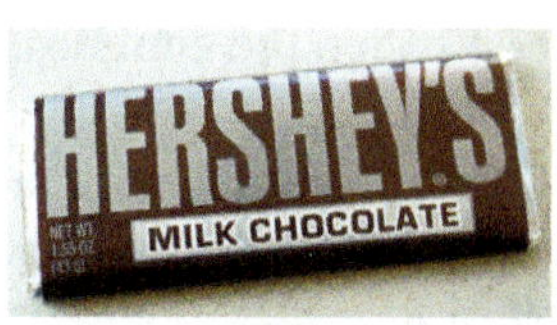

102. Reducing the Size of a Candy Bar Rework Problem 101 if the reduction is to be 20%.

103. Constructing a Border around a Pool A circular pool measures 10 feet across. One cubic yard of concrete is to be used to create a circular border of uniform width around the pool. If the border is to have a depth of 3 inches, how wide will the border be? (1 cubic yard = 27 cubic feet) See the illustration.

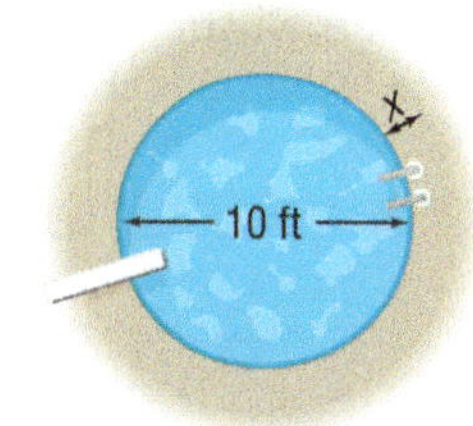

104. Constructing a Border around a Pool Rework Problem 103 if the depth of the border is 4 inches.

105. Constructing a Border around a Garden A landscaper, who just completed a rectangular flower garden measuring 6 feet by 10 feet, orders 1 cubic yard of premixed cement, all of which is to be used to create a border of uniform width around the garden. If the border is to have a depth of 3 inches, how wide will the border be?

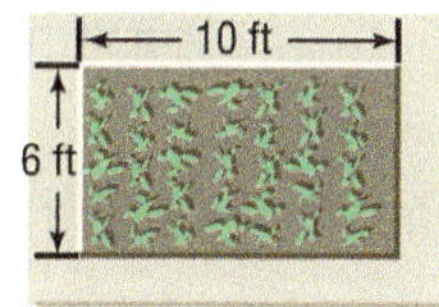

106. Dimensions of a Patio A contractor orders 8 cubic yards of premixed cement, all of which is to be used to pour a patio that will be 4 inches thick. If the length of the patio is specified to be twice the width, what will be the patio dimensions? (1 cubic yard = 27 cubic feet)

107. The sum of the consecutive integers $1, 2, 3, \ldots, n$ is given by the formula $\frac{1}{2}n(n + 1)$. How many consecutive integers, starting with 1, must be added to get a sum of 666?

108. Geometry If a polygon of n sides has $\frac{1}{2}n(n - 3)$ diagonals, how many sides will a polygon with 65 diagonals have? Is there a polygon with 80 diagonals?

109. Show that the sum of the roots of a quadratic equation is $-\frac{b}{a}$.

110. Show that the product of the roots of a quadratic equation is $\frac{c}{a}$.

111. Find k such that the equation $kx^2 + x + k = 0$ has a repeated real solution.

112. Find k such that the equation $x^2 - kx + 4 = 0$ has a repeated real solution.

113. Show that the real solutions of the equation $ax^2 + bx + c = 0$ are the negatives of the real solutions of the equation $ax^2 - bx + c = 0$. Assume that $b^2 - 4ac \geq 0$.

114. Show that the real solutions of the equation $ax^2 + bx + c = 0$ are the reciprocals of the real solutions of the equation $cx^2 + bx + a = 0$. Assume that $b^2 - 4ac \geq 0$.

Discussion and Writing

115. Which of the following pairs of equations are equivalent? Explain.
(a) $x^2 = 9$; $x = 3$
(b) $x = \sqrt{9}$; $x = 3$
(c) $(x - 1)(x - 2) = (x - 1)^2$; $x - 2 = x - 1$

116. Describe three ways that you might solve a quadratic equation. State your preferred method; explain why you chose it.

117. Explain the benefits of evaluating the discriminant of a quadratic equation before attempting to solve it.

118. Create three quadratic equations: one having two distinct solutions, one having no real solution, and one having exactly one real solution.

119. The word *quadratic* seems to imply four (*quad*), yet a quadratic equation is an equation that involves a polynomial of degree 2. Investigate the origin of the term *quadratic* as it is used in the expression *quadratic equation*. Write a brief essay on your findings.

'Are You Prepared?' Answers

1. $(x - 6)(x + 1)$ **2.** $(2x - 3)(x + 1)$ **3.** $\left\{-\frac{5}{3}, 3\right\}$ **4.** $2\sqrt{10}$

1.4 Complex Numbers; Quadratic Equations in the Complex Number System

PREPARING FOR THIS SECTION *Before getting started, review the following:*

- Classification of Numbers (Section R.1, pp. 4–5)
- Rationalizing Denominators (Section R.8, p. 74)

Now Work the 'Are You Prepared?' problems on page 131.

OBJECTIVES
1 Add, Subtract, Multiply, and Divide Complex Numbers (p. 125)
2 Solve Quadratic Equations in the Complex Number System (p. 129)

Complex Numbers

One property of a real number is that its square is nonnegative (greater than or equal to 0). For example, there is no real number x for which

$$x^2 = -1$$

To remedy this situation, we introduce a number called the **imaginary unit,** which we denote by ***i***, whose square is -1; that is,

$$i^2 = -1$$

This should not surprise you. If our universe were to consist only of integers, there would be no number x for which $2x = 1$. This unfortunate circumstance was remedied by introducing numbers such as $\frac{1}{2}$ and $\frac{2}{3}$, the *rational numbers*. If our universe were to consist only of rational numbers, there would be no x whose square equals 2. That is, there would be no number x for which $x^2 = 2$. To remedy this, we introduced numbers such as $\sqrt{2}$ and $\sqrt[3]{5}$, the *irrational numbers*. The *real numbers,* you will recall, consist of the rational numbers and the irrational numbers. Now, if our universe were to consist only of real numbers, then there would be no number x whose square is -1. To remedy this, we introduce a number i, whose square is -1.

In the progression outlined, each time that we encountered a situation that was unsuitable, we introduced a new number system to remedy this situation. And each new number system contained the earlier number system as a subset. The number system that results from introducing the number i is called the **complex number system.**

Complex numbers are numbers of the form $\boldsymbol{a + bi}$, where a and b are real numbers. The real number a is called the **real part** of the number $a + bi$; the real number b is called the **imaginary part** of $a + bi$.

For example, the complex number $-5 + 6i$ has the real part -5 and the imaginary part 6.

When a complex number is written in the form $a + bi$, where a and b are real numbers, we say it is in **standard form.** However, if the imaginary part of a complex number is negative, such as in the complex number $3 + (-2)i$, we agree to write it instead in the form $3 - 2i$.

Also, the complex number $a + 0i$ is usually written simply as a. This serves to remind us that the real numbers are a subset of the complex numbers. The complex number $0 + bi$ is usually written as bi. Sometimes the complex number bi is called a **pure imaginary number.**

1 Add, Subtract, Multiply, and Divide Complex Numbers

Equality, addition, subtraction, and multiplication of complex numbers are defined so as to preserve the familiar rules of algebra for real numbers. Thus, two complex numbers are equal if and only if their real parts are equal and their imaginary parts are equal. That is,

Equality of Complex Numbers

$$a + bi = c + di \quad \text{if and only if} \quad a = c \text{ and } b = d \tag{1}$$

Two complex numbers are added by forming the complex number whose real part is the sum of the real parts and whose imaginary part is the sum of the imaginary parts. That is,

Sum of Complex Numbers

$$(a + bi) + (c + di) = (a + c) + (b + d)i \tag{2}$$

To subtract two complex numbers, we use this rule:

Difference of Complex Numbers

$$(a + bi) - (c + di) = (a - c) + (b - d)i \tag{3}$$

EXAMPLE 1 Adding and Subtracting Complex Numbers

(a) $(3 + 5i) + (-2 + 3i) = [3 + (-2)] + (5 + 3)i = 1 + 8i$

(b) $(6 + 4i) - (3 + 6i) = (6 - 3) + (4 - 6)i = 3 + (-2)i = 3 - 2i$

Figure 40

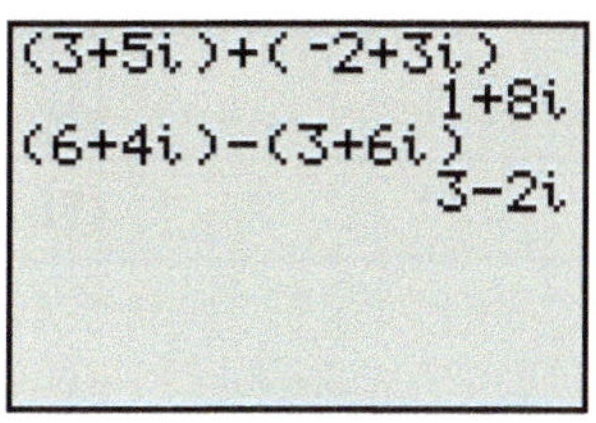

Some graphing calculators have the capability of handling complex numbers. For example, Figure 40 shows the results of Example 1 using a TI-84 Plus graphing calculator.

Now Work PROBLEM 13

Products of complex numbers are calculated as illustrated in Example 2.

EXAMPLE 2 Multiplying Complex Numbers

$$\begin{aligned}
(5 + 3i)\cdot(2 + 7i) &= 5\cdot(2 + 7i) + 3i(2 + 7i) && \text{Distributive Property}\\
&= 10 + 35i + 6i + 21i^2 && \text{Distributive Property}\\
&= 10 + 41i + 21(-1) && i^2 = -1\\
&= -11 + 41i
\end{aligned}$$

Based on the procedure of Example 2, we define the **product** of two complex numbers by the following formula:

Product of Complex Numbers

$$(a + bi) \cdot (c + di) = (ac - bd) + (ad + bc)i \qquad \textbf{(4)}$$

Figure 41

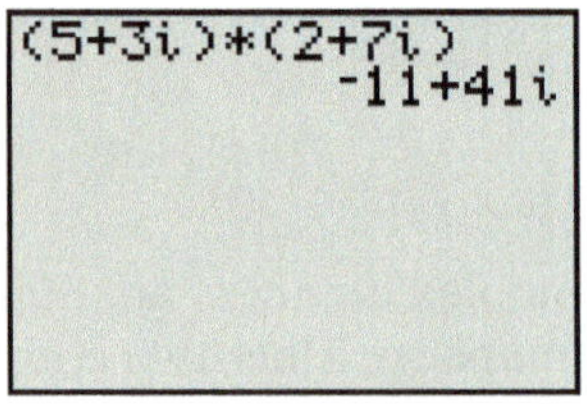

Do not bother to memorize formula (4). Instead, whenever it is necessary to multiply two complex numbers, follow the usual rules for multiplying two binomials, as in Example 2, remembering that $i^2 = -1$. For example,

$$(2i)(2i) = 4i^2 = -4$$

$$(2 + i)(1 - i) = 2 - 2i + i - i^2 = 3 - i$$

Graphing calculators may also be used to multiply complex numbers. Figure 41 shows the result obtained in Example 2 using a TI-84 Plus graphing calculator.

Now Work PROBLEM 19

Algebraic properties for addition and multiplication, such as the Commutative, Associative, and Distributive Properties, hold for complex numbers. However, the property that every nonzero complex number has a multiplicative inverse, or reciprocal, requires a closer look.

Conjugates

If $z = a + bi$ is a complex number, then its **conjugate,** denoted by $\bar{z}$, is defined as

$$\bar{z} = \overline{a + bi} = a - bi$$

For example, $\overline{2 + 3i} = 2 - 3i$ and $\overline{-6 - 2i} = -6 + 2i$.

EXAMPLE 3 Multiplying a Complex Number by Its Conjugate

Find the product of the complex number $z = 3 + 4i$ and its conjugate $\bar{z}$.

Solution Since $\bar{z} = 3 - 4i$, we have

$$z\bar{z} = (3 + 4i)(3 - 4i) = 9 - 12i + 12i - 16i^2 = 9 + 16 = 25$$

The result obtained in Example 3 has an important generalization.

THEOREM The product of a complex number and its conjugate is a nonnegative real number. That is, if $z = a + bi$, then

$$z\bar{z} = a^2 + b^2 \qquad \textbf{(5)}$$

Proof If $z = a + bi$, then

$$z\bar{z} = (a + bi)(a - bi) = a^2 - abi + abi - (bi)^2 = a^2 - b^2i^2 = a^2 + b^2 \quad \blacksquare$$

To express the reciprocal of a nonzero complex number z in standard form, multiply the numerator and denominator of $\frac{1}{z}$ by its conjugate $\bar{z}$. That is, if $z = a + bi$ is a nonzero complex number, then

$$\frac{1}{a + bi} = \frac{1}{z} = \frac{1}{z} \cdot \frac{\bar{z}}{\bar{z}} = \frac{\bar{z}}{z\bar{z}} \underset{\uparrow \text{ Use (5).}}{=} \frac{a - bi}{a^2 + b^2} = \frac{a}{a^2 + b^2} - \frac{b}{a^2 + b^2}i$$

EXAMPLE 4 Writing the Reciprocal of a Complex Number in Standard Form

Write $\frac{1}{3 + 4i}$ in standard form $a + bi$; that is, find the reciprocal of $3 + 4i$.

Solution Multiply the numerator and denominator of $\frac{1}{3 + 4i}$ by the conjugate of $3 + 4i$, the complex number $3 - 4i$. The result is

$$\frac{1}{3 + 4i} = \frac{1}{3 + 4i} \cdot \frac{3 - 4i}{3 - 4i} = \frac{3 - 4i}{9 + 16} = \frac{3}{25} - \frac{4}{25}i$$

Figure 42

A graphing calculator can be used to verify the result of Example 4. See Figure 42.

To express the quotient of two complex numbers in standard form, we multiply the numerator and denominator of the quotient by the conjugate of the denominator.

EXAMPLE 5 Writing the Quotient of a Complex Number in Standard Form

Write each of the following in standard form.

(a) $\frac{1 + 4i}{5 - 12i}$ (b) $\frac{2 - 3i}{4 - 3i}$

Solution (a) $$\frac{1 + 4i}{5 - 12i} = \frac{1 + 4i}{5 - 12i} \cdot \frac{5 + 12i}{5 + 12i} = \frac{5 + 12i + 20i + 48i^2}{25 + 144}$$
$$= \frac{-43 + 32i}{169} = \frac{-43}{169} + \frac{32}{169}i$$

(b) $$\frac{2 - 3i}{4 - 3i} = \frac{2 - 3i}{4 - 3i} \cdot \frac{4 + 3i}{4 + 3i} = \frac{8 + 6i - 12i - 9i^2}{16 + 9}$$
$$= \frac{17 - 6i}{25} = \frac{17}{25} - \frac{6}{25}i$$

Now Work PROBLEM 27

EXAMPLE 6 Writing Other Expressions in Standard Form

If $z = 2 - 3i$ and $w = 5 + 2i$, write each of the following expressions in standard form.

(a) $\frac{z}{w}$ (b) $\overline{z + w}$ (c) $z + \bar{z}$

Solution (a) $$\frac{z}{w} = \frac{z \cdot \bar{w}}{w \cdot \bar{w}} = \frac{(2 - 3i)(5 - 2i)}{(5 + 2i)(5 - 2i)} = \frac{10 - 4i - 15i + 6i^2}{25 + 4}$$
$$= \frac{4 - 19i}{29} = \frac{4}{29} - \frac{19}{29}i$$

(b) $\overline{z + w} = \overline{(2 - 3i) + (5 + 2i)} = \overline{7 - i} = 7 + i$

(c) $z + \bar{z} = (2 - 3i) + (2 + 3i) = 4$

The conjugate of a complex number has certain general properties that we shall find useful later.

For a real number $a = a + 0i$, the conjugate is $\bar{a} = \overline{a + 0i} = a - 0i = a$. That is,

THEOREM The conjugate of a real number is the real number itself.

Other properties that are direct consequences of the definition of the conjugate are given next. In each statement, z and w represent complex numbers.

THEOREM The conjugate of the conjugate of a complex number is the complex number itself.

$$(\bar{\bar{z}}) = z \tag{6}$$

The conjugate of the sum of two complex numbers equals the sum of their conjugates.

$$\overline{z + w} = \bar{z} + \bar{w} \tag{7}$$

The conjugate of the product of two complex numbers equals the product of their conjugates.

$$\overline{z \cdot w} = \bar{z} \cdot \bar{w} \tag{8}$$

We leave the proofs of equations (6), (7), and (8) as exercises. See Problems 88–90.

Powers of i

The powers of i follow a pattern that is useful to know.

$i^1 = i$	$i^5 = i^4 \cdot i = 1 \cdot i = i$
$i^2 = -1$	$i^6 = i^4 \cdot i^2 = -1$
$i^3 = i^2 \cdot i = -i$	$i^7 = i^4 \cdot i^3 = -i$
$i^4 = i^2 \cdot i^2 = (-1)(-1) = 1$	$i^8 = i^4 \cdot i^4 = 1$

And so on. The powers of i repeat with every fourth power.

EXAMPLE 7 Evaluating Powers of i

(a) $i^{27} = i^{24} \cdot i^3 = (i^4)^6 \cdot i^3 = 1^6 \cdot i^3 = -i$

(b) $i^{101} = i^{100} \cdot i^1 = (i^4)^{25} \cdot i = 1^{25} \cdot i = i$

EXAMPLE 8 Writing the Power of a Complex Number in Standard Form

Write $(2 + i)^3$ in standard form.

Solution We use the special product formula for $(x + a)^3$.

$$(x + a)^3 = x^3 + 3ax^2 + 3a^2x + a^3$$

Using this special product formula,

$$\begin{aligned}(2 + i)^3 &= 2^3 + 3 \cdot i \cdot 2^2 + 3 \cdot i^2 \cdot 2 + i^3 \\ &= 8 + 12i + 6(-1) + (-i) \\ &= 2 + 11i\end{aligned}$$

NOTE If you did not remember the special product formula for $(x + a)^3$, you could find $(2 + i)^3$ by simplifying $(2 + i)^2(2 + i)$.

Now Work PROBLEMS 33 AND 41

2 Solve Quadratic Equations in the Complex Number System

Quadratic equations with a negative discriminant have no real number solution. However, if we extend our number system to allow complex numbers, quadratic equations will always have a solution. Since the solution to a quadratic equation involves the square root of the discriminant, we begin with a discussion of square roots of negative numbers.

DEFINITION

If N is a positive real number, we define the **principal square root of $-N$,** denoted by $\sqrt{-N}$, as

$$\sqrt{-N} = \sqrt{N}i$$

where i is the imaginary unit and $i^2 = -1$.

EXAMPLE 9 Evaluating the Square Root of a Negative Number

(a) $\sqrt{-1} = \sqrt{1}i = i$ (b) $\sqrt{-16} = \sqrt{16}i = 4i$

(c) $\sqrt{-8} = \sqrt{8}i = 2\sqrt{2}i$

EXAMPLE 10 Using the Square Root Method in the Complex Number System

Solve each equation in the complex number system.

(a) $x^2 = 4$ (b) $x^2 = -9$

Solution

(a) $x^2 = 4$

$$x = \pm\sqrt{4} = \pm 2$$

The equation has the solution set $\{-2, 2\}$.

(b) $x^2 = -9$

$$x = \pm\sqrt{-9} = \pm\sqrt{9}i = \pm 3i$$

The equation has the solution set $\{-3i, 3i\}$.

WARNING When working with square roots of negative numbers, do not set the square root of a product equal to the product of the square roots (which can be done with positive numbers). To see why, look at this calculation: We know that $\sqrt{100} = 10$. However, it is also true that $100 = (-25)(-4)$, so

$$\begin{aligned} 10 &= \sqrt{100} \\ &= \sqrt{(-25)(-4)} \\ &\neq \sqrt{-25}\sqrt{-4} \end{aligned}$$

because

$$\begin{aligned} \sqrt{-25}\cdot\sqrt{-4} &= \left(\sqrt{25}i\right)\left(\sqrt{4}i\right) \\ &= (5i)(2i) \\ &= 10i^2 = -10 \end{aligned}$$

Now Work PROBLEM 53

Because we have defined the square root of a negative number, we can now restate the quadratic formula without restriction.

THEOREM

In the complex number system, the solutions of the quadratic equation $ax^2 + bx + c = 0$, where a, b, and c are real numbers and $a \neq 0$, are given by the formula

$$x = \frac{-b \pm \sqrt{b^2 - 4ac}}{2a} \qquad \textbf{(9)}$$

EXAMPLE 11 Solving a Quadratic Equation in the Complex Number System

Solve the equation $x^2 - 4x + 8 = 0$ in the complex number system.

Solution Here $a = 1, b = -4, c = 8$, and $b^2 - 4ac = 16 - 4(1)(8) = -16$. Using equation (9), we find that

$$x = \frac{-(-4) \pm \sqrt{-16}}{2(1)} = \frac{4 \pm \sqrt{16}i}{2} = \frac{4 \pm 4i}{2} = 2 \pm 2i$$

The equation has the solution set $\{2 - 2i, 2 + 2i\}$.

Check:

$$2 + 2i: \quad (2 + 2i)^2 - 4(2 + 2i) + 8 = 4 + 8i + 4i^2 - 8 - 8i + 8 = 4 - 4 = 0$$

$$2 - 2i: \quad (2 - 2i)^2 - 4(2 - 2i) + 8 = 4 - 8i + 4i^2 - 8 + 8i + 8 = 4 - 4 = 0$$

Figure 43

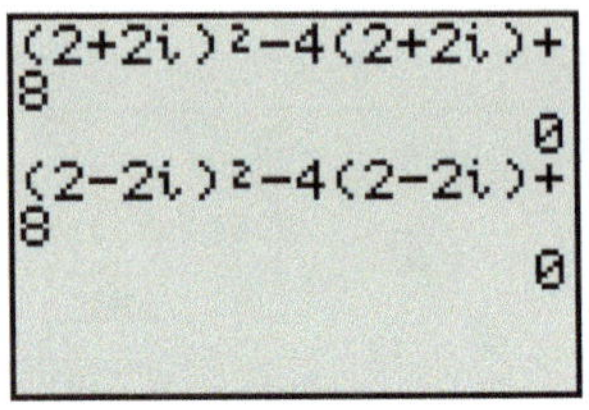

Figure 43 shows the check of the solution using a TI-84 Plus graphing calculator. Graph $Y_i = x^2 - 4x + 8$. How many x-intercepts are there?

Now Work PROBLEM 59

The discriminant $b^2 - 4ac$ of a quadratic equation still serves as a way to determine the character of the solutions.

Character of the Solutions of a Quadratic Equation

In the complex number system, consider a quadratic equation $ax^2 + bx + c = 0$ with real coefficients.

1. If $b^2 - 4ac > 0$, the equation has two unequal real solutions.
2. If $b^2 - 4ac = 0$, the equation has a repeated real solution, a double root.
3. If $b^2 - 4ac < 0$, the equation has two complex solutions that are not real. The solutions are conjugates of each other.

The third conclusion in the display is a consequence of the fact that if $b^2 - 4ac = -N < 0$ then, by the quadratic formula, the solutions are

$$x = \frac{-b + \sqrt{b^2 - 4ac}}{2a} = \frac{-b + \sqrt{-N}}{2a} = \frac{-b + \sqrt{N}i}{2a} = \frac{-b}{2a} + \frac{\sqrt{N}}{2a}i$$

and

$$x = \frac{-b - \sqrt{b^2 - 4ac}}{2a} = \frac{-b - \sqrt{-N}}{2a} = \frac{-b - \sqrt{N}i}{2a} = \frac{-b}{2a} - \frac{\sqrt{N}}{2a}i$$

which are conjugates of each other.

EXAMPLE 12 Determining the Character of the Solutions of a Quadratic Equation

Without solving, determine the character of the solutions of each equation.

(a) $3x^2 + 4x + 5 = 0$ (b) $2x^2 + 4x + 1 = 0$

(c) $9x^2 - 6x + 1 = 0$

Solution (a) Here $a = 3, b = 4$, and $c = 5$, so $b^2 - 4ac = 4^2 - 4(3)(5) = -44$. The solutions are two complex numbers that are not real and are conjugates of each other.

(b) Here $a = 2, b = 4$, and $c = 1$, so $b^2 - 4ac = 4^2 - 4(2)(1) = 8$. The solutions are two unequal real numbers.

(c) Here $a = 9, b = -6$, and $c = 1$, so $b^2 - 4ac = (-6)^2 - 4(9)(1) = 0$. The solution is a repeated real number, that is, a double root.

Now Work PROBLEM 73

1.4 Assess Your Understanding

'Are You Prepared?' *Answers are given at the end of these exercises. If you get a wrong answer, read the pages listed in red.*

1. Name the integers and the rational numbers in the set $\left\{-3, 0, \sqrt{2}, \frac{6}{5}, \pi\right\}$. (pp. 4–5)
2. ***True or False*** Rational numbers and irrational numbers are in the set of real numbers. (pp. 4–5)
3. Rationalize the denominator of $\frac{3}{2+\sqrt{3}}$. (p. 74)

Concepts and Vocabulary

4. In the complex number $5 + 2i$, the number 5 is called the _____ part; the number 2 is called the _____ part; the number i is called the _____ _____.
5. The equation $x^2 = -4$ has the solution set _____.
6. ***True or False*** The conjugate of $2 + 5i$ is $-2 - 5i$.
7. ***True or False*** All real numbers are complex numbers.
8. ***True or False*** If $2 - 3i$ is a solution of a quadratic equation with real coefficients, then $-2 + 3i$ is also a solution.

Skill Building

In Problems 9–46, write each expression in the standard form $a + bi$. Verify your results using a graphing utility.

9. $(2 - 3i) + (6 + 8i)$
10. $(4 + 5i) + (-8 + 2i)$
11. $(-3 + 2i) - (4 - 4i)$
12. $(3 - 4i) - (-3 - 4i)$
13. $(2 - 5i) - (8 + 6i)$
14. $(-8 + 4i) - (2 - 2i)$
15. $3(2 - 6i)$
16. $-4(2 + 8i)$
17. $2i(2 - 3i)$
18. $3i(-3 + 4i)$
19. $(3 - 4i)(2 + i)$
20. $(5 + 3i)(2 - i)$
21. $(-6 + i)(-6 - i)$
22. $(-3 + i)(3 + i)$
23. $\frac{10}{3 - 4i}$
24. $\frac{13}{5 - 12i}$
25. $\frac{2 + i}{i}$
26. $\frac{2 - i}{-2i}$
27. $\frac{6 - i}{1 + i}$
28. $\frac{2 + 3i}{1 - i}$
29. $\left(\frac{1}{2} + \frac{\sqrt{3}}{2}i\right)^2$
30. $\left(\frac{\sqrt{3}}{2} - \frac{1}{2}i\right)^2$
31. $(1 + i)^2$
32. $(1 - i)^2$
33. i^{23}
34. i^{14}
35. i^{-15}
36. i^{-23}
37. $i^6 - 5$
38. $4 + i^3$
39. $6i^3 - 4i^5$
40. $4i^3 - 2i^2 + 1$
41. $(1 + i)^3$
42. $(3i)^4 + 1$
43. $i^7(1 + i^2)$
44. $2i^4(1 + i^2)$
45. $i^6 + i^4 + i^2 + 1$
46. $i^7 + i^5 + i^3 + i$

In Problems 47–52, perform the indicated operations and express your answer in the form $a + bi$.

47. $\sqrt{-4}$

48. $\sqrt{-9}$

49. $\sqrt{-25}$

50. $\sqrt{-64}$

51. $\sqrt{(3 + 4i)(4i - 3)}$

52. $\sqrt{(4 + 3i)(3i - 4)}$

In Problems 53–72, solve each equation in the complex number system. Check your results using a graphing utility.

53. $x^2 + 4 = 0$

54. $x^2 - 4 = 0$

55. $x^2 - 16 = 0$

56. $x^2 + 25 = 0$

57. $x^2 - 6x + 13 = 0$

58. $x^2 + 4x + 8 = 0$

59. $x^2 - 6x + 10 = 0$

60. $x^2 - 2x + 5 = 0$

61. $8x^2 - 4x + 1 = 0$

62. $10x^2 + 6x + 1 = 0$

63. $5x^2 + 2x + 1 = 0$

64. $13x^2 + 6x + 1 = 0$

65. $x^2 + x + 1 = 0$

66. $x^2 - x + 1 = 0$

67. $x^3 - 8 = 0$

68. $x^3 + 27 = 0$

69. $x^4 - 16 = 0$

70. $x^4 - 1 = 0$

71. $x^4 + 13x^2 + 36 = 0$

72. $x^4 + 3x^2 - 4 = 0$

In Problems 73–78, without solving, determine the character of the solutions of each equation. Verify your answer using a graphing utility.

73. $3x^2 - 3x + 4 = 0$

74. $2x^2 - 4x + 1 = 0$

75. $2x^2 + 3x - 4 = 0$

76. $x^2 + 2x + 6 = 0$

77. $9x^2 - 12x + 4 = 0$

78. $4x^2 + 12x + 9 = 0$

Applications and Extensions

79. $2 + 3i$ is a solution of a quadratic equation with real coefficients. Find the other solution.

80. $4 - i$ is a solution of a quadratic equation with real coefficients. Find the other solution.

In Problems 81–84, $z = 3 - 4i$ and $w = 8 + 3i$. Write each expression in the standard form $a + bi$.

81. $z + \bar{z}$

82. $w - \bar{w}$

83. $z\bar{z}$

84. $\overline{z - w}$

85. **Electrical Circuits** The impedance Z, in ohms, of a circuit element is defined as the ratio of the phasor voltage V, in volts, across the element to the phasor current I, in amperes, through the elements. That is, $Z = \frac{V}{I}$. If the voltage across a circuit element is $18 + i$ volts and the current through the element is $3-4i$ amperes, determine the impedance.

86. **Parallel Circuits** In an ac circuit with two parallel pathways, the total impedance Z, in ohms, satisfies the formula $\frac{1}{Z} = \frac{1}{Z_1} + \frac{1}{Z_2}$, where Z_1 is the impedance of the first pathway and Z_2 is the impedance of the second pathway. Determine the total impedance if the impedances of the two pathways are $Z_1 = 2 + i$ ohms and $Z_2 = 4 - 3i$ ohms.

87. Use $z = a + bi$ to show that $z + \bar{z} = 2a$ and that $z - \bar{z} = 2bi$.

88. Use $z = a + bi$ to show that $(\bar{\bar{z}}) = z$.

89. Use $z = a + bi$ and $w = c + di$ to show that $\overline{z + w} = \bar{z} + \bar{w}$.

90. Use $z = a + bi$ and $w = c + di$ to show that $\overline{z \cdot w} = \bar{z} \cdot \bar{w}$.

Discussion and Writing

91. Explain to a friend how you would add two complex numbers and how you would multiply two complex numbers. Explain any differences in the two explanations.

92. Write a brief paragraph that compares the method used to rationalize denominators and the method used to write the quotient of two complex numbers in standard form.

93. Use an Internet search engine to investigate the origins of complex numbers. Write a paragraph describing what you find and present it to the class.

94. Explain how the method of multiplying two complex numbers is related to multiplying two binomials.

95. **What Went Wrong?** A student multiplied $\sqrt{-9}$ and $\sqrt{-9}$ as follows:

$$\sqrt{-9} \cdot \sqrt{-9} = \sqrt{(-9)(-9)} = \sqrt{81} = 9$$

The instructor marked the problem incorrect. Why?

'Are You Prepared?' Answers

1. Integers: $\{-3, 0\}$; rational numbers: $\left\{-3, 0, \frac{6}{5}\right\}$

2. True

3. $3(2 - \sqrt{3})$

1.5 Radical Equations; Equations Quadratic in Form; Absolute Value Equations; Factorable Equations

PREPARING FOR THIS SECTION *Before getting started, review the following:*

- Absolute Value (Section R.2, p. 19)
- Square Roots (Section R.2, pp. 23–24)
- Factoring Polynomials (Section R.5, pp. 49–55)
- nth Roots; Rational Exponents (Section R.8, pp. 72–76)

Now Work the 'Are You Prepared?' problems on page 138.

OBJECTIVES
1. Solve Radical Equations (p. 133)
2. Solve Equations Quadratic in Form (p. 135)
3. Solve Absolute Value Equations (p. 136)
4. Solve Equations by Factoring (p. 137)

1 Solve Radical Equations

When the variable in an equation occurs in a square root, cube root, and so on, that is, when it occurs in a radical, the equation is called a **radical equation.** Sometimes a suitable operation will change a radical equation to one that is linear or quadratic. A commonly used procedure is to isolate the most complicated radical on one side of the equation and then eliminate it by raising each side to a power equal to the index of the radical. Care must be taken, however, because apparent solutions that are not, in fact, solutions of the original equation may result. Recall that these are called extraneous solutions. In radical equations, extraneous solutions may occur when the index of the radical is even. Therefore, we need to check all answers when working with radical equations.

EXAMPLE 1 Solving a Radical Equation

Find the real solutions of the equation: $\sqrt[3]{2x-4}-2=0$

Algebraic Solution

The equation contains a radical whose index is 3. We isolate it on the left side:

$$\sqrt[3]{2x-4}-2=0$$
$$\sqrt[3]{2x-4}=2$$

Because the index of the radical is 3, raise each side to the third power and solve.

$$\left(\sqrt[3]{2x-4}\right)^3=2^3 \quad \text{Raise each side to the power 3.}$$
$$2x-4=8 \quad \text{Simplify.}$$
$$2x=12 \quad \text{Add 4 to both sides.}$$
$$x=6 \quad \text{Divide both sides by 2.}$$

Graphing Solution

Figure 44 shows the graph of the equation $Y_1=\sqrt[3]{2x-4}-2$. From the graph, we see one x-intercept near 6. Using ZERO (or ROOT), we find that the x-intercept is 6. The only solution is $x=6$.

Figure 44

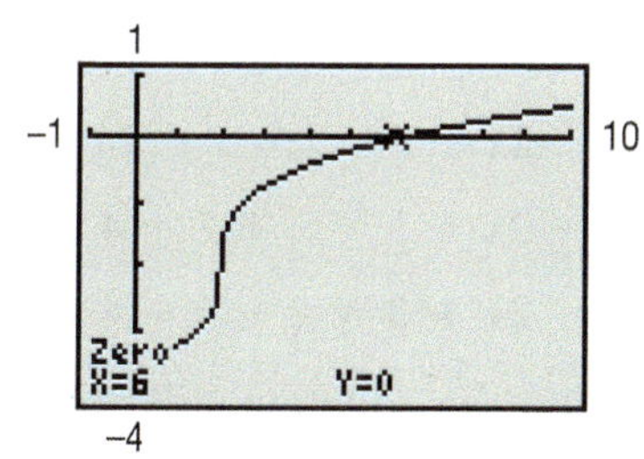

Check: $\sqrt[3]{2(6)-4}-2=\sqrt[3]{12-4}-2=\sqrt[3]{8}-2=2-2=0$

The solution set is $\{6\}$.

Now Work PROBLEM 11

EXAMPLE 2 Solving a Radical Equation

Find the real solutions of the equation: $\sqrt{x-1} = x - 7$

Algebraic Solution

We square both sides since the index of a square root is 2.

$$\sqrt{x-1} = x - 7$$

$$\left(\sqrt{x-1}\right)^2 = (x-7)^2 \qquad \text{Square both sides.}$$

$$x - 1 = x^2 - 14x + 49 \qquad \text{Remove parentheses.}$$

$$x^2 - 15x + 50 = 0 \qquad \text{Put in standard form.}$$

$$(x-10)(x-5) = 0 \qquad \text{Factor.}$$

$$x = 10 \quad \text{or} \quad x = 5 \qquad \text{Apply the Zero-Product Property and solve.}$$

Graphing Solution

Graph $Y_1 = \sqrt{x-1}$ and $Y_2 = x - 7$. See Figure 45. From the graph, there is one point of intersection. Using INTERSECT, the point of intersection is $(10, 3)$, so the solution is $x = 10$.

Figure 45

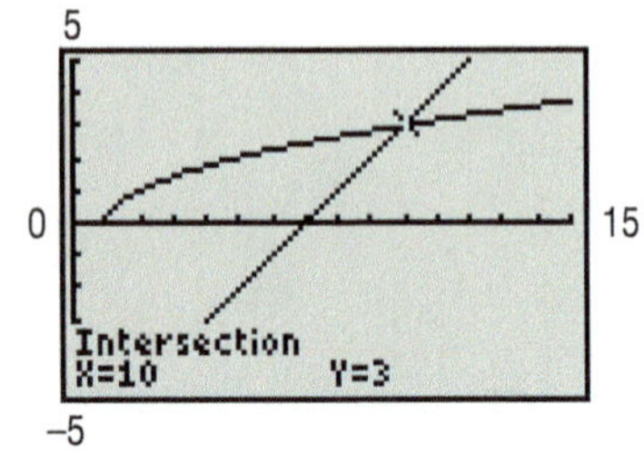

There is a discrepancy between the algebraic solution and graphing solution. Let's check the results of our algebraic solution.

Check: $x = 10$: $\sqrt{x-1} = \sqrt{10-1} = \sqrt{9} = 3$ and $x - 7 = 10 - 7 = 3$

$x = 5$: $\sqrt{x-1} = \sqrt{5-1} = \sqrt{4} = 2$ and $x - 7 = 5 - 7 = -2$

The apparent algebraic solution $x = 5$ is extraneous; the only solution of the equation is $x = 10$. The solution set is $\{10\}$.

Now Work PROBLEM 23

Sometimes, we need to raise each side to a power more than once in order to solve a radical equation algebraically.

EXAMPLE 3 Solving a Radical Equation

Find the real solutions of the equation: $\sqrt{2x+3} - \sqrt{x+2} = 2$

Algebraic Solution

First, we choose to isolate the more complicated radical expression (in this case, $\sqrt{2x+3}$) on the left side:

$$\sqrt{2x+3} = \sqrt{x+2} + 2$$

Now square both sides (the index of the radical is 2).

$$\left(\sqrt{2x+3}\right)^2 = \left(\sqrt{x+2} + 2\right)^2$$

$$2x + 3 = \left(\sqrt{x+2}\right)^2 + 4\sqrt{x+2} + 4 \qquad \text{Remove parentheses.}$$

$$2x + 3 = x + 2 + 4\sqrt{x+2} + 4 \qquad \text{Simplify.}$$

$$2x + 3 = x + 6 + 4\sqrt{x+2} \qquad \text{Combine like terms.}$$

Because the equation still contains a radical, we isolate the remaining radical on the right side and again square both sides.

$$x - 3 = 4\sqrt{x+2} \qquad \text{Isolate the radical on the right side.}$$

$$(x-3)^2 = 16(x+2) \qquad \text{Square both sides.}$$

$$x^2 - 6x + 9 = 16x + 32 \qquad \text{Remove parentheses.}$$

$$x^2 - 22x - 23 = 0 \qquad \text{Put in standard form.}$$

$$(x-23)(x+1) = 0 \qquad \text{Factor.}$$

$$x = 23 \quad \text{or} \quad x = -1$$

Graphing Solution

Graph $Y_1 = \sqrt{2x+3} - \sqrt{x+2}$ and $Y_2 = 2$. See Figure 46. From the graph there is one point of intersection. Using INTERSECT, the point of intersection is $(23, 2)$, so the solution is $x = 23$.

Figure 46

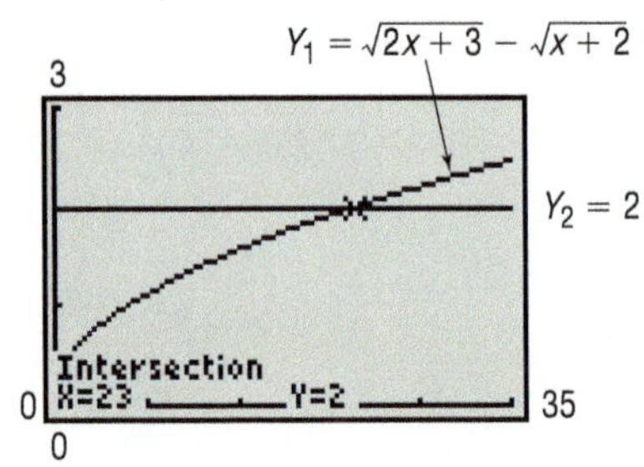

Check: $x = 23$: $\sqrt{2(23) + 3} - \sqrt{23 + 2} = \sqrt{49} - \sqrt{25} = 7 - 5 = 2$

$x = -1$: $\sqrt{2(-1) + 3} - \sqrt{-1 + 2} = \sqrt{1} - \sqrt{1} = 1 - 1 = 0$

The apparent solution $x = -1$ is extraneous; the only solution is $x = 23$. The solution set of the equation is $\{23\}$.

Now Work PROBLEM 31

2 Solve Equations Quadratic in Form

The equation $x^4 + x^2 - 12 = 0$ is not quadratic in x, but it is quadratic in x^2. That is, if we let $u = x^2$, we get $u^2 + u - 12 = 0$, a quadratic equation. This equation can be solved for u and, in turn, by using $u = x^2$, we can find the solutions x of the original equation.

In general, if an appropriate substitution u transforms an equation into one of the form

$$au^2 + bu + c = 0, \qquad a \neq 0$$

then the original equation is called an **equation of the quadratic type** or an **equation quadratic in form.**

The difficulty of solving such an equation lies in the determination that the equation is, in fact, quadratic in form. After you are told an equation is quadratic in form, it is easy enough to see it, but some practice is needed to enable you to recognize them on your own.

EXAMPLE 4 Solving Equations That Are Quadratic in Form

Find the real solutions of the equation: $(x + 2)^2 + 11(x + 2) - 12 = 0$

Solution For this equation, let $u = x + 2$. Then $u^2 = (x + 2)^2$, and the original equation,

$$(x + 2)^2 + 11(x + 2) - 12 = 0$$

becomes

$$u^2 + 11u - 12 = 0 \quad \text{Let } u = x + 2.$$

$$(u + 12)(u - 1) = 0 \quad \text{Factor.}$$

$$u = -12 \quad \text{or} \quad u = 1 \quad \text{Solve.}$$

But we want to solve for x. Because $u = x + 2$, we have

$$x + 2 = -12 \quad \text{or} \quad x + 2 = 1$$

$$x = -14 \qquad\qquad x = -1$$

Check: $x = -14$: $(-14 + 2)^2 + 11(-14 + 2) - 12$

$$= (-12)^2 + 11(-12) - 12 = 144 - 132 - 12 = 0$$

$x = -1$: $(-1 + 2)^2 + 11(-1 + 2) - 12 = 1 + 11 - 12 = 0$

The original equation has the solution set $\{-14, -1\}$.

Check: Verify the solution of Example 4 using a graphing utility.

EXAMPLE 5 **Solving Equations That Are Quadratic in Form**

Find the real solutions of the equation: $x + 2\sqrt{x} - 3 = 0$

Solution For the equation $x + 2\sqrt{x} - 3 = 0$, let $u = \sqrt{x}$. Then $u^2 = x$, and the original equation,

$$x + 2\sqrt{x} - 3 = 0$$

becomes

$$u^2 + 2u - 3 = 0 \quad \text{Let } u = \sqrt{x}.$$

$$(u + 3)(u - 1) = 0 \quad \text{Factor.}$$

$$u = -3 \quad \text{or} \quad u = 1 \quad \text{Solve.}$$

Since $u = \sqrt{x}$, we have $\sqrt{x} = -3$ or $\sqrt{x} = 1$. The first of these, $\sqrt{x} = -3$, has no real solution, since the square root of a real number is never negative. The second, $\sqrt{x} = 1$, has the solution $x = 1$.

Check: $1 + 2\sqrt{1} - 3 = 1 + 2 - 3 = 0$

The solution set of the original equation is $\{1\}$.

Check: Verify the solution to Example 5 using a graphing utility.

Another method for solving Example 5 would be to treat it as a radical equation. Solve it this way for practice.

The idea should now be clear. If an equation contains an expression and that same expression squared, make a substitution for the expression. You may get a quadratic equation.

Now Work PROBLEM 51

3 Solve Absolute Value Equations

Recall that, on the real number line, the absolute value of a equals the distance from the origin to the point whose coordinate is a. For example, there are two points whose distance from the origin is 5 units, -5 and 5. Thus the equation $|x| = 5$ will have the solution set $\{-5, 5\}$.

Another way to obtain this result is to use the algebraic definition of absolute value, $|a| = a$ if $a \geq 0$, $|a| = -a$ if $a < 0$. The equation $|u| = a$ leads to two equations depending on whether u is nonnegative (greater than or equal to zero) or negative.

$$|u| = a$$

If $u < 0$

$$|u| = a$$

$$-u = a \quad |u| = -u \text{ when } u < 0$$

$$u = -a \quad \text{Multiply both sides by } -1.$$

If $u \geq 0$

$$|u| = a$$

$$u = a \quad |u| = u \text{ when } u \geq 0$$

So we have the following result.

THEOREM

If a is a positive real number and if u is any algebraic expression, then

$$|u| = a \quad \text{is equivalent to} \quad u = a \quad \text{or} \quad u = -a \qquad \textbf{(1)}$$

EXAMPLE 6 **Solving an Equation Involving Absolute Value**

Solve the equation: $|2x - 3| + 2 = 7$

Algebraic Solution

$$|2x - 3| + 2 = 7$$

$$|2x - 3| = 5 \quad \text{Subtract 2 from each side.}$$

$$2x - 3 = 5 \quad \text{or} \quad 2x - 3 = -5 \quad \text{Apply statement (1).}$$

$$2x = 8 \quad \text{or} \quad 2x = -2$$

$$x = 4 \quad \text{or} \quad x = -1$$

The solution set is $\{-1, 4\}$.

Graphing Solution

For this equation, graph $Y_1 = |2x - 3| + 2$ and $Y_2 = 7$ on the same screen and find their point(s) of intersection, if any. See Figure 47. Using the INTERSECT command (twice), we find the points of intersection to be $(-1, 7)$ and $(4, 7)$. The solution set is $\{-1, 4\}$.

Figure 47

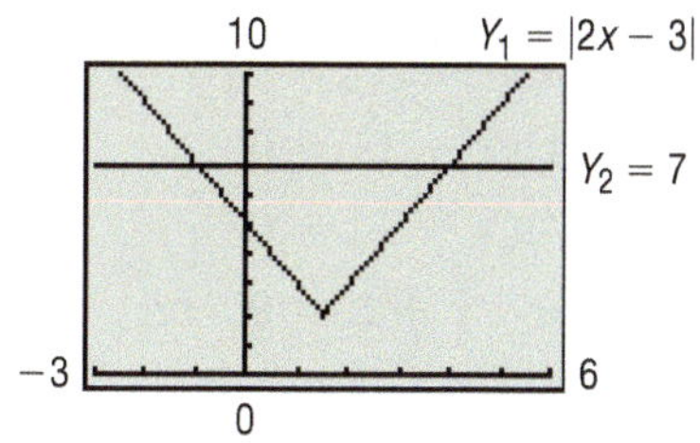

Now Work PROBLEM 67

4 Solve Equations by Factoring

We have already solved certain quadratic equations using factoring. Let's look at examples of other kinds of equations that can be solved by factoring.

EXAMPLE 7 **Solving Equations by Factoring**

Solve the equation: $x^4 = 4x^2$

Algebraic Solution

We begin by collecting all terms on one side. This results in 0 on one side and an expression to be factored on the other.

$$x^4 = 4x^2$$

$$x^4 - 4x^2 = 0$$

$$x^2(x^2 - 4) = 0 \quad \text{Factor.}$$

$$x^2 = 0 \quad \text{or} \quad x^2 - 4 = 0 \quad \text{Apply the Zero-Product Property.}$$

$$x^2 = 4$$

$$x = 0 \quad \text{or} \quad x = -2 \quad \text{or} \quad x = 2 \quad \text{Use the Square Root Method.}$$

Graphing Solution

Graph $Y_1 = x^4$ and $Y_2 = 4x^2$ on the same screen and find their point(s) of intersection, if any. See Figure 48. Using the INTERSECT command (three times), we find the points of intersection to be $(-2, 16)$, $(0, 0)$, and $(2, 16)$. The solutions are $x = -2, x = 0$, or $x = 2$.

Figure 48

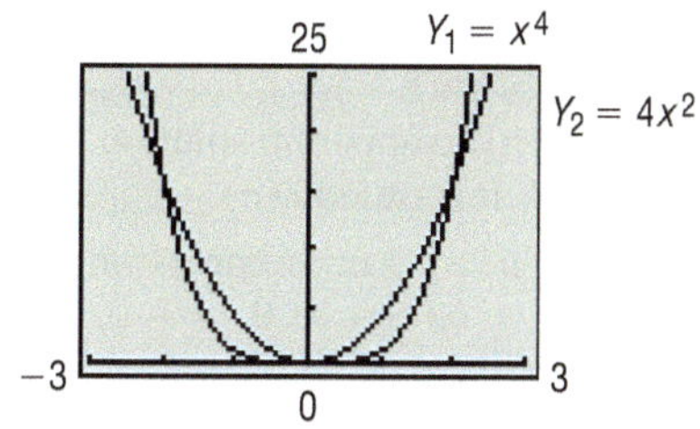

Check: $x = -2$: $(-2)^4 = 16$ and $4(-2)^2 = 16$ So −2 is a solution.

$x = 0$: $0^4 = 0$ and $4 \cdot 0^2 = 0$ So 0 is a solution.

$x = 2$: $2^4 = 16$ and $4 \cdot 2^2 = 16$ So 2 is a solution.

The solution set is $\{-2, 0, 2\}$.

EXAMPLE 8 Solving Equations by Factoring

Solve the equation: $x^3 - x^2 - 4x + 4 = 0$

Algebraic Solution

Do you recall the method of factoring by grouping? (If not, review pp. 53–54.) We group the terms of $x^3 - x^2 - 4x + 4 = 0$ as follows:

$$(x^3 - x^2) - (4x - 4) = 0$$

Factor out x^2 from the first grouping and 4 from the second.

$$x^2(x - 1) - 4(x - 1) = 0$$

This reveals the common factor $(x - 1)$, so we have

$$(x^2 - 4)(x - 1) = 0$$

$$(x - 2)(x + 2)(x - 1) = 0 \quad \text{Factor again.}$$

$$x - 2 = 0 \quad \text{or} \quad x + 2 = 0 \quad \text{or} \quad x - 1 = 0 \quad \text{Set each factor equal to 0.}$$

$$x = 2 \qquad x = -2 \qquad x = 1 \quad \text{Solve.}$$

Graphing Solution

Graph $Y_1 = x^3 - x^2 - 4x + 4$. See Figure 49. Using ZERO (three times), the values of x for which $y = 0$ are -2, 1, and 2.

Figure 49

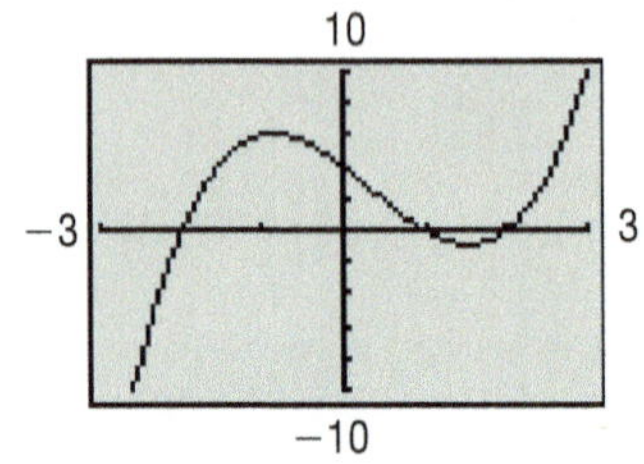

Check:

$x = -2$: $(-2)^3 - (-2)^2 - 4(-2) + 4 = -8 - 4 + 8 + 4 = 0$ −2 is a solution.

$x = 1$: $1^3 - 1^2 - 4(1) + 4 = 1 - 1 - 4 + 4 = 0$ 1 is a solution.

$x = 2$: $2^3 - 2^2 - 4(2) + 4 = 8 - 4 - 8 + 4 = 0$ 2 is a solution.

The solution set is $\{-2, 1, 2\}$.

Now Work PROBLEM 89

1.5 Assess Your Understanding

'Are You Prepared?' *Answers are given at the end of these exercises. If you get a wrong answer, read the pages listed in red.*

1. ***True or False*** The principal square root of any nonnegative real number is always nonnegative. (p. 23)
2. $\left(\sqrt[3]{x}\right)^3 =$ _____. (p. 75)
3. Factor $2x^2 - 7x - 4$. (pp. 54–55)
4. Factor $x^3 + 4x^2 - 9x - 36$. (pp. 53–54)
5. Use a real number line to describe why $|-4| = 4$. (p. 19)

Concepts and Vocabulary

6. When an apparent solution does not satisfy the original equation, it is called a(n) _____ solution.
7. If u is an expression involving x, the equation $au^2 + bu + c = 0$, $a \neq 0$, is called a(n) equation _____ _____ _____.
8. ***True or False*** Radical equations sometimes have no real solution.

Skill Building

In Problems 9–40, find the real solutions of each equation. Verify your results using a graphing utility.

9. $\sqrt{y + 3} = 5$
10. $\sqrt{t - 3} = 7$
11. $\sqrt{2t - 1} = 1$
12. $\sqrt{3t + 4} = 2$
13. $\sqrt{3t + 4} = -6$
14. $\sqrt{5t + 3} = -2$
15. $\sqrt[3]{1 - 2x} - 3 = 0$
16. $\sqrt[3]{1 - 2x} - 1 = 0$
17. $\sqrt[4]{5x - 4} = 2$

18. $\sqrt[5]{2x-3}=-1$
19. $\sqrt[5]{x^2+2x}=-1$
20. $\sqrt[4]{x^2+16}=\sqrt{5}$
21. $x=8\sqrt{x}$
22. $x=3\sqrt{x}$
23. $\sqrt{15-2x}=x$
24. $\sqrt{12-x}=x$
25. $\sqrt{x^2-x-4}=x+2$
26. $\sqrt{x^2-x-8}=x+5$
27. $3+\sqrt{3x+1}=x$
28. $2+\sqrt{12-2x}=x$
29. $\sqrt{2x+3}-\sqrt{x+1}=1$
30. $\sqrt{3x+7}+\sqrt{x+2}=1$
31. $\sqrt{3x+1}-\sqrt{x-1}=2$
32. $\sqrt{3x-5}-\sqrt{x+7}=2$
33. $\sqrt{3-2\sqrt{x}}=\sqrt{x}$
34. $\sqrt{10+3\sqrt{x}}=\sqrt{x}$
35. $(3x+1)^{1/2}=4$
36. $(3x-5)^{1/2}=2$
37. $(5x-2)^{1/3}=2$
38. $(2x+1)^{1/3}=-1$
39. $(x^2+9)^{1/2}=5$
40. $(x^2-16)^{1/2}=9$

In Problems 41–66, find the real solutions of each equation. Verify your results using a graphing utility.

41. $t^4-16=0$
42. $y^4-4=0$
43. $x^4-5x^2+4=0$
44. $x^4-10x^2+24=0$
45. $3x^4-2x^2-1=0$
46. $2x^4-5x^2-12=0$
47. $x^6+7x^3-8=0$
48. $x^6-7x^3-8=0$
49. $(x+2)^2+7(x+2)+12=0$
50. $(2x+5)^2-(2x+5)-6=0$
51. $2(s+1)^2-5(s+1)=3$
52. $3(1-y)^2+5(1-y)+2=0$
53. $x-4\sqrt{x}=0$
54. $x-8\sqrt{x}=0$
55. $x+\sqrt{x}=20$
56. $x+\sqrt{x}=6$
57. $t^{1/2}-2t^{1/4}+1=0$
58. $z^{1/2}-4z^{1/4}+4=0$
59. $4x^{1/2}-9x^{1/4}+4=0$
60. $x^{1/2}-3x^{1/4}+2=0$
61. $\dfrac{1}{(x+1)^2}=\dfrac{1}{x+1}+2$
62. $\dfrac{1}{(x-1)^2}+\dfrac{1}{x-1}=12$
63. $3x^{-2}-7x^{-1}-6=0$
64. $2x^{-2}-3x^{-1}-4=0$
65. $2x^{2/3}-5x^{1/3}-3=0$
66. $3x^{4/3}+5x^{2/3}-2=0$

In Problems 67–84, solve each equation. Verify your results using a graphing utility.

67. $|2x+3|=5$
68. $|3x-1|=2$
69. $|1-4t|+8=13$
70. $|1-2z|+6=9$
71. $|-2x|=8$
72. $|-x|=1$
73. $4-|2x|=3$
74. $5-\left|\frac{1}{2}x\right|=3$
75. $\frac{2}{3}|x|=9$
76. $\frac{3}{4}|x|=9$
77. $\left|\frac{x}{3}+\frac{2}{5}\right|=2$
78. $\left|\frac{x}{2}-\frac{1}{3}\right|=1$
79. $|u-2|=-\frac{1}{2}$
80. $|2-v|=-1$
81. $|x^2-9|=0$
82. $|x^2-16|=0$
83. $|x^2-2x|=3$
84. $|x^2+x|=12$

In Problems 85–94, find the real solutions of each equation by factoring. Verify your results using a graphing utility.

85. $x^3-9x=0$
86. $x^4-81x^2=0$
87. $x^3+x^2-20x=0$
88. $x^3+6x^2-7x=0$
89. $x^3+x^2-x-1=0$
90. $x^3+4x^2-x-4=0$
91. $x^3-3x^2-4x+12=0$
92. $x^3-3x^2-x+3=0$
93. $2x^3+4=x^2+8x$
94. $3x^3+4x^2=27x+36$

In Problems 95–98, find the real solutions of each equation. Use a calculator to express any solutions rounded to two decimal places.

95. $x-4x^{1/2}+2=0$
96. $x^{2/3}+4x^{1/3}+2=0$
97. $x^4+\sqrt{3}x^2-3=0$
98. $x^4+\sqrt{2}x^2-2=0$

Mixed Practice

In Problems 99–118, find the real solutions of each equation. Verify your results using a graphing utility.

99. $3x^2+7x-20=0$
100. $2x^2-13x+21=0$
101. $5a^3-45a=-2a^2+18$
102. $3z^3-12z=-5z^2+20$
103. $-3|5x-2|+9=0$
104. $\frac{1}{4}|2x-3|=\frac{3}{2}$
105. $4(w-3)=w+3$
106. $6(k+3)-2k=12$

107. $\left(\frac{v}{v+1}\right)^2 + \frac{2v}{v+1} = 8$ **108.** $\left(\frac{y}{y-1}\right)^2 = \frac{6y}{y-1} + 7$ **109.** $|-3x+2| = x+10$ **110.** $|4x-3| = x+2$

111. $\sqrt{2x+5} - x = 1$ **112.** $\sqrt{3x+1} - 2x = -6$ **113.** $3m^2 + 6m = -1$ **114.** $4y^2 - 8y = 3$

115. $|x^2 + x - 1| = 1$ **116.** $|x^2 + 3x - 2| = 2$ **117.** $\sqrt[4]{5x^2 - 6} = x$ **118.** $\sqrt[4]{4 - 3x^2} = x$

In Problems 119–122, find all complex solutions of each equation.

119. $t^4 - 16 = 0$ **120.** $y^4 - 81 = 0$ **121.** $x^6 - 9x^3 + 8 = 0$ **122.** $z^6 + 28z^3 + 27 = 0$

Applications and Extensions

123. If $k = \frac{x+3}{x-3}$ and $k^2 - k = 12$, find x.

124. If $k = \frac{x+3}{x-4}$ and $k^2 - 3k = 28$, find x.

125. Find all points having an x-coordinate of 2 whose distance from the point $(-2, -1)$ is 5.

126. Find all points having a y-coordinate of -3 whose distance from the point $(1, 2)$ is 13.

127. Find all points on the x-axis that are 5 units from the point $(4, -3)$.

128. Find all points on the y-axis that are 5 units from the point $(4, 4)$.

129. Physics: Using Sound to Measure Distance The distance to the surface of the water in a well can sometimes be found by dropping an object into the well and measuring the time elapsed until a sound is heard. If t_1 is the time (measured in seconds) that it takes for the object to strike the water, then t_1 will obey the equation $s = 16t_1^2$, where s is the distance (measured in feet). It follows that $t_1 = \frac{\sqrt{s}}{4}$. Suppose that t_2 is the time that it takes for the sound of the impact to reach your ears. Because sound travels at a speed of approximately 1100 feet per second, the time t_2 for the sound to travel the distance s will be $t_2 = \frac{s}{1100}$. See the illustration. Now $t_1 + t_2$ is the total time that elapses from the moment that the object is dropped to the moment that a sound is heard. We have the equation

$$\text{Total time elapsed} = \frac{\sqrt{s}}{4} + \frac{s}{1100}$$

Find the distance to the water's surface if the total time elapsed from dropping a rock to hearing it hit water is 4 seconds.

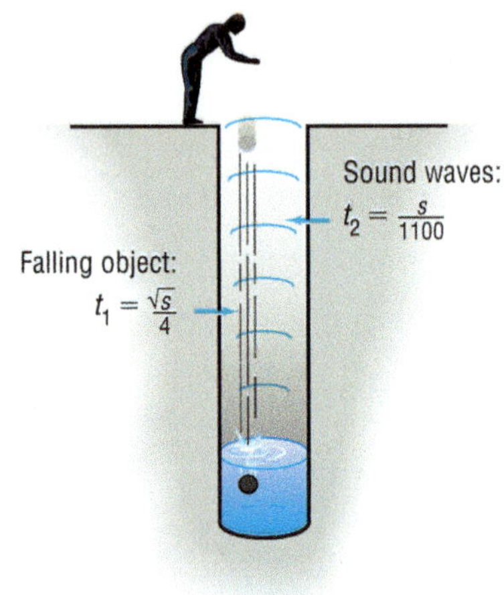

130. Crushing Load A civil engineer relates the thickness T, in inches, and height H, in feet, of a square wooden pillar to its crushing load L, in tons, using the model $T = \sqrt[4]{\frac{LH^2}{25}}$. If a square wooden pillar is 4 inches thick and 10 feet high, what is its crushing load?

131. Foucault's Pendulum The period of a pendulum is the time it takes the pendulum to make one full swing back and forth. The period T, in seconds, is given by the formula $T = 2\pi\sqrt{\frac{l}{32}}$, where l is the length, in feet, of the pendulum. In 1851, Jean-Bernard-Leon Foucault demonstrated the axial rotation of Earth using a large pendulum that he hung in the Pantheon in Paris. The period of Foucault's pendulum was approximately 16.5 seconds. What was its length?

Discussion and Writing

132. Make up a radical equation that has no solution.

133. Make up a radical equation that has an extraneous solution.

134. Discuss the step in the solving process for radical equations that leads to the possibility of extraneous solutions. Why is there no such possibility for linear and quadratic equations?

135. The equation $|x| = -2$ has no real solution. Why?

136. What Went Wrong? On an exam, Jane solved the equation $\sqrt{2x+3} - x = 0$ and wrote that the solution set was $\{-1, 3\}$. Jane received 3 out of 5 points for the problem. Jane asks you why she received 3 out of 5 points. Provide an explanation.

'Are You Prepared?' Answers

1. True **2.** x **3.** $(2x+1)(x-4)$ **4.** $(x-3)(x+3)(x+4)$

5. The distance from the origin to -4 on a real number line is 4 units.

1.6 Problem Solving: Interest, Mixture, Uniform Motion, Constant Rate Jobs

OBJECTIVES

Applied (word) problems do not come in the form "Solve the equation. . . . " Instead, they supply information using words, a verbal description of the real problem. So, to solve applied problems, we must be able to translate the verbal description into the language of mathematics. We do this by using variables to represent unknown quantities and then finding relationships (such as equations) that involve these variables. The process of doing all this is called **mathematical modeling.** The equation or inequality that represents the relationship among the variables is called a **model.**

Any solution to the mathematical problem must be checked against the mathematical problem, the verbal description, and the real problem. See Figure 50 for an illustration of the **modeling process.**

Figure 50

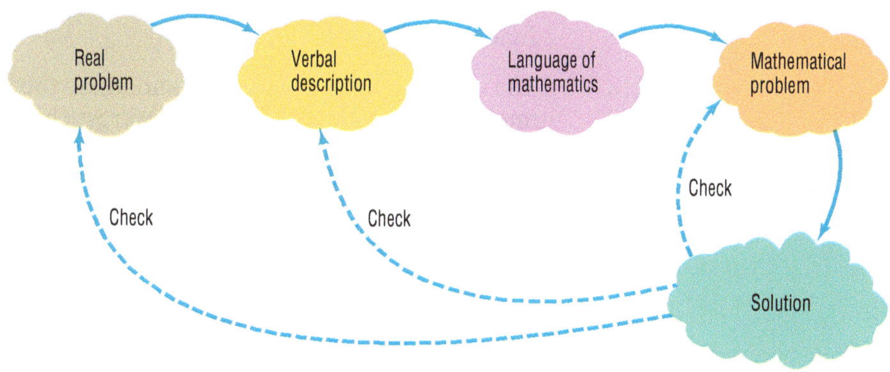

1 Translate Verbal Descriptions into Mathematical Expressions

Let's look at a few examples that will help you to translate certain words into mathematical symbols.

EXAMPLE 1 **Translating Verbal Descriptions into Mathematical Expressions**

(a) For uniform motion, the speed (rate) of an object equals the distance traveled divided by the time required.

Translation: If r is the speed (or rate), d the distance, and t the time, then $r = \frac{d}{t}$.

(b) Let x denote a number.

The number 5 times as large as x is $5x$.
The number 3 less than x is $x - 3$.
The number that exceeds x by 4 is $x + 4$.
The number that, when added to x, gives 5 is $5 - x$.

Now Work PROBLEM 7

Always check the units used to measure the variables of an applied problem. In Example 1(a), if r is measured in miles per hour, then the distance d must be expressed in miles and the time t must be expressed in hours. It is a good practice to check units to be sure that they are consistent and make sense.

2 Solve Interest Problems

Interest is money paid for the use of money. The total amount borrowed (whether by an individual from a bank in the form of a loan or by a bank from an individual in the form of a savings account) is called the **principal.** The **rate of interest,** expressed as a percent, is the amount charged for the use of the principal for a given period of time, usually on a yearly (that is, per annum) basis.

Simple Interest Formula

If a principal of P dollars is borrowed for a period of t years at a per annum interest rate r, expressed as a decimal, the interest I charged is

$$I = Prt \qquad (1)$$

Interest charged according to formula (1) is called **simple interest**. When using formula (1), be sure to express r as a decimal.

EXAMPLE 2 Using the Simple Interest Formula: Computing Interest on a Loan

Suppose that Juanita borrows \$500 for 6 months at the simple interest rate of 9% per annum. What is the interest that Juanita will be charged on this loan? How much does Juanita owe after 6 months?

Solution The rate of interest is given per annum, so the actual time that the money is borrowed must be expressed in years. The interest charged would be the principal, \$500, times the rate of interest ($9\% = 0.09$) times the time in years, $\frac{1}{2}$:

$$\text{Interest charged} = I = Prt = (500)(0.09)\left(\frac{1}{2}\right) = \$22.50$$

After 6 months, Juanita will owe what she borrowed plus the interest: $\$500 + \$22.50 = \$522.50$.

In the next example we use the steps for solving applied problems, given on page 108.

EXAMPLE 3 Financial Planning

Candy has \$70,000 to invest and requires an overall rate of return of 9%. She can invest in a safe, government-insured certificate of deposit, but it only pays 8%. To obtain 9%, she agrees to invest some of her money in noninsured corporate bonds paying 12%. How much should be placed in each investment to achieve her goal?

Solution

STEP 1: The question is asking for two dollar amounts: the principal to invest in the corporate bonds and the principal to invest in the certificate of deposit.

STEP 2: We let b represent the amount (in dollars) to be invested in the bonds. Then $70{,}000 - b$ is the amount that will be invested in the certificate. (Do you see why?)

NOTE Don't forget to choose variables that remind you of what they represent, as in b for bonds.

STEP 3: We set up a table:

	Principal ($)	Rate	Time (yr)	Interest ($)
Bonds	b	12% = 0.12	1	$0.12b$
Certificate	$70{,}000 - b$	8% = 0.08	1	$0.08(70{,}000 - b)$
Total	70,000	9% = 0.09	1	$0.09(70{,}000) = 6300$

Since the total interest from the investments is equal to $0.09(70{,}000) = 6300$, we have the equation

$$0.12b + 0.08(70{,}000 - b) = 6300 \quad \text{The Model}$$

(Note that the units are consistent: the unit is dollars on each side.)

STEP 4:

$$\begin{aligned} 0.12b + 5600 - 0.08b &= 6300 \\ 0.04b &= 700 \\ b &= 17{,}500 \end{aligned}$$

Candy should place \$17,500 in the bonds and \$70,000 − \$17,500 = \$52,500 in the certificate.

STEP 5: The interest on the bonds after 1 year is 0.12(\$17,500) = \$2100; the interest on the certificate after 1 year is 0.08(\$52,500) = \$4200. The total annual interest is \$6300, the required amount.

Now Work PROBLEM 17

3 Solve Mixture Problems

Oil refineries sometimes produce gasoline that is a blend of two or more types of fuel; bakeries occasionally blend two or more types of flour for their bread. These problems are referred to as **mixture problems** because they combine two or more quantities to form a mixture.

EXAMPLE 4 Blending Coffees

The manager of a Starbucks store decides to experiment with a new blend of coffee. She will mix some B grade Colombian coffee that sells for \$5 per pound with some A grade Arabica coffee that sells for \$10 per pound to get 100 pounds of the new blend. The selling price of the new blend is to be \$7 per pound, and there is to be no difference in revenue from selling the new blend versus selling the other types. How many pounds of the B grade Colombian and A grade Arabica coffees are required?

Solution Let c represent the number of pounds of the B grade Colombian coffee. Then $100 - c$ equals the number of pounds of the A grade Arabica coffee. See Figure 51.

Figure 51

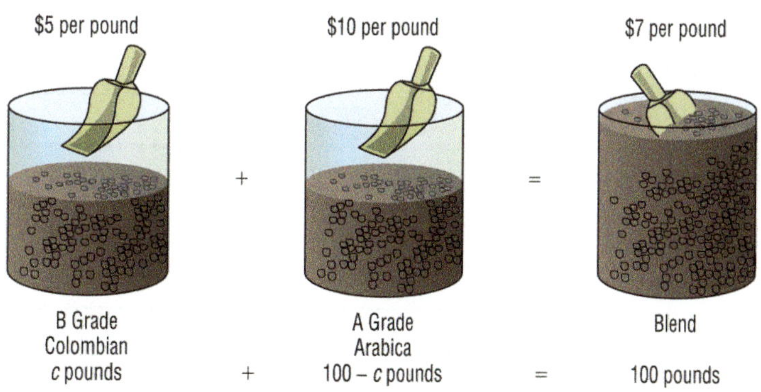

Since there is to be no difference in revenue between selling the A and B grades separately versus the blend, we have

$$\begin{Bmatrix}\text{Price per pound}\\ \text{of B grade}\end{Bmatrix}\begin{Bmatrix}\text{No. pounds}\\ \text{B grade}\end{Bmatrix} + \begin{Bmatrix}\text{Price per pound}\\ \text{of A grade}\end{Bmatrix}\begin{Bmatrix}\text{No. pounds}\\ \text{A grade}\end{Bmatrix} = \begin{Bmatrix}\text{Price per pound}\\ \text{of blend}\end{Bmatrix}\begin{Bmatrix}\text{No. pounds}\\ \text{blend}\end{Bmatrix}$$

$$\$5 \cdot c + \$10 \cdot (100 - c) = \$7 \cdot 100 \quad \text{The Model}$$

We have the equation

$$\begin{aligned} 5c + 10(100 - c) &= 7 \cdot (100) \\ 5c + 1000 - 10c &= 700 \\ -5c &= -300 \\ c &= 60 \end{aligned}$$

The manager should blend 60 pounds of B grade Colombian coffee with $100 - 60 = 40$ pounds of A grade Arabica coffee to get the desired blend.

Check: The 60 pounds of B grade coffee would sell for $(\$5)(60) = \$300$, and the 40 pounds of A grade coffee would sell for $(\$10)(40) = \$400$; the total revenue, \$700, equals the revenue obtained from selling the blend, as desired.

Now Work PROBLEM 21

4 Solve Uniform Motion Problems

Objects that move at a constant speed are said to be in **uniform motion.** When the average speed of an object is known, it can be interpreted as its constant speed. For example, a bicyclist traveling at an average speed of 25 miles per hour is in uniform motion.

Uniform Motion Formula

If an object moves at an average speed (or rate) r, the distance d covered in time t is given by the formula

$$d = r \cdot t \tag{2}$$

That is, Distance = Rate · Time.

EXAMPLE 5 Physics: Uniform Motion

Tanya, who is a long-distance runner, runs at an average speed of 8 miles per hour (mi/hr). Two hours after Tanya leaves your house, you leave in your Honda and follow the same route. If your average speed is 40 mi/hr, how long will it be before you catch up to Tanya? How far will each of you be from your home?

Solution Refer to Figure 52. We use t to represent the time (in hours) that it takes the Honda to catch up to Tanya. When this occurs, the total time elapsed for Tanya is $t + 2$ hours.

Figure 52

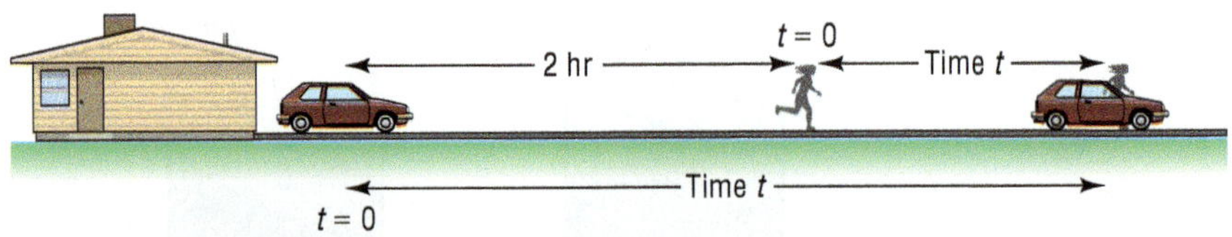

Set up the following table:

	Rate mi/hr	Time hr	Distance mi
Tanya	8	$t + 2$	$8(t + 2)$
Honda	40	t	$40t$

Since the distance traveled is the same, we are led to the following equation:

$$8(t + 2) = 40t \quad \text{The Model}$$
$$8t + 16 = 40t$$
$$32t = 16$$
$$t = \frac{1}{2} \text{ hour}$$

It will take the Honda $\frac{1}{2}$ hour to catch up to Tanya. Each will have gone 20 miles.

Check: In 2.5 hours, Tanya travels a distance of $(2.5)(8) = 20$ miles. In $\frac{1}{2}$ hour, the Honda travels a distance of $\left(\frac{1}{2}\right)(40) = 20$ miles.

EXAMPLE 6 Physics: Uniform Motion

A motorboat heads upstream a distance of 24 miles on a river whose current is running at 3 miles per hour (mi/hr). The trip up and back takes 6 hours. Assuming that the motorboat maintained a constant speed relative to the water, what was its speed?

Solution See Figure 53. We use r to represent the constant speed of the motorboat relative to the water. Then the true speed going upstream is $r - 3$ mi/hr, and the true speed going downstream is $r + 3$ mi/hr. Since Distance = Rate × Time, then Time $= \frac{\text{Distance}}{\text{Rate}}$. We set up a table.

Figure 53

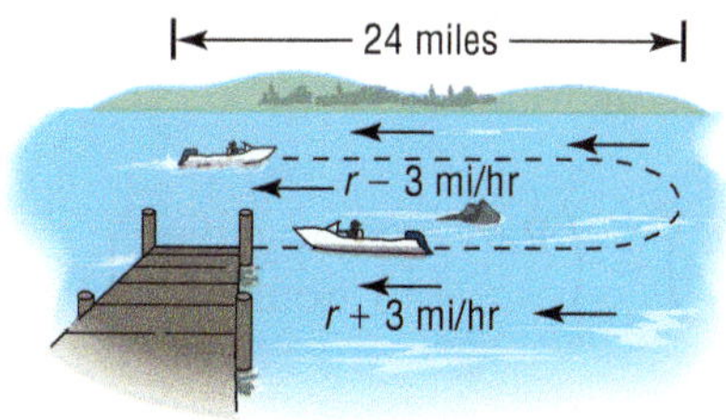

	Rate mi/hr	Distance mi	Time = $\frac{\text{Distance}}{\text{Velocity}}$ hr
Upstream	$r - 3$	24	$\frac{24}{r-3}$
Downstream	$r + 3$	24	$\frac{24}{r+3}$

Since the total time up and back is 6 hours, we have

$$\frac{24}{r-3} + \frac{24}{r+3} = 6 \quad \text{The Model}$$

$$\frac{24(r+3) + 24(r-3)}{(r-3)(r+3)} = 6 \quad \text{Add the quotients on the left.}$$

$$\frac{48r}{r^2 - 9} = 6 \quad \text{Simplify.}$$

$$48r = 6(r^2 - 9)$$

$$6r^2 - 48r - 54 = 0 \quad \text{Place in standard form.}$$

$$r^2 - 8r - 9 = 0 \quad \text{Divide by 6.}$$

$$(r - 9)(r + 1) = 0 \quad \text{Factor.}$$

$$r = 9 \quad \text{or} \quad r = -1 \quad \text{Apply the Zero-Product Property and solve.}$$

We discard the solution $r = -1$ mi/hr, so the speed of the motorboat relative to the water is 9 mi/hr.

Now Work PROBLEM 27

5 Solve Constant Rate Job Problems

This section involves jobs that are performed at a **constant rate.** Our assumption is that, if a job can be done in t units of time, $\frac{1}{t}$ of the job is done in 1 unit of time. So, if a job takes 4 hours, then $\frac{1}{4}$ of the job is done in 1 hour.

EXAMPLE 7 Working Together to Do a Job

At 10 AM Danny is asked by his father to weed the garden. From past experience, Danny knows that this will take him 4 hours, working alone. His older brother, Mike, when it is his turn to do this job, requires 6 hours. Since Mike wants to go golfing with Danny and has a reservation for 1 PM, he agrees to help Danny. Assuming no gain or loss of efficiency, when will they finish if they work together? Can they make the tee time?

Solution We set up Table 4. If it takes Danny 4 hours to do the entire job, and he works at a constant rate, then in 1 hour, Danny does $\frac{1}{4}$ of the job. In 1 hour, Mike does $\frac{1}{6}$ of the job. Let t be the time (in hours) that it takes them to do the job together. In 1 hour, then, $\frac{1}{t}$ of the job is completed. We reason as follows:

Table 4

	Hours to Do Job	Part of Job Done in 1 Hour
Danny	4	$\frac{1}{4}$
Mike	6	$\frac{1}{6}$
Together	t	$\frac{1}{t}$

$$\begin{pmatrix}\text{Part done by Danny}\\ \text{in 1 hour}\end{pmatrix} + \begin{pmatrix}\text{Part done by Mike}\\ \text{in 1 hour}\end{pmatrix} = \begin{pmatrix}\text{Part done together}\\ \text{in 1 hour}\end{pmatrix}$$

From Table 4,

$$\frac{1}{4} + \frac{1}{6} = \frac{1}{t} \qquad \text{The Model}$$

$$\frac{3}{12} + \frac{2}{12} = \frac{1}{t} \qquad \text{LCD} = 12$$

$$\frac{5}{12} = \frac{1}{t}$$

$$5t = 12 \qquad \text{Multiply both sides by } 12t.$$

$$t = \frac{12}{5} \qquad \text{Divide both sides by } 5.$$

Working together, the job can be done in $\frac{12}{5}$ hours, or 2 hours, 24 minutes. They should make the tee time, since they will finish at 12:24 PM.

Now Work PROBLEM 29

1.6 Assess Your Understanding

Concepts and Vocabulary

1. Using variables to represent unknown quantities and then finding relationships that involve these variables is referred to as _____ _____.
2. The money paid for the use of money is _____.
3. Objects that move at a constant speed are said to be in _____ _____.
4. ***True or False*** The amount charged for the use of principal for a given period of time is called the rate of interest.
5. ***True or False*** If an object moves at an average rate r, the distance d covered in time t is given by the formula $d = r \cdot t$.
6. Suppose that you want to mix two coffees to obtain 100 pounds of the blend. If x represents the number of pounds of coffee A, write an algebraic expression that represents the number of pounds of coffee B.

Applications and Extensions

In Problems 7–16, translate each sentence into a mathematical equation. Be sure to identify the meaning of all symbols.

7. **Geometry** The area of a circle is the product of the number π and the square of the radius.

8. **Geometry** The circumference of a circle is the product of the number π and twice the radius.

9. **Geometry** The area of a square is the square of the length of a side.

10. **Geometry** The perimeter of a square is four times the length of a side.

11. **Physics** Force equals the product of mass and acceleration.

12. **Physics** Pressure is force per unit area.

13. **Physics** Work equals force times distance.

14. **Physics** Kinetic energy is one-half the product of the mass and the square of the velocity.

15. **Business** The total variable cost of manufacturing x dishwashers is \$150 per dishwasher times the number of dishwashers manufactured.

16. **Business** The total revenue derived from selling x dishwashers is \$250 per dishwasher times the number of dishwashers sold.

17. **Financial Planning** Betsy, a recent retiree, requires \$6000 per year in extra income. She has \$84,000 to invest and can invest in B-rated bonds paying 8% per year or in a certificate of deposit (CD) paying 6% per year. How much money should be invested in each to realize exactly \$6000 in interest per year?

18. **Financial Planning** After 2 years, Betsy (see Problem 17) finds that she will now require \$6400 per year. Assuming that the remaining information is the same, how should the money be reinvested?

19. **Banking** A bank loaned out \$120,000, part of it at the rate of 8% per year and the rest at the rate of 12% per year. If the interest received totaled \$11,000, how much was loaned at 8%?

20. **Banking** Wendy, a loan officer at a bank, has \$1,000,000 to lend and is required to obtain an average return of 8% per year. If she can lend at the rate of 7% or at the rate of 11%, how much can she lend at the 7% rate and still meet her requirement?

21. **Blending Teas** The manager of a store that specializes in selling tea decides to experiment with a new blend. She will mix some Earl Grey tea that sells for \$5 per pound with some Orange Pekoe tea that sells for \$3 per pound to get 100 pounds of the new blend. The selling price of the new blend is to be \$4.50 per pound, and there is to be no difference in revenue from selling the new blend versus selling the other types. How many pounds of the Earl Grey tea and Orange Pekoe tea are required?

22. **Business: Blending Coffee** A coffee manufacturer wants to market a new blend of coffee that sells for \$3.90 per pound by mixing two coffees that sell for \$2.75 and \$5 per pound, respectively. What amounts of each coffee should be blended to obtain the desired mixture?

    [**Hint:** Assume that the total weight of the desired blend is 100 pounds.]

23. **Business: Mixing Nuts** A nut store normally sells cashews for \$9.00 per pound and almonds for \$3.50 per pound. But at the end of the month the almonds had not sold well, so, in order to sell 60 pounds of almonds, the manager decided to mix the 60 pounds of almonds with some cashews and sell the mixture for \$7.50 per pound. How many pounds of cashews should be mixed with the almonds to ensure no change in the profit?

24. **Business: Mixing Candy** A candy store sells boxes of candy containing caramels and cremes. Each box sells for \$12.50 and holds 30 pieces of candy (all pieces are the same size). If the caramels cost \$0.25 to produce and the cremes cost \$0.45 to produce, how many of each should be in a box to make a profit of \$3?

25. **Physics: Uniform Motion** A motorboat can maintain a constant speed of 16 miles per hour relative to the water. The boat makes a trip upstream to a certain point in 20 minutes; the return trip takes 15 minutes. What is the speed of the current? (See the figure.)

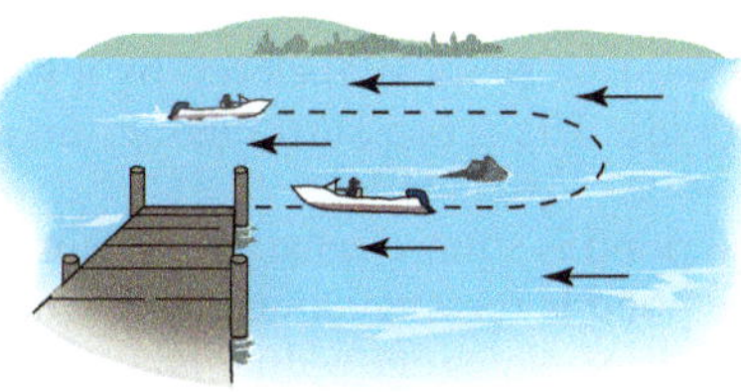

26. **Physics: Uniform Motion** A motorboat heads upstream on a river that has a current of 3 miles per hour. The trip upstream takes 5 hours, and the return trip takes 2.5 hours. What is the speed of the motorboat? (Assume that the motorboat maintains a constant speed relative to the water.)

27. **Physics: Uniform Motion** A motorboat maintained a constant speed of 15 miles per hour relative to the water in going 10 miles upstream and then returning. The total time for the trip was 1.5 hours. Use this information to find the speed of the current.

28. **Physics: Uniform Motion** Two cars enter the Florida Turnpike at Commercial Boulevard at 8:00 AM, each heading for Wildwood. One car's average speed is 10 miles per hour more than the other's. The faster car arrives at Wildwood at 11:00 AM, $\frac{1}{2}$ hour before the other car. What is the average speed of each car? How far did each travel?

29. **Working Together on a Job** Trent can deliver his newspapers in 30 minutes. It takes Lois 20 minutes to do the same route. How long will it take them to deliver the newspapers if they work together?

30. **Working Together on a Job** Patrice, by himself, can paint four rooms in 10 hours. If he hires April to help, they can do the same job together in 6 hours. If he lets April work alone, how long will it take her to paint four rooms?

Mixed Practice

31. Enclosing a Garden A gardener has 46 feet of fencing to be used to enclose a rectangular garden that has a border 2 feet wide surrounding it (see the figure).

(a) If the length of the garden is to be twice its width, what are the dimensions of the garden?
(b) What is the area of the garden?
(c) If the length and width of the garden are to be the same, what are the dimensions of the garden?
(d) What is the area of the square garden?

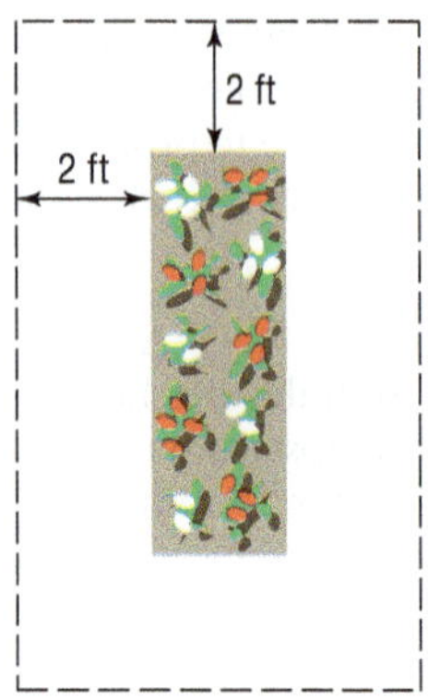

32. Construction A pond is enclosed by a wooden deck that is 3 feet wide. The fence surrounding the deck is 100 feet long.

(a) If the pond is square, what are its dimensions?
(b) If the pond is rectangular and the length of the pond is to be three times its width, what are its dimensions?
(c) If the pond is circular, what is its diameter?
(d) Which pond has the most area?

33. Football A tight end can run the 100-yard dash in 12 seconds. A defensive back can do it in 10 seconds. The tight end catches a pass at his own 20-yard line with the defensive back at the 15-yard line. (See the figure.) If no other players are nearby, at what yard line will the defensive back catch up to the tight end?

[**Hint:** At time $t = 0$, the defensive back is 5 yards behind the tight end.]

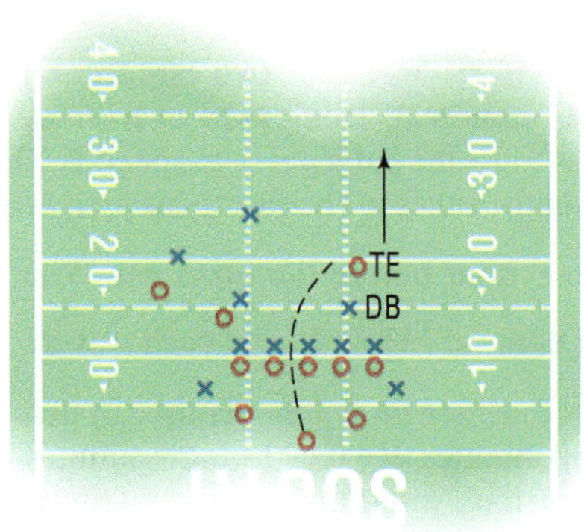

34. Computing Business Expense Therese, an outside salesperson, uses her car for both business and pleasure. Last year she traveled 30,000 miles, using 900 gallons of gasoline. Her car gets 40 miles per gallon on the highway and 25 in the city. She can deduct all highway travel, but no city travel, on her taxes. How many miles should Therese be allowed as a business expense?

35. Mixing Water and Antifreeze How much water should be added to 1 gallon of pure antifreeze to obtain a solution that is 60% antifreeze?

36. Mixing Water and Antifreeze The cooling system of a certain foreign-made car has a capacity of 15 liters. If the system is filled with a mixture that is 40% antifreeze, how much of this mixture should be drained and replaced by pure antifreeze so that the system is filled with a solution that is 60% antifreeze?

37. Moving Walkways The speed of a moving walkway is typically about 2.5 feet per second. Walking on such a moving walkway, it takes Karen a total of 40 seconds to travel 50 feet with the movement of the walkway and then back again against the movement of the walkway. What is Karen's normal walking speed?

Source: Answers.com

38. Moving Walkways The Gare Montparnasse train station in Paris has a high-speed version of a moving walkway. If Jean Claude walks while riding this moving walkway, he can travel 200 meters in 30 seconds less time than if he stands still on the moving walkway. If he walks at a normal rate of 1.5 meters per second, what is the speed of the Gare Montparnasse walkway?

Source: Answers.com

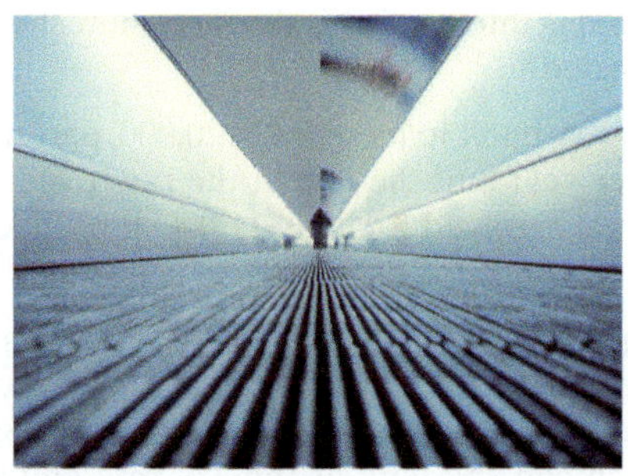

39. Laser Printers It takes an HP LaserJet 1300 laser printer 10 minutes longer to complete a 600-page print job by itself than it takes an HP LaserJet 2420 to complete the same job by itself. Together the two printers can complete the job in 12 minutes. How long does it take each printer to complete the print job alone? What is the speed of each printer?

Source: Hewlett-Packard

40. Tennis Anyone? A regulation doubles tennis court has an area of 2808 square feet. If it is 6 feet longer than twice its width, determine the dimensions of the court.

Source: United States Tennis Association

41. Chemistry: Salt Solutions How much water must be evaporated from 32 ounces of a 4% salt solution to make a 6% salt solution?

42. Chemistry: Salt Solutions How much water must be evaporated from 240 gallons of a 3% salt solution to produce a 5% salt solution?

43. Purity of Gold The purity of gold is measured in karats, with pure gold being 24 karats. Other purities of gold are expressed as proportional parts of pure gold. Thus, 18-karat gold is $\frac{18}{24}$, or 75% pure gold; 12-karat gold is $\frac{12}{24}$, or 50% pure gold; and so on. How much 12-karat gold should be mixed with pure gold to obtain 60 grams of 16-karat gold?

44. Chemistry: Sugar Molecules A sugar molecule has twice as many atoms of hydrogen as it does oxygen and one more atom of carbon than oxygen. If a sugar molecule has a total of 45 atoms, how many are oxygen? How many are hydrogen?

45. Running a Race Mike can run the mile in 6 minutes, and Dan can run the mile in 9 minutes. If Mike gives Dan a head start of 1 minute, how far from the start will Mike pass Dan? How long does it take?

46. Range of an Airplane An air rescue plane averages 300 miles per hour in still air. It carries enough fuel for 5 hours of flying time. If, upon takeoff, it encounters a head wind of 30 mi/hr, how far can it fly and return safely? (Assume that the wind remains constant.)

47. Emptying Oil Tankers An oil tanker can be emptied by the main pump in 4 hours. An auxiliary pump can empty the tanker in 9 hours. If the main pump is started at 9 AM, when should the auxiliary pump be started so that the tanker is emptied by noon?

48. Cement A 20-pound bag of Economy brand cement mix contains 25% cement and 75% sand. How much pure cement must be added to produce a cement mix that is 40% cement?

49. Emptying a Tub A bathroom tub fills in 15 minutes with both faucets open and the stopper in place. With both faucets closed and the stopper removed, the tub empties in 20 minutes. How long will it take for the tub to fill if both faucets are open and the stopper is removed?

50. Using Two Pumps A 5-horsepower (hp) pump can empty a pool in 5 hours. A smaller, 2-hp pump empties the same pool in 8 hours. The pumps are used together to begin emptying this pool. After 2 hours, the 2-hp pump breaks down. How long will it take the larger pump to empty the pool?

51. A Biathlon Suppose that you have entered an 87-mile biathlon that consists of a run and a bicycle race. During your run, your average speed is 6 miles per hour, and during your bicycle race, your average speed is 25 miles per hour. You finish the race in 5 hours. What is the distance of the run? What is the distance of the bicycle race?

52. Constructing a Coffee Can A 39-ounce can of Hills Bros.® coffee requires 188.5 square inches of aluminum. If its height is 7 inches, what is its radius? (The surface area S of a right cylinder is $S = 2\pi r^2 + 2\pi rh$, where r is the radius and h is the height.)

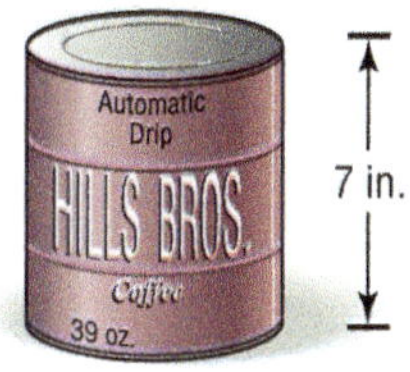

53. Comparing Olympic Heroes In the 1984 Olympics, Carl Lewis of the United States won the gold medal in the 100-meter race with a time of 9.99 seconds. In the 1896 Olympics, Thomas Burke, also of the United States, won the gold medal in the 100-meter race in 12.0 seconds. If they ran in the same race repeating their respective times, by how many meters would Lewis beat Burke?

Discussion and Writing

54. Critical Thinking You are the manager of a clothing store and have just purchased 100 dress shirts for $20.00 each. After 1 month of selling the shirts at the regular price, you plan to have a sale giving 40% off the original selling price. However, you still want to make a profit of $4 on each shirt at the sale price. What should you price the shirts at initially to ensure this? If, instead of 40% off at the sale, you give 50% off, by how much is your profit reduced?

55. Critical Thinking Make up a word problem that requires solving a linear equation as part of its solution. Exchange problems with a friend. Write a critique of your friend's problem.

56. Critical Thinking Without solving, explain what is wrong with the following mixture problem: How many liters of 25% ethanol should be added to 20 liters of 48% ethanol to obtain a solution of 58% ethanol? Now go through an algebraic solution. What happens?

57. Computing Average Speed In going from Chicago to Atlanta, a car averages 45 miles per hour, and in going from Atlanta to Miami, it averages 55 miles per hour. If Atlanta is halfway between Chicago and Miami, what is the average speed from Chicago to Miami? Discuss an intuitive solution. Write a paragraph defending your intuitive solution. Then solve the problem algebraically. Is your intuitive solution the same as the algebraic one? If not, find the flaw.

58. Speed of a Plane On a recent flight from Phoenix to Kansas City, a distance of 919 nautical miles, the plane arrived 20 minutes early. On leaving the aircraft, I asked the captain, "What was our tail wind?" He replied, "I don't know, but our ground speed was 550 knots." How can you determine if enough information is provided to find the tail wind? If possible, find the tail wind. (1 knot = 1 nautical mile per hour)

1.7 Solving Inequalities

PREPARING FOR THIS SECTION *Before getting started, review the following:*

- Inequalities (Section R.2, pp. 18–19)
- Absolute Value (Section R.2, pp. 19–20)
- Union of Sets (Section R.1, pp. 2–3)

Now Work the 'Are You Prepared?' problems on page 158.

OBJECTIVES
1 Use Interval Notation (p. 150)
2 Use Properties of Inequalities (p. 151)
3 Solve Linear Inequalities Algebraically and Graphically (p. 153)
4 Solve Combined Inequalities Algebraically and Graphically (p. 155)
5 Solve Absolute Value Inequalities Algebraically and Graphically (p. 155)

Suppose that a and b are two real numbers and $a < b$. We shall use the notation $a < x < b$ to mean that x is a number *between* a and b. So, the expression $a < x < b$ is equivalent to the two inequalities $a < x$ and $x < b$. Similarly, the expression $a \le x \le b$ is equivalent to the two inequalities $a \le x$ and $x \le b$. The remaining two possibilities, $a \le x < b$ and $a < x \le b$, are defined similarly.

Although it is acceptable to write $3 \ge x \ge 2$, it is preferable to reverse the inequality symbols and write instead $2 \le x \le 3$ so that, as you read from left to right, the values go from smaller to larger.

A statement such as $2 \le x \le 1$ is false because there is no number x for which $2 \le x$ and $x \le 1$. Finally, we never mix inequality symbols, as in $2 \le x \ge 3$.

1 Use Interval Notation

Let a and b represent two real numbers with $a < b$:

> **In Words**
> The notation $[a, b]$ represents all real numbers between a and b, inclusive. The notation (a, b) represents all real numbers between a and b, not including either a or b.

A **closed interval,** denoted by $[a, b]$, consists of all real numbers x for which $a \le x \le b$.
An **open interval,** denoted by (a, b), consists of all real numbers x for which $a < x < b$.
The **half-open,** or **half-closed, intervals** are $(a, b]$, consisting of all real numbers x for which $a < x \le b$, and $[a, b)$, consisting of all real numbers x for which $a \le x < b$.

In each of these definitions, a is called the **left endpoint** and b the **right endpoint** of the interval.

The symbol ∞ (read as "infinity") is not a real number but a notational device used to indicate unboundedness in the positive direction. The symbol $-\infty$ (read as "minus infinity" or "negative infinity") also is not a real number, but a notational device used to indicate unboundedness in the negative direction. Using the symbols ∞ and $-\infty$, we can define five other kinds of intervals:

$[a, \infty)$	consists of all real numbers x for which $x \ge a$
(a, ∞)	consists of all real numbers x for which $x > a$
$(-\infty, a]$	consists of all real numbers x for which $x \le a$
$(-\infty, a)$	consists of all real numbers x for which $x < a$
$(-\infty, \infty)$	consists of all real numbers x

Note that ∞ and $-\infty$ are never included as endpoints since they are not real numbers.

Table 5 summarizes interval notation, corresponding inequality notation, and their graphs.

Table 5

Interval	Inequality	Graph
The open interval (a, b)	$a < x < b$	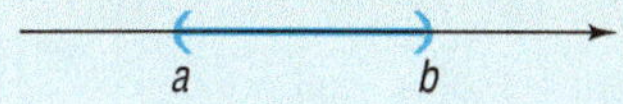
The closed interval $[a, b]$	$a \le x \le b$	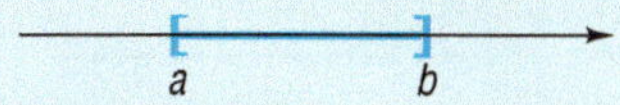
The half-open interval $[a, b)$	$a \le x < b$	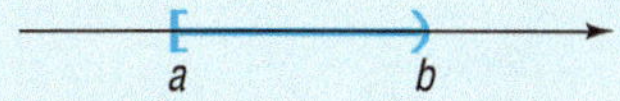
The half-open interval $(a, b]$	$a < x \le b$	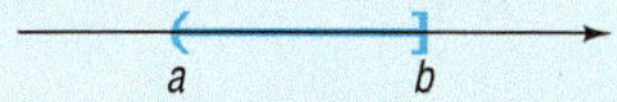
The interval $[a, \infty)$	$x \ge a$	
The interval (a, ∞)	$x > a$	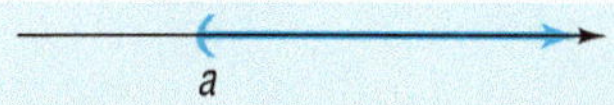
The interval $(-\infty, a]$	$x \le a$	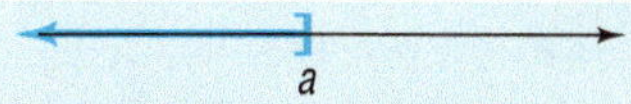
The interval $(-\infty, a)$	$x < a$	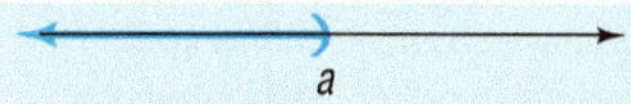
The interval $(-\infty, \infty)$	All real numbers	

EXAMPLE 1 Writing Inequalities Using Interval Notation

Write each inequality using interval notation.

(a) $1 \le x \le 3$
(b) $-4 < x < 0$
(c) $x > 5$
(d) $x \le 1$

Solution

(a) $1 \le x \le 3$ represents all numbers x between 1 and 3, inclusive. In interval notation, we write $[1, 3]$.
(b) In interval notation, $-4 < x < 0$ is written $(-4, 0)$.
(c) $x > 5$ consists of all numbers x greater than 5. In interval notation, we write $(5, \infty)$.
(d) In interval notation, $x \le 1$ is written $(-\infty, 1]$.

EXAMPLE 2 Writing Intervals Using Inequality Notation

Write each interval as an inequality involving x.

(a) $[1, 4)$ (b) $(2, \infty)$ (c) $[2, 3]$ (d) $(-\infty, -3]$

Solution

(a) $[1, 4)$ consists of all numbers x for which $1 \le x < 4$.
(b) $(2, \infty)$ consists of all numbers x for which $x > 2$.
(c) $[2, 3]$ consists of all numbers x for which $2 \le x \le 3$.
(d) $(-\infty, -3]$ consists of all numbers x for which $x \le -3$.

Now Work PROBLEMS 11, 21, AND 29

2 Use Properties of Inequalities

The product of two positive real numbers is positive, the product of two negative real numbers is positive, and the product of 0 and 0 is 0. For any real number a, the value of a^2 is 0 or positive; that is, a^2 is nonnegative. This is called the **nonnegative property.**

In Words
The square of a real number is never negative.

Nonnegative Property

For any real number a,

$$a^2 \geq 0 \quad (1)$$

If we add the same number to both sides of an inequality, we obtain an equivalent inequality. For example, since $3 < 5$, then $3 + 4 < 5 + 4$ or $7 < 9$. This is called the **addition property** of inequalities.

In Words
The addition property states that the sense, or direction, of an inequality remains unchanged if the same number is added to each side.

Addition Property of Inequalities

For real numbers a, b, and c:

$$\text{If } a < b, \text{ then } a + c < b + c. \quad \textbf{(2a)}$$

$$\text{If } a > b, \text{ then } a + c > b + c. \quad \textbf{(2b)}$$

Figure 54 illustrates the addition property (2a). In Figure 54(a), we see that a lies to the left of b. If c is positive, then $a + c$ and $b + c$ each lie c units to the right of a and b, respectively. Consequently, $a + c$ must lie to the left of $b + c$; that is, $a + c < b + c$. Figure 54(b) illustrates the situation if c is negative.

Figure 54

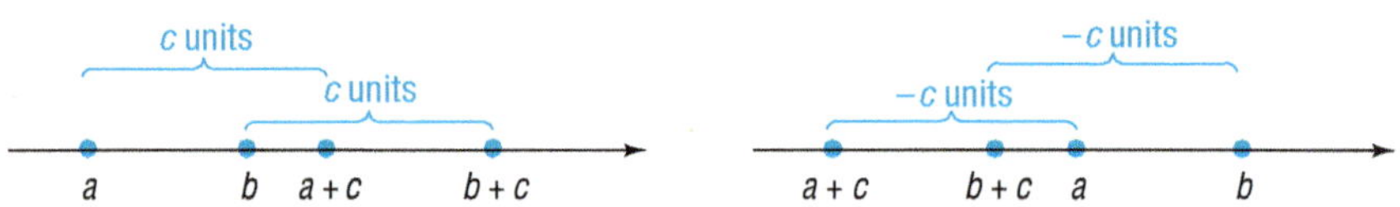

(a) If $a < b$ and $c > 0$, then $a + c < b + c$.
(b) If $a < b$ and $c < 0$, then $a + c < b + c$.

Draw an illustration similar to Figure 54 that illustrates the addition property (2b).

EXAMPLE 3 Addition Property of Inequalities

(a) If $x < -5$, then $x + 5 < -5 + 5$ or $x + 5 < 0$.
(b) If $x > 2$, then $x + (-2) > 2 + (-2)$ or $x - 2 > 0$.

Now Work PROBLEM 37

We will use two examples to arrive at our next property.

EXAMPLE 4 Multiplying an Inequality by a Positive Number

Express as an inequality the result of multiplying each side of the inequality $3 < 7$ by 2.

Solution We begin with

$$3 < 7$$

Multiplying each side by 2 yields the numbers 6 and 14, so we have

$$6 < 14$$

EXAMPLE 5 Multiplying an Inequality by a Negative Number

Express as an inequality the result of multiplying each side of the inequality $9 > 2$ by -4.

Solution We begin with

$$9 > 2$$

Multiplying each side by -4 yields the numbers -36 and -8, so we have

$$-36 < -8$$

Note that the effect of multiplying both sides of $9 > 2$ by the negative number -4 is that the direction of the inequality symbol is reversed.

Examples 4 and 5 illustrate the following general **multiplication properties** for inequalities:

In Words
The multiplication properties state that the sense, or direction, of an inequality remains the same if each side is multiplied by a positive real number, whereas the direction is reversed if each side is multiplied by a negative real number.

Multiplication Properties for Inequalities

For real numbers a, b, and c:

If $a < b$ and if $c > 0$, then $ac < bc$.
If $a < b$ and if $c < 0$, then $ac > bc$. **(3a)**

If $a > b$ and if $c > 0$, then $ac > bc$.
If $a > b$ and if $c < 0$, then $ac < bc$. **(3b)**

EXAMPLE 6 Multiplication Property of Inequalities

(a) If $2x < 6$, then $\frac{1}{2}(2x) < \frac{1}{2}(6)$ or $x < 3$.

(b) If $\frac{x}{-3} > 12$, then $-3\left(\frac{x}{-3}\right) < -3(12)$ or $x < -36$.

(c) If $-4x > -8$, then $\frac{-4x}{-4} < \frac{-8}{-4}$ or $x < 2$.

Now Work PROBLEM 43

3 Solve Linear Inequalities Algebraically and Graphically

An **inequality in one variable** is a statement involving two expressions, at least one containing the variable, separated by one of the inequality symbols, $<$, $\leq$, $>$, or $\geq$. To **solve an inequality** means to find all values of the variable for which the statement is true. These values are called **solutions** of the inequality.

For example, the following are all inequalities involving one variable, x:

$$x + 5 < 8 \qquad 2x - 3 \geq 4 \qquad x^2 - 1 \leq 3 \qquad \frac{x+1}{x-2} > 0$$

As with equations, one method for solving an inequality is to replace it by a series of equivalent inequalities until an inequality with an obvious solution, such as $x < 3$, is obtained. We obtain equivalent inequalities by applying some of the same operations as those used to find equivalent equations. The addition property and the multiplication properties form the basis for the following procedures.

Procedures That Leave the Inequality Symbol Unchanged

1. Simplify both sides of the inequality by combining like terms and eliminating parentheses:

   Replace $x + 2 + 6 > 2x + 5(x + 1)$
   by $x + 8 > 7x + 5$

2. Add or subtract the same expression on both sides of the inequality:

   Replace $3x - 5 < 4$
   by $(3x - 5) + 5 < 4 + 5$

3. Multiply or divide both sides of the inequality by the same *positive* expression:

   Replace $4x > 16$ by $\dfrac{4x}{4} > \dfrac{16}{4}$

Procedures That Reverse the Sense or Direction of the Inequality Symbol

1. Interchange the two sides of the inequality:

   Replace $3 < x$ by $x > 3$

2. Multiply or divide both sides of the inequality by the same *negative* expression:

   Replace $-2x > 6$ by $\dfrac{-2x}{-2} < \dfrac{6}{-2}$

To solve an inequality using a graphing utility, we follow these steps:

Steps for Solving Inequalities Graphically

Step 1: Write the inequality in one of the following forms:

$$Y_1 < Y_2, \quad Y_1 > Y_2, \quad Y_1 \le Y_2, \quad Y_1 \ge Y_2$$

Step 2: Graph Y_1 and Y_2 on the same screen.

Step 3: If the inequality is of the form $Y_1 < Y_2$, determine on what interval Y_1 is below Y_2.

If the inequality is of the form $Y_1 > Y_2$, determine on what interval Y_1 is above Y_2.

If the inequality is not strict ($\le$ or $\ge$), include the x-coordinates of the points of intersection in the solution.

As the examples that follow illustrate, we solve linear inequalities using many of the same steps that we would use to solve linear equations. The ultimate goal is to get the variable on one side of the inequality, and the constant on the other. In writing the solution of an inequality, we may use either set notation or interval notation, whichever is more convenient.

EXAMPLE 7 Solving an Inequality Algebraically and Graphically

Solve the inequality $4x + 7 \ge 2x - 3$, and graph the solution set.

Algebraic Solution

$4x + 7 \ge 2x - 3$	
$4x + 7 - 7 \ge 2x - 3 - 7$	Subtract 7 from both sides.
$4x \ge 2x - 10$	Simplify.
$4x - 2x \ge 2x - 10 - 2x$	Subtract 2x from both sides.
$2x \ge -10$	Simplify.
$\dfrac{2x}{2} \ge \dfrac{-10}{2}$	Divide both sides by 2. (The direction of the inequality symbol is unchanged.)
$x \ge -5$	Simplify.

The solution set is $\{x \mid x \ge -5\}$ or, using interval notation, all numbers in the interval $[-5, \infty)$.

Graphing Solution

We graph $Y_1 = 4x + 7$ and $Y_2 = 2x - 3$ on the same screen. See Figure 55. Using the INTERSECT command, we find that Y_1 and Y_2 intersect at $x = -5$. The graph of Y_1 is above that of Y_2, $Y_1 > Y_2$, to the right of the point of intersection. Since the inequality is not strict, the solution set is $\{x \mid x \ge -5\}$ or, using interval notation, $[-5, \infty)$.

Figure 55

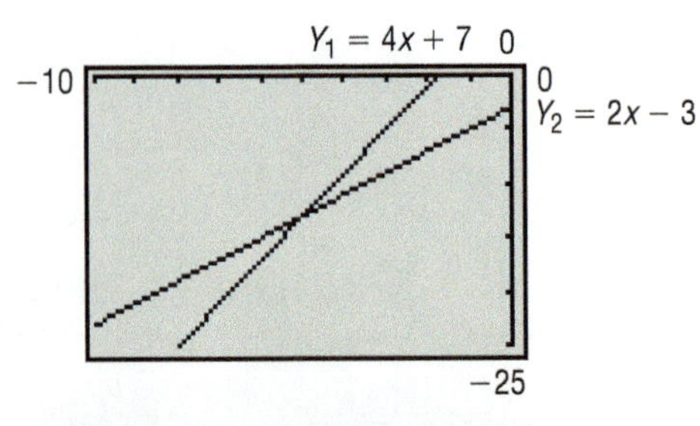

Figure 56

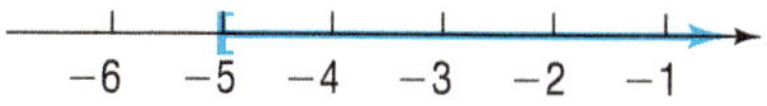

See Figure 56 for the graph of the solution set.

Now Work PROBLEM 49

4 Solve Combined Inequalities Algebraically and Graphically

Now let's look at how to solve combined inequalities.

EXAMPLE 8 Solving a Combined Inequality Algebraically and Graphically

Solve the inequality: $-1 \leq \dfrac{3-5x}{2} \leq 9$

Algebraic Solution

$$-1 \leq \frac{3-5x}{2} \leq 9$$

$$2(-1) \leq 2\left(\frac{3-5x}{2}\right) \leq 2(9)$$ Multiply each part by 2 to remove the denominator.

$$-2 \leq 3-5x \leq 18$$ Simplify.

$$-2-3 \leq 3-5x-3 \leq 18-3$$ Subtract 3 from each part to isolate the term containing x.

$$-5 \leq -5x \leq 15$$ Simplify.

$$\frac{-5}{-5} \geq \frac{-5x}{-5} \geq \frac{15}{-5}$$ Divide each part by −5 (reverse the sense of each inequality symbol).

$$1 \geq x \geq -3$$ Simplify.

$$-3 \leq x \leq 1$$ Reverse the order so the numbers get larger as you read from left to right.

The solution set is $\{x \mid -3 \leq x \leq 1\}$. In interval notation, the solution is $[-3, 1]$.

Graphing Solution

To solve a combined inequality graphically, we graph each part: $Y_1 = -1$, $Y_2 = \dfrac{3-5x}{2}$, $Y_3 = 9$. We seek the values of x for which the graph of Y_2 is between the graphs of Y_1 and Y_3. See Figure 57. The point of intersection of Y_1 and Y_2 is $(1, -1)$, and the point of intersection of Y_2 and Y_3 is $(-3, 9)$. The inequality is true for all values of x between these two intersection points. Since the inequality is nonstrict, the solution set is $\{x \mid -3 \leq x \leq 1\}$ or, using interval notation, $[-3, 1]$.

Figure 57

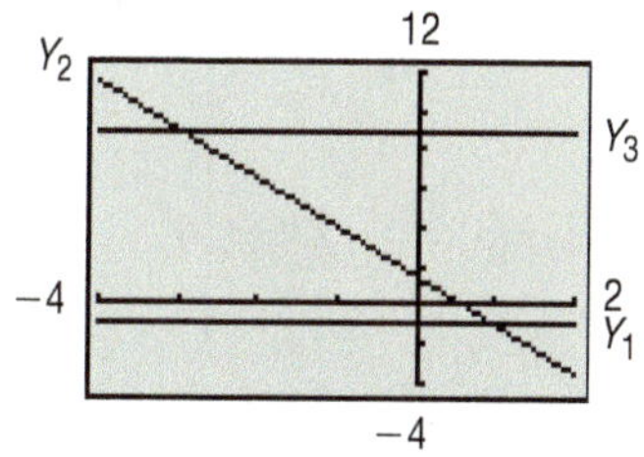

See Figure 58 for the graph of the solution set.

Figure 58

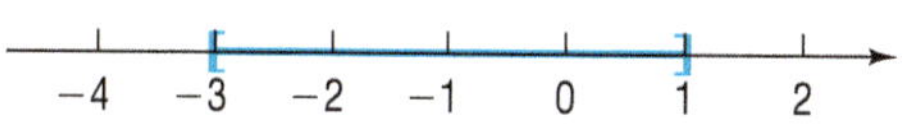

Now Work PROBLEM 65

5 Solve Absolute Value Inequalities Algebraically and Graphically

Let's look at an inequality involving absolute value.

EXAMPLE 9 Solving an Inequality Involving Absolute Value

Solve the inequality: $|x| < 4$

Algebraic Solution

We are looking for all points whose coordinate x is a distance less than 4 units from the origin. See Figure 59 for an illustration.

Figure 59

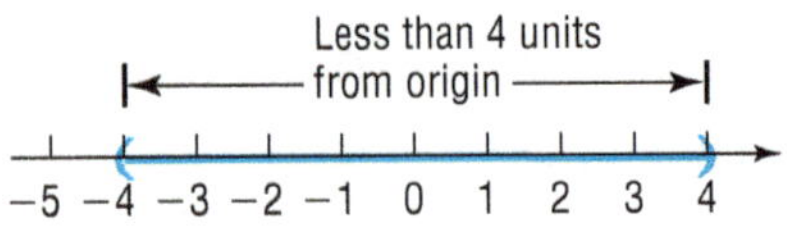

Because any x between -4 and 4 satisfies the condition $|x| < 4$, the solution set consists of all numbers x for which $-4 < x < 4$, that is, all x in the interval $(-4, 4)$.

Graphing Solution

We graph $Y_1 = |x|$ and $Y_2 = 4$ on the same screen. See Figure 60. Using the INTERSECT command (twice), we find that Y_1 and Y_2 intersect at $x = -4$ and at $x = 4$. The graph of Y_1 is below that of Y_2, $Y_1 < Y_2$, between the points of intersection. Since the inequality is strict, the solution set is $\{x \mid -4 < x < 4\}$ or, using interval notation, $(-4, 4)$.

Figure 60

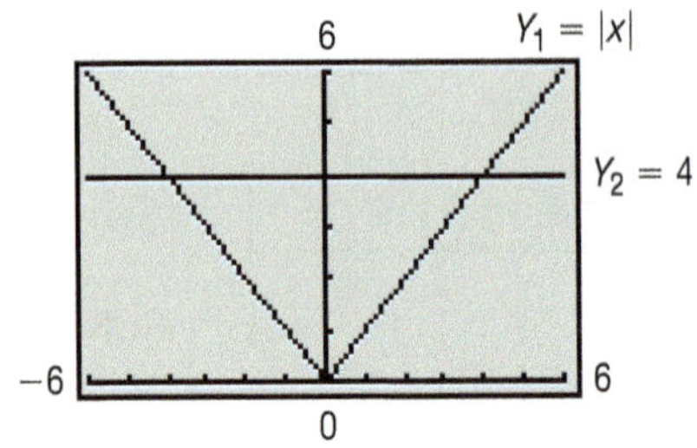

We are led to the following results:

THEOREM

If a is any positive number and if u is any algebraic expression, then

$\lvert u\rvert < a$	is equivalent to	$-a < u < a$	**(4)**
$\lvert u\rvert \le a$	is equivalent to	$-a \le u \le a$	**(5)**

In other words, $|u| < a$ is equivalent to $-a < u$ and $u < a$.

EXAMPLE 10 Solving an Inequality Involving Absolute Value

Solve the inequality $|2x + 4| \le 3$, and graph the solution set.

Algebraic Solution

$$|2x + 4| \le 3$$ This follows the form of statement (5); the expression $u = 2x + 4$ is inside the absolute value bars.

$$-3 \le 2x + 4 \le 3$$ Apply statement (5).

$$-3 - 4 \le 2x + 4 - 4 \le 3 - 4$$ Subtract 4 from each part.

$$-7 \le 2x \le -1$$ Simplify.

$$\frac{-7}{2} \le \frac{2x}{2} \le \frac{-1}{2}$$ Divide each part by 2.

$$-\frac{7}{2} \le x \le -\frac{1}{2}$$ Simplify.

The solution set is $\left\{x \middle| -\frac{7}{2} \le x \le -\frac{1}{2}\right\}$, that is, all x in the interval $\left[-\frac{7}{2}, -\frac{1}{2}\right]$.

Figure 62

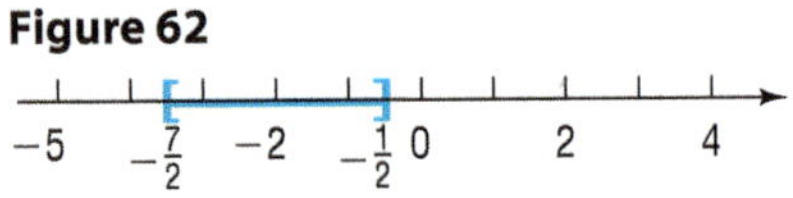

Graphing Solution

We graph $Y_1 = |2x + 4|$ and $Y_2 = 3$ on the same screen. See Figure 61. Using the INTERSECT command (twice), we find that Y_1 and Y_2 intersect at $x = -3.5$ and at $x = -0.5$. The graph of Y_1 is below that of Y_2, $Y_1 < Y_2$, between the points of intersection. Since the inequality is not strict, the solution set is $\{x \mid -3.5 \le x \le -0.5\}$ or, using interval notation, $[-3.5, -0.5]$.

Figure 61

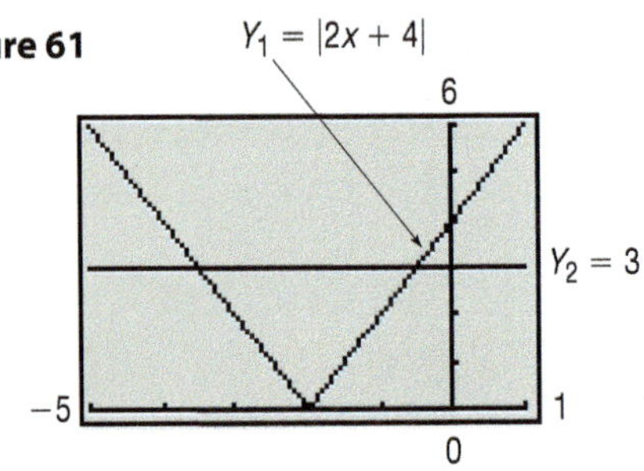

See Figure 62 for a graph of the solution set.

Now Work PROBLEM 77

EXAMPLE 11 Solving an Inequality Involving Absolute Value

Solve the inequality $|x| > 3$.

Algebraic Solution

We are looking for all points whose coordinate x is a distance greater than 3 units from the origin. Figure 63 illustrates the situation.

Figure 63

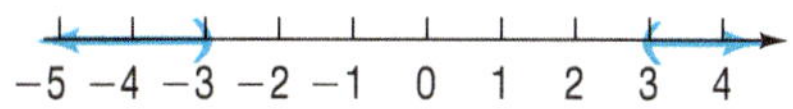

We conclude that any x less than -3 or greater than 3 satisfies the condition $|x| > 3$. Consequently, the solution set is $\{x \mid x < -3 \text{ or } x > 3\}$. Using interval notation, the solution is $(-\infty, -3) \cup (3, \infty)$.*

Graphing Solution

We graph $Y_1 = |x|$ and $Y_2 = 3$ on the same screen. See Figure 64. Using the INTERSECT command (twice), we find that Y_1 and Y_2 intersect at $x = -3$ and at $x = 3$. The graph of Y_1 is above that of Y_2, $Y_1 > Y_2$, to the left of $x = -3$ and to the right of $x = 3$. Since the inequality is strict, the solution set is $\{x \mid x < -3 \text{ or } x > 3\}$. Using interval notation, the solution is $(-\infty, -3) \cup (3, \infty)$.*

Figure 64

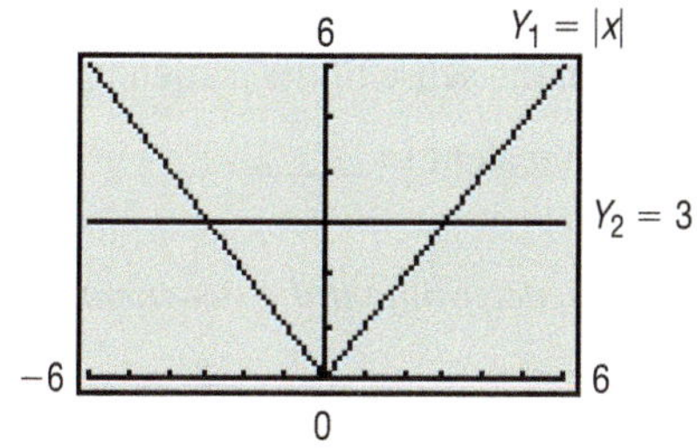

THEOREM

If a is any positive number and u is any algebraic expression, then

$\lvert u\rvert > a$	is equivalent to	$u < -a$ or $u > a$	**(6)**
$\lvert u\rvert \ge a$	is equivalent to	$u \le -a$ or $u \ge a$	**(7)**

WARNING A common error to be avoided is to attempt to write the solution $x < 1$ or $x > 4$ as $1 > x > 4$, which is incorrect, since there are no numbers x for which $1 > x$ *and* $x > 4$. Another common error is to "mix" the symbols and write $1 < x > 4$, which makes no sense.

EXAMPLE 12 Solving an Inequality Involving Absolute Value

Solve the inequality $|2x - 5| > 3$, and graph the solution set.

Algebraic Solution

$|2x - 5| > 3$ — This follows the form of statement (6); the expression $u = 2x - 5$ is inside the absolute value bars.

$2x - 5 < -3$ or $2x - 5 > 3$ — Apply statement (6).

$2x - 5 + 5 < -3 + 5$ or $2x - 5 + 5 > 3 + 5$ — Add 5 to each part.

$2x < 2$ or $2x > 8$ — Simplify.

$\dfrac{2x}{2} < \dfrac{2}{2}$ or $\dfrac{2x}{2} > \dfrac{8}{2}$ — Divide each part by 2.

$x < 1$ or $x > 4$ — Simplify.

The solution set is $\{x \mid x < 1 \text{ or } x > 4\}$. Using interval notation, the solution is $(-\infty, 1) \cup (4, \infty)$.

Graphing Solution

We graph $Y_1 = |2x - 5|$ and $Y_2 = 3$ on the same screen. See Figure 65. Using the INTERSECT command (twice), we find that Y_1 and Y_2 intersect at $x = 1$ and at $x = 4$. The graph of Y_1 is above that of Y_2, $Y_1 > Y_2$, to the left of $x = 1$ and to the right of $x = 4$. Since the inequality is strict, the solution set is $\{x \mid x < 1 \text{ or } x > 4\}$. Using interval notation, the solution is $(-\infty, 1) \cup (4, \infty)$.

Figure 65

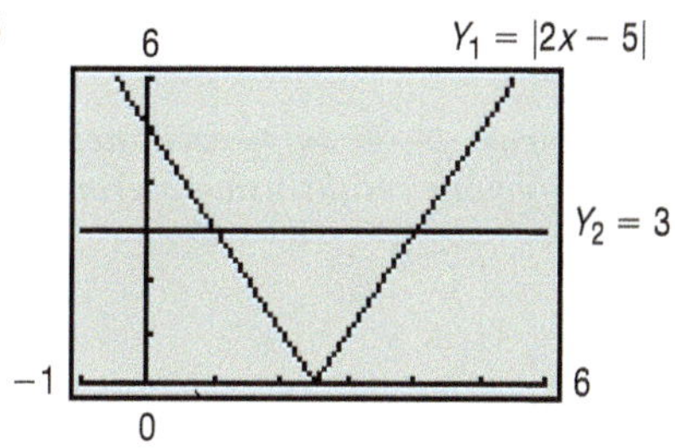

Figure 66

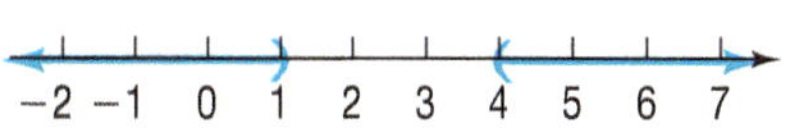

See Figure 66 for a graph of the solution set.

Now Work PROBLEM 79

*Recall that the symbol ∪ represents the union of two sets and means "or".

1.7 Assess Your Understanding

'Are You Prepared?'

Answers are given at the end of these exercises. If you get a wrong answer, read the pages listed in red.

1. Graph the inequality: $x \geq -2$. (p. 19)

2. ***True or False*** The absolute value of a negative number is positive. (p. 19)

3. If $A = \{a, e, i, o, u\}$ and $B = \{a, b, c, d, e\}$, what is $A \cup B$? (pp. 2–3)

Concepts and Vocabulary

4. If each side of an inequality is multiplied by a(n) _____ number, then the sense of the inequality symbol is reversed.

5. A(n) _____ _____, denoted $[a, b]$, consists of all real numbers x for which $a \leq x \leq b$.

6. $|u| \leq a$ is equivalent to _____.

In Problems 7–10, determine if the statement is True or False if $a < b$ and $c < 0$.

7. $a + c < b + c$ **8.** $a - c < b - c$ **9.** $ac > bc$ **10.** $\frac{a}{c} < \frac{b}{c}$

Skill Building

In Problems 11–16, express the graph shown in color using interval notation. Also express each as an inequality involving x.

11. 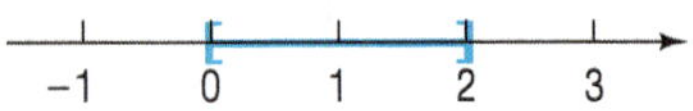

12. 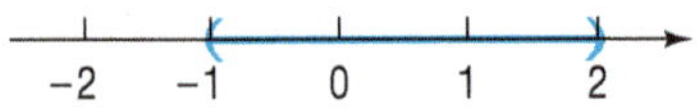

13. 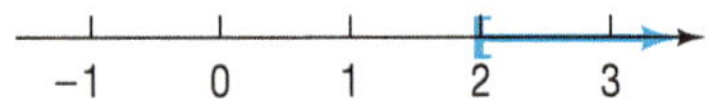

14. 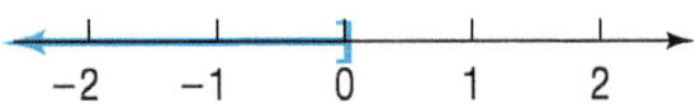

15. 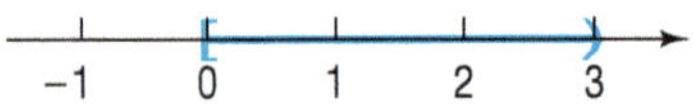

16. 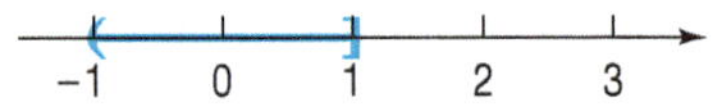

In Problems 17–20, an inequality is given. Write the inequality obtained by:

(a) Adding 3 to each side of the given inequality.

(b) Subtracting 5 from each side of the given inequality.

(c) Multiplying each side of the given inequality by 3.

(d) Multiplying each side of the given inequality by −2.

17. $3 < 5$ **18.** $2 > 1$ **19.** $2x + 1 < 2$ **20.** $1 - 2x > 5$

In Problems 21–28, write each inequality using interval notation, and illustrate each inequality using the real number line.

21. $0 \leq x \leq 4$ **22.** $-1 < x < 5$ **23.** $4 \leq x < 6$ **24.** $-2 < x \leq 0$

25. $x \geq 4$ **26.** $x \leq 5$ **27.** $x < -4$ **28.** $x > 1$

In Problems 29–36, write each interval as an inequality involving x, and illustrate each inequality using the real number line.

29. $[2, 5]$ **30.** $(1, 2)$ **31.** $(-3, -2)$ **32.** $[0, 1)$

33. $[4, \infty)$ **34.** $(-\infty, 2]$ **35.** $(-\infty, -3)$ **36.** $(-8, \infty)$

In Problems 37–48, fill in the blank with the correct inequality symbol.

37. If $x < 5$, then $x - 5$ _____ 0.

38. If $x < -4$, then $x + 4$ _____ 0.

39. If $x > -4$, then $x + 4$ _____ 0.

40. If $x > 6$, then $x - 6$ _____ 0.

41. If $x \geq -4$, then $3x$ _____ −12.

42. If $x \leq 3$, then $2x$ _____ 6.

43. If $x > 6$, then $-2x$ _____ −12.

44. If $x > -2$, then $-4x$ _____ 8.

45. If $2x < 6$, then x _____ 3.

46. If $3x \leq 12$, then x _____ 4.

47. If $-\frac{1}{2}x \leq 3$, then x _____ −6.

48. If $-\frac{1}{4}x > 1$, then x _____ −4.

In Problems 49–86, solve each inequality algebraically. Express your answer using set notation or interval notation. Graph the solution set. Verify your results using a graphing utility.

49. $3x - 7 > 2$ **50.** $2x + 5 > 1$ **51.** $1 - 2x \leq 3$

52. $2 - 3x \leq 5$ **53.** $3x - 1 \geq 3 + x$ **54.** $2x - 2 \geq 3 + x$

55. $-2(x + 3) < 8$ **56.** $-3(1 - x) < 12$ **57.** $4 - 3(1 - x) \leq 3$

58. $8 - 4(2 - x) \leq -2x$ **59.** $\frac{1}{2}(x - 4) > x + 8$ **60.** $3x + 4 > \frac{1}{3}(x - 2)$

61. $0 \leq 2x - 6 \leq 4$ **62.** $4 \leq 2x + 2 \leq 10$ **63.** $-5 \leq 4 - 3x \leq 2$

64. $-3 \leq 3 - 2x \leq 9$ **65.** $-3 < \frac{2x - 1}{4} < 0$ **66.** $0 < \frac{3x + 2}{2} < 4$

67. $1 < 1 - \frac{1}{2}x < 4$

68. $0 < 1 - \frac{1}{3}x < 1$

69. $(x + 2)(x - 3) > (x - 1)(x + 1)$

70. $(x - 1)(x + 1) > (x - 3)(x + 4)$

71. $|2x| < 8$

72. $|3x| < 15$

73. $|3x| > 12$

74. $|2x| > 6$

75. $|3t - 2| \leq 4$

76. $|2u + 5| \leq 7$

77. $|x - 2| + 2 < 3$

78. $|x + 4| + 3 < 5$

79. $|x - 3| \geq 2$

80. $|x + 4| \geq 2$

81. $|1 - 2x| > |-3|$

82. $|2 - 3x| > |-1|$

83. $|1 - 4x| - 7 < -2$

84. $|1 - 2x| - 4 < -1$

85. $|2x + 1| < -1$

86. $|3x - 4| \geq 0$

Mixed Practice

In Problems 87–106, solve each inequality algebraically. Express your answer using set notation or interval notation. Graph the solution set. Verify your results using a graphing utility.

87. $3 - 4x < 11$

88. $1 - 3x \leq 7$

89. $|2x + 1| - 5 \geq -1$

90. $|5x + 2| - 3 > 9$

91. $\frac{x}{2} \geq 1 - \frac{x}{4}$

92. $\frac{x}{3} \geq 2 + \frac{x}{6}$

93. $-\frac{1}{3} \leq \frac{x + 1}{6} < \frac{4}{3}$

94. $-\frac{3}{2} < \frac{x - 3}{4} \leq \frac{5}{4}$

95. $x(4x + 3) \leq (2x + 1)^2$

96. $x(9x - 5) \leq (3x - 1)^2$

97. $|(3x - 2) - 7| < \frac{1}{2}$

98. $|(4x - 1) - 11)| < \frac{1}{4}$

99. $-3 < 5 - 2x \leq 11$

100. $2 \leq 3 - 2(x + 1) < 8$

101. $7 - |x - 1| > 4$

102. $9 - |x + 3| \geq 5$

103. $-3 < x + 5 < 2x$

104. $2 < x - 3 < 2x$

105. $x + 2 < 2x - 1 < 5x$

106. $2x - 1 < 3x + 5 < 5x - 7$

Applications and Extensions

107. Express the fact that x differs from 2 by less than $\frac{1}{2}$ as an inequality involving an absolute value. Solve for x.

108. Express the fact that x differs from -1 by less than 1 as an inequality involving an absolute value. Solve for x.

109. Express the fact that x differs from -3 by more than 2 as an inequality involving an absolute value. Solve for x.

110. Express the fact that x differs from 2 by more than 3 as an inequality involving an absolute value. Solve for x.

111. A young adult may be defined as someone older than 21, but less than 30 years of age. Express this statement using inequalities.

112. Middle-aged may be defined as being 40 or more and less than 60. Express this statement using inequalities.

113. Body Temperature Normal human body temperature is 98.6°F. If a temperature x that differs from normal by at least 1.5° is considered unhealthy, write the condition for an unhealthy temperature x as an inequality involving an absolute value, and solve for x.

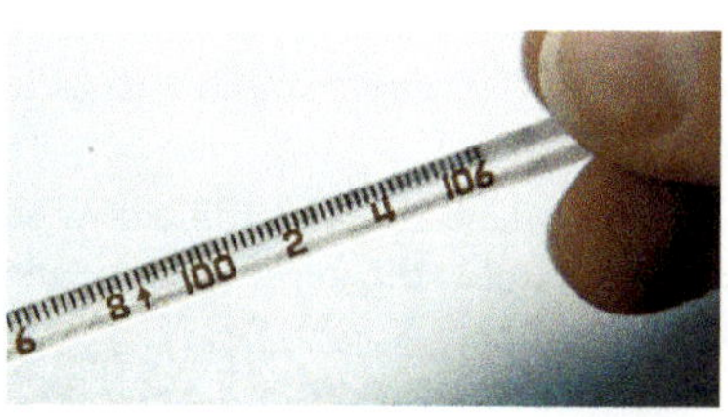

114. Household Voltage In the United States, normal household voltage is 115 volts. However, it is not uncommon for actual voltage to differ from normal voltage by at most 5 volts. Express this situation as an inequality involving an absolute value. Use x as the actual voltage and solve for x.

115. Life Expectancy The Social Security Administration determined that an average 30-year-old male in 2005 could expect to live at least 49.66 more years and an average 30-year-old female in 2005 could expect to live at least 53.58 more years.

(a) To what age can an average 30-year-old male expect to live? Express your answer as an inequality.

(b) To what age can an average 30-year-old female expect to live? Express your answer as an inequality.

(c) Who can expect to live longer, a male or a female? By how many years?

Source: Actuarial Study No. 120, August 2005

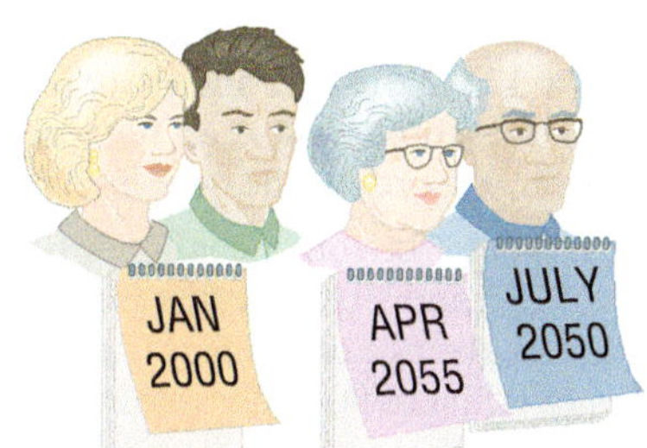

116. General Chemistry For a certain ideal gas, the volume V (in cubic centimeters) equals 20 times the temperature T in kelvins (K). If the temperature varies from 353 to 393 K, inclusive, what is the corresponding range of the volume of the gas?

117. Real Estate A real estate agent agrees to sell a large apartment complex according to the following commission schedule: \$45,000 plus 25% of the selling price in excess of \$900,000. Assuming that the complex will sell at some price between \$900,000 and \$1,100,000, inclusive, over what range does the agent's commission vary? How does the commission vary as a percent of selling price?

118. Sales Commission A used car salesperson is paid a commission of \$25 plus 40% of the selling price in excess of owner's cost. The owner claims that used cars typically sell for at least owner's cost plus \$70 and at most owner's cost plus \$300. For each sale made, over what range can the salesperson expect the commission to vary?

119. Federal Tax Withholding The percentage method of withholding for federal income tax (2006) states that a single person whose weekly wages, after subtracting withholding allowances, are over \$620, but not over \$1409, shall have \$78.30 plus 25% of the excess over \$620 withheld. Over what range does the amount withheld vary if the weekly wages vary from \$700 to \$900, inclusive?

Source: *Employer's Tax Guide.* Department of the Treasury, Internal Revenue Service, 2006.

120. Exercising Sue wants to lose weight. For healthy weight loss, the American College of Sports Medicine (ACSM) recommends 200 to 300 minutes of exercise per week. For the first six days of the week, Sue exercised 40, 45, 0, 50, 25, and 35 minutes. How long should Sue exercise on the seventh day in order to stay within the ACSM guidelines?

121. Electricity Rates Commonwealth Edison Company's summer charge for electricity is 8.275¢ per kilowatt-hour. In addition, each monthly bill contains a customer charge of \$7.13. If last summer's bills ranged from a low of \$65.96 to a high of \$217.02, over what range did usage vary (in kilowatt-hours)?

Source: Commonwealth Edison Co., Chicago, Illinois, 2006.

122. Water Bills The Village of Oak Lawn charges homeowners \$28.84 per quarter-year plus \$2.28 per 1000 gallons for water usage in excess of 12,000 gallons. In 2006, one homeowner's quarterly bill ranged from a high of \$74.44 to a low of \$42.52. Over what range did water usage vary?

Source: Village of Oak Lawn, Illinois, 2006.

123. Markup of a New Car The markup over dealer's cost of a new car ranges from 12% to 18%. If the sticker price is \$8800, over what range will the dealer's cost vary?

124. IQ Tests A standard intelligence test has an average score of 100. According to statistical theory, of the people who take the test, the 2.5% with the highest scores will have scores of more than 1.96σ above the average, where σ (sigma, a number called the *standard deviation*) depends on the nature of the test. If $\sigma = 12$ for this test and there is (in principle) no upper limit to the score possible on the test, write the interval of possible test scores of the people in the top 2.5%.

125. Computing Grades In your Economics 101 class, you have scores of 68, 82, 87, and 89 on the first four of five tests. To get a grade of B, the average of the first five test scores must be greater than or equal to 80 and less than 90.

(a) Find the range of the score that you need on the last test to get a B.
(b) What score do you need if the fifth test counts double?

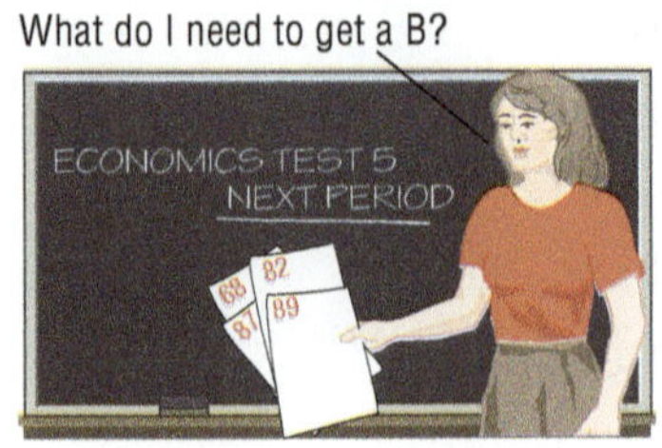

126. "Light" Foods For food products to be labeled "light," the U.S. Food and Drug Administration requires that the altered product must either contain one-third or fewer calories than the regular product or it must contain one-half or less fat than the regular product. If a serving of Miracle Whip® Light contains 20 calories and 1.5 grams of fat, then what must be true about either the number of calories or the grams of fat in a serving of regular Miracle Whip®?

127. Reading Books A Gallup poll conducted May 20–22, 2005, found that Americans read an average of 13.4 books per year. Gallup is 99% confident that the result from this poll is off by fewer than 1.35 books from the actual average x. Express this situation as an inequality involving absolute value, and solve the inequality for x to determine the interval in which the actual average is likely to fall.
[**Note:** In statistics, this interval is called a 99% **confidence interval.**]

128. Speed of Sound According to data from the Hill Aerospace Museum (Hill Air Force Base, Utah), the speed of sound varies depending on altitude, barometric pressure, and temperature. For example, at 20,000 feet, 13.75 inches of mercury, and $-12.3°$F, the speed of sound is about 707 miles per hour, but the speed can vary from this result by as much as 55 miles per hour as conditions change.
(a) Express this situation as an inequality involving an absolute value.
(b) Using x for the speed of sound, solve for x to find an interval for the speed of sound.

129. Arithmetic Mean If $a < b$, show that $a < \frac{a+b}{2} < b$. The number $\frac{a+b}{2}$ is called the **arithmetic mean** of a and b.

130. Refer to Problem 129. Show that the arithmetic mean of a and b is equidistant from a and b.

131. Geometric Mean If $0 < a < b$, show that $a < \sqrt{ab} < b$. The number $\sqrt{ab}$ is called the **geometric mean** of a and b.

132. Refer to Problems 129 and 131. Show that the geometric mean of a and b is less than the arithmetic mean of a and b.

133. Harmonic Mean For $0 < a < b$, let h be defined by

$$\frac{1}{h} = \frac{1}{2}\left(\frac{1}{a} + \frac{1}{b}\right)$$

Show that $a < h < b$. The number h is called the **harmonic mean** of a and b.

134. Refer to Problems 129, 131, and 133. Show that the harmonic mean of a and b equals the geometric mean squared divided by the arithmetic mean.

Discussion and Writing

135. Make up an inequality that has no solution. Make up one that has exactly one solution.

136. How would you explain to a fellow student the underlying reason for the multiplication properties for inequalities (page 153); that is, the sense or direction of an inequality remains the same if each side is multiplied by a positive real number, while the direction is reversed if each side is multiplied by a negative real number.

137. The inequality $x^2 + 1 < -5$ has no solution. Explain why.

138. Do you prefer to use inequality notation or interval notation to express the solution to an inequality? Give your reasons. Are there particular circumstances when you prefer one to the other? Cite examples.

'Are You Prepared?' Answers

1. 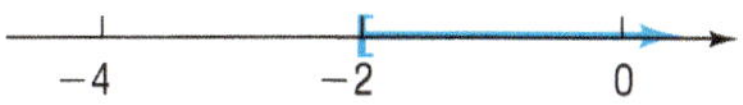

2. True

3. $A \cup B = \{a, b, c, d, e, i, o, u\}$

CHAPTER REVIEW

Things to Know

Formulas

Distance formula (p. 89) $\quad d = \sqrt{(x_2 - x_1)^2 + (y_2 - y_1)^2}$

Midpoint formula (p. 91) $\quad (x, y) = \left(\dfrac{x_1 + x_2}{2}, \dfrac{y_1 + y_2}{2}\right)$

Quadratic equation and quadratic formula

The real solutions of the equation $ax^2 + bx + c = 0$, $a \neq 0$, are given by $x = \dfrac{-b \pm \sqrt{b^2 - 4ac}}{2a}$, provided $b^2 - 4ac \geq 0$. If $b^2 - 4ac < 0$, there are no real solutions. (p. 118)

In the complex number system, the solutions of the equation $ax^2 + bx + c = 0$, $a \neq 0$, are given by $x = \dfrac{-b \pm \sqrt{b^2 - 4ac}}{2a}$. (p. 129)

Discriminant (pp. 118 and 130)

If $b^2 - 4ac > 0$, there are two unequal real solutions.

If $b^2 - 4ac = 0$, there is one repeated real solution, a root of multiplicity 2.

If $b^2 - 4ac < 0$, there are no real solutions, but there are two distinct complex solutions that are not real; the complex solutions are conjugates of each other.

Interval notation (pp. 150–151)

$[a, b]$	$\{x \mid a \leq x \leq b\}$	(a, b)	$\{x \mid a < x < b\}$	$(-\infty, a]$	$\{x \mid x \leq a\}$
$[a, b)$	$\{x \mid a \leq x < b\}$	$[a, \infty)$	$\{x \mid x \geq a\}$	$(-\infty, a)$	$\{x \mid x < a\}$
$(a, b]$	$\{x \mid a < x \leq b\}$	(a, ∞)	$\{x \mid x > a\}$	$(-\infty, \infty)$	All real numbers

Properties of inequalities

Addition property (p. 152)
If $a < b$, then $a + c < b + c$.
If $a > b$, then $a + c > b + c$.

Multiplication properties (p. 153)
(a) If $a < b$ and if $c > 0$, then $ac < bc$.
If $a < b$ and if $c < 0$, then $ac > bc$.
(b) If $a > b$ and if $c > 0$, then $ac > bc$.
If $a > b$ and if $c < 0$, then $ac < bc$.

Absolute value

If $|u| = a$, $a > 0$, then $u = -a$ or $u = a$. (p. 136)

If $|u| \leq a$, $a > 0$, then $-a \leq u \leq a$. (p. 156)

If $|u| \geq a$, $a > 0$, then $u \leq -a$ or $u \geq a$. (p. 157)

Objectives

Section		You should be able to	Example(s)	Review Exercises
1.1	1	Use the distance formula (p. 88)	2, 3, 4	91(a)–96(a); 105, 106, 107
	2	Use the midpoint formula (p. 91)	5	91(b)–96(b)
	3	Graph equations by hand by plotting points (p. 91)	6, 7, 8	99–104
	4	Graph equations using a graphing utility (p. 94)	9, 10	97, 99–104
	5	Use a graphing utility to create tables (p. 96)	11	97
	6	Find intercepts from a graph (p. 96)	12	98
	7	Use a graphing utility to approximate intercepts (p. 97)	13	99–104
1.2	1	Solve equations using a graphing utility (p. 103)	1, 2	51–54
	2	Solve linear equations (p. 105)	3, 4	1–6
	3	Solve rational equations (p. 106)	5, 6	7, 8, 36, 47, 48
	4	Solve problems that can be modeled by linear equations (p. 108)	7, 8	129
1.3	1	Solve quadratic equations by factoring (p. 113)	1, 2	9, 10, 13, 33–35
	2	Solve quadratic equations using the Square Root Method (p. 114)	3	49, 50
	3	Solve quadratic equations by completing the square (p. 115)	4, 5	9, 10, 13–16, 19, 33–35
	4	Solve quadratic equations using the quadratic formula (p. 117)	6, 7	9, 10, 13–16, 19, 33–35
	5	Solve problems that can be modeled by quadratic equations (p. 120)	8	113, 119, 126, 127, 131, 132
1.4	1	Add, subtract, multiply, and divide complex numbers (p. 125)	1–8	73–82
	2	Solve quadratic equations in the complex number system (p. 129)	10, 11	83–90
1.5	1	Solve radical equations (p. 133)	1–3	17, 18, 23–28, 37, 38
	2	Solve equations quadratic in form (p. 135)	4, 5	21, 22, 31, 32
	3	Solve absolute value equations (p. 136)	6	39–42
	4	Solve equations by factoring (p. 137)	7, 8	43–46
1.6	1	Translate verbal descriptions into mathematical expressions (p. 141)	1	108, 109
	2	Solve interest problems (p. 142)	2, 3	110, 111
	3	Solve mixture problems (p. 143)	4	122–125
	4	Solve uniform motion problems (p. 144)	5, 6	112, 114–118, 128, 134
	5	Solve constant rate job problems (p. 146)	7	120, 121, 130, 133
1.7	1	Use interval notation (p. 150)	1, 2	55–68
	2	Use properties of inequalities (p. 151)	3–6	55–68
	3	Solve linear inequalities algebraically and graphically (p. 153)	7	55–56
	4	Solve combined inequalities algebraically and graphically (p. 155)	8	57–60
	5	Solve absolute value inequalities algebraically and graphically (p. 155)	9–12	61–68

Review Exercises

In Problems 1–50, find all the real solutions, if any, of each equation. (Where they appear, a, b, m, and n are positive constants.) When possible, verify your results using a graphing utility.

1. $2 - \frac{x}{3} = 8$

2. $\frac{x}{4} - 2 = 4$

3. $-2(5 - 3x) + 8 = 4 + 5x$

4. $(6 - 3x) - 2(1 + x) = 6x$

5. $\frac{3x}{4} - \frac{x}{3} = \frac{1}{12}$

6. $\frac{4 - 2x}{3} + \frac{1}{6} = 2x$

7. $\frac{x}{x - 1} = \frac{6}{5}, \quad x \neq 1$

8. $\frac{4x - 5}{3 - 7x} = 2, \quad x \neq \frac{3}{7}$

9. $x(1 - x) = 6$

10. $x(1 + x) = 6$

11. $\frac{1}{2}\left(x - \frac{1}{3}\right) = \frac{3}{4} - \frac{x}{6}$

12. $\frac{1 - 3x}{4} = \frac{x + 6}{3} + \frac{1}{2}$

13. $(x - 1)(2x + 3) = 3$

14. $x(2 - x) = 3(x - 4)$

15. $2x + 3 = 4x^2$

16. $1 + 6x = 4x^2$

17. $\sqrt[3]{x^2 - 1} = 2$

18. $\sqrt{1 + x^3} = 3$

19. $x(x + 1) + 2 = 0$

20. $3x^2 - x + 1 = 0$

21. $x^4 - 5x^2 + 4 = 0$

22. $3x^4 + 4x^2 + 1 = 0$

23. $\sqrt{2x - 3} + x = 3$

24. $\sqrt{2x - 1} = x - 2$

25. $\sqrt[4]{2x+3}=2$
26. $\sqrt[5]{3x+1}=-1$
27. $\sqrt{x+1}+\sqrt{x-1}=\sqrt{2x+1}$
28. $\sqrt{2x-1}-\sqrt{x-5}=3$
29. $2x^{1/2}-3=0$
30. $3x^{1/4}-2=0$
31. $x^{-6}-7x^{-3}-8=0$
32. $6x^{-1}-5x^{-1/2}+1=0$
33. $x^2+m^2=2mx+(nx)^2, \quad n\neq 1, n\neq -1$
34. $b^2x^2+2ax=x^2+a^2, \quad b\neq 1, b\neq -1$
35. $10a^2x^2-2abx-36b^2=0$
36. $\frac{1}{x-m}+\frac{1}{x-n}=\frac{2}{x}, \quad x\neq 0, x\neq m, x\neq n$
37. $\sqrt{x^2+3x+7}-\sqrt{x^2-3x+9}+2=0$
38. $\sqrt{x^2+3x+7}-\sqrt{x^2+3x+9}=2$
39. $|2x+3|=7$
40. $|3x-1|=5$
41. $|2-3x|+2=9$
42. $|1-2x|+1=4$
43. $2x^3=3x^2$
44. $5x^4=9x^3$
45. $2x^3+5x^2-8x-20=0$
46. $3x^3+5x^2-3x-5=0$
47. $\frac{1}{x-1}+\frac{3}{x+2}=\frac{11}{x^2+x-2}$
48. $\frac{5}{x+5}-\frac{2}{x-3}=\frac{-1}{x^2+2x-15}$
49. $(x-2)^2=9$
50. $(2x+1)^2=25$

In Problems 51–54, use a graphing utility to approximate the solutions of each equation rounded to two decimal places. All solutions lie between −10 and 10.

51. $x^3-5x+3=0$
52. $-x^3+3x+1=0$
53. $x^4-3=2x+1$
54. $-x^4+7=x^2-2$

In Problems 55–68, solve each inequality. Express your answer using set notation or interval notation. Graph the solution set. Verify your results using a graphing utility.

55. $\frac{2x-3}{5}+2\leq\frac{x}{2}$
56. $\frac{5-x}{3}\leq 6x-4$
57. $-9\leq\frac{2x+3}{-4}\leq 7$
58. $-4<\frac{2x-2}{3}<6$
59. $2<\frac{3-3x}{12}<6$
60. $-3\leq\frac{5-3x}{2}\leq 6$
61. $|3x+4|<\frac{1}{2}$
62. $|1-2x|<\frac{1}{3}$
63. $|2x-5|\geq 9$
64. $|3x+1|\geq 10$
65. $2+|2-3x|\leq 4$
66. $\frac{1}{2}+\left|\frac{2x-1}{3}\right|\leq 1$
67. $1-|2-3x|<-4$
68. $1-\left|\frac{2x-1}{3}\right|<-2$

In Problems 69–72, what number should be added to complete the square of each expression?

69. x^2+6x
70. x^2-10x
71. $x^2-\frac{4}{3}x$
72. $x^2+\frac{4}{5}x$

In Problems 73–82, use the complex number system and write each expression in the standard form a + bi.

73. $(6+3i)-(2-4i)$
74. $(8-3i)+(-6+2i)$
75. $4(3-i)+3(-5+2i)$
76. $2(1+i)-3(2-3i)$
77. $\frac{3}{3+i}$
78. $\frac{4}{2-i}$
79. i^{50}
80. i^{29}
81. $(2+3i)^3$
82. $(3-2i)^3$

In Problems 83–90, solve each equation in the complex number system.

83. $x^2+x+1=0$
84. $x^2-x+1=0$
85. $2x^2+x-2=0$
86. $3x^2-2x-1=0$
87. $x^2+3=x$
88. $2x^2+1=2x$
89. $x(1-x)=6$
90. $x(1+x)=2$

In Problems 91–96, find the following for each pair of points:
(a) The distance between the points
(b) The midpoint of the line segment connecting the points

91. (0, 0); (4, 2)
92. (0, 0); (−4, 6)
93. (1, −1); (−2, 3)
94. (−2, 2); (1, 4)
95. (4, −4); (4, 8)
96. (−3, 4); (2, 4)

97. Graph $y=-x^2+15$ using a graphing utility. Create a table of values to determine a good initial viewing window.

98. List the intercepts of the following graph.

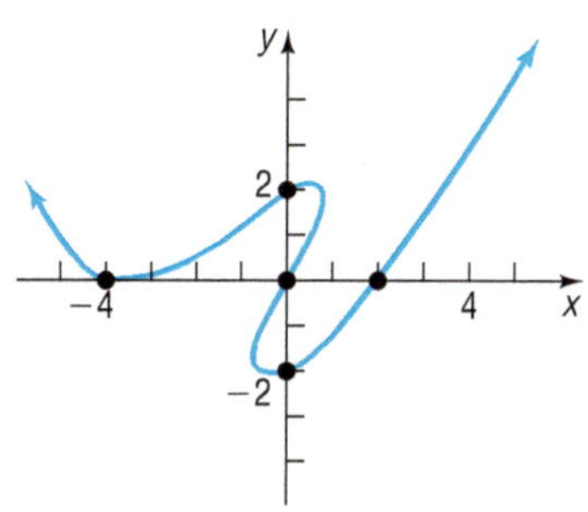

In Problems 99–104, graph each equation by hand by plotting points. Verify your results using a graphing utility. Approximate the intercepts using a graphing utility and label them on the graph.

99. $2x - 3y = 6$

100. $x + 2y = 4$

101. $y = x^2 - 9$

102. $y = x^2 + 4$

103. $x^2 + 2y = 16$

104. $2x^2 - 4y = 24$

---

105. Show that the points $A = (3, 4)$, $B = (1, 1)$, and $C = (-2, 3)$ are the vertices of an isosceles triangle.

106. Show that the points $A = (-2, 0)$, $B = (-4, 4)$, and $C = (8, 5)$ are the vertices of a right triangle.

107. Find two numbers y such that the distance from $(-3, 2)$ to $(5, y)$ is 10.

108. Translate the following statement into a mathematical expression: The perimeter p of a rectangle is the sum of two times the length l and two times the width w.

109. Translate the following statement into a mathematical expression: The total cost C of manufacturing x bicycles in one day is \$50,000 plus \$95 times the number of bicycles manufactured.

110. Banking A bank lends out \$9000 at 7% simple interest. At the end of 1 year, how much interest is owed on the loan?

111. Financial Planning Steve, a recent retiree, requires \$5000 per year in extra income. He has \$70,000 to invest and can invest in A-rated bonds paying 8% per year or in a certificate of deposit (CD) paying 5% per year. How much money should be invested in each to realize exactly \$5000 in interest per year?

112. Lightning and Thunder A flash of lightning is seen, and the resulting thunderclap is heard 3 seconds later. If the speed of sound averages 1100 feet per second, how far away is the storm?

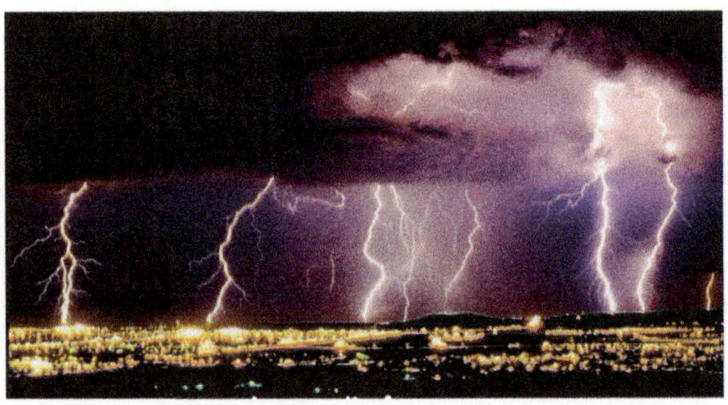

113. Physics: Intensity of Light The intensity I (in candlepower) of a certain light source obeys the equation $I = \dfrac{900}{x^2}$, where x is the distance (in meters) from the light. Over what range of distances can an object be placed from this light source so that the range of intensity of light is from 1600 to 3600 candlepower, inclusive?

114. Extent of Search and Rescue A search plane has a cruising speed of 250 miles per hour and carries enough fuel for at most 5 hours of flying. If there is a wind that averages 30 miles per hour and the direction of the search is with the wind one way and against it the other, how far can the search plane travel before it has to turn back?

115. Extent of Search and Rescue If the search plane described in Problem 114 is able to add a supplementary fuel tank that allows for an additional 2 hours of flying, how much farther can the plane extend its search?

116. Rescue at Sea A life raft, set adrift from a sinking ship 150 miles offshore, travels directly toward a Coast Guard station at the rate of 5 miles per hour. At the time that the raft is set adrift, a rescue helicopter is dispatched from the Coast Guard station. If the helicopter's average speed is 90 miles per hour, how long will it take the helicopter to reach the life raft?

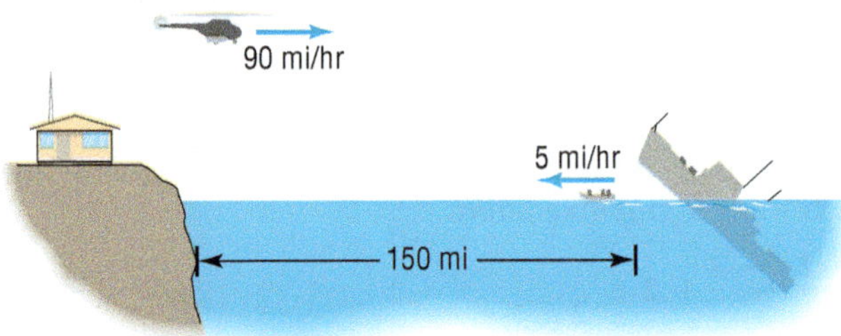

117. Physics: Uniform Motion Two bees leave two locations 150 meters apart and fly, without stopping, back and forth between these two locations at average speeds of 3 meters per second and 5 meters per second, respectively. How long is it until the bees meet for the first time? How long is it until they meet for the second time?

118. Physics: Uniform Motion A Metra commuter train leaves Union Station in Chicago at 12 noon. Two hours later, an Amtrak train leaves on the same track, traveling at an average speed that is 50 miles per hour faster than the Metra train. At 3 PM the Amtrak train is 10 miles behind the commuter train. How fast is each going?

119. Physics An object is thrown down from the top of a building 1280 feet tall with an initial velocity of 32 feet per second. The distance s (in feet) of the object from the ground after t seconds is $s = 1280 - 32t - 16t^2$.
(a) When will the object strike ground?
(b) What is the height of the object after 4 seconds?

120. Working Together to Get a Job Done Clarissa and Shawna, working together, can paint the exterior of a house in 6 days. Clarissa by herself can complete this job in 5 days less than Shawna. How long will it take Clarissa to complete the job by herself?

121. Emptying a Tank Two pumps of different sizes, working together, can empty a fuel tank in 5 hours. The larger pump can empty this tank in 4 hours less than the smaller. If the larger pump is out of order, how long will it take the smaller one to do the job alone?

122. Chemistry: Mixing Acids A laboratory has 60 cubic centimeters (cm^3) of a solution that is 40% HCl acid. How many cubic centimeters of a 15% solution of HCl acid should be mixed with the 60 cm^3 of 40% acid to obtain a solution of 25% HCl? How much of the 25% solution is there?

123. Business: Blending Coffee A coffee house has 20 pounds of a coffee that sells for \$4 per pound. How many pounds of a coffee that sells for \$8 per pound should be mixed with the 20 pounds of \$4-per-pound coffee to obtain a blend that will sell for \$5 per pound? How much of the \$5-per-pound coffee is there to sell?

124. Chemistry: Salt Solutions How much water should be added to 64 ounces of a 10% salt solution to make a 2% salt solution?

125. Chemistry: Salt Solutions How much water must be evaporated from 64 ounces of a 2% salt solution to make a 10% salt solution?

126. Geometry The hypotenuse of a right triangle measures 13 centimeters. Find the lengths of the legs if their sum is 17 centimeters.

127. Geometry The diagonal of a rectangle measures 10 inches. If the length is 2 inches more than the width, find the dimensions of the rectangle.

128. Physics: Uniform Motion A man is walking at an average speed of 4 miles per hour alongside a railroad track. A freight train, going in the same direction at an average speed of 30 miles per hour, requires 5 seconds to pass the man. How long is the freight train? Give your answer in feet.

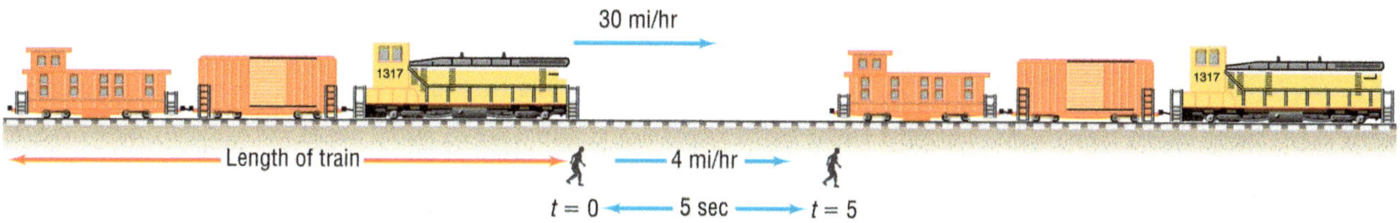

129. Framing a Painting An artist has 50 inches of oak trim to frame a painting. The frame is to have a border 3 inches wide surrounding the painting.

(a) If the painting is square, what are its dimensions? What are the dimensions of the frame?

(b) If the painting is rectangular with a length twice its width, what are the dimensions of the painting? What are the dimensions of the frame?

130. Using Two Pumps An 8-horsepower (hp) pump can fill a tank in 8 hours. A smaller, 3-hp pump fills the same tank in 12 hours. The pumps are used together to begin filling this tank. After 4 hours, the 8-hp pump breaks down. How long will it take the smaller pump to fill the tank?

131. Pleasing Proportion One formula stating the relationship between the length l and width w of a rectangle of "pleasing proportion" is $l^2 = w(l + w)$. How should a 4 foot by 8 foot sheet of plasterboard be cut so that the result is a rectangle of "pleasing proportion" with a width of 4 feet?

132. Business: Determining the Cost of a Charter A group of 20 senior citizens can charter a bus for a one-day excursion trip for \$15 per person. The charter company agrees to reduce the price of each ticket by 10¢ for each additional passenger in excess of 20 who goes on the trip, up to a maximum of 44 passengers (the capacity of the bus). If the final bill from the charter company was \$482.40, how many seniors went on the trip, and how much did each pay?

133. Utilizing Copying Machines A new copying machine can do a certain job in 1 hour less than an older copier. Together they can do this job in 72 minutes. How long would it take the older copier by itself to do the job?

134. Evening Up a Race In a 100-meter race, Todd crosses the finish line 5 meters ahead of Scott. To even things up, Todd suggests to Scott that they race again, this time with Todd lining up 5 meters behind the start.

(a) Assuming that Todd and Scott run at the same pace as before, does the second race end in a tie?

(b) If not, who wins?

(c) By how many meters does he win?

(d) How far back should Todd start so that the race ends in a tie?

After running the race a second time, Scott, to even things up, suggests to Todd that he (Scott) line up 5 meters in front of the start.

(e) Assuming again that they run at the same pace as in the first race, does the third race result in a tie?

(f) If not, who wins?

(g) By how many meters?

(h) How far up should Scott start so that the race ends in a tie?

135. Explain the differences among the following three problems. Are there any similarities in their solution?

(a) Write the expression as a single quotient:

$$\frac{x}{x-2} + \frac{x}{x^2-4}$$

(b) Solve: $\frac{x}{x-2} + \frac{x}{x^2-4} = 0$

(c) Solve: $\frac{x}{x-2} + \frac{x}{x^2-4} < 0$

CHAPTER TEST

1. Suppose the points $(-2, -3)$ and $(4, 5)$ are the endpoints of a line segment.
   (a) Find the distance between the two points.
   (b) Find the midpoint of the line segment connecting the two points.

In Problems 2–9, solve each equation algebraically in the real number system. Answers should be exact.

2. $2x^2 + 6x = x - 3$
3. $x + 1 = \sqrt{x + 7}$
4. $2 - \frac{3}{m} = \frac{2}{m + 2}$
5. $5x - 8 = -4(x - 1) + 6$
6. $5|3 - 2b| - 7 = 8$
7. $x^4 + x^2 = 3x^2 + 8$
8. $x^2 - 4x + 2 = 0$
9. $2x^2 + x - 1 = x(x + 7) + 2$

In Problems 10 and 11, graph each equation by hand by plotting points. Use a graphing utility to approximate the intercepts and label them on the graph.

10. $2x - 7y = 21$
11. $y = x^2 - 5$

In Problems 12–14, use a graphing utility to approximate the real solutions of each equation rounded to two decimal places. All solutions lie between −10 and 10.

12. $2x^3 - x^2 - 2x + 1 = 0$
13. $x^4 - 5x^2 - 8 = 0$
14. $-x^3 + 7x - 2 = x^2 + 3x - 3$

In Problems 15–18, solve each inequality. Express your answer in set-builder notation or interval notation. Graph the solution set.

15. $\frac{2x + 3}{4} < -2$
16. $|2x + 3| - 4 \geq 3$
17. $-7 < 3 - 5x \leq 8$
18. $|3x + 4| < 8$

In Problems 19–21, write each expression in the standard form $a + bi$.

19. $2(3 - 7i) - (4 + 11i)$
20. $(3 + 10i)(8 + i)$
21. $\frac{2 + i}{5 - 3i}$
22. Solve the equation $4x^2 - 4x + 5 = 0$ in the complex number system.
23. Jamie is a cashier at a local supermarket. On average, she can check out a customer in 5 minutes. Scott, a cashier trainee, takes an average of 8.5 minutes to check out customers. How long would it take Jamie and Scott to check out 65 customers if they work together at different registers? Round to two decimal places if necessary.
24. A health food retailer sells a mixture of dried cherries, cranberries, and pecans for \$15.00 per pound and dried banana chips for \$2.25 per pound. The retailer decides to create a new mix by adding banana chips to the original mix. How many pounds of banana chips must be mixed with 40 pounds of the original mix to obtain a new mixture that sells for \$10.25 per pound with no loss in revenue? Round to two decimal places if necessary.
25. A Mountainsmith Auspex 4000 backpack originally sold for \$275.00 but is advertised at 42% off. What is the sale price of the backpack?
26. Glenn invests \$10,000 in a certificate of deposit (CD) that pays simple interest of 4% per annum. How much interest will Glenn earn after 3 months?

CHAPTER PROJECTS

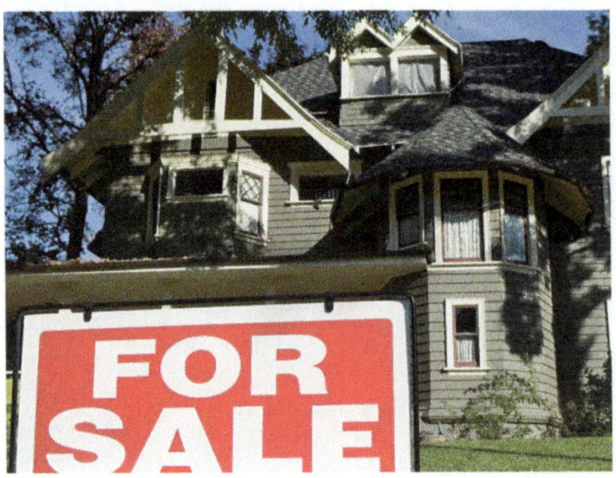

I. **Home Mortgages** While you may not be in the market for a home right now, it is probably an event that will occur for you within the next few years. Formula (1) gives the monthly payment P required to pay off a loan L at an annual rate of interest r, expressed as a decimal, but usually given as a percent. The time t, measured in months, is the duration of the loan, so a 15-year mortgage requires $t = 12 \times 15 = 180$ monthly payments.

$$P = L\left[\frac{\frac{r}{12}}{1 - \left(1 + \frac{r}{12}\right)^{-t}}\right] \quad (1)$$

P = monthly payment
L = loan amount
r = annual rate of interest, expressed as a decimal
t = length of loan, in months

On May 12, 2006, average interest rates were:

(a) 6.58% on 30-year mortgages
(b) 6.17% on 15-year mortgages

1. For each of these loans, calculate the monthly payment for a loan of \$200,000.
2. Then compute the total amount paid over the term of the loan.
3. Finally, calculate the interest paid.

On June 6, 2007, average interest rates were:

(a) 6.09% on 30-year mortgages

(b) 5.78% on 15-year mortgages

4. For each of these loans, calculate the monthly payment for a loan of $200,000.
5. Then compute the total amount paid over the term of the loan.
6. Finally, calculate the interest paid.
7. Solve equation (1) for L.
8. If you can afford to pay $1000 per month for a mortgage payment, calculate the amount you can borrow on June 6, 2007.
   (a) On a 30-year mortgage
   (b) On a 15-year mortgage
9. Repeat Problem 8 for the rates on May 12, 2006.
10. Check with your local lending institution for current rates for 30-year and 15-year mortgages. Calculate how much you can borrow with a $1000 per month payment.
11. Repeat Problem 10 if you can afford a payment of $1300 per month.
12. Do you think that the interest rate plays an important role in determining how much you can afford to pay for a house?
13. Comment on the two types of mortgages: 30-year and 15-year. Which would you take? Why?

The following project is also available on the Instructor's Resource Center (IRC):

II. **Project at Motorola How Many Cellular Phones Can I Make?** An industrial engineer uses a model involving equations to be sure production levels meet customer demand.

Graphs

2

The First Modern Olympics: Athens, 1896

The birth of the modern Olympic Games

By John Gettings—"I hereby proclaim the opening of the first International Olympic Games at Athens." With these words on April 6, 1896, King George I of Greece welcomed the crowd that had gathered in the newly reconstructed Panathenean Stadium to the modern-day Olympic Summer Games.

The event was the idea of Baron Pierre de Coubertin of France, who traveled the world to gather support for his dream to have nations come together and overcome national disputes, all in the name of sport.

The program for the Games included track and field, fencing, weightlifting, rifle and pistol shooting, tennis, cycling, swimming, gymnastics, and wrestling. Although 14 nations participated, most of the athletes were Greek.

The Games reached their high point on Day 11 with the first modern-day marathon. The idea to hold an event to commemorate the Ancient Olympic games was suggested by a friend of de Coubertin and was met with great anticipation. The race was run from Marathon to Athens (estimated at 22–26 miles), watched by more than 100,000 people, and won by a Greek runner, Spiridon Louis.

Source: Fact Monster™ Database, © 2006 Pearson Education, Inc.

—See Chapter Project I—

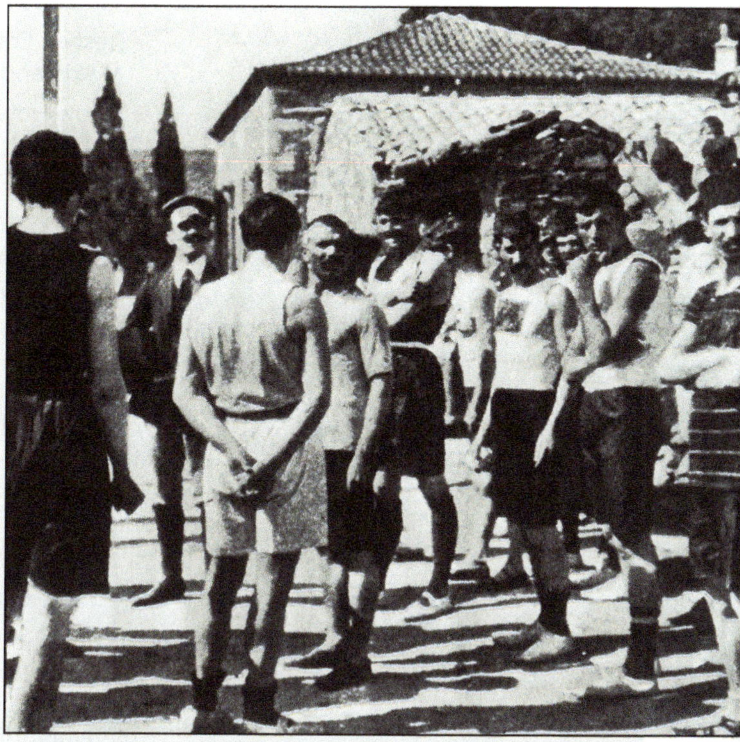

Source: Adapted from John Gettings, "The First Modern Olympics: Athens, 1896," accessed at www.infoplease.com. ©2006 Pearson Education, Inc. Reprinted with permission.

A Look Back

In Chapter R we reviewed algebra essentials and geometry essentials. In Chapter 1 we introduced the Cartesian plane to graph equations in two variables and studied equations in one variable.

A Look Ahead

Now we will continue our discussion of the Cartesian plane by introducing additional techniques that can be used to obtain the graph of an equation with two variables. We also introduce two specific graphs: lines and circles. We conclude the chapter by looking at variation.

Outline

2.1 Intercepts; Symmetry; Graphing Key Equations

PREPARING FOR THIS SECTION *Before getting started, review the following:*

- Introduction to Graphing Equations (Section 1.1, pp. 91–95)
- Solving Quadratic Equations (Section 1.3, pp. 113–120)
- Finding Intercepts Using a Graphing Utility (Section 1.1, p. 97)
- Finding Intercepts from a Graph (Section 1.1, pp. 96–97)

Now Work the 'Are You Prepared?' problems on page 175.

OBJECTIVES
1. Find Intercepts Algebraically from an Equation (p. 170)
2. Test an Equation for Symmetry with Respect to (a) the x-Axis, (b) the y-Axis, and (c) the Origin (p. 171)
3. Know How to Graph Key Equations (p. 173)

1 Find Intercepts Algebraically from an Equation

Figure 1

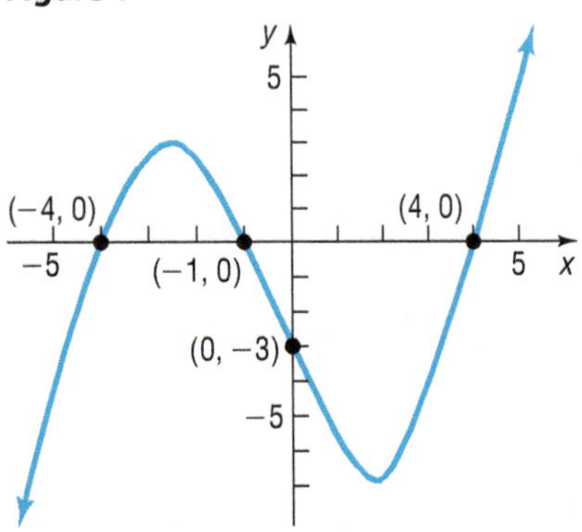

In Section 1.1, we discussed how to find intercepts from a graph and how to approximate intercepts from an equation using a graphing utility. Now we discuss how to find intercepts from an equation algebraically. To help understand the procedure, we present Figure 1. From the graph, we can see that the intercepts are $(-4, 0)$, $(-1, 0)$, $(4, 0)$, and $(0, -3)$. The x-intercepts are -4, -1, and 4. The y-intercept is -3. Notice that x-intercepts have y-coordinates that equal 0; y-intercepts have x-coordinates that equal 0. This leads to the following procedure for finding intercepts.

Procedure for Finding Intercepts

1. To find the x-intercept(s), if any, of the graph of an equation, let $y = 0$ in the equation and solve for x, where x is a real number.
2. To find the y-intercept(s), if any, of the graph of an equation, let $x = 0$ in the equation and solve for y, where y is a real number.

EXAMPLE 1 **Finding Intercepts from an Equation**

Find the x-intercept(s) and the y-intercept(s) of the graph of $y = x^2 - 4$. Then graph $y = x^2 - 4$ by plotting points.

Solution To find the x-intercept(s), we let $y = 0$ and obtain the equation

$$x^2 - 4 = 0 \quad y = x^2 - 4 \text{ with } y = 0$$
$$(x + 2)(x - 2) = 0 \quad \text{Factor.}$$
$$x + 2 = 0 \quad \text{or} \quad x - 2 = 0 \quad \text{Zero-Product Property}$$
$$x = -2 \quad \text{or} \quad x = 2 \quad \text{Solve.}$$

The equation has two solutions, -2 and 2. The x-intercepts are -2 and 2.

To find the y-intercept(s), we let $x = 0$ in the equation.

$$y = x^2 - 4$$
$$= 0^2 - 4 = -4$$

The y-intercept is -4.

Since $x^2 \geq 0$ for all x, we deduce from the equation $y = x^2 - 4$ that $y \geq -4$ for all x. This information, the intercepts, and the points from Table 1 enable us to graph $y = x^2 - 4$ by hand. See Figure 2.

Table 1

x	$y = x^2 - 4$	(x, y)
-3	$(-3)^2 - 4 = 5$	$(-3, 5)$
-1	-3	$(-1, -3)$
1	-3	$(1, -3)$
3	5	$(3, 5)$

Figure 2

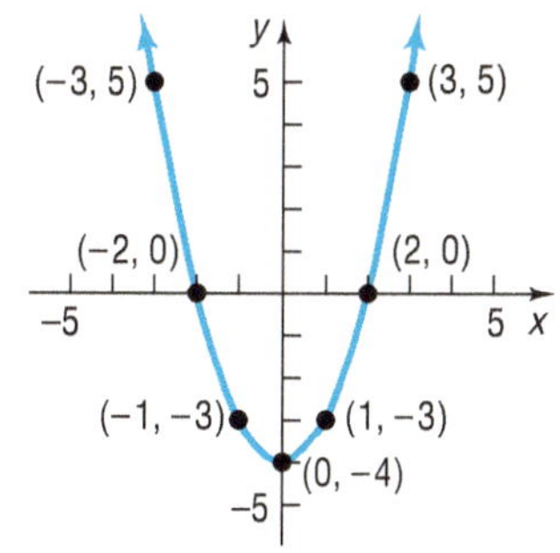

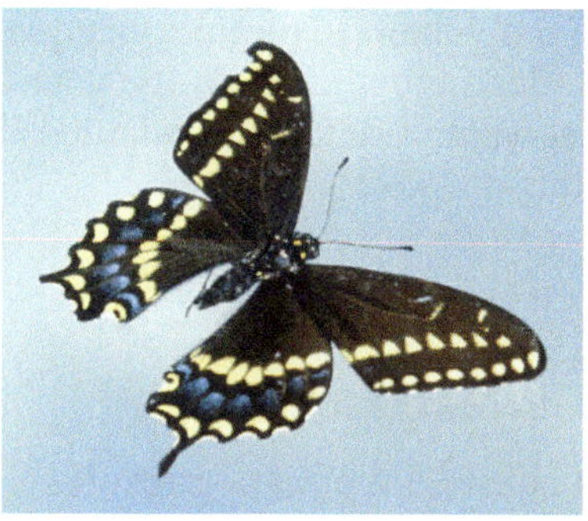

Now Work PROBLEM 13

2 Test an Equation for Symmetry

Another helpful tool for graphing equations by hand involves *symmetry,* particularly symmetry with respect to the x-axis, the y-axis, and the origin.

Symmetry often occurs in nature. Consider the picture of the butterfly. Do you see the symmetry?

DEFINITION

A graph is said to be **symmetric with respect to the x-axis** if, for every point (x, y) on the graph, the point $(x, -y)$ is also on the graph.

A graph is said to be **symmetric with respect to the y-axis** if, for every point (x, y) on the graph, the point $(-x, y)$ is also on the graph.

A graph is said to be **symmetric with respect to the origin** if, for every point (x, y) on the graph, the point $(-x, -y)$ is also on the graph.

Figure 3 illustrates the definition. Notice that, when a graph is symmetric with respect to the x-axis, the part of the graph above the x-axis is a reflection or mirror image of the part below it, and vice versa. When a graph is symmetric with respect to the y-axis, the part of the graph to the right of the y-axis is a reflection of the part to the left of it, and vice versa. Symmetry with respect to the origin may be viewed in two ways:

1. As a reflection about the y-axis, followed by a reflection about the x-axis
2. As a projection along a line through the origin so that the distances from the origin are equal

Figure 3

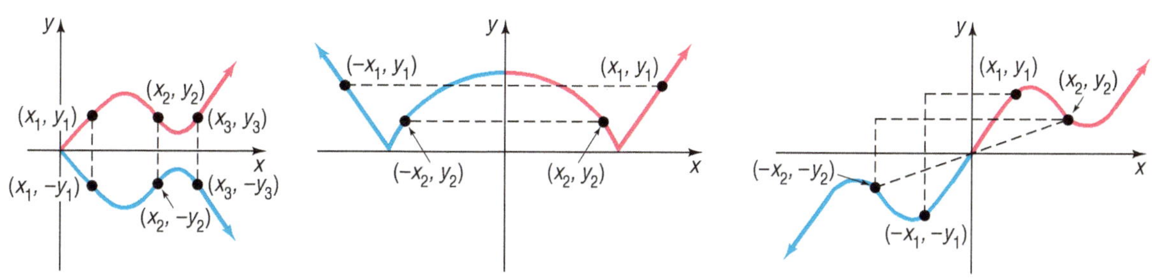

Symmetry with respect to the x-axis | Symmetry with respect to the y-axis | Symmetry with respect to the origin

EXAMPLE 2 **Symmetric Points**

(a) If a graph is symmetric with respect to the x-axis and the point $(4, 2)$ is on the graph, then the point $(4, -2)$ is also on the graph.

(b) If a graph is symmetric with respect to the y-axis and the point $(4, 2)$ is on the graph, then the point $(-4, 2)$ is also on the graph.

(c) If a graph is symmetric with respect to the origin and the point $(4, 2)$ is on the graph, then the point $(-4, -2)$ is also on the graph.

Now Work PROBLEM 21

When the graph of an equation is symmetric with respect to the x-axis, the y-axis, or the origin, the number of points that you need to plot in order to see the pattern is reduced. For example, if the graph of an equation is symmetric with respect to the y-axis, then, once points to the right of the y-axis are plotted, an equal number of points on the graph can be obtained by reflecting them about the y-axis. Because of this, before we graph an equation, we first want to determine whether it has any symmetry. The following tests are used for this purpose.

Tests for Symmetry

To test the graph of an equation for symmetry with respect to the

x-Axis Replace y by $-y$ in the equation. If an equivalent equation results, the graph of the equation is symmetric with respect to the x-axis.

y-Axis Replace x by $-x$ in the equation. If an equivalent equation results, the graph of the equation is symmetric with respect to the y-axis.

Origin Replace x by $-x$ and y by $-y$ in the equation. If an equivalent equation results, the graph of the equation is symmetric with respect to the origin.

EXAMPLE 3 Finding Intercepts and Testing an Equation for Symmetry

For the equation $y = \dfrac{x^2 - 4}{x^2 + 1}$ (a) find the intercepts and (b) test for symmetry.

Solution (a) To obtain the x-intercept(s), let $y = 0$ in the equation and solve for x.

$$\frac{x^2 - 4}{x^2 + 1} = 0 \quad \text{Let } y = 0.$$

$$x^2 - 4 = 0 \quad \text{Multiply both sides by } x^2 + 1.$$

$$x = -2 \quad \text{or} \quad x = 2 \quad \text{Factor and use the Zero-product Property.}$$

To obtain the y-intercept(s), let $x = 0$ in the equation and solve for y.

$$y = \frac{x^2 - 4}{x^2 + 1} = \frac{0^2 - 4}{0^2 + 1} = \frac{-4}{1} = -4$$

The x-intercepts are -2 and 2; the y-intercept is -4.

(b) We now test the equation for symmetry with respect to the x-axis, the y-axis, and the origin.

x-Axis: To test for symmetry with respect to the x-axis, replace y by $-y$. Since

$$-y = \frac{x^2 - 4}{x^2 + 1} \text{ is not equivalent to } y = \frac{x^2 - 4}{x^2 + 1}$$

the graph of the equation is not symmetric with respect to the x-axis.

y-Axis: To test for symmetry with respect to the y-axis, replace x by $-x$. Since

$$y = \frac{(-x)^2 - 4}{(-x)^2 + 1} = \frac{x^2 - 4}{x^2 + 1} \text{ is equivalent to } y = \frac{x^2 - 4}{x^2 + 1}$$

the graph of the equation is symmetric with respect to the y-axis.

Origin: To test for symmetry with respect to the origin, replace x by $-x$ and y by $-y$.

$$-y = \frac{(-x)^2 - 4}{(-x)^2 + 1} \quad \text{Replace } x \text{ by } -x \text{ and } y \text{ by } -y.$$

$$-y = \frac{x^2 - 4}{x^2 + 1} \quad \text{Simplify.}$$

Since the result is not equivalent to the original equation, the graph of the equation $y = \frac{x^2 - 4}{x^2 + 1}$ is not symmetric with respect to the origin.

Seeing the Concept

Figure 4

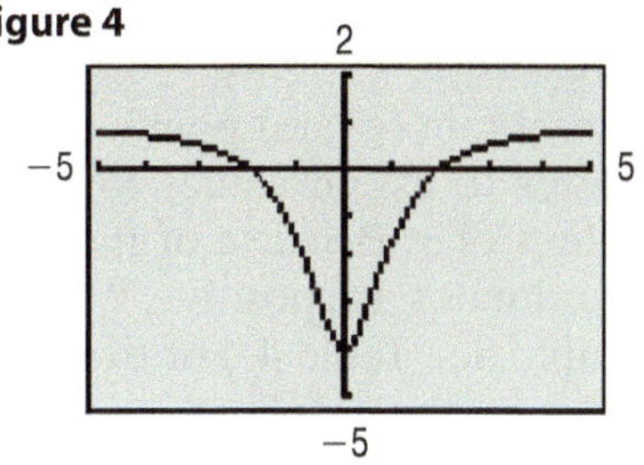

Figure 4 shows the graph of $y = \frac{x^2 - 4}{x^2 + 1}$ using a graphing utility. Do you see the symmetry with respect to the y-axis? Also, did you notice that the point (2, 0) is on the graph along with (−2, 0)? How could we have found the second x-intercept using symmetry?

Now Work PROBLEM 51

3 Know How to Graph Key Equations

There are certain equations whose graphs we should be able to easily visualize in our mind's eye. We have already discussed the graph of $y = x^2$ in Example 8 in Section 1.1. The next three examples use intercepts, symmetry, and point plotting to obtain the graphs of additional key equations. It is important to know the graphs of these key equations because we use them later.

EXAMPLE 4 Graphing the Equation $y = x^3$ by Finding Intercepts and Checking for Symmetry

Graph the equation $y = x^3$ by hand by plotting points. Find any intercepts and check for symmetry first.

Solution First, we seek the intercepts of $y = x^3$. When $x = 0$, then $y = 0$; and when $y = 0$, then $x = 0$. The origin $(0, 0)$ is the only intercept. Now we test $y = x^3$ for symmetry.

x-Axis: Replace y by $-y$. Since $-y = x^3$ is not equivalent to $y = x^3$, the graph is not symmetric with respect to the x-axis.

y-Axis: Replace x by $-x$. Since $y = (-x)^3 = -x^3$ is not equivalent to $y = x^3$, the graph is not symmetric with respect to the y-axis.

Origin: Replace x by $-x$ and y by $-y$. Since $-y = (-x)^3 = -x^3$ is equivalent to $y = x^3$ (multiply both sides by -1), the graph is symmetric with respect to the origin.

Figure 5

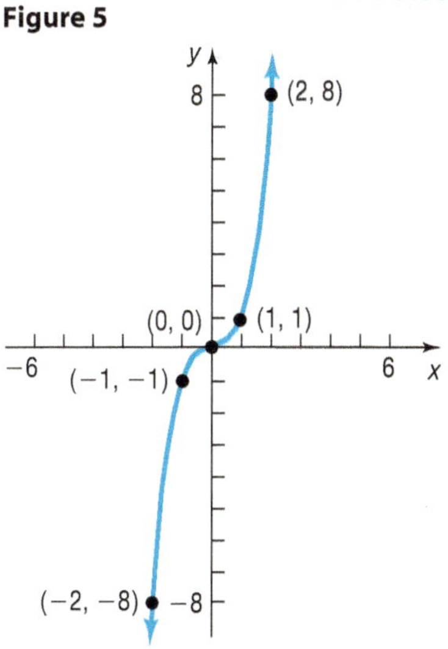

To graph by hand, we use the equation to obtain several points on the graph. Because of the symmetry with respect to the origin, we only need to locate points on the graph for which $x \geq 0$. See Table 2 on page 174. Points on the graph could also be obtained using the TABLE feature on a graphing utility. See Table 3. Do you see the symmetry with respect to the origin from the table? Figure 5 shows the graph.

Table 2

x	$y = x^3$	(x, y)
0	0	(0, 0)
1	1	(1, 1)
2	8	(2, 8)
3	27	(3, 27)

Table 3

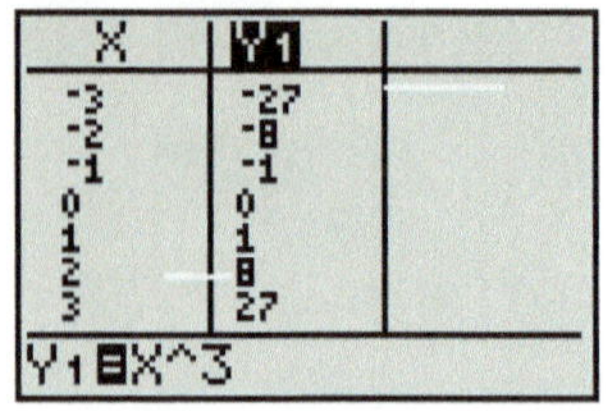

X	Y1
-3	-27
-2	-8
-1	-1
0	0
1	1
2	8
3	27

Y1=X^3

EXAMPLE 5 Graphing the Equation $x = y^2$

Graph the equation $x = y^2$. Find any intercepts and check for symmetry first.

Solution

The lone intercept is $(0, 0)$. The graph is symmetric with respect to the x-axis since $x = (-y)^2$ is equivalent to $x = y^2$. The graph is not symmetric with respect to the y-axis or the origin.

To graph $x = y^2$ by hand, we use the equation to obtain several points on the graph. Because the equation is solved for x, it is easier to assign values to y and use the equation to determine the corresponding values of x. Because of the symmetry, we can restrict ourselves to points whose y-coordinates are positive. We then use the symmetry to find additional points on the graph. See Table 4. For example, since $(1, 1)$ is on the graph, so is $(1, -1)$. Since $(4, 2)$ is on the graph, so is $(4, -2)$, and so on. We plot these points and connect them with a smooth curve to obtain Figure 6.

Table 4

y	$x = y^2$	(x, y)
0	0	(0, 0)
1	1	(1, 1)
2	4	(4, 2)
3	9	(9, 3)

Figure 6

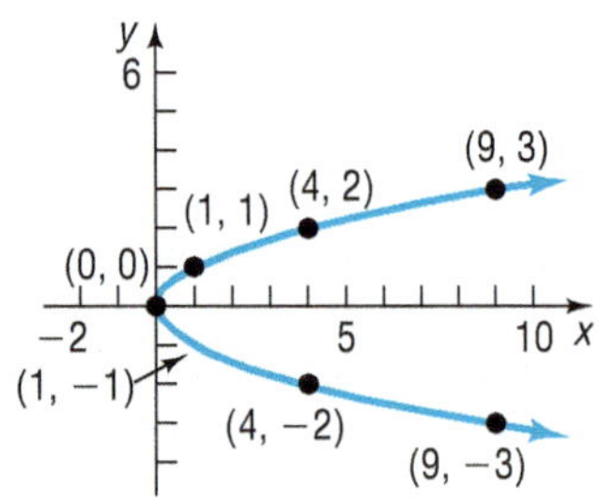

To graph the equation $x = y^2$ using a graphing utility, we must write the equation in the form $y = \{\text{expression in } x\}$. We proceed to solve for y.

$$x = y^2$$
$$y^2 = x$$
$$y = \pm\sqrt{x} \quad \text{Square Root Method}$$

To graph $x = y^2$, we need to graph both $Y_1 = \sqrt{x}$ and $Y_2 = -\sqrt{x}$ on the same screen. Figure 7 shows the result. Table 5 shows various values of y for a given value of x when $Y_1 = \sqrt{x}$ and $Y_2 = -\sqrt{x}$. Notice that when $x < 0$ we get an error. Can you explain why?

Figure 7

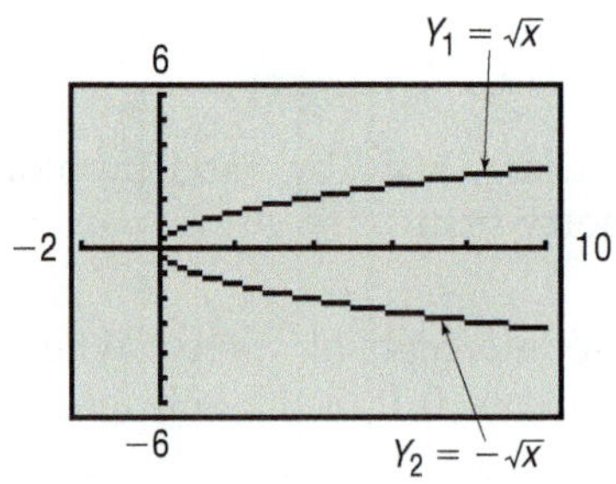

Table 5

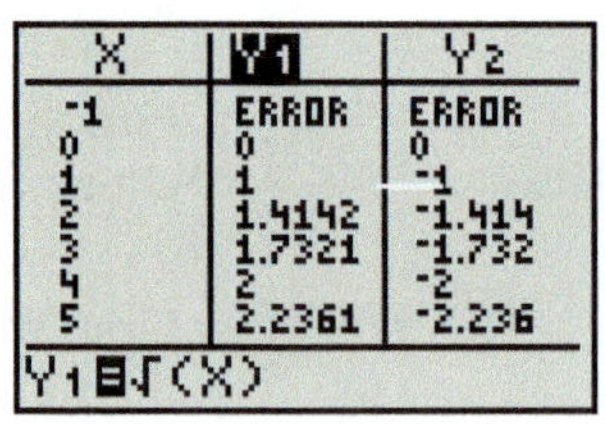

X	Y1	Y2
-1	ERROR	ERROR
0	0	0
1	1	-1
2	1.4142	-1.414
3	1.7321	-1.732
4	2	-2
5	2.2361	-2.236

Y1=√(X)

Look again at either Figure 6 or 7. If we restrict y so that $y \geq 0$, the equation $x = y^2$, $y \geq 0$, may be written as $y = \sqrt{x}$. The portion of the graph of $x = y^2$ in quadrant I plus the origin is the graph of $y = \sqrt{x}$. See Figure 8.

Figure 8

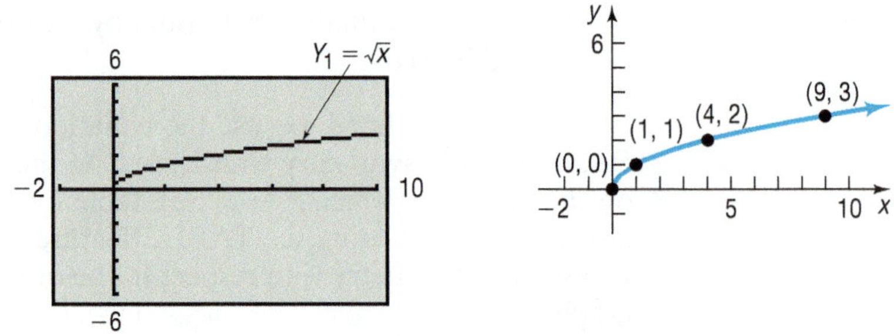

EXAMPLE 6

Graphing the Equation $y = \frac{1}{x}$

Graph the equation $y = \frac{1}{x}$. Find any intercepts and check for symmetry first.

Solution We check for intercepts first. If we let $x = 0$, we obtain a 0 in the denominator, which is not defined. We conclude that there is no y-intercept. If we let $y = 0$, we get the equation $\frac{1}{x} = 0$, which has no solution. We conclude that there is no x-intercept. The graph of $y = \frac{1}{x}$ does not cross or touch the coordinate axes.

Next we check for symmetry.

x-Axis: Replacing y by $-y$ yields $-y = \frac{1}{x}$, which is not equivalent to $y = \frac{1}{x}$.

y-Axis: Replacing x by $-x$ yields $y = \frac{1}{-x} = -\frac{1}{x}$, which is not equivalent to $y = \frac{1}{x}$.

Origin: Replacing x by $-x$ and y by $-y$ yields $-y = -\frac{1}{x}$, which is equivalent to $y = \frac{1}{x}$. The graph is symmetric only with respect to the origin.

Table 6

x	$y = \frac{1}{x}$	(x, y)
$\frac{1}{10}$	10	$\left(\frac{1}{10}, 10\right)$
$\frac{1}{3}$	3	$\left(\frac{1}{3}, 3\right)$
$\frac{1}{2}$	2	$\left(\frac{1}{2}, 2\right)$
1	1	$(1, 1)$
2	$\frac{1}{2}$	$\left(2, \frac{1}{2}\right)$
3	$\frac{1}{3}$	$\left(3, \frac{1}{3}\right)$
10	$\frac{1}{10}$	$\left(10, \frac{1}{10}\right)$

We can use the equation to form Table 6 and obtain some points on the graph. Because of symmetry, we only find points (x, y) for which x is positive. From Table 6 we infer that, if x is a large and positive number, then $y = \frac{1}{x}$ is a positive number close to 0. We also infer that if x is a positive number close to 0 then $y = \frac{1}{x}$ is a large and positive number. Armed with this information, we can graph the equation. Figure 9 illustrates some of these points and the graph of $y = \frac{1}{x}$. Observe how the absence of intercepts and the existence of symmetry with respect to the origin were utilized. Figure 10 confirms our algebraic analysis using a TI-84 Plus.

Figure 9

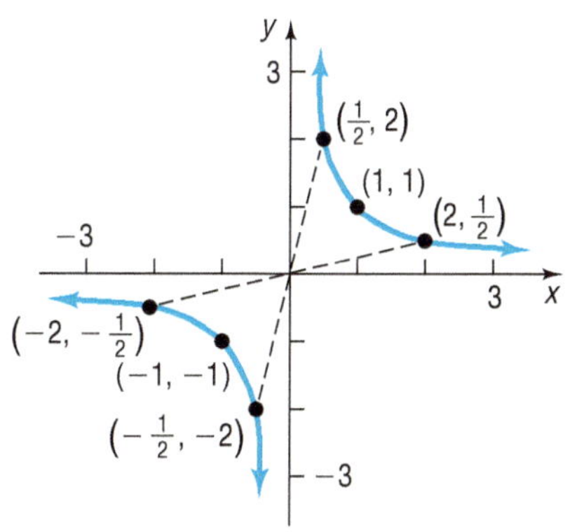

Figure 10

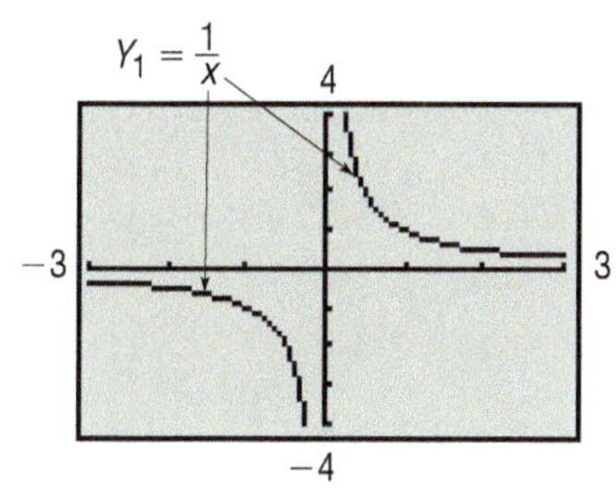

2.1 Assess Your Understanding

'Are You Prepared?' *Answers are given at the end of these exercises. If you get a wrong answer, read the pages listed in red.*

1. Graph $y = 2x - 4$ by plotting points. Based on the graph, determine the intercepts. (pp. 91–93)

2. Solve: $x^2 - 4x - 12 = 0$ (pp. 113–114)

Concepts and Vocabulary

3. The x-intercepts of the graph of an equation are those x-values for which _______.

4. If for every point (x, y) on the graph of an equation the point $(-x, y)$ is also on the graph, then the graph is symmetric with respect to the _______.

5. If the graph of an equation is symmetric with respect to the y-axis and -4 is an x-intercept of this graph, then _______ is also an x-intercept.

6. If the graph of an equation is symmetric with respect to the origin and $(3, -4)$ is a point on the graph, then _______ is also a point on the graph.

7. ***True or False*** To find the y-intercepts of the graph of an equation, let $x = 0$ and solve for y.

8. ***True or False*** If a graph is symmetric with respect to the x-axis, then it cannot be symmetric with respect to the y-axis.

Skill Building

In Problems 9–20, find the intercepts and graph each equation by plotting points. Be sure to label the intercepts.

9. $y = x + 2$ **10.** $y = x - 6$ **11.** $y = 2x + 8$ **12.** $y = 3x - 9$

13. $y = x^2 - 1$ **14.** $y = x^2 - 9$ **15.** $y = -x^2 + 4$ **16.** $y = -x^2 + 1$

17. $2x + 3y = 6$ **18.** $5x + 2y = 10$ **19.** $9x^2 + 4y = 36$ **20.** $4x^2 + y = 4$

In Problems 21–30, plot each point. Then plot the point that is symmetric to it with respect to (a) the x-axis; (b) the y-axis; (c) the origin.

21. $(3, 4)$ **22.** $(5, 3)$ **23.** $(-2, 1)$ **24.** $(4, -2)$ **25.** $(5, -2)$

26. $(-1, -1)$ **27.** $(-3, -4)$ **28.** $(4, 0)$ **29.** $(0, -3)$ **30.** $(-3, 0)$

In Problems 31–38, the graph of an equation is given. Indicate whether the graph is symmetric with respect to the x-axis, the y-axis, or the origin.

31.
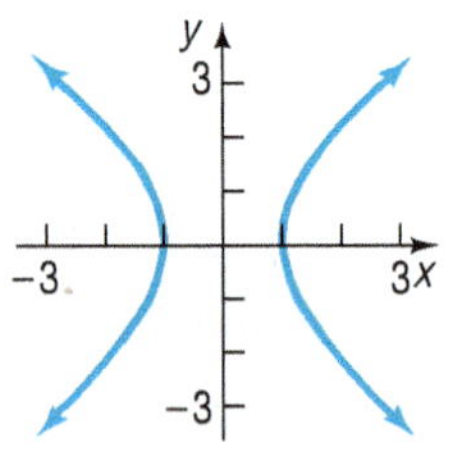

32.
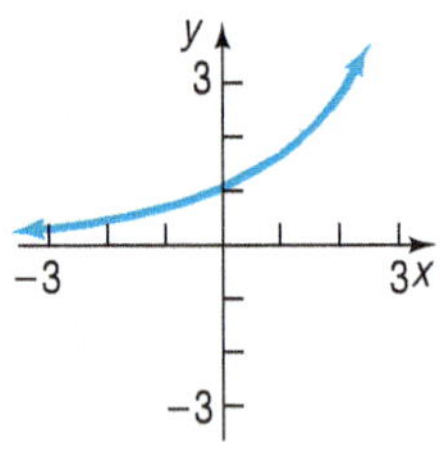

33.
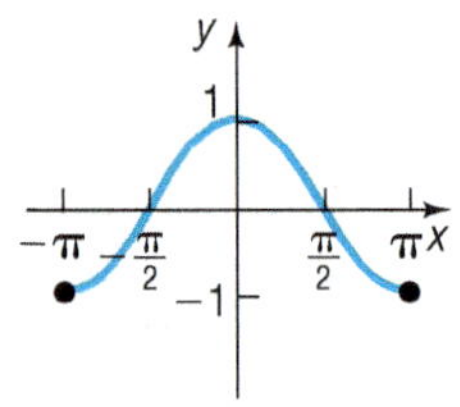

34.
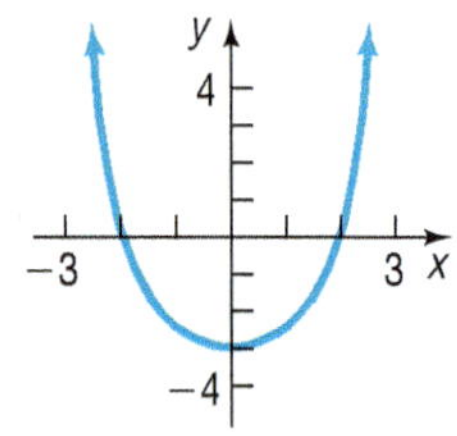

35.
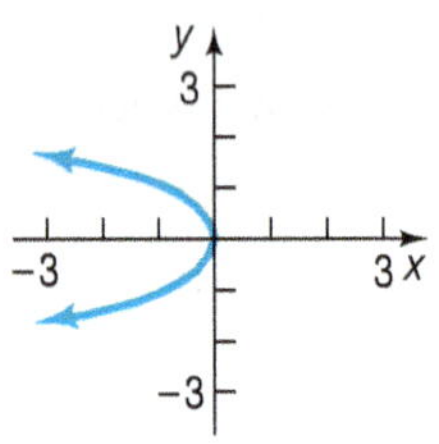

36.
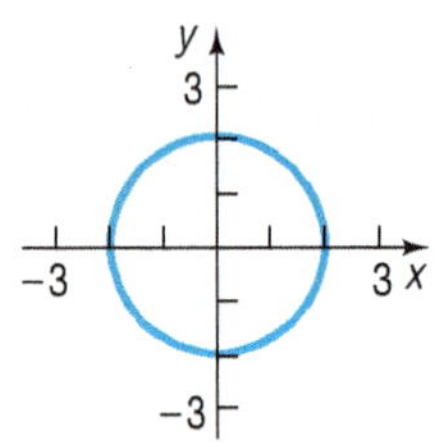

37.
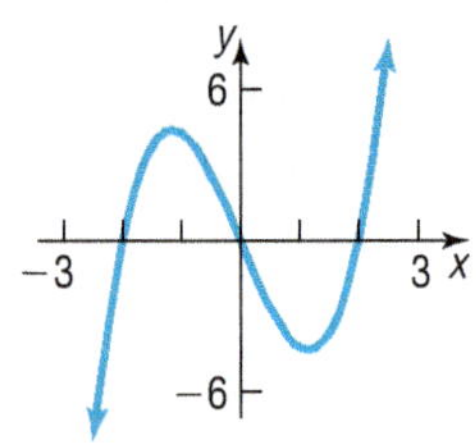

38.
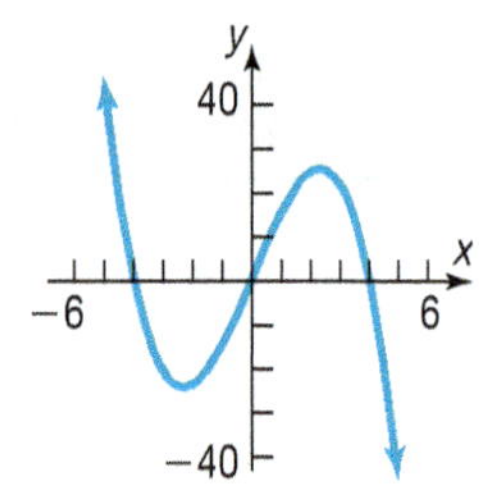

In Problems 39–42, draw a complete graph so that it has the type of symmetry indicated.

39. y-axis
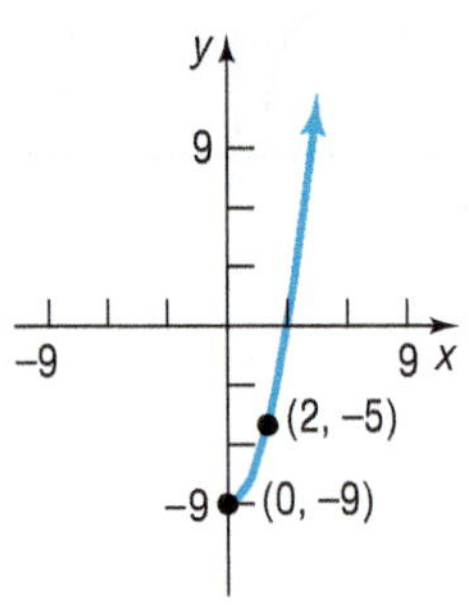

40. x-axis
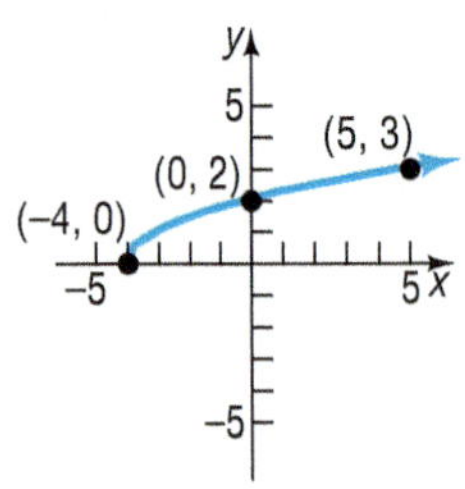

41. Origin
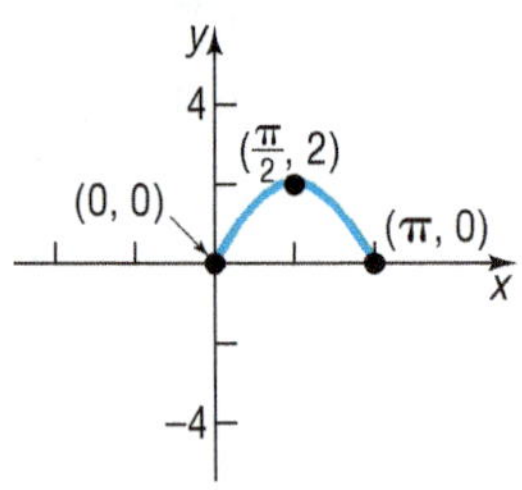

42. y-axis
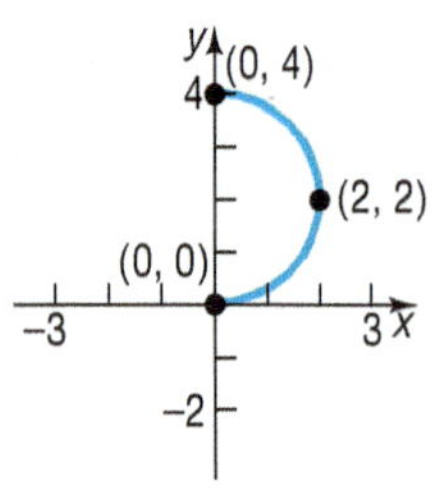

In Problems 43–46, draw a quick sketch of the equation by hand. Be sure to label all intercepts.

43. $y = x^3$ **44.** $x = y^2$ **45.** $y = \sqrt{x}$ **46.** $y = \dfrac{1}{x}$

In Problems 47–62, list the intercepts and test for symmetry.

47. $y^2 = x + 4$ **48.** $y^2 = x + 9$ **49.** $y = \sqrt[3]{x}$ **50.** $y = \sqrt[5]{x}$

51. $y = x^4 - 8x^2 - 9$ **52.** $y = x^4 - 2x^2 - 8$ **53.** $9x^2 + 4y^2 = 36$ **54.** $4x^2 + y^2 = 4$

55. $y = x^3 + x^2 - 9x - 9$ **56.** $y = x^3 + 2x^2 - 4x - 8$ **57.** $y = |x| - 4$ **58.** $y = |x| - 2$

59. $y = \dfrac{3x}{x^2 + 9}$ **60.** $y = \dfrac{x^2 - 4}{2x}$ **61.** $y = \dfrac{-x^3}{x^2 - 9}$ **62.** $y = \dfrac{x^4 + 1}{2x^5}$

Mixed Practice

In Problems 63–70, (a) find the intercepts of each equation, (b) test each equation for symmetry with respect to the x-axis, the y-axis, and the origin, and (c) graph each equation by hand by plotting points. Be sure to label the intercepts on the graph and use any symmetry to assist in drawing the graph. Verify your results using a graphing utility.

63. $y = x^2 - 5$ **64.** $y = x^2 - 8$ **65.** $x - y^2 = -9$ **66.** $x + y^2 = 4$

67. $x^2 + y^2 = 9$ **68.** $x^2 + y^2 = 16$ **69.** $y = x^3 - 4x$ **70.** $y = x^3 - x$

Applications and Extensions

71. Given that the point $(1, 2)$ is on the graph of an equation that is symmetric with respect to the origin, what other point is on the graph?

72. If the graph of an equation is symmetric with respect to the y-axis and 6 is an x-intercept of this graph, name another x-intercept.

73. If the graph of an equation is symmetric with respect to the origin and -4 is an x-intercept of this graph, name another x-intercept.

74. If the graph of an equation is symmetric with respect to the x-axis and 2 is a y-intercept, name another y-intercept.

75. Microphones In studios and on stages, cardioid microphones are often preferred for the richness they add to voices and for their ability to reduce the level of sound from the sides and rear of the microphone. Suppose one such cardioid pattern is given by the equation $(x^2 + y^2 - x)^2 = x^2 + y^2$.

(a) Find the intercepts of the graph of the equation.

(b) Test for symmetry with respect to the x-axis, y-axis, and origin.

Source: *www.notaviva.com*

76. Solar Energy The solar electric generating systems at Kramer Junction, California, use parabolic troughs to heat a heat-transfer fluid to a high temperature. This fluid is used to generate steam that drives a power conversion system to produce electricity. For troughs 7.5 feet wide, an equation for the cross section is $16y^2 = 120x - 225$.

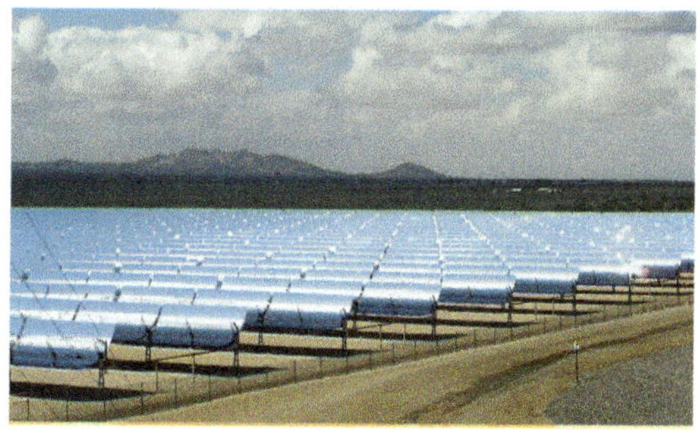

(a) Find the intercepts of the graph of the equation.

(b) Test for symmetry with respect to the x-axis, y-axis, and origin.

Source: *U.S. Department of Energy*

Discussion and Writing

77. An equation is being tested for symmetry with respect to the x-axis, the y-axis, and the origin. Explain why, if two of these symmetries are present, the remaining one must also be present.

78. Draw a graph that contains the points $(-2, 5)$, $(-1, 3)$, and $(0, 2)$ that is symmetric with respect to the y-axis. Compare your graph with those of other students; comment on any similarities. Can a graph contain these points and be symmetric with respect to the x-axis? the origin? Why or why not?

'Are You Prepared?' Answers

1.

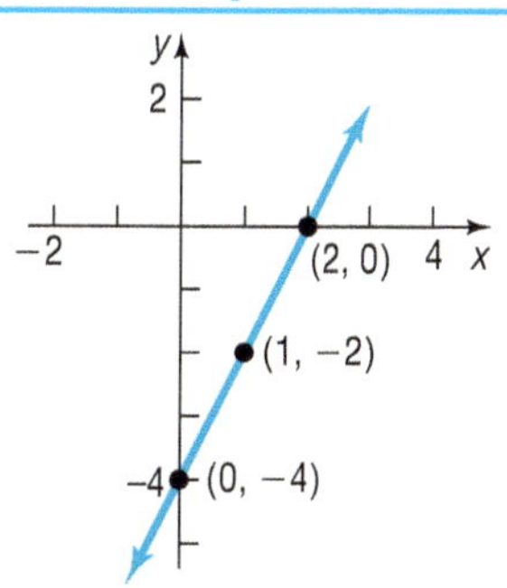

x-intercept: 2; y-intercept: -4

2. $\{-2, 6\}$

2.2 Lines

OBJECTIVES

In this section we study a certain type of equation that contains two variables, called a *linear equation,* and its graph, a *line.*

1 Calculate and Interpret the Slope of a Line

Figure 11

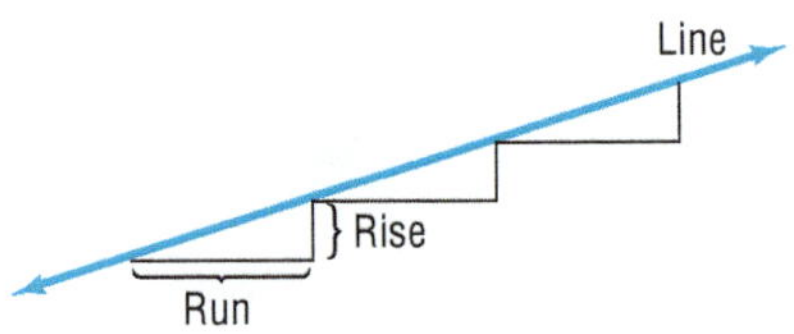

Consider the staircase illustrated in Figure 11. Each step contains exactly the same horizontal **run** and the same vertical **rise.** The ratio of the rise to the run, called the *slope,* is a numerical measure of the steepness of the staircase. For example, if the run is increased and the rise remains the same, the staircase becomes less steep. If the run is kept the same, but the rise is increased, the staircase becomes more steep. This important characteristic of a line is best defined using rectangular coordinates.

DEFINITION

Let $P = (x_1, y_1)$ and $Q = (x_2, y_2)$ be two distinct points. If $x_1 \neq x_2$, the **slope *m*** of the nonvertical line L containing P and Q is defined by the formula

$$m = \frac{y_2 - y_1}{x_2 - x_1} \qquad x_1 \neq x_2 \tag{1}$$

If $x_1 = x_2$, L is a **vertical line** and the slope m of L is **undefined** (since this results in division by 0).

Figure 12(a) provides an illustration of the slope of a nonvertical line; Figure 12(b) illustrates a vertical line.

Figure 12

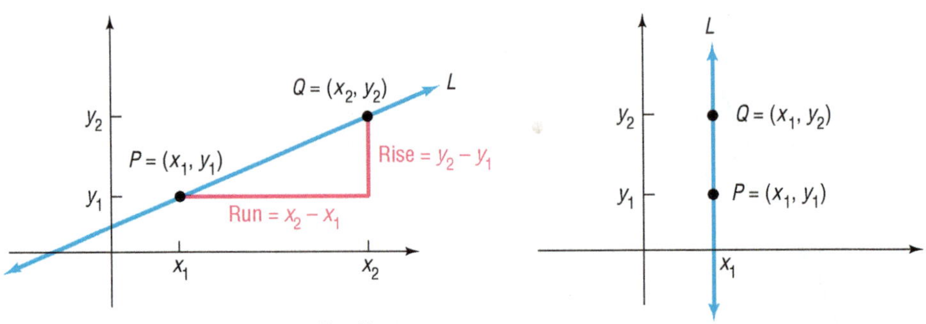

(a) Slope of L is $m = \frac{y_2 - y_1}{x_2 - x_1}$

(b) Slope is undefined; L is vertical

As Figure 12(a) illustrates, the slope m of a nonvertical line may be viewed as

$$m = \frac{y_2 - y_1}{x_2 - x_1} = \frac{\text{Rise}}{\text{Run}} \quad \text{or} \quad m = \frac{y_2 - y_1}{x_2 - x_1} = \frac{\text{Change in } y}{\text{Change in } x} = \frac{\Delta y}{\Delta x}$$

The slope m of a nonvertical line L measures the amount that y changes as x changes from x_1 to x_2. We call this the **average rate of change** of y with respect to x.

In Words

The symbol Δ is the Greek letter delta. In mathematics, we read Δ as "change in," so $\frac{\Delta y}{\Delta x}$ is read, "change in y divided by change in x."

Two comments about computing the slope of a nonvertical line may prove helpful:

1. Any two distinct points on the line can be used to compute the slope of the line. (See Figure 13 for justification.)

Figure 13

Triangles ABC and PQR are similar (equal angles). Hence, ratios of corresponding sides are proportional so that

$$\text{Slope using } P \text{ and } Q = \frac{y_2 - y_1}{x_2 - x_1} = \frac{d(B, C)}{d(A, C)} = \text{Slope using } A \text{ and } B$$

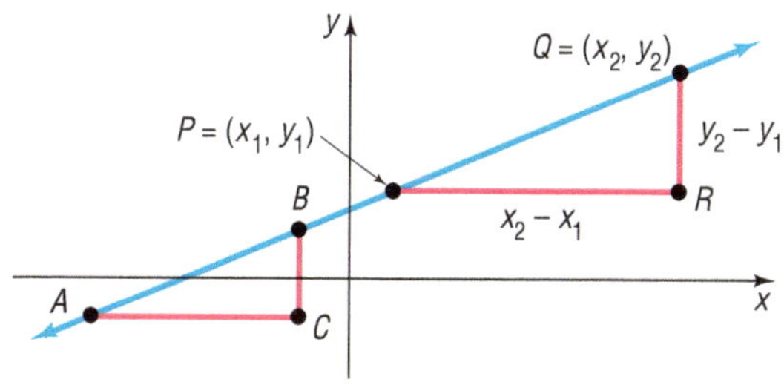

2. The slope of a line may be computed from $P = (x_1, y_1)$ to $Q = (x_2, y_2)$ or from Q to P because

$$\frac{y_2 - y_1}{x_2 - x_1} = \frac{y_1 - y_2}{x_1 - x_2}$$

EXAMPLE 1 Finding and Interpreting the Slope of a Line Containing Two Points

The slope m of the line containing the points $(1, 2)$ and $(5, -3)$ may be computed as

$$m = \frac{-3 - 2}{5 - 1} = \frac{-5}{4} = -\frac{5}{4} \quad \text{or as} \quad m = \frac{2 - (-3)}{1 - 5} = \frac{5}{-4} = -\frac{5}{4}$$

For every 4-unit change in x, y will change by -5 units. That is, if x increases by 4 units, then y will decrease by 5 units. The average rate of change of y with respect to x is $-\frac{5}{4}$.

Now Work PROBLEMS 11 AND 17

Square Screens

To get an undistorted view of slope, the same scale must be used on each axis. However, most graphing utilities have a rectangular screen. Because of this, using the same interval for both x and y will result in a distorted view. For example, Figure 14 shows the graph of the line $y = x$ connecting the points $(-4, -4)$ and $(4, 4)$. We expect the line to bisect the first and third quadrants, but it doesn't. We need to adjust the selections for Xmin, Xmax, Ymin, and Ymax so that a **square screen** results. On most graphing utilities, this is accomplished by setting the ratio of x to y at 3:2.*

Figure 14

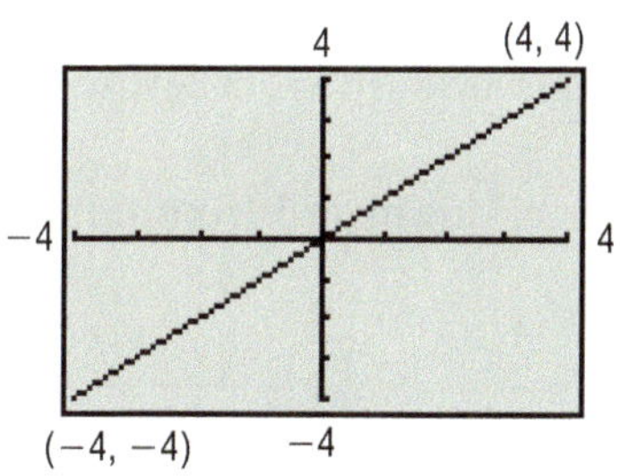

Figure 15 shows the graph of the line $y = x$ on a square screen using a TI-84 Plus. Notice that the line now bisects the first and third quadrants. Compare this illustration to Figure 14.

Figure 15

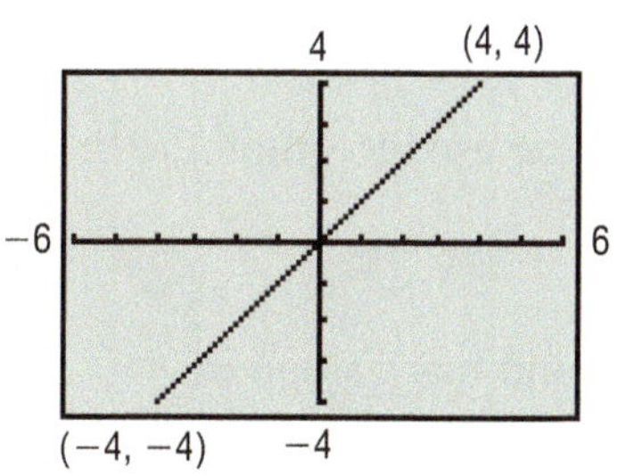

*Most graphing utilities have a feature that automatically squares the viewing window. Consult your owner's manual for the appropriate keystrokes.

To get a better idea of the meaning of the slope m of a line, consider the following:

Exploration

On the same square screen, graph the following equations:

$Y_1 = 0$ Slope of line is 0.

$Y_2 = \frac{1}{4}x$ Slope of line is $\frac{1}{4}$.

$Y_3 = \frac{1}{2}x$ Slope of line is $\frac{1}{2}$.

$Y_4 = x$ Slope of line is 1.

$Y_5 = 2x$ Slope of line is 2.

$Y_6 = 6x$ Slope of line is 6.

See Figure 16.

Figure 16

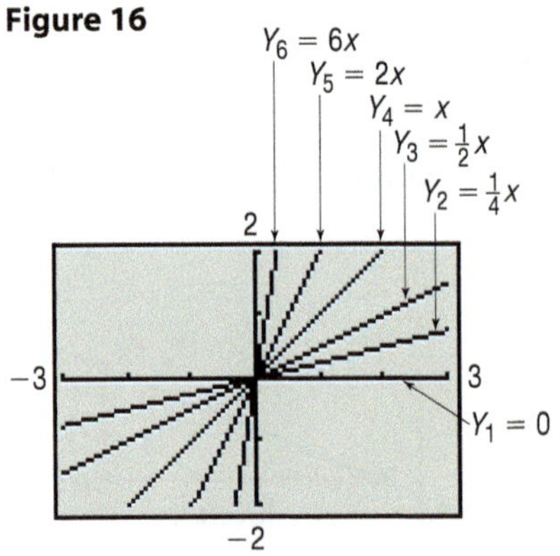

Exploration

On the same square screen, graph the following equations:

$Y_1 = 0$ Slope of line is 0.

$Y_2 = -\frac{1}{4}x$ Slope of line is $-\frac{1}{4}$.

$Y_3 = -\frac{1}{2}x$ Slope of line is $-\frac{1}{2}$.

$Y_4 = -x$ Slope of line is −1.

$Y_5 = -2x$ Slope of line is −2.

$Y_6 = -6x$ Slope of line is −6.

See Figure 17.

Figure 17

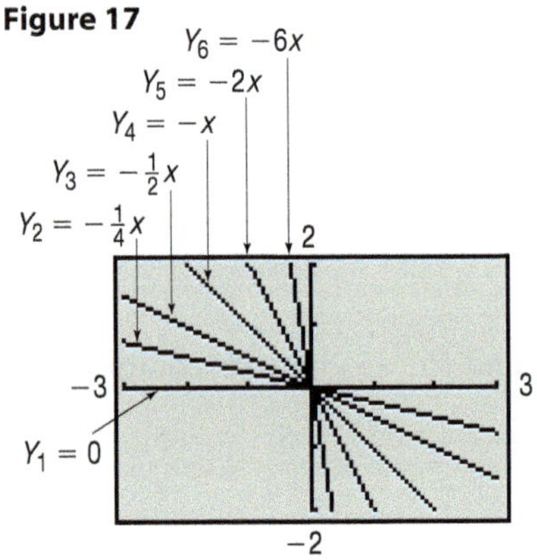

Figures 16 and 17 illustrate the following facts:

1. When the slope of a line is positive, the line slants upward from left to right.
2. When the slope of a line is negative, the line slants downward from left to right.
3. When the slope is 0, the line is horizontal.

Figures 16 and 17 also illustrate that the closer the line is to the vertical position, the greater the magnitude of the slope. So, a line with slope 6 will be steeper than a line whose slope is 3.

2 Graph Lines Given a Point and the Slope

The next example illustrates how the slope of a line can be used to graph the line.

EXAMPLE 2 **Graphing a Line Given a Point and a Slope**

Draw a graph of the line that contains the point $(3, 2)$ and has a slope of:

(a) $\frac{3}{4}$ (b) $-\frac{4}{5}$

Solution (a) Slope $= \dfrac{\text{Rise}}{\text{Run}}$. The fact that the slope is $\dfrac{3}{4}$ means that for every horizontal movement (run) of 4 units to the right there will be a vertical movement (rise) of 3 units. If we start at the given point $(3, 2)$ and move 4 units to the right and 3 units up, we reach the point $(7, 5)$. By drawing the line through this point and the point $(3, 2)$, we have the graph. See Figure 18.

Figure 18

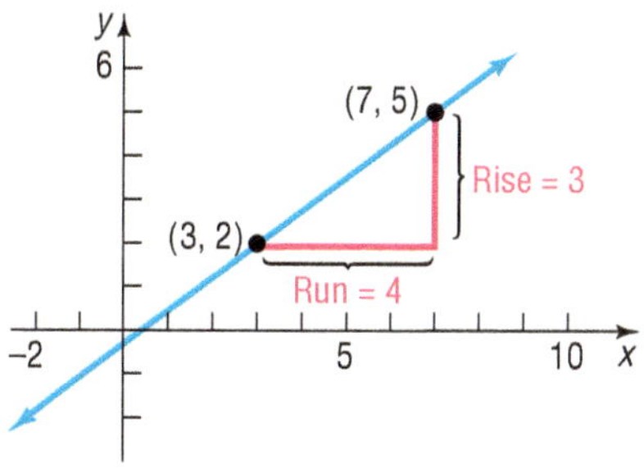

(b) The fact that the slope is

$$-\frac{4}{5} = \frac{-4}{5} = \frac{\text{Rise}}{\text{Run}}$$

means that for every horizontal movement of 5 units to the right there will be a corresponding vertical movement of -4 units (a downward movement of 4 units). If we start at the given point $(3, 2)$ and move 5 units to the right and then 4 units down, we arrive at the point $(8, -2)$. By drawing the line through these points, we have the graph. See Figure 19.

Alternatively, we can set

$$-\frac{4}{5} = \frac{4}{-5} = \frac{\text{Rise}}{\text{Run}}$$

so that for every horizontal movement of -5 units (a movement to the left of 5 units) there will be a corresponding vertical movement of 4 units (upward). This approach brings us to the point $(-2, 6)$, which is also on the graph shown in Figure 19.

Figure 19

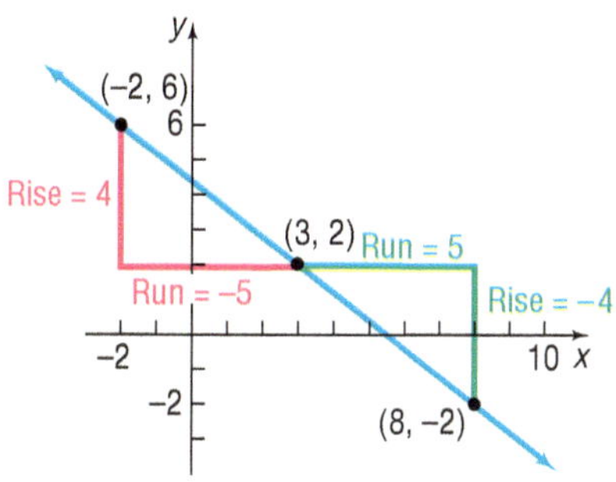

Now Work PROBLEM 23

3 Find the Equation of a Vertical Line

Now that we have discussed the slope of a line, we are ready to derive equations of lines. As we shall see, there are several forms of the equation of a line. Let's start with an example.

EXAMPLE 3 Graphing a Line

Graph the equation: $x = 3$

Solution To graph $x = 3$ by hand, we are looking for all points (x, y) in the plane for which $x = 3$. No matter what y-coordinate is used, the corresponding x-coordinate always equals 3. Consequently, the graph of the equation $x = 3$ is a vertical line with x-intercept 3 and undefined slope. See Figure 20(a).

To use a graphing utility, we need to express the equation in the form $y = \{\text{expression in } x\}$. But $x = 3$ cannot be put into this form, so an alternative method must be used. Consult your manual to determine the methodology required to draw vertical lines. Figure 20(b) shows the graph that you should obtain.

Figure 20
$x = 3$

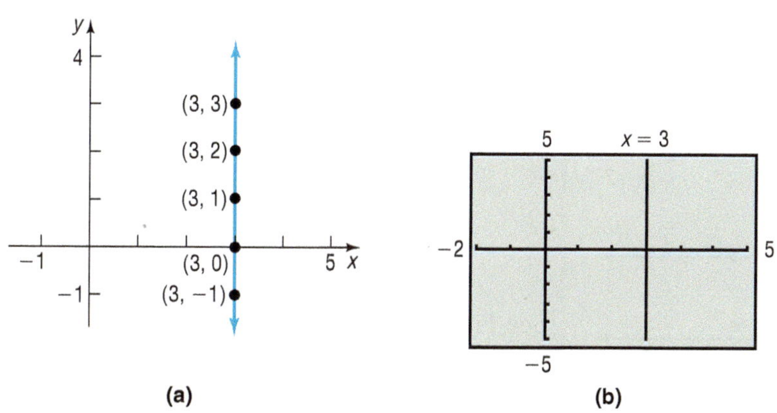

As suggested by Example 3, we have the following result:

THEOREM

Equation of a Vertical Line

A vertical line is given by an equation of the form

$$x = a$$

where a is the x-intercept.

Figure 21

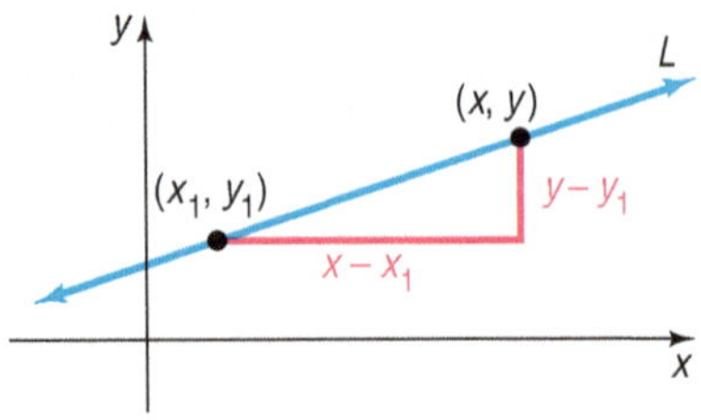

4 Use the Point–Slope Form of a Line; Identify Horizontal Lines

Now let L be a nonvertical line with slope m and containing the point (x_1, y_1). See Figure 21. For any other point (x, y) on L, we have

$$m = \frac{y - y_1}{x - x_1} \quad \text{or} \quad y - y_1 = m(x - x_1)$$

THEOREM

Point–Slope Form of an Equation of a Line

An equation of a nonvertical line of slope m that contains the point (x_1, y_1) is

$$y - y_1 = m(x - x_1) \qquad \textbf{(2)}$$

EXAMPLE 4 **Using the Point–Slope Form of a Line**

An equation of the line with slope 4 and containing the point $(1, 2)$ can be found by using the point–slope form with $m = 4$, $x_1 = 1$, and $y_1 = 2$.

$$\begin{aligned} y - y_1 &= m(x - x_1) && \text{Point-slope form} \\ y - 2 &= 4(x - 1) && m = 4, x_1 = 1, y_1 = 2 \\ y &= 4x - 2 \end{aligned}$$

See Figure 22 for the graph.

Figure 22

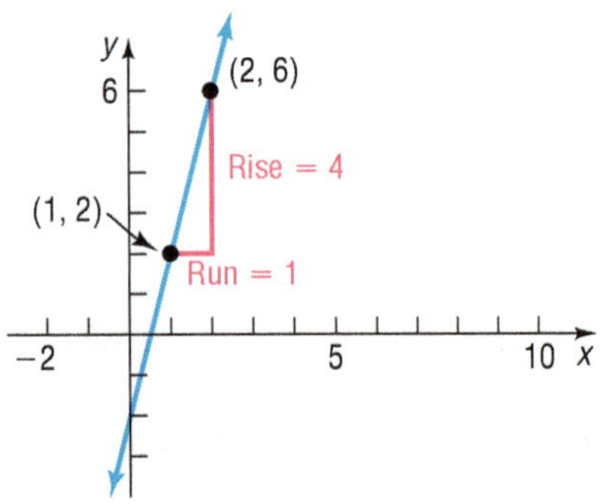

EXAMPLE 5 **Finding the Equation of a Horizontal Line**

Find an equation of the horizontal line containing the point $(3, 2)$.

Solution Because all the y-values are the same on a horizontal line, the slope of a horizontal line is 0. To get an equation, we use the point–slope form with $m = 0$, $x_1 = 3$, and $y_1 = 2$.

$$\begin{aligned} y - y_1 &= m(x - x_1) && \text{Point-slope form} \\ y - 2 &= 0 \cdot (x - 3) && m = 0, x_1 = 3, \text{and } y_1 = 2 \\ y - 2 &= 0 \\ y &= 2 \end{aligned}$$

See Figure 23 for the graph.

Figure 23
$y = 2$

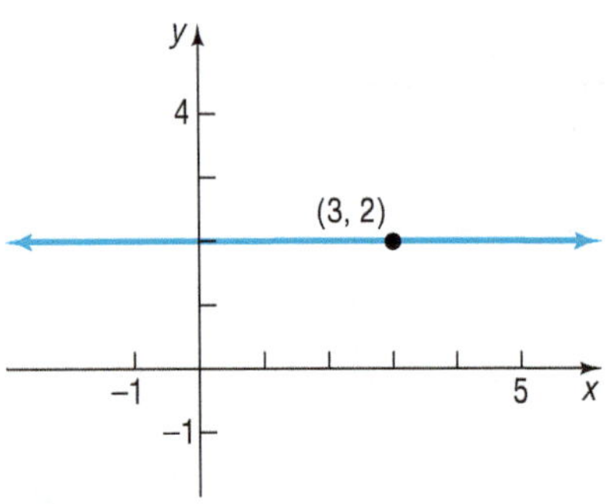

As suggested by Example 5, we have the following result:

THEOREM

Equation of a Horizontal Line

A horizontal line is given by an equation of the form

$$y = b$$

where b is the y-intercept.

5 Find the Equation of a Line Given Two Points

We use the slope formula and the point–slope form of a line to find an equation given two points.

EXAMPLE 6 **Finding an Equation of a Line Given Two Points**

Find an equation of the line containing the points $(2, 3)$ and $(-4, 5)$. Graph the line.

Solution We first compute the slope of the line with $(x_1, y_1) = (2, 3)$ and $(x_2, y_2) = (-4, 5)$.

$$m = \frac{5 - 3}{-4 - 2} = \frac{2}{-6} = -\frac{1}{3} \qquad m = \frac{y_2 - y_1}{x_2 - x_1}$$

We use the point $(x_1, y_1) = (2, 3)$ and the slope $m = -\frac{1}{3}$ to get the point–slope form of the equation of the line.

$$y - 3 = -\frac{1}{3}(x - 2) \qquad y - y_1 = m(x - x_1)$$

See Figure 24 for the graph.

Figure 24

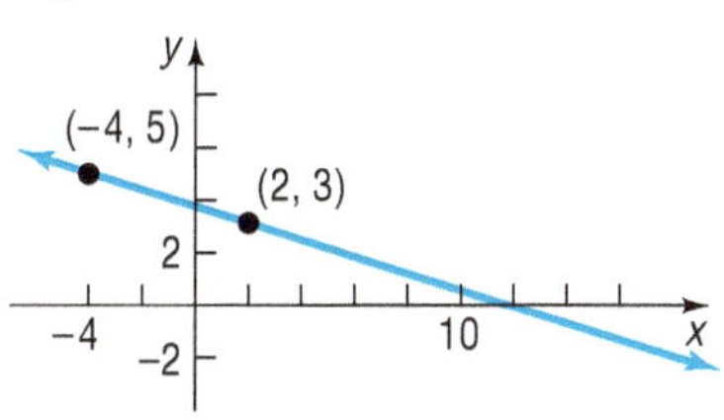

In the solution to Example 6, we could have used the other point, $(-4, 5)$, instead of the point $(2, 3)$. The equation that results, although it looks different, is equivalent to the equation that we obtained in the example. (Try it for yourself.)

Now Work PROBLEM 37

6 Write the Equation of a Line in Slope–Intercept Form

Another useful equation of a line is obtained when the slope m and y-intercept b are known. In this event, we know both the slope m of the line and a point $(0, b)$ on the line; thus, we may use the point–slope form, equation (2), to obtain the following equation:

$$y - b = m(x - 0) \quad \text{or} \quad y = mx + b$$

THEOREM

Slope–Intercept Form of an Equation of a Line

An equation of a line with slope m and y-intercept b is

$$y = mx + b \tag{3}$$

Now Work PROBLEM 45
(EXPRESS ANSWER IN SLOPE–INTERCEPT FORM)

Seeing the Concept

Figure 25
$y = mx + 2$

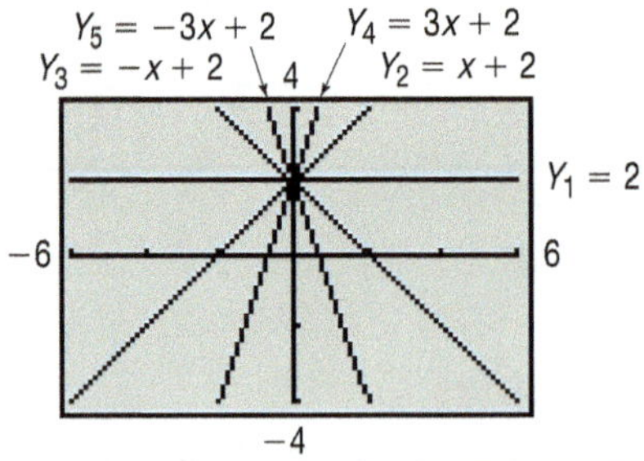

To see the role that the slope m plays, graph the following lines on the same square screen.

$$Y_1 = 2 \quad m = 0$$
$$Y_2 = x + 2 \quad m = 1$$
$$Y_3 = -x + 2 \quad m = -1$$
$$Y_4 = 3x + 2 \quad m = 3$$
$$Y_5 = -3x + 2 \quad m = -3$$

See Figure 25. What do you conclude about the lines $y = mx + 2$?

Seeing the Concept

Figure 26
$y = 2x + b$

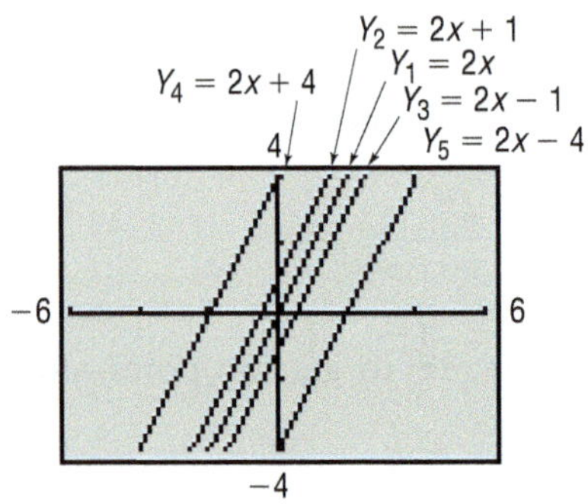

To see the role of the y-intercept b, graph the following lines on the same square screen.

$$Y_1 = 2x \quad b = 0$$
$$Y_2 = 2x + 1 \quad b = 1$$
$$Y_3 = 2x - 1 \quad b = -1$$
$$Y_4 = 2x + 4 \quad b = 4$$
$$Y_5 = 2x - 4 \quad b = -4$$

See Figure 26. What do you conclude about the lines $y = 2x + b$?

7 Identify the Slope and y-Intercept of a Line from Its Equation

When the equation of a line is written in slope–intercept form, it is easy to find the slope m and y-intercept b of the line. For example, suppose that the equation of a line is

$$y = -2x + 3$$

Compare it to $y = mx + b$.

$$y = -2x + 3$$
$$\uparrow \quad \uparrow$$
$$y = mx + b$$

The slope of this line is -2 and its y-intercept is 3.

Now Work PROBLEM 71

EXAMPLE 7 **Finding the Slope and y-Intercept**

Find the slope m and y-intercept b of the equation $2x + 4y = 8$. Graph the equation.

Solution To obtain the slope and y-intercept, we transform the equation into its slope–intercept form by solving for y.

$$2x + 4y = 8$$
$$4y = -2x + 8$$
$$y = -\frac{1}{2}x + 2 \qquad y = mx + b$$

Figure 27

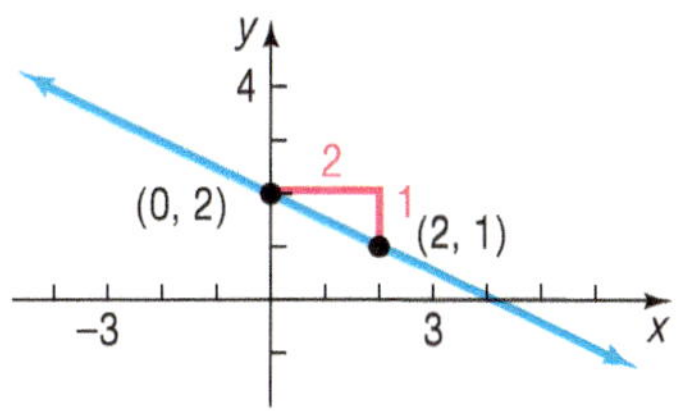

The coefficient of x, $-\frac{1}{2}$, is the slope, and the y-intercept is 2. We can graph the line using the fact that the y-intercept is 2 and the slope is $-\frac{1}{2}$. Starting at the point $(0, 2)$, go to the right 2 units and then down 1 unit to the point $(2, 1)$. See Figure 27.

Now Work PROBLEM 77

8 Graph Lines Written in General Form Using Intercepts

The form of the equation of the line in Example 7, $2x + 4y = 8$, is called the *general form*.

DEFINITION

The equation of a line is in **general form** when it is written as

$$Ax + By = C \tag{4}$$

where A, B, and C are real numbers and A and B are not both 0.

When we want to graph a linear equation that is written in general form, we can solve the equation for y and write the equation in slope–intercept form as we did in Example 7. Another approach to graphing the equation would be to find its intercepts. Remember, the intercepts of the graph of an equation are the points where the graph crosses or touches a coordinate axis.

EXAMPLE 8 Graphing a Linear Equation Using Its Intercepts

Graph the linear equation $2x + 4y = 8$ by finding its intercepts.

Solution To obtain the x-intercept, let $y = 0$ in the equation and solve for x.

$$2x + 4y = 8$$
$$2x + 4(0) = 8 \quad \text{Let } y = 0.$$
$$2x = 8$$
$$x = 4 \quad \text{Divide both sides by 2.}$$

The x-intercept is 4 and the point $(4, 0)$ is on the graph of the equation.

To obtain the y-intercept, let $x = 0$ in the equation and solve for y.

$$2x + 4y = 8$$
$$2(0) + 4y = 8 \quad \text{Let } x = 0.$$
$$4y = 8$$
$$y = 2 \quad \text{Divide both sides by 4.}$$

The y-intercept is 2 and the point $(0, 2)$ is on the graph of the equation.

We plot the points $(4, 0)$ and $(0, 2)$ in a Cartesian plane and draw a line through the points. See Figure 28.

Figure 28

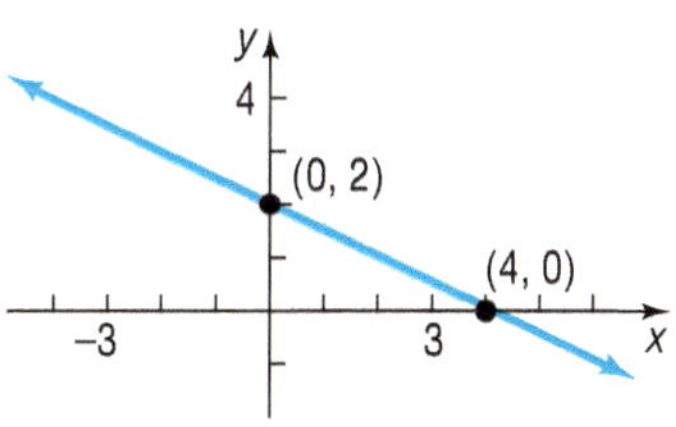

Now Work PROBLEM 91

Every line has an equation that is equivalent to an equation written in general form. For example, a vertical line whose equation is

$$x = a$$

can be written in the general form

$$1 \cdot x + 0 \cdot y = a \quad A = 1, B = 0, C = a$$

A horizontal line whose equation is

$$y = b$$

can be written in the general form

$$0 \cdot x + 1 \cdot y = b \quad A = 0, B = 1, C = b$$

Lines that are neither vertical nor horizontal have general equations of the form

$$Ax + By = C \quad A \neq 0 \text{ and } B \neq 0$$

Because the equation of every line can be written in general form, any equation equivalent to equation (4) is called a **linear equation.**

Figure 29
The distinct lines are parallel if and only if their slopes are equal.

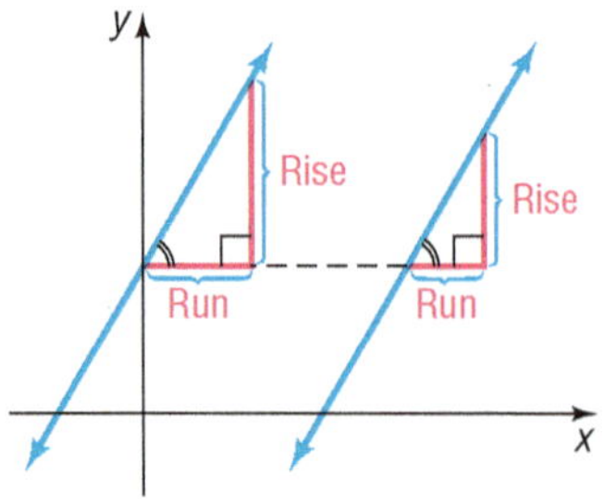

9 Find Equations of Parallel Lines

When two lines (in the plane) do not intersect (that is, they have no points in common), they are said to be **parallel.** Look at Figure 29. There we have drawn two lines and have constructed two right triangles by drawing sides parallel to the coordinate axes. These lines are parallel if and only if the right triangles are similar. (Do you see why? Two angles are equal.) And the triangles are similar if and only if the ratios of corresponding sides are equal.

This suggests the following result:

THEOREM

Criterion for Parallel Lines

Two nonvertical lines are parallel if and only if their slopes are equal and they have different y-intercepts.

The use of the words "if and only if" in the preceding theorem means that actually two statements are being made, one the converse of the other.

If two nonvertical lines are parallel, then their slopes are equal and they have different y-intercepts.

If two nonvertical lines have equal slopes and they have different y-intercepts, then they are parallel.

EXAMPLE 9 Showing That Two Lines Are Parallel

Show that the lines given by the following equations are parallel:

$$L_1: \; 2x + 3y = 6, \qquad L_2: \; 4x + 6y = 0$$

Solution We need to determine whether these lines have equal slopes and different y-intercepts, so we write each equation in slope–intercept form:

$$L_1: \; 2x + 3y = 6 \qquad\qquad L_2: \; 4x + 6y = 0$$
$$3y = -2x + 6 \qquad\qquad 6y = -4x$$
$$y = -\frac{2}{3}x + 2 \qquad\qquad y = -\frac{2}{3}x$$

$$\text{Slope} = -\frac{2}{3}; \; y\text{-intercept} = 2 \qquad \text{Slope} = -\frac{2}{3}; \; y\text{-intercept} = 0$$

Because these lines have the same slope, $-\frac{2}{3}$, but different y-intercepts, the lines are parallel. See Figure 30.

Figure 30
Parallel lines

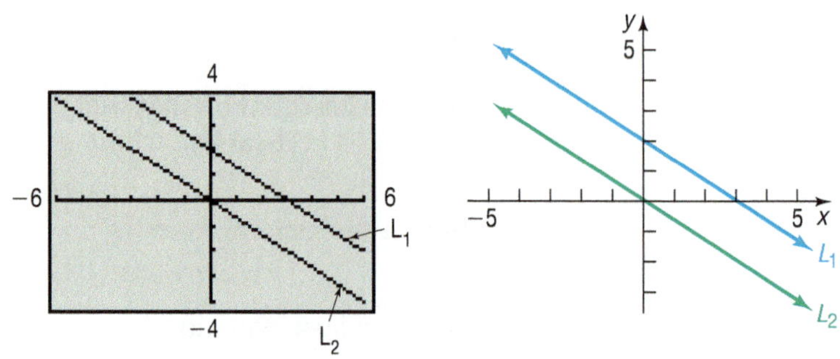

EXAMPLE 10 Finding a Line That Is Parallel to a Given Line

Find an equation for the line that contains the point $(2, -3)$ and is parallel to the line $2x + y = 6$. By hand, graph the two lines.

Solution Since the two lines are to be parallel, the slope of the line that we seek equals the slope of the line $2x + y = 6$. We begin by writing the equation of the line $2x + y = 6$ in slope–intercept form.

$$2x + y = 6$$
$$y = -2x + 6$$

The slope is -2. Since the line that we seek contains the point $(2, -3)$, we use the point–slope form to obtain

$$y - y_1 = m(x - x_1) \quad \text{Point–slope form}$$
$$y - (-3) = -2(x - 2) \quad m = -2, x_1 = 2, y_1 = -3$$
$$y + 3 = -2x + 4 \quad \text{Simplify.}$$
$$y = -2x + 1 \quad \text{Slope–intercept form}$$
$$2x + y = 1 \quad \text{General form}$$

This line is parallel to the line $2x + y = 6$ and contains the point $(2, -3)$. See Figure 31.

Figure 31

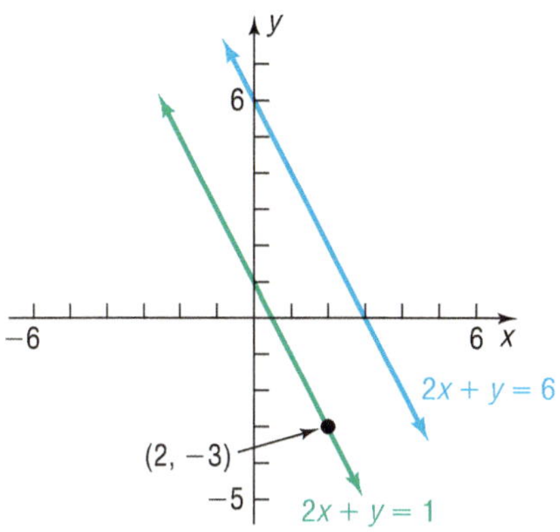

Now Work PROBLEM 59

10 Find Equations of Perpendicular Lines

When two lines intersect at a right angle (90°), they are said to be **perpendicular.** See Figure 32.

Figure 32 Perpendicular lines

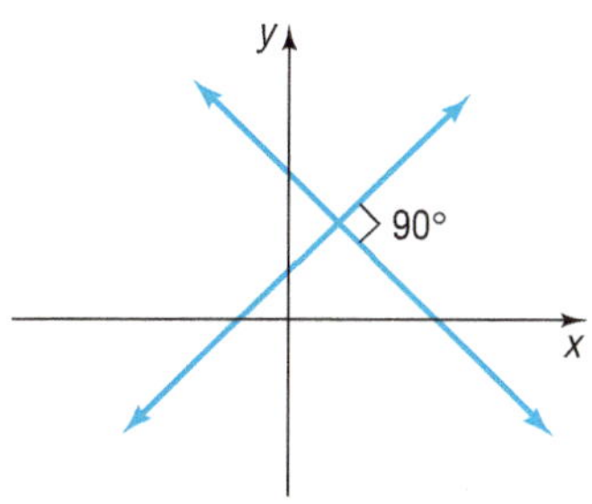

The following result gives a condition, in terms of their slopes, for two lines to be perpendicular.

THEOREM

Criterion for Perpendicular Lines

Two nonvertical lines are perpendicular if and only if the product of their slopes is -1.

Here we shall prove the "only if" part of the statement:

If two nonvertical lines are perpendicular, then the product of their slopes is -1.

In Problem 128, you are asked to prove the "if" part of the theorem; that is:

If two nonvertical lines have slopes whose product is -1, then the lines are perpendicular.

Figure 33

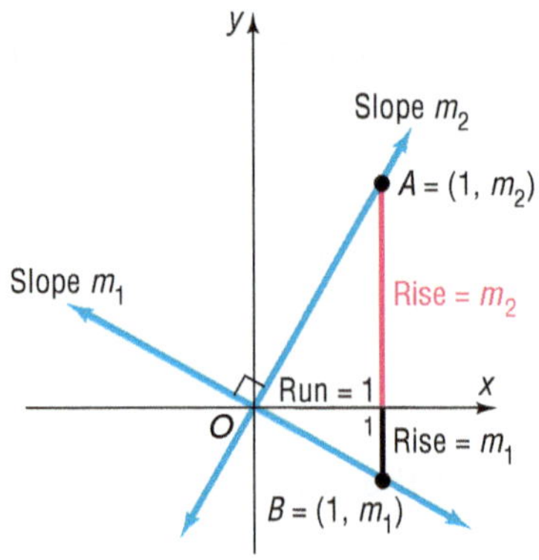

Proof Let m_1 and m_2 denote the slopes of the two lines. There is no loss in generality (that is, neither the angle nor the slopes are affected) if we situate the lines so that they meet at the origin. See Figure 33. The point $A = (1, m_2)$ is on the line having slope m_2, and the point $B = (1, m_1)$ is on the line having slope m_1. (Do you see why this must be true?)

Suppose that the lines are perpendicular. Then triangle OAB is a right triangle. As a result of the Pythagorean Theorem, it follows that

$$[d(O, A)]^2 + [d(O, B)]^2 = [d(A, B)]^2 \qquad \textbf{(5)}$$

By the distance formula, we can write the squares of these distances as

$$[d(O, A)]^2 = (1 - 0)^2 + (m_2 - 0)^2 = 1 + m_2^2$$
$$[d(O, B)]^2 = (1 - 0)^2 + (m_1 - 0)^2 = 1 + m_1^2$$
$$[d(A, B)]^2 = (1 - 1)^2 + (m_2 - m_1)^2 = m_2^2 - 2m_1m_2 + m_1^2$$

Using these facts in equation (5), we get

$$(1 + m_2^2) + (1 + m_1^2) = m_2^2 - 2m_1m_2 + m_1^2$$

which, upon simplification, can be written as

$$m_1m_2 = -1$$

If the lines are perpendicular, the product of their slopes is -1. ■

You may find it easier to remember the condition for two nonvertical lines to be perpendicular by observing that the equality $m_1m_2 = -1$ means that m_1 and m_2 are negative reciprocals of each other; that is, either $m_1 = -\frac{1}{m_2}$ or $m_2 = -\frac{1}{m_1}$.

EXAMPLE 11 Finding the Slope of a Line Perpendicular to Another Line

If a line has slope $\frac{3}{2}$, any line having slope $-\frac{2}{3}$ is perpendicular to it.

EXAMPLE 12 Finding the Equation of a Line Perpendicular to a Given Line

Find an equation of the line that contains the point $(1, -2)$ and is perpendicular to the line $x + 3y = 6$. Graph the two lines.

Solution We first write the equation of the given line in slope–intercept form to find its slope.

$$x + 3y = 6$$
$$3y = -x + 6 \qquad \text{Proceed to solve for } y.$$
$$y = -\frac{1}{3}x + 2 \qquad \text{Place in the form } y = mx + b.$$

The given line has slope $-\frac{1}{3}$. Any line perpendicular to this line will have slope 3.

Because we require the point $(1, -2)$ to be on this line with slope 3, we use the point–slope form of the equation of a line.

$$y - y_1 = m(x - x_1) \qquad \text{Point–slope form}$$
$$y - (-2) = 3(x - 1) \qquad m = 3, x_1 = 1, y_1 = -2$$

To obtain other forms of the equation, we proceed as follows:

$$y + 2 = 3(x - 1)$$
$$y + 2 = 3x - 3 \qquad \text{Simplify}$$
$$y = 3x - 5 \qquad \text{Slope–intercept form}$$
$$3x - y = 5 \qquad \text{General form}$$

Figure 34 shows the graphs.

Figure 34

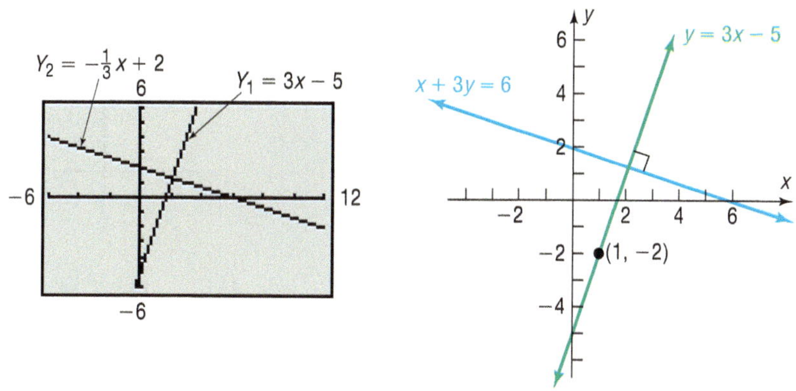

WARNING Be sure to use a square screen when you graph perpendicular lines. Otherwise, the angle between the two lines will appear distorted. ■

Now Work PROBLEM 65

2.2 Assess Your Understanding

Concepts and Vocabulary

1. The slope of a vertical line is __________; the slope of a horizontal line is __________.
2. For the line $2x + 3y = 6$, the x-intercept is __________ and the y-intercept is __________.
3. A horizontal line is given by an equation of the form __________, where b is the __________.
4. ***True or False*** Vertical lines have an undefined slope.
5. ***True or False*** The slope of the line $2y = 3x + 5$ is 3.
6. ***True or False*** The point (1, 2) is on the line $2x + y = 4$.
7. Two nonvertical lines have slopes m_1 and m_2, respectively. The lines are parallel if __________ and the __________ are unequal; the lines are perpendicular if __________.
8. The lines $y = 2x + 3$ and $y = ax + 5$ are parallel if $a =$ __________.
9. The lines $y = 2x - 1$ and $y = ax + 2$ are perpendicular if $a =$ __________.
10. ***True or False*** Perpendicular lines have slopes that are reciprocals of one another.

Skill Building

In Problems 11–14, (a) find the slope of the line and (b) interpret the slope.

11.
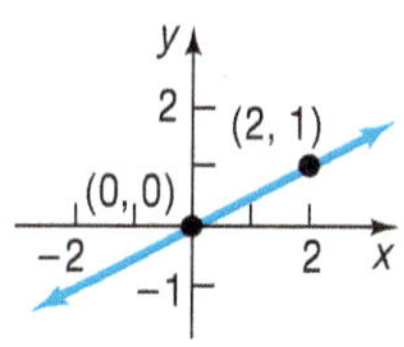

12.
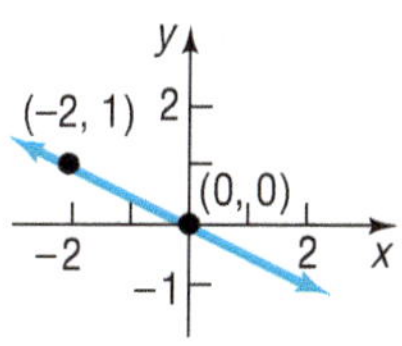

13.
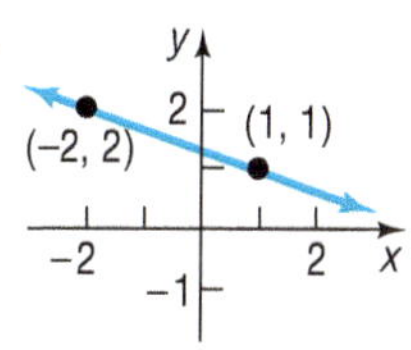

14.
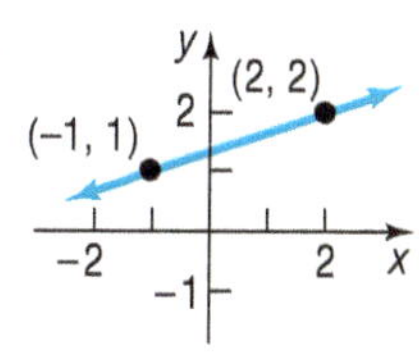

In Problems 15–22, plot each pair of points and determine the slope of the line containing them. Graph the line.

15. (2, 3); (4, 0)
16. (4, 2); (3, 4)
17. (−2, 3); (2, 1)
18. (−1, 1); (2, 3)
19. (−3, −1); (2, −1)
20. (4, 2); (−5, 2)
21. (−1, 2); (−1, −2)
22. (2, 0); (2, 2)

In Problems 23–30, graph the line containing the point P and having slope m.

23. $P = (1, 2); m = 3$
24. $P = (2, 1); m = 4$
25. $P = (2, 4); m = -\frac{3}{4}$
26. $P = (1, 3); m = -\frac{2}{5}$
27. $P = (-1, 3); m = 0$
28. $P = (2, -4); m = 0$
29. $P = (0, 3)$; slope undefined
30. $P = (-2, 0)$; slope undefined

In Problems 31–36, the slope and a point on a line are given. Use this information to locate three additional points on the line. Answers may vary.
[**Hint:** *It is not necessary to find the equation of the line. See Example 2.*]

31. Slope 4; point (1, 2)
32. Slope 2; point (−2, 3)
33. Slope $-\frac{3}{2}$; point (2, −4)
34. Slope $\frac{4}{3}$; point (−3, 2)
35. Slope −2; point (−2, −3)
36. Slope −1; point (4, 1)

In Problems 37–44, find an equation of the line L.

37.

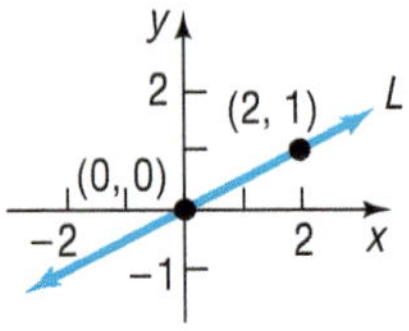

38.

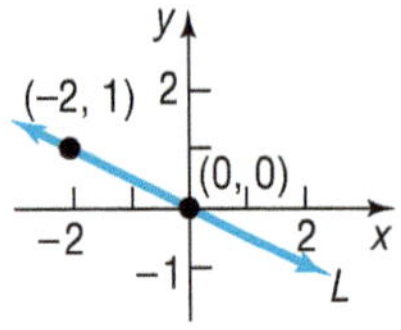

39.

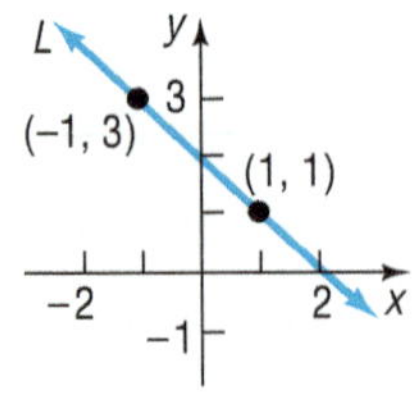

40.

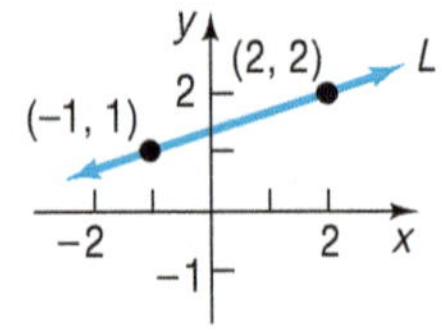

41.

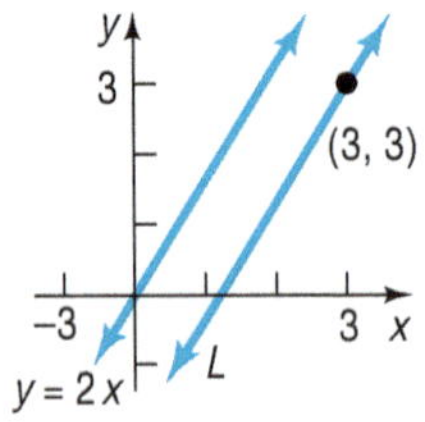

L is parallel to y = 2x

42.

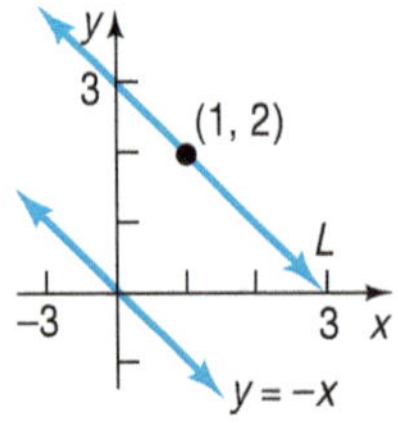

L is parallel to y = −x

43.

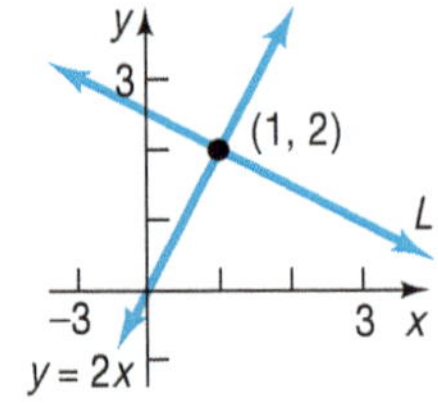

L is perpendicular to y = 2x

44.

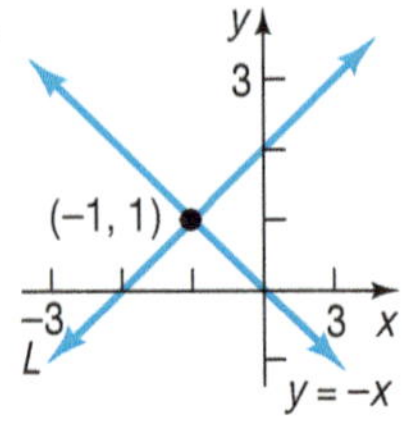

L is perpendicular to y = −x

In Problems 45–70, find an equation for the line with the given properties. Express your answer using either the general form or the slope–intercept form of the equation of a line, whichever you prefer.

45. Slope $= 3$; containing the point $(-2, 3)$

46. Slope $= 2$; containing the point $(4, -3)$

47. Slope $= -\frac{2}{3}$; containing the point $(1, -1)$

48. Slope $= \frac{1}{2}$; containing the point $(3, 1)$

49. Containing the points $(1, 3)$ and $(-1, 2)$

50. Containing the points $(-3, 4)$ and $(2, 5)$

51. Slope $= -3$; y-intercept $= 3$

52. Slope $= -2$; y-intercept $= -2$

53. x-intercept $= 2$; y-intercept $= -1$

54. x-intercept $= -4$; y-intercept $= 4$

55. Slope undefined; containing the point $(2, 4)$

56. Slope undefined; containing the point $(3, 8)$

57. Horizontal; containing the point $(-3, 2)$

58. Vertical; containing the point $(4, -5)$

59. Parallel to the line $y = 2x$; containing the point $(-1, 2)$

60. Parallel to the line $y = -3x$; containing the point $(-1, 2)$

61. Parallel to the line $2x - y = -2$; containing the point $(0, 0)$

62. Parallel to the line $x - 2y = -5$; containing the point $(0, 0)$

63. Parallel to the line $x = 5$; containing the point $(4, 2)$

64. Parallel to the line $y = 5$; containing the point $(4, 2)$

65. Perpendicular to the line $y = \frac{1}{2}x + 4$; containing the point $(1, -2)$

66. Perpendicular to the line $y = 2x - 3$; containing the point $(1, -2)$

67. Perpendicular to the line $2x + y = 2$; containing the point $(-3, 0)$

68. Perpendicular to the line $x - 2y = -5$; containing the point $(0, 4)$

69. Perpendicular to the line $x = 8$; containing the point $(3, 4)$

70. Perpendicular to the line $y = 8$; containing the point $(3, 4)$

In Problems 71–90, find the slope and y-intercept of each line. Graph the line by hand. Check your graph using a graphing utility.

71. $y = 2x + 3$ **72.** $y = -3x + 4$ **73.** $\frac{1}{2}y = x - 1$ **74.** $\frac{1}{3}x + y = 2$ **75.** $y = \frac{1}{2}x + 2$

76. $y = 2x + \frac{1}{2}$ **77.** $x + 2y = 4$ **78.** $-x + 3y = 6$ **79.** $2x - 3y = 6$ **80.** $3x + 2y = 6$

81. $x + y = 1$ **82.** $x - y = 2$ **83.** $x = -4$ **84.** $y = -1$ **85.** $y = 5$

86. $x = 2$ **87.** $y - x = 0$ **88.** $x + y = 0$ **89.** $2y - 3x = 0$ **90.** $3x + 2y = 0$

In Problems 91–100, (a) find the intercepts of the graph of each equation and (b) graph the equation.

91. $2x + 3y = 6$

92. $3x - 2y = 6$

93. $-4x + 5y = 40$

94. $6x - 4y = 24$

95. $7x + 2y = 21$

96. $5x + 3y = 18$

97. $\frac{1}{2}x + \frac{1}{3}y = 1$

98. $x - \frac{2}{3}y = 4$

99. $0.2x - 0.5y = 1$

100. $-0.3x + 0.4y = 1.2$

101. Find an equation of the x-axis.

102. Find an equation of the y-axis.

In Problems 103–106, the equations of two lines are given. Determine if the lines are parallel, perpendicular, or neither.

103. $y = 2x - 3$
$y = 2x + 4$

104. $y = \frac{1}{2}x - 3$
$y = -2x + 4$

105. $y = 4x + 5$
$y = -4x + 2$

106. $y = -2x + 3$
$y = -\frac{1}{2}x + 2$

In Problems 107–110, match each graph with the correct equation:

(a) $y = x$ (b) $y = 2x$ (c) $y = \frac{x}{2}$ (d) $y = 4x$

107.

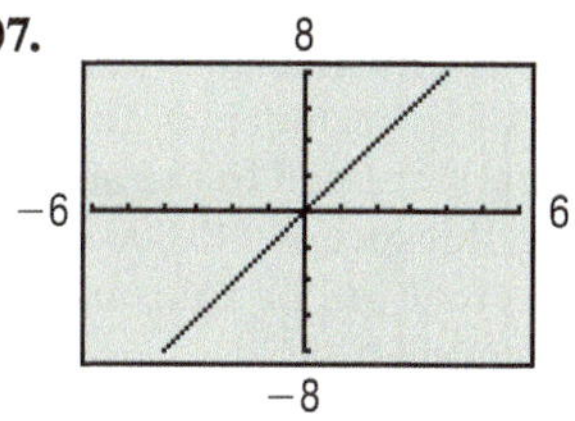

108.

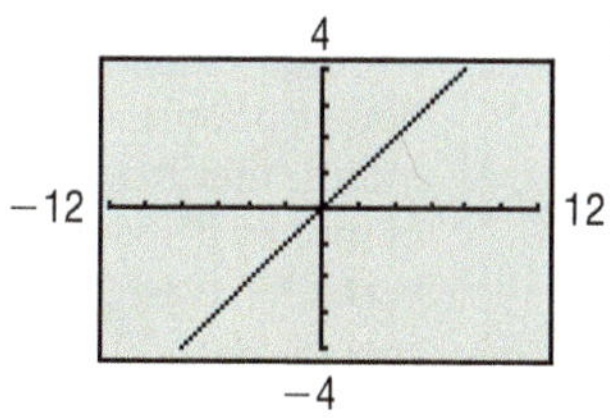

109.

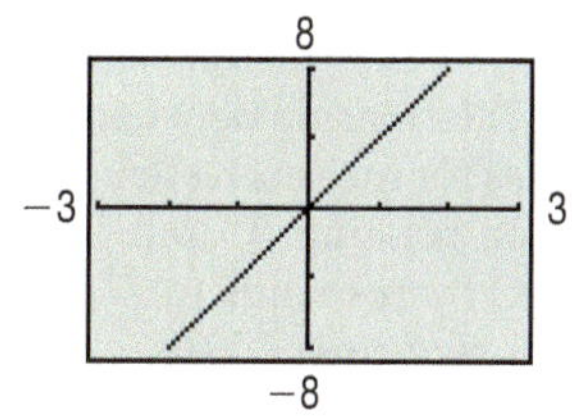

110.

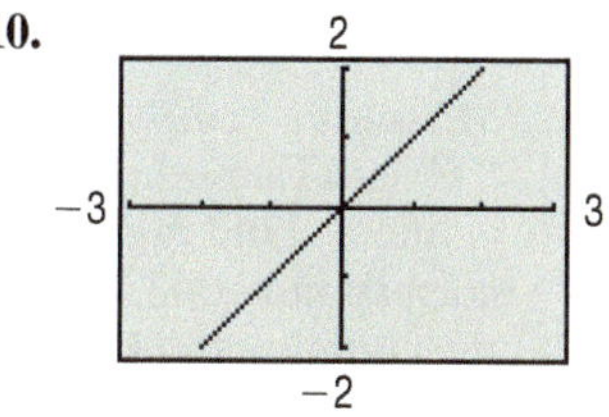

Applications and Extensions

111. Geometry Use slopes to show that the triangle whose vertices are $(-2, 5)$, $(1, 3)$, and $(-1, 0)$ is a right triangle.

112. Geometry Use slopes to show that the quadrilateral whose vertices are $(1, -1)$, $(4, 1)$, $(2, 2)$, and $(5, 4)$ is a parallelogram.

113. Geometry Use slopes to show that the quadrilateral whose vertices are $(-1, 0)$, $(2, 3)$, $(1, -2)$, and $(4, 1)$ is a rectangle.

114. Geometry Use slopes and the distance formula to show that the quadrilateral whose vertices are $(0, 0)$, $(1, 3)$, $(4, 2)$, and $(3, -1)$ is a square.

115. Truck Rentals A truck rental company rents a moving truck for one day by charging \$29 plus \$0.20 per mile. Write a linear equation that relates the cost C, in dollars, of renting the truck to the number x of miles driven. What is the cost of renting the truck if the truck is driven 110 miles? 230 miles?

116. Cost Equation The **fixed costs** of operating a business are the costs incurred regardless of the level of production. Fixed costs include rent, fixed salaries, and costs of leasing machinery. The **variable costs** of operating a business are the costs that change with the level of output. Variable costs include raw materials, hourly wages, and electricity. Suppose that a manufacturer of jeans has fixed daily costs of \$500 and variable costs of \$8 for each pair of jeans manufactured. Write a linear equation that relates the daily cost C, in dollars, of manufacturing the jeans to the number x of jeans manufactured. What is the cost of manufacturing 400 pairs of jeans? 740 pairs?

117. Cost of Sunday Home Delivery The cost to the *Chicago Tribune* for Sunday home delivery is approximately \$0.53 per newspaper with fixed costs of \$1,070,000. Write a linear equation that relates the cost C and the number x of copies delivered.

Source: *Chicago Tribune*, 2002.

118. Wages of a Car Salesperson Dan receives \$375 per week for selling new and used cars at a car dealership in Oak Lawn, Illinois. In addition, he receives 5% of the profit on any sales that he generates. Write a linear equation that relates Dan's weekly salary S when he has sales that generate a profit of x dollars.

119. Electricity Rates in Illinois Commonwealth Edison Company supplies electricity to residential customers for a monthly customer charge of \$7.58 plus 8.275 cents per kilowatt-hour for up to 400 kilowatt-hours.

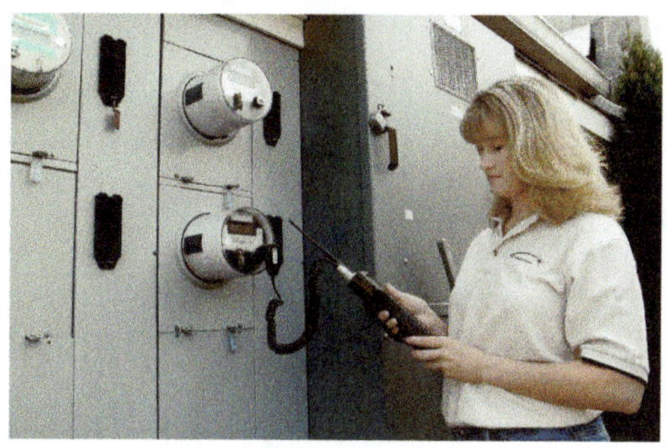

(a) Write a linear equation that relates the monthly charge C, in dollars, to the number x of kilowatt-hours used in a month, $0 \le x \le 400$.

(b) Graph this equation.
(c) What is the monthly charge for using 100 kilowatt-hours?
(d) What is the monthly charge for using 300 kilowatt-hours?
(e) Interpret the slope of the line.

Source: *Commonwealth Edison Company,* April 2006.

120. Electricity Rates in Florida Florida Power & Light Company supplies electricity to residential customers for a monthly customer charge of \$5.17 plus 10.07 cents per kilowatt-hour for up to 1000 kilowatt-hours.
(a) Write a linear equation that relates the monthly charge C, in dollars, to the number x of kilowatt-hours used in a month, $0 \le x \le 1000$.
(b) Graph this equation.
(c) What is the monthly charge for using 200 kilowatt-hours?
(d) What is the monthly charge for using 500 kilowatt-hours?
(e) Interpret the slope of the line.

Source: *Florida Power & Light Company,* April 2006.

121. Measuring Temperature The relationship between Celsius (°C) and Fahrenheit (°F) degrees for measuring temperature is linear. Find a linear equation relating °C and °F if 0°C corresponds to 32°F and 100°C corresponds to 212°F. Use the equation to find the Celsius measure of 70°F.

122. Measuring Temperature The Kelvin (K) scale for measuring temperature is obtained by adding 273 to the Celsius temperature.
(a) Write a linear equation relating K and °C.
(b) Write a linear equation relating K and °F (see Problem 121).

123. Access Ramp A wooden access ramp is being built to reach a platform that sits 30 inches above the floor. The proposed ramp drops 2 inches for every 25-inch run.

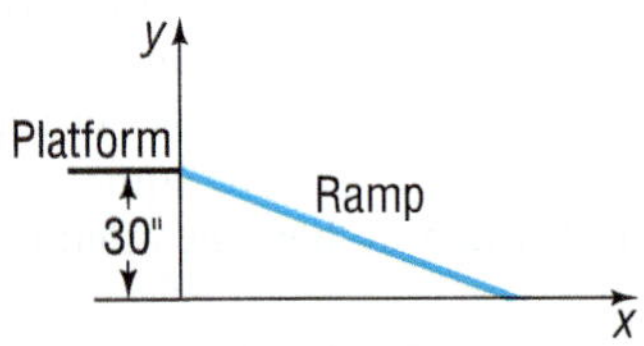

(a) Write a linear equation that relates the height y above the floor to the horizontal distance x from the platform.
(b) Find and interpret the x-intercept of the graph of your equation.
(c) Design requirements stipulate that the maximum run be 30 feet and that the maximum slope be a drop of 1 inch for each 12 inches of run. Will this ramp meet the requirements? Explain.
(d) What slopes could be used to obtain the 30-inch rise and still meet design requirements?

Source: *www.adaptiveaccess.com/wood_ramps.php*

124. Cigarette Use A study by the Partnership for a Drug-Free America indicated that, in 1998, 42% of teens in grades 7 through 12 had recently used cigarettes. A similar study in 2005 indicated that 22% of such teens had recently used cigarettes.
(a) Write a linear equation that relates the percent of teens that recently used cigarettes y to the number of years after 1998, x.
(b) Find the intercepts of the graph of your equation.
(c) Do the intercepts have any meaningful interpretation?
(d) Use your equation to predict the percent for the year 2019. Is this result reasonable?

Source: *www.drugfree.org*

125. Product Promotion A cereal company finds that the number of people who will buy one of its products in the first month that it is introduced is linearly related to the amount of money it spends on advertising. If it spends \$40,000 on advertising, then 100,000 boxes of cereal will be sold, and if it spends \$60,000, then 200,000 boxes will be sold.
(a) Write a linear equation that relates the amount A spent on advertising to the number x of boxes the company aims to sell.
(b) How much advertising is needed to sell 300,000 boxes of cereal?
(c) Interpret the slope.

126. Show that the line containing the points (a, b) and (b, a), $a \ne b$, is perpendicular to the line $y = x$. Also show that the midpoint of (a, b) and (b, a) lies on the line $y = x$.

127. The equation $2x - y = C$ defines a **family of lines,** one line for each value of C. On one set of coordinate axes, graph the members of the family when $C = -4$, $C = 0$, and $C = 2$. Can you draw a conclusion from the graph about each member of the family?

128. Prove that if two nonvertical lines have slopes whose product is -1 then the lines are perpendicular. **[Hint:** Refer to Figure 33 and use the converse of the Pythagorean Theorem.]

Discussion and Writing

129. Which of the following equations might have the graph shown? (More than one answer is possible.)
(a) $2x + 3y = 6$
(b) $-2x + 3y = 6$
(c) $3x - 4y = -12$
(d) $x - y = 1$
(e) $x - y = -1$
(f) $y = 3x - 5$
(g) $y = 2x + 3$
(h) $y = -3x + 3$

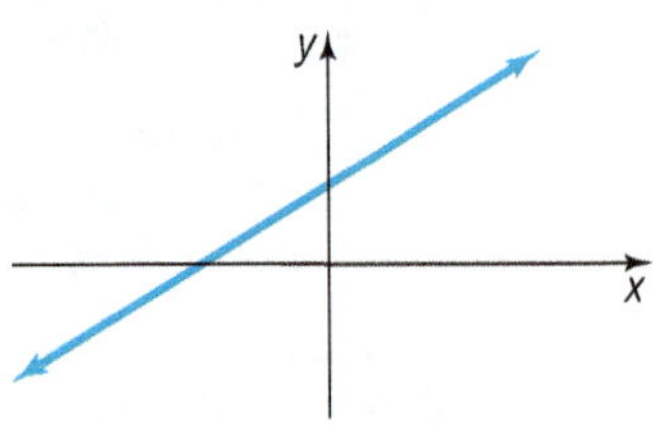

130. Which of the following equations might have the graph shown? (More than one answer is possible.)
(a) $2x + 3y = 6$
(b) $2x - 3y = 6$
(c) $3x + 4y = 12$
(d) $x - y = 1$
(e) $x - y = -1$
(f) $y = -2x - 1$
(g) $y = -\frac{1}{2}x + 10$
(h) $y = x + 4$

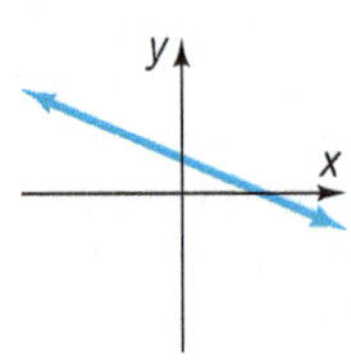

131. The figure shows the graph of two parallel lines. Which of the following pairs of equations might have such a graph?

(a) $x - 2y = 3$
$x + 2y = 7$
(b) $x + y = 2$
$x + y = -1$
(c) $x - y = -2$
$x - y = 1$
(d) $x - y = -2$
$2x - 2y = -4$
(e) $x + 2y = 2$
$x + 2y = -1$

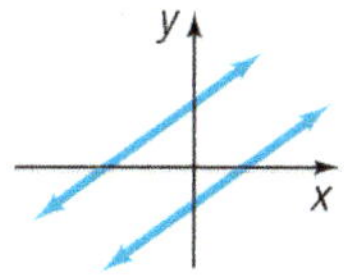

132. The figure shows the graph of two perpendicular lines. Which of the following pairs of equations might have such a graph?

(a) $y - 2x = 2$
$y + 2x = -1$
(b) $y - 2x = 0$
$2y + x = 0$
(c) $2y - x = 2$
$2y + x = -2$
(d) $y - 2x = 2$
$x + 2y = -1$
(e) $2x + y = -2$
$2y + x = -2$

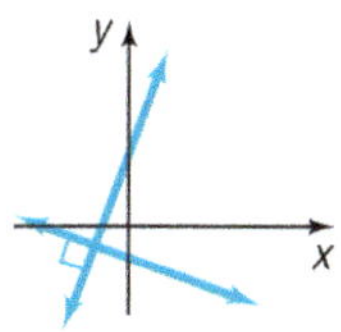

133. ***m* Is for Slope** The accepted symbol used to denote the slope of a line is the letter *m*. Investigate the origin of this symbolism. Begin by consulting a French dictionary and looking up the French word *monter*. Write a brief essay on your findings.

134. **Grade of a Road** The term *grade* is used to describe the inclination of a road. How does this term relate to the notion of slope of a line? Is a 4% grade very steep? Investigate the grades of some mountainous roads and determine their slopes. Write a brief essay on your findings.

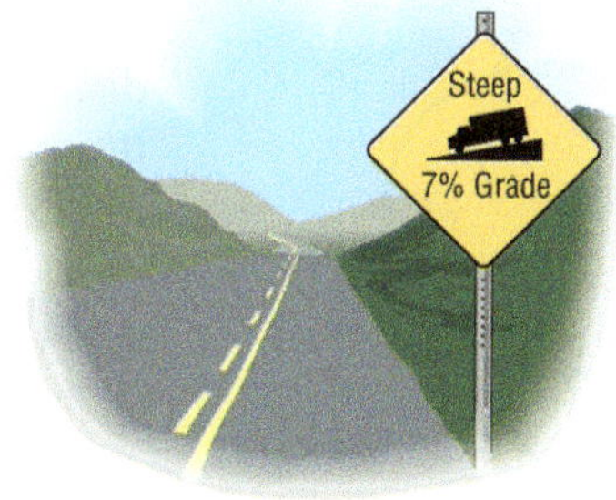

135. **Carpentry** Carpenters use the term *pitch* to describe the steepness of staircases and roofs. How does pitch relate to slope? Investigate typical pitches used for stairs and for roofs. Write a brief essay on your findings.

136. Can the equation of every line be written in slope–intercept form? Why?

137. Does every line have exactly one x-intercept and one y-intercept? Are there any lines that have no intercepts?

138. What can you say about two lines that have equal slopes and equal y-intercepts?

139. What can you say about two lines with the same x-intercept and the same y-intercept? Assume that the x-intercept is not 0.

140. If two distinct lines have the same slope, but different x-intercepts, can they have the same y-intercept?

141. If two distinct lines have the same y-intercept, but different slopes, can they have the same x-intercept?

142. **What Went Wrong?** A student is asked to find the slope of the line joining $(-3, 2)$ and $(1, -4)$. He states that the slope is $\frac{3}{2}$. Is he correct? If not, what went wrong?

2.3 Circles

PREPARING FOR THIS SECTION *Before getting started, review the following:*

- Completing the Square (Section 1.3, pp. 115–116)
- Square Root Method (Section 1.3, pp. 114–115)

Now Work the 'Are You Prepared?' problems on page 198

OBJECTIVES
1 Write the Standard Form of the Equation of a Circle (p. 193)
2 Graph a Circle by Hand and by Using a Graphing Utility (p. 194)
3 Work with the General Form of the Equation of a Circle (p. 196)

1 Write the Standard Form of the Equation of a Circle

One advantage of a coordinate system is that it enables us to translate a geometric statement into an algebraic statement, and vice versa. Consider, for example, the following geometric statement that defines a circle.

DEFINITION A **circle** is a set of points in the xy-plane that are a fixed distance r from a fixed point (h, k). The fixed distance r is called the **radius**, and the fixed point (h, k) is called the **center** of the circle.

Figure 35

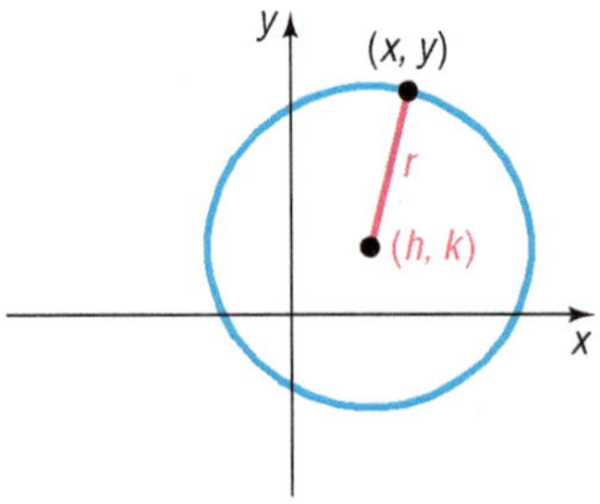

Figure 35 shows the graph of a circle. To find the equation, we let (x, y) represent the coordinates of any point on a circle with radius r and center (h, k). Then the distance between the points (x, y) and (h, k) must always equal r. That is, by the distance formula

$$\sqrt{(x - h)^2 + (y - k)^2} = r$$

After squaring both sides, we obtain

$$(x - h)^2 + (y - k)^2 = r^2$$

DEFINITION

The **standard form of an equation of a circle** with radius r and center (h, k) is

$$(x - h)^2 + (y - k)^2 = r^2 \qquad \textbf{(1)}$$

THEOREM

The standard form of an equation of a circle of radius r with center at the origin $(0, 0)$ is

$$x^2 + y^2 = r^2$$

DEFINITION

If the radius $r = 1$, the circle whose center is at the origin is called the **unit circle** and has the equation

$$x^2 + y^2 = 1$$

See Figure 36.

Figure 36
Unit circle $x^2 + y^2 = 1$

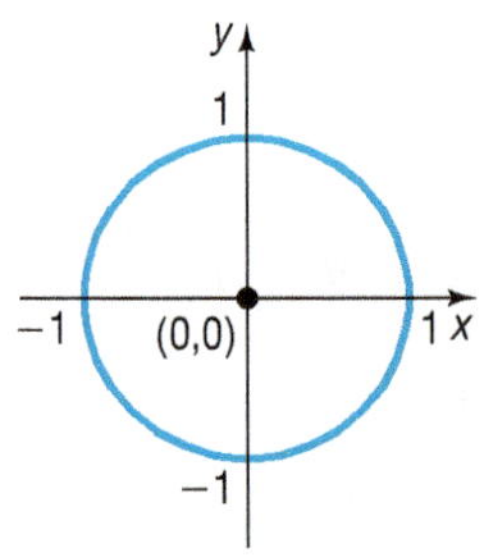

EXAMPLE 1 Writing the Standard Form of the Equation of a Circle

Write the standard form of the equation of the circle with radius 5 and center $(-3, 6)$.

Solution Using the form of equation (1) and substituting the values $r = 5$, $h = -3$, and $k = 6$, we have

$$(x - h)^2 + (y - k)^2 = r^2$$
$$(x - (-3))^2 + (y - 6)^2 = 5^2$$
$$(x + 3)^2 + (y - 6)^2 = 25$$

Now Work PROBLEM 7

2 Graph a Circle by Hand and by Using a Graphing Utility

The graph of any equation of the form $(x - h)^2 + (y - k)^2 = r^2$ is that of a circle with radius r and center (h, k).

EXAMPLE 2 Graphing a Circle by Hand and by Using a Graphing Utility

Graph the equation: $(x + 3)^2 + (y - 2)^2 = 16$

Solution Since the equation is in the form of equation (1), its graph is a circle. To graph the equation by hand, we first compare the given equation to the standard form of the equation of a circle. The comparison yields information about the circle.

$$(x + 3)^2 + (y - 2)^2 = 16$$
$$(x - (-3))^2 + (y - 2)^2 = 4^2$$
$$\uparrow \qquad \uparrow \qquad \uparrow$$
$$(x - h)^2 + (y - k)^2 = r^2$$

Figure 37

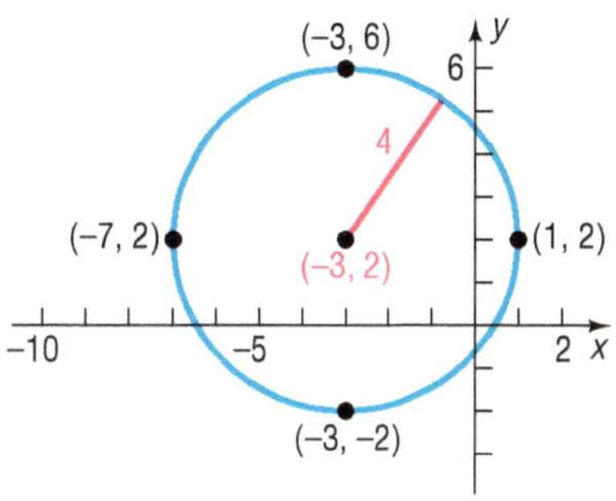

We see that $h = -3$, $k = 2$, and $r = 4$. The circle has center $(-3, 2)$ and a radius of 4 units. To graph this circle, we first plot the center $(-3, 2)$. Since the radius is 4, we can locate four points on the circle by plotting points 4 units to the left, to the right, up, and down from the center. These four points can then be used as guides to obtain the graph. See Figure 37.

To graph a circle on a graphing utility, we must write the equation in the form $y = \{\text{expression involving } x\}$.* We must solve for y in the equation

$$(x + 3)^2 + (y - 2)^2 = 16$$
$$(y - 2)^2 = 16 - (x + 3)^2 \qquad \text{Subtract } (x + 3)^2 \text{ from both sides.}$$
$$y - 2 = \pm\sqrt{16 - (x + 3)^2} \qquad \text{Use the Square Root Method.}$$
$$y = 2 \pm \sqrt{16 - (x + 3)^2} \qquad \text{Add 2 to both sides.}$$

To graph the circle, we graph the top half

$$Y_1 = 2 + \sqrt{16 - (x + 3)^2}$$

and the bottom half

$$Y_2 = 2 - \sqrt{16 - (x + 3)^2}$$

Figure 38

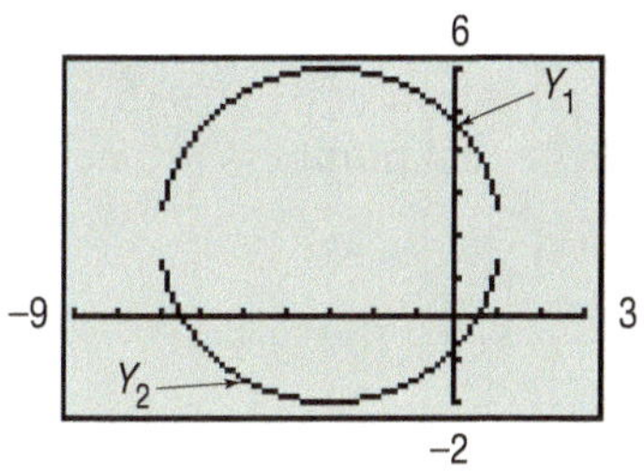

Also, be sure to use a square screen. Otherwise, the circle will appear distorted. Figure 38 shows the graph on a TI-84 Plus.

Now Work PROBLEMS 23(a) AND (b)

EXAMPLE 3 Finding the Intercepts of a Circle

For the circle $(x + 3)^2 + (y - 2)^2 = 16$, find the intercepts, if any, of its graph.

Solution This is the equation discussed and graphed in Example 2. To find the x-intercepts, if any, let $y = 0$ and solve for x. Then

$$(x + 3)^2 + (y - 2)^2 = 16$$
$$(x + 3)^2 + (0 - 2)^2 = 16 \qquad y = 0$$
$$(x + 3)^2 + 4 = 16 \qquad \text{Simplify.}$$
$$(x + 3)^2 = 12 \qquad \text{Subtract 4 from both sides.}$$
$$x + 3 = \pm\sqrt{12} \qquad \text{Apply the Square Root Method.}$$
$$x = -3 \pm 2\sqrt{3} \qquad \text{Solve for } x.$$

The x-intercepts are $-3 - 2\sqrt{3} \approx -6.46$ and $-3 + 2\sqrt{3} \approx 0.46$.

*Some graphing utilities (e.g., TI-83, TI-84, and TI-86) have a CIRCLE function that allows the user to enter only the coordinates of the center of the circle and its radius to graph the circle.

To find the y-intercepts, if any, we let $x = 0$ and solve for y. Then

$$\begin{aligned} (x+3)^2 + (y-2)^2 &= 16 \\ (0+3)^2 + (y-2)^2 &= 16 && x = 0 \\ 9 + (y-2)^2 &= 16 \\ (y-2)^2 &= 7 \\ y - 2 &= \pm\sqrt{7} && \text{Apply the Square Root Method.} \\ y &= 2 \pm \sqrt{7} && \text{Solve for } y. \end{aligned}$$

The y-intercepts are $2 - \sqrt{7} \approx -0.65$ and $2 + \sqrt{7} \approx 4.65$.

Look back at Figure 37 to verify the approximate locations of the intercepts.

Now Work PROBLEM 23(c)

3 Work with the General Form of the Equation of a Circle

If we eliminate the parentheses from the standard form of the equation of the circle given in Example 3, we get

$$(x+3)^2 + (y-2)^2 = 16$$
$$x^2 + 6x + 9 + y^2 - 4y + 4 = 16$$

which we find, upon simplifying, is equivalent to

$$x^2 + y^2 + 6x - 4y - 3 = 0$$

It can be shown that any equation of the form

$$x^2 + y^2 + ax + by + c = 0$$

has a graph that is a circle, or a point, or has no graph at all. For example, the graph of the equation $x^2 + y^2 = 0$ is the single point $(0, 0)$. The equation $x^2 + y^2 + 5 = 0$, or $x^2 + y^2 = -5$, has no graph, because sums of squares of real numbers are never negative.

DEFINITION When its graph is a circle, the equation

$$x^2 + y^2 + ax + by + c = 0$$

is referred to as the **general form of the equation of a circle.**

If an equation of a circle is in the general form, we use the method of completing the square to put the equation in standard form so that we can identify its center and radius.

EXAMPLE 4 Graphing a Circle Whose Equation Is in General Form

Graph the equation $x^2 + y^2 + 4x - 6y + 12 = 0$

Solution We complete the square in both x and y to put the equation in standard form. Group the expression involving x, group the expression involving y, and put the constant on the right side of the equation. The result is

$$(x^2 + 4x) + (y^2 - 6y) = -12$$

Next, complete the square of each expression in parentheses. Remember that any number added on the left side of the equation must be added on the right.

$$(x^2 + 4x + 4) + (y^2 - 6y + 9) = -12 + 4 + 9$$

$$\left(\frac{4}{2}\right)^2 = 4 \qquad \left(\frac{-6}{2}\right)^2 = 9$$

$$(x+2)^2 + (y-3)^2 = 1 \qquad \text{Factor}$$

We recognize this equation as the standard form of the equation of a circle with radius 1 and center $(-2, 3)$. To graph the equation by hand, use the center $(-2, 3)$ and the radius 1. See Figure 39(a).

To graph the equation using a graphing utility, we need to solve for y.

$$(y - 3)^2 = 1 - (x + 2)^2$$

$$y - 3 = \pm\sqrt{1 - (x + 2)^2} \quad \text{Use the Square Root Method.}$$

$$y = 3 \pm \sqrt{1 - (x + 2)^2} \quad \text{Add 3 to both sides.}$$

Figure 39(b) illustrates the graph on a TI-84 Plus graphing calculator.

Figure 39

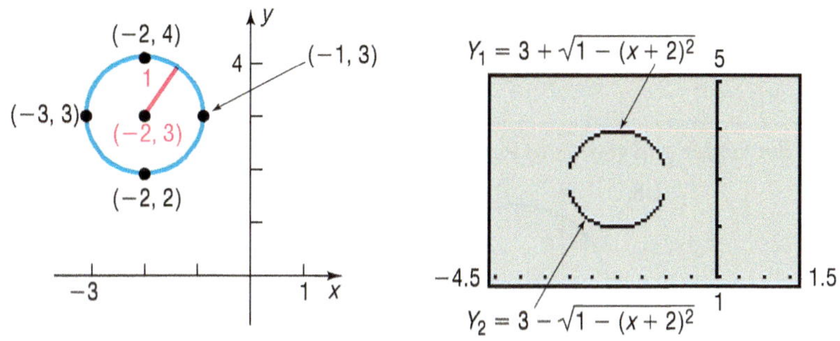

(a) (b)

Now Work PROBLEM 27

EXAMPLE 5 Finding the General Equation of a Circle

Find the general equation of the circle whose center is $(1, -2)$ and whose graph contains the point $(4, -2)$.

Solution To find the equation of a circle, we need to know its center and its radius. Here, we know that the center is $(1, -2)$. Since the point $(4, -2)$ is on the graph, the radius r will equal the distance from $(4, -2)$ to the center $(1, -2)$. See Figure 40. Thus,

Figure 40

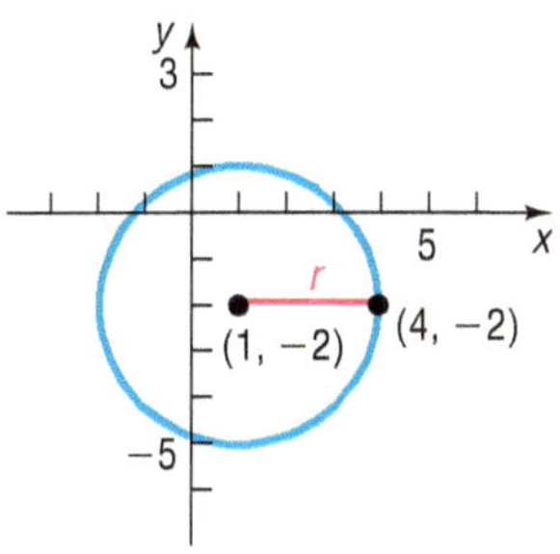

$$r = \sqrt{(4 - 1)^2 + [-2 - (-2)]^2}$$
$$= \sqrt{9} = 3$$

The standard form of the equation of the circle is

$$(x - 1)^2 + (y + 2)^2 = 9$$

Eliminating the parentheses and rearranging terms, we get the general equation

$$x^2 + y^2 - 2x + 4y - 4 = 0$$

Now Work PROBLEM 13

Overview

The discussion in Sections 2.2 and 2.3 about lines and circles dealt with two main types of problems that can be generalized as follows:

1. Given an equation, classify it and graph it.
2. Given a graph, or information about a graph, find its equation.

This text deals with both types of problems. We shall study various equations, classify them, and graph them. The second type of problem is usually more difficult to solve than the first. In many instances a graphing utility can be used to solve problems when information about the problem (such as data) is given.

2.3 Assess Your Understanding

Are You Prepared?

Answers are given at the end of these exercises. If you get a wrong answer, read the pages listed in red.

1. To complete the square of $x^2 + 10x$, you would (*add/subtract*) the number ____. (pp. 115–116)

2. Use the Square Root Method to solve the equation $(x - 2)^2 = 9$. (pp. 114–115)

Concepts and Vocabulary

3. ***True or False*** Every equation of the form

$$x^2 + y^2 + ax + by + c = 0$$

has a circle as its graph.

4. For a circle, the ____ is the distance from the center to any point on the circle.

5. ***True or False*** The radius of the circle $x^2 + y^2 = 9$ is 3.

6. ***True or False*** The center of the circle

$$(x + 3)^2 + (y - 2)^2 = 13$$

is $(3, -2)$.

Skill Building

In Problems 7–10, find the center and radius of each circle. Write the standard form of the equation.

7.

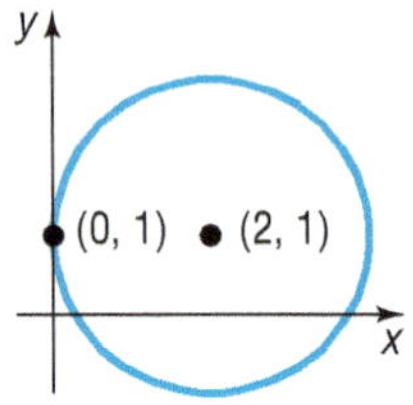

8.

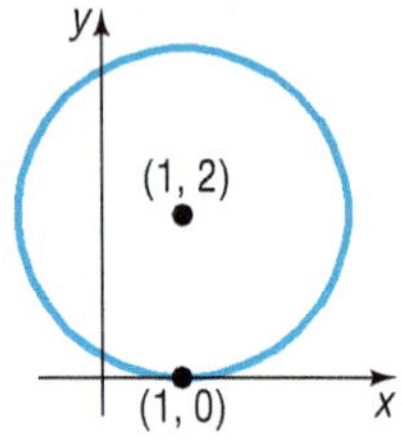

9.

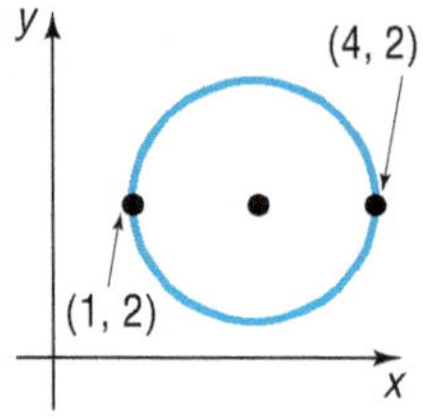

10.

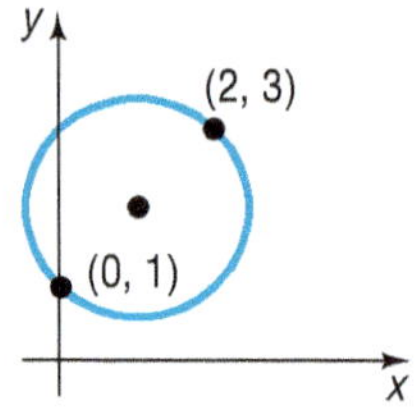

In Problems 11–20, write the standard form of the equation and the general form of the equation of each circle of radius r and center (h, k). By hand, graph each circle.

11. $r = 2;\quad (h, k) = (0, 0)$

12. $r = 3;\quad (h, k) = (0, 0)$

13. $r = 2;\quad (h, k) = (0, 2)$

14. $r = 3;\quad (h, k) = (1, 0)$

15. $r = 5;\quad (h, k) = (4, -3)$

16. $r = 4;\quad (h, k) = (2, -3)$

17. $r = 4;\quad (h, k) = (-2, 1)$

18. $r = 7;\quad (h, k) = (-5, -2)$

19. $r = \frac{1}{2};\quad (h, k) = \left(\frac{1}{2}, 0\right)$

20. $r = \frac{1}{2};\quad (h, k) = \left(0, -\frac{1}{2}\right)$

In Problems 21–34, (a) find the center (h, k) and radius r of each circle; (b) by hand, graph each circle; (c) find the intercepts, if any.

21. $x^2 + y^2 = 4$

22. $x^2 + (y - 1)^2 = 1$

23. $2(x - 3)^2 + 2y^2 = 8$

24. $3(x + 1)^2 + 3(y - 1)^2 = 6$

25. $x^2 + y^2 - 2x - 4y - 4 = 0$

26. $x^2 + y^2 + 4x + 2y - 20 = 0$

27. $x^2 + y^2 + 4x - 4y - 1 = 0$

28. $x^2 + y^2 - 6x + 2y + 9 = 0$

29. $x^2 + y^2 - x + 2y + 1 = 0$

30. $x^2 + y^2 + x + y - \frac{1}{2} = 0$

31. $2x^2 + 2y^2 - 12x + 8y - 24 = 0$

32. $2x^2 + 2y^2 + 8x + 7 = 0$

33. $2x^2 + 8x + 2y^2 = 0$

34. $3x^2 + 3y^2 - 12y = 0$

In Problems 35–42, find the standard form of the equation of each circle.

35. Center at the origin and containing the point $(-2, 3)$

36. Center $(1, 0)$ and containing the point $(-3, 2)$

37. Center $(2, 3)$ and tangent to the x-axis

38. Center $(-3, 1)$ and tangent to the y-axis

39. With endpoints of a diameter at $(1, 4)$ and $(-3, 2)$

40. With endpoints of a diameter at $(4, 3)$ and $(0, 1)$

41. Center $(-1, 3)$ and tangent to the line $y = 2$

42. Center $(4, -2)$ and tangent to the line $x = 1$

In Problems 43–46, match each graph with the correct equation.

(a) $(x - 3)^2 + (y + 3)^2 = 9$ (b) $(x + 1)^2 + (y - 2)^2 = 4$ (c) $(x - 1)^2 + (y + 2)^2 = 4$ (d) $(x + 3)^2 + (y - 3)^2 = 9$

43.

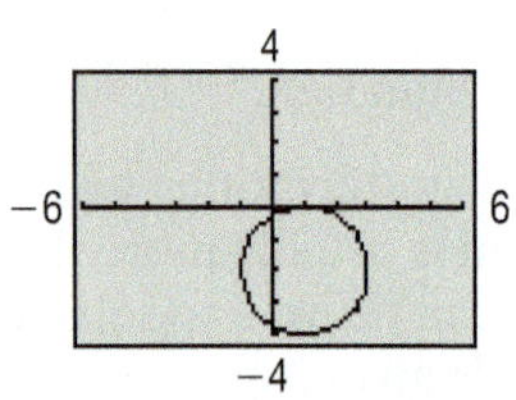

44.

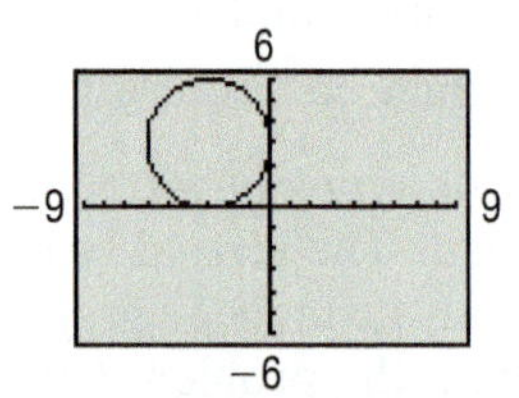

45.

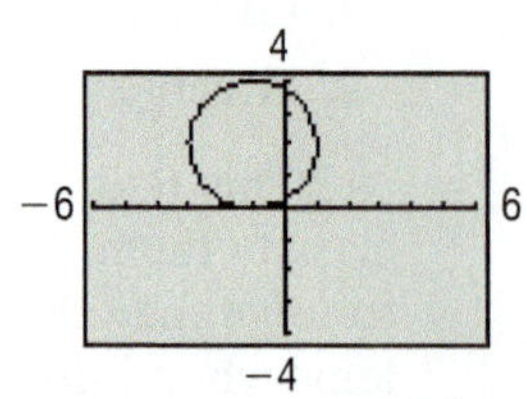

46.

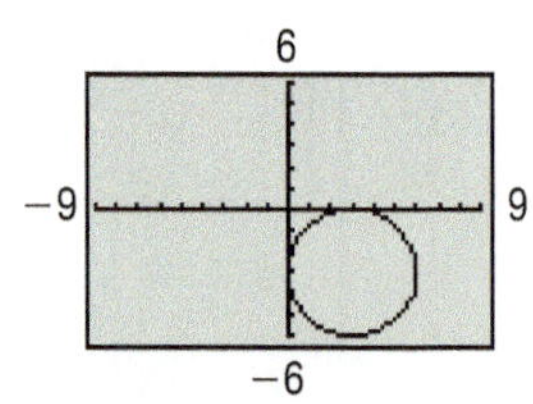

Applications and Extensions

47. Find the area of the square in the figure.

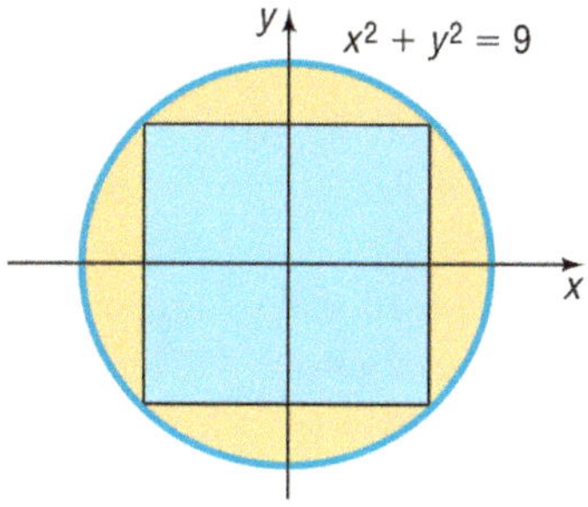

48. Find the area of the blue shaded region in the figure, assuming the quadrilateral inside the circle is a square.

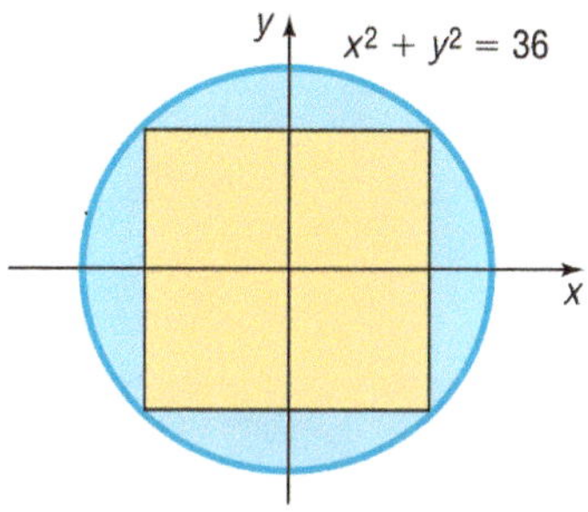

49. Ferris Wheel The original Ferris wheel was built in 1893 by Pittsburg, Pennsylvania bridge builder George W. Ferris. The Ferris wheel was originally built for the 1893 World's Fair in Chicago, but was also later reconstructed for the 1904 World's Fair in St. Louis. It had a maximum height of 264 feet and a wheel diameter of 250 feet. Find an equation for the wheel if the center of the wheel is on the y-axis.

Source: inventors.about.com

50. Ferris Wheel In 2006, the Star of Nanchang (in the Jiangxi province) opened as the world's largest Ferris wheel. It has a maximum height of 160 meters and a diameter of 153 meters, with one full rotation taking approximately 30 minutes. Find an equation for the wheel if the center of the wheel is on the y-axis.

Source: AsiaOne Travel

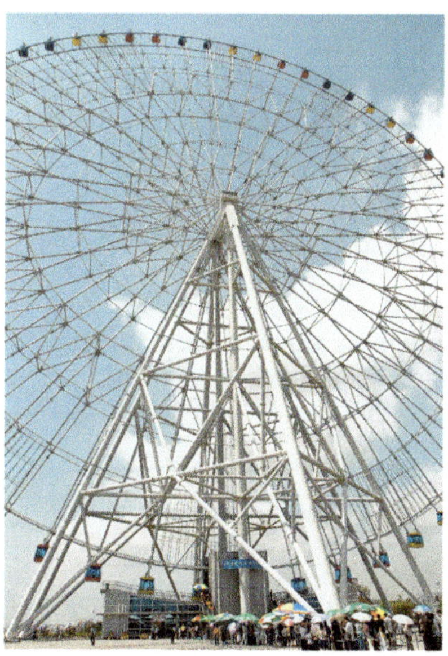

Note: Two other even larger Ferris wheels are reportedly to be completed in Asia by 2008 in time for the 2008 summer Olympics.

51. Weather Satellites Earth is represented on a map of a portion of the solar system so that its surface is the circle with equation $x^2 + y^2 + 2x + 4y - 4091 = 0$. A weather satellite circles 0.6 unit above Earth with the center of its circular orbit at the center of Earth. Find the equation for the orbit of the satellite on this map.

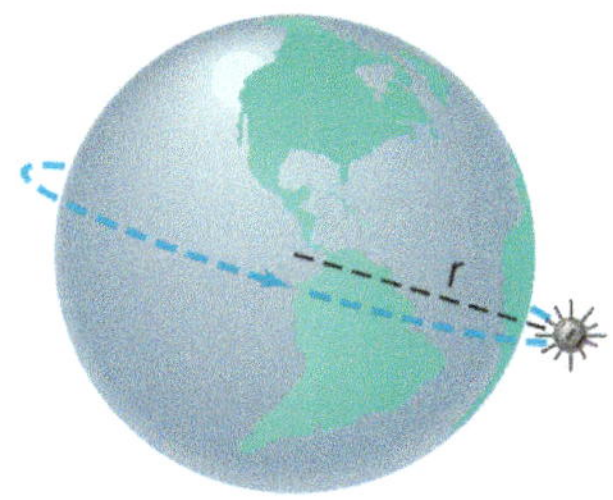

52. The **tangent line** to a circle may be defined as the line that intersects the circle in a single point, called the **point of tangency.** See the figure.

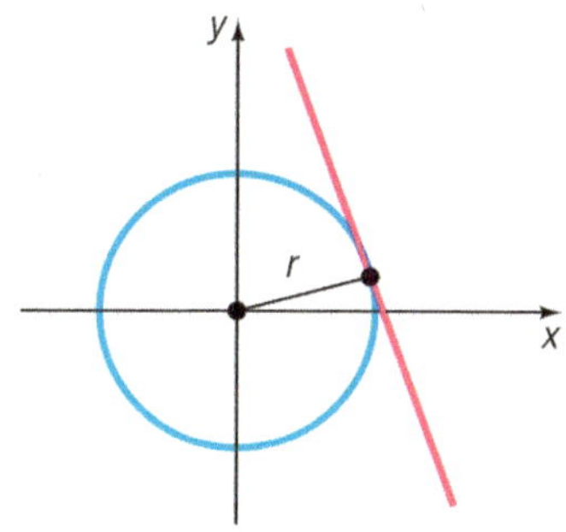

If the equation of the circle is $x^2 + y^2 = r^2$ and the equation of the tangent line is $y = mx + b$, show that:

(a) $r^2(1 + m^2) = b^2$

[**Hint:** The quadratic equation $x^2 + (mx + b)^2 = r^2$ has exactly one solution.]

(b) The point of tangency is $\left(\frac{-r^2m}{b}, \frac{r^2}{b}\right)$.

(c) The tangent line is perpendicular to the line containing the center of the circle and the point of tangency.

53. The Greek Method The Greek method for finding the equation of the tangent line to a circle uses the fact that at any point on a circle the lines containing the center and the tangent line are perpendicular (see Problem 52). Use this method to find an equation of the tangent line to the circle $x^2 + y^2 = 9$ at the point $(1, 2\sqrt{2})$.

54. Use the Greek method described in Problem 53 to find an equation of the tangent line to the circle $x^2 + y^2 - 4x + 6y + 4 = 0$ at the point $(3, 2\sqrt{2} - 3)$.

55. Refer to Problem 52. The line $x - 2y + 4 = 0$ is tangent to a circle at $(0, 2)$. The line $y = 2x - 7$ is tangent to the same circle at $(3, -1)$. Find the center of the circle.

56. Find an equation of the line containing the centers of the two circles

$$x^2 + y^2 - 4x + 6y + 4 = 0$$

and

$$x^2 + y^2 + 6x + 4y + 9 = 0$$

57. If a circle of radius 2 is made to roll along the x-axis, what is an equation for the path of the center of the circle?

58. If the circumference of a circle is 6π, what is its radius?

Discussion and Writing

59. Which of the following equations might have the graph shown? (More than one answer is possible.)
(a) $(x - 2)^2 + (y + 3)^2 = 13$
(b) $(x - 2)^2 + (y - 2)^2 = 8$
(c) $(x - 2)^2 + (y - 3)^2 = 13$
(d) $(x + 2)^2 + (y - 2)^2 = 8$
(e) $x^2 + y^2 - 4x - 9y = 0$
(f) $x^2 + y^2 + 4x - 2y = 0$
(g) $x^2 + y^2 - 9x - 4y = 0$
(h) $x^2 + y^2 - 4x - 4y = 4$

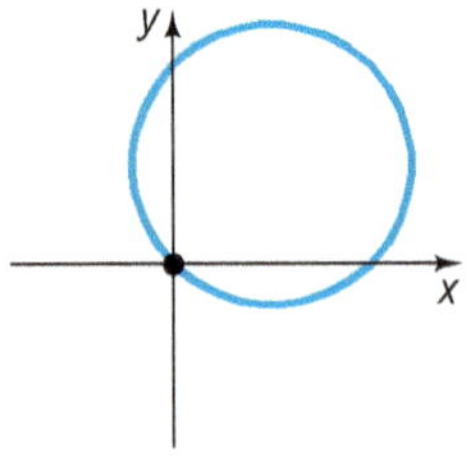

60. Which of the following equations might have the graph shown? (More than one answer is possible.)
(a) $(x - 2)^2 + y^2 = 3$
(b) $(x + 2)^2 + y^2 = 3$
(c) $x^2 + (y - 2)^2 = 3$
(d) $(x + 2)^2 + y^2 = 4$
(e) $x^2 + y^2 + 10x + 16 = 0$
(f) $x^2 + y^2 + 10x - 2y = 1$
(g) $x^2 + y^2 + 9x + 10 = 0$
(h) $x^2 + y^2 - 9x - 10 = 0$

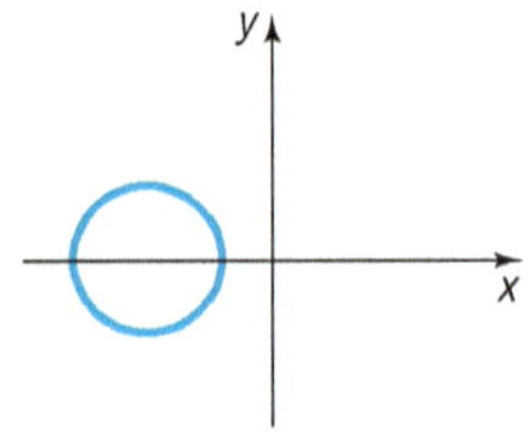

61. Explain how the center and radius of a circle can be used to graph a circle.

62. **What Went Wrong?** A student stated that the center and radius of the graph whose equation is $(x + 3)^2 + (y - 2)^2 = 16$ are $(3, -2)$ and 4, respectively. Why is this incorrect?

'Are You Prepared?' Answers

1. add; 25 **2.** $\{-1, 5\}$

2.4 Variation

OBJECTIVES
1 Construct a Model Using Direct Variation (p. 200)
2 Construct a Model Using Inverse Variation (p. 201)
3 Construct a Model Using Joint or Combined Variation (p. 202)

When a mathematical model is developed for a real-world problem, it often involves relationships between quantities that are expressed in terms of proportionality:

Force is proportional to acceleration.

When an ideal gas is held at a constant temperature, pressure and volume are inversely proportional.

The force of attraction between two objects is inversely proportional to the square of the distance between them.

Revenue is directly proportional to sales.

Each of these statements illustrates the idea of **variation,** or how one quantity varies in relation to another quantity. Quantities may vary *directly, inversely,* or *jointly.*

1 Construct a Model Using Direct Variation

DEFINITION

Let x and y denote two quantities. Then y **varies directly** with x, or y is **directly proportional to** x, if there is a nonzero number k such that

$$y = kx$$

The number k is called the **constant of proportionality.**

Figure 41
$y = kx; k > 0, x \geq 0$

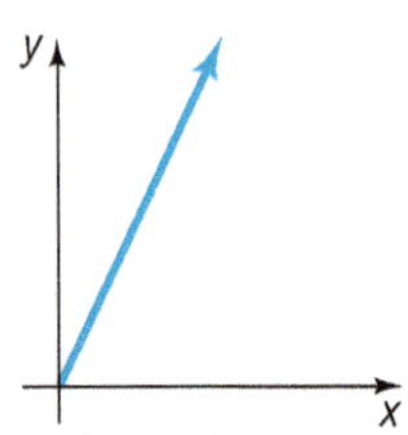

The graph in Figure 41 illustrates the relationship between y and x if y varies directly with x and $k > 0$, $x \geq 0$. Note that the constant of proportionality is, in fact, the slope of the line.

If we know that two quantities vary directly, then knowing the value of each quantity in one instance enables us to write a formula that is true in all cases.

EXAMPLE 1 Mortgage Payments

The monthly payment p on a mortgage varies directly with the amount B borrowed. If the monthly payment on a 30-year mortgage is \$6.65 for every \$1000 borrowed, find a formula that relates the monthly payment p to the amount B borrowed for a mortgage with these terms. Then find the monthly payment p when the amount B borrowed is \$120,000.

Solution Because p varies directly with B, we know that

$$p = kB$$

for some constant k. Because $p = 6.65$ when $B = 1000$, it follows that

$$6.65 = k(1000)$$

$$k = 0.00665 \quad \text{Solve for } k.$$

Since $p = kB$, we have

$$p = 0.00665B$$

In particular, when $B = \$120{,}000$, we find that

$$p = 0.00665(\$120{,}000) = \$798$$

Figure 42

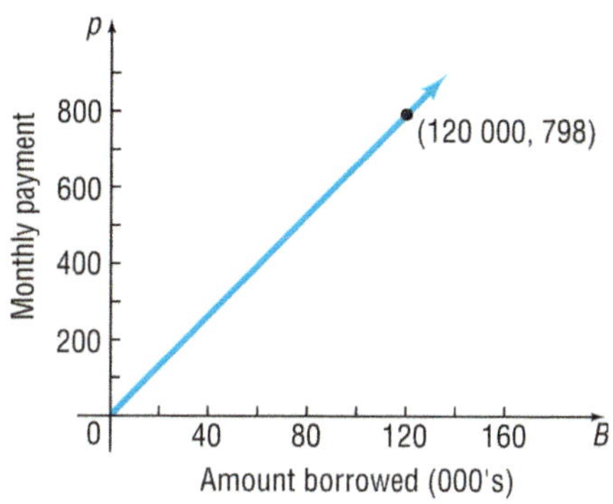

Figure 42 illustrates the relationship between the monthly payment p and the amount B borrowed.

Now Work PROBLEMS 3 AND 21

2 Construct a Model Using Inverse Variation

DEFINITION

Let x and y denote two quantities. Then y **varies inversely** with x, or y is **inversely proportional to** x, if there is a nonzero constant k such that

$$y = \frac{k}{x}$$

Figure 43

$y = \frac{k}{x}; k > 0, x > 0$

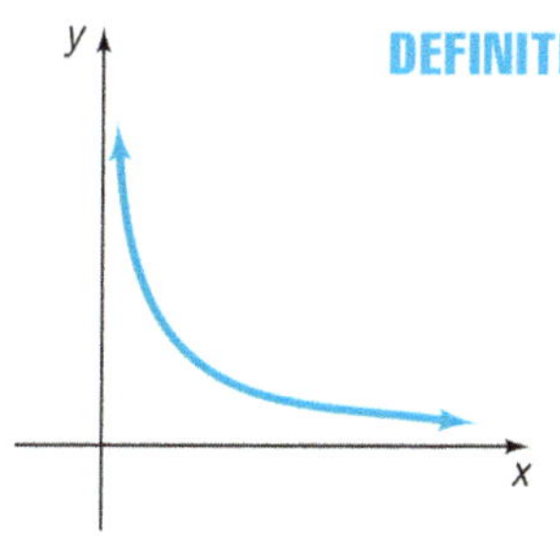

The graph in Figure 43 illustrates the relationship between y and x if y varies inversely with x and $k > 0$, $x > 0$.

EXAMPLE 2 Maximum Weight That Can Be Supported by a Piece of Pine

Figure 44

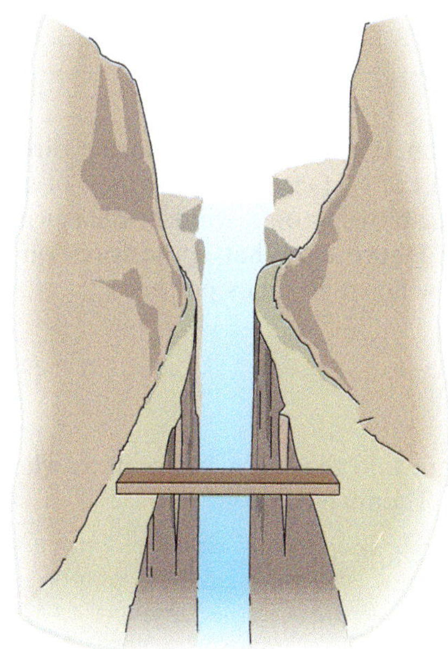

See Figure 44. The maximum weight W that can be safely supported by a 2-inch by 4-inch piece of pine varies inversely with its length l. Experiments indicate that the maximum weight that a 10-foot-long 2-by-4 piece of pine can support is 500 pounds. Write a general formula relating the maximum weight W (in pounds) to length l (in feet). Find the maximum weight W that can be safely supported by a length of 25 feet.

Solution Because W varies inversely with l, we know that

$$W = \frac{k}{l}$$

for some constant k. Because $W = 500$ when $l = 10$, we have

$$500 = \frac{k}{10}$$

$$k = 5000$$

Since $W = \frac{k}{l}$, we have

$$W = \frac{5000}{l}$$

Figure 45

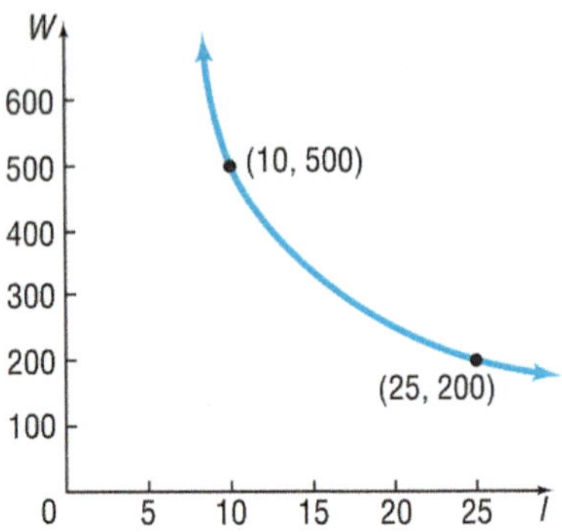

In particular, the maximum weight W that can be safely supported by a piece of pine 25 feet in length is

$$W = \frac{5000}{25} = 200 \text{ pounds}$$

Figure 45 illustrates the relationship between the weight W and the length l.

Now Work PROBLEM 31

3 Construct a Model Using Joint Variation or Combined Variation

When a variable quantity Q is proportional to the product of two or more other variables, we say that Q **varies jointly** with these quantities. Combinations of direct and/or inverse variation may also occur. This is usually referred to as **combined variation.**

Let's look at an example.

EXAMPLE 3 Loss of Heat Through a Wall

The loss of heat through a wall varies jointly with the area of the wall and the difference between the inside and outside temperatures and varies inversely with the thickness of the wall. Write an equation that relates these quantities.

Solution We begin by assigning symbols to represent the quantities:

L = Heat loss T = Temperature difference
A = Area of wall d = Thickness of wall

Then

$$L = k\frac{AT}{d}$$

where k is the constant of proportionality.

In direct or inverse variation, the quantities that vary may be raised to powers. For example, in the early seventeenth century, Johannes Kepler (1571–1630) discovered that the square of the period of revolution T around the Sun varies directly with the cube of its mean distance a from the Sun. That is, $T^2 = ka^3$, where k is the constant of proportionality.

EXAMPLE 4 Force of the Wind on a Window

The force F of the wind on a flat surface positioned at a right angle to the direction of the wind varies jointly with the area A of the surface and the square of the speed v of the wind. A wind of 30 miles per hour blowing on a window measuring 4 feet by 5 feet has a force of 150 pounds. See Figure 46. What is the force on a window measuring 3 feet by 4 feet caused by a wind of 50 miles per hour?

Solution Since F varies jointly with A and v^2, we have

$$F = kAv^2$$

where k is the constant of proportionality. We are told that $F = 150$ pounds when $A = 4 \cdot 5 = 20$ feet^2 and $v = 30$ miles per hour. Then we have

Figure 46

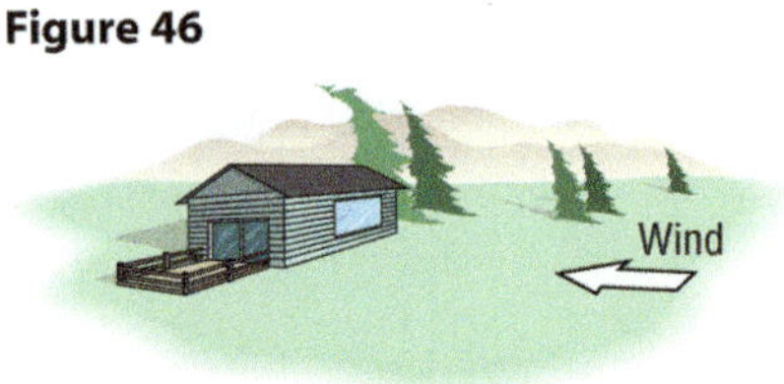

$$150 = k(20)(900) \quad F = kAv^2, F = 150, A = 20, v = 30$$

$$k = \frac{1}{120}$$

Since $F = kAv^2$, we have

$$F = \frac{1}{120}Av^2$$

For a wind of 50 miles per hour blowing on a window whose area is $A = 3 \cdot 4 = 12$ square feet, the force F is

$$F = \frac{1}{120}(12)(2500) = 250 \text{ pounds}$$

Now Work PROBLEM 39

2.4 Assess Your Understanding

Concepts and Vocabulary

1. If x and y are two quantities, then y is directly proportional to x if there is a nonzero number k such that __________.

2. ***True or False*** If y varies directly with x, then $y = \frac{k}{x}$, where k is a constant.

Skill Building

In Problems 3–14, write a general formula to describe each variation.

3. y varies directly with x; $y = 2$ when $x = 10$

4. v varies directly with t; $v = 16$ when $t = 2$

5. A varies directly with x^2; $A = 4\pi$ when $x = 2$

6. V varies directly with x^3; $V = 36\pi$ when $x = 3$

7. F varies inversely with d^2; $F = 10$ when $d = 5$

8. y varies inversely with $\sqrt{x}$; $y = 4$ when $x = 9$

9. z varies directly with the sum of the squares of x and y; $z = 5$ when $x = 3$ and $y = 4$

10. T varies jointly with the cube root of x and the square of d; $T = 18$ when $x = 8$ and $d = 3$

11. M varies directly with the square of d and inversely with the square root of x; $M = 24$ when $x = 9$ and $d = 4$

12. z varies directly with the sum of the cube of x and the square of y; $z = 1$ when $x = 2$ and $y = 3$

13. The square of T varies directly with the cube of a and inversely with the square of d; $T = 2$ when $a = 2$ and $d = 4$

14. The cube of z varies directly with the sum of the squares of x and y; $z = 2$ when $x = 9$ and $y = 4$

Applications and Extensions

In Problems 15–20, write an equation that relates the quantities.

15. **Geometry** The volume V of a sphere varies directly with the cube of its radius r. The constant of proportionality is $\frac{4\pi}{3}$.

16. **Geometry** The square of the length of the hypotenuse c of a right triangle varies jointly with the sum of the squares of the lengths of its legs a and b. The constant of proportionality is 1.

17. **Geometry** The area A of a triangle varies jointly with the lengths of the base b and the height h. The constant of proportionality is $\frac{1}{2}$.

18. **Geometry** The perimeter p of a rectangle varies jointly with the sum of the lengths of its sides l and w. The constant of proportionality is 2.

19. **Physics: Newton's Law** The force F (in newtons) of attraction between two objects varies jointly with their masses m and M (in kilograms) and inversely with the square of the distance d (in meters) between them. The constant of proportionality is $G = 6.67 \times 10^{-11}$.

20. **Physics: Simple Pendulum** The **period** of a pendulum is the time required for one oscillation; the pendulum is usually referred to as **simple** when the angle made to the vertical is less than 5°. The period T of a simple pendulum (in seconds) varies directly with the square root of its length l (in feet). The constant of proportionality is $\frac{2\pi}{\sqrt{32}}$.

21. **Mortgage Payments** The monthly payment p on a mortgage varies directly with the amount B borrowed. If the monthly payment on a 30-year mortgage is \$6.49 for every \$1000 borrowed, find a linear equation that relates the monthly payment p to the amount B borrowed for a mortgage with the same terms. Then find the monthly payment p when the amount B borrowed is \$145,000.

22. **Mortgage Payments** The monthly payment p on a mortgage varies directly with the amount B borrowed. If the monthly payment on a 15-year mortgage is \$8.99 for every \$1000 borrowed, find a linear equation that relates the monthly payment p to the amount B borrowed for a mortgage with the same terms. Then find the monthly payment p when the amount B borrowed is \$175,000.

23. Physics: Falling Objects The distance s that an object falls is directly proportional to the square of the time t of the fall. If an object falls 16 feet in 1 second, how far will it fall in 3 seconds? How long will it take an object to fall 64 feet?

24. Physics: Falling Objects The velocity v of a falling object is directly proportional to the time t of the fall. If, after 2 seconds, the velocity of the object is 64 feet per second, what will its velocity be after 3 seconds?

25. Physics: Stretching a Spring The elongation E of a spring balance varies directly with the applied weight W (see the figure). If $E = 3$ when $W = 20$, find E when $W = 15$.

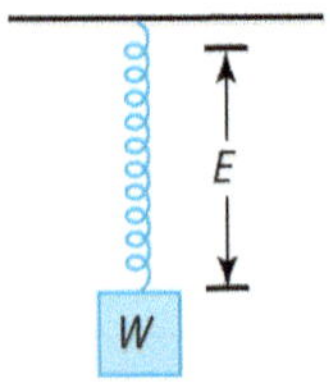

26. Physics: Vibrating String The rate of vibration of a string under constant tension varies inversely with the length of the string. If a string is 48 inches long and vibrates 256 times per second, what is the length of a string that vibrates 576 times per second?

27. Revenue Equation At the corner Shell station, the revenue R varies directly with the number g of gallons of gasoline sold. If the revenue is \$47.40 when the number of gallons sold is 12, find a linear equation that relates revenue R to the number g of gallons of gasoline. Then find the revenue R when the number of gallons of gasoline sold is 10.5.

28. Cost Equation The cost C of chocolate-covered almonds varies directly with the number A of pounds of almonds purchased. If the cost is \$23.75 when the number of pounds of chocolate-covered almonds purchased is 5, find a linear equation that relates the cost C to the number A of pounds of almonds purchased. Then find the cost C when the number of pounds of almonds purchased is 3.5.

29. Demand Suppose that the demand D for candy at the movie theater is inversely related to the price p.

(a) When the price of candy is \$2.75 per bag, the theater sells 156 bags of candy. Express the demand for candy in terms of its price.

(b) Determine the number of bags of candy that will be sold if the price is raised to \$3 a bag.

30. Driving to School The time t that it takes to get to school varies inversely with your average speed s.

(a) Suppose that it takes you 40 minutes to get to school when your average speed is 30 miles per hour. Express the driving time to school in terms of average speed.

(b) Suppose that your average speed to school is 40 miles per hour. How long will it take you to get to school?

31. Pressure The volume V of a gas held at a constant temperature in a closed container varies inversely with its pressure P. If the volume of a gas is 600 cubic centimeters (cm^3) when the pressure is 150 millimeters of mercury (mm Hg), find the volume when the pressure is 200 mm Hg.

32. Resistance The current i in a circuit is inversely proportional to its resistance Z measured in ohms. Suppose that when the current in a circuit is 30 amperes the resistance is 8 ohms. Find the current in the same circuit when the resistance is 10 ohms.

33. Weight The weight of an object above the surface of Earth varies inversely with the square of the distance from the center of Earth. If Maria weighs 125 pounds when she is on the surface of Earth (3960 miles from the center), determine Maria's weight if she is at the top of Mount McKinley (3.8 miles from the surface of Earth).

34. Intensity of Light The intensity I of light (measured in foot-candles) varies inversely with the square of the distance from the bulb. Suppose that the intensity of a 100-watt light bulb at a distance of 2 meters is 0.075 foot-candle. Determine the intensity of the bulb at a distance of 5 meters.

35. Geometry The volume V of a right circular cylinder varies jointly with the square of its radius r and its height h. The constant of proportionality is π. See the figure. Write an equation for V.

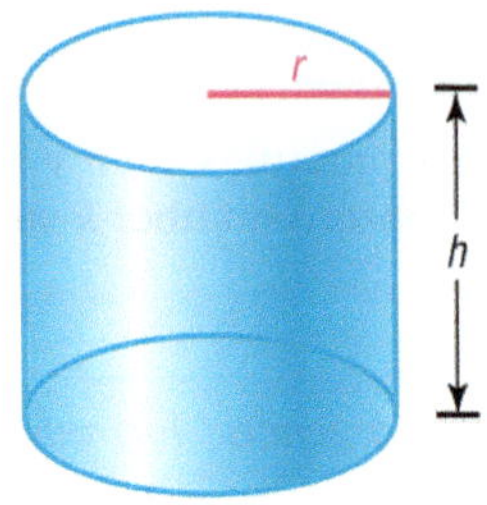

36. Geometry The volume V of a right circular cone varies jointly with the square of its radius r and its height h. The constant of proportionality is $\frac{\pi}{3}$. See the figure. Write an equation for V.

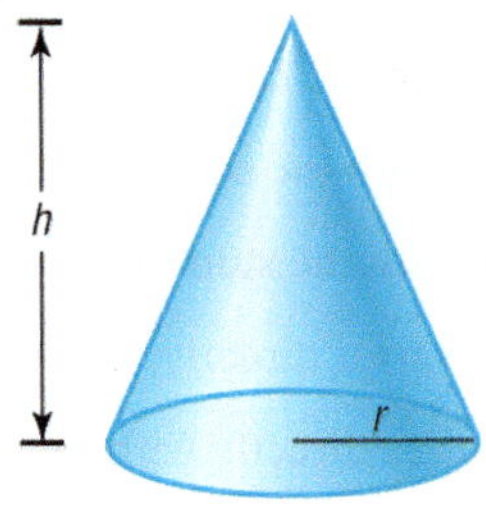

37. Weight of a Body The weight of a body above the surface of Earth varies inversely with the square of the distance from the center of Earth. If a certain body weighs 55 pounds when it is 3960 miles from the center of Earth, how much will it weigh when it is 3965 miles from the center?

38. Force of the Wind on a Window The force exerted by the wind on a plane surface varies jointly with the area of the surface and the square of the velocity of the wind. If the force on an area of 20 square feet is 11 pounds when the wind velocity is 22 miles per hour, find the force on a surface area of 47.125 square feet when the wind velocity is 36.5 miles per hour.

39. Horsepower The horsepower (hp) that a shaft can safely transmit varies jointly with its speed (in revolutions per minute, rpm) and the cube of its diameter. If a shaft of a certain material 2 inches in diameter can transmit 36 hp at 75 rpm, what diameter must the shaft have in order to transmit 45 hp at 125 rpm?

40. Chemistry: Gas Laws The volume V of an ideal gas varies directly with the temperature T and inversely with the pressure P. Write an equation relating V, T, and P using k as the constant of proportionality. If a cylinder contains oxygen at a temperature of 300 K and a pressure of 15 atmospheres in a volume of 100 liters, what is the constant of proportionality k?

If a piston is lowered into the cylinder, decreasing the volume occupied by the gas to 80 liters and raising the temperature to 310 K, what is the gas pressure?

41. **Physics: Kinetic Energy** The kinetic energy K of a moving object varies jointly with its mass m and the square of its velocity v. If an object weighing 25 kilograms and moving with a velocity of 10 meters per second has a kinetic energy of 1250 joules, find its kinetic energy when the velocity is 15 meters per second.

42. **Electrical Resistance of a Wire** The electrical resistance of a wire varies directly with the length of the wire and inversely with the square of the diameter of the wire. If a wire 432 feet long and 4 millimeters in diameter has a resistance of 1.24 ohms, find the length of a wire of the same material whose resistance is 1.44 ohms and whose diameter is 3 millimeters.

43. **Measuring the Stress of Materials** The stress in the material of a pipe subject to internal pressure varies jointly with the internal pressure and the internal diameter of the pipe and inversely with the thickness of the pipe. The stress is 100 pounds per square inch when the diameter is 5 inches, the thickness is 0.75 inch, and the internal pressure is 25 pounds per square inch. Find the stress when the internal pressure is 40 pounds per square inch if the diameter is 8 inches and the thickness is 0.50 inch.

44. **Safe Load for a Beam** The maximum safe load for a horizontal rectangular beam varies jointly with the width of the beam and the square of the thickness of the beam and inversely with its length. If an 8-foot beam will support up to 750 pounds when the beam is 4 inches wide and 2 inches thick, what is the maximum safe load in a similar beam 10 feet long, 6 inches wide, and 2 inches thick?

Discussion and Writing

45. In the early 17th century, Johannes Kepler discovered that the square of the period T for a planet to orbit the Sun varies directly with the cube of its mean distance a from the Sun. Go to the library and research this law and Kepler's other two laws. Write a brief paper about these laws and Kepler's place in history.

46. Using a situation that has not been discussed in the text, write a real-world problem that you think involves two variables that vary directly. Exchange your problem with another student's to solve and critique.

47. Using a situation that has not been discussed in the text, write a real-world problem that you think involves two variables that vary inversely. Exchange your problem with another student's to solve and critique.

48. Using a situation that has not been discussed in the text, write a real-world problem that you think involves three variables that vary jointly. Exchange your problem with another student's to solve and critique.

CHAPTER REVIEW

Things to Know

Formulas

Slope (p. 178)	$m = \dfrac{y_2 - y_1}{x_2 - x_1}$ if $x_1 \neq x_2$; undefined if $x_1 = x_2$
Parallel lines (p. 186)	Equal slopes ($m_1 = m_2$) and different y-intercepts ($b_1 \neq b_2$)
Perpendicular lines (p.187)	Product of slopes is -1 ($m_1 \cdot m_2 = -1$); slopes are negative reciprocals of each other $\left(m_1 = -\dfrac{1}{m_2}\right)$
Direct variation (p. 200)	$y = kx, k \neq 0$
Inverse variation (p. 201)	$y = \dfrac{k}{x}, k \neq 0$

Equations of Lines and Circles

Vertical line (p. 182)	$x = a$; a is the x-intercept
Horizontal line (p. 183)	$y = b$; b is the y-intercept
Point–slope form of the equation of a line (p. 182)	$y - y_1 = m(x - x_1)$; m is the slope of the line, (x_1, y_1) is a point on the line
Slope–intercept form of the equation of a line (p. 183)	$y = mx + b$; m is the slope of the line, b is the y-intercept
General form of the equation of a line (p. 185)	$Ax + By = C$; A, B not both 0
Standard form of the equation of a circle (p. 194)	$(x - h)^2 + (y - k)^2 = r^2$; r is the radius of the circle, (h, k) is the center of the circle
Equation of the unit circle (p. 194)	$x^2 + y^2 = 1$
General form of the equation of a circle (p. 196)	$x^2 + y^2 + ax + by + c = 0$, with restrictions on a, b, and c

Objectives

Section		You should be able to ...	Example(s)	Review Exercises
2.1	1	Find intercepts algebraically from an equation (p. 170)	1	7–14
	2	Test an equation for symmetry with respect to the x-axis, the y-axis, and the origin (p. 171)	2, 3	7–14
	3	Know how to graph key equations (p. 173)	4–6	43, 44
2.2	1	Calculate and interpret the slope of a line (p. 178)	1	1(a)–6(a), 1(b)–6(b), 47
	2	Graph lines given a point and the slope (p. 180)	2	45
	3	Find the equation of a vertical line (p. 181)	3	27
	4	Use the point–slope form of a line; identify horizontal lines (p. 182)	4, 5	25, 26
	5	Find the equation of a line given two points (p. 183)	6	28–30
	6	Write the equation of a line in slope–intercept form (p. 183)	pg. 183	25, 26, 28–34
	7	Identify the slope and y-intercept of a line from its equation (p. 184)	7	35–38
	8	Graph lines written in general form using intercepts (p. 185)	8	39–42
	9	Find equations of parallel lines (p. 186)	9, 10	31, 32
	10	Find equations of perpendicular lines (p. 187)	11, 12	33, 34
2.3	1	Write the standard form of the equation of a circle (p. 193)	1	15–18, 46
	2	Graph a circle by hand and by using a graphing utility (p. 194)	2, 3	19–24
	3	Work with the general form of the equation of a circle (p. 196)	4, 5	21–24
2.4	1	Construct a model using direct variation (p. 200)	1	48, 49, 51
	2	Construct a model using inverse variation (p. 201)	2	50
	3	Construct a model using joint or combined variation (p. 202)	3, 4	51

Review Exercises

In Problems 1–6, for each pair of points:
(a) Find the slope of the line containing the points
(b) Interpret the slope found in part (a)

1. $(0, 0)$; $(4, 2)$ **2.** $(0, 0)$; $(-4, 6)$ **3.** $(1, -1)$; $(-2, 3)$
4. $(-2, 2)$; $(1, 4)$ **5.** $(4, -4)$; $(4, 8)$ **6.** $(-3, 4)$; $(2, 4)$

In Problems 7–14, list the intercepts and test for symmetry with respect to the x-axis, the y-axis, and the origin.

7. $2x = 3y^2$ **8.** $y = 5x$ **9.** $x^2 + 4y^2 = 16$ **10.** $9x^2 - y^2 = 9$
11. $y = x^4 + 2x^2 + 1$ **12.** $y = x^3 - x$ **13.** $x^2 + x + y^2 + 2y = 0$ **14.** $x^2 + 4x + y^2 - 2y = 0$

In Problems 15–18, find the standard form of the equation of the circle whose center and radius are given.

15. $(h, k) = (-2, 3)$; $r = 4$ **16.** $(h, k) = (3, 4)$; $r = 4$ **17.** $(h, k) = (-1, -2)$; $r = 1$ **18.** $(h, k) = (2, -4)$; $r = 3$

In Problems 19–24, find the center and radius of each circle. Graph each circle. Find the intercepts, if any, of each circle.

19. $x^2 + (y - 1)^2 = 4$ **20.** $(x + 2)^2 + y^2 = 9$ **21.** $x^2 + y^2 - 2x + 4y - 4 = 0$
22. $x^2 + y^2 + 4x - 4y - 1 = 0$ **23.** $3x^2 + 3y^2 - 6x + 12y = 0$ **24.** $2x^2 + 2y^2 - 4x = 0$

In Problems 25–34, find an equation of the line having the given characteristics. Express your answer using either the general form or the slope–intercept form of the equation of a line, whichever you prefer.

25. Slope $= -2$; containing the point $(3, -1)$
26. Slope $= 0$; containing the point $(-5, 4)$
27. Vertical; containing the point $(-3, 4)$
28. x-intercept $= 2$; containing the point $(4, -5)$
29. y-intercept $= -2$; containing the point $(5, -3)$
30. Containing the points $(3, -4)$ and $(2, 1)$
31. Parallel to the line $2x - 3y = -4$; containing the point $(-5, 3)$
32. Parallel to the line $x + y = 2$; containing the point $(1, -3)$
33. Perpendicular to the line $x + y = 2$; containing the point $(4, -3)$
34. Perpendicular to the line $3x - y = -4$; containing the point $(-2, 4)$

In Problems 35–38, find the slope and y-intercept of each line. Graph the line by hand, labeling any intercepts.

35. $4x - 5y = -20$ **36.** $3x + 4y = 12$ **37.** $\frac{1}{2}x - \frac{1}{3}y = -\frac{1}{6}$ **38.** $-\frac{3}{4}x + \frac{1}{2}y = 0$

In Problems 39–42, find the intercepts and graph each line by hand.

39. $2x - 3y = 12$ **40.** $x - 2y = 8$ **41.** $\frac{1}{2}x + \frac{1}{3}y = 2$ **42.** $\frac{1}{3}x - \frac{1}{4}y = 1$

43. Sketch a graph of $y = x^3$.

44. Sketch a graph of $y = \sqrt{x}$.

45. Graph the line with slope $\frac{2}{3}$ containing the point $(1, 2)$.

46. The endpoints of the diameter of a circle are $(-3, 2)$ and $(5, -6)$. Find the center and radius of the circle. Write the standard equation of this circle.

47. Show that the points $A = (2, 5)$, $B = (6, 1)$, and $C = (8, -1)$ lie on a line by using slopes.

48. Mortgage Payments The monthly payment p on a mortgage varies directly with the amount B borrowed. If the monthly payment on a 30-year mortgage is \$854.00 when \$130,000 is borrowed, find an equation that relates the monthly payment p to the amount B borrowed for a mortgage with the same terms. Then find the monthly payment p when the amount borrowed is \$165,000.

49. Revenue Function At the corner Esso station, the revenue R varies directly with the number g of gallons of gasoline sold. If the revenue is \$46.67 when the number of gallons sold is 13, find an equation that relates revenue R to the number g of gallons of gasoline. Then find the revenue R when the number of gallons of gasoline sold is 11.2.

50. Weight of a Body The weight of a body varies inversely with the square of its distance from the center of Earth. Assuming that the radius of Earth is 3960 miles, how much would a man weigh at an altitude of 1 mile above Earth's surface if he weighs 200 pounds on Earth's surface?

51. Kepler's Third Law of Planetary Motion Kepler's Third Law of Planetary Motion states that the square of the period of revolution T of a planet varies directly with the cube of its mean distance a from the Sun. If the mean distance of Earth from the Sun is 93 million miles, what is the mean distance of the planet Mercury from the Sun, given that Mercury has a "year" of 88 days?

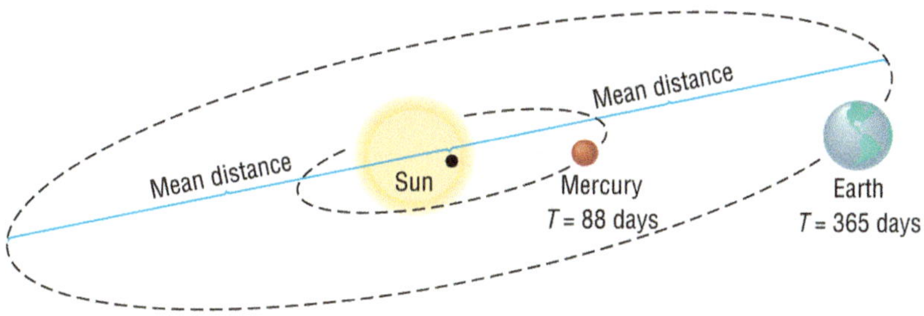

52. Create four problems that you might be asked to do given the two points $(-3, 4)$ and $(6, 1)$. Each problem should involve a different concept. Be sure that your directions are clearly stated.

53. Describe each of the following graphs in the xy-plane. Give justification.

(a) $x = 0$ (b) $y = 0$ (c) $x + y = 0$ (d) $xy = 0$ (e) $x^2 + y^2 = 0$

54. Suppose that you have a rectangular field that requires watering. Your watering system consists of an arm of variable length that rotates so that the watering pattern is a circle. Decide where to position the arm and what length it should be so that the entire field is watered most efficiently. When does it become desirable to use more than one arm?

[**Hint:** Use a rectangular coordinate system positioned as shown in the figures. Write equations for the circle(s) swept out by the watering arm(s).]

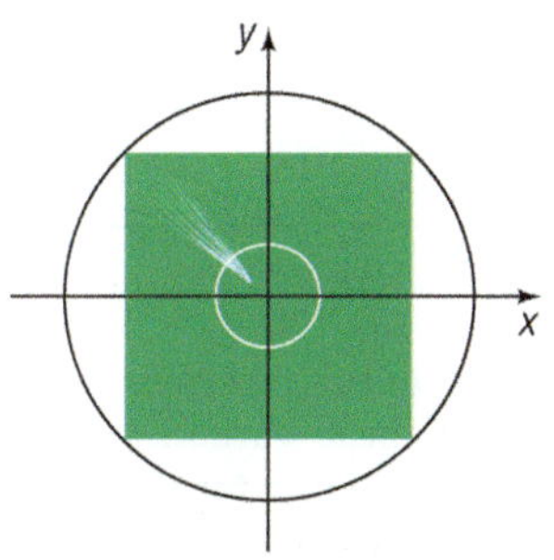

Square field

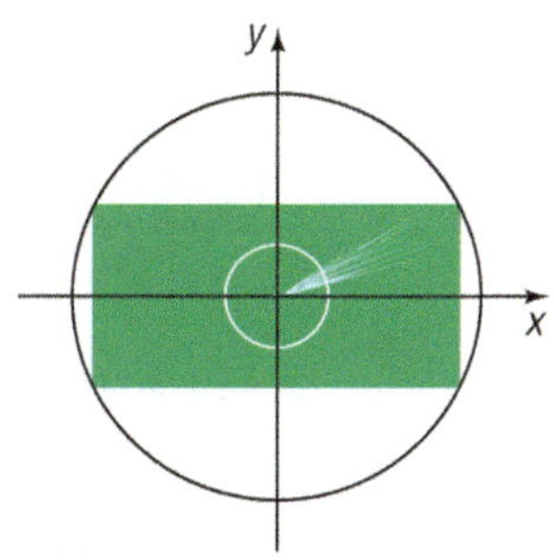

Rectangular field, one arm

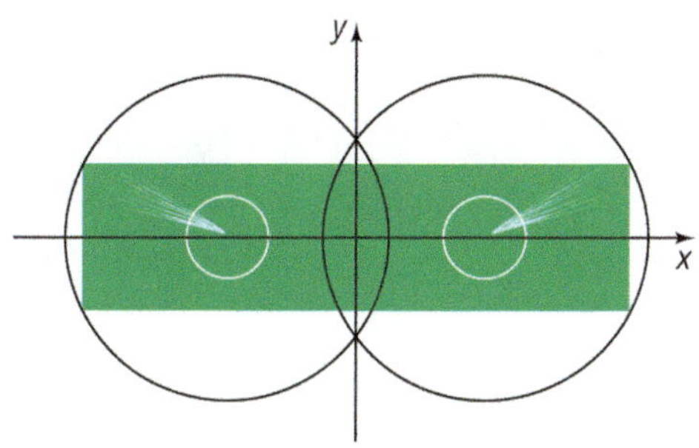

Rectangular field, two arms

CHAPTER TEST

1. Use $P_1 = (-1, 3)$ and $P_2 = (5, -1)$.
   (a) Find the slope of the line containing P_1 and P_2.
   (b) Interpret this slope.
2. Sketch the graph of $y^2 = x$.
3. List the intercepts and test for symmetry: $x^2 + y = 9$.
4. Write the slope–intercept form of the line with slope -2 containing the point $(3, -4)$. Graph the line.
5. Write the general form of the circle with center $(4, -3)$ and radius 5.
6. Find the center and radius of the circle $x^2 + y^2 + 4x - 2y - 4 = 0$. Graph this circle.
7. For the line $2x + 3y = 6$, find a line parallel to it containing the point $(1, -1)$. Also find a line perpendicular to it containing the point $(0, 3)$.
8. **Resistance due to a Conductor** The resistance (in ohms) of a circular conductor varies directly with the length of the conductor and inversely with the square of the radius of the conductor. If 50 feet of wire with a radius of 6×10^{-3} inch has a resistance of 10 ohms, what would be the resistance of 100 feet of the same wire if the radius is increased to 7×10^{-3} inch?

CUMULATIVE REVIEW

In Problems 1–8, find the real solution(s) of each equation.

1. $3x - 5 = 0$
2. $x^2 - x - 12 = 0$
3. $2x^2 - 5x - 3 = 0$
4. $x^2 - 2x - 2 = 0$
5. $x^2 + 2x + 5 = 0$
6. $\sqrt{2x + 1} = 3$
7. $|x - 2| = 1$
8. $\sqrt{x^2 + 4x} = 2$

In Problems 9 and 10, solve each equation in the complex number system.

9. $x^2 = -9$
10. $x^2 - 2x + 5 = 0$

In Problems 11–14, solve each inequality. Graph the solution set.

11. $2x - 3 \le 7$
12. $-1 < x + 4 < 5$
13. $|x - 2| \le 1$
14. $|2 + x| > 3$
15. Find the distance between the points $P = (-1, 3)$ and $Q = (4, -2)$. Find the midpoint of the line segment from P to Q.
16. Which of the following points are on the graph of $y = x^3 - 3x + 1$?
    (a) $(-2, -1)$ (b) $(2, 3)$ (c) $(3, 1)$
17. Sketch the graph of $y = x^3$.
18. Find the equation of the line containing the points $(-1, 4)$ and $(2, -2)$. Express your answer in slope–intercept form.
19. Find the equation of the line perpendicular to the line $y = 2x + 1$ and containing the point $(3, 5)$. Express your answer in slope–intercept form and graph the line.
20. Graph the equation $x^2 + y^2 - 4x + 8y - 5 = 0$.

CHAPTER PROJECTS

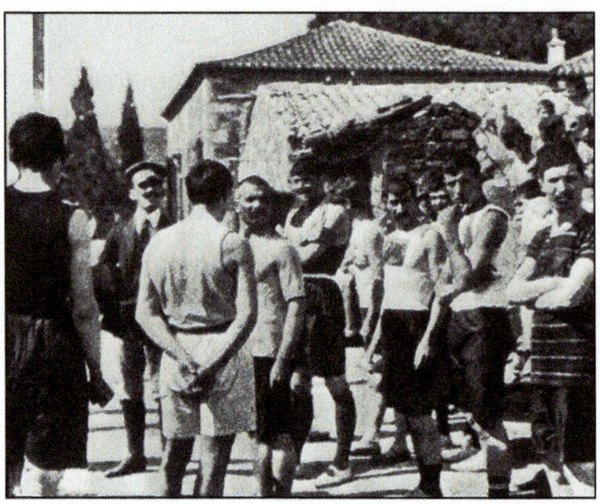

I. **Predicting Olympic Performance** Measurements of human performance over time sometimes follow a strong linear relationship for reasonably short periods. In 2004, the Summer Olympic Games returned to Greece, the home of both the ancient Olympics and the first modern Olympics. The following data represent the winning times (in hours) for men and women in the Olympic marathon.

Year	Men	Women
1984	2.16	2.41
1988	2.18	2.43
1992	2.22	2.54
1996	2.21	2.43
2000	2.17	2.39

Source: www.hickoksports.com/history/olmtandf.shtml

1. Treating year as x and the winning time as y, find linear equations relating these variables (separately for men and women) using the data for the years 1992 and 1996. Compare the equations and comment on any similarities or differences.

2. Interpret the slopes in your equations from part 1. Do the y-intercepts have a reasonable interpretation? Why or why not?
3. Use your lines to predict the winning time in the 2004 Olympics. Compare your predictions to the actual results (2.18 hours for men and 2.44 hours for women). How well did your equations do in predicting the winning times?
4. Repeat parts 1 to 3 using the data for the years 1996 and 2000. How do your results compare?
5. Would your equations be useful in predicting the winning marathon times in the 2112 Summer Olympics? Why or why not?
6. Pick your favorite Winter Olympics event and find the winning value (that is, distance, time, or the like) in two Winter Olympics prior to 2006. Repeat parts 1 to 3 using your selected event and years and compare to the actual results of the 2006 Winter Olympics in Torino, Italy.

The following projects are available on the Instructors Resource Center (IRC):

II. Project at Motorola ***How Many Cellular Phones Can I Make?*** An industrial engineer uses a model involving equations to be sure that production levels meet customer demand.

III. Isocost Lines An isocost line is a linear equation used in economics to represent various combinations of two inputs while maintaining a constant level of expenditure. Explore various input options for labor and capital.

Functions and Their Graphs

3

U.S. Consumers Make Wireless Top Choice

July 11, 2005—WASHINGTON, D.C., CTIA—The Wireless Association President and CEO Steve Largent highlights the Federal Communications Commission (FCC) report on local telephone competition, which indicates for the first time there are more wireless subscribers in the United States than wireline service lines. According to the FCC report, as of December 31, 2004, there were 181.1 million wireless subscribers in the U.S., compared to a combined 178 million incumbent local exchange carrier switched access lines and competitive local exchange carrier switched access lines. Also according to the report, there was a 15% increase in wireless subscribers for the 12-month period ending December 31, 2004.

"This significant milestone reflects consumers' approval of a vibrant, ultracompetitive industry that lets us communicate how we want, where we want and with whom we want," said Largent.

"Wireless is so popular because it lets people communicate on their terms. And it's not just about talking anymore. You can browse the Web, take pictures and video, download music, play games or conduct business. Wireless can satisfy a lot of different communication needs and desires and that's extremely popular to millions of consumers."

Source: "U.S. Consumers Make Wireless Top Choice," July 11, 2005 http://ctia.org/news_media/press/body.cfm?record_id=1533. Used with the permission of CTIA—The Wireless Association®.

—See Chapter Project I—

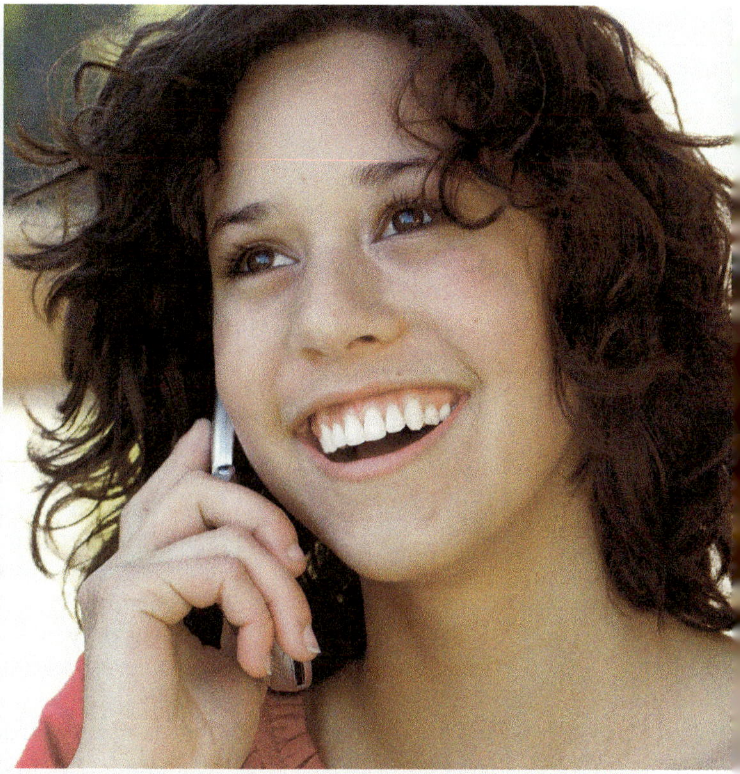

A Look Back

Up to now, our discussion has focused on equations. We have solved equations containing one variable and developed techniques for graphing equations containing two variables.

A Look Ahead

In this chapter, we look at a special type of equation involving two variables called a *function*. This chapter deals with what a function is, how to graph functions, properties of functions, and how functions are used in applications. The word *function* apparently was introduced by René Descartes in 1637. For him, a function simply meant any positive integral power of a variable *x*. Gottfried Wilhelm Leibniz (1646–1716), who always emphasized the geometric side of mathematics, used the word *function* to denote any quantity associated with a curve, such as the coordinates of a point on the curve. Leonhard Euler (1707–1783) employed the word to mean any equation or formula involving variables and constants. His idea of a function is similar to the one most often seen in courses that precede calculus. Later, the use of functions in investigating heat flow equations led to a very broad definition, due to Lejeune Dirichlet (1805–1859), which describes a function as a rule or correspondence between two sets. It is his definition that we use here.

Outline

3.1 Functions

PREPARING FOR THIS SECTION *Before getting started, review the following:*

- Intervals (Section 1.7, pp. 150–151)
- Solving Inequalities (Section 1.7, pp. 153–155)
- Evaluating Algebraic Expressions, Domain of a Variable (Section R.2, pp. 20–21)

Now Work the 'Are You Prepared?' problems on page 223.

OBJECTIVES
1 Determine Whether a Relation Represents a Function (p. 212)
2 Find the Value of a Function (p. 216)
3 Find the Domain of a Function Defined by an Equation (p. 219)
4 Form the Sum, Difference, Product, and Quotient of Two Functions (p. 221)

1 Determine Whether a Relation Represents a Function

We often see situations where one variable is somehow linked to the value of another variable. For example, an individual's level of education is linked to annual income. Engine size is linked to gas mileage. When the value of one variable is related to the value of a second variable, we have a *relation*. A **relation** is a correspondence between two sets. If x and y are two elements in these sets and if a relation exists between x and y, then we say that x **corresponds** to y or that y **depends on** x, and we write $x \rightarrow y$.

We have a number of ways to express relations between two sets. For example, the equation $y = 3x - 1$ shows a relation between x and y. It says that if we take some number x, multiply it by 3, and then subtract 1 we obtain the corresponding value of y. In this sense, x serves as the **input** to the relation and y is the **output** of the relation. We can also express this relation as a graph, as shown in Figure 1.

Figure 1

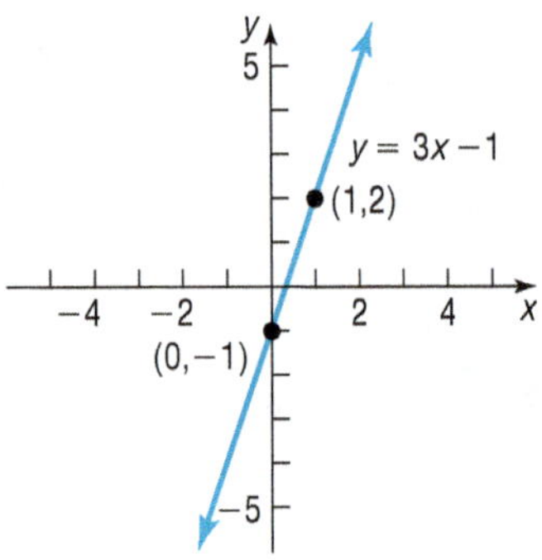

Not only can a relation be expressed through an equation or graph, but we can also express a relation through a technique called *mapping*. A **map** illustrates a relation by using a set of inputs and drawing arrows to the corresponding element in the set of outputs. Ordered pairs can also be used to represent $x \rightarrow y$ as (x, y). We illustrate these two concepts in Example 1.

EXAMPLE 1 **Maps and Ordered Pairs as Relations**

Figure 2 shows a relation between states and the number of representatives each state has in the House of Representatives. The relation might be named "number of representatives."

Figure 2

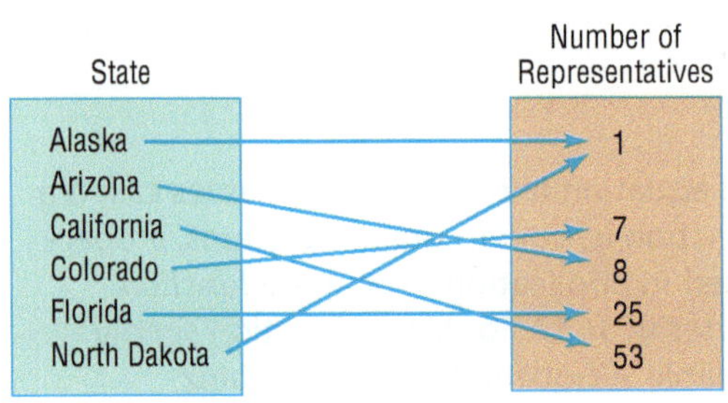

In this relation, Alaska corresponds to 1, Arizona corresponds to 8, and so on. Using ordered pairs, this relation would be expressed as

{(Alaska, 1), (Arizona, 8), (California, 53), (Colorado, 7), (Florida, 25), (North Dakota, 1)}

We now present one of the most important concepts in algebra—the *function*. A function is a special type of relation. To understand the idea behind a function, let's revisit the relation presented in Example 1. If we were to ask, "How many representatives does Alaska have?," you would respond 1. In fact, each input *state* corresponds to a single output *number of representatives*.

Let's consider a second relation where we have a correspondence between the weight of a brain (in grams) and IQ. See Figure 3. Notice the brain that weighs 1342 grams corresponds to two different IQs: 104 and 111. So, if someone asked, "What is the IQ of an individual whose brain weighs 1342 grams?," you cannot provide a single answer.

Figure 3

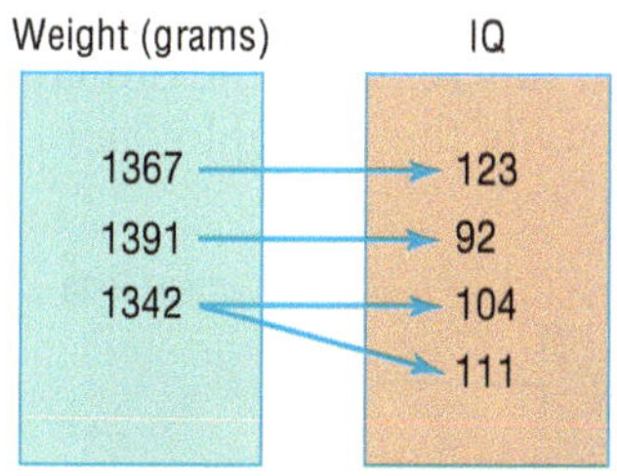

Let's look at one more relation. Figure 4 is a relation that shows a correspondence between *animals* and *life expectancy*. If asked to determine the life expectancy of a dog, we would all respond "11 years." If asked to determine the life expectancy of a rabbit, we would all respond "7 years."

Figure 4

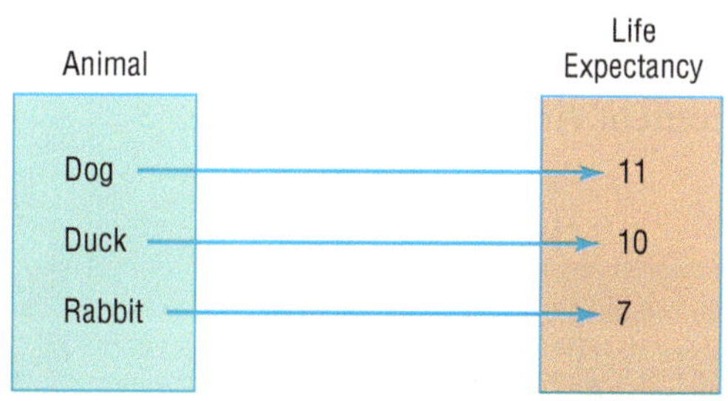

Notice that the relations presented in Figures 2 and 4 have something in common. What is it? The common link between these two relations is that each input corresponds to exactly one output. This leads to the definition of a *function*.

DEFINITION

> Let X and Y be two nonempty sets.* A **function** from X into Y is a relation that associates with each element of X exactly one element of Y.

The set X is called the **domain** of the function. For each element x in X, the corresponding element y in Y is called the **value** of the function at x, or the **image** of x. The set of all images of the elements in the domain is called the **range** of the function. See Figure 5.

Figure 5

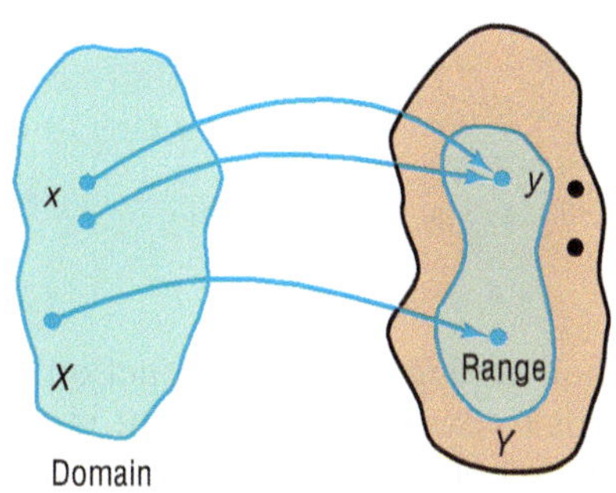

*The sets X and Y will usually be sets of real numbers, in which case a (real) function results. The two sets can also be sets of complex numbers, and then we have defined a complex function. In the broad definition (due to Lejeune Dirichlet), X and Y can be any two sets.

Since there may be some elements in Y that are not the image of some x in X, it follows that the range of a function may be a subset of Y, as shown in Figure 5.

Not all relations between two sets are functions. The next example shows how to determine whether a relation is a function.

EXAMPLE 2 Determining Whether a Relation Represents a Function

Determine which of the following relations represent a function. If the relation is a function, then state its domain and range.

(a) See Figure 6. For this relation, the domain represents the level of education and the range represents the unemployment rate.

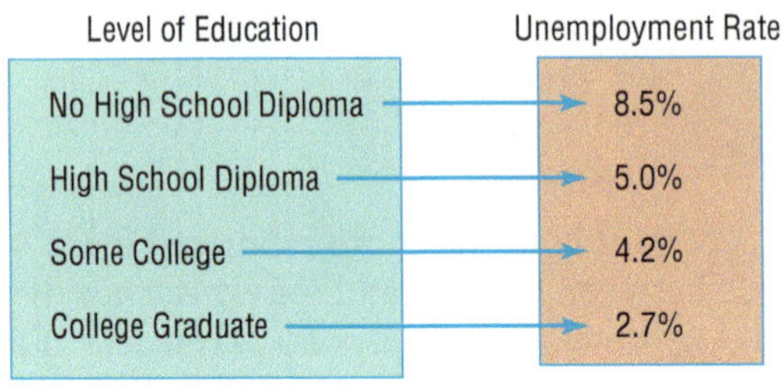

Figure 6
Source: *Statistical Abstract of the United States, 2006*

(b) See Figure 7. For this relation, the domain represents the number of calories in a sandwich from a fast-food restaurant and the range represents the fat content (in grams).

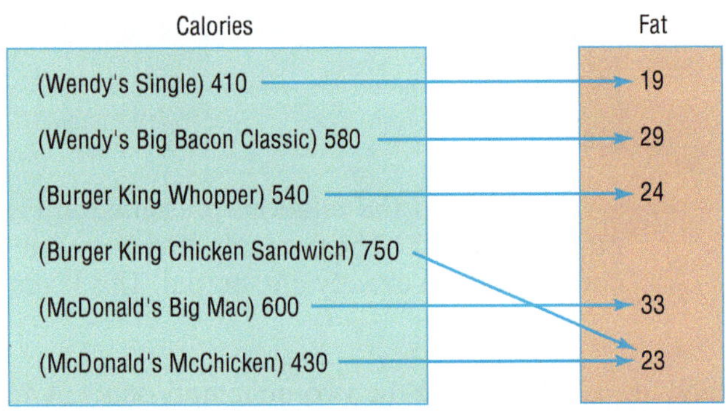

Figure 7
Source: Each company's Web site

(c) See Figure 8. For this relation, the domain represents the weight (in carats) of pear-cut diamonds and the range represents the price (in dollars).

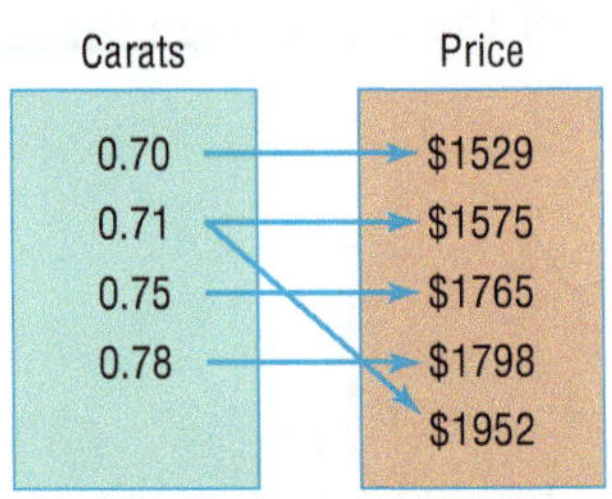

Figure 8
Source: diamonds.com

Solution (a) The relation in Figure 6 is a function because each element in the domain corresponds to exactly one element in the range. The domain of the relation is {No High School Diploma, High School Diploma, Some College, College Graduate}, and the range of the relation is $\{8.5\%, 5.0\%, 4.2\%, 2.7\%\}$.

(b) The relation in Figure 7 is a function because each element in the domain corresponds to exactly one element in the range. The domain of the relation is $\{410, 580, 540, 750, 600, 430\}$. The range of the relation is $\{19, 29, 24, 33, 23\}$.

Notice that it is okay for more than one element in the domain to correspond to the same element in the range (McDonald's and Burger King's chicken sandwich both have 23 grams of fat).

(c) The relation in Figure 8 is not a function because each element in the domain does not correspond to exactly one element in the range. If a 0.71-carat diamond is chosen from the domain, a single price cannot be assigned to it.

Now Work PROBLEM 15

> **In Words**
> For a function, no input has more than one output.

The idea behind a function is its predictability. If the input is known, we can use the function to determine the output. With "nonfunctions," we don't have this predictability. Look back at Figure 7. The inputs are $\{410, 580, 540, 750, 600, 430\}$. The correspondence is *number of fat grams,* and the outputs are $\{19, 29, 24, 33, 23\}$. If asked, "How many grams of fat are in a 410-calorie sandwich?," we can use the correspondence to answer "19." Now consider Figure 8. If asked, "What is the price of a 0.71-carat diamond?," we could not give a single response because two outputs result from the single input "0.71." For this reason, the relation in Figure 8 is not a function.

> **In Words**
> For a function, the domain is the set of inputs, and the range is the set of outputs.

We may also think of a function as a set of ordered pairs (x, y) in which no ordered pairs have the same first element and different second elements. The set of all first elements x is the domain of the function, and the set of all second elements y is its range. Each element x in the domain corresponds to exactly one element y in the range.

EXAMPLE 3 Determining Whether a Relation Represents a Function

Determine whether each relation represents a function. If it is a function, state the domain and range.

(a) $\{(1, 4), (2, 5), (3, 6), (4, 7)\}$

(b) $\{(1, 4), (2, 4), (3, 5), (6, 10)\}$

(c) $\{(-3, 9), (-2, 4), (0, 0), (1, 1), (-3, 8)\}$

Solution

(a) This relation is a function because there are no ordered pairs with the same first element and different second elements. The domain of this function is $\{1, 2, 3, 4\}$, and its range is $\{4, 5, 6, 7\}$.

(b) This relation is a function because there are no ordered pairs with the same first element and different second elements. The domain of this function is $\{1, 2, 3, 6\}$, and its range is $\{4, 5, 10\}$.

(c) This relation is not a function because there are two ordered pairs, $(-3, 9)$ and $(-3, 8)$, that have the same first element and different second elements.

In Example 3(b), notice that 1 and 2 in the domain each have the same image in the range. This does not violate the definition of a function; two different first elements can have the same second element. A violation of the definition occurs when two ordered pairs have the same first element and different second elements, as in Example 3(c).

Now Work PROBLEM 19

Up to now we have shown how to identify when a relation is a function for relations defined by mappings (Example 2) and ordered pairs (Example 3). We know that relations can also be expressed as equations. We discuss next the circumstances under which equations are functions.

To determine whether an equation, where y depends on x, is a function, it is often easiest to solve the equation for y. If any value of x in the domain corresponds to more than one y, the equation does not define a function; otherwise, it does define a function.

EXAMPLE 4 Determining Whether an Equation Is a Function

Determine if the equation $y = 2x - 5$ defines y as a function of x.

Solution The equation tells us to take an input x, multiply it by 2, and then subtract 5. For any input x, these operations yield only one output y. For example, if $x = 1$, then $y = 2(1) - 5 = -3$. If $x = 3$, then $y = 2(3) - 5 = 1$. For this reason, the equation is a function.

EXAMPLE 5 Determining Whether an Equation Is a Function

Determine if the equation $x^2 + y^2 = 1$ defines y as a function of x.

Solution To determine whether the equation $x^2 + y^2 = 1$, which defines the unit circle, is a function, we need to solve the equation for y.

$$x^2 + y^2 = 1$$
$$y^2 = 1 - x^2$$
$$y = \pm\sqrt{1 - x^2}$$

For values of x between -1 and 1, two values of y result. For example, if $x = 0$, then $y = \pm 1$, so two different outputs result from the same input. This means that the equation $x^2 + y^2 = 1$ does not define a function.

Now Work PROBLEM 33

2 Find the Value of a Function

Functions are often denoted by letters such as f, F, g, G, and others. If f is a function, then for each number x in its domain the corresponding image in the range is designated by the symbol $f(x)$, read as "f of x" or as "f at x." We refer to $f(x)$ as the **value of f at the number x**; $f(x)$ is the number that results when x is given and the function f is applied; $f(x)$ is the output corresponding to x or the image of x; $f(x)$ does *not* mean "f times x." For example, the function given in Example 4 may be written as $y = f(x) = 2x - 5$. Then $f\left(\frac{3}{2}\right) = -2$.

Figure 9 illustrates some other functions. Notice that, in every function, for each x in the domain there is one value in the range.

Figure 9

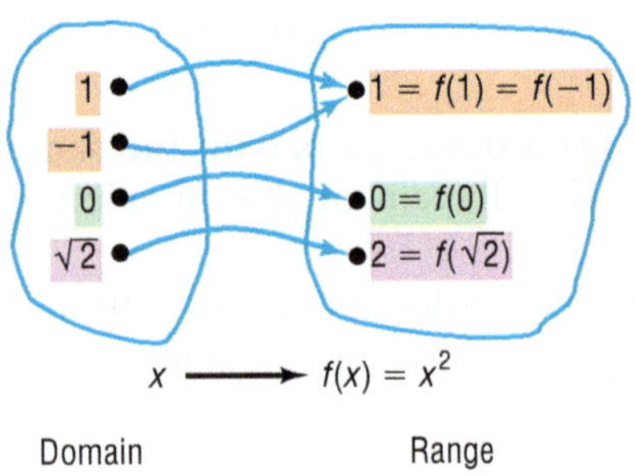

(a) $f(x) = x^2$

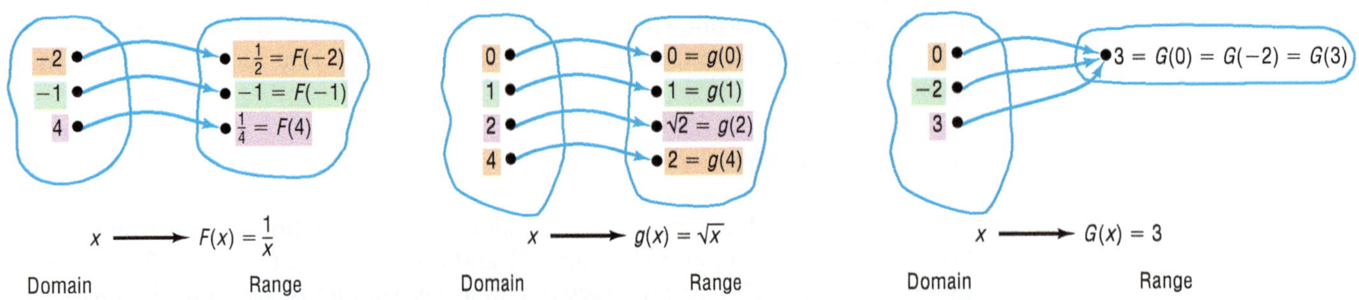

(b) $F(x) = \frac{1}{x}$ (c) $g(x) = \sqrt{x}$ (d) $G(x) = 3$

Figure 10

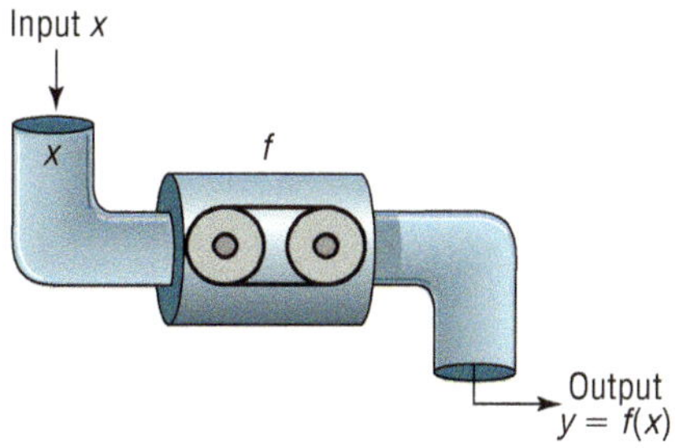

Sometimes it is helpful to think of a function f as a machine that receives as input a number from the domain, manipulates it, and outputs the value. See Figure 10.

The restrictions on this input/output machine are as follows:

1. It only accepts numbers from the domain of the function.
2. For each input, there is exactly one output (which may be repeated for different inputs).

For a function $y = f(x)$, the variable x is called the **independent variable,** because it can be assigned any of the permissible numbers from the domain. The variable y is called the **dependent variable,** because its value depends on x.

Any symbol can be used to represent the independent and dependent variables. For example, if f is the *cube function,* then f can be given by $f(x) = x^3$ or $f(t) = t^3$ or $f(z) = z^3$. All three functions are the same. Each tells us to cube the independent variable to get the output. In practice, the symbols used for the independent and dependent variables are based on common usage, such as using C for cost in business.

The independent variable is also called the **argument** of the function. Thinking of the independent variable as an argument can sometimes make it easier to find the value of a function. For example, if f is the function defined by $f(x) = x^3$, then f tells us to cube the argument. Thus, $f(2)$ means to cube 2, $f(a)$ means to cube the number a, and $f(x + h)$ means to cube the quantity $x + h$.

EXAMPLE 6 Illustrating Language Used with Functions

Consider the function $f(p) = p^2$. The name of the function is f. The *independent variable* or *argument* of f is p. If $q = f(p)$, then q is the *dependent variable.* Since $f(3) = 9$, we say that 9 is the *value* of f at 3 or 9 is the *image* of 3. Because we can square any real number, the *domain* of the function is the set of all real numbers, or $(-\infty, \infty)$ using interval notation. Because the square of any real number is greater than or equal to zero, the *range* of f is $[0, \infty)$. In words, the function f tells us to square the argument.

EXAMPLE 7 Finding Values of a Function

For the function f defined by $f(x) = 2x^2 - 3x$, evaluate

(a) $f(3)$ (b) $f(x) + f(3)$ (c) $3f(x)$

(d) $f(-x)$ (e) $-f(x)$ (f) $f(3x)$

(g) $f(x + 3)$ (h) $\dfrac{f(x+h) - f(x)}{h}$ $h \neq 0$

Solution (a) We substitute 3 for x in the equation for f, $f(x) = 2x^2 - 3x$, to get

$$f(3) = 2(3)^2 - 3(3) = 18 - 9 = 9$$

The image of 3 is 9.

(b) $f(x) + f(3) = (2x^2 - 3x) + (9) = 2x^2 - 3x + 9$

(c) We multiply the equation for f by 3.

$$3f(x) = 3(2x^2 - 3x) = 6x^2 - 9x$$

(d) We substitute $-x$ for x in the equation for f and simplify.

$$f(-x) = 2(-x)^2 - 3(-x) = 2x^2 + 3x$$

(e) $-f(x) = -(2x^2 - 3x) = -2x^2 + 3x$

(f) We substitute $3x$ for x in the equation for f and simplify.

$$f(3x) = 2(3x)^2 - 3(3x) = 2(9x^2) - 9x = 18x^2 - 9x$$

(g) We substitute $x + 3$ for x in the equation for f and simplify.

$$\begin{aligned} f(x+3) &= 2(x+3)^2 - 3(x+3) \quad \text{Notice the use of parentheses here.} \\ &= 2(x^2 + 6x + 9) - 3x - 9 \\ &= 2x^2 + 12x + 18 - 3x - 9 \\ &= 2x^2 + 9x + 9 \end{aligned}$$

(h) $$\frac{f(x+h) - f(x)}{h} = \frac{[2(x+h)^2 - 3(x+h)] - [2x^2 - 3x]}{h}$$

↑ $f(x + h) = 2(x + h)^2 - 3(x + h)$

$$\begin{aligned} &= \frac{2(x^2 + 2xh + h^2) - 3x - 3h - 2x^2 + 3x}{h} \quad \text{Simplify.} \\ &= \frac{2x^2 + 4xh + 2h^2 - 3h - 2x^2}{h} \quad \text{Distribute and combine like terms.} \\ &= \frac{4xh + 2h^2 - 3h}{h} \quad \text{Combine like terms.} \\ &= \frac{h(4x + 2h - 3)}{h} \quad \text{Factor out } h. \\ &= 4x + 2h - 3 \quad \text{Divide out the } h\text{'s.} \end{aligned}$$

Notice in Example 7 that $f(x + 3) \neq f(x) + f(3)$, $f(-x) \neq -f(x)$, and $3f(x) \neq f(3x)$.

The expression in part (h) is called the **difference quotient** of f, an important expression in calculus.

Now Work PROBLEMS 39 AND 73

Most calculators have special keys that allow you to find the value of certain commonly used functions. For example, you should be able to find the square function $f(x) = x^2$, the square root function $f(x) = \sqrt{x}$, the reciprocal function $f(x) = \dfrac{1}{x} = x^{-1}$, and many others that will be discussed later in this book (such as $\ln x$ and $\log x$). Verify the results of Example 7, which follows, on your calculator.

EXAMPLE 8 Finding Values of a Function on a Calculator

(a) $f(x) = x^2$ $\quad f(1.234) = 1.234^2 = 1.522756$

(b) $F(x) = \dfrac{1}{x}$ $\quad F(1.234) = \dfrac{1}{1.234} \approx 0.8103727715$

(c) $g(x) = \sqrt{x}$ $\quad g(1.234) = \sqrt{1.234} \approx 1.110855526$

COMMENT Graphing calculators can be used to evaluate any function that you wish. Figure 11 shows the result obtained in Example 7(a) on a TI-84 Plus graphing calculator with the function to be evaluated, $f(x) = 2x^2 - 3x$, in Y_1.

Figure 11

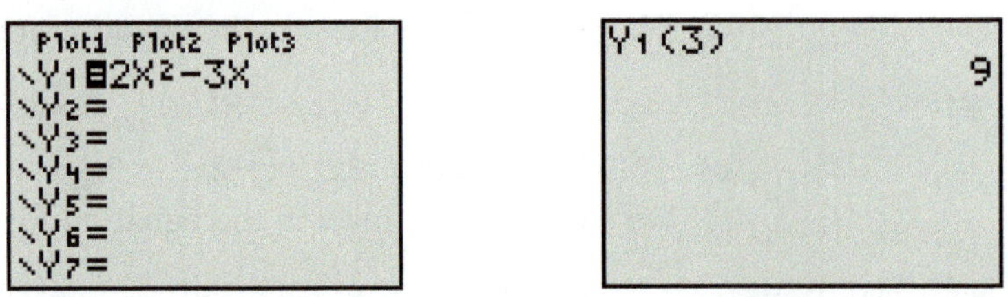

Implicit Form of a Function

In general, when a function f is defined by an equation in x and y, we say that the function f is given **implicitly.** If it is possible to solve the equation for y in terms of x, then we write $y = f(x)$ and say that the function is given **explicitly.** For example,

Implicit Form	Explicit Form
$3x + y = 5$	$y = f(x) = -3x + 5$
$x^2 - y = 6$	$y = f(x) = x^2 - 6$
$xy = 4$	$y = f(x) = \dfrac{4}{x}$

COMMENT The explicit form of a function is the form required by a graphing calculator. ■

We list next a summary of some important facts to remember about a function f.

SUMMARY Important Facts about Functions

(a) For each x in a domain of a function f, there is exactly one image $f(x)$ in the range; however, an element in the range can result from more than one x in the domain.
(b) f is the symbol that we use to denote the function. It is symbolic of the equation that we use to get from an x in the domain to $f(x)$ in the range.
(c) If $y = f(x)$, then x is called the independent variable or argument of f, and y is called the dependent variable or the value of f at x.

3 Find the Domain of a Function Defined by an Equation

Often the domain of a function f is not specified; instead, only the equation defining the function is given. In such cases, we agree that the **domain of f** is the largest set of real numbers for which the value $f(x)$ is a real number. The domain of a function f is the same as the domain of the variable x in the expression $f(x)$.

EXAMPLE 9 Finding the Domain of a Function

Find the domain of each of the following functions:

(a) $f(x) = x^2 + 5x$ (b) $g(x) = \dfrac{3x}{x^2 - 4}$ (c) $h(t) = \sqrt{4 - 3t}$

Solution

(a) The function tells us to square a number and then add five times the number. Since these operations can be performed on any real number, we conclude that the domain of f is the set of all real numbers.

(b) The function g tells us to divide $3x$ by $x^2 - 4$. Since division by 0 is not defined, the denominator $x^2 - 4$ can never be 0, so x can never equal -2 or 2. The domain of the function g is $\{x|x \neq -2, x \neq 2\}$.

(c) The function h tells us to take the square root of $4 - 3t$. But only nonnegative numbers have real square roots, so the expression under the square root (the radicand) must be nonnegative (greater than or equal to zero). This requires that

$$4 - 3t \geq 0$$
$$-3t \geq -4$$
$$t \leq \frac{4}{3}$$

The domain of h is $\left\{t \middle| t \leq \frac{4}{3}\right\}$ or the interval $\left(-\infty, \frac{4}{3}\right]$. ■

In Words

The domain of g found in Example 9(b) is $\{x | x \neq -2, x \neq 2\}$. This notation is read, "The domain of the function g is the set of all real numbers x such that x does not equal -2 and x does not equal 2."

For the functions that we will encounter in this book, the following steps may prove helpful for finding the domain of a function that is defined by an equation whose domain is a subset of real numbers.

Finding the Domain of a Function Defined by an Equation

1. Start with the domain as the set of real numbers.
2. If the equation has a denominator, exclude any numbers that give a zero denominator.
3. If the equation has a radical of even index, exclude any numbers that cause the expression inside the radical to be negative.

Now Work PROBLEM 51

If x is in the domain of a function f, we shall say that **f is defined at x,** or **$f(x)$ exists.** If x is not in the domain of f, we say that **f is not defined at x,** or **$f(x)$ does not exist.** For example, if $f(x) = \frac{x}{x^2 - 1}$, then $f(0)$ exists, but $f(1)$ and $f(-1)$ do not exist. (Do you see why?)

We have not said much about finding the range of a function. The reason is that when a function is defined by an equation it is often difficult to find the range.* Therefore, we shall usually be content to find just the domain of a function when the function is defined by an equation. We shall express the domain of a function using inequalities, interval notation, set notation, or words, whichever is most convenient.

When we use functions in applications, the domain may be restricted by physical or geometric considerations. For example, the domain of the function f defined by $f(x) = x^2$ is the set of all real numbers. However, if f is used to obtain the area of a square when the length x of a side is known, then we must restrict the domain of f to the positive real numbers, since the length of a side can never be 0 or negative.

EXAMPLE 10 Finding the Domain in an Application

Express the area of a circle as a function of its radius. Find the domain.

Solution See Figure 12. We know that the formula for the area A of a circle of radius r is $A = \pi r^2$. If we use r to represent the independent variable and A to represent the dependent variable, the function expressing this relationship is

Figure 12

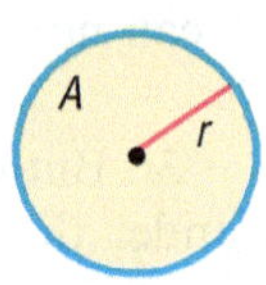

$$A(r) = \pi r^2$$

In this setting, the domain is $\{r | r > 0\}$. (Do you see why?)

Observe in the solution to Example 10 that we used the symbol A in two ways: It is used to name the function, and it is used to symbolize the dependent variable. This double use is common in applications and should not cause any difficulty.

Now Work PROBLEM 87

*In Section 6.2, we discuss a way to find the range for a special class of functions.

4 Form the Sum, Difference, Product, and Quotient of Two Functions

Next we introduce some operations on functions. We shall see that functions, like numbers, can be added, subtracted, multiplied, and divided. For example, if $f(x) = x^2 + 9$ and $g(x) = 3x + 5$, then

$$f(x) + g(x) = (x^2 + 9) + (3x + 5) = x^2 + 3x + 14$$

The new function $y = x^2 + 3x + 14$ is called the *sum function* $f + g$. Similarly,

$$f(x) \cdot g(x) = (x^2 + 9)(3x + 5) = 3x^3 + 5x^2 + 27x + 45$$

The new function $y = 3x^3 + 5x^2 + 27x + 45$ is called the *product function* $f \cdot g$.
The general definitions are given next.

DEFINITION

If f and g are functions:
The **sum** $\boldsymbol{f + g}$ is the function defined by

$$(f + g)(x) = f(x) + g(x)$$

In Words

Remember, the symbol ∩ means intersection. It means you should find the elements that are common to two sets.

The domain of $f + g$ consists of the numbers x that are in the domains of both f and g. That is, domain of $f + g$ = domain of $f \cap$ domain of g.

DEFINITION

The **difference** $\boldsymbol{f - g}$ is the function defined by

$$(f - g)(x) = f(x) - g(x)$$

The domain of $f - g$ consists of the numbers x that are in the domains of both f and g. That is, domain of $f - g$ = domain of $f \cap$ domain of g.

DEFINITION

The **product** $\boldsymbol{f \cdot g}$ is the function defined by

$$(f \cdot g)(x) = f(x) \cdot g(x)$$

The domain of $f \cdot g$ consists of the numbers x that are in the domains of both f and g. That is, domain of $f \cdot g$ = domain of $f \cap$ domain of g.

DEFINITION

The **quotient** $\dfrac{f}{g}$ is the function defined by

$$\left(\frac{f}{g}\right)(x) = \frac{f(x)}{g(x)} \qquad g(x) \neq 0$$

The domain of $\dfrac{f}{g}$ consists of the numbers x for which $g(x) \neq 0$ that are in the domains of both f and g. That is,

$$\text{domain of } \frac{f}{g} = \{x \mid g(x) \neq 0\} \cap \text{domain of } f \cap \text{domain of } g.$$

EXAMPLE 11 **Operations on Functions**

Let f and g be two functions defined as

$$f(x) = \frac{1}{x+2} \quad \text{and} \quad g(x) = \frac{x}{x-1}$$

Find the following, and determine the domain in each case.

(a) $(f+g)(x)$ (b) $(f-g)(x)$

(c) $(f \cdot g)(x)$ (d) $\left(\frac{f}{g}\right)(x)$

Solution The domain of f is $\{x | x \neq -2\}$ and the domain of g is $\{x | x \neq 1\}$.

(a) $(f+g)(x) = f(x) + g(x) = \frac{1}{x+2} + \frac{x}{x-1}$

$$= \frac{x-1}{(x+2)(x-1)} + \frac{x(x+2)}{(x+2)(x-1)} = \frac{x^2+3x-1}{(x+2)(x-1)}$$

The domain of $f+g$ consists of those numbers x that are in the domains of both f and g. Therefore, the domain of $f+g$ is $\{x | x \neq -2, x \neq 1\}$.

(b) $(f-g)(x) = f(x) - g(x) = \frac{1}{x+2} - \frac{x}{x-1}$

$$= \frac{x-1}{(x+2)(x-1)} - \frac{x(x+2)}{(x+2)(x-1)} = \frac{-(x^2+x+1)}{(x+2)(x-1)}$$

The domain of $f-g$ consists of those numbers x that are in the domains of both f and g. Therefore, the domain of $f-g$ is $\{x | x \neq -2, x \neq 1\}$.

(c) $(f \cdot g)(x) = f(x) \cdot g(x) = \frac{1}{x+2} \cdot \frac{x}{x-1} = \frac{x}{(x+2)(x-1)}$

The domain of $f \cdot g$ consists of those numbers x that are in the domains of both f and g. Therefore, the domain of $f \cdot g$ is $\{x | x \neq -2, x \neq 1\}$.

(d) $\left(\frac{f}{g}\right)(x) = \frac{f(x)}{g(x)} = \frac{\frac{1}{x+2}}{\frac{x}{x-1}} = \frac{1}{x+2} \cdot \frac{x-1}{x} = \frac{x-1}{x(x+2)}$

The domain of $\frac{f}{g}$ consists of the numbers x for which $g(x) \neq 0$ that are in the domains of both f and g. Since $g(x) = 0$ when $x = 0$, we exclude 0 as well as -2 and 1 from the domain. The domain of $\frac{f}{g}$ is $\{x | x \neq -2, x \neq 0, x \neq 1\}$.

Now Work PROBLEM 61

In calculus, it is sometimes helpful to view a complicated function as the sum, difference, product, or quotient of simpler functions. For example,

$F(x) = x^2 + \sqrt{x}$ is the sum of $f(x) = x^2$ and $g(x) = \sqrt{x}$.

$H(x) = \frac{x^2-1}{x^2+1}$ is the quotient of $f(x) = x^2 - 1$ and $g(x) = x^2 + 1$.

SUMMARY

We list here some of the important vocabulary introduced in this section, with a brief description of each term.

Function A relation between two sets of real numbers so that each number x in the first set, the domain, has corresponding to it exactly one number y in the second set

A set of ordered pairs (x, y) or $(x, f(x))$ in which no first element is paired with two different second elements

The range is the set of y values of the function that are the images of the x values in the domain.

A function f may be defined implicitly by an equation involving x and y or explicitly by writing $y = f(x)$.

Unspecified domain If a function f is defined by an equation and no domain is specified, then the domain will be taken to be the largest set of real numbers for which the equation defines a real number.

Function notation $y = f(x)$

f is a symbol for the function.

x is the independent variable or argument.

y is the dependent variable.

$f(x)$ is the value of the function at x, or the image of x.

3.1 Assess Your Understanding

'Are You Prepared?' *Answers are given at the end of these exercises. If you get a wrong answer, read the pages listed in red.*

1. The inequality $-1 < x < 3$ can be written in interval notation as _____. (pp. 150–151)
2. If $x = -2$, the value of the expression $3x^2 - 5x + \frac{1}{x}$ is _____. (pp. 20–21)
3. The domain of the variable in the expression $\frac{x-3}{x+4}$ is _____. (p. 21)
4. Solve the inequality: $3 - 2x > 5$. Graph the solution set. (pp. 153–155)

Concepts and Vocabulary

5. If f is a function defined by the equation $y = f(x)$, then x is called the _____ variable and y is the _____ variable.
6. The set of all images of the elements in the domain of a function is called the _____.
7. If the domain of f is all real numbers in the interval $[0, 7]$ and the domain of g is all real numbers in the interval $[-2, 5]$, the domain of $f + g$ is all real numbers in the interval _____.
8. If $f(5) = 17$, then 17 is the _____ of 5.
9. If $f(x) = x + 1$ and $g(x) = x^3$, then _____ $= x^3 - (x + 1)$.
10. ***True or False*** Every relation is a function.
11. ***True or False*** The domain of $(f \cdot g)(x)$ consists of the numbers x that are in the domains of both f and g.
12. ***True or False*** The independent variable is sometimes referred to as the argument of the function.
13. ***True or False*** If no domain is specified for a function f, then the domain of f is taken to be the set of real numbers.
14. ***True or False*** The domain of the function $f(x) = \frac{x^2 - 4}{x}$ is $\{x | x \neq \pm 2\}$.

Skill Building

In Problems 15–26, determine whether each relation represents a function. For each function, state the domain and range.

15.

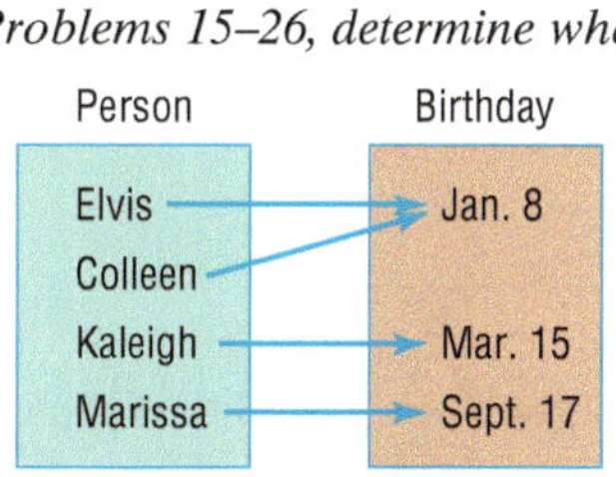

16.

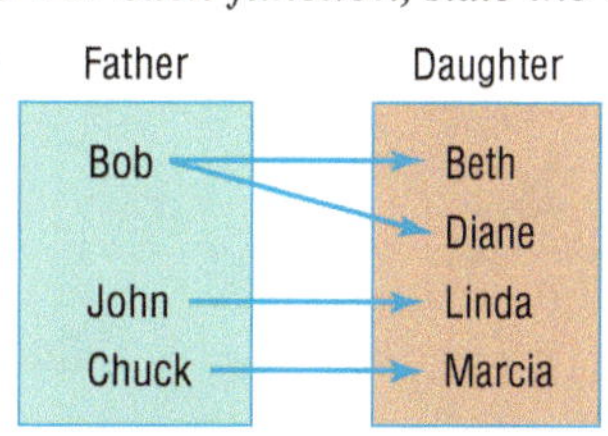

17.

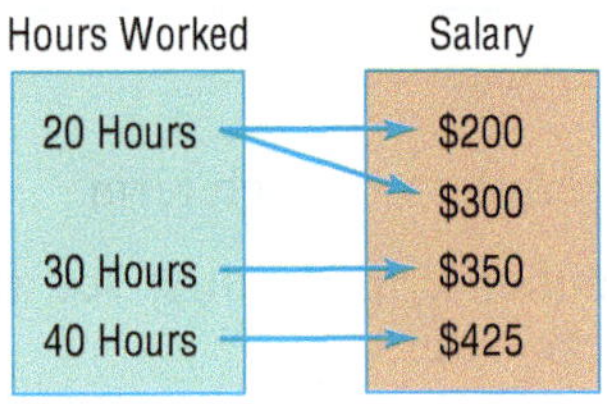

18.

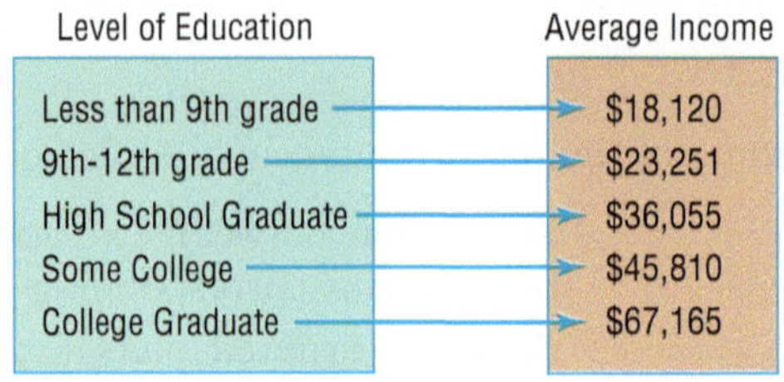

19. $\{(2, 6), (-3, 6), (4, 9), (2, 10)\}$ **20.** $\{(-2, 5), (-1, 3), (3, 7), (4, 12)\}$ **21.** $\{(1, 3), (2, 3), (3, 3), (4, 3)\}$

22. $\{(0, -2), (1, 3), (2, 3), (3, 7)\}$ **23.** $\{(-2, 4), (-2, 6), (0, 3), (3, 7)\}$ **24.** $\{(-4, 4), (-3, 3), (-2, 2), (-1, 1), (-4, 0)\}$

25. $\{(-2, 4), (-1, 1), (0, 0), (1, 1)\}$ **26.** $\{(-2, 16), (-1, 4), (0, 3), (1, 4)\}$

In Problems 27–38, determine whether the equation defines y as a function of x.

27. $y = x^2$ **28.** $y = x^3$ **29.** $y = \frac{1}{x}$ **30.** $y = |x|$

31. $y = \pm\sqrt{4 - x^2}$ **32.** $y = \pm\sqrt{1 - 2x}$ **33.** $x = y^2$ **34.** $x + y^2 = 1$

35. $y = 2x^2 - 3x + 4$ **36.** $y = \frac{3x - 1}{x + 2}$ **37.** $2x^2 + 3y^2 = 1$ **38.** $x^2 - 4y^2 = 1$

In Problems 39–46, find the following for each function:

(a) $f(0)$ *(b)* $f(1)$ *(c)* $f(-1)$ *(d)* $f(-x)$ *(e)* $-f(x)$ *(f)* $f(x + 1)$ *(g)* $f(2x)$ *(h)* $f(x + h)$

39. $f(x) = 3x^2 + 2x - 4$ **40.** $f(x) = -2x^2 + x - 1$ **41.** $f(x) = \frac{x}{x^2 + 1}$ **42.** $f(x) = \frac{x^2 - 1}{x + 4}$

43. $f(x) = |x| + 4$ **44.** $f(x) = \sqrt{x^2 + x}$ **45.** $f(x) = \frac{2x + 1}{3x - 5}$ **46.** $f(x) = 1 - \frac{1}{(x + 2)^2}$

In Problems 47–60, find the domain of each function.

47. $f(x) = -5x + 4$ **48.** $f(x) = x^2 + 2$ **49.** $f(x) = \frac{x}{x^2 + 1}$ **50.** $f(x) = \frac{x^2}{x^2 + 1}$

51. $g(x) = \frac{x}{x^2 - 16}$ **52.** $h(x) = \frac{2x}{x^2 - 4}$ **53.** $F(x) = \frac{x - 2}{x^3 + x}$ **54.** $G(x) = \frac{x + 4}{x^3 - 4x}$

55. $h(x) = \sqrt{3x - 12}$ **56.** $G(x) = \sqrt{1 - x}$ **57.** $f(x) = \frac{4}{\sqrt{x - 9}}$ **58.** $f(x) = \frac{x}{\sqrt{x - 4}}$

59. $p(x) = \sqrt{\frac{2}{x - 1}}$ **60.** $q(x) = \sqrt{-x - 2}$

In Problems 61–70, for the given functions f and g, find the following. For parts (a)–(d), also find the domain.

(a) $(f + g)(x)$ *(b)* $(f - g)(x)$ *(c)* $(f \cdot g)(x)$ *(d)* $\left(\frac{f}{g}\right)(x)$

(e) $(f + g)(3)$ *(f)* $(f - g)(4)$ *(g)* $(f \cdot g)(2)$ *(h)* $\left(\frac{f}{g}\right)(1)$

61. $f(x) = 3x + 4;\ g(x) = 2x - 3$ **62.** $f(x) = 2x + 1;\ g(x) = 3x - 2$

63. $f(x) = x - 1;\ g(x) = 2x^2$ **64.** $f(x) = 2x^2 + 3;\ g(x) = 4x^3 + 1$

65. $f(x) = \sqrt{x};\ g(x) = 3x - 5$ **66.** $f(x) = |x|;\ g(x) = x$

67. $f(x) = 1 + \frac{1}{x};\ g(x) = \frac{1}{x}$ **68.** $f(x) = \sqrt{x - 1};\ g(x) = \sqrt{4 - x}$

69. $f(x) = \frac{2x + 3}{3x - 2};\ g(x) = \frac{4x}{3x - 2}$ **70.** $f(x) = \sqrt{x + 1};\ g(x) = \frac{2}{x}$

71. Given $f(x) = 3x + 1$ and $(f + g)(x) = 6 - \frac{1}{2}x$, find the function g.

72. Given $f(x) = \frac{1}{x}$ and $\left(\frac{f}{g}\right)(x) = \frac{x + 1}{x^2 - x}$, find the function g.

In Problems 73–80, find the difference quotient of f; that is, find $\dfrac{f(x+h)-f(x)}{h}$, $h \neq 0$, *for each function. Be sure to simplify.*

73. $f(x) = 4x + 3$
74. $f(x) = -3x + 1$
75. $f(x) = x^2 - x + 4$
76. $f(x) = x^2 + 5x - 1$
77. $f(x) = 3x^2 - 2x + 6$
78. $f(x) = 4x^2 + 5x - 7$
79. $f(x) = x^3 - 2$
80. $f(x) = \dfrac{1}{x+3}$

Applications and Extensions

81. If $f(x) = 2x^3 + Ax^2 + 4x - 5$ and $f(2) = 5$, what is the value of A?

82. If $f(x) = 3x^2 - Bx + 4$ and $f(-1) = 12$, what is the value of B?

83. If $f(x) = \dfrac{3x+8}{2x-A}$ and $f(0) = 2$, what is the value of A?

84. If $f(x) = \dfrac{2x-B}{3x+4}$ and $f(2) = \dfrac{1}{2}$, what is the value of B?

85. If $f(x) = \dfrac{2x-A}{x-3}$ and $f(4) = 0$, what is the value of A? Where is f not defined?

86. If $f(x) = \dfrac{x-B}{x-A}$, $f(2) = 0$ and $f(1)$ is undefined, what are the values of A and B?

87. **Geometry** Express the area A of a rectangle as a function of the length x if the length of the rectangle is twice its width.

88. **Geometry** Express the area A of an isosceles right triangle as a function of the length x of one of the two equal sides.

89. **Constructing Functions** Express the gross salary G of a person who earns \$10 per hour as a function of the number x of hours worked.

90. **Constructing Functions** Tiffany, a commissioned salesperson, earns \$100 base pay plus \$10 per item sold. Express her gross salary G as a function of the number x of items sold.

91. **Population as a Function of Age** The function

$$P(a) = 0.015a^2 - 4.962a + 290.580$$

represents the population P (in millions) of Americans in 2005 that are a years of age or older.

Source: U.S. Census Bureau

(a) Identify the dependent and independent variable.
(b) Evaluate $P(20)$. Provide a verbal explanation of the meaning of $P(20)$.
(c) Evaluate $P(0)$. Provide a verbal explanation of the meaning of $P(0)$.

92. **Number of Rooms** The function

$$N(r) = -1.44r^2 + 14.52r - 14.96$$

represents the number N of housing units (in millions) in 2005 that have r rooms, where r is an integer and $2 \leq r \leq 9$.

Source: U.S. Census Bureau

(a) Identify the dependent and independent variable.
(b) Evaluate $N(3)$. Provide a verbal explanation of the meaning of $N(3)$.

93. **Effect of Gravity on Earth** If a rock falls from a height of 20 meters on Earth, the height H (in meters) after x seconds is approximately

$$H(x) = 20 - 4.9x^2$$

(a) What is the height of the rock when $x = 1$ second? $x = 1.1$ seconds? $x = 1.2$ seconds? $x = 1.3$ seconds?
(b) When is the height of the rock 15 meters? When is it 10 meters? When is it 5 meters?
(c) When does the rock strike the ground?

94. **Effect of Gravity on Jupiter** If a rock falls from a height of 20 meters on the planet Jupiter, its height H (in meters) after x seconds is approximately

$$H(x) = 20 - 13x^2$$

(a) What is the height of the rock when $x = 1$ second? $x = 1.1$ seconds? $x = 1.2$ seconds?
(b) When is the height of the rock 15 meters? When is it 10 meters? When is it 5 meters?
(c) When does the rock strike the ground?

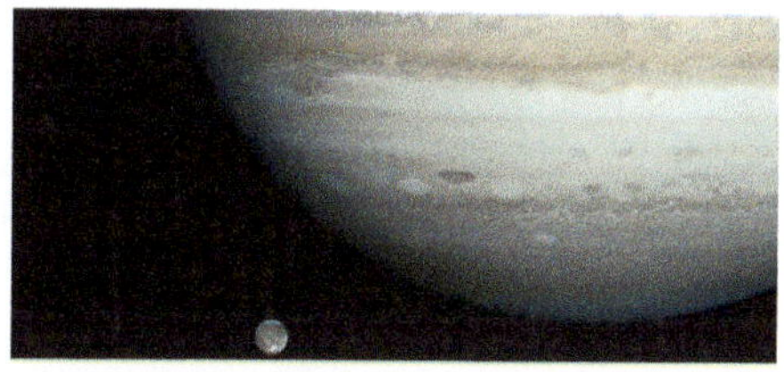

95. **Cost of Trans-Atlantic Travel** A Boeing 747 crosses the Atlantic Ocean (3000 miles) with an airspeed of 500 miles per hour. The cost C (in dollars) per passenger is given by

$$C(x) = 100 + \frac{x}{10} + \frac{36{,}000}{x}$$

where x is the ground speed (airspeed $\pm$ wind).

(a) What is the cost per passenger for quiescent (no wind) conditions?
(b) What is the cost per passenger with a head wind of 50 miles per hour?
(c) What is the cost per passenger with a tail wind of 100 miles per hour?
(d) What is the cost per passenger with a head wind of 100 miles per hour?

96. **Cross-sectional Area** The cross-sectional area of a beam cut from a log with radius 1 foot is given by the function $A(x) = 4x\sqrt{1-x^2}$, where x represents the length, in feet, of half the base of the beam. See the figure. Determine the cross-sectional area of the beam if the length of half the base of the beam is as follows:

(a) One-third of a foot
(b) One-half of a foot
(c) Two-thirds of a foot

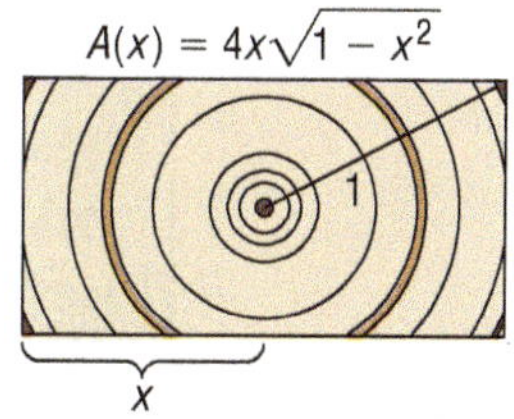

97. Economics The **participation rate** is the number of people in the labor force divided by the civilian population (excludes military). Let $L(x)$ represent the size of the labor force in year x and $P(x)$ represent the civilian population in year x. Determine a function that represents the participation rate R as a function of x.

98. Crimes Suppose that $V(x)$ represents the number of violent crimes committed in year x and $P(x)$ represents the number of property crimes committed in year x. Determine a function T that represents the combined total of violent crimes and property crimes in year x.

99. Health Care Suppose that $P(x)$ represents the percentage of income spent on health care in year x and $I(x)$ represents income in year x. Determine a function H that represents total health care expenditures in year x.

100. Income Tax Suppose that $I(x)$ represents the income of an individual in year x before taxes and $T(x)$ represents the individual's tax bill in year x. Determine a function N that represents the individual's net income (income after taxes) in year x.

101. Profit Function Suppose that the revenue R, in dollars, from selling x cell phones, in hundreds, is $R(x) = -1.2x^2 + 220x$. The cost C, in dollars, of selling x cell phones, in hundreds, is $C(x) = 0.05x^3 - 2x^2 + 65x + 500$.
(a) Find the profit function, $P(x) = R(x) - C(x)$.
(b) Find the profit if $x = 15$ hundred cell phones are sold.
(c) Interpret $P(15)$.

102. Profit Function Suppose that the revenue R, in dollars, from selling x clocks is $R(x) = 30x$. The cost C, in dollars, of selling x clocks is $C(x) = 0.1x^2 + 7x + 400$.
(a) Find the profit function, $P(x) = R(x) - C(x)$.
(b) Find the profit if $x = 30$ clocks are sold.
(c) Interpret $P(30)$.

103. Some functions f have the property that $f(a + b) = f(a) + f(b)$ for all real numbers a and b. Which of the following functions have this property?
(a) $h(x) = 2x$ (b) $g(x) = x^2$
(c) $F(x) = 5x - 2$ (d) $G(x) = \frac{1}{x}$

Discussion and Writing

104. Are the functions $f(x) = x - 1$ and $g(x) = \frac{x^2 - 1}{x + 1}$ the same? Explain.

105. Investigate when, historically, the use of the function notation $y = f(x)$ first appeared.

106. Write the equation of the following function f. The square root of three times a number x increased by 5.

107. Write the following function in words: $g(t) = t^2 - 2$.

'Are You Prepared?' Answers

1. $(-1, 3)$ **2.** 21.5 **3.** $\{x \mid x \neq -4\}$ **4.** $\{x \mid x < -1\}$ 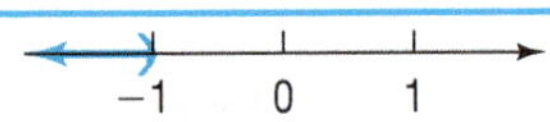

3.2 The Graph of a Function

PREPARING FOR THIS SECTION *Before getting started, review the following:*

- Graphs of Equations (Section 1.1, pp. 91–95)
- Intercepts (Section 2.1, pp. 170–171)

Now Work the 'Are You Prepared?' problems on page 230.

OBJECTIVES
1 Identify the Graph of a Function (p. 227)
2 Obtain Information from or about the Graph of a Function (p. 228)

In applications, a graph often demonstrates more clearly the relationship between two variables than, say, an equation or table would. For example, Table 1 shows the average price of gasoline in California adjusted for inflation (based on 2005 dollars) for the years 1979–2006. If we plot these data and then connect the points, we obtain Figure 13.

Table 1

Year	Price	Year	Price	Year	Price	Year	Price
1979	1.9829	1986	1.3459	1993	1.5189	2000	1.8249
1980	2.4929	1987	1.3274	1994	1.4709	2001	1.7536
1981	2.4977	1988	1.3111	1995	1.4669	2002	1.5955
1982	2.1795	1989	1.3589	1996	1.5397	2003	1.8950
1983	1.8782	1990	1.4656	1997	1.5329	2004	2.1521
1984	1.8310	1991	1.4973	1998	1.3246	2005	2.4730
1985	1.7540	1992	1.3969	1999	1.5270	2006	2.8919

Source: Statistical Abstract of the United States

Figure 13

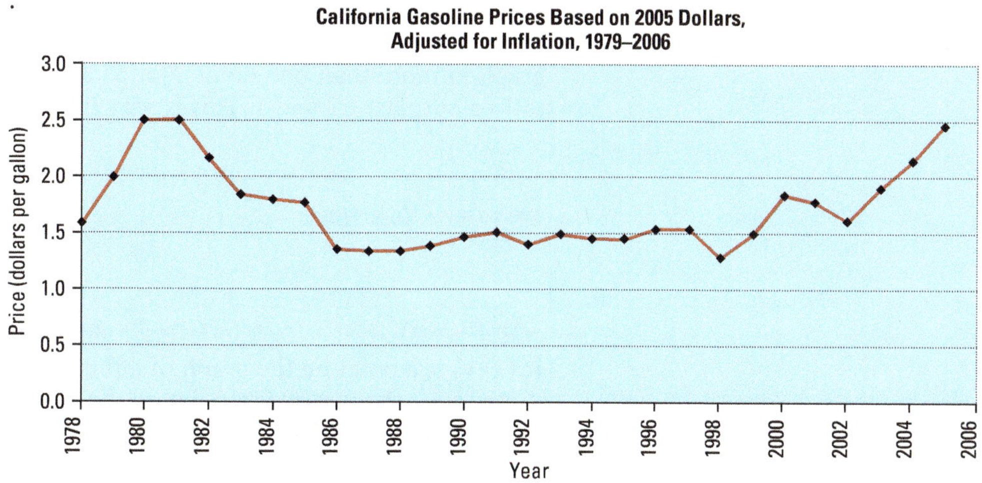

SOURCE: Statistical Abstact of the United States.

We can see from the graph that the price of gasoline (adjusted for inflation) was falling from 2000 to 2002 and has been rising rapidly since 2002. The graph also shows that the lowest price occurred in 1998. To learn information such as this from an equation requires that some calculations be made.

Look again at Figure 13. The graph shows that for each date on the horizontal axis there is only one price on the vertical axis. The graph represents a function, although the exact rule for getting from date to price is not given.

When a function is defined by an equation in x and y, the **graph of the function** is the graph of the equation, that is, the set of points (x, y) in the xy-plane that satisfies the equation.

1 Identify the Graph of a Function

Not every collection of points in the xy-plane represents the graph of a function. Remember, for a function, each number x in the domain has exactly one image y in the range. This means that the graph of a function cannot contain two points with the same x-coordinate and different y-coordinates. Therefore, the graph of a function must satisfy the following **vertical-line test.**

THEOREM

Vertical-line Test

A set of points in the xy-plane is the graph of a function if and only if every vertical line intersects the graph in at most one point.

In other words, if any vertical line intersects a graph at more than one point, the graph is not the graph of a function.

EXAMPLE 1 Identifying the Graph of a Function

Which of the graphs in Figure 14 are graphs of functions?

Figure 14

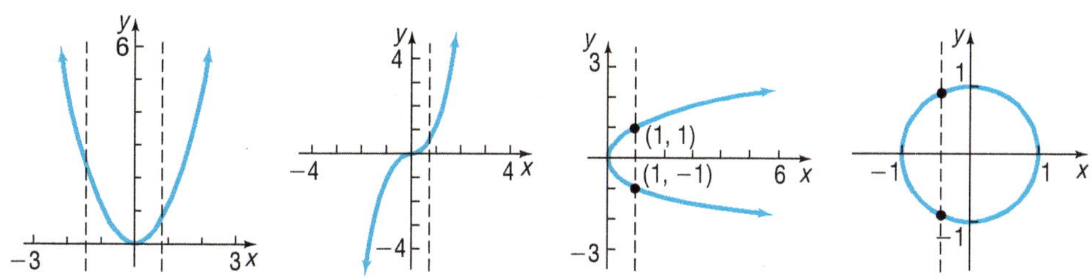

(a) $y = x^2$ **(b)** $y = x^3$ **(c)** $x = y^2$ **(d)** $x^2 + y^2 = 1$

Solution The graphs in Figures 14(a) and 14(b) are graphs of functions, because every vertical line intersects each graph in at most one point. The graphs in Figures 14(c) and 14(d) are not graphs of functions, because there is a vertical line that intersects each graph in more than one point. Notice in Figure 14(c) that the input 1 corresponds to two outputs, -1 and 1. This is why the graph does not represent a function.

Now Work PROBLEM 15

2 Obtain Information from or about the Graph of a Function

If (x, y) is a point on the graph of a function f, then y is the value of f at x. That is, if $y = f(x)$, then (x, y) is a point on the graph of f. The next example illustrates how to obtain information about a function if its graph is given.

EXAMPLE 2 Obtaining Information from the Graph of a Function

Figure 15

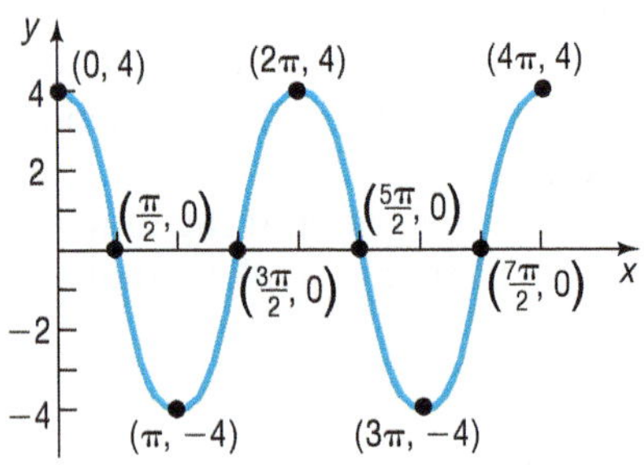

Let f be the function whose graph is given in Figure 15. (The graph of f might represent the distance that the bob of a pendulum is from its *at-rest* position. Negative values of y mean that the pendulum is to the left of the at-rest position, and positive values of y mean that the pendulum is to the right of the at-rest position.)

(a) What are $f(0)$, $f\left(\frac{3\pi}{2}\right)$, and $f(3\pi)$?

(b) What is the domain of f?

(c) What is the range of f?

(d) List the intercepts. (Recall that these are the points, if any, where the graph crosses or touches the coordinate axes.)

(e) How often does the line $y = 2$ intersect the graph?

(f) For what values of x does $f(x) = -4$?

(g) For what values of x is $f(x) > 0$?

Solution (a) Since $(0, 4)$ is on the graph of f, the y-coordinate 4 is the value of f at the x-coordinate 0; that is, $f(0) = 4$. In a similar way, we find that when $x = \frac{3\pi}{2}$ then $y = 0$, so $f\left(\frac{3\pi}{2}\right) = 0$. When $x = 3\pi$, then $y = -4$, so $f(3\pi) = -4$.

(b) To determine the domain of f, we notice that the points on the graph of f have x-coordinates between 0 and 4π, inclusive; and for each number x between 0 and 4π, there is a point $(x, f(x))$ on the graph. The domain of f is $\{x | 0 \le x \le 4\pi\}$ or the interval $[0, 4\pi]$.

(c) The points on the graph all have y-coordinates between -4 and 4, inclusive; and for each such number y, there is at least one number x in the domain. The range of f is $\{y | -4 \le y \le 4\}$ or the interval $[-4, 4]$.

(d) The intercepts are

$$(0, 4), \left(\frac{\pi}{2}, 0\right), \left(\frac{3\pi}{2}, 0\right), \left(\frac{5\pi}{2}, 0\right), \quad \text{and} \quad \left(\frac{7\pi}{2}, 0\right)$$

(e) If we draw the horizontal line $y = 2$ on the graph in Figure 15, we find that it intersects the graph four times.

(f) Since $(\pi, -4)$ and $(3\pi, -4)$ are the only points on the graph for which $y = f(x) = -4$, we have $f(x) = -4$ when $x = \pi$ and $x = 3\pi$.

(g) To determine where $f(x) > 0$, we look at Figure 15 and determine the x-values from 0 to 4π for which the y-coordinate is positive. This occurs on $\left[0, \frac{\pi}{2}\right) \cup \left(\frac{3\pi}{2}, \frac{5\pi}{2}\right) \cup \left(\frac{7\pi}{2}, 4\pi\right]$. Using inequality notation, $f(x) > 0$ for $0 \le x < \frac{\pi}{2}$ or $\frac{3\pi}{2} < x < \frac{5\pi}{2}$ or $\frac{7\pi}{2} < x \le 4\pi$.

When the graph of a function is given, its domain may be viewed as the shadow created by the graph on the x-axis by vertical beams of light. Its range can be viewed as the shadow created by the graph on the y-axis by horizontal beams of light. Try this technique with the graph given in Figure 15.

Now Work PROBLEMS 9 AND 13

EXAMPLE 3 Obtaining Information about the Graph of a Function

Consider the function: $f(x) = \frac{x+1}{x+2}$

(a) Is the point $\left(1, \frac{1}{2}\right)$ on the graph of f?
(b) If $x = 2$, what is $f(x)$? What point is on the graph of f?
(c) If $f(x) = 2$, what is x? What point is on the graph of f?
(d) What are the x-intercepts of the graph of f (if any)? What point(s) are on the graph of f?

Solution

(a) When $x = 1$, then

$$f(x) = \frac{x+1}{x+2}$$
$$f(1) = \frac{1+1}{1+2} = \frac{2}{3}$$

The point $\left(1, \frac{2}{3}\right)$ is on the graph of f; the point $\left(1, \frac{1}{2}\right)$ is not.

(b) If $x = 2$, then

$$f(x) = \frac{x+1}{x+2}$$
$$f(2) = \frac{2+1}{2+2} = \frac{3}{4}$$

The point $\left(2, \frac{3}{4}\right)$ is on the graph of f.

(c) If $f(x) = 2$, then

$$f(x) = 2$$
$$\frac{x+1}{x+2} = 2$$
$$x + 1 = 2(x + 2) \quad \text{Multiply both sides by } x + 2.$$
$$x + 1 = 2x + 4 \quad \text{Remove parentheses.}$$
$$x = -3 \quad \text{Solve for } x.$$

If $f(x) = 2$, then $x = -3$. The point $(-3, 2)$ is on the graph of f.

(d) The x-intercepts of the graph of f are the real solutions of the equation $f(x) = 0$. The only real solution of the equation $f(x) = \frac{x+1}{x+2} = 0$ is $x = -1$, so -1 is the only x-intercept. Since $f(-1) = 0$, the point $(-1, 0)$ is on the graph of f.

Now Work PROBLEM 25

EXAMPLE 4 **Average Cost Function**

The average cost $\overline{C}$ of manufacturing x computers per day is given by the function

$$\overline{C}(x) = 0.56x^2 - 34.39x + 1212.57 + \frac{20{,}000}{x}$$

Determine the average cost of manufacturing:

(a) 30 computers per day
(b) 40 computers per day
(c) 50 computers per day
(d) Graph the function $\overline{C} = \overline{C}(x), 0 < x \le 80$.
(e) Create a TABLE with TblStart = 1 and ΔTbl = 1. Which value of x minimizes the average cost?

Solution (a) The average cost of manufacturing $x = 30$ computers is

$$\overline{C}(30) = 0.56(30)^2 - 34.39(30) + 1212.57 + \frac{20{,}000}{30} = \$1351.54$$

(b) The average cost of manufacturing $x = 40$ computers is

$$\overline{C}(40) = 0.56(40)^2 - 34.39(40) + 1212.57 + \frac{20{,}000}{40} = \$1232.97$$

(c) The average cost of manufacturing $x = 50$ computers is

$$\overline{C}(50) = 0.56(50)^2 - 34.39(50) + 1212.57 + \frac{20{,}000}{50} = \$1293.07$$

(d) See Figure 16 for the graph of $\overline{C} = \overline{C}(x)$.
(e) With the function $\overline{C} = \overline{C}(x)$ in Y_1, we create Table 2. We scroll down until we find a value of x for which Y_1 is smallest. Table 3 shows that manufacturing $x = 41$ computers minimizes the average cost at \$1231.74 per computer.

Figure 16

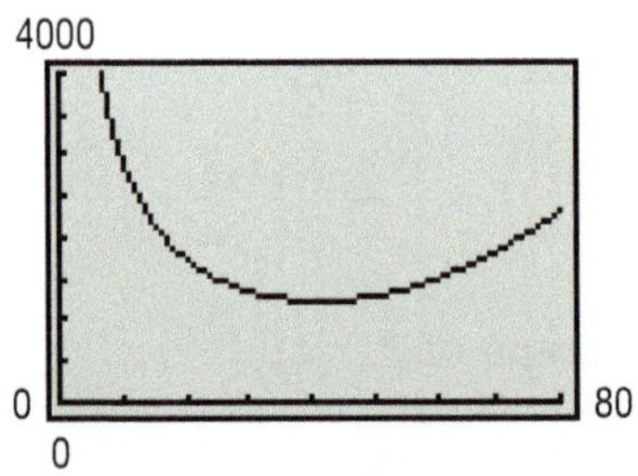

Table 2

X	Y1
1	21179
2	11146
3	7781.1
4	6084
5	5054.6
6	4359.7
7	3856.4

Y1 = .56X²−34.39X...

Table 3

X	Y1
38	1240.7
39	1235.9
40	1233
41	1231.7
42	1232.2
43	1234.4
44	1238.1

Y1=1231.74487805

Now Work PROBLEM 31

SUMMARY

Graph of a function The collection of points (x, y) that satisfies the equation $y = f(x)$.

Vertical-line test A collection of points is the graph of a function provided that every vertical line intersects the graph in at most one point.

3.2 Assess Your Understanding

'Are You Prepared?' *Answers are given at the end of these exercises. If you get a wrong answer, read the pages listed in red.*

1. The intercepts of the equation $x^2 + 4y^2 = 16$ are _____. (pp. 170–171)

2. ***True or False*** The point $(-2, -6)$ is on the graph of the equation $x = 2y - 2$. (pp. 91–93)

Concepts and Vocabulary

3. A set of points in the xy-plane is the graph of a function if and only if every _____ line intersects the graph in at most one point.
4. If the point $(5, -3)$ is a point on the graph of f, then $f(\_\_\_\_) = \_\_\_\_$.
5. If $g(-2) = 7$, then the point _____ is on the graph of g.
6. ***True or False*** A function can have more than one y-intercept.
7. ***True or False*** The graph of a function $y = f(x)$ always crosses the y-axis.
8. ***True or False*** The y-intercept of the graph of the function $y = f(x)$, whose domain is all real numbers, is $f(0)$.

Skill Building

9. Use the graph below of the function f to answer parts (a)–(n).

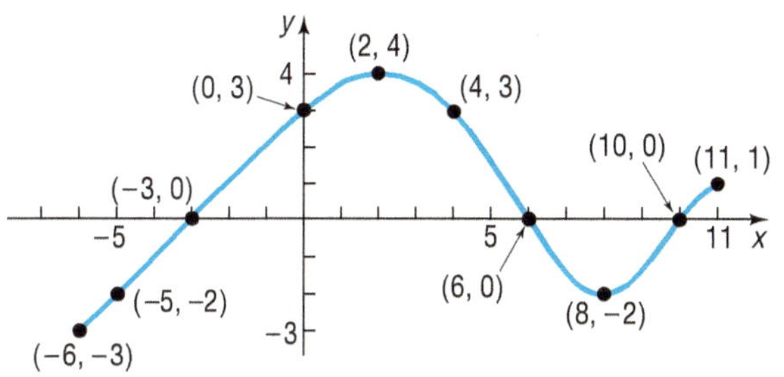

(a) Find $f(0)$ and $f(-6)$.
(b) Find $f(6)$ and $f(11)$.
(c) Is $f(3)$ positive or negative?
(d) Is $f(-4)$ positive or negative?
(e) For what values of x is $f(x) = 0$?
(f) For what values of x is $f(x) > 0$?
(g) What is the domain of f?
(h) What is the range of f?
(i) What are the x-intercepts?
(j) What is the y-intercept?
(k) How often does the line $y = \frac{1}{2}$ intersect the graph?
(l) How often does the line $x = 5$ intersect the graph?
(m) For what values of x does $f(x) = 3$?
(n) For what values of x does $f(x) = -2$?

10. Use the graph below of the function f to answer parts (a)–(n).

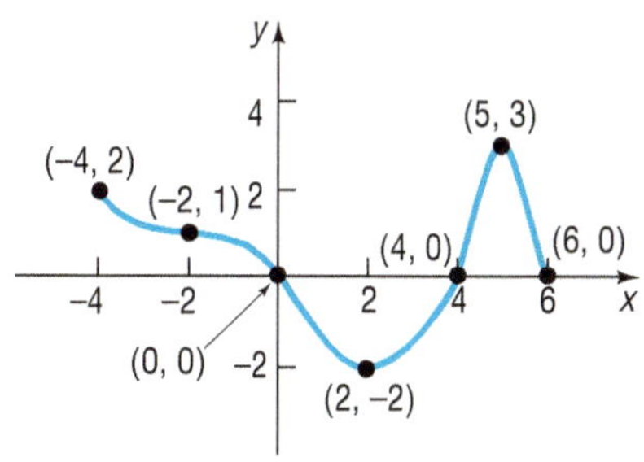

(a) Find $f(0)$ and $f(6)$.
(b) Find $f(2)$ and $f(-2)$.
(c) Is $f(3)$ positive or negative?
(d) Is $f(-1)$ positive or negative?
(e) For what values of x is $f(x) = 0$?
(f) For what values of x is $f(x) < 0$?
(g) What is the domain of f?
(h) What is the range of f?
(i) What are the x-intercepts?
(j) What is the y-intercept?
(k) How often does the line $y = -1$ intersect the graph?
(l) How often does the line $x = 1$ intersect the graph?
(m) For what value of x does $f(x) = 3$?
(n) For what value of x does $f(x) = -2$?

In Problems 11–22, determine whether the graph is that of a function by using the vertical-line test. If it is, use the graph to find:
(a) The domain and range
(b) The intercepts, if any
(c) Any symmetry with respect to the x-axis, the y-axis, or the origin

11.
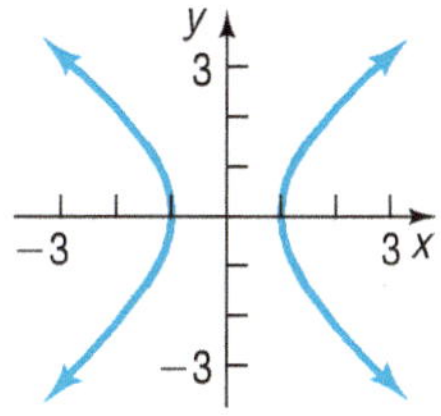

12.
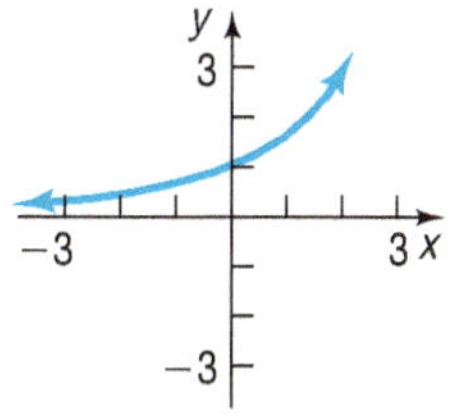

13.
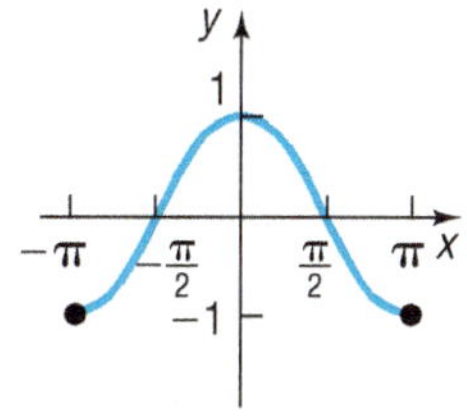

14.
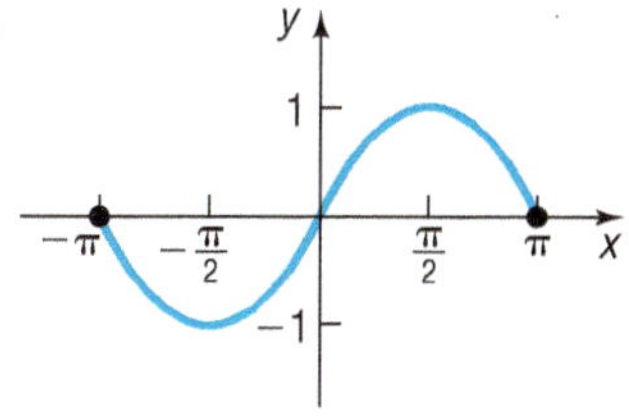

15.
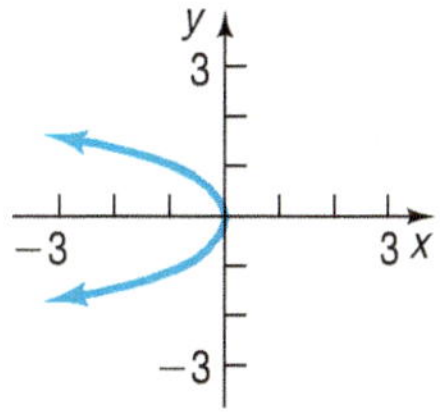

16.
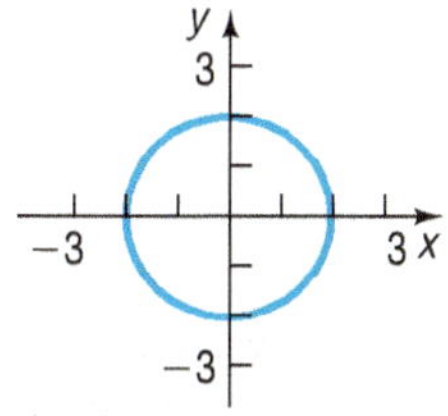

17.
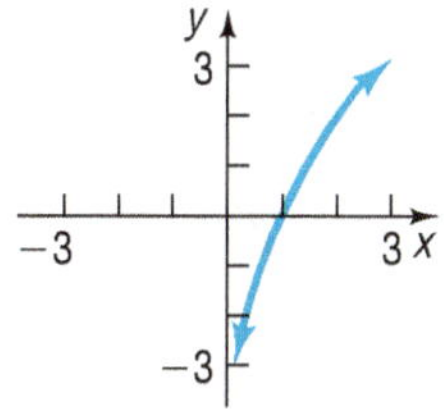

18.
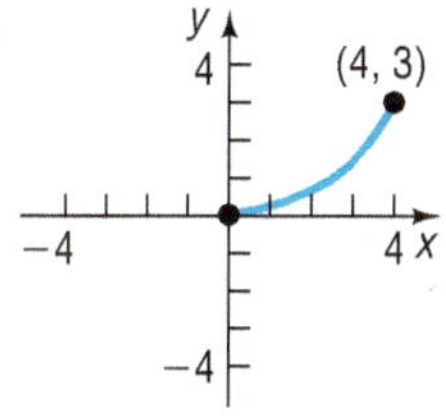

19.

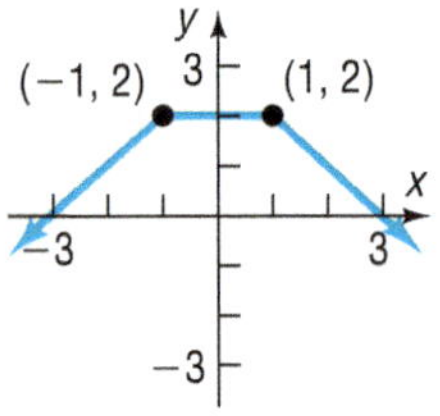

20.

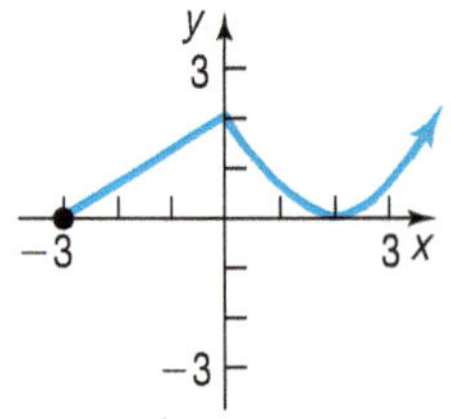

21.

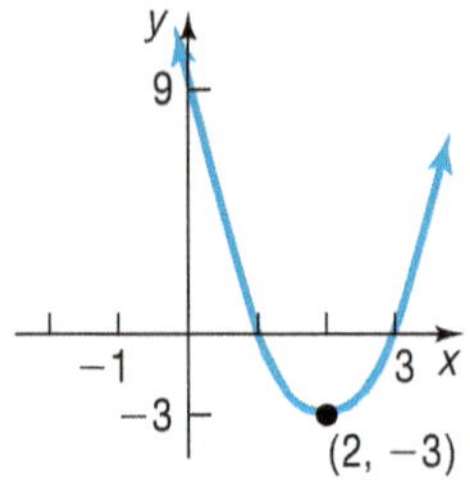

22.

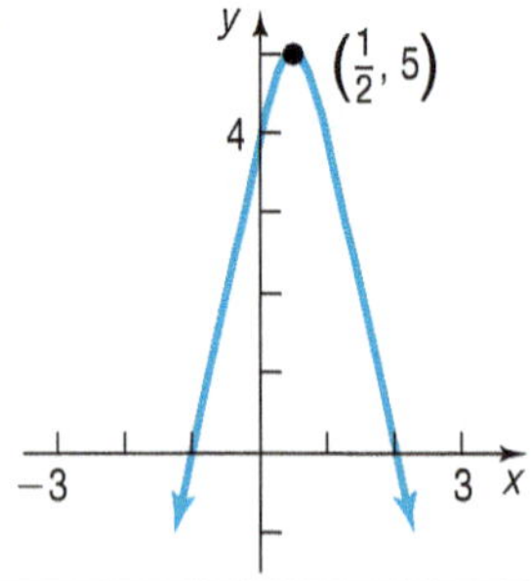

In Problems 23–28, answer the questions about the given function.

23. $f(x) = 2x^2 - x - 1$
(a) Is the point $(-1, 2)$ on the graph of f?
(b) If $x = -2$, what is $f(x)$? What point is on the graph of f?
(c) If $f(x) = -1$, what is x? What point(s) are on the graph of f?
(d) What is the domain of f?
(e) List the x-intercepts, if any, of the graph of f.
(f) List the y-intercept, if there is one, of the graph of f.

24. $f(x) = -3x^2 + 5x$
(a) Is the point $(-1, 2)$ on the graph of f?
(b) If $x = -2$, what is $f(x)$? What point is on the graph of f?
(c) If $f(x) = -2$, what is x? What point(s) are on the graph of f?
(d) What is the domain of f?
(e) List the x-intercepts, if any, of the graph of f.
(f) List the y-intercept, if there is one, of the graph of f.

25. $f(x) = \dfrac{x + 2}{x - 6}$
(a) Is the point $(3, 14)$ on the graph of f?
(b) If $x = 4$, what is $f(x)$? What point is on the graph of f?
(c) If $f(x) = 2$, what is x? What point(s) are on the graph of f?
(d) What is the domain of f?
(e) List the x-intercepts, if any, of the graph of f.
(f) List the y-intercept, if there is one, of the graph of f.

26. $f(x) = \dfrac{x^2 + 2}{x + 4}$
(a) Is the point $\left(1, \dfrac{3}{5}\right)$ on the graph of f?
(b) If $x = 0$, what is $f(x)$? What point is on the graph of f?
(c) If $f(x) = \dfrac{1}{2}$, what is x? What point(s) are on the graph of f?
(d) What is the domain of f?
(e) List the x-intercepts, if any, of the graph of f.
(f) List the y-intercept, if there is one, of the graph of f.

27. $f(x) = \dfrac{2x^2}{x^4 + 1}$
(a) Is the point $(-1, 1)$ on the graph of f?
(b) If $x = 2$, what is $f(x)$? What point is on the graph of f?
(c) If $f(x) = 1$, what is x? What point(s) are on the graph of f?
(d) What is the domain of f?
(e) List the x-intercepts, if any, of the graph of f.
(f) List the y-intercept, if there is one, of the graph of f.

28. $f(x) = \dfrac{2x}{x - 2}$
(a) Is the point $\left(\dfrac{1}{2}, -\dfrac{2}{3}\right)$ on the graph of f?
(b) If $x = 4$, what is $f(x)$? What point is on the graph of f?
(c) If $f(x) = 1$, what is x? What point(s) are on the graph of f?
(d) What is the domain of f?
(e) List the x-intercepts, if any, of the graph of f.
(f) List the y-intercept, if there is one, of the graph of f.

Applications and Extensions

29. Free-throw Shots According to physicist Peter Brancazio, the key to a successful foul shot in basketball lies in the arc of the shot. Brancazio determined the optimal angle of the arc from the free-throw line to be 45 degrees. The arc also depends on the velocity with which the ball is shot. If a player shoots a foul shot, releasing the ball at a 45-degree angle from a position 6 feet above the floor, then the path of the ball can be modeled by the function

$$h(x) = -\frac{44x^2}{v^2} + x + 6$$

where h is the height of the ball above the floor, x is the forward distance of the ball in front of the foul line, and v is the initial velocity with which the ball is shot in feet per second. Suppose a player shoots a ball with an initial velocity of 28 feet per second.
(a) Determine the height of the ball after it has traveled 8 feet in front of the foul line.
(b) Determine the height of the ball after it has traveled 12 feet in front of the foul line.
(c) Find additional points and graph the path of the basketball by hand.
(d) The center of the hoop is 10 feet above the floor and 15 feet in front of the foul line. Will the ball go through the hoop? Why or why not? If not, with what initial velocity must the ball be shot in order for the ball to go through the hoop?

Source: *The Physics of Foul Shots,* Discover, *Vol. 21, No. 10, October 2000*

30. Granny Shots The last player in the NBA to use an underhand foul shot (a "granny" shot) was Hall of Fame forward Rick Barry who retired in 1980. Barry believes that current NBA players could increase their free-throw percentage if they were to use an underhand shot. Since underhand shots are released from a lower position, the angle of the shot must be increased. If a player shoots an underhand foul shot, releasing the ball at a 70-degree angle from a position 3.5 feet above the floor, then the path of the ball can be modeled by the function $h(x) = -\dfrac{136x^2}{v^2} + 2.7x + 3.5$, where h is the height of the ball above the floor, x is the forward distance of the ball in front of the foul line, and v is the initial velocity with which the ball is shot in feet per second.

(a) The center of the hoop is 10 feet above the floor and 15 feet in front of the foul line. Determine the initial velocity with which the ball must be shot in order for the ball to go through the hoop.
(b) Write the function for the path of the ball using the velocity found in part (a).
(c) Determine the height of the ball after it has traveled 9 feet in front of the foul line.
(d) Find additional points and graph the path of the basketball by hand.

Source: *The Physics of Foul Shots,* Discover, *Vol. 21, No. 10, October 2000*

31. Cost of Trans-Atlantic Travel A Boeing 747 crosses the Atlantic Ocean (3000 miles) with an airspeed of 500 miles per hour. The cost C (in dollars) per passenger is given by

$$C(x) = 100 + \frac{x}{10} + \frac{36{,}000}{x}$$

where x is the ground speed (airspeed $\pm$ wind).

(a) Use a graphing utility to graph the function $C = C(x)$.
(b) Create a TABLE with TblStart = 0 and ΔTbl = 50.
(c) To the nearest 50 miles per hour, what ground speed minimizes the cost per passenger?

32. Cross-sectional Area The cross-sectional area of a beam cut from a log with radius 1 foot is given by the function $A(x) = 4x\sqrt{1 - x^2}$, where x represents the length, in feet, of half the base of the beam. See the figure.

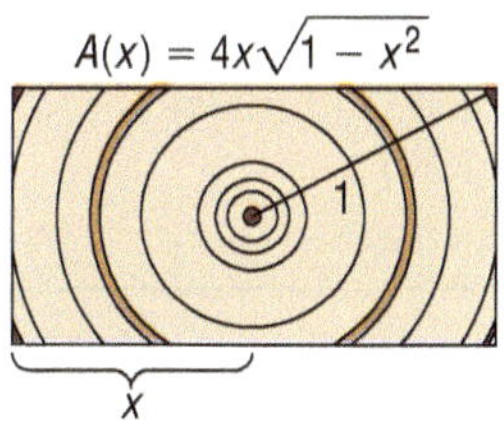

(a) Find the domain of A.
(b) Use a graphing utility to graph the function $A = A(x)$.
(c) Create a TABLE with TblStart = 0 and ΔTbl = 0.1 for $0 \le x \le 1$. Which value of x maximizes the cross-sectional area? What should be the length of the base of the beam to maximize the cross-sectional area?

33. The graphs of two functions, f and g, are illustrated. Use the graphs to answer parts (a)–(f).

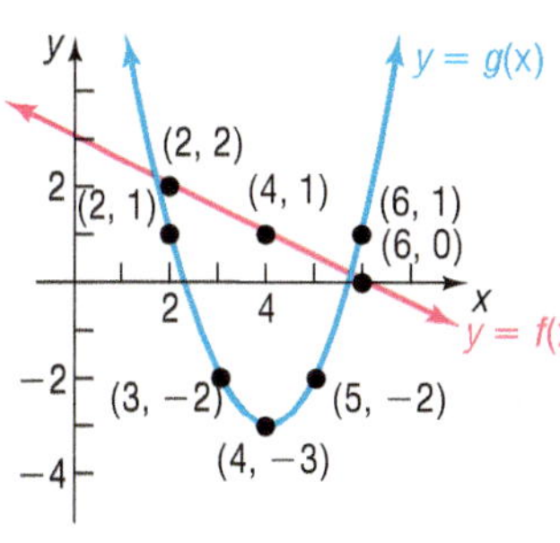

(a) $(f + g)(2)$ (b) $(f + g)(4)$ (c) $(f - g)(6)$
(d) $(g - f)(6)$ (e) $(f \cdot g)(2)$ (f) $\left(\dfrac{f}{g}\right)(4)$

34. Reading and Interpreting Graphs Let C be the function whose graph is given below. This graph represents the cost C of manufacturing q computers in a day.

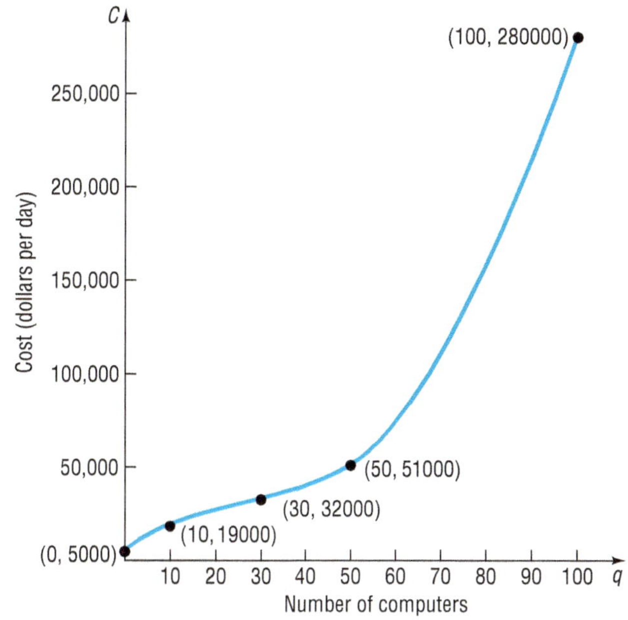

(a) Determine $C(0)$. Interpret this value.
(b) Determine $C(10)$. Interpret this value.
(c) Determine $C(50)$. Interpret this value.
(d) What is the domain of C? What does this domain imply in terms of daily production?
(e) Describe the shape of the graph.
(f) The point (30, 32000) is called an *inflection point*. Describe the behavior of the graph around the inflection point.

35. Reading and Interpreting Graphs Let C be the function whose graph is given below. This graph represents the cost C of using m anytime cell phone minutes in a month for a five-person family plan.

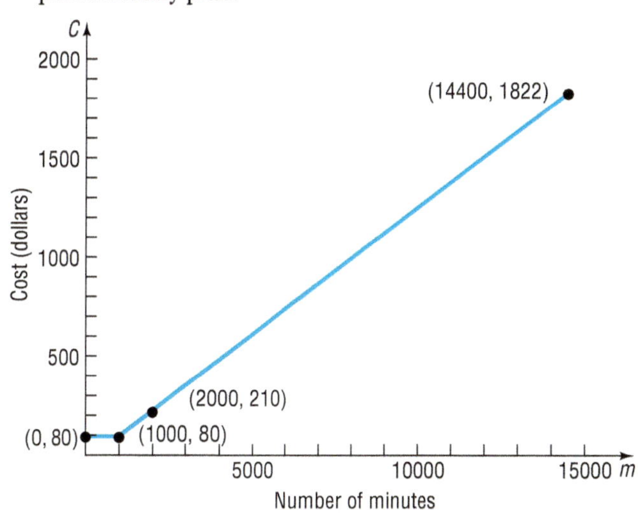

(a) Determine $C(0)$. Interpret this value.
(b) Determine $C(1000)$. Interpret this value.
(c) Determine $C(2000)$. Interpret this value.
(d) What is the domain of C? What does this domain imply in terms of the number of anytime minutes?
(e) Describe the shape of the graph.

Discussion and Writing

36. Describe how you would proceed to find the domain and range of a function if you were given its graph. How would your strategy change if you were given the equation defining the function instead of its graph?

37. How many x-intercepts can the graph of a function have? How many y-intercepts can the graph of a function have?

38. Is a graph that consists of a single point the graph of a function? Can you write the equation of such a function?

39. Match each of the following functions with the graph that best describes the situation.
    (a) The cost of building a house as a function of its square footage
    (b) The height of an egg dropped from a 300-foot building as a function of time
    (c) The height of a human as a function of time
    (d) The demand for Big Macs as a function of price
    (e) The height of a child on a swing as a function of time

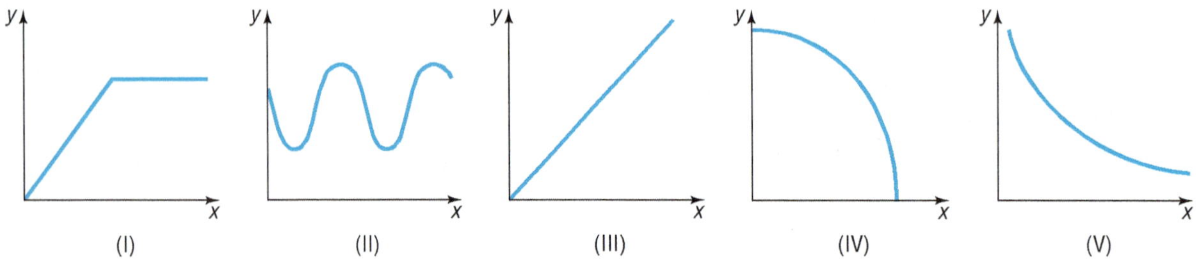

40. Match each of the following functions with the graph that best describes the situation.
    (a) The temperature of a bowl of soup as a function of time
    (b) The number of hours of daylight per day over a 2-year period
    (c) The population of Florida as a function of time
    (d) The distance traveled by a car going at a constant velocity as a function of time
    (e) The height of a golf ball hit with a 7-iron as a function of time

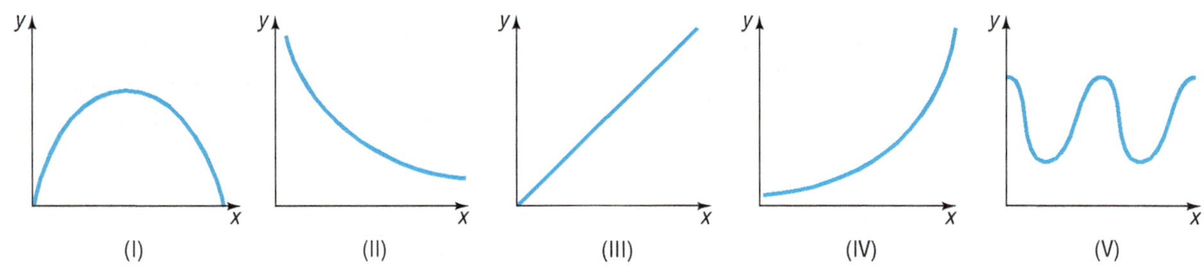

41. Consider the following scenario: Barbara decides to take a walk. She leaves home, walks 2 blocks in 5 minutes at a constant speed, and realizes that she forgot to lock the door. So Barbara runs home in 1 minute. While at her doorstep, it takes her 1 minute to find her keys and lock the door. Barbara walks 5 blocks in 15 minutes and then decides to jog home. It takes her 7 minutes to get home. Draw a graph of Barbara's distance from home (in blocks) as a function of time.

42. Consider the following scenario: Jayne enjoys riding her bicycle through the woods. At the forest preserve, she gets on her bicycle and rides up a 2000-foot incline in 10 minutes. She then travels down the incline in 3 minutes. The next 5000 feet are level terrain and she covers the distance in 20 minutes. She rests for 15 minutes. Jayne then travels 10,000 feet in 30 minutes. Draw a graph of Jayne's distance traveled (in feet) as a function of time.

43. The sketch to the right represents the distance d (in miles) that Kevin is from home as a function of time t (in hours). Answer the questions based on the graph. In parts (a)–(g), how many hours elapsed and how far was Kevin from home during this time?

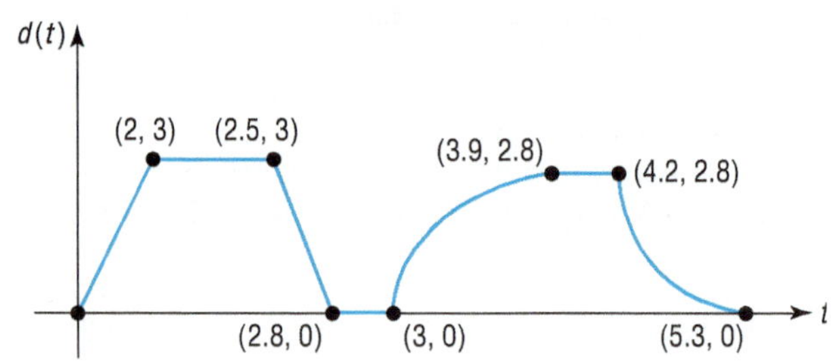

(a) From $t = 0$ to $t = 2$
(b) From $t = 2$ to $t = 2.5$
(c) From $t = 2.5$ to $t = 2.8$
(d) From $t = 2.8$ to $t = 3$
(e) From $t = 3$ to $t = 3.9$
(f) From $t = 3.9$ to $t = 4.2$
(g) From $t = 4.2$ to $t = 5.3$
(h) What is the farthest distance that Kevin is from home?
(i) How many times did Kevin return home?

44. The following sketch represents the speed v (in miles per hour) of Michael's car as a function of time t (in minutes).

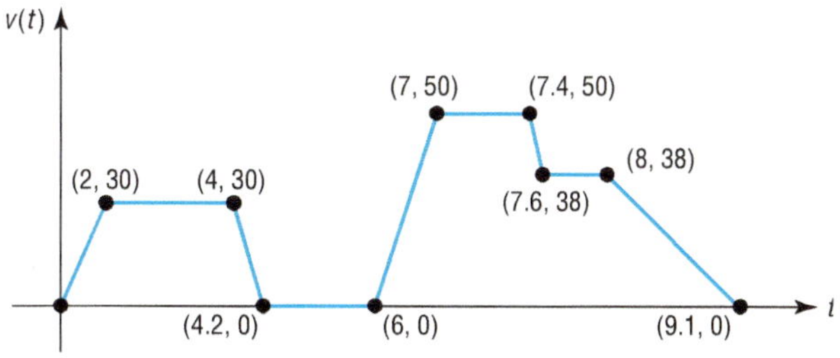

(a) Over what interval of time is Michael traveling fastest?
(b) Over what interval(s) of time is Michael's speed zero?
(c) What is Michael's speed between 0 and 2 minutes?
(d) What is Michael's speed between 4.2 and 6 minutes?
(e) What is Michael's speed between 7 and 7.4 minutes?
(f) When is Michael's speed constant?

45. Draw the graph of a function whose domain is $\{x|-3 \le x \le 8,\ x \ne 5\}$ and whose range is $\{y|-1 \le y \le 2,\ y \ne 0\}$. What point(s) in the rectangle $-3 \le x \le 8, -1 \le y \le 2$ cannot be on the graph? Compare your graph with those of other students. What differences do you see?

46. Is there a function whose graph is symmetric with respect to the x-axis? Explain.

47. Explain why the vertical-line test works.

'Are You Prepared?' Answers

1. $(-4, 0), (4, 0), (0, -2), (0, 2)$ **2.** False

3.3 Properties of Functions

PREPARING FOR THIS SECTION *Before getting started, review the following:*

- Intervals (Section 1.7, pp. 150–151)
- Intercepts (Section 2.1, pp. 170–171)
- Slope of a Line (Section 2.2, pp. 178–179)
- Point–Slope Form of a Line (Section 2.2, p. 182)
- Symmetry (Section 2.1, pp. 171–173)

Now Work the 'Are You Prepared?' problems on page 242.

OBJECTIVES
1 Determine Even and Odd Functions from a Graph (p. 235)
2 Identify Even and Odd Functions from the Equation (p. 236)
3 Use a Graph to Determine Where a Function Is Increasing, Decreasing, or Constant (p. 237)
4 Use a Graph to Locate Local Maxima and Local Minima (p. 238)
5 Use a Graphing Utility to Approximate Local Maxima and Local Minima and to Determine Where a Function Is Increasing or Decreasing (p. 239)
6 Find the Average Rate of Change of a Function (p. 240)

To obtain the graph of a function $y = f(x)$, it is often helpful to know certain properties that the function has and the impact of these properties on the way that the graph will look.

1 Determine Even and Odd Functions from a Graph

The words *even* and *odd*, when applied to a function f, describe the symmetry that exists for the graph of the function.

A function f is even, if and only if, whenever the point (x, y) is on the graph of f then the point $(-x, y)$ is also on the graph. Using function notation, we define an even function as follows:

DEFINITION

A function f is **even** if, for every number x in its domain, the number $-x$ is also in the domain and

$$f(-x) = f(x)$$

A function f is odd, if and only if, whenever the point (x, y) is on the graph of f then the point $(-x, -y)$ is also on the graph. Using function notation, we define an odd function as follows:

DEFINITION

A function f is **odd** if, for every number x in its domain, the number $-x$ is also in the domain and

$$f(-x) = -f(x)$$

Refer to page 172, where the tests for symmetry are listed. The following results are then evident.

THEOREM

A function is even if and only if its graph is symmetric with respect to the y-axis. A function is odd if and only if its graph is symmetric with respect to the origin.

EXAMPLE 1 Determining Even and Odd Functions from the Graph

Determine whether each graph given in Figure 17 is the graph of an even function, an odd function, or a function that is neither even nor odd.

Figure 17

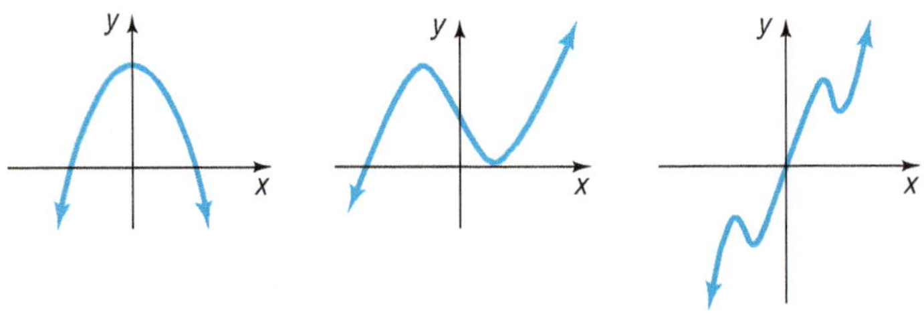

Solution

(a) The graph in Figure 17(a) is that of an even function, because the graph is symmetric with respect to the y-axis.

(b) The function whose graph is given in Figure 17(b) is neither even nor odd, because the graph is neither symmetric with respect to the y-axis nor symmetric with respect to the origin.

(c) The function whose graph is given in Figure 17(c) is odd, because its graph is symmetric with respect to the origin.

Now Work PROBLEMS 21(a), (b), AND (d)

2 Identify Even and Odd Functions from the Equation

A graphing utility can be used to conjecture whether a function is even, odd, or neither. Remember that, when the graph of an even function contains the point (x, y), it must also contain the point $(-x, y)$. Therefore, if the graph shows evidence of symmetry with respect to the y-axis, we would conjecture that the function is even. In addition, if the graph shows evidence of symmetry with respect to the origin, we would conjecture that the function is odd.

In the next example, we use a graphing utility to conjecture whether a function is even, odd, or neither. Then we verify our conjecture algebraically.

EXAMPLE 2 Identifying Even and Odd Functions

Use a graphing utility to conjecture whether each of the following functions is even, odd, or neither. Verify the conjecture algebraically. Then state whether the graph is symmetric with respect to the y-axis or with respect to the origin.

(a) $f(x) = x^2 - 5$ (b) $g(x) = x^3 - 1$ (c) $h(x) = 5x^3 - x$

Solution (a) Graph the function. See Figure 18. It appears that the graph is symmetric with respect to the y-axis. We conjecture that the function is even.

To algebraically verify the conjecture, we replace x by $-x$ in $f(x) = x^2 - 5$. Then

$$f(-x) = (-x)^2 - 5 = x^2 - 5 = f(x)$$

Since $f(-x) = f(x)$, we conclude that f is an even function, and the graph is symmetric with respect to the y-axis.

Figure 18

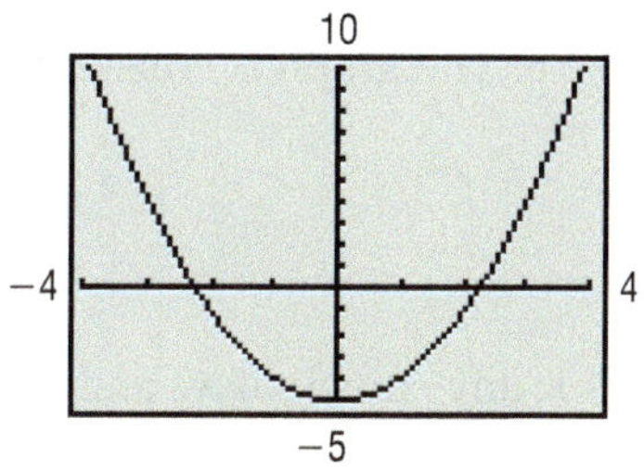

(b) Graph the function. See Figure 19. It appears that there is no symmetry. We conjecture that the function is neither even nor odd.

To algebraically verify that the function is not even, we find $g(-x)$ and compare the result with $g(x)$.

$$g(-x) = (-x)^3 - 1 = -x^3 - 1; \quad g(x) = x^3 - 1$$

Since $g(-x) \neq g(x)$, the function is not even.

To algebraically verify that the function is not odd, we find $-g(x)$ and compare the result with $g(-x)$.

$$-g(x) = -(x^3 - 1) = -x^3 + 1; \quad g(-x) = -x^3 - 1$$

Since $g(-x) \neq -g(x)$, the function is not odd. The graph is not symmetric with respect to the y-axis nor is it symmetric with respect to the origin.

Figure 19

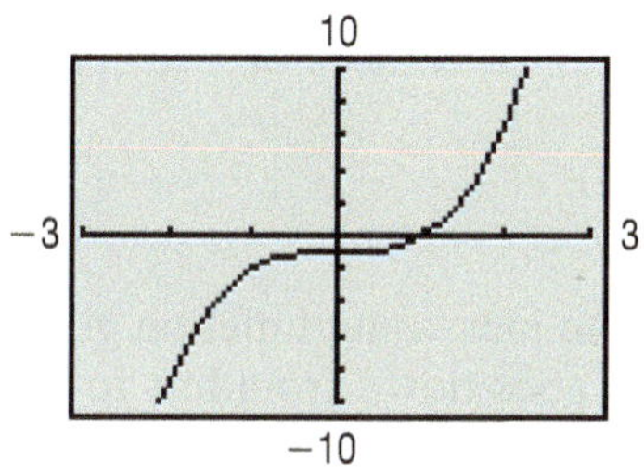

(c) Graph the function. See Figure 20. It appears that there is symmetry with respect to the origin. We conjecture that the function is odd.

To algebraically verify the conjecture, we replace x by $-x$ in $h(x) = 5x^3 - x$. Then

$$h(-x) = 5(-x)^3 - (-x) = -5x^3 + x = -(5x^3 - x) = -h(x)$$

Since $h(-x) = -h(x)$, h is an odd function, and the graph of h is symmetric with respect to the origin.

Figure 20

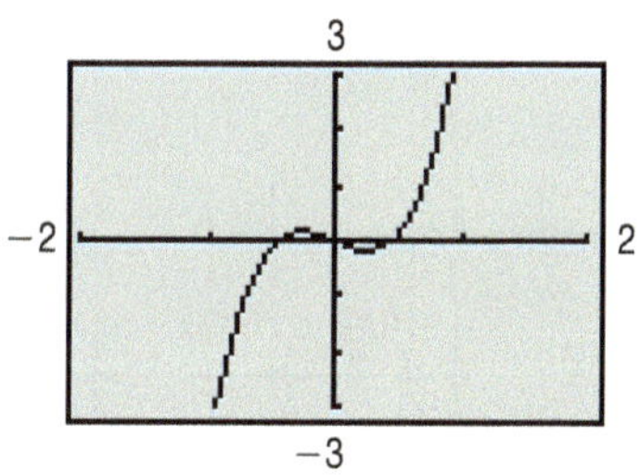

Now Work PROBLEM 33

3 Use a Graph to Determine Where a Function Is Increasing, Decreasing, or Constant

Consider the graph given in Figure 21. If you look from left to right along the graph of the function, you will notice that parts of the graph are rising, parts are falling, and parts are horizontal. In such cases, the function is described as *increasing, decreasing,* or *constant,* respectively.

Figure 21

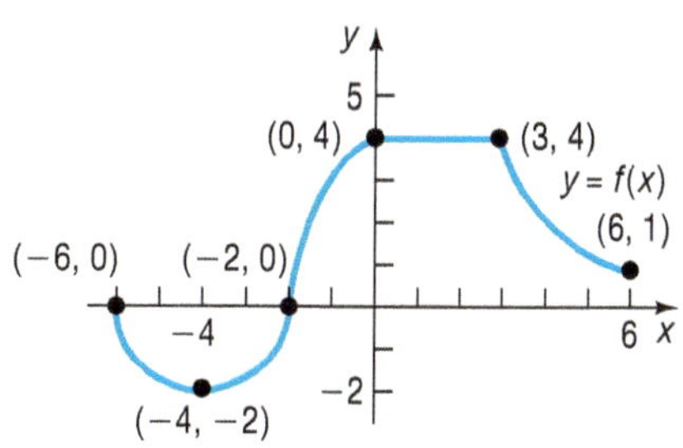

EXAMPLE 3 Determining Where a Function Is Increasing, Decreasing, or Constant from Its Graph

Where is the function in Figure 21 increasing? Where is it decreasing? Where is it constant?

Solution To answer the question of where a function is increasing, where it is decreasing, and where it is constant, we use strict inequalities involving the independent variable x, or we use open intervals* of x-coordinates. The graph in Figure 21 is rising (increasing) from the point $(-4, -2)$ to the point $(0, 4)$, so we conclude that it is

*The open interval (a, b) consists of all real numbers x for which $a < x < b$.

WARNING We describe the behavior of a graph in terms of its x-values. Do not say the graph in Figure 21 is increasing from the point $(-4, -2)$ to $(0, 4)$. Rather, say it is increasing on the interval $(-4, 0)$.

increasing on the open interval $(-4, 0)$ or for $-4 < x < 0$. The graph is falling (decreasing) from the point $(-6, 0)$ to the point $(-4, -2)$ and from the point $(3, 4)$ to the point $(6, 1)$. We conclude that the graph is decreasing on the open intervals $(-6, -4)$ and $(3, 6)$ or for $-6 < x < -4$ and $3 < x < 6$. The graph is constant on the open interval $(0, 3)$ or for $0 < x < 3$.

More precise definitions follow:

DEFINITIONS

A function f is **increasing** on an open interval I if, for any choice of x_1 and x_2 in I, with $x_1 < x_2$, we have $f(x_1) < f(x_2)$.

A function f is **decreasing** on an open interval I if, for any choice of x_1 and x_2 in I, with $x_1 < x_2$, we have $f(x_1) > f(x_2)$.

A function f is **constant** on an open interval I if, for all choices of x in I, the values $f(x)$ are equal.

In Words

If a function is decreasing, then, as the values of x get bigger, the values of the function get smaller. If a function is increasing, then, as the values of x get bigger, the values of the function also get bigger. If a function is constant, then, as the values of x get bigger, the values of the function remain unchanged.

Figure 22 illustrates the definitions. The graph of an increasing function goes up from left to right, the graph of a decreasing function goes down from left to right, and the graph of a constant function remains at a fixed height.

Figure 22

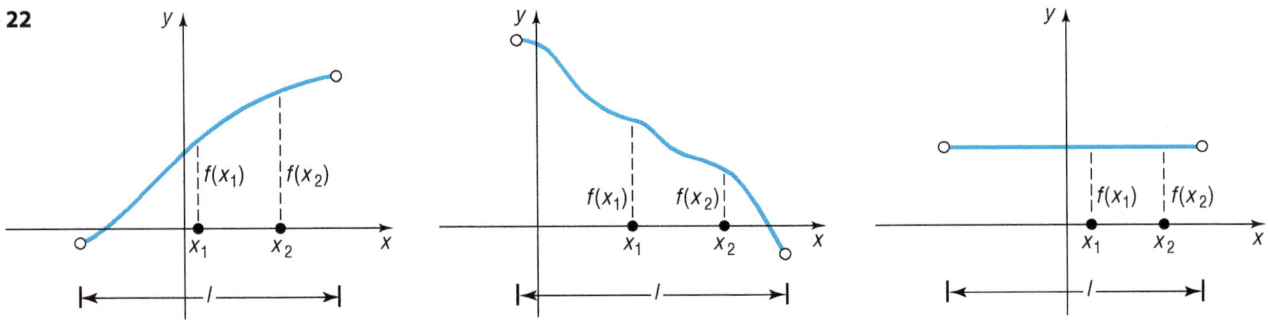

(a) For $x_1 < x_2$ in I, $f(x_1) < f(x_2)$; f is increasing on I

(b) For $x_1 < x_2$ in I, $f(x_1) > f(x_2)$; f is decreasing on I

(c) For all x in I, the values of f are equal; f is constant on I

Now Work PROBLEMS 11, 13, 15, AND 21(c)

4 Use a Graph to Locate Local Maxima and Local Minima

When the graph of a function is increasing to the left of $x = c$ and decreasing to the right of $x = c$, then at c the value of f is largest. This value is called a *local maximum* of f.* See Figure 23(a).

Figure 23

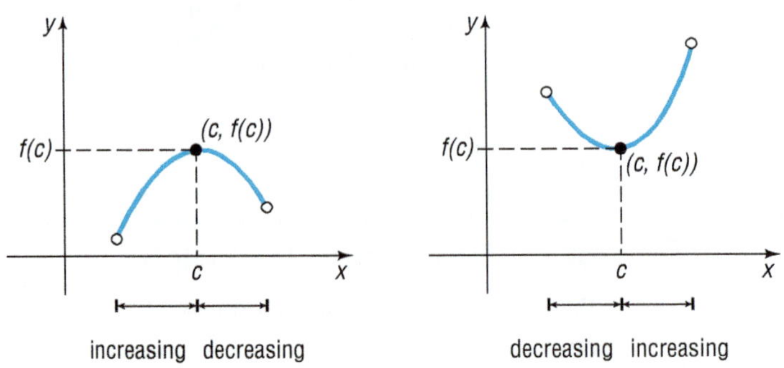

The local maximum is $f(c)$ and occurs at $x = c$.

The local minimum is $f(c)$ and occurs at $x = c$.

(a) **(b)**

*Some texts will use the term *relative* instead of *local*.

When the graph of a function is decreasing to the left of $x = c$ and is increasing to the right of $x = c$, then at c the value of f is the smallest. This value is called a *local minimum* of f. See Figure 23(b).

DEFINITIONS

A function f has a **local maximum** at c if there is an open interval I containing c so that, for all x not equal to c in I, $f(x) < f(c)$. We call $f(c)$ a **local maximum of f.**

A function f has a **local minimum** at c if there is an open interval I containing c so that, for all x not equal to c in I, $f(x) > f(c)$. We call $f(c)$ a **local minimum of f.**

If f has a local maximum at c, then the value of f at c is greater than the values of f near c. If f has a local minimum at c, then the value of f at c is less than the values of f near c. The word *local* is used to suggest that it is only near c that the value $f(c)$ is largest or smallest.

EXAMPLE 4

Finding Local Maxima and Local Minima from the Graph of a Function and Determining Where the Function Is Increasing, Decreasing, or Constant

Figure 24

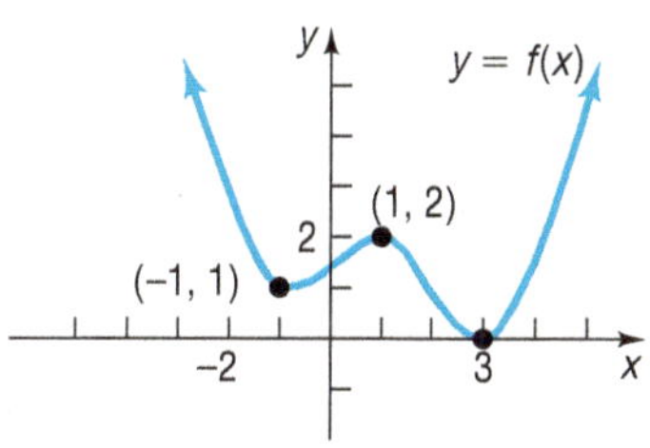

Figure 24 shows the graph of a function f.

(a) At what value(s) of x, if any, does f have a local maximum?
(b) What are the local maxima?
(c) At what value(s) of x, if any, does f have a local minimum?
(d) What are the local minima?
(e) List the intervals on which f is increasing. List the intervals on which f is decreasing.

Solution The domain of f is the set of real numbers.

(a) f has a local maximum at 1, since for all x close to 1, $x \neq 1$, we have $f(x) < f(1)$.
(b) The local maximum is $f(1) = 2$.
(c) f has a local minimum at -1 and at 3.
(d) The local minima are $f(-1) = 1$ and $f(3) = 0$.
(e) The function whose graph is given in Figure 24 is increasing for all values of x between -1 and 1 and for all values of x greater than 3. That is, the function is increasing on the intervals $(-1, 1)$ and $(3, \infty)$ or for $-1 < x < 1$ and $x > 3$. The function is decreasing for all values of x less than -1 and for all values of x between 1 and 3. That is, the function is decreasing on the intervals $(-\infty, -1)$ and $(1, 3)$ or for $x < -1$ and $1 < x < 3$.

WARNING The y-value is the local maximum or local minimum and it occurs at some x-value. For Figure 24, we say the local maximum is 2 and that the local maximum occurs at $x = 1$. ■

Now Work PROBLEMS 17 AND 19

5 Use a Graphing Utility to Approximate Local Maxima and Local Minima and to Determine Where a Function Is Increasing or Decreasing

To locate the exact value at which a function f has a local maximum or a local minimum usually requires calculus. However, a graphing utility may be used to approximate these values by using the MAXIMUM and MINIMUM features.

EXAMPLE 5

Using a Graphing Utility to Approximate Local Maxima and Minima and to Determine Where a Function Is Increasing or Decreasing

(a) Use a graphing utility to graph $f(x) = 6x^3 - 12x + 5$ for $-2 < x < 2$. Approximate where f has a local maximum and where f has a local minimum.

(b) Determine where f is increasing and where it is decreasing.

Solution

(a) Graphing utilities have a feature that finds the maximum or minimum point of a graph within a given interval. Graph the function f for $-2 < x < 2$. The MAXIMUM and MINIMUM commands require us to first determine the open interval I. The graphing utility will then approximate the maximum or minimum value in the interval. Using MAXIMUM we find that the local maximum is 11.53 and it occurs at $x = -0.82$, rounded to two decimal places. See Figure 25(a). Using MINIMUM, we find that the local minimum is -1.53 and it occurs at $x = 0.82$, rounded to two decimal places. See Figure 25(b).

Figure 25

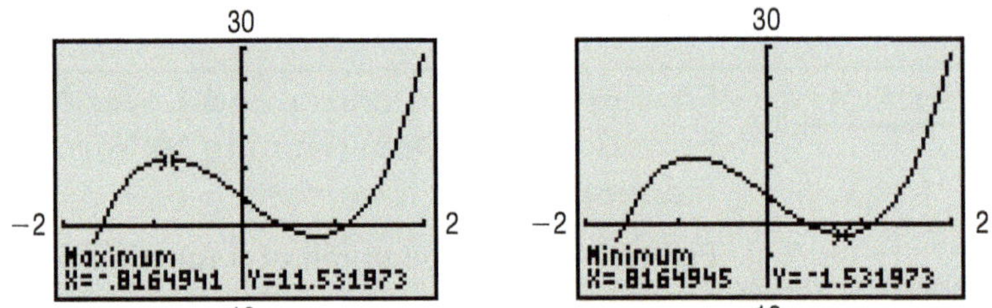

(a) (b)

(b) Looking at Figures 25(a) and (b), we see that the graph of f is increasing from $x = -2$ to $x = -0.82$ and from $x = 0.82$ to $x = 2$, so f is increasing on the intervals $(-2, -0.82)$ and $(0.82, 2)$ or for $-2 < x < -0.82$ and $0.82 < x < 2$. The graph is decreasing from $x = -0.82$ to $x = 0.82$, so f is decreasing on the interval $(-0.82, 0.82)$ or for $-0.82 < x < 0.82$.

Now Work PROBLEM 45

6 Find the Average Rate of Change of a Function

In Section 2.2, we said that the slope of a line could be interpreted as the average rate of change of the line. To find the average rate of change of a function between any two points on its graph, we calculate the slope of the line containing the two points.

DEFINITION

If a and b, $a \neq b$, are in the domain of a function $y = f(x)$, the **average rate of change of f** from a to b is defined as

$$\text{Average rate of change} = \frac{\Delta y}{\Delta x} = \frac{f(b) - f(a)}{b - a} \qquad a \neq b \qquad \textbf{(1)}$$

The symbol Δy in (1) is the "change in y," and Δx is the "change in x." The average rate of change of f is the change in y divided by the change in x.

EXAMPLE 6

Finding the Average Rate of Change

Find the average rate of change of $f(x) = 3x^2$:

(a) From 1 to 3 (b) From 1 to 5 (c) From 1 to 7

Solution (a) The average rate of change of $f(x) = 3x^2$ from 1 to 3 is

$$\frac{\Delta y}{\Delta x} = \frac{f(3) - f(1)}{3 - 1} = \frac{27 - 3}{3 - 1} = \frac{24}{2} = 12$$

(b) The average rate of change of $f(x) = 3x^2$ from 1 to 5 is

$$\frac{\Delta y}{\Delta x} = \frac{f(5) - f(1)}{5 - 1} = \frac{75 - 3}{5 - 1} = \frac{72}{4} = 18$$

(c) The average rate of change of $f(x) = 3x^2$ from 1 to 7 is

$$\frac{\Delta y}{\Delta x} = \frac{f(7) - f(1)}{7 - 1} = \frac{147 - 3}{7 - 1} = \frac{144}{6} = 24$$

Figure 26

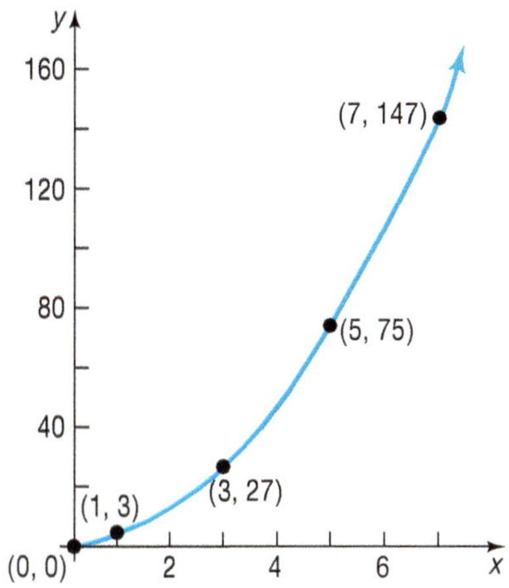

See Figure 26 for a graph of $f(x) = 3x^2$. The function f is increasing for $x > 0$. The fact that the average rates of change are getting larger indicates that the graph is getting steeper; that is, it is increasing at an increasing rate.

Now Work PROBLEM 53

The Secant Line

The average rate of change of a function has an important geometric interpretation. Look at the graph of $y = f(x)$ in Figure 27. We have labeled two points on the graph: $(a, f(a))$ and $(b, f(b))$. The line containing these two points is called the **secant line;** its slope is

$$m_{\text{sec}} = \frac{f(b) - f(a)}{b - a}$$

Figure 27

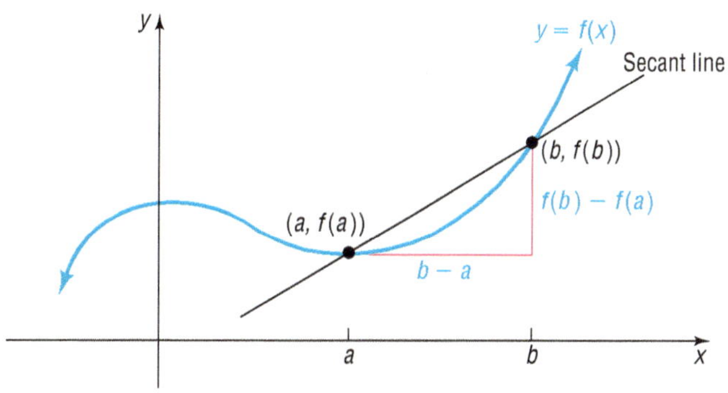

THEOREM

Slope of the Secant Line

The average rate of change of a function from a to b equals the slope of the secant line containing two points $(a, f(a))$ and $(b, f(b))$ on its graph.

EXAMPLE 7 **Finding the Equation of a Secant Line**

Suppose that $g(x) = 3x^2 - 2x + 3$.

(a) Find the average rate of change of g from -2 to 1.
(b) Find an equation of the secant line containing $(-2, g(-2))$ and $(1, g(1))$.
(c) Using a graphing utility, draw the graph of g and the secant line obtained in part (b) on the same screen.

Solution (a) The average rate of change of $g(x) = 3x^2 - 2x + 3$ from -2 to 1 is

$$\text{Average rate of change} = \frac{g(1) - g(-2)}{1 - (-2)}$$
$$= \frac{4 - 19}{3} \qquad g(1) = 3(1^2) - 2(1) + 3 = 4 \quad g(-2) = 3(-2)^2 - 2(-2) + 3 = 19$$
$$= -\frac{15}{3} = -5$$

Figure 28

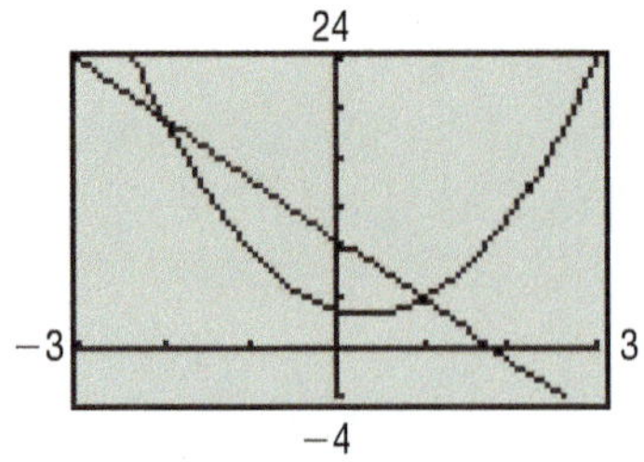

(b) The slope of the secant line containing $(-2, g(-2)) = (-2, 19)$ and $(1, g(1)) = (1, 4)$ is $m_{sec} = -5$. We use the point–slope form to find an equation of the secant line.

$$y - y_1 = m_{sec}(x - x_1) \qquad \text{Point–slope form of the secant line}$$
$$y - 19 = -5(x - (-2)) \qquad x_1 = -2, y_1 = g(-2) = 19, m_{sec} = -5$$
$$y - 19 = -5x - 10 \qquad \text{Simplify.}$$
$$y = -5x + 9 \qquad \text{Slope–intercept form of the secant line}$$

(c) Figure 28 shows the graph of g along with the secant line $y = -5x + 9$.

Now Work PROBLEM 59

3.3 Assess Your Understanding

'Are You Prepared?' *Answers are given at the end of these exercises. If you get a wrong answer, read the pages listed in red.*

1. The interval $(2, 5)$ can be written as the inequality _____. (pp. 150–151)
2. The slope of the line containing the points $(-2, 3)$ and $(3, 8)$ is _____. (pp. 178–179)
3. Test the equation $y = 5x^2 - 1$ for symmetry with respect to the x-axis, the y-axis, and the origin. (pp. 171–173)
4. Write the point–slope form of the line with slope 5 containing the point $(3, -2)$. (p. 182)
5. The intercepts of the equation $y = x^2 - 9$ are _____. (pp. 170–171)

Concepts and Vocabulary

6. A function f is _____ on an open interval I if, for any choice of x_1 and x_2 in I, with $x_1 < x_2$, we have $f(x_1) < f(x_2)$.
7. An _____ function f is one for which $f(-x) = f(x)$ for every x in the domain of f; an _____ function f is one for which $f(-x) = -f(x)$ for every x in the domain of f.
8. ***True or False*** A function f is decreasing on an open interval I if, for any choice of x_1 and x_2 in I, with $x_1 < x_2$, we have $f(x_1) > f(x_2)$.
9. ***True or False*** A function f has a local maximum at c if there is an open interval I containing c so that, for all x not equal to c in I, $f(x) < f(c)$.
10. ***True or False*** Even functions have graphs that are symmetric with respect to the origin.

Skill Building

In Problems 11–20, use the given graph of the function f.

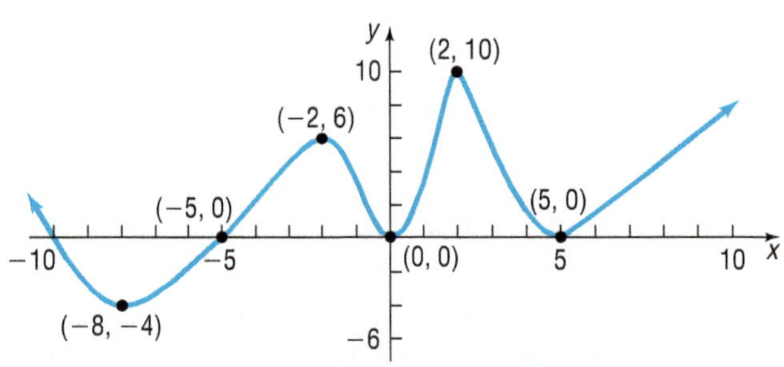

11. Is f increasing on the interval $(-8, -2)$?

12. Is f decreasing on the interval $(-8, -4)$?

13. Is f increasing on the interval $(2, 10)$?

14. Is f decreasing on the interval $(2, 5)$?

15. List the interval(s) on which f is increasing.

16. List the interval(s) on which f is decreasing.

17. Is there a local maximum at 2? If yes, what is it?

18. Is there a local maximum at 5? If yes, what is it?

19. List the numbers at which f has a local maximum. What are these local maxima?

20. List the numbers at which f has a local minimum. What are these local minima?

In Problems 21–28, the graph of a function is given. Use the graph to find:
(a) The intercepts, if any
(b) The domain and range
(c) The intervals on which it is increasing, decreasing, or constant
(d) Whether it is even, odd, or neither

21.
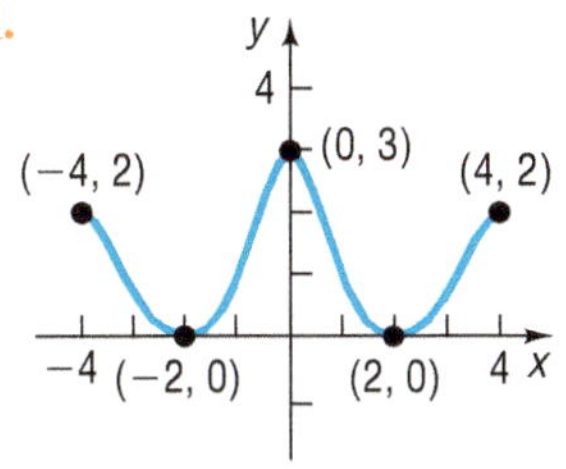

22.
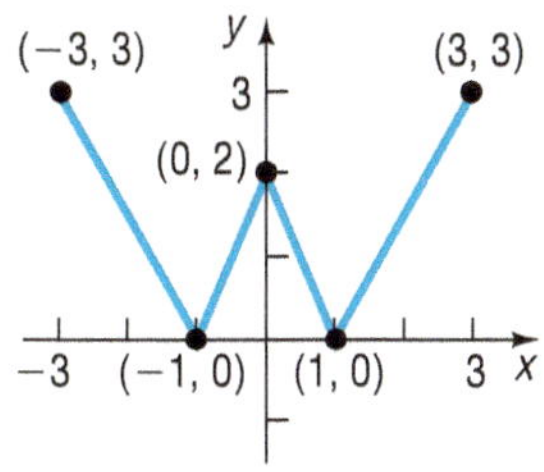

23.
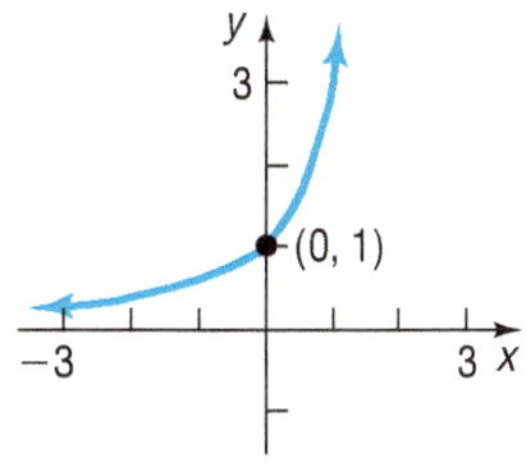

24.
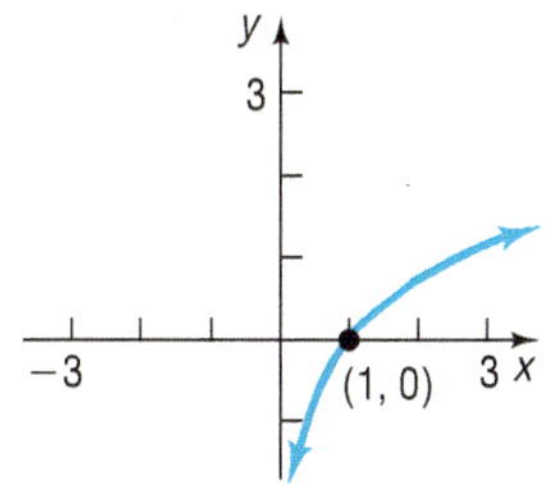

25.
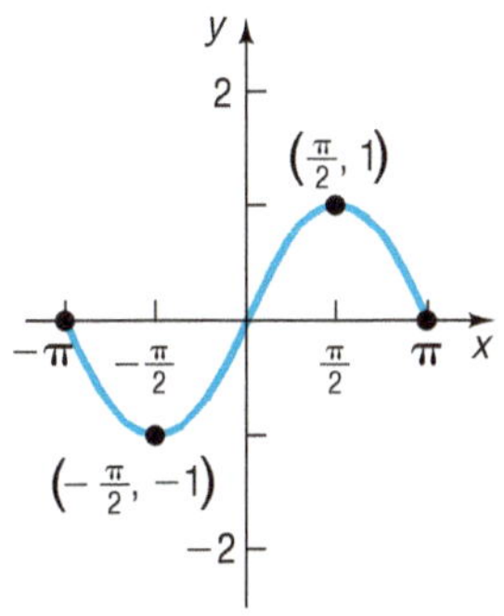

26.
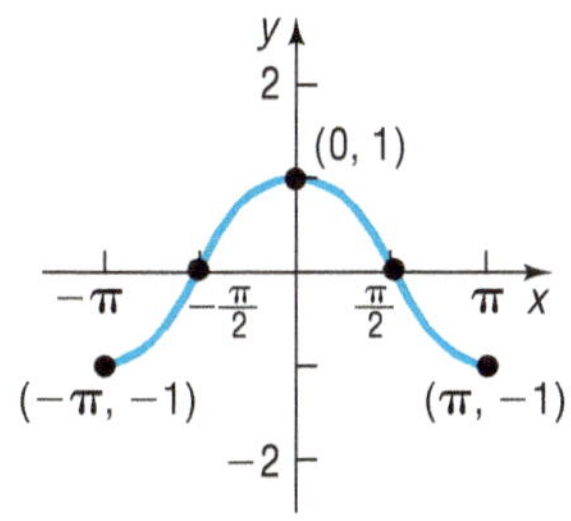

27.
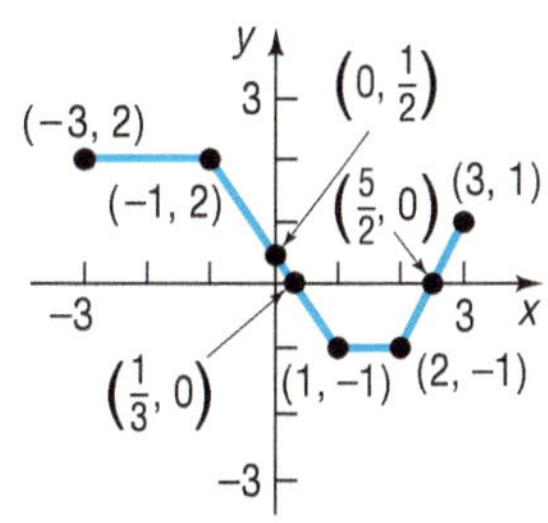

28.
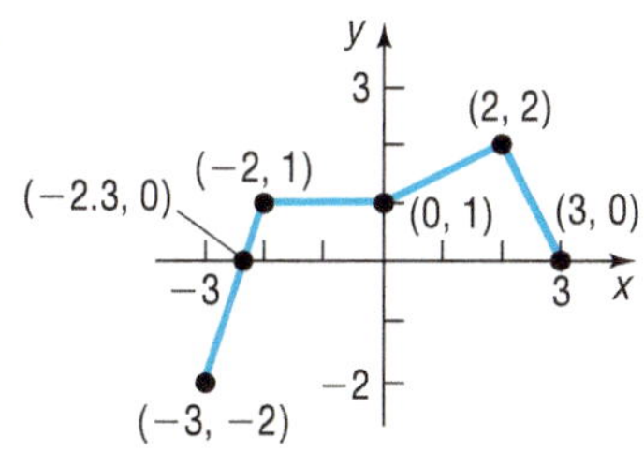

In Problems 29–32, the graph of a function f is given. Use the graph to find:
(a) The values, if any, at which f has a local maximum. What are these local maxima?
(b) The values, if any, at which f has a local minimum. What are these local minima?

29.
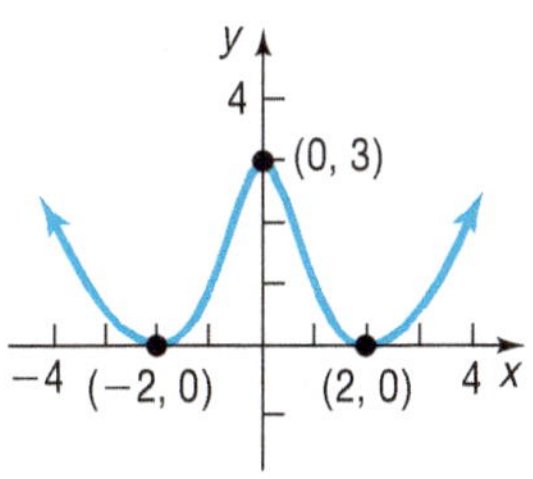

30.
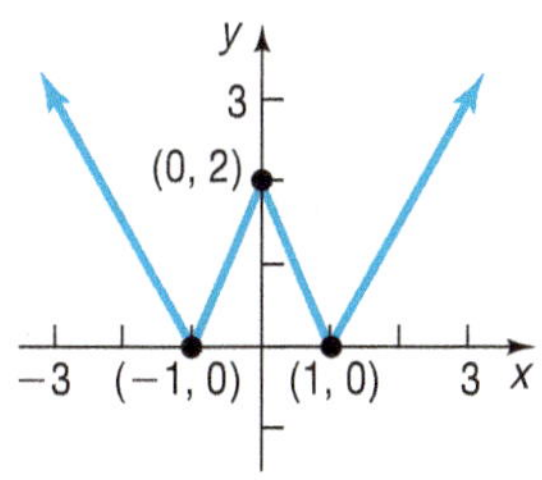

31.
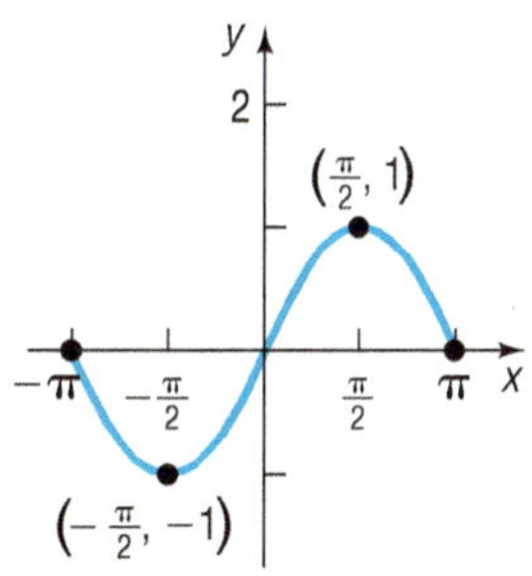

32.
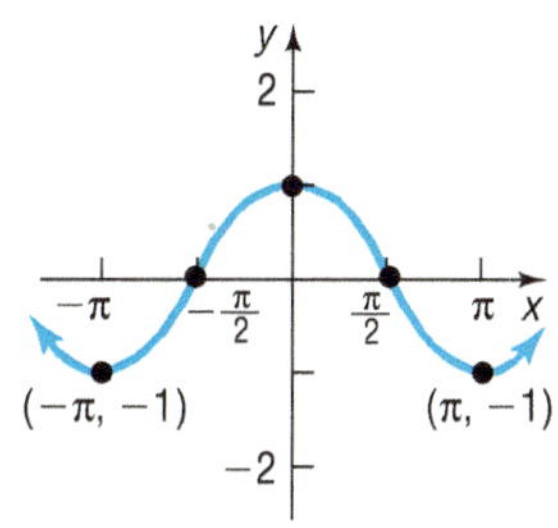

In Problems 33–44, determine algebraically whether each function is even, odd, or neither.

33. $F(x) = 4x^3 + x$

34. $f(x) = 2x^4 - x^2$

35. $g(x) = -3x^2 - 5$

36. $h(x) = 3x^5 + 5x$

37. $F(x) = \sqrt[3]{x}$

38. $G(x) = \sqrt{x}$

39. $f(x) = x + |x|$

40. $f(x) = \sqrt[3]{2x^2 + 1}$

41. $g(x) = \dfrac{1}{x^2}$

42. $h(x) = \dfrac{x}{x^2 - 1}$

43. $h(x) = \dfrac{-x^3}{3x^2 - 9}$

44. $F(x) = \dfrac{2x}{|x|}$

In Problems 45–52, use a graphing utility to graph each function over the indicated interval and approximate any local maxima and local minima. Determine where the function is increasing and where it is decreasing. Round answers to two decimal places.

45. $f(x) = x^3 - 3x + 2 \quad (-2, 2)$

46. $f(x) = x^3 - 3x^2 + 5 \quad (-1, 3)$

47. $f(x) = x^5 - x^3 \quad (-2, 2)$

48. $f(x) = x^4 - x^2 \quad (-2, 2)$

49. $f(x) = -0.2x^3 - 0.6x^2 + 4x - 6 \quad (-6, 4)$

50. $f(x) = -0.4x^3 + 0.6x^2 + 3x - 2 \quad (-4, 5)$

51. $f(x) = 0.25x^4 + 0.3x^3 - 0.9x^2 + 3 \quad (-3, 2)$

52. $f(x) = -0.4x^4 - 0.5x^3 + 0.8x^2 - 2 \quad (-3, 2)$

53. Find the average rate of change of $f(x) = -2x^2 + 4$
(a) From 0 to 2
(b) From 1 to 3
(c) From 1 to 4

54. Find the average rate of change of $f(x) = -x^3 + 1$
(a) From 0 to 2
(b) From 1 to 3
(c) From −1 to 1

55. Find the average rate of change of $g(x) = x^3 - 2x + 1$
(a) From −3 to −2
(b) From −1 to 1
(c) From 1 to 3

56. Find the average rate of change of $h(x) = x^2 - 2x + 3$
(a) From −1 to 1
(b) From 0 to 2
(c) From 2 to 5

57. $f(x) = 5x - 2$
(a) Find the average rate of change from 1 to 3.
(b) Find an equation of the secant line containing $(1, f(1))$ and $(3, f(3))$.

58. $f(x) = -4x + 1$
(a) Find the average rate of change from 2 to 5.
(b) Find an equation of the secant line containing $(2, f(2))$ and $(5, f(5))$.

59. $g(x) = x^2 - 2$
(a) Find the average rate of change from −2 to 1.
(b) Find an equation of the secant line containing $(-2, g(-2))$ and $(1, g(1))$.

60. $g(x) = x^2 + 1$
(a) Find the average rate of change from −1 to 2.
(b) Find an equation of the secant line containing $(-1, g(-1))$ and $(2, g(2))$.

61. $h(x) = x^2 - 2x$
(a) Find the average rate of change from 2 to 4.
(b) Find an equation of the secant line containing $(2, h(2))$ and $(4, h(4))$.

62. $h(x) = -2x^2 + x$
(a) Find the average rate of change from 0 to 3.
(b) Find an equation of the secant line containing $(0, h(0))$ and $(3, h(3))$.

Mixed Practice

63. $g(x) = x^3 - 4x$
(a) Determine whether g is even, odd, or neither.
(b) Using a graphing utility, graph g.
(c) There is a local minimum in quadrant IV. Approximate the local minimum.
(d) Using the results from part (a), determine the local maximum in quadrant II without using your graphing utility. Verify your result using your graphing utility.

64. $f(x) = -x^3 + 6x$
(a) Determine whether f is even, odd, or neither.
(b) Using a graphing utility, graph f.
(c) There is a local maximum in quadrant I. Approximate the local maximum.
(d) Using the results from part (a), determine the local minimum in quadrant III without using your graphing utility. Verify your result using your graphing utility.

65. $F(x) = -x^4 + 8x^2 + 8$
(a) Determine whether F is even, odd, or neither.
(b) Using a graphing utility, graph F.
(c) There is a local maximum in quadrant I. Approximate the local maximum.
(d) Using the results from part (a), determine the local maximum in quadrant II without using your graphing utility.
(e) Suppose the area under the graph of F between $x = 0$ and $x = 3$ that is bounded below by the x-axis is 47.4 square units. Using the results from part (a), determine the area under the graph of F between $x = -3$ and $x = 0$ bounded below by the x-axis.

66. $G(x) = -x^4 + 32x^2 + 144$
(a) Determine whether G is even, odd, or neither.
(b) Using a graphing utility, graph G.
(c) There is a local maximum in quadrant I. Approximate the local maximum.
(d) Using the results from part (a), determine the local maximum in quadrant II without using your graphing utility.
(e) Suppose the area under the graph of G between $x = 0$ and $x = 6$ that is bounded below by the x-axis is 1612.8 square units. Using the results from part (a), determine the area under the graph of F between $x = -6$ and $x = 0$ bounded below by the x-axis.

Applications and Extensions

67. **Minimum Average Cost** The average cost per hour in dollars, $\overline{C}$, of producing x riding lawn mowers can be modeled by the function

$$\overline{C}(x) = 0.3x^2 + 21x - 251 + \frac{2500}{x}$$

(a) Use a graphing utility to graph $\overline{C} = \overline{C}(x)$.
(b) Determine the number of riding lawn mowers to produce in order to minimize average cost.
(c) What is the minimum average cost?

68. **Medicine Concentration** The concentration C of a medication in the bloodstream t hours after being administered is modeled by the function

$$C(t) = -0.002t^4 + 0.039t^3 - 0.285t^2 + 0.766t + 0.085$$

(a) After how many hours will the concentration be highest?
(b) A woman nursing a child must wait until the concentration is below 0.5 before she can feed him. After taking the medication, how long must she wait before feeding her child?

69. E-coli Growth A strain of E-coli Beu 397-recA441 is placed into a nutrient broth at 30° Celsius and allowed to grow. The data shown in the table are collected. The population is measured in grams and the time in hours. Since population P depends on time t and each input corresponds to exactly one output, we can say that population is a function of time; so $P(t)$ represents the population at time t.

(a) Find the average rate of change of the population from 0 to 2.5 hours.
(b) Find the average rate of change of the population from 4.5 to 6 hours.
(c) What is happening to the average rate of change as time passes?

Time (hours), t	Population (grams), P
0	0.09
2.5	0.18
3.5	0.26
4.5	0.35
6	0.50

70. e-Filing Tax Returns The Internal Revenue Service Restructuring and Reform Act (RRA) was signed into law by President Bill Clinton in 1998. A major objective of the RRA was to promote electronic filing of tax returns. The data in the table show the percentage of individual income tax returns filed electronically for filing years 1998–2006. Since the percentage P of returns filed electronically depends on the filing year y and each input corresponds to exactly one output, the percentage of returns filed electronically is a function of the filing year; so $P(y)$ represents the percentage of returns filed electronically for filing year y.

(a) Find the average rate of change of the percentage of e-filed returns from 1998 to 2000.
(b) Find the average rate of change of the percentage of e-filed returns from 2001 to 2003.
(c) Find the average rate of change of the percentage of e-filed returns from 2004 to 2006.
(d) What is happening to the average rate of change as time passes?

Year	Percentage of returns e-filed
1998	20.7
1999	23.5
2000	27.6
2001	30.7
2002	35.6
2003	40.2
2004	46.5
2005	51.1
2006	57.1

Source: Internal Revenue Service

71. For the function $f(x) = x^2$, compute each average rate of change:

(a) From 0 to 1
(b) From 0 to 0.5
(c) From 0 to 0.1
(d) From 0 to 0.01
(e) From 0 to 0.001
(f) Use a graphing utility to graph each of the secant lines along with f.
(g) What do you think is happening to the secant lines?
(h) What is happening to the slopes of the secant lines? Is there some number that they are getting closer to? What is that number?

72. For the function $f(x) = x^2$, compute each average rate of change:

(a) From 1 to 2
(b) From 1 to 1.5
(c) From 1 to 1.1
(d) From 1 to 1.01
(e) From 1 to 1.001
(f) Use a graphing utility to graph each of the secant lines along with f.
(g) What do you think is happening to the secant lines?
(h) What is happening to the slopes of the secant lines? Is there some number that they are getting closer to? What is that number?

Problems 73–80 require the following discussion of a secant line. The slope of the secant line containing the two points $(x, f(x))$ and $(x + h, f(x + h))$ on the graph of a function $y = f(x)$ may be given as

$$m_{\text{sec}} = \frac{f(x+h) - f(x)}{(x+h) - x} = \frac{f(x+h) - f(x)}{h} \qquad h \neq 0$$

In calculus, this expression is called the **difference quotient of f.**

(a) Express the slope of the secant line of each function in terms of x and h. Be sure to simplify your answer.
(b) Find m_{sec} for $h = 0.5, 0.1,$ and 0.01 at $x = 1$. What value does m_{sec} approach as h approaches 0?
(c) Find the equation for the secant line at $x = 1$ with $h = 0.01$.
(d) Use a graphing utility to graph f and the secant line found in part (c) on the same viewing window.

73. $f(x) = 2x + 5$

74. $f(x) = -3x + 2$

75. $f(x) = x^2 + 2x$

76. $f(x) = 2x^2 + x$

77. $f(x) = 2x^2 - 3x + 1$

78. $f(x) = -x^2 + 3x - 2$

79. $f(x) = \frac{1}{x}$

80. $f(x) = \frac{1}{x^2}$

Discussion and Writing

81. Draw the graph of a function that has the following properties: domain: all real numbers; range: all real numbers; intercepts: $(0, -3)$ and $(3, 0)$; a local maximum of -2 is at -1; a local minimum of -6 is at 2. Compare your graph with those of others. Comment on any differences.

82. Redo Problem 81 with the following additional information: increasing on $(-\infty, -1)$, $(2, \infty)$; decreasing on $(-1, 2)$. Again compare your graph with others and comment on any differences.

83. How many x-intercepts can a function defined on an interval have if it is increasing on that interval? Explain.

84. Suppose that a friend of yours does not understand the idea of increasing and decreasing functions. Provide an explanation, complete with graphs, that clarifies the idea.

85. Can a function be both even and odd? Explain.

86. Using a graphing utility, graph $y = 5$ on the interval $(-3, 3)$. Use MAXIMUM to find the local maxima on $(-3, 3)$. Comment on the result provided by the calculator.

'Are You Prepared?' Answers

1. $2 < x < 5$ **2.** 1 **3.** symmetric with respect to the y-axis **4.** $y + 2 = 5(x - 3)$ **5.** $(-3, 0), (3, 0), (0, -9)$

3.4 Library of Functions; Piecewise-defined Functions

PREPARING FOR THIS SECTION *Before getting started, review the following:*

- Intercepts (Section 2.1, pp. 170–171)
- Graphs of Key Equations (Section 1.1: Example 8, p. 93; Section 2.1: Example 4, p. 173; Example 5, p. 174; Example 6, p. 175)

Now Work the 'Are You Prepared?' problems on page 252.

OBJECTIVES 1 Graph the Functions Listed in the Library of Functions (p. 246)
2 Graph Piecewise-defined Functions (p. 251)

1 Graph the Functions Listed in the Library of Functions

We now introduce a few more functions to add to our list of important functions. We begin with the *square root function.*

On page 174, we graphed the equation $y = \sqrt{x}$. Figure 29 shows a graph of the function $f(x) = \sqrt{x}$. Based on the graph, we have the following properties:

Figure 29

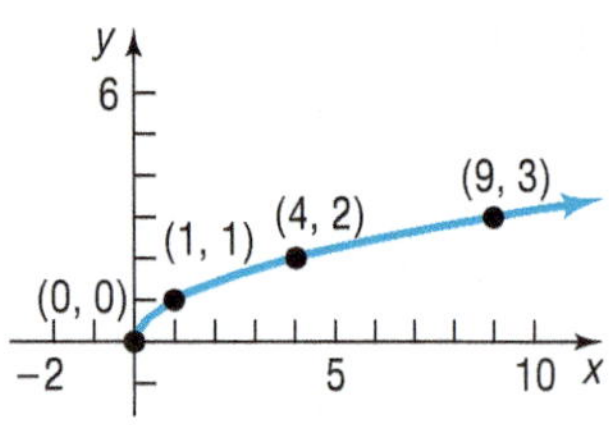

Properties of $f(x) = \sqrt{x}$

1. The domain and the range are the set of nonnegative real numbers.
2. The x-intercept of the graph of $f(x) = \sqrt{x}$ is 0. The y-intercept of the graph of $f(x) = \sqrt{x}$ is also 0.
3. The function is neither even nor odd.
4. It is increasing on the interval $(0, \infty)$.
5. It has a minimum value of 0 at $x = 0$.

EXAMPLE 1 **Graphing the Cube Root Function**

(a) Determine whether $f(x) = \sqrt[3]{x}$ is even, odd, or neither. State whether the graph of f is symmetric with respect to the y-axis or symmetric with respect to the origin.

(b) Determine the intercepts, if any, of the graph of $f(x) = \sqrt[3]{x}$.

(c) Graph $f(x) = \sqrt[3]{x}$.

Solution (a) Because

$$f(-x) = \sqrt[3]{-x} = -\sqrt[3]{x} = -f(x)$$

the function is odd. The graph of f is symmetric with respect to the origin.

(b) The y-intercept is $f(0) = \sqrt[3]{0} = 0$. The x-intercept is found by solving the equation $f(x) = 0$.

$$f(x) = 0$$
$$\sqrt[3]{x} = 0 \quad f(x) = \sqrt[3]{x}$$
$$x = 0 \quad \text{Cube both sides of the equation.}$$

The x-intercept is also 0.

(c) We use the function to form Table 4 and obtain some points on the graph. Because of the symmetry with respect to the origin, we need to find only points (x, y) for which $x \geq 0$. Figure 30 shows the graph of $f(x) = \sqrt[3]{x}$.

Table 4

x	$y = f(x) = \sqrt[3]{x}$	(x, y)
0	0	(0, 0)
$\frac{1}{8}$	$\frac{1}{2}$	$\left(\frac{1}{8}, \frac{1}{2}\right)$
1	1	(1, 1)
2	$\sqrt[3]{2} \approx 1.26$	$(2, \sqrt[3]{2})$
8	2	(8, 2)

Figure 30

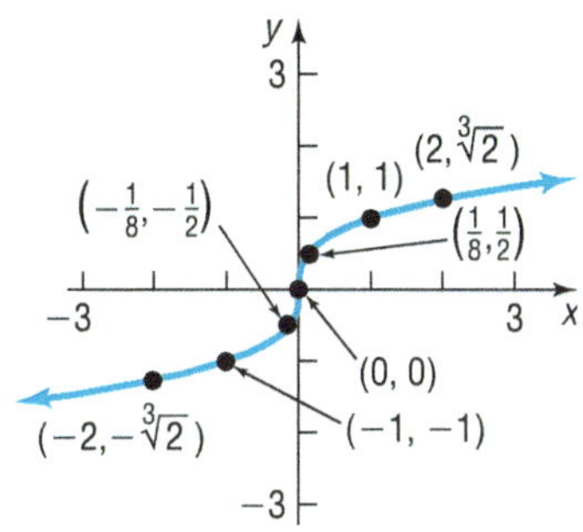

From the results of Example 1 and Figure 30, we have the following properties of the cube root function.

Properties of $f(x) = \sqrt[3]{x}$

1. The domain and the range are the set of all real numbers.
2. The x-intercept of the graph of $f(x) = \sqrt[3]{x}$ is 0. The y-intercept of the graph of $f(x) = \sqrt[3]{x}$ is also 0.
3. The graph is symmetric with respect to the origin. The function is odd.
4. It is increasing on the interval $(-\infty, \infty)$.
5. It does not have a local minimum or a local maximum.

EXAMPLE 2 Graphing the Absolute Value Function

(a) Determine whether $f(x) = |x|$ is even, odd, or neither. State whether the graph of f is symmetric with respect to the y-axis or symmetric with respect to the origin.

(b) Determine the intercepts, if any, of the graph of $f(x) = |x|$.

(c) Graph $f(x) = |x|$.

Solution (a) Because

$$f(-x) = |-x|$$
$$= |x| = f(x)$$

the function is even. The graph of f is symmetric with respect to the y-axis.

(b) The y-intercept is $f(0) = |0| = 0$. The x-intercept is found by solving the equation $f(x) = 0$ or $|x| = 0$. So the x-intercept is 0.

(c) We use the function to form Table 5 and obtain some points on the graph. Because of the symmetry with respect to the y-axis, we need to find only points (x, y) for which $x \geq 0$. Figure 31 shows the graph of $f(x) = |x|$.

Table 5

x	$y = f(x) = \lvert x \rvert$	(x, y)
0	0	(0, 0)
1	1	(1, 1)
2	2	(2, 2)
3	3	(3, 3)

Figure 31

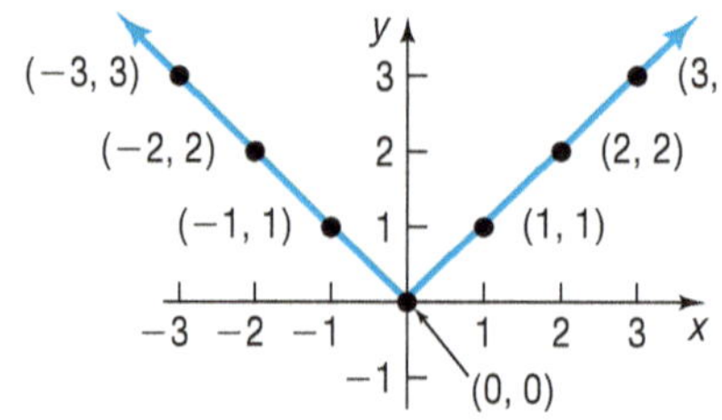

From the results of Example 2 and Figure 31, we have the following properties of the absolute value function.

Properties of $f(x) = |x|$

1. The domain is the set of all real numbers. The range of f is $\{y|y \geq 0\}$.
2. The x-intercept of the graph of $f(x) = |x|$ is 0. The y-intercept of the graph of $f(x) = |x|$ is also 0.
3. The graph is symmetric with respect to the y-axis. The function is even.
4. It is decreasing on the interval $(-\infty, 0)$. It is increasing on the interval $(0, \infty)$.
5. It has a local minimum of 0 at $x = 0$.

Seeing the Concept

Graph $y = |x|$ on a square screen and compare what you see with Figure 31. Note that some graphing calculators use abs(x) for absolute value.

We now provide a summary of the key functions that we have encountered. In going through this list, pay special attention to the properties of each function, particularly to the shape of each graph. Knowing these graphs along with key points on the graph will lay the foundation for later graphing techniques.

Constant Function

$$f(x) = b \qquad b \text{ is a real number}$$

Figure 32 Constant Function

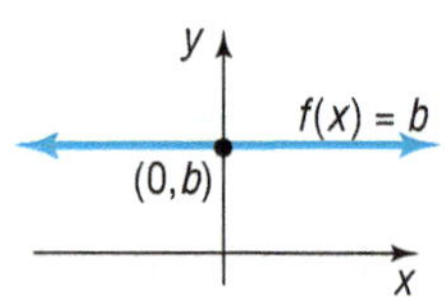

See Figure 32.

The domain of a **constant function** is the set of all real numbers; its range is the set consisting of a single number b. Its graph is a horizontal line whose y-intercept is b. The constant function is an even function whose graph is a horizontal line over its domain.

Identity Function

$$f(x) = x$$

Figure 33 Identity Function

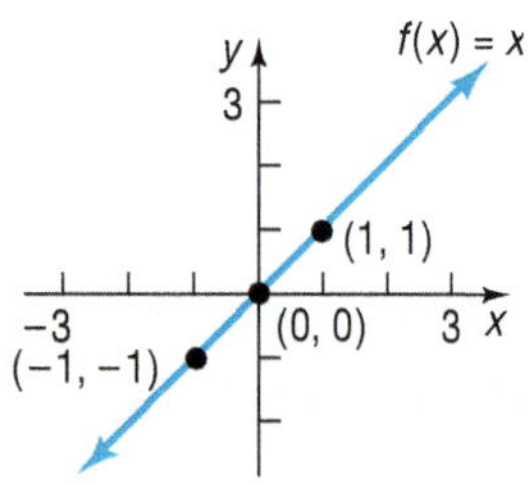

See Figure 33.

The domain and the range of the **identity function** are the set of all real numbers. Its graph is a line whose slope is $m = 1$ and whose y-intercept is 0. The line consists of all points for which the x-coordinate equals the y-coordinate. The identity function is an odd function that is increasing over its domain. Note that the graph bisects quadrants I and III.

Figure 34
Square Function

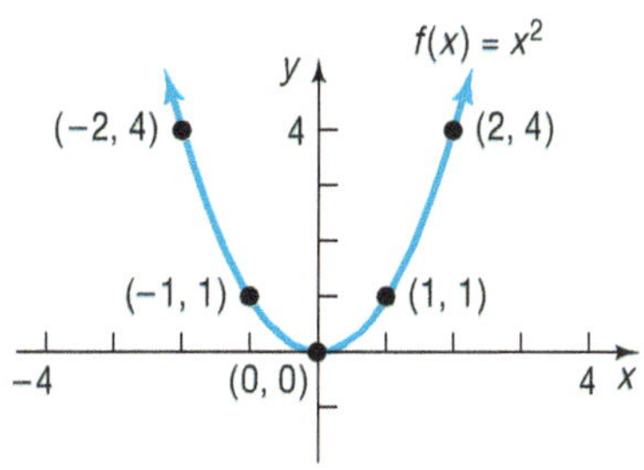

Square Function

$$f(x) = x^2$$

See Figure 34.

The domain of the **square function** f is the set of all real numbers; its range is the set of nonnegative real numbers. The graph of this function is a parabola whose intercept is at $(0, 0)$. The square function is an even function that is decreasing on the interval $(-\infty, 0)$ and increasing on the interval $(0, \infty)$.

Figure 35
Cube Function

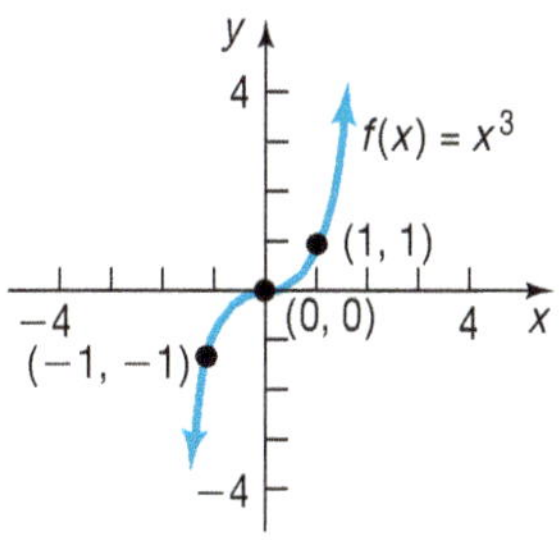

Cube Function

$$f(x) = x^3$$

See Figure 35.

The domain and the range of the **cube function** are the set of all real numbers. The intercept of the graph is at $(0, 0)$. The cube function is odd and is increasing on the interval $(-\infty, \infty)$.

Figure 36
Square Root Function

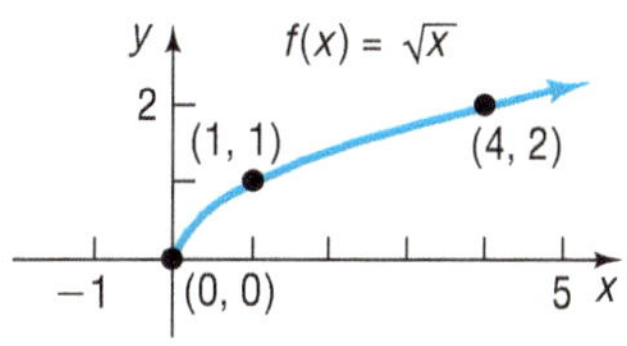

Square Root Function

$$f(x) = \sqrt{x}$$

See Figure 36.

The domain and the range of the **square root function** are the set of nonnegative real numbers. The intercept of the graph is at $(0, 0)$. The square root function is neither even nor odd and is increasing on the interval $(0, \infty)$.

Figure 37
Cube Root Function

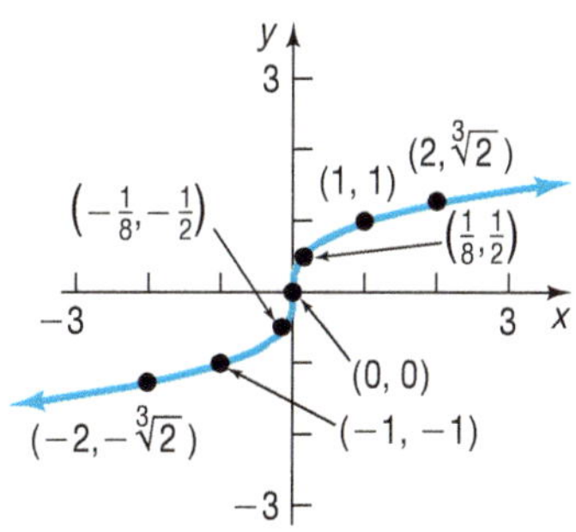

Cube Root Function

$$f(x) = \sqrt[3]{x}$$

See Figure 37.

The domain and the range of the **cube root function** are the set of all real numbers. The intercept of the graph is at $(0, 0)$. The cube root function is an odd function that is increasing on the interval $(-\infty, \infty)$.

Figure 38
Reciprocal Function

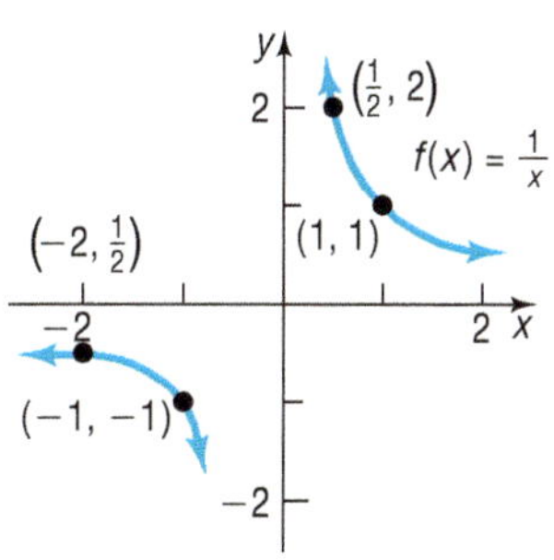

Reciprocal Function

$$f(x) = \frac{1}{x}$$

Refer to Example 6, page 175, for a discussion of the equation $y = \dfrac{1}{x}$. See Figure 38.

The domain and the range of the **reciprocal function** are the set of all nonzero real numbers. The graph has no intercepts. The reciprocal function is decreasing on the intervals $(-\infty, 0)$ and $(0, \infty)$ and is an odd function.

Absolute Value Function

$$f(x) = |x|$$

Figure 39
Absolute Value Function

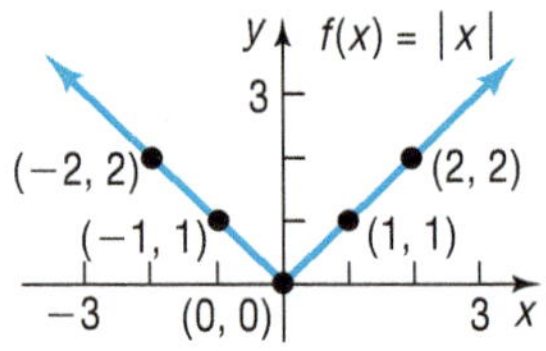

See Figure 39.

The domain of the **absolute value function** is the set of all real numbers; its range is the set of nonnegative real numbers. The intercept of the graph is at $(0, 0)$. If $x \geq 0$, then $f(x) = x$, and the graph of f is part of the line $y = x$; if $x < 0$, then $f(x) = -x$, and the graph of f is part of the line $y = -x$. The absolute value function is an even function; it is decreasing on the interval $(-\infty, 0)$ and increasing on the interval $(0, \infty)$.

The notation $\text{int}(x)$ stands for the largest integer less than or equal to x. For example,

$$\text{int}(1) = 1, \quad \text{int}(2.5) = 2, \quad \text{int}\left(\frac{1}{2}\right) = 0, \quad \text{int}\left(-\frac{3}{4}\right) = -1, \quad \text{int}(\pi) = 3$$

This type of correspondence occurs frequently enough in mathematics that we give it a name.

DEFINITION

Greatest Integer Function

$$f(x) = \text{int}(x)^* = \text{greatest integer less than or equal to } x$$

Table 6

x	$y = f(x) = \text{int}(x)$	(x, y)
-1	-1	$(-1, -1)$
$-\frac{1}{2}$	-1	$\left(-\frac{1}{2}, -1\right)$
$-\frac{1}{4}$	-1	$\left(-\frac{1}{4}, -1\right)$
0	0	$(0, 0)$
$\frac{1}{4}$	0	$\left(\frac{1}{4}, 0\right)$
$\frac{1}{2}$	0	$\left(\frac{1}{2}, 0\right)$
$\frac{3}{4}$	0	$\left(\frac{3}{4}, 0\right)$

We obtain the graph of $f(x) = \text{int}(x)$ by plotting several points. See Table 6. For values of x, $-1 \leq x < 0$, the value of $f(x) = \text{int}(x)$ is -1; for values of x, $0 \leq x < 1$, the value of f is 0. See Figure 40 for the graph.

Figure 40
Greatest Integer Function

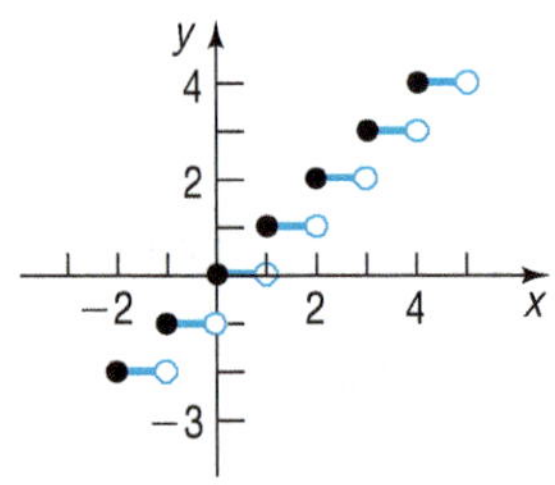

The domain of the **greatest integer function** is the set of all real numbers; its range is the set of integers. The y-intercept of the graph is 0. The x-intercepts lie in the interval $[0, 1)$. The greatest integer function is neither even nor odd. It is constant on every interval of the form $[k, k + 1)$, for k an integer. In Figure 40, we use a solid dot to indicate, for example, that at $x = 1$ the value of f is $f(1) = 1$; we use an open circle to illustrate that the function does not assume the value of 0 at $x = 1$.

Although a precise definition requires the idea of a limit, discussed in calculus, in a rough sense, a function is said to be **continuous** if its graph has no gaps or holes and can be drawn without lifting a pencil from the paper on which the graph is drawn. We contrast this with a *discontinuous* function. A function is **discontinuous** if its graph has gaps or holes so that its graph cannot be drawn without lifting a pencil from the paper.

From the graph of the greatest integer function, we can see why it is also called a **step function.** At $x = 0$, $x = \pm 1$, $x = \pm 2$, and so on, this function is discontinuous because, at integer values, the graph suddenly "steps" from one value to another without taking on any of the intermediate values. For example, to the immediate left of $x = 3$, the y-coordinates of the points on the graph are 2, and to the immediate right of $x = 3$, the y-coordinates of the points on the graph are 3. So, the graph has gaps in it.

COMMENT When graphing a function using a graphing utility, you can choose either the **connected mode,** in which points plotted on the screen are connected, making the graph appear without any breaks, or the **dot mode,** in which only the points plotted appear. When graphing the greatest integer function with a graphing utility, it is necessary to be in the dot mode. This is to prevent the utility from "connecting the dots" when $f(x)$ changes from one integer value to the next. See Figure 41.

*Some books use the notation $f(x) = [\![x]\!]$ instead of $\text{int}(x)$.

Figure 41
$f(x) = \text{int}(x)$

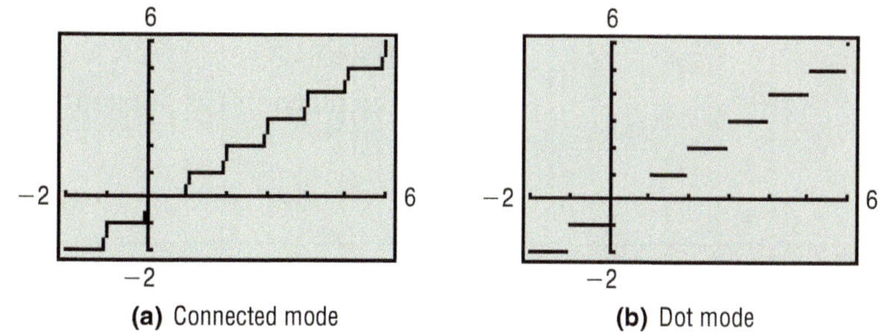

The functions that we have discussed so far are basic. Whenever you encounter one of them, you should see a mental picture of its graph. For example, if you encounter the function $f(x) = x^2$, you should see in your mind's eye a picture like Figure 34.

Now Work PROBLEMS 9 THROUGH 16

2 Graph Piecewise-defined Functions

Sometimes a function is defined differently on different parts of its domain. For example, the absolute value function $f(x) = |x|$ is actually defined by two equations: $f(x) = x$ if $x \geq 0$ and $f(x) = -x$ if $x < 0$. For convenience, we generally combine these equations into one expression as

$$f(x) = |x| = \begin{cases} x & \text{if } x \geq 0 \\ -x & \text{if } x < 0 \end{cases}$$

When functions are defined by more than one equation, they are called **piecewise-defined** functions.

Let's look at another example of a piecewise-defined function.

EXAMPLE 3 Analyzing a Piecewise-defined Function

The function f is defined as

$$f(x) = \begin{cases} -x + 1 & \text{if } -3 \leq x < 1 \\ 2 & \text{if } x = 1 \\ x^2 & \text{if } x > 1 \end{cases}$$

(a) Find $f(0)$, $f(1)$, and $f(2)$.
(b) Determine the domain of f.
(c) Graph f.
(d) Use the graph to find the range of f.
(e) Is f continuous on its domain?

Solution (a) To find $f(0)$, we observe that when $x = 0$ the equation for f is given by $f(x) = -x + 1$. So we have

$$f(0) = -0 + 1 = 1$$

When $x = 1$, the equation for f is $f(x) = 2$. Thus,

$$f(1) = 2$$

When $x = 2$, the equation for f is $f(x) = x^2$. So

$$f(2) = 2^2 = 4$$

(b) To find the domain of f, we look at its definition. Since f is defined for all x greater than or equal to -3, the domain of f is $\{x | x \geq -3\}$, or the interval $[-3, \infty)$.

(c) To graph f, we graph "each piece." First we graph the line $y = -x + 1$ and keep only the part for which $-3 \leq x < 1$. Then we plot the point $(1, 2)$ because, when $x = 1$, $f(x) = 2$. Finally, we graph the parabola $y = x^2$ and keep only the part for which $x > 1$. See Figure 42.

Figure 42

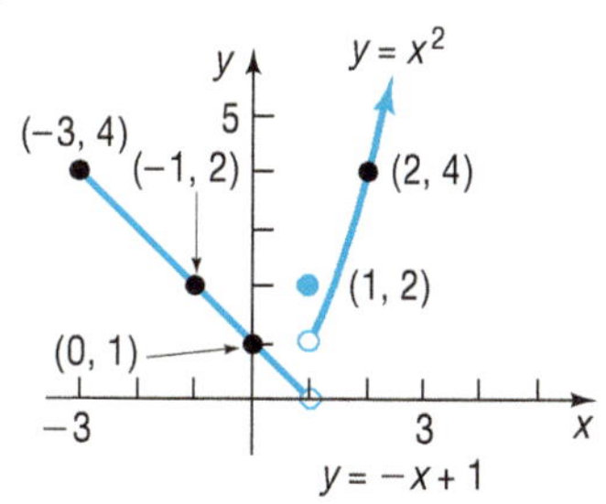

(d) From the graph, we conclude that the range of f is $\{y | y > 0\}$, or the interval $(0, \infty)$.

(e) The function f is not continuous because there is a "jump" in the graph at $x = 1$.

Now Work PROBLEM 29

EXAMPLE 4 Cost of Electricity

In May 2006, Commonwealth Edison Company supplied electricity to residences for a monthly customer charge of \$7.58 plus 8.275¢ per kilowatt-hour (kWhr) for the first 400 kWhr supplied in the month and 6.208¢ per kWhr for all usage over 400 kWhr in the month.

(a) What is the charge for using 300 kWhr in a month?

(b) What is the charge for using 700 kWhr in a month?

(c) If C is the monthly charge for x kWhr, develop a model relating the monthly charge and kilowatt-hours used. That is, express C as a function of x.

Source: Commonwealth Edison Co., Chicago, Illinois, 2006

Solution (a) For 300 kWhr, the charge is \$7.58 plus 8.275¢ = \$0.08275 per kWhr. That is,

$$\text{Charge} = \$7.58 + \$0.08275(300) = \$32.41$$

(b) For 700 kWhr, the charge is \$7.58 plus 8.275¢ per kWhr for the first 400 kWhr plus 6.208¢ per kWhr for the 300 kWhr in excess of 400. That is,

$$\text{Charge} = \$7.58 + \$0.08275(400) + \$0.06208(300) = \$59.30$$

(c) Let x represent the number of kilowatt-hours used. If $0 \le x \le 400$, the monthly charge C (in dollars) can be found by multiplying x times \$0.08275 and adding the monthly customer charge of \$7.58. So, if $0 \le x \le 400$, then $C(x) = 0.08275x + 7.58$.

For $x > 400$, the charge is $0.08275(400) + 7.58 + 0.06208(x - 400)$, since $x - 400$ equals the usage in excess of 400 kWhr, which costs \$0.06208 per kWhr. That is, if $x > 400$, then

$$\begin{aligned} C(x) &= 0.08275(400) + 7.58 + 0.06208(x - 400) \\ &= 40.68 + 0.06208(x - 400) \\ &= 0.06208x + 15.848 \end{aligned}$$

The rule for computing C follows two equations:

$$C(x) = \begin{cases} 0.08275x + 7.58 & \text{if } 0 \le x \le 400 \\ 0.06208x + 15.848 & \text{if } x > 400 \end{cases}$$

The Model

See Figure 43 for the graph.

Figure 43

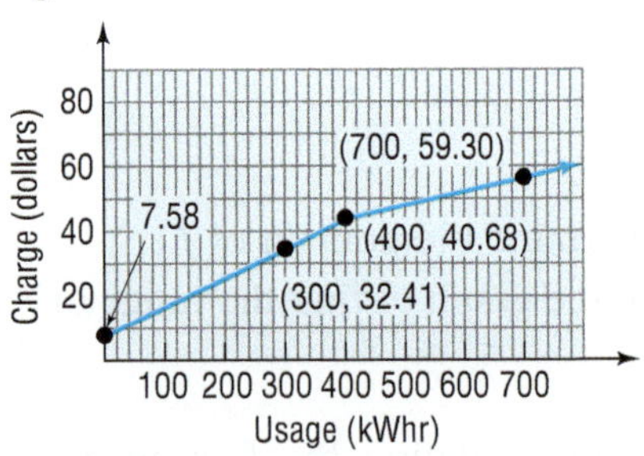

3.4 Assess Your Understanding

'Are You Prepared?' *Answers are given at the end of these exercises. If you get a wrong answer, read the pages listed in red.*

1. Sketch the graph of $y = \sqrt{x}$. (p. 174)
2. Sketch the graph of $y = \frac{1}{x}$. (p. 175)
3. List the intercepts of the equation $y = x^3 - 8$. (pp. 170–171)

Concepts and Vocabulary

4. The function $f(x) = x^2$ is decreasing on the interval ______.
5. When functions are defined by more than one equation, they are called ______ functions.
6. ***True or False*** The cube function is odd and is increasing on the interval $(-\infty, \infty)$.
7. ***True or False*** The cube root function is odd and is decreasing on the interval $(-\infty, \infty)$.
8. ***True or False*** The domain and the range of the reciprocal function are the set of all real numbers.

Skill Building

In Problems 9–16, match each graph to its function.

A. *Constant function* B. *Identity function* C. *Square function*
D. *Cube function* E. *Square root function* F. *Reciprocal function*
G. *Absolute value function* H. *Cube root function*

9. 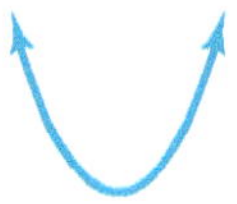

10.

11. 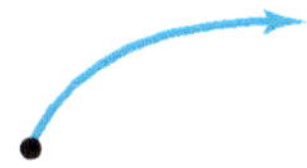

12. 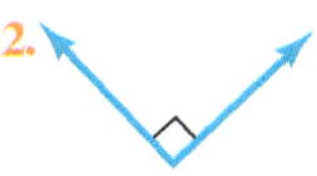

13. 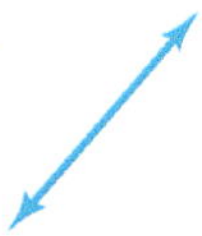

14. 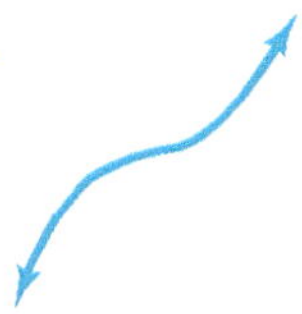

15. 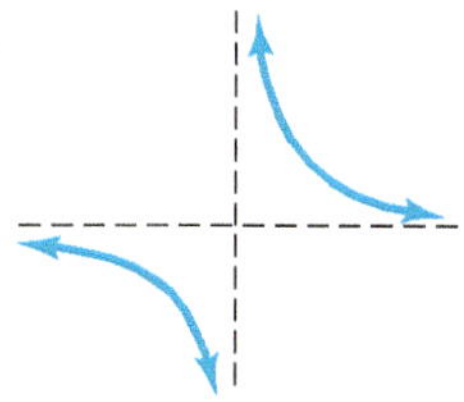

16. 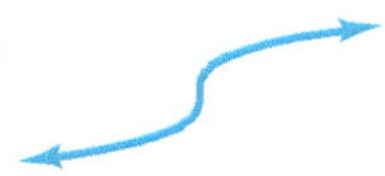

In Problems 17–24, sketch the graph of each function. Be sure to label three points on the graph.

17. $f(x) = x$ **18.** $f(x) = x^2$ **19.** $f(x) = x^3$ **20.** $f(x) = \sqrt{x}$

21. $f(x) = \dfrac{1}{x}$ **22.** $f(x) = |x|$ **23.** $f(x) = \sqrt[3]{x}$ **24.** $f(x) = 3$

25. If $f(x) = \begin{cases} x^2 & \text{if } x < 0 \\ 2 & \text{if } x = 0 \\ 2x + 1 & \text{if } x > 0 \end{cases}$

find: (a) $f(-2)$ (b) $f(0)$ (c) $f(2)$

26. If $f(x) = \begin{cases} -3x & \text{if } x < -1 \\ 0 & \text{if } x = -1 \\ 2x^2 + 1 & \text{if } x > -1 \end{cases}$

find: (a) $f(-2)$ (b) $f(-1)$ (c) $f(0)$

27. If $f(x) = \begin{cases} 2x - 4 & \text{if } -1 \le x \le 2 \\ x^3 - 2 & \text{if } 2 < x \le 3 \end{cases}$

find: (a) $f(0)$ (b) $f(1)$ (c) $f(2)$ (d) $f(3)$

28. If $f(x) = \begin{cases} x^3 & \text{if } -2 \le x < 1 \\ 3x + 2 & \text{if } 1 \le x \le 4 \end{cases}$

find: (a) $f(-1)$ (b) $f(0)$ (c) $f(1)$ (d) $f(3)$

In Problems 29–40:

(a) Find the domain of each function. *(b) Locate any intercepts.* *(c) Graph each function.*
(d) Based on the graph, find the range. *(e) Is f continuous on its domain?*

29. $f(x) = \begin{cases} 2x & \text{if } x \ne 0 \\ 1 & \text{if } x = 0 \end{cases}$

30. $f(x) = \begin{cases} 3x & \text{if } x \ne 0 \\ 4 & \text{if } x = 0 \end{cases}$

31. $f(x) = \begin{cases} -2x + 3 & x < 1 \\ 3x - 2 & x \ge 1 \end{cases}$

32. $f(x) = \begin{cases} x + 3 & \text{if } x < -2 \\ -2x - 3 & \text{if } x \ge -2 \end{cases}$

33. $f(x) = \begin{cases} x + 3 & \text{if } -2 \le x < 1 \\ 5 & \text{if } x = 1 \\ -x + 2 & \text{if } x > 1 \end{cases}$

34. $f(x) = \begin{cases} 2x + 5 & \text{if } -3 \le x < 0 \\ -3 & \text{if } x = 0 \\ -5x & \text{if } x > 0 \end{cases}$

35. $f(x) = \begin{cases} 1 + x & \text{if } x < 0 \\ x^2 & \text{if } x \ge 0 \end{cases}$

36. $f(x) = \begin{cases} \dfrac{1}{x} & \text{if } x < 0 \\ \sqrt[3]{x} & \text{if } x \ge 0 \end{cases}$

37. $f(x) = \begin{cases} |x| & \text{if } -2 \le x < 0 \\ x^3 & \text{if } x > 0 \end{cases}$

38. $f(x) = \begin{cases} 2 - x & \text{if } -3 \le x < 1 \\ \sqrt{x} & \text{if } x > 1 \end{cases}$

39. $f(x) = 2\,\text{int}(x)$

40. $f(x) = \text{int}(2x)$

In Problems 41–44, the graph of a piecewise-defined function is given. Write a definition for each function.

41. 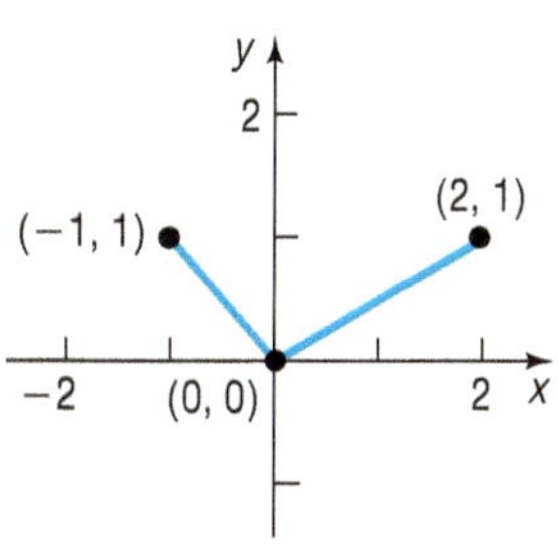

42. 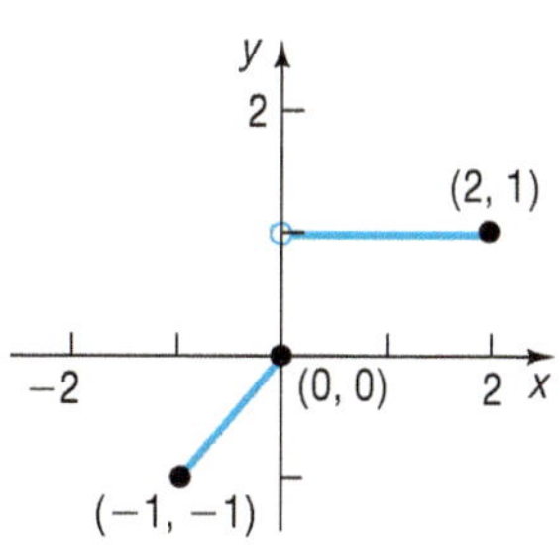

43. 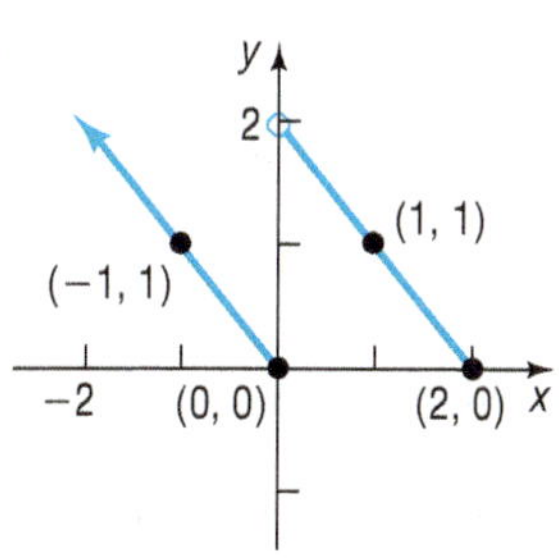

44. 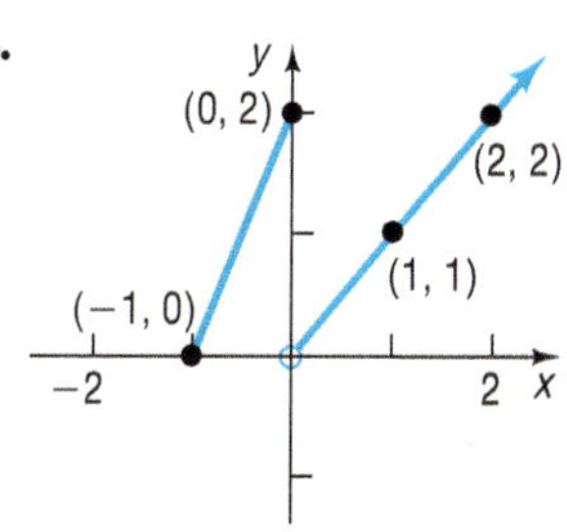

45. If $f(x) = \text{int}(2x)$, find (a) $f(1.2)$ (b) $f(1.6)$ (c) $f(-1.8)$

46. If $f(x) = \text{int}\left(\frac{x}{2}\right)$, find (a) $f(1.2)$ (b) $f(1.6)$ (c) $f(-1.8)$

Applications and Extensions

47. Cell Phone Service Sprint PCS offers a monthly cellular phone plan for \$39.99. It includes 450 anytime minutes and charges \$0.45 per minute for additional minutes. The following function is used to compute the monthly cost for a subscriber:

$$C(x) = \begin{cases} 39.99 & \text{if } 0 < x \le 450 \\ 0.45x - 162.51 & \text{if } x > 450 \end{cases}$$

where x is the number of anytime minutes used. Compute the monthly cost of the cellular phone for use of the following anytime minutes:

(a) 200 (b) 465 (c) 451

Source: *Sprint PCS*

48. Parking at O'Hare International Airport The short-term (no more than 24 hours) parking fee F (in dollars) for parking x hours at O'Hare International Airport's main parking garage can be modeled by the function

$$F(x) = \begin{cases} 3 & \text{if } 0 < x \le 3 \\ 5\,\text{int}(x + 1) + 1 & \text{if } 3 < x < 9 \\ 50 & \text{if } 9 \le x \le 24 \end{cases}$$

Determine the fee for parking in the short-term parking garage for
(a) 2 hours
(b) 7 hours
(c) 15 hours
(d) 8 hours and 24 minutes

Source: *O'Hare International Airport*

49. Cost of Natural Gas In February 2007, Peoples Energy had the following rate schedule for natural gas usage in single-family residences:

Monthly service charge	\$9.45
Per therm service charge	
1st 50 therms	\$0.36375/therm
Over 50 therms	\$0.11445/therm
Gas charge	\$0.8738/therm

(a) What is the charge for using 50 therms in a month?
(b) What is the charge for using 500 therms in a month?
(c) Develop a model that relates the monthly charge C for x therms of gas.
(d) Graph the function found in part (c).

Source: *Peoples Energy, Chicago, Illinois, 2007*

50. Cost of Natural Gas In February 2007, Nicor Gas had the following rate schedule for natural gas usage in single-family residences:

Monthly customer charge	\$8.85
Distribution charge	
1st 20 therms	\$0.1473/therm
Next 30 therms	\$0.0579/therm
Over 50 therms	\$0.0519/therm
Gas supply charge	\$0.67/therm

(a) What is the charge for using 40 therms in a month?
(b) What is the charge for using 202 therms in a month?
(c) Develop a model that gives the monthly charge C for x therms of gas.
(d) Graph the function found in part (c).

Source: *Nicor Gas, Aurora, Illinois, 2007*

51. Federal Income Tax Two 2007 Tax Rate Schedules are given in the accompanying table. If x equals taxable income and y equals the tax due, construct a function $y = f(x)$ for Schedule X.

REVISED 2007 TAX RATE SCHEDULES

Schedule X—Single						Schedule Y-1—Married Filing jointly or qualifying Widow(er)					
If Taxable Income Is over	But Not over	The Tax Is This Amount		Plus This %	Of the Excess over	If Taxable Income Is over	But Not over	The Tax Is This Amount		Plus This %	Of the Excess over
0	7825	0.00	+	10%	\$0	\$0	15,650	0.00	+	10%	\$0
7825	31,850	782.50	+	15%	7825	15,650	63,700	1565.00	+	15%	15,650
31,850	77,100	4386.25	+	25%	31,850	63,700	128,500	8772.50	+	25%	63,700
77,100	160,850	15,698.75	+	28%	77,100	128,500	195,850	24,972.50	+	28%	128,500
160,850	349,700	39,148.75	+	33%	160,850	195,850	349,700	43,830.50	+	33%	195,850
349,700		101,469.25	+	35%	349,700	349,700	—	94,601.00	+	35%	349,700

Source: *Internal Revenue Service*

52. Federal Income Tax Refer to the revised 2007 tax rate schedules. If x equals taxable income and y equals the tax due, construct a function $y = f(x)$ for Schedule Y-1.

53. Cost of Transporting Goods A trucking company transports goods between Chicago and New York, a distance of 960 miles. The company's policy is to charge, for each pound, \$0.50 per mile for the first 100 miles, \$0.40 per mile for the next 300 miles, \$0.25 per mile for the next 400 miles, and no charge for the remaining 160 miles.
(a) Graph the relationship between the cost of transportation in dollars and mileage over the entire 960-mile route.
(b) Find the cost as a function of mileage for hauls between 100 and 400 miles from Chicago.
(c) Find the cost as a function of mileage for hauls between 400 and 800 miles from Chicago.

54. **Car Rental Costs** An economy car rented in Florida from National Car Rental® on a weekly basis costs \$95 per week. Extra days cost \$24 per day until the day rate exceeds the weekly rate, in which case the weekly rate applies. Also, any part of a day used counts as a full day. Find the cost C of renting an economy car as a piecewise-defined function of the number x of days used, where $7 \le x \le 14$. Graph this function.

55. **Minimum Payments for Credit Cards** Holders of credit cards issued by banks, department stores, oil companies, and so on, receive bills each month that state minimum amounts that must be paid by a certain due date. The minimum due depends on the total amount owed. One such credit card company uses the following rules: For a bill of less than \$10, the entire amount is due. For a bill of at least \$10 but less than \$500, the minimum due is \$10. A minimum of \$30 is due on a bill of at least \$500 but less than \$1000, a minimum of \$50 is due on a bill of at least \$1000 but less than \$1500, and a minimum of \$70 is due on bills of \$1500 or more. Find the function f that describes the minimum payment due on a bill of x dollars. Graph f.

56. **Interest Payments for Credit Cards** Refer to Problem 55. The card holder may pay any amount between the minimum due and the total owed. The organization issuing the card charges the card holder interest of 1.5% per month for the first \$1000 owed and 1% per month on any unpaid balance over \$1000. Find the function g that gives the amount of interest charged per month on a balance of x dollars. Graph g.

57. **Wind Chill** The wind chill factor represents the equivalent air temperature at a standard wind speed that would produce the same heat loss as the given temperature and wind speed.

One formula for computing the equivalent temperature is

$$W = \begin{cases} t & 0 \le v < 1.79 \\ 33 - \dfrac{(10.45 + 10\sqrt{v} - v)(33 - t)}{22.04} & 1.79 \le v \le 20 \\ 33 - 1.5958(33 - t) & v > 20 \end{cases}$$

where v represents the wind speed (in meters per second) and t represents the air temperature (°C). Compute the wind chill for the following:

(a) An air temperature of 10°C and a wind speed of 1 meter per second (m/sec)
(b) An air temperature of 10°C and a wind speed of 5 m/sec
(c) An air temperature of 10°C and a wind speed of 15 m/sec
(d) An air temperature of 10°C and a wind speed of 25 m/sec
(e) Explain the physical meaning of the equation corresponding to $0 \le v < 1.79$.
(f) Explain the physical meaning of the equation corresponding to $v > 20$.

58. **Wind Chill** Redo Problem 57(a)–(d) for an air temperature of −10°C.

59. **First-class Mail** In 2007 the U.S. Postal Service charged \$0.80 postage for first-class mail retail flats (such as an 8.5" by 11" envelope) weighing up to 1 ounce, plus \$0.17 for each additional ounce up to 13 ounces. First-class rates do not apply to flats weighing more than 13 ounces. Develop a model that relates C, the first-class postage charged, for a flat weighing x ounces. Graph the function.

Source: United States Postal Service

Discussion and Writing

In Problems 60–67, use a graphing utility.

60. **Exploration** Graph $y = x^2$. Then on the same screen graph $y = x^2 + 2$, followed by $y = x^2 + 4$, followed by $y = x^2 - 2$. What pattern do you observe? Can you predict the graph of $y = x^2 - 4$? Of $y = x^2 + 5$?

61. **Exploration** Graph $y = x^2$. Then on the same screen graph $y = (x - 2)^2$, followed by $y = (x - 4)^2$, followed by $y = (x + 2)^2$. What pattern do you observe? Can you predict the graph of $y = (x + 4)^2$? Of $y = (x - 5)^2$?

62. **Exploration** Graph $y = |x|$. Then on the same screen graph $y = 2|x|$, followed by $y = 4|x|$, followed by $y = \frac{1}{2}|x|$. What pattern do you observe? Can you predict the graph of $y = \frac{1}{4}|x|$? Of $y = 5|x|$?

63. **Exploration** Graph $y = x^2$. Then on the same screen graph $y = -x^2$. What pattern do you observe? Now try $y = |x|$ and $y = -|x|$. What do you conclude?

64. **Exploration** Graph $y = \sqrt{x}$. Then on the same screen graph $y = \sqrt{-x}$. What pattern do you observe? Now try $y = 2x + 1$ and $y = 2(-x) + 1$. What do you conclude?

65. **Exploration** Graph $y = x^3$. Then on the same screen graph $y = (x - 1)^3 + 2$. Could you have predicted the result?

66. **Exploration** Graph $y = x^2$, $y = x^4$, and $y = x^6$ on the same screen. What do you notice is the same about each graph? What do you notice that is different?

67. **Exploration** Graph $y = x^3$, $y = x^5$, and $y = x^7$ on the same screen. What do you notice is the same about each graph? What do you notice that is different?

68. **Consider the equation**

$$y = \begin{cases} 1 & \text{if } x \text{ is rational} \\ 0 & \text{if } x \text{ is irrational} \end{cases}$$

Is this a function? What is its domain? What is its range? What is its y-intercept, if any? What are its x-intercepts, if any? Is it even, odd, or neither? How would you describe its graph?

69. Define some functions that pass through $(0, 0)$ and $(1, 1)$ and are increasing for $x \ge 0$. Begin your list with $y = \sqrt{x}$, $y = x$, and $y = x^2$. Can you propose a general result about such functions?

'Are You Prepared' Answers

1.

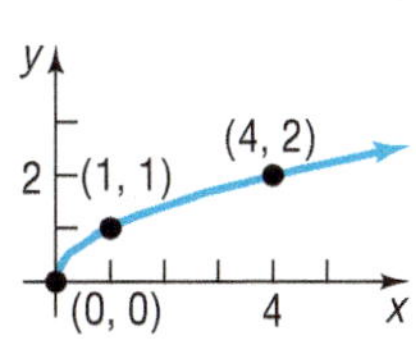

2.

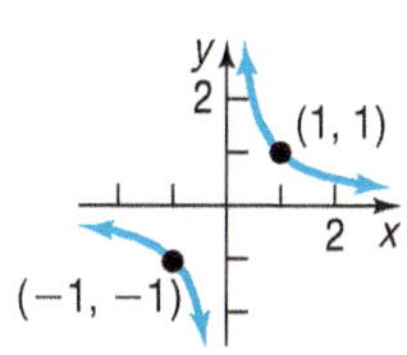

3. $(0, -8), (2, 0)$

3.5 Graphing Techniques: Transformations

OBJECTIVES 1 Graph Functions Using Vertical and Horizontal Shifts (p. 256)
2 Graph Functions Using Compressions and Stretches (p. 258)
3 Graph Functions Using Reflections about the x-Axis or y-Axis (p. 260)

At this stage, if you were asked to graph any of the functions defined by $y = x$, $y = x^2$, $y = x^3$, $y = \sqrt{x}$, $y = \sqrt[3]{x}$, $y = |x|$, or $y = \frac{1}{x}$, your response should be, "Yes, I recognize these functions and know the general shapes of their graphs." (If this is not your answer, review the previous section, Figures 33 through 39.)

Sometimes we are asked to graph a function that is "almost" like one that we already know how to graph. In this section, we look at some of these functions and develop techniques for graphing them. Collectively, these techniques are referred to as **transformations.** We introduce the method of transformations because it is a more efficient method of graphing than point-plotting.

1 Graph Functions Using Vertical and Horizontal Shifts

Exploration

On the same screen, graph each of the following functions:

$$Y_1 = x^2$$
$$Y_2 = x^2 + 2$$
$$Y_3 = x^2 - 2$$

What do you observe? Now create a table of values for Y_1, Y_2, and Y_3. What do you observe?

Result Figure 44 illustrates the graphs. You should have observed a general pattern. With $Y_1 = x^2$ on the screen, the graph of $Y_2 = x^2 + 2$ is identical to that of $Y_1 = x^2$, except that it is shifted vertically up 2 units. The graph of $Y_3 = x^2 - 2$ is identical to that of $Y_1 = x^2$, except that it is shifted vertically down 2 units. From Table 7(a), we see that the y-coordinates on $Y_2 = x^2 + 2$ are 2 units larger than the y-coordinates on $Y_1 = x^2$ for any given x-coordinate. From Table 7(b), we see that the y-coordinates on $Y_3 = x^2 - 2$ are 2 units smaller than the y-coordinates on $Y_1 = x^2$ for any given x-coordinate.

Figure 44

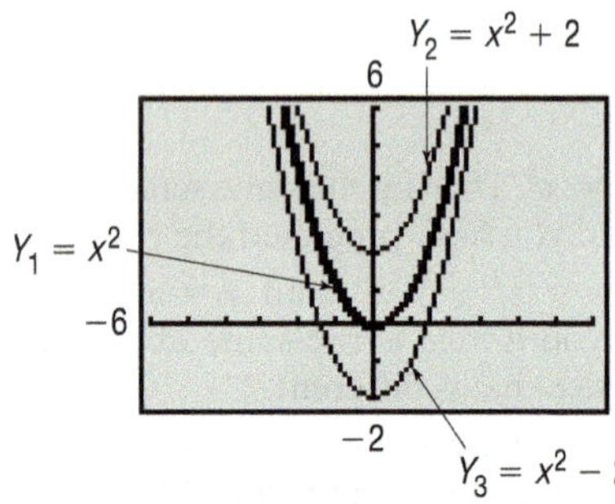

Table 7

X	Y1	Y2
-2	4	6
-1	1	3
0	0	2
1	1	3
2	4	6
3	9	11
4	16	18

Y2=X²+2

(a)

X	Y1	Y3
-2	4	2
-1	1	-1
0	0	-2
1	1	-1
2	4	2
3	9	7
4	16	14

Y3=X²-2

(b)

We are led to the following conclusions:

If a positive real number k is added to the outputs of a function $y = f(x)$, the graph of the new function $y = f(x) + k$ is the graph of f **shifted vertically up** k units.

If a positive real number k is subtracted from the outputs of a function $y = f(x)$, the graph of the new function $y = f(x) - k$ is the graph of f **shifted vertically down** k units.

EXAMPLE 1 **Vertical Shift Down**

Use the graph of $f(x) = x^2$ to obtain the graph of $h(x) = x^2 - 4$.

Solution Table 8 lists some points on the graphs of $Y_1 = f(x) = x^2$ and $Y_2 = h(x) = f(x) - 4 = x^2 - 4$. Notice that each y-coordinate of h is 4 units less than the corresponding y-coordinate of f.

Table 8

X	Y_1	Y_2
-3	9	5
-2	4	0
-1	1	-3
0	0	-4
1	1	-3
2	4	0
3	9	5

$Y_2 = X^2 - 4$

To obtain the graph of h from the graph of f, subtract 4 from each y-coordinate on the graph of f. The graph of h is identical to that of f, except that it is shifted down 4 units. See Figure 45.

Figure 45

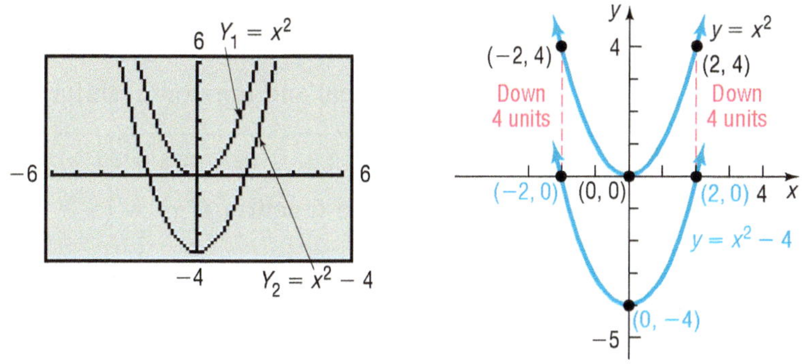

Now Work PROBLEM 39

Exploration

On the same screen, graph each of the following functions:

$$Y_1 = x^2$$
$$Y_2 = (x - 3)^2$$
$$Y_3 = (x + 2)^2$$

What do you observe?

Figure 46

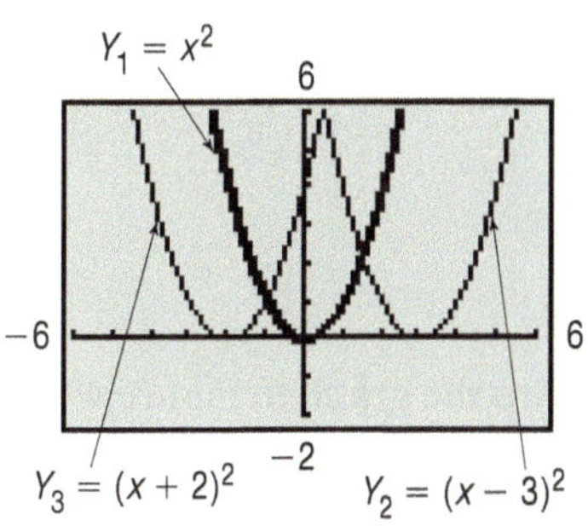

Result Figure 46 illustrates the graphs. You should have observed the following pattern. With the graph of $Y_1 = x^2$ on the screen, the graph of $Y_2 = (x - 3)^2$ is identical to that of $Y_1 = x^2$, except it is shifted horizontally to the right 3 units. The graph of $Y_3 = (x + 2)^2$ is identical to that of $Y_1 = x^2$, except it is shifted horizontally to the left 2 units.

From Table 9(a), we see the x-coordinates on $Y_2 = (x - 3)^2$ are 3 units larger than they are for $Y_1 = x^2$ for any given y-coordinate. For example, when $Y_1 = 0$, then $x = 0$, and when $Y_2 = 0$, then $x = 3$. Also, when $Y_1 = 1$, then $x = -1$ or 1, and when $Y_2 = 1$, then $x = 2$ or 4. From Table 9(b), we see the x-coordinates on $Y_3 = (x + 2)^2$ are 2 units smaller than they are for $Y_1 = x^2$ for any given y-coordinate. For example, when $Y_1 = 0$, then $x = 0$, and when $Y_3 = 0$, then $x = -2$. Also, when $Y_1 = 4$, then $x = -2$ or 2, and when $Y_3 = 4$, then $x = -4$ or 0.

Table 9

X	Y_1	Y_2
-1	1	16
0	0	9
1	1	4
2	4	1
3	9	0
4	16	1
5	25	4

$Y_2 = (X-3)^2$

(a)

X	Y_1	Y_3
-4	16	4
-3	9	1
-2	4	0
-1	1	1
0	0	4
1	1	9
2	4	16

$Y_3 = (X+2)^2$

(b)

We are led to the following conclusions:

If the argument x of a function f is replaced by $x - h, h > 0$, the graph of the new function $y = f(x - h)$ is the graph of f **shifted horizontally right** h units. If the argument x of a function f is replaced by $x + h, h > 0$, the graph of the new function $y = f(x + h)$ is the graph of f **shifted horizontally left** h units.

Now Work PROBLEM 43

NOTE Vertical shifts result when adding or subtracting a real number k after performing the operation suggested by the basic function, while horizontal shifts result when adding or subtracting a real number h to or from x before performing the operation suggested by the basic function. For example, the graph of $f(x) = \sqrt{x} + 3$ is obtained by shifting the graph of $y = \sqrt{x}$ up 3 units, because we evaluate the square root function first and then add 3. The graph of $g(x) = \sqrt{x + 3}$ is obtained by shifting the graph of $y = \sqrt{x}$ left 3 units, because we first add 3 to x before we evaluate the square root function. ■

Vertical and horizontal shifts are sometimes combined.

EXAMPLE 2 Combining Vertical and Horizontal Shifts

Graph the function $f(x) = (x + 3)^2 - 5$.

Solution We graph f in steps. First, we note that the rule for f is basically a square function, so we begin with the graph of $y = x^2$ as shown in Figure 47(a). To get the graph of $y = (x + 3)^2$, we shift the graph of $y = x^2$ horizontally 3 units to the left. See Figure 47(b). Finally, to get the graph of $y = (x + 3)^2 - 5$, we shift the graph of $y = (x + 3)^2$ vertically down 5 units. See Figure 47(c). Note the points plotted on each graph. Using key points can be helpful in keeping track of the transformation that has taken place.

Figure 47

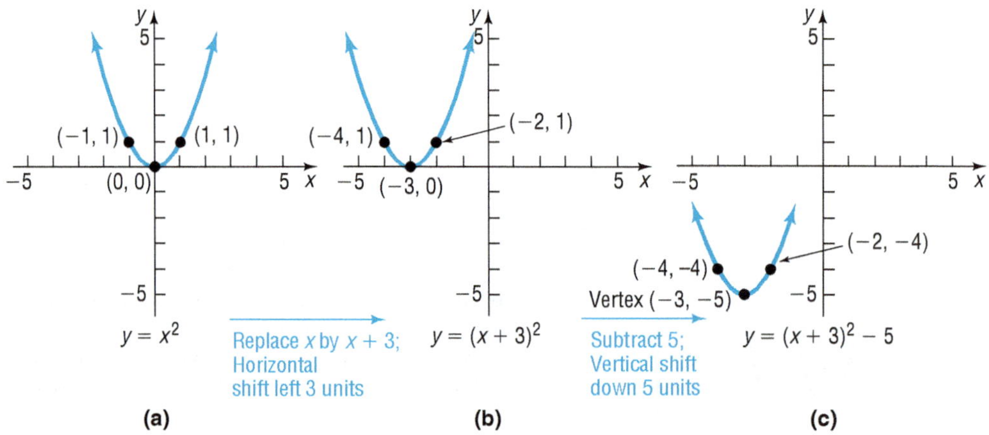

Check: Graph $Y_1 = f(x) = (x + 3)^2 - 5$ and compare the graph to Figure 47(c).

In Example 2, if the vertical shift had been done first, followed by the horizontal shift, the final graph would have been the same. (Try it for yourself.)

Now Work PROBLEMS 45 AND 77

2 Graph Functions Using Compressions and Stretches

Exploration

On the same screen, graph each of the following functions:

$$Y_1 = |x|$$
$$Y_2 = 2|x|$$
$$Y_3 = \frac{1}{2}|x|$$

Figure 48

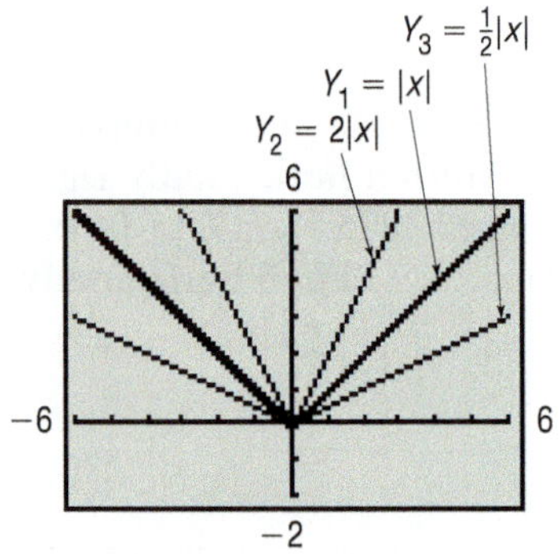

Then create a table of values and compare the y-coordinates for any given x-coordinate.

Result Figure 48 illustrates the graphs.

Now look at Table 10, where $Y_1 = |x|$ and $Y_2 = 2|x|$. Notice that the values for Y_2 are two times the values of Y_1 for each x-value. This means that the graph of $Y_2 = 2|x|$ can be obtained from the graph of $Y_1 = |x|$ by multiplying each y-coordinate of $Y_1 = |x|$ by 2. Therefore, the graph of Y_2 will be vertically *stretched* by a factor of 2.

Look at Table 11 where $Y_1 = |x|$ and $Y_3 = \frac{1}{2}|x|$. The values of Y_3 are half the values of Y_1 for each x-value. So, the graph of $Y_3 = \frac{1}{2}|x|$ can be obtained from the graph of $Y_1 = |x|$ by multiplying each y-coordinate by $\frac{1}{2}$. Therefore, the graph of Y_3 will be vertically *compressed* by a factor of $\frac{1}{2}$.

Table 10

X	Y1	Y2
-2	2	4
-1	1	2
0	0	0
1	1	2
2	2	4
3	3	6
4	4	8

Y2=2abs(X)

Table 11

X	Y1	Y3
-2	2	1
-1	1	.5
0	0	0
1	1	.5
2	2	1
3	3	1.5
4	4	2

Y3=.5abs(X)

Based on the Exploration, we have the following result:

When the right side of a function $y = f(x)$ is multiplied by a positive number a, the graph of the new function $y = af(x)$ is obtained by multiplying each y-coordinate on the graph of $y = f(x)$ by a. The new graph is a **vertically compressed** (if $0 < a < 1$) or a **vertically stretched** (if $a > 1$) version of the graph of $y = f(x)$.

Now Work PROBLEM 47

What happens if the argument x of a function $y = f(x)$ is multiplied by a positive number a, creating a new function $y = f(ax)$? To find the answer, we look at the following Exploration.

Exploration

On the same screen, graph each of the following functions:

$$Y_1 = f(x) = \sqrt{x} \qquad Y_2 = f(2x) = \sqrt{2x} \qquad Y_3 = f\left(\frac{1}{2}x\right) = \sqrt{\frac{1}{2}x} = \sqrt{\frac{x}{2}}$$

Create a table of values to explore the relation between the x- and y-coordinates of each function.

Result You should have obtained the graphs in Figure 49. Look at Table 12(a). Notice that (1, 1), (4, 2), and (9, 3) are points on the graph of $Y_1 = \sqrt{x}$. Also, (0.5, 1), (2, 2), and (4.5, 3) are points on the graph of $Y_2 = \sqrt{2x}$. For a given y-coordinate, the x-coordinate on the graph of Y_2 is $\frac{1}{2}$ of the x-coordinate on Y_1. We conclude that the graph of $Y_2 = \sqrt{2x}$ is obtained by multiplying the x-coordinate of each point on the graph of $Y_1 = \sqrt{x}$ by $\frac{1}{2}$. The graph of $Y_2 = \sqrt{2x}$ is the graph of $Y_1 = \sqrt{x}$ *compressed* horizontally.

Figure 49

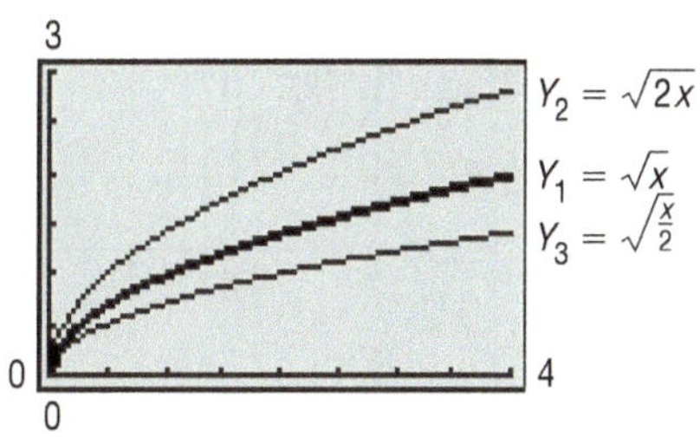

Look at Table 12(b). Notice that (1, 1), (4, 2), and (9, 3) are points on the graph of $Y_1 = \sqrt{x}$. Also notice that (2, 1), (8, 2), and (18, 3) are points on the graph of $Y_3 = \sqrt{\frac{x}{2}}$. For a given y-coordinate, the x-coordinate on the graph of Y_3 is 2 times the x-coordinate on Y_1. We conclude that the graph of $Y_3 = \sqrt{\frac{x}{2}}$ is obtained by multiplying the x-coordinate of each point on the graph of $Y_1 = \sqrt{x}$ by 2. The graph of $Y_3 = \sqrt{\frac{x}{2}}$ is the graph of $Y_1 = \sqrt{x}$ *stretched* horizontally.

Table 12

X	Y1	Y2
0	0	0
.5	.70711	1
1	1	1.4142
2	1.4142	2
4	2	2.8284
4.5	2.1213	3
9	3	4.2426

Y2=√(2X)

(a)

X	Y1	Y3
0	0	0
1	1	.70711
2	1.4142	1
4	2	1.4142
8	2.8284	2
9	3	2.1213
18	4.2426	3

Y3=√(X/2)

(b)

Based on the results of the Exploration, we have the following result:

If the argument x of a function $y = f(x)$ is multiplied by a positive number a, the graph of the new function $y = f(ax)$ is obtained by multiplying each x-coordinate of $y = f(x)$ by $\frac{1}{a}$. A **horizontal compression** results if $a > 1$, and a **horizontal stretch** occurs if $0 < a < 1$.

Let's look at an example.

EXAMPLE 3 Graphing Using Stretches and Compressions

The graph of $y = f(x)$ is given in Figure 50. Use this graph to find the graphs of

(a) $y = 2f(x)$ (b) $y = f(3x)$

Figure 50

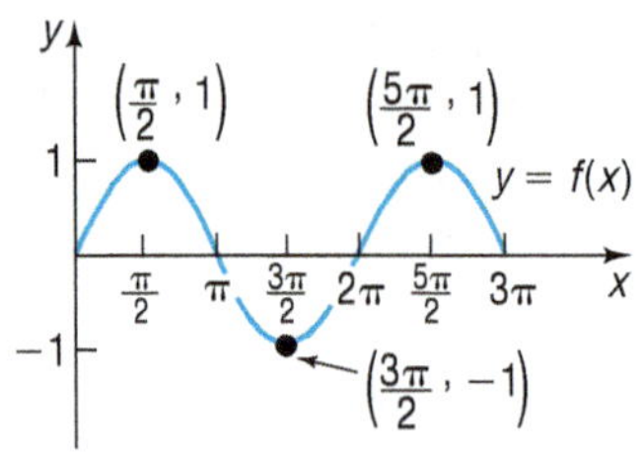

Solution (a) The graph of $y = 2f(x)$ is obtained by multiplying each y-coordinate of $y = f(x)$ by 2. See Figure 51.

(b) The graph of $y = f(3x)$ is obtained from the graph of $y = f(x)$ by multiplying each x-coordinate of $y = f(x)$ by $\frac{1}{3}$. See Figure 52.

Figure 51

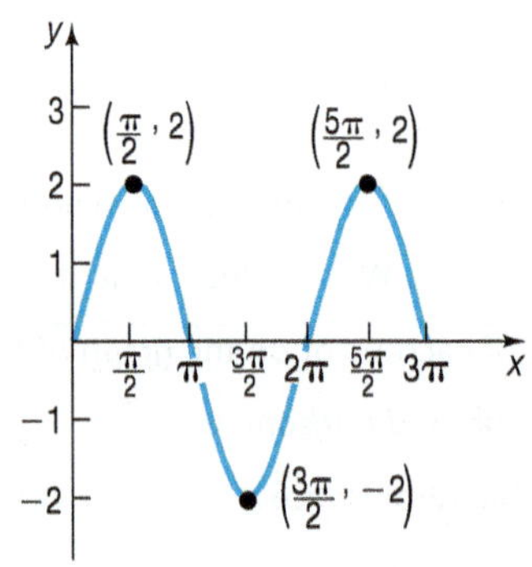

$y = 2f(x)$

Figure 52

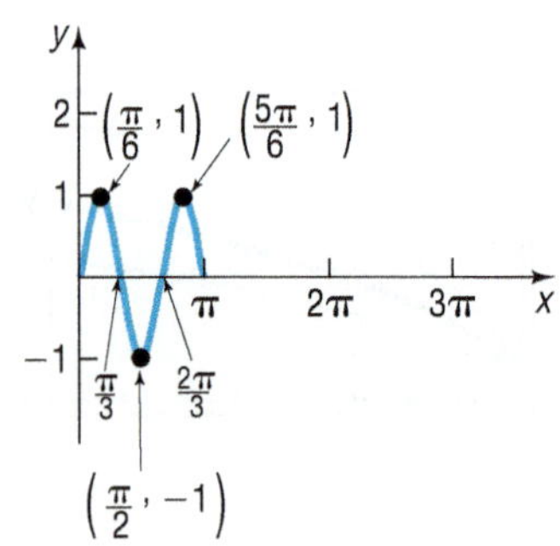

$y = f(3x)$

Now Work PROBLEMS 69(e) AND (g)

3 Graph Functions Using Reflections about the *x*-Axis or *y*-Axis

Exploration

Reflection about the x-axis:

(a) Graph and create a table of $Y_1 = x^2$ and $Y_2 = -x^2$.

(b) Graph and create a table of $Y_1 = |x|$ and $Y_2 = -|x|$.

(c) Graph and create a table of $Y_1 = x^2 - 4$ and $Y_2 = -(x^2 - 4) = -x^2 + 4$.

Result See Tables 13(a), (b), and (c) and Figures 53(a), (b), and (c). For each point (x, y) on the graph of Y_1, the point $(x, -y)$ is on the graph of Y_2. Put another way, Y_2 is the reflection about the x-axis of Y_1.

Table 13

(a)

X	Y1	Y2
-3	9	-9
-2	4	-4
-1	1	-1
0	0	0
1	1	-1
2	4	-4
3	9	-9

Y2 = -X²

(b)

X	Y1	Y2
-3	3	-3
-2	2	-2
-1	1	-1
0	0	0
1	1	-1
2	2	-2
3	3	-3

Y2 = -abs(X)

(c)

X	Y1	Y2
-3	5	-5
-2	0	0
-1	-3	3
0	-4	4
1	-3	3
2	0	0
3	5	-5

Y2 = -X²+4

Figure 53

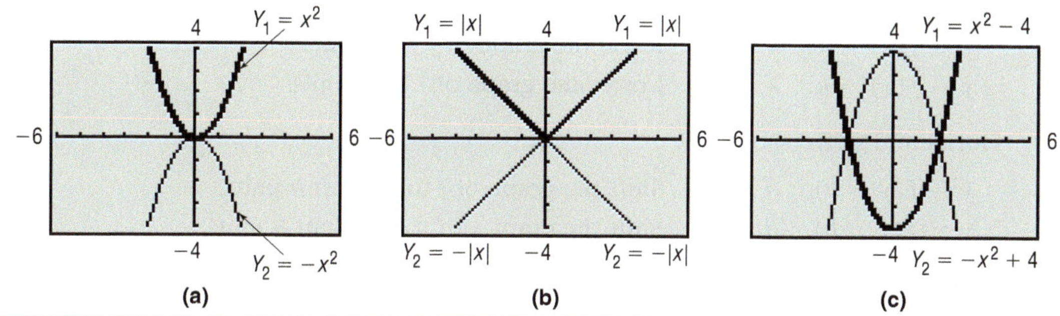

The results of the previous Exploration lead to the following result.

When the right side of the function $y = f(x)$ is multiplied by -1, the graph of the new function $y = -f(x)$ is the **reflection about the x-axis** of the graph of the function $y = f(x)$.

Now Work PROBLEM 51

Exploration

Reflection about the y-axis:

(a) Graph $Y_1 = \sqrt{x}$, followed by $Y_2 = \sqrt{-x}$.

(b) Graph $Y_1 = x + 1$, followed by $Y_2 = -x + 1$.

(c) Graph $Y_1 = x^4 + x$, followed by $Y_2 = (-x)^4 + (-x) = x^4 - x$.

Result See Tables 14(a), (b), and (c) and Figures 54(a), (b), and (c). For each point (x, y) on the graph of Y_1, the point $(-x, y)$ is on the graph of Y_2. Put another way, Y_2 is the reflection about the y-axis of Y_1.

Table 14

(a)

X	Y1	Y2
-3	ERROR	1.7321
-2	ERROR	1.4142
-1	ERROR	1
0	0	0
1	1	ERROR
2	1.4142	ERROR
3	1.7321	ERROR

Y2 = √(-X)

(b)

X	Y1	Y2
-3	-2	4
-2	-1	3
-1	0	2
0	1	1
1	2	0
2	3	-1
3	4	-2

Y2 = -X+1

(c)

X	Y1	Y2
-3	78	84
-2	14	18
-1	0	2
0	0	0
1	2	0
2	18	14
3	84	78

Y2 = X^4-X

Figure 54

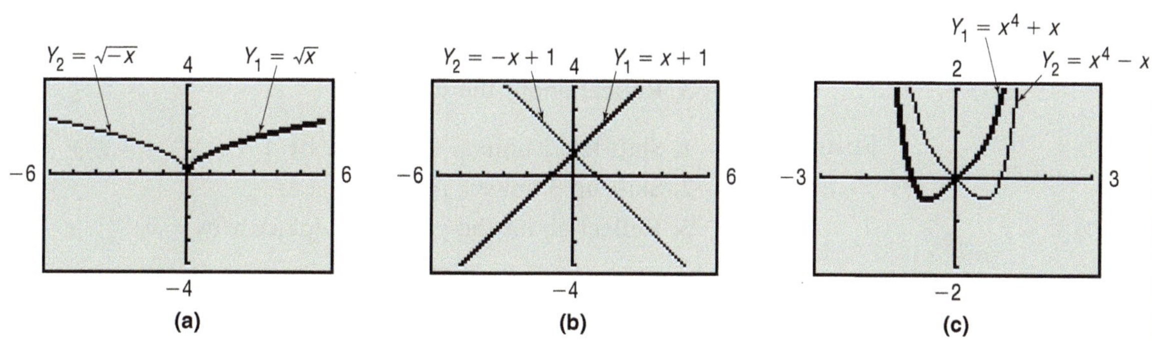

The results of the previous Exploration lead to the following result.

When the graph of the function $y = f(x)$ is known, the graph of the new function $y = f(-x)$ is the **reflection about the y-axis** of the graph of the function $y = f(x)$.

SUMMARY OF GRAPHING TECHNIQUES

To Graph:	Draw the Graph of f and:	Functional Change to $f(x)$
Vertical shifts		
$y = f(x) + k, \quad k > 0$	Raise the graph of f by k units.	Add k to $f(x)$.
$y = f(x) - k, \quad k > 0$	Lower the graph of f by k units.	Subtract k from $f(x)$.
Horizontal shifts		
$y = f(x + h), \quad h > 0$	Shift the graph of f to the left h units.	Replace x by $x + h$.
$y = f(x - h), \quad h > 0$	Shift the graph of f to the right h units.	Replace x by $x - h$.
Compressing or stretching		
$y = af(x), \quad a > 0$	Multiply each y-coordinate of $y = f(x)$ by a. Stretch the graph of f vertically if $a > 1$. Compress the graph of f vertically if $0 < a < 1$.	Multiply $f(x)$ by a.
$y = f(ax), \quad a > 0$	Multiply each x-coordinate of $y = f(x)$ by $\frac{1}{a}$ Stretch the graph of f horizontally if $0 < a < 1$. Compress the graph of f horizontally if $a > 1$.	Replace x by ax.
Reflection about the x-axis		
$y = -f(x)$	Reflect the graph of f about the x-axis.	Multiply $f(x)$ by -1.
Reflection about the y-axis		
$y = f(-x)$	Reflect the graph of f about the y-axis.	Replace x by $-x$.

The examples that follow combine some of the procedures outlined in this section to get the required graph.

EXAMPLE 4 Determining the Function Obtained from a Series of Transformations

Find the function that is finally graphed after the following three transformations are applied to the graph of $y = |x|$.

1. Shift left 2 units.
2. Shift up 3 units.
3. Reflect about the y-axis.

Solution

1. Shift left 2 units: Replace x by $x + 2$. $\quad y = |x + 2|$
2. Shift up 3 units: Add 3. $\quad y = |x + 2| + 3$
3. Reflect about the y-axis: Replace x by $-x$. $\quad y = |-x + 2| + 3$

Now Work PROBLEM 27

EXAMPLE 5 Combining Graphing Procedures

Graph the function $f(x) = \dfrac{3}{x-2} + 1$. Find the domain and the range of f.

Solution It is helpful to write f as $f(x) = 3\left(\dfrac{1}{x-2}\right) + 1$. Now we use the following steps to obtain the graph of f:

STEP 1: $y = \dfrac{1}{x}$ — Reciprocal function

STEP 2: $y = 3 \cdot \left(\dfrac{1}{x}\right) = \dfrac{3}{x}$ — Multiply by 3; vertical stretch of the graph of $y = \dfrac{1}{x}$ by a factor of 3.

STEP 3: $y = \dfrac{3}{x-2}$ — Replace x by $x - 2$; horizontal shift to the right 2 units.

STEP 4: $y = \dfrac{3}{x-2} + 1$ — Add 1; vertical shift up 1 unit.

See Figure 55.

Figure 55

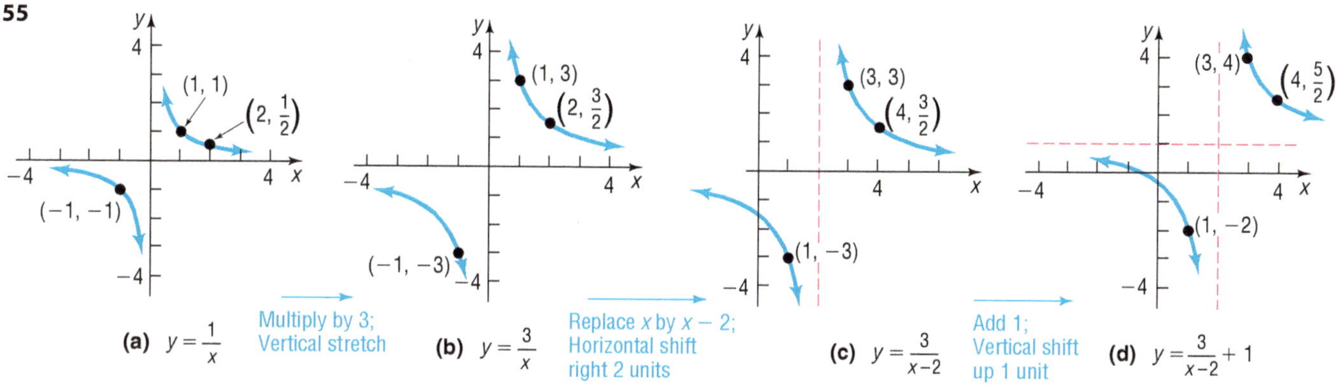

The domain of $y = \dfrac{1}{x}$ is $\{x \mid x \neq 0\}$ and its range is $\{y \mid y \neq 0\}$. Because we shifted right 2 units and up 1 unit to obtain f, the domain of f is $\{x \mid x \neq 2\}$ and its range is $\{y \mid y \neq 1\}$.

Other ordering of the steps shown in Example 5 would also result in the graph of f. For example, try this one:

STEP 1: $y = \dfrac{1}{x}$ — Reciprocal function

STEP 2: $y = \dfrac{1}{x-2}$ — Replace x by $x - 2$; horizontal shift to the right 2 units.

STEP 3: $y = \dfrac{3}{x-2}$ — Multiply by 3; vertical stretch of the graph of $y = \dfrac{1}{x-2}$ by a factor of 3.

STEP 4: $y = \dfrac{3}{x-2} + 1$ — Add 1; vertical shift up 1 unit.

HINT Although the order in which transformations are performed can be altered, you may consider using the following order for consistency:

1. Reflections
2. Compressions and stretches
3. Shifts

EXAMPLE 6 Combining Graphing Procedures

Graph the function $f(x) = \sqrt{1-x} + 2$. Find the domain and the range of f.

Solution Because horizontal shifts require the form $x - h$, we begin by rewriting $f(x)$ as $f(x) = \sqrt{1-x} + 2 = \sqrt{-(x-1)} + 2$. Now use the following steps:

STEP 1: $y = \sqrt{x}$ — Square root function

STEP 2: $y = \sqrt{-x}$ — Replace x by $-x$; reflect about the y-axis.

Step 3: $y = \sqrt{-(x - 1)} = \sqrt{1 - x}$ Replace x by x − 1; horizontal shift to the right 1 unit.

Step 4: $y = \sqrt{1 - x} + 2$ Add 2; vertical shift up 2 units.

See Figure 56.

Figure 56

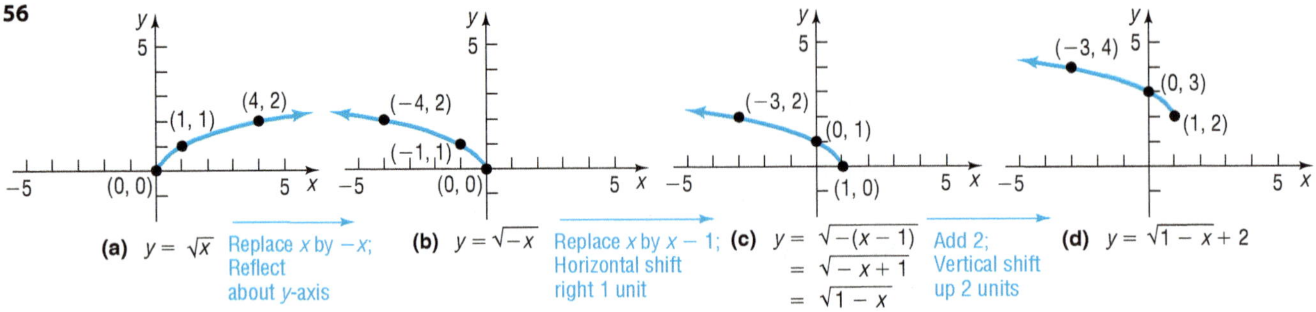

The domain of f is $(-\infty, 1]$ and its range is $[2, \infty)$.

Now Work PROBLEM 61

3.5 Assess Your Understanding

Concepts and Vocabulary

1. Suppose that the graph of a function f is known. Then the graph of $y = f(x - 2)$ may be obtained by a(n) _____ shift of the graph of f to the _____ a distance of 2 units.
2. Suppose that the graph of a function f is known. Then the graph of $y = f(-x)$ may be obtained by a reflection about the _____-axis of the graph of the function $y = f(x)$.
3. Suppose that the x-intercepts of the graph of $y = f(x)$ are -2, 1, and 5. The x-intercepts of $y = f(x + 3)$ are _____, _____, and _____.
4. ***True or False*** The graph of $y = -f(x)$ is the reflection about the x-axis of the graph of $y = f(x)$.
5. ***True or False*** To obtain the graph of $y = f(x + 2) - 3$, shift the graph of $y = f(x)$ horizontally to the right 2 units and vertically down 3 units.
6. ***True or False*** Suppose that the x-intercepts of the graph of $y = f(x)$ are -3 and 2. Then the x-intercepts of the graph of $y = 2f(x)$ are also -3 and 2.

Skill Building

In Problems 7–18, match each graph to one of the following functions:

A. $y = x^2 + 2$	B. $y = -x^2 + 2$	C. $y = \lvert x\rvert + 2$	D. $y = -\lvert x\rvert + 2$
E. $y = (x - 2)^2$	F. $y = -(x + 2)^2$	G. $y = \lvert x - 2\rvert$	H. $y = -\lvert x + 2\rvert$
I. $y = 2x^2$	J. $y = -2x^2$	K. $y = 2\lvert x\rvert$	L. $y = -2\lvert x\rvert$

7.
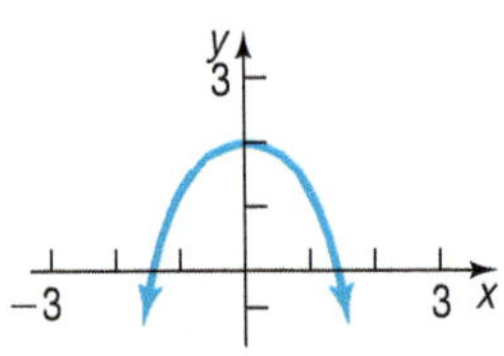

8.
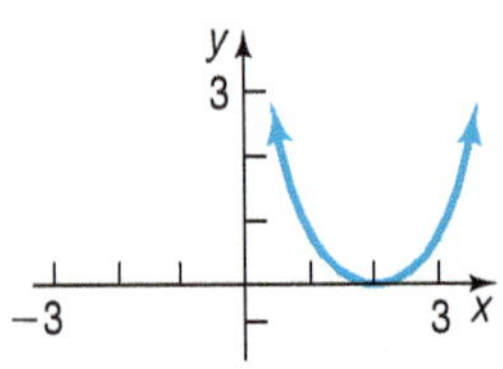

9.
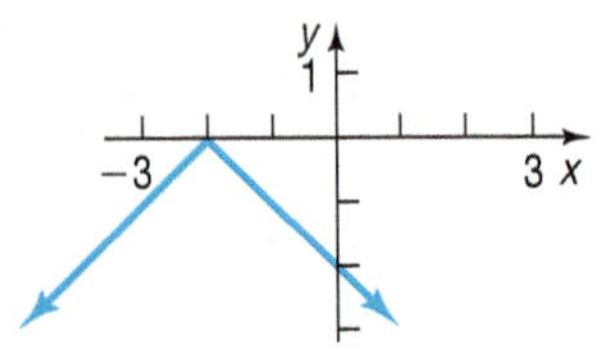

10.
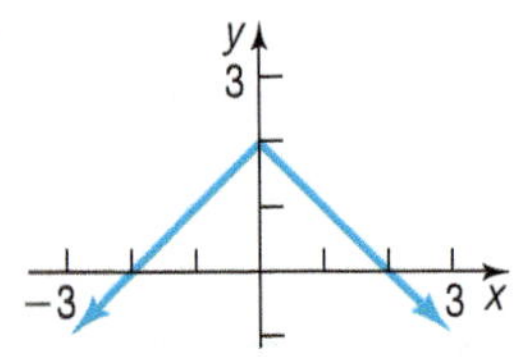

11.
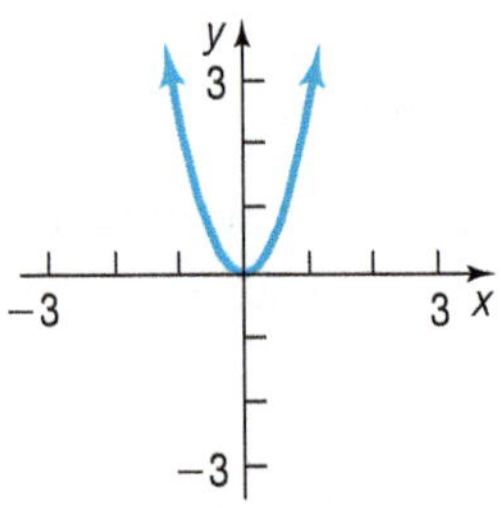

12.
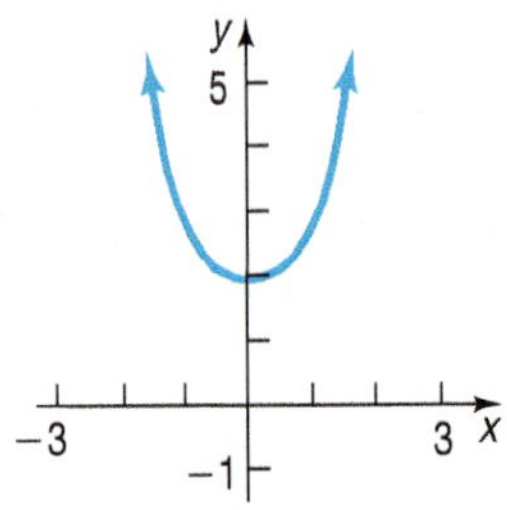

13.
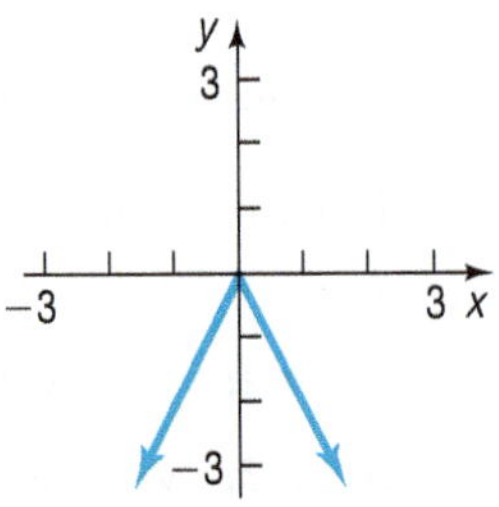

14.
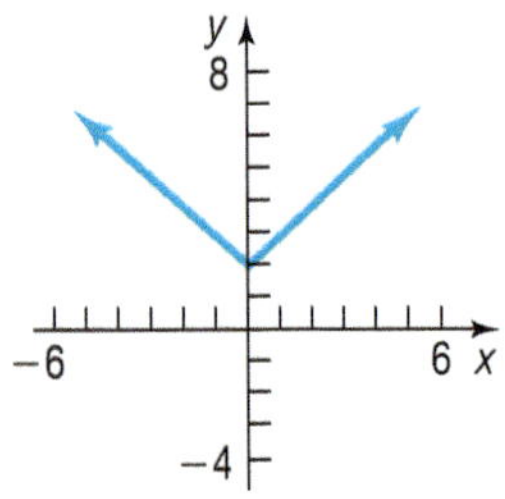

15.

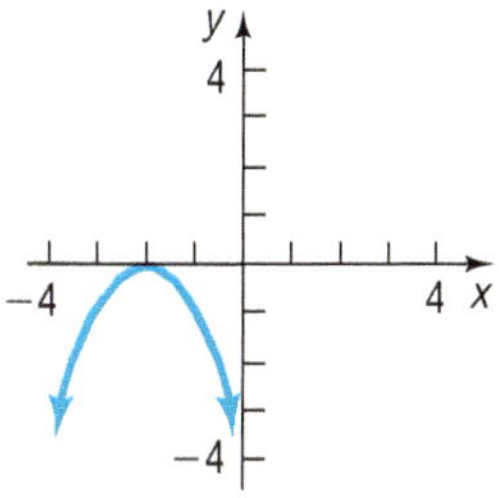

16.

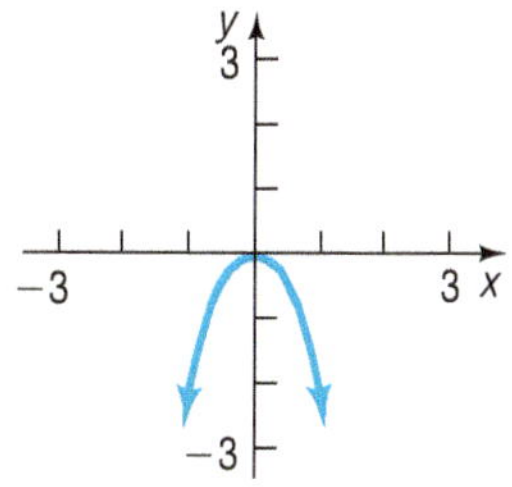

17.

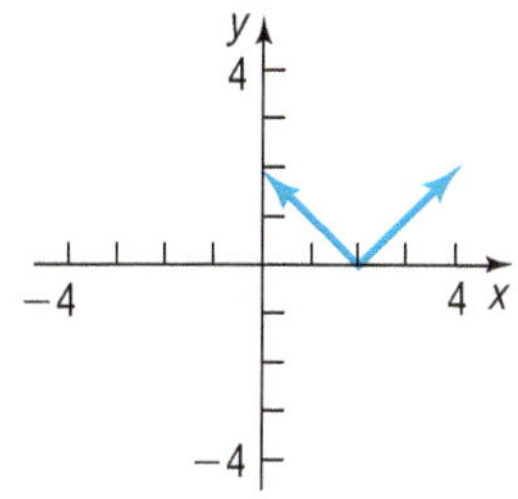

18.

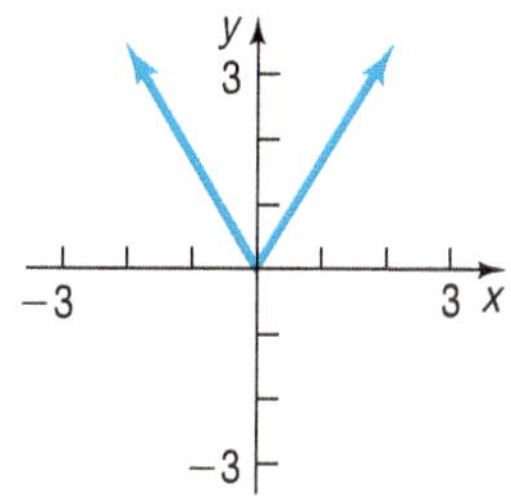

In Problems 19–26, write the function whose graph is the graph of $y = x^3$, *but is:*

19. Shifted to the right 4 units

20. Shifted to the left 4 units

21. Shifted up 4 units

22. Shifted down 4 units

23. Reflected about the y-axis

24. Reflected about the x-axis

25. Vertically stretched by a factor of 4

26. Horizontally stretched by a factor of 4

In Problems 27–30, find the function that is finally graphed after the following transformations are applied to the graph of $y = \sqrt{x}$.

27. (1) Shift up 2 units
(2) Reflect about the x-axis
(3) Reflect about the y-axis

28. (1) Reflect about the x-axis
(2) Shift right 3 units
(3) Shift down 2 units

29. (1) Reflect about the x-axis
(2) Shift up 2 units
(3) Shift left 3 units

30. (1) Shift up 2 units
(2) Reflect about the y-axis
(3) Shift left 3 units

31. If $(3, 6)$ is a point on the graph of $y = f(x)$, which of the following points must be on the graph of $y = -f(x)$?
(a) $(6, 3)$
(b) $(6, -3)$
(c) $(3, -6)$
(d) $(-3, 6)$

32. If $(3, 6)$ is a point on the graph of $y = f(x)$, which of the following points must be on the graph of $y = f(-x)$?
(a) $(6, 3)$
(b) $(6, -3)$
(c) $(3, 6)$
(d) $(-3, 6)$

33. If $(1, 3)$ is a point on the graph of $y = f(x)$, which of the following points must be on the graph of $y = 2f(x)$?
(a) $(1, 3)$
(b) $(2, 3)$
(c) $(1, 6)$
(d) $\left(\frac{1}{2}, 3\right)$

34. If $(4, 2)$ is a point on the graph of $y = f(x)$, which of the following points must be on the graph of $y = f(2x)$?
(a) $(4, 1)$
(b) $(8, 2)$
(c) $(2, 2)$
(d) $(4, 4)$

35. Suppose that the x-intercepts of the graph of $y = f(x)$ are -5 and 3.
(a) What are the x-intercepts of the graph of $y = f(x + 2)$?
(b) What are the x-intercepts of the graph of $y = f(x - 2)$?
(c) What are the x-intercepts of the graph of $y = 4f(x)$?
(d) What are the x-intercepts of the graph of $y = f(-x)$?

36. Suppose that the x-intercepts of the graph of $y = f(x)$ are -8 and 1.
(a) What are the x-intercepts of the graph of $y = f(x + 4)$?
(b) What are the x-intercepts of the graph of $y = f(x - 3)$?
(c) What are the x-intercepts of the graph of $y = 2f(x)$?
(d) What are the x-intercepts of the graph of $y = f(-x)$?

37. Suppose that the function $y = f(x)$ is increasing on the interval $(-1, 5)$.
(a) Over what interval is the graph of $y = f(x + 2)$ increasing?
(b) Over what interval is the graph of $y = f(x - 5)$ increasing?
(c) What can be said about the graph of $y = -f(x)$?
(d) What can be said about the graph of $y = f(-x)$?

38. Suppose that the function $y = f(x)$ is decreasing on the interval $(-2, 7)$.
(a) Over what interval is the graph of $y = f(x + 2)$ decreasing?
(b) Over what interval is the graph of $y = f(x - 5)$ decreasing?
(c) What can be said about the graph of $y = -f(x)$?
(d) What can be said about the graph of $y = f(-x)$?

In Problems 39–68, graph each function using the techniques of shifting, compressing, stretching, and/or reflecting. Start with the graph of the basic function (for example, $y = x^2$*) and show all stages. Be sure to show at least three key points. Find the domain and the range of each function. Verify your results using a graphing utility.*

39. $f(x) = x^2 - 1$

40. $f(x) = x^2 + 4$

41. $g(x) = x^3 + 1$

42. $g(x) = x^3 - 1$

43. $h(x) = \sqrt{x - 2}$

44. $h(x) = \sqrt{x + 1}$

45. $f(x) = (x - 1)^3 + 2$

46. $f(x) = (x + 2)^3 - 3$

47. $g(x) = 4\sqrt{x}$

48. $g(x) = \frac{1}{2}\sqrt{x}$

49. $h(x) = \frac{1}{2x}$

50. $h(x) = \sqrt[3]{2x}$

51. $f(x) = -\sqrt[3]{x}$

52. $f(x) = -\sqrt{x}$

53. $g(x) = \sqrt[3]{-x}$

54 $g(x) = \frac{1}{-x}$

55. $h(x) = -x^3 + 2$

56. $h(x) = \frac{1}{-x} + 2$

57. $f(x) = 2(x + 1)^2 - 3$

58. $f(x) = 3(x - 2)^2 + 1$

59. $g(x) = 2\sqrt{x - 2} + 1$

60. $g(x) = 3|x + 1| - 3$

61. $h(x) = \sqrt{-x} - 2$

62. $h(x) = \frac{4}{x} + 2$

63. $f(x) = -(x + 1)^3 - 1$

64. $f(x) = -4\sqrt{x - 1}$

65. $g(x) = 2|1 - x|$

66. $g(x) = 4\sqrt{2 - x}$

67. $h(x) = 2\,\text{int}(x - 1)$

68. $h(x) = \text{int}(-x)$

In Problems 69–72, the graph of a function f is illustrated. Use the graph of f as the first step toward graphing each of the following functions:

(a) $F(x) = f(x) + 3$ (b) $G(x) = f(x + 2)$ (c) $P(x) = -f(x)$ (d) $H(x) = f(x + 1) - 2$

(e) $Q(x) = \frac{1}{2}f(x)$ (f) $g(x) = f(-x)$ (g) $h(x) = f(2x)$

69.

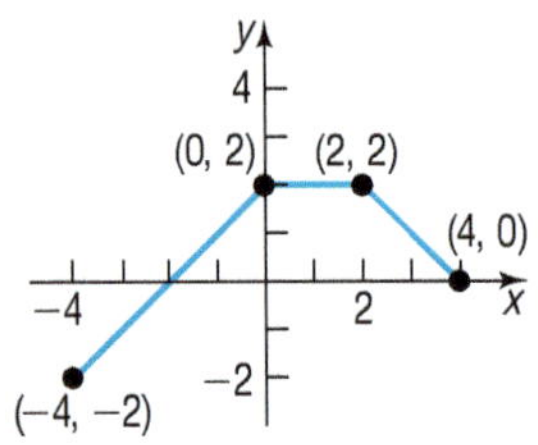

70.

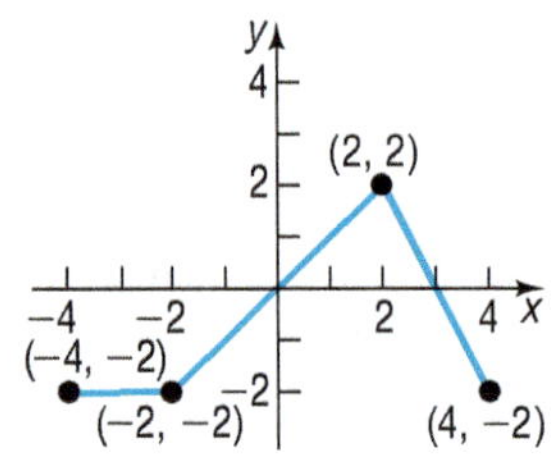

71.

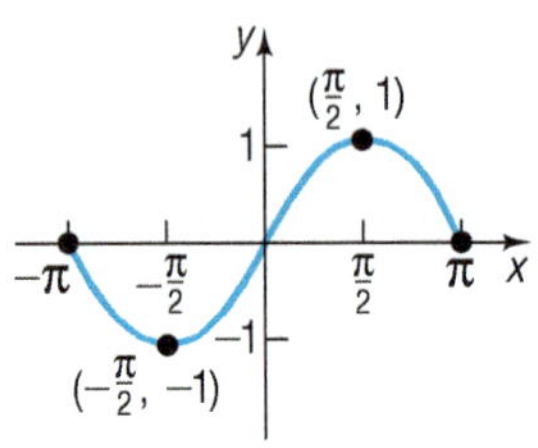

72.

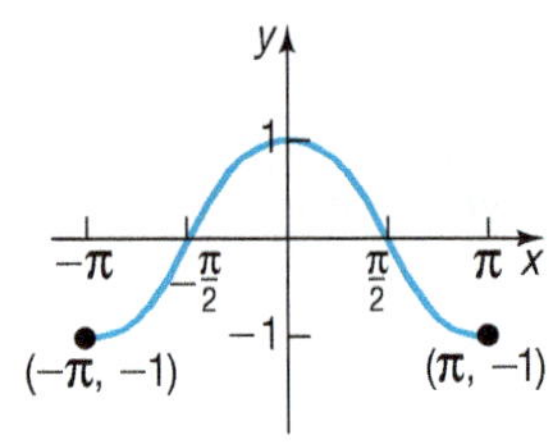

Mixed Practice

73. (a) Using a graphing utility, graph $f(x) = x^3 - 9x$ for $-4 < x < 4$.
(b) Find the x-intercepts of the graph of f.
(c) Approximate any local maxima and local minima.
(d) Determine where f is increasing and where it is decreasing.
(e) Without using a graphing utility, repeat parts (b)–(d) for $y = f(x + 2)$.
(f) Without using a graphing utility, repeat parts (b)–(d) for $y = 2f(x)$.
(g) Without using a graphing utility, repeat parts (b)–(d) for $y = f(-x)$.

74. (a) Using a graphing utility, graph $f(x) = x^3 - 4x$ for $-3 < x < 3$.
(b) Find the x-intercepts of the graph of f.
(c) Approximate any local maxima and local minima.
(d) Determine where f is increasing and where it is decreasing.
(e) Without using a graphing utility, repeat parts (b)–(d) for $y = f(x - 4)$.
(f) Without using a graphing utility, repeat parts (b)–(d) for $y = f(2x)$.
(g) Without using a graphing utility, repeat parts (b)–(d) for $y = -f(x)$.

In Problems 75–84, complete the square of each quadratic expression. Then graph each function using the technique of shifting. (If necessary, refer to Section 1.3 to review completing the square.)

75. $f(x) = x^2 + 2x$

76. $f(x) = x^2 - 6x$

77. $f(x) = x^2 - 8x + 1$

78. $f(x) = x^2 + 4x + 2$

79. $f(x) = x^2 + x + 1$

80. $f(x) = x^2 - x + 1$

81. $f(x) = 2x^2 - 12x + 19$

82. $f(x) = 3x^2 + 6x + 1$

83. $f(x) = -3x^2 - 12x - 17$

84. $f(x) = -2x^2 - 12x - 13$

Applications and Extensions

85. The equation $y = (x - c)^2$ defines a *family of parabolas*, one parabola for each value of c. On one set of coordinate axes, graph the members of the family for $c = 0$, $c = 3$, and $c = -2$.

86. Repeat Problem 85 for the family of parabolas $y = x^2 + c$.

87. Thermostat Control Energy conservation experts estimate that homeowners can save 5% to 10% on winter heating bills by programming their thermostats 5 to 10 degrees lower while sleeping. In the given graph, the temperature T (in degrees Fahrenheit) of a home is given as a function of time t (in hours after midnight) over a 24-hour period.

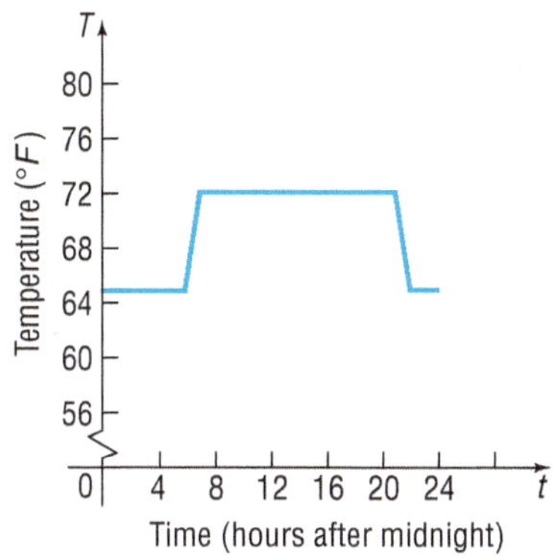

(a) At what temperature is the thermostat set during daytime hours? At what temperature is the thermostat set overnight?

(b) The homeowner reprograms the thermostat to $y = T(t) - 2$. Explain how this affects the temperature in the house. Graph this new function.

(c) The homeowner reprograms the thermostat to $y = T(t + 1)$. Explain how this affects the temperature in the house. Graph this new function.

Source: Roger Albright, *547 Ways to be Fuel Smart*, 2000

88. Digital Music Revenues The total projected worldwide digital music revenues R, in millions of dollars, for the years 2005 through 2010 can be estimated by the function

$$R(x) = 170.7x^2 + 1373x + 1080$$

where x is the number of years after 2005.

(a) Find $R(0)$, $R(3)$, and $R(5)$ and explain what each value represents.

(b) Find $r = R(x - 5)$.

(c) Find $r(5)$, $r(8)$, and $r(10)$ and explain what each value represents.

(d) In the model r, what does x represent?

(e) Would there be an advantage in using the model r when estimating the projected revenues for a given year instead of the model R?

Source: *eMarketer.com, May 2006*

89. Temperature Measurements The relationship between the Celsius (°C) and Fahrenheit (°F) scales for measuring temperature is given by the equation

$$F = \frac{9}{5}C + 32$$

The relationship between the Celsius (°C) and Kelvin (K) scales is $K = C + 273$. Graph the equation $F = \frac{9}{5}C + 32$ using degrees Fahrenheit on the y-axis and degrees Celsius on the x-axis. Use the techniques introduced in this section to obtain the graph showing the relationship between Kelvin and Fahrenheit temperatures.

90. Period of a Pendulum The period T (in seconds) of a simple pendulum is a function of its length l (in feet) defined by the equation

$$T = 2\pi\sqrt{\frac{l}{g}}$$

where $g \approx 32.2$ feet per second per second is the acceleration of gravity.

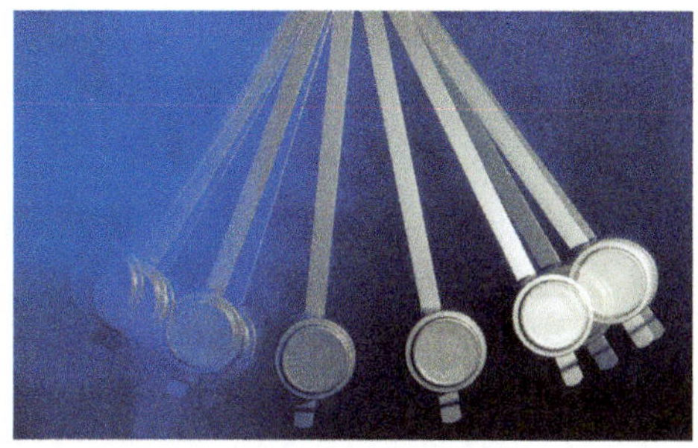

(a) Use a graphing utility to graph the function $T = T(l)$.

(b) Now graph the functions $T = T(l + 1)$, $T = T(l + 2)$, and $T = T(l + 3)$.

(c) Discuss how adding to the length l changes the period T.

(d) Now graph the functions $T = T(2l)$, $T = T(3l)$, and $T = T(4l)$.

(e) Discuss how multiplying the length l by factors of 2, 3, and 4 changes the period T.

91. Cigar Company Profits The daily profits of a cigar company from selling x cigars are given by

$$p(x) = -0.05x^2 + 100x - 2000$$

The government wishes to impose a tax on cigars (sometimes called a *sin tax*) that gives the company the option of either paying a flat tax of \$10,000 per day or a tax of 10% on profits. As chief financial officer (CFO) of the company, you need to decide which tax is the better option for the company.

(a) On the same screen, graph $Y_1 = p(x) - 10{,}000$ and $Y_2 = (1 - 0.10)p(x)$.

(b) Based on the graph, which option would you select? Why?

(c) Using the terminology learned in this section, describe each graph in terms of the graph of $p(x)$.

(d) Suppose that the government offered the options of a flat tax of \$4800 or a tax of 10% on profits. Which would you select? Why?

92. The graph of a function f is illustrated in the figure.

(a) Draw the graph of $y = |f(x)|$.

(b) Draw the graph of $y = f(|x|)$.

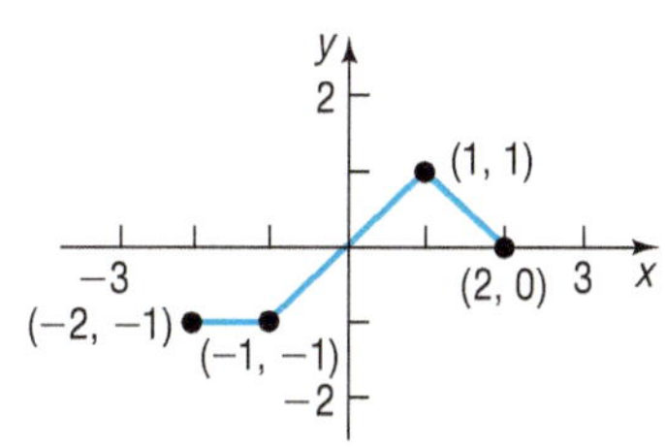

93. The graph of a function f is illustrated in the figure.
(a) Draw the graph of $y = |f(x)|$.
(b) Draw the graph of $y = f(|x|)$.

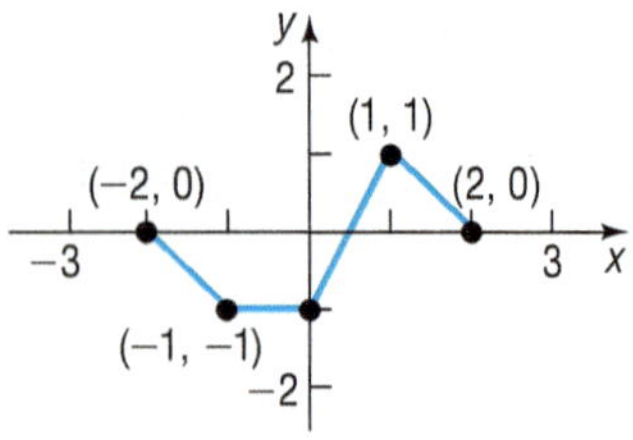

94. Suppose $(1, 3)$ is a point on the graph of $y = f(x)$.
(a) What point is on the graph of $y = f(x + 3) - 5$?
(b) What point is on the graph of $y = -2f(x - 2) + 1$?
(c) What point is on the graph of $y = f(2x + 3)$?

95. Suppose $(-3, 5)$ is a point on the graph of $y = f(x)$.
(a) What point is on the graph of $y = f(x + 1) - 3$?
(b) What point is on the graph of $y = -3f(x - 4) + 3$?
(c) What point is on the graph of $y = f(3x + 9)$?

Discussion and Writing

96. Suppose that the graph of a function f is known. Explain how the graph of $y = 4f(x)$ differs from the graph of $y = f(4x)$.

97. The area under the curve $y = \sqrt{x}$ bounded below by the x-axis, on the left by $x = 0$, and on the right by $x = 4$ is $\frac{16}{3}$ square units. Using the ideas presented in this section, what do you think is the area under the curve of $y = \sqrt{-x}$ bounded below by the x-axis, on the left by $x = -4$, and on the right by $x = 0$? Justify your answer.

3.6 Mathematical Models: Building Functions

OBJECTIVE 1 Build and Analyze Functions (p. 268)

1 Build and Analyze Functions

Real-world problems often result in mathematical models that involve functions. These functions need to be constructed or built based on the information given. In building functions, we must be able to translate the verbal description into the language of mathematics. We do this by assigning symbols to represent the independent and dependent variables and then by finding the function or rule that relates these variables.

EXAMPLE 1 Finding the Distance from the Origin to a Point on a Graph

Let $P = (x, y)$ be a point on the graph of $y = x^2 - 1$.

(a) Express the distance d from P to the origin O as a function of x.
(b) What is d if $x = 0$?
(c) What is d if $x = 1$?
(d) What is d if $x = \frac{\sqrt{2}}{2}$?
(e) Use a graphing utility to graph the function $d = d(x)$, $x \geq 0$. Rounded to two decimal places, find the value(s) of x at which d has a local minimum. [This gives the point(s) on the graph of $y = x^2 - 1$ closest to the origin.]

Solution (a) Figure 57 illustrates the graph of $y = x^2 - 1$. The distance d from P to O is

$$d = \sqrt{(x - 0)^2 + (y - 0)^2} = \sqrt{x^2 + y^2}$$

Since P is a point on the graph of $y = x^2 - 1$, we substitute $x^2 - 1$ for y. Then

$$d(x) = \sqrt{x^2 + (x^2 - 1)^2} = \sqrt{x^4 - x^2 + 1}$$

We have expressed the distance d as a function of x.

Figure 57

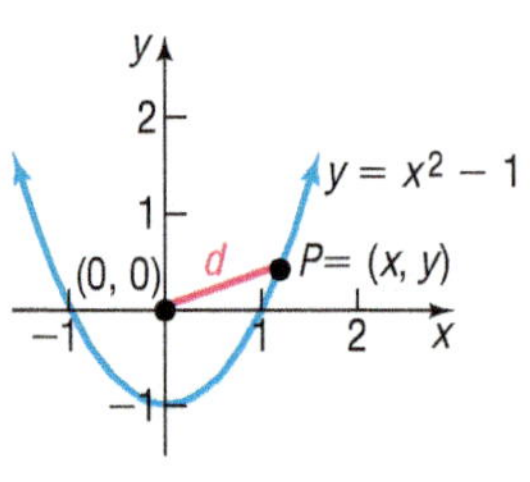

(b) If $x = 0$, the distance d is

$$d(0) = \sqrt{0^4 - 0^2 + 1} = \sqrt{1} = 1$$

(c) If $x = 1$, the distance d is

$$d(1) = \sqrt{1^4 - 1^2 + 1} = 1$$

(d) If $x = \frac{\sqrt{2}}{2}$, the distance d is

$$d\left(\frac{\sqrt{2}}{2}\right) = \sqrt{\left(\frac{\sqrt{2}}{2}\right)^4 - \left(\frac{\sqrt{2}}{2}\right)^2 + 1} = \sqrt{\frac{1}{4} - \frac{1}{2} + 1} = \frac{\sqrt{3}}{2}$$

Figure 58

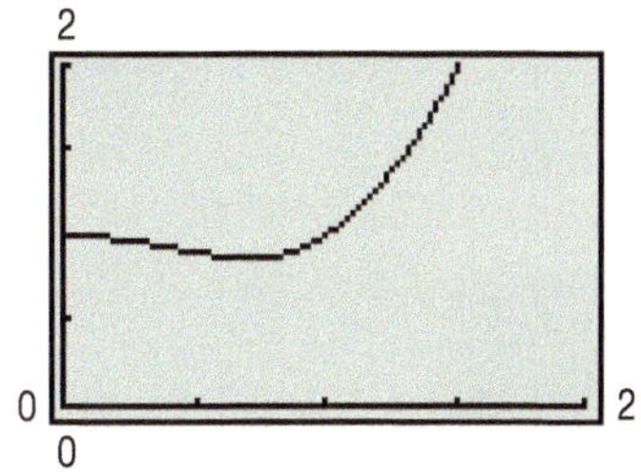

(e) Figure 58 shows the graph of $Y_1 = \sqrt{x^4 - x^2 + 1}$. Using the MINIMUM feature on a graphing utility, we find that when $x \approx 0.71$ the value of d is smallest. The local minimum is $d \approx 0.87$ rounded to two decimal places. Since $d(x)$ is even, by symmetry, it follows that when $x \approx -0.71$ the value of d is also a local minimum. Since $(\pm 0.71)^2 - 1 = -0.50$, the points $(-0.71, -0.50)$ and $(0.71, 0.50)$ on the graph of $y = x^2 - 1$ are closest to the origin.

Now Work PROBLEM 1

EXAMPLE 2 Area of a Rectangle

Figure 59

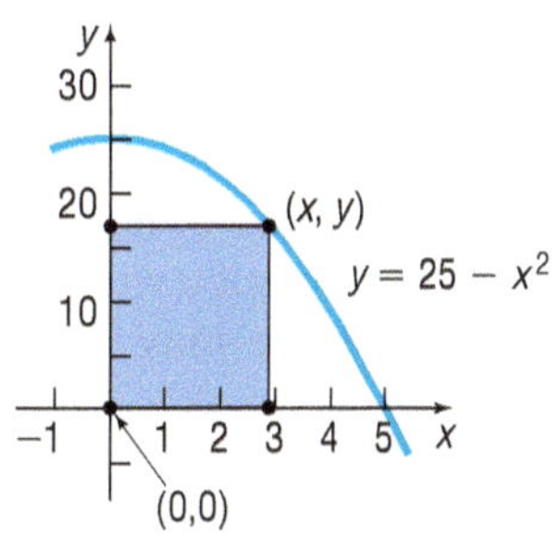

A rectangle has one corner in quadrant I on the graph of $y = 25 - x^2$, another at the origin, a third on the positive y-axis, and the fourth on the positive x-axis. See Figure 59.

(a) Express the area A of the rectangle as a function of x.
(b) What is the domain of A?
(c) Graph $A = A(x)$.
(d) For what value of x is the area largest?

Solution

(a) The area A of the rectangle is $A = xy$, where $y = 25 - x^2$. Substituting this expression for y, we obtain $A(x) = x(25 - x^2) = 25x - x^3$.
(b) Since (x, y) is in quadrant I, we have $x > 0$. Also, $y = 25 - x^2 > 0$, which implies that $x^2 < 25$, so $-5 < x < 5$. Combining these restrictions, we have the domain of A as $\{x | 0 < x < 5\}$, or $(0, 5)$ using interval notation.
(c) See Figure 60 for the graph of $A = A(x)$.
(d) Using MAXIMUM, we find that the maximum area is 48.11 square units at $x = 2.89$, each rounded to two decimal places. See Figure 61.

Figure 60

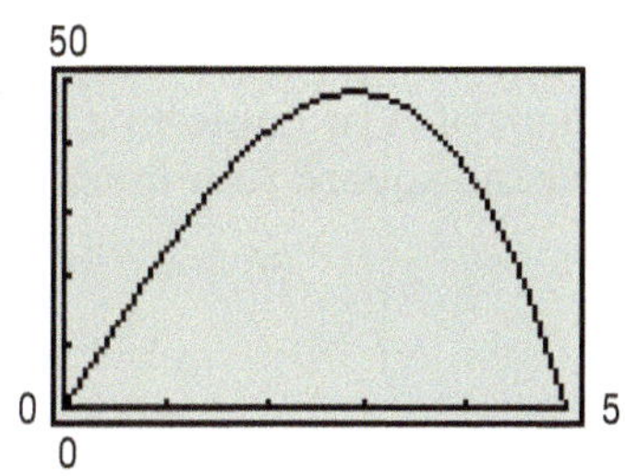

Figure 61

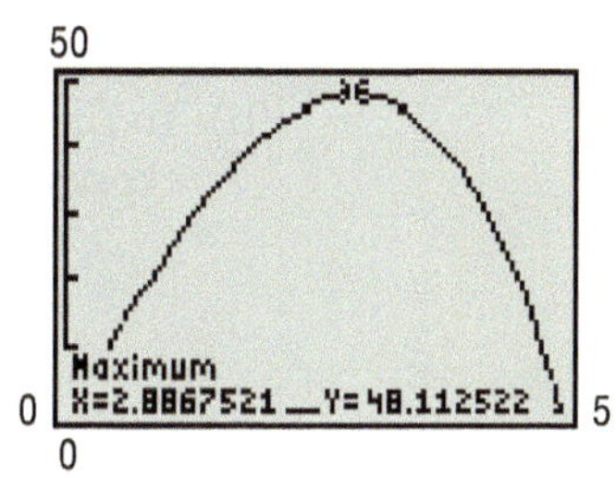

Now Work PROBLEM 7

EXAMPLE 3 Making a Playpen*

A manufacturer of children's playpens makes a square model that can be opened at one corner and attached at right angles to a wall or, perhaps, the side of a house. If each side is 3 feet in length, the open configuration doubles the available area in which the child can play from 9 square feet to 18 square feet. See Figure 62.

Now suppose that we place hinges at the outer corners to allow for a configuration like the one shown in Figure 63.

Figure 62

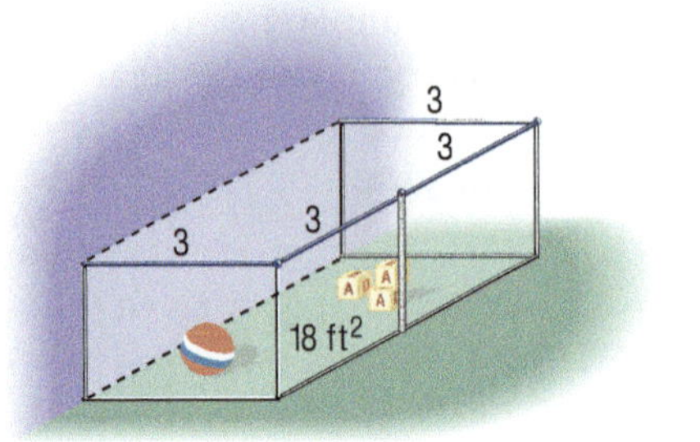

Figure 63

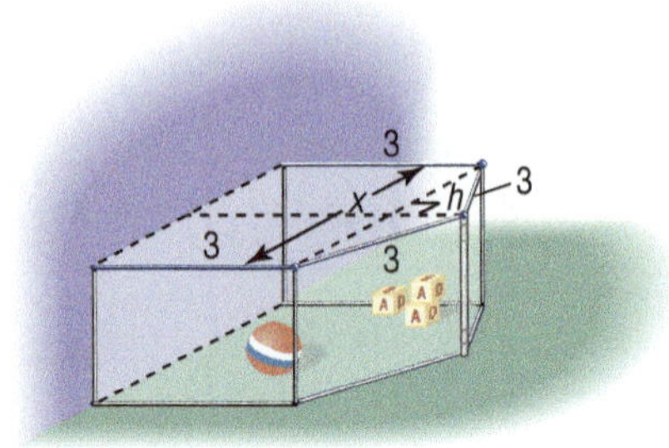

(a) Build a model that expresses the area A of the configuration shown in Figure 63 as a function of the distance x between the two parallel sides.

(b) Find the domain of A.

(c) Find A if $x = 5$.

(d) Graph $A = A(x)$. For what value of x is the area largest? What is the maximum area?

Solution (a) Refer to Figure 63. The area A that we seek consists of the area of a rectangle (with width 3 and length x) and the area of an isosceles triangle (with base x and two equal sides of length 3). The height h of the triangle may be found using the Pythagorean Theorem.

$$h^2 + \left(\frac{x}{2}\right)^2 = 3^2$$

$$h^2 = 3^2 - \left(\frac{x}{2}\right)^2 = 9 - \frac{x^2}{4} = \frac{36 - x^2}{4}$$

$$h = \frac{1}{2}\sqrt{36 - x^2}$$

The area A enclosed by the playpen is

$$A = \text{area of rectangle} + \text{area of triangle} = 3x + \frac{1}{2}x\left(\frac{1}{2}\sqrt{36 - x^2}\right)$$

The area A expressed as a function of x is

$$A(x) = 3x + \frac{x\sqrt{36 - x^2}}{4} \quad \text{The Model}$$

(b) To find the domain of A, we note first that $x > 0$, since x is a length. Also, the expression under the square root must be positive, so

$$36 - x^2 > 0$$
$$x^2 < 36$$
$$-6 < x < 6$$

Combining these restrictions, we find that the domain of A is $0 < x < 6$, or $(0, 6)$ using interval notation.

*Adapted from *Proceedings, Summer Conference for College Teachers on Applied Mathematics* (University of Missouri, Rolla), 1971.

Figure 64

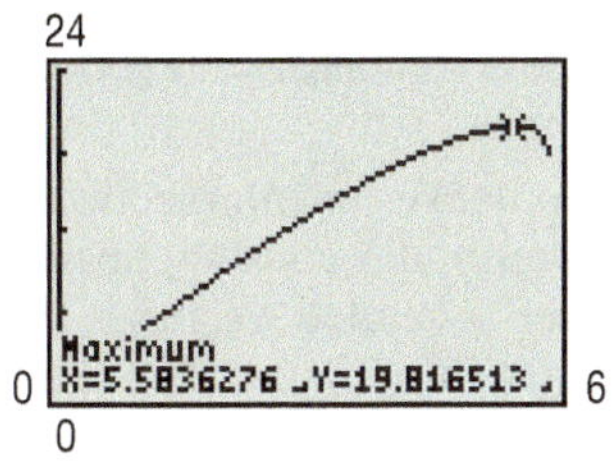

(c) If $x = 5$, the area is

$$A(5) = 3(5) + \frac{5}{4}\sqrt{36 - (5)^2} \approx 19.15 \text{ square feet}$$

If the length of the playpen is 5 feet, its area is 19.15 square feet.

(d) See Figure 64. The maximum area is about 19.82 square feet, obtained when x is about 5.58 feet.

3.6 Assess Your Understanding

Applications and Extensions

1. Let $P = (x, y)$ be a point on the graph of $y = x^2 - 8$.
   (a) Express the distance d from P to the origin as a function of x.
   (b) What is d if $x = 0$?
   (c) What is d if $x = 1$?
   (d) Use a graphing utility to graph $d = d(x)$.
   (e) For what values of x is d smallest?

2. Let $P = (x, y)$ be a point on the graph of $y = x^2 - 8$.
   (a) Express the distance d from P to the point $(0, -1)$ as a function of x.
   (b) What is d if $x = 0$?
   (c) What is d if $x = -1$?
   (d) Use a graphing utility to graph $d = d(x)$.
   (e) For what values of x is d smallest?

3. Let $P = (x, y)$ be a point on the graph of $y = \sqrt{x}$.
   (a) Express the distance d from P to the point $(1, 0)$ as a function of x.
   (b) Use a graphing utility to graph $d = d(x)$.
   (c) For what values of x is d smallest?

4. Let $P = (x, y)$ be a point on the graph of $y = \frac{1}{x}$.
   (a) Express the distance d from P to the origin as a function of x.
   (b) Use a graphing utility to graph $d = d(x)$.
   (c) For what values of x is d smallest?

5. A right triangle has one vertex on the graph of $y = x^3$, $x > 0$, at (x, y), another at the origin, and the third on the positive y-axis at $(0, y)$, as shown in the figure. Express the area A of the triangle as a function of x.

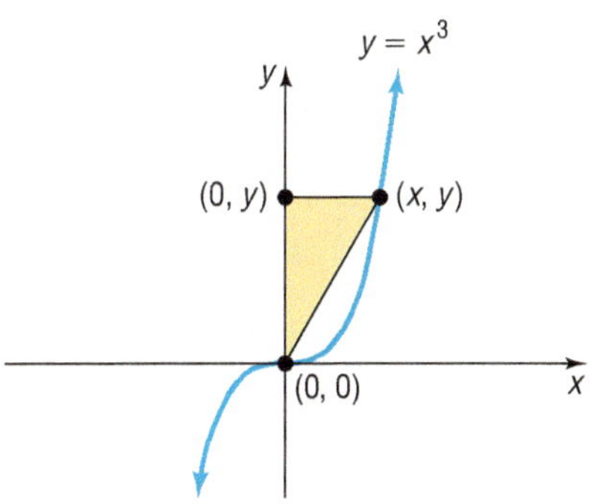

6. A right triangle has one vertex on the graph of $y = 9 - x^2$, $x > 0$, at (x, y), another at the origin, and the third on the positive x-axis at $(x, 0)$. Express the area A of the triangle as a function of x.

7. A rectangle has one corner in quadrant I on the graph of $y = 16 - x^2$, another at the origin, a third on the positive y-axis, and the fourth on the positive x-axis. See the figure.

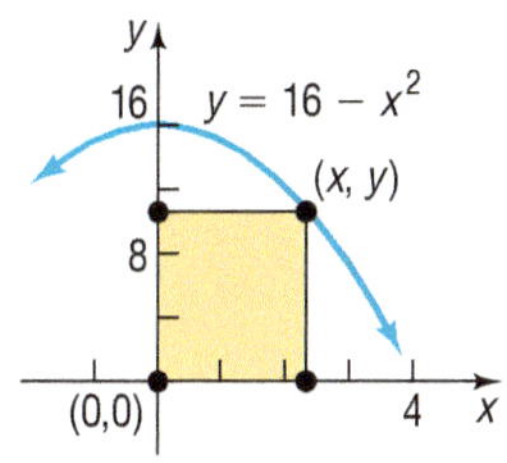

   (a) Express the area A of the rectangle as a function of x.
   (b) What is the domain of A?
   (c) Graph $A = A(x)$. For what value of x is A largest?

8. A rectangle is inscribed in a semicircle of radius 2. See the figure. Let $P = (x, y)$ be the point in quadrant I that is a vertex of the rectangle and is on the circle.

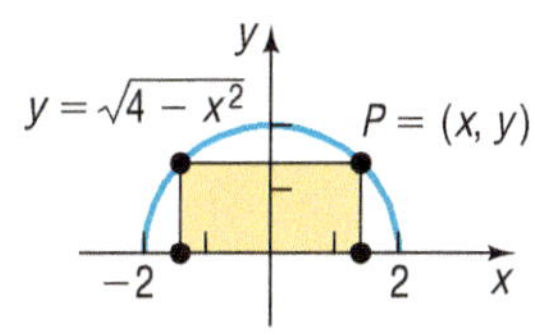

   (a) Express the area A of the rectangle as a function of x.
   (b) Express the perimeter p of the rectangle as a function of x.
   (c) Graph $A = A(x)$. For what value of x is A largest?
   (d) Graph $p = p(x)$. For what value of x is p largest?

9. A rectangle is inscribed in a circle of radius 2. See the figure. Let $P = (x, y)$ be the point in quadrant I that is a vertex of the rectangle and is on the circle.

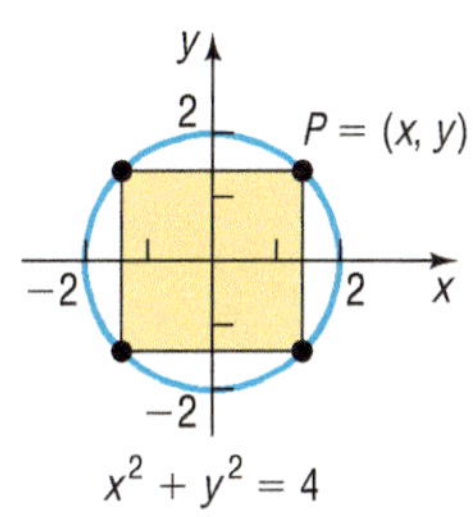

   (a) Express the area A of the rectangle as a function of x.
   (b) Express the perimeter p of the rectangle as a function of x.
   (c) Graph $A = A(x)$. For what value of x is A largest?
   (d) Graph $p = p(x)$. For what value of x is p largest?

10. A circle of radius r is inscribed in a square. See the figure.

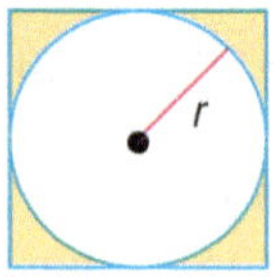

(a) Express the area A of the square as a function of the radius r of the circle.
(b) Express the perimeter p of the square as a function of r.

11. Geometry A wire 10 meters long is to be cut into two pieces. One piece will be shaped as a square, and the other piece will be shaped as a circle. See the figure.

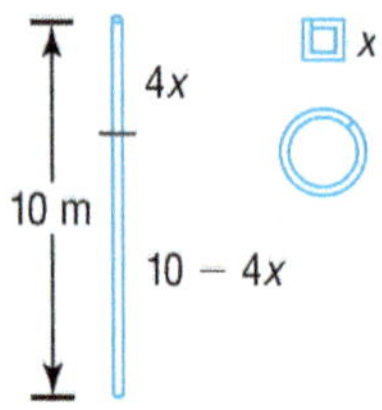

(a) Express the total area A enclosed by the pieces of wire as a function of the length x of a side of the square.
(b) What is the domain of A?
(c) Graph $A = A(x)$. For what value of x is A smallest?

12. Geometry A wire 10 meters long is to be cut into two pieces. One piece will be shaped as an equilateral triangle, and the other piece will be shaped as a circle.
(a) Express the total area A enclosed by the pieces of wire as a function of the length x of a side of the equilateral triangle.
(b) What is the domain of A?
(c) Graph $A = A(x)$. For what value of x is A smallest?

13. A wire of length x is bent into the shape of a circle.
(a) Express the circumference of the circle as a function of x.
(b) Express the area of the circle as a function of x.

14. A wire of length x is bent into the shape of a square.
(a) Express the perimeter of the square as a function of x.
(b) Express the area of the square as a function of x.

15. Geometry A semicircle of radius r is inscribed in a rectangle so that the diameter of the semicircle is the length of the rectangle. See the figure.

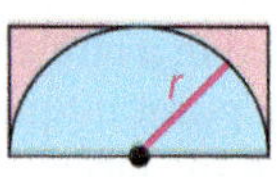

(a) Express the area A of the rectangle as a function of the radius r of the semicircle.
(b) Express the perimeter p of the rectangle as a function of r.

16. Geometry An equilateral triangle is inscribed in a circle of radius r. See the figure. Express the circumference C of the circle as a function of the length x of a side of the triangle.

[**Hint:** First show that $r^2 = \frac{x^2}{3}$.]

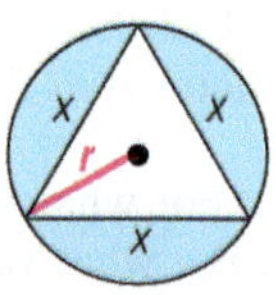

17. Geometry An equilateral triangle is inscribed in a circle of radius r. See the figure in Problem 16. Express the area A within the circle, but outside the triangle, as a function of the length x of a side of the triangle.

18. Uniform Motion Two cars leave an intersection at the same time. One is headed south at a constant speed of 30 miles per hour, and the other is headed west at a constant speed of 40 miles per hour (see the figure). Build a model that expresses the distance d between the cars as a function of the time t.

[**Hint:** At $t = 0$, the cars leave the intersection.]

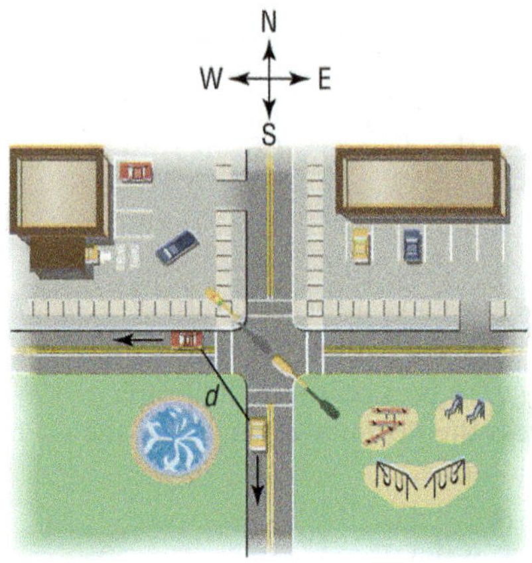

19. Uniform Motion Two cars are approaching an intersection. One is 2 miles south of the intersection and is moving at a constant speed of 30 miles per hour. At the same time, the other car is 3 miles east of the intersection and is moving at a constant speed of 40 miles per hour.
(a) Build a model that expresses the distance d between the cars as a function of time t.

[**Hint:** At $t = 0$, the cars are 2 miles south and 3 miles east of the intersection, respectively.]

(b) Use a graphing utility to graph $d = d(t)$. For what value of t is d smallest?

20. Inscribing a Cylinder in a Sphere Inscribe a right circular cylinder of height h and radius r in a sphere of fixed radius R. See the illustration. Express the volume V of the cylinder as a function of h.

[**Hint:** $V = \pi r^2 h$. Note also the right triangle.]

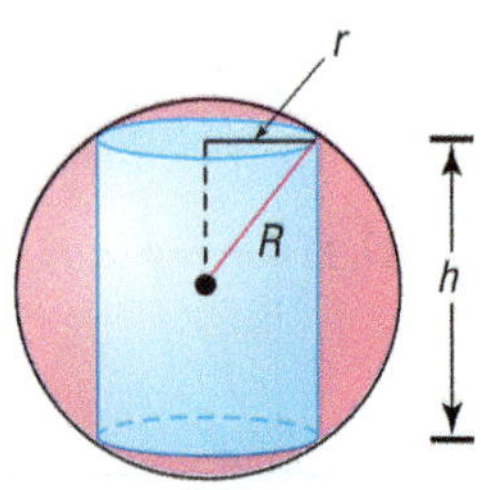

Sphere

21. Inscribing a Cylinder in a Cone Inscribe a right circular cylinder of height h and radius r in a cone of fixed radius R and fixed height H. See the illustration. Express the volume V of the cylinder as a function of r.

[**Hint:** $V = \pi r^2 h$. Note also the similar triangles.]

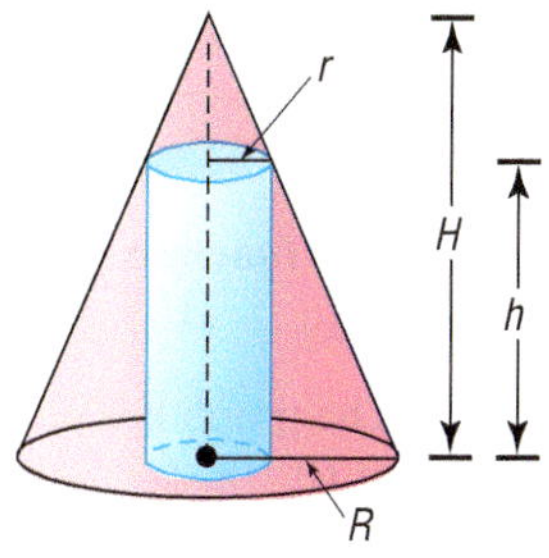

Cone

22. Installing Cable TV MetroMedia Cable is asked to provide service to a customer whose house is located 2 miles from the road along which the cable is buried. The nearest connection box for the cable is located 5 miles down the road. See the figure.

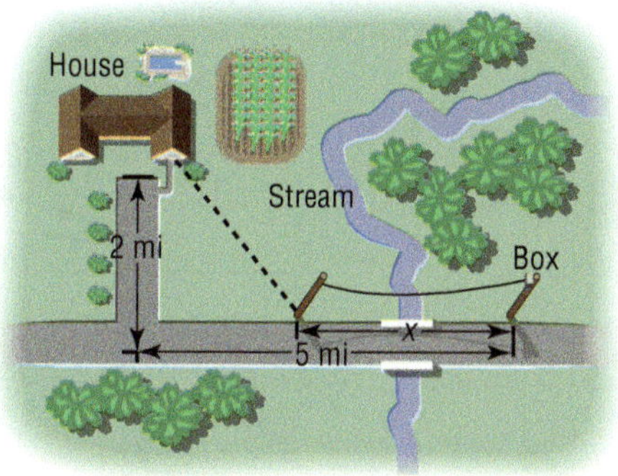

(a) If the installation cost is \$500 per mile along the road and \$700 per mile off the road, build a model that expresses the total cost C of installation as a function of the distance x (in miles) from the connection box to the point where the cable installation turns off the road. Give the domain.
(b) Compute the cost if $x = 1$ mile.
(c) Compute the cost if $x = 3$ miles.
(d) Graph the function $C = C(x)$. Use TRACE to see how the cost C varies as x changes from 0 to 5.
(e) What value of x results in the least cost?

23. Time Required to Go from an Island to a Town An island is 2 miles from the nearest point P on a straight shoreline. A town is 12 miles down the shore from P. See the illustration.

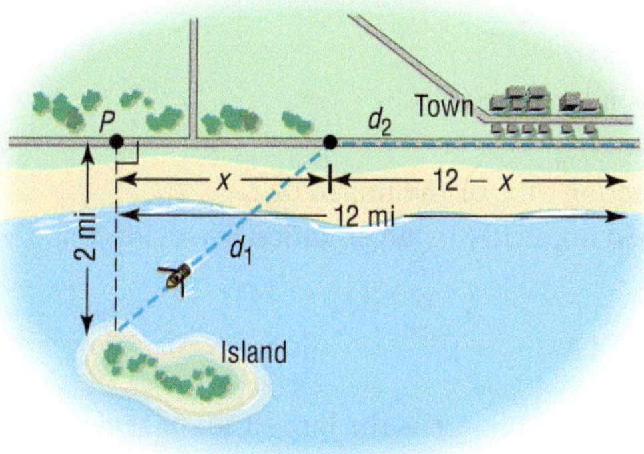

(a) If a person can row a boat at an average speed of 3 miles per hour and the same person can walk 5 miles per hour, build a model that expresses the time T that it takes to go from the island to town as a function of the distance x from P to where the person lands the boat.
(b) What is the domain of T?
(c) How long will it take to travel from the island to town if the person lands the boat 4 miles from P?
(d) How long will it take if the person lands the boat 8 miles from P?

24. Filling a Conical Tank Water is poured into a container in the shape of a right circular cone with radius 4 feet and height 16 feet. See the figure. Express the volume V of the water in the cone as a function of the height h of the water.

[**Hint:** The volume V of a cone of radius r and height h is $V = \frac{1}{3}\pi r^2 h$.]

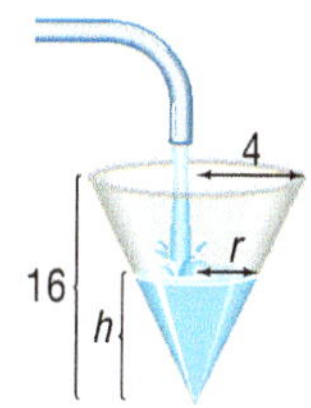

25. Constructing an Open Box An open box with a square base is to be made from a square piece of cardboard 24 inches on a side by cutting out a square from each corner and turning up the sides. See the figure.

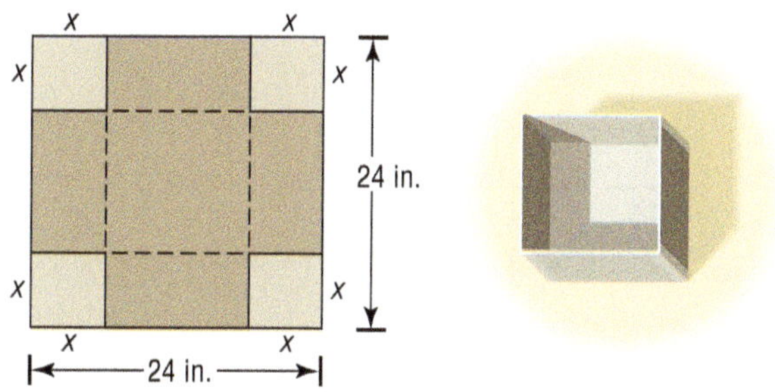

(a) Express the volume V of the box as a function of the length x of the side of the square cut from each corner.
(b) What is the volume if a 3-inch square is cut out?
(c) What is the volume if a 10-inch square is cut out?
(d) Graph $V = V(x)$. For what value of x is V largest?

26. Constructing an Open Box An open box with a square base is required to have a volume of 10 cubic feet.
(a) Express the amount A of material used to make such a box as a function of the length x of a side of the square base.
(b) How much material is required for a base 1 foot by 1 foot?
(c) How much material is required for a base 2 feet by 2 feet?
(d) Use a graphing utility to graph $A = A(x)$. For what value of x is A smallest?

CHAPTER REVIEW

Library of Functions

Constant function (p. 248)

$f(x) = b$

The graph is a horizontal line with y-intercept b.

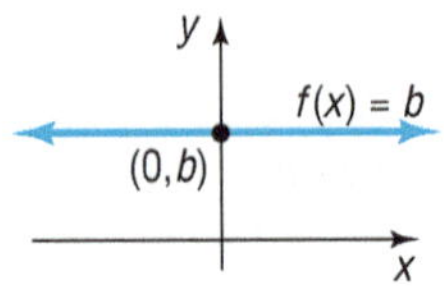

Identity function (p. 248)

$f(x) = x$

The graph is a line with slope 1 and y-intercept 0.

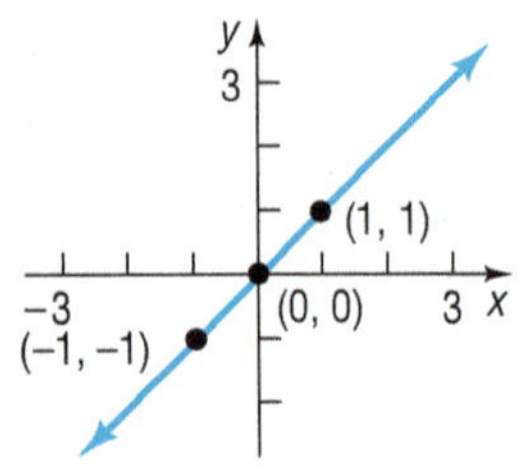

Square function (p. 249)

$f(x) = x^2$

The graph is a parabola with intercept at (0, 0).

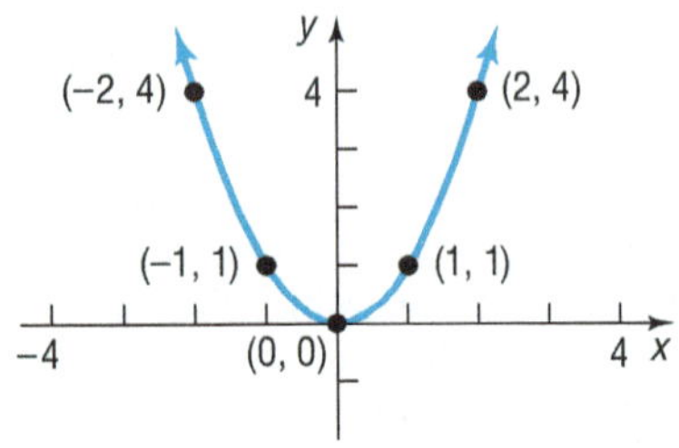

Cube function (p. 249)

$f(x) = x^3$

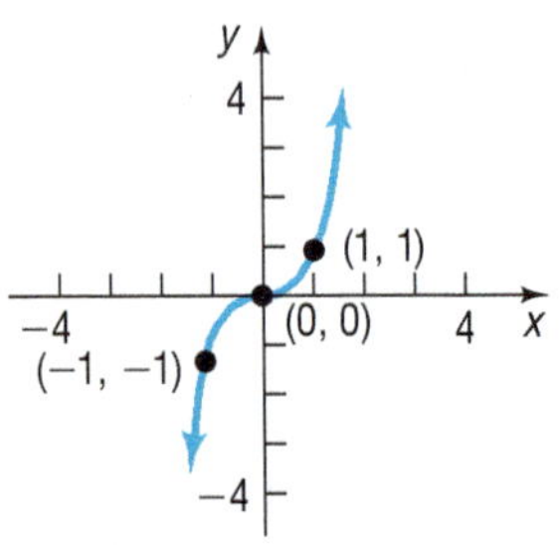

Square root function (p. 249)

$f(x) = \sqrt{x}$

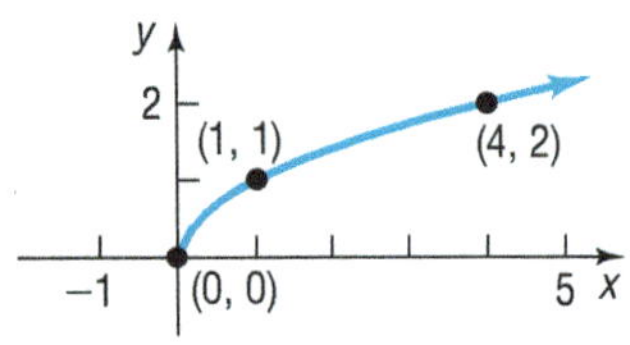

Cube root function (p. 249)

$f(x) = \sqrt[3]{x}$

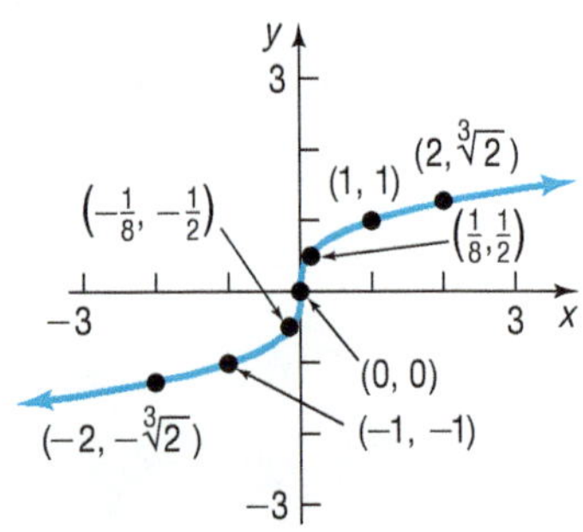

Reciprocal function (p. 249)

$f(x) = \dfrac{1}{x}$

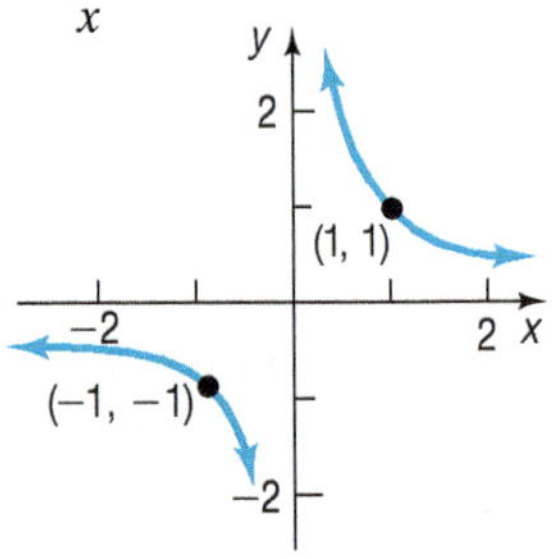

Absolute value function (pp. 250)

$f(x) = |x|$

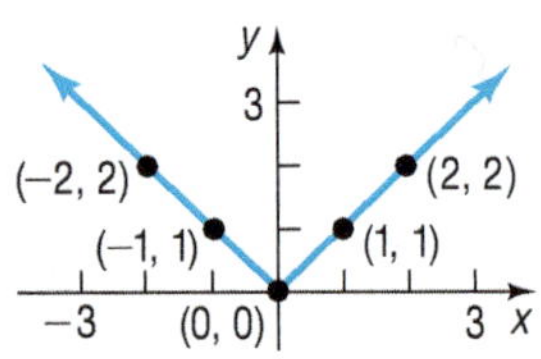

Greatest integer function (p. 250)

$f(x) = \text{int}(x)$

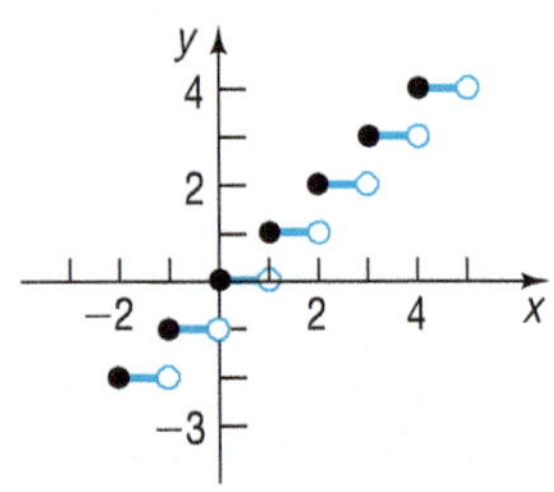

Things to Know

Function (pp. 212–216)

A relation between two sets so that each element x in the first set, the domain, has corresponding to it exactly one element y in the second set. The range is the set of y values of the function for the x values in the domain.

A function can also be characterized as a set of ordered pairs (x, y) in which no first element is paired with two different second elements.

Function notation (pp. 216–219)

$y = f(x)$

f is a symbol for the function.

x is the argument, or independent variable.

y is the dependent variable.

$f(x)$ is the value of the function at x, or the image of x.

A function f may be defined implicitly by an equation involving x and y or explicitly by writing $y = f(x)$.

Difference quotient of f (pp. 218 and 245)

$$\frac{f(x + h) - f(x)}{h} \qquad h \neq 0$$

Domain (pp. 213 and 219)

If unspecified, the domain of a function f is the largest set of real numbers for which $f(x)$ is a real number.

Vertical-line test (p. 227)	A set of points in the plane is the graph of a function if and only if every vertical line intersects the graph in at most one point.
Even function f (p. 235)	$f(-x) = f(x)$ for every x in the domain ($-x$ must also be in the domain).
Odd function f (p. 236)	$f(-x) = -f(x)$ for every x in the domain ($-x$ must also be in the domain).
Increasing function (p. 238)	A function f is increasing on an open interval I if, for any choice of x_1 and x_2 in I, with $x_1 < x_2$, we have $f(x_1) < f(x_2)$.
Decreasing function (p. 238)	A function f is decreasing on an open interval I if, for any choice of x_1 and x_2 in I, with $x_1 < x_2$, we have $f(x_1) > f(x_2)$.
Constant function (p. 238)	A function f is constant on an open interval I if, for all choices of x in I, the values of $f(x)$ are equal.
Local maximum (p. 239)	A function f has a local maximum at c if there is an open interval I containing c so that, for all x not equal to c in I, $f(x) < f(c)$.
Local minimum (p. 239)	A function f has a local minimum at c if there is an open interval I containing c so that, for all x not equal to c in I, $f(x) > f(c)$.
Average rate of change of a function (p. 240)	The average rate of change of f from a to b is $$\frac{\Delta y}{\Delta x} = \frac{f(b) - f(a)}{b - a} \qquad a \neq b$$

Objectives

Section		You should be able to . . .	Examples	Review Exercises
3.1	1	Determine whether a relation represents a function (p. 212)	1–5	1, 2
	2	Find the value of a function (p. 216)	6–8	3–8, 23, 24, 71, 72
	3	Find the domain of a function defined by an equation (p. 219)	9, 10	9–16
	4	Form the sum, difference, product, and quotient of two functions (p. 221)	11	17–22
3.2	1	Identify the graph of a function (p. 227)	1	47–50
	2	Obtain information from or about the graph of a function (p. 228)	2–4	25(a)–(e), 26(a)–(e), 27(a), 27(d), 27(f), 28(a), 28(d), 28(f)
3.3	1	Determine even and odd functions from a graph (p. 235)	1	27(e), 28(e)
	2	Identify even and odd functions from the equation (p. 236)	2	29–36
	3	Use a graph to determine where a function is increasing, decreasing, or constant (p. 237)	3	27(b), 28(b)
	4	Use a graph to locate local maxima and local minima (p. 238)	4	27(c), 28(c)
	5	Use a graphing utility to approximate local maxima and local minima and to determine where a function is increasing or decreasing (p. 239)	5	37–40, 74(d), 75(b)
	6	Find the average rate of change of a function (p. 240)	6, 7	41–46
3.4	1	Graph the functions listed in the library of functions (p. 246)	1, 2	51–54
	2	Graph piecewise-defined functions (p. 251)	3, 4	67–70
3.5	1	Graph functions using vertical and horizontal shifts (p. 256)	1, 2	25(f), 26(f), 26(g) 55, 56, 59–66
	2	Graph functions using compressions and stretches (p. 258)	3	25(g), 26(h), 57, 58, 65, 66
	3	Graph functions using reflections about the x-axis or y-axis (p. 260)	4–6	25(h), 57, 61, 62, 66
3.6	1	Build and analyze functions (p. 268)	1–3	73–75

Review Exercises

In Problems 1 and 2, determine whether each relation represents a function. For each function, state the domain and the range.

1. $\{(-1, 0), (2, 3), (4, 0)\}$ **2.** $\{(4, -1), (2, 1), (4, 2)\}$

In Problems 3–8, find the following for each function:

(a) $f(2)$ *(b)* $f(-2)$ *(c)* $f(-x)$ *(d)* $-f(x)$ *(e)* $f(x - 2)$ *(f)* $f(2x)$

3. $f(x) = \dfrac{3x}{x^2 - 1}$ **4.** $f(x) = \dfrac{x^2}{x + 1}$ **5.** $f(x) = \sqrt{x^2 - 4}$

6. $f(x) = |x^2 - 4|$ **7.** $f(x) = \dfrac{x^2 - 4}{x^2}$ **8.** $f(x) = \dfrac{x^3}{x^2 - 9}$

In Problems 9–16, find the domain of each function.

9. $f(x) = \dfrac{x}{x^2 - 9}$ **10.** $f(x) = \dfrac{3x^2}{x - 2}$ **11.** $f(x) = \sqrt{2 - x}$ **12.** $f(x) = \sqrt{x + 2}$

13. $h(x) = \dfrac{\sqrt{x}}{|x|}$ **14.** $g(x) = \dfrac{|x|}{x}$ **15.** $f(x) = \dfrac{x}{x^2 + 2x - 3}$ **16.** $F(x) = \dfrac{1}{x^2 - 3x - 4}$

In Problems 17–22, find $f + g$, $f - g$, $f \cdot g$, and $\dfrac{f}{g}$ for each pair of functions. State the domain of each.

17. $f(x) = 2 - x$; $g(x) = 3x + 1$ **18.** $f(x) = 2x - 1$; $g(x) = 2x + 1$ **19.** $f(x) = 3x^2 + x + 1$; $g(x) = 3x$

20. $f(x) = 3x$; $g(x) = 1 + x + x^2$ **21.** $f(x) = \dfrac{x + 1}{x - 1}$; $g(x) = \dfrac{1}{x}$ **22.** $f(x) = \dfrac{1}{x - 3}$; $g(x) = \dfrac{3}{x}$

In Problems 23 and 24, find the difference quotient of each function f; that is, find

$$\frac{f(x + h) - f(x)}{h} \qquad h \neq 0$$

23. $f(x) = -2x^2 + x + 1$ **24.** $f(x) = 3x^2 - 2x + 4$

25. Using the graph of the function f shown:

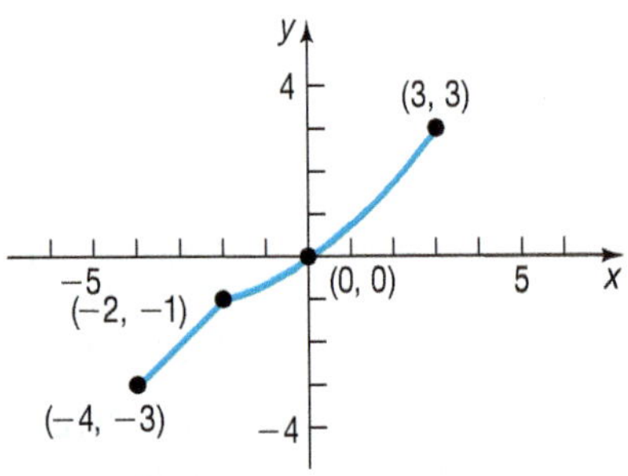

(a) Find the domain and the range of f.
(b) List the intercepts.
(c) Find $f(-2)$.
(d) For what value of x does $f(x) = -3$?
(e) Solve $f(x) > 0$.
(f) Graph $y = f(x - 3)$.
(g) Graph $y = f\left(\dfrac{1}{2}x\right)$.
(h) Graph $y = -f(x)$.

26. Using the graph of the function g shown:

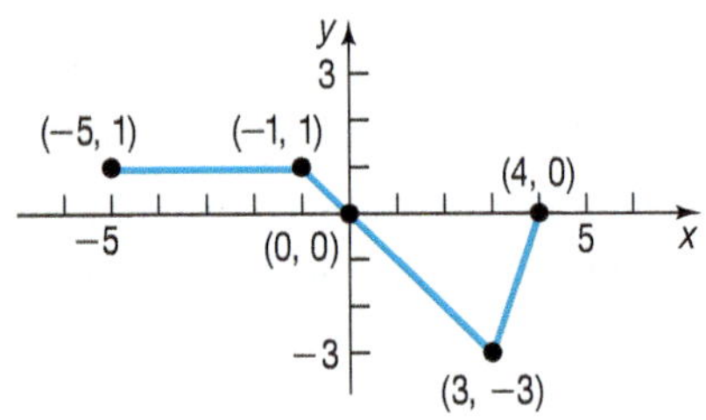

(a) Find the domain and the range of g.
(b) Find $g(-1)$.
(c) List the intercepts.
(d) For what value of x does $g(x) = -3$?
(e) Solve $g(x) > 0$.
(f) Graph $y = g(x - 2)$.
(g) Graph $y = g(x) + 1$.
(h) Graph $y = 2g(x)$.

In Problems 27 and 28, use the graph of the function f to find:
(a) The domain and the range of f.
(b) The intervals on which f is increasing, decreasing, or constant.
(c) The local minima and local maxima.
(d) Whether the graph is symmetric with respect to the x-axis, the y-axis, or the origin.
(e) Whether the function is even, odd, or neither.
(f) The intercepts, if any.

27.

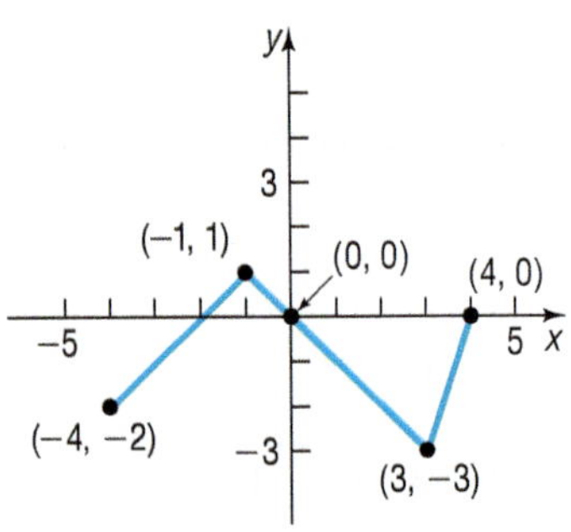

28.

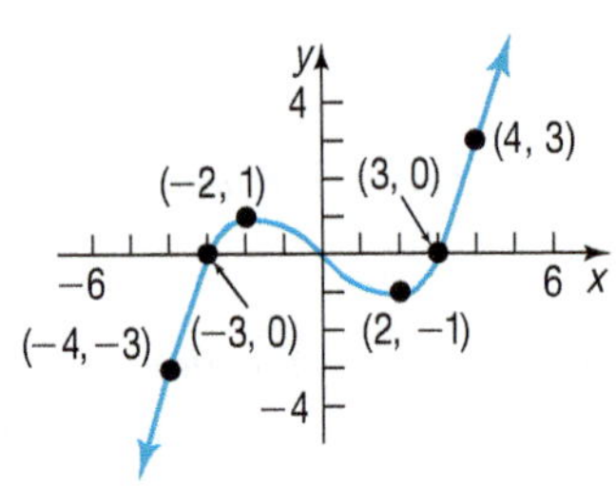

In Problems 29–36, determine (algebraically) whether the given function is even, odd, or neither.

29. $f(x) = x^3 - 4x$

30. $g(x) = \dfrac{4 + x^2}{1 + x^4}$

31. $h(x) = \dfrac{1}{x^4} + \dfrac{1}{x^2} + 1$

32. $F(x) = \sqrt{1 - x^3}$

33. $G(x) = 1 - x + x^3$

34. $H(x) = 1 + x + x^2$

35. $f(x) = \dfrac{x}{1 + x^2}$

36. $g(x) = \dfrac{1 + x^2}{x^3}$

In Problems 37–40, use a graphing utility to graph each function over the indicated interval. Approximate any local maxima and local minima. Determine where the function is increasing and where it is decreasing.

37. $f(x) = 2x^3 - 5x + 1 \quad (-3, 3)$

38. $f(x) = -x^3 + 3x - 5 \quad (-3, 3)$

39. $f(x) = 2x^4 - 5x^3 + 2x + 1 \quad (-2, 3)$

40. $f(x) = -x^4 + 3x^3 - 4x + 3 \quad (-2, 3)$

In Problems 41 and 42, find the average rate of change of f:

(a) From 1 to 2 *(b) From 0 to 1* *(c) From 2 to 4*

41. $f(x) = 8x^2 - x$

42. $f(x) = 2x^3 + x$

In Problems 43–46, find the average rate of change from 2 to 3 for each function f. Be sure to simplify.

43. $f(x) = 2 - 5x$

44. $f(x) = 2x^2 + 7$

45. $f(x) = 3x - 4x^2$

46. $f(x) = x^2 - 3x + 2$

In Problems 47–50, is the graph shown the graph of a function?

47.

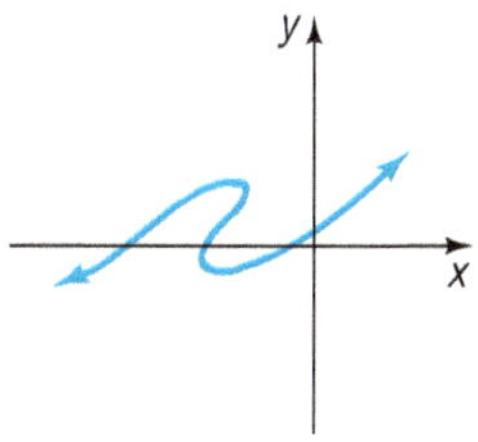

48.

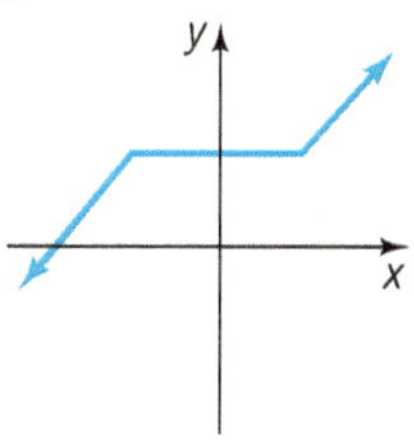

49.

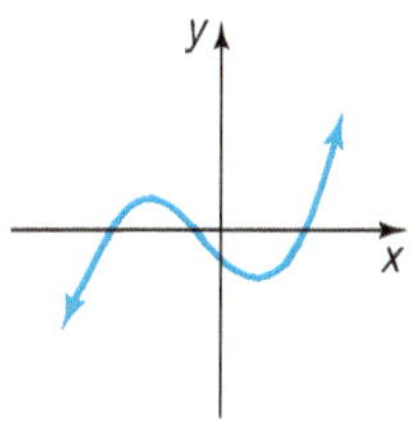

50.

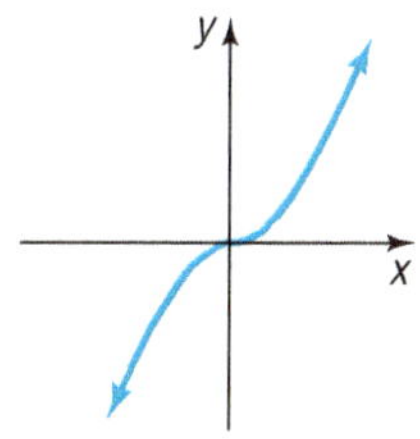

In Problems 51–54, sketch the graph of each function. Be sure to label at least three points.

51. $f(x) = |x|$

52. $f(x) = \sqrt[3]{x}$

53. $f(x) = \sqrt{x}$

54. $f(x) = \dfrac{1}{x}$

In Problems 55–66, graph each function using the techniques of shifting, compressing or stretching, and reflections. Identify any intercepts on the graph. State the domain and, based on the graph, find the range.

55. $F(x) = |x| - 4$

56. $f(x) = |x| + 4$

57. $g(x) = -2|x|$

58. $g(x) = \dfrac{1}{2}|x|$

59. $h(x) = \sqrt{x - 1}$

60. $h(x) = \sqrt{x} - 1$

61. $f(x) = \sqrt{1 - x}$

62. $f(x) = -\sqrt{x + 3}$

63. $h(x) = (x - 1)^2 + 2$

64. $h(x) = (x + 2)^2 - 3$

65. $g(x) = 3(x - 1)^3 + 1$

66. $g(x) = -2(x + 2)^3 - 8$

In Problems 67–70,

(a) Find the domain of each function.
(b) Locate any intercepts.
(c) Graph each function.
(d) Based on the graph, find the range.
(e) Is f continuous on its domain?

67. $f(x) = \begin{cases} 3x & \text{if } -2 < x \le 1 \\ x + 1 & \text{if } x > 1 \end{cases}$

68. $f(x) = \begin{cases} x - 1 & \text{if } -3 < x < 0 \\ 3x - 1 & \text{if } x \ge 0 \end{cases}$

69. $f(x) = \begin{cases} x & \text{if } -4 \le x < 0 \\ 1 & \text{if } x = 0 \\ 3x & \text{if } x > 0 \end{cases}$

70. $f(x) = \begin{cases} x^2 & \text{if } -2 \le x \le 2 \\ 2x - 1 & \text{if } x > 2 \end{cases}$

71. A function f is defined by

$$f(x) = \frac{Ax + 5}{6x - 2}$$

If $f(1) = 4$, find A.

72. A function g is defined by

$$g(x) = \frac{A}{x} + \frac{8}{x^2}$$

If $g(-1) = 0$, find A.

73. Page Design A page with dimensions of $8\frac{1}{2}$ inches by 11 inches has a border of uniform width x surrounding the printed matter of the page, as shown in the figure.

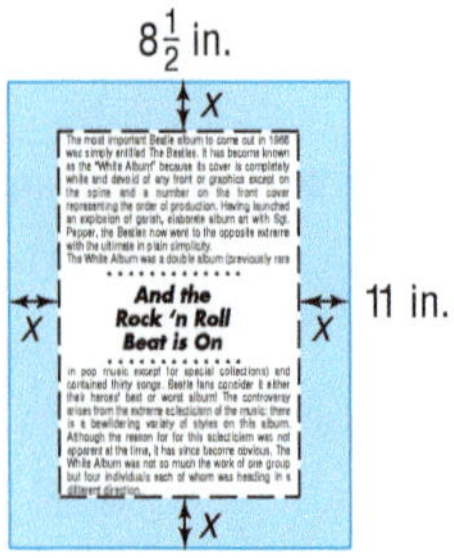

(a) Develop a model that expresses the area A of the printed part of the page as a function of the width x of the border.
(b) Give the domain and the range of A.
(c) Find the area of printed page for borders of widths 1 inch, 1.2 inches, and 1.5 inches.
(d) Graph the function $A = A(x)$.
(e) Use a graphing utility to determine what margin should be used to obtain areas of 70 square inches and 50 square inches by solving $A(x) = 70$ and $A(x) = 50$.

74. Constructing a Closed Box A closed box with a square base is required to have a volume of 10 cubic feet.
(a) Build a model that expresses the amount A of material used to make such a box as a function of the length x of a side of the square base.
(b) How much material is required for a base 1 foot by 1 foot?
(c) How much material is required for a base 2 feet by 2 feet?
(d) Graph $A = A(x)$. For what value of x is A smallest?

75. A rectangle has one vertex in quadrant I on the graph of $y = 10 - x^2$, another at the origin, one on the positive x-axis, and one on the positive y-axis.
(a) Express the area A of the rectangle as a function of x.
(b) Find the largest area A that can be enclosed by the rectangle.

CHAPTER TEST

1. Determine whether each relation represents a function. For each function, state the domain and the range.
(a) $\{(2, 5), (4, 6), (6, 7), (8, 8)\}$
(b) $\{(1, 3), (4, -2), (-3, 5), (1, 7)\}$
(c)

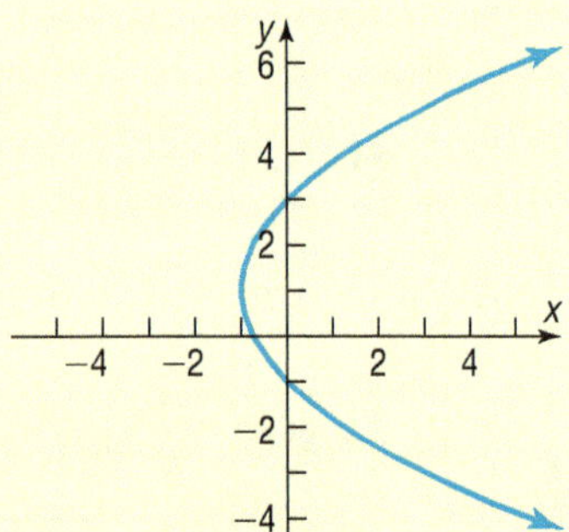

(d)

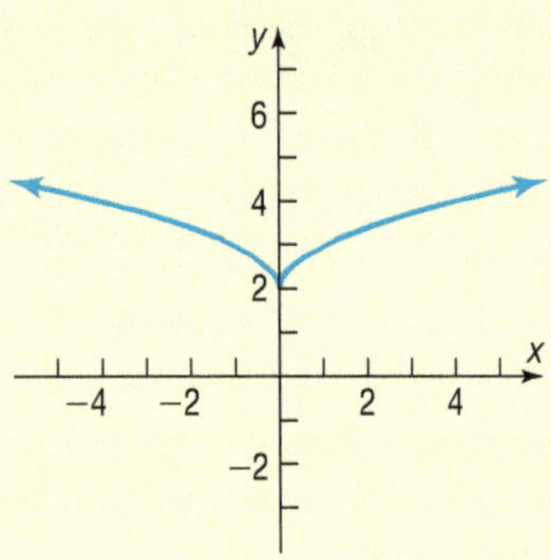

In Problems 2–4, find the domain of each function and evaluate each function at $x = -1$.

2. $f(x) = \sqrt{4 - 5x}$

3. $g(x) = \dfrac{x + 2}{|x + 2|}$

4. $h(x) = \dfrac{x - 4}{x^2 + 5x - 36}$

5. Using the graph of the function f:

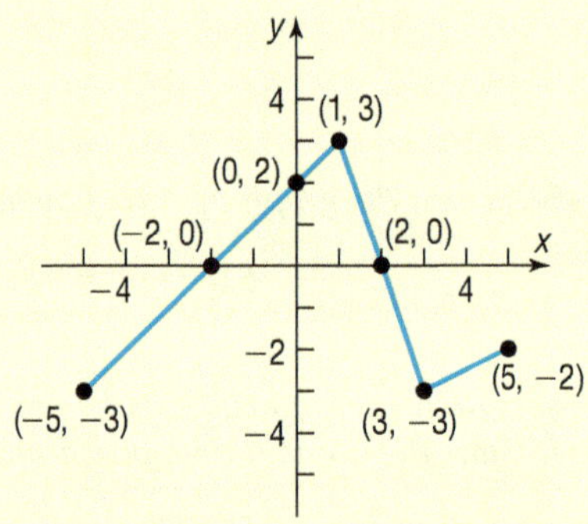

(a) Find the domain and the range of f.
(b) List the intercepts.
(c) Find $f(1)$.
(d) For what value(s) of x does $f(x) = -3$?
(e) Solve $f(x) < 0$.

6. Use a graphing utility to graph the function $f(x) = -x^4 + 2x^3 + 4x^2 - 2$ on the interval $(-5, 5)$. Approximate any local maxima and local minima rounded to two decimal places. Determine where the function is increasing and where it is decreasing.

7. Consider the function $g(x) = \begin{cases} 2x + 1 & \text{if } x < -1 \\ x - 4 & \text{if } x \geq -1 \end{cases}$
(a) Graph the function.
(b) List the intercepts.
(c) Find $g(-5)$.
(d) Find $g(2)$.

8. For the function $f(x) = 3x^2 - 2x + 4$, find the average rate of change of f from 3 to 4.

9. For the functions $f(x) = 2x^2 + 1$ and $g(x) = 3x - 2$, find the following and simplify:
(a) $f - g$
(b) $f \cdot g$
(c) $f(x + h) - f(x)$

10. Graph each function using the techniques of shifting, compressing or stretching, and reflections. Start with the graph of the basic function and show all stages.

(a) $h(x) = -2(x + 1)^3 + 3$

(b) $g(x) = |x + 4| + 2$

11. The variable interest rate on a student loan changes each July 1 based on the bank prime loan rate. For the years 1992–2007, this rate can be approximated by the model $r(x) = -0.115x^2 + 1.183x + 5.623$, where x is the number of years since 1992 and r is the interest rate as a percent.

(a) Use a graphing utility to estimate the highest rate during this time period. During which year was the interest rate the highest?

(b) Use the model to estimate the rate in 2010. Does this value seem reasonable?

Source: U.S. Federal Reserve

12. A community skating rink is in the shape of a rectangle with semicircles attached at the ends. The length of the rectangle is 20 feet less than twice the width. The thickness of the ice is 0.75 inch.

(a) Build a model that expresses the ice volume, V, as a function of the width, x.

(b) How much ice is in the rink if the width is 90 feet?

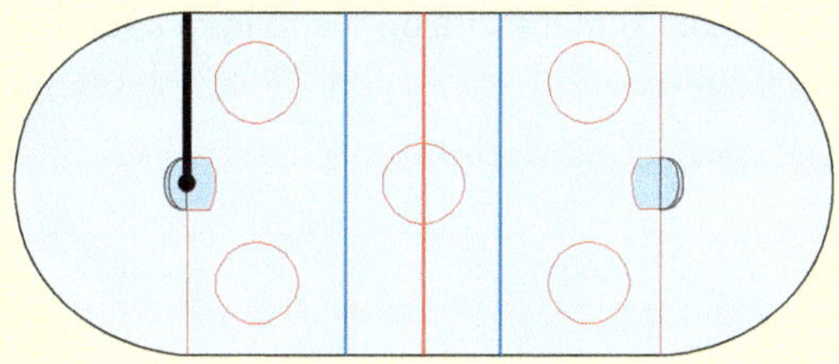

CUMULATIVE REVIEW

In Problems 1–8, find the real solutions of each equation.

1. $-5x + 4 = 0$

2. $x^2 - 7x + 12 = 0$

3. $3x^2 - 5x - 2 = 0$

4. $4x^2 + 4x + 1 = 0$

5. $4x^2 - 2x + 4 = 0$

6. $\sqrt[3]{1 - x} = 2$

7. $\sqrt[5]{1 - x} = 2$

8. $|2 - 3x| = 1$

9. In the complex number system, solve $4x^2 - 2x + 4 = 0$.

10. Solve the inequality $-2 < 3x - 5 < 7$. Graph the solution set.

In Problems 11–14, graph each equation.

11. $-3x + 4y = 12$

12. $y = 3x + 12$

13. $x^2 + y^2 + 2x - 4y + 4 = 0$

14. $y = (x + 1)^2 - 3$

15. For the graph of the function f:

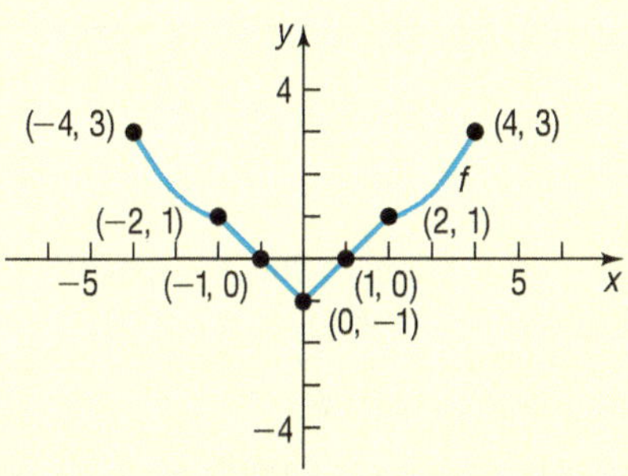

(a) Find the domain and the range of f.

(b) Find the intercepts.

(c) Is the graph of f symmetric with respect to the x-axis, the y-axis, or the origin?

(d) Find $f(2)$.

(e) For what value(s) of x is $f(x) = 3$?

(f) Solve $f(x) < 0$.

(g) Graph $y = f(x) + 2$

(h) Graph $y = f(-x)$.

(i) Graph $y = 2f(x)$.

(j) Is f even, odd, or neither?

(k) Find the interval(s) on which f is increasing.

(l) Find the interval(s) on which f is decreasing.

(m) Find the local maxima and local minima.

(n) Find the average rate of change of f from 1 to 4.

16. Find the distance between the points $P = (-1, 3)$ and $Q = (4, -2)$.

17. Which of the following points are on the graph of $y = x^3 - 3x + 1$?

(a) $(-2, -1)$ (b) $(2, 3)$ (c) $(3, 1)$

18. Determine the intercepts of $y = 3x^2 + 14x - 5$.

19. Use a graphing utility to find the solution(s) of the equation $x^4 - 3x^3 + 4x - 1 = 0$.

[**Hint:** All solutions are between $x = -10$ and $x = 10$.]

20. Find the equation of the line perpendicular to the line $y = 2x + 1$ and containing the point $(3, 5)$. Express your answer in slope–intercept form and graph both lines.

21. Determine whether the following relation represents a function: $\{(-3, 8), (1, 3), (2, 5), (3, 8)\}$.

22. For the function f defined by $f(x) = x^2 - 4x + 1$, evaluate:

(a) $f(2)$

(b) $f(x) + f(2)$

(c) $f(-x)$

(d) $-f(x)$

(e) $f(x + 2)$

(f) $\dfrac{f(x + h) - f(x)}{h}$ $\quad h \neq 0$

23. Find the domain of $h(z) = \dfrac{3z - 1}{z^2 - 6z - 7}$.

24. Determine whether the following graph is the graph of a function:

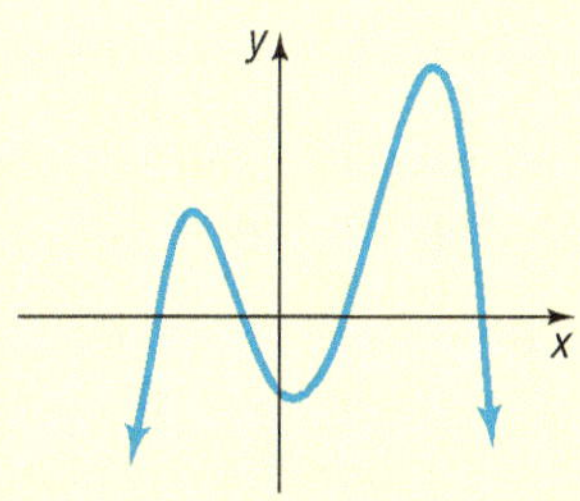

25. Consider the function $f(x) = \dfrac{x}{x + 4}$.

(a) Is the point $\left(1, \dfrac{1}{4}\right)$ on the graph of f?

(b) If $x = -2$, what is $f(x)$? What point is on the graph of f?

(c) If $f(x) = 2$, what is x? What point is on the graph of f?

26. Factor: $6x^2(x^2 + 1)^{1/2} + 2(x^2 + 1)^{3/2}$

27. Factor the expression and write it with only positive exponents: $2(x + 5)^{1/2} + x(x + 5)^{-1/2}$

CHAPTER PROJECTS

I. Cell Phone Service This month, as you were paying your bills, you noticed that your cell phone service contract had expired. Since cell phone numbers are now portable, you decided to investigate different companies. Since you regularly travel outside your local calling area, you decide to look only at plans with no roaming charges. Here are your findings:

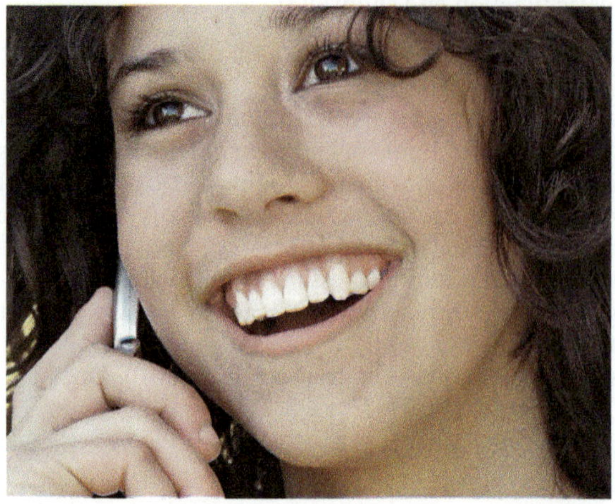

	Anytime Minutes Included	Charge for Each Extra Minute	Mobile-to-Mobile Minutes	National Long Distance	Nights and Weekends
Company A:					
Plan A1: $39.99	450 with rollover	$0.45	Unlimited	Included	5000 (9 PM–6 AM)
Plan A2: $59.99	900 with rollover	$0.40	Unlimited	Included	Unlimited (9 PM–6 AM)
Company B:					
Plan B1: $39.99	600	$0.40	None	Included	Unlimited (9 PM–7 AM)
Plan B2: $49.99	1000	$0.40	None	Included	Unlimited (9 PM–7 AM)
Company C:					
Plan C1: $59.99	550	551–1050, $5 for each 50 minutes; above 1050, $0.10 each minute	Unlimited	Included	Unlimited (7 PM–7 AM)
Plan C2: $69.99	800	801–1300, $5 for each 50 minutes; above 1300, $0.10 each minute	Unlimited	Included	Unlimited (7 PM–7 AM)

Source: Based on rates from the Web sites of the companies Cingular, T Mobile, and Sprint PCS for the zip code 76201 on May 20, 2006. (*www.cingular.com, www.t-mobile.com, www.sprintpcs.com*)

Each plan requires a 2-year contract.

1. Determine the total cost of each plan for the life of the contract, assuming that you stay within the allotted anytime minutes provided by each contract.
2. If you expect to use 400 anytime, 200 mobile-to-mobile, and 4500 night and weekend minutes per month, which plan provides the best deal? If you expect to use 400 anytime, 200 mobile-to-mobile, and 5500 night and weekend minutes per month, which plan provides the best deal? If you expect to use 500 anytime, and 1000 mobile-to-mobile, and 2000 night and weekend minutes per month, which plan provides the best deal?
3. Ignoring any mobile-to-mobile and night and weekend usage, if you expect to use 850 anytime minutes each month, which option provides the best deal? What if you use 1050 anytime minutes per month?
4. Each monthly charge includes a specific number of peak time minutes in the monthly fee. Write a function for each option, where C is the monthly cost and x is the number of anytime minutes used.
5. Graph each function from part 4.
6. For each of the companies A, B, and C, determine the average price per minute for each plan, based on no extra minutes used. For each company, which plan is better?
7. Looking at the three plans that you found to be the best for companies A, B, and C in part 6, which of these three seems to be the best deal?
8. Based on your own cell phone usage, which plan would be the best for you?

The following projects are available on the Instructor's Resource Center (IRC):

II. Project at Motorola: ***Wireless Internet Service*** Use functions and their graphs to analyze the total cost of various wireless Internet service plans.

III. Cost of Cable When government regulations and customer preference influence the path of a new cable line, the Pythagorean Theorem can be used to assess the cost of installation.

IV. Oil Spill Functions are used to analyze the size and spread of an oil spill from a leaking tanker.

Linear and Quadratic Functions

4

Valuing Hewlett-Packard

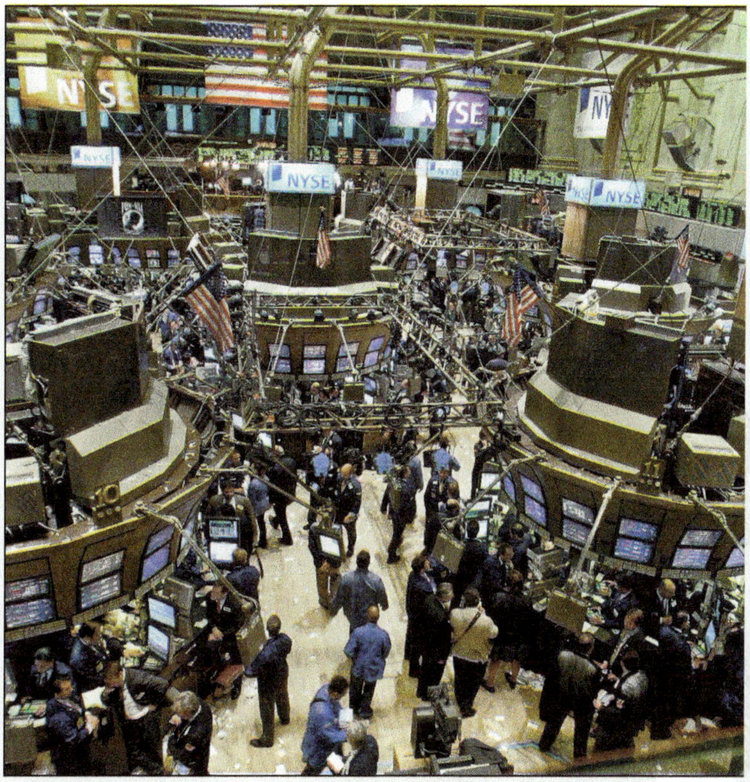

17 MAY 2006—With the market focused on Hewlett-Packard Co. (HPQ) taking market share from Dell Inc. (DELL), the casual observer might be fooled into mistaking it as a computer hardware company. It's not; while the Compaq acquisition of 2002 expanded the company's hardware business, 53.4 percent of HPQ's profits still come from its imaging and printing division.

When we value stocks, we use the capital asset pricing model to price shares, which examines the volatility of the stock through beta and then projects how much earnings growth would be needed to justify the current stock price. Historically, HPQ has a beta of 2.05. The computer peripherals industry average, though, is 1.98. But the industry average beta for the computer hardware companies is 1.58.

A weighted average beta for the two industries seems appropriate for a company that walks the line between them. In this case, a beta weighted between the computer peripherals and computer hardware industries by the percentages of HPQ's profits yields an expected beta of 1.8. If we use this beta in a capital-asset-pricing model, assuming that Hewlett-Packard will maintain a price-to-earnings-to-growth ratio of 1.3 and that full fiscal year 2006 earnings per share will be at the low end of management's guidance, \$2.04, then a new position taken at the current price would require an annualized, per-share earnings growth rate of 9.9% over the next five years to break even. The average long-term growth rate among the 12 analysts who provide one is 12.3% for Hewlett-Packard; this is well above what our analysis indicates that we require, implying a strong undervaluation even after the price spike.

Source: Adapted from *Paul DeMartino, "Valuing Hewlett-Packard," May 17, 2006, www.reuters.com.* Adapted with permission.

—See Chapter Project I—

A Look Back

Up to now, our discussion has focused on graphs of equations and the idea of a function. We have learned how to graph equations using the point-plotting method, intercepts, and the tests for symmetry. In addition, we have learned what a function is and how to identify whether a relation represents a function. We also introduced properties of functions, such as domain and range, increasing and decreasing, even and odd, and average rate of change.

A Look Ahead

Going forward, we will look at specific categories of functions. In this chapter, we focus on linear and quadratic functions, their properties and applications.

Outline

4.1 Linear Functions, Their Properties, and Linear Models

PREPARING FOR THIS SECTION *Before getting started, review the following:*

- Lines (Section 2.2, pp. 178–186)
- Solve Linear Equations (Section 1.2, pp. 105–106)
- Functions (Section 3.1, pp. 212–220)
- The Graph of a Function (Section 3.2, pp. 226–230)
- Properties of Functions (Section 3.3, pp. 235–242)

Now Work the 'Are You Prepared?' problems on page 288.

OBJECTIVES
1. Graph Linear Functions (p. 282)
2. Use Average Rate of Change to Identify Linear Functions (p. 282)
3. Determine Whether a Linear Function Is Increasing, Decreasing, or Constant (p. 285)
4. Build Linear Models from Verbal Descriptions (p. 286)

1 Graph Linear Functions

In Section 2.2, we discussed lines. In particular, we developed the slope–intercept form of the equation of a line $y = mx + b$. When we write the slope–intercept form of a line using function notation, we have a *linear function*.

DEFINITION A **linear function** is a function of the form

$$f(x) = mx + b$$

The graph of a linear function is a line with slope m and y-intercept b. Its domain is the set of all real numbers. If $m \neq 0$, the range of a linear function is the set of all real numbers. If $m = 0$, the range is $\{y \mid y = b\}$.

EXAMPLE 1 Graphing a Linear Function

Graph the linear function $f(x) = -3x + 7$. What is the domain and the range of f?

Solution This is a linear function with slope $m = -3$ and y-intercept $b = 7$. To graph this function, we start by plotting the point $(0, 7)$, the y-intercept, and use the slope to find an additional point by moving right 1 unit and down 3 units. See Figure 1. The domain and the range of f are the set of all real numbers.

Figure 1

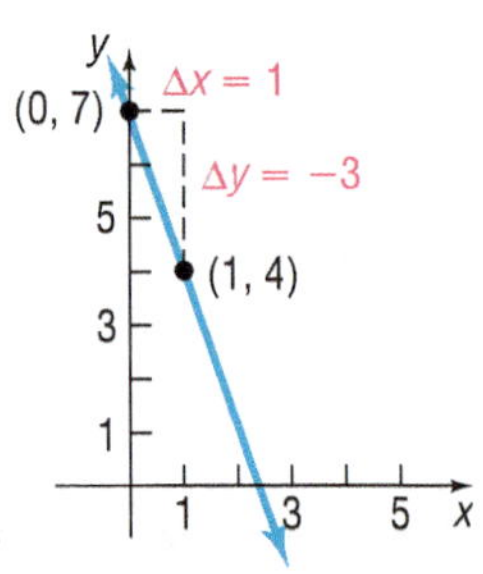

Alternatively, we could have found an additional point by evaluating the function at some $x \neq 0$. For $x = 1$, we find $f(1) = -3(1) + 7 = 4$ and obtain the point $(1, 4)$ on the graph.

Now Work PROBLEMS 13(a), (b), and (c)

2 Use Average Rate of Change to Identify Linear Functions

Look at Table 1, which shows certain values of the independent variable x and corresponding values of the dependent variable y for the function $y = f(x) = -3x + 7$. Notice that as the value of the independent variable x increases by 1, the value of the dependent variable y decreases by 3. That is, the average rate of change of y with respect to x is a constant, -3.

Table 1

x	$y = f(x) = -3x + 7$	Average Rate of Change $= \dfrac{\Delta y}{\Delta x}$
-2	13	
		$\dfrac{10 - 13}{-1 - (-2)} = \dfrac{-3}{1} = -3$
-1	10	
		$\dfrac{7 - 10}{0 - (-1)} = \dfrac{-3}{1} = -3$
0	7	
		-3
1	4	
		-3
2	1	
		-3
3	-2	

It is not a coincidence that the average rate of change of the linear function $f(x) = -3x + 7$ is the slope of the linear function. That is, $\dfrac{\Delta y}{\Delta x} = m = -3$. The following theorem states this fact.

THEOREM

Average Rate of Change of a Linear Function

Linear functions have a constant average rate of change. That is, the average rate of change of a linear function $f(x) = mx + b$ is

$$\frac{\Delta y}{\Delta x} = m$$

Proof The average rate of change of $f(x) = mx + b$ from x_1 to x_2, $x_1 \neq x_2$, is

$$\frac{\Delta y}{\Delta x} = \frac{f(x_2) - f(x_1)}{x_2 - x_1} = \frac{(mx_2 + b) - (mx_1 + b)}{x_2 - x_1}$$

$$= \frac{mx_2 - mx_1}{x_2 - x_1} = \frac{m(x_2 - x_1)}{x_2 - x_1} = m$$

■

Based on the theorem just proved, the average rate of change of the function $g(x) = -\frac{2}{5}x + 5$ is $-\frac{2}{5}$.

Now Work PROBLEM 13(d)

As it turns out, only linear functions have a constant average rate of change. Because of this, we can use the average rate of change to determine whether a particular function is linear or not. Functions that are not linear are said to be **nonlinear.**

EXAMPLE 2 Using the Average Rate of Change to Identify Linear Functions

(a) A strain of E-coli Beu 397-recA441 is placed into a Petri dish at 30° Celsius and allowed to grow. The data shown in Table 2 on the following page are collected. The population is measured in grams and the time in hours. Plot the ordered pairs (x, y) in the Cartesian plane and use the average rate of change to determine whether the function is linear.

(b) The data in Table 3 represent the maximum number of heartbeats that a healthy individual should have during a 15-second interval of time while exercising for different ages. Plot the ordered pairs (x, y) in the Cartesian plane and use the average rate of change to determine whether the function is linear.

Table 2

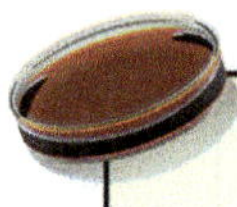

Time (hours), x	Population (grams), y	(x, y)
0	0.09	(0, 0.09)
1	0.12	(1, 0.12)
2	0.16	(2, 0.16)
3	0.22	(3, 0.22)
4	0.29	(4, 0.29)
5	0.39	(5, 0.39)

Table 3

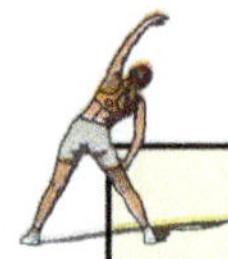

Age, x	Maximum Number of Heart Beats, y	(x, y)
20	50	(20, 50)
30	47.5	(30, 47.5)
40	45	(40, 45)
50	42.5	(50, 42.5)
60	40	(60, 40)
70	37.5	(70, 37.5)

Source: American Heart Association

Solution We compute the average rate of change of each function. If the average rate of change is constant, the function is linear. If the average rate of change is not constant, the function is nonlinear.

(a) Figure 2 shows the points listed in Table 2 plotted in the Cartesian plane. Notice that it is impossible to draw a straight line that contains all the points. Table 4 displays the average rate of change of population.

Figure 2

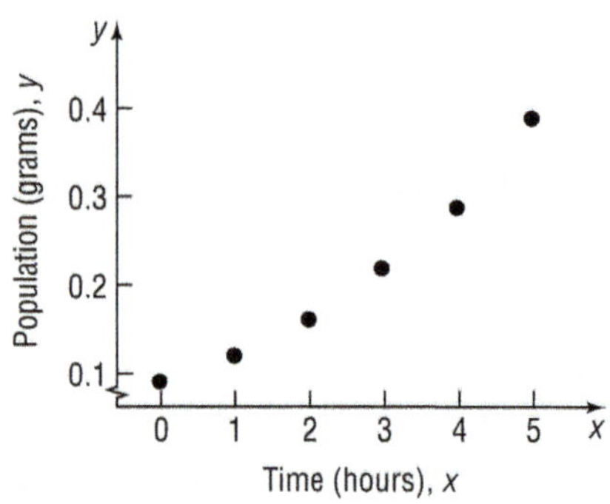

Table 4

Time (hours), x	Population (grams), y	Average Rate of Change $= \dfrac{\Delta y}{\Delta x}$
0	0.09	
		$\dfrac{0.12 - 0.09}{1 - 0} = 0.03$
1	0.12	
		0.04
2	0.16	
		0.06
3	0.22	
		0.07
4	0.29	
		0.10
5	0.39	

Because the average rate of change is not constant, we know that the function is not linear. In fact, because the average rate of change is increasing as the value of the independent variable increases, we say that the function is increasing at an increasing rate. So not only is the population increasing over time, but it is also growing more rapidly as time passes.

(b) Figure 3 shows the points listed in Table 3 plotted in the Cartesian plane. We can see that the data in Figure 3 lie on a straight line. Table 5 shows the average rate of change of the maximum number of heartbeats. The average rate of change of the heartbeat data is constant, -0.25 beat per year, so the function is linear. We can find the linear function using the point–slope formula with $x_1 = 20$, $y_1 = 50$, and $m = -0.25$.

$$\begin{aligned} y - y_1 &= m(x - x_1) \\ y - 50 &= -0.25(x - 20) \\ y - 50 &= -0.25x + 5 \\ y &= -0.25x + 55 \end{aligned}$$

Figure 3

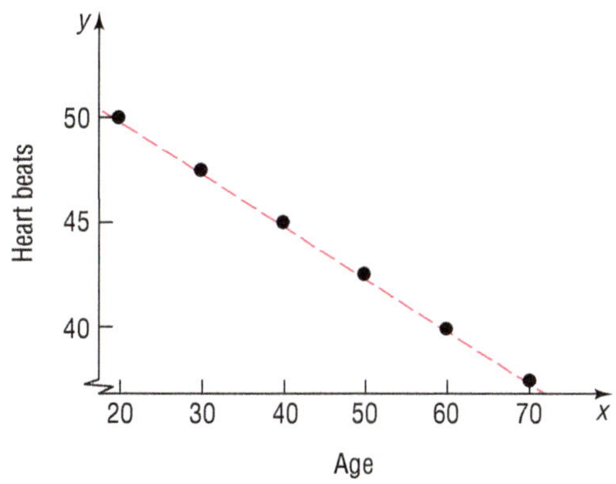

Table 5

Age, x	Maximum Number of Heartbeats, y	Average Rate of Change $= \frac{\Delta y}{\Delta x}$
20	50	
		$\frac{47.5 - 50}{30 - 20} = -0.25$
30	47.5	
		-0.25
40	45	
		-0.25
50	42.5	
		-0.25
60	40	
		-0.25
70	37.5	

Now Work PROBLEM 21

3 Determine Whether a Linear Function Is Increasing, Decreasing, or Constant

Look back at the Explorations on page 180. We know that when the slope m of a linear function is positive $(m > 0)$ the line slants upward from left to right. When the slope m of a linear function is negative $(m < 0)$, the line slants downward from left to right. When the slope m of a linear function is zero $(m = 0)$, the line is horizontal.

Based on these results, we have the following theorem.

THEOREM

Increasing, Decreasing, and Constant Linear Functions

A linear function $f(x) = mx + b$ is increasing over its domain if its slope, m, is positive. It is decreasing over its domain if its slope, m, is negative. It is constant over its domain if its slope, m, is zero.

EXAMPLE 3 Determining Whether a Linear Function Is Increasing, Decreasing, or Constant

Determine whether the following linear functions are increasing, decreasing, or constant.

(a) $f(x) = 5x - 2$ (b) $g(x) = -2x + 8$

(c) $s(t) = \frac{3}{4}t - 4$ (d) $h(z) = 7$

Solution

(a) For the linear function $f(x) = 5x - 2$, the slope is 5, which is positive. The function f is increasing on the interval $(-\infty, \infty)$.

(b) For the linear function $g(x) = -2x + 8$, the slope is -2, which is negative. The function g is decreasing on the interval $(-\infty, \infty)$.

(c) For the linear function $s(t) = \frac{3}{4}t - 4$, the slope is $\frac{3}{4}$, which is positive. The function s is increasing on the interval $(-\infty, \infty)$.

(d) We can write the linear function h as $h(z) = 0z + 7$. Because the slope is 0, the function h is constant on the interval $(-\infty, \infty)$.

Now Work PROBLEM 13(e)

4 Build Linear Models from Verbal Descriptions

When the average rate of change of a function is constant, we can use a linear function to model the relation between the two variables. For example, if your phone company charges you \$0.07 per minute to talk regardless of the number of minutes used, we can model the relation between the cost C and minutes used x as the linear function $C(x) = 0.07x$, with slope $m = \frac{0.07 \text{ dollar}}{1 \text{ minute}}$.

Modeling with a Linear Function

If the average rate of change of a function is a constant m, a linear function f can be used to model the relation between the two variables as follows:

$$f(x) = mx + b$$

where b is the value of f at 0, that is, $f(0)$.

We illustrate how to develop a linear model in the next example.

EXAMPLE 4 Modeling Straight-line Depreciation

Book value is the value of an asset that a company uses to create its balance sheet. Some companies depreciate their assets using straight-line depreciation so that the value of the asset declines by a fixed amount each year. The amount of the decline depends on the useful life that the company places on the asset. Suppose that a company just purchased a fleet of new cars for its sales force at a cost of \$28,000 per car. The company chooses to depreciate each vehicle using the straight-line method over 7 years. This means that each car will depreciate by $\frac{\$28{,}000}{7} = \$4000$ per year.

(a) Build a linear model that expresses the book value V of each car as a function of its age, x.
(b) What is the implied domain of the function found in part (a)?
(c) Graph the linear function.
(d) What is the book value of each car after 3 years?
(e) Interpret the slope.
(f) When will the book value of each car be \$8000?
[**Hint:** Solve the equation $V(x) = 8000$.]

Solution

(a) If we let $V(x)$ represent the value of each car after x years, then $V(0)$ represents the original value of each car, so $V(0) = \$28{,}000$. The y-intercept of the linear function is \$28,000. Because each car depreciates by \$4000 per year, the slope of the linear function is -4000. The linear function that represents the book value V of each car after x years is

$$V(x) = -4000x + 28{,}000 \quad \text{The Model}$$

(b) Because x represents the age of the car, we have $x \ge 0$. Also, the value of the car cannot be negative. Since $V(7) = 0$, we have $x \le 7$. Therefore, the domain of the function V is $\{x \mid 0 \le x \le 7\}$ or $[0, 7]$, using interval notation.

(c) Figure 4 shows the graph of V.

Figure 4

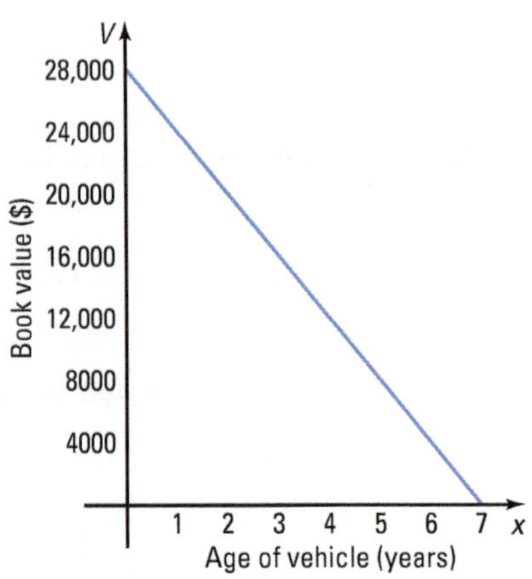

(d) The book value of each car after 3 years is

$$V(3) = -4000(3) + 28{,}000$$
$$= \$16{,}000$$

(e) Since the slope of $V(x) = -4000x + 28{,}000$ is -4000, the average rate of change of book value is $-\$4000$/year. So for each additional year that passes the book value of the car decreases by \$4000.

(f) To find when the book value is \$8000, we solve the equation

$$\begin{aligned} V(x) &= 8000 \\ -4000x + 28{,}000 &= 8000 \\ -4000x &= -20{,}000 && \text{Subtract 28,000 from each side.} \\ x &= \frac{-20{,}000}{-4000} = 5 && \text{Divide by } -4000. \end{aligned}$$

The car will have a book value of \$8000 when it is 5 years old.

Now Work PROBLEM 45

Next we look at an application from economics.

EXAMPLE 5 Supply and Demand

The **quantity supplied** of a good is the amount of a product that a company is willing to make available for sale at a given price. The **quantity demanded** of a good is the amount of a product that consumers are willing to purchase at a given price. Suppose that the quantity supplied, S, and quantity demanded, D, of cellular telephones each month are given by the following functions:

$$S(p) = 60p - 900$$
$$D(p) = -15p + 2850$$

where p is the price (in dollars) of the telephone.

(a) The **equilibrium price** of a product is defined as the price at which quantity supplied equals quantity demanded. That is, the equilibrium price is the price at which $S(p) = D(p)$. Find the equilibrium price of cellular telephones. What is the **equilibrium quantity,** the amount demanded (or supplied) at the equilibrium price?

(b) Determine the prices for which quantity supplied is greater than quantity demanded. That is, solve the inequality $S(p) > D(p)$.

(c) Graph $S = S(p)$, $D = D(p)$ and label the equilibrium price.

Solution (a) To find the equilibrium price, we solve the equation $S(p) = D(p)$.

$$\begin{aligned} 60p - 900 &= -15p + 2850 && S(p) = 60p - 900;\ D(p) = -15p + 2850 \\ 60p &= -15p + 3750 && \text{Add 900 to each side.} \\ 75p &= 3750 && \text{Add } 15p \text{ to each side.} \\ p &= 50 && \text{Divide each side by 75.} \end{aligned}$$

The equilibrium price is \$50 per cellular phone. To find the equilibrium quantity, we evaluate either $S(p)$ or $D(p)$ at $p = 50$.

$$S(50) = 60(50) - 900 = 2100$$

The equilibrium quantity is 2100 cellular phones. At a price of \$50 per phone, the company will produce and sell 2100 phones each month and have no shortages or excess inventory.

(b) We solve the inequality $S(p) > D(p)$.

$$\begin{aligned} 60p - 900 &> -15p + 2850 && S(p) > D(p) \\ 60p &> -15p + 3750 && \text{Add 900 to each side.} \\ 75p &> 3750 && \text{Add } 15p \text{ to each side.} \\ p &> 50 && \text{Divide each side by 75.} \end{aligned}$$

If the company charges more than \$50 per phone, quantity supplied will exceed quantity demanded. In this case the company will have excess phones in inventory.

(c) Figure 5 shows the graphs of $S = S(p)$ and $D = D(p)$ with the equilibrium point labeled. Notice that we only graph each function in quadrant I. Do you know why?

Figure 5

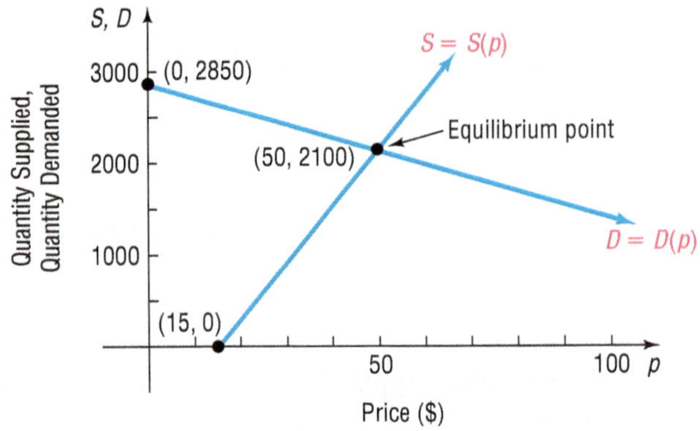

Now Work PROBLEM 39

4.1 Assess Your Understanding

'Are You Prepared?' *Answers are given at the end of these exercises. If you get a wrong answer, read the pages listed in red.*

1. Graph $y = 2x - 3$. (pp. 184–185)
2. Find the slope of the line joining the points $(2, 5)$ and $(-1, 3)$. (pp. 178–179)
3. Find the average rate of change of $f(x) = 3x^2 - 2$ from 2 to 4. (pp. 240–242)
4. Solve: $60x - 900 = -15x + 2850$. (pp. 105–106)
5. If $f(x) = x^2 - 4$, find $f(-2)$. (pp. 217–218)
6. ***True or False*** The graph of the function $f(x) = x^2$ is increasing on the interval $(0, \infty)$. (pp. 237–240)

Concepts and Vocabulary

7. For the graph of the linear function $f(x) = mx + b$, m is the _____ and b is the _____.
8. For the graph of the linear function $H(z) = -4z + 3$, the slope is _____ and the y-intercept is _____.
9. If the slope m of the graph of a linear function is _____, the function is increasing over its domain.
10. ***True or False*** The slope of a line is the average rate of change of the linear function.
11. ***True or False*** If the average rate of change of a linear function is $\frac{2}{3}$ then, if y increases by 3, x will increase by 2.
12. ***True or False*** The average rate of change of $f(x) = 2x + 8$ is 8.

Skill Building

In Problems 13–20, a linear function is given.
(a) Determine the slope and y-intercept of each function.
(b) Use the slope and y-intercept to graph the linear function.
(c) Determine the domain and the range of each function.
(d) Determine the average rate of change of each function.
(e) Determine whether the linear function is increasing, decreasing, or constant.

13. $f(x) = 2x + 3$
14. $g(x) = 5x - 4$
15. $h(x) = -3x + 4$
16. $p(x) = -x + 6$
17. $f(x) = \frac{1}{4}x - 3$
18. $h(x) = -\frac{2}{3}x + 4$
19. $F(x) = 4$
20. $G(x) = -2$

In Problems 21–28, determine whether the given function is linear or nonlinear. If it is linear, determine the equation of the line.

21.

x	y
−2	4
−1	1
0	−2
1	−5
2	−8

22.

x	y
−2	$\frac{1}{4}$
−1	$\frac{1}{2}$
0	1
1	2
2	4

23.

x	y
−20	−8
−10	−3
0	0
10	1
20	0

24.

x	y
−20	−40
−10	0
0	40
10	80
20	120

25.

x	y
−2	−26
−1	−4
0	2
1	−2
2	−10

26.

x	y
−2	−4
−1	−3.5
0	−3
1	−2.5
2	−2

27.

x	y
−10	8
−5	8
0	8
5	8
10	8

28.

x	y
−10	0
−5	1
0	4
5	9
10	16

Applications and Extensions

29. Suppose that $f(x) = 4x - 1$ and $g(x) = -2x + 5$.
(a) Solve $f(x) = 0$. (b) Solve $f(x) > 0$.
(c) Solve $f(x) = g(x)$. (d) Solve $f(x) \le g(x)$.
(e) Graph $y = f(x)$ and $y = g(x)$ and label the point that represents the solution to the equation $f(x) = g(x)$.

30. Suppose that $f(x) = 3x + 5$ and $g(x) = -2x + 15$.
(a) Solve $f(x) = 0$. (b) Solve $f(x) < 0$.
(c) Solve $f(x) = g(x)$. (d) Solve $f(x) \ge g(x)$.
(e) Graph $y = f(x)$ and $y = g(x)$ and label the point that represents the solution to the equation $f(x) = g(x)$.

31. In parts (a)–(f), use the following figure.

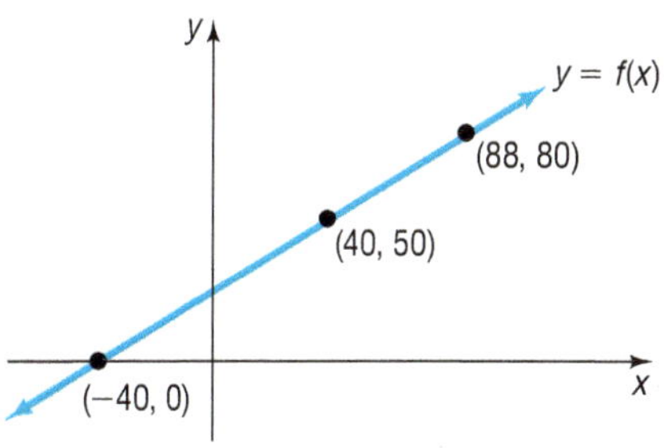

(a) Solve $f(x) = 50$. (b) Solve $f(x) = 80$.
(c) Solve $f(x) = 0$. (d) Solve $f(x) > 50$.
(e) Solve $f(x) \le 80$. (f) Solve $0 < f(x) < 80$.

32. In parts (a)–(f), use the following figure.

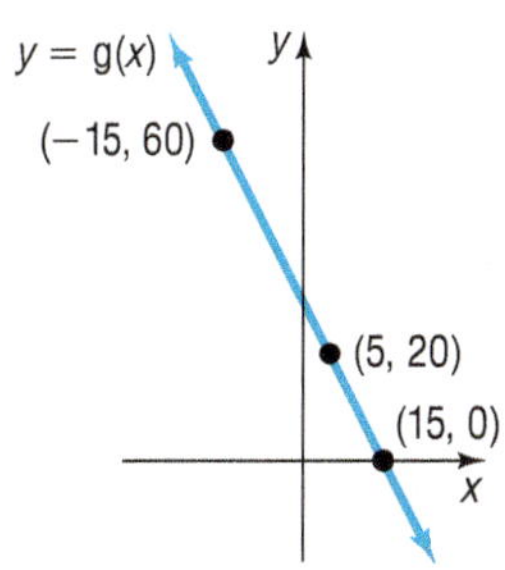

(a) Solve $g(x) = 20$. (b) Solve $g(x) = 60$.
(c) Solve $g(x) = 0$. (d) Solve $g(x) > 20$.
(e) Solve $g(x) \le 60$. (f) Solve $0 < g(x) < 60$.

33. In parts (a) and (b) use the following figure.

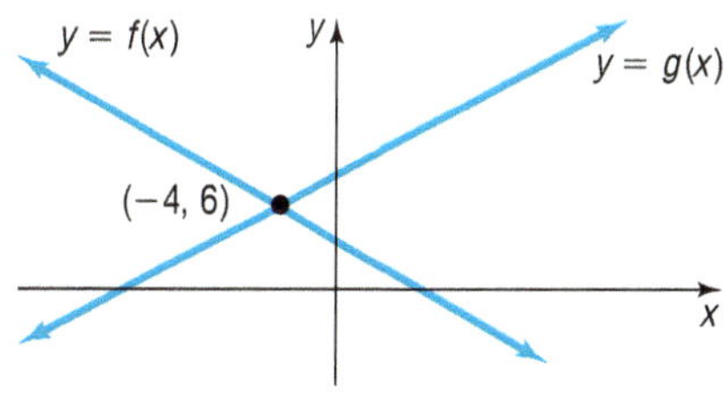

(a) Solve the equation: $f(x) = g(x)$.
(b) Solve the inequality: $f(x) > g(x)$.

34. In parts (a) and (b), use the following figure.

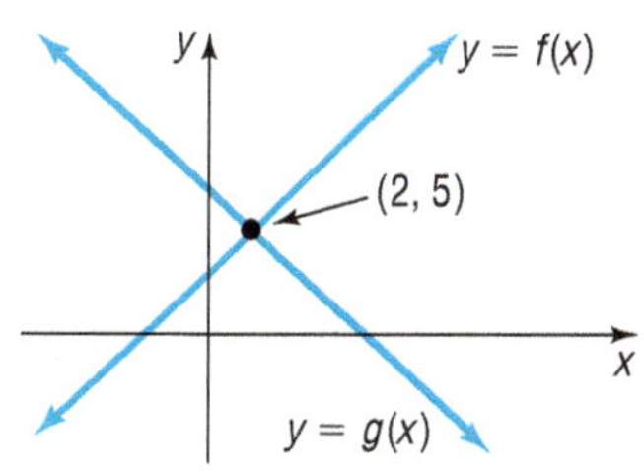

(a) Solve the equation: $f(x) = g(x)$.
(b) Solve the inequality: $f(x) \le g(x)$.

35. In parts (a) and (b), use the following figure.

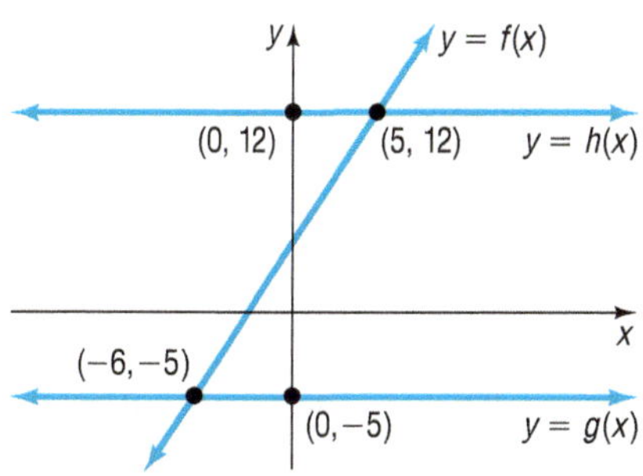

(a) Solve the equation: $f(x) = g(x)$.
(b) Solve the inequality: $g(x) \le f(x) < h(x)$.

36. In parts (a) and (b), use the following figure.

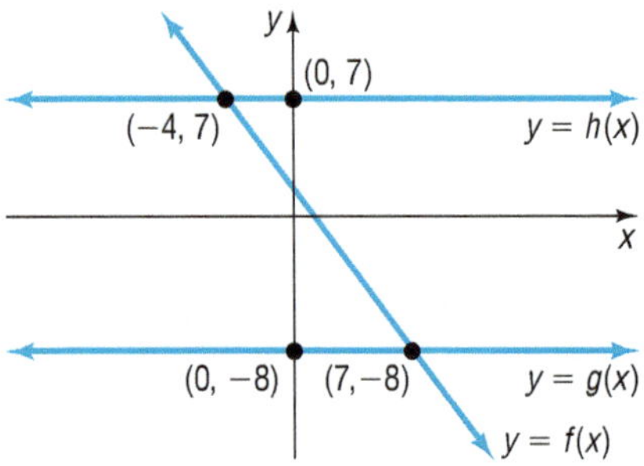

(a) Solve the equation: $f(x) = g(x)$.
(b) Solve the inequality: $g(x) < f(x) \le h(x)$.

37. Car Rentals The cost C, in dollars, of renting a moving truck for a day is modeled by the function $C(x) = 0.25x + 35$, where x is the number of miles driven.
(a) What is the cost if you drive $x = 40$ miles?
(b) If the cost of renting the moving truck is \$80, how many miles did you drive?
(c) Suppose that you want the cost to be no more than \$100. What is the maximum number of miles that you can drive?
(d) What is the implied domain of C?

38. Phone Charges The monthly cost C, in dollars, for international calls on a certain cellular phone plan is modeled by the function $C(x) = 0.38x + 5$, where x is the number of minutes used.
(a) What is the cost if you talk on the phone for $x = 50$ minutes?
(b) Suppose that your monthly bill is \$29.32. How many minutes did you use the phone?
(c) Suppose that you budget yourself \$60 per month for the phone. What is the maximum number of minutes that you can talk?
(d) What is the implied domain of C if there are 30 days in the month?

39. Supply and Demand Suppose that the quantity supplied S and quantity demanded D of T-shirts at a concert are given by the following functions:

$$S(p) = -200 + 50p$$
$$D(p) = 1000 - 25p$$

where p is the price of a T-shirt.
(a) Find the equilibrium price for T-shirts at this concert. What is the equilibrium quantity?
(b) Determine the prices for which quantity demanded is greater than quantity supplied.
(c) What do you think will eventually happen to the price of T-shirts if quantity demanded is greater than quantity supplied?

40. Supply and Demand Suppose that the quantity supplied S and quantity demanded D of hot dogs at a baseball game are given by the following functions:

$$S(p) = -2000 + 3000p$$
$$D(p) = 10{,}000 - 1000p$$

where p is the price of a hot dog.
(a) Find the equilibrium price for hot dogs at the baseball game. What is the equilibrium quantity?
(b) Determine the prices for which quantity demanded is less than quantity supplied.
(c) What do you think will eventually happen to the price of hot dogs if quantity demanded is less than quantity supplied?

41. Taxes The function $T(x) = 0.15(x - 7825) + 782.5$ represents the tax bill T of a single person whose adjusted gross income is x dollars for income between \$7825 and \$31,850, inclusive in 2007.

Source: Internal Revenue Service
(a) What is the domain of this linear function?
(b) What is a single filer's tax bill if adjusted gross income is \$18,000?
(c) Which variable is independent and which is dependent?
(d) Graph the linear function over the domain specified in part (a).
(e) What is a single filer's adjusted gross income if the tax bill is \$3358.75?

42. Luxury Tax In 2002, major league baseball signed a labor agreement with the players. In this agreement, any team whose payroll exceeds \$136.5 million in 2006 will have to pay a luxury tax of 40% (for second offenses). The linear function $T(p) = 0.40(p - 136.5)$ describes the luxury tax T of a team whose payroll is p (in millions of dollars).

Source: Major League Baseball
(a) What is the implied domain of this linear function?
(b) What is the luxury tax for the New York Yankees whose 2006 payroll was \$171.1 million?
(c) Graph the linear function.
(d) What is the payroll of a team that pays a luxury tax of \$11.7 million?

The point at which a company's profits equal zero is called the company's ***break-even point.*** *For Problems 43 and 44, let R represent a company's revenue, let C represent the company's costs, and let x represent the number of units produced and sold each day.*
(a) Find the firm's break-even point; that is, find x so that $R = C$.
(b) Find the values of x such that $R(x) > C(x)$. This represents the number of units that the company must sell to earn a profit.

43. $R(x) = 8x$
$C(x) = 4.5x + 17{,}500$

44. $R(x) = 12x$
$C(x) = 10x + 15{,}000$

45. Straight-line Depreciation Suppose that a company has just purchased a new computer for \$3000. The company chooses to depreciate the computer using the straight-line method over 3 years.
(a) Write a linear model that expresses the book value V of the computer as a function of its age x.
(b) What is the implied domain of the function found in part (a)?
(c) Graph the linear function.
(d) What is the book value of the computer after 2 years?
(e) When will the computer have a book value of \$2000?

46. Straight-line Depreciation Suppose that a company has just purchased a new machine for its manufacturing facility for \$120,000. The company chooses to depreciate the machine using the straight-line method over 10 years.
(a) Write a linear model that expresses the book value V of the machine as a function of its age x.
(b) What is the implied domain of the function found in part (a)?
(c) Graph the linear function.

(d) What is the book value of the machine after 4 years?
(e) When will the machine have a book value of \$72,000?

47. Cost Function The simplest cost function is the linear cost function, $C(x) = mx + b$, where the y-intercept b represents the fixed costs of operating a business and the slope m represents the cost of each item produced. Suppose that a small bicycle manufacturer has daily fixed costs of \$1800 and each bicycle costs \$90 to manufacture.
(a) Write a linear model that expresses the cost C of manufacturing x bicycles in a day.
(b) Graph the model.
(c) What is the cost of manufacturing 14 bicycles in a day?
(d) How many bicycles could be manufactured for \$3780?

48. Cost Function Refer to Problem 47. Suppose that the landlord of the building increases the bicycle manufacturer's rent by \$100 per month.
(a) Assuming that the manufacturer is open for business 20 days per month, what are the new daily fixed costs?
(b) Write a linear model that expresses the cost C of manufacturing x bicycles in a day with the higher rent.
(c) Graph the model.
(d) What is the cost of manufacturing 14 bicycles in a day?
(e) How many bicycles can be manufactured for \$3780?

49. Truck Rentals A truck rental company rents a truck for one day by charging \$29 plus \$0.07 per mile.
(a) Write a linear model that relates the cost C, in dollars, of renting the truck to the number x of miles driven.
(b) What is the cost of renting the truck if the truck is driven 110 miles? 230 miles?

50. Long Distance A phone company offers a domestic long distance package by charging \$5 plus \$0.05 per minute.
(a) Write a linear model that relates the cost C, in dollars, of talking x minutes.
(b) What is the cost of talking 105 minutes? 180 minutes?

51. Developing a Linear Model from Data The following data represent the price p and quantity demanded per day q of an MP3 player.

Price, p (in dollars)	Quantity Demanded, q
150	100
200	80
250	60
300	40

(a) Plot the ordered pairs (p, q) in a Cartesian plane.
(b) Show that quantity demanded q is a linear function of the price p.
(c) Determine the linear function that describes the relation between p and q.
(d) What is the implied domain of the linear function?
(e) Graph the linear function in the Cartesian plane drawn in part (a).
(f) Interpret the slope.
(g) Interpret the value of the intercepts.

52. Developing a Linear Model from Data The following data represent the various combinations of soda and hot dogs that Yolanda can buy at a baseball game with \$60.

Soda, s	Hot Dogs, h
20	0
15	3
10	6
5	9

(a) Plot the ordered pairs (s, h) in a Cartesian plane.
(b) Show that the number of hot dogs purchased h is a linear function of the number of sodas purchased s.
(c) Determine the linear function that describes the relation between s and h.
(d) What is the implied domain of the linear function?
(e) Graph the linear function in the Cartesian plane drawn in part (a).
(f) Interpret the slope.
(g) Interpret the value of the intercepts.

Discussion and Writing

53. Which of the following functions might have the graph shown? (More than one answer is possible.)
(a) $f(x) = 2x - 7$
(b) $g(x) = -3x + 4$
(c) $H(x) = 5$
(d) $F(x) = 3x + 4$
(e) $G(x) = \frac{1}{2}x + 2$

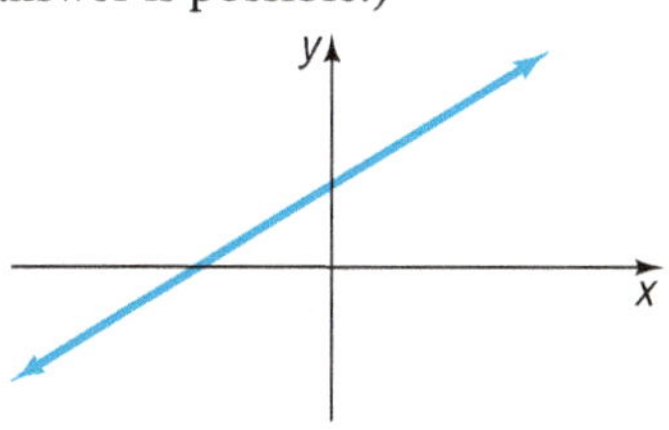

54. Which of the following functions might have the graph shown? (More than one answer is possible.)
(a) $f(x) = 3x + 1$
(b) $g(x) = -2x + 3$
(c) $H(x) = 3$
(d) $F(x) = -4x - 1$
(e) $G(x) = -\frac{2}{3}x + 3$

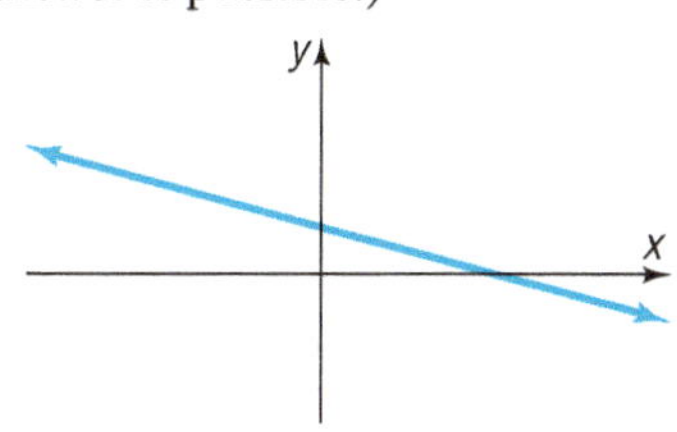

55. Under what circumstances is a linear function $f(x) = mx + b$ odd? Can a linear function ever be even?

'Are You Prepared?' Answers

1.

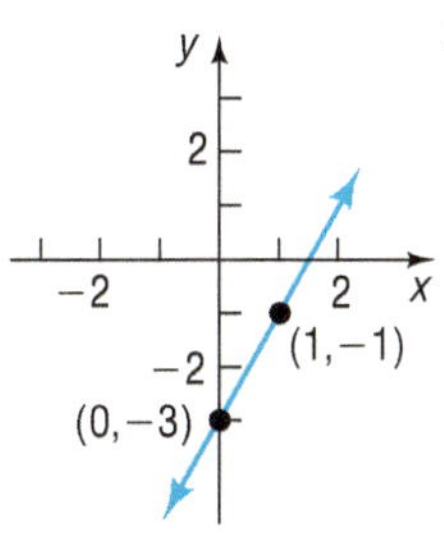

2. $\frac{2}{3}$ **3.** 18 **4.** $\{50\}$ **5.** 0 **6.** True

4.2 Building Linear Models from Data

PREPARING FOR THIS SECTION *Before getting started, review the following:*

- Rectangular Coordinates (Section 1.1, pp. 86–88)
- Lines (Section 2.2, pp. 178–185)
- Functions (Section 3.1, pp. 212–218)

Now Work the 'Are You Prepared?' problems on page 295.

OBJECTIVES
1. Draw and Interpret Scatter Diagrams (p. 292)
2. Distinguish between Linear and Nonlinear Relations (p. 293)
3. Use a Graphing Utility to Find the Line of Best Fit (p. 294)

1 Draw and Interpret Scatter Diagrams

In Section 4.1, we built linear models from verbal descriptions. Linear models can also be constructed by fitting a linear function to data. The first step in finding this relation is to plot the ordered pairs using rectangular coordinates. The resulting graph is called a **scatter diagram.**

EXAMPLE 1 Drawing and Interpreting a Scatter Diagram

In baseball, the on-base percentage for a team represents the percentage of time that the players safely reach base. The data given in Table 6 represent the number of runs scored y and the on-base percentage x for teams in the National League during the 2006 baseball season.

Table 6

Team	On-Base Percentage, x	Runs Scored, y	(x, y)
Atlanta	33.7	849	(33.7, 849)
St. Louis	33.7	781	(33.7, 781)
Colorado	34.1	813	(34.1, 813)
Houston	33.2	735	(33.2, 735)
Philadelphia	34.7	865	(34.7, 865)
San Francisco	32.4	746	(32.4, 746)
Pittsburgh	32.7	691	(32.7, 691)
Florida	33.1	758	(33.1, 758)
Chicago Cubs	31.9	716	(31.9, 716)
Arizona	33.1	773	(33.1, 773)
Milwaukee	32.7	730	(32.7, 730)
Washington	33.8	746	(33.8, 746)
Cincinnati	33.6	749	(33.6, 749)
San Diego	33.2	731	(33.2, 731)
NY Mets	33.4	834	(33.4, 834)
Los Angeles	34.8	820	(34.8, 820)

Source: espn.com

(a) Draw a scatter diagram of the data, treating on-base percentage as the independent variable.

(b) Use a graphing utility to draw a scatter diagram.

(c) Describe what happens to runs scored as the on-base percentage increases.

Solution (a) To draw a scatter diagram, we plot the ordered pairs listed in Table 6, with the on-base percentage as the x-coordinate and the runs scored as the y-coordinate. See Figure 6(a). Notice that the points in the scatter diagram are not connected.

(b) Figure 6(b) shows a scatter diagram using a TI-84 Plus graphing calculator.

(c) We see from the scatter diagrams that, as the on-base percentage increases, the trend is that the number of runs scored also increases.

Figure 6

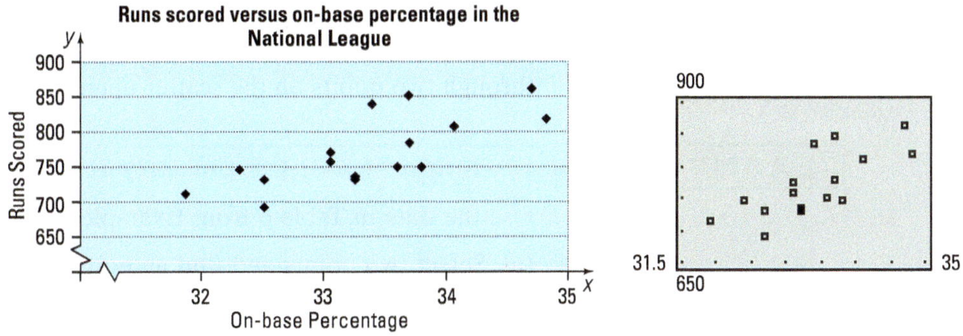

(a) (b)

Now Work PROBLEM 11(a)

2 Distinguish between Linear and Nonlinear Relations

You should notice that the points in Figure 6 do not follow a perfect linear relation (as they do in Figure 3 in Section 4.1). However, the data do exhibit a linear pattern. There are numerous explanations as to why the data are not perfectly linear, but one easy explanation is the fact that other variables besides on-base percentage play a role in determining runs scored, such as number of home runs hit.

Scatter diagrams are used to help us to see the type of relation that exists between two variables. In this text, we will discuss a variety of different relations that may exist between two variables. For now, we concentrate on distinguishing between linear and nonlinear relations. See Figure 7.

Figure 7

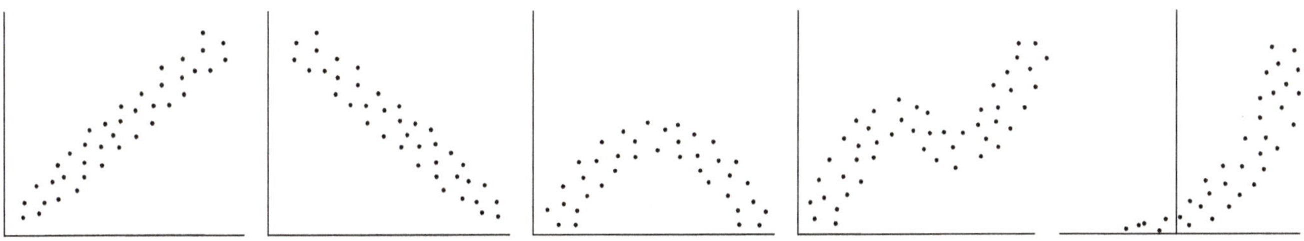

(a) Linear $y = mx + b, m > 0$ **(b)** Linear $y = mx + b, m < 0$ **(c)** Nonlinear **(d)** Nonlinear **(e)** Nonlinear

EXAMPLE 2 Distinguishing between Linear and Nonlinear Relations

Determine whether the relation between the two variables in Figure 8 is linear or nonlinear.

Figure 8

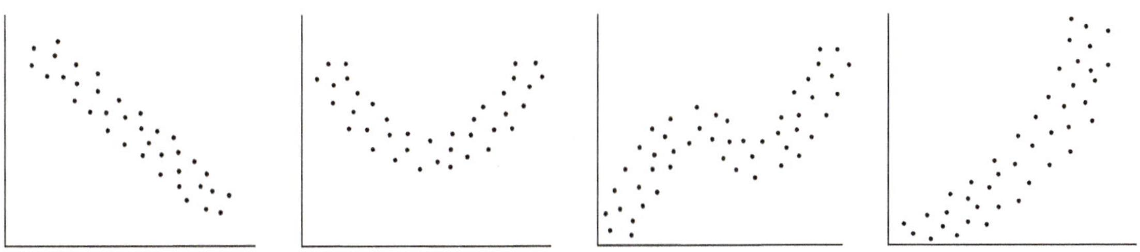

(a) (b) (c) (d)

Solution (a) Linear (b) Nonlinear (c) Nonlinear (d) Nonlinear

Now Work PROBLEM 5

In this section we will study data whose scatter diagrams imply that a linear relation exists between the two variables.

Suppose that the scatter diagram of a set of data appears to be linearly related as in Figure 7(a) or (b). We might wish to find an equation of a line (a model) that relates the two variables. One way to obtain a model for such data is to draw a line through two points on the scatter diagram and determine the equation of the line.

EXAMPLE 3 **Finding a Model for Linearly Related Data**

Use the data in Table 6 from Example 1 to:

(a) Select two points and find an equation of the line containing the points.
(b) Graph the line on the scatter diagram obtained in Example 1(b).

Solution (a) Select two points, say (32.7, 730) and (34.7, 865). The slope of the line joining the points (32.7, 730) and (34.7, 865) is

$$m = \frac{865 - 730}{34.7 - 32.7} = \frac{135}{2} = 67.5$$

The equation of the line with slope 67.5 and passing through (32.7, 730) is found using the point–slope form with $m = 67.5$, $x_1 = 32.7$, and $y_1 = 730$.

$$\begin{aligned} y - y_1 &= m(x - x_1) && \text{Point–slope form of a line} \\ y - 730 &= 67.5(x - 32.7) && x_1 = 32.7,\ y_1 = 730,\ m = 67.5 \\ y - 730 &= 67.5x - 2207.25 \\ y &= 67.5x - 1477.25 && \text{The Model} \end{aligned}$$

Figure 9

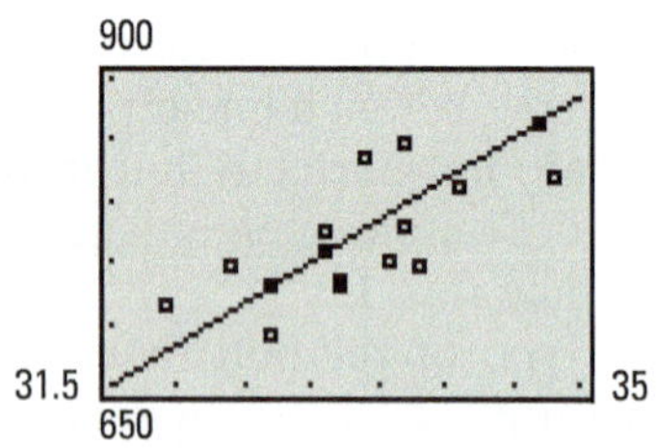

(b) Figure 9 shows the scatter diagram with the graph of the line found in part (a).

Select two other points and complete the solution. Graph the line on the scatter diagram obtained in Figure 9.

Now Work PROBLEMS 11(b) AND (c)

3 Use a Graphing Utility to Find the Line of Best Fit

The model obtained in Example 3 depends on the selection of points, which will vary from person to person. So the model that we found might be different from the model you found. Although the model we found in Example 3 appears to fit the data well, there may be a model that "fits it better." Do you think your model fits the data better? Is there a *line of best fit*? As it turns out, there is a method for finding a model that best fits linearly related data (called the **line of best fit**).*

EXAMPLE 4 **Finding a Model for Linearly Related Data**

Use the data in Table 6 from Example 1.

(a) Use a graphing utility to find the line of best fit that models the relation between on-base percentage and runs scored.
(b) Graph the line of best fit on the scatter diagram obtained in Example 1(b).
(c) Interpret the slope.
(d) Use the line of best fit to predict the number of runs a team will score if their on-base percentage is 34.1.

* We shall not discuss the underlying mathematics of lines of best fit in this book.

Solution (a) Graphing utilities contain built-in programs that find the line of best fit for a collection of points in a scatter diagram. Upon executing the LINear REGression program, we obtain the results shown in Figure 10. The output that the utility provides shows us the equation $y = ax + b$, where a is the slope of the line and b is the y-intercept. The line of best fit that relates on-base percentage to runs scored may be expressed as the line

$$y = 49.19x - 870.94 \quad \text{The Model}$$

(b) Figure 11 shows the graph of the line of best fit, along with the scatter diagram.

Figure 10

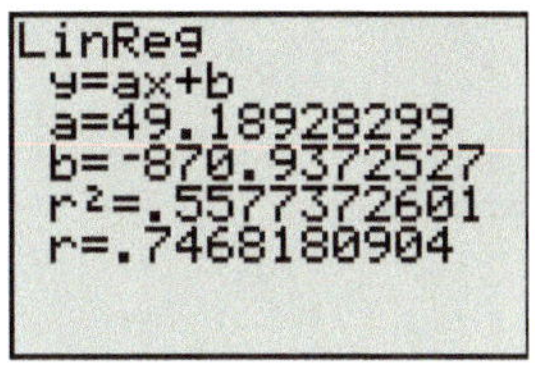

Figure 11

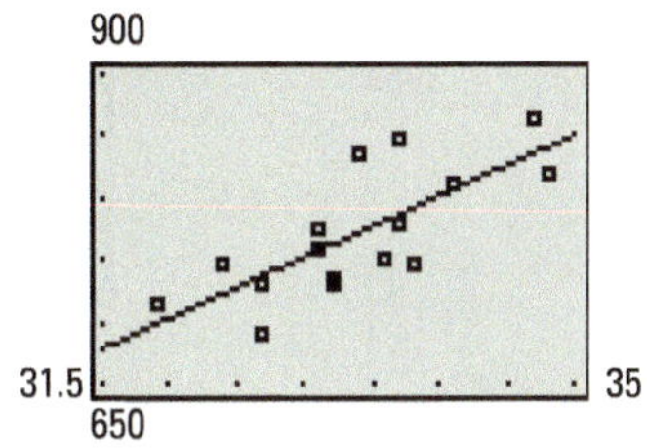

(c) The slope of the line of best fit is 49.19, which means that, for every 1 percent increase in the on-base percentage, runs scored increase 49.19, on average.

(d) Letting $x = 34.1$ in the equation of the line of best fit, we obtain $y = 49.19(34.1) - 870.94 \approx 806$ runs.

Now Work PROBLEMS 11(d) AND (e)

Does the line of best fit appear to be a good fit? In other words, does the line appear to accurately describe the relation between on-base percentage and runs scored?

And just how "good" is this line of best fit? Look again at Figure 10. The last line of output is $r = 0.747$. This number, called the **correlation coefficient,** $r, -1 \leq r \leq 1$, is a measure of the strength of the linear relation that exists between two variables. The closer that $|r|$ is to 1, the more perfect the linear relationship is. If r is close to 0, there is little or no linear relationship between the variables. A negative value of r, $r < 0$, indicates that as x increases y decreases; a positive value of r, $r > 0$, indicates that as x increases y does also. The data given in Table 6, having a correlation coefficient of 0.747, are indicative of a relatively strong linear relationship with positive slope.

4.2 Assess Your Understanding

'Are You Prepared?' *Answers are given at the end of these exercises. If you get a wrong answer, read the pages listed in red.*

1. Plot the points (1, 5), (2, 6), (3, 9), (1, 12) in the Cartesian plane. Is the relation {(1, 5), (2, 6), (3, 9), (1, 12)} a function? Why? (pp. 86 and 215)

2. Find the equation of the line containing the points (1, 4) and (3, 8). (p. 183)

Concepts and Vocabulary

3. A _____ _____ is used to help us to see the type of relation, if any, that may exist between two variables.

4. ***True or False*** The correlation coefficient is a measure of the strength of a linear relation between two variables and must lie between −1 and 1, inclusive.

Skill Building

In Problems 5–10, examine the scatter diagram and determine whether the type of relation is linear or nonlinear.

5.
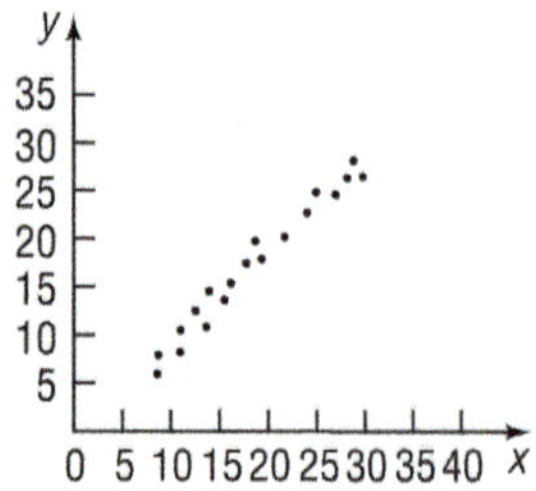

6.
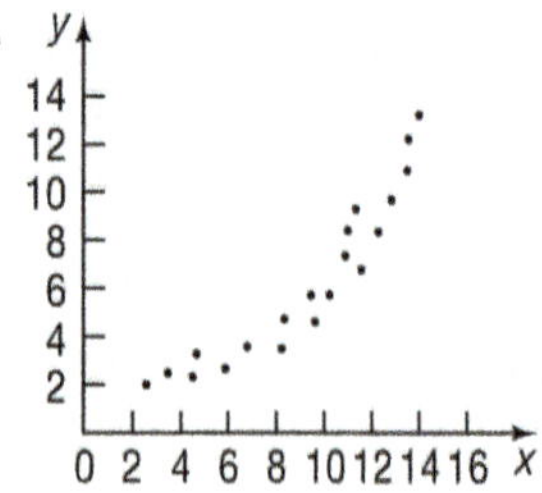

7.
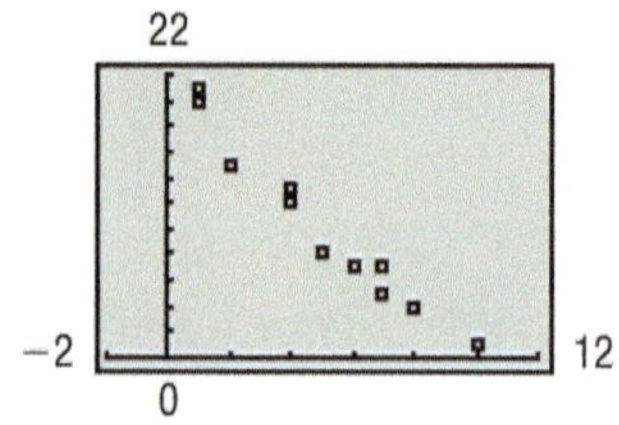

8.
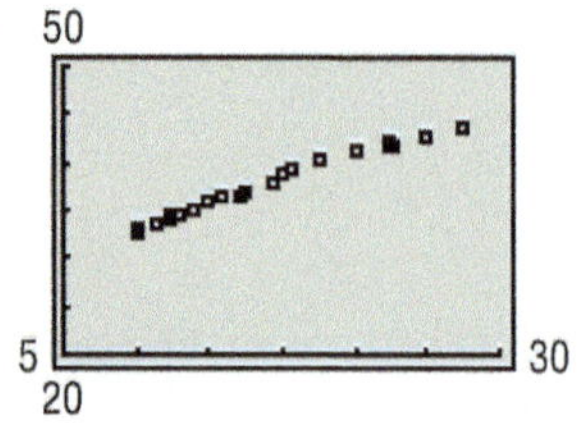

9.
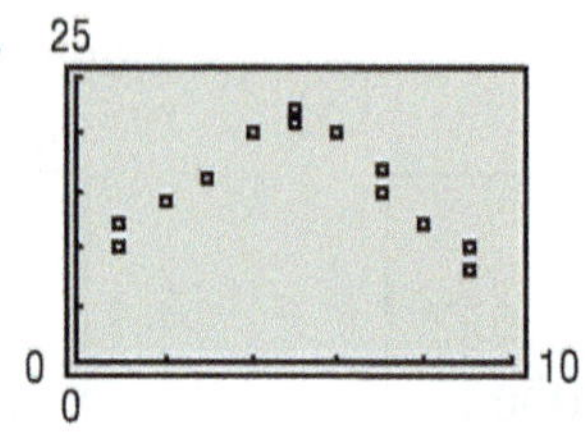

10.
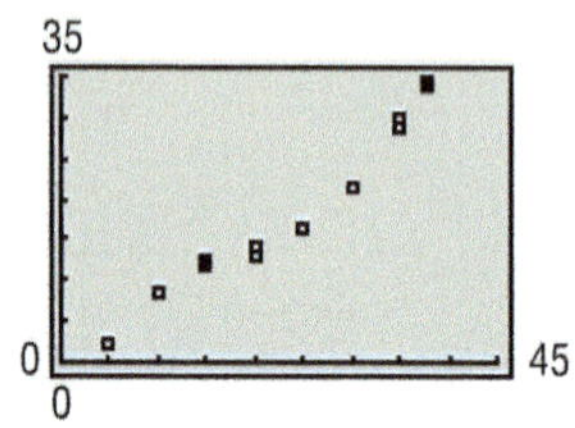

In Problems 11–16,

(a) Draw a scatter diagram.
(b) Select two points from the scatter diagram and find the equation of the line containing the points selected.
(c) Graph the line found in part (b) on the scatter diagram.
(d) Use a graphing utility to find the line of best fit.
(e) Use a graphing utility to draw the scatter diagram and graph the line of best fit on it.

11.

x	3	4	5	6	7	8	9
y	4	6	7	10	12	14	16

12.

x	3	5	7	9	11	13
y	0	2	3	6	9	11

13.

x	−2	−1	0	1	2
y	−4	0	1	4	5

14.

x	−2	−1	0	1	2
y	7	6	3	2	0

15.

x	−20	−17	−15	−14	−10
y	100	120	118	130	140

16.

x	−30	−27	−25	−20	−14
y	10	12	13	13	18

Applications and Extensions

17. **Candy** The following data represent the weight (in grams) of various candy bars and the corresponding number of calories.

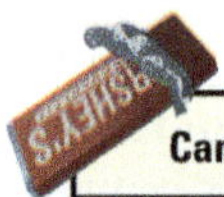

Candy Bar	Weight, x	Calories, y
Hershey's Milk Chocolate®	44.28	230
Nestle's Crunch®	44.84	230
Butterfinger®	61.30	270
Baby Ruth®	66.45	280
Almond Joy®	47.33	220
Twix® (with Caramel)	58.00	280
Snickers®	61.12	280
Heath®	39.52	210

Source: Megan Pocius, Student at Joliet Junior College

(a) Draw a scatter diagram of the data treating weight as the independent variable.
(b) What type of relation appears to exist between the weight of a candy bar and the number of calories?
(c) Select two points and find a linear model that contains the points.
(d) Graph the line on the scatter diagram drawn in part (a).
(e) Use the linear model to predict the number of calories in a candy bar that weighs 62.3 grams.
(f) Interpret the slope of the line found in part (c).

18. **Raisins** The following data represent the weight (in grams) of a box of raisins and the number of raisins in the box.

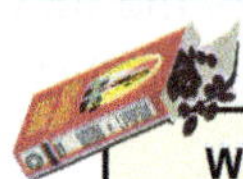

Weight (in grams), w	Number of Raisins, N
42.3	87
42.7	91
42.8	93
42.4	87
42.6	89
42.4	90
42.3	82
42.5	86
42.7	86
42.5	86

Source: Jennifer Maxwell, Student at Joliet Junior College

(a) Does the relation defined by the set of ordered pairs (w, N) represent a function?
(b) Draw a scatter diagram of the data treating weight as the independent variable.
(c) Select two points and find a linear model that contains the points.
(d) Graph the line on the scatter diagram drawn in part (b).
(e) Express the relationship found in part (c) using function notation.
(f) Use the linear model to predict the number of raisins in a box that weighs 42.5 grams.
(g) Interpret the slope of the line found in part (c).

19. Height versus Head Circumference A pediatrician wanted to find a linear model that relates a child's height, H, to head circumference, C. She randomly selects nine children from her practice, measures their height and head circumference, and obtains the data shown. Let H represent the independent variable and C the dependent variable.

(a) Use a graphing utility to draw a scatter diagram.
(b) Use a graphing utility to find the line of best fit that models the relation between height and head circumference. Express the solution using function notation.
(c) Interpret the slope.
(d) Predict the head circumference of a child that is 26 inches tall.
(e) What is the height of a child whose head circumference is 17.4 inches?

Height, H (inches)	Head Circumference, C (inches)
25.25	16.4
25.75	16.9
25	16.9
27.75	17.6
26.5	17.3
27	17.5
26.75	17.3
26.75	17.5
27.5	17.5

Source: Denise Slucki, Student at Joliet Junior College

20. Gestation Period versus Life Expectancy A researcher would like to develop a linear model relating the gestation period of an animal, G, and its life expectancy, L. She collects the following data. Let G represent the independent variable and L the dependent variable.

Animal	Gestation (or incubation) Period, G (days)	Life Expectancy, L (years)
Cat	63	11
Chicken	22	7.5
Dog	63	11
Duck	28	10
Goat	151	12
Lion	108	10
Parakeet	18	8
Pig	115	10
Rabbit	31	7
Squirrel	44	9

Source: Time Almanac 2007

(a) Use a graphing utility to draw a scatter diagram.
(b) Use a graphing utility to find the line of best fit that models the relation between gestation period and life expectancy. Express the solution using function notation.
(c) Interpret the slope.
(d) Predict the life expectancy of an animal whose gestation period is 89 days.

Mixed Practice

21. Demand for Jeans The marketing manager at Levi-Strauss wishes to find a function that relates the demand D for men's jeans and p, the price of the jeans. The following data were obtained based on a price history of the jeans.

Price (\$/Pair), p	Demand (Pairs of Jeans Sold per Day), D
20	60
22	57
23	56
23	53
27	52
29	49
30	44

(a) Does the relation defined by the set of ordered pairs (p, D) represent a function?
(b) Draw a scatter diagram of the data.
(c) Using a graphing utility, find the line of best fit that models the relation between price and quantity demanded.
(d) Interpret the slope.
(e) Express the relationship found in part (c) using function notation.
(f) What is the domain of the function?
(g) How many jeans will be demanded if the price is \$28 a pair?

22. Advertising and Sales Revenue A marketing firm wishes to find a function that relates the sales S of a product and A, the amount spent on advertising the product. The data are obtained from past experience. Advertising and sales are measured in thousands of dollars.

Advertising Expenditures, A	Sales, S
20	335
22	339
22.5	338
24	343
24	341
27	350
28.3	351

(a) Does the relation defined by the set of ordered pairs (A, S) represent a function?
(b) Draw a scatter diagram of the data.
(c) Using a graphing utility, find the line of best fit that models the relation between advertising expenditures and sales.
(d) Interpret the slope.
(e) Express the relationship found in part (c) using function notation.
(f) What is the domain of the function?
(g) Predict sales if advertising expenditures are \$25,000.

Discussion and Writing

23. **Maternal Age versus Down Syndrome** A biologist would like to know how the age of the mother affects the incidence rate of Down syndrome. The data to the right represent the age of the mother and the incidence rate of Down syndrome per 1000 pregnancies.

Draw a scatter diagram treating age of the mother as the independent variable. Would it make sense to find the line of best fit for these data? Why or why not?

24. Find the line of best fit for the ordered pairs (1, 5) and (3, 8). What is the correlation coefficient for these data? Why is this result reasonable?

25. What does a correlation coefficient of 0 imply?

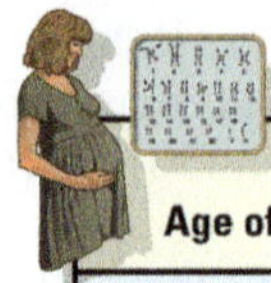

Age of Mother, x	Incidence of Down Syndrome, y
33	2.4
34	3.1
35	4
36	5
37	6.7
38	8.3
39	10
40	13.3
41	16.7
42	22.2
43	28.6
44	33.3
45	50

Source: Hook, E.B., *Journal of the American Medical Association*, 249, 2034-2038, 1983.

'Are You Prepared?' Answers

1.

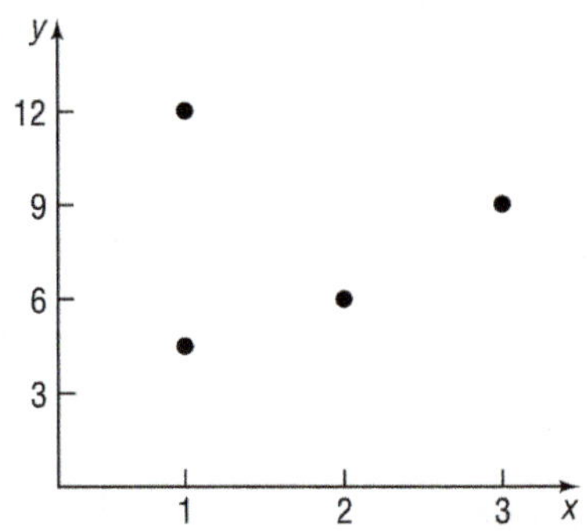

No, because an input, 1, corresponds to two different outputs.

2. $y = 2x + 2$

4.3 Quadratic Functions and Their Properties

PREPARING FOR THIS SECTION *Before getting started, review the following:*

- Intercepts (Section 2.1, pp. 170–171)
- Graphing Techniques: Transformations (Section 3.5, pp. 256–264)
- Completing the Square (Section 1.3, pp. 115–117)
- Quadratic Equations (Section 1.3, pp. 113–120)

Now Work the 'Are You Prepared?' problems on page 307.

OBJECTIVES
1 Graph a Quadratic Function Using Transformations (p. 300)
2 Identify the Vertex and Axis of Symmetry of a Quadratic Function (p. 302)
3 Graph a Quadratic Function Using Its Vertex, Axis, and Intercepts (p. 302)
4 Find the Maximum or Minimum Value of a Quadratic Function (p. 306)

Quadratic Functions

Here are some examples of quadratic functions.

$$F(x) = 3x^2 - 5x + 1 \qquad g(x) = -6x^2 + 1 \qquad H(x) = \frac{1}{2}x^2 + \frac{2}{3}x$$

Do you see what is similar among these three functions?

DEFINITION

A **quadratic function** is a function of the form

$$f(x) = ax^2 + bx + c$$

where a, b, and c are real numbers and $a \neq 0$. The domain of a quadratic function is the set of all real numbers.

In Words

A quadratic function is a function defined by a second-degree polynomial in one variable.

Many applications require a knowledge of quadratic functions. For example, suppose that Texas Instruments collects the data shown in Table 7, which relate the number of calculators sold at the price p (in dollars) per calculator. Since the price of a product determines the quantity that will be purchased, we treat price as the independent variable. The relationship between the number x of calculators sold and the price p per calculator can be modeled by the linear equation

$$x = 21{,}000 - 150p$$

Table 7

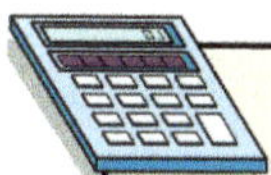

Price per Calculator, p (Dollars)	Number of Calculators, x
60	12,000
65	11,250
70	10,500
75	9,750
80	9,000
85	8,250
90	7,500

Then the revenue R derived from selling x calculators at the price p per calculator is equal to the unit selling price p of the calculator times the number x of units actually sold. That is,

$$\begin{aligned} R &= xp \\ R(p) &= (21{,}000 - 150p)p \quad x = 21{,}000 - 150p \\ &= -150p^2 + 21{,}000p \end{aligned}$$

Figure 12

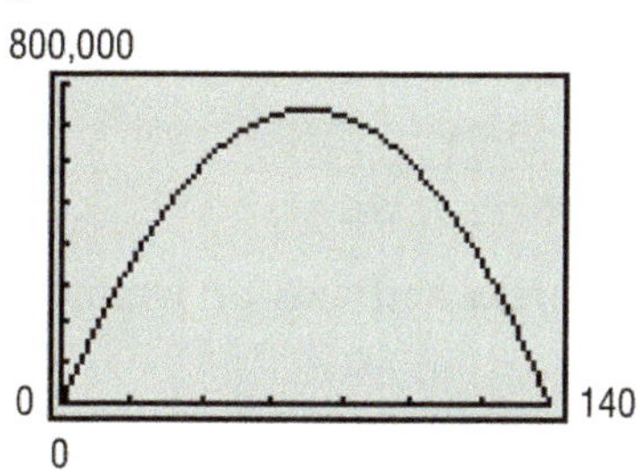

So the revenue R is a quadratic function of the price p. Figure 12 illustrates the graph of this revenue function, whose domain is $0 \le p \le 140$, since both x and p must be nonnegative. In the next section we shall determine the price p that maximizes revenue.

A second situation in which a quadratic function appears involves the motion of a projectile. Based on Newton's Second Law of Motion (force equals mass times acceleration, $F = ma$), it can be shown that, ignoring air resistance, the path of a projectile propelled upward at an inclination to the horizontal is the graph of a quadratic function. See Figure 13 for an illustration. In the next section we shall analyze the path of a projectile.

Figure 13
Path of a cannonball

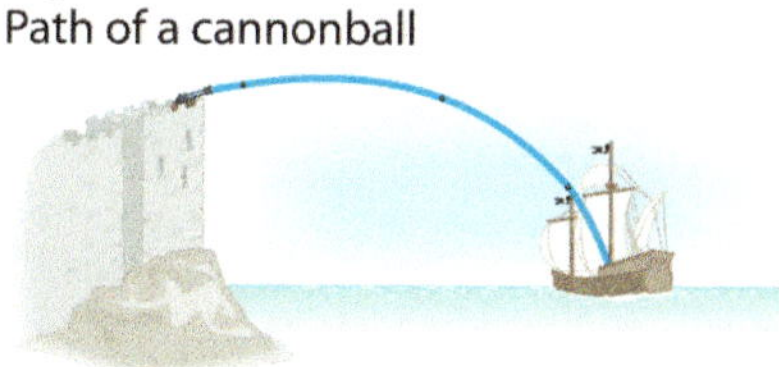

1 Graph a Quadratic Function Using Transformations

We know how to graph the square function $f(x) = x^2$. Figure 14 shows the graph of three functions of the form $f(x) = ax^2, a > 0$, for $a = 1, a = \frac{1}{2}$, and $a = 3$. Notice that the larger the value of a, the "narrower" the graph is, and the smaller the value of a, the "wider" the graph is.

Figure 15 shows graphs of $f(x) = ax^2$ for $a < 0$. Notice that these graphs are reflections about the x-axis of the graphs in Figure 15. Based on the results of these two figures, we can draw some general conclusions about the graph of $f(x) = ax^2$. First, as $|a|$ increases, the graph becomes "taller" (a vertical stretch), and as $|a|$ gets closer to zero, the graph gets "shorter" (a vertical compression). Second, if a is positive, the graph opens "up," and if a is negative, the graph opens "down."

Figure 14

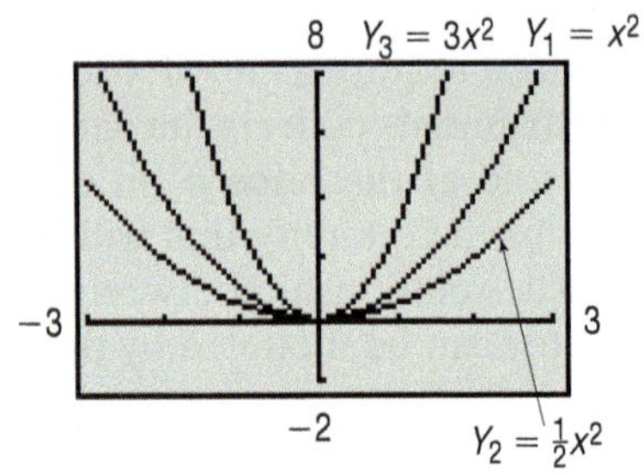

Figure 15

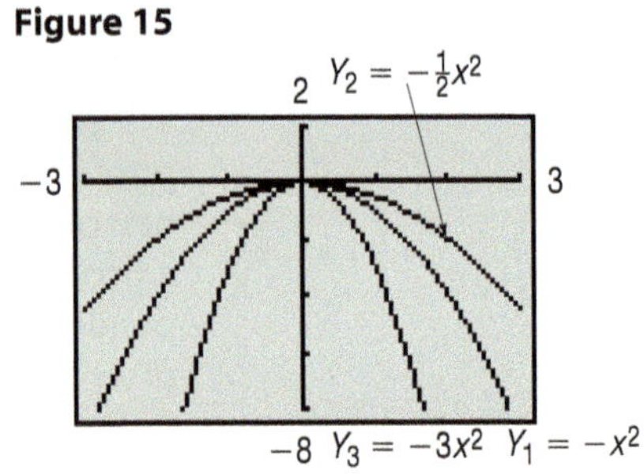

Figure 16
Graphs of a quadratic function, $f(x) = ax^2 + bx + c, a \neq 0$

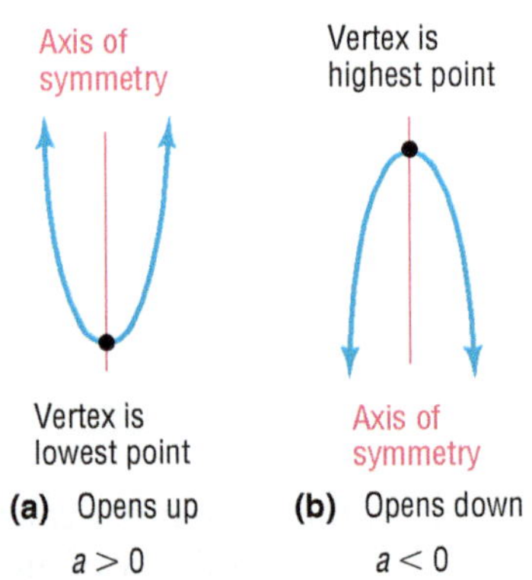

The graphs in Figures 14 and 15 are typical of the graphs of all quadratic functions, which we call **parabolas.*** Refer to Figure 16, where two parabolas are pictured. The one on the left **opens up** and has a lowest point; the one on the right **opens down** and has a highest point. The lowest or highest point of a parabola is called the **vertex.** The vertical line passing through the vertex in each parabola in Figure 16 is called the **axis of symmetry** (usually abbreviated to **axis**) of the parabola. Because the parabola is symmetric about its axis, the axis of symmetry of a parabola can be used to find additional points on the parabola.

The parabolas shown in Figure 16 are the graphs of a quadratic function $f(x) = ax^2 + bx + c, a \neq 0$. Notice that the coordinate axes are not included in the figure. Depending on the values of a, b, and c, the axes could be placed anywhere. The important fact is that the shape of the graph of a quadratic function will look like one of the parabolas in Figure 16.

In the following example, we use techniques from Section 3.5 to graph a quadratic function $f(x) = ax^2 + bx + c, a \neq 0$. In so doing, we shall complete the square and write the function f in the form $f(x) = a(x - h)^2 + k$.

EXAMPLE 1 Graphing a Quadratic Function Using Transformations

Graph the function $f(x) = 2x^2 + 8x + 5$. Find the vertex and axis of symmetry.

Solution We begin by completing the square on the right side.

$$\begin{aligned} f(x) &= 2x^2 + 8x + 5 \\ &= 2(x^2 + 4x) + 5 && \text{Factor out the 2 from } 2x^2 + 8x. \\ &= 2(x^2 + 4x + 4) + 5 - 8 && \text{Complete the square of } x^2 + 4x \text{ by adding 4. Notice that the factor of 2 requires that 8 be added and subtracted.} \\ &= 2(x + 2)^2 - 3 \end{aligned}$$

*We shall study parabolas using a geometric definition later in this book.

The graph of f can be obtained in three stages, as shown in Figure 17. Now compare this graph to the graph in Figure 16(a). The graph of $f(x) = 2x^2 + 8x + 5$ is a parabola that opens up and has its vertex (lowest point) at $(-2, -3)$. Its axis of symmetry is the line $x = -2$.

Figure 17

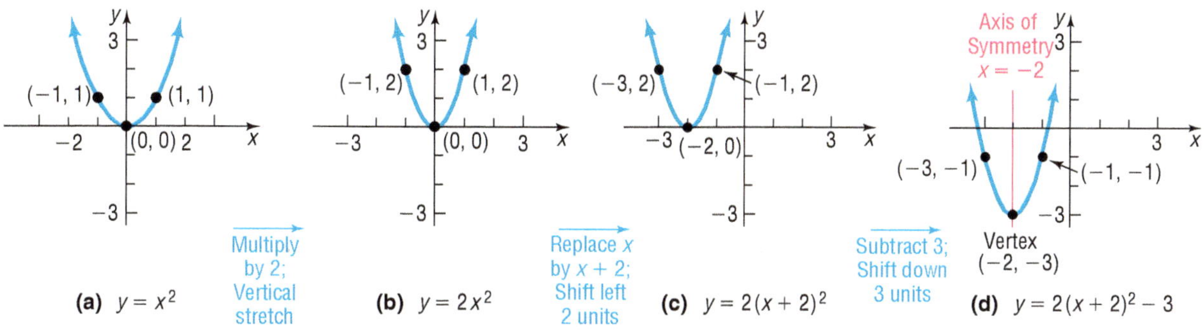

(a) $y = x^2$ (b) $y = 2x^2$ (c) $y = 2(x+2)^2$ (d) $y = 2(x+2)^2 - 3$

Check: Use a graphing utility to graph $Y_1 = f(x) = 2x^2 + 8x + 5$ and use the MINIMUM command to locate its vertex.

Now Work PROBLEM 21

The method used in Example 1 can be used to graph any quadratic function $f(x) = ax^2 + bx + c$, $a \neq 0$, as follows:

$$\begin{aligned} f(x) &= ax^2 + bx + c \\ &= a\left(x^2 + \frac{b}{a}x\right) + c && \text{Factor out } a \text{ from } ax^2 + bx. \\ &= a\left(x^2 + \frac{b}{a}x + \frac{b^2}{4a^2}\right) + c - a\left(\frac{b^2}{4a^2}\right) && \text{Complete the square by adding } \frac{b^2}{4a^2}. \text{ Look closely at this step!} \\ &= a\left(x + \frac{b}{2a}\right)^2 + c - \frac{b^2}{4a} && \text{Factor; Simplify} \\ &= a\left(x + \frac{b}{2a}\right)^2 + \frac{4ac - b^2}{4a} && c - \frac{b^2}{4a} = c \cdot \frac{4a}{4a} - \frac{b^2}{4a} = \frac{4ac - b^2}{4a} \\ &= a\left(x - \left(-\frac{b}{2a}\right)\right)^2 + \frac{4ac - b^2}{4a} \end{aligned}$$

Based on these results, we conclude the following:

If $h = -\dfrac{b}{2a}$ and $k = \dfrac{4ac - b^2}{4a}$, then

$$f(x) = ax^2 + bx + c = a(x - h)^2 + k \qquad \textbf{(1)}$$

The graph of $f(x) = a(x - h)^2 + k$ is the parabola $y = ax^2$ shifted horizontally h units (replace x by $x - h$) and vertically k units (add k). As a result, the vertex is at (h, k), and the graph opens up if $a > 0$ and down if $a < 0$. The axis of symmetry is the vertical line $x = h$.

For example, compare equation (1) with the solution given in Example 1.

$$\begin{aligned} f(x) &= 2(x + 2)^2 - 3 \\ &= 2(x - (-2))^2 + (-3) \\ &\ \ \uparrow \qquad\quad \uparrow \qquad\quad \uparrow \\ &= a(x - \quad h)^2 + \quad k \end{aligned}$$

We conclude that $a = 2$, so the graph opens up. Also, we find that $h = -2$ and $k = -3$, so its vertex is at $(-2, -3)$.

2 Identify the Vertex and Axis of Symmetry of a Quadratic Function

We do not need to complete the square to obtain the vertex. In almost every case, it is easier to obtain the vertex of a quadratic function f by remembering that its x-coordinate is $h = -\frac{b}{2a}$. The y-coordinate k can then be found by evaluating f at $-\frac{b}{2a}$. That is, $k = f\left(-\frac{b}{2a}\right)$.

We summarize these remarks as follows:

Properties of the Graph of a Quadratic Function

$$f(x) = ax^2 + bx + c \qquad a \neq 0$$

$$\text{Vertex} = \left(-\frac{b}{2a}, f\left(-\frac{b}{2a}\right)\right) \qquad \text{Axis of symmetry: the line } x = -\frac{b}{2a} \qquad \textbf{(2)}$$

Parabola opens up if $a > 0$; the vertex is a minimum point.
Parabola opens down if $a < 0$; the vertex is a maximum point.

EXAMPLE 2 **Locating the Vertex without Graphing**

Without graphing, locate the vertex and axis of symmetry of the parabola defined by $f(x) = -3x^2 + 6x + 1$. Does it open up or down?

Solution For this quadratic function, $a = -3$, $b = 6$, and $c = 1$. The x-coordinate of the vertex is

$$h = -\frac{b}{2a} = -\frac{6}{-6} = 1$$

The y-coordinate of the vertex is

$$k = f\left(-\frac{b}{2a}\right) = f(1) = -3(1)^2 + 6(1) + 1 = 4$$

The vertex is located at the point $(1, 4)$. The axis of symmetry is the line $x = 1$. Because $a = -3 < 0$, the parabola opens down.

Check: Use a graphing utility to graph $Y_1 = f(x) = -3x^2 + 6x + 1$ and use the MAXIMUM command to locate its vertex.

3 Graph a Quadratic Function Using Its Vertex, Axis, and Intercepts

The location of the vertex and the location of the intercepts, along with knowledge as to whether the graph of the quadratic function opens up or down, usually provide enough information to graph $f(x) = ax^2 + bx + c$, $a \neq 0$.

The y-intercept is the value of f at $x = 0$; that is, the y-intercept is $f(0) = c$.

The x-intercepts, if there are any, are found by solving the quadratic equation

$$ax^2 + bx + c = 0$$

This equation has two, one, or no real solutions, depending on whether the discriminant $b^2 - 4ac$ is positive, 0, or negative. Depending on the value of the discriminant, the graph of f has x-intercepts, as follows:

The x-Intercepts of a Quadratic Function

1. If the discriminant $b^2 - 4ac > 0$, the graph of $f(x) = ax^2 + bx + c$ has two distinct x-intercepts, so it crosses the x-axis in two places.
2. If the discriminant $b^2 - 4ac = 0$, the graph of $f(x) = ax^2 + bx + c$ has one x-intercept, so it touches the x-axis at its vertex.
3. If the discriminant $b^2 - 4ac < 0$, the graph of $f(x) = ax^2 + bx + c$ has no x-intercept, so it does not cross or touch the x-axis.

Figure 18 illustrates these possibilities for parabolas that open up.

Figure 18
$f(x) = ax^2 + bx + c, a > 0$

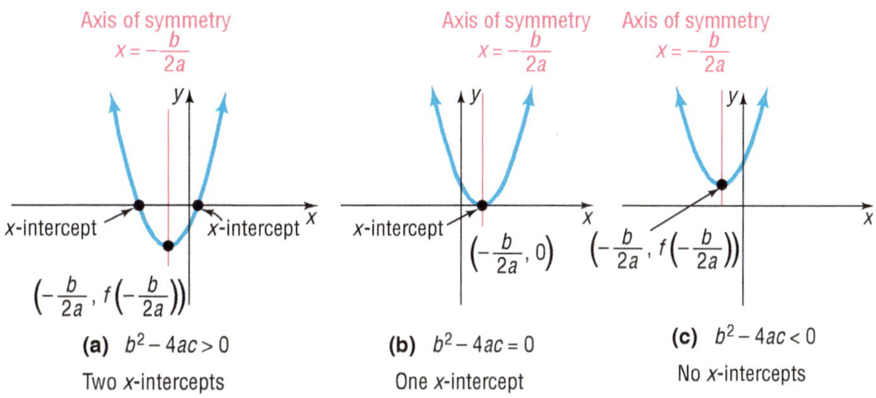

EXAMPLE 3 How to Graph a Quadratic Function by Hand Using Its Properties

Graph $f(x) = -3x^2 + 6x + 1$ using its properties. Determine the domain and the range of f. Determine where f is increasing and where it is decreasing.

Step-by-Step Solution

STEP 1 Determine whether the graph of f opens up or down.

In Example 2, we determined that the graph of $f(x) = -3x^2 + 6x + 1$ opens down because $a = -3 < 0$.

STEP 2 Determine the vertex and axis of symmetry of the graph of f.

In Example 2, we found the vertex to be at the point whose coordinates are $(1, 4)$. The axis of symmetry is the line $x = 1$.

STEP 3 Determine the intercepts of the graph of f.

The y-intercept is found by letting $x = 0$. The y-intercept is $f(0) = 1$. The x-intercepts are found by solving the equation $f(x) = 0$.

$$f(x) = 0$$
$$-3x^2 + 6x + 1 = 0 \qquad a = -3, b = 6, c = 1$$

The discriminant $b^2 - 4ac = 6^2 - 4(-3)(1) = 36 + 12 = 48 > 0$, so the equation has two real solutions and the graph has two x-intercepts. Using the quadratic formula, we find that

$$x = \frac{-b + \sqrt{b^2 - 4ac}}{2a} = \frac{-6 + \sqrt{48}}{2(-3)} = \frac{-6 + 4\sqrt{3}}{-6} \approx -0.15$$

and

$$x = \frac{-b - \sqrt{b^2 - 4ac}}{2a} = \frac{-6 - \sqrt{48}}{2(-3)} = \frac{-6 - 4\sqrt{3}}{-6} \approx 2.15$$

The x-intercepts are approximately -0.15 and 2.15.

STEP 4 Use the information in Steps 1 through 3 to graph f.

The graph is illustrated in Figure 19. Notice how we use the y-intercept and the axis of symmetry, $x = 1$, to obtain the additional point $(2, 1)$ on the graph.

Figure 19

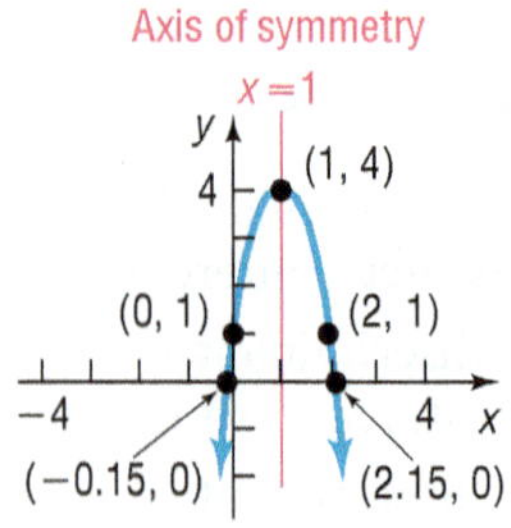

The domain of f is the set of all real numbers. Based on the graph, the range of f is the interval $(-\infty, 4]$. The function f is increasing on the interval $(-\infty, 1)$ and decreasing on the interval $(1, \infty)$.

Graph the function in Example 3 by completing the square and using transformations. Which method do you prefer?

Now Work PROBLEM 29

If the graph of a quadratic function has only one x-intercept or no x-intercepts, it is usually necessary to plot an additional point to obtain the graph.

EXAMPLE 4 Graphing a Quadratic Function Using Its Vertex, Axis, and Intercepts

(a) Graph $f(x) = x^2 - 6x + 9$ by determining whether the graph opens up or down and by finding its vertex, axis of symmetry, y-intercept, and x-intercepts, if any.

(b) Determine the domain and the range of f.

(c) Determine where f is increasing and where it is decreasing.

Solution

(a) **STEP 1:** For $f(x) = x^2 - 6x + 9$, we have $a = 1$, $b = -6$, and $c = 9$. Since $a = 1 > 0$, the parabola opens up.

STEP 2: The x-coordinate of the vertex is

$$h = -\frac{b}{2a} = -\frac{-6}{2(1)} = 3$$

The y-coordinate of the vertex is

$$k = f(3) = (3)^2 - 6(3) + 9 = 0$$

So the vertex is at $(3, 0)$. The axis of symmetry is the line $x = 3$.

STEP 3: The y-intercept is $f(0) = 9$. Since the vertex $(3, 0)$ lies on the x-axis, the graph touches the x-axis at the x-intercept.

STEP 4: By using the axis of symmetry and the y-intercept at $(0, 9)$, we can locate the additional point $(6, 9)$ on the graph. See Figure 20.

Figure 20

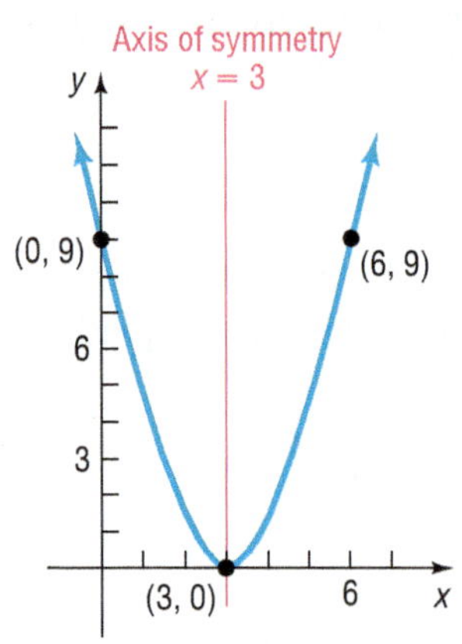

(b) The domain of f is the set of all real numbers. Based on the graph, the range of f is the interval $[0, \infty)$.

(c) The function f is decreasing on the interval $(-\infty, 3)$ and increasing on the interval $(3, \infty)$.

Now Work PROBLEM 35

EXAMPLE 5 **Graphing a Quadratic Function Using Its Vertex, Axis, and Intercepts**

(a) Graph $f(x) = 2x^2 + x + 1$ by determining whether the graph opens up or down and by finding its vertex, axis of symmetry, y-intercept, and x-intercepts, if any.

(b) Determine the domain and the range of f.

(c) Determine where f is increasing and where it is decreasing.

Solution (a) **Step 1:** For $f(x) = 2x^2 + x + 1$, we have $a = 2, b = 1$, and $c = 1$. Since $a = 2 > 0$, the parabola opens up.

Step 2: The x-coordinate of the vertex is

$$h = -\frac{b}{2a} = -\frac{1}{4}$$

The y-coordinate of the vertex is

$$k = f\left(-\frac{1}{4}\right) = 2\left(\frac{1}{16}\right) + \left(-\frac{1}{4}\right) + 1 = \frac{7}{8}$$

So the vertex is at $\left(-\frac{1}{4}, \frac{7}{8}\right)$. The axis of symmetry is the line $x = -\frac{1}{4}$.

Step 3: The y-intercept is $f(0) = 1$. The x-intercept(s), if any, obey the equation $2x^2 + x + 1 = 0$. Since the discriminant $b^2 - 4ac = (1)^2 - 4(2)(1) = -7 < 0$, this equation has no real solutions, and therefore the graph has no x-intercepts.

Step 4: We use the point $(0, 1)$ and the axis of symmetry $x = -\frac{1}{4}$ to locate the additional point $\left(-\frac{1}{2}, 1\right)$ on the graph. See Figure 21.

NOTE In Example 5, since the vertex is above the x-axis and the parabola opens up, we can conclude that the graph of the quadratic function will have no x-intercepts. ■

Figure 21

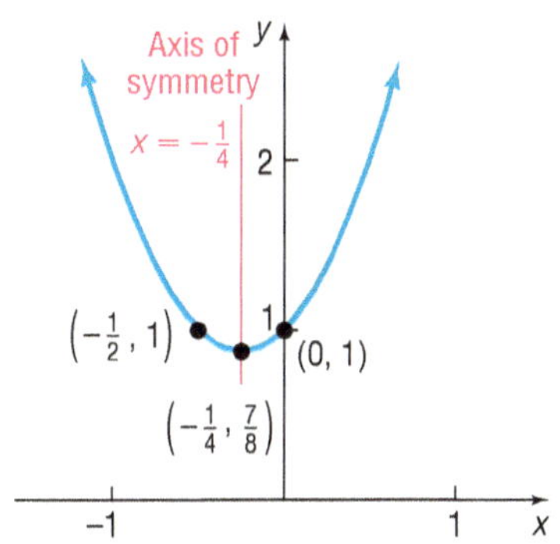

(b) The domain of f is the set of all real numbers. Based on the graph, the range of f is the interval $\left[\frac{7}{8}, \infty\right)$.

(c) The function f is decreasing on the interval $\left(-\infty, -\frac{1}{4}\right)$ and is increasing on the interval $\left(-\frac{1}{4}, \infty\right)$.

Now Work PROBLEM 39

SUMMARY **Steps for Graphing a Quadratic Function $f(x) = ax^2 + bx + c, a \neq 0$, by Hand**

Option 1

Step 1: Complete the square in x to write the quadratic function in the form $f(x) = a(x - h)^2 + k$.

Step 2: Graph the function in stages using transformations.

Option 2

Step 1: Determine whether the graph of f opens up or down.

Step 2: Determine the vertex $\left(-\frac{b}{2a}, f\left(-\frac{b}{2a}\right)\right)$ and the axis of symmetry, $x = -\frac{b}{2a}$.

Step 3: Determine the y-intercept, $f(0)$. Determine the x-intercept(s), if any.

(a) If $b^2 - 4ac > 0$, then the graph of the quadratic function has two x-intercepts, which are found by solving the equation $ax^2 + bx + c = 0$.

(b) If $b^2 - 4ac = 0$, the vertex is the x-intercept.

(c) If $b^2 - 4ac < 0$, there are no x-intercepts.

Step 4: Determine an additional point by using the y-intercept and the axis of symmetry. Plot the points and draw the graph.

Given the vertex (h, k) and one additional point on the graph of a quadratic function $f(x) = ax^2 + bx + c$, $a \neq 0$, we can use

$$f(x) = a(x - h)^2 + k \qquad (3)$$

to obtain the quadratic function.

EXAMPLE 6 Finding the Quadratic Function Given Its Vertex and One Other Point

Determine the quadratic function whose vertex is $(1, -5)$ and whose y-intercept is -3. The graph of the parabola is shown in Figure 22.

Figure 22

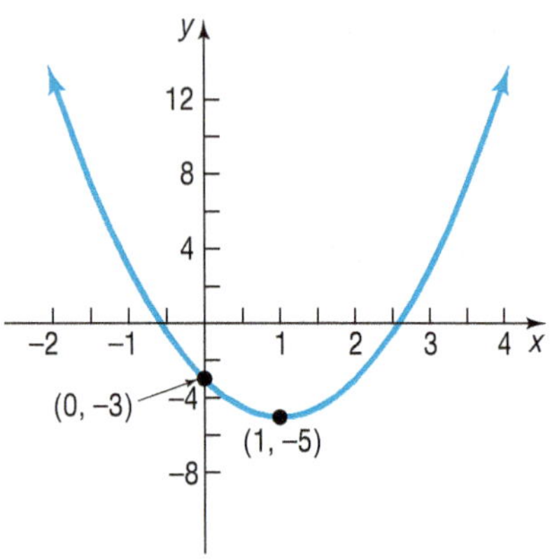

Solution The vertex is $(1, -5)$, so $h = 1$ and $k = -5$. Substitute these values into equation (3).

$$f(x) = a(x - h)^2 + k \quad \text{Equation (3)}$$
$$f(x) = a(x - 1)^2 - 5 \quad h = 1, k = -5$$

To determine the value of a, we use the fact that $f(0) = -3$ (the y-intercept).

$$f(x) = a(x - 1)^2 - 5$$
$$-3 = a(0 - 1)^2 - 5 \quad x = 0, y = f(0) = -3$$
$$-3 = a - 5$$
$$a = 2$$

The quadratic function whose graph is shown in Figure 22 is

$$f(x) = a(x - h)^2 + k = 2(x - 1)^2 - 5 = 2x^2 - 4x - 3$$

Check: Figure 23 shows the graph $f(x) = 2x^2 - 4x - 3$ using a graphing utility.

Figure 23

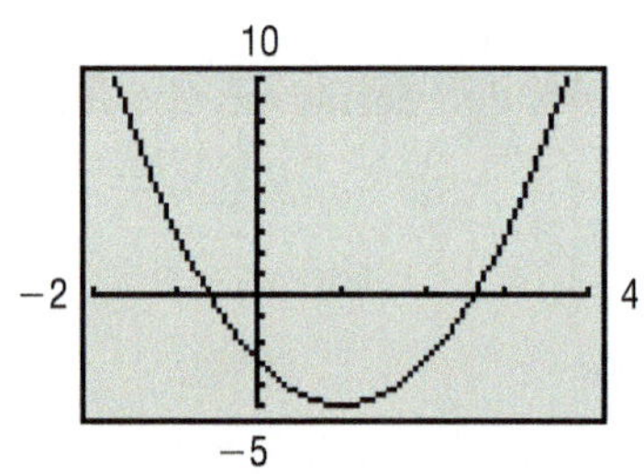

Now Work PROBLEM 43

4 Find the Maximum or Minimum Value of a Quadratic Function

The graph of a quadratic function

$$f(x) = ax^2 + bx + c, \qquad a \neq 0$$

is a parabola with vertex at $\left(-\dfrac{b}{2a}, f\left(-\dfrac{b}{2a}\right)\right)$. This vertex is the highest point on the graph if $a < 0$ and the lowest point on the graph if $a > 0$. If the vertex is the highest point $(a < 0)$, then $f\left(-\dfrac{b}{2a}\right)$ is the **maximum value** of f. If the vertex is the lowest point $(a > 0)$, then $f\left(-\dfrac{b}{2a}\right)$ is the **minimum value** of f.

EXAMPLE 7 **Finding the Maximum or Minimum Value of a Quadratic Function**

Determine whether the quadratic function

$$f(x) = x^2 - 4x - 5$$

has a maximum or minimum value. Then find the maximum or minimum value.

Solution We compare $f(x) = x^2 - 4x - 5$ to $f(x) = ax^2 + bx + c$. We conclude that $a = 1, b = -4$, and $c = -5$. Since $a > 0$, the graph of f opens up, so the vertex is a minimum point. The minimum value occurs at

$$x = -\frac{b}{2a} \underset{\uparrow}{=} -\frac{-4}{2(1)} = \frac{4}{2} = 2$$

$a = 1, b = -4$

The minimum value is

$$f\left(-\frac{b}{2a}\right) = f(2) = 2^2 - 4(2) - 5 = 4 - 8 - 5 = -9$$

Table 8

X	Y1
-1	0
0	-5
1	-8
2	-9
3	-8
4	-5
5	0

Y1=X²-4X-5

Check: We can support the algebraic solution using the TABLE feature on a graphing utility. Create Table 8 by letting $Y_1 = x^2 - 4x - 5$. From the table we see that the smallest value of y occurs when $x = 2$, leading us to investigate the number 2 further. The symmetry about $x = 2$ confirms that 2 is the x-coordinate of the vertex of the parabola. We conclude that the minimum value is -9 and occurs at $x = 2$.

Now Work PROBLEM 51

4.3 Assess Your Understanding

'Are You Prepared?' *Answers are given at the end of these exercises. If you get a wrong answer, read the pages listed in red.*

1. List the intercepts of the graph of $y = x^2 - 9$. (pp. 170–171)
2. Find the real solutions of the equation $2x^2 + 7x - 4 = 0$. (pp. 113–114)
3. To complete the square of $x^2 - 5x$, you add the number ______. (pp. 115–116)
4. To graph $y = (x - 4)^2$, you shift the graph of $y = x^2$ to the ______ a distance of ______ units. (pp. 256–257)

Concepts and Vocabulary

5. The graph of a quadratic function is called a(n) ______.
6. The vertical line passing through the vertex of a parabola is called the ______.
7. The x-coordinate of the vertex of $f(x) = ax^2 + bx + c$, $a \neq 0$, is ______.
8. ***True or False*** The graph of $f(x) = 2x^2 + 3x - 4$ opens up.
9. ***True or False*** The y-coordinate of the vertex of $f(x) = -x^2 + 4x + 5$ is $f(2)$.
10. ***True or False*** If the discriminant $b^2 - 4ac = 0$, the graph of $f(x) = ax^2 + bx + c$, $a \neq 0$, will touch the x-axis at its vertex.

Skill Building

In Problems 11–18, match graphs A–H to one the following functions.

11. $f(x) = x^2 - 1$
12. $f(x) = -x^2 - 1$
13. $f(x) = x^2 - 2x + 1$
14. $f(x) = x^2 + 2x + 1$
15. $f(x) = x^2 - 2x + 2$
16. $f(x) = x^2 + 2x$
17. $f(x) = (x - 1)^2 - 1$
18. $f(x) = (x + 1)^2 + 1$

A.
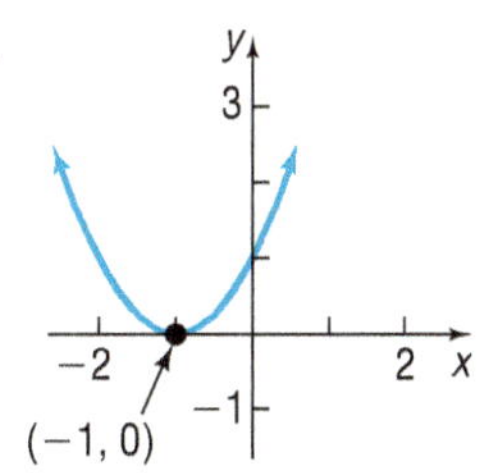

B.
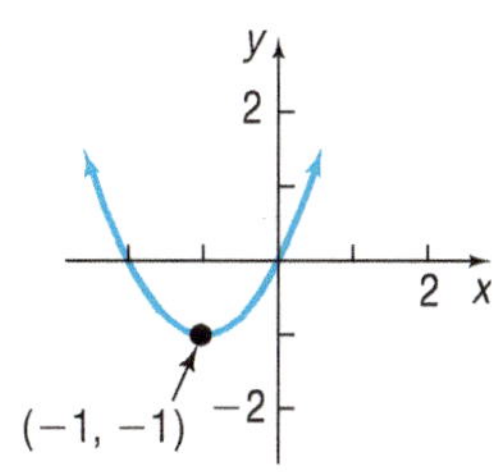

C.
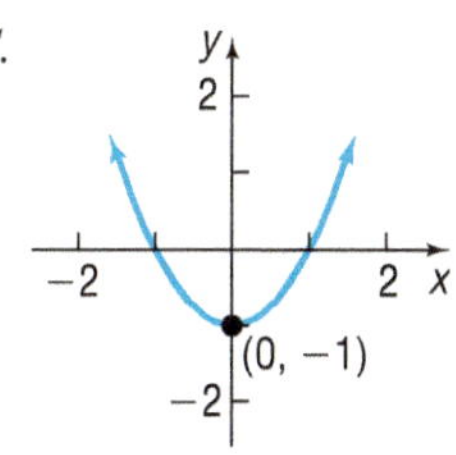

D.
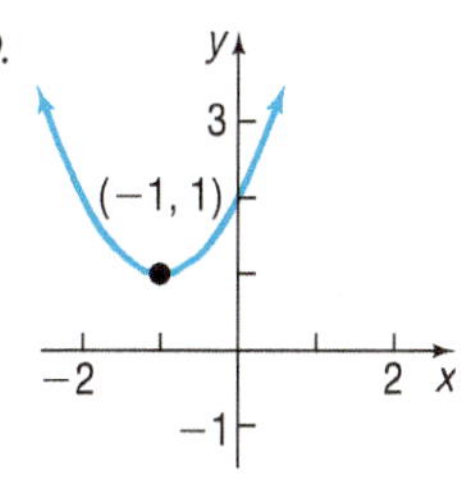

E.
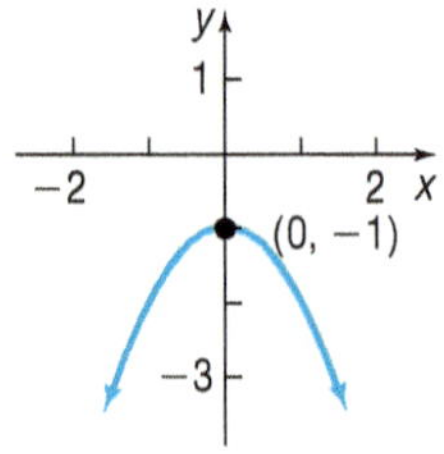

F.
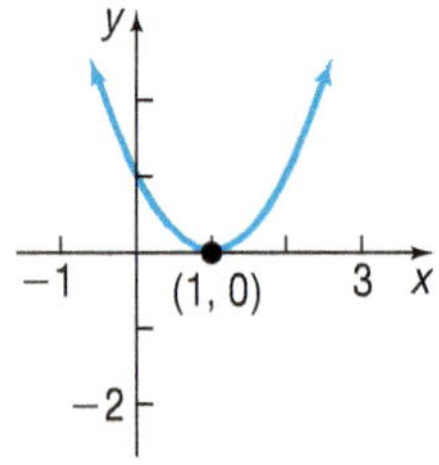

G.
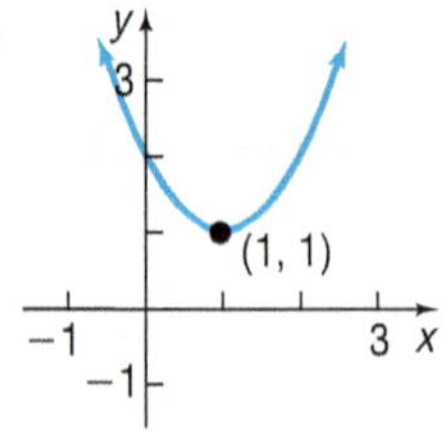

H.
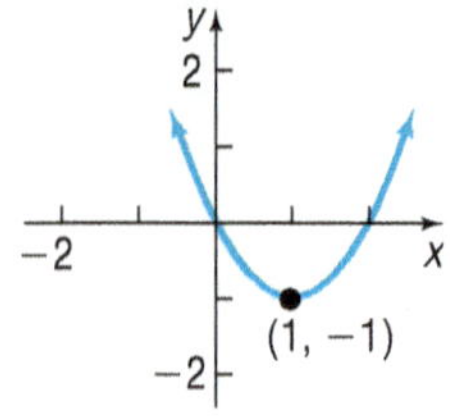

In Problems 19–28, graph the function f by starting with the graph of $y = x^2$ and using transformations (shifting, compressing, stretching, and/or reflection). Verify your results using a graphing utility.
[**Hint:** If necessary, write f in the form $f(x) = a(x - h)^2 + k$.]

19. $f(x) = \frac{1}{4}x^2 - 2$ **20.** $f(x) = 2x^2 + 4$ **21.** $f(x) = (x + 2)^2 - 2$ **22.** $f(x) = (x - 3)^2 - 10$

23. $f(x) = 2(x - 1)^2 - 1$ **24.** $f(x) = 3(x + 1)^2 - 3$ **25.** $f(x) = -x^2 - 2x$ **26.** $f(x) = -2x^2 + 6x + 2$

27. $f(x) = 2x^2 + 4x - 2$ **28.** $f(x) = 3x^2 + 12x + 5$

In Problems 29–42, (a) graph each quadratic function by determining whether its graph opens up or down and by finding its vertex, axis of symmetry, y-intercept, and x-intercepts, if any. (b) Determine the domain and the range of the function. (c) Determine where the function is increasing and where it is decreasing. Verify your results using a graphing utility.

29. $f(x) = x^2 + 2x$ **30.** $f(x) = x^2 - 4x$ **31.** $f(x) = -x^2 - 6x$ **32.** $f(x) = -x^2 + 4x$

33. $f(x) = x^2 + 2x - 8$ **34.** $f(x) = x^2 - 2x - 3$ **35.** $f(x) = x^2 + 2x + 1$ **36.** $f(x) = x^2 + 6x + 9$

37. $f(x) = 2x^2 - x + 2$ **38.** $f(x) = 4x^2 - 2x + 1$ **39.** $f(x) = -2x^2 + 2x - 3$ **40.** $f(x) = -3x^2 + 3x - 2$

41. $f(x) = -4x^2 - 6x + 2$ **42.** $f(x) = 3x^2 - 8x + 2$

In Problems 43–48, determine the quadratic function whose graph is given. Verify your results using a graphing utility.

43.
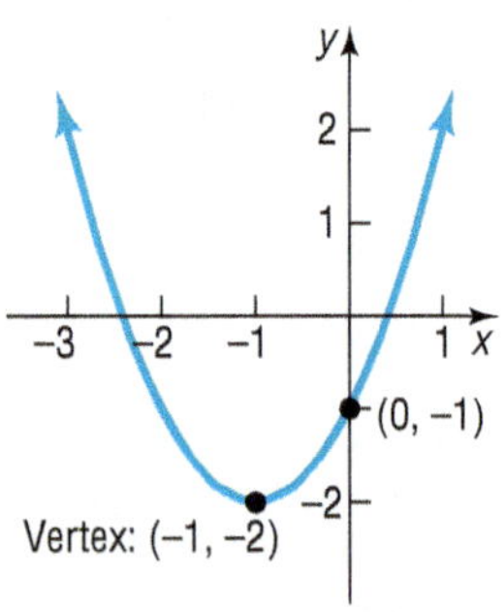

44.
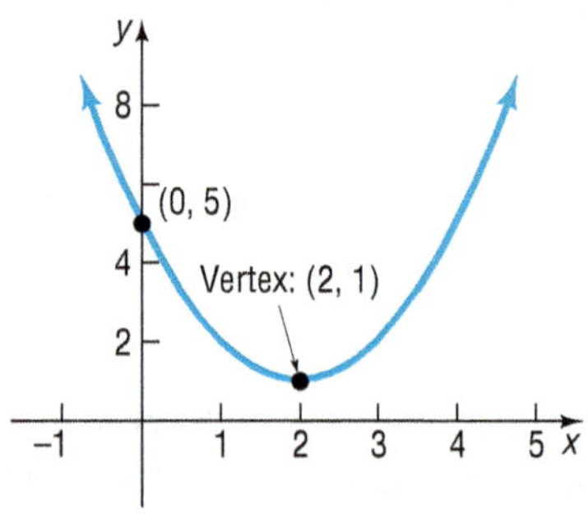

45.
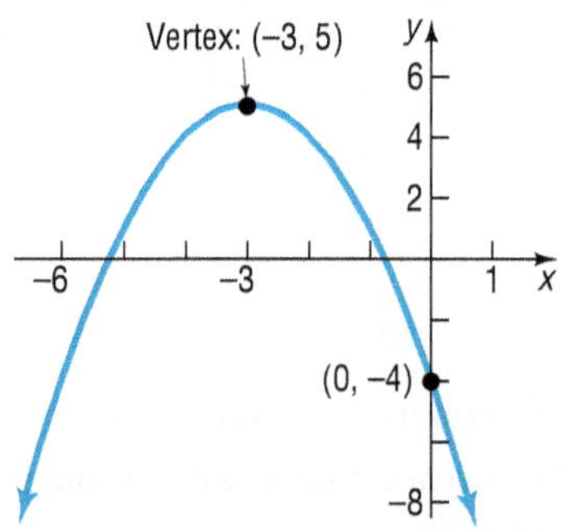

46.
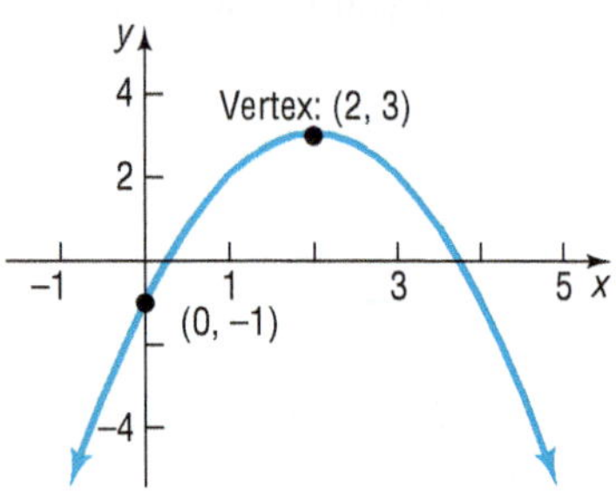

47.
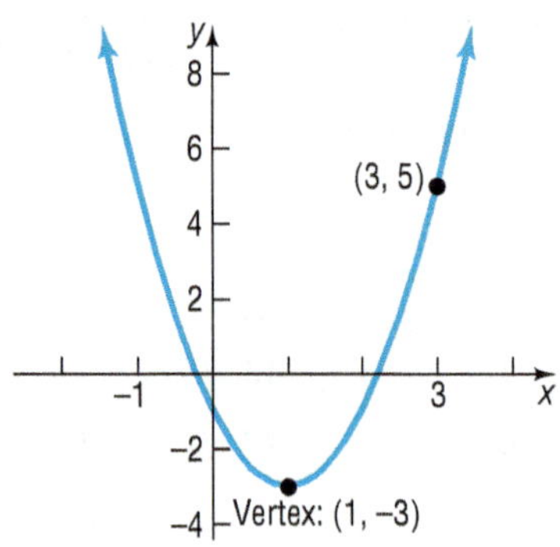

48.
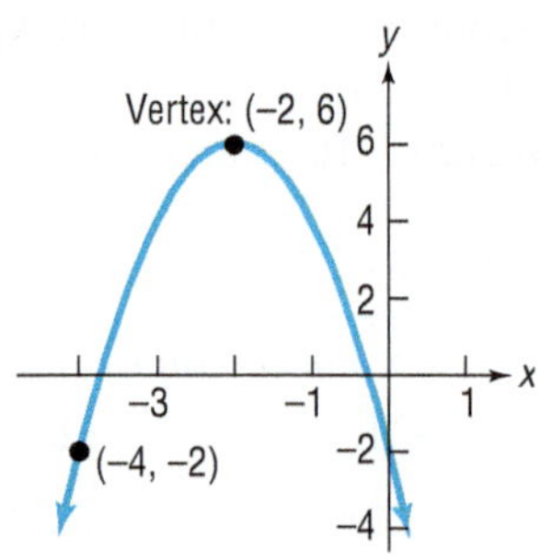

In Problems 49–56, determine, without graphing, whether the given quadratic function has a maximum value or a minimum value and then find the value. Use a graphing utility to verify your result.

49. $f(x) = 2x^2 + 12x$ **50.** $f(x) = -2x^2 + 12x$ **51.** $f(x) = 2x^2 + 12x - 3$ **52.** $f(x) = 4x^2 - 8x + 3$

53. $f(x) = -x^2 + 10x - 4$ **54.** $f(x) = -2x^2 + 8x + 3$ **55.** $f(x) = -3x^2 + 12x + 1$ **56.** $f(x) = 4x^2 - 4x$

Mixed Practice

In Problems 57–68, (a) graph each function. (b) Determine the domain and the range of the function. (c) Determine where the function is increasing and where it is decreasing.

57. $f(x) = x^2 - 2x - 15$
58. $g(x) = x^2 - 2x - 8$
59. $F(x) = 2x - 5$
60. $f(x) = \frac{3}{2}x - 2$
61. $g(x) = -2(x - 3)^2 + 2$
62. $h(x) = -3(x + 1)^2 + 4$
63. $f(x) = 2x^2 + x + 1$
64. $G(x) = 3x^2 + 2x + 5$
65. $h(x) = -\frac{2}{5}x + 4$
66. $f(x) = -3x + 2$
67. $H(x) = -4x^2 - 4x - 1$
68. $F(x) = -4x^2 + 20x - 25$

Applications and Extensions

69. The graph of the function $f(x) = ax^2 + bx + c$ has vertex at $(0, 2)$ and passes through the point $(1, 8)$. Find a, b, and c.

70. The graph of the function $f(x) = ax^2 + bx + c$ has vertex at $(1, 4)$ and passes through the point $(-1, -8)$. Find a, b, and c.

In Problems 71–76, for the given functions f and g,
(a) Graph f and g on the same Cartesian plane.
(b) Solve $f(x) = g(x)$.
(c) Use the result of part (b) to label the points of intersection of the graphs of f and g.
(d) Shade the region above f and below g between the points of intersection.

71. $f(x) = 2x - 1;\quad g(x) = x^2 - 4$
72. $f(x) = -2x - 1;\quad g(x) = x^2 - 9$
73. $f(x) = -x^2 + 4;\quad g(x) = -2x + 1$
74. $f(x) = -x^2 + 9;\quad g(x) = 2x + 1$
75. $f(x) = -x^2 + 5x;\quad g(x) = x^2 + 3x - 4$
76. $f(x) = -x^2 + 7x - 6;\quad g(x) = x^2 + x - 6$

Answer Problems 77 and 78 using the following: A quadratic function of the form $f(x) = ax^2 + bx + c$ with $b^2 - 4ac > 0$ may also be written in the form $f(x) = a(x - r_1)(x - r_2)$, where r_1 and r_2 are the x-intercepts of the graph of the quadratic function.

77. (a) Find a quadratic function whose x-intercepts are -3 and 1 with $a = 1$; $a = 2$; $a = -2$; $a = 5$.
(b) How does the value of a affect the intercepts?
(c) How does the value of a affect the axis of symmetry?
(d) How does the value of a affect the vertex?
(e) Compare the x-coordinate of the vertex with the midpoint of the x-intercepts. What might you conclude?

78. (a) Find a quadratic function whose x-intercepts are -5 and 3 with $a = 1$; $a = 2$; $a = -2$; $a = 5$.
(b) How does the value of a affect the intercepts?
(c) How does the value of a affect the axis of symmetry?
(d) How does the value of a affect the vertex?
(e) Compare the x-coordinate of the vertex with the midpoint of the x-intercepts. What might you conclude?

79. Suppose that $f(x) = x^2 + 4x - 21$.
(a) What is the vertex of f?
(b) What are the x-intercepts of the graph of f?
(c) Solve $f(x) = -21$ for x. What points are on the graph of f?
(d) Use the information obtained in parts (a)–(c) to graph $f(x) = x^2 + 4x - 21$.

80. Suppose that $f(x) = x^2 + 2x - 8$.
(a) What is the vertex of f?
(b) What are the x-intercepts of the graph of f?
(c) Solve $f(x) = -8$ for x. What points are on the graph of f?
(d) Use the information obtained in parts (a)–(c) to graph $f(x) = x^2 + 2x - 8$.

81. Find the point on the line $y = x$ that is closest to the point $(3, 1)$.
[**Hint:** Express the distance d from the point to the line as a function of x, and then find the minimum value of $[d(x)]^2$.

82. Find the point on the line $y = x + 1$ that is closest to the point $(4, 1)$.

83. Maximizing Revenue Suppose that the manufacturer of a gas clothes dryer has found that, when the unit price is p dollars, the revenue R (in dollars) is

$$R(p) = -4p^2 + 4000p$$

What unit price should be established for the dryer to maximize revenue? What is the maximum revenue?

84. Maximizing Revenue The John Deere company has found that the revenue, in dollars, from sales of heavy-duty tractors is a function of the unit price p, in dollars, that it charges. If the revenue R is

$$R(p) = -\frac{1}{2}p^2 + 1900p$$

what unit price p should be charged to maximize revenue? What is the maximum revenue?

85. Minimizing Marginal Cost The **marginal cost** of a product can be thought of as the cost of producing one additional unit of output. For example, if the marginal cost of producing the 50th product is \$6.20, it cost \$6.20 to increase production from 49 to 50 units of output. Suppose the marginal cost C (in dollars) to produce x thousand mp3 players is given by the function

$$C(x) = x^2 - 140x + 7400$$

(a) How many players should be produced to minimize the marginal cost?
(b) What is the minimum marginal cost?

86. Minimizing Marginal Cost (See Problem 85.) The marginal cost C (in dollars) of manufacturing x cell phones (in thousands) is given by

$$C(x) = 5x^2 - 200x + 4000$$

(a) How many cell phones should be manufactured to minimize the marginal cost?
(b) What is the minimum marginal cost?

87. Hunting The function

$$H(x) = -3.24x^2 + 242.1x - 738.4$$

models the number of individuals whose age is x and who engage in hunting activities.

(a) What is the age at which there are the most hunters? Approximately how many hunters are this age?

(b) Are the number of hunters increasing or decreasing for individuals who are between 40 and 45 years of age?

Source: National Sporting Goods Association

88. Male Murder Victims The function

$$M(x) = 1.00x^2 - 136.74x + 4764.89$$

models the number of male murder victims who are x years of age $(20 \leq x < 90)$.

(a) Use the model to approximate the number of male murder victims who are $x = 23$ years of age.

(b) At what age is the number of male murder victims 1456?

(c) Describe what happens to the number of male murder victims as age increases from 20 to 65.

Source: Federal Bureau of Investigation

89. Business The monthly revenue R achieved by selling x wristwatches is modeled by $R(x) = 75x - 0.2x^2$. The monthly cost C of selling x wristwatches is $C(x) = 32x + 1750$.

(a) How many wristwatches must the firm sell to maximize revenue? What is the maximum revenue?

(b) Profit is given as $P(x) = R(x) - C(x)$. What is the profit function?

(c) How many wristwatches must the firm sell to maximize profit? What is the maximum profit?

(d) Provide a reasonable explanation as to why the answers found in parts (a) and (c) differ. Explain why a quadratic function is a reasonable model for revenue.

90. Business The daily revenue R achieved by selling x boxes of candy is modeled by $R(x) = 9.5x - 0.04x^2$. The daily cost C of selling x boxes of candy is $C(x) = 1.25x + 250$.

(a) How many boxes of candy must the firm sell to maximize revenue? What is the maximum revenue?

(b) Profit is given as $P(x) = R(x) - C(x)$. What is the profit function?

(c) How many boxes of candy must the firm sell to maximize profit? What is the maximum profit?

(d) Provide a reasonable explanation as to why the answers found in parts (a) and (c) differ. Explain why a quadratic function is a reasonable model for revenue.

91. Let $f(x) = ax^2 + bx + c$, where a, b, and c are odd integers. If x is an integer, show that $f(x)$ must be an odd integer.
[**Hint:** x is either an even integer or an odd integer.]

Discussion and Writing

92. Make up a quadratic function that opens down and has only one x-intercept. Compare yours with others in the class. What are the similarities? What are the differences?

93. On one set of coordinate axes, graph the family of parabolas $f(x) = x^2 + 2x + c$ for $c = -3$, $c = 0$, and $c = 1$. Describe the characteristics of a member of this family.

94. On one set of coordinate axes, graph the family of parabolas $f(x) = x^2 + bx + 1$ for $b = -4$, $b = 0$, and $b = 4$. Describe the general characteristics of this family.

95. State the circumstances that cause the graph of a quadratic function $f(x) = ax^2 + bx + c$ to have no x-intercepts.

96. Why does the graph of a quadratic function open up if $a > 0$ and down if $a < 0$?

97. Can a quadratic function have a range of $(-\infty, \infty)$? Justify your answer.

98. What are the possibilities for the number of times the graphs of two different quadratic functions intersect?

'Are You Prepared?' Answers

1. $(0, -9), (-3, 0), (3, 0)$ **2.** $\left\{-4, \frac{1}{2}\right\}$ **3.** $\frac{25}{4}$ **4.** right; 4

4.4 Building Quadratic Models from Verbal Descriptions and from Data

PREPARING FOR THIS SECTION *Before getting started, review the following:*

- Problem Solving (Section 1.6, p. 141)

Now Work the 'Are You Prepared?' problems on page 316.

OBJECTIVES
1 Build Quadratic Models from Verbal Descriptions (p. 311)
2 Build Quadratic Models from Data (p. 314)

In this section we will first discuss models that lead to a quadratic function when a verbal description of the problem is given. We end the section by fitting a quadratic function to data, which is another form of modeling.

When a mathematical model leads to a quadratic function, the properties of the graph of the quadratic function can provide important information about the model. In particular, we can use the quadratic function to determine the maximum or minimum value of the function. The fact that the graph of a quadratic function has a maximum or minimum value enables us to answer questions involving **optimization,** that is, finding the maximum or minimum values in models involving quadratic functions.

1 Build Quadratic Models from Verbal Descriptions

In economics, revenue R, in dollars, is defined as the amount of money received from the sale of an item and is equal to the unit selling price p, in dollars, of the item times the number x of units actually sold. That is,

$$R = xp$$

In economics, the Law of Demand states that p and x are related: As one increases, the other decreases. The equation that relates p and x is called the **demand equation.**

EXAMPLE 1 Maximizing Revenue

The marketing department at Texas Instruments has found that, when certain calculators are sold at a price of p dollars per unit, the number x of calculators sold is given by the demand equation

$$x = 21{,}000 - 150p$$

(a) Find a model that expresses the revenue R as a function of the price p.

(b) What is the domain of R?

(c) What unit price should be used to maximize revenue?

(d) If this price is charged, what is the maximum revenue?

(e) How many units are sold at this price?

(f) Graph R.

Solution

(a) The revenue R is $R = xp$, where $x = 21{,}000 - 150p$.

$$R = xp = (21{,}000 - 150p)p = -150p^2 + 21{,}000p \quad \text{The Model}$$

(b) Because x represents the number of calculators sold, we have $x \geq 0$, so $21{,}000 - 150p \geq 0$. Solving this linear inequality, we find that $p \leq 140$. In addition, Texas Instruments will only charge a positive price for the calculator, so $p > 0$. Combining these inequalities, we have the domain of R is $\{p \mid 0 < p \leq 140\}$.

(c) The function R is a quadratic function with $a = -150, b = 21{,}000$, and $c = 0$. Because $a < 0$, the vertex is the highest point on the parabola. The revenue R is therefore a maximum when the price p is

$$p = -\frac{b}{2a} \underset{\substack{\uparrow \\ a = -150,\ b = 21{,}000}}{=} -\frac{21{,}000}{2(-150)} = -\frac{21{,}000}{-300} = \$70.00$$

(d) The maximum revenue R is

$$R(70) = -150(70)^2 + 21{,}000(70) = \$735{,}000$$

(e) The number of calculators sold is given by the demand equation $x = 21{,}000 - 150p$. At a price of $p = \$70$,

$$x = 21{,}000 - 150(70) = 10{,}500$$

calculators are sold.

(f) See Figure 24 for the graph of R.

Figure 24

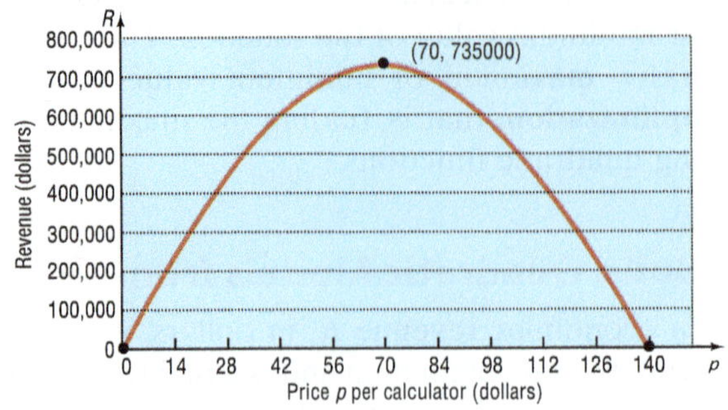

Now Work PROBLEM 3

EXAMPLE 2 Maximizing the Area Enclosed by a Fence

A farmer has 2000 yards of fence to enclose a rectangular field. What are the dimensions of the rectangle that encloses the most area?

Solution Figure 25 illustrates the situation. The available fence represents the perimeter of the rectangle. If x is the length and w is the width, then

Figure 25

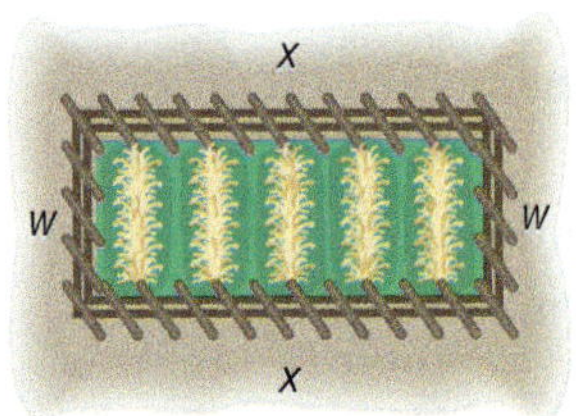

$$2x + 2w = 2000 \qquad (1)$$

The area A of the rectangle is

$$A = xw$$

To express A in terms of a single variable, we solve equation (1) for w and substitute the result in $A = xw$. Then A involves only the variable x. [You could also solve equation (1) for x and express A in terms of w alone. Try it!]

$$2x + 2w = 2000$$
$$2w = 2000 - 2x$$
$$w = \frac{2000 - 2x}{2} = 1000 - x$$

Then the area A is

$$A = xw = x(1000 - x) = -x^2 + 1000x$$

Now, A is a quadratic function of x.

$$A(x) = -x^2 + 1000x \quad a = -1, b = 1000, c = 0$$

Figure 26

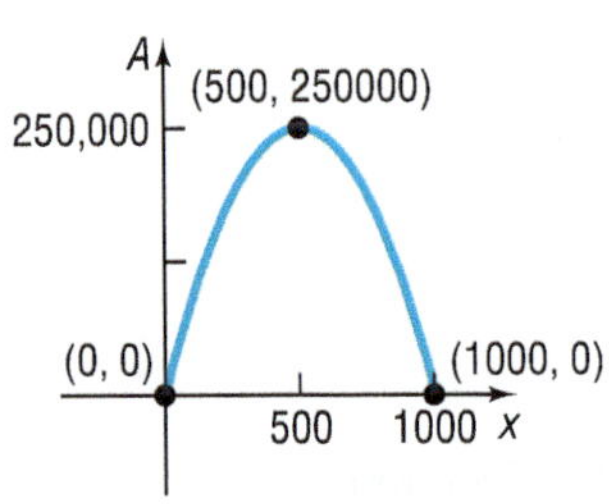

Figure 26 shows a graph of $A(x) = -x^2 + 1000x$. Since $a < 0$, the vertex is a maximum point on the graph of A. The maximum value occurs at

$$x = -\frac{b}{2a} = -\frac{1000}{2(-1)} = 500$$

The maximum value of A is

$$A\left(-\frac{b}{2a}\right) = A(500) = -500^2 + 1000(500) = -250{,}000 + 500{,}000 = 250{,}000$$

The largest rectangle that can be enclosed by 2000 yards of fence has an area of 250,000 square yards. Its dimensions are 500 yards by 500 yards.

Now Work PROBLEM 7

EXAMPLE 3 Analyzing the Motion of a Projectile

A projectile is fired from a cliff 500 feet above the water at an inclination of 45° to the horizontal, with a muzzle velocity of 400 feet per second. In physics, it is established that the height h of the projectile above the water can be modeled by

$$h(x) = \frac{-32x^2}{(400)^2} + x + 500$$

where x is the horizontal distance of the projectile from the base of the cliff. See Figure 27.

Figure 27

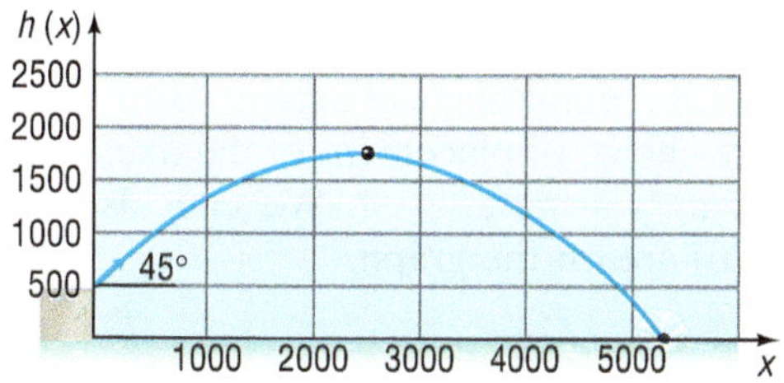

(a) Find the maximum height of the projectile.

(b) How far from the base of the cliff will the projectile strike the water?

Solution (a) The height of the projectile is given by a quadratic function.

$$h(x) = \frac{-32x^2}{(400)^2} + x + 500 = \frac{-1}{5000}x^2 + x + 500$$

We are looking for the maximum value of h. Since $a < 0$, the maximum value is obtained at the vertex. We compute

$$x = -\frac{b}{2a} = -\frac{1}{2\left(-\frac{1}{5000}\right)} = \frac{5000}{2} = 2500$$

The maximum height of the projectile is

$$h(2500) = \frac{-1}{5000}(2500)^2 + 2500 + 500 = -1250 + 2500 + 500 = 1750 \text{ ft}$$

(b) The projectile will strike the water when the height is zero. To find the distance x traveled, we need to solve the equation

$$h(x) = \frac{-1}{5000}x^2 + x + 500 = 0$$

We find the discriminant first.

$$b^2 - 4ac = 1^2 - 4\left(\frac{-1}{5000}\right)(500) = 1.4$$

Then

$$x = \frac{-b \pm \sqrt{b^2 - 4ac}}{2a} = \frac{-1 \pm \sqrt{1.4}}{2\left(-\frac{1}{5000}\right)} \approx \begin{cases} -458 \\ 5458 \end{cases}$$

We discard the negative solution and find that the projectile will strike the water at a distance of about 5458 feet from the base of the cliff.

Seeing the Concept

Graph

$$h(x) = \frac{-1}{5000}x^2 + x + 500$$
$$0 \le x \le 5500$$

Use MAXIMUM to find the maximum height of the projectile, and use ROOT or ZERO to find the distance from the base of the cliff to where it strikes the water. Compare your results with those obtained in Example 3.

Now Work PROBLEM 11

EXAMPLE 4 **The Golden Gate Bridge**

The Golden Gate Bridge, a suspension bridge, spans the entrance to San Francisco Bay. Its 746-foot-tall towers are 4200 feet apart. The bridge is suspended from two huge cables more than 3 feet in diameter; the 90-foot-wide roadway is 220 feet above the water. The cables are parabolic in shape* and touch the road surface at the center of the bridge. Find the height of the cable above the road at a distance of 1000 feet from the center.

Solution See Figure 28. We begin by choosing the placement of the coordinate axes so that the x-axis coincides with the road surface and the origin coincides with the center of the bridge. As a result, the twin towers will be vertical (height $746 - 220 = 526$ feet above the road) and located 2100 feet from the center. Also, the cable, which has the shape of a parabola, will extend from the towers, open up, and have its vertex at $(0, 0)$. The choice of placement of the axes enables us to identify the equation of the parabola as $y = ax^2$, $a > 0$. We can also see that the points $(-2100, 526)$ and $(2100, 526)$ are on the graph.

Figure 28

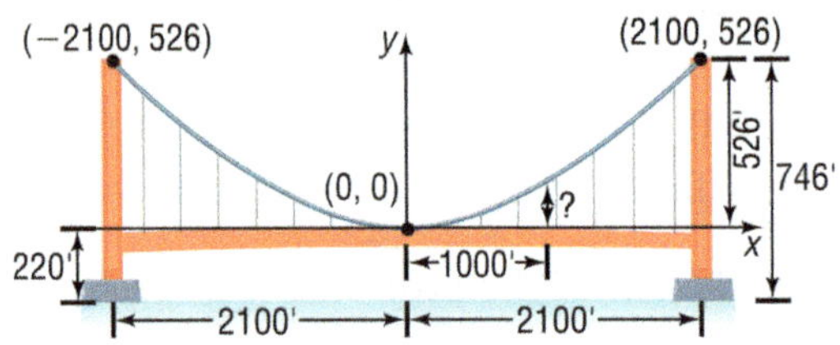

Based on these facts, we can find the value of a in $y = ax^2$.

$$y = ax^2$$

$$526 = a(2100)^2 \qquad x = 2100, y = 526$$

$$a = \frac{526}{(2100)^2}$$

The equation of the parabola is therefore

$$y = \frac{526}{(2100)^2}x^2$$

The height of the cable when $x = 1000$ is

$$y = \frac{526}{(2100)^2}(1000)^2 \approx 119.3 \text{ feet}$$

The cable is 119.3 feet above the road at a distance of 1000 feet from the center of the bridge.

Now Work PROBLEM 13

2 Build Quadratic Models from Data

In Section 4.2, we found the line of best fit for data that appeared to be linearly related. It was noted that data may also follow a nonlinear relation. Figures 29(a) and (b) show scatter diagrams of data that follow a quadratic relation.

*A cable suspended from two towers is in the shape of a **catenary,** but when a horizontal roadway is suspended from the cable, the cable takes the shape of a parabola.

Figure 29

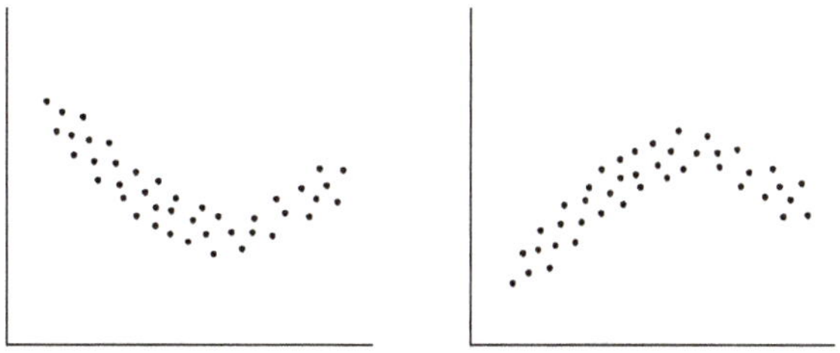

$y = ax^2 + bx + c, a > 0$

(a)

$y = ax^2 + bx + c, a < 0$

(b)

EXAMPLE 5 Fitting a Quadratic Function to Data

A farmer collected the data given in Table 9, which shows crop yields Y for various amounts x of fertilizer used.

(a) Use a graphing utility to draw a scatter diagram of the data. Comment on the type of relation that may exist between the two variables.

(b) Use a graphing utility to find the quadratic function of best fit that models the relation between amount of fertilizer and crop yield.

(c) Use the function found in part (b) to determine the optimal amount of fertilizer to apply.

(d) Use the function found in part (b) to predict crop yield when the optimal amount of fertilizer is applied.

(e) Draw the quadratic function of best fit on the scatter diagram.

Table 9

Plot	Fertilizer, x (Pounds/100 ft²)	Yield, Y (Bushels)
1	0	4
2	0	6
3	5	10
4	5	7
5	10	12
6	10	10
7	15	15
8	15	17
9	20	18
10	20	21
11	25	20
12	25	21
13	30	21
14	30	22
15	35	21
16	35	20
17	40	19
18	40	19

Solution

(a) Figure 30 shows the scatter diagram, from which it appears that the data follow a quadratic relation, with $a < 0$.

(b) Upon executing the QUADratic REGression program, we obtain the results shown in Figure 31. The output that the utility provides shows us the equation $y = ax^2 + bx + c$. The quadratic function of best fit is

$$Y(x) = -0.0171x^2 + 1.0765x + 3.8939 \quad \text{The Model}$$

where x represents the amount of fertilizer used and Y represents crop yield.

Figure 30

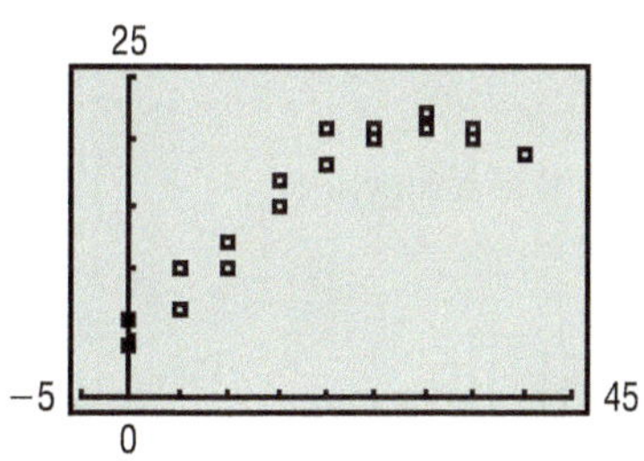

Figure 31

(c) Based on the quadratic function of best fit, the optimal amount of fertilizer to apply is

$$x = -\frac{b}{2a} = -\frac{1.0765}{2(-0.0171)} \approx 31.5 \text{ pounds of fertilizer per 100 square feet}$$

(d) We evaluate the function $Y(x)$ for $x = 31.5$.

$$Y(31.5) = -0.0171(31.5)^2 + 1.0765(31.5) + 3.8939 \approx 20.8 \text{ bushels}$$

If we apply 31.5 pounds of fertilizer per 100 square feet, the crop yield will be 20.8 bushels according to the quadratic function of best fit.

(e) Figure 32 shows the graph of the quadratic function found in part (b) drawn on the scatter diagram.

Figure 32

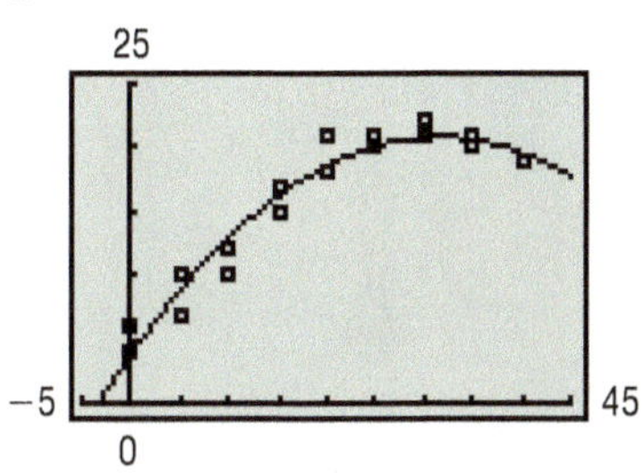

Look again at Figure 31. Notice that the output given by the graphing calculator does not include r, the correlation coefficient. Recall that the correlation coefficient is a measure of the strength of a linear relation that exists between two variables. The graphing calculator does not provide an indication of how well the function fits the data in terms of r since a quadratic function cannot be expressed as a linear function.

Now Work PROBLEM 27

4.4 Assess Your Understanding

'Are You Prepared?' *Answers are given at the end of these exercises. If you get a wrong answer, read the pages listed in red.*

1. Translate the following sentence to a mathematical equation: The total revenue R from selling x hot dogs is \$3 times the number of hot dogs sold. (p. 141)

2. Translate the following sentence to a mathematical equation: The perimeter P of a rectangle is the sum of twice the length l and twice the width w. (p. 141)

Applications and Extensions

3. **Maximizing Revenue** The price p (in dollars) and the quantity x sold of a certain product obey the demand equation

$$p = -\frac{1}{6}x + 100$$

(a) Find a model that expresses the revenue R as a function of x. (Remember, $R = xp$.)
(b) What is the domain of R?
(c) What is the revenue if 200 units are sold?
(d) What quantity x maximizes revenue? What is the maximum revenue?
(e) What price should the company charge to maximize revenue?

4. **Maximizing Revenue** The price p (in dollars) and the quantity x sold of a certain product obey the demand equation

$$p = -\frac{1}{3}x + 100$$

(a) Find a model that expresses the revenue R as a function of x.
(b) What is the domain of R?
(c) What is the revenue if 100 units are sold?
(d) What quantity x maximizes revenue? What is the maximum revenue?
(e) What price should the company charge to maximize revenue?

5. **Maximizing Revenue** The price p (in dollars) and the quantity x sold of a certain product obey the demand equation

$$x = -5p + 100 \qquad 0 < p \le 20$$

(a) Express the revenue R as a function of x.
(b) What is the revenue if 15 units are sold?
(c) What quantity x maximizes revenue? What is the maximum revenue?
(d) What price should the company charge to maximize revenue?

6. **Maximizing Revenue** The price p (in dollars) and the quantity x sold of a certain product obey the demand equation

$$x = -20p + 500 \qquad 0 < p \le 25$$

(a) Express the revenue R as a function of x.
(b) What is the revenue if 20 units are sold?
(c) What quantity x maximizes revenue? What is the maximum revenue?
(d) What price should the company charge to maximize revenue?

7. **Enclosing a Rectangular Field** David has 400 yards of fencing and wishes to enclose a rectangular area.
(a) Express the area A of the rectangle as a function of the width w of the rectangle.
(b) For what value of w is the area largest?
(c) What is the maximum area?

8. **Enclosing a Rectangular Field** Beth has 3000 feet of fencing available to enclose a rectangular field.
(a) Express the area A of the rectangle as a function of x, where x is the length of the rectangle.
(b) For what value of x is the area largest?
(c) What is the maximum area?

9. **Enclosing the Most Area with a Fence** A farmer with 4000 meters of fencing wants to enclose a rectangular plot that borders on a river. If the farmer does not fence the side along the river, what is the largest area that can be enclosed? (See the figure.)

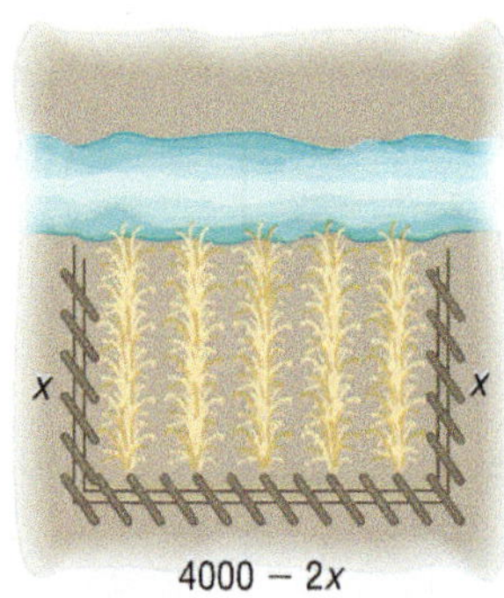

10. Enclosing the Most Area with a Fence A farmer with 2000 meters of fencing wants to enclose a rectangular plot that borders on a straight highway. If the farmer does not fence the side along the highway, what is the largest area that can be enclosed?

11. Analyzing the Motion of a Projectile A projectile is fired from a cliff 200 feet above the water at an inclination of 45° to the horizontal, with a muzzle velocity of 50 feet per second. The height h of the projectile above the water is modeled by

$$h(x) = \frac{-32x^2}{(50)^2} + x + 200$$

where x is the horizontal distance of the projectile from the face of the cliff.

(a) At what horizontal distance from the face of the cliff is the height of the projectile a maximum?
(b) Find the maximum height of the projectile.
(c) At what horizontal distance from the face of the cliff will the projectile strike the water?
(d) Using a graphing utility, graph the function h, $0 \le x \le 200$.
(e) Use a graphing utility to verify the solutions found in parts (b) and (c).
(f) When the height of the projectile is 100 feet above the water, how far is it from the cliff?

12. Analyzing the Motion of a Projectile A projectile is fired at an inclination of 45° to the horizontal, with a muzzle velocity of 100 feet per second. The height h of the projectile is modeled by

$$h(x) = \frac{-32x^2}{(100)^2} + x$$

where x is the horizontal distance of the projectile from the firing point.

(a) At what horizontal distance from the firing point is the height of the projectile a maximum?
(b) Find the maximum height of the projectile.
(c) At what horizontal distance from the firing point will the projectile strike the ground?
(d) Using a graphing utility, graph the function h, $0 \le x \le 350$.
(e) Use a graphing utility to verify the results obtained in parts (b) and (c).
(f) When the height of the projectile is 50 feet above the ground, how far has it traveled horizontally?

13. Suspension Bridge A suspension bridge with weight uniformly distributed along its length has twin towers that extend 75 meters above the road surface and are 400 meters apart. The cables are parabolic in shape and are suspended from the tops of the towers. The cables touch the road surface at the center of the bridge. Find the height of the cables at a point 100 meters from the center. (Assume that the road is level.)

14. Architecture A parabolic arch has a span of 120 feet and a maximum height of 25 feet. Choose suitable rectangular coordinate axes and find the equation of the parabola. Then calculate the height of the arch at points 10 feet, 20 feet, and 40 feet from the center.

15. Constructing Rain Gutters A rain gutter is to be made of aluminum sheets that are 12 inches wide by turning up the edges 90°. See the illustration. What depth will provide maximum cross-sectional area and hence allow the most water to flow?

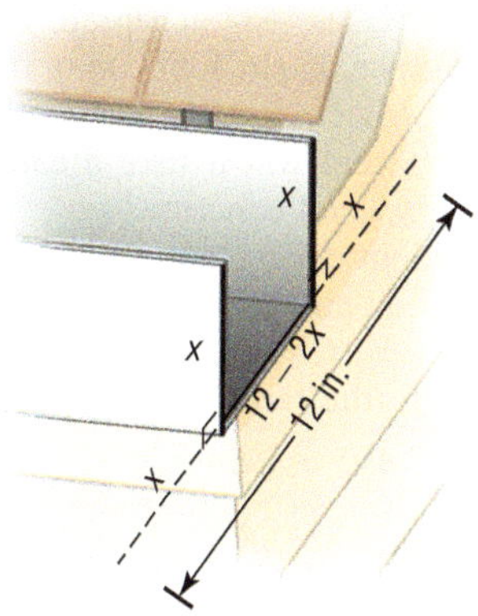

16. Norman Windows A **Norman window** has the shape of a rectangle surmounted by a semicircle of diameter equal to the width of the rectangle. See the figure. If the perimeter of the window is 20 feet, what dimensions will admit the most light (maximize the area)?

[**Hint:** Circumference of a circle $= 2\pi r$; area of a circle $= \pi r^2$, where r is the radius of the circle.]

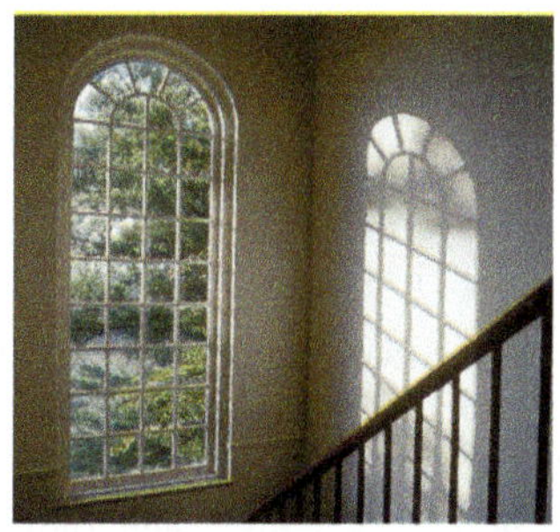

17. Constructing a Stadium A track and field playing area is in the shape of a rectangle with semicircles at each end. See the figure. The inside perimeter of the track is to be 1500 meters. What should the dimensions of the rectangle be so that the area of the rectangle is a maximum?

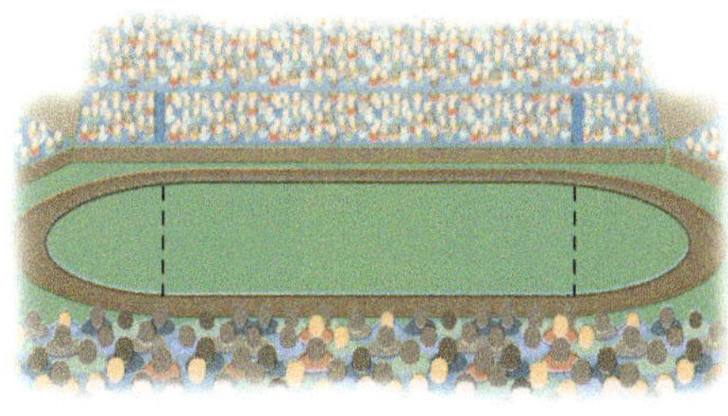

18. Architecture A special window has the shape of a rectangle surmounted by an equilateral triangle. See the figure. If the perimeter of the window is 16 feet, what dimensions will admit the most light?

[**Hint:** Area of an equilateral triangle $= \left(\frac{\sqrt{3}}{4}\right)x^2$, where x is the length of a side of the triangle.]

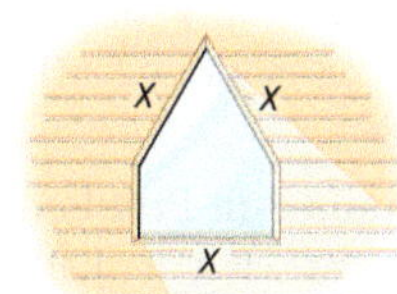

19. Stopping Distance An accepted relationship between stopping distance, d (in feet), and the speed of a car, v (in mph), is $d = 1.1v + 0.06v^2$ on dry, level concrete.

(a) How many feet will it take a car traveling 45 mph to stop on dry, level concrete?

(b) If an accident occurs 200 feet ahead of you, what is the maximum speed you can be traveling to avoid being involved?

(c) What might the term $1.1v$ represent?

Source: www2.nsta.org/Energy/fn_braking.html

20. Insurance Claims The years 1999 to 2005 were particularly costly for insurance companies, with 7 of the 10 most costly catastrophes in U.S. history (as of 2005). For the United States Automobile Association (USAA) and its affiliates, the total cost of claims for catastrophic losses, in millions, can be modeled by $C(x) = 34.87x^2 - 98.1x + 258.3$, where $x = 0$ for 1999, $x = 1$ for 2000, $x = 2$ for 2001, and so on.

(a) Estimate the total cost of claims for the year 2003.

(b) According to the model, during which year were catastrophic loss claims at a minimum?

(c) Would $C(x)$ be useful for predicting total catastrophic loss claims for the year 2015? Why or why not?

Source: USAA Report to Members 2005

21. Chemical Reactions A self-catalytic chemical reaction results in the formation of a compound that causes the formation ratio to increase. If the reaction rate V is modeled by

$$V(x) = kx(a - x), \qquad 0 \le x \le a$$

where k is a positive constant, a is the initial amount of the compound, and x is the variable amount of the compound, for what value of x is the reaction rate a maximum?

22. Calculus: Simpson's Rule The figure shows the graph of $y = ax^2 + bx + c$. Suppose that the points $(-h, y_0)$, $(0, y_1)$, and (h, y_2) are on the graph. It can be shown that the area enclosed by the parabola, the x-axis, and the lines $x = -h$ and $x = h$ is

$$\text{Area} = \frac{h}{3}(2ah^2 + 6c)$$

Show that this area may also be given by

$$\text{Area} = \frac{h}{3}(y_0 + 4y_1 + y_2)$$

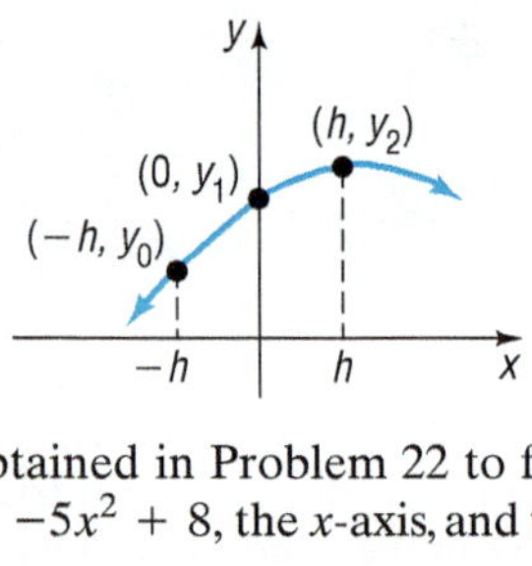

23. Use the result obtained in Problem 22 to find the area enclosed by $f(x) = -5x^2 + 8$, the x-axis, and the lines $x = -1$ and $x = 1$.

24. Use the result obtained in Problem 22 to find the area enclosed by $f(x) = 2x^2 + 8$, the x-axis, and the lines $x = -2$ and $x = 2$.

25. Use the result obtained in Problem 22 to find the area enclosed by $f(x) = x^2 + 3x + 5$, the x-axis, and the lines $x = -4$ and $x = 4$.

26. Use the result obtained in Problem 22 to find the area enclosed by $f(x) = -x^2 + x + 4$, the x-axis, and the lines $x = -1$ and $x = 1$.

27. Life Cycle Hypothesis An individual's income varies with his or her age. The following table shows the median income I of individuals of different age groups within the United States for 2003. For each age group, let the class midpoint represent the independent variable, x. For the class "65 years and older," we will assume that the class midpoint is 69.5.

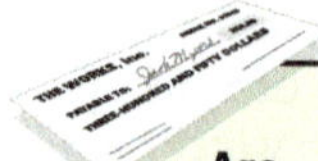

Age	Class Midpoint, x	Median Income, I
15–24 years	19.5	$9,961
25–34 years	29.5	$30,562
35–44 years	39.5	$39,195
45–54 years	49.5	$42,079
55–64 years	59.5	$38,915
65 years and older	69.5	$20,363

Source: U.S. Census Bureau

(a) Use a graphing utility to draw a scatter diagram of the data. Comment on the type of relation that may exist between the two variables.

(b) Use a graphing utility to find the quadratic function of best fit that models the relation between age and median income.

(c) Use the function found in part (b) to determine the age at which an individual can expect to earn the most income.

(d) Use the function found in part (b) to predict the peak income earned.

(e) With a graphing utility, graph the quadratic function of best fit on the scatter diagram.

28. Height of a Ball A shot-putter throws a ball at an inclination of 45° to the horizontal. The following data represent the height of the ball h at the instant that it has traveled x feet horizontally.

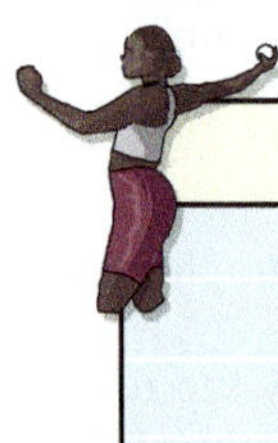

Distance, x	Height, h
20	25
40	40
60	55
80	65
100	71
120	77
140	77
160	75
180	71
200	64

(a) Use a graphing utility to draw a scatter diagram of the data. Comment on the type of relation that may exist between the two variables.

(b) Use a graphing utility to find the quadratic function of best fit that models the relation between distance and height.

(c) Use the function found in part (b) to determine how far the ball will travel before it reaches its maximum height.

(d) Use the function found in part (b) to find the maximum height of the ball.

(e) With a graphing utility, graph the quadratic function of best fit on the scatter diagram.

Mixed Practice

29. Which Model? The following data represent the square footage and rents (dollars per month) for apartments in the Del Mar area of San Diego, California.

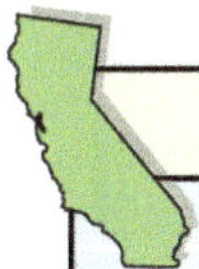

Square Footage, x	Rent per Month, R
686	1600
770	1665
817	1750
800	1685
809	1700
901	1770
803	1725

Source: apartments.com

(a) Using a graphing utility, draw a scatter diagram of the data treating square footage as the independent variable. What type of relation appears to exist between square footage and rent?

(b) Based on your response to part (a), find either a linear or quadratic model that describes the relation between square footage and rent.

(c) Use your model to predict the rent of an apartment in San Diego that is 850 square feet.

30. Which Model? An engineer collects the following data showing the speed s of a Toyota Camry and its average miles per gallon, M.

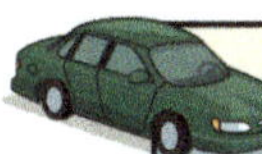

Speed, s	Miles per Gallon, M
30	18
35	20
40	23
40	25
45	25
50	28
55	30
60	29
65	26
65	25
70	25

(a) Using a graphing utility, draw a scatter diagram of the data treating speed as the independent variable. What type of relation appears to exist between speed and miles per gallon?

(b) Based on your response to part (a), find either a linear or quadratic model that describes the relation between speed and miles per gallon.

(c) Use your model to predict the miles per gallon for a Camry that is traveling 63 miles per hour.

31. Which Model? The following data represent the percentage of the U.S. population aged 25 years or older whose age is x who do not have a high school diploma as of March 2005.

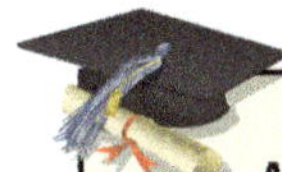

Age, a	Percentage without a High School Diploma, P
30	13.3
40	11.6
50	10.9
60	13.7
70	22.3
80	30.2

Source: U.S. Census Bureau

(a) Using a graphing utility, draw a scatter diagram of the data treating age as the independent variable. What type of relation appears to exist between age and percentage of the population without a high school diploma?

(b) Based on your response to part (a), find either a linear or quadratic model that describes the relation between age and percentage of the population that do not have a high school diploma.

(c) Use your model to predict the percentage of 35-year-olds that do not have a high school diploma.

32. Which Model? Crickets make a chirping noise by sliding their wings rapidly over each other. Perhaps you have noticed that the number of chirps seems to increase with the temperature. The following data list the temperature (in degrees Fahrenheit) and the number of chirps per second for the striped ground cricket.

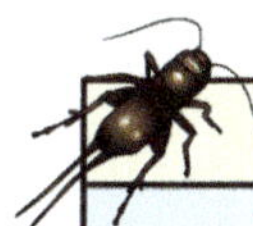

Temperature, x	Chirps per Second, C
88.6	20.0
93.3	19.8
80.6	17.1
69.7	14.7
69.4	15.4
79.6	15.0
80.6	16.0
76.3	14.4
75.2	15.5

Source: Pierce, George W. *The Songs of Insects*. Cambridge, MA Harvard University Press, 1949, pp. 12 – 21

(a) Using a graphing utility, draw a scatter diagram of the data treating temperature as the independent variable. What type of relation appears to exist between temperature and chirps per second?

(b) Based on your response to part (a), find either a linear or quadratic model that best describes the relation between temperature and chirps per second.

(c) Use your model to predict the chirps per second if the temperature is 80°F.

'Are You Prepared?' Answers

1. $R = 3x$ **2.** $P = 2l + 2w$

4.5 Inequalities Involving Quadratic Functions

PREPARING FOR THIS SECTION *Before getting started, review the following:*

- Solve Inequalities (Section 1.7, pp. 153–155)
- Use Interval Notation (Section 1.7, pp. 150–151)

Now Work the 'Are You Prepared?' problems on page 322.

OBJECTIVE 1 Solve Inequalities Involving a Quadratic Function (p. 320)

1 Solve Inequalities Involving a Quadratic Function

In this section we solve inequalities that involve quadratic functions. We will accomplish this by using their graphs. For example, to solve the inequality

$$ax^2 + bx + c > 0 \qquad a \neq 0$$

we graph the function $f(x) = ax^2 + bx + c$ and, from the graph, determine where it is above the x-axis, that is, where $f(x) > 0$. To solve the inequality $ax^2 + bx + c < 0, a \neq 0$, we graph the function $f(x) = ax^2 + bx + c$ and determine where the graph is below the x-axis. If the inequality is not strict, we include the x-intercepts in the solution.

Let's look at an example.

EXAMPLE 1 Solving an Inequality

Solve the inequality $x^2 - 4x - 12 \leq 0$ and graph the solution set.

By Hand Solution

We graph the function $f(x) = x^2 - 4x - 12$. The intercepts are

y-intercept: $f(0) = -12$ *Evaluate f at 0.*

x-intercepts (if any):

$$x^2 - 4x - 12 = 0 \quad \text{Solve } f(x) = 0.$$
$$(x - 6)(x + 2) = 0 \quad \text{Factor.}$$
$$x - 6 = 0 \text{ or } x + 2 = 0 \quad \text{Apply the Zero-Product Property.}$$
$$x = 6 \text{ or } x = -2$$

The y-intercept is -12; the x-intercepts are -2 and 6.

The vertex is at $x = -\frac{b}{2a} = -\frac{-4}{2(1)} = 2$. Since $f(2) = -16$, the vertex is $(2, -16)$. See Figure 33 for the graph.

Figure 33

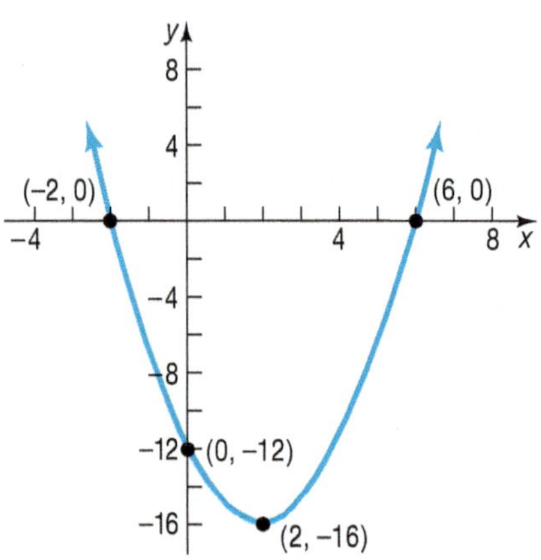

The graph is below the x-axis for $-2 < x < 6$. Since the original inequality is not strict, we include the x-intercepts. The solution set is $\{x | -2 \leq x \leq 6\}$ or, using interval notation, $[-2, 6]$.

Graphing Utility Solution

We graph $Y_1 = x^2 - 4x - 12$. See Figure 34. Using the ZERO command, we find that the x-intercepts of Y_1 are -2 and 6. Since the inequality is not strict, the solution set is $\{x | -2 \leq x \leq 6\}$ or, using interval notation, $[-2, 6]$.

Figure 34

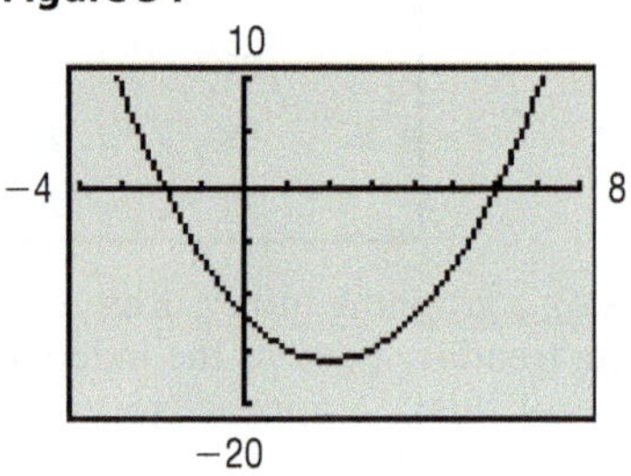

Figure 35

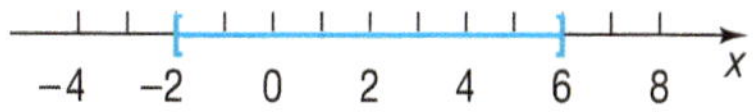

See Figure 35 for the graph of the solution set.

Now Work PROBLEM 9

EXAMPLE 2 Solving an Inequality

Solve the inequality $2x^2 < x + 10$ and graph the solution set.

Solution *Option 1* We rearrange the inequality so that 0 is on the right side.

$$2x^2 < x + 10$$

$$2x^2 - x - 10 < 0 \qquad \text{Subtract } x + 10 \text{ from both sides.}$$

This inequality is equivalent to the one that we wish to solve.

Next we graph the function $f(x) = 2x^2 - x - 10$. We want to determine where the graph is below the x-axis. The intercepts are

y-intercept: $f(0) = -10$ — Evaluate f at 0.

x-intercepts (if any):

$$2x^2 - x - 10 = 0 \qquad \text{Solve } f(x) = 0.$$

$$(2x - 5)(x + 2) = 0 \qquad \text{Factor.}$$

$$2x - 5 = 0 \quad \text{or} \quad x + 2 = 0 \qquad \text{Apply the Zero-Product Property.}$$

$$x = \frac{5}{2} \quad \text{or} \quad x = -2$$

Figure 36

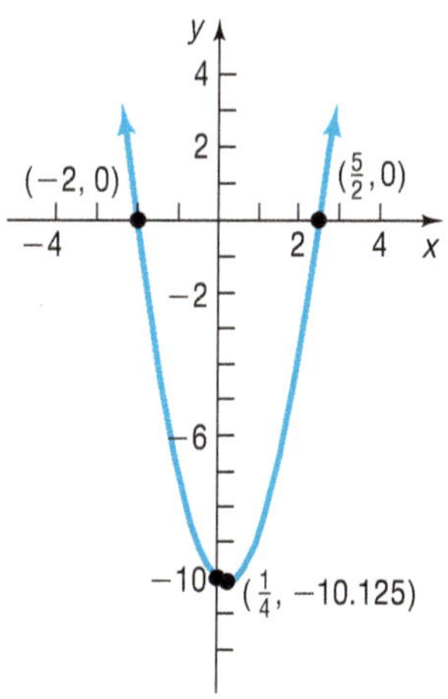

The y-intercept is -10; the x-intercepts are -2 and $\frac{5}{2}$.

The vertex is at $x = -\frac{b}{2a} = -\frac{-1}{2(2)} = \frac{1}{4}$. Since $f\left(\frac{1}{4}\right) = -10.125$, the vertex is $\left(\frac{1}{4}, -10.125\right)$. See Figure 36 for the graph.

The graph is below the x-axis between $x = -2$ and $x = \frac{5}{2}$. Since the inequality is strict, the solution set is $\left\{x \middle| -2 < x < \frac{5}{2}\right\}$ or, using interval notation, $\left(-2, \frac{5}{2}\right)$.

Check: Verify the result of Method 1 by graphing $Y_1 = 2x^2 - x - 10$ and using the ZERO feature.

Option 2 If $f(x) = 2x^2$ and $g(x) = x + 10$, the inequality that we wish to solve is $f(x) < g(x)$. We graph the functions $f(x) = 2x^2$ and $g(x) = x + 10$. See Figure 37. The graphs intersect where $f(x) = g(x)$. Then

Figure 37

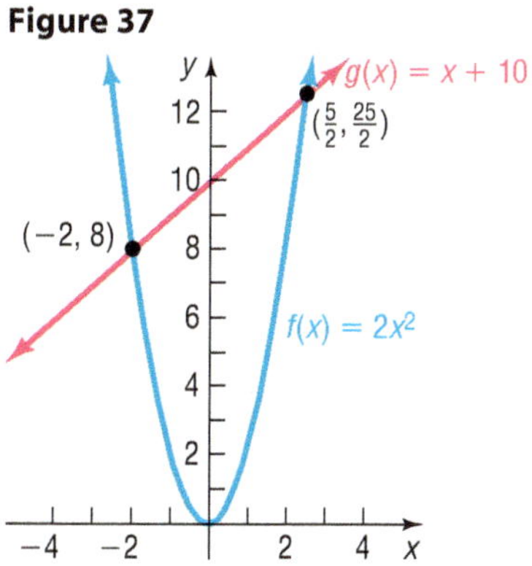

$$2x^2 = x + 10 \qquad f(x) = g(x)$$

$$2x^2 - x - 10 = 0$$

$$(2x - 5)(x + 2) = 0 \qquad \text{Factor.}$$

$$2x - 5 = 0 \quad \text{or} \quad x + 2 = 0 \qquad \text{Apply the Zero-Product Property.}$$

$$x = \frac{5}{2} \quad \text{or} \quad x = -2$$

The graphs intersect at the points $(-2, 8)$ and $\left(\frac{5}{2}, \frac{25}{2}\right)$. The inequality that we wish to solve is $f(x) < g(x)$, so we seek where the graph of f is below that of g.

This happens between the points of intersection. Since the inequality is strict, the solution set is $\left\{x \middle| -2 < x < \frac{5}{2}\right\}$ or, using interval notation, $\left(-2, \frac{5}{2}\right)$.

Figure 38

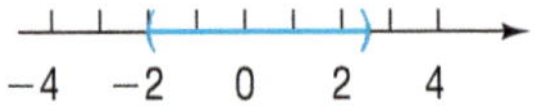

See Figure 38 for the graph of the solution set.

Check: Verify the result of Option 2 by graphing $Y_1 = 2x^2$ and $Y_2 = x + 10$ on the same screen. Use INTERSECT to find the x-coordinates of the points of intersection. Determine where Y_1 is below Y_2 to solve $Y_1 < Y_2$.

Now Work PROBLEMS 5 AND 13

4.5 Assess Your Understanding

'Are You Prepared?' *Answers are given at the end of these exercises. If you get a wrong answer, read the pages listed in red.*

1. Solve the inequality $-3x - 2 < 7$. (pp. 153–155)

2. Write $(-2, 7]$ using inequality notation and graph the inequality. (pp. 150–151)

Skill Building

In Problems 3–6, use the figure to solve each inequality.

3.

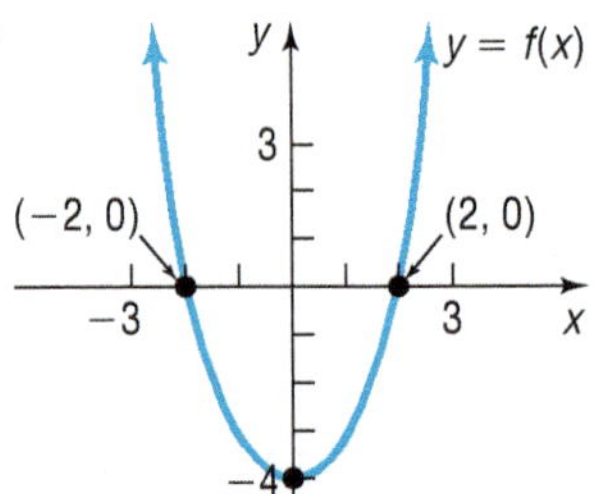

(a) $f(x) > 0$
(b) $f(x) \le 0$

4.

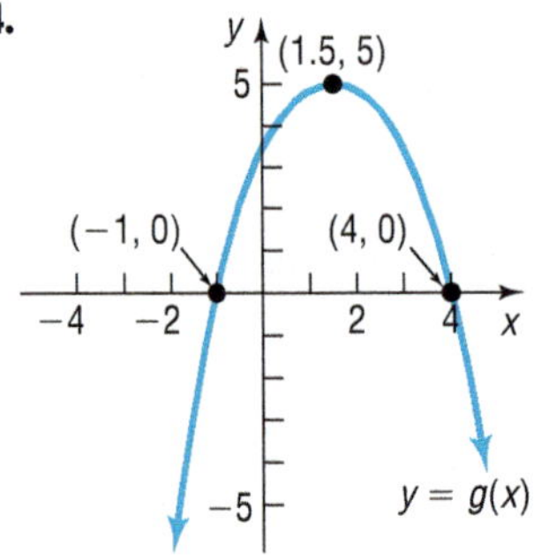

(a) $g(x) < 0$
(b) $g(x) \ge 0$

5.

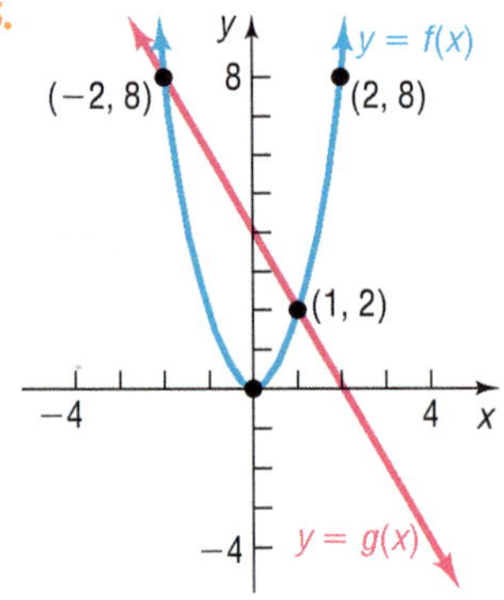

(a) $g(x) \ge f(x)$
(b) $f(x) > g(x)$

6.

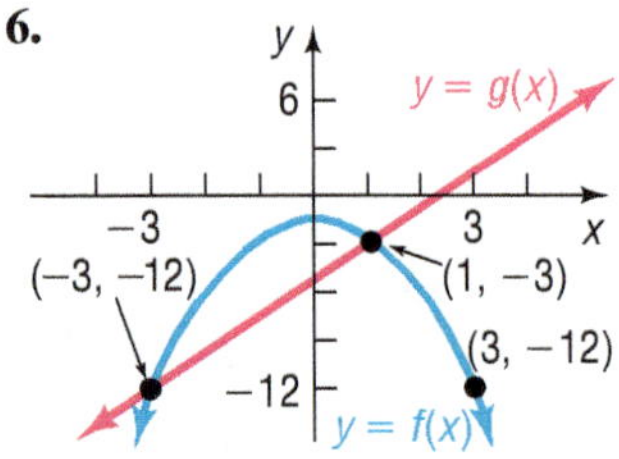

(a) $f(x) < g(x)$
(b) $f(x) \ge g(x)$

In Problems 7–22, solve each inequality. Verify your results using a graphing utility.

7. $x^2 - 3x - 10 \le 0$
8. $x^2 + 3x - 10 > 0$
9. $x^2 - 4x > 0$
10. $x^2 + 8x \le 0$
11. $x^2 - 9 < 0$
12. $x^2 - 1 < 0$
13. $x^2 + x > 12$
14. $x^2 + 7x < -12$
15. $2x^2 \le 5x + 3$
16. $6x^2 \le 6 + 5x$
17. $x(x - 7) > 8$
18. $x(x + 1) > 20$
19. $4x^2 + 9 < 6x$
20. $25x^2 + 16 < 40x$
21. $6(x^2 + 1) \ge 5x$
22. $2(2x^2 - 3x) \ge -9$

Mixed Practice

23. What is the domain of the function $f(x) = \sqrt{x^2 - 16}$?

24. What is the domain of the function $f(x) = \sqrt{x - 3x^2}$?

In Problems 25–32, use the given functions f and g.

(a) Solve $f(x) = 0$. *(b) Solve $g(x) = 0$.* *(c) Solve $f(x) = g(x)$.* *(d) Solve $f(x) > 0$.*
(e) Solve $g(x) \le 0$. *(f) Solve $f(x) > g(x)$.* *(g) Solve $f(x) \ge 1$.*

25. $f(x) = x^2 - 1$
$g(x) = 3x + 3$

26. $f(x) = -x^2 + 3$
$g(x) = -3x + 3$

27. $f(x) = -x^2 + 1$
$g(x) = 4x + 1$

28. $f(x) = -x^2 + 4$
$g(x) = -x - 2$

29. $f(x) = x^2 - 4$
$g(x) = -x^2 + 4$

30. $f(x) = x^2 - 2x + 1$
$g(x) = -x^2 + 1$

31. $f(x) = x^2 - x - 2$
$g(x) = x^2 + x - 2$

32. $f(x) = -x^2 - x + 1$
$g(x) = -x^2 + x + 6$

Applications and Extensions

33. Physics A ball is thrown vertically upward with an initial velocity of 80 feet per second. The distance s (in feet) of the ball from the ground after t seconds is $s(t) = 80t - 16t^2$.

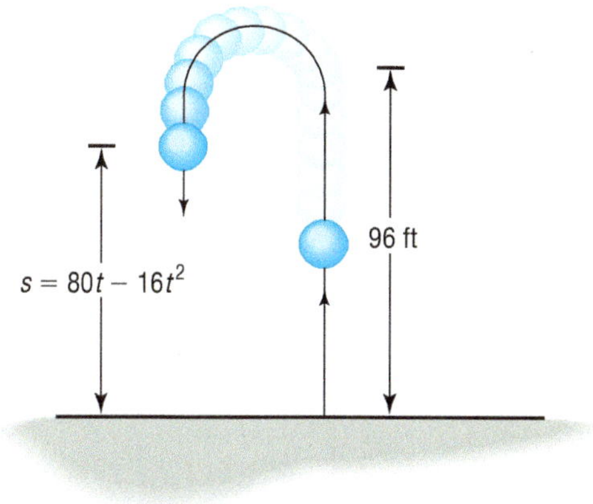

(a) At what time t will the ball strike the ground?
(b) For what time t is the ball more than 96 feet above the ground?

34. Physics A ball is thrown vertically upward with an initial velocity of 96 feet per second. The distance s (in feet) of the ball from the ground after t seconds is $s(t) = 96t - 16t^2$.
(a) At what time t will the ball strike the ground?
(b) For what time t is the ball more than 128 feet above the ground?

35. Revenue Suppose that the manufacturer of a gas clothes dryer has found that, when the unit price is p dollars, the revenue R (in dollars) is

$$R(p) = -4p^2 + 4000p$$

(a) At what prices p is revenue zero?
(b) For what range of prices will revenue exceed \$800,000?

36. Revenue The John Deere company has found that the revenue from sales of heavy-duty tractors is a function of the unit price p, in dollars, that it charges. The revenue R, in dollars, is

$$R(p) = -\frac{1}{2}p^2 + 1900p$$

(a) At what prices p is revenue zero?
(b) For what range of prices will revenue exceed \$1,200,000?

37. Artillery A projectile fired from the point $(0, 0)$ at an angle to the positive x-axis has a trajectory given by

$$y = cx - (1 + c^2)\left(\frac{g}{2}\right)\left(\frac{x}{v}\right)^2$$

where
x = horizontal distance in meters
y = height in meters
v = initial muzzle velocity in meters per second (m/sec)
g = acceleration due to gravity = 9.81 meters per second squared (m/sec^2)
$c > 0$ is a constant determined by the angle of elevation.

A howitzer fires an artillery round with a muzzle velocity of 897 m/sec.
(a) If the round must clear a hill 200 meters high at a distance of 2000 meters in front of the howitzer, what values of c are permitted in the trajectory equation?
(b) If the goal in part (a) is to hit a target on the ground 75 kilometers away, is it possible to do so? If so, for what values of c? If not, what is the maximum distance the round will travel?

Source: www.answers.com

38. Runaway Car Using Hooke's Law, we can show that the *work* done in compressing a spring a distance of x feet from its at-rest position is $W = \frac{1}{2}kx^2$, where k is a stiffness constant depending on the spring. It can also be shown that the work done by a body in motion before it comes to rest is given by $\tilde{W} = \frac{w}{2g}v^2$, where w = weight of the object (lb), g = acceleration due to gravity (32.2 ft/sec^2), and v = object's velocity (in ft/sec). A parking garage has a spring shock absorber at the end of a ramp to stop runaway cars. The spring has a stiffness constant $k = 9450$ lb/ft and must be able to stop a 4000-lb car traveling at 25 mph. What is the least compression required of the spring? Express your answer using feet to the nearest tenth. [**Hint:** Solve $W > \tilde{W}$, $x \geq 0$].

Source: www.sciforums.com

Discussion and Writing

39. Show that the inequality $(x - 4)^2 \leq 0$ has exactly one solution.

40. Show that the inequality $(x - 2)^2 > 0$ has one real number that is not a solution.

41. Explain why the inequality $x^2 + x + 1 > 0$ has all real numbers as the solution set.

42. Explain why the inequality $x^2 - x + 1 < 0$ has the empty set as solution set.

'Are You Prepared?' Answers

1. $\{x|x > -3\}$ or $(-3, \infty)$ **2.** $-2 < x \leq 7$

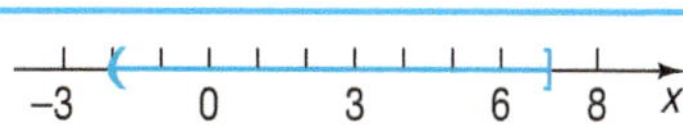

CHAPTER REVIEW

Things to Know

Linear function (p. 282)

$f(x) = mx + b$

Average rate of change $= m$
The graph is a line with slope m and y-intercept b.

Quadratic function (pp. 299–302)

$f(x) = ax^2 + bx + c, a \neq 0$

The graph is a parabola that opens up if $a > 0$ and opens down if $a < 0$.

Vertex: $\left(-\frac{b}{2a}, f\left(-\frac{b}{2a}\right)\right)$

Axis of symmetry: $x = -\frac{b}{2a}$

y-intercept: $f(0) = c$

x-intercept(s): If any, found by finding the real solutions of the equation $ax^2 + bx + c = 0$

Objectives

Section		You should be able to . . .	Example(s)	Review Exercises
4.1	1	Graph linear functions (p. 282)	1	1(a)–6(a), 1(b)–6(b)
	2	Use average rate of change to identify linear functions (p. 282)	2	7, 8
	3	Determine whether a linear function is increasing, decreasing, or constant (p. 285)	3	1(e)–6(e)
	4	Build linear models from verbal descriptions (p. 286)	4, 5	35, 36
4.2	1	Draw and interpret scatter diagrams (p. 292)	1	45(a), 46(a)
	2	Distinguish between linear and nonlinear relations (p. 293)	2	45(b), 46(a)
	3	Use a graphing utility to find the line of best fit (p. 294)	4	45(c)
4.3	1	Graph a quadratic function using transformations (p. 300)	1	9–14
	2	Identify the vertex and axis of symmetry of a quadratic function (p. 302)	2	15–24
	3	Graph a quadratic function using its vertex, axis, and intercepts (p. 302)	3–5	15–24
	4	Find the maximum or minimum value of a quadratic function (p. 306)	7	25–30, 37–42
4.4	1	Building quadratic models from verbal descriptions (p. 311)	1–4	37–44
	2	Build quadratic models from data (p. 314)	5	46
4.5	1	Solve inequalities involving a quadratic function (p. 320)	1, 2	31–34

Review Exercises

In Problems 1–6:

(a) Determine the slope and y-intercept of each linear function.
(b) Graph each function. Label the intercepts.
(c) Determine the domain and the range of each function.
(d) Determine the average rate of change of each function.
(e) Determine whether the function is increasing, decreasing, or constant.

1. $f(x) = 2x - 5$

2. $g(x) = -4x + 7$

3. $h(x) = \frac{4}{5}x - 6$

4. $F(x) = -\frac{1}{3}x + 1$

5. $G(x) = 4$

6. $H(x) = -3$

In Problems 7 and 8, determine whether the function is linear or nonlinear. If the function is linear, determine the equation of the line.

7.

x	y
−1	−2
0	3
1	8
2	13
3	18

8.

x	y
−1	−3
0	4
1	7
2	6
3	1

In Problems 9–14, graph each quadratic function using transformations (shifting, compressing, stretching, and/or reflecting). Verify your results using a graphing utility.

9. $f(x) = (x - 2)^2 + 2$ **10.** $f(x) = (x + 1)^2 - 4$ **11.** $f(x) = -(x - 4)^2$

12. $f(x) = (x - 1)^2 - 3$ **13.** $f(x) = 2(x + 1)^2 + 4$ **14.** $f(x) = -3(x + 2)^2 + 1$

In Problems 15–24, (a) graph each quadratic function by determining whether its graph opens up or down and by finding its vertex, axis of symmetry, y-intercept, and x-intercepts, if any. (b) Determine the domain and the range of the function. (c) Determine where the function is increasing and where it is decreasing. Verify your results using a graphing utility.

15. $f(x) = (x - 2)^2 + 2$ **16.** $f(x) = (x + 1)^2 - 4$ **17.** $f(x) = \frac{1}{4}x^2 - 16$ **18.** $f(x) = -\frac{1}{2}x^2 + 2$

19. $f(x) = -4x^2 + 4x$ **20.** $f(x) = 9x^2 - 6x + 3$ **21.** $f(x) = \frac{9}{2}x^2 + 3x + 1$ **22.** $f(x) = -x^2 + x + \frac{1}{2}$

23. $f(x) = 3x^2 + 4x - 1$ **24.** $f(x) = -2x^2 - x + 4$

In Problems 25–30, determine whether the given quadratic function has a maximum value or a minimum value, and then find the value. Verify your results using a graphing utility.

25. $f(x) = 3x^2 - 6x + 4$ **26.** $f(x) = 2x^2 + 8x + 5$ **27.** $f(x) = -x^2 + 8x - 4$

28. $f(x) = -x^2 - 10x - 3$ **29.** $f(x) = -3x^2 + 12x + 4$ **30.** $f(x) = -2x^2 + 4$

In Problems 31–34, solve each quadratic inequality. Verify your results using a graphing utility.

31. $x^2 + 6x - 16 < 0$ **32.** $3x^2 - 2x - 1 \geq 0$ **33.** $3x^2 \geq 14x + 5$ **34.** $4x^2 < 13x - 3$

---

35. Comparing Phone Companies Marissa must decide between one of two companies as her long-distance phone provider. Company A charges a monthly fee of \$7.00 plus \$0.06 per minute, while Company B does not have a monthly fee, but charges \$0.08 per minute.
(a) Find a linear function that relates cost, C, to total minutes on the phone, x, for each company.
(b) Determine the number of minutes for which the bill from Company A will equal the bill from Company B.
(c) Over what interval of minutes will the bill from Company B be less than the bill from Company A?

36. Sales Commissions Bill was just offered a sales position for a computer company. His salary would be \$15,000 per year plus 1% of his total annual sales.
(a) Find a linear function that relates Bill's annual salary, S, to his total annual sales, x.
(b) In 2005, Bill had total annual sales of \$1,000,000. What was Bill's salary?
(c) What would Bill have to sell to earn \$100,000?
(d) Determine the sales required of Bill for his salary to exceed \$150,000.

37. Demand Equation The price p (in dollars) and the quantity x sold of a certain product obey the demand equation

$$p = -\frac{1}{10}x + 150 \qquad 0 \leq x \leq 1500$$

(a) Express the revenue R as a function of x.
(b) What is the revenue if 100 units are sold?
(c) What quantity x maximizes revenue? What is the maximum revenue?
(d) What price should the company charge to maximize revenue?

38. Landscaping A landscape engineer has 200 feet of border to enclose a rectangular pond. What dimensions will result in the largest pond?

39. Enclosing the Most Area with a Fence A farmer with 10,000 meters of fencing wants to enclose a rectangular field and then divide it into two plots with a fence parallel to one of the sides. See the figure. What is the largest area that can be enclosed?

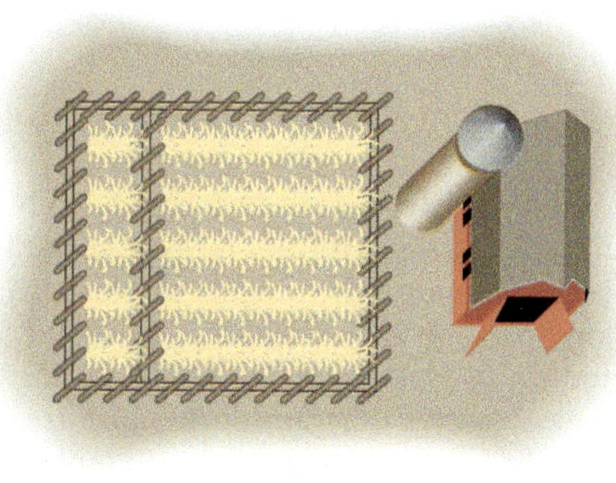

40. Architecture A special window in the shape of a rectangle with semicircles at each end is to be constructed so that the outside dimensions are 100 feet in length. See the illustration. Find the dimensions of the rectangle that maximizes its area.

41. Minimizing Marginal Cost Callaway Golf Company has determined that the marginal cost C of manufacturing x Big Bertha drivers may be expressed by the quadratic function

$$C(x) = 4.9x^2 - 617.4x + 19{,}600$$

(a) How many drivers should be manufactured to minimize the marginal cost?
(b) At this level of production, what is the marginal cost?

42. Violent Crimes The function

$$V(t) = 5.0t^2 - 87.3t + 1761.1$$

models the number V (in thousands) of violent crimes committed in the United States t years after 1995, based on data obtained from the Federal Bureau of Investigation. So $t = 0$ represents 1995, $t = 1$ represents 1996, and so on.
(a) Determine the year in which the fewest violent crimes were committed.
(b) Approximately how many violent crimes were committed during this year?
(c) Using a graphing utility, graph $V = V(t)$. Were the number of violent crimes increasing or decreasing during the years 1995 to 2003?

43. A rectangle has one vertex on the line $y = 10 - x$, $x > 0$, another at the origin, one on the positive x-axis, and one on the positive y-axis. Express the area A of the rectangle as a function of x. Find the largest area A that can be enclosed by the rectangle.

44. Parabolic Arch Bridge A horizontal bridge is in the shape of a parabolic arch. Given the information shown in the figure, what is the height h of the arch 2 feet from shore?

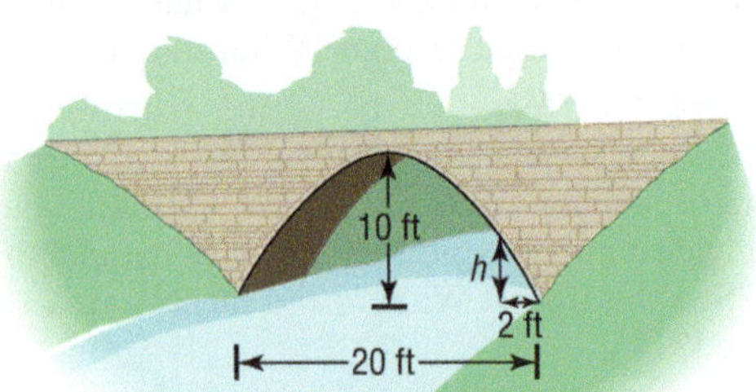

45. Bone Length Research performed at NASA, led by Dr. Emily R. Morey-Holton, measured the lengths of the right humerus and right tibia in 11 rats that were sent to space on Spacelab Life Sciences 2. The following data were collected.
(a) Draw a scatter diagram of the data treating length of the right humerus as the independent variable.

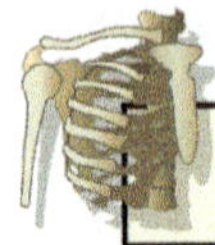

Right Humerus (mm), x	Right Tibia (mm), y
24.80	36.05
24.59	35.57
24.59	35.57
24.29	34.58
23.81	34.20
24.87	34.73
25.90	37.38
26.11	37.96
26.63	37.46
26.31	37.75
26.84	38.50

Source: NASA Life Sciences Data Archive

(b) Based on the scatter diagram, do you think that there is a linear relation between the length of the right humerus and the length of the right tibia?
(c) Use a graphing utility to find the line of best fit relating length of the right humerus and length of the right tibia.
(d) Predict the length of the right tibia on a rat whose right humerus is 26.5 millimeters (mm).

46. Advertising A small manufacturing firm collected the following data on advertising expenditures A (in thousands of dollars) and total revenue R (in thousands of dollars).

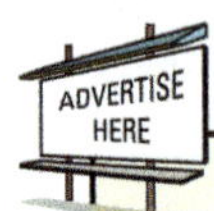

Advertising	Total Revenue
20	\$6101
22	\$6222
25	\$6350
25	\$6378
27	\$6453
28	\$6423
29	\$6360
31	\$6231

(a) Draw a scatter diagram of the data. Comment on the type of relation that may exist between the two variables.
(b) Use a graphing utility to find the quadratic function of best fit that models the relation between advertising and total revenue.
(c) Use the function found in part (b) to determine the optimal level of advertising.
(d) Use the function found in part (b) to predict the total revenue when the optimal level of advertising is spent.
(e) Use a graphing utility to draw a scatter diagram of the data, and then graph the quadratic function of best fit on the scatter diagram.

CHAPTER TEST

1. For the linear function $f(x) = -4x + 3$,
   (a) Find the slope and y-intercept.
   (b) Determine whether f is increasing, decreasing, or constant.
   (c) Graph f.

2. Determine whether the given function is linear or nonlinear If it is linear, determine the equation of the line.

x	y
−2	12
−1	7
0	2
1	−3
2	−8

3. Graph $f(x) = (x - 3)^2 - 2$ using transformations.

In Problems 4 and 5,
(a) Determine whether the graph opens up or down.
(b) Determine the vertex of the graph of the quadratic function.
(c) Determine the axis of symmetry of the graph of the quadratic function.
(d) Determine the intercepts of the graph of the quadratic function.
(e) Use the information in parts (a)–(d) to graph the quadratic function.
(f) Based on the graph, determine the domain and the range of the quadratic function.
(g) Based on the graph, determine where the function is increasing and where it is decreasing.

4. $f(x) = 3x^2 - 12x + 4$
5. $g(x) = -2x^2 + 4x - 5$
6. Determine the quadratic function for the given graph.

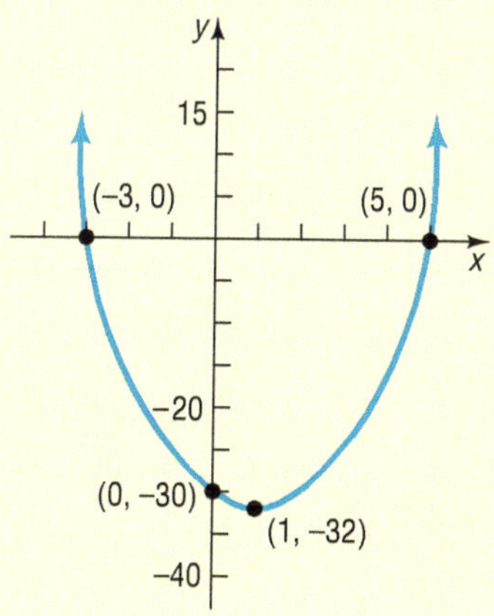

7. Determine whether $f(x) = -2x^2 + 12x + 3$ has a maximum or minimum value. Then find the maximum or minimum value.
8. Solve $x^2 - 10x + 24 \geq 0$.
9. The weekly rental cost of a 20-foot recreational vehicle is \$129.50 plus \$0.15 per mile.
   (a) Find a linear function that expresses the cost C as a function of miles driven m.
   (b) What is the rental cost if 860 miles are driven?
   (c) How many miles were driven if the rental cost is \$213.80?
10. The price p (in dollars) and the quantity x sold of a certain product obey the demand equation $p = -\frac{1}{10}x + 1000$.
   (a) Find a model that expresses the revenue R as a function of x.
   (b) What is the revenue if 400 units are sold?
   (c) What quantity x maximizes revenue? What is the maximum revenue?
   (d) What price should the company charge to maximize revenue?
11. Consider these two data sets:

Set A

x	−2	−1	0	0	1	2	2
y	5	2	1	−3	−8	−12	−10

Set B

x	−2	−1	0	0	1	2	2
y	10	4	2	3	5	10	12

One data set follows a linear pattern and one data set follows a quadratic relation.
   (a) Draw a scatter diagram of each data set. Determine which is linear and which is quadratic. For the linear data, indicate whether the relation shows a positive or negative slope. For the quadratic relation, indicate whether the quadratic function of best fit will open up or down.
   (b) For the linear data set, find the line of best fit.
   (c) For the quadratic data set, find the quadratic function of best fit.

CUMULATIVE REVIEW

1. For the function f defined by $f(x) = x^2 - 4x + 1$, find:
   (a) $f(2)$ (b) $f(x) + f(2)$ (c) $f(-x)$
   (d) $-f(x)$ (e) $f(x + 2)$ (f) $\dfrac{f(x+h) - f(x)}{h}$ $h \neq 0$
2. Find the domain of $h(z) = \dfrac{3z - 1}{6z - 7}$.
3. Is the following graph the graph of a function?

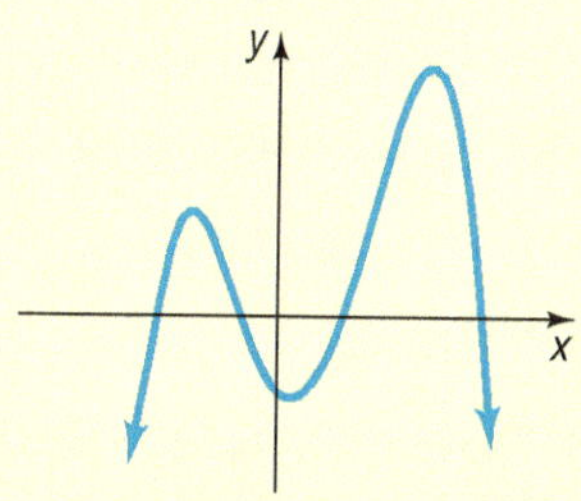

4. Consider the function $f(x) = \frac{x}{x + 4}$.

(a) Is the point $\left(1, \frac{1}{4}\right)$ on the graph of f?

(b) If $x = -2$, what is $f(x)$? What point is on the graph of f?

(c) If $f(x) = 2$, what is x? What point is on the graph of f?

5. Is the function $f(x) = \frac{x^2}{2x + 1}$ even, odd, or neither?

6. Approximate the local maxima and local minima of $f(x) = x^3 - 5x + 1$ on $(-4, 4)$. Determine where the function is increasing and where it is decreasing.

7. If $f(x) = 3x + 5$ and $g(x) = 2x + 1$,

(a) Solve $f(x) = g(x)$.

(b) Solve $f(x) > g(x)$.

8. Given that $f(x) = x^2 + 3x$ and $g(x) = 5x + 3$,

(a) Solve $f(x) = g(x)$. (b) Solve $f(x) > g(x)$.

9. For the graph of the function f to the right,

(a) Find the domain and the range of f.

(b) Find the intercepts.

(c) Is the graph of f symmetric with respect to the x-axis, the y-axis, or the origin?

(d) Find $f(2)$.

(e) For what value(s) of x is $f(x) = 3$?

(f) Solve $f(x) < 0$.

(g) Graph $y = f(x) + 2$

(h) Graph $y = f(-x)$.

(i) Graph $y = 2f(x)$.

(j) Is f even, odd, or neither?

(k) Find the interval(s) on which f is increasing.

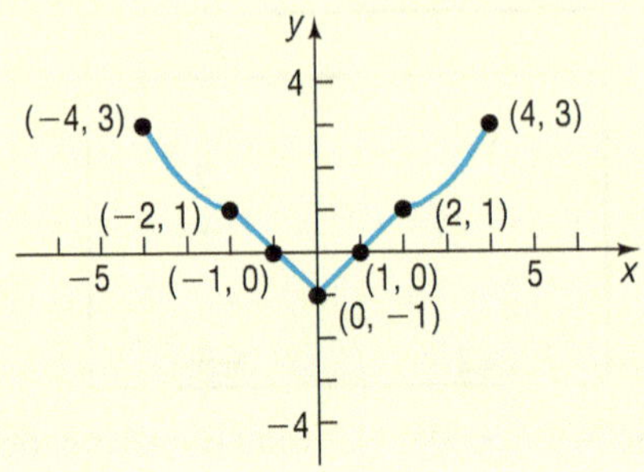

CHAPTER PROJECTS

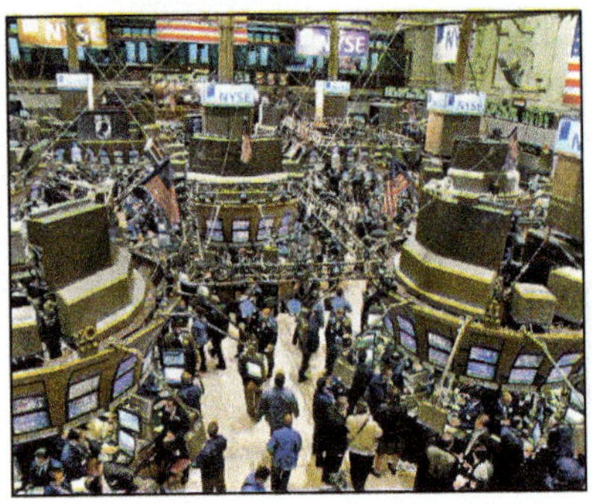

I. **Analyzing a Stock** The beta, β, of a stock represents the relative risk of a stock compared with a market basket of stocks, such as Standard and Poor's 500 Index of stocks. Beta is computed by finding the slope of the line of best fit between the rate of return of the stock and rate of return of the S&P 500. The rates of return are computed on a weekly basis.

1. Find the weekly closing price of your favorite stock and the S&P 500 for 20 weeks. One good source is on the Internet at *finance.yahoo.com.*
2. Compute the rate of return by computing the weekly percentage change in the closing price of your stock and the weekly percentage change in the S&P 500 using the following formula:

$$\text{Weekly \% change} = \frac{P_2 - P_1}{P_1}$$

where P_1 is last week's closing price and P_2 is this week's closing price.

3. Using a graphing utility, find the line of best fit, treating the weekly rate of return of the S&P 500 as the independent variable and the weekly rate of return of your stock as the dependent variable.
4. What is the beta of your stock?
5. Compare your result with that of the Value Line Investment Survey found in your library. What might account for any differences?

The following projects are available on the Instructor's Resource Center (IRC):

II. **Cannons** A battery commander uses the weight of a missile, its initial velocity, and the position of its gun to determine where the missile will travel.

III. **First and Second Differences** Finite differences provide a numerical method that is used to estimate the graph of an unknown function.

IV. **CBL Experiment** Computer simulation is used to study the physical properties of a bouncing ball.

Polynomial and Rational Functions

5

We Don't Want to Be Number One Again—in Tree Pollen

LEXINGTON, Ky. (March 29, 2006)—Were your allergies especially bothersome last spring? It's no wonder. In the spring of 2005, Lexington, Ky., was the tree pollen capital of the U.S. No allergy sufferer is rooting for the Bluegrass to earn that title again in 2006. Either way, allergy sufferers should be prepared. The spring allergy season is here.

If you can't stop sneezing, blowing your nose, and rubbing your itchy eyes, you may have allergic rhinitis, as do 40 million other Americans. The pollens of trees such as maple, oak and walnut trigger much of the allergic rhinitis, commonly referred to as hay fever, in the early spring. Pollinating grasses and weeds are responsible for summer hay fever. Perennial allergic rhinitis can be triggered any time of year by indoor allergens, such as animal dander, mold or dust mites.

"Spring can be the start of misery for many allergy sufferers because they have multiple pollen allergies," said Dr. Beth Miller, director of the UK Asthma, Allergy and Sinus Clinics.

Source: Adapted from *Melanie Jackson, "We Don't Want to Be Number One Again—In Tree Pollen," University of Kentucky News, March 29, 2006, http://news.uky.edu.*

—See Chapter Project I—

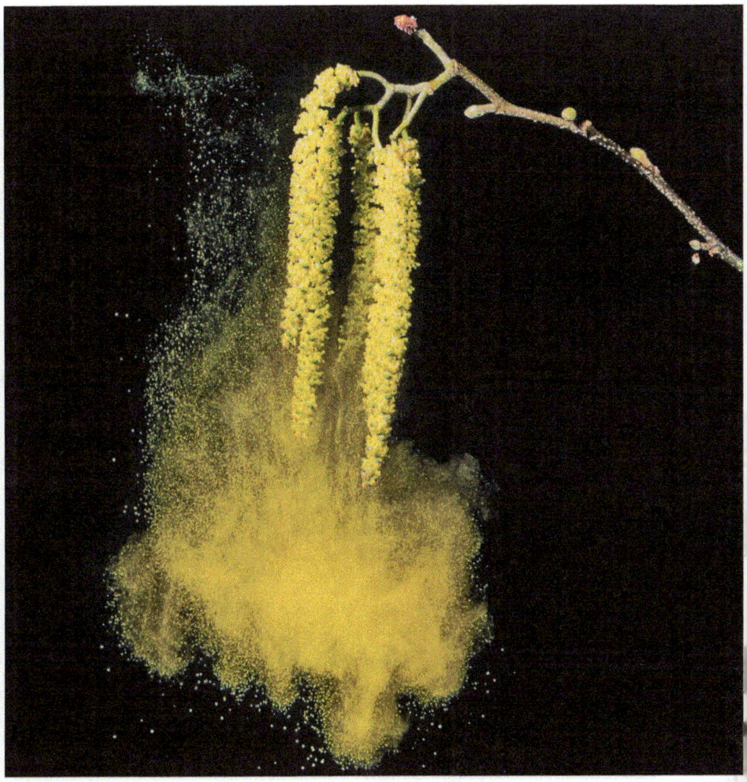

A Look Back

In Chapter 3, we began our discussion of functions. We defined domain and range and independent and dependent variables; we found the value of a function and graphed functions. We continued our study of functions by listing the properties that a function might have, such as being even or odd, and we created a library of functions, naming key functions and listing their properties, including their graphs.

In Chapter 4, we discussed linear functions and quadratic functions, which belong to the class of *polynomial functions*.

A Look Ahead

In this chapter, we look at two general classes of functions, polynomial functions and rational functions, and examine their properties. Polynomial functions are arguably the simplest expressions in algebra. For this reason, they are often used to approximate other, more complicated functions. Rational functions are ratios of polynomial functions.

Outline

5.1 Polynomial Functions and Models

PREPARING FOR THIS SECTION *Before getting started, review the following:*

- Polynomials (Section R.4, pp. 39–41)
- Using a Graphing Utility to Approximate Local Maxima and Local Minima (Section 3.3, pp. 239–240)
- Intercepts of a Function (Section 3.2, pp. 228–229)
- Graphing Techniques: Transformations (Section 3.5, pp. 256–264)
- Intercepts (Section 2.1, pp. 170–171)

Now Work the 'Are You Prepared?' problems on page 343.

OBJECTIVES
1. Identify Polynomial Functions and Their Degree (p. 330)
2. Graph Polynomial Functions Using Transformations (p. 333)
3. Identify the Real Zeros of a Polynomial Function and Their Multiplicity (p. 334)
4. Analyze the Graph of a Polynomial Function (p. 339)
5. Build Cubic Models from Data (p. 342)

1 Identify Polynomial Functions and Their Degree

In Chapter 4 we studied the linear function $f(x) = mx + b$, which we can write as

$$f(x) = a_1x + a_0$$

and we studied the quadratic function $f(x) = ax^2 + bx + c,\ a \neq 0$, which we can write as

$$f(x) = a_2x^2 + a_1x + a_0, \qquad a_2 \neq 0$$

Each of these functions is an example of a *polynomial function.*

DEFINITION

A **polynomial function** is a function of the form

$$f(x) = a_nx^n + a_{n-1}x^{n-1} + \cdots + a_1x + a_0 \qquad \textbf{(1)}$$

where $a_n, a_{n-1}, \ldots, a_1, a_0$ are real numbers and n is a nonnegative integer. The domain of a polynomial function is the set of all real numbers.

In Words

A polynomial function is the sum of monomials.

A polynomial function is a function whose rule is given by a polynomial in one variable. The **degree** of a polynomial function is the largest power of x that appears. The zero polynomial function $f(x) = 0 + 0x + 0x^2 + \cdots + 0x^n$ is not assigned a degree.

Polynomial functions are among the simplest expressions in algebra. They are easy to evaluate: only addition and repeated multiplication are required. Because of this, they are often used to approximate other, more complicated functions. In this section, we investigate properties of this important class of functions.

EXAMPLE 1 Identifying Polynomial Functions

Determine which of the following are polynomial functions. For those that are, state the degree; for those that are not, tell why not.

(a) $f(x) = 2 - 3x^4$ (b) $g(x) = \sqrt{x}$ (c) $h(x) = \dfrac{x^2 - 2}{x^3 - 1}$

(d) $F(x) = 0$ (e) $G(x) = 8$ (f) $H(x) = -2x^3(x - 1)^2$

Solution (a) f is a polynomial function of degree 4.

(b) g is not a polynomial function because $g(x) = \sqrt{x} = x^{\frac{1}{2}}$, so the variable x is raised to the $\frac{1}{2}$ power, which is not a nonnegative integer.

(c) h is not a polynomial function. It is the ratio of two polynomials, and the polynomial in the denominator is of positive degree.

(d) F is the zero polynomial function; it is not assigned a degree.

(e) G is a nonzero constant function. It is a polynomial function of degree 0 since $G(x) = 8 = 8x^0$.

(f) $H(x) = -2x^3(x-1)^2 = -2x^3(x^2 - 2x + 1) = -2x^5 + 4x^4 - 2x^3$. So H is a polynomial function of degree 5. Do you see a way to find the degree of H without multiplying out?

Now Work PROBLEMS 11 AND 15

We have already discussed in detail polynomial functions of degrees 0, 1, and 2. See Table 1 for a summary of the properties of the graphs of these polynomial functions.

Table 1

Degree	Form	Name	Graph
No degree	$f(x) = 0$	Zero function	The x-axis
0	$f(x) = a_0, \quad a_0 \neq 0$	Constant function	Horizontal line with y-intercept a_0
1	$f(x) = a_1x + a_0, \quad a_1 \neq 0$	Linear function	Nonvertical, nonhorizontal line with slope a_1 and y-intercept a_0
2	$f(x) = a_2x^2 + a_1x + a_0, \quad a_2 \neq 0$	Quadratic function	Parabola: graph opens up if $a_2 > 0$; graph opens down if $a_2 < 0$

One objective of this section is to analyze the graph of a polynomial function. If you take a course in calculus, you will learn that the graph of every polynomial function is both smooth and continuous. By **smooth,** we mean that the graph contains no sharp corners or cusps; by **continuous,** we mean that the graph has no gaps or holes and can be drawn without lifting pencil from paper. See Figures 1(a) and (b).

Figure 1

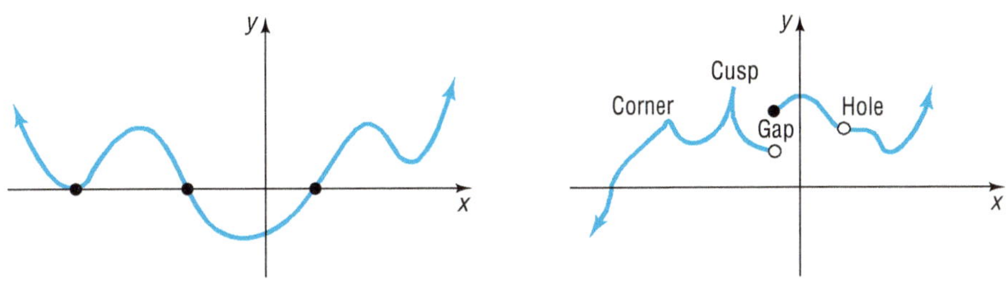

(a) Graph of a polynomial function: smooth, continuous

(b) Cannot be the graph of a polynomial function

Power Functions

We begin the analysis of the graph of a polynomial function by discussing *power functions,* a special kind of polynomial function.

DEFINITION

A **power function of degree n** is a monomial of the form

$$f(x) = ax^n \tag{2}$$

where a is a real number, $a \neq 0$, and $n > 0$ is an integer.

In Words

A power function is defined by a single monomial.

Examples of power functions are

$f(x) = 3x$ (degree 1) $\quad f(x) = -5x^2$ (degree 2) $\quad f(x) = 8x^3$ (degree 3) $\quad f(x) = -5x^4$ (degree 4)

The graph of a power function of degree 1, $f(x) = ax$, is a straight line, with slope a, that passes through the origin. The graph of a power function of degree 2, $f(x) = ax^2$, is a parabola, with vertex at the origin, that opens up if $a > 0$ and down if $a < 0$.

If we know how to graph a power function of the form $f(x) = x^n$, a compression or stretch and, perhaps, a reflection about the x-axis will enable us to obtain the graph of $g(x) = ax^n$. Consequently, we shall concentrate on graphing power functions of the form $f(x) = x^n$.

We begin with power functions of even degree of the form $f(x) = x^n$, $n \geq 2$ and *n even*.

Exploration

Using your graphing utility and the viewing window $-2 \leq x \leq 2$, $-4 \leq y \leq 16$, graph the function $Y_1 = f(x) = x^4$. On the same screen, graph $Y_2 = g(x) = x^8$. Now, also on the same screen, graph $Y_3 = h(x) = x^{12}$. What do you notice about the graphs as the magnitude of the exponent increases? Repeat this procedure for the viewing window $-1 \leq x \leq 1$, $0 \leq y \leq 1$. What do you notice?

Result See Figures 2(a) and (b).

Figure 2

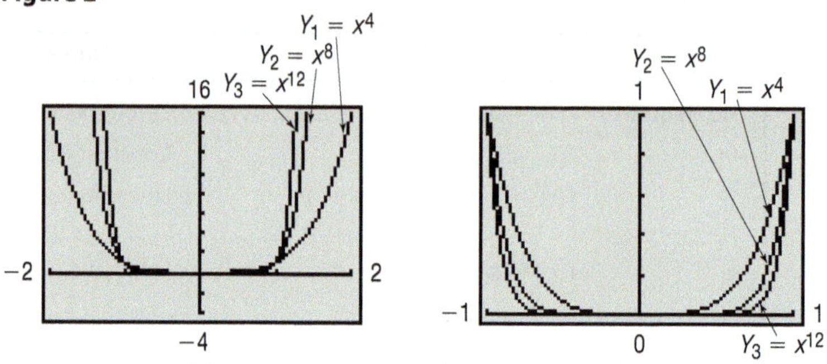

(a) (b)

Table 2

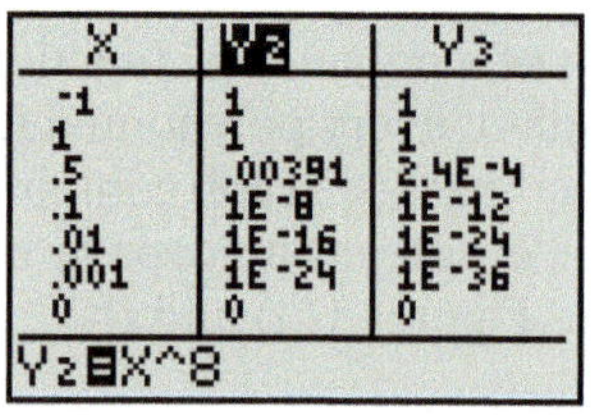

X	Y2	Y3
-1	1	1
1	1	1
.5	.00391	2.4E-4
.1	1E-8	1E-12
.01	1E-16	1E-24
.001	1E-24	1E-36
0	0	0

Y2=X^8

NOTE Don't forget how graphing calculators express scientific notation. In Table 2, 1E −8 means 1×10^{-8}. ■

The domain of $f(x) = x^n$, $n \geq 2$ and n even, is the set of all real numbers, and the range is the set of nonnegative real numbers. Such a power function is an even function (do you see why?), so its graph is symmetric with respect to the y-axis. Its graph always contains the origin (0, 0) and the points (−1, 1) and (1, 1).

If $n = 2$, the graph is the parabola $y = x^2$ that opens up, with vertex at the origin. For large n, it appears that the graph coincides with the x-axis near the origin, but it does not; the graph actually touches the x-axis only at the origin. See Table 2, where $Y_2 = x^8$ and $Y_3 = x^{12}$. For x close to 0, the values of y are positive and close to 0. Also, for large n, it may appear that for $x < -1$ or for $x > 1$ the graph is vertical, but it is not; it is only increasing very rapidly. If you TRACE along one of the graphs, these distinctions will be clear.

To summarize:

Properties of Power Functions, $f(x) = x^n$, n Is an Even Integer

1. The graph is symmetric with respect to the y-axis, so f is even.
2. The domain is the set of all real numbers. The range is the set of nonnegative real numbers.
3. The graph always contains the points $(0, 0)$, $(1, 1)$, and $(-1, 1)$.
4. As the exponent n increases in magnitude, the graph becomes more vertical when $x < -1$ or $x > 1$; but for x near the origin, the graph tends to flatten out and lie closer to the x-axis.

Now we consider power functions of odd degree of the form $f(x) = x^n$, *n odd*.

Exploration

Using your graphing utility and the viewing window $-2 \leq x \leq 2$, $-16 \leq y \leq 16$, graph the function $Y_1 = f(x) = x^3$. On the same screen, graph $Y_2 = g(x) = x^7$ and $Y_3 = h(x) = x^{11}$. What do you notice about the graphs as the magnitude of the exponent increases? Repeat this procedure for the viewing window $-1 \leq x \leq 1$, $-1 \leq y \leq 1$. What do you notice?

Result The graphs on your screen should look like Figures 3(a) and (b).

Figure 3

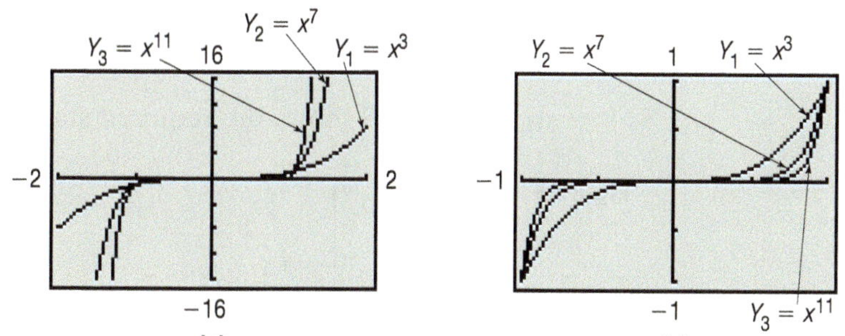

(a) (b)

The domain and the range of $f(x) = x^n$, $n \geq 3$ and n odd, are the set of real numbers. Such a power function is an odd function (do you see why?), so its graph is symmetric with respect to the origin. Its graph always contains the origin (0, 0) and the points (−1, −1) and (1, 1).

It appears that the graph coincides with the x-axis near the origin, but it does not; the graph actually crosses the x-axis only at the origin. Also, it appears that as x increases the graph is vertical, but it is not; it is increasing very rapidly. TRACE along the graphs to verify these distinctions.

To summarize:

Properties of Power Functions, $f(x) = x^n$, n Is an Odd Integer

1. The graph is symmetric with respect to the origin, so f is odd.
2. The domain and the range are the set of all real numbers.
3. The graph always contains the points $(0, 0)$, $(1, 1)$, and $(-1, -1)$.
4. As the exponent n increases in magnitude, the graph becomes more vertical when $x < -1$ or $x > 1$, but for x near the origin, the graph tends to flatten out and lie closer to the x-axis.

2 Graph Polynomial Functions Using Transformations

The methods of shifting, compression, stretching, and reflection studied in Section 3.5, when used with the facts just presented, will enable us to graph polynomial functions that are transformations of power functions.

EXAMPLE 2 Graphing Polynomial Functions Using Transformations

Graph: $f(x) = 1 - x^5$

Solution It is helpful to rewrite f as $f(x) = -x^5 + 1$. Figure 4 shows the required stages.

Figure 4

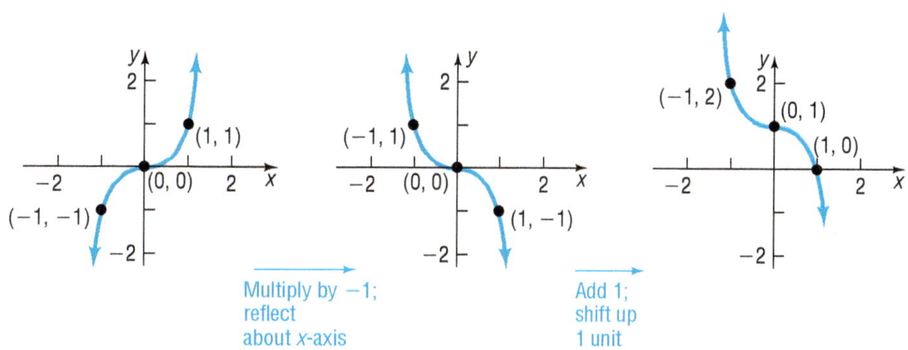

(a) $y = x^5$ (b) $y = -x^5$ (c) $y = -x^5 + 1 = 1 - x^5$

Check: Verify the graph of f by graphing $Y_1 = 1 - x^5$.

EXAMPLE 3 Graphing Polynomial Functions Using Transformations

Graph: $f(x) = \frac{1}{2}(x-1)^4$

Solution Figure 5 shows the required stages.

Figure 5

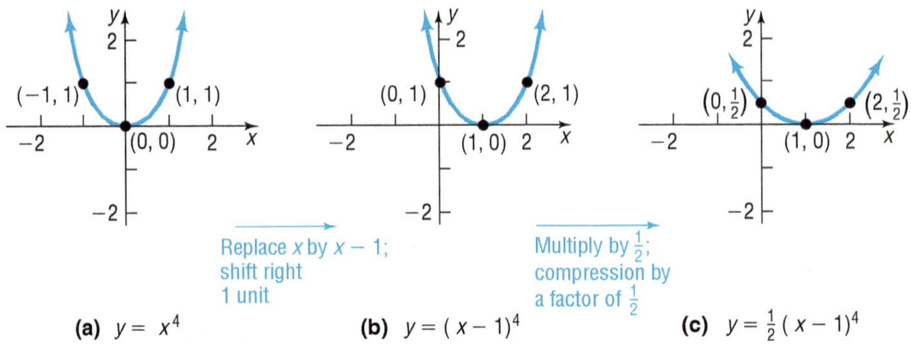

Check: Verify the graph of f by graphing $Y_1 = \frac{1}{2}(x-1)^4$.

Now Work PROBLEMS 23 AND 29

3 Identify the Real Zeros of a Polynomial Function and Their Multiplicity

Figure 6 shows the graph of a polynomial function with four x-intercepts. Notice that at the x-intercepts the graph must either cross the x-axis or touch the x-axis. Consequently, between consecutive x-intercepts the graph is either above the x-axis or below the x-axis. We will make use of this property of the graph of a polynomial function shortly.

Figure 6

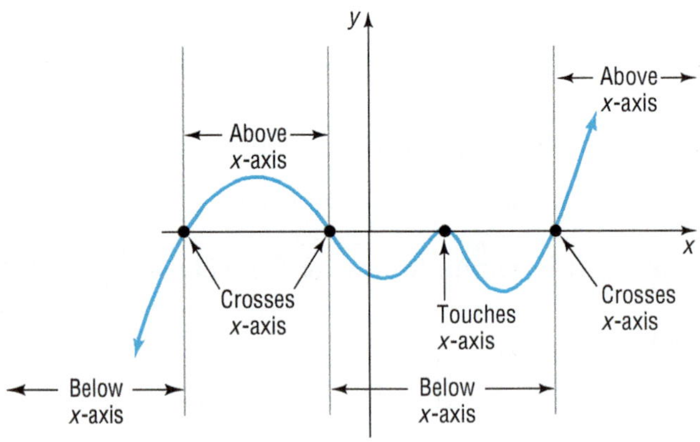

If a polynomial function f is factored completely, it is easy to solve the equation $f(x) = 0$ using the Zero-Product Property and locate the x-intercepts of the graph. For example, if $f(x) = (x-1)^2(x+3)$, then the solutions of the equation

$$f(x) = (x-1)^2(x+3) = 0$$

are 1 and −3. So, if $x = 1$ or $x = -3$, the value of the function is zero. This leads to the following definition.

DEFINITION

If f is a function and r is a real number for which $f(r) = 0$, then r is called a **real zero of** f.

As a consequence of this definition, the following statements are equivalent.

1. r is a real zero of a polynomial function f.
2. r is an x-intercept of the graph of f.
3. $x - r$ is a factor of f.
4. r is a solution to the equation $f(x) = 0$.

So the real zeros of a polynomial function are the x-intercepts of its graph, and they are found by solving the equation $f(x) = 0$.

EXAMPLE 4 Finding a Polynomial from Its Zeros

(a) Find a polynomial of degree 3 whose zeros are -3, 2, and 5.

(b) Use a graphing utility to graph the polynomial found in part (a) to verify your result.

Solution (a) If r is a real zero of a polynomial f, then $x - r$ is a factor of f. This means that $x - (-3) = x + 3$, $x - 2$, and $x - 5$ are factors of f. As a result, any polynomial of the form

$$f(x) = a(x + 3)(x - 2)(x - 5)$$

where a is any nonzero real number, qualifies. The value of a causes a stretch, compression, or reflection, but does not affect the x-intercepts. Do you know why?

(b) We choose to graph f with $a = 1$. Then

$$f(x) = (x + 3)(x - 2)(x - 5) = x^3 - 4x^2 - 11x + 30$$

Figure 7 shows the graph of f. Notice that the x-intercepts are -3, 2, and 5.

Figure 7

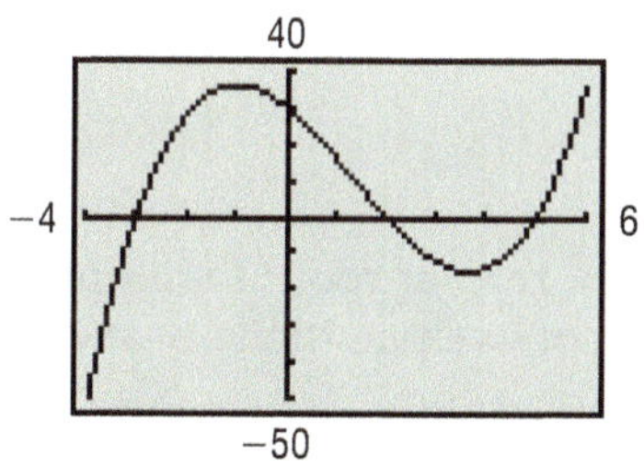

Seeing the Concept

Graph the function found in Example 4 for $a = 2$ and $a = -1$. Does the value of a affect the zeros of f? How does the value of a affect the graph of f?

Now Work PROBLEM 37

If the same factor $x - r$ occurs more than once, r is called a **repeated,** or **multiple, zero of f.** More precisely, we have the following definition.

DEFINITION If $(x - r)^m$ is a factor of a polynomial f and $(x - r)^{m+1}$ is not a factor of f, then r is called a **zero of multiplicity m of f.***

EXAMPLE 5 Identifying Zeros and Their Multiplicities

For the polynomial

$$f(x) = 5(x - 2)(x + 3)^2\left(x - \frac{1}{2}\right)^4$$

2 is a zero of multiplicity 1 because the exponent on the factor $x - 2$ is 1.

-3 is a zero of multiplicity 2 because the exponent on the factor $x + 3$ is 2.

$\frac{1}{2}$ is a zero of multiplicity 4 because the exponent on the factor $x - \frac{1}{2}$ is 4.

Now Work PROBLEM 45(a)

*Some books use the terms **multiple root** and **root of multiplicity m.**

In Example 5 notice that, if you add the multiplicities $(1 + 2 + 4 = 7)$, you obtain the degree of the polynomial.

Suppose that it is possible to factor completely a polynomial function and, as a result, locate all the x-intercepts of its graph (the real zeros of the function). The following example illustrates the role that the multiplicity of an x-intercept plays.

EXAMPLE 6 Investigating the Role of Multiplicity

For the polynomial $f(x) = x^2(x - 2)$:

(a) Find the x- and y-intercepts of the graph of f.
(b) Using a graphing utility, graph the polynomial.
(c) For each x-intercept, determine whether it is of odd or even multiplicity.

Solution (a) The y-intercept is $f(0) = 0^2(0 - 2) = 0$. The x-intercepts satisfy the equation

$$f(x) = x^2(x - 2) = 0$$

from which we find that

$$x^2 = 0 \quad \text{or} \quad x - 2 = 0$$
$$x = 0 \quad \text{or} \quad x = 2$$

The x-intercepts are 0 and 2.

(b) See Figure 8 for the graph of f.

(c) We can see from the factored form of f that 0 is a zero or root of multiplicity 2, and 2 is a zero or root of multiplicity 1; so 0 is of even multiplicity and 2 is of odd multiplicity.

Figure 8

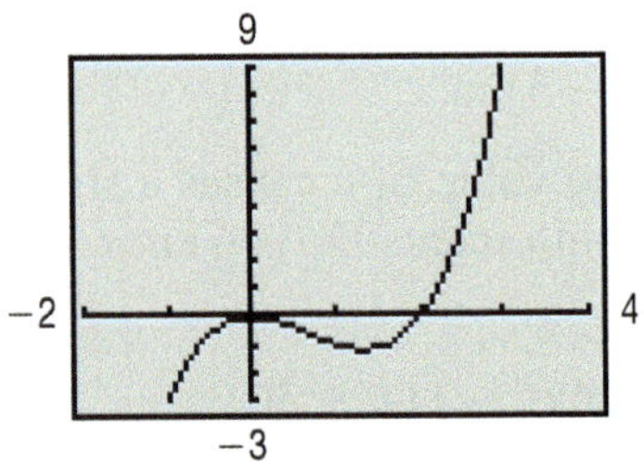

We can use a TABLE to further analyze the graph. See Table 3. The sign of $f(x)$ is the same on each side of $x = 0$, and the graph of f just *touches* the x-axis at $x = 0$ (a zero of *even* multiplicity). The sign of $f(x)$ changes from one side of $x = 2$ to the other, and the graph of f *crosses* the x-axis at $x = 2$ (a zero of *odd* multiplicity). These observations suggest the following result:

Table 3

X	Y1
-2	-16
-1	-3
0	0
1	-1
2	0
3	9
4	32

Y1 = X²(X−2)

If r Is a Zero of Even Multiplicity

Sign of $f(x)$ does not change from one side of r to the other side of r. — Graph **touches** x-axis at r.

If r Is a Zero of Odd Multiplicity

Sign of $f(x)$ changes from one side of r to the other side of r. — Graph **crosses** x-axis at r.

Now Work PROBLEM 45(b)

Figure 9

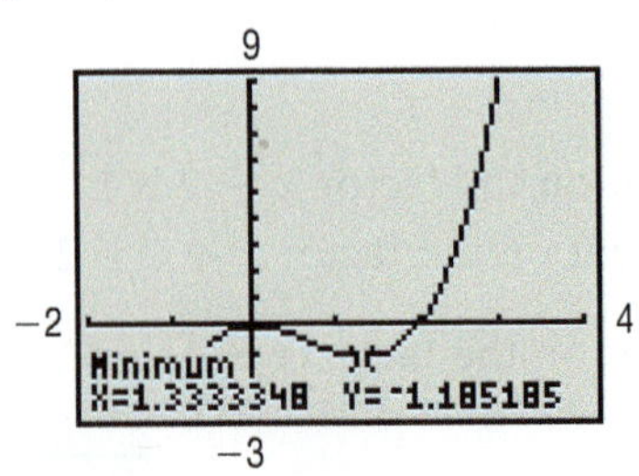

Turning Points

Points on the graph where the graph changes from an increasing function to a decreasing function, or vice versa, are called **turning points.***

Look at Figure 9. The graph of $f(x) = x^2(x - 2) = x^3 - 2x^2$ has a turning point at $(0, 0)$. After utilizing MINIMUM, we find that the graph also has a turning point at $(1.33, -1.19)$, rounded to two decimal places.

*Graphing utilities can be used to approximate turning points. Calculus is needed to find the exact location of turning points.

Exploration

Graph $Y_1 = x^3$, $Y_2 = x^3 - x$ and $Y_3 = x^3 + 3x^2 + 4$. How many turning points do you see? Graph $Y_1 = x^4$, $Y_2 = x^4 - \frac{4}{3}x^3$, and $Y_3 = x^4 - 2x^2$. How many turning points do you see? How does the number of turning points compare to the degree?

The following theorem from calculus supplies the answer to the question posed in the Exploration.

THEOREM

Turning Points

If f is a polynomial function of degree n, then f has at most $n - 1$ turning points.

If the graph of a polynomial function f has $n - 1$ turning points, the degree of f is at least n.

For example, a polynomial function of degree 5 will have at most $5 - 1 = 4$ turning points. If a polynomial function has 3 turning points, then its degree must be at least 4.

EXAMPLE 7 Identifying the Graph of a Polynomial Function

Which of the graphs in Figure 10 could be the graph of a polynomial function? For those that could, list the real zeros and state the least degree the polynomial can have. For those that could not, say why not.

Figure 10

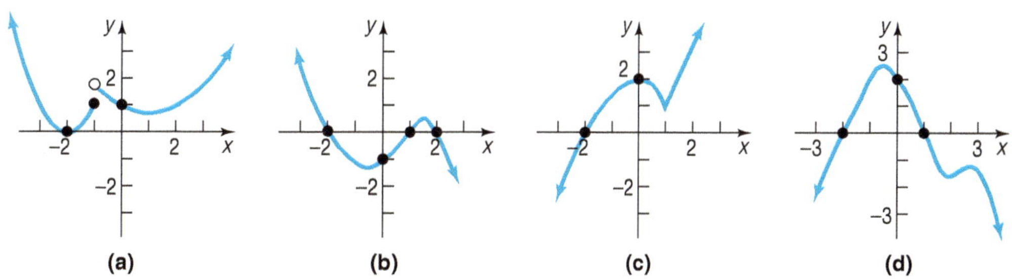

Solution

(a) The graph in Figure 10(a) cannot be the graph of a polynomial function because of the gap that occurs at $x = -1$. Remember, the graph of a polynomial function is continuous—no gaps or holes.

(b) The graph in Figure 10(b) could be the graph of a polynomial function. It has three real zeros, at -2, 1, and 2. Since the graph has two turning points, the degree of the polynomial function must be at least 3.

(c) The graph in Figure 10(c) cannot be the graph of a polynomial function because of the cusp at $x = 1$. Remember, the graph of a polynomial function is smooth.

(d) The graph in Figure 10(d) could be the graph of a polynomial function. It has two real zeros, at -2 and 1. Since the graph has three turning points, the degree of the polynomial function is at least 4.

Now Work PROBLEMS 45(c) AND 57

End Behavior

One last remark about Figure 8. For very large values of x, either positive or negative, the graph of $f(x) = x^2(x - 2)$ looks like the graph of $y = x^3$. To see why, we write f in the form

$$f(x) = x^2(x - 2) = x^3 - 2x^2 = x^3\left(1 - \frac{2}{x}\right)$$

Now, for large values of x, either positive or negative, the term $\frac{2}{x}$ is close to 0, so for large values of x

$$f(x) = x^3 - 2x^2 = x^3\left(1 - \frac{2}{x}\right) \approx x^3$$

The behavior of the graph of a function for large values of x, either positive or negative, is referred to as its **end behavior.** The end behavior of $f(x) = x^2(x - 2)$ is $y = x^3$.

THEOREM

End Behavior

For large values of x, either positive or negative, the graph of the polynomial

$$f(x) = a_nx^n + a_{n-1}x^{n-1} + \cdots + a_1x + a_0$$

resembles the graph of the power function

$$y = a_nx^n$$

Seeing the Concept

For each pair of functions Y_1 and Y_2 given in parts (a), (b), and (c), graph Y_1 and Y_2 on the same viewing window. Create a TABLE or TRACE for large positive and large negative values of x. What do you notice about the graphs of Y_1 and Y_2 as x becomes very large and positive or very large and negative?

(a) $Y_1 = x^2(x - 2); \quad Y_2 = x^3$

(b) $Y_1 = x^4 - 3x^3 + 7x - 3; \quad Y_2 = x^4$

(c) $Y_1 = -2x^3 + 4x^2 - 8x + 10; \quad Y_2 = -2x^3$

Look back at Figures 2(a) and 3(a). Based on the preceding theorem and the previous discussion on power functions, the end behavior of a polynomial can only be of four types. See Figure 11.

Figure 11
End Behavior of $f(x) = a_nx^n + a_{n-1}x^{n-1} + \cdots + a_1x + a_0$

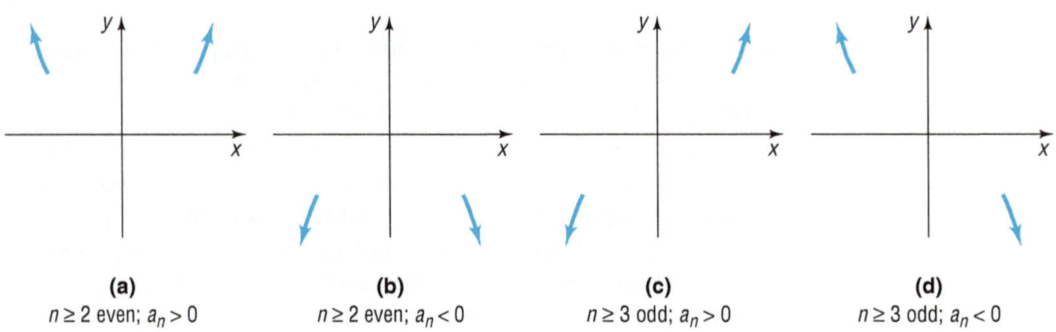

For example, if $f(x) = -2x^4 + x^3 + 4x^2 - 7x + 1$, the graph of f will resemble the graph of the power function $y = -2x^4$ for large $|x|$. Therefore, the graph of f will behave like Figure 11(b) for large $|x|$.

Now Work PROBLEM 45(d)

EXAMPLE 8 **Identifying the Graph of a Polynomial Function**

Which of the graphs in Figure 12 could be the graph of

$$f(x) = x^4 + 5x^3 + 5x^2 - 5x - 6?$$

Figure 12

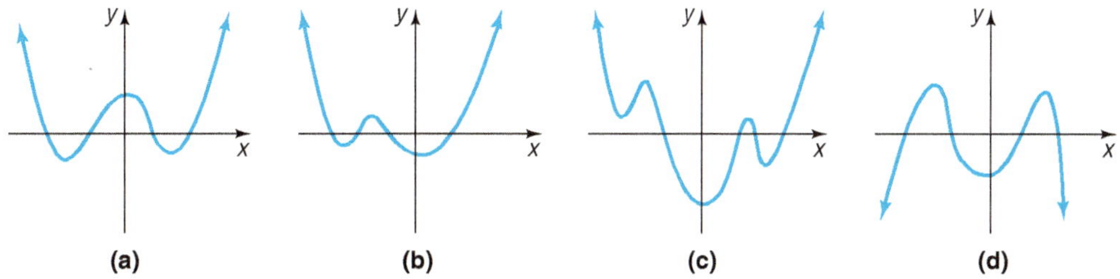

Solution The y-intercept of f is $f(0) = -6$. We can eliminate the graph in Figure 12(a), whose y-intercept is positive.

We don't have any methods for finding the x-intercepts of f, so we move on to investigate the turning points of each graph. Since f is of degree 4, the graph of f has at most 3 turning points. We eliminate the graph in Figure 12(c) since that graph has 5 turning points.

Now we look at end behavior. For large values of x, the graph of f will behave like the graph of $y = x^4$. This eliminates the graph in Figure 12(d), whose end behavior is like the graph of $y = -x^4$.

Only the graph in Figure 12(b) could be (and, in fact, is) the graph of $f(x) = x^4 + 5x^3 + x^2 - 5x - 6$.

Now Work PROBLEM 61

SUMMARY **Graph of a Polynomial Function** $f(x) = a_n x^n + a_{n-1}x^{n-1} + \cdots + a_1 x + a_0 \quad a_n \neq 0$

Degree of the polynomial f: n

Maximum number of turning points: $n - 1$

At a zero of even multiplicity: The graph of f touches the x-axis.

At a zero of odd multiplicity: The graph of f crosses the x-axis.

Between zeros, the graph of f is either above or below the x-axis.

End behavior: For large $|x|$, the graph of f behaves like the graph of $y = a_n x^n$.

4 Analyze the Graph of a Polynomial Function

EXAMPLE 9 **How to Analyze the Graph of a Polynomial Function**

Analyze the graph of the polynomial function $f(x) = (2x + 1)(x - 3)^2$.

Step-by-Step Solution

STEP 1 *Determine the end behavior of the graph of the function.*

First, we expand the polynomial to write it in the form

$$f(x) = a_n x^n + a_{n-1}x^{n-1} + \cdots + a_1 x + a_0$$

$$\begin{aligned} f(x) &= (2x + 1)(x - 3)^2 \\ &= (2x + 1)(x^2 - 6x + 9) \\ &= 2x^3 - 12x^2 + 18x + x^2 - 6x + 9 && \text{Multiply.} \\ &= 2x^3 - 11x^2 + 12x + 9 && \text{Combine like terms.} \end{aligned}$$

The polynomial function f is of degree 3. The graph of f behaves like $y = 2x^3$ for large values of $|x|$.

Step 2 Find the x- and y-intercepts of the graph of the function.

The y-intercept is $f(0) = 9$. To find the x-intercepts, we solve $f(x) = 0$.

$$f(x) = 0$$
$$(2x + 1)(x - 3)^2 = 0$$
$$2x + 1 = 0 \quad \text{or} \quad (x - 3)^2 = 0$$
$$x = -\frac{1}{2} \quad \text{or} \quad x - 3 = 0$$
$$x = 3$$

The x-intercepts are $-\frac{1}{2}$ and 3.

Step 3 Determine the zeros of the function and their multiplicity. Use this information to determine whether the graph crosses or touches the x-axis at each x-intercept.

The zeros of f are $-\frac{1}{2}$ and 3. The zero $-\frac{1}{2}$ is a zero of multiplicity 1, so the graph of f crosses the x-axis at $x = -\frac{1}{2}$. The zero 3 is a zero of multiplicity 2, so the graph of f touches the x-axis at $x = 3$.

Step 4 Use a graphing utility to graph the function.

See Figure 13 for the graph of f.

Figure 13

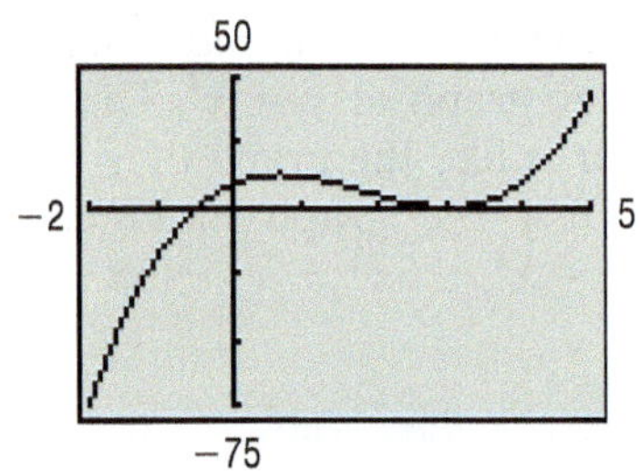

Step 5 Approximate the turning points of the graph.

From the graph of f shown in Figure 13, we can see that f has two turning points. Using MAXIMUM, one turning point is at (0.67, 12.70), rounded to two decimal places. Using MINIMUM, the other turning point is at (3, 0).

Step 6 Use the information in Steps 1 to 5 to draw a complete graph of the function by hand.

Figure 14 shows a graph of f using the information in Steps 1 through 5.

Figure 14

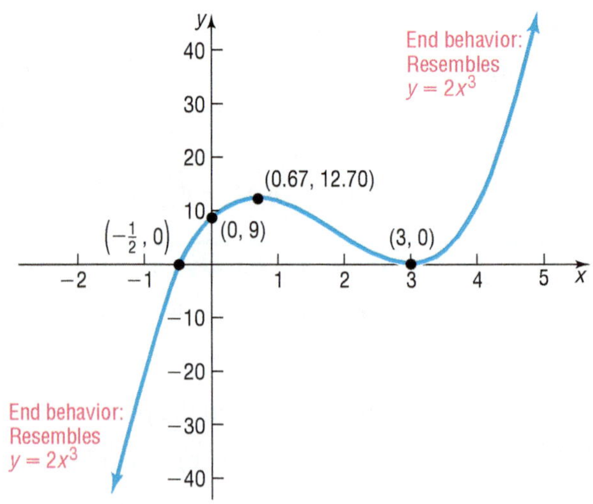

Step 7 Find the domain and the range of the function.

The domain and the range of f is the set of all real numbers.

Step 8 Use the graph to determine where the function is increasing and where it is decreasing.

Based on the graph, f is increasing on the intervals $(-\infty, 0.67)$ and $(3, \infty)$. Also, f is decreasing on the interval $(0.67, 3)$.

SUMMARY Analyzing the Graph of a Polynomial Function

STEP 1: Determine the end behavior of the graph of the function.

STEP 2: Find the x- and y-intercepts of the graph of the function.

STEP 3: Determine the zeros of the function and their multiplicity. Use this information to determine whether the graph crosses or touches the x-axis at each x-intercept.

STEP 4: Use a graphing utility to graph the function.

STEP 5: Approximate the turning points of the graph.

STEP 6: Use the information in Steps 1 through 5 to draw a complete graph of the function by hand.

STEP 7: Find the domain and the range of the function.

STEP 8: Use the graph to determine where the function is increasing and where it is decreasing.

Now Work PROBLEM 65

For polynomial functions that have noninteger coefficients and for polynomials that are not easily factored, we utilize the graphing utility early in the analysis. This is because the amount of information that can be obtained from algebraic analysis is limited.

EXAMPLE 10 How to Use a Graphing Utility to Analyze the Graph of a Polynomial Function

Analyze the graph of the polynomial function

$$f(x) = x^3 + 2.48x^2 - 4.3155x + 2.484406$$

Step-by-Step Solution

STEP 1 Determine the end behavior of the graph of the function.

The polynomial function f is of degree 3. The graph of f behaves like $y = x^3$ for large values of $|x|$.

STEP 2 Graph the function using a graphing utility.

See Figure 15 for the graph of f.

Figure 15

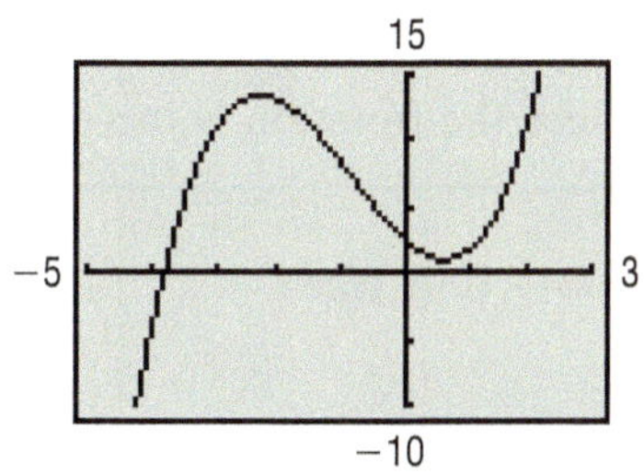

STEP 3 Use a graphing utility to approximate the x- and y-intercepts of the graph.

The y-intercept is $f(0) = 2.484406$. In Example 9 the polynomial function was factored, so it was easy to find the x-intercepts algebraically. However, it is not readily apparent how to factor f in this example. Therefore, we use a graphing utility's ZERO (or ROOT or SOLVE) feature and find the lone x-intercept to be -3.79, rounded to two decimal places.

STEP 4 Use a graphing utility to create a TABLE to find points on the graph around each x-intercept.

Table 4 shows values of x around the x-intercept. The points $(-4, -4.57)$ and $(-2, 13.04)$ are on the graph.

Table 4

X	Y1
-4	-4.574
-2	13.035

Y1=X^3+2.48X²-4...

STEP 5 Approximate the turning points of the graph.

From the graph of f shown in Figure 15, we can see that f has two turning points. Using MAXIMUM, one turning point is at $(-2.28, 13.36)$, rounded to two decimal places. Using MINIMUM, the other turning point is at $(0.63, 1)$, rounded to two decimal places.

STEP 6 Use the information in Steps 1 through 5 to draw a complete graph of the function by hand.

Figure 16 shows a graph of f using the information in Steps 1 to 5.

Figure 16

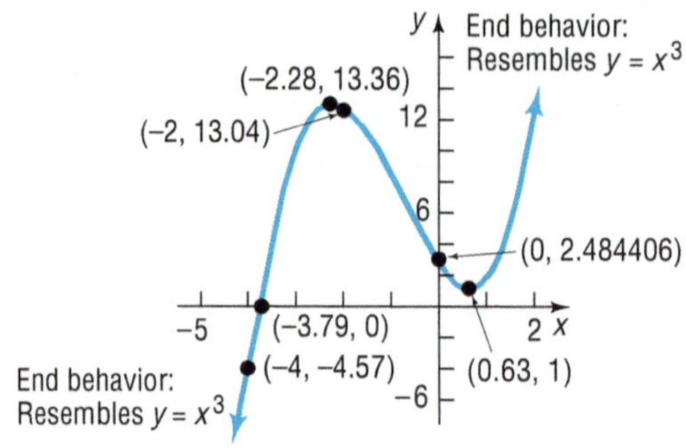

STEP 7 Find the domain and the range of the function.

The domain and the range of f are the set of all real numbers.

STEP 8 Use the graph to determine where the function is increasing and where it is decreasing.

Based on the graph, f is increasing on the intervals $(-\infty, -2.28)$ and $(0.63, \infty)$. Also, f is decreasing on the interval $(-2.28, 0.63)$.

SUMMARY Using a Graphing Utility to Analyze the Graph of a Polynomial Function

STEP 1: Determine the end behavior of the graph of the function.

STEP 2: Graph the function using a graphing utility.

STEP 3: Use a graphing utility to approximate the x- and y-intercepts of the graph.

STEP 4: Use a graphing utility to create a TABLE to find points on the graph around each x-intercept.

STEP 5: Approximate the turning points of the graph.

STEP 6: Use the information in Steps 1 through 5 to draw a complete graph of the function by hand.

STEP 7: Find the domain and the range of the function.

STEP 8: Use the graph to determine where the function is increasing and where it is decreasing.

Now Work PROBLEM 83

5 Build Cubic Models from Data

In Section 4.2 we found the line of best fit from data, and in Section 4.4 we found the quadratic function of best fit. It is also possible to find polynomial functions of best fit. However, most statisticians do not recommend finding polynomials of best fit of degree higher than 3.

Data that follow a cubic relation should look like Figure 17(a) or (b).

Figure 17

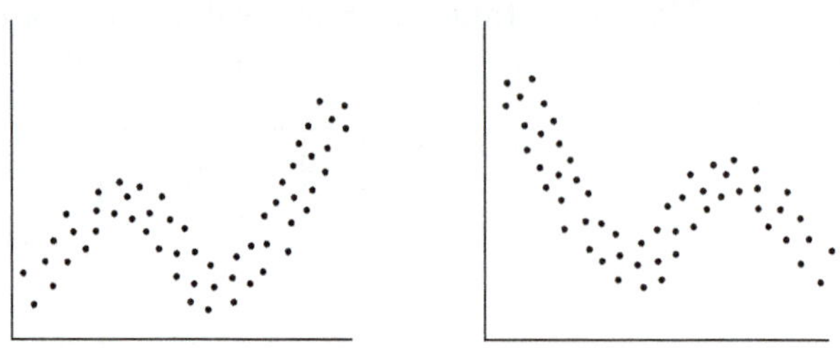

$y = ax^3 + bx^2 + cx + d, a > 0$
(a)

$y = ax^3 + bx^2 + cx + d, a < 0$
(b)

EXAMPLE 11 A Cubic Function of Best Fit

The data in Table 5 represent the weekly cost C (in thousands of dollars) of printing x thousand textbooks.

Table 5

Number of Text Books, x	Cost, C
0	100
5	128.1
10	144
13	153.5
17	161.2
18	162.6
20	166.3
23	178.9
25	190.2
27	221.8

(a) Draw a scatter diagram of the data using x as the independent variable and C as the dependent variable. Comment on the type of relation that may exist between the two variables x and C.

(b) Using a graphing utility, find the cubic function of best fit $C = C(x)$ that models the relation between number of texts and cost.

(c) Graph the cubic function of best fit on your scatter diagram.

(d) Use the function found in part (b) to predict the cost of printing 22 thousand texts per week.

Solution

(a) Figure 18 shows the scatter diagram. A cubic relation may exist between the two variables.

(b) Upon executing the CUBIC REGression program, we obtain the results shown in Figure 19. The output the utility provides shows us the equation $y = ax^3 + bx^2 + cx + d$. The cubic function of best fit to the data is $C(x) = 0.0179x^3 - 0.7279x^2 + 11.3079x + 88.3169$.

(c) Figure 20 shows the graph of the cubic function of best fit on the scatter diagram. The function fits the data reasonably well.

Figure 18

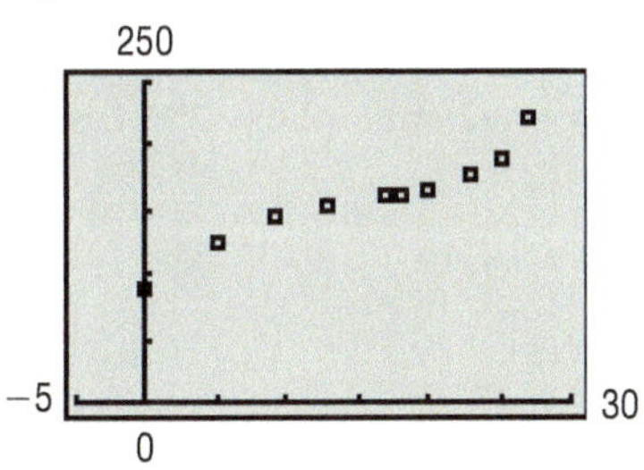

Figure 19

Figure 20

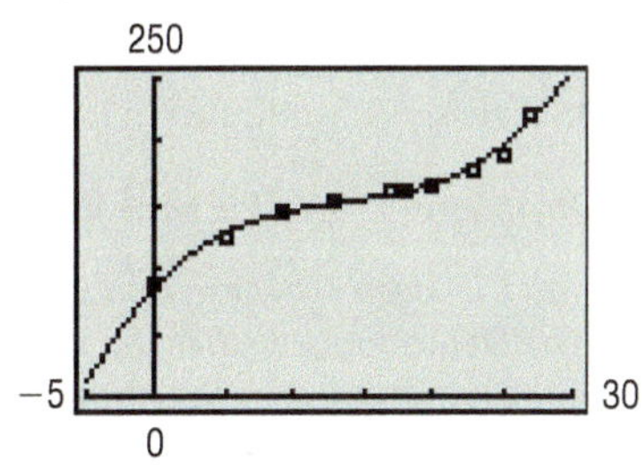

(d) We evaluate the function $C(x)$ at $x = 22$.

$$C(22) = 0.0179(22)^3 - 0.7279(22)^2 + 11.3079(22) + 88.3169 \approx 175.4$$

The model predicts that the cost of printing 22 thousand textbooks in a week will be 175.4 thousand dollars, that is \$175,400.

5.1 Assess Your Understanding

'Are You Prepared?' *Answers are given at the end of these exercises. If you get a wrong answer, read the pages listed in red.*

1. The intercepts of the equation $9x^2 + 4y = 36$ are _____. (pp. 170–171)

2. ***True or False*** The expression $4x^3 - 3.6x^2 - \sqrt{2}$ is a polynomial. (pp. 40–41)

3. To graph $y = x^2 - 4$, you would shift the graph of $y = x^2$ _____ a distance of _____ units. (pp. 256–257)

4. Use a graphing utility to approximate (rounded to two decimal places) the local maxima and local minima of $f(x) = x^3 - 2x^2 - 4x + 5$, for $-3 < x < 3$. (pp. 239–240)

Concepts and Vocabulary

5. The graph of every polynomial function is both _____ and _____.

6. A real number r for which $f(r) = 0$ is called a(n) _____ of the function f.

7. If r is a real zero of even multiplicity of a function f, the graph of f _____ the x-axis at r.

8. ***True or False*** The graph of $f(x) = x^2(x - 3)(x + 4)$ has exactly three x-intercepts.

9. ***True or False*** The x-intercepts of the graph of a polynomial function are called turning points.

10. ***True or False*** End behavior: the graph of the function $f(x) = 3x^4 - 6x^2 + 2x + 5$ resembles $y = x^4$ for large values of $|x|$.

Skill Building

In Problems 11–22, determine which functions are polynomial functions. For those that are, state the degree. For those that are not, tell why not.

11. $f(x) = 4x + x^3$
12. $f(x) = 5x^2 + 4x^4$
13. $g(x) = \dfrac{1 - x^2}{2}$
14. $h(x) = 3 - \dfrac{1}{2}x$
15. $f(x) = 1 - \dfrac{1}{x}$
16. $f(x) = x(x - 1)$
17. $g(x) = x^{3/2} - x^2 + 2$
18. $h(x) = \sqrt{x}(\sqrt{x} - 1)$
19. $F(x) = 5x^4 - \pi x^3 + \dfrac{1}{2}$
20. $F(x) = \dfrac{x^2 - 5}{x^3}$
21. $G(x) = 2(x - 1)^2(x^2 + 1)$
22. $G(x) = -3x^2(x + 2)^3$

In Problems 23–36, use transformations of the graph of $y = x^4$ or $y = x^5$ to graph each function. Verify your results using a graphing utility.

23. $f(x) = (x + 1)^4$
24. $f(x) = (x - 2)^5$
25. $f(x) = x^5 - 3$
26. $f(x) = x^4 + 2$
27. $f(x) = \dfrac{1}{2}x^4$
28. $f(x) = 3x^5$
29. $f(x) = -x^5$
30. $f(x) = -x^4$
31. $f(x) = (x - 1)^5 + 2$
32. $f(x) = (x + 2)^4 - 3$
33. $f(x) = 2(x + 1)^4 + 1$
34. $f(x) = \dfrac{1}{2}(x - 1)^5 - 2$
35. $f(x) = 4 - (x - 2)^5$
36. $f(x) = 3 - (x + 2)^4$

In Problems 37–44, form a polynomial whose real zeros and degree are given. Answers will vary depending on the choice of a leading coefficient. Graph the function using a graphing utility to verify your results.

37. Zeros: $-1, 1, 3$; degree 3
38. Zeros: $-2, 2, 3$; degree 3
39. Zeros: $-3, 0, 4$; degree 3
40. Zeros: $-4, 0, 2$; degree 3
41. Zeros: $-4, -1, 2, 3$; degree 4
42. Zeros: $-3, -1, 2, 5$; degree 4
43. Zeros: -1, multiplicity 1; 3, multiplicity 2; degree 3
44. Zeros: -2, multiplicity 2; 4, multiplicity 1; degree 3

In Problems 45–56, for each polynomial function:

(a) List each real zero and its multiplicity.
(b) Determine whether the graph crosses or touches the x-axis at each x-intercept.
(c) Determine the maximum number of turning points on the graph.
(d) Determine the end behavior; that is, find the power function that the graph of f resembles for large values of $|x|$.

45. $f(x) = 3(x - 7)(x + 3)^2$
46. $f(x) = 4(x + 4)(x + 3)^3$
47. $f(x) = 4(x^2 + 1)(x - 2)^3$
48. $f(x) = 2(x - 3)(x^2 + 4)^3$
49. $f(x) = -2\left(x + \dfrac{1}{2}\right)^2(x + 4)^3$
50. $f(x) = \left(x - \dfrac{1}{3}\right)^2(x - 1)^3$
51. $f(x) = (x - 5)^3(x + 4)^2$
52. $f(x) = \left(x + \sqrt{3}\right)^2(x - 2)^4$
53. $f(x) = 3(x^2 + 8)(x^2 + 9)^2$
54. $f(x) = -2(x^2 + 3)^3$
55. $f(x) = -2x^2(x^2 - 2)$
56. $f(x) = 4x(x^2 - 3)$

In Problems 57–60, identify which of the graphs could be the graph of a polynomial function. For those that could, list the real zeros and state the least degree the polynomial can have. For those that could not, say why not.

57. 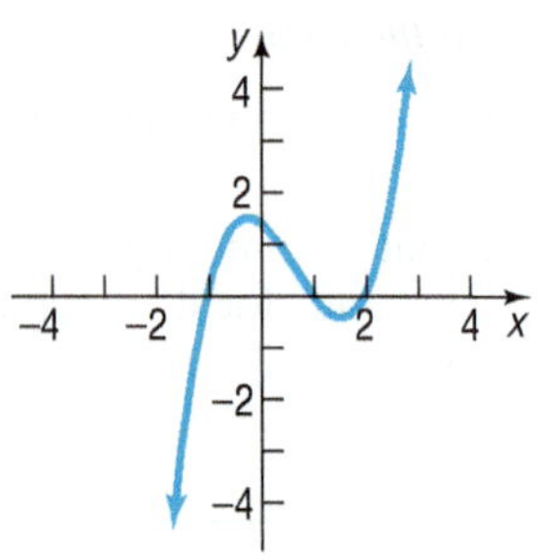

58. 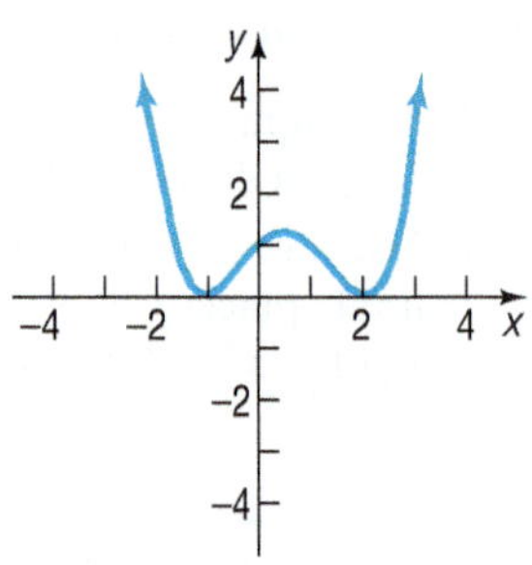

59. 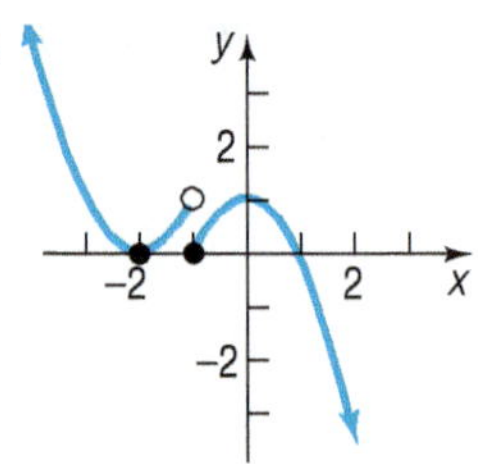

60. 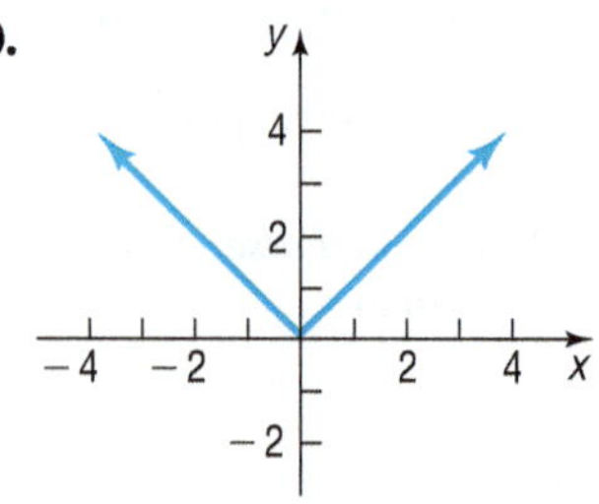

In Problems 61–64, construct a polynomial function that might have the given graph. (More than one answer may be possible.)

61. 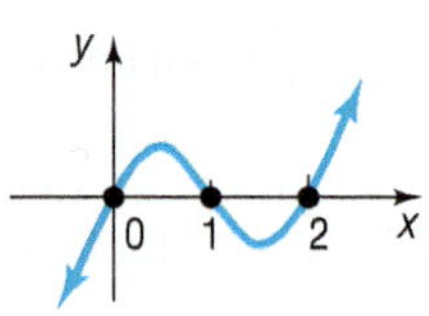

62. 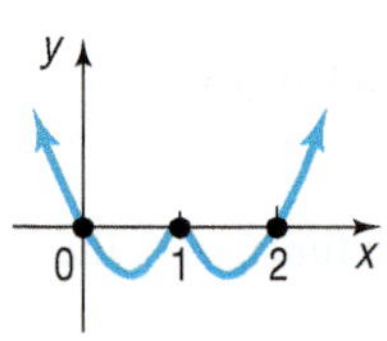

63.

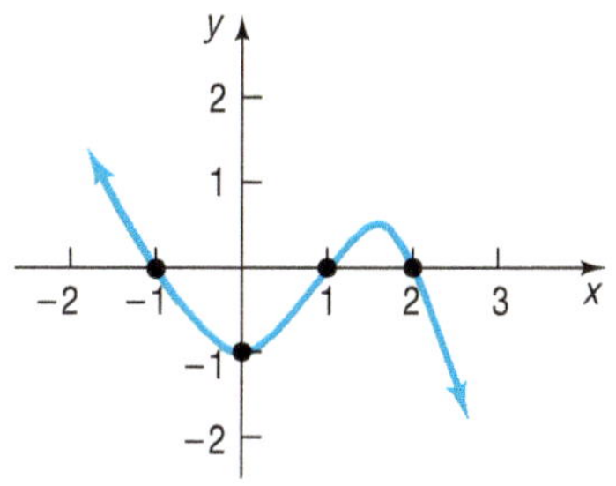

64.

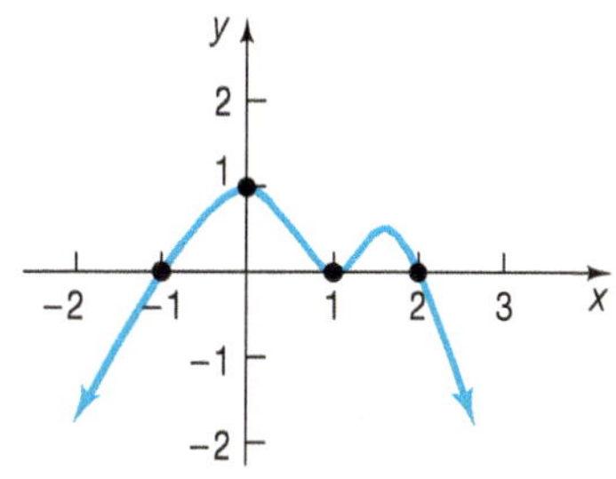

In Problems 65–82, analyze each polynomial function by following Steps 1 through 8 on page 341.

65. $f(x) = x^2(x - 3)$
66. $f(x) = x(x + 2)^2$
67. $f(x) = (x + 4)(x - 2)^2$
68. $f(x) = (x - 1)(x + 3)^2$
69. $f(x) = -2(x + 2)(x - 2)^3$
70. $f(x) = -\frac{1}{2}(x + 4)(x - 1)^3$
71. $f(x) = (x + 1)(x - 2)(x + 4)$
72. $f(x) = (x - 1)(x + 4)(x - 3)$
73. $f(x) = x^2(x - 2)(x + 2)$
74. $f(x) = x^2(x - 3)(x + 4)$
75. $f(x) = (x + 1)^2(x - 2)^2$
76. $f(x) = (x + 1)^3(x - 3)$
77. $f(x) = x^2(x - 3)(x + 1)$
78. $f(x) = x^2(x - 3)(x - 1)$
79. $f(x) = (x + 2)^2(x - 4)^2$
80. $f(x) = (x - 2)^2(x + 2)(x + 4)$
81. $f(x) = x^2(x - 2)(x^2 + 3)$
82. $f(x) = x^2(x^2 + 1)(x + 4)$

In Problems 83–90, analyze each polynomial function f by following Steps 1 through 8 on page 342.

83. $f(x) = x^3 + 0.2x^2 - 1.5876x - 0.31752$
84. $f(x) = x^3 - 0.8x^2 - 4.6656x + 3.73248$
85. $f(x) = x^3 + 2.56x^2 - 3.31x + 0.89$
86. $f(x) = x^3 - 2.91x^2 - 7.668x - 3.8151$
87. $f(x) = x^4 - 2.5x^2 + 0.5625$
88. $f(x) = x^4 - 18.5x^2 + 50.2619$
89. $f(x) = 2x^4 - \pi x^3 + \sqrt{5}x - 4$
90. $f(x) = -1.2x^4 + 0.5x^2 - \sqrt{3}x + 2$

Mixed Practice

In Problems 91–98, analyze each polynomial function by following Steps 1 through 8 on page 341.
[**Hint:** You will need to first factor the polynomial.]

91. $f(x) = 4x - x^3$
92. $f(x) = x - x^3$
93. $f(x) = x^3 + x^2 - 12x$
94. $f(x) = x^3 + 2x^2 - 8x$
95. $f(x) = 2x^4 + 12x^3 - 8x^2 - 48x$
96. $f(x) = 4x^3 + 10x^2 - 4x - 10$
97. $f(x) = -x^5 - x^4 + x^3 + x^2$
98. $f(x) = -x^5 + 5x^4 + 4x^3 - 20x^2$

Applications and Extensions

99. **Future Value of Money** Suppose you make deposits of \$500 at the beginning of every year into an Individual Retirement Account (IRA). At the beginning of the first year, the value of the account will be 500 dollars; at the beginning of the second year, the value of the account, in dollars, will be

$$\underbrace{500 + 500r}_{\text{Value of 1st deposit}} + \underbrace{500}_{\text{Value of 2nd deposit}} = 500(1 + r) + 500 = 500r + 1000$$

(a) Verify that the value of the account at the beginning of the third year is $T(r) = 500r^2 + 1500r + 1500$.
(b) The account value at the beginning of the fourth year is $F(r) = 500r^3 + 2000r^2 + 3000r + 2000$. If the annual rate of interest is $5\% = 0.05$, what will be the value of the account at the beginning of the fourth year?

100. **A Geometric Series** In calculus, you will learn that certain functions can be approximated by polynomial functions. We will explore one such function now.

(a) Using a graphing utility, create a table of values with $Y_1 = f(x) = \frac{1}{1 - x}$ and $Y_2 = g_2(x) = 1 + x + x^2 + x^3$ for $-1 < x < 1$ with $\Delta\text{Tbl} = 0.1$.

(b) Using a graphing utility, create a table of values with $Y_1 = f(x) = \frac{1}{1 - x}$ and $Y_3 = g_3(x) = 1 + x + x^2 + x^3 + x^4$ for $-1 < x < 1$ with $\Delta\text{Tbl} = 0.1$.

(c) Using a graphing utility, create a table of values with $Y_1 = f(x) = \frac{1}{1 - x}$ and $Y_4 = g_4(x) = 1 + x + x^2 + x^3 + x^4 + x^5$ for $-1 < x < 1$ with $\Delta\text{Tbl} = 0.1$.

(d) What do you notice about the values of the function as more terms are added to the polynomial? Are there some values of x for which the approximation is better?

101. Hurricanes In 2005, hurricane Katrina struck the Gulf Coast of the United States, killing 1289 people and causing an estimated $200 billion in damage. The following data represent the number of major hurricane strikes in the United States (category 3, 4, or 5) each decade from 1921 to 2000.

Decade, x	Major Hurricanes Striking United States, H
1921–1930, 1	5
1931–1940, 2	8
1941–1950, 3	10
1951–1960, 4	8
1961–1970, 5	6
1971–1980, 6	4
1981–1990, 7	5
1991–2000, 8	5

Source: National Oceanic & Atmospheric Administration

(a) Draw a scatter diagram of the data. Comment on the type of relation that may exist between the two variables.
(b) Use a graphing utility to find the cubic function of best fit that models the relation between decade and number of major hurricanes.
(c) Use the model found in part (b) to predict the number of major hurricanes that struck the United States between 1961 and 1970.
(d) With a graphing utility, draw a scatter diagram of the data and then graph the cubic function of best fit on the scatter diagram.
(e) Concern has risen about the increase in the number and intensity of hurricanes, but some scientists believe this is just a natural fluctuation that could last another decade or two. Use your model to predict the number of major hurricanes that will strike the United States between 2001 and 2010. Does your result appear to agree with what these scientists believe?
(f) From 2001 through 2005, five major hurricanes struck the United States. Does this support or contradict your prediction in part (e)?

102. Active Duty The following data represent the number of active duty military personnel (in millions) in the four major branches of the U.S. military for the years 1998 to 2005, where 1 represents 1998, 2 represents 1999, and so on.

Year, x	Active Duty Personnel, N
1998, 1	1.41
1999, 2	1.39
2000, 3	1.38
2001, 4	1.39
2002, 5	1.41
2003, 6	1.42
2004, 7	1.41
2005, 8	1.39

Source: infoplease.com

(a) Draw a scatter diagram of the data. Comment on the type of relation that may exist between the two variables.
(b) Use a graphing utility to find the cubic function of best fit that models the relation between year and active duty personnel.
(c) Use the model found in part (b) to predict the number of active duty military personnel in 2002.
(d) With a graphing utility, draw a scatter diagram of the data and then graph the cubic function of best fit on the scatter diagram.
(e) Do you think that the function given in part (b) will be useful in predicting the number of active duty military personnel in 2010? Explain.

Mixed Practice

103. Temperature The following data represent the temperature T (°Fahrenheit) in Kansas City, Missouri, x hours after midnight on May 15, 2005.

Hours after Midnight, x	Temperature (°F), T
3	45.0
6	44.1
9	51.1
12	57.9
15	63.0
18	63.0
21	59.0
24	54.0

Source: The Weather Underground

(a) Draw a scatter diagram of the data. Comment on the type of relation that may exist between the two variables.
(b) Find the average rate of change in temperature from 9 AM to 12 noon.
(c) What is the average rate of change in temperature from 3 PM to 6 PM?
(d) Decide on a function of best fit to these data (linear, quadratic, or cubic) and use this function to predict the temperature at 5 PM.
(e) With a graphing utility, draw a scatter diagram of the data and then graph the function of best fit on the scatter diagram.
(f) Interpret the y-intercept.

Discussion and Writing

104. Can the graph of a polynomial function have no y-intercept? Can it have no x-intercepts? Explain.

105. Write a few paragraphs that provide a general strategy for graphing a polynomial function. Be sure to mention the following: degree, intercepts, end behavior, and turning points.

106. Make up a polynomial that has the following characteristics: crosses the x-axis at -1 and 4, touches the x-axis at 0 and 2, and is above the x-axis between 0 and 2. Give your polynomial to a fellow classmate and ask for a written critique of your polynomial.

107. Make up two polynomials, not of the same degree, with the following characteristics: crosses the x-axis at -2, touches the x-axis at 1, and is above the x-axis between -2 and 1. Give your polynomials to a fellow classmate and ask for a written critique of your polynomials.

108. The graph of a polynomial function is always smooth and continuous. Name a function studied earlier that is smooth and not continuous. Name one that is continuous, but not smooth.

109. Which of the following statements are true regarding the graph of the cubic polynomial $f(x) = x^3 + bx^2 + cx + d$? (Give reasons for your conclusions.)

(a) It intersects the y-axis in one and only one point.
(b) It intersects the x-axis in at most three points.
(c) It intersects the x-axis at least once.
(d) For $|x|$ very large, it behaves like the graph of $y = x^3$.
(e) It is symmetric with respect to the origin.
(f) It passes through the origin.

110. The illustration shows the graph of a polynomial function.

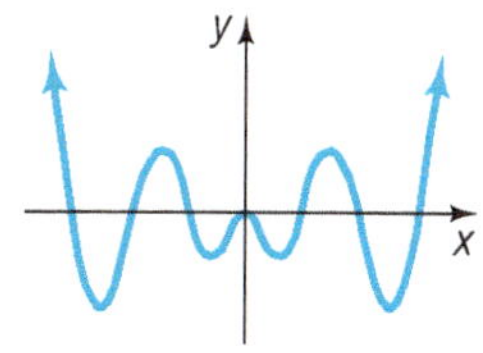

(a) Is the degree of the polynomial even or odd?
(b) Is the leading coefficient positive or negative?
(c) Is the function even, odd, or neither?
(d) Why is x^2 necessarily a factor of the polynomial?
(e) What is the minimum degree of the polynomial?
(f) Formulate five different polynomials whose graphs could look like the one shown. Compare yours to those of other students. What similarities do you see? What differences?

111. Design a polynomial function with the following characteristics: degree 6; four distinct real zeros, one of multiplicity 3; y-intercept 3; behaves like $y = -5x^6$ for large values of $|x|$. Is this polynomial unique? Compare your polynomial with those of other students. What terms will be the same as everyone else's? Add some more characteristics, such as symmetry or naming the real zeros. How does this modify the polynomial?

'Are You Prepared?' Answers

1. $(-2,0), (2,0), (0,9)$ **2.** True **3.** down; 4 **4.** Local maximum 6.48 at $x = -0.67$; local minimum -3 at $x = 2$

5.2 Properties of Rational Functions

PREPARING FOR THIS SECTION *Before getting started, review the following:*

- Rational Expressions (Section R.7, pp. 61–67)
- Polynomial Division (Section R.4, pp. 44–47)
- Graph of $f(x) = \dfrac{1}{x}$ (Section 2.1, Example 6, p. 175)
- Graphing Techniques: Transformations (Section 3.5, pp. 256–264)

Now Work the 'Are You Prepared?' problems on page 356.

OBJECTIVES
1 Find the Domain of a Rational Function (p. 348)
2 Find the Vertical Asymptotes of a Rational Function (p. 351)
3 Find the Horizontal or Oblique Asymptotes of a Rational Function (p. 352)

Ratios of integers are called *rational numbers*. Similarly, ratios of polynomial functions are called *rational functions*. Examples of rational functions are

$$R(x) = \frac{x^2 - 4}{x^2 + x + 1} \qquad F(x) = \frac{x^3}{x^2 - 4} \qquad G(x) = \frac{3x^2}{x^4 - 1}$$

DEFINITION

A **rational function** is a function of the form

$$R(x) = \frac{p(x)}{q(x)}$$

where p and q are polynomial functions and q is not the zero polynomial. The domain of a rational function is the set of all real numbers except those for which the denominator q is 0.

1 Find the Domain of a Rational Function

EXAMPLE 1 Finding the Domain of a Rational Function

(a) The domain of $R(x) = \frac{2x^2 - 4}{x + 5}$ is the set of all real numbers x except -5; that is, $\{x | x \neq -5\}$.

(b) The domain of $R(x) = \frac{1}{x^2 - 4}$ is the set of all real numbers x except -2 and 2; that is, $\{x | x \neq -2, x \neq 2\}$.

(c) The domain of $R(x) = \frac{x^3}{x^2 + 1}$ is the set of all real numbers.

(d) The domain of $R(x) = \frac{-x^2 + 2}{3}$ is the set of all real numbers.

(e) The domain of $R(x) = \frac{x^2 - 1}{x - 1}$ is the set of all real numbers x except 1; that is, $\{x | x \neq 1\}$.

It is important to observe that the functions

$$R(x) = \frac{x^2 - 1}{x - 1} \quad \text{and} \quad f(x) = x + 1$$

are not equal, since the domain of R is $\{x | x \neq 1\}$ and the domain of f is the set of all real numbers.

Now Work PROBLEM 13

If $R(x) = \frac{p(x)}{q(x)}$ is a rational function and if p and q have no common factors, then the rational function R is said to be in **lowest terms.** For a rational function $R(x) = \frac{p(x)}{q(x)}$ in lowest terms, the zeros, if any, of the numerator are the x-intercepts of the graph of R and so will play a major role in the graph of R. The zeros of the denominator of R [that is, the numbers x, if any, for which $q(x) = 0$], although not in the domain of R, also play a major role in the graph of R. We will discuss this role shortly.

We have already discussed the properties of the rational function $f(x) = \frac{1}{x}$. (Refer to Example 6, page 175.) The next rational function that we take up is $H(x) = \frac{1}{x^2}$.

EXAMPLE 2 Graphing $y = \frac{1}{x^2}$

Analyze the graph of $H(x) = \frac{1}{x^2}$.

Solution The domain of $H(x) = \dfrac{1}{x^2}$ consists of all real numbers x except 0. The graph has no y-intercept, because x can never equal 0. The graph has no x-intercept because the equation $H(x) = 0$ has no solution. Therefore, the graph of H will not cross either coordinate axis.

Because

$$H(-x) = \frac{1}{(-x)^2} = \frac{1}{x^2} = H(x)$$

H is an even function, so its graph is symmetric with respect to the y-axis.

See Figure 21. Notice that the graph confirms the conclusions just reached. But what happens to the graph as the values of x get closer and closer to 0? We use a TABLE to answer the question. See Table 6. The first four rows show that as x approaches 0 the values of $H(x)$ become larger and larger positive numbers. When this happens, we say that $H(x)$ is **unbounded in the positive direction.** We symbolize this by writing $H(x) \to \infty$ [read as "$H(x)$ **approaches infinity**"]. In calculus, **limits** are used to convey these ideas. There we use the symbolism $\lim_{x \to 0} H(x) = \infty$, read as "the limit of $H(x)$ as x approaches 0 is infinity" to mean that $H(x) \to \infty$ as $x \to 0$.

Look at the last three rows of Table 6. As $x \to \infty$, the values of $H(x)$ approach 0 (the end behavior of the graph). This is symbolized in calculus by writing $\lim_{x \to \infty} H(x) = 0$.

Figure 21

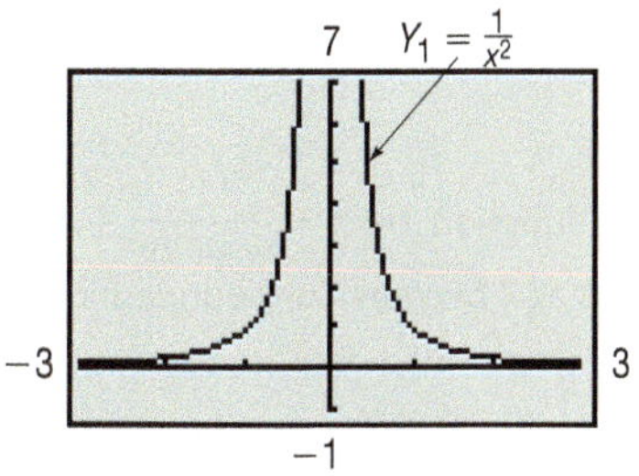

Table 6

X	Y1
.1	100
.01	10000
.001	1E6
1E-4	1E8
10	.01
100	1E-4
1000	1E-6

Y1=1/X²

Figure 22 shows the graph of $H(x) = \dfrac{1}{x^2}$ drawn by hand.

Figure 22

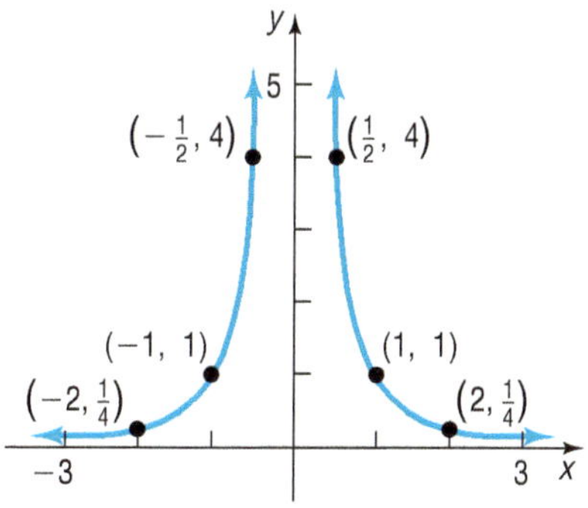

Sometimes transformations (shifting, compressing, stretching, and reflection) can be used to graph a rational function.

EXAMPLE 3 Using Transformations to Graph a Rational Function

Graph the rational function: $R(x) = \dfrac{1}{(x-2)^2} + 1$

Solution First, notice that the domain of R is the set of all real numbers except $x = 2$. To graph R, we start with the graph of $y = \dfrac{1}{x^2}$. See Figure 23 for the stages.

Figure 23

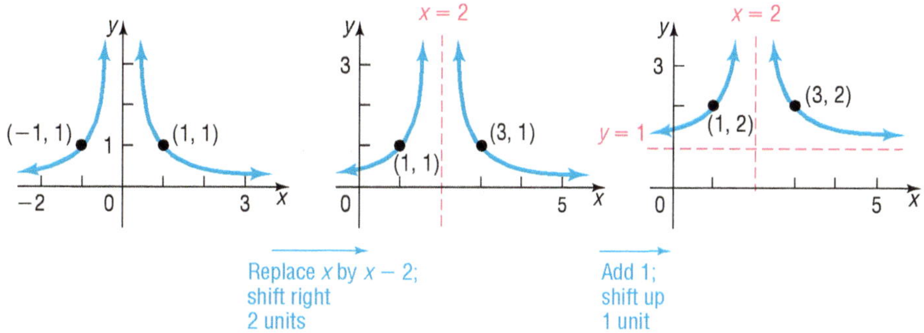

(a) $y = \dfrac{1}{x^2}$ (b) $y = \dfrac{1}{(x-2)^2}$ (c) $y = \dfrac{1}{(x-2)^2} + 1$

Check: Graph $Y_1 = \frac{1}{(x-2)^2} + 1$ using a graphing utility to verify the graph obtained in Figure 23(c).

Now Work PROBLEM 31

Asymptotes

Notice that the y-axis in Figure 23(a) is transformed into the vertical line $x = 2$ in Figure 23(c), and the x-axis in Figure 23(a) is transformed into the horizontal line $y = 1$ in Figure 23(c). The **Exploration** that follows will help us analyze the role of these lines.

Exploration

(a) Using a graphing utility and the TABLE feature, evaluate the function $H(x) = \frac{1}{(x-2)^2} + 1$ at $x = 10, 100, 1000,$ and $10{,}000$. What happens to the values of H as x becomes unbounded in the positive direction, symbolized by $\lim_{x\to\infty} H(x)$?

(b) Evaluate H at $x = -10, -100, -1000,$ and $-10{,}000$. What happens to the values of H as x becomes unbounded in the negative direction, symbolized by $\lim_{x\to-\infty} H(x)$?

(c) Evaluate H at $x = 1.5, 1.9, 1.99, 1.999,$ and 1.9999. What happens to the values of H as x approaches 2, $x < 2$, symbolized by $\lim_{x\to 2^-} H(x)$?

(d) Evaluate H at $x = 2.5, 2.1, 2.01, 2.001,$ and 2.0001. What happens to the values of H as x approaches 2, $x > 2$, symbolized by $\lim_{x\to 2^+} H(x)$?

Result

(a) Table 7 shows the values of $Y_1 = H(x)$ as x approaches ∞. Notice that the values of H are approaching 1, so $\lim_{x\to\infty} H(x) = 1$.

(b) Table 8 shows the values of $Y_1 = H(x)$ as x approaches $-\infty$. Again the values of H are approaching 1, so $\lim_{x\to-\infty} H(x) = 1$.

(c) From Table 9 we see that, as x approaches 2, $x < 2$, the values of H are increasing without bound, so $\lim_{x\to 2^-} H(x) = \infty$.

(d) Finally, Table 10 reveals that, as x approaches 2, $x > 2$, the values of H are increasing without bound, so $\lim_{x\to 2^+} H(x) = \infty$.

Table 7

X	Y1
10	1.0156
100	1.0001
1000	1
10000	1

Y1=1/(X−2)²+1

Table 8

X	Y1
-10	1.0069
-100	1.0001
-1000	1
-10000	1

Y1=1/(X−2)²+1

Table 9

X	Y1
1.5	5
1.9	101
1.99	10001
1.999	1E6
1.9999	1E8

Y1=1/(X−2)²+1

Table 10

X	Y1
2.5	5
2.1	101
2.01	10001
2.001	1E6
2.0001	1E8

Y1=1/(X−2)²+1

The results of the Exploration reveal an important property of rational functions. The vertical line $x = 2$ and the horizontal line $y = 1$ are called *asymptotes* of the graph of H, which we define as follows:

DEFINITION

Let R denote a function:

If, as $x \to -\infty$ or as $x \to \infty$, the values of $R(x)$ approach some fixed number L, then the line $y = L$ is a **horizontal asymptote** of the graph of R. [Refer to Figures 24(a) and (b).]

If, as x approaches some number c, the values $|R(x)| \to \infty$, then the line $x = c$ is a **vertical asymptote** of the graph of R. The graph of R never intersects a vertical asymptote. [Refer to Figures 24(c) and (d).]

Figure 24

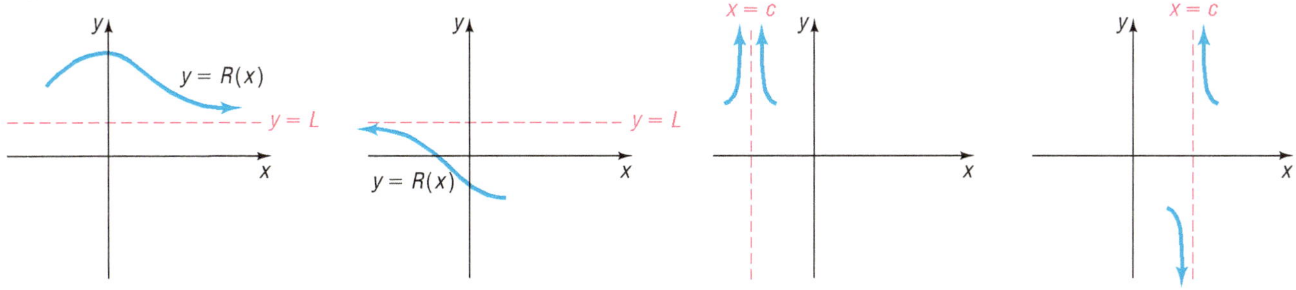

(a) End behavior: As $x \to \infty$, the values of $R(x)$ approach L [symbolized by $\lim_{x \to \infty} R(x) = L$]. That is, the points on the graph of R are getting closer to the line $y = L$; $y = L$ is a horizontal asymptote.

(b) End behavior: As $x \to -\infty$, the values of $R(x)$ approach L [symbolized by $\lim_{x \to -\infty} R(x) = L$]. That is, the points on the graph of R are getting closer to the line $y = L$; $y = L$ is a horizontal asymptote.

(c) As x approaches c, the values of $R(x) \to \infty$ [for $x < c$, this is symbolized by $\lim_{x \to c^-} R(x) = \infty$; for $x > c$, this is symbolized by $\lim_{x \to c^+} R(x) = \infty$]. That is, the points on the graph of R are getting closer to the line $x = c$; $x = c$ is a vertical asymptote.

(d) As x approaches c, the values of $|R(x)| \to \infty$ [for $x < c$, this is symbolized by $\lim_{x \to c^-} R(x) = -\infty$; for $x > c$, this is symbolized by $\lim_{x \to c^+} R(x) = \infty$]. That is, the points on the graph of R are getting closer to the line $x = c$; $x = c$ is a vertical asymptote.

Figure 25

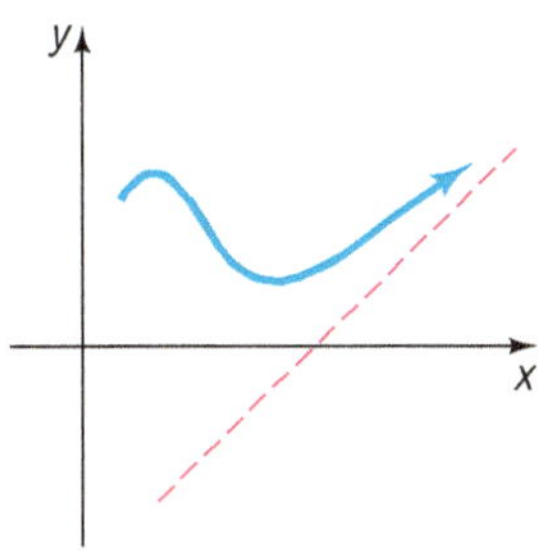

A horizontal asymptote, when it occurs, describes the end behavior of the graph as $x \to \infty$ or as $x \to -\infty$. **The graph of a function may intersect a horizontal asymptote.**

A vertical asymptote, when it occurs, describes the behavior of the graph when x is close to some number c. **The graph of a function will never intersect a vertical asymptote.**

There is a third possibility. If, as $x \to -\infty$ or as $x \to \infty$, the value of a rational function $R(x)$ approaches a linear expression $ax + b$, $a \neq 0$, then the line $y = ax + b, a \neq 0$, is an **oblique asymptote** of R. Figure 25 shows an oblique asymptote. An oblique asymptote, when it occurs, describes the end behavior of the graph. **The graph of a function may intersect an oblique asymptote.**

Now Work PROBLEM 23

2 Find the Vertical Asymptotes of a Rational Function

The vertical asymptotes of a rational function $R(x) = \dfrac{p(x)}{q(x)}$, in lowest terms, are located at the real zeros of the denominator of $q(x)$. Suppose that r is a real zero of q, so $x - r$ is a factor of q. As x approaches r, symbolized as $x \to r$, the values of $x - r$ approach 0, causing the ratio to become unbounded; that is, $|R(x)| \to \infty$. Based on the definition, we conclude that the line $x = r$ is a vertical asymptote.

THEOREM

Locating Vertical Asymptotes

A rational function $R(x) = \dfrac{p(x)}{q(x)}$, *in lowest terms,* will have a vertical asymptote $x = r$ if r is a real zero of the denominator q. That is, if $x - r$ is a factor of the denominator q of a rational function $R(x) = \dfrac{p(x)}{q(x)}$, in lowest terms, R will have the vertical asymptote $x = r$.

WARNING If a rational function is not in lowest terms, an application of this theorem may result in an incorrect listing of vertical asymptotes. ■

EXAMPLE 4 **Finding Vertical Asymptotes**

Find the vertical asymptotes, if any, of the graph of each rational function.

(a) $R(x) = \dfrac{x}{x^2 - 4}$ (b) $F(x) = \dfrac{x + 3}{x - 1}$

(c) $H(x) = \dfrac{x^2}{x^2 + 1}$ (d) $G(x) = \dfrac{x^2 - 9}{x^2 + 4x - 21}$

Solution

(a) R is in lowest terms, and the zeros of the denominator $x^2 - 4$ are -2 and 2. The lines $x = -2$ and $x = 2$ are the vertical asymptotes of the graph of R.

WARNING In Example 4(a), the vertical asymptotes are $x = -2$ and $x = 2$. Do not say that the vertical asymptotes are -2 and 2. ■

(b) F is in lowest terms, and the only zero of the denominator is 1. The line $x = 1$ is the vertical asymptote of the graph of F.

(c) H is in lowest terms, and the denominator has no real zeros, because the equation $x^2 + 1$ has no real solutions. The graph of H has no vertical asymptotes.

(d) Factor $G(x)$ to determine if it is in lowest terms.

$$G(x) = \frac{x^2 - 9}{x^2 + 4x - 21} = \frac{(x + 3)(x - 3)}{(x + 7)(x - 3)} = \frac{x + 3}{x + 7} \qquad x \neq 3$$

The only zero of the denominator of $G(x)$ in lowest terms is -7. The line $x = -7$ is the only vertical asymptote of the graph of G.

As Example 4 points out, rational functions can have no vertical asymptotes, one vertical asymptote, or more than one vertical asymptote. However, the graph of a rational function will never intersect any of its vertical asymptotes. (Do you know why?)

Exploration

Graph each of the following rational functions:

$$R(x) = \frac{1}{x - 1} \quad R(x) = \frac{1}{(x - 1)^2} \quad R(x) = \frac{1}{(x - 1)^3} \quad R(x) = \frac{1}{(x - 1)^4}$$

Each has the vertical asymptote $x = 1$. What happens to the value of $R(x)$ as x approaches 1 from the right side of the vertical asymptote; that is, what is $\lim_{x \to 1^+} R(x)$? What happens to the value of $R(x)$ as x approaches 1 from the left side of the vertical asymptote; that is, what is $\lim_{x \to 1^-} R(x)$? How does the multiplicity of the zero in the denominator affect the graph of R?

Now Work PROBLEM 45 (FIND THE VERTICAL ASYMPTOTES, IF ANY)

3 Find the Horizontal or Oblique Asymptotes of a Rational Function

The procedure for finding horizontal and oblique asymptotes is somewhat more involved. To find such asymptotes, we need to know how the value of a function behaves as $x \to -\infty$ or as $x \to \infty$.

If a rational function $R(x)$ is **proper,** that is, if the degree of the numerator is less than the degree of the denominator, then as $x \to -\infty$ or as $x \to \infty$ the value of $R(x)$ approaches 0. Consequently, the line $y = 0$ (the x-axis) is a horizontal asymptote of the graph.

THEOREM

If a rational function is proper, the line $y = 0$ is a horizontal asymptote of its graph.

EXAMPLE 5 **Finding Horizontal Asymptotes**

Find the horizontal asymptotes, if any, of the graph of

$$R(x) = \frac{x - 12}{4x^2 + x + 1}$$

Solution Since the degree of the numerator, 1, is less than the degree of the denominator, 2, the rational function R is proper. We conclude that the line $y = 0$ is a horizontal asymptote of the graph of R.

To see why $y = 0$ is a horizontal asymptote of the function R in Example 5, we need to investigate the behavior of R as $x \to -\infty$ and $x \to \infty$. When $|x|$ is unbounded, the numerator of R, which is $x - 12$, can be approximated by the power function $y = x$, while the denominator of R, which is $4x^2 + x + 1$, can be approximated by the power function $y = 4x^2$. Applying these ideas to $R(x)$, we find

$$R(x) = \frac{x - 12}{4x^2 + x + 1} \underset{\substack{\uparrow \\ \text{For } |x| \text{ unbounded}}}{\approx} \frac{x}{4x^2} = \frac{1}{4x} \underset{\substack{\uparrow \\ \text{As } x \to -\infty \text{ or } x \to \infty}}{\to} 0$$

This shows that the line $y = 0$ is a horizontal asymptote of the graph of R.

We verify these results in Tables 11(a) and (b). Notice, as $x \to -\infty$ [Table 11(a)] or $x \to \infty$ [Table 11(b)], respectively, that the value of $R(x)$ approaches 0.

Table 11

X	Y1
-10	-.0563
-100	-.0028
-1000	-3E-4
-10000	-3E-5
-1E5	-3E-6
-1E6	-3E-7
-1E7	-3E-8

Y1=(X-12)/(4X²+...

(a)

X	Y1
10	-.0049
100	.00219
1000	2.5E-4
10000	2.5E-5
100000	2.5E-6
1E6	2.5E-7
1E7	2.5E-8

Y1=(X-12)/(4X²+...

(b)

If a rational function $R(x) = \dfrac{p(x)}{q(x)}$ is **improper,** that is, if the degree of the numerator is greater than or equal to the degree of the denominator, we must use long division to write the rational function as the sum of a polynomial $f(x)$ plus a proper rational function $\dfrac{r(x)}{q(x)}$ using long division. That is, we write

$$R(x) = \frac{p(x)}{q(x)} = f(x) + \frac{r(x)}{q(x)}$$

where $f(x)$ is a polynomial and $\dfrac{r(x)}{q(x)}$ is a proper rational function. Since $\dfrac{r(x)}{q(x)}$ is proper, then $\dfrac{r(x)}{q(x)} \to 0$ as $x \to -\infty$ or as $x \to \infty$. As a result,

$$R(x) = \frac{p(x)}{q(x)} \to f(x), \qquad \text{as } x \to -\infty \text{ or as } x \to \infty$$

The possibilities are listed next.

1. If $f(x) = b$, a constant, then the line $y = b$ is a horizontal asymptote of the graph of R.
2. If $f(x) = ax + b$, $a \neq 0$, then the line $y = ax + b$ is an oblique asymptote of the graph of R.
3. In all other cases, the graph of R approaches the graph of f, and there are no horizontal or oblique asymptotes.

We illustrate each of these possibilities in Examples 6 through 8.

EXAMPLE 6 Finding Horizontal or Oblique Asymptotes

Find the horizontal or oblique asymptotes, if any, of the graph of

$$H(x) = \frac{3x^4 - x^2}{x^3 - x^2 + 1}$$

Solution Since the degree of the numerator, 4, is larger than the degree of the denominator, 3, the rational function H is improper. To find any horizontal or oblique asymptotes, we use long division.

$$\begin{array}{r} 3x \quad + 3 \qquad\qquad \\ x^3 - x^2 + 1 \overline{)\, 3x^4 \qquad\qquad - x^2 \qquad\quad} \\ \underline{3x^4 - 3x^3 \qquad + 3x} \quad \\ 3x^3 - x^2 - 3x \quad \\ \underline{3x^3 - 3x^2 \qquad + 3} \\ 2x^2 - 3x - 3 \end{array}$$

As a result,

$$H(x) = \frac{3x^4 - x^2}{x^3 - x^2 + 1} = 3x + 3 + \frac{2x^2 - 3x - 3}{x^3 - x^2 + 1}$$

As $x \to -\infty$ or as $x \to \infty$,

$$\frac{2x^2 - 3x - 3}{x^3 - x^2 + 1} \approx \frac{2x^2}{x^3} = \frac{2}{x} \to 0$$

As $x \to -\infty$ or as $x \to \infty$, we have $H(x) \to 3x + 3$. We conclude that the graph of the rational function H has an oblique asymptote $y = 3x + 3$.

We verify these results in Tables 12(a) and (b) with $Y_1 = H(x)$ and $Y_2 = 3x + 3$. As $x \to -\infty$ or $x \to \infty$, the difference in the values between Y_1 and Y_2 is indistinguishable.

Table 12

X	Y1	Y2
-10	-27.21	-27
-100	-297	-297
-1000	-2997	-2997
-10000	-29997	-29997
-1E5	-3E5	-3E5
-1E6	-3E6	-3E6
-1E7	-3E7	-3E7

Y1=(3X^4-X²)/(X...

(a)

X	Y1	Y2
10	33.185	33
100	303.02	303
1000	3003	3003
10000	30003	30003
100000	300003	300003
1E6	3E6	3E6
1E7	3E7	3E7

Y1=(3X^4-X²)/(X...

(b)

EXAMPLE 7 Finding Horizontal or Oblique Asymptotes

Find the horizontal or oblique asymptotes, if any, of the graph of

$$R(x) = \frac{8x^2 - x + 2}{4x^2 - 1}$$

Solution Since the degree of the numerator, 2, equals the degree of the denominator, 2, the rational function R is improper. To find any horizontal or oblique asymptotes, we use long division.

$$\begin{array}{r} 2 \\ 4x^2 - 1 \overline{\smash{)}\,8x^2 - x + 2} \\ \underline{8x^2 - 2} \\ -x + 4 \end{array}$$

As a result,

$$R(x) = \frac{8x^2 - x + 2}{4x^2 - 1} = 2 + \frac{-x + 4}{4x^2 - 1}$$

Then, as $x \to -\infty$ or as $x \to \infty$,

$$\frac{-x + 4}{4x^2 - 1} \approx \frac{-x}{4x^2} = \frac{-1}{4x} \to 0$$

As $x \to -\infty$ or as $x \to \infty$, we have $R(x) \to 2$. We conclude that $y = 2$ is a horizontal asymptote of the graph.

✓**Check:** Verify the results by creating a TABLE with $Y_1 = R(x)$ and $Y_2 = 2$.

In Example 7, we note that the quotient 2 obtained by long division is the quotient of the leading coefficients of the numerator polynomial and the denominator polynomial $\left(\frac{8}{4}\right)$. This means that we can avoid the long division process for rational functions whose numerator and denominator *are of the same degree* and conclude that the quotient of the leading coefficients will give us the horizontal asymptote.

Now Work PROBLEMS 41 AND 43

EXAMPLE 8 **Finding Horizontal or Oblique Asymptotes**

Find the horizontal or oblique asymptotes, if any, of the graph of

$$G(x) = \frac{2x^5 - x^3 + 2}{x^3 - 1}$$

Solution Since the degree of the numerator, 5, is larger than the degree of the denominator, 3, the rational function G is improper. To find any horizontal or oblique asymptotes, we use long division.

$$\begin{array}{r} 2x^2 - 1 \\ x^3 - 1 \overline{) 2x^5 - x^3 + 2} \\ \underline{2x^5 - 2x^2 } \\ -x^3 + 2x^2 + 2 \\ \underline{-x^3 + 1} \\ 2x^2 + 1 \end{array}$$

As a result,

$$G(x) = \frac{2x^5 - x^3 + 2}{x^3 - 1} = 2x^2 - 1 + \frac{2x^2 + 1}{x^3 - 1}$$

Then, as $x \to -\infty$ or as $x \to \infty$,

$$\frac{2x^2 + 1}{x^3 - 1} \approx \frac{2x^2}{x^3} = \frac{2}{x} \to 0$$

As $x \to -\infty$ or as $x \to \infty$, we have $G(x) \to 2x^2 - 1$. We conclude that, for large values of $|x|$, the graph of G approaches the graph of $y = 2x^2 - 1$. That is, the graph of G will look like the graph of $y = 2x^2 - 1$ as $x \to -\infty$ or $x \to \infty$. Since $y = 2x^2 - 1$ is not a linear function, G has no horizontal or oblique asymptotes.

We now summarize the procedure for finding horizontal and oblique asymptotes.

SUMMARY **Finding Horizontal and Oblique Asymptotes of a Rational Function R**

Consider the rational function

$$R(x) = \frac{p(x)}{q(x)} = \frac{a_n x^n + a_{n-1}x^{n-1} + \cdots + a_1 x + a_0}{b_m x^m + b_{m-1}x^{m-1} + \cdots + b_1 x + b_0}$$

in which the degree of the numerator is n and the degree of the denominator is m.

1. If $n < m$ (the degree of the numerator is less than the degree of the denominator), then R is a proper rational function, and the graph of R will have the horizontal asymptote $y = 0$ (the x-axis).
2. If $n \geq m$ (the degree of the numerator is greater than or equal to the degree of the denominator), then R is improper. Here long division is used.
   (a) If $n = m$ (the degree of the numerator equals the degree of the denominator), the quotient obtained will be the number $\frac{a_n}{b_m}$, and the line $y = \frac{a_n}{b_m}$ is a horizontal asymptote.
   (b) If $n = m + 1$ (the degree of the numerator is one more than the degree of the denominator), the quotient obtained is of the form $ax + b$ (a polynomial of degree 1), and the line $y = ax + b$ is an oblique asymptote.
   (c) If $n \geq m + 2$ (the degree of the numerator is two or more greater than the degree of the denominator), the quotient obtained is a polynomial of degree 2 or higher, and R has neither a horizontal nor an oblique asymptote. In this case, for $|x|$ unbounded, the graph of R will behave like the graph of the quotient.

NOTE *The graph of a rational function either has one horizontal or one oblique asymptote or else has no horizontal and no oblique asymptote.*

5.2 Assess Your Understanding

'Are You Prepared?' *Answers are given at the end of these exercises. If you get a wrong answer, read the pages listed in red.*

1. ***True or False*** The quotient of two polynomial expressions is a rational expression. (p. 61)
2. What is the quotient and remainder when $3x^4 - x^2$ is divided by $x^3 - x^2 + 1$? (pp. 44–47)
3. Graph $y = \dfrac{1}{x}$. (p. 175)
4. Graph $y = 2(x + 1)^2 - 3$ using transformations. (pp. 256–264)

Concepts and Vocabulary

5. The line _____ is a horizontal asymptote of $R(x) = \dfrac{x^3 - 1}{x^3 + 1}$.
6. The line _____ is a vertical asymptote of $R(x) = \dfrac{x - 1}{x + 1}$.
7. For a rational function R, if the degree of the numerator is less than the degree of the denominator, then R is _____.
8. ***True or False*** The domain of every rational function is the set of all real numbers.
9. ***True or False*** If an asymptote is neither horizontal nor vertical, it is called oblique.
10. ***True or False*** If the degree of the numerator of a rational function equals the degree of the denominator, then the ratio of the leading coefficients gives rise to the horizontal asymptote.

Skill Building

In Problems 11–22, find the domain of each rational function.

11. $R(x) = \dfrac{4x}{x - 3}$
12. $R(x) = \dfrac{5x^2}{3 + x}$
13. $H(x) = \dfrac{-4x^2}{(x - 2)(x + 4)}$
14. $G(x) = \dfrac{6}{(x + 3)(4 - x)}$
15. $F(x) = \dfrac{3x(x - 1)}{2x^2 - 5x - 3}$
16. $Q(x) = \dfrac{-x(1 - x)}{3x^2 + 5x - 2}$
17. $R(x) = \dfrac{x}{x^3 - 8}$
18. $R(x) = \dfrac{x}{x^4 - 1}$
19. $H(x) = \dfrac{3x^2 + x}{x^2 + 4}$
20. $G(x) = \dfrac{x - 3}{x^4 + 1}$
21. $R(x) = \dfrac{3(x^2 - x - 6)}{4(x^2 - 9)}$
22. $F(x) = \dfrac{-2(x^2 - 4)}{3(x^2 + 4x + 4)}$

In Problems 23–28, use the graph shown to find:
(a) The domain and range of each function
(b) The intercepts, if any
(c) Horizontal asymptotes, if any
(d) Vertical asymptotes, if any
(e) Oblique asymptotes, if any

23.
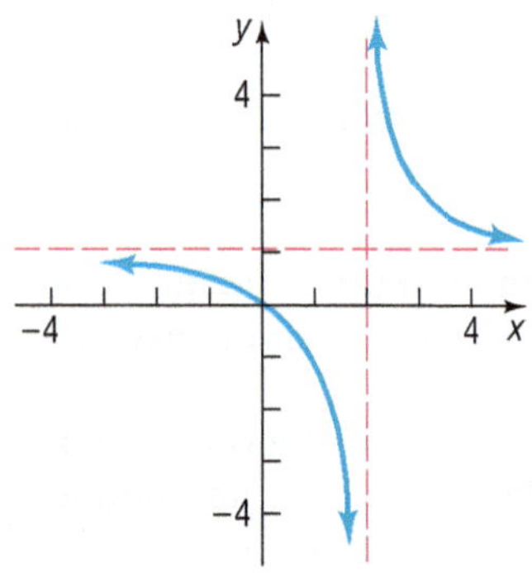

24.
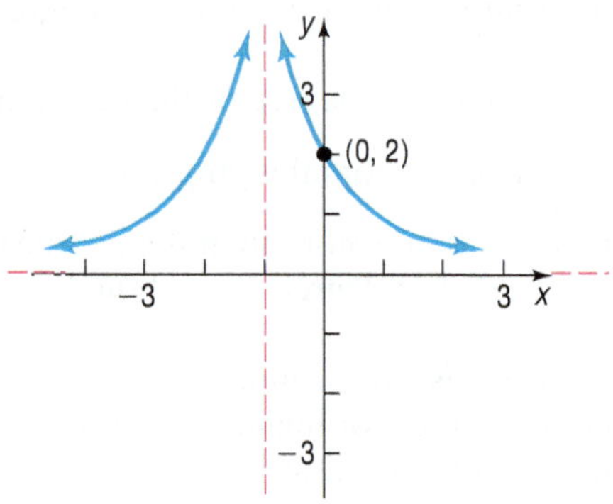

25.
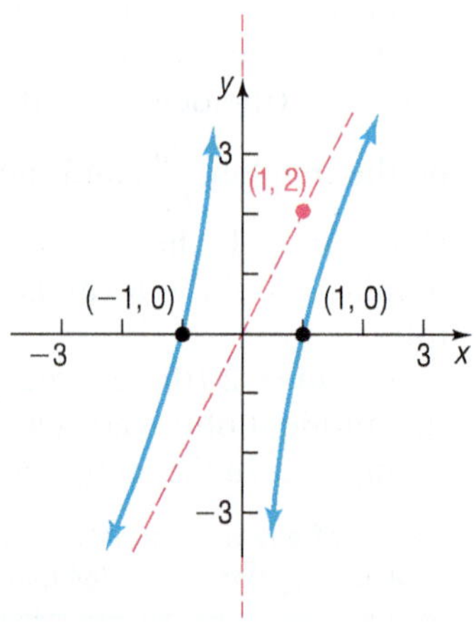

26.
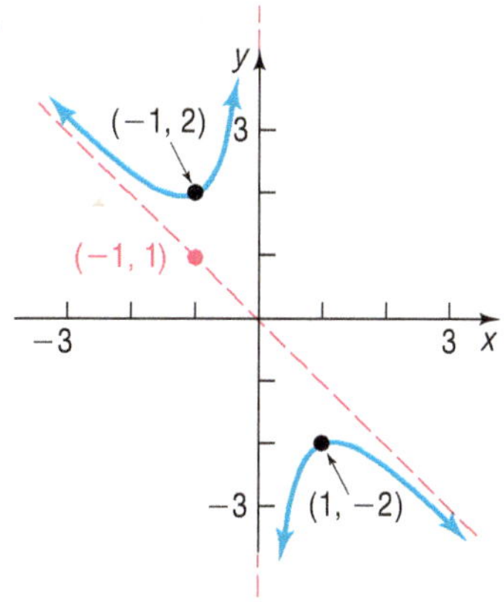

27.
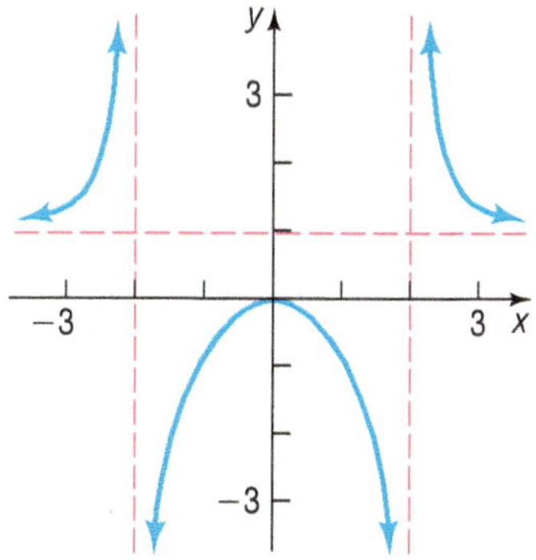

28.
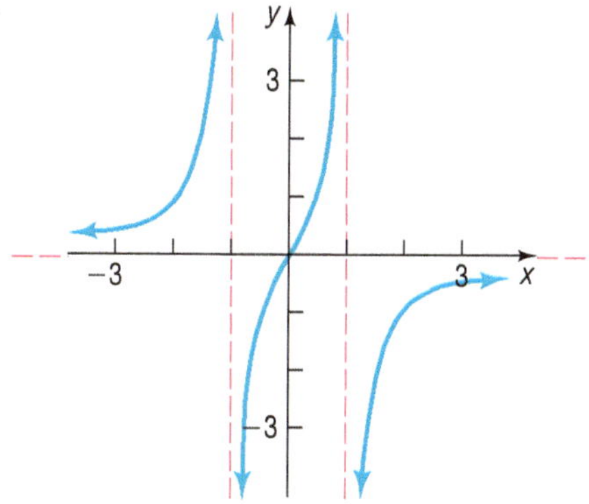

In Problems 29–40, graph each rational function using transformations. Verify your results using a graphing utility.

29. $F(x) = 2 + \dfrac{1}{x}$

30. $Q(x) = 3 + \dfrac{1}{x^2}$

31. $R(x) = \dfrac{1}{(x-1)^2}$

32. $R(x) = \dfrac{3}{x}$

33. $H(x) = \dfrac{-2}{x+1}$

34. $G(x) = \dfrac{2}{(x+2)^2}$

35. $R(x) = \dfrac{-1}{x^2+4x+4}$

36. $R(x) = \dfrac{1}{x-1} + 1$

37. $G(x) = 1 + \dfrac{2}{(x-3)^2}$

38. $F(x) = 2 - \dfrac{1}{x+1}$

39. $R(x) = \dfrac{x^2-4}{x^2}$

40. $R(x) = \dfrac{x-4}{x}$

In Problems 41–52, find the vertical, horizontal, and oblique asymptotes, if any, of each rational function.

41. $R(x) = \dfrac{3x}{x+4}$

42. $R(x) = \dfrac{3x+5}{x-6}$

43. $H(x) = \dfrac{x^3-8}{x^2-5x+6}$

44. $G(x) = \dfrac{x^3+1}{x^2-5x+6}$

45. $T(x) = \dfrac{x^3}{x^4-1}$

46. $P(x) = \dfrac{4x^2}{x^3-1}$

47. $Q(x) = \dfrac{2x^2-5x-12}{3x^2-11x-4}$

48. $F(x) = \dfrac{x^2+6x+5}{2x^2+7x+5}$

49. $R(x) = \dfrac{6x^2+7x-5}{3x+5}$

50. $R(x) = \dfrac{8x^2+26x-7}{4x-1}$

51. $G(x) = \dfrac{x^4-1}{x^2-x}$

52. $F(x) = \dfrac{x^4-16}{x^2-2x}$

Applications and Extensions

53. Gravity In physics, it is established that the acceleration due to gravity, g (in meters/sec^2), at a height h meters above sea level is given by

$$g(h) = \frac{3.99 \times 10^{14}}{(6.374 \times 10^6 + h)^2}$$

where 6.374×10^6 is the radius of Earth in meters.

(a) What is the acceleration due to gravity at sea level?

(b) The Sears Tower in Chicago, Illinois, is 443 meters tall. What is the acceleration due to gravity at the top of the Sears Tower?

(c) The peak of Mount Everest is 8848 meters above sea level. What is the acceleration due to gravity on the peak of Mount Everest?

(d) Find the horizontal asymptote of $g(h)$.

(e) Solve $g(h) = 0$. How do you interpret your answer?

54. Population Model A rare species of insect was discovered in the Amazon Rain Forest. To protect the species, environmentalists declare the insect endangered and transplant the insect into a protected area. The population P of the insect t months after being transplanted is

$$P(t) = \frac{50(1 + 0.5t)}{(2 + 0.01t)}$$

(a) How many insects were discovered? In other words, what was the population when $t = 0$?

(b) What will the population be after 5 years?

(c) Determine the horizontal asymptote of $P(t)$. What is the largest population that the protected area can sustain?

55. Resistance in Parallel Circuits From Ohm's law for circuits, it follows that the total resistance R_{tot} of two components hooked in parallel is given by the equation

$$R_{tot} = \frac{R_1 R_2}{R_1 + R_2}$$

where R_1 and R_2 are the individual resistances.

(a) Let $R_1 = 10$ ohms, and graph R_{tot} as a function of R_2.
(b) Find and interpret any asymptotes of the graph obtained in part (a).
(c) If $R_2 = 2\sqrt{R_1}$, what value of R_1 will yield an R_{tot} of 17 ohms?

Source: *en.wikipedia.org/wiki/Series_and_parallel_circuits*

56. Newton's Method In calculus you will learn that, if

$$p(x) = a_n x^n + a_{n-1}x^{n-1} + \cdots + a_1 x + a_0$$

is a polynomial, then the *derivative* of $p(x)$ is

$$p'(x) = na_n x^{n-1} + (n-1)a_{n-1}x^{n-2} + \cdots + 2a_2 x + a_1$$

Newton's Method is an efficient method for finding the x-intercepts (or real zeros) of a function, such as $p(x)$. The following steps outline Newton's Method.

Step 1: Select an initial value x_0 that is somewhat close to the x-intercept being sought.

Step 2: Find values for x using the relation

$$x_{n+1} = x_n - \frac{p(x_n)}{p'(x_n)} \qquad n = 1, 2, \ldots$$

until you get two consecutive values x_n and x_{n+1} that agree to whatever decimal place accuracy you desire.

Step 3: The approximate zero will be x_{n+1}.

Consider the polynomial $p(x) = x^3 - 7x - 40$.
(a) Evaluate $p(5)$ and $p(-3)$.
(b) What might we conclude about a zero of p? Explain.
(c) Use Newton's Method to approximate an x-intercept, r, $-3 < r < 5$, of $p(x)$ to four decimal places.
(d) Use a graphing utility to graph $p(x)$ and verify your answer in part (c).
(e) Using a graphing utility, evaluate $p(r)$ to verify your result.

Discussion and Writing

57. If the graph of a rational function R has the vertical asymptote $x = 4$, the factor $x - 4$ must be present in the denominator of R. Explain why.

58. If the graph of a rational function R has the horizontal asymptote $y = 2$, the degree of the numerator of R equals the degree of the denominator of R. Explain why.

59. Can the graph of a rational function have both a horizontal and an oblique asymptote? Explain.

60. Make up a rational function that has $y = 2x + 1$ as an oblique asymptote. Explain the methodology that you used.

'Are You Prepared?' Answers

1. True **2.** Quotient: $3x + 3$; remainder: $2x^2 - 3x - 3$ **3.**

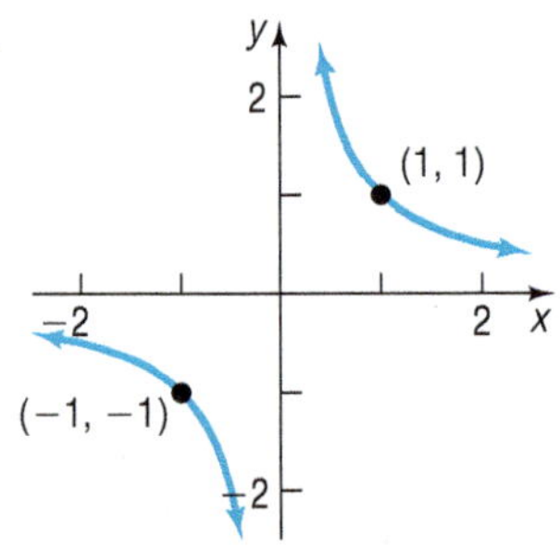

4.

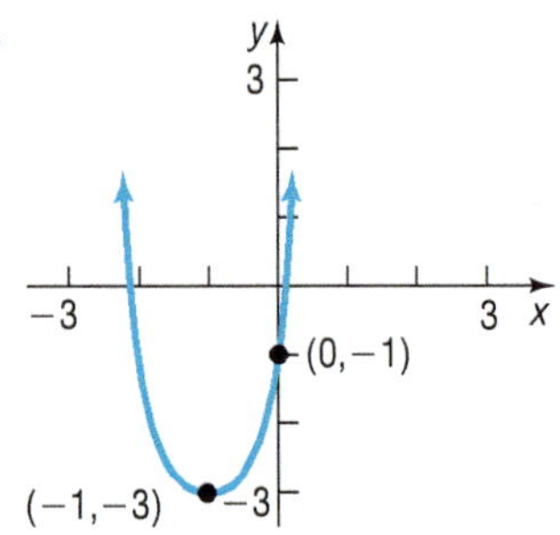

5.3 The Graph of a Rational Function

PREPARING FOR THIS SECTION *Before getting started, review the following:*

- Finding Intercepts (Section 2.1, pp. 170–171)
- Even or Odd Functions (Section 3.3, pp. 235–237)
- Solving Rational Equations (Section 1.2, pp. 106–108)

Now Work the 'Are You Prepared?' problems on page 366.

OBJECTIVES 1 Analyze the Graph of a Rational Function (p. 358)
2 Solve Applied Problems Involving Rational Functions (p. 365)

1 Analyze the Graph of a Rational Function

Graphing utilities make the task of graphing rational functions less time consuming. However, the results of algebraic analysis must be taken into account before drawing conclusions based on the graph provided by the utility. In the next example we illustrate how to use the information collected in the last section in conjunction with the graphing utility to analyze the graph of a rational function $R(x) = \frac{p(x)}{q(x)}$.

EXAMPLE 1 How to Analyze the Graph of a Rational Function

Analyze the graph of the rational function: $R(x) = \frac{x - 1}{x^2 - 4}$

Step-by-Step Solution

STEP 1 Factor the numerator and denominator of R. Find the domain of the rational function.

$$R(x) = \frac{x - 1}{x^2 - 4} = \frac{x - 1}{(x + 2)(x - 2)}$$

The domain of R is $\{x \mid x \neq -2, x \neq 2\}$.

STEP 2 Write R in lowest terms.

Because there are no common factors between the numerator and denominator, R is in lowest terms.

STEP 3 Locate the intercepts of the graph.

The x-intercepts are found by determining the real zeros of the numerator of R written in lowest terms. By solving $x - 1 = 0$, we find the only real zero of the numerator is 1, so the only x-intercept of the graph of R is 1. Since 0 is in the domain of R, the y-intercept is $R(0) = \frac{1}{4}$.

STEP 4 Test for symmetry. If $R(-x) = R(x)$, the function is even and its graph will be symmetric with respect to the y-axis. If $R(-x) = -R(x)$, the function is odd and its graph will be symmetric with respect to the origin.

Because

$$R(-x) = \frac{-x - 1}{(-x)^2 - 4} = \frac{-x - 1}{x^2 - 4} = -\frac{x + 1}{x^2 - 4}$$

we conclude that $R(-x) \neq R(x)$, so R is not even. Because $R(-x) \neq -R(x)$, R is not odd. So there is no symmetry with respect to the y-axis or the origin.

STEP 5 Locate the vertical asymptotes.

We locate the vertical asymptotes by finding the zeros of the denominator with the rational function in lowest terms. With R written in lowest terms, we find that the graph of R has two vertical asymptotes: the lines $x = -2$ and $x = 2$.

STEP 6 Locate the horizontal or oblique asymptotes. Determine points, if any, at which the graph of R intersects these asymptotes.

Because the degree of the numerator is less than the degree of the denominator, R is proper and the line $y = 0$ (the x-axis) is a horizontal asymptote of the graph. To determine if the graph of R intersects the horizontal asymptote, we solve the equation $R(x) = 0$:

$$\frac{x - 1}{x^2 - 4} = 0$$
$$x - 1 = 0$$
$$x = 1$$

The only solution is $x = 1$, so the graph of R intersects the horizontal asymptote at $(1, 0)$.

STEP 7 Graph R using a graphing utility.

The analysis in Steps 1 through 6 helps us to determine an appropriate viewing window to obtain a complete graph. Figure 26(a) shows the graph of $R(x) = \frac{x - 1}{x^2 - 4}$ in connected mode, and Figure 26(b) shows it in dot mode. Notice in Figure 26(a) that the graph has vertical lines at $x = -2$ and $x = 2$. This is due to the fact that, when the graphing utility is in connected mode, it will "connect the dots" between consecutive pixels. We know that the graph of R does not cross the lines $x = -2$ and $x = 2$, since R is not defined at $x = -2$ or $x = 2$. When graphing rational functions, dot mode should be used if extraneous vertical lines appear. You should confirm that all the algebraic conclusions that we came to in Steps 1 through 6 are part of the graph.

For example, the graph has a horizontal asymptote at $y = 0$ and vertical asymptotes at $x = -2$ and $x = 2$. The y-intercept is $\frac{1}{4}$ and the x-intercept is 1.

Figure 26

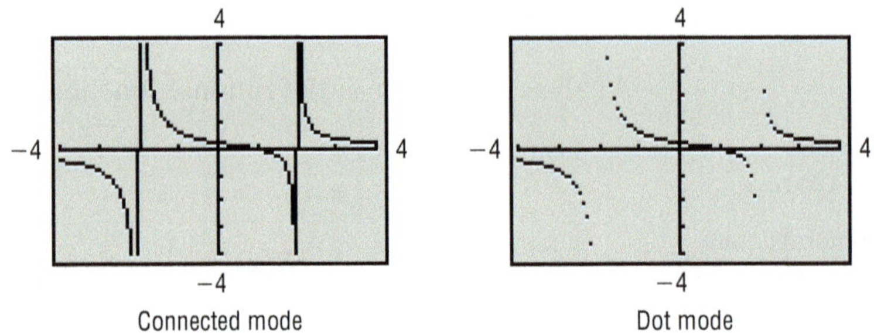

STEP 8 Use the results obtained in Steps 1 through 7 to graph R by hand.

Figure 27 shows the graph of R.

Figure 27

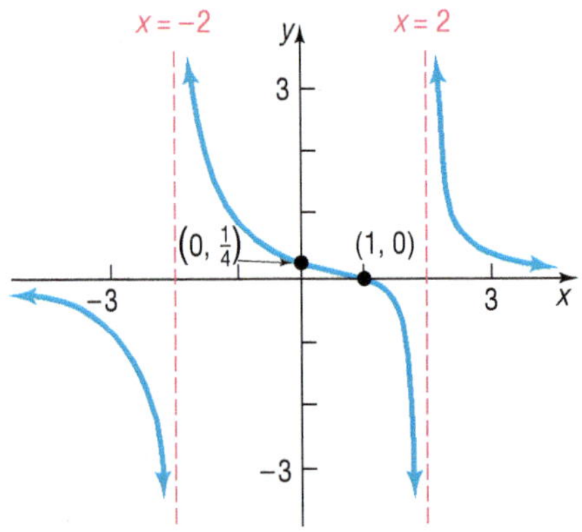

Below we summarize the steps to be used when analyzing a rational function.

Analyzing the Graph of a Rational Function

STEP 1: Factor the numerator and denominator of R. Find the domain of the rational function.

STEP 2: Write R in lowest terms.

STEP 3: Locate the intercepts of the graph. The x-intercepts, if any, of $R(x) = \frac{p(x)}{q(x)}$ in lowest terms satisfy the equation $p(x) = 0$. The y-intercept, if there is one, is $R(0)$.

STEP 4: Test for symmetry. Replace x by $-x$ in $R(x)$. If $R(-x) = R(x)$, there is symmetry with respect to the y-axis; if $R(-x) = -R(x)$, there is symmetry with respect to the origin.

STEP 5: Locate the vertical asymptotes. The vertical asymptotes, if any, of $R(x) = \frac{p(x)}{q(x)}$ in lowest terms are found by identifying the real zeros of $q(x)$. Each zero of the denominator gives rise to a vertical asymptote.

STEP 6: Locate the horizontal or oblique asymptotes, if any, using the procedure given in Section 5.2. Determine points, if any, at which the graph of R intersects these asymptotes.

STEP 7: Graph R using a graphing utility.

STEP 8: Use the results obtained in Steps 1 through 7 to graph R by hand.

Now Work PROBLEM 7

EXAMPLE 2 Analyzing the Graph of a Rational Function

Analyze the graph of the rational function: $R(x) = \dfrac{x^2 - 1}{x}$

Solution

STEP 1: $R(x) = \dfrac{x^2 - 1}{x} = \dfrac{(x + 1)(x - 1)}{x}$
The domain of R is $\{x | x \neq 0\}$.

STEP 2: R is in lowest terms.

STEP 3: The graph has two x-intercepts: -1 and 1. There is no y-intercept, since x cannot equal 0.

STEP 4: Since $R(-x) = -R(x)$, the function is odd and the graph is symmetric with respect to the origin.

STEP 5: The real zero of the denominator with R in lowest terms is 0, so the graph of $R(x)$ has the line $x = 0$ (the y-axis) as a vertical asymptote.

STEP 6: Since the degree of the numerator, 2, is one more than the degree of the denominator, 1, the rational function R is improper and will have an oblique asymptote. To find the oblique asymptote, we use long division.

NOTE Because the denominator of the rational function is a monomial, we can also find the oblique asymptote as follows:

$$\frac{x^2 - 1}{x} = \frac{x^2}{x} - \frac{1}{x} = x - \frac{1}{x}$$

Since $\frac{1}{x} \rightarrow 0$ as $x \rightarrow \infty$, $y = x$ is an oblique asymptote. ■

$$\begin{array}{r} x \\ x\overline{)x^2 - 1} \\ \underline{x^2 } \\ -1 \end{array}$$

The quotient is x, so the line $y = x$ is an oblique asymptote of the graph. To determine whether the graph of R intersects the asymptote $y = x$, we solve the equation $R(x) = x$.

$$\begin{aligned} R(x) = \frac{x^2 - 1}{x} &= x \\ x^2 - 1 &= x^2 \\ -1 &= 0 \quad \text{Impossible.} \end{aligned}$$

We conclude that the equation $\dfrac{x^2 - 1}{x} = x$ has no solution, so the graph of $R(x)$ does not intersect the line $y = x$.

STEP 7: See Figure 28. We see from the graph that there is no y-intercept and there are two x-intercepts, -1 and 1. The symmetry with respect to the origin is also evident. We can also see that there is a vertical asymptote at $x = 0$. Finally, it is not necessary to graph this function in dot mode since no extraneous vertical lines are present.

Figure 28

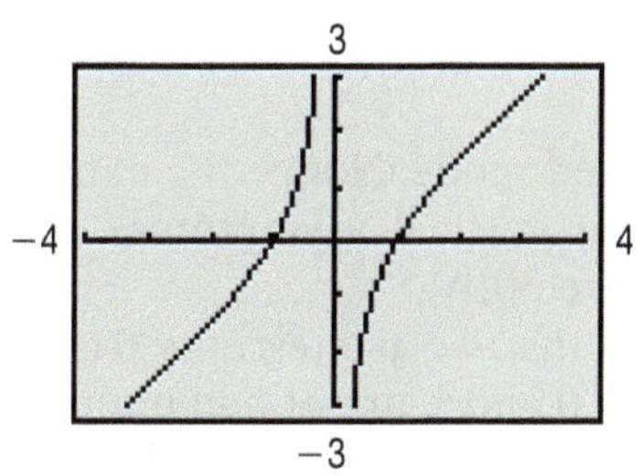

STEP 8: Using the information gathered in Steps 1 through 7, we obtain the graph of R shown in Figure 29. Notice how the oblique asymptote is used as a guide in graphing the rational function by hand.

Figure 29

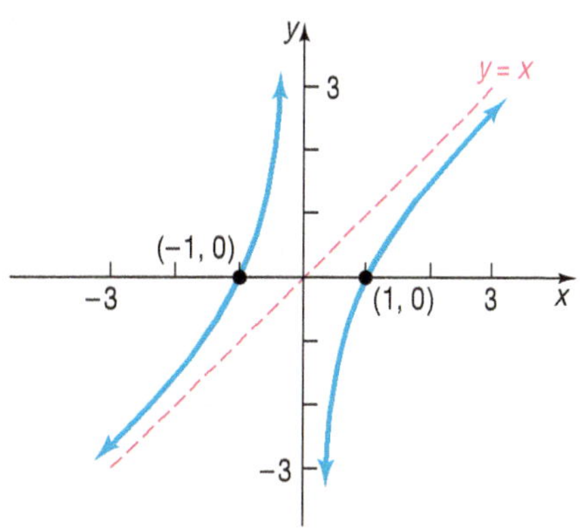

Now Work PROBLEM 15

EXAMPLE 3 Analyzing the Graph of a Rational Function

Analyze the graph of the rational function: $R(x) = \dfrac{3x^2 - 3x}{x^2 + x - 12}$

Solution

Step 1: We factor R to get

$$R(x) = \frac{3x(x - 1)}{(x + 4)(x - 3)}$$

The domain of R is $\{x \mid x \neq -4, x \neq 3\}$.

Step 2: R is in lowest terms.

Step 3: The graph has two x-intercepts: 0 and 1. The y-intercept is $R(0) = 0$.

Step 4: Because

$$R(-x) = \frac{3(-x)^2 - 3(-x)}{(-x)^2 + (-x) - 12} = \frac{3x^2 + 3x}{x^2 - x - 12}$$

we conclude that R is neither even nor odd. There is no symmetry with respect to the y-axis or the origin.

Step 5: The real zeros of the denominator of R with R in lowest terms are -4 and 3. So the graph of R has two vertical asymptotes: $x = -4$ and $x = 3$.

Step 6: Since the degree of the numerator equals the degree of the denominator, the graph has a horizontal asymptote. To find it, we form the quotient of the leading coefficient of the numerator, 3, and the leading coefficient of the denominator, 1. The graph of R has the horizontal asymptote $y = 3$. To find out whether the graph of R intersects the asymptote, we solve the equation $R(x) = 3$.

$$\begin{aligned} R(x) = \frac{3x^2 - 3x}{x^2 + x - 12} &= 3 \\ 3x^2 - 3x &= 3x^2 + 3x - 36 \\ -6x &= -36 \\ x &= 6 \end{aligned}$$

The graph intersects the line $y = 3$ at $x = 6$, and $(6, 3)$ is a point on the graph of R.

Step 7: Figure 30(a) shows the graph of R in connected mode. Notice the extraneous vertical lines at $x = -4$ and $x = 3$ (the vertical asymptotes). As a result, we also graph R in dot mode. See Figure 30(b).

Step 8: Figure 30 does not display the graph between the two x-intercepts, 0 and 1. However, because the zeros in the numerator, 0 and 1, are of odd multiplicity (both are multiplicity 1), we know that the graph of R crosses the x-axis at 0 and 1. Therefore, the graph of R is above the x-axis for $0 < x < 1$. To see this part better, we graph R for $-1 \leq x \leq 2$ in Figure 31. Using MAXIMUM, we approximate the turning point to be $(0.52, 0.07)$, rounded to two decimal places.

Figure 30

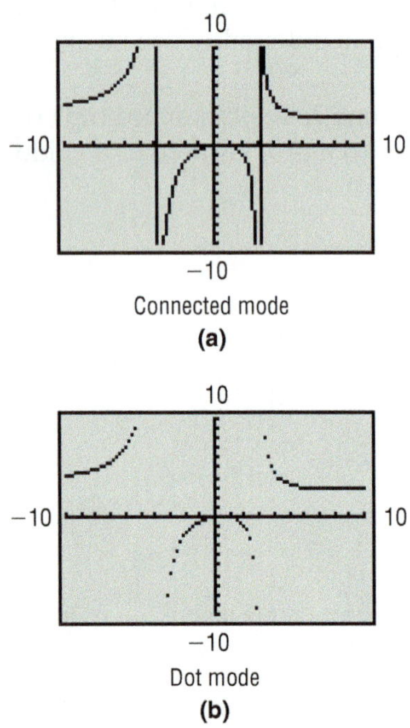

Connected mode
(a)

Dot mode
(b)

Figure 31

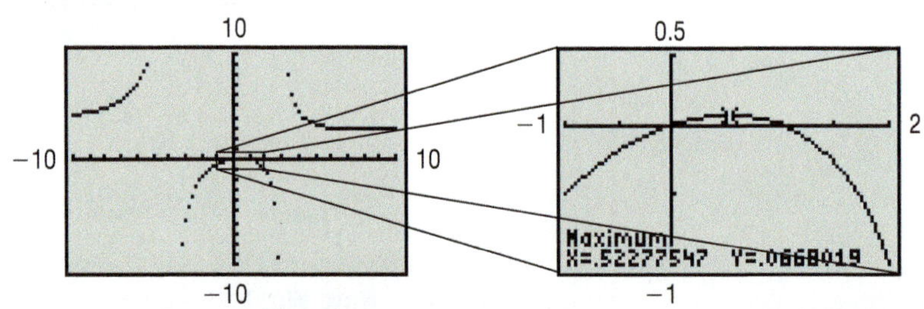

Figure 30 also does not display the graph of R crossing the horizontal asymptote at $(6, 3)$. To see this part better, we graph R for $4 \le x \le 60$ in Figure 32. Using MINIMUM, we approximate the turning point to be $(11.48, 2.75)$, rounded to two decimal places.

Figure 32

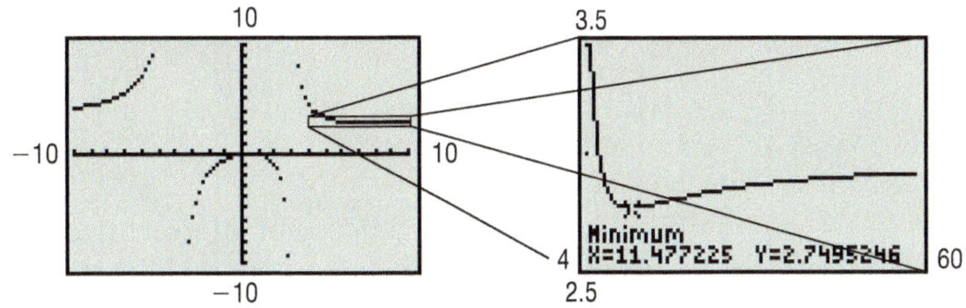

Using this information along with the information gathered in Steps 1 through 7, we obtain the graph of R shown in Figure 33.

Figure 33

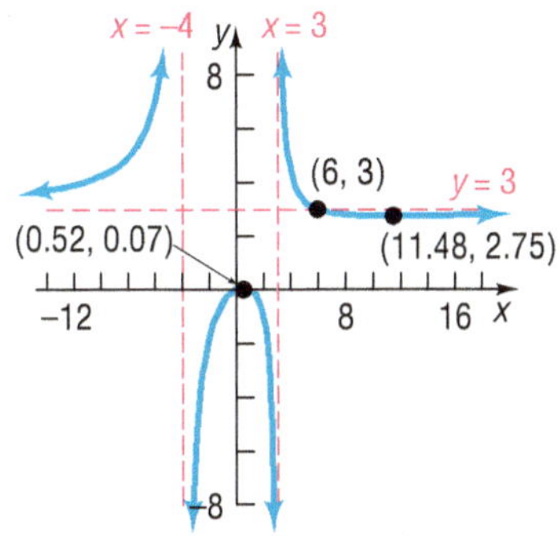

EXAMPLE 4 Analyzing the Graph of a Rational Function with a Hole

Analyze the graph of the rational function: $R(x) = \dfrac{2x^2 - 5x + 2}{x^2 - 4}$

Solution

STEP 1: We factor R and obtain

$$R(x) = \frac{(2x - 1)(x - 2)}{(x + 2)(x - 2)}$$

The domain of R is $\{x \mid x \neq -2, x \neq 2\}$.

STEP 2: In lowest terms,

$$R(x) = \frac{2x - 1}{x + 2} \qquad x \neq 2$$

STEP 3: The graph has one x-intercept: 0.5. The y-intercept is $R(0) = -0.5$.

STEP 4: Because

$$R(-x) = \frac{2x^2 + 5x + 2}{x^2 - 4}$$

we conclude that R is neither even nor odd. There is no symmetry with respect to the y-axis or the origin.

STEP 5: Since $x + 2$ is the only factor of the denominator of $R(x)$ *in lowest terms,* the graph has one vertical asymptote, $x = -2$. However, the rational function is undefined at both $x = 2$ and $x = -2$.

STEP 6: Since the degree of the numerator equals the degree of the denominator, the graph has a horizontal asymptote. To find it, we form the quotient of the leading coefficient of the numerator, 2, and the leading coefficient of the denominator, 1. The graph of R has the horizontal asymptote $y = 2$. To find whether the graph of R intersects the asymptote, we solve the equation $R(x) = 2$.

$$
\begin{aligned}
R(x) = \frac{2x - 1}{x + 2} &= 2 \\
2x - 1 &= 2(x + 2) \\
2x - 1 &= 2x + 4 \\
-1 &= 4 \qquad \text{Impossible}
\end{aligned}
$$

The graph does not intersect the line $y = 2$.

Step 7: Figure 34 shows the graph of $R(x)$. Notice that the graph has one vertical asymptote at $x = -2$. Also, the function appears to be continuous at $x = 2$.

Figure 34

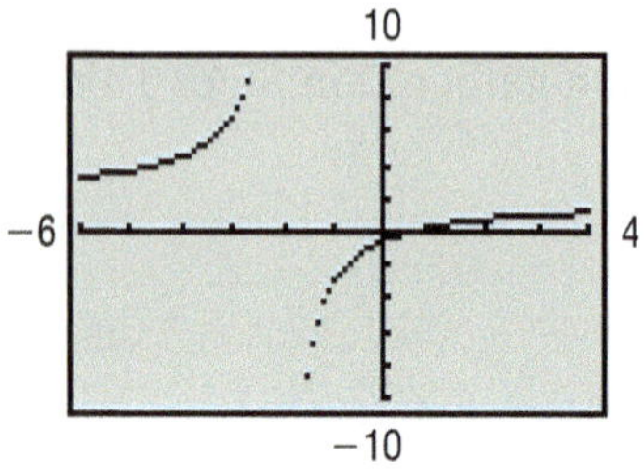

Step 8: The analysis presented thus far does not explain the behavior of the graph at $x = 2$. We use the TABLE feature of our graphing utility to determine the behavior of the graph of R as x approaches 2. See Table 13. From the table, we conclude that the value of R approaches 0.75 as x approaches 2. This result is further verified by evaluating R in lowest terms at $x = 2$. We conclude that there is a hole in the graph at $(2, 0.75)$. Using the information gathered in Steps 1 through 7, we obtain the graph of R shown in Figure 35.

Table 13

X	Y1
1.99	.74687
1.999	.74969
1.9999	.74997
2	ERROR
2.0001	.75003
2.001	.75031
2.01	.75312

Y1=(2X²−5X+2)/(...

Figure 35

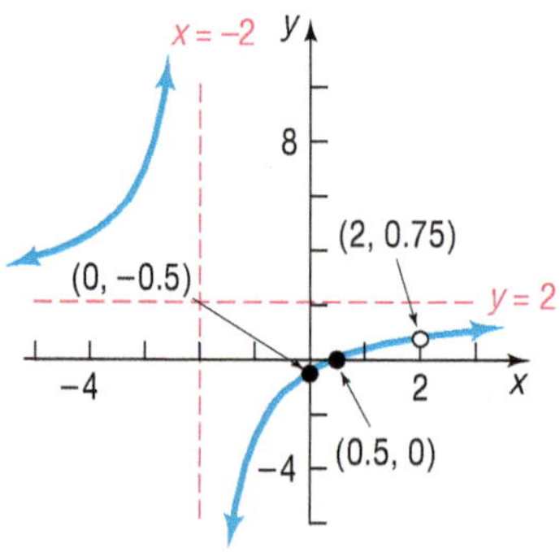

As Example 4 shows, **the values excluded from the domain of a rational function give rise to either vertical asymptotes or holes.**

Now Work PROBLEM 33

EXAMPLE 5 Constructing a Rational Function from Its Graph

Make up a rational function that might have the graph shown in Figure 36.

Figure 36

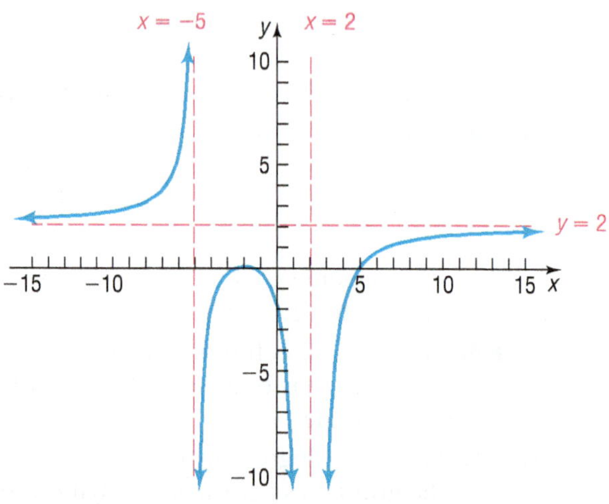

Solution The numerator of a rational function $R(x) = \dfrac{p(x)}{q(x)}$ in lowest terms determines the x-intercepts of its graph. The graph shown in Figure 36 has x-intercepts -2 (even multiplicity; graph touches the x-axis) and 5 (odd multiplicity; graph crosses the x-axis). So one possibility for the numerator is $p(x) = (x + 2)^2(x - 5)$.

The denominator of a rational function in lowest terms determines the vertical asymptotes of its graph. The vertical asymptotes of the graph are $x = -5$ and $x = 2$. Since $R(x)$ approaches ∞ from the left of $x = -5$ and $R(x)$ approaches $-\infty$ from the right of $x = -5$, we know that $(x + 5)$ is a factor of odd multiplicity in $q(x)$. Also, $R(x)$ approaches $-\infty$ from both sides of $x = 2$, so $(x - 2)$ is a factor of even multiplicity in $q(x)$. A possibility for the denominator is $q(x) = (x + 5)(x - 2)^2$.

So far we have $R(x) = \dfrac{(x + 2)^2(x - 5)}{(x + 5)(x - 2)^2}$. However, the horizontal asymptote of the graph given in Figure 36 is $y = 2$, so we know that the degree of the numerator must equal the degree of the denominator, and the quotient of leading coefficients must be $\dfrac{2}{1}$. This leads to $R(x) = \dfrac{2(x + 2)^2(x - 5)}{(x + 5)(x - 2)^2}$. Figure 37 shows the graph of R drawn on a graphing utility. Since Figure 37 looks similar to Figure 36, we have found a rational function R for the graph in Figure 36.

Figure 37

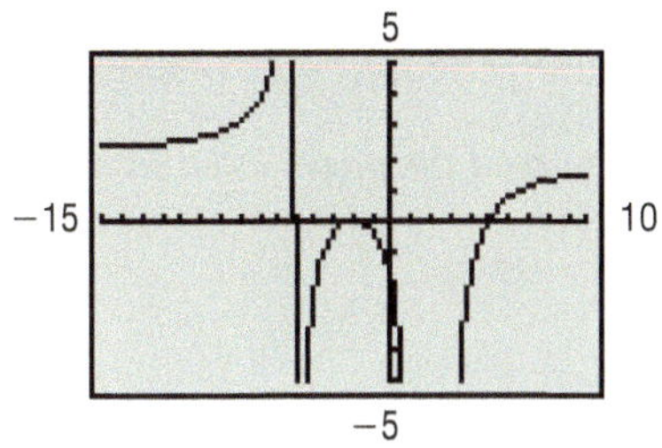

Now Work PROBLEM 51

2 Solve Applied Problems Involving Rational Functions

EXAMPLE 6 Finding the Least Cost of a Can

Reynolds Metal Company manufactures aluminum cans in the shape of a cylinder with a capacity of 500 cubic centimeters (cm^3), or $\frac{1}{2}$ liter. The top and bottom of the can are made of a special aluminum alloy that costs 0.05¢/per square centimeter (cm^2). The sides of the can are made of material that costs 0.02¢/cm^2.

(a) Express the cost of material for the can as a function of the radius r of the can.
(b) Use a graphing utility to graph the function $C = C(r)$.
(c) What value of r will result in the least cost?
(d) What is this least cost?

Solution (a) Figure 38 illustrates the situation. Notice that the material required to produce a cylindrical can of height h and radius r consists of a rectangle of area $2\pi rh$ and two circles, each of area πr^2. The total cost C (in cents) of manufacturing the can is therefore

Figure 38

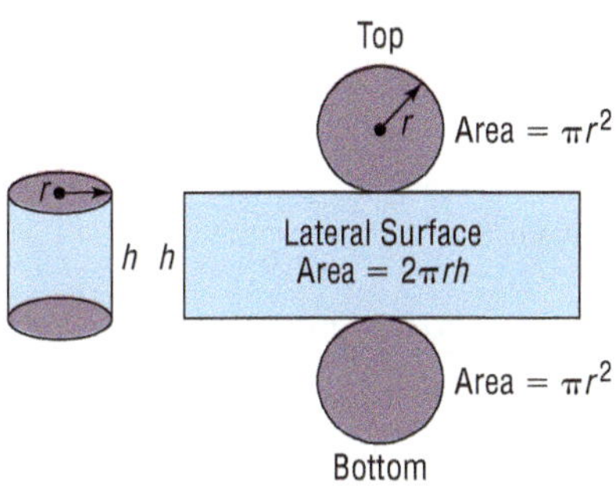

$$\begin{aligned} C &= \text{Cost of the top and bottom} + \text{Cost of the side} \\ &= \underbrace{2(\pi r^2)}_{\substack{\text{Total area}\\\text{of top and}\\\text{bottom}}}\ \underbrace{(0.05)}_{\substack{\text{Cost/unit}\\\text{area}}} + \underbrace{(2\pi rh)}_{\substack{\text{Total}\\\text{area of}\\\text{side}}}\ \underbrace{(0.02)}_{\substack{\text{Cost/unit}\\\text{area}}} \\ &= 0.10\pi r^2 + 0.04\pi rh \end{aligned}$$

But we have the additional restriction that the height h and radius r must be chosen so that the volume V of the can is 500 cm^3. Since $V = \pi r^2 h$, we have

$$500 = \pi r^2 h \quad \text{or} \quad h = \frac{500}{\pi r^2}$$

Figure 39

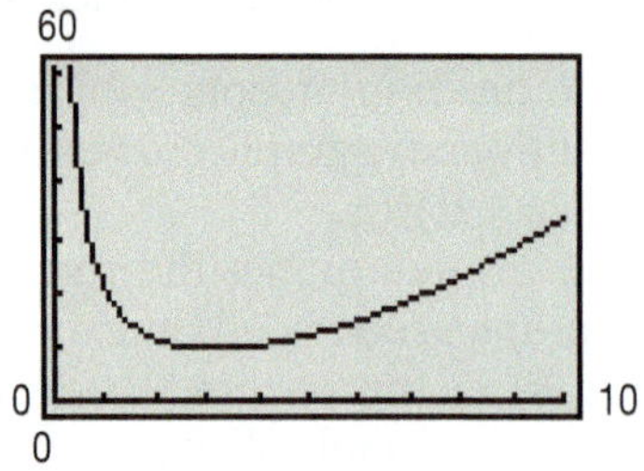

Substituting this expression for h, the cost C, in cents, as a function of the radius r, is

$$C(r) = 0.10\pi r^2 + 0.04\pi r\frac{500}{\pi r^2} = 0.10\pi r^2 + \frac{20}{r} = \frac{0.10\pi r^3 + 20}{r}$$

(b) See Figure 39 for the graph of $C = C(r)$.
(c) Using the MINIMUM command, the cost is least for a radius of about 3.17 cm.
(d) The least cost is $C(3.17) \approx 9.47$¢.

Now Work PROBLEM 61

5.3 Assess Your Understanding

'Are You Prepared?' *The answer is given at the end of these exercises. If you get a wrong answer, read the pages listed in red.*

1. (a) Find the intercepts of the graph of the function $f(x) = \frac{x^2 - 1}{x^2 - 4}$. (pp. 170–171; 229)
(b) Is the function in part (a) even, odd, or neither? (pp. 235–237)

2. Solve $\frac{x - 3}{x^2 + 1} = -2$. (pp. 106–108)

Concepts and Vocabulary

3. If the numerator and the denominator of a rational function have no common factors, the rational function is ________.

4. ***True or False*** The graph of a rational function never intersects a horizontal asymptote.

5. ***True or False*** The graph of a rational function sometimes intersects a vertical asymptote.

6. ***True or False*** The graph of a rational function sometimes has a hole.

Skill Building

In Problems 7–44, follow Steps 1 through 8 on page 360 to analyze the graph of each function.

7. $R(x) = \frac{x + 1}{x(x + 4)}$ **8.** $R(x) = \frac{x}{(x - 1)(x + 2)}$ **9.** $R(x) = \frac{3x + 3}{2x + 4}$ **10.** $R(x) = \frac{2x + 4}{x - 1}$

11. $R(x) = \frac{3}{x^2 - 4}$ **12.** $R(x) = \frac{6}{x^2 - x - 6}$ **13.** $P(x) = \frac{x^4 + x^2 + 1}{x^2 - 1}$ **14.** $Q(x) = \frac{x^4 - 1}{x^2 - 4}$

15. $H(x) = \frac{x^3 - 1}{x^2 - 9}$ **16.** $G(x) = \frac{x^3 + 1}{x^2 + 2x}$ **17.** $R(x) = \frac{x^2}{x^2 + x - 6}$ **18.** $R(x) = \frac{x^2 + x - 12}{x^2 - 4}$

19. $G(x) = \frac{x}{x^2 - 4}$ **20.** $G(x) = \frac{3x}{x^2 - 1}$ **21.** $R(x) = \frac{3}{(x - 1)(x^2 - 4)}$ **22.** $R(x) = \frac{-4}{(x + 1)(x^2 - 9)}$

23. $H(x) = \frac{4(x^2 - 1)}{x^4 - 16}$ **24.** $H(x) = \frac{x^2 + 4}{x^4 - 1}$ **25.** $F(x) = \frac{x^2 - 3x - 4}{x + 2}$ **26.** $F(x) = \frac{x^2 + 3x + 2}{x - 1}$

27. $R(x) = \frac{x^2 + x - 12}{x - 4}$ **28.** $R(x) = \frac{x^2 - x - 12}{x + 5}$ **29.** $F(x) = \frac{x^2 + x - 12}{x + 2}$ **30.** $G(x) = \frac{x^2 - x - 12}{x + 1}$

31. $R(x) = \frac{x(x - 1)^2}{(x + 3)^3}$ **32.** $R(x) = \frac{(x - 1)(x + 2)(x - 3)}{x(x - 4)^2}$ **33.** $R(x) = \frac{x^2 + x - 12}{x^2 - x - 6}$

34. $R(x) = \frac{x^2 + 3x - 10}{x^2 + 8x + 15}$ **35.** $R(x) = \frac{6x^2 - 7x - 3}{2x^2 - 7x + 6}$ **36.** $R(x) = \frac{8x^2 + 26x + 15}{2x^2 - x - 15}$

37. $R(x) = \frac{x^2 + 5x + 6}{x + 3}$ **38.** $R(x) = \frac{x^2 + x - 30}{x + 6}$ **39.** $f(x) = x + \frac{1}{x}$

40. $f(x) = 2x + \frac{9}{x}$ **41.** $f(x) = x^2 + \frac{1}{x}$ **42.** $f(x) = 2x^2 + \frac{9}{x}$

43. $f(x) = x + \frac{1}{x^3}$ **44.** $f(x) = 2x + \frac{9}{x^3}$

In Problems 45–50, graph each function using a graphing utility; then use MINIMUM to obtain the minimum value, rounded to two decimal places.

45. $f(x) = x + \dfrac{1}{x}, \quad x > 0$

46. $f(x) = 2x + \dfrac{9}{x}, \quad x > 0$

47. $f(x) = x^2 + \dfrac{1}{x}, \quad x > 0$

48. $f(x) = 2x^2 + \dfrac{9}{x}, \quad x > 0$

49. $f(x) = x + \dfrac{1}{x^3}, \quad x > 0$

50. $f(x) = 2x + \dfrac{9}{x^3}, \quad x > 0$

In Problems 51–54, find a rational function that might have the given graph. (More than one answer might be possible.)

51.

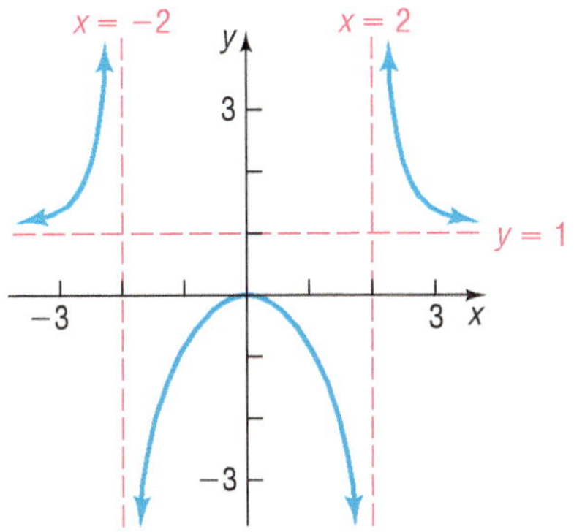

52.

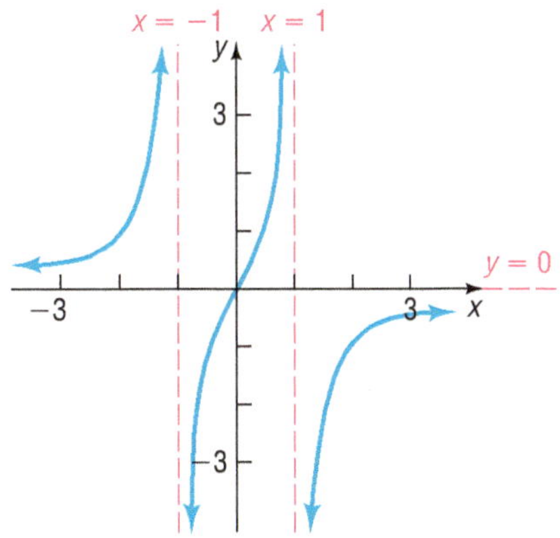

53.

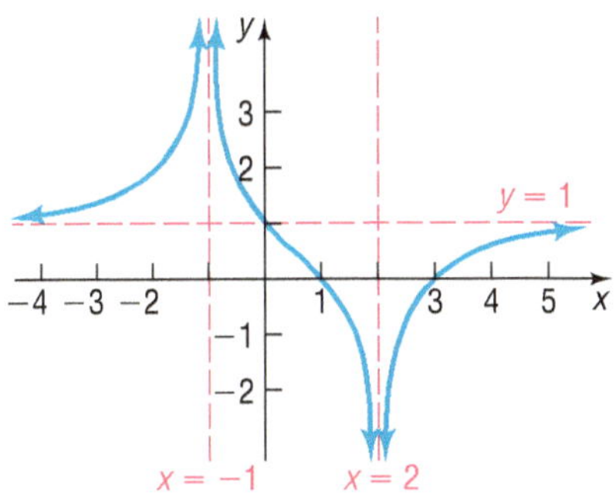

54.

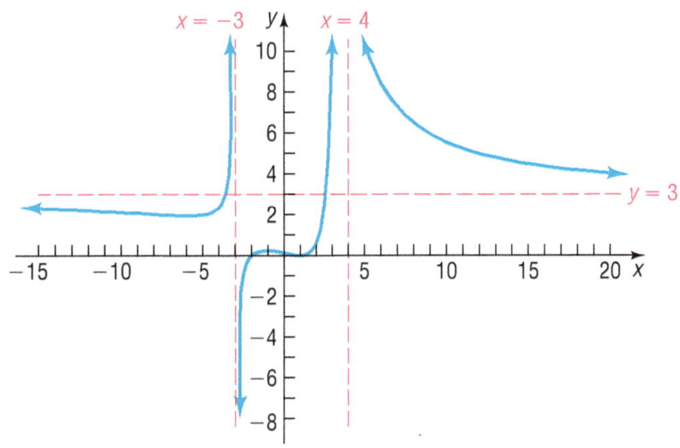

Applications and Extensions

55. Drug Concentration The concentration C of a certain drug in a patient's bloodstream t hours after injection is given by

$$C(t) = \frac{t}{2t^2 + 1}$$

(a) Find the horizontal asymptote of $C(t)$. What happens to the concentration of the drug as t increases?
(b) Using your graphing utility, graph $C = C(t)$.
(c) Determine the time at which the concentration is highest.

56. Drug Concentration The concentration C of a certain drug in a patient's bloodstream t minutes after injection is given by

$$C(t) = \frac{50t}{t^2 + 25}$$

(a) Find the horizontal asymptote of $C(t)$. What happens to the concentration of the drug as t increases?
(b) Using your graphing utility, graph $C = C(t)$.
(c) Determine the time at which the concentration is highest.

57. Minimum Cost A rectangular area adjacent to a river is to be fenced in; no fence is needed on the river side. The enclosed area is to be 1000 square feet. Fencing for the side parallel to the river is \$5 per linear foot, and fencing for the other two sides is \$8 per linear foot; the four corner posts are \$25 apiece. Let x be the length of one of the sides perpendicular to the river.

(a) Write a function $C(x)$ that describes the cost of the project.
(b) What is the domain of C?
(c) Use a graphing utility to graph $C = C(x)$.
(d) Find the dimensions of the cheapest enclosure.

Source: *www.uncwil.edu/courses/math111hb/PandR/rational/rational.html*

58. Doppler Effect The Doppler effect (named after Christian Doppler) is the change in the pitch (frequency) of the sound from a source (s) as heard by an observer (o) when one or both are in motion. If we assume both the source and the observer are moving in the same direction, the relationship is

$$f' = f_a\left(\frac{v - v_o}{v - v_s}\right)$$

where
f' = perceived pitch by the observer
f_a = actual pitch of the source
v = speed of sound in air (assume 772.4 mph)
v_o = speed of the observer
v_s = speed of the source

Suppose you are traveling down the road at 45 mph and you hear an ambulance (with siren) coming toward you from the rear. The actual pitch of the siren is 600 hertz (Hz).

(a) Write a function $f'(v_s)$ that describes this scenario.
(b) If $f' = 620$ Hz, find the speed of the ambulance.
(c) Use a graphing utility to graph the function.
(d) Verify your answer from part (b).

Source: www.kettering.edu/~drussell/

59. **Minimizing Surface Area** United Parcel Service has contracted you to design a closed box with a square base that has a volume of 10,000 cubic inches. See the illustration.

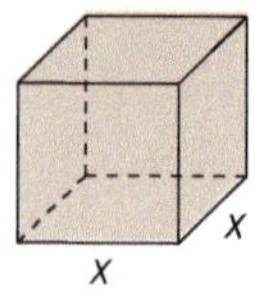

(a) Express the surface area S of the box as a function of x.
(b) Using a graphing utility, graph the function found in part (a).
(c) What is the minimum amount of cardboard that can be used to construct the box?
(d) What are the dimensions of the box that minimize the surface area?
(e) Why might UPS be interested in designing a box that minimizes the surface area?

60. **Minimizing Surface Area** United Parcel Service has contracted you to design an open box with a square base that has a volume of 5000 cubic inches. See the illustration.

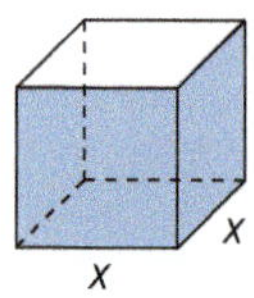

(a) Express the surface area S of the box as a function of x.
(b) Using a graphing utility, graph the function found in part (a).
(c) What is the minimum amount of cardboard that can be used to construct the box?
(d) What are the dimensions of the box that minimize the surface area?
(e) Why might UPS be interested in designing a box that minimizes the surface area?

61. **Cost of a Can** A can in the shape of a right circular cylinder is required to have a volume of 500 cubic centimeters. The top and bottom are made of material that costs 6¢ per square centimeter, while the sides are made of material that costs 4¢ per square centimeter.
(a) Express the total cost C of the material as a function of the radius r of the cylinder. (Refer to Figure 38.)
(b) Graph $C = C(r)$. For what value of r is the cost C a minimum?

62. **Material Needed to Make a Drum** A steel drum in the shape of a right circular cylinder is required to have a volume of 100 cubic feet.

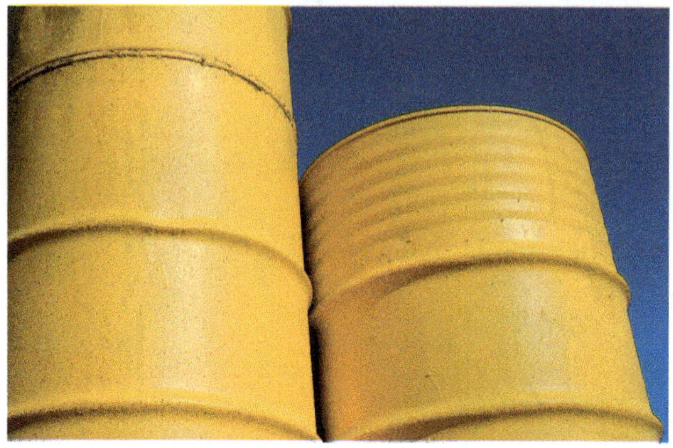

(a) Express the amount A of material required to make the drum as a function of the radius r of the cylinder.
(b) How much material is required if the drum's radius is 3 feet?
(c) How much material is required if the drum's radius is 4 feet?
(d) How much material is required if the drum's radius is 5 feet?
(e) Graph $A = A(r)$. For what value of r is A smallest?

Discussion and Writing

63. Graph each of the following functions:

$$y = \frac{x^2 - 1}{x - 1} \qquad y = \frac{x^3 - 1}{x - 1}$$

$$y = \frac{x^4 - 1}{x - 1} \qquad y = \frac{x^5 - 1}{x - 1}$$

Is $x = 1$ a vertical asymptote? Why not? What is happening for $x = 1$? What do you conjecture about $y = \frac{x^n - 1}{x - 1}$, $n \geq 1$ an integer, for $x = 1$?

64. Graph each of the following functions:

$$y = \frac{x^2}{x - 1} \qquad y = \frac{x^4}{x - 1} \qquad y = \frac{x^6}{x - 1} \qquad y = \frac{x^8}{x - 1}$$

What similarities do you see? What differences?

65. Write a few paragraphs that provide a general strategy for graphing a rational function. Be sure to mention the following: proper, improper, intercepts, and asymptotes.

66. Create a rational function that has the following characteristics: crosses the x-axis at 2; touches the x-axis at -1; one vertical asymptote at $x = -5$ and another at $x = 6$; and one horizontal asymptote, $y = 3$. Compare your function to a fellow classmate's. How do they differ? What are their similarities?

67. Create a rational function that has the following characteristics: crosses the x-axis at 3; touches the x-axis at -2; one vertical asymptote, $x = 1$; and one horizontal asymptote, $y = 2$. Give your rational function to a fellow classmate and ask for a written critique of your rational function.

68. Create a rational function with the following characteristics: three real zeros, one of multiplicity 2; y-intercept 1; vertical asymptotes, $x = -2$ and $x = 3$; oblique asymptote, $y = 2x + 1$. Is this rational function unique? Compare your function with those of other students. What will be the same as everyone else's? Add some more characteristics, such as symmetry or naming the real zeros. How does this modify the rational function?

69. Explain the circumstances under which the graph of a rational function will have a hole.

'Are You Prepared?' Answers

1. (a) $\left(0, \frac{1}{4}\right), (1, 0), (-1, 0)$ (b) Even   2. $\left\{-1, \frac{1}{2}\right\}$

5.4 Polynomial and Rational Inequalities

PREPARING FOR THIS SECTION *Before getting started, review the following:*

- Solving Linear Inequalities (Section 1.7, pp. 153–155)
- Solving Quadratic Inequalities (Section 4.5, pp. 320–322)

Now Work the 'Are You Prepared?' problem on page 373.

OBJECTIVES
1 Solve Polynomial Inequalities Algebraically and Graphically (p. 369)
2 Solve Rational Inequalities Algebraically and Graphically (p. 371)

1 Solve Polynomial Inequalities Algebraically and Graphically

In this section we solve inequalities that involve polynomials of degree 3 and higher, along with inequalities that involve rational functions. To help understand the algebraic procedure for solving such inequalities, we use the information obtained in the previous three sections about the graphs of polynomial and rational functions. The approach follows the same logic as we used to solve inequalities involving quadratic functions.

EXAMPLE 1 **Solving a Polynomial Inequality Using Its Graph**

Solve $(x + 3)(x - 1)^2 > 0$ by graphing $f(x) = (x + 3)(x - 1)^2$.

By Hand Graphical Solution

We graph $f(x) = (x + 3)(x - 1)^2$ and determine the intervals of x for which the graph is above the x-axis. Do you see why? These values of x result in $f(x)$ being positive. Using Steps 1 through 8 on page 341, we obtain the graph shown in Figure 40.

Figure 40

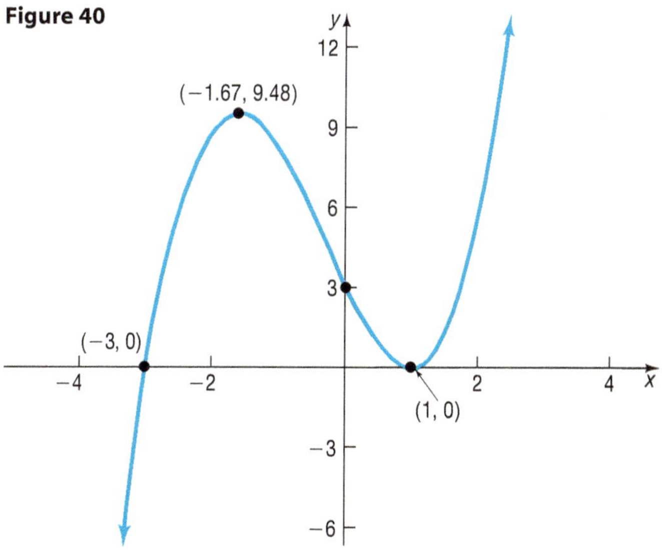

From the graph, we can see that $f(x) > 0$ for $-3 < x < 1$ or $x > 1$. The solution set is $\{x \mid -3 < x < 1 \text{ or } x > 1\}$ or, using interval notation, $(-3, 1) \cup (1, \infty)$.

Graphing Utility Solution

We graph $Y_1 = (x + 3)(x - 1)^2$. See Figure 41. Using the ZERO command, we find that the x-intercepts of the graph of Y_1 are -3 and 1. The graph of Y_1 is above the x-axis (and therefore f is positive) for $-3 < x < 1$ or $x > 1$. Therefore, the solution set is $\{x \mid -3 < x < 1 \text{ or } x > 1\}$ or, using interval notation, $(-3, 1) \cup (1, \infty)$.

Figure 41

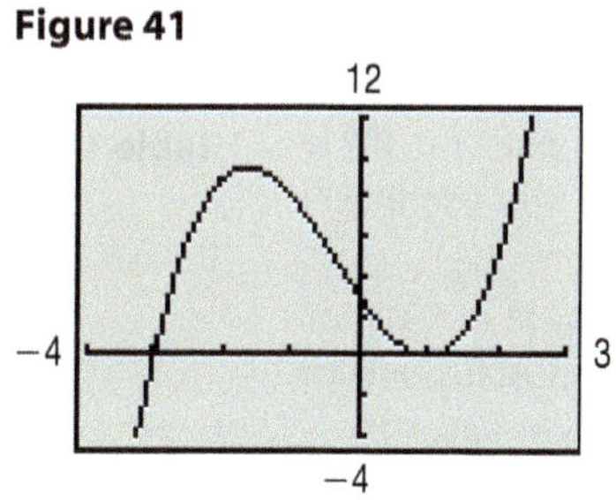

Now Work PROBLEM 9

The results of Example 1 lead to the following approach to solving polynomial inequalities algebraically. Suppose that the polynomial or rational inequality is in one of the forms

$$f(x) < 0 \qquad f(x) > 0 \qquad f(x) \le 0 \qquad f(x) \ge 0$$

Locate the zeros (x-intercepts of the graph) of the polynomial function f. We know that the sign of f can change on either side of an x-intercept, so we use these zeros to divide the real number line into intervals. Then we know that on each interval the graph of f is either above $[f(x) > 0]$ or below $[f(x) < 0]$ the x-axis.

EXAMPLE 2 How to Solve a Polynomial Inequality Algebraically

Solve the inequality $x^4 > x$ algebraically, and graph the solution set.

Step-by-Step Solution

STEP 1 Write the inequality so that a polynomial expression f is on the left side and zero is on the right side.

Rearrange the inequality so that 0 is on the right side.

$$x^4 > x$$

$$x^4 - x > 0 \quad \text{Subtract } x \text{ from both sides of the inequality.}$$

This inequality is equivalent to the one we wish to solve.

STEP 2 Determine the real zeros (x-intercepts of the graph) of f.

Find the zeros of $f(x) = x^4 - x$ by solving $x^4 - x = 0$.

$$x^4 - x = 0$$

$$x(x^3 - 1) = 0 \quad \text{Factor out } x.$$

$$x(x - 1)(x^2 + x + 1) = 0 \quad \text{Factor the difference of two cubes.}$$

$$x = 0 \quad \text{or} \quad x - 1 = 0 \quad \text{or} \quad x^2 + x + 1 = 0 \quad \text{Set each factor to zero and solve.}$$

$$x = 0 \quad \text{or} \quad x = 1$$

The equation $x^2 + x + 1 = 0$ has no real solutions. Do you see why?

STEP 3 Use the zeros found in Step 2 to divide the real number line into intervals.

We use the zeros to separate the real number line into three intervals:

$$(-\infty, 0) \qquad (0, 1) \qquad (1, \infty)$$

STEP 4 Select a number in each interval, evaluate f at the number, and determine whether f is positive or negative. If f is positive, all values of x in the interval are positive. If f is negative, all values of x in the interval are negative.

Note: If the inequality is not strict ($\le$ or $\ge$), include the solutions of $f(x) = 0$ in the solution set.

We select a test number in each interval found in Step 3 and evaluate $f(x) = x^4 - x$ at each number to determine if $f(x)$ is positive or negative. See Table 14.

Table 14

	0	1	→ x
Interval	$(-\infty, 0)$	$(0, 1)$	$(1, \infty)$
Number chosen	-1	$\frac{1}{2}$	2
Value of f	$f(-1) = 2$	$f\left(\frac{1}{2}\right) = -\frac{7}{16}$	$f(2) = 14$
Conclusion	Positive	Negative	Positive

Since we want to know where $f(x)$ is positive, we conclude that $f(x) > 0$ for all numbers x for which $x < 0$ or $x > 1$. Because the original inequality is strict, numbers x that satisfy the equation $x^4 = x$ are not solutions. The solution set of the inequality $x^4 > x$ is $\{x \mid x < 0 \text{ or } x > 1\}$ or, using interval notation, $(-\infty, 0) \cup (1, \infty)$.

Figure 42 shows the graph of the solution set.

Figure 42

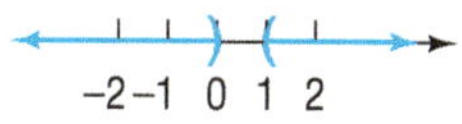

Check: Graph $Y_1 = x^4$ and $Y_2 = x$ on the same screen. Use INTERSECT to find where $Y_1 = Y_2$. Then determine where $Y_1 > Y_2$.

Now Work PROBLEM 21

2 Solve Rational Inequalities Algebraically and Graphically

Just as we presented a graphical approach to help understand the algebraic procedure for solving inequalities involving polynomials, we present a graphical approach to help understand the algebraic procedure for solving inequalities involving rational expressions.

EXAMPLE 3 Solving a Rational Inequality Using Its Graph

Solve $\frac{x-1}{x^2-4} \geq 0$ by graphing $R(x) = \frac{x-1}{x^2-4}$.

By Hand Graphical Solution

We graph $R(x) = \frac{x-1}{x^2-4}$ and determine the intervals of x such that the graph is above or on the x-axis. Do you see why? These values of x result in $R(x)$ being positive or zero. We graphed $R(x) = \frac{x-1}{x^2-4}$ in Example 1 from Section 5.3 (p. 359). We reproduce the graph in Figure 43.

Figure 43

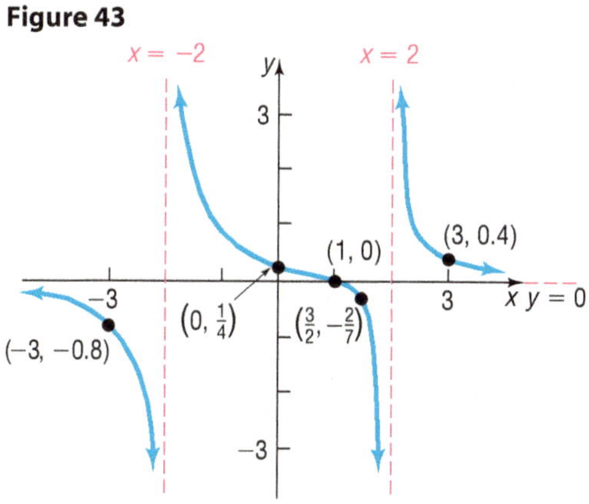

From the graph, we can see that $R(x) \geq 0$ for $-2 < x \leq 1$ or $x > 2$. The solution set is $\{x \mid -2 < x \leq 1 \text{ or } x > 2\}$ or, using interval notation, $(-2, 1] \cup (2, \infty)$.

Graphing Utility Solution

We graph $Y_1 = \frac{x-1}{x^2-4}$. See Figure 44. Using the ZERO command, we find that the x-intercept of the graph of Y_1 is 1. The graph of Y_1 is above the x-axis (and therefore R is positive) for $-2 < x < 1$ or $x > 2$. Since the inequality is not strict, we include 1 in the solution set. Therefore, the solution set is $\{x \mid -2 < x \leq 1 \text{ or } x > 2\}$ or, using interval notation, $(-2, 1] \cup (2, \infty)$.

Figure 44

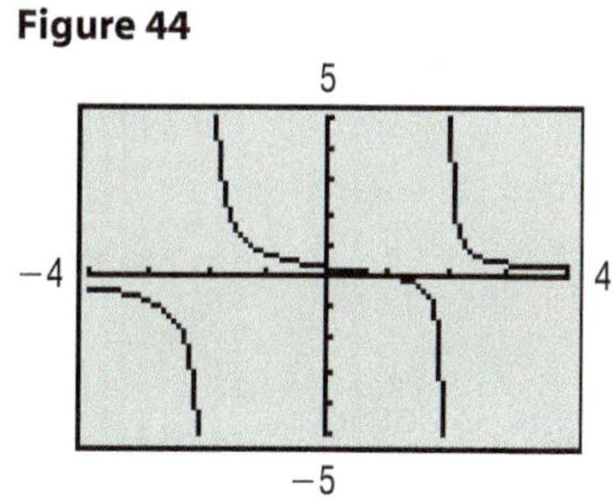

Now Work PROBLEM 33

To solve a rational inequality algebraically, we follow the same approach that we used to solve a polynomial inequality graphically. However, we must also identify the zeros of the denominator of the rational function, because the sign of a rational function may change on either side of a vertical asymptote. Convince yourself of this by looking at Figure 43. Notice the function values are negative for $x < -2$ and are positive for $x > -2$ (but less than 1).

EXAMPLE 4 How to Solve a Rational Inequality Algebraically

Solve the inequality $\frac{4x + 5}{x + 2} \geq 3$ algebraically, and graph the solution set.

Step-by-Step Solution

Step 1 Write the inequality so that a rational expression f is on the left side and zero is on the right side.

Rearrange the inequality so that 0 is on the right side.

$$\frac{4x + 5}{x + 2} \geq 3$$

$$\frac{4x + 5}{x + 2} - 3 \geq 0 \quad \text{Subtract 3 from both sides of the inequality.}$$

$$\frac{4x + 5}{x + 2} - 3 \cdot \frac{x + 2}{x + 2} \geq 0 \quad \text{Multiply 3 by } \frac{x + 2}{x + 2}$$

$$\frac{4x + 5 - 3x - 6}{x + 2} \geq 0 \quad \text{Write as a single quotient.}$$

$$\frac{x - 1}{x + 2} \geq 0 \quad \text{Combine like terms.}$$

Step 2 Determine the real zeros (x-intercepts of the graph) of f and the real numbers for which f is undefined.

The zero of $f(x) = \frac{x - 1}{x + 2}$ is 1. Also, f is undefined for $x = -2$.

Step 3 Use the zeros and undefined values found in Step 2 to divide the real number line into intervals.

We use the zeros and undefined values to separate the real number line into three intervals:

$$(-\infty, -2) \qquad (-2, 1) \qquad (1, \infty)$$

Step 4 Select a number in each interval, evaluate f at the number, and determine whether f is positive or negative. If f is positive, all values of f in the interval are positive. If f is negative, all values of f in the interval are negative.

Note: If the inequality is not strict ($\leq$ or $\geq$), include the solutions of $f(x) = 0$ in the solution set.

We select a test number in each interval found in Step 3 and evaluate $f(x) = \frac{x - 1}{x + 2}$ at each number to determine if $f(x)$ is positive or negative. See Table 15.

Table 15

(number line: −2, 1, x)			
Interval	$(-\infty, -2)$	$(-2, 1)$	$(1, \infty)$
Number chosen	-3	0	2
Value of f	$f(-3) = 4$	$f(0) = -\frac{1}{2}$	$f(2) = \frac{1}{4}$
Conclusion	Positive	Negative	Positive

Since we want to know where $f(x)$ is positive or zero, we conclude that $f(x) \geq 0$ for all numbers x for which $x < -2$ or $x \geq 1$. Notice we do not include -2 in the solution because -2 is not in the domain of f. The solution set of the inequality $\frac{4x + 5}{x + 2} \geq 3$ is $\{x \mid x < -2 \text{ or } x \geq 1\}$ or, using interval notation, $(-\infty, -2) \cup [1, \infty)$.

Figure 45 shows the graph of the solution set.

Figure 45

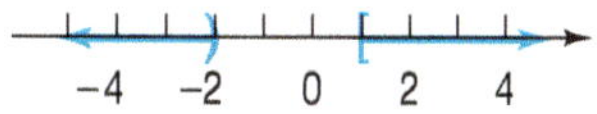

Check: Graph $Y_1 = \dfrac{4x+5}{x+2}$ and $Y_2 = 3$ on the same screen. Use INTERSECT to find where $Y_1 = Y_2$. Then determine where $Y_1 \geq Y_2$.

Now Work PROBLEM 39

Next we summarize the steps for solving a polynomial or rational inequality algebraically.

SUMMARY Steps for Solving Polynomial and Rational Inequalities Algebraically

STEP 1: Write the inequality so that a polynomial or rational expression f is on the left side and zero is on the right side in one of the following forms:

$$f(x) > 0 \qquad f(x) \geq 0 \qquad f(x) < 0 \qquad f(x) \leq 0$$

For rational expressions, be sure that the left side is written as a single quotient and find the domain of f.

STEP 2: Determine the real numbers at which the expression f equals zero and, if the expression is rational, the real numbers at which the expression f is undefined.

STEP 3: Use the numbers found in Step 2 to separate the real number line into intervals.

STEP 4: Select a number in each interval and evaluate f at the number.
(a) If the value of f is positive, then $f(x) > 0$ for all numbers x in the interval.
(b) If the value of f is negative, then $f(x) < 0$ for all numbers x in the interval.

If the inequality is not strict ($\geq$ or $\leq$), include the solutions of $f(x) = 0$ in the solution set. Be careful to exclude values of x where f is undefined.

5.4 Assess Your Understanding

'Are You Prepared?' *Answers are given at the end of these exercises. If you get a wrong answer, read the pages listed in red.*

1. Solve the inequality $3 - 4x > 5$. Graph the solution set. (pp. 153–155)

2. Solve the inequality $x^2 - 5x \leq 24$. Graph the solution set. (pp. 320–322)

Concepts and Vocabulary

3. ***True or False*** A test number for the interval $-2 < x < 5$ could be 4.

4. ***True or False*** The graph of $f(x) = \dfrac{x}{x-3}$ is above the x-axis for $x < 0$ or $x > 3$, so the solution set of the inequality $\dfrac{x}{x-3} \geq 0$ is $\{x \mid x \leq 0 \text{ or } x \geq 3\}$.

Skill Building

In Problems 5–8, use the graph of the function f to solve the inequality.

5. (a) $f(x) > 0$
(b) $f(x) \leq 0$

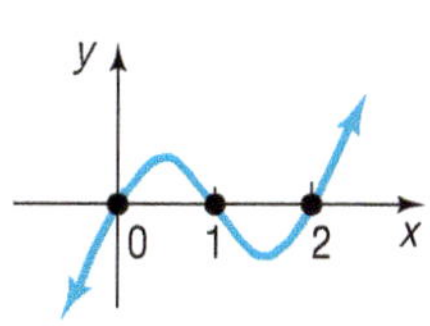

6. (a) $f(x) < 0$
(b) $f(x) \geq 0$

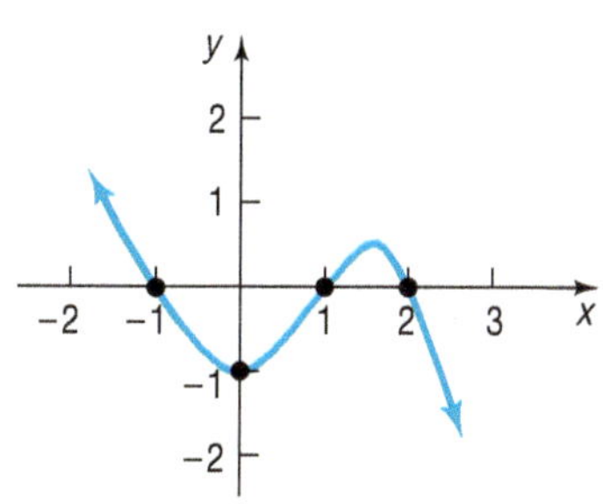

7. **(a)** $f(x) < 0$
(b) $f(x) \geq 0$

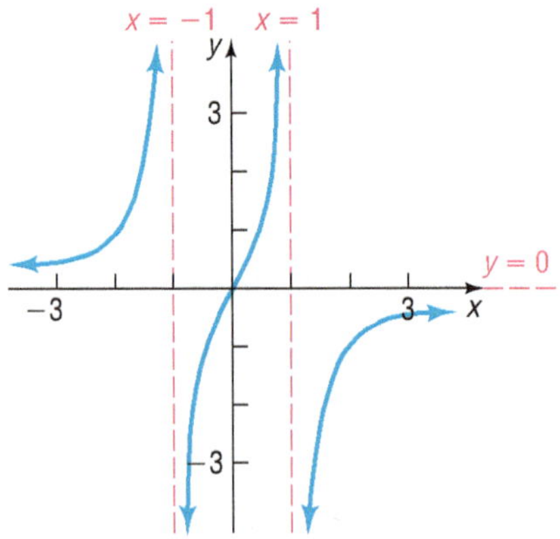

8. **(a)** $f(x) > 0$
(b) $f(x) \leq 0$

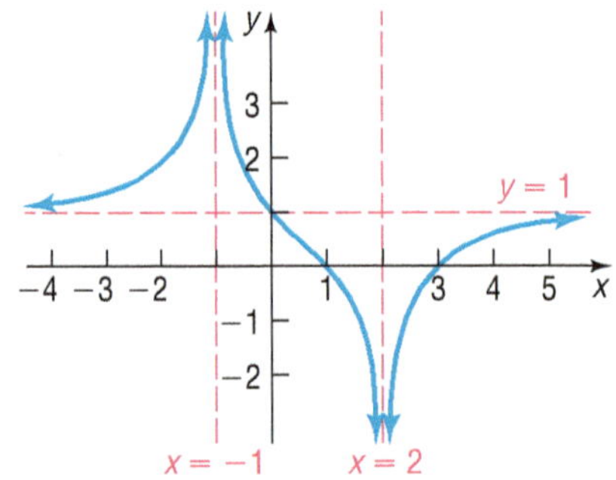

In Problems 9–14, solve the inequality by using the graph of the function.
[**Hint:** The graphs were drawn in Problems 65–70 of Section 5.1.]

9. Solve $f(x) < 0$, where $f(x) = x^2(x - 3)$.

10. Solve $f(x) \leq 0$, where $f(x) = x(x + 2)^2$.

11. Solve $f(x) \geq 0$, where $f(x) = (x + 4)(x - 2)^2$.

12. Solve $f(x) > 0$, where $f(x) = (x - 1)(x + 3)^2$.

13. Solve $f(x) \leq 0$, where $f(x) = -2(x + 2)(x - 2)^3$.

14. Solve $f(x) < 0$, where $f(x) = -\frac{1}{2}(x + 4)(x - 1)^3$.

In Problems 15–18, solve the inequality by using the graph of the function.
[**Hint:** The graphs were drawn in Problems 7–10 of Section 5.3.]

15. Solve $R(x) > 0$, where $R(x) = \dfrac{x + 1}{x(x + 4)}$.

16. Solve $R(x) < 0$, where $R(x) = \dfrac{x}{(x - 1)(x + 2)}$.

17. Solve $R(x) \leq 0$, where $R(x) = \dfrac{3x + 3}{2x + 4}$.

18. Solve $R(x) \geq 0$, where $R(x) = \dfrac{2x + 4}{x - 1}$.

In Problems 19–48, solve each inequality algebraically. Verify your results using a graphing utility.

19. $(x - 5)^2(x + 2) < 0$

20. $(x - 5)(x + 2)^2 > 0$

21. $x^3 - 4x^2 > 0$

22. $x^3 + 8x^2 < 0$

23. $2x^3 > -8x^2$

24. $3x^3 < -15x^2$

25. $(x - 1)(x - 2)(x - 3) \leq 0$

26. $(x + 1)(x + 2)(x + 3) \leq 0$

27. $x^3 - 2x^2 - 3x > 0$

28. $x^3 + 2x^2 - 3x > 0$

29. $x^4 > x^2$

30. $x^4 < 9x^2$

31. $x^4 > 1$

32. $x^3 > 1$

33. $\dfrac{x + 1}{x - 1} > 0$

34. $\dfrac{x - 3}{x + 1} > 0$

35. $\dfrac{(x - 1)(x + 1)}{x} \leq 0$

36. $\dfrac{(x - 3)(x + 2)}{x - 1} \leq 0$

37. $\dfrac{(x - 2)^2}{x^2 - 1} \geq 0$

38. $\dfrac{(x + 5)^2}{x^2 - 4} \geq 0$

39. $\dfrac{x + 4}{x - 2} \leq 1$

40. $\dfrac{x + 2}{x - 4} \geq 1$

41. $\dfrac{3x - 5}{x + 2} \leq 2$

42. $\dfrac{x - 4}{2x + 4} \geq 1$

43. $\dfrac{1}{x - 2} < \dfrac{2}{3x - 9}$

44. $\dfrac{5}{x - 3} > \dfrac{3}{x + 1}$

45. $\dfrac{x^2(3 + x)(x + 4)}{(x + 5)(x - 1)} \geq 0$

46. $\dfrac{x(x^2 + 1)(x - 2)}{(x - 1)(x + 1)} \geq 0$

47. $\dfrac{(3 - x)^3(2x + 1)}{x^3 - 1} < 0$

48. $\dfrac{(2 - x)^3(3x - 2)}{x^3 + 1} < 0$

Mixed Practice

In Problems 49–60, solve each inequality algebraically. Verify your results using a graphing utility.

49. $(x + 1)(x - 3)(x - 5) > 0$

50. $(2x - 1)(x + 2)(x + 5) < 0$

51. $7x - 4 \geq -2x^2$

52. $x^2 + 3x \geq 10$

53. $\dfrac{x + 1}{x - 3} \leq 2$

54. $\dfrac{x - 1}{x + 2} \geq -2$

55. $3(x^2 - 2) < 2(x - 1)^2 + x^2$

56. $(x - 3)(x + 2) < x^2 + 3x + 5$

57. $6x - 5 < \dfrac{6}{x}$

58. $x + \dfrac{12}{x} < 7$

59. $x^3 - 9x \leq 0$

60. $x^3 - x \geq 0$

Applications and Extensions

61. For what positive numbers will the cube of a number exceed four times its square?

62. For what positive numbers will the cube of a number be less than the number?

63. What is the domain of the function $f(x) = \sqrt{x^4 - 16}$?

64. What is the domain of the function $f(x) = \sqrt{x^3 - 3x^2}$?

65. What is the domain of the function $f(x) = \sqrt{\dfrac{x - 2}{x + 4}}$?

66. What is the domain of the function $f(x) = \sqrt{\dfrac{x - 1}{x + 4}}$?

In Problems 67–70, determine where the graph of f is below the graph of g by solving the inequality $f(x) \le g(x)$. Graph f and g together.

67. $f(x) = x^4 - 1$
$g(x) = -2x^2 + 2$

68. $f(x) = x^4 - 1$
$g(x) = x - 1$

69. $f(x) = x^4 - 4$
$g(x) = 3x^2$

70. $f(x) = x^4$
$g(x) = 2 - x^2$

71. Average Cost Suppose that the daily cost C of manufacturing bicycles is given by $C(x) = 80x + 5000$. Then the average daily cost $\overline{C}$ is given by $\overline{C}(x) = \dfrac{80x + 5000}{x}$. How many bicycles must be produced each day for the average cost to be no more than \$100?

72. Average Cost See Problem 71. Suppose that the government imposes a \$1000 per day tax on the bicycle manufacturer so that the daily cost C of manufacturing x bicycles is now given by $C(x) = 80x + 6000$. Now the average daily cost $\overline{C}$ is given by $\overline{C}(x) = \dfrac{80x + 6000}{x}$. How many bicycles must be produced each day for the average cost to be no more than \$100?

73. Bungee Jumping Originating on Pentecost Island in the Pacific, the practice of a person jumping from a high place harnessed to a flexible attachment was introduced to Western culture in 1979 by the Oxford University Dangerous Sport Club. One important parameter to know before attempting a bungee jump is the amount the cord will stretch at the bottom of the fall. The stiffness of the cord is related to the amount of stretch by the equation

$$K = \frac{2W(S + L)}{S^2}$$

where W = weight of the jumper (pounds)
K = cord's stiffness (pounds per foot)
L = free length of the cord (feet)
S = stretch (feet)

(a) A 150-pound person plans to jump off a ledge attached to a cord of length 42 feet. If the stiffness of the cord is no less than 16 pounds per foot, how much will the cord stretch?

(b) If safety requirements will not permit the jumper to get any closer than 3 feet to the ground, what is the minimum height required for the ledge in part (a)?

Source: American Institute of Physics, Physics News Update, No. 150, November 5, 1993

74. Gravitational Force According to Newton's Law of universal gravitation, the attractive force F between two bodies is given by

$$F = G\frac{m_1 m_2}{r^2}$$

where m_1, m_2 = the masses of the two bodies
r = distance between the two bodies
G = gravitational constant 6.6742×10^{-11} newtons meter^2 kilogram^{-2}

Suppose an object is traveling directly from Earth to the moon. The mass of Earth is 5.9742×10^{24} kilograms, the mass of the moon is 7.349×10^{22} kilograms, and the mean distance from Earth to the moon is 384,400 kilometers. For an object between Earth and the moon, how far from Earth is the force on the object due to the moon greater than the force on the object due to Earth?

Source: www.solarviews.com;en.wikipedia.org

75. Field Trip Mrs. West has decided to take her fifth grade class to a play. The manager of the theater agreed to discount the regular \$40 price of the ticket by \$0.20 for each ticket sold. The cost of the bus, \$500, will be split equally among each of the students. How many students must attend to keep the cost per student at or below \$40?

Discussion and Writing

76. Make up an inequality that has no solution. Make up one that has exactly one solution.

77. The inequality $x^4 + 1 < -5$ has no solution. Explain why.

78. A student attempted to solve the inequality $\dfrac{x + 4}{x - 3} \le 0$ by multiplying both sides of the inequality by $x - 3$ to get $x + 4 \le 0$. This led to a solution of $\{x \mid x \le -4\}$. Is the student correct? Explain.

79. Write a rational inequality whose solution set is $\{x \mid -3 < x \le 5\}$.

'Are You Prepared?'

1. $\left\{x \,\middle|\, x < -\dfrac{1}{2}\right\}$ or $\left(-\infty, -\dfrac{1}{2}\right)$

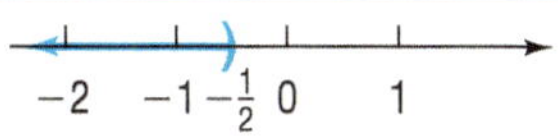

2. $\{x \mid -3 \le x \le 8\}$ or $[-3, 8]$

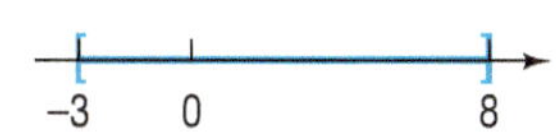

5.5 The Real Zeros of a Polynomial Function

PREPARING FOR THIS SECTION *Before getting started, review the following:*

- Classification of Numbers (Section R.1, pp. 4–5)
- Factoring Polynomials (Section R.5, pp. 49–55)
- Synthetic Division (Section R.6, pp. 57–60)
- Polynomial Division (Section R.4, pp. 44–47)
- Quadratic Formula (Section 1.3, pp. 117–120)
- Evaluate Functions (Section 3.1, pp. 216–218)

Now Work the 'Are You Prepared?' problems on page 387.

OBJECTIVES
1. Use the Remainder and Factor Theorems (p. 376)
2. Use the Rational Zeros Theorem to List the Potential Rational Zeros of a Polynomial Function (p. 379)
3. Find the Real Zeros of a Polynomial Function (p. 380)
4. Solve Polynomial Equations (p. 382)
5. Use the Theorem for Bounds on Zeros (p. 383)
6. Use the Intermediate Value Theorem (p. 386)

In Section 5.1, we were able to identify the real zeros of a polynomial function because either the polynomial function was in factored form or it could be easily factored. But how do we find the real zeros of a polynomial function if it is not factored or cannot be easily factored?

Recall that if r is a real zero of a polynomial function f then $f(r) = 0$, r is an x-intercept of the graph of f, $x - r$ is a factor of f, and r is a solution of the equation $f(x) = 0$. For example, if $x - 4$ is a factor of f, then 4 is a real zero of f and 4 is a solution to the equation $f(x) = 0$. For polynomial and rational functions, we have seen the importance of the real zeros for graphing. In most cases, however, the real zeros of a polynomial function are difficult to find using algebraic methods. No nice formulas like the quadratic formula are available to help us find zeros for polynomials of degree 3 or higher. Formulas do exist for solving any third- or fourth-degree polynomial equation, but they are somewhat complicated. No general formulas exist for polynomial equations of degree 5 or higher. Refer to the Historical Feature at the end of this section for more information.

1 Use the Remainder and Factor Theorems

When we divide one polynomial (the dividend) by another (the divisor), we obtain a quotient polynomial and a remainder, the remainder being either the zero polynomial or a polynomial whose degree is less than the degree of the divisor. To check our work, we verify that

$$(\text{Quotient})(\text{Divisor}) + \text{Remainder} = \text{Dividend}$$

This checking routine is the basis for a famous theorem called the **division algorithm* for polynomials,** which we now state without proof.

THEOREM

Division Algorithm for Polynomials

If $f(x)$ and $g(x)$ denote polynomial functions and if $g(x)$ is a polynomial whose degree is greater than zero, then there are unique polynomial functions $q(x)$ and $r(x)$ such that

$$\frac{f(x)}{g(x)} = q(x) + \frac{r(x)}{g(x)} \quad \text{or} \quad \underset{\text{dividend}}{f(x)} = \underset{\text{quotient}}{q(x)}\underset{\text{divisor}}{g(x)} + \underset{\text{remainder}}{r(x)} \qquad \textbf{(1)}$$

where $r(x)$ is either the zero polynomial or a polynomial of degree less than that of $g(x)$.

*A systematic process in which certain steps are repeated a finite number of times is called an **algorithm.** For example, long division is an algorithm.

In equation (1), $f(x)$ is the **dividend,** $g(x)$ is the **divisor,** $q(x)$ is the **quotient,** and $r(x)$ is the **remainder.**

If the divisor $g(x)$ is a first-degree polynomial of the form

$$g(x) = x - c \qquad c \text{ a real number}$$

then the remainder $r(x)$ is either the zero polynomial or a polynomial of degree 0. As a result, for such divisors, the remainder is some number, say R, and we may write

$$f(x) = (x - c)q(x) + R \qquad \textbf{(2)}$$

This equation is an identity in x and is true for all real numbers x. Suppose that $x = c$. Then equation (2) becomes

$$f(c) = (c - c)q(c) + R$$
$$f(c) = R$$

Substitute $f(c)$ for R in equation (2) to obtain

$$f(x) = (x - c)q(x) + f(c) \qquad \textbf{(3)}$$

We have now proved the **Remainder Theorem.**

REMAINDER THEOREM

Let f be a polynomial function. If $f(x)$ is divided by $x - c$, then the remainder is $f(c)$.

EXAMPLE 1 Using the Remainder Theorem

Find the remainder if $f(x) = x^3 - 4x^2 - 5$ is divided by

(a) $x - 3$ (b) $x + 2$

Solution

(a) We could use long division or synthetic division, but it is easier to use the Remainder Theorem, which says that the remainder is $f(3)$.

$$f(3) = (3)^3 - 4(3)^2 - 5 = 27 - 36 - 5 = -14$$

The remainder is -14.

(b) To find the remainder when $f(x)$ is divided by $x + 2 = x - (-2)$, we evaluate $f(-2)$.

$$f(-2) = (-2)^3 - 4(-2)^2 - 5 = -8 - 16 - 5 = -29$$

The remainder is -29.

Figure 46

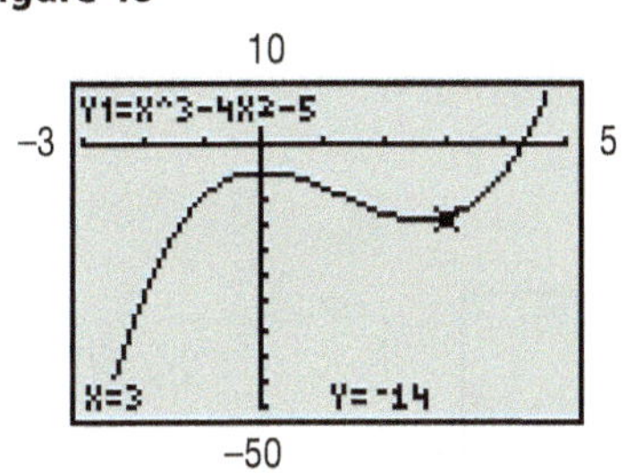

Compare the method used in Example 1(a) with the method used in Example 1 on page 59 (synthetic division). Which method do you prefer? Give reasons.

NOTE A graphing utility provides another way to find the value of a function, using the eVALUEate feature. Consult your manual for details. See Figure 46 for the result of Example 1(a).

An important and useful consequence of the Remainder Theorem is the **Factor Theorem.**

FACTOR THEOREM

Let f be a polynomial function. Then $x - c$ is a factor of $f(x)$ if and only if $f(c) = 0$.

The Factor Theorem actually consists of two separate statements:

1. If $f(c) = 0$, then $x - c$ is a factor of $f(x)$.
2. If $x - c$ is a factor of $f(x)$, then $f(c) = 0$.

The proof requires two parts.

Proof

1. Suppose that $f(c) = 0$. Then, by equation (3), we have

$$\begin{aligned} f(x) &= (x - c)q(x) + f(c) \\ &= (x - c)q(x) + 0 \\ &= (x - c)q(x) \end{aligned}$$

for some polynomial $q(x)$. That is, $x - c$ is a factor of $f(x)$.

2. Suppose that $x - c$ is a factor of $f(x)$. Then there is a polynomial function q such that

$$f(x) = (x - c)q(x)$$

Replacing x by c, we find that

$$f(c) = (c - c)q(c) = 0 \cdot q(c) = 0$$

This completes the proof. ■

One use of the Factor Theorem is to determine whether a polynomial has a particular factor.

EXAMPLE 2 Using the Factor Theorem

Use the Factor Theorem to determine whether the function

$$f(x) = 2x^3 - x^2 + 2x - 3$$

has the factor

(a) $x - 1$ (b) $x + 2$

Solution The Factor Theorem states that if $f(c) = 0$ then $x - c$ is a factor.

(a) Because $x - 1$ is of the form $x - c$ with $c = 1$, we find the value of $f(1)$. We choose to use substitution.

$$f(1) = 2(1)^3 - (1)^2 + 2(1) - 3 = 2 - 1 + 2 - 3 = 0$$

See also Figure 47(a). By the Factor Theorem, $x - 1$ is a factor of $f(x)$.

(b) We first need to write $x + 2$ in the form $x - c$. Since $x + 2 = x - (-2)$, we find the value of $f(-2)$. See Figure 47(b). Because $f(-2) = -27 \neq 0$, we conclude from the Factor Theorem that $x - (-2) = x + 2$ is not a factor of $f(x)$.

Figure 47

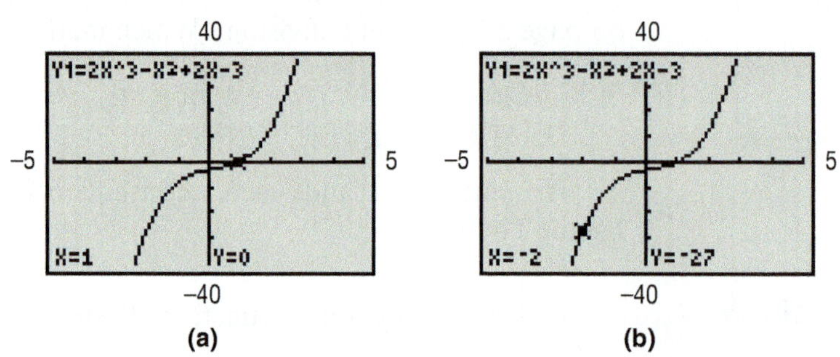

In Example 2, we found that $x - 1$ was a factor of f. To write f in factored form, we can use long division or synthetic division. Using synthetic division, we find that

$$\begin{array}{r|rrrr} 1 & 2 & -1 & 2 & -3 \\ & & 2 & 1 & 3 \\ \hline & 2 & 1 & 3 & 0 \end{array}$$

The quotient is $q(x) = 2x^2 + x + 3$ with a remainder of 0, as expected. We can write f in factored form as

$$f(x) = 2x^3 - x^2 + 2x - 3 = (x - 1)(2x^2 + x + 3)$$

Now Work PROBLEM 11

The next theorem concerns the number of real zeros that a polynomial function may have. In counting the zeros of a polynomial, we count each zero as many times as its multiplicity.

THEOREM

Number of Real Zeros

A polynomial function of degree n, $n \geq 1$, has at most n real zeros.

Proof The proof is based on the Factor Theorem. If r is a zero of a polynomial function f, then $f(r) = 0$ and, hence, $x - r$ is a factor of $f(x)$. Each zero corresponds to a factor of degree 1. Because f cannot have more first-degree factors than its degree, the result follows. ■

2 Use the Rational Zeros Theorem to List the Potential Rational Zeros of a Polynomial Function

The next result, called the **Rational Zeros Theorem,** provides information about the rational zeros of a polynomial *with integer coefficients*.

THEOREM

Rational Zeros Theorem

Let f be a polynomial function of degree 1 or higher of the form

$$f(x) = a_n x^n + a_{n-1}x^{n-1} + \cdots + a_1 x + a_0 \qquad a_n \neq 0, a_0 \neq 0$$

where each coefficient is an integer. If $\frac{p}{q}$, in lowest terms, is a rational zero of f, then p must be a factor of a_0, and q must be a factor of a_n.

EXAMPLE 3 **Listing Potential Rational Zeros**

List the potential rational zeros of

$$f(x) = 2x^3 + 11x^2 - 7x - 6$$

Solution Because f has integer coefficients, we may use the Rational Zeros Theorem. First, we list all the integers p that are factors of $a_0 = -6$ and all the integers q that are factors of the leading coefficient $a_3 = 2$.

$$p: \quad \pm 1, \pm 2, \pm 3, \pm 6$$
$$q: \quad \pm 1, \pm 2$$

Now we form all possible quotients $\frac{p}{q}$.

$$\frac{p}{q}: \quad \pm 1, \pm 2, \pm 3, \pm 6, \pm\frac{1}{2}, \pm\frac{3}{2}$$

If f has a rational zero, it will be found in this list, which contains 12 possibilities.

Now Work PROBLEM 21

In Words

For the polynomial function

$f(x) = 2x^3 + 11x^2 - 7x - 6$

we know 5 is not a zero, because 5 is not in the list of potential rational zeros. However, -1 may or may not be a zero.

Be sure that you understand what the Rational Zeros Theorem says: For a polynomial with integer coefficients, *if* there is a rational zero, it is one of those listed. It may be the case that the function does not have any rational zeros.

The Rational Zeros Theorem provides a list of potential rational zeros of a function f. If we graph f, we can get a better sense of the location of the x-intercepts and test to see if they are rational. We can also use the potential rational zeros to select our initial viewing window to graph f and then adjust the window based on the results. The graphs shown throughout the text will be those obtained after setting the final viewing window.

3 Find the Real Zeros of a Polynomial Function

EXAMPLE 4 How to Find the Real Zeros of a Polynomial Function

Find the real zeros of the polynomial function $f(x) = 2x^3 + 11x^2 - 7x - 6$. Write f in factored form.

Step-by-Step Solution

STEP 1 Use the degree of the polynomial to determine the maximum number of zeros.

Since f is a polynomial of degree 3, there are at most three real zeros.

STEP 2 If the polynomial has integer coefficients, use the Rational Zeros Theorem to identify those rational numbers that potentially can be zeros.

We list the potential rational zeros obtained in Example 3:

$$\pm 1, \pm 2, \pm 3, \pm 6, \pm\frac{1}{2}, \pm\frac{3}{2}$$

STEP 3 Using a graphing utility, graph the polynomial function.

Figure 48 shows the graph of f. We see that f has three zeros: one near -6, one between -1 and 0, and one near 1.

Figure 48

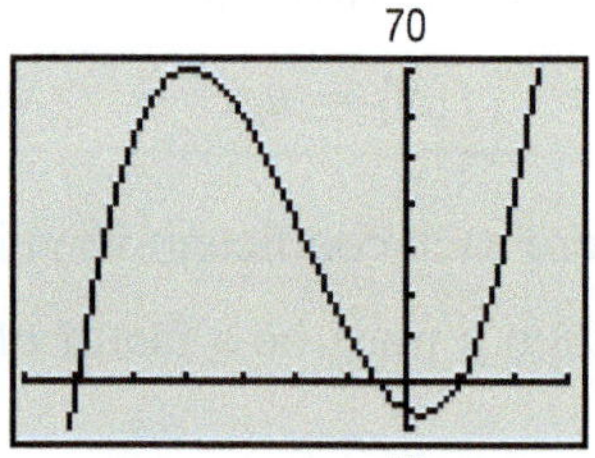

STEP 4 Use the Factor Theorem to determine if the potential rational zero is a zero. If it is, use synthetic division or long division to factor the polynomial function. Repeat Step 4 until all the zeros of the polynomial function have been identified and the polynomial function is completely factored.

From our list of potential rational zeros, we will test -6 to determine if it is a zero of f. Because

$$\begin{aligned} f(-6) &= 2(-6)^3 + 11(-6)^2 - 7(-6) - 6 \\ &= 2(-216) + 11(36) + 42 - 6 \\ &= -432 + 396 + 36 \\ &= 0 \end{aligned}$$

we know that -6 is a zero and $x - (-6) = x + 6$ is a factor of f. We can use long division or synthetic division to factor f. (We will not show the division here, but you are encouraged to verify the results shown.) After dividing f by $x + 6$, we find that the quotient is $2x^2 - x - 1$, so

$$\begin{aligned} f(x) &= 2x^3 + 11x^2 - 7x - 6 \\ &= (x + 6)(2x^2 - x - 1) \end{aligned}$$

Now any solution of the equation $2x^2 - x - 1 = 0$ will be a zero of f. Because of this, we call the equation $2x^2 - x - 1 = 0$ a **depressed equation** of f. Because any zero of f is a solution to the equation $2x^2 - x - 1 = 0$, we work with the depressed equation to find the zeros of f.

The depressed equation $2x^2 - x - 1 = 0$ is a quadratic equation with discriminant $b^2 - 4ac = (-1)^2 - 4(2)(-1) = 9 > 0$. The equation has two real solutions, which can be found by factoring.

$$2x^2 - x - 1 = (2x + 1)(x - 1) = 0$$
$$2x + 1 = 0 \quad \text{or} \quad x - 1 = 0$$
$$x = -\frac{1}{2} \quad \text{or} \quad x = 1$$

The zeros of f are -6, $-\frac{1}{2}$, and 1.

We completely factor f as follows:

$$f(x) = 2x^3 + 11x^2 - 7x - 6 = (x + 6)(2x^2 - x - 1) = (x + 6)(2x + 1)(x - 1)$$

Notice that the three zeros of f are in the list of potential rational zeros in Step 2.

We summarize the steps used in Example 4 next.

SUMMARY Steps for Finding the Real Zeros of a Polynomial Function

Step 1: Use the degree of the polynomial to determine the maximum number of zeros.

Step 2: If the polynomial has integer coefficients, use the Rational Zeros Theorem to identify those rational numbers that potentially can be zeros.

Step 3: Using a graphing utility, graph the polynomial function.

Step 4: Use the Factor Theorem to determine if the potential rational zero is a zero. If it is, use synthetic division or long division to factor the polynomial function. Each time that a zero (and thus a factor) is found, repeat Step 4 on the depressed equation. In attempting to find the zeros, remember to use (if possible) the factoring techniques that you already know (special products, factoring by grouping, and so on).

EXAMPLE 5 Finding the Real Zeros of a Polynomial Function

Find the real zeros of $f(x) = x^5 - x^4 - 4x^3 + 8x^2 - 32x + 48$. Write f in factored form.

Solution

Step 1: There are at most five real zeros.

Step 2: To obtain the list of potential rational zeros, we write the factors p of $a_0 = 48$ and the factors q of the leading coefficient $a_5 = 1$.

$$p: \quad \pm1, \pm2, \pm3, \pm4, \pm6, \pm8, \pm12, \pm16, \pm24, \pm48$$
$$q: \quad \pm1$$

The potential rational zeros consist of all possible quotients $\frac{p}{q}$:

$$\frac{p}{q}: \quad \pm1, \pm2, \pm3, \pm4, \pm6, \pm8, \pm12, \pm16, \pm24, \pm48$$

Figure 49

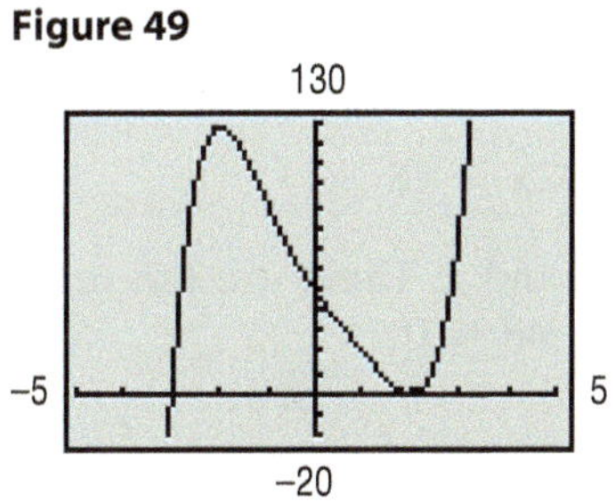

Step 3: Figure 49 shows the graph of f. The graph has the characteristics that we expect of the given polynomial of degree 5: no more than four turning points, y-intercept 48, and it behaves like $y = x^5$ for large $|x|$.

STEP 4: Since -3 appears to be a zero and -3 is a potential rational zero, we evaluate $f(-3)$ and find that $f(-3) = 0$. By the Factor Theorem, $x + 3$ is a factor of f. We use synthetic division to factor f.

$$\begin{array}{r|rrrrrr} -3 & 1 & -1 & -4 & 8 & -32 & 48 \\ & & -3 & 12 & -24 & 48 & -48 \\ \hline & 1 & -4 & 8 & -16 & 16 & 0 \end{array}$$

We can factor f as

$$f(x) = x^5 - x^4 - 4x^3 + 8x^2 - 32x + 48 = (x + 3)(x^4 - 4x^3 + 8x^2 - 16x + 16)$$

We now work with the first depressed equation:

$$q_1(x) = x^4 - 4x^3 + 8x^2 - 16x + 16 = 0$$

Repeat Step 4: In looking back at Figure 49, it appears that 2 might be a zero of even multiplicity. We check the potential rational zero 2 and find that $f(2) = 0$. Using synthetic division,

$$\begin{array}{r|rrrrr} 2 & 1 & -4 & 8 & -16 & 16 \\ & & 2 & -4 & 8 & -16 \\ \hline & 1 & -2 & 4 & -8 & 0 \end{array}$$

Now we can factor f as

$$f(x) = (x + 3)(x - 2)(x^3 - 2x^2 + 4x - 8)$$

Repeat Step 4: The depressed equation $q_2(x) = x^3 - 2x^2 + 4x - 8 = 0$ can be factored by grouping.

$$\begin{aligned} x^3 - 2x^2 + 4x - 8 = (x^3 - 2x^2) + (4x - 8) &= x^2(x - 2) + 4(x - 2) \\ &= (x - 2)(x^2 + 4) = 0 \\ x - 2 = 0 \quad &\text{or} \quad x^2 + 4 = 0 \\ x = 2 \quad & \end{aligned}$$

Since $x^2 + 4 = 0$ has no real solutions, the real zeros of f are -3 and 2, with 2 being a zero of multiplicity 2. The factored form of f is

$$\begin{aligned} f(x) &= x^5 - x^4 - 4x^3 + 8x^2 - 32x + 48 \\ &= (x + 3)(x - 2)^2(x^2 + 4) \end{aligned}$$

Now Work PROBLEM 39

4 Solve Polynomial Equations

EXAMPLE 6 Solving a Polynomial Equation

Solve the equation: $x^5 - x^4 - 4x^3 + 8x^2 - 32x + 48 = 0$

Solution The solutions of this equation are the zeros of the polynomial function

$$f(x) = x^5 - x^4 - 4x^3 + 8x^2 - 32x + 48$$

Using the result of Example 5, the real zeros are -3 and 2. These are the real solutions of the equation $x^5 - x^4 - 4x^3 + 8x^2 - 32x + 48 = 0$.

Now Work PROBLEM 63

In Example 5, the quadratic factor $x^2 + 4$ that appears in the factored form of $f(x)$ is called *irreducible,* because the polynomial $x^2 + 4$ cannot be factored over the real numbers. In general, we say that a quadratic factor $ax^2 + bx + c$ is **irreducible** if it cannot be factored over the real numbers, that is, if it is prime over the real numbers.

Refer back to Examples 4 and 5. The polynomial function of Example 4 has three real zeros, and its factored form contains three linear factors. The polynomial function of Example 5 has two distinct real zeros, and its factored form contains two distinct linear factors and one irreducible quadratic factor.

THEOREM

Every polynomial function (with real coefficients) can be uniquely factored into a product of linear factors and/or irreducible quadratic factors.

We shall prove this result in Section 5.6, and, in fact, we shall draw several additional conclusions about the zeros of a polynomial function. One conclusion is worth noting now. If a polynomial (with real coefficients) is of odd degree, then it must contain at least one linear factor. (Do you see why?) This means that it must have at least one real zero.

COROLLARY

A polynomial function (with real coefficients) of odd degree has at least one real zero.

5 Use the Theorem for Bounds on Zeros

One challenge in using a graphing utility is to set the viewing window so that a complete graph is obtained. The next theorem is a tool that can be used to find bounds on the zeros. This will assure that the function does not have any zeros outside these bounds. Then using these bounds to set *X*min and *X*max assures that all the x-intercepts appear in the viewing window.

A positive number M is a **bound** on the zeros of a polynomial if every zero r lies between $-M$ and M, inclusive. That is, M is a bound to the zeros of a polynomial f if

$$-M \leq \text{any zero of } f \leq M$$

THEOREM

Bounds on Zeros

Let f denote a polynomial function whose leading coefficient is 1.

$$f(x) = x^n + a_{n-1}x^{n-1} + \cdots + a_1x + a_0$$

A bound M on the zeros of f is the smaller of the two numbers

$$\text{Max}\{1, |a_0| + |a_1| + \cdots + |a_{n-1}|\}, \quad 1 + \text{Max}\{|a_0|, |a_1|, \ldots, |a_{n-1}|\} \qquad \textbf{(4)}$$

where Max { } means "choose the largest entry in { }."

An example will help to make the theorem clear.

EXAMPLE 7 **Using the Theorem for Finding Bounds on Zeros**

Find a bound to the zeros of each polynomial.

(a) $f(x) = x^5 + 3x^3 - 9x^2 + 5$ (b) $g(x) = 4x^5 - 2x^3 + 2x^2 + 1$

Solution (a) The leading coefficient of f is 1.

$$f(x) = x^5 + 3x^3 - 9x^2 + 5 \quad a_4 = 0, a_3 = 3, a_2 = -9, a_1 = 0, a_0 = 5$$

We evaluate the expressions in formula (4).

$$\begin{aligned}\text{Max}\{1, |a_0| + |a_1| + \cdots + |a_{n-1}|\} &= \text{Max}\{1, |5| + |0| + |-9| + |3| + |0|\}\\ &= \text{Max}\{1, 17\} = 17\\ 1 + \text{Max}\{|a_0|, |a_1|, \ldots, |a_{n-1}|\} &= 1 + \text{Max}\{|5|, |0|, |-9|, |3|, |0|\}\\ &= 1 + 9 = 10\end{aligned}$$

The smaller of the two numbers, 10, is the bound. Every zero of f lies between -10 and 10.

(b) First we write g so that its leading coefficient is 1.

$$g(x) = 4x^5 - 2x^3 + 2x^2 + 1 = 4\left(x^5 - \frac{1}{2}x^3 + \frac{1}{2}x^2 + \frac{1}{4}\right)$$

Next we evaluate the two expressions in formula (4) with $a_4 = 0$, $a_3 = -\frac{1}{2}$, $a_2 = \frac{1}{2}$, $a_1 = 0$, and $a_0 = \frac{1}{4}$.

$$\begin{aligned}\text{Max}\{1, |a_0| + |a_1| + \cdots + |a_{n-1}|\} &= \text{Max}\left\{1, \left|\frac{1}{4}\right| + |0| + \left|\frac{1}{2}\right| + \left|-\frac{1}{2}\right| + |0|\right\}\\ &= \text{Max}\left\{1, \frac{5}{4}\right\} = \frac{5}{4}\\ 1 + \text{Max}\{|a_0|, |a_1|, \ldots, |a_{n-1}|\} &= 1 + \text{Max}\left\{\left|\frac{1}{4}\right|, |0|, \left|\frac{1}{2}\right|, \left|-\frac{1}{2}\right|, |0|\right\}\\ &= 1 + \frac{1}{2} = \frac{3}{2}\end{aligned}$$

The smaller of the two numbers, $\frac{5}{4}$, is the bound. Every zero of g lies between $-\frac{5}{4}$ and $\frac{5}{4}$.

EXAMPLE 8 **Obtaining Graphs Using Bounds on Zeros**

Obtain a graph for each polynomial.

(a) $f(x) = x^5 + 3x^3 - 9x^2 + 5$ (b) $g(x) = 4x^5 - 2x^3 + 2x^2 + 1$

Solution (a) Based on Example 7(a), every zero lies between -10 and 10. Using $X\text{min} = -10$ and $X\text{max} = 10$, we graph $Y_1 = f(x) = x^5 + 3x^3 - 9x^2 + 5$. Figure 50(a) shows the graph obtained using ZOOM-FIT. Figure 50(b) shows the graph after adjusting the viewing window to improve the graph.

Figure 50

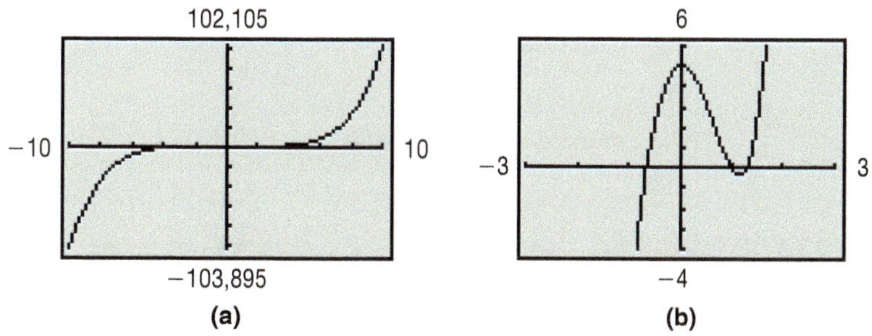

Figure 51

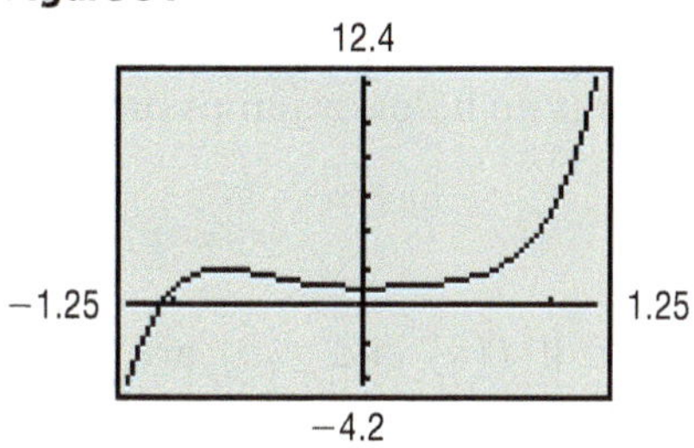

(b) Based on Example 7(b), every zero lies between $-\frac{5}{4}$ and $\frac{5}{4}$. Using $X\text{min} = -\frac{5}{4}$ and $X\text{max} = \frac{5}{4}$, we graph $Y_1 = g(x) = 4x^5 - 2x^3 + 2x^2 + 1$. Figure 51 shows the graph after using ZOOM-FIT. Here no adjustment of the viewing window is needed.

Now Work PROBLEM 33

The next example shows how to proceed when some of the coefficients of the polynomial are not integers.

EXAMPLE 9 Finding the Real Zeros of a Polynomial

Find all the real zeros of the polynomial function

$$f(x) = x^5 - 1.8x^4 - 17.79x^3 + 31.672x^2 + 37.95x - 8.7121$$

Round answers to two decimal places.

Solution

STEP 1: There are at most five real zeros.

STEP 2: Since there are noninteger coefficients, the Rational Zeros Theorem does not apply.

STEP 3: We determine the bounds of f. The leading coefficient of f is 1 with $a_4 = -1.8$, $a_3 = -17.79$, $a_2 = 31.672$, $a_1 = 37.95$, and $a_0 = -8.7121$. We evaluate the expressions using formula (4).

Figure 52

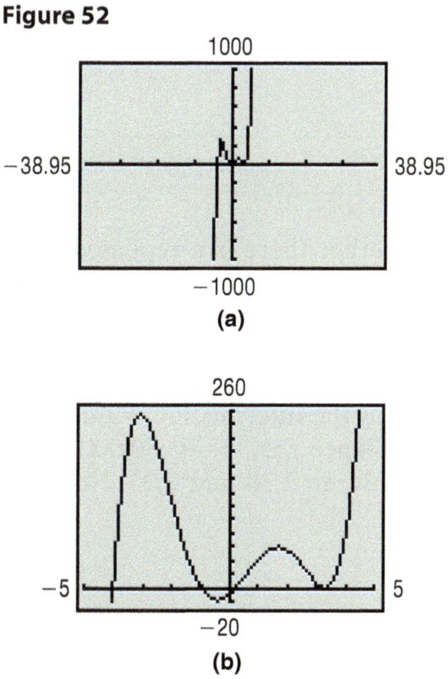

$$\begin{aligned}\text{Max}\{1, |-8.7121| + |37.95| + |31.672| + |-17.79| + |-1.8|\} &= \text{Max}\{1, 97.9241\}\\ &= 97.9241\end{aligned}$$

$$\begin{aligned}1 + \text{Max}\{|-8.7121|, |37.95|, |31.672|, |-17.79|, |-1.8|\} &= 1 + 37.95\\ &= 38.95\end{aligned}$$

The smaller of the two numbers, 38.95, is the bound. Every real zero of f lies between -38.95 and 38.95. Figure 52(a) shows the graph of f with $X\text{min} = -38.95$ and $X\text{max} = 38.95$. Figure 52(b) shows a graph of f after adjusting the viewing window to improve the graph.

STEP 4: From Figure 52(b), we see that f appears to have four x-intercepts: one near -4, one near -1, one between 0 and 1, and one near 3. The x-intercept near 3 might be a zero of even multiplicity since the graph seems to touch the x-axis at that point.

We use the Factor Theorem to determine if -4 and -1 are zeros. We find that $f(-4) = f(-1) = 0$. As a result -4 and -1 are zeros. Using ZERO (or ROOT), we find that the remaining zeros are 0.20 and 3.30, rounded to two decimal places.

There are no real zeros on the graph that have not already been identified. So either 3.30 is a zero of multiplicity 2 or there are two distinct zeros, each of which is 3.30, rounded to two decimal places. (Example 10 will explain how to determine which is true.)

Now Work PROBLEM 57

6 Use the Intermediate Value Theorem

The Intermediate Value Theorem requires that the function be *continuous*. Although calculus is needed to explain the meaning precisely, we have already said that, very basically, a function f is continuous when its graph can be drawn without lifting pencil from paper, that is, when the graph contains no holes or jumps or gaps. Every polynomial function is continuous.

INTERMEDIATE VALUE THEOREM

Let f denote a continuous function. If $a < b$ and if $f(a)$ and $f(b)$ are of opposite sign, then f has at least one zero between a and b.

Although the proof of this result requires advanced methods in calculus, it is easy to "see" why the result is true. Look at Figure 53.

Figure 53
If $f(a) < 0$ and $f(b) > 0$ and if f is continuous, there is at least one zero between a and b.

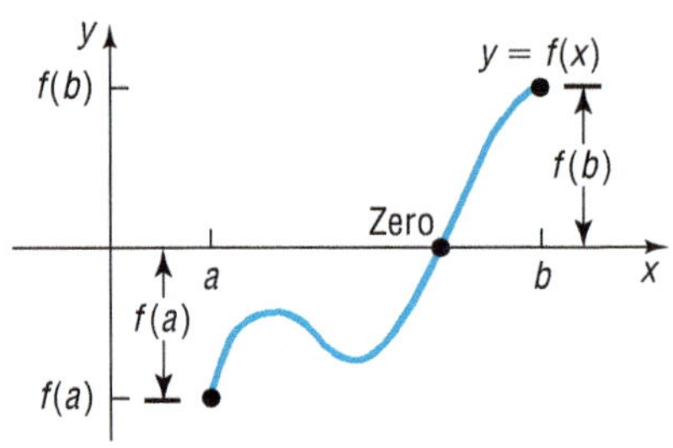

The Intermediate Value Theorem together with the TABLE feature of a graphing utility provides a basis for finding zeros.

EXAMPLE 10 Using the Intermediate Value Theorem and a Graphing Utility to Locate Zeros

Continue working with Example 9 to determine whether there is a repeated zero or two distinct zeros near 3.30.

Solution We use the TABLE feature of a graphing utility. See Table 16. Since $f(3.29) = 0.00956 > 0$ and $f(3.30) = -0.0001 < 0$, by the Intermediate Value Theorem there is a zero between 3.29 and 3.30. Similarly, since $f(3.30) = -0.0001 < 0$ and $f(3.31) = 0.0097 > 0$, there is another zero between 3.30 and 3.31. Now we know that the five zeros of f are distinct.

Table 16

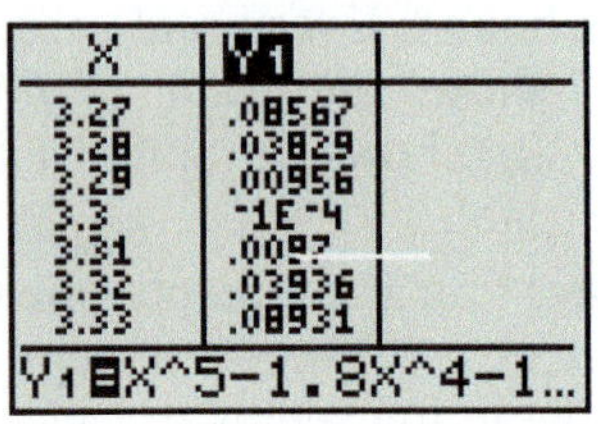

X	Y1
3.27	.08567
3.28	.03829
3.29	.00956
3.3	-1E-4
3.31	.0097
3.32	.03936
3.33	.08931

Y1=X^5-1.8X^4-1…

Now Work PROBLEM 75

Historical Feature

Formulas for the solution of third- and fourth-degree polynomial equations exist, and, while not very practical, they do have an interesting history.

In the 1500s in Italy, mathematical contests were a popular pastime, and persons possessing methods for solving problems kept them secret. (Solutions that were published were already common knowledge.) Niccolo of Brescia (1499–1557), commonly referred to as Tartaglia ("the stammerer"), had the secret for solving cubic (third-degree) equations, which gave him a decided advantage in the contests. Girolamo Cardano (1501–1576) found out that Tartaglia had the secret, and, being interested in cubics, he requested it from Tartaglia. The reluctant Tartaglia hesitated for some time, but finally, swearing Cardano to secrecy with midnight oaths by candlelight, told him the secret. Cardano then published the solution in his book *Ars Magna* (1545), giving Tartaglia the credit but rather compromising the secrecy. Tartaglia exploded into bitter recriminations, and each wrote pamphlets that reflected on the other's mathematics, moral character, and ancestry.

The quartic (fourth-degree) equation was solved by Cardano's student Lodovico Ferrari, and this solution also was included, with credit and this time with permission, in the *Ars Magna*.

Attempts were made to solve the fifth-degree equation in similar ways, all of which failed. In the early 1800s, P. Ruffini, Niels Abel, and Evariste Galois all found ways to show that it is not possible to solve fifth-degree equations by formula, but the proofs required the introduction of new methods. Galois's methods eventually developed into a large part of modern algebra.

Historical Problems

Problems 1–8 develop the Tartaglia–Cardano solution of the cubic equation and show why it is not altogether practical.

1. Show that the general cubic equation $y^3 + by^2 + cy + d = 0$ can be transformed into an equation of the form $x^3 + px + q = 0$ by using the substitution $y = x - \frac{b}{3}$.
2. In the equation $x^3 + px + q = 0$, replace x by $H + K$. Let $3HK = -p$, and show that $H^3 + K^3 = -q$.
3. Based on Problem 2, we have the two equations
$$3HK = -p \quad \text{and} \quad H^3 + K^3 = -q$$
Solve for K in $3HK = -p$ and substitute into $H^3 + K^3 = -q$. Then show that
$$H = \sqrt[3]{\frac{-q}{2} + \sqrt{\frac{q^2}{4} + \frac{p^3}{27}}}$$
[**Hint:** Look for an equation that is quadratic in form.]
4. Use the solution for H from Problem 3 and the equation $H^3 + K^3 = -q$ to show that
$$K = \sqrt[3]{\frac{-q}{2} - \sqrt{\frac{q^2}{4} + \frac{p^3}{27}}}$$
5. Use the results from Problems 2 to 4 to show that the solution of $x^3 + px + q = 0$ is
$$x = \sqrt[3]{\frac{-q}{2} + \sqrt{\frac{q^2}{4} + \frac{p^3}{27}}} + \sqrt[3]{\frac{-q}{2} - \sqrt{\frac{q^2}{4} + \frac{p^3}{27}}}$$
6. Use the result of Problem 5 to solve the equation $x^3 - 6x - 9 = 0$.
7. Use a calculator and the result of Problem 5 to solve the equation $x^3 + 3x - 14 = 0$.
8. Use the methods of this chapter to solve the equation $x^3 + 3x - 14 = 0$.

5.5 Assess Your Understanding

'Are You Prepared?'

Answers are given at the end of these exercises. If you get a wrong answer, read the pages listed in red.

1. Find $f(-1)$ if $f(x) = 2x^2 - x$. (pp. 216–218)
2. Factor the expression $6x^2 + x - 2$. (pp. 54–55)
3. Find the quotient and remainder if $3x^4 - 5x^3 + 7x - 4$ is divided by $x - 3$. (pp. 45–47 or 59)
4. Solve the equation $x^2 + x - 3 = 0$. (pp. 117–120)

Concepts and Vocabulary

5. In the process of polynomial division, (Divisor)(Quotient) + __________ = __________.
6. When a polynomial function f is divided by $x - c$, the remainder is __________.
7. If a function f, whose domain is all real numbers, is even and if 4 is a zero of f, then __________ is also a zero.
8. ***True or False*** Every polynomial function of degree 3 with real coefficients has exactly three real zeros.
9. ***True or False*** The only potential rational zeros of $f(x) = 2x^5 - x^3 + x^2 - x + 1$ are $\pm 1, \pm 2$.
10. ***True or False*** If f is a polynomial function of degree 4 and if $f(2) = 5$, then
$$\frac{f(x)}{x - 2} = p(x) + \frac{5}{x - 2}$$
where $p(x)$ is a polynomial of degree 3.

Skill Building

In Problems 11–20, use the Remainder Theorem to find the remainder when $f(x)$ is divided by $x - c$. Then use the Factor Theorem to determine whether $x - c$ is a factor of $f(x)$. If $x - c$ is a factor of f, write f in factored form; that is, write f in the form $f(x) = (x - c)$ (quotient).

11. $f(x) = 4x^3 - 3x^2 - 8x + 4;\quad x - 2$
12. $f(x) = -4x^3 + 5x^2 + 8;\quad x + 3$
13. $f(x) = 3x^4 - 6x^3 - 5x + 10;\quad x - 2$
14. $f(x) = 4x^4 - 15x^2 - 4;\quad x - 2$
15. $f(x) = 3x^6 + 82x^3 + 27;\quad x + 3$
16. $f(x) = 2x^6 - 18x^4 + x^2 - 9;\quad x + 3$
17. $f(x) = 4x^6 - 64x^4 + x^2 - 15;\quad x + 4$
18. $f(x) = x^6 - 16x^4 + x^2 - 16;\quad x + 4$
19. $f(x) = 2x^4 - x^3 + 2x - 1;\quad x - \frac{1}{2}$
20. $f(x) = 3x^4 + x^3 - 3x + 1;\quad x + \frac{1}{3}$

In Problems 21–32, determine the maximum number of real zeros that each polynomial function may have. Then list the potential rational zeros of each polynomial function. Do not attempt to find the zeros.

21. $f(x) = 3x^4 - 3x^3 + x^2 - x + 1$
22. $f(x) = x^5 - x^4 + 2x^2 + 3$
23. $f(x) = x^5 - 6x^2 + 9x - 3$
24. $f(x) = 2x^5 - x^4 - x^2 + 1$
25. $f(x) = -4x^3 - x^2 + x + 2$
26. $f(x) = 6x^4 - x^2 + 2$
27. $f(x) = 6x^4 - x^2 + 9$
28. $f(x) = -4x^3 + x^2 + x + 6$
29. $f(x) = 2x^5 - x^3 + 2x^2 + 12$
30. $f(x) = 3x^5 - x^2 + 2x + 18$
31. $f(x) = 6x^4 + 2x^3 - x^2 + 20$
32. $f(x) = -6x^3 - x^2 + x + 10$

In Problems 33–38, find the bounds to the zeros of each polynomial function. Use the bounds to obtain a complete graph of f.

33. $f(x) = 2x^3 + x^2 - 1$
34. $f(x) = 3x^3 - 2x^2 + x + 4$
35. $f(x) = x^3 - 5x^2 - 11x + 11$
36. $f(x) = 2x^3 - x^2 - 11x - 6$
37. $f(x) = x^4 + 3x^3 - 5x^2 + 9$
38. $f(x) = 4x^4 - 12x^3 + 27x^2 - 54x + 81$

In Problems 39–56, find the real zeros of f. Use the real zeros to factor f.

39. $f(x) = x^3 + 2x^2 - 5x - 6$
40. $f(x) = x^3 + 8x^2 + 11x - 20$
41. $f(x) = 2x^3 - 13x^2 + 24x - 9$
42. $f(x) = 2x^3 - 5x^2 - 4x + 12$
43. $f(x) = 3x^3 + 4x^2 + 4x + 1$
44. $f(x) = 3x^3 - 7x^2 + 12x - 28$
45. $f(x) = x^3 - 8x^2 + 17x - 6$
46. $f(x) = x^3 + 6x^2 + 6x - 4$
47. $f(x) = x^4 + x^3 - 3x^2 - x + 2$
48. $f(x) = x^4 - x^3 - 6x^2 + 4x + 8$
49. $f(x) = 2x^4 + 17x^3 + 35x^2 - 9x - 45$
50. $f(x) = 4x^4 - 15x^3 - 8x^2 + 15x + 4$
51. $f(x) = 2x^4 - 3x^3 - 21x^2 - 2x + 24$
52. $f(x) = 2x^4 + 11x^3 - 5x^2 - 43x + 35$
53. $f(x) = 4x^4 + 7x^2 - 2$
54. $f(x) = 4x^4 + 15x^2 - 4$
55. $f(x) = 4x^5 - 8x^4 - x + 2$
56. $f(x) = 4x^5 + 12x^4 - x - 3$

In Problems 57–62, find the real zeros of f. If necessary, round to two decimal places.

57. $f(x) = x^3 + 3.2x^2 - 16.83x - 5.31$
58. $f(x) = x^3 + 3.2x^2 - 7.25x - 6.3$
59. $f(x) = x^4 - 1.4x^3 - 33.71x^2 + 23.94x + 292.41$
60. $f(x) = x^4 + 1.2x^3 - 7.46x^2 - 4.692x + 15.2881$
61. $f(x) = x^3 + 19.5x^2 - 1021x + 1000.5$
62. $f(x) = x^3 + 42.2x^2 - 664.8x + 1490.4$

In Problems 63–72, find the real solutions of each equation.

63. $x^4 - x^3 + 2x^2 - 4x - 8 = 0$
64. $2x^3 + 3x^2 + 2x + 3 = 0$
65. $3x^3 + 4x^2 - 7x + 2 = 0$
66. $2x^3 - 3x^2 - 3x - 5 = 0$
67. $3x^3 - x^2 - 15x + 5 = 0$
68. $2x^3 - 11x^2 + 10x + 8 = 0$
69. $x^4 + 4x^3 + 2x^2 - x + 6 = 0$
70. $x^4 - 2x^3 + 10x^2 - 18x + 9 = 0$
71. $x^3 - \frac{2}{3}x^2 + \frac{8}{3}x + 1 = 0$
72. $x^3 - \frac{2}{3}x^2 + 3x - 2 = 0$

In Problems 73–78, use the Intermediate Value Theorem to show that each function has a zero in the given interval. Approximate the zero rounded to two decimal places.

73. $f(x) = 8x^4 - 2x^2 + 5x - 1$; $[0, 1]$

74. $f(x) = x^4 + 8x^3 - x^2 + 2$; $[-1, 0]$

75. $f(x) = 2x^3 + 6x^2 - 8x + 2$; $[-5, -4]$

76. $f(x) = 3x^3 - 10x + 9$; $[-3, -2]$

77. $f(x) = x^5 - x^4 + 7x^3 - 7x^2 - 18x + 18$; $[1.4, 1.5]$

78. $f(x) = x^5 - 3x^4 - 2x^3 + 6x^2 + x + 2$; $[1.7, 1.8]$

Mixed Practice

In Problems 79–84, analyze each polynomial function using Steps 1 through 8 on page 341 in Section 5.1.

79. $f(x) = x^3 + 2x^2 - 5x - 6$
[**Hint:** See Problem 39.]

80. $f(x) = x^3 + 8x^2 + 11x - 20$
[**Hint:** See Problem 40.]

81. $f(x) = x^4 + x^3 - 3x^2 - x + 2$
[**Hint:** See Problem 47.]

82. $f(x) = x^4 - x^3 - 6x^2 + 4x + 8$
[**Hint:** See Problem 48.]

83. $f(x) = 4x^5 - 8x^4 - x + 2$
[**Hint:** See Problem 55.]

84. $f(x) = 4x^5 + 12x^4 - x - 3$
[**Hint:** See Problem 56.]

Applications and Extensions

85. Find k such that $f(x) = x^3 - kx^2 + kx + 2$ has the factor $x - 2$.

86. Find k such that $f(x) = x^4 - kx^3 + kx^2 + 1$ has the factor $x + 2$.

87. What is the remainder when $f(x) = 2x^{20} - 8x^{10} + x - 2$ is divided by $x - 1$?

88. What is the remainder when $f(x) = -3x^{17} + x^9 - x^5 + 2x$ is divided by $x + 1$?

89. Use the Factor Theorem to prove that $x - c$ is a factor of $x^n - c^n$ for any positive integer n.

90. Use the Factor Theorem to prove that $x + c$ is a factor of $x^n + c^n$ if $n \geq 1$ is an odd integer.

91. One solution of the equation $x^3 - 8x^2 + 16x - 3 = 0$ is 3. Find the sum of the remaining solutions.

92. One solution of the equation $x^3 + 5x^2 + 5x - 2 = 0$ is -2. Find the sum of the remaining solutions.

93. **Geometry** What is the length of the edge of a cube if, after a slice 1 inch thick is cut from one side, the volume remaining is 294 cubic inches?

94. **Geometry** What is the length of the edge of a cube if its volume could be doubled by an increase of 6 centimeters in one edge, an increase of 12 centimeters in a second edge, and a decrease of 4 centimeters in the third edge?

95. Let $f(x)$ be a polynomial function whose coefficients are integers. Suppose that r is a real zero of f and that the leading coefficient of f is 1. Use the Rational Zeros Theorem to show that r is either an integer or an irrational number.

96. Prove the Rational Zeros Theorem.

[**Hint:** Let $\frac{p}{q}$, where p and q have no common factors except 1 and -1, be a zero of the polynomial function

$$f(x) = a_nx^n + a_{n-1}x^{n-1} + \cdots + a_1x + a_0$$

whose coefficients are all integers. Show that

$$a_np^n + a_{n-1}p^{n-1}q + \cdots + a_1pq^{n-1} + a_0q^n = 0$$

Now, because p is a factor of the first n terms of this equation, p must also be a factor of the term a_0q^n. Since p is not a factor of q (why?), p must be a factor of a_0. Similarly, q must be a factor of a_n.]

97. **Bisection Method for Approximating Zeros of a Function f** We begin with two consecutive integers, a and $a + 1$, such that $f(a)$ and $f(a + 1)$ are of opposite sign. Evaluate f at the midpoint m_1 of a and $a + 1$. If $f(m_1) = 0$, then m_1 is the zero of f, and we are finished. Otherwise, $f(m_1)$ is of opposite sign to either $f(a)$ or $f(a + 1)$. Suppose that it is $f(a)$ and $f(m_1)$ that are of opposite sign. Now evaluate f at the midpoint m_2 of a and m_1. Repeat this process until the desired degree of accuracy is obtained. Note that each iteration places the zero in an interval whose length is half that of the previous interval. Use the bisection method to approximate the zero of $f(x) = 8x^4 - 2x^2 + 5x - 1$ in the interval $[0, 1]$, correct to three decimal places. Verify your result using a graphing utility.

[**Hint:** The process ends when both endpoints agree to the desired number of decimal places.]

Discussion and Writing

98. Is $\frac{1}{3}$ a zero of $f(x) = 2x^3 + 3x^2 - 6x + 7$? Explain.

99. Is $\frac{1}{3}$ a zero of $f(x) = 4x^3 - 5x^2 - 3x + 1$? Explain.

100. Is $\frac{3}{5}$ a zero of $f(x) = 2x^6 - 5x^4 + x^3 - x + 1$? Explain.

101. Is $\frac{2}{3}$ a zero of $f(x) = x^7 + 6x^5 - x^4 + x + 2$? Explain.

'Are You Prepared?' Answers

1. 3 **2.** $(3x + 2)(2x - 1)$ **3.** Quotient: $3x^3 + 4x^2 + 12x + 43$; remainder: 125 **4.** $\left\{\frac{-1 - \sqrt{13}}{2}, \frac{-1 + \sqrt{13}}{2}\right\}$

5.6 Complex Zeros; Fundamental Theorem of Algebra

PREPARING FOR THIS SECTION *Before getting started, review the following:*

- Complex Numbers (Section 1.4, pp. 124–128)
- Quadratic Equations in the Complex Number System (Section 1.4, pp. 129–131)

Now Work the 'Are You Prepared?' problems on page 394.

OBJECTIVES
1 Use the Conjugate Pairs Theorem (p. 391)
2 Find a Polynomial Function with Specified Zeros (p. 392)
3 Find the Complex Zeros of a Polynomial Function (p. 393)

In Section 1.3, we found the real solutions of a quadratic equation. That is, we found the real zeros of a polynomial function of degree 2. Then, in Section 1.4 we found the solutions of a quadratic equation in the complex number system; that is, we found the *complex zeros* of a polynomial function of degree 2.

In Section 5.5 we found the real zeros of polynomial functions of degree 3 or higher. In this section we will find the *complex zeros* of polynomial functions of degree 3 or higher.

DEFINITION

A variable in the complex number system is referred to as a **complex variable.** A **complex polynomial function** f of degree n is a function of the form

$$f(x) = a_n x^n + a_{n-1}x^{n-1} + \cdots + a_1 x + a_0 \quad \textbf{(1)}$$

where $a_n, a_{n-1}, \ldots, a_1, a_0$ are complex numbers, $a_n \neq 0$, n is a nonnegative integer, and x is a complex variable. As before, a_n is called the **leading coefficient** of f. A complex number r is called a **complex zero** of f if $f(r) = 0$.

In most of our work the coefficients in (1) will be real numbers.

We have learned that some quadratic equations have no real solutions, but that in the complex number system every quadratic equation has a solution, either real or complex. The next result, proved by Karl Friedrich Gauss (1777–1855) when he was 22 years old,* gives an extension to complex polynomials. In fact, this result is so important and useful that it has become known as the **Fundamental Theorem of Algebra.**

FUNDAMENTAL THEOREM OF ALGEBRA

Every complex polynomial function $f(x)$ of degree $n \geq 1$ has at least one complex zero.

We shall not prove this result, as the proof is beyond the scope of this book. However, using the Fundamental Theorem of Algebra and the Factor Theorem, we can prove the following result:

THEOREM

Every complex polynomial function $f(x)$ of degree $n \geq 1$ can be factored into n linear factors (not necessarily distinct) of the form

$$f(x) = a_n(x - r_1)(x - r_2) \cdot \cdots \cdot (x - r_n) \quad \textbf{(2)}$$

where $a_n, r_1, r_2, \ldots, r_n$ are complex numbers. That is, every complex polynomial function of degree $n \geq 1$ has exactly n complex zeros, some of which may repeat.

*In all, Gauss gave four different proofs of this theorem, the first one in 1799 being the subject of his doctoral dissertation.

Proof Let

$$f(x) = a_n x^n + a_{n-1}x^{n-1} + \cdots + a_1 x + a_0$$

By the Fundamental Theorem of Algebra, f has at least one zero, say r_1. Then, by the Factor Theorem, $x - r_1$ is a factor, and

$$f(x) = (x - r_1)q_1(x)$$

where $q_1(x)$ is a complex polynomial of degree $n - 1$ whose leading coefficient is a_n. Repeating this argument n times, we arrive at

$$f(x) = (x - r_1)(x - r_2) \cdot \cdots \cdot (x - r_n)q_n(x)$$

where $q_n(x)$ is a complex polynomial of degree $n - n = 0$ whose leading coefficient is a_n. That is, $q_n(x) = a_n x^0 = a_n$, and so

$$f(x) = a_n(x - r_1)(x - r_2) \cdot \cdots \cdot (x - r_n)$$

We conclude that every complex polynomial function $f(x)$ of degree $n \geq 1$ has exactly n (not necessarily distinct) zeros. ■

1 Use the Conjugate Pairs Theorem

We can use the Fundamental Theorem of Algebra to obtain valuable information about the complex zeros of polynomials whose coefficients are real numbers.

CONJUGATE PAIRS THEOREM

Let $f(x)$ be a polynomial whose coefficients are real numbers. If $r = a + bi$ is a zero of f, the complex conjugate $\bar{r} = a - bi$ is also a zero of f.

In other words, for polynomials whose coefficients are real numbers, the complex zeros occur in conjugate pairs. This result should not be all that surprising since the complex zeros of a quadratic function occurred in conjugate pairs.

Proof Let

$$f(x) = a_n x^n + a_{n-1}x^{n-1} + \cdots + a_1 x + a_0$$

where $a_n, a_{n-1}, \ldots, a_1, a_0$ are real numbers and $a_n \neq 0$. If $r = a + bi$ is a zero of f, then $f(r) = f(a + bi) = 0$, so

$$a_n r^n + a_{n-1}r^{n-1} + \cdots + a_1 r + a_0 = 0$$

We take the conjugate of both sides to get

$$\overline{a_n r^n + a_{n-1}r^{n-1} + \cdots + a_1 r + a_0} = \bar{0}$$

$$\overline{a_n r^n} + \overline{a_{n-1}r^{n-1}} + \cdots + \overline{a_1 r} + \overline{a_0} = \bar{0}$$ The conjugate of a sum equals the sum of the conjugates (see Section 1.4).

$$\overline{a_n}(\bar{r})^n + \overline{a_{n-1}}(\bar{r})^{n-1} + \cdots + \overline{a_1}\bar{r} + \overline{a_0} = \bar{0}$$ The conjugate of a product equals the product of the conjugates.

$$a_n(\bar{r})^n + a_{n-1}(\bar{r})^{n-1} + \cdots + a_1\bar{r} + a_0 = 0$$ The conjugate of a real number equals the real number.

This last equation states that $f(\bar{r}) = 0$; that is, $\bar{r} = a - bi$ is a zero of f. ■

The importance of this result should be clear. Once we know that, say, $3 + 4i$ is a zero of a polynomial with real coefficients, then we know that $3 - 4i$ is also a zero. This result has an important corollary.

COROLLARY

A polynomial f of odd degree with real coefficients has at least one real zero.

Proof Because complex zeros occur as conjugate pairs in a polynomial with real coefficients, there will always be an even number of zeros that are not real numbers. Consequently, since f is of odd degree, one of its zeros has to be a real number. ■

For example, the polynomial $f(x) = x^5 - 3x^4 + 4x^3 - 5$ has at least one zero that is a real number, since f is of degree 5 (odd) and has real coefficients.

EXAMPLE 1 **Using the Conjugate Pairs Theorem**

A polynomial f of degree 5 whose coefficients are real numbers has the zeros 1, $5i$, and $1 + i$. Find the remaining two zeros.

Solution Since f has coefficients that are real numbers, complex zeros appear as conjugate pairs. It follows that $-5i$, the conjugate of $5i$, and $1 - i$, the conjugate of $1 + i$, are the two remaining zeros.

Now Work PROBLEM 7

2 Find a Polynomial Function with Specified Zeros

EXAMPLE 2 **Finding a Polynomial Function Whose Zeros Are Given**

(a) Find a polynomial f of degree 4 whose coefficients are real numbers and that has the zeros 1, 1, and $-4 + i$.

(b) Graph the polynomial found in part (a) to verify your result.

Solution (a) Since $-4 + i$ is a zero, by the Conjugate Pairs Theorem, $-4 - i$ must also be a zero of f. Because of the Factor Theorem, if $f(c) = 0$, then $x - c$ is a factor of $f(x)$. So we can now write f as

$$f(x) = a(x - 1)(x - 1)[x - (-4 + i)][x - (-4 - i)]$$

where a is any real number. If we let $a = 1$, we obtain

$$\begin{aligned} f(x) &= (x - 1)(x - 1)[x - (-4 + i)][x - (-4 - i)] \\ &= (x^2 - 2x + 1)[x^2 - (-4 + i)x - (-4 - i)x + (-4 + i)(-4 - i)] \\ &= (x^2 - 2x + 1)(x^2 + 4x - ix + 4x + ix + 16 + 4i - 4i - i^2) \\ &= (x^2 - 2x + 1)(x^2 + 8x + 17) \\ &= x^4 + 8x^3 + 17x^2 - 2x^3 - 16x^2 - 34x + x^2 + 8x + 17 \\ &= x^4 + 6x^3 + 2x^2 - 26x + 17 \end{aligned}$$

(b) A quick analysis of the polynomial f tells us what to expect:

At most three turning points

For large $|x|$, the graph will behave like $y = x^4$.

A repeated real zero at 1 so that the graph will touch the x-axis at 1

The only x-intercept is at 1.

Figure 54 shows the complete graph. (Do you see why? The graph has exactly three turning points and the degree of the polynomial is 4.)

Figure 54

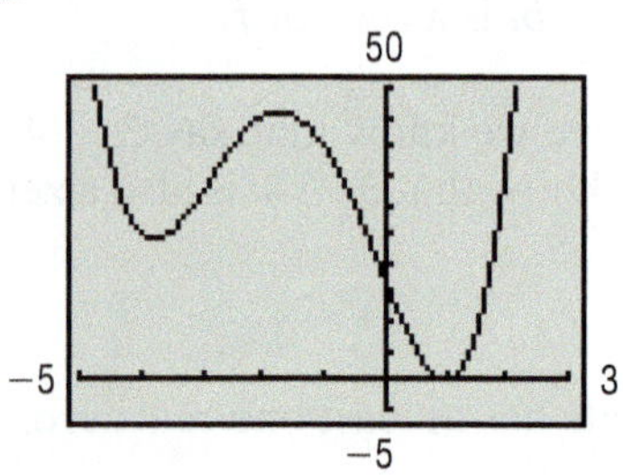

Now Work PROBLEM 17

Exploration

Graph the function found in Example 2 for $a = 2$ and $a = -1$. Does the value of a affect the zeros of f? How does the value of a affect the graph of f?

Now we can prove the theorem we conjectured earlier in Section 5.5.

THEOREM

Every polynomial function with real coefficients can be uniquely factored over the real numbers into a product of linear factors and/or irreducible quadratic factors.

Proof Every complex polynomial f of degree n has exactly n zeros and can be factored into a product of n linear factors. If its coefficients are real, then those zeros that are complex numbers will always occur as conjugate pairs. As a result, if $r = a + bi$ is a complex zero, then so is $\bar{r} = a - bi$. Consequently, when the linear factors $x - r$ and $x - \bar{r}$ of f are multiplied, we have

$$(x - r)(x - \bar{r}) = x^2 - (r + \bar{r})x + r\bar{r} = x^2 - 2ax + a^2 + b^2$$

This second-degree polynomial has real coefficients and is irreducible (over the real numbers). Thus, the factors of f are either linear or irreducible quadratic factors. ■

3 Find the Complex Zeros of a Polynomial Function

The steps for finding the complex zeros of a polynomial function are the same as those for finding the real zeros.

EXAMPLE 3 **Finding the Complex Zeros of a Polynomial Function**

Find the complex zeros of the polynomial function

$$f(x) = 3x^4 + 5x^3 + 25x^2 + 45x - 18$$

Write f in factored form.

Solution

STEP 1: The degree of f is 4. So f will have four complex zeros.

STEP 2: The Rational Zeros Theorem provides information about the potential rational zeros of polynomials with integer coefficients. For this polynomial (which has integer coefficients), the potential rational zeros are

$$\pm\frac{1}{3}, \pm\frac{2}{3}, \pm 1, \pm 2, \pm 3, \pm 6, \pm 9, \pm 18$$

Figure 55

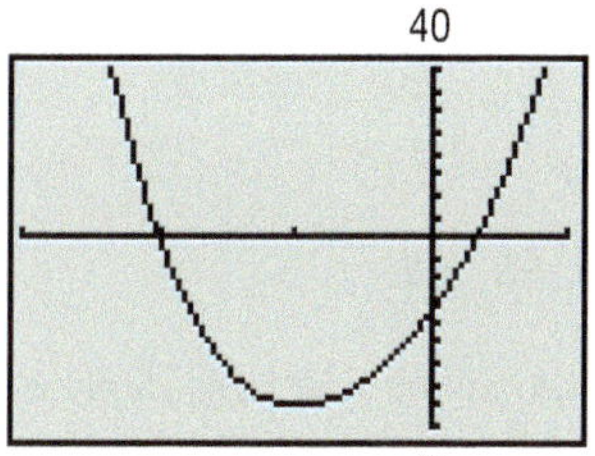

STEP 3: Figure 55 shows the graph of f. The graph has the characteristics that we expect of this polynomial of degree 4: It behaves like $y = 3x^4$ for large $|x|$ and has y-intercept -18. There are x-intercepts near -2 and between 0 and 1.

STEP 4: Because $f(-2) = 0$, we know that -2 is a zero of f and $x - (-2) = x + 2$ is a factor of f. We can use long division or synthetic division to factor f. Using synthetic division, we factor f.

$$\begin{array}{r|rrrrr} -2 & 3 & 5 & 25 & 45 & -18 \\ & & -6 & 2 & -54 & 18 \\ \hline & 3 & -1 & 27 & -9 & 0 \end{array}$$

So $f(x) = (x + 2)(3x^3 - x^2 + 27x - 9)$. The depressed equation is

$$q_1(x) = 3x^3 - x^2 + 27x - 9 = 0$$

Repeat Step 4: The depressed equation $3x^3 - x^2 + 27x - 9 = 0$ can be factored by grouping.

$$3x^3 - x^2 + 27x - 9 = 0$$

$$x^2(3x - 1) + 9(3x - 1) = 0 \quad \text{Factor } x^2 \text{ from } 3x^3 - x^2 \text{ and 9 from } 27x - 9.$$

$$(x^2 + 9)(3x - 1) = 0 \quad \text{Factor out the common factor } 3x - 1.$$

$$x^2 + 9 = 0 \quad \text{or} \quad 3x - 1 = 0 \quad \text{Apply the Zero-Product Property.}$$

$$x^2 = -9 \quad \text{or} \quad 3x = 1$$

$$x = -3i, \quad x = 3i \quad \text{or} \quad x = \frac{1}{3}$$

The four complex zeros of f are $\left\{-3i, 3i, -2, \frac{1}{3}\right\}$.

The factored form of f is

$$\begin{aligned} f(x) &= 3x^4 + 5x^3 + 25x^2 + 45x - 18 \\ &= (x + 3i)(x - 3i)(x + 2)(3x - 1) \\ &= 3(x + 3i)(x - 3i)(x + 2)\left(x - \frac{1}{3}\right) \end{aligned}$$

Now Work PROBLEM 33

5.6 Assess Your Understanding

'Are You Prepared?' *Answers are given at the end of these exercises. If you get the wrong answer, read the pages listed in red.*

1. Find the sum and the product of the complex numbers $3 - 2i$ and $-3 + 5i$. (pp. 125–126)

2. In the complex number system, solve $x^2 + 2x + 2 = 0$. (pp. 129–130)

Concepts and Vocabulary

3. Every polynomial function of odd degree with real coefficients will have at least ________ real zero(s).

4. If $3 + 4i$ is a zero of a polynomial function of degree 5 with real coefficients, then so is ________.

5. ***True or False*** A polynomial function of degree n with real coefficients has exactly n complex zeros. At most n of them are real zeros.

6. ***True or False*** A polynomial function of degree 4 with real coefficients could have $-3, 2 + i, 2 - i$, and $-3 + 5i$ as its zeros.

Skill Building

In Problems 7–16, information is given about a polynomial $f(x)$ whose coefficients are real numbers. Find the remaining zeros of f.

7. Degree 3; zeros: $3, 4 - i$
8. Degree 3; zeros: $4, 3 + i$
9. Degree 4; zeros: $i, 1 + i$
10. Degree 4; zeros: $1, 2, 2 + i$
11. Degree 5; zeros: $1, i, 2i$
12. Degree 5; zeros: $0, 1, 2, i$
13. Degree 4; zeros: $i, 2, -2$
14. Degree 4; zeros: $2 - i, -i$
15. Degree 6; zeros: $2, 2 + i, -3 - i, 0$
16. Degree 6; zeros: $i, 3 - 2i, -2 + i$

In Problems 17–22, form a polynomial $f(x)$ with real coefficients having the given degree and zeros. Answers will vary depending on the choice of leading coefficient.

17. Degree 4; zeros: $3 + 2i$; 4, multiplicity 2
18. Degree 4; zeros: $i, 1 + 2i$
19. Degree 5; zeros: $2; -i; 1 + i$
20. Degree 6; zeros: $i, 4 - i; 2 + i$
21. Degree 4; zeros: 3, multiplicity 2; $-i$
22. Degree 5; zeros: 1, multiplicity 3; $1 + i$

In Problems 23–30, use the given zero to find the remaining zeros of each function.

23. $f(x) = x^3 - 4x^2 + 4x - 16$; zero: $2i$

24. $g(x) = x^3 + 3x^2 + 25x + 75$; zero: $-5i$

25. $f(x) = 2x^4 + 5x^3 + 5x^2 + 20x - 12$; zero: $-2i$

26. $h(x) = 3x^4 + 5x^3 + 25x^2 + 45x - 18$; zero: $3i$

27. $h(x) = x^4 - 9x^3 + 21x^2 + 21x - 130$; zero: $3 - 2i$

28. $f(x) = x^4 - 7x^3 + 14x^2 - 38x - 60$; zero: $1 + 3i$

29. $h(x) = 3x^5 + 2x^4 + 15x^3 + 10x^2 - 528x - 352$; zero: $-4i$

30. $g(x) = 2x^5 - 3x^4 - 5x^3 - 15x^2 - 207x + 108$; zero: $3i$

In Problems 31–40, find the complex zeros of each polynomial function. Write f in factored form.

31. $f(x) = x^3 - 1$

32. $f(x) = x^4 - 1$

33. $f(x) = x^3 - 8x^2 + 25x - 26$

34. $f(x) = x^3 + 13x^2 + 57x + 85$

35. $f(x) = x^4 + 5x^2 + 4$

36. $f(x) = x^4 + 13x^2 + 36$

37. $f(x) = x^4 + 2x^3 + 22x^2 + 50x - 75$

38. $f(x) = x^4 + 3x^3 - 19x^2 + 27x - 252$

39. $f(x) = 3x^4 - x^3 - 9x^2 + 159x - 52$

40. $f(x) = 2x^4 + x^3 - 35x^2 - 113x + 65$

Discussion and Writing

In Problems 41 and 42, explain why the facts given are contradictory.

41. $f(x)$ is a polynomial of degree 3 whose coefficients are real numbers; its zeros are $4 + i$, $4 - i$, and $2 + i$.

42. $f(x)$ is a polynomial of degree 3 whose coefficients are real numbers; its zeros are 2, i, and $3 + i$.

43. $f(x)$ is a polynomial of degree 4 whose coefficients are real numbers; three of its zeros are 2, $1 + 2i$, and $1 - 2i$. Explain why the remaining zero must be a real number.

44. $f(x)$ is a polynomial of degree 4 whose coefficients are real numbers; two of its zeros are -3 and $4 - i$. Explain why one of the remaining zeros must be a real number. Write down one of the missing zeros.

'Are You Prepared?' Answers

1. Sum: $3i$; product: $1 + 21i$

2. $-1 - i, -1 + i$

CHAPTER REVIEW

Things to Know

Power function (pp. 331–333)

$f(x) = x^n$, $n \geq 2$ even — Domain: all real numbers Range: nonnegative real numbers
Passes through $(-1, 1)$, $(0, 0)$, $(1, 1)$
Even function
Decreasing on $(-\infty, 0)$, increasing on $(0, \infty)$

$f(x) = x^n$, $n \geq 3$ odd — Domain: all real numbers Range: all real numbers
Passes through $(-1, -1)$, $(0, 0)$, $(1, 1)$
Odd function
Increasing on $(-\infty, \infty)$

Polynomial function (pp. 330, 336–341)

$f(x) = a_nx^n + a_{n-1}x^{n-1} + \cdots + a_1x + a_0$, $a_n \neq 0$ — Domain: all real numbers
At most $n - 1$ turning points
End behavior: Behaves like $y = a_nx^n$ for large $|x|$

Real zeros of a polynomial function f (pp. 334–335) — Real number r for which $f(r) = 0$; if r is a real zero of f, then r is an x-intercept of the graph of f and $x - r$ is a factor of f.

Rational function (pp. 347–355)

$R(x) = \dfrac{p(x)}{q(x)}$ Domain: $\{x | q(x) \neq 0\}$

p, q are polynomial functions. Vertical asymptotes: With $R(x)$ in lowest terms, if $q(r) = 0$ for some real number, then $x = r$ is a vertical asymptote.

Horizontal or oblique asymptotes: See the summary on page 355.

Remainder Theorem (p. 377) If a polynomial function $f(x)$ is divided by $x - c$, then the remainder is $f(c)$.

Factor Theorem (p. 377) $x - c$ is a factor of a polynomial function $f(x)$ if and only if $f(c) = 0$.

Rational Zeros Theorem (p. 379) Let f be a polynomial function of degree 1 or higher of the form

$$f(x) = a_n x^n + a_{n-1}x^{n-1} + \cdots + a_1 x + a_0 \qquad a_n \neq 0, a_0 \neq 0$$

where each coefficient is an integer. If $\frac{p}{q}$, in lowest terms, is a rational zero of f, then p must be a factor of a_0, and q must be a factor of a_n.

Bounds on Zeros (p. 383) Let f denote a polynomial function whose leading coefficient is 1, so $f(x) = x^n + a_{n-1}x^{n-1} + \cdots + a_1 x + a_0$. A bound M on the zeros of f is the smaller of the two numbers

$$\text{Max}\{1, |a_0| + |a_1| + \cdots + |a_{n-1}|\}, \quad 1 + \text{Max}\{|a_0|, |a_1|, \ldots, |a_{n-1}|\}$$

Intermediate Value Theorem (p. 386) Let f be a continuous function. If $a < b$ and $f(a)$ and $f(b)$ are of opposite sign, then there is at least one real zero of f between a and b.

Fundamental Theorem of Algebra (p. 390) Every complex polynomial function $f(x)$ of degree $n \geq 1$ has at least one complex zero.

Conjugate Pairs Theorem (p. 391) Let f be a polynomial whose coefficients are real numbers. If $r = a + bi$ is a zero of f, then its complex conjugate $\bar{r} = a - bi$ is also a zero of f.

Objectives

Section		You should be able to . . .	Examples	Review Exercises
5.1	1	Identify polynomial functions and their degree (p. 330)	1	1–4
	2	Graph polynomial functions using transformations (p. 333)	2, 3	5–10
	3	Identify the real zeros of a polynomial function and their multiplicity (p. 334)	4–6	11–18
	4	Analyze the graph of a polynomial function (p. 339)	9, 10	11–18
	5	Build cubic models from data (p. 342)	11	88
5.2	1	Find the domain of a rational function (p. 348)	1	19–22
	2	Find the vertical asymptotes of a rational function (p. 351)	4	19–22
	3	Find the horizontal or oblique asymptotes of a rational function (p. 352)	5–8	19–22
5.3	1	Analyze the graph of a rational function (p. 358)	1–5	23–34
	2	Solve applied problems involving rational functions (p. 365)	6	87
5.4	1	Solve polynomial inequalities algebraically and graphically (p. 369)	1, 2	35, 36
	2	Solve rational inequalities algebraically and graphically (p. 371)	3, 4	37–44
5.5	1	Use the Remainder and Factor Theorems (p. 376)	1, 2	45–50
	2	Use the Rational Zeros Theorem to list the potential rational zeros of a polynomial function (p. 379)	3	51, 52
	3	Find the real zeros of a polynomial function (p. 380)	4, 5, 9	53–62
	4	Solve polynomial equations (p. 382)	6	63–66
	5	Use the Theorem for Bounds on Zeros (p. 383)	7, 8	67–70
	6	Use the Intermediate Value Theorem (p. 386)	10	71–74
5.6	1	Use the Conjugate Pairs Theorem (p. 391)	1	75–78
	2	Find a polynomial function with specified zeros (p. 392)	2	75–78
	3	Find the complex zeros of a polynomial function (p. 393)	3	79–86

Review Exercises

In Problems 1–4, determine which functions are polynomial functions. For those that are, state the degree. For those that are not, tell why not.

1. $f(x) = 4x^5 - 3x^2 + 5x - 2$
2. $f(x) = \dfrac{3x^5}{2x + 1}$
3. $f(x) = 3x^2 + 5x^{1/2} - 1$
4. $f(x) = 3$

In Problems 5–10, graph each function using transformations (shifting, compressing, stretching, and reflection). Show all the stages.

5. $f(x) = (x + 2)^3$
6. $f(x) = -x^3 + 3$
7. $f(x) = -(x - 1)^4$
8. $f(x) = (x - 1)^4 - 2$
9. $f(x) = (x - 1)^4 + 2$
10. $f(x) = (1 - x)^3$

In Problems 11–18, analyze the graph of each polynomial function by following Steps 1 through 8 on page 341.

11. $f(x) = x(x + 2)(x + 4)$
12. $f(x) = x(x - 2)(x - 4)$
13. $f(x) = (x - 2)^2(x + 4)$
14. $f(x) = (x - 2)(x + 4)^2$
15. $f(x) = x^3 - 4x^2$
16. $f(x) = x^3 + 4x$
17. $f(x) = (x - 1)^2(x + 3)(x + 1)$
18. $f(x) = (x - 4)(x + 2)^2(x - 2)$

In Problems 19–22, find the domain of each rational function. Find any horizontal, vertical, or oblique asymptotes.

19. $R(x) = \dfrac{x + 2}{x^2 - 9}$
20. $R(x) = \dfrac{x^2 + 4}{x - 2}$
21. $R(x) = \dfrac{2x^2 + 3x + 2}{(x + 2)^2}$
22. $R(x) = \dfrac{-4x^3}{x^3 - 1}$

In Problems 23–34, analyze each rational function following Steps 1 through 8 on page 360.

23. $R(x) = \dfrac{2x - 6}{x}$
24. $R(x) = \dfrac{4 - x}{x}$
25. $H(x) = \dfrac{x + 2}{x(x - 2)}$
26. $H(x) = \dfrac{x}{x^2 - 1}$
27. $R(x) = \dfrac{x^2 + x - 6}{x^2 - x - 6}$
28. $R(x) = \dfrac{x^2 - 6x + 9}{x^2}$
29. $F(x) = \dfrac{x^3}{x^2 - 4}$
30. $F(x) = \dfrac{3x^3}{(x - 1)^2}$
31. $R(x) = \dfrac{2x^4}{(x - 1)^2}$
32. $R(x) = \dfrac{x^4}{x^2 - 9}$
33. $G(x) = \dfrac{x^2 - 4}{x^2 - x - 2}$
34. $F(x) = \dfrac{(x - 1)^2}{x^2 - 1}$

In Problems 35–44, solve each inequality. Graph the solution set.

35. $x^3 + x^2 < 4x + 4$
36. $x^3 + 4x^2 \geq x + 4$
37. $\dfrac{6}{x + 3} \geq 1$
38. $\dfrac{-2}{1 - 3x} < 1$
39. $\dfrac{2x - 6}{1 - x} < 2$
40. $\dfrac{3 - 2x}{2x + 5} \geq 2$
41. $\dfrac{(x - 2)(x - 1)}{x - 3} \geq 0$
42. $\dfrac{x + 1}{x(x - 5)} \leq 0$
43. $\dfrac{x^2 - 8x + 12}{x^2 - 16} > 0$
44. $\dfrac{x(x^2 + x - 2)}{x^2 + 9x + 20} \leq 0$

In Problems 45–48, find the remainder R when $f(x)$ is divided by $g(x)$. Is g a factor of f?

45. $f(x) = 8x^3 - 3x^2 + x + 4;\ \ g(x) = x - 1$
46. $f(x) = 2x^3 + 8x^2 - 5x + 5;\ \ g(x) = x - 2$
47. $f(x) = x^4 - 2x^3 + 15x - 2;\ \ g(x) = x + 2$
48. $f(x) = x^4 - x^2 + 2x + 2;\ \ g(x) = x + 1$

49. Find the value of $f(x) = 12x^6 - 8x^4 + 1$ at $x = 4$.
50. Find the value of $f(x) = -16x^3 + 18x^2 - x + 2$ at $x = -2$.

In Problems 51 and 52, determine the maximum number of zeros each polynomial function may have. Then list the potential rational zeros of each polynomial function. Do not attempt to find the zeros.

51. $f(x) = 12x^8 - x^7 + 8x^4 - 2x^3 + x + 3$
52. $f(x) = -6x^5 + x^4 + 5x^3 + x + 1$

In Problems 53–58, use the Rational Zeros Theorem to find all the real zeros of each polynomial function. Use the zeros to factor f over the real numbers.

53. $f(x) = x^3 - 3x^2 - 6x + 8$
54. $f(x) = x^3 - x^2 - 10x - 8$
55. $f(x) = 4x^3 + 4x^2 - 7x + 2$
56. $f(x) = 4x^3 - 4x^2 - 7x - 2$
57. $f(x) = x^4 - 4x^3 + 9x^2 - 20x + 20$
58. $f(x) = x^4 + 6x^3 + 11x^2 + 12x + 18$

In Problems 59–62, determine the real zeros of the polynomial function. Approximate all irrational zeros rounded to two decimal places.

59. $f(x) = 2x^3 - 11.84x^2 - 9.116x + 82.46$
60. $f(x) = 12x^3 + 39.8x^2 - 4.4x - 3.4$
61. $g(x) = 15x^4 - 21.5x^3 - 1718.3x^2 + 5308x + 3796.8$
62. $g(x) = 3x^4 + 67.93x^3 + 486.265x^2 + 1121.32x + 412.195$

In Problems 63–66, solve each equation in the real number system.

63. $2x^4 + 2x^3 - 11x^2 + x - 6 = 0$

64. $3x^4 + 3x^3 - 17x^2 + x - 6 = 0$

65. $2x^4 + 7x^3 + x^2 - 7x - 3 = 0$

66. $2x^4 + 7x^3 - 5x^2 - 28x - 12 = 0$

In Problems 67–70, find bounds to the real zeros of each polynomial function. Obtain a complete graph of f using a graphing utility.

67. $f(x) = x^3 - x^2 - 4x + 2$

68. $f(x) = x^3 + x^2 - 10x - 5$

69. $f(x) = 2x^3 - 7x^2 - 10x + 35$

70. $f(x) = 3x^3 - 7x^2 - 6x + 14$

In Problems 71–74, use the Intermediate Value Theorem to show that each polynomial has a zero in the given interval.

71. $f(x) = 3x^3 - x - 1;\ [0, 1]$

72. $f(x) = 2x^3 - x^2 - 3;\ [1, 2]$

73. $f(x) = 8x^4 - 4x^3 - 2x - 1;\ [0, 1]$

74. $f(x) = 3x^4 + 4x^3 - 8x - 2;\ [1, 2]$

In Problems 75–78, information is given about a complex polynomial $f(x)$ whose coefficients are real numbers. Find the remaining zeros of f. Then find a polynomial with real coefficients that has the zeros.

75. Degree 3; zeros: $4 + i, 6$

76. Degree 3; zeros: $3 + 4i, 5$

77. Degree 4; zeros: $i, 1 + i$

78. Degree 4; zeros: $1, 2, 1 + i$

In Problems 79–86, find the complex zeros of each polynomial function $f(x)$. Write f in factored form.

79. $f(x) = x^3 - 3x^2 - 6x + 8$

80. $f(x) = x^3 - x^2 - 10x - 8$

81. $f(x) = 4x^3 + 4x^2 - 7x + 2$

82. $f(x) = 4x^3 - 4x^2 - 7x - 2$

83. $f(x) = x^4 - 4x^3 + 9x^2 - 20x + 20$

84. $f(x) = x^4 + 6x^3 + 11x^2 + 12x + 18$

85. $f(x) = 2x^4 + 2x^3 - 11x^2 + x - 6$

86. $f(x) = 3x^4 + 3x^3 - 17x^2 + x - 6$

87. Making a Can A can in the shape of a right circular cylinder is required to have a volume of 250 cubic centimeters.

(a) Express the amount A of material to make the can as a function of the radius r of the cylinder.

(b) How much material is required if the can is of radius 3 centimeters?

(c) How much material is required if the can is of radius 5 centimeters?

(d) Graph $A = A(r)$. For what value of r is A smallest?

88. Model It: Poverty Rates The following data represent the percentage of families in the United States whose income is below the poverty level.

Year, *t*	Percent below Poverty Level, *p*
1990, 1	10.9
1991, 2	11.5
1992, 3	11.9
1993, 4	12.3
1994, 5	11.6
1995, 6	10.8
1996, 7	11.0
1997, 8	10.3
1998, 9	10.0
1999, 10	9.3
2000, 11	8.7
2001, 12	9.2
2002, 13	9.6
2003, 14	10.0
2004, 15	10.2

Source: U.S. Census Bureau

(a) With a graphing utility, draw a scatter diagram of the data. Comment on the type of relation that appears to exist between the two variables.

(b) Decide on a function of best fit to these data (linear, quadratic, or cubic), and use this function to predict the percentage of U.S. families that were below the poverty level in 2005 ($t = 16$).

(c) Draw the function of best fit on the scatter diagram drawn in part (a).

89. Design a polynomial function with the following characteristics: degree 6; four real zeros, one of multiplicity 3; y-intercept 3; behaves like $y = -5x^6$ for large values of $|x|$. Is this polynomial unique? Compare your polynomial with those of other students. What terms will be the same as everyone else's? Add some more characteristics, such as symmetry or naming the real zeros. How does this modify the polynomial?

90. Design a rational function with the following characteristics: three real zeros, one of multiplicity 2; y-intercept 1; vertical asymptotes $x = -2$ and $x = 3$; oblique asymptote $y = 2x + 1$. Is this rational function unique? Compare yours with those of other students. What will be the same as everyone else's? Add some more characteristics, such as symmetry or naming the real zeros. How does this modify the rational function?

91. The illustration shows the graph of a polynomial function.

(a) Is the degree of the polynomial even or odd?

(b) Is the leading coefficient positive or negative?

(c) Is the function even, odd, or neither?

(d) Why is x^2 necessarily a factor of the polynomial?

(e) What is the minimum degree of the polynomial?

(f) Formulate five different polynomials whose graphs could look like the one shown. Compare yours to those of other students. What similarities do you see? What differences?

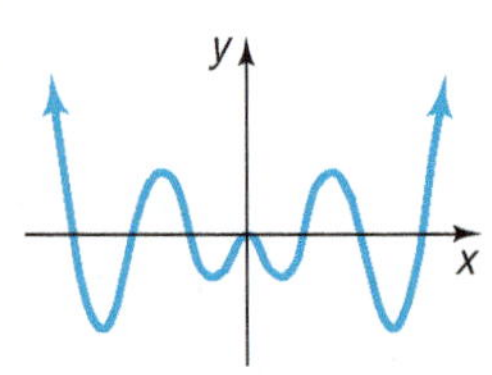

CHAPTER TEST

1. Graph $f(x) = (x - 3)^4 - 2$ using transformations.
2. For the polynomial function $g(x) = 2x^3 + 5x^2 - 28x - 15$,
   (a) Determine the maximum number of real zeros that the function may have.
   (b) Find bounds to the zeros of the function.
   (c) List the potential rational zeros.
   (d) Determine the real zeros of g. Factor g over the reals.
   (e) Find the x- and y-intercepts of the graph of g.
   (f) Determine whether the graph crosses or touches the x-axis at each x-intercept.
   (g) Find the power function that the graph of g resembles for large values of $|x|$.
   (h) Approximate the turning points on the graph of g.
   (i) Put all the information together to obtain the graph of g.
3. Find the complex zeros of $f(x) = x^3 - 4x^2 + 25x - 100$.
4. Solve $3x^3 + 2x - 1 = 8x^2 - 4$ in the complex number system.

In Problems 5 and 6, find the domain of each function. Find any horizontal, vertical, or oblique asymptotes.

5. $g(x) = \dfrac{2x^2 - 14x + 24}{x^2 + 6x - 40}$
6. $r(x) = \dfrac{x^2 + 2x - 3}{x + 1}$
7. Sketch the graph of the function in Problem 6. Label all intercepts, vertical asymptotes, horizontal asymptotes, and oblique asymptotes.

In Problems 8 and 9, write a function that meets the given conditions.

8. Fourth-degree polynomial with real coefficients; zeros: $-2, 0, 3 + i$
9. Rational function; asymptotes: $y = 2$, $x = 4$; domain: $\{x \mid x \neq 4, x \neq 9\}$
10. Use the Intermediate Value Theorem to show that the function $f(x) = -2x^2 - 3x + 8$ has at least one real zero on the interval $[0, 4]$.
11. Solve: $\dfrac{x + 2}{x - 3} < 2$

CUMULATIVE REVIEW

1. Find the distance between the points $P = (1, 3)$ and $Q = (-4, 2)$.
2. Solve the inequality $x^2 \geq x$ and graph the solution set.
3. Solve the inequality $x^2 - 3x < 4$ and graph the solution set.
4. Find a linear function with slope -3 that contains the point $(-1, 4)$. Graph the function.
5. Find the equation of the line parallel to the line $y = 2x + 1$ and containing the point $(3, 5)$. Express your answer in slope–intercept form and graph the line.
6. Graph the equation $y = x^3$.
7. Does the relation $\{(3, 6), (1, 3), (2, 5), (3, 8)\}$ represent a function? Why or why not?
8. Solve the equation $x^3 - 6x^2 + 8x = 0$.
9. Solve the inequality $3x + 2 \leq 5x - 1$ and graph the solution set.
10. Find the center and radius of the circle $x^2 + 4x + y^2 - 2y - 4 = 0$. Graph the circle.
11. For the equation $y = x^3 - 9x$, determine the intercepts and test for symmetry.
12. Find an equation of the line perpendicular to $3x - 2y = 7$ that contains the point $(1, 5)$.
13. Is the following graph the graph of a function? Why or why not?

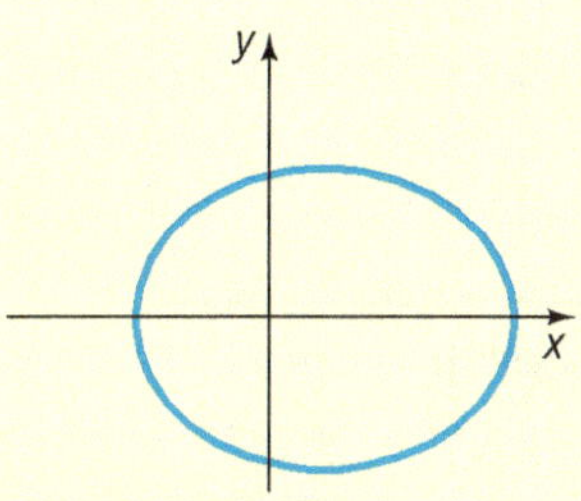

14. For the function $f(x) = x^2 + 5x - 2$, find
    (a) $f(3)$
    (b) $f(-x)$
    (c) $-f(x)$
    (d) $f(3x)$
    (e) $\dfrac{f(x + h) - f(x)}{h}$ $\quad h \neq 0$
15. Answer the following questions regarding the function
    $$f(x) = \frac{x + 5}{x - 1}$$
    (a) What is the domain of f?
    (b) Is the point $(2, 6)$ on the graph of f?
    (c) If $x = 3$, what is $f(x)$? What point is on the graph of f?
    (d) If $f(x) = 9$, what is x? What point is on the graph of f?
16. Graph the function $f(x) = -3x + 7$.
17. Graph $f(x) = 2x^2 - 4x + 1$ by determining whether its graph opens up or down and by finding its vertex, axis of symmetry, y-intercept, and x-intercepts, if any.
18. Find the average rate of change of $f(x) = x^2 + 3x + 1$ from 1 to 2. Use this result to find the equation of the secant line containing $(1, f(1))$ and $(2, f(2))$.
19. In parts (a) to (f), use the following graph.

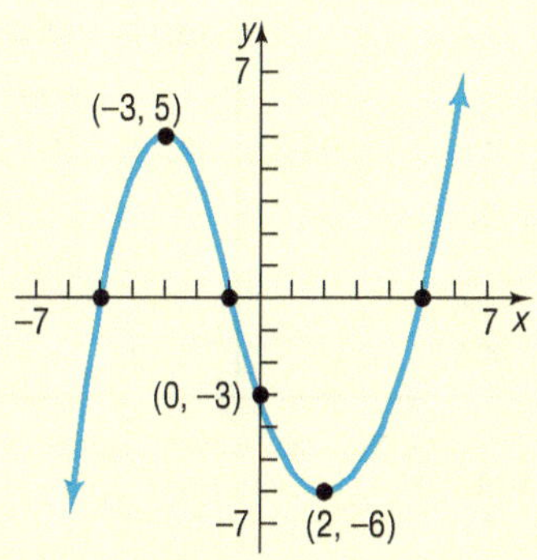

(a) Determine the intercepts.
(b) Based on the graph, tell whether the graph is symmetric with respect to the x-axis, the y-axis, and/or the origin.
(c) Based on the graph, tell whether the function is even, odd, or neither.
(d) List the intervals on which f is increasing. List the intervals on which f is decreasing.
(e) List the numbers, if any, at which f has a local maximum. What are these local maxima?
(f) List the numbers, if any, at which f has a local minimum. What are these local minima?

20. Determine algebraically whether the function

$$f(x) = \frac{5x}{x^2 - 9}$$

is even, odd, or neither.

21. For the function $f(x) = \begin{cases} 2x + 1 & \text{if } -3 < x < 2 \\ -3x + 4 & \text{if } x \geq 2 \end{cases}$

(a) Find the domain of f.
(b) Locate any intercepts.
(c) Graph the function.
(d) Based on the graph, find the range.

22. Graph the function $f(x) = -3(x + 1)^2 + 5$ using transformations.

23. Suppose that $f(x) = x^2 - 5x + 1$ and $g(x) = -4x - 7$.
(a) Find $f + g$ and state its domain.
(b) Find $\frac{f}{g}$ and state its domain.

24. The price p (in dollars) and the quantity x sold of a certain product obey the demand equation

$$p = -\frac{1}{10}x + 150, \qquad 0 \leq x \leq 1500$$

(a) Express the revenue R as a function of x.
(b) What is the revenue if 100 units are sold?
(c) What quantity x maximizes revenue? What is the maximum revenue?
(d) What price should the company charge to maximize revenue?

CHAPTER PROJECTS

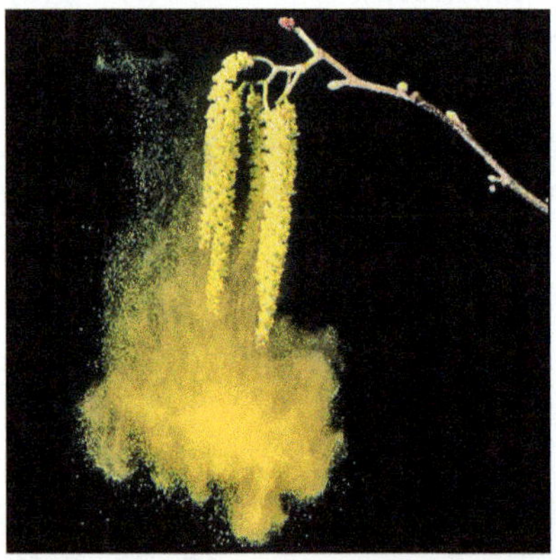

I. Weed Pollen The table shows the tree pollen count for Lexington, Kentucky, from April 1, 2006, to May 16, 2006, during the prime season for tree pollen.

Date	Count	Date	Count	Date	Count	Date	Count	Date	Count
4/1	76	4/10	19	4/19	1162	4/29	359	5/8	63
4/2	43	4/11	207	4/20	924	4/30	164	5/9	91
4/3	114	4/12	903	4/21	1030	5/1	65	5/10	166
4/4	73	4/13	1058	4/22	82	5/2	69	5/11	80
4/5	140	4/14	981	4/23	1514	5/3	30	5/12	153
4/6	138	4/15	1832	4/25	724	5/4	181	5/13	71
4/7	153	4/16	2058	4/26	102	5/5	95	5/14	63
4/8	264	4/17	1387	4/27	110	5/6	300	5/15	13
4/9	203	4/18	1188	4/28	320	5/7	91	5/16	9

Source: National Allergy Bureau at the American Academy of Allergy, Asthma and Immunology Web site. Used with permission.

1. Let the independent variable D represent the date, where $D = 1$ on 4/1, $D = 2$ on 4/2, $D = 3$ on 4/3, and $D = 46$ on 5/16. Let the dependent variable P represent the number of grains per cubic meter of weed pollen. Draw a scatter diagram of the data using your graphing utility and by hand on graph paper.
2. In the hand-drawn scatter diagram, sketch a smooth curve that fits the data. Try to have the curve pass through as many of the points as closely as possible. How many turning points are there? What does this tell you about the degree of the polynomial that could model the data? If the ends of the graph on the left and on the right are extended downward, how many real zeros are indicated by the graph? How many complex? Why?
3. Use a graphing utility to determine the quartic function of best fit. Graph it on your graphing utility. Does it look like what you expected from your hand-drawn graph? Explain.
4. Use a graphing utility to determine the cubic function of best fit. Graph it on your graphing utility. Does it look like what you expected from your hand-drawn graph? Explain.
5. Use a graphing utility to determine the quadratic function of best fit. Graph it on your graphing utility. Does it look like what you expected from your hand-drawn graph? Explain.
6. Which of the three functions seems to fit the best? Explain your reasoning.
7. Investigate weed pollen counts at other times of the year. (Go to the National Allergy Bureau at www.aaaai.org.) Would the polynomial you found in part (3) work for other 6-week periods? Why or why not? What about for the same period of time in other years?

The following project is available at the Instructor's Resource Center (IRC):

II. Theory of Equations The coefficients of a polynomial function can be found if its zeros are known, an advantage of using polynomials in modeling.

Exponential and Logarithmic Functions

6

The McDonald's Scalding Coffee Case

April 3, 1996—There is a lot of hype about the McDonald's scalding coffee case. No one is in favor of frivolous cases or outlandish results; however, it is important to understand some points that were not reported in most of the stories about the case. McDonald's coffee was not only hot, it was scalding, capable of almost instantaneous destruction of skin, flesh, and muscle.

Plaintiff's expert, a scholar in thermodynamics applied to human skin burns, testified that liquids at 180 degrees will cause a full thickness burn to human skin in two to seven seconds. Other testimony showed that, as the temperature decreases toward 155 degrees, the extent of the burn relative to that temperature decreases exponentially. Thus, if (the) spill had involved coffee at 155 degrees, the liquid would have cooled and given her time to avoid a serious burn.

Source: ATLA fact sheet © 1995, 1996 Consumer Attorneys of CA. Used with permission of the Association of Trial Lawyers of America.

—See Chapter Project I—

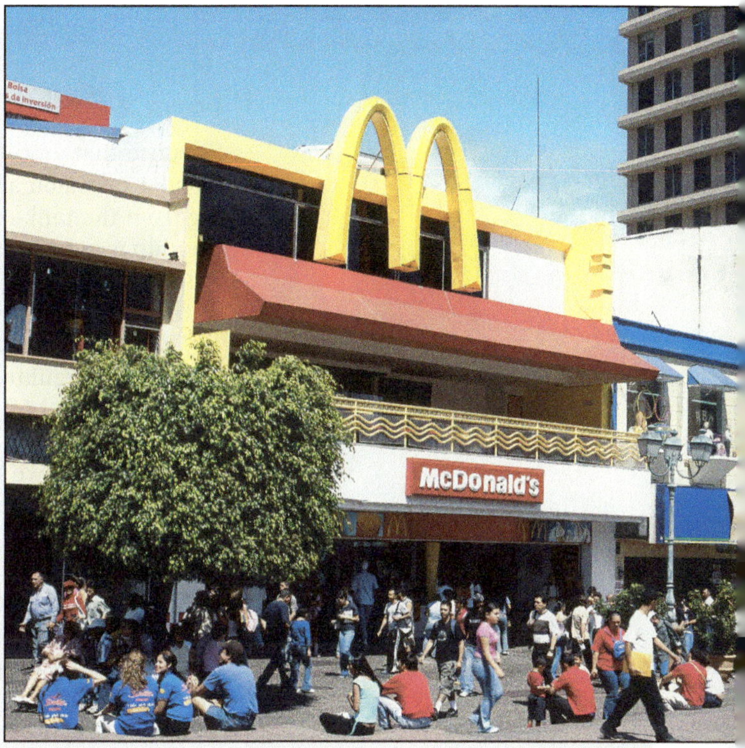

A Look Back

Until now, our study of functions has concentrated on polynomial and rational functions. These functions belong to the class of **algebraic functions,** that is, functions that can be expressed in terms of sums, differences, products, quotients, powers, or roots of polynomials. Functions that are not algebraic are termed **transcendental** (they transcend, or go beyond, algebraic functions).

A Look Ahead

In this chapter, we study two transcendental functions: the exponential function and the logarithmic function. These functions occur frequently in a wide variety of applications, such as biology, chemistry, economics, and psychology.

The chapter begins with a discussion of composite, one-to-one, and inverse functions, concepts needed to see the relationship between exponential and logarithmic functions.

Outline

6.1 Composite Functions

PREPARING FOR THIS SECTION *Before getting started, review the following:*

- Finding Values of a Function (Section 3.1, pp. 216–218)
- Domain of a Function (Section 3.1, pp. 219–220)

Now Work the 'Are You Prepared?' problems on page 409.

OBJECTIVES 1 Form a Composite Function (p. 404)
2 Find the Domain of a Composite Function (p. 405)

1 Form a Composite Function

Figure 1

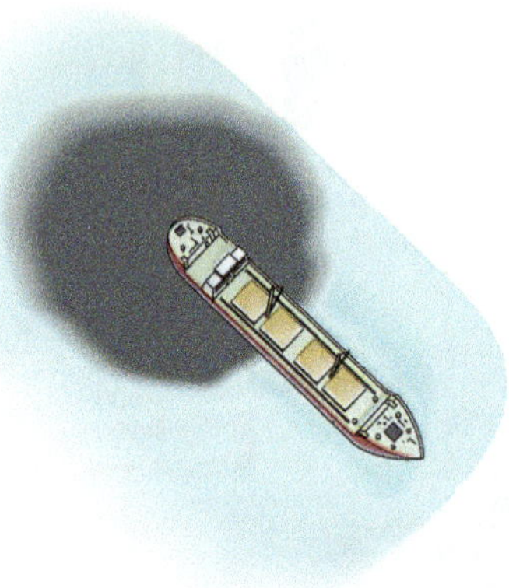

Suppose that an oil tanker is leaking oil and you want to determine the area of the circular oil patch around the ship. See Figure 1. It is determined that the oil is leaking from the tanker in such a way that the radius of the circular patch of oil around the ship is increasing at a rate of 3 feet per minute. Therefore, the radius r of the oil patch at any time t, in minutes, is given by $r(t) = 3t$. So after 20 minutes the radius of the oil patch is $r(20) = 3(20) = 60$ feet.

The area A of a circle as a function of the radius r is given by $A(r) = \pi r^2$. The area of the circular patch of oil after 20 minutes is $A(60) = \pi(60)^2 = 3600\pi$ square feet. Notice that $60 = r(20)$, so $A(60) = A(r(20))$. The argument of the function A is itself a function!

In general, we can find the area of the oil patch as a function of time t by evaluating $A(r(t))$. The function $A(r(t))$ is a special type of function called a *composite function.*

As another example, consider the function $y = (2x + 3)^2$. If we write $y = f(u) = u^2$ and $u = g(x) = 2x + 3$, then, by a substitution process, we can obtain the original function: $y = f(u) = f(g(x)) = (2x + 3)^2$.

In general, suppose that f and g are two functions and that x is a number in the domain of g. By evaluating g at x, we get $g(x)$. If $g(x)$ is in the domain of f, then we may evaluate f at $g(x)$ and obtain the expression $f(g(x))$. The correspondence from x to $f(g(x))$ is called a *composite function* $f \circ g$.

DEFINITION

Given two functions f and g, the **composite function,** denoted by $f \circ g$ (read as "f composed with g"), is defined by

$$(f \circ g)(x) = f(g(x))$$

The domain of $f \circ g$ is the set of all numbers x in the domain of g such that $g(x)$ is in the domain of f.

Look carefully at Figure 2. Only those x's in the domain of g for which $g(x)$ is in the domain of f can be in the domain of $f \circ g$. The reason is that if $g(x)$ is not in the domain of f then $f(g(x))$ is not defined. Because of this, the domain of $f \circ g$ is a subset of the domain of g; the range of $f \circ g$ is a subset of the range of f.

Figure 2

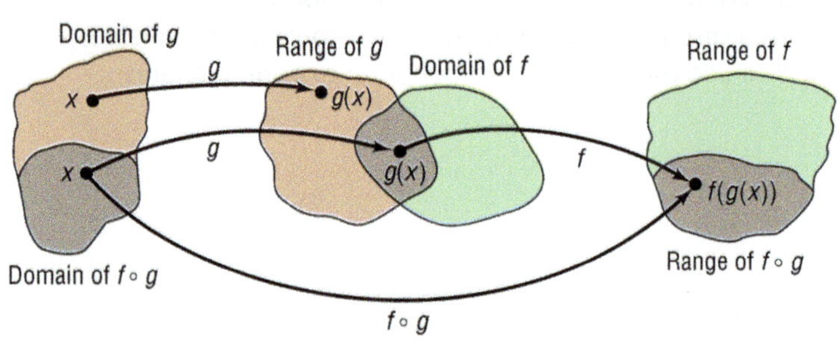

Figure 3 provides a second illustration of the definition. Here x is the input to the function g, yielding $g(x)$. Then $g(x)$ is the input to the function f, yielding $f(g(x))$. Notice that the "inside" function g in $f(g(x))$ is done first.

Figure 3

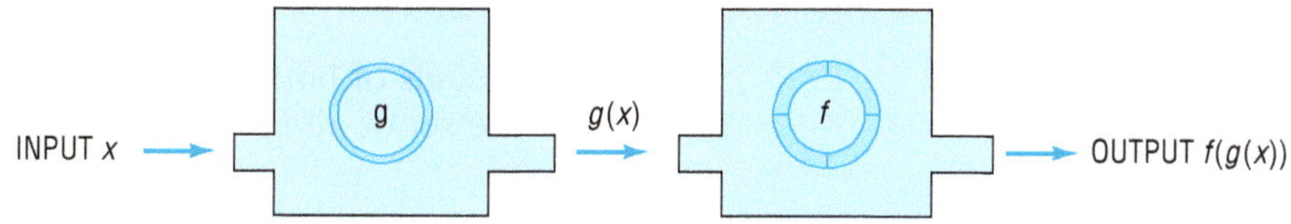

Let's look at some examples.

EXAMPLE 1 **Evaluating a Composite Function**

Suppose that $f(x) = 2x^2 - 3$ and $g(x) = 4x$. Find:

(a) $(f \circ g)(1)$ (b) $(g \circ f)(1)$ (c) $(f \circ f)(-2)$ (d) $(g \circ g)(-1)$

Solution

(a) $(f \circ g)(1) = f(g(1)) = f(4) = 2 \cdot 4^2 - 3 = 29$

$g(x) = 4x$, $g(1) = 4$; $f(x) = 2x^2 - 3$

(b) $(g \circ f)(1) = g(f(1)) = g(-1) = 4 \cdot (-1) = -4$

$f(x) = 2x^2 - 3$, $f(1) = -1$; $g(x) = 4x$

(c) $(f \circ f)(-2) = f(f(-2)) = f(5) = 2 \cdot 5^2 - 3 = 47$

$f(-2) = 2(-2)^2 - 3 = 5$

(d) $(g \circ g)(-1) = g(g(-1)) = g(-4) = 4 \cdot (-4) = -16$

$g(-1) = -4$

Figure 4

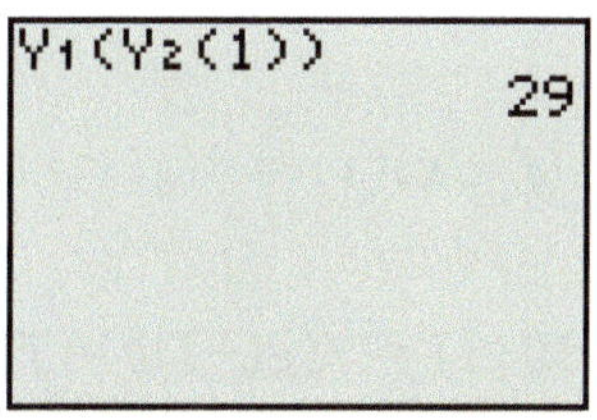

COMMENT Graphing calculators can be used to evaluate composite functions.* Let $Y_1 = f(x) = 2x^2 - 3$ and $Y_2 = g(x) = 4x$. Then, using a TI-84 Plus graphing calculator, $(f \circ g)(1)$ is found as shown in Figure 4. Notice that this is the result obtained in Example 1(a).

Now Work PROBLEM 11

2 Find the Domain of a Composite Function

EXAMPLE 2 **Finding a Composite Function and Its Domain**

Suppose that $f(x) = x^2 + 3x - 1$ and $g(x) = 2x + 3$.

Find: (a) $f \circ g$ (b) $g \circ f$

Then find the domain of each composite function.

Solution The domain of f and the domain of g are the set of all real numbers.

(a) $(f \circ g)(x) = f(g(x)) = f(2x + 3) = (2x + 3)^2 + 3(2x + 3) - 1$

$f(x) = x^2 + 3x - 1$

$= 4x^2 + 12x + 9 + 6x + 9 - 1 = 4x^2 + 18x + 17$

Since the domains of both f and g are the set of all real numbers, the domain of $f \circ g$ is the set of all real numbers.

*Consult your owner's manual for the appropriate keystrokes.

(b) $(g \circ f)(x) = g(f(x)) = g(x^2 + 3x - 1) = 2(x^2 + 3x - 1) + 3$

$\uparrow$ $g(x) = 2x + 3$

$= 2x^2 + 6x - 2 + 3 = 2x^2 + 6x + 1$

Since the domains of both f and g are the set of all real numbers, the domain of $g \circ f$ is the set of all real numbers.

Look back at Figure 2. In determining the domain of the composite function $(f \circ g)(x) = f(g(x))$, keep the following two thoughts in mind about the input x.

1. $g(x)$ must be defined so that any x not in the domain of g must be excluded.
2. $f(g(x))$ must be defined so that any x for which $g(x)$ is not in the domain of f must be excluded.

EXAMPLE 3 Finding the Domain of $f \circ g$

Find the domain of $(f \circ g)(x)$ if $f(x) = \dfrac{1}{x+2}$ and $g(x) = \dfrac{4}{x-1}$.

Solution For $(f \circ g)(x) = f(g(x))$, we first note that the domain of g is $\{x | x \neq 1\}$, so we exclude 1 from the domain of $f \circ g$. Next we note that the domain of f is $\{x | x \neq -2\}$, which means that $g(x)$ cannot equal -2. We solve the equation $g(x) = -2$ to determine what values of x to exclude.

$$\begin{aligned} \frac{4}{x-1} &= -2 \qquad g(x) = -2 \\ 4 &= -2(x-1) \\ 4 &= -2x + 2 \\ 2x &= -2 \\ x &= -1 \end{aligned}$$

We also exclude -1 from the domain of $f \circ g$. The domain of $f \circ g$ is $\{x | x \neq -1, x \neq 1\}$.

Check: For $x = 1$, $g(x) = \dfrac{4}{x-1}$ is not defined, so $(f \circ g)(x) = f(g(x))$ is not defined.

For $x = -1$, $g(-1) = \dfrac{4}{-2} = -2$, and $(f \circ g)(-1) = f(g(-1)) = f(-2)$ is not defined.

Now Work PROBLEM 21

EXAMPLE 4 Finding a Composite Function and Its Domain

Suppose that $f(x) = \dfrac{1}{x+2}$ and $g(x) = \dfrac{4}{x-1}$.

Find: (a) $f \circ g$ (b) $f \circ f$

Then find the domain of each composite function.

Solution The domain of f is $\{x | x \neq -2\}$ and the domain of g is $\{x | x \neq 1\}$.

(a) $(f \circ g)(x) = f(g(x)) = f\left(\dfrac{4}{x-1}\right) = \dfrac{1}{\dfrac{4}{x-1} + 2} = \dfrac{x-1}{4 + 2(x-1)} = \dfrac{x-1}{2x+2} = \dfrac{x-1}{2(x+1)}$

$\uparrow$ $f(x) = \dfrac{1}{x+2}$ $\qquad$ $\uparrow$ Multiply by $\dfrac{x-1}{x-1}$.

In Example 3, we found the domain of $f \circ g$ to be $\{x | x \neq -1, x \neq 1\}$.

We could also find the domain of $f \circ g$ by first looking at the domain of g: $\{x|x \neq 1\}$. We exclude 1 from the domain of $f \circ g$ as a result. Then we look at $f \circ g$ and notice that x cannot equal -1, since $x = -1$ results in division by 0. So we also exclude -1 from the domain of $f \circ g$. Therefore, the domain of $f \circ g$ is $\{x|x \neq -1, x \neq 1\}$.

(b) $$(f \circ f)(x) = f(f(x)) = f\left(\frac{1}{x+2}\right) = \frac{1}{\frac{1}{x+2} + 2} = \frac{x+2}{1 + 2(x+2)} = \frac{x+2}{2x+5}$$

$f(x) = \frac{1}{x+2}$ ↑ ↑ Multiply by $\frac{x+2}{x+2}$.

The domain of $f \circ f$ consists of those x in the domain of f, $\{x|x \neq -2\}$, for which

$$f(x) = \frac{1}{x+2} \neq -2$$

$$\frac{1}{x+2} = -2$$
$$1 = -2(x+2)$$
$$1 = -2x - 4$$
$$2x = -5$$
$$x = -\frac{5}{2}$$

or, equivalently,

$$x \neq -\frac{5}{2}$$

The domain of $f \circ f$ is $\left\{x \middle| x \neq -\frac{5}{2}, x \neq -2\right\}$.

We could also find the domain of $f \circ f$ by recognizing that -2 is not in the domain of f and so should be excluded from the domain of $f \circ f$. Then, looking at $f \circ f$, we see that x cannot equal $-\frac{5}{2}$. Do you see why? Therefore, the domain of $f \circ f$ is $\left\{x \middle| x \neq -\frac{5}{2}, x \neq -2\right\}$.

Now Work PROBLEMS 33 AND 35

Look back at Example 2, which illustrates that, in general, $f \circ g \neq g \circ f$. Sometimes $f \circ g$ does equal $g \circ f$, as shown in the next example.

EXAMPLE 5 Showing That Two Composite Functions Are Equal

If $f(x) = 3x - 4$ and $g(x) = \frac{1}{3}(x + 4)$, show that

$$(f \circ g)(x) = (g \circ f)(x) = x$$

for every x in the domain of $f \circ g$ and $g \circ f$.

Solution

$$\begin{aligned}(f \circ g)(x) &= f(g(x)) \\ &= f\left(\frac{x+4}{3}\right) && g(x) = \frac{1}{3}(x+4) = \frac{x+4}{3} \\ &= 3\left(\frac{x+4}{3}\right) - 4 && \text{Substitute } g(x) \text{ into the rule for } f,\ f(x) = 3x - 4. \\ &= x + 4 - 4 = x\end{aligned}$$

Seeing the Concept

Using a graphing calculator, let

$$Y_1 = f(x) = 3x - 4$$
$$Y_2 = g(x) = \frac{1}{3}(x + 4)$$
$$Y_3 = f \circ g, Y_4 = g \circ f$$

Using the viewing window $-3 \le x \le 3$, $-2 \le y \le 2$, graph only Y_3 and Y_4. What do you see? TRACE to verify that $Y_3 = Y_4$.

$$\begin{aligned}(g \circ f)(x) &= g(f(x)) \\ &= g(3x - 4) && f(x) = 3x - 4 \\ &= \frac{1}{3}[(3x - 4) + 4] && \text{Substitute } f(x) \text{ into the rule for } g, g(x) = \frac{1}{3}(x + 4). \\ &= \frac{1}{3}(3x) = x\end{aligned}$$

We conclude that $(f \circ g)(x) = (g \circ f)(x) = x$.

In Section 6.2, we shall see that there is an important relationship between functions f and g for which $(f \circ g)(x) = (g \circ f)(x) = x$.

Now Work PROBLEM 45

Calculus Application

Some techniques in calculus require that we be able to determine the components of a composite function. For example, the function $H(x) = \sqrt{x + 1}$ is the composition of the functions f and g, where $f(x) = \sqrt{x}$ and $g(x) = x + 1$, because $H(x) = (f \circ g)(x) = f(g(x)) = f(x + 1) = \sqrt{x + 1}$.

EXAMPLE 6 Finding the Components of a Composite Function

Find functions f and g such that $f \circ g = H$ if $H(x) = (x^2 + 1)^{50}$.

Solution The function H takes $x^2 + 1$ and raises it to the power 50. A natural way to decompose H is to raise the function $g(x) = x^2 + 1$ to the power 50. If we let $f(x) = x^{50}$ and $g(x) = x^2 + 1$, then

Figure 5

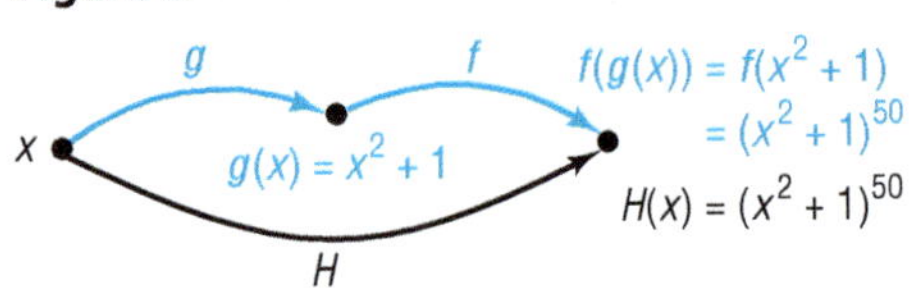

$$\begin{aligned}(f \circ g)(x) &= f(g(x)) \\ &= f(x^2 + 1) \\ &= (x^2 + 1)^{50} = H(x)\end{aligned}$$

See Figure 5.

Other functions f and g may be found for which $f \circ g = H$ in Example 6. For example, if $f(x) = x^2$ and $g(x) = (x^2 + 1)^{25}$, then

$$(f \circ g)(x) = f(g(x)) = f((x^2 + 1)^{25}) = [(x^2 + 1)^{25}]^2 = (x^2 + 1)^{50}$$

Although the functions f and g found as a solution to Example 6 are not unique, there is usually a "natural" selection for f and g that comes to mind first.

EXAMPLE 7 Finding the Components of a Composite Function

Find functions f and g such that $f \circ g = H$ if $H(x) = \dfrac{1}{x + 1}$.

Solution Here H is the reciprocal of $g(x) = x + 1$. If we let $f(x) = \dfrac{1}{x}$ and $g(x) = x + 1$, we find that

$$(f \circ g)(x) = f(g(x)) = f(x + 1) = \frac{1}{x + 1} = H(x)$$

Now Work PROBLEM 53

6.1 Assess Your Understanding

'Are You Prepared?' *Answers are given at the end of these exercises. If you get a wrong answer, read the pages listed in red.*

1. Find $f(3)$ if $f(x) = -4x^2 + 5x$. (pp. 216–218)

2. Find $f(3x)$ if $f(x) = 4 - 2x^2$. What is the argument of f? (p. 217)

3. Find the domain of the function $f(x) = \frac{x^2 - 1}{x^2 - 25}$. (pp. 219–220)

Concepts and Vocabulary

4. If $f(x) = x + 1$ and $g(x) = x^3$, then _____ $= (x + 1)^3$.

5. ***True or False*** $f(g(x)) = f(x) \cdot g(x)$.

6. ***True or False*** The domain of the composite function $(f \circ g)(x)$ is the same as the domain of $g(x)$.

Skill Building

In Problems 7 and 8, evaluate each expression using the values given in the table.

7.

x	−3	−2	−1	0	1	2	3
f(x)	−7	−5	−3	−1	3	5	5
g(x)	8	3	0	−1	0	3	8

(a) $(f \circ g)(1)$ (b) $(f \circ g)(-1)$ (c) $(g \circ f)(-1)$ (d) $(g \circ f)(0)$ (e) $(g \circ g)(-2)$ (f) $(f \circ f)(-1)$

8.

x	−3	−2	−1	0	1	2	3
f(x)	11	9	7	5	3	1	−1
g(x)	−8	−3	0	1	0	−3	−8

(a) $(f \circ g)(1)$ (b) $(f \circ g)(2)$ (c) $(g \circ f)(2)$ (d) $(g \circ f)(3)$ (e) $(g \circ g)(1)$ (f) $(f \circ f)(3)$

In Problems 9 and 10, evaluate each expression using the graphs of $y = f(x)$ and $y = g(x)$ shown in the figure.

9. (a) $(g \circ f)(-1)$ (b) $(g \circ f)(0)$
(c) $(f \circ g)(-1)$ (d) $(f \circ g)(4)$

10. (a) $(g \circ f)(1)$ (b) $(g \circ f)(5)$
(c) $(f \circ g)(0)$ (d) $(f \circ g)(2)$

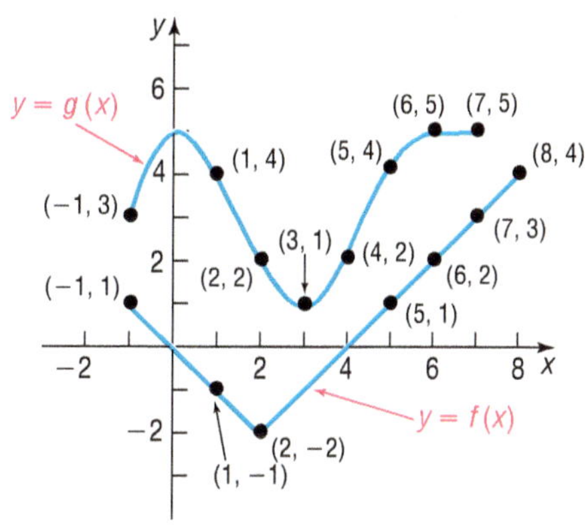

In Problems 11–20, for the given functions f and g, find:

(a) $(f \circ g)(4)$ *(b)* $(g \circ f)(2)$ *(c)* $(f \circ f)(1)$ *(d)* $(g \circ g)(0)$

11. $f(x) = 2x;\quad g(x) = 3x^2 + 1$

12. $f(x) = 3x + 2;\quad g(x) = 2x^2 - 1$

13. $f(x) = 4x^2 - 3;\quad g(x) = 3 - \frac{1}{2}x^2$

14. $f(x) = 2x^2;\quad g(x) = 1 - 3x^2$

15. $f(x) = \sqrt{x};\quad g(x) = 2x$

16. $f(x) = \sqrt{x + 1};\quad g(x) = 3x$

17. $f(x) = |x|;\quad g(x) = \frac{1}{x^2 + 1}$

18. $f(x) = |x - 2|;\quad g(x) = \frac{3}{x^2 + 2}$

19. $f(x) = \frac{3}{x + 1};\quad g(x) = \sqrt[3]{x}$

20. $f(x) = x^{3/2};\quad g(x) = \frac{2}{x + 1}$

In Problems 21–28, find the domain of the composite function $f \circ g$.

21. $f(x) = \frac{3}{x - 1};\quad g(x) = \frac{2}{x}$

22. $f(x) = \frac{1}{x + 3};\quad g(x) = -\frac{2}{x}$

23. $f(x) = \frac{x}{x-1}$; $g(x) = -\frac{4}{x}$

24. $f(x) = \frac{x}{x+3}$; $g(x) = \frac{2}{x}$

25. $f(x) = \sqrt{x}$; $g(x) = 2x + 3$

26. $f(x) = x - 2$; $g(x) = \sqrt{1-x}$

27. $f(x) = x^2 + 1$; $g(x) = \sqrt{x-1}$

28. $f(x) = x^2 + 4$; $g(x) = \sqrt{x-2}$

In Problems 29–44, for the given functions f and g, find:

(a) $f \circ g$ *(b)* $g \circ f$ *(c)* $f \circ f$ *(d)* $g \circ g$

State the domain of each composite function.

29. $f(x) = 2x + 3$; $g(x) = 3x$

30. $f(x) = -x$; $g(x) = 2x - 4$

31. $f(x) = 3x + 1$; $g(x) = x^2$

32. $f(x) = x + 1$; $g(x) = x^2 + 4$

33. $f(x) = x^2$; $g(x) = x^2 + 4$

34. $f(x) = x^2 + 1$; $g(x) = 2x^2 + 3$

35. $f(x) = \frac{3}{x-1}$; $g(x) = \frac{2}{x}$

36. $f(x) = \frac{1}{x+3}$; $g(x) = -\frac{2}{x}$

37. $f(x) = \frac{x}{x-1}$; $g(x) = -\frac{4}{x}$

38. $f(x) = \frac{x}{x+3}$; $g(x) = \frac{2}{x}$

39. $f(x) = \sqrt{x}$; $g(x) = 2x + 3$

40. $f(x) = \sqrt{x-2}$; $g(x) = 1 - 2x$

41. $f(x) = x^2 + 1$; $g(x) = \sqrt{x-1}$

42. $f(x) = x^2 + 4$; $g(x) = \sqrt{x-2}$

43. $f(x) = \frac{x-5}{x+1}$; $g(x) = \frac{x+2}{x-3}$

44. $f(x) = \frac{2x-1}{x-2}$; $g(x) = \frac{x+4}{2x-5}$

In Problems 45–52, show that $(f \circ g)(x) = (g \circ f)(x) = x$.

45. $f(x) = 2x$; $g(x) = \frac{1}{2}x$

46. $f(x) = 4x$; $g(x) = \frac{1}{4}x$

47. $f(x) = x^3$; $g(x) = \sqrt[3]{x}$

48. $f(x) = x + 5$; $g(x) = x - 5$

49. $f(x) = 2x - 6$; $g(x) = \frac{1}{2}(x + 6)$

50. $f(x) = 4 - 3x$; $g(x) = \frac{1}{3}(4 - x)$

51. $f(x) = ax + b$; $g(x) = \frac{1}{a}(x - b)$ $a \neq 0$

52. $f(x) = \frac{1}{x}$; $g(x) = \frac{1}{x}$

In Problems 53–58, find functions f and g so that $f \circ g = H$.

53. $H(x) = (2x + 3)^4$

54. $H(x) = (1 + x^2)^3$

55. $H(x) = \sqrt{x^2 + 1}$

56. $H(x) = \sqrt{1 - x^2}$

57. $H(x) = |2x + 1|$

58. $H(x) = |2x^2 + 3|$

Applications and Extensions

59. If $f(x) = 2x^3 - 3x^2 + 4x - 1$ and $g(x) = 2$, find $(f \circ g)(x)$ and $(g \circ f)(x)$.

60. If $f(x) = \frac{x+1}{x-1}$, find $(f \circ f)(x)$.

61. If $f(x) = 2x^2 + 5$ and $g(x) = 3x + a$, find a so that the graph of $f \circ g$ crosses the y-axis at 23.

62. If $f(x) = 3x^2 - 7$ and $g(x) = 2x + a$, find a so that the graph of $f \circ g$ crosses the y-axis at 68.

In Problems 63 and 64, use the functions f and g to find:

(a) $f \circ g$

(b) $g \circ f$

(c) the domain of $f \circ g$ *and of* $g \circ f$

(d) the conditions for which $f \circ g = g \circ f$

63. $f(x) = ax + b$; $g(x) = cx + d$

64. $f(x) = \frac{ax+b}{cx+d}$; $g(x) = mx$

65. Surface Area of a Balloon The surface area S (in square meters) of a hot-air balloon is given by

$$S(r) = 4\pi r^2$$

where r is the radius of the balloon (in meters). If the radius r is increasing with time t (in seconds) according to the formula $r(t) = \frac{2}{3}t^3$, $t \geq 0$, find the surface area S of the balloon as a function of the time t.

66. Volume of a Balloon The volume V (in cubic meters) of the hot-air balloon described in Problem 65 is given by

$V(r) = \frac{4}{3}\pi r^3$. If the radius r is the same function of t as in Problem 65, find the volume V as a function of the time t.

67. Automobile Production The number N of cars produced at a certain factory in 1 day after t hours of operation is given by $N(t) = 100t - 5t^2, 0 \le t \le 10$. If the cost C (in dollars) of producing N cars is $C(N) = 15{,}000 + 8000N$, find the cost C as a function of the time t of operation of the factory.

68. Environmental Concerns The spread of oil leaking from a tanker is in the shape of a circle. If the radius r (in feet) of the spread after t hours is $r(t) = 200\sqrt{t}$, find the area A of the oil slick as a function of the time t.

69. Production Cost The price p, in dollars, of a certain product and the quantity x sold obey the demand equation

$$p = -\frac{1}{4}x + 100 \quad 0 \le x \le 400$$

Suppose that the cost C, in dollars, of producing x units is

$$C = \frac{\sqrt{x}}{25} + 600$$

Assuming that all items produced are sold, find the cost C as a function of the price p.

[**Hint:** Solve for x in the demand equation and then form the composite.]

70. Cost of a Commodity The price p, in dollars, of a certain commodity and the quantity x sold obey the demand equation

$$p = -\frac{1}{5}x + 200 \quad 0 \le x \le 1000$$

Suppose that the cost C, in dollars, of producing x units is

$$C = \frac{\sqrt{x}}{10} + 400$$

Assuming that all items produced are sold, find the cost C as a function of the price p.

71. Volume of a Cylinder The volume V of a right circular cylinder of height h and radius r is $V = \pi r^2 h$. If the height is twice the radius, express the volume V as a function of r.

72. Volume of a Cone The volume V of a right circular cone is $V = \frac{1}{3}\pi r^2 h$. If the height is twice the radius, express the volume V as a function of r.

73. Foreign Exchange Traders often buy foreign currency in hope of making money when the currency's value changes. For example, on April 3, 2007, one U.S. dollar could purchase 0.7480 Euros, and one Euro could purchase 158.7160 yen. Let $f(x)$ represent the number of Euros you can buy with x dollars, and let $g(x)$ represent the number of yen you can buy with x Euros.

(a) Find a function that relates dollars to Euros.
(b) Find a function that relates Euros to yen.
(c) Use the results of parts (a) and (b) to find a function that relates dollars to yen. That is, find $(g \circ f)(x) = g(f(x))$.
(d) What is $g(f(1000))$?

74. Temperature Conversion The function $C(F) = \frac{5}{9}(F - 32)$ converts a temperature in degrees Fahrenheit, F, to a temperature in degrees Celsius, C. The function $K(C) = C + 273$, converts a temperature in degrees Celsius to a temperature in kelvins, K.

(a) Find a function that converts a temperature in degrees Fahrenheit to a temperature in kelvins.
(b) Determine 80 degrees Fahrenheit in kelvins.

75. Discounts The manufacturer of a computer is offering two discounts on last year's model computer. The first discount is a \$200 rebate and the second discount is 20% off the regular price, p.

(a) Write a function f that represents the sale price if only the rebate applies.
(b) Write a function g that represents the sale price if only the 20% discount applies.
(c) Find $f \circ g$ and $g \circ f$. What does each of these functions represent? Which combination of discounts represents a better deal for the consumer? Why?

76. If f and g are odd functions, show that the composite function $f \circ g$ is also odd.

77. If f is an odd function and g is an even function, show that the composite functions $f \circ g$ and $g \circ f$ are both even.

'Are You Prepared?' Answers

1. -21 **2.** $4 - 18x^2$; $3x$ **3.** $\{x \mid x \neq -5, x \neq 5\}$

6.2 One-to-One Functions; Inverse Functions

PREPARING FOR THIS SECTION *Before getting started, review the following:*

- Functions (Section 3.1, pp. 212–220)
- Increasing/Decreasing Functions (Section 3.3, pp. 237–238)

Now Work the 'Are You Prepared?' problems on page 421.

OBJECTIVES
1. Determine Whether a Function Is One-to-One (p. 412)
2. Determine the Inverse of a Function Defined by a Map or a Set of Ordered Pairs (p. 414)
3. Obtain the Graph of the Inverse Function from the Graph of the Function (p. 416)
4. Find the Inverse of a Function Defined by an Equation (p. 417)

1 Determine Whether a Function Is One-to-One

In Section 3.1, we presented four different ways to represent a function: as (1) a map, (2) a set of ordered pairs, (3) a graph, and (4) an equation. For example, Figures 6 and 7 illustrate two different functions represented as mappings. The function in Figure 6 shows the correspondence between states and their population (in millions). The function in Figure 7 shows a correspondence between animals and life expectancy (in years).

Figure 6

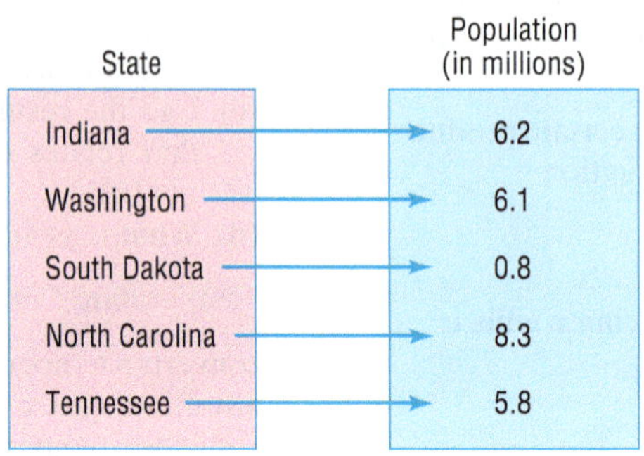

Figure 7

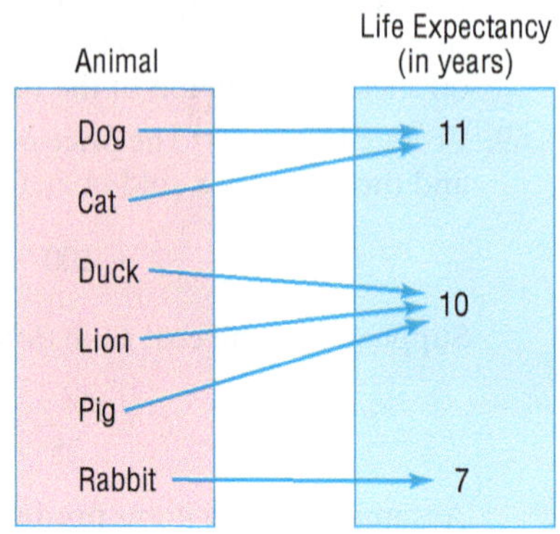

Suppose we asked a group of people to name the state that has a population of 0.8 million based on the function in Figure 6. Everyone in the group would respond South Dakota. Now, if we asked the same group of people to name the animal whose life expectancy is 11 years based on the function in Figure 7, some would respond dog, while others would respond cat. What is the difference between the functions in Figures 6 and 7? In Figure 6, we can see that no two elements in the domain corresponds to the same element in the range. In Figure 7, this is not the case: two different elements in the domain correspond to the same element in the range. We give functions such as the one in Figure 6 a special name.

DEFINITION

A function is **one-to-one** if any two different inputs in the domain correspond to two different outputs in the range. That is, if x_1 and x_2 are two different inputs of a function f, then f is one-to-one if $f(x_1) \neq f(x_2)$.

In Words

A function is not one-to-one if two different inputs correspond to the same output.

Put another way, a function f is one-to-one if no y in the range is the image of more than one x in the domain. A function is not one-to-one if two different elements in the domain correspond to the same element in the range. So the function in Figure 7 is not one-to-one because two different elements in the domain, *dog* and *cat,* both correspond to 11. Figure 8 illustrates the distinction of one-to-one functions, functions that are not one-to-one, and nonfunctions.

Figure 8

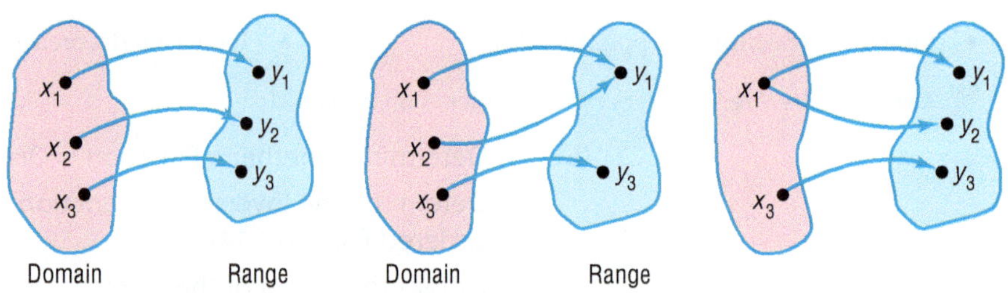

(a) One-to-one function: Each x in the domain has one and only one image in the range

(b) Not a one-to-one function: y_1 is the image of both x_1 and x_2.

(c) Not a function: x_1 has two images, y_1 and y_2.

EXAMPLE 1 Determining Whether a Function Is One-to-One

Determine whether the following functions are one-to-one.

(a) For the following function, the domain represents the age of five males and the range represents their HDL (good) cholesterol (mg/dL).

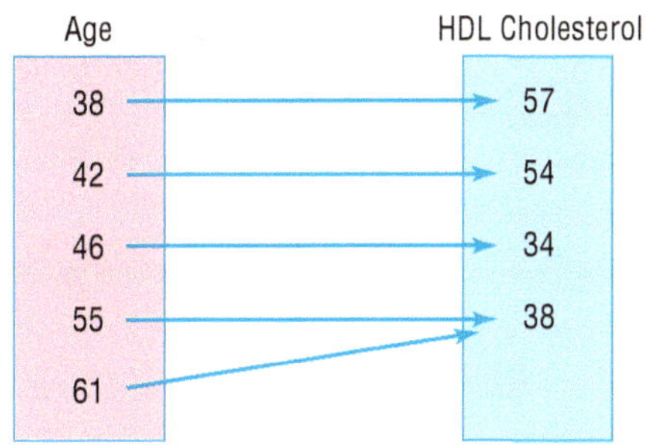

(b) $\{(-2, 6), (-1, 3), (0, 2), (1, 5), (2, 8)\}$

Solution (a) The function is not one-to-one because there are two different inputs, 55 and 61, that correspond to the same output, 38.

(b) The function is one-to-one because there are no two distinct inputs that correspond to the same output.

Now Work PROBLEMS 9 AND 13

For functions defined by an equation $y = f(x)$ and for which the graph of f is known, there is a simple test, called the **horizontal-line test,** to determine whether f is one-to-one.

THEOREM

Horizontal-line Test

If every horizontal line intersects the graph of a function f in at most one point, then f is one-to-one.

Figure 9
$f(x_1) = f(x_2) = h$ and $x_1 \neq x_2$; f is not a one-to-one function.

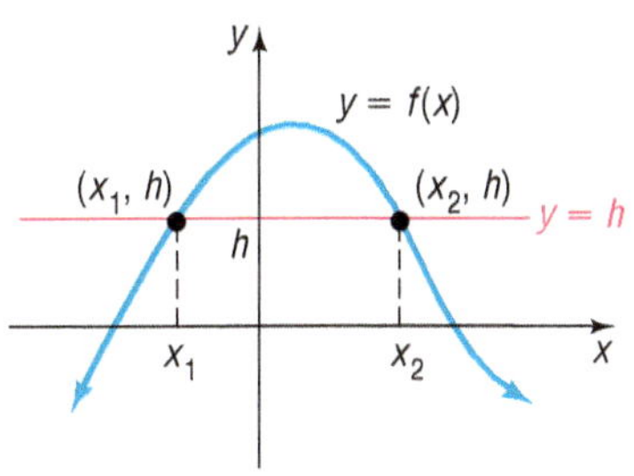

The reason that this test works can be seen in Figure 9, where the horizontal line $y = h$ intersects the graph at two distinct points, (x_1, h) and (x_2, h). Since h is the image of both x_1 and x_2 and $x_1 \neq x_2$, f is not one-to-one. Based on Figure 9, we can state the horizontal-line test in another way: If the graph of any horizontal line intersects the graph of a function f at more than one point, then f is not one-to-one.

EXAMPLE 2 Using the Horizontal-line Test

For each function, use its graph to determine whether the function is one-to-one.

(a) $f(x) = x^2$ (b) $g(x) = x^3$

Solution (a) Figure 10(a) on page 414 illustrates the horizontal-line test for $f(x) = x^2$. The horizontal line $y = 1$ intersects the graph of f twice, at $(1, 1)$ and at $(-1, 1)$, so f is not one-to-one.

(b) Figure 10(b) illustrates the horizontal-line test for $g(x) = x^3$. Because every horizontal line intersects the graph of g exactly once, it follows that g is one-to-one.

Figure 10

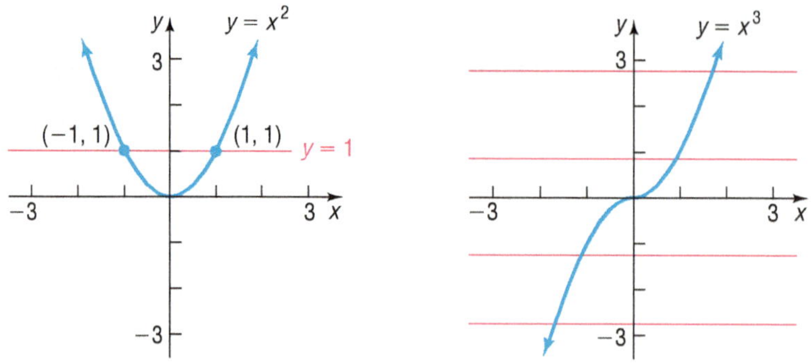

(a) A horizontal line intersects the graph twice; f is not one-to-one

(b) Every horizontal line intersects the graph exactly once; g is one-to-one

Now Work PROBLEM 17

Let's look more closely at the one-to-one function $g(x) = x^3$. This function is an increasing function. Because an increasing (or decreasing) function will always have different y-values for unequal x-values, it follows that a function that is increasing (or decreasing) over its domain is also a one-to-one function.

THEOREM

A function that is increasing on an interval I is a one-to-one function on I.
A function that is decreasing on an interval I is a one-to-one function on I.

2 Determine the Inverse of a Function Defined by a Map or a Set of Ordered Pairs

We begin with the definition.

DEFINITION

Suppose f is a one-to-one function. Then, to each x in the domain of f, there is exactly one y in the range (because f is a function); and to each y in the range of f, there is exactly one x in the domain (because f is one-to-one). The correspondence from the range of f back to the domain of f is called the **inverse function of f.** We use the symbol f^{-1} to denote the inverse of f.

In Words

Suppose we have a one-to-one function f where the input 5 corresponds to the output 10. In the inverse function f^{-1}, the input 10 would correspond to the output 5.

Recall that we have a variety of ways of representing functions. We will discuss how to find inverses for all four representations of functions: (1) maps, (2) sets of ordered pairs (3) graphs, and (4) equations. We begin with finding inverses of functions represented by maps or sets of ordered pairs.

EXAMPLE 3 **Finding the Inverse of a Function Defined by a Map**

Find the inverse of the following function. Let the domain of the function represent certain states, and let the range represent the state's population (in millions). State the domain and the range of the inverse function.

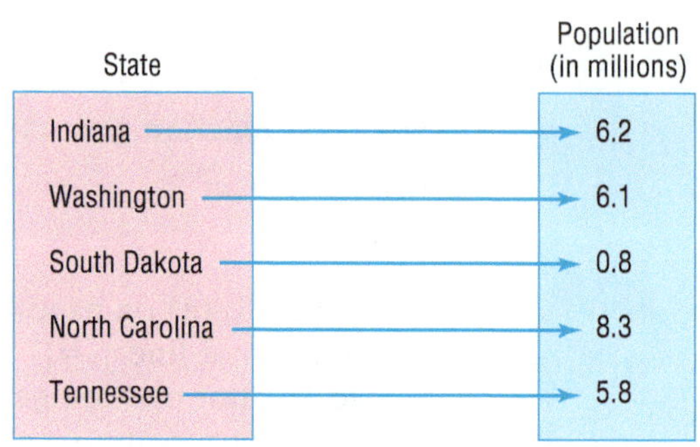

Solution The function is one-to-one. To find the inverse function, we interchange the elements in the domain with the elements in the range. For example, the function

receives as input Indiana and outputs 6.2 million. So the inverse receives as input 6.2 million and outputs Indiana. The inverse function is shown next.

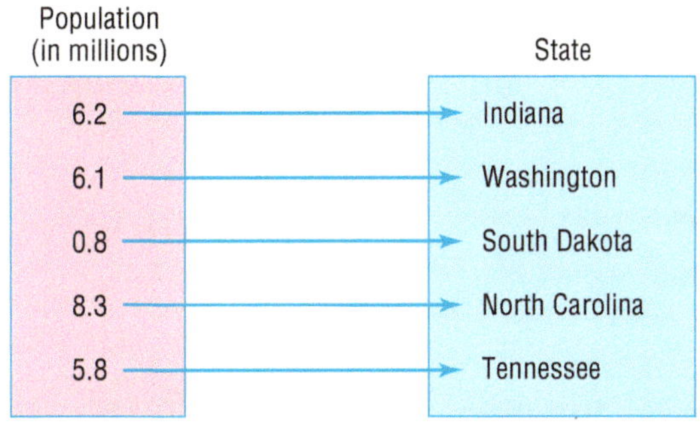

The domain of the inverse function is $\{6.2, 6.1, 0.8, 8.3, 5.8\}$. The range of the inverse function is {Indiana, Washington, South Dakota, North Carolina, Tennessee}.

If the function f is a set of ordered pairs (x, y), then the inverse of f, denoted f^{-1}, is the set of ordered pairs (y, x).

EXAMPLE 4 Finding the Inverse of a Function Defined by a Set of Ordered Pairs

Find the inverse of the following one-to-one function:

$$\{(-3, -27), (-2, -8), (-1, -1), (0, 0), (1, 1), (2, 8), (3, 27)\}$$

State the domain and the range of the function and its inverse.

Solution The inverse of the given function is found by interchanging the entries in each ordered pair and so is given by

$$\{(-27, -3), (-8, -2), (-1, -1), (0, 0), (1, 1), (8, 2), (27, 3)\}$$

The domain of the function is $\{-3, -2, -1, 0, 1, 2, 3\}$. The range of the function is $\{-27, -8, -1, 0, 1, 8, 27\}$. The domain of the inverse function is $\{-27, -8, -1, 0, 1, 8, 27\}$. The range of the inverse function is $\{-3, -2, -1, 0, 1, 2, 3\}$.

Now Work PROBLEMS 23 AND 27

Figure 11

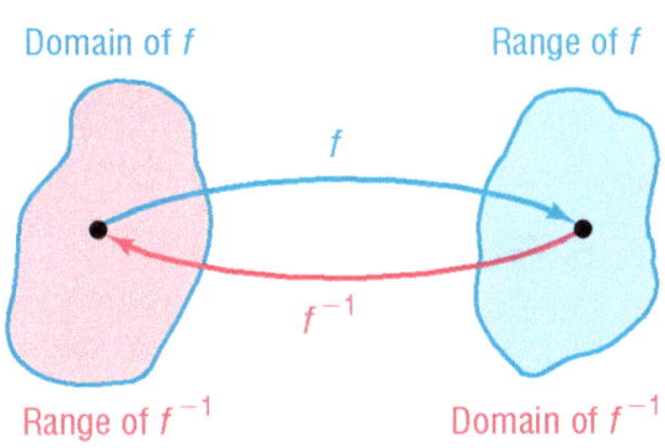

WARNING Be careful! f^{-1} is a symbol for the inverse function of f. The -1 used in f^{-1} is not an exponent. That is, f^{-1} does not mean the reciprocal of f; $f^{-1}(x)$ is not equal to $\dfrac{1}{f(x)}$.

Remember, if f is a one-to-one function, it has an inverse function, f^{-1}. See Figure 11.

Based on the results of Example 4 and Figure 11, two facts are now apparent about a one-to-one function f and its inverse f^{-1}.

$$\text{Domain of } f = \text{Range of } f^{-1} \qquad \text{Range of } f = \text{Domain of } f^{-1}$$

Look again at Figure 11 to visualize the relationship. If we start with x, apply f, and then apply f^{-1}, we get x back again. If we start with x, apply f^{-1}, and then apply f, we get the number x back again. To put it simply, what f does, f^{-1} undoes, and vice versa. See the illustration that follows.

Input x from domain of f $\xrightarrow{\text{Apply } f}$ $f(x)$ $\xrightarrow{\text{Apply } f^{-1}}$ $f^{-1}(f(x)) = x$

Input x from domain of f^{-1} $\xrightarrow{\text{Apply } f^{-1}}$ $f^{-1}(x)$ $\xrightarrow{\text{Apply } f}$ $f(f^{-1}(x)) = x$

In other words,

$$f^{-1}(f(x)) = x \quad \text{where } x \text{ is in the domain of } f$$
$$f(f^{-1}(x)) = x \quad \text{where } x \text{ is in the domain of } f^{-1}$$

Consider the function $f(x) = 2x$, which multiplies the argument x by 2. The inverse function f^{-1} undoes whatever f does. So the inverse function of f is $f^{-1}(x) = \frac{1}{2}x$, which divides the argument by 2. For example, $f(3) = 2(3) = 6$ and $f^{-1}(6) = \frac{1}{2}(6) = 3$, so f^{-1} undoes what f did. We can verify this by showing that

$$f^{-1}(f(x)) = f^{-1}(2x) = \frac{1}{2}(2x) = x \quad \text{and} \quad f(f^{-1}(x)) = f\left(\frac{1}{2}x\right) = 2\left(\frac{1}{2}x\right) = x$$

See Figure 12.

Figure 12

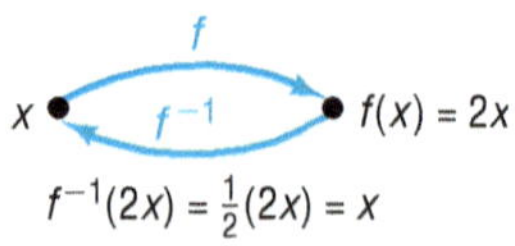

EXAMPLE 5 Verifying Inverse Functions

Exploration

Simultaneously graph $Y_1 = x$, $Y_2 = x^3$, and $Y_3 = \sqrt[3]{x}$ on a square screen with $-3 \le x \le 3$. What do you observe about the graphs of $Y_2 = x^3$, its inverse $Y_3 = \sqrt[3]{x}$, and the line $Y_1 = x$?

Repeat this experiment by simultaneously graphing $Y_1 = x$, $Y_2 = 2x + 3$, and $Y_3 = \frac{1}{2}(x - 3)$ on a square screen with $-6 \le x \le 3$. Do you see the symmetry of the graph of Y_2 and its inverse Y_3 with respect to the line $Y_1 = x$?

(a) We verify that the inverse of $g(x) = x^3$ is $g^{-1}(x) = \sqrt[3]{x}$ by showing that

$$g^{-1}(g(x)) = g^{-1}(x^3) = \sqrt[3]{x^3} = x \quad \text{for all } x \text{ in the domain of } g$$
$$g(g^{-1}(x)) = g(\sqrt[3]{x}) = (\sqrt[3]{x})^3 = x \quad \text{for all } x \text{ in the domain of } g^{-1}$$

(b) We verify that the inverse of $f(x) = 2x + 3$ is $f^{-1}(x) = \frac{1}{2}(x - 3)$ by showing that

$$f^{-1}(f(x)) = f^{-1}(2x + 3) = \frac{1}{2}[(2x + 3) - 3] = \frac{1}{2}(2x) = x \quad \text{for all } x \text{ in the domain of } f$$

$$f(f^{-1}(x)) = f\left(\frac{1}{2}(x - 3)\right) = 2\left[\frac{1}{2}(x - 3)\right] + 3 = (x - 3) + 3 = x \quad \text{for all } x \text{ in the domain of } f^{-1}$$

EXAMPLE 6 Verifying Inverse Functions

Verify that the inverse of $f(x) = \dfrac{1}{x - 1}$ is $f^{-1}(x) = \dfrac{1}{x} + 1$. For what values of x is $f^{-1}(f(x)) = x$? For what values of x is $f(f^{-1}(x)) = x$?

Solution The domain of f is $\{x | x \ne 1\}$ and the domain of f^{-1} is $\{x | x \ne 0\}$. Now

$$f^{-1}(f(x)) = f^{-1}\left(\frac{1}{x - 1}\right) = \frac{1}{\frac{1}{x - 1}} + 1 = x - 1 + 1 = x \quad \text{provided } x \ne 1$$

$$f(f^{-1}(x)) = f\left(\frac{1}{x} + 1\right) = \frac{1}{\frac{1}{x} + 1 - 1} = \frac{1}{\frac{1}{x}} = x \quad \text{provided } x \ne 0$$

Now Work PROBLEM 31

3 Obtain the Graph of the Inverse Function from the Graph of the Function

For the functions in Example 5(b), we list points on the graph of $f = Y_1$ and on the graph of $f^{-1} = Y_2$ in Table 1. We notice that whenever (a, b) is on the graph of f then (b, a) is on the graph of f^{-1}. Figure 13 shows these points plotted. Also shown is the graph of $y = x$, which you should observe is a line of symmetry of the points.

Table 1

X	Y1
-5	-7
-4	-5
-3	-3
-2	-1
-1	1
0	3
1	5

Y1■2X+3

X	Y2
-7	-5
-5	-4
-3	-3
-1	-2
1	-1
3	0
5	1

Y2■(1/2)(X−3)

Figure 13

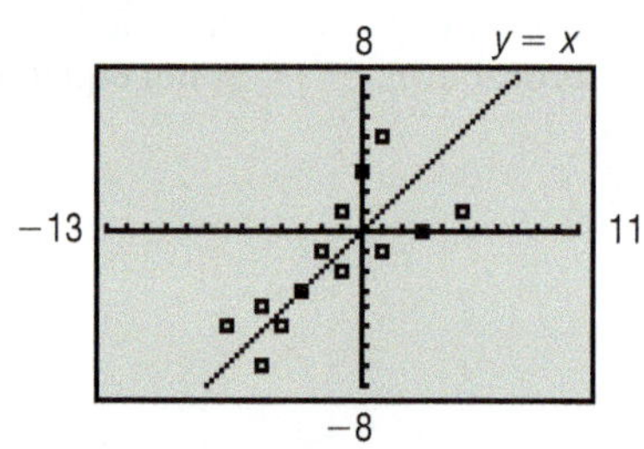

Figure 14

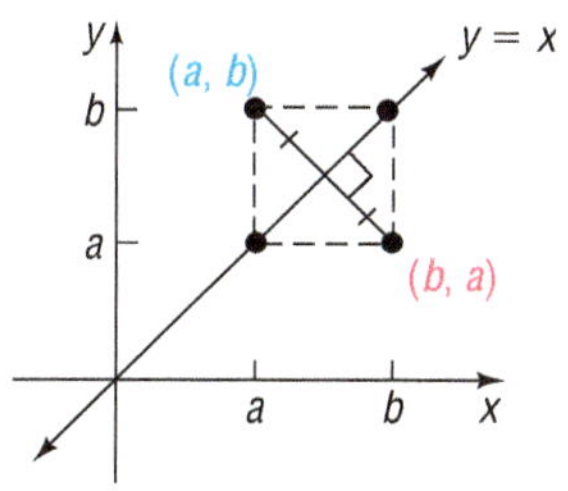

Suppose that (a, b) is a point on the graph of a one-to-one function f defined by $y = f(x)$. Then $b = f(a)$. This means that $a = f^{-1}(b)$, so (b, a) is a point on the graph of the inverse function f^{-1}. The relationship between the point (a, b) on f and the point (b, a) on f^{-1} is shown in Figure 14. The line segment containing (a, b) and (b, a) is perpendicular to the line $y = x$ and is bisected by the line $y = x$. (Do you see why?) It follows that the point (b, a) on f^{-1} is the reflection about the line $y = x$ of the point (a, b) on f.

THEOREM

The graph of a function f and the graph of its inverse f^{-1} are symmetric with respect to the line $y = x$.

Figure 15 illustrates this result. Notice that, once the graph of f is known, the graph of f^{-1} may be obtained by reflecting the graph of f about the line $y = x$.

Figure 15

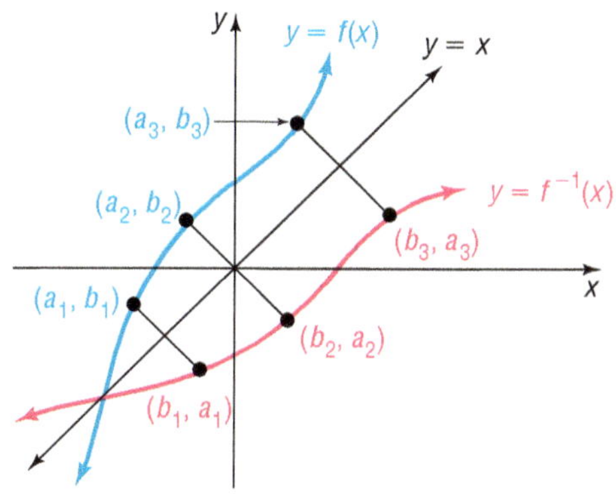

EXAMPLE 7 Graphing the Inverse Function

The graph in Figure 16(a) is that of a one-to-one function $y = f(x)$. Draw the graph of its inverse.

Solution We begin by adding the graph of $y = x$ to Figure 16(a). Since the points $(-2, -1)$, $(-1, 0)$, and $(2, 1)$ are on the graph of f, we know that the points $(-1, -2)$, $(0, -1)$, and $(1, 2)$ must be on the graph of f^{-1}. Keeping in mind that the graph of f^{-1} is the reflection about the line $y = x$ of the graph of f, we can draw f^{-1}. See Figure 16(b).

Figure 16

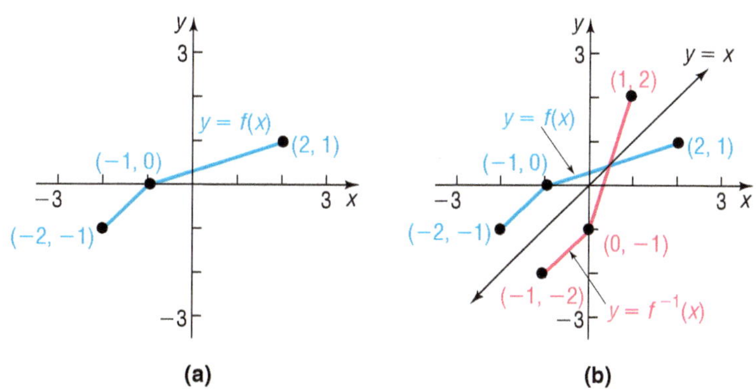

Now Work PROBLEM 41

4 Find the Inverse of a Function Defined by an Equation

The fact that the graphs of a one-to-one function f and its inverse function f^{-1} are symmetric with respect to the line $y = x$ tells us more. It says that we can obtain f^{-1}

by interchanging the roles of x and y in f. Look again at Figure 11. If f is defined by the equation

$$y = f(x)$$

then f^{-1} is defined by the equation

$$x = f(y)$$

The equation $x = f(y)$ defines f^{-1} *implicitly*. If we can solve this equation for y, we will have the *explicit* form of f^{-1}, that is,

$$y = f^{-1}(x)$$

Let's use this procedure to find the inverse of $f(x) = 2x + 3$. (Since f is a linear function and is increasing, we know that f is one-to-one and so has an inverse function.)

EXAMPLE 8 How to Find the Inverse Function

Find the inverse of $f(x) = 2x + 3$. Graph f and f^{-1} on the same coordinate axes.

Step-by-Step Solution

STEP 1 Replace $f(x)$ with y. In $y = f(x)$, interchange the variables x and y to obtain $x = f(y)$. This equation defines the inverse function f^{-1} implicitly.

We replace $f(x)$ with y in $f(x) = 2x + 3$ and obtain $y = 2x + 3$. Now we interchange the variables x and y to obtain

$$x = 2y + 3$$

This equation defines the inverse f^{-1} implicitly.

STEP 2 If possible, solve the implicit equation for y in terms of x to obtain the explicit form of f^{-1}, $y = f^{-1}(x)$.

To find the explicit form of the inverse, we solve $x = 2y + 3$ for y.

$$x = 2y + 3$$

$$2y + 3 = x \quad \text{Reflexive Property; If } a = b, \text{ then } b = a.$$

$$2y = x - 3 \quad \text{Subtract 3 from both sides.}$$

$$y = \frac{1}{2}(x - 3) \quad \text{Divide both sides by 2.}$$

The explicit form of the inverse f^{-1} is therefore

$$f^{-1}(x) = \frac{1}{2}(x - 3)$$

STEP 3 Check the result by showing that $f^{-1}(f(x)) = x$ and $f(f^{-1}(x)) = x$.

We verified that f and f^{-1} are inverses in Example 5(b) on page 416.

The graphs of $f(x) = 2x + 3$ and its inverse $f^{-1}(x) = \frac{1}{2}(x - 3)$ are shown in Figure 17. Note the symmetry of the graphs with respect to the line $y = x$.

Figure 17

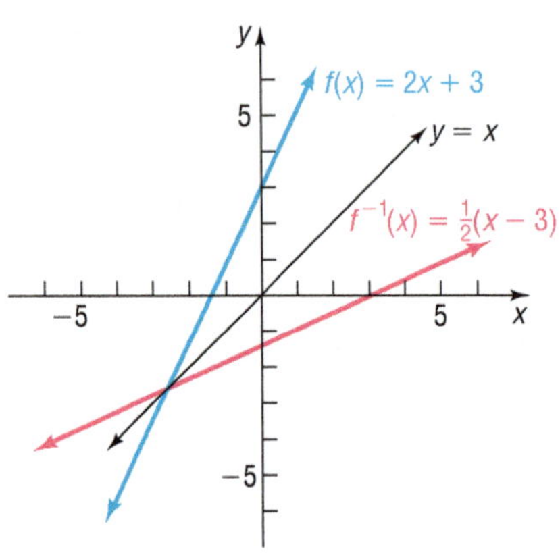

We now list the steps to follow for finding the inverse of a one-to-one function.

Procedure for Finding the Inverse of a One-to-One Function

Step 1: In $y = f(x)$, interchange the variables x and y to obtain

$$x = f(y)$$

This equation defines the inverse function f^{-1} implicitly.

Step 2: If possible, solve the implicit equation for y in terms of x to obtain the explicit form of f^{-1}:

$$y = f^{-1}(x)$$

Step 3: Check the result by showing that

$$f^{-1}(f(x)) = x \quad \text{and} \quad f(f^{-1}(x)) = x$$

EXAMPLE 9 Finding the Inverse Function

The function

$$f(x) = \frac{2x + 1}{x - 1} \qquad x \neq 1$$

is one-to-one. Find its inverse and check the result.

Solution **Step 1:** Replace $f(x)$ with y and interchange the variables x and y in

$$y = \frac{2x + 1}{x - 1}$$

to obtain

$$x = \frac{2y + 1}{y - 1}$$

Step 2: Solve for y.

$$x = \frac{2y + 1}{y - 1}$$

$$x(y - 1) = 2y + 1 \quad \text{Multiply both sides by } y - 1.$$

$$xy - x = 2y + 1 \quad \text{Apply the Distributive Property.}$$

$$xy - 2y = x + 1 \quad \text{Subtract 2y from both sides; add x to both sides.}$$

$$(x - 2)y = x + 1 \quad \text{Factor.}$$

$$y = \frac{x + 1}{x - 2} \quad \text{Divide by } x - 2.$$

The inverse is

$$f^{-1}(x) = \frac{x + 1}{x - 2} \quad x \neq 2 \quad \text{Replace } y \text{ by } f^{-1}(x).$$

Step 3: ✓ Check:

$$f^{-1}(f(x)) = f^{-1}\left(\frac{2x + 1}{x - 1}\right) = \frac{\frac{2x + 1}{x - 1} + 1}{\frac{2x + 1}{x - 1} - 2} = \frac{2x + 1 + x - 1}{2x + 1 - 2(x - 1)} = \frac{3x}{3} = x \quad x \neq 1$$

$$f(f^{-1}(x)) = f\left(\frac{x + 1}{x - 2}\right) = \frac{2\left(\frac{x + 1}{x - 2}\right) + 1}{\frac{x + 1}{x - 2} - 1} = \frac{2(x + 1) + x - 2}{x + 1 - (x - 2)} = \frac{3x}{3} = x \quad x \neq 2$$

Exploration

In Example 9, we found that, if $f(x) = \dfrac{2x+1}{x-1}$, then $f^{-1}(x) = \dfrac{x+1}{x-2}$. Compare the vertical and horizontal asymptotes of f and f^{-1}.

Result The vertical asymptote of f is $x = 1$, and the horizontal asymptote is $y = 2$. The vertical asymptote of f^{-1} is $x = 2$, and the horizontal asymptote is $y = 1$.

Now Work PROBLEM 49

We said in Chapter 3 that finding the range of a function f is not easy. However, if f is one-to-one, we can find its range by finding the domain of the inverse function f^{-1}.

EXAMPLE 10 Finding the Range of a Function

Find the domain and the range of

$$f(x) = \frac{2x+1}{x-1}$$

Solution The domain of f is $\{x|x \neq 1\}$. To find the range of f, we use the fact that the domain of f^{-1} equals the range of f. Based on Example 9, we have

$$f^{-1}(x) = \frac{x+1}{x-2}$$

The domain of f^{-1} is $\{x|x \neq 2\}$, so the range of f is $\{y|y \neq 2\}$. Also, because the domain of f is $\{x|x \neq 1\}$, the range of f^{-1} is $\{y|y \neq 1\}$.

Now Work PROBLEM 63

If a function is not one-to-one, it has no inverse function. Sometimes, though, an appropriate restriction on the domain of such a function will yield a new function that is one-to-one. Then the function defined on the restricted domain has an inverse function. Let's look at an example of this common practice.

EXAMPLE 11 Finding the Inverse of a Domain-restricted Function

Find the inverse of $y = f(x) = x^2$ if $x \geq 0$.

Solution The function $y = x^2$ is not one-to-one. [Refer to Example 2(a).] However, if we restrict the domain of this function to $x \geq 0$, as indicated, we have a new function that is increasing and therefore is one-to-one. As a result, the function defined by $y = f(x) = x^2$, $x \geq 0$, has an inverse function, f^{-1}.

We follow the steps given previously to find f^{-1}.

STEP 1: In the equation $y = x^2$, $x \geq 0$, interchange the variables x and y. The result is

$$x = y^2 \qquad y \geq 0$$

This equation defines (implicitly) the inverse function.

STEP 2: We solve for y to get the explicit form of the inverse. Since $y \geq 0$, only one solution for y is obtained: $y = \sqrt{x}$. So $f^{-1}(x) = \sqrt{x}$.

STEP 3: Check: $f^{-1}(f(x)) = f^{-1}(x^2) = \sqrt{x^2} = |x| = x$ since $x \geq 0$

$$f(f^{-1}(x)) = f(\sqrt{x}) = (\sqrt{x})^2 = x$$

Figure 18 illustrates the graphs of $f(x) = x^2$, $x \geq 0$, and $f^{-1}(x) = \sqrt{x}$.

Figure 18

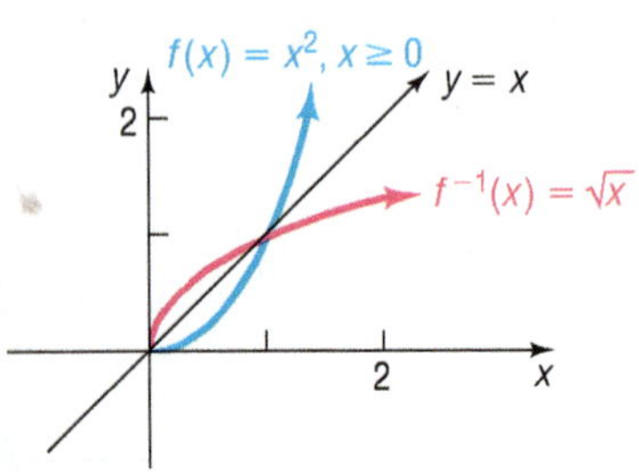

SUMMARY

1. If a function f is one-to-one, then it has an inverse function f^{-1}.
2. Domain of f = Range of f^{-1}; Range of f = Domain of f^{-1}.
3. To verify that f^{-1} is the inverse of f, show that $f^{-1}(f(x)) = x$ for every x in the domain of f and $f(f^{-1}(x)) = x$ for every x in the domain of f^{-1}.
4. The graphs of f and f^{-1} are symmetric with respect to the line $y = x$.
5. To find the range of a one-to-one function f, find the domain of its inverse function f^{-1}.

6.2 Assess Your Understanding

'Are You Prepared?'

Answers are given at the end of these exercises. If you get a wrong answer, read the pages listed in red.

1. Is the set of ordered pairs $\{(1, 3), (2, 3), (-1, 2)\}$ a function? Why or why not? (pp. 212–216)
2. Where is the function $f(x) = x^2$ increasing? Where is it decreasing? (pp. 237–238)
3. What is the domain of $f(x) = \dfrac{x + 5}{x^2 + 3x - 18}$? (pp. 219–220)

Concepts and Vocabulary

4. If every horizontal line intersects the graph of a function f at no more than one point, f is a(n) _____ function.
5. If f^{-1} denotes the inverse of a function f, then the graphs of f and f^{-1} are symmetric with respect to the line _____.
6. If the domain of a one-to-one function f is $[4, \infty)$, the range of its inverse, f^{-1}, is _____.
7. ***True or False*** If f and g are inverse functions, the domain of f is the same as the domain of g.
8. ***True or False*** If f and g are inverse functions, their graphs are symmetric with respect to the line $y = x$.

Skill Building

In Problems 9–16, determine whether the function is one-to-one.

9.

10.
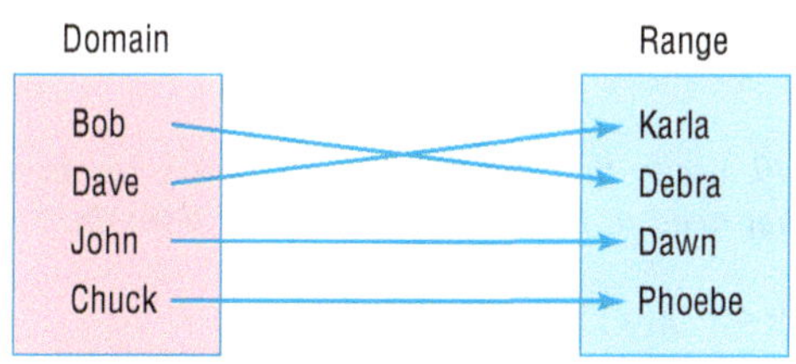

11.
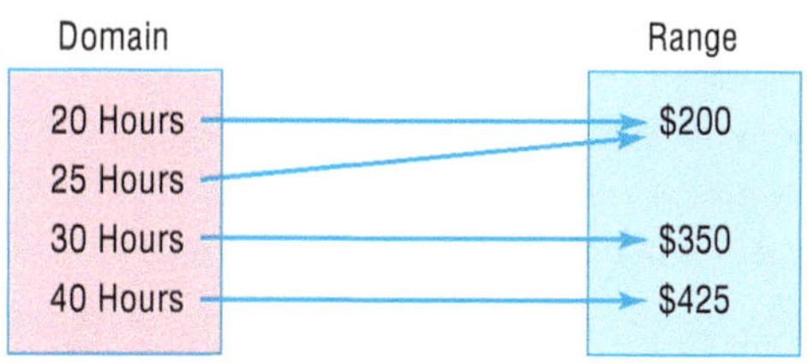

12.
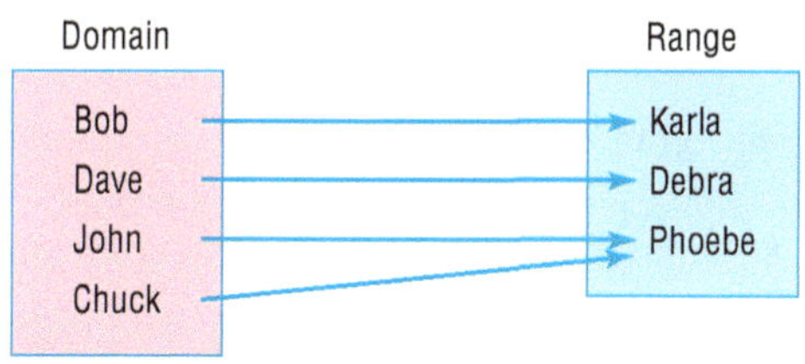

13. $\{(2, 6), (-3, 6), (4, 9), (1, 10)\}$
14. $\{(-2, 5), (-1, 3), (3, 7), (4, 12)\}$
15. $\{(0, 0), (1, 1), (2, 16), (3, 81)\}$
16. $\{(1, 2), (2, 8), (3, 18), (4, 32)\}$

In Problems 17–22, the graph of a function f is given. Use the horizontal-line test to determine whether f is one-to-one.

17.
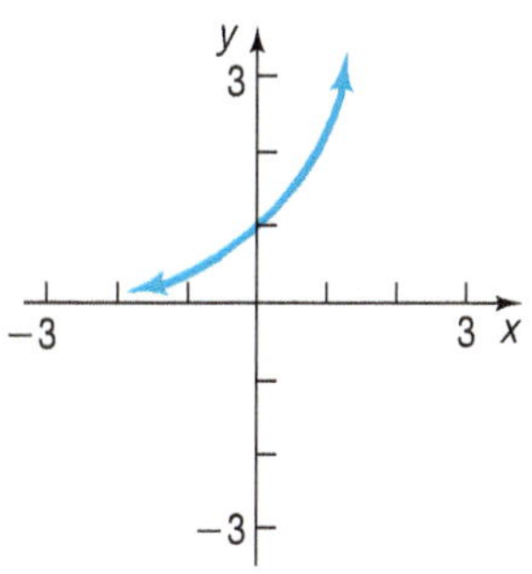

18.
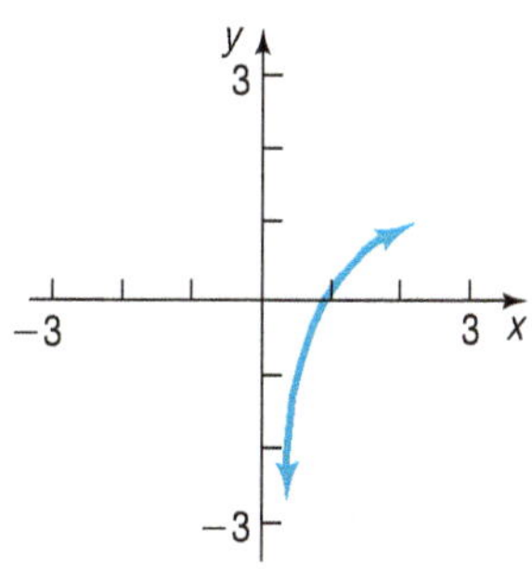

19.
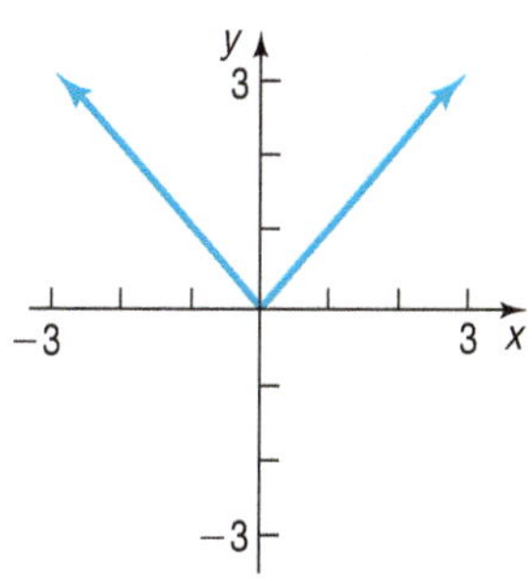

20.

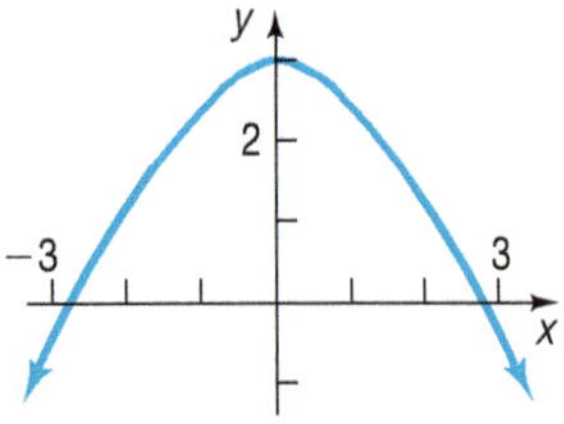

21.

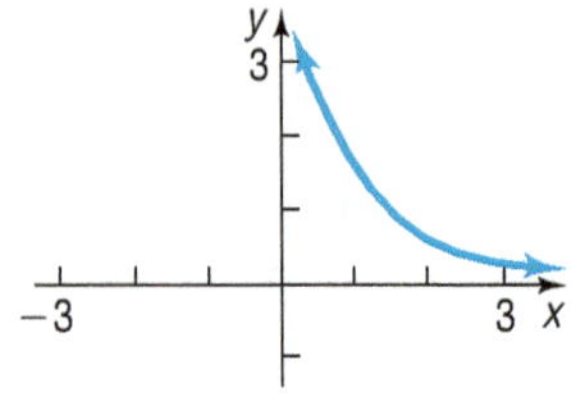

22.

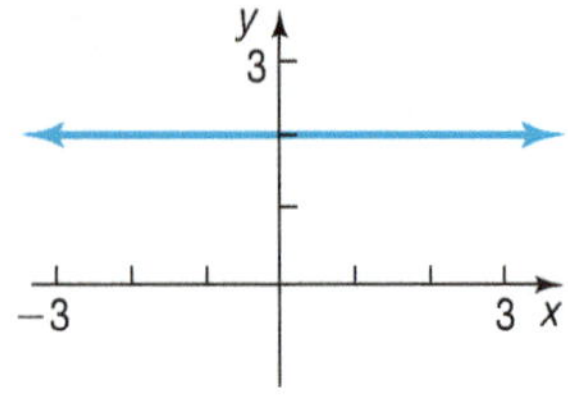

In Problems 23–30, find the inverse of each one-to-one function. State the domain and the range of each inverse function.

23.

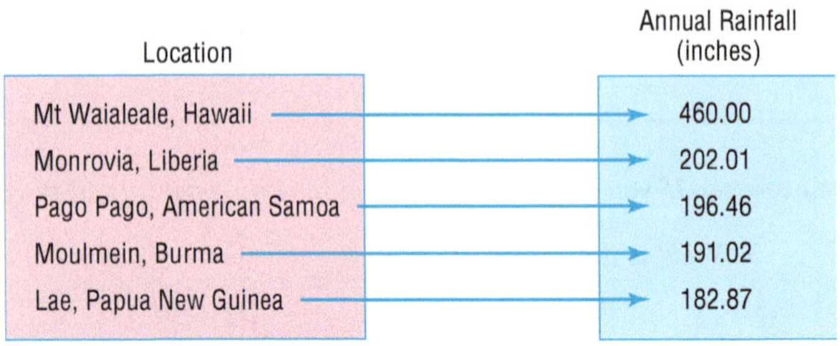

Source: *Information Please Almanac*

24.

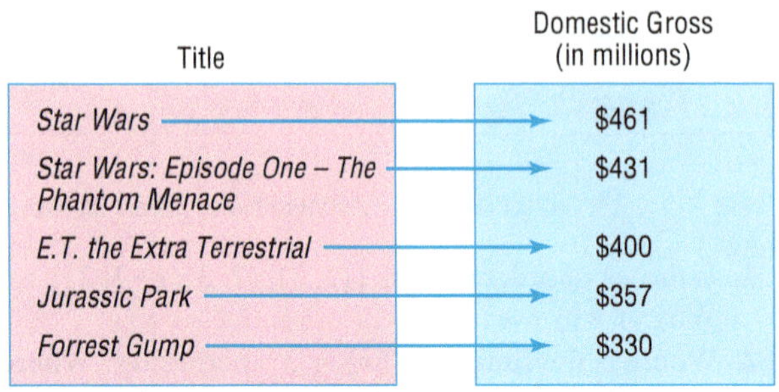

Source: *Information Please Almanac*

25.

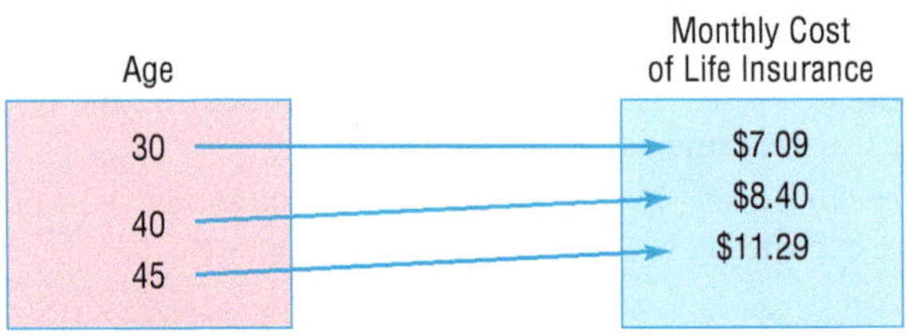

Source: eterm.com

26.

Source: *United States Statistical Abstract*

27. $\{(-3, 5), (-2, 9), (-1, 2), (0, 11), (1, -5)\}$

28. $\{(-2, 2), (-1, 6), (0, 8), (1, -3), (2, 9)\}$

29. $\{(-2, 1), (-3, 2), (-10, 0), (1, 9), (2, 4)\}$

30. $\{(-2, -8), (-1, -1), (0, 0), (1, 1), (2, 8)\}$

In Problems 31–40, verify that the functions f and g are inverses of each other by showing that $f(g(x)) = x$ and $g(f(x)) = x$. Give any values of x that need to be excluded from the domain.

31. $f(x) = 3x + 4; \quad g(x) = \frac{1}{3}(x - 4)$

32. $f(x) = 3 - 2x; \quad g(x) = -\frac{1}{2}(x - 3)$

33. $f(x) = 4x - 8; \quad g(x) = \frac{x}{4} + 2$

34. $f(x) = 2x + 6; \quad g(x) = \frac{1}{2}x - 3$

35. $f(x) = x^3 - 8; \quad g(x) = \sqrt[3]{x + 8}$

36. $f(x) = (x - 2)^2, x \geq 2; \quad g(x) = \sqrt{x} + 2$

37. $f(x) = \frac{1}{x}; \quad g(x) = \frac{1}{x}$

38. $f(x) = x; \quad g(x) = x$

39. $f(x) = \frac{2x + 3}{x + 4}; \quad g(x) = \frac{4x - 3}{2 - x}$

40. $f(x) = \frac{x - 5}{2x + 3}; \quad g(x) = \frac{3x + 5}{1 - 2x}$

In Problems 41–46, the graph of a one-to-one function f is given. Draw the graph of the inverse function f^{-1}. For convenience (and as a hint), the graph of $y = x$ is also given.

41.

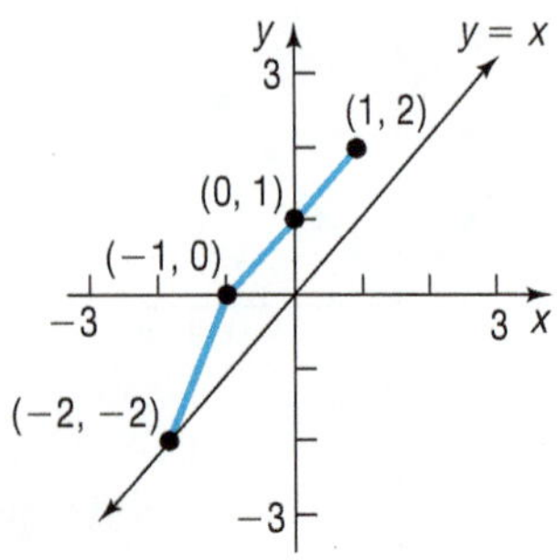

42.

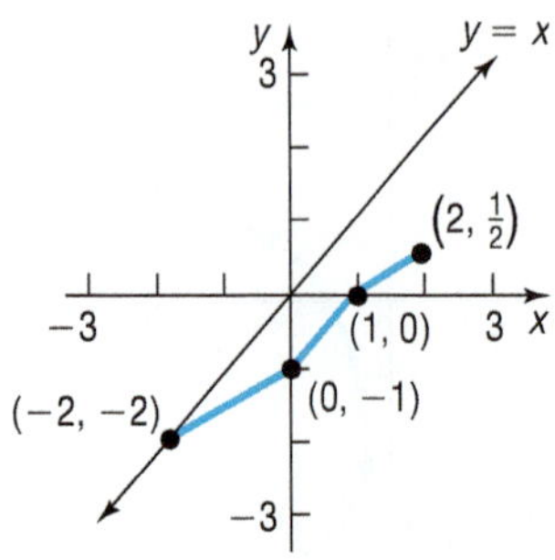

43.

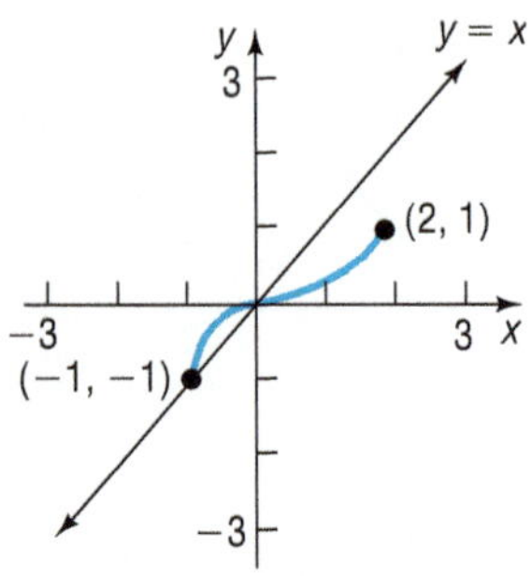

44.

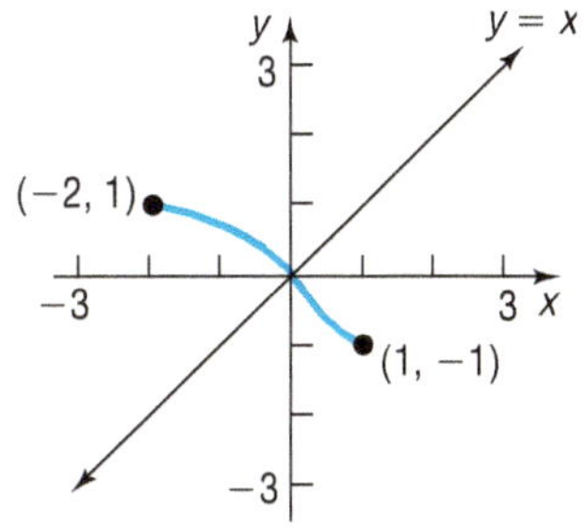

45.

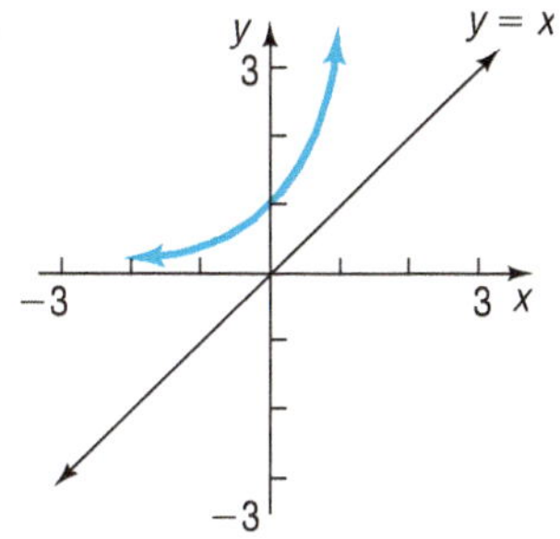

46.

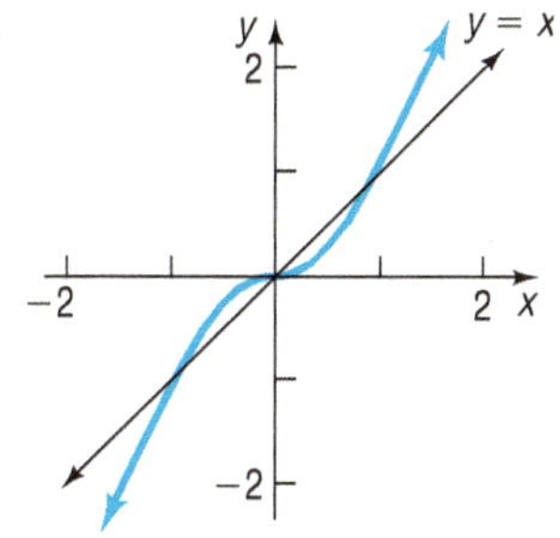

In Problems 47–58, the function f is one-to-one. Find its inverse and check your answer. Graph f, f^{-1}, and $y = x$ on the same coordinate axes.

47. $f(x) = 3x$

48. $f(x) = -4x$

49. $f(x) = 4x + 2$

50. $f(x) = 1 - 3x$

51. $f(x) = x^3 - 1$

52. $f(x) = x^3 + 1$

53. $f(x) = x^2 + 4 \quad x \geq 0$

54. $f(x) = x^2 + 9 \quad x \geq 0$

55. $f(x) = \dfrac{4}{x}$

56. $f(x) = -\dfrac{3}{x}$

57. $f(x) = \dfrac{1}{x - 2}$

58. $f(x) = \dfrac{4}{x + 2}$

In Problems 59–70, the function f is one-to-one. Find its inverse and check your answer. State the domain and the range of f and f^{-1}.

59. $f(x) = \dfrac{2}{3 + x}$

60. $f(x) = \dfrac{4}{2 - x}$

61. $f(x) = \dfrac{3x}{x + 2}$

62. $f(x) = -\dfrac{2x}{x - 1}$

63. $f(x) = \dfrac{2x}{3x - 1}$

64. $f(x) = -\dfrac{3x + 1}{x}$

65. $f(x) = \dfrac{3x + 4}{2x - 3}$

66. $f(x) = \dfrac{2x - 3}{x + 4}$

67. $f(x) = \dfrac{2x + 3}{x + 2}$

68. $f(x) = \dfrac{-3x - 4}{x - 2}$

69. $f(x) = \dfrac{x^2 - 4}{2x^2} \quad x > 0$

70. $f(x) = \dfrac{x^2 + 3}{3x^2} \quad x > 0$

Applications and Extensions

71. Use the graph of $y = f(x)$ given in Problem 41 to evaluate the following:
(a) $f(-1)$ (b) $f(1)$ (c) $f^{-1}(1)$ (d) $f^{-1}(2)$

72. Use the graph of $y = f(x)$ given in Problem 42 to evaluate the following:
(a) $f(2)$ (b) $f(1)$ (c) $f^{-1}(0)$ (d) $f^{-1}(-1)$

73. If $f(7) = 13$ and f is one-to-one, what is $f^{-1}(13)$?

74. If $g(-5) = 3$ and g is one-to-one, what is $g^{-1}(3)$?

75. The domain of a one-to-one function f is $[5, \infty)$, and its range is $[-2, \infty)$. State the domain and the range of f^{-1}.

76. The domain of a one-to-one function f is $[0, \infty)$, and its range is $[5, \infty)$. State the domain and the range of f^{-1}.

77. The domain of a one-to-one function g is the set of all real numbers, and its range is $[0, \infty)$. State the domain and the range of g^{-1}.

78. The domain of a one-to-one function g is $[0, 15]$, and its range is $(0, 8)$. State the domain and the range of g^{-1}.

79. A function $y = f(x)$ is increasing on the interval $(0, 5)$. What conclusions can you draw about the graph of $y = f^{-1}(x)$?

80. A function $y = f(x)$ is decreasing on the interval $(0, 5)$. What conclusions can you draw about the graph of $y = f^{-1}(x)$?

81. Find the inverse of the linear function

$$f(x) = mx + b \quad m \neq 0$$

82. Find the inverse of the function

$$f(x) = \sqrt{r^2 - x^2} \quad 0 \leq x \leq r$$

83. A function f has an inverse function. If the graph of f lies in quadrant I, in which quadrant does the graph of f^{-1} lie?

84. A function f has an inverse function. If the graph of f lies in quadrant II, in which quadrant does the graph of f^{-1} lie?

85. The function $f(x) = |x|$ is not one-to-one. Find a suitable restriction on the domain of f so that the new function that results is one-to-one. Then find the inverse of f.

86. The function $f(x) = x^4$ is not one-to-one. Find a suitable restriction on the domain of f so that the new function that results is one-to-one. Then find the inverse of f.

In applications, the symbols used for the independent and dependent variables are often based on common usage. So, rather than using $y = f(x)$ to represent a function, an applied problem might use $C = C(q)$ to represent the cost C of manufacturing q units of a good since, in economics, q is used for output. Because of this, the inverse notation f^{-1} used in a pure mathematics problem is not used when finding inverses of applied problems. Rather, the inverse of a function such as $C = C(q)$ will be $q = q(C)$. So $C = C(q)$ is a function that represents the cost C as a function of the output q, while $q = q(C)$ is a function that represents the output q as a function of the cost C. Problems 87–90 illustrate this idea.

87. Vehicle Stopping Distance Taking into account reaction time, the distance d (in feet) that a car requires to come to a complete stop while traveling r miles per hour is given by the function

$$d(r) = 6.97r - 90.39$$

(a) Express the speed r at which the car is traveling as a function of the distance d required to come to a complete stop.
(b) Verify that $r = r(d)$ is the inverse of $d = d(r)$ by showing that $r(d(r)) = r$ and $d(r(d)) = d$.
(c) Predict the speed that a car was traveling if the distance required to stop was 300 feet.

88. Height and Head Circumference The head circumference C of a child is related to the height H of the child (both in inches) through the function

$$H(C) = 2.15C - 10.53$$

(a) Express the head circumference C as a function of height H.
(b) Verify that $C = C(H)$ is the inverse of $H = H(C)$ by showing that $H(C(H)) = H$ and $C(H(C)) = C$.
(c) Predict the head circumference of a child who is 26 inches tall.

89. Ideal Body Weight One model for the ideal body weight W for men (in kilograms) as a function of height h (in inches) is given by the function

$$W(h) = 50 + 2.3(h - 60)$$

(a) What is the ideal weight of a 6-foot male?
(b) Express the height h as a function of weight W.
(c) Verify that $h = h(W)$ is the inverse of $W = W(h)$ by showing that $h(W(h)) = h$ and $W(h(W)) = W$.
(d) What is the height of a male who is at his ideal weight of 80 kilograms?

[**Note:** The ideal body weight W for women (in kilograms) as a function of height h (in inches) is given by $W(h) = 45.5 + 2.3(h - 60)$.]

90. Temperature Conversion The function $F(C) = \frac{9}{5}C + 32$ converts a temperature from C degrees Celsius to F degrees Fahrenheit.

(a) Express the temperature in degrees Celsius C as a function of the temperature in degrees Fahrenheit F.
(b) Verify that $C = C(F)$ is the inverse of $F = F(C)$ by showing that $C(F(C)) = C$ and $F(C(F)) = F$.
(c) What is the temperature in degrees Celsius if it is 70 degrees Fahrenheit?

91. Income Taxes The function

$$T(g) = 4386.25 + 0.25(g - 31{,}850)$$

represents the 2007 federal income tax T (in dollars) due for a "single" filer whose modified adjusted gross income is g dollars, where $31{,}850 \le g \le 77{,}100$.

(a) What is the domain of the function T?
(b) Given that the tax due T is an increasing linear function of modified adjusted gross income g, find the range of the function T.
(c) Find adjusted gross income g as a function of federal income tax T. What are the domain and the range of this function?

92. Income Taxes The function

$$T(g) = 1565 + 0.15(g - 15{,}650)$$

represents the 2007 federal income tax T (in dollars) due for a "married filing jointly" filer whose modified adjusted gross income is g dollars, where $15{,}650 \le g \le 63{,}700$.

(a) What is the domain of the function T?
(b) Given that the tax due T is an increasing linear function of modified adjusted gross income g, find the range of the function T.
(c) Find adjusted gross income g as a function of federal income tax T. What are the domain and the range of this function?

93. Gravity on Earth If a rock falls from a height of 100 meters on Earth, the height H (in meters) after t seconds is approximately

$$H(t) = 100 - 4.9t^2$$

(a) In general, quadratic functions are not one-to-one. However, the function $H(t)$ is one-to-one. Why?
(b) Find the inverse of H and verify your result.
(c) How long will it take a rock to fall 80 meters?

94. Period of a Pendulum The period T (in seconds) of a simple pendulum as a function of its length l (in feet) is given by

$$T(l) = 2\pi\sqrt{\frac{l}{32.2}}$$

(a) Express the length l as a function of the period T.
(b) How long is a pendulum whose period is 3 seconds?

95. Given

$$f(x) = \frac{ax + b}{cx + d}$$

find $f^{-1}(x)$. If $c \neq 0$, under what conditions on a, b, c, and d is $f = f^{-1}$?

Discussion and Writing

96. Can a one-to-one function and its inverse be equal? What must be true about the graph of f for this to happen? Give some examples to support your conclusion.

97. Draw the graph of a one-to-one function that contains the points $(-2, -3)$, $(0, 0)$, and $(1, 5)$. Now draw the graph of its inverse. Compare your graph to those of other students. Discuss any similarities. What differences do you see?

98. Give an example of a function whose domain is the set of real numbers and that is neither increasing nor decreasing on its domain, but is one-to-one.

[**Hint:** Use a piecewise-defined function.]

99. Is every odd function one-to-one? Explain.

100. Suppose $C(g)$ represents the cost C, in dollars, of manufacturing g cars. Explain what $C^{-1}(800{,}000)$ represents.

'Are You Prepared?' Answers

1. Yes; for each input x there is one output y.

2. Increasing on $(0, \infty)$; decreasing on $(-\infty, 0)$

3. $\{x | x \neq -6, x \neq 3\}$

6.3 Exponential Functions

PREPARING FOR THIS SECTION *Before getting started, review the following:*

- Exponents (Section R.2, pp. 21–23, and Section R.8, pp. 75–76)
- Graphing Techniques: Transformations (Section 3.5, pp. 256–264)
- Average Rate of Change (Section 3.3, pp. 240–242)
- Solving Linear and Quadratic Equations (Sections 1.2 and 1.3, pp. 105–106 and 113–114)
- Linear Functions (Section 4.1, pp. 282–285)
- Horizontal Asymptotes (Section 5.2, pp. 352–355)

Now Work the 'Are You Prepared?' problems on page 435.

OBJECTIVES

1 Evaluate Exponential Functions

In Chapter R, Section R.8, we give a definition for raising a real number a to a rational power. Based on that discussion, we gave meaning to expressions of the form

$$a^r$$

where the base a is a positive real number and the exponent r is a rational number.

But what is the meaning of a^x, where the base a is a positive real number and the exponent x is an irrational number? Although a rigorous definition requires methods discussed in calculus, the basis for the definition is easy to follow: Select a rational number r that is formed by truncating (removing) all but a finite number of digits from the irrational number x. Then it is reasonable to expect that

$$a^x \approx a^r$$

For example, take the irrational number $\pi = 3.14159\ldots$. Then an approximation to a^π is

$$a^\pi \approx a^{3.14}$$

where the digits after the hundredths position have been removed from the value for π. A better approximation would be

$$a^\pi \approx a^{3.14159}$$

where the digits after the hundred-thousandths position have been removed. Continuing in this way, we can obtain approximations to a^π to any desired degree of accuracy.

Most calculators have an $\boxed{x^y}$ key or a caret key $\boxed{\wedge}$ for working with exponents. To evaluate expressions of the form a^x, enter the base a, then press the $\boxed{x^y}$ key (or the $\boxed{\wedge}$ key), enter the exponent x, and press $\boxed{=}$ (or $\boxed{\text{ENTER}}$).

EXAMPLE 1 Using a Calculator to Evaluate Powers of 2

Using a calculator, evaluate:

(a) $2^{1.4}$ (b) $2^{1.41}$ (c) $2^{1.414}$ (d) $2^{1.4142}$ (e) $2^{\sqrt{2}}$

Solution Figure 19 shows the solution to part (a) using a TI-84 Plus graphing calculator.

Figure 19

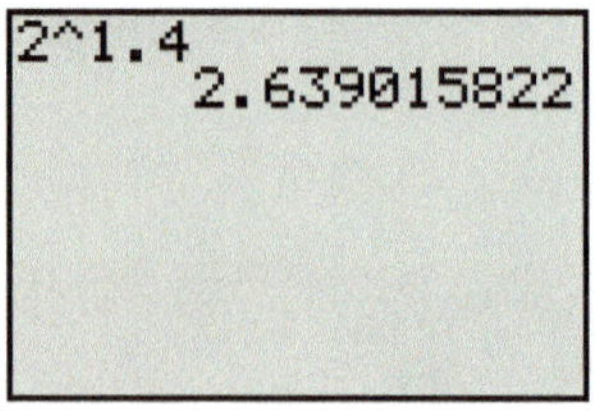

(a) $2^{1.4} \approx 2.639015822$ (b) $2^{1.41} \approx 2.657371628$

(c) $2^{1.414} \approx 2.66474965$ (d) $2^{1.4142} \approx 2.665119089$

(e) $2^{\sqrt{2}} \approx 2.665144143$

Now Work PROBLEM 11

It can be shown that the familiar laws for rational exponents hold for real exponents.

THEOREM

Laws of Exponents

If s, t, a, and b are real numbers with $a > 0$ and $b > 0$, then

$$a^s \cdot a^t = a^{s+t} \qquad (a^s)^t = a^{st} \qquad (ab)^s = a^s \cdot b^s$$

$$1^s = 1 \qquad a^{-s} = \frac{1}{a^s} = \left(\frac{1}{a}\right)^s \qquad a^0 = 1 \tag{1}$$

Introduction to Exponential Growth

Let's examine a function f that has the following two properties:

1. The value of f doubles with every 1-unit increase in the independent variable x.
2. The value of f at $x = 0$ is 5, so $f(0) = 5$.

Table 2 shows values of the function f for $x = 0, 1, 2, 3,$ and 4.

Table 2

x	$f(x)$
0	5
1	10
2	20
3	40
4	80

We seek an equation $y = f(x)$ that describes this function f. The key is the fact that the value of f doubles for every 1-unit increase in x.

$f(0) = 5$

$f(1) = 2f(0) = 2 \cdot 5 = 5 \cdot 2^1$ We double the value of f at 0 to get the value at 1.

$f(2) = 2f(1) = 2(5 \cdot 2) = 5 \cdot 2^2$ We double the value of f at 1 to get the value at 2.

$f(3) = 2f(2) = 2(5 \cdot 2^2) = 5 \cdot 2^3$

$f(4) = 2f(3) = 2(5 \cdot 2^3) = 5 \cdot 2^4$

The pattern leads us to

$$f(x) = 2f(x-1) = 2(5 \cdot 2^{x-1}) = 5 \cdot 2^x$$

We are now ready for the following definition.

DEFINITION

An **exponential function** is a function of the form

$$f(x) = Ca^x$$

where a is a positive real number $(a > 0)$ and $a \neq 1$, and $C \neq 0$ is a real number. The domain of f is the set of all real numbers. The base a is the **growth factor**, and because $f(0) = Ca^0 = C$, we call C the **initial value.**

WARNING It is important to distinguish a power function, $g(x) = ax^n$, $n \geq 2$, an integer, from an exponential function, $f(x) = C \cdot a^x$, $a \neq 1$, $a > 0$. In a power function, the base is a variable and the exponent is a constant. In an exponential function, the base is a constant and the exponent is a variable. ■

In the definition of an exponential function, we exclude the base $a = 1$ because this function is simply the constant function $f(x) = C \cdot 1^x = C$. We also need to exclude bases that are negative; otherwise, we would have to exclude many values of x from the domain, such as $x = \frac{1}{2}$ and $x = \frac{3}{4}$. [Recall that $(-2)^{1/2} = \sqrt{-2}$, $(-3)^{3/4} = \sqrt[4]{(-3)^3} = \sqrt[4]{-27}$, and so on, are not defined in the set of real numbers.]

Some examples of exponential functions are

$$f(x) = 2^x \qquad F(x) = \left(\frac{1}{3}\right)^x \qquad G(x) = 2 \cdot 3^x$$

Notice for each function that the base is a constant and the exponent is a variable.

In the function $f(x) = 5 \cdot 2^x$, notice that the ratio of consecutive outputs is constant for 1-unit increases in the input. This ratio equals the constant 2, the base of the exponential function. In other words,

$$\frac{f(1)}{f(0)} = \frac{5 \cdot 2^1}{5} = 2 \qquad \frac{f(2)}{f(1)} = \frac{5 \cdot 2^2}{5 \cdot 2^1} = 2 \qquad \frac{f(3)}{f(2)} = \frac{5 \cdot 2^3}{5 \cdot 2^2} = 2 \quad \text{and so on}$$

This leads to the following result.

THEOREM

For an exponential function $f(x) = C \cdot a^x$, $a > 0$, $a \neq 1$, if x is any real number, then

$$\frac{f(x+1)}{f(x)} = a \quad \text{or} \quad f(x+1) = af(x)$$

In Words

For 1-unit changes in the input x of an exponential function $f(x) = C \cdot a^x$, the ratio of consecutive outputs is the constant a.

Proof

$$\frac{f(x+1)}{f(x)} = \frac{Ca^{x+1}}{Ca^x} = a^{x+1-x} = a^1 = a \quad ■$$

EXAMPLE 2 Identifying Linear or Exponential Functions

Determine whether the given function is linear, exponential, or neither. For those that are linear, find a linear function that models the data. For those that are exponential, find an exponential function that models the data.

(a)

x	y
−1	5
0	2
1	−1
2	−4
3	−7

(b)

x	y
−1	2
0	4
1	7
2	11
3	15

(c)

x	y
−1	32
0	16
1	8
2	4
3	2

Solution For each function we compute the average rate of change of y with respect to x. If the average rate of change is constant, then the function is linear. Also, we compute the ratio of consecutive outputs. If the ratio is constant, then the function is exponential.

(a) See Table 3(a). The average rate of change for every 1-unit increase in x is –3. Therefore, the function is a linear function. In a linear function the average rate of change is the slope m, so $m = -3$. The y-intercept b is the value of the function at $x = 0$, so $b = 2$. The linear function that models the data is $f(x) = mx + b = -3x + 2$.

Table 3

x	y	Average Rate of Change	Ratio of Consecutive Outputs
−1	5		
		$\frac{\Delta y}{\Delta x} = \frac{2-5}{0-(-1)} = -3$	$\frac{2}{5}$
0	2		
		−3	$-\frac{1}{2}$
1	−1		
		−3	4
2	−4		
		−3	$\frac{7}{4}$
3	−7		

(a)

x	y	Average Rate of Change	Ratio of Consecutive Outputs
−1	2		
		$\frac{\Delta y}{\Delta x} = \frac{4-2}{0-(-1)} = 2$	2
0	4		
		3	$\frac{7}{4}$
1	7		
		4	$\frac{11}{7}$
2	11		
		5	$\frac{16}{11}$
3	16		

(b)

x	y	Average Rate of Change	Ratio of Consecutive Outputs
−1	32		
		$\frac{\Delta y}{\Delta x} = \frac{16-32}{0-(-1)} = -16$	$\frac{16}{32} = \frac{1}{2}$
0	16		
		−8	$\frac{8}{16} = \frac{1}{2}$
1	8		
		−4	$\frac{4}{8} = \frac{1}{2}$
2	4		
		−2	$\frac{2}{4} = \frac{1}{2}$
3	2		

(c)

(b) See Table 3(b). In this function the average rate of change from -1 to 0 is 2, and the average rate of change from 0 to 1 is 3. Because the average rate of change is not constant, the function is not a linear function. In this function, the ratio of consecutive outputs from -1 to 0 is 2, and the ratio of consecutive outputs from 0 to 1 is $\frac{7}{4}$. Because the ratio of consecutive outputs is not a constant, the function is not an exponential function.

(c) See Table 3(c). In this function the average rate of change from -1 to 0 is -16, and the average rate of change from 0 to 1 is -8. Because the average rate of change is not constant, the function is not a linear function. The ratio of consecutive outputs for a 1-unit increase in the inputs is a constant, $\frac{1}{2}$. Because the ratio of consecutive outputs is constant, the function is an exponential function with growth factor $a = \frac{1}{2}$. The initial value of the exponential function is $C = 16$. Therefore, the exponential function that models the data is $g(x) = Ca^x = 16 \cdot \left(\frac{1}{2}\right)^x$.

Now Work PROBLEM 21

2 Graph Exponential Functions

If we know how to graph an exponential function of the form $f(x) = a^x$, then a compression, stretch, and perhaps reflection about the x-axis will enable us to obtain the graph of $f(x) = C \cdot a^x$.

First, we graph the exponential function $f(x) = 2^x$.

EXAMPLE 3 Graphing an Exponential Function

Graph the exponential function: $f(x) = 2^x$

Solution The domain of $f(x) = 2^x$ is the set of all real numbers. We begin by locating some points on the graph of $f(x) = 2^x$, as listed in Table 4.

Table 4

X	Y1
-10	9.8E-4
-3	.125
-1	.5
0	1
1	2
3	8
10	1024

Y1=2^X

Since $2^x > 0$ for all x, the range of f is $(0, \infty)$. From this, we conclude that the graph has no x-intercepts, and, in fact, the graph will lie above the x-axis for all x. As Table 4 indicates, the y-intercept is 1. Table 4 also indicates that as $x \to -\infty$ the value of $f(x) = 2^x$ gets closer and closer to 0. We conclude that the x-axis ($y = 0$) is a horizontal asymptote to the graph as $x \to -\infty$. This gives us the end behavior for x large and negative.

To determine the end behavior for x large and positive, look again at Table 4. As $x \to \infty$, $f(x) = 2^x$ grows very quickly, causing the graph of $f(x) = 2^x$ to rise very rapidly. It is apparent that f is an increasing function and hence is one-to-one.

Figure 20 shows the graph of $f(x) = 2^x$. Notice that all the conclusions given earlier are confirmed by the graph.

Figure 20

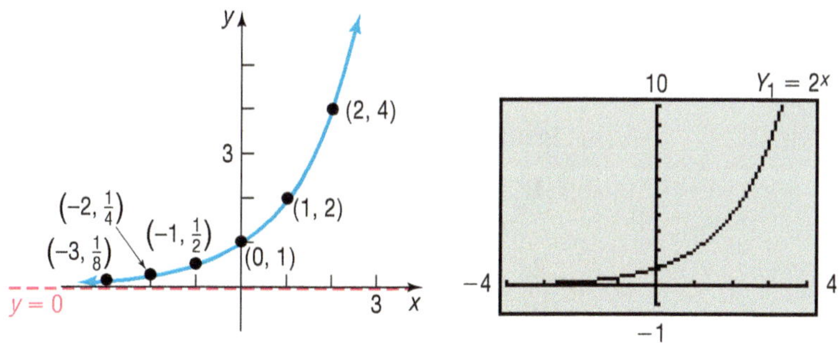

As we shall see, graphs that look like the one in Figure 20 occur very frequently in a variety of situations. For example, look at the graph in Figure 21, which illustrates the number of cellular telephone subscribers at the end of each year from 1985 to 2006. We might conclude from this graph that the number of cellular telephone subscribers is growing *exponentially*.

Figure 21

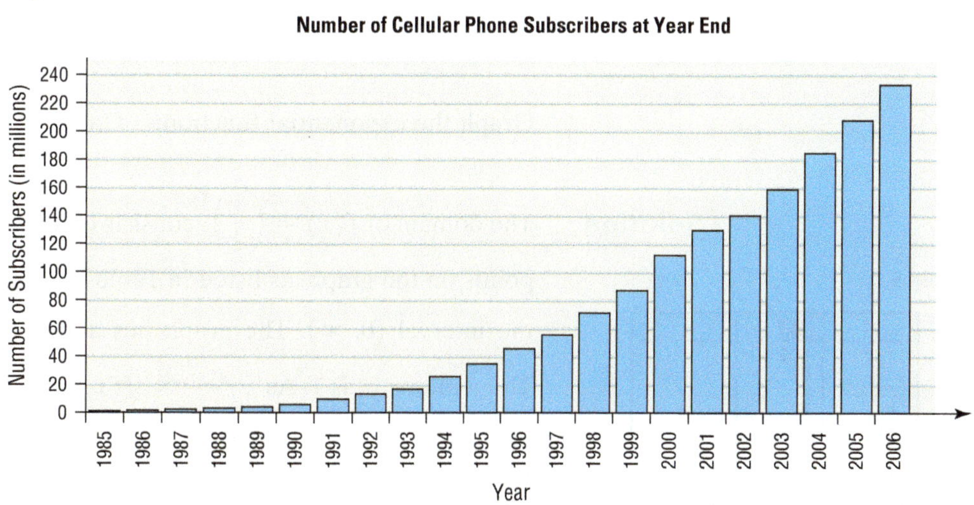

Source: ©2006 CTIA–The Wireless Association®. All Rights Reserved.

We shall have more to say about situations that lead to exponential growth later in this chapter. For now, we continue to seek properties of the exponential functions.

The graph of $f(x) = 2^x$ in Figure 20 is typical of all exponential functions of the form $f(x) = a^x$ with $a > 1$. Such functions are increasing functions and hence are one-to-one. Their graphs lie above the x-axis, pass through the point $(0, 1)$, and thereafter rise rapidly as $x \to \infty$. As $x \to -\infty$, the x-axis $(y = 0)$ is a horizontal asymptote. There are no vertical asymptotes. Finally, the graphs are smooth and continuous, with no corners or gaps.

Figures 22 and 23 illustrate the graphs of two more exponential functions whose bases are larger than 1. Notice the larger the base, the steeper the graph is when $x > 0$ and when $x < 0$, the larger the base, the closer the graph of the equation is to the x-axis.

Figure 22

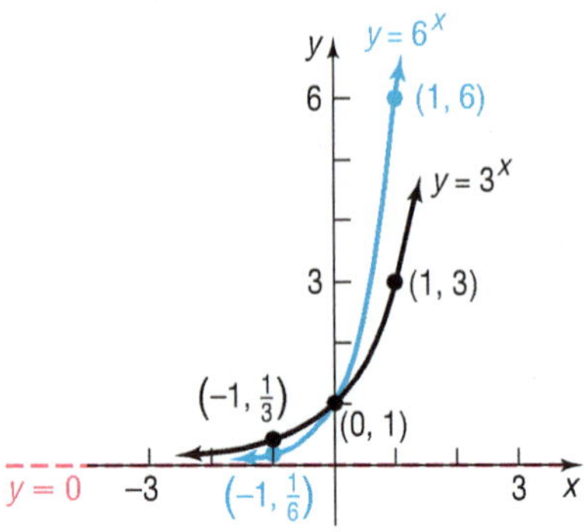

Figure 23

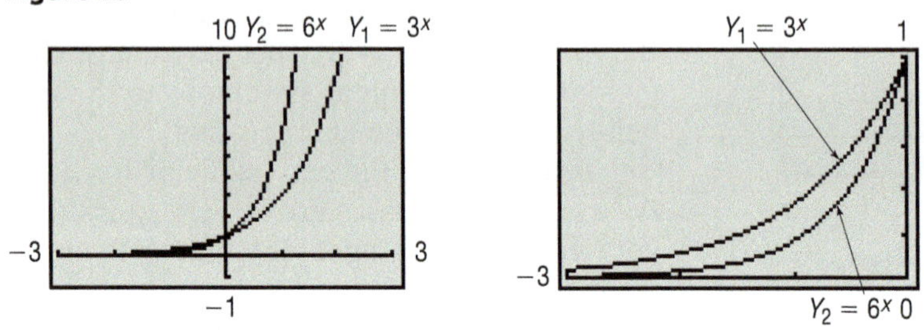

(a) (b)

The following display summarizes the information that we have about $f(x) = a^x, a > 1$.

Properties of the Exponential Function $f(x) = a^x, a > 1$

1. The domain is the set of all real numbers; the range is the set of positive real numbers.
2. There are no x-intercepts; the y-intercept is 1.
3. The x-axis $(y = 0)$ is a horizontal asymptote as $x \to -\infty$.
4. $f(x) = a^x, a > 1$, is an increasing function and is one-to-one.
5. The graph of f contains the points $(0, 1)$, $(1, a)$, and $\left(-1, \frac{1}{a}\right)$.
6. The graph of f is smooth and continuous, with no corners or gaps. See Figure 24.

Figure 24
$f(x) = a^x, a > 1$

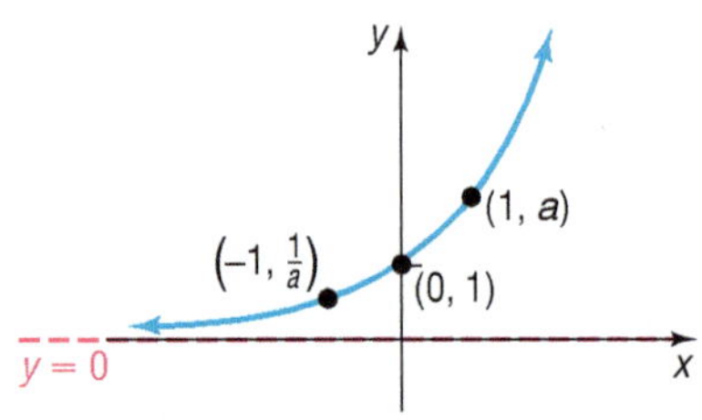

Now we consider $f(x) = a^x$ when $0 < a < 1$.

EXAMPLE 4 Graphing an Exponential Function

Graph the exponential function: $f(x) = \left(\frac{1}{2}\right)^x$

Solution The domain of $f(x) = \left(\frac{1}{2}\right)^x$ consists of all real numbers. As before, we locate some points on the graph, as listed in Table 5. Since $\left(\frac{1}{2}\right)^x > 0$ for all x, the range of f is the interval $(0, \infty)$. The graph lies above the x-axis and so has no x-intercepts. The y-intercept is 1. As $x \to -\infty$, $f(x) = \left(\frac{1}{2}\right)^x$ grows very quickly. As $x \to \infty$, the values of $f(x)$ approach 0. The x-axis $(y = 0)$ is a horizontal asymptote as $x \to \infty$. It is apparent that f is a decreasing function and hence is one-to-one. Figure 25 illustrates the graph.

Table 5

X	Y1
-10	1024
-3	8
-1	2
0	1
1	.5
3	.125
10	9.8E-4

Y1=(1/2)^X

Figure 25

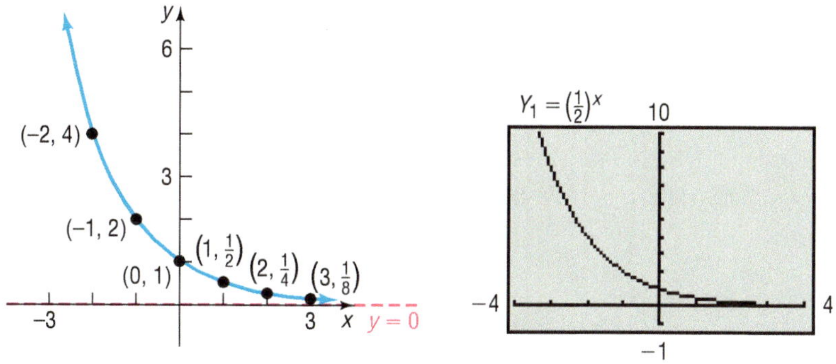

Seeing the Concept

Using a graphing utility, simultaneously graph:

(a) $Y_1 = 3^x, Y_2 = \left(\frac{1}{3}\right)^x$

(b) $Y_1 = 6^x, Y_2 = \left(\frac{1}{6}\right)^x$

Conclude that the graph of $Y_2 = \left(\frac{1}{a}\right)^x$, for $a > 0$, is the reflection about the y-axis of the graph of $Y_1 = a^x$.

We could have obtained the graph of $y = \left(\frac{1}{2}\right)^x$ from the graph of $y = 2^x$ using transformations. The graph of $y = \left(\frac{1}{2}\right)^x = 2^{-x}$ is a reflection about the y-axis of the graph of $y = 2^x$. Compare Figures 20 and 25.

The graph of $f(x) = \left(\frac{1}{2}\right)^x$ in Figure 25 is typical of all exponential functions of the form $f(x) = a^x$ with $0 < a < 1$. Such functions are decreasing and one-to-one. Their graphs lie above the x-axis and pass through the point $(0, 1)$. The graphs rise rapidly as $x \to -\infty$. As $x \to \infty$, the x-axis $(y = 0)$ is a horizontal asymptote. There are no vertical asymptotes. Finally, the graphs are smooth and continuous, with no corners or gaps.

Figures 26 and 27 illustrate the graphs of two more exponential functions whose bases are between 0 and 1. Notice the smaller the base, the steeper the graph when $x < 0$ and when $x > 0$, the smaller the base, the closer the graph of the equation is to the x-axis.

The following display summarizes the information that we have about the function $f(x) = a^x, 0 < a < 1$.

Figure 26

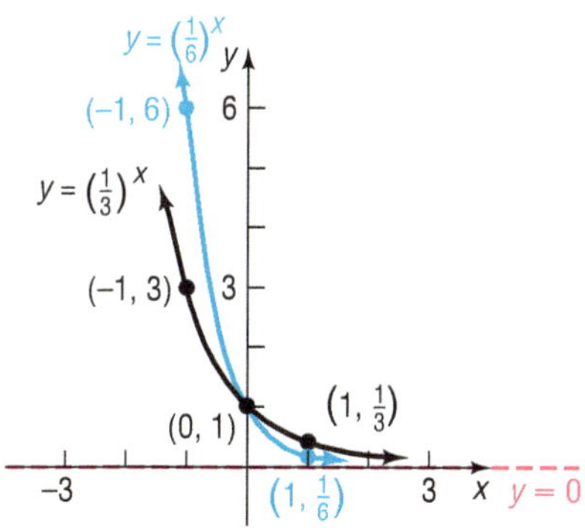

Figure 27

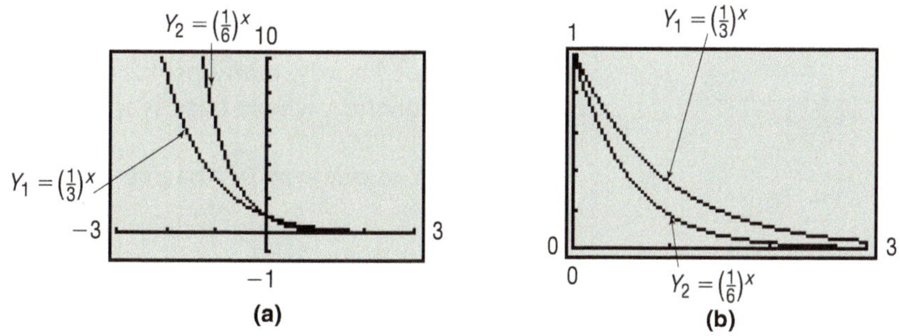

Figure 28

$f(x) = a^x, 0 < a < 1$

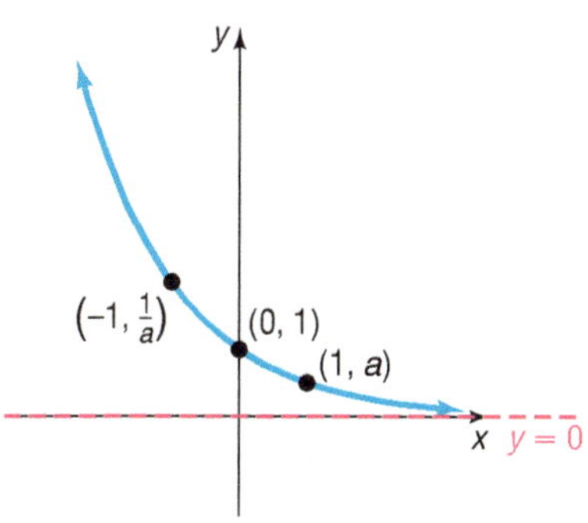

Properties of the Exponential Function $f(x) = a^x$, $0 < a < 1$

1. The domain is the set of all real numbers; the range is the set of positive real numbers.
2. There are no x-intercepts; the y-intercept is 1.
3. The x-axis $(y = 0)$ is a horizontal asymptote as $x \to \infty$.
4. $f(x) = a^x, 0 < a < 1$, is a decreasing function and is one-to-one.
5. The graph of f contains the points $(0, 1)$, $(1, a)$, and $\left(-1, \frac{1}{a}\right)$.
6. The graph of f is smooth and continuous, with no corners or gaps. See Figure 28.

EXAMPLE 5 Graphing Exponential Functions Using Transformations

Graph $f(x) = 2^{-x} - 3$ and determine the domain, range, and horizontal asymptote of f.

Solution We begin with the graph of $y = 2^x$. Figure 29 shows the stages.

Figure 29

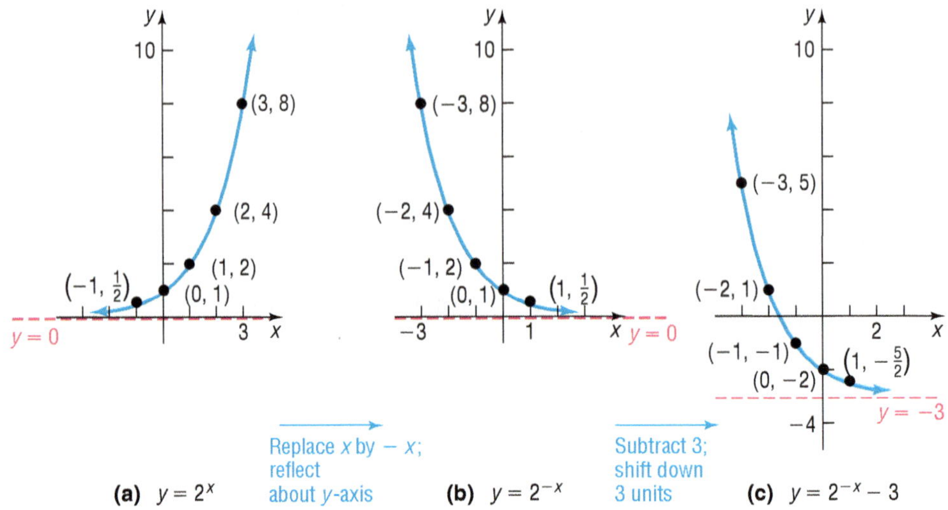

As Figure 29(c) illustrates, the domain of $f(x) = 2^{-x} - 3$ is the interval $(-\infty, \infty)$, and the range is the interval $(-3, \infty)$. The horizontal asymptote of f is the line $y = -3$.

Check: Graph $Y_1 = 2^{-x} - 3$ to verify the graph obtained in Figure 29(c).

Now Work PROBLEM 37

3 Define the Number *e*

As we shall see shortly, many problems that occur in nature require the use of an exponential function whose base is a certain irrational number, symbolized by the letter *e*.

Let's look at one way of arriving at this important number *e*.

DEFINITION

The **number *e*** is defined as the number that the expression

$$\left(1 + \frac{1}{n}\right)^n \quad (2)$$

approaches as $n \to \infty$. In calculus, this is expressed using limit notation as

$$e = \lim_{n\to\infty}\left(1 + \frac{1}{n}\right)^n$$

Table 6, illustrates what happens to the defining expression (2) as n takes on increasingly large values. The last number in the right column in the table is correct to nine decimal places and is the same as the entry given for e on your calculator (if expressed correctly to nine decimal places).

Table 6

n	$\frac{1}{n}$	$1 + \frac{1}{n}$	$\left(1 + \frac{1}{n}\right)^n$
1	1	2	2
2	0.5	1.5	2.25
5	0.2	1.2	2.48832
10	0.1	1.1	2.59374246
100	0.01	1.01	2.704813829
1,000	0.001	1.001	2.716923932
10,000	0.0001	1.0001	2.718145927
100,000	0.00001	1.00001	2.718268237
1,000,000	0.000001	1.000001	2.718280469
1,000,000,000	10^{-9}	$1 + 10^{-9}$	2.718281827

Table 7

The exponential function $f(x) = e^x$, whose base is the number e, occurs with such frequency in applications that it is usually referred to as *the* exponential function. Graphing calculators have the key $\boxed{e^x}$ or $\boxed{\exp(x)}$, which may be used to evaluate the exponential function for a given value of x. Use your calculator to find e^x for $x = -2, x = -1, x = 0, x = 1$, and $x = 2$, as we have done to create Table 7. The graph of the exponential function $f(x) = e^x$ is given in Figures 30(a) and 30(b). Since $2 < e < 3$, the graph of $y = e^x$ is increasing and lies between the graphs of $y = 2^x$ and $y = 3^x$. [See Figure 30(c).]

Figure 30

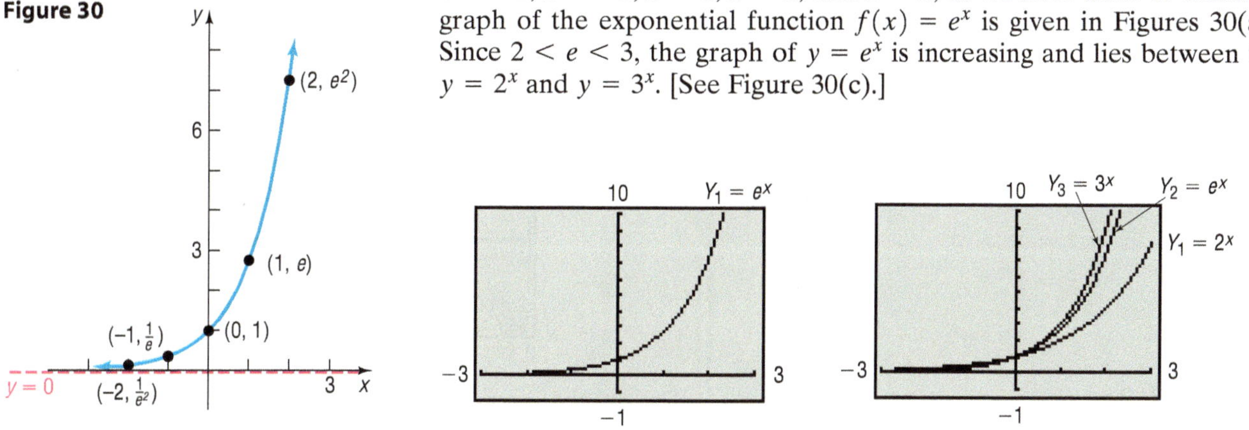

(a) (b) (c)

EXAMPLE 6 Graphing Exponential Functions Using Transformations

Graph $f(x) = -e^{x-3}$ and determine the domain, range, and horizontal asymptote of f.

Solution We begin with the graph of $y = e^x$. Figure 31 shows the stages.

Figure 31

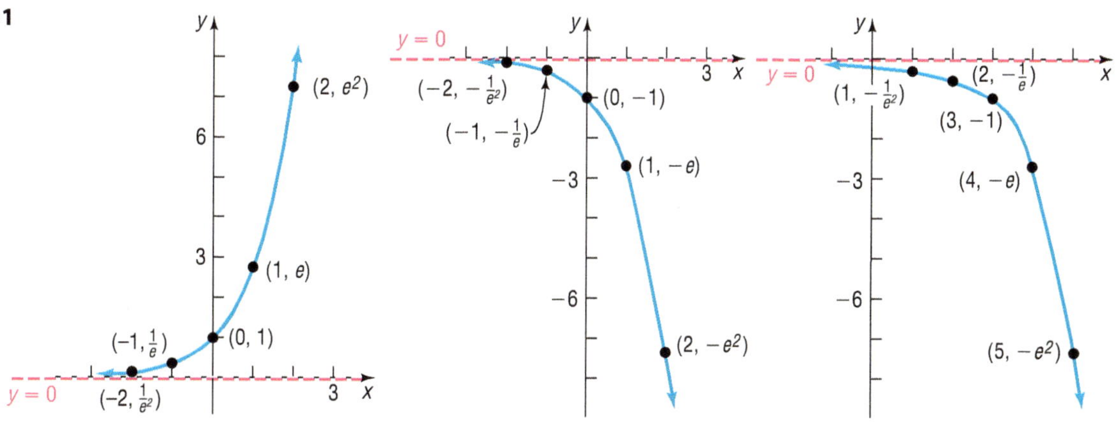

(a) $y = e^x$ Multiply by −1; Reflect about x-axis. (b) $y = -e^x$ Replace x by $x - 3$; Shift right 3 units. (c) $y = -e^{x-3}$

As Figure 31(c) illustrates, the domain of $f(x) = -e^{x-3}$ is the interval $(-\infty, \infty)$, and the range is the interval $(-\infty, 0)$. The horizontal asymptote is the line $y = 0$.

✓Check: Graph $Y_1 = -e^{x-3}$ to verify the graph obtained in Figure 31(c).

Now Work PROBLEM 49

4 Solve Exponential Equations

Equations that involve terms of the form a^x, $a > 0$, $a \neq 1$, are referred to as **exponential equations.** Such equations can sometimes be solved by appropriately applying the Laws of Exponents and property (3):

$$\text{If} \quad a^u = a^v, \quad \text{then} \quad u = v \qquad \textbf{(3)}$$

In Words
When two exponential expressions with the same base are equal, then their exponents are equal.

Property (3) is a consequence of the fact that exponential functions are one-to-one. To use property (3), each side of the equality must be written with the same base.

EXAMPLE 7 Solving an Exponential Equation

Solve: $3^{x+1} = 81$

Algebraic Solution

$$3^{x+1} = 81$$
$$3^{x+1} = 3^4 \qquad 81 = 3^4$$

Now we have the same base, 3, on each side of the equation, so we can set the exponents equal to each other to obtain

$$x + 1 = 4$$
$$x = 3$$

The solution set is $\{3\}$.

Graphing Solution

Graph $Y_1 = 3^{x+1}$ and $Y_2 = 81$. Use INTERSECT to determine the point of intersection. See Figure 32.

Figure 32

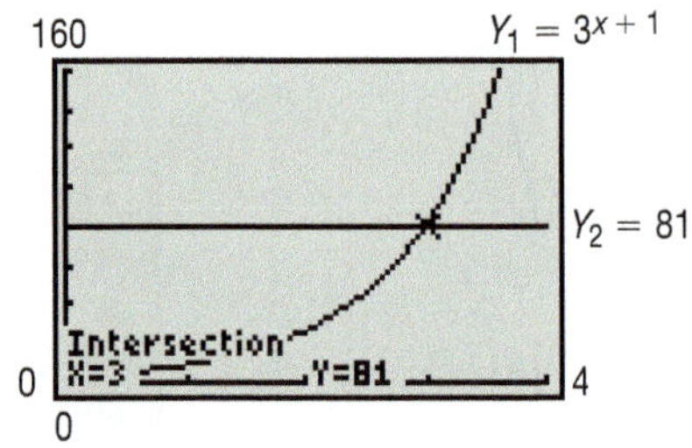

The graphs intersect at (3, 81), so the solution set is $\{3\}$.

Now Work PROBLEM 59

EXAMPLE 8 Solving an Exponential Equation

Solve: $e^{-x^2} = (e^x)^2 \cdot \dfrac{1}{e^3}$

Solution We use the Laws of Exponents first to get the base e on the right side.

$$(e^x)^2 \cdot \frac{1}{e^3} = e^{2x} \cdot e^{-3} = e^{2x-3}$$

As a result,

$$e^{-x^2} = e^{2x-3}$$
$$-x^2 = 2x - 3 \qquad \text{Apply Property (3).}$$
$$x^2 + 2x - 3 = 0 \qquad \text{Place the quadratic equation in standard form.}$$
$$(x+3)(x-1) = 0 \qquad \text{Factor.}$$
$$x = -3 \quad \text{or} \quad x = 1 \qquad \text{Use the Zero-Product Property.}$$

The solution set is $\{-3, 1\}$.

✓Check: Verify these solutions using a graphing utility.

Many applications involve the exponential function. Let's look at one.

EXAMPLE 9 **Exponential Probability**

Between 9:00 PM and 10:00 PM cars arrive at Burger King's drive-thru at the rate of 12 cars per hour (0.2 car per minute). The following formula from statistics can be used to determine the probability that a car will arrive within t minutes of 9:00 PM.

$$F(t) = 1 - e^{-0.2t}$$

(a) Determine the probability that a car will arrive within 5 minutes of 9 PM (that is, before 9:05 PM).

(b) Determine the probability that a car will arrive within 30 minutes of 9 PM (before 9:30 PM).

(c) Graph F using your graphing utility.

(d) What value does F approach as t becomes unbounded in the positive direction?

Solution (a) The probability that a car will arrive within 5 minutes is found by evaluating $F(t)$ at $t = 5$.

$$F(5) = 1 - e^{-0.2(5)} \approx 0.63212$$

Figure 33

We evaluate this expression in Figure 33. We conclude that there is a 63% probability that a car will arrive within 5 minutes.

(b) The probability that a car will arrive within 30 minutes is found by evaluating $F(t)$ at $t = 30$.

$$F(30) = 1 - e^{-0.2(30)} \approx 0.9975$$

↑ Use a calculator.

There is a 99.75% probability that a car will arrive within 30 minutes.

Figure 34

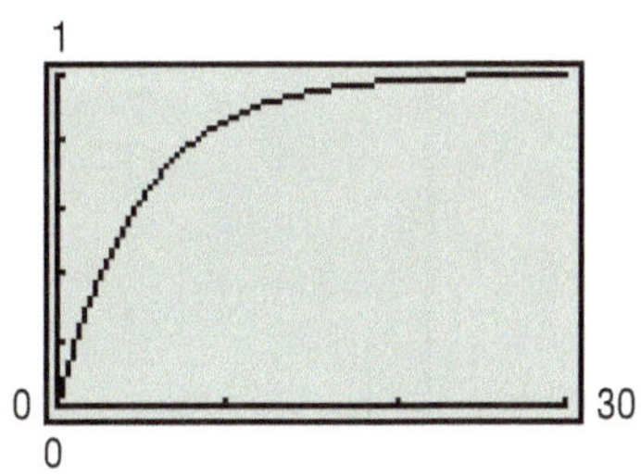

(c) See Figure 34 for the graph of F.

(d) As time passes, the probability that a car will arrive increases. The value that F approaches can be found by letting $t \to \infty$. Since $e^{-0.2t} = \frac{1}{e^{0.2t}}$, it follows that $e^{-0.2t} \to 0$ as $t \to \infty$. We conclude that F approaches 1 as t gets large.

Now Work PROBLEM 101

SUMMARY Properties of the Exponential Function

$f(x) = a^x, \quad a > 1$

Domain: the interval $(-\infty, \infty)$; Range: the interval $(0, \infty)$
x-intercepts: none; y-intercept: 1
Horizontal asymptote: x-axis $(y = 0)$ as $x \to -\infty$
Increasing; one-to-one; smooth; continuous
See Figure 24 for a typical graph.

$f(x) = a^x, \quad 0 < a < 1$

Domain: the interval $(-\infty, \infty)$; Range: the interval $(0, \infty)$
x-intercepts: none; y-intercept: 1
Horizontal asymptote: x-axis $(y = 0)$ as $x \to \infty$
Decreasing; one-to-one; smooth; continuous
See Figure 28 for a typical graph.

If $a^u = a^v$, then $u = v$.

6.3 Assess Your Understanding

'Are You Prepared?' *Answers are given at the end of these exercises. If you get a wrong answer, read the pages listed in red.*

1. (a) $4^3 =$ _____ (b) $8^{2/3} =$ _____ (c) $3^{-2} =$ _____
(d) $(x^2)^3 =$ _____ (e) $a^3 \cdot a^5 =$ _____ (pp. 21–23 and p. 75)

2. Solve: (a) $5x - 2 = 3$ (pp. 105–106)
(b) $2x^2 + 3x = 14$ (pp. 113–114)

3. ***True or False*** To graph $y = (x - 2)^3$, shift the graph of $y = x^3$ to the left 2 units. (pp. 256–258)
4. Find the average rate of change of $f(x) = 3x - 5$ from $x = 0$ to $x = 4$. (pp. 240–241 and 282–285)
5. ***True or False*** The function $f(x) = \dfrac{2x}{x - 3}$ has $y = 2$ as a horizontal asymptote. (pp. 352–355)

Concepts and Vocabulary

6. The graph of every exponential function $f(x) = a^x$, $a > 0$, $a \neq 1$, passes through three points: ______, ______, and ______.
7. If the graph of the exponential function $f(x) = a^x$, $a > 0$, $a \neq 1$, is decreasing, then a must be less than ______.
8. If $3^x = 3^4$, then $x =$ ______.
9. ***True or False*** The graphs of $y = 3^x$ and $y = \left(\dfrac{1}{3}\right)^x$ are identical.
10. ***True or False*** The range of the exponential function $f(x) = a^x$, $a > 0$, $a \neq 1$, is the set of all real numbers.

Skill Building

In Problems 11–20, approximate each number using a calculator. Express your answer rounded to three decimal places.

11. (a) $3^{2.2}$ (b) $3^{2.23}$ (c) $3^{2.236}$ (d) $3^{\sqrt{5}}$
12. (a) $5^{1.7}$ (b) $5^{1.73}$ (c) $5^{1.732}$ (d) $5^{\sqrt{3}}$
13. (a) $2^{3.14}$ (b) $2^{3.141}$ (c) $2^{3.1415}$ (d) 2^{π}
14. (a) $2^{2.7}$ (b) $2^{2.71}$ (c) $2^{2.718}$ (d) 2^{e}
15. (a) $3.1^{2.7}$ (b) $3.14^{2.71}$ (c) $3.141^{2.718}$ (d) π^{e}
16. (a) $2.7^{3.1}$ (b) $2.71^{3.14}$ (c) $2.718^{3.141}$ (d) e^{π}
17. $e^{1.2}$
18. $e^{-1.3}$
19. $e^{-0.85}$
20. $e^{2.1}$

In Problems 21–28, determine whether the given function is linear, exponential, or neither. For those that are linear functions, find a linear function that models the data; for those that are exponential, find an exponential function that models the data.

21.

x	$f(x)$
−1	3
0	6
1	12
2	18
3	30

22.

x	$g(x)$
−1	2
0	5
1	8
2	11
3	14

23.

x	$H(x)$
−1	$\frac{1}{4}$
0	1
1	4
2	16
3	64

24.

x	$F(x)$
−1	$\frac{2}{3}$
0	1
1	$\frac{3}{2}$
2	$\frac{9}{4}$
3	$\frac{27}{8}$

25.

x	$f(x)$
−1	$\frac{3}{2}$
0	3
1	6
2	12
3	24

26.

x	$g(x)$
−1	6
0	1
1	0
2	3
3	10

27.

x	$H(x)$
−1	2
0	4
1	6
2	8
3	10

28.

x	$F(x)$
−1	$\frac{1}{2}$
0	$\frac{1}{4}$
1	$\frac{1}{8}$
2	$\frac{1}{16}$
3	$\frac{1}{32}$

In Problems 29–36, the graph of an exponential function is given. Match each graph to one of the following functions.

(a) $y = 3^x$ (b) $y = 3^{-x}$ (c) $y = -3^x$ (d) $y = -3^{-x}$
(e) $y = 3^x - 1$ (f) $y = 3^{x-1}$ (g) $y = 3^{1-x}$ (h) $y = 1 - 3^x$

29.

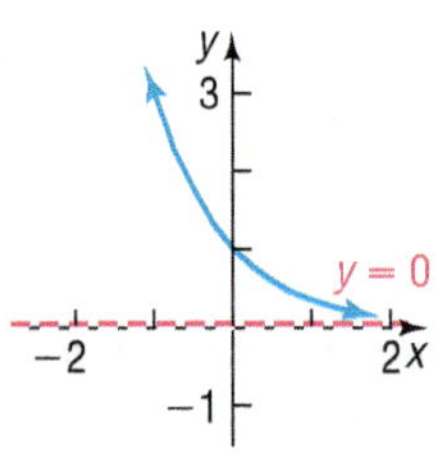

30.

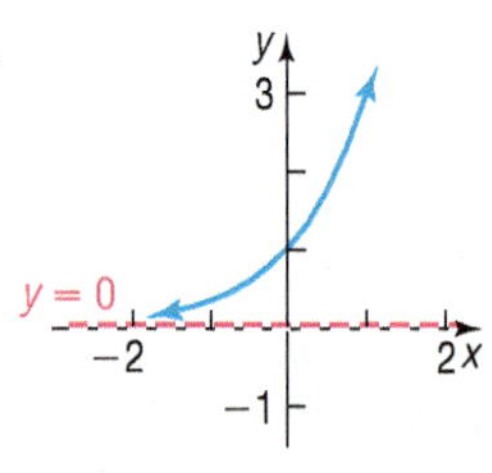

31.

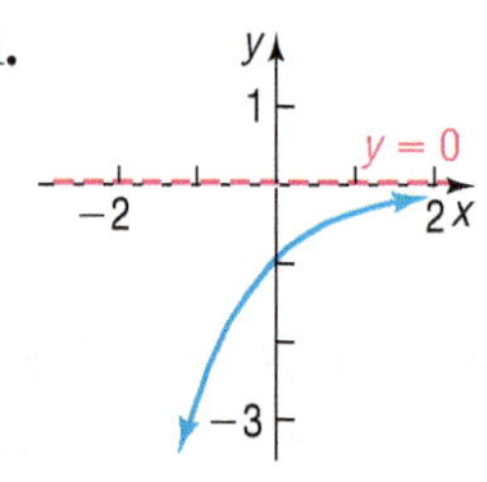

32.

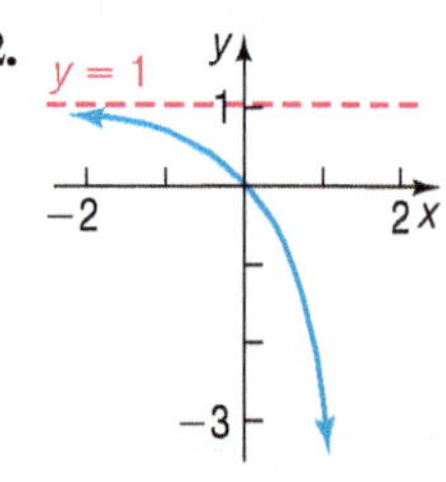

33.

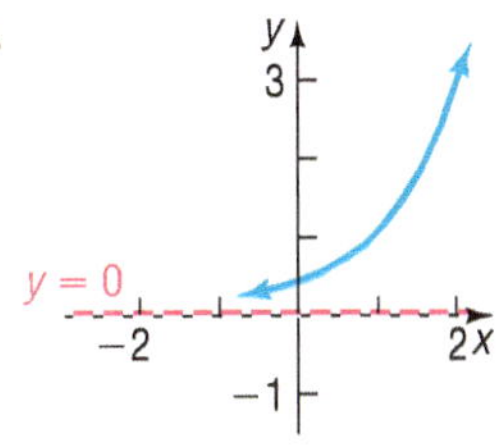

34.

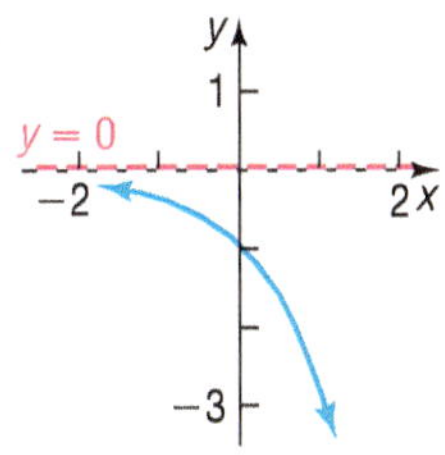

35.

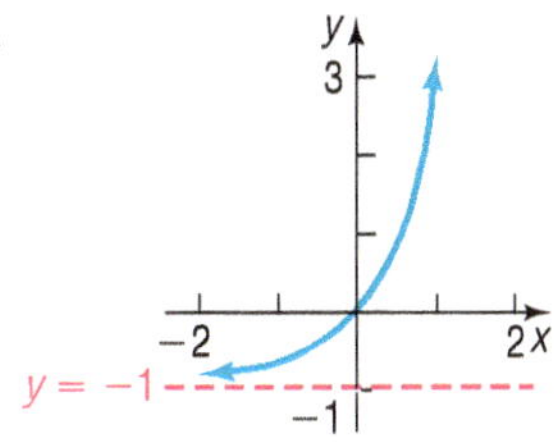

36.

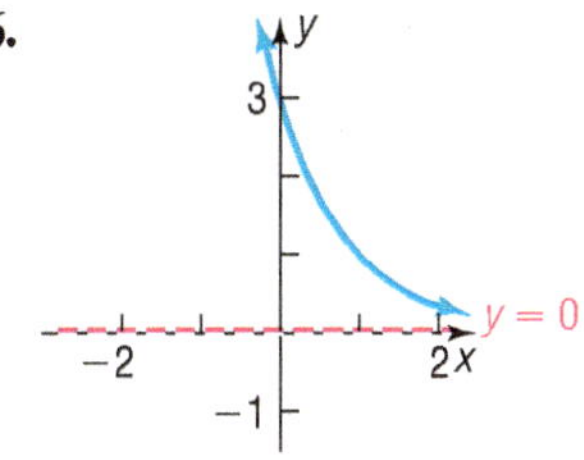

In Problems 37–48, use transformations to graph each function. Determine the domain, range, and horizontal asymptote of each function.

37. $f(x) = 2^x + 1$ **38.** $f(x) = 3^x - 2$ **39.** $f(x) = 3^{x-1}$ **40.** $f(x) = 2^{x+2}$

41. $f(x) = 3 \cdot \left(\frac{1}{2}\right)^x$ **42.** $f(x) = 4 \cdot \left(\frac{1}{3}\right)^x$ **43.** $f(x) = 3^{-x} - 2$ **44.** $f(x) = -3^x + 1$

45. $f(x) = 2 + 4^{x-1}$ **46.** $f(x) = 1 - 2^{x+3}$ **47.** $f(x) = 2 + 3^{x/2}$ **48.** $f(x) = 1 - 2^{-x/3}$

In Problems 49–56, begin with the graph of $y = e^x$ (Figure 30(a)) and use transformations to graph each function. Determine the domain, range, and horizontal asymptote of each function.

49. $f(x) = e^{-x}$ **50.** $f(x) = -e^x$ **51.** $f(x) = e^{x+2}$ **52.** $f(x) = e^x - 1$

53. $f(x) = 5 - e^{-x}$ **54.** $f(x) = 9 - 3e^{-x}$ **55.** $f(x) = 2 - e^{-x/2}$ **56.** $f(x) = 7 - 3e^{2x}$

In Problems 57–76, solve each equation. Verify your results using a graphing utility.

57. $7^x = 7^3$ **58.** $5^x = 5^{-6}$ **59.** $2^{-x} = 16$ **60.** $3^{-x} = 81$

61. $\left(\frac{1}{5}\right)^x = \frac{1}{25}$ **62.** $\left(\frac{1}{4}\right)^x = \frac{1}{64}$ **63.** $2^{2x-1} = 4$ **64.** $5^{x+3} = \frac{1}{5}$

65. $3^{x^3} = 9^x$ **66.** $4^{x^2} = 2^x$ **67.** $8^{-x+14} = 16^x$ **68.** $9^{-x+15} = 27^x$

69. $3^{x^2-7} = 27^{2x}$ **70.** $5^{x^2+8} = 125^{2x}$ **71.** $4^x \cdot 2^{x^2} = 16^2$ **72.** $9^{2x} \cdot 27^{x^2} = 3^{-1}$

73. $e^x = e^{3x+8}$ **74.** $e^{3x} = e^{2-x}$ **75.** $e^{x^2} = e^{3x} \cdot \frac{1}{e^2}$ **76.** $(e^4)^x \cdot e^{x^2} = e^{12}$

77. If $4^x = 7$, what does 4^{-2x} equal? **78.** If $2^x = 3$, what does 4^{-x} equal?

79. If $3^{-x} = 2$, what does 3^{2x} equal? **80.** If $5^{-x} = 3$, what does 5^{3x} equal?

In Problems 81–84, determine the exponential function whose graph is given.

81.

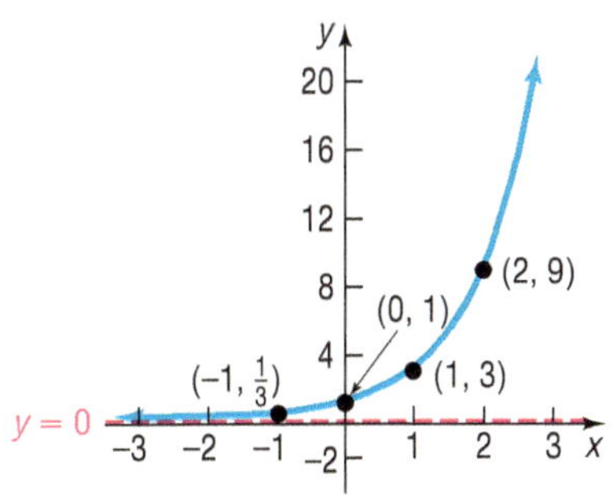

82.

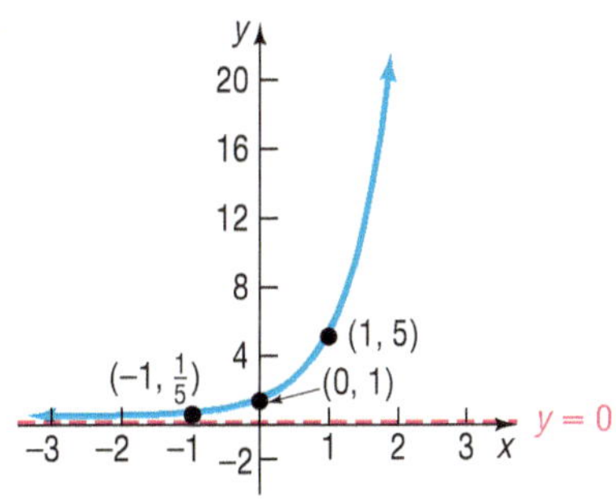

83.

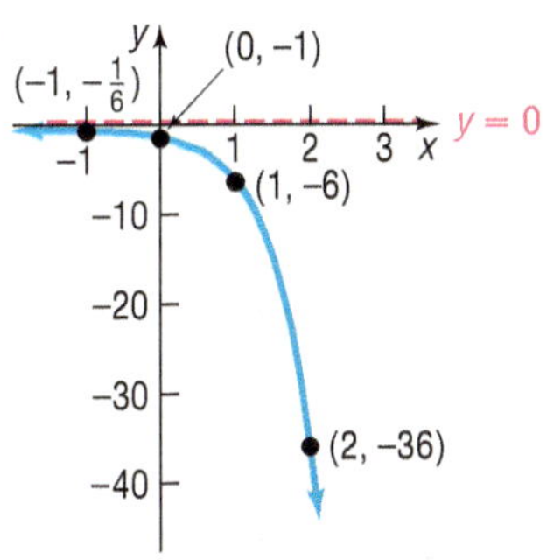

84.

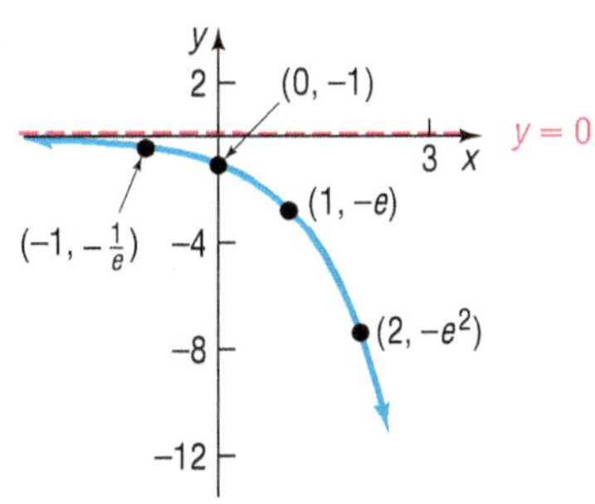

Mixed Practice

85. Suppose that $f(x) = 2^x$.
(a) What is $f(4)$? What point is on the graph of f?
(b) If $f(x) = \frac{1}{16}$, what is x? What point is on the graph of f?

86. Suppose that $f(x) = 3^x$.
(a) What is $f(4)$? What point is on the graph of f?
(b) If $f(x) = \frac{1}{9}$, what is x? What point is on the graph of f?

87. Suppose that $g(x) = 4^x + 2$.
(a) What is $g(-1)$? What point is on the graph of g?
(b) If $g(x) = 66$, what is x? What point is on the graph of g?

88. Suppose that $g(x) = 5^x - 3$.
(a) What is $g(-1)$? What point is on the graph of g?
(b) If $g(x) = 122$, what is x? What point is on the graph of g?

89. Suppose that $H(x) = \left(\frac{1}{2}\right)^x - 4$.
(a) What is $H(-6)$? What point is on the graph of H?
(b) If $H(x) = 12$, what is x? What point is on the graph of H?
(c) Find the zero of H.

90. Suppose that $F(x) = \left(\frac{1}{3}\right)^x - 3$.
(a) What is $F(-5)$? What point is on the graph of F?
(b) If $F(x) = 24$, what is x? What point is on the graph of F?
(c) Find the zero of F.

In Problems 91–94, graph each function. Based on the graph, state the domain and the range and find any intercepts.

91. $f(x) = \begin{cases} e^{-x} & \text{if } x < 0 \\ e^{x} & \text{if } x \geq 0 \end{cases}$

92. $f(x) = \begin{cases} e^{x} & \text{if } x < 0 \\ e^{-x} & \text{if } x \geq 0 \end{cases}$

93. $f(x) = \begin{cases} -e^{x} & \text{if } x < 0 \\ -e^{-x} & \text{if } x \geq 0 \end{cases}$

94. $f(x) = \begin{cases} -e^{-x} & \text{if } x < 0 \\ -e^{x} & \text{if } x \geq 0 \end{cases}$

Applications and Extensions

95. Optics If a single pane of glass obliterates 3% of the light passing through it, the percent p of light that passes through n successive panes is given approximately by the function

$$p(n) = 100(0.97)^n$$

(a) What percent of light will pass through 10 panes?
(b) What percent of light will pass through 25 panes?

96. Atmospheric Pressure The atmospheric pressure p on a balloon or plane decreases with increasing height. This pressure, measured in millimeters of mercury, is related to the height h (in kilometers) above sea level by the function

$$p(h) = 760e^{-0.145h}$$

(a) Find the atmospheric pressure at a height of 2 kilometers (over a mile).
(b) What is it at a height of 10 kilometers (over 30,000 feet)?

97. Depreciation The price p, in dollars, of a Honda Civic DX Sedan that is x years old is modeled by

$$p(x) = 16{,}630(0.90)^x$$

(a) How much does a 3-year-old Civic DX Sedan cost?
(b) How much does a 9-year-old Civic DX Sedan cost?

98. Healing of Wounds The normal healing of wounds can be modeled by an exponential function. If A_0 represents the original area of the wound and if A equals the area of the wound, then the function

$$A(n) = A_0e^{-0.35n}$$

describes the area of a wound after n days following an injury when no infection is present to retard the healing. Suppose that a wound initially had an area of 100 square millimeters.
(a) If healing is taking place, how large will the area of the wound be after 3 days?
(b) How large will it be after 10 days?

99. Drug Medication The function

$$D(h) = 5e^{-0.4h}$$

can be used to find the number of milligrams D of a certain drug that is in a patient's bloodstream h hours after the drug has been administered. How many milligrams will be present after 1 hour? After 6 hours?

100. Spreading of Rumors A model for the number N of people in a college community who have heard a certain rumor is

$$N = P(1 - e^{-0.15d})$$

where P is the total population of the community and d is the number of days that have elapsed since the rumor began. In a community of 1000 students, how many students will have heard the rumor after 3 days?

101. Exponential Probability Between 12:00 PM and 1:00 PM, cars arrive at Citibank's drive-thru at the rate of 6 cars per hour (0.1 car per minute). The following formula from probability can be used to determine the probability that a car will arrive within t minutes of 12:00 PM:

$$F(t) = 1 - e^{-0.1t}$$

(a) Determine the probability that a car will arrive within 10 minutes of 12:00 PM (that is, before 12:10 PM).
(b) Determine the probability that a car will arrive within 40 minutes of 12:00 PM (before 12:40 PM).
(c) What value does F approach as t becomes unbounded in the positive direction?
(d) Graph F using a graphing utility.
(e) Using INTERSECT, determine how many minutes are needed for the probability to reach 50%.

102. Exponential Probability Between 5:00 PM and 6:00 PM, cars arrive at Jiffy Lube at the rate of 9 cars per hour (0.15 car per minute). The following formula from probability can be used to determine the probability that a car will arrive within t minutes of 5:00 PM:

$$F(t) = 1 - e^{-0.15t}$$

(a) Determine the probability that a car will arrive within 15 minutes of 5:00 PM (that is, before 5:15 PM).
(b) Determine the probability that a car will arrive within 30 minutes of 5:00 PM (before 5:30 PM).
(c) What value does F approach as t becomes unbounded in the positive direction?
(d) Graph F using a graphing utility.
(e) Using INTERSECT, determine how many minutes are needed for the probability to reach 60%.

103. Poisson Probability Between 5:00 PM and 6:00 PM, cars arrive at McDonald's drive-thru at the rate of 20 cars per hour. The following formula from probability can be used to determine the probability that x cars will arrive between 5:00 PM and 6:00 PM.

$$P(x) = \frac{20^x e^{-20}}{x!}$$

where

$$x! = x \cdot (x-1) \cdot (x-2) \cdot \cdots \cdot 3 \cdot 2 \cdot 1$$

(a) Determine the probability that $x = 15$ cars will arrive between 5:00 PM and 6:00 PM.
(b) Determine the probability that $x = 20$ cars will arrive between 5:00 PM and 6:00 PM.

104. Poisson Probability People enter a line for the *Demon Roller Coaster* at the rate of 4 per minute. The following formula from probability can be used to determine the probability that x people will arrive within the next minute.

$$P(x) = \frac{4^x e^{-4}}{x!}$$

where

$$x! = x \cdot (x-1) \cdot (x-2) \cdot \cdots \cdot 3 \cdot 2 \cdot 1$$

(a) Determine the probability that $x = 5$ people will arrive within the next minute.
(b) Determine the probability that $x = 8$ people will arrive within the next minute.

105. Relative Humidity The relative humidity is the ratio (expressed as a percent) of the amount of water vapor in the air to the maximum amount that it can hold at a specific temperature. The relative humidity, R, is found using the following formula:

$$R = 10^{\left(\frac{4221}{T+459.4} - \frac{4221}{D+459.4} + 2\right)}$$

where T is the air temperature (in °F) and D is the dew point temperature (in °F).

(a) Determine the relative humidity if the air temperature is 50° Fahrenheit and the dew point temperature is 41° Fahrenheit.
(b) Determine the relative humidity if the air temperature is 68° Fahrenheit and the dew point temperature is 59° Fahrenheit.
(c) What is the relative humidity if the air temperature and the dew point temperature are the same?

106. Learning Curve Suppose that a student has 500 vocabulary words to learn. If the student learns 15 words after 5 minutes, the function

$$L(t) = 500(1 - e^{-0.0061t})$$

approximates the number of words L that the student will learn after t minutes.

(a) How many words will the student learn after 30 minutes?
(b) How many words will the student learn after 60 minutes?

107. Current in a *RL* Circuit The equation governing the amount of current I (in amperes) after time t (in seconds) in a single RL circuit consisting of a resistance R (in ohms), an inductance L (in henrys), and an electromotive force E (in volts) is

$$I = \frac{E}{R}\left[1 - e^{-(R/L)t}\right]$$

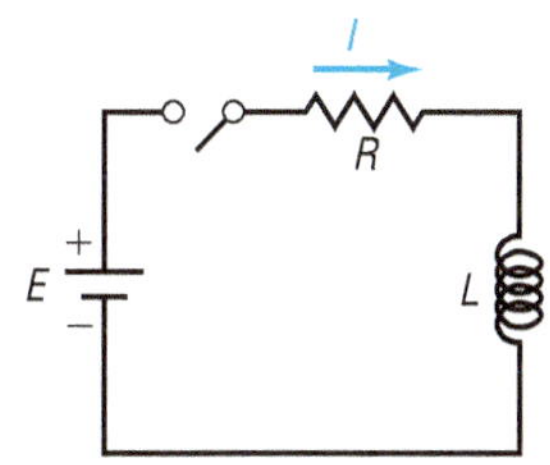

(a) If $E = 120$ volts, $R = 10$ ohms, and $L = 5$ henrys, how much current I_1 is flowing after 0.3 second? After 0.5 second? After 1 second?
(b) What is the maximum current?
(c) Graph the function $I = I_1(t)$, measuring I along the y-axis and t along the x-axis.
(d) If $E = 120$ volts, $R = 5$ ohms, and $L = 10$ henrys, how much current I_2 is flowing after 0.3 second? After 0.5 second? After 1 second?
(e) What is the maximum current?
(f) Graph the function $I = I_2(t)$ on the same coordinate axes as $I_1(t)$.

108. Current in a *RC* Circuit The equation governing the amount of current I (in amperes) after time t (in microseconds) in a single RC circuit consisting of a resistance R (in ohms), a capacitance C (in microfarads), and an electromotive force E (in volts) is

$$I = \frac{E}{R}e^{-t/(RC)}$$

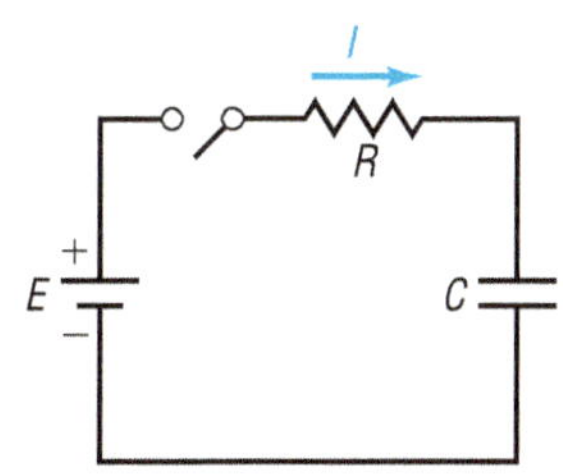

(a) If $E = 120$ volts, $R = 2000$ ohms, and $C = 1.0$ microfarad, how much current I_1 is flowing initially ($t = 0$)? After 1000 microseconds? After 3000 microseconds?
(b) What is the maximum current?
(c) Graph the function $I = I_1(t)$, measuring I along the y-axis and t along the x-axis.
(d) If $E = 120$ volts, $R = 1000$ ohms, and $C = 2.0$ microfarads, how much current I_2 is flowing initially? After 1000 microseconds? After 3000 microseconds?
(e) What is the maximum current?
(f) Graph the function $I = I_2(t)$ on the same coordinate axes as $I_1(t)$.

109. Another Formula for *e* Use a calculator to compute the values of

$$2 + \frac{1}{2!} + \frac{1}{3!} + \cdots + \frac{1}{n!}$$

for $n = 4, 6, 8$, and 10. Compare each result with e.
[**Hint:** $1! = 1, 2! = 2 \cdot 1, 3! = 3 \cdot 2 \cdot 1,$
$n! = n(n-1) \cdot \cdots \cdot (3)(2)(1)$.]

110. Another Formula for *e* Use a calculator to compute the various values of the expression. Compare the values to e.

$$2 + \cfrac{1}{1 + \cfrac{1}{2 + \cfrac{2}{3 + \cfrac{3}{4 + \cfrac{4}{\text{etc.}}}}}}$$

111. Difference Quotient If $f(x) = a^x$, show that

$$\frac{f(x+h) - f(x)}{h} = a^x \cdot \frac{a^h - 1}{h} \quad h \neq 0$$

112. If $f(x) = a^x$, show that $f(A + B) = f(A) \cdot f(B)$.

113. If $f(x) = a^x$, show that $f(-x) = \dfrac{1}{f(x)}$.

114. If $f(x) = a^x$, show that $f(\alpha x) = [f(x)]^\alpha$.

Problems 115 and 116 provide definitions for two other transcendental functions.

115. The **hyperbolic sine function,** designated by $\sinh x$, is defined as

$$\sinh x = \frac{1}{2}(e^x - e^{-x})$$

(a) Show that $f(x) = \sinh x$ is an odd function.
(b) Graph $f(x) = \sinh x$ using a graphing utility.

116. The **hyperbolic cosine function,** designated by $\cosh x$, is defined as

$$\cosh x = \frac{1}{2}(e^x + e^{-x})$$

(a) Show that $f(x) = \cosh x$ is an even function.
(b) Graph $f(x) = \cosh x$ using a graphing utility.
(c) Refer to Problem 115. Show that, for every x,

$$(\cosh x)^2 - (\sinh x)^2 = 1$$

117. Historical Problem Pierre de Fermat (1601–1665) conjectured that the function

$$f(x) = 2^{(2^x)} + 1$$

for $x = 1, 2, 3, \ldots$, would always have a value equal to a prime number. But Leonhard Euler (1707–1783) showed that this formula fails for $x = 5$. Use a calculator to determine the prime numbers produced by f for $x = 1, 2, 3, 4$. Then show that $f(5) = 641 \times 6{,}700{,}417$, which is not prime.

Discussion and Writing

118. The bacteria in a 4-liter container double every minute. After 60 minutes the container is full. How long did it take to fill half the container?

119. Explain in your own words what the number e is. Provide at least two applications that use this number.

120. Do you think that there is a power function that increases more rapidly than an exponential function whose base is greater than 1? Explain.

121. As the base a of an exponential function $f(x) = a^x, a > 1$, increases, what happens to the behavior of its graph for $x > 0$? What happens to the behavior of its graph for $x < 0$?

122. The graphs of $y = a^{-x}$ and $y = \left(\dfrac{1}{a}\right)^x$ are identical. Why?

'Are You Prepared?' Answers

1. (a) 64 (b) 4 (c) $\frac{1}{9}$ (d) x^6 (e) a^8 **2.** (a) $\{1\}$ (b) $\left\{-\frac{7}{2}, 2\right\}$ **3.** False **4.** 3 **5.** True

6.4 Logarithmic Functions

PREPARING FOR THIS SECTION *Before getting started, review the following:*

- Solving Inequalities (Section 1.7, pp. 153–155)
- Quadratic Inequalities (Section 4.5, pp. 320–322)
- Polynomial and Rational Inequalities (Section 5.4, pp. 369–373)

Now Work the 'Are You Prepared?' problems on page 449.

OBJECTIVES
1 Change Exponential Statements to Logarithmic Statements and Logarithmic Statements to Exponential Statements (p. 441)
2 Evaluate Logarithmic Expressions (p. 441)
3 Determine the Domain of a Logarithmic Function (p. 442)
4 Graph Logarithmic Functions (p. 443)
5 Solve Logarithmic Equations (p. 447)

Recall that a one-to-one function $y = f(x)$ has an inverse function that is defined (implicitly) by the equation $x = f(y)$. In particular, the exponential function $y = f(x) = a^x$, $a > 0$, $a \neq 1$, is one-to-one and hence has an inverse function that is defined implicitly by the equation

$$x = a^y, \qquad a > 0, \qquad a \neq 1$$

This inverse function is so important that it is given a name, the *logarithmic function.*

DEFINITION

The **logarithmic function to the base *a*,** where $a > 0$ and $a \neq 1$, is denoted by $y = \log_a x$ (read as "y is the logarithm to the base a of x") and is defined by

$$y = \log_a x \quad \text{if and only if} \quad x = a^y$$

The domain of the logarithmic function $y = \log_a x$ is $x > 0$.

As this definition illustrates, a *logarithm* is a name for a certain exponent.

EXAMPLE 1 Relating Logarithms to Exponents

(a) If $y = \log_3 x$, then $x = 3^y$. For example, $4 = \log_3 81$ is equivalent to $81 = 3^4$.

(b) If $y = \log_5 x$, then $x = 5^y$. For example, $-1 = \log_5\left(\frac{1}{5}\right)$ is equivalent to $\frac{1}{5} = 5^{-1}$.

1 Change Exponential Statements to Logarithmic Statements and Logarithmic Statements to Exponential Statements

We can use the definition of a logarithm to convert from exponential form to logarithmic form, and vice versa, as the following two examples illustrate.

EXAMPLE 2 Changing Exponential Statements to Logarithmic Statements

Change each exponential statement to an equivalent statement involving a logarithm.

(a) $1.2^3 = m$ (b) $e^b = 9$ (c) $a^4 = 24$

Solution We use the fact that $y = \log_a x$ and $x = a^y$, $a > 0$, $a \neq 1$, are equivalent.

(a) If $1.2^3 = m$, then $3 = \log_{1.2} m$. (b) If $e^b = 9$, then $b = \log_e 9$.

(c) If $a^4 = 24$, then $4 = \log_a 24$.

Now Work PROBLEM 9

EXAMPLE 3 Changing Logarithmic Statements to Exponential Statements

Change each logarithmic statement to an equivalent statement involving an exponent.

(a) $\log_a 4 = 5$ (b) $\log_e b = -3$ (c) $\log_3 5 = c$

Solution (a) If $\log_a 4 = 5$, then $a^5 = 4$. (b) If $\log_e b = -3$, then $e^{-3} = b$.

(c) If $\log_3 5 = c$, then $3^c = 5$.

Now Work PROBLEM 17

2 Evaluate Logarithmic Expressions

To find the exact value of a logarithm, we write the logarithm in exponential notation and use the fact that if $a^u = a^v$ then $u = v$.

EXAMPLE 4 **Finding the Exact Value of a Logarithmic Expression**

Find the exact value of:

(a) $\log_2 16$ (b) $\log_3 \frac{1}{27}$

Solution

> **In Words**
> The expression $\log_2 16$ means, "give me the exponent y such that $2^y = 16$."

(a)
$$y = \log_2 16$$
$$2^y = 16 \quad \text{Change to exponential form.}$$
$$2^y = 2^4 \quad 16 = 2^4$$
$$y = 4 \quad \text{Equate exponents.}$$

Therefore, $\log_2 16 = 4$.

(b)
$$y = \log_3 \frac{1}{27}$$
$$3^y = \frac{1}{27} \quad \text{Change to exponential form.}$$
$$3^y = 3^{-3} \quad \frac{1}{27} = \frac{1}{3^3} = 3^{-3}$$
$$y = -3 \quad \text{Equate exponents.}$$

Therefore, $\log_3 \frac{1}{27} = -3$.

Now Work PROBLEM 25

3 Determine the Domain of a Logarithmic Function

The logarithmic function $y = \log_a x$ has been defined as the inverse of the exponential function $y = a^x$. That is, if $f(x) = a^x$, then $f^{-1}(x) = \log_a x$. Based on the discussion given in Section 6.2 on inverse functions, for a function f and its inverse f^{-1}, we have

$$\text{Domain of } f^{-1} = \text{Range of } f \quad \text{and} \quad \text{Range of } f^{-1} = \text{Domain of } f$$

Consequently, it follows that

> Domain of the logarithmic function = Range of the exponential function = $(0, \infty)$
>
> Range of the logarithmic function = Domain of the exponential function = $(-\infty, \infty)$

In the next box, we summarize some properties of the logarithmic function:

> $y = \log_a x$ (defining equation: $x = a^y$)
> Domain: $0 < x < \infty$ Range: $-\infty < y < \infty$

The domain of a logarithmic function consists of the *positive* real numbers, so the argument of a logarithmic function must be greater than zero.

EXAMPLE 5 **Finding the Domain of a Logarithmic Function**

Find the domain of each logarithmic function.

(a) $F(x) = \log_2(x + 3)$ (b) $g(x) = \log_5\left(\frac{1 + x}{1 - x}\right)$ (c) $h(x) = \log_{1/2}|x|$

Solution

(a) The domain of F consists of all x for which $x + 3 > 0$, that is, $x > -3$. Using interval notation, the domain of f is $(-3, \infty)$.

(b) The domain of g is restricted to

$$\frac{1 + x}{1 - x} > 0$$

Solving this inequality, we find that the domain of g consists of all x between -1 and 1, that is, $-1 < x < 1$ or, using interval notation, $(-1, 1)$.

(c) Since $|x| > 0$, provided that $x \neq 0$, the domain of h consists of all real numbers except zero or, using interval notation, $(-\infty, 0) \cup (0, \infty)$.

Now Work PROBLEMS 39 AND 45

4 Graph Logarithmic Functions

Since exponential functions and logarithmic functions are inverses of each other, the graph of the logarithmic function $y = \log_a x$ is the reflection about the line $y = x$ of the graph of the exponential function $y = a^x$, as shown in Figure 35.

Figure 35

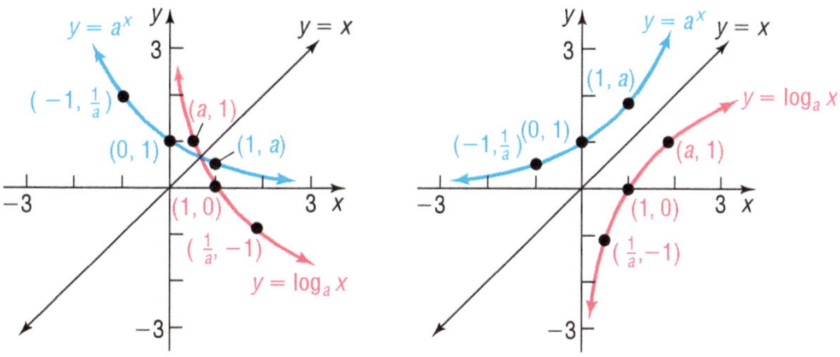

(a) $0 < a < 1$ **(b)** $a > 1$

For example, to graph $y = \log_2 x$, graph $y = 2^x$ and reflect it about the line $y = x$. See Figure 36. To graph $y = \log_{1/3} x$, graph $y = \left(\frac{1}{3}\right)^x$ and reflect it about the line $y = x$. See Figure 37.

Figure 36

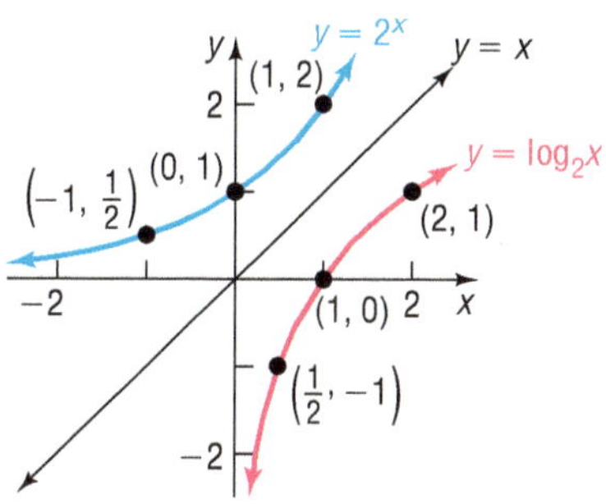

Figure 37

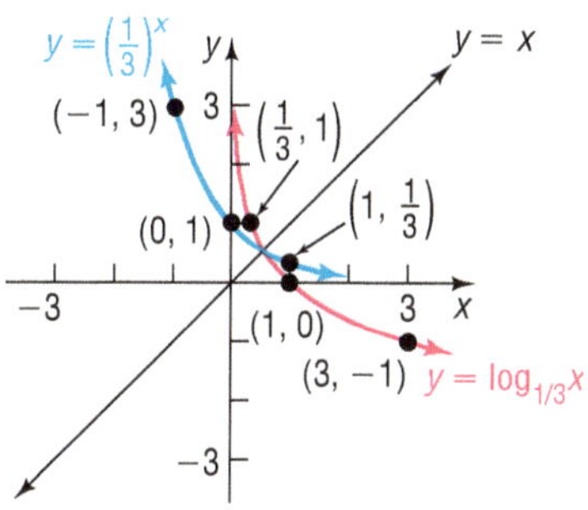

Now Work PROBLEM 59

The graphs of $y = \log_a x$ in Figures 35(a) and (b) lead to the following properties.

Properties of the Logarithmic Function $f(x) = \log_a x$

1. The domain is the set of positive real numbers; the range is the set of all real numbers.
2. The x-intercept of the graph is 1. There is no y-intercept.
3. The y-axis ($x = 0$) is a vertical asymptote of the graph.
4. A logarithmic function is decreasing if $0 < a < 1$ and increasing if $a > 1$.
5. The graph of f contains the points $(1, 0)$, $(a, 1)$, and $\left(\frac{1}{a}, -1\right)$.
6. The graph is smooth and continuous, with no corners or gaps.

If the base of a logarithmic function is the number e, then we have the **natural logarithm function.** This function occurs so frequently in applications that it is given a special symbol, **ln** (from the Latin, *logarithmus naturalis*). That is,

In Words
$y = \log_e x$ is written $y = \ln x$

$$y = \ln x \quad \text{if and only if} \quad x = e^y \tag{1}$$

Since $y = \ln x$ and the exponential function $y = e^x$ are inverse functions, we can obtain the graph of $y = \ln x$ by reflecting the graph of $y = e^x$ about the line $y = x$. See Figure 38.

Table 8 displays other points on the graph of $f(x) = \ln x$. Notice for $x \le 0$ that we obtain an error message. Do you recall why?

Figure 38

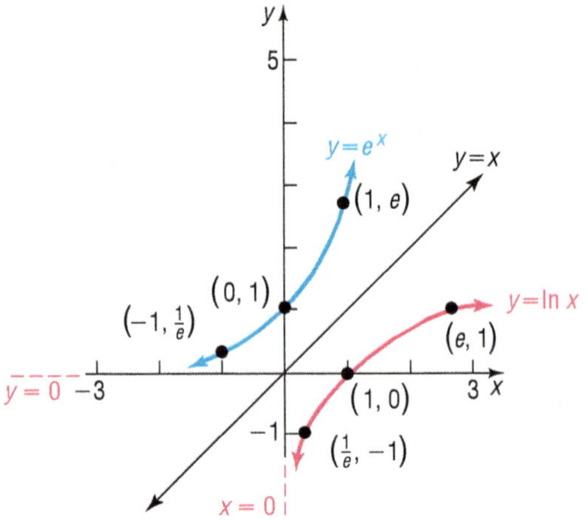

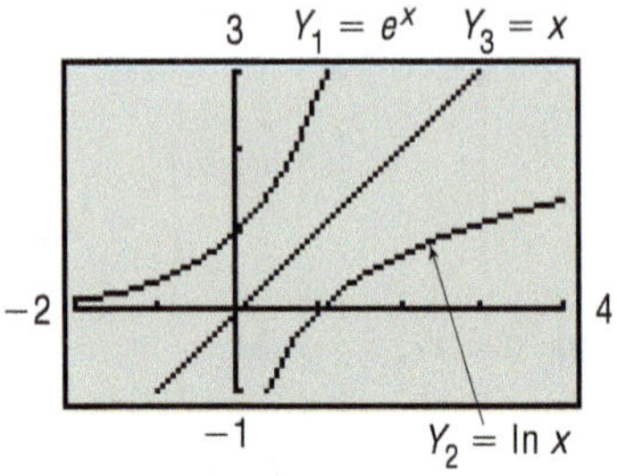

Table 8

X	Y2
-1	ERROR
0.5	-.6931
1	0
2	.69315
2.7183	1
3	1.0986

Y2=ln(X)

EXAMPLE 6 Graphing a Logarithmic Function and Its Inverse

(a) Find the domain of the logarithmic function $f(x) = -\ln(x - 2)$.
(b) Graph f.
(c) From the graph, determine the range and vertical asymptote of f.
(d) Find f^{-1}, the inverse of f.
(e) Use f^{-1} to confirm the range of f found in part (c). From the domain of f, find the range of f^{-1}.
(f) Graph f^{-1}.

Solution

(a) The domain of f consists of all x for which $x - 2 > 0$ or, equivalently, $x > 2$. The domain of f is $\{x | x > 2\}$ or $(2, \infty)$.
(b) To obtain the graph of $y = -\ln(x - 2)$, we begin with the graph of $y = \ln x$ and use transformations. See Figure 39.

Figure 39

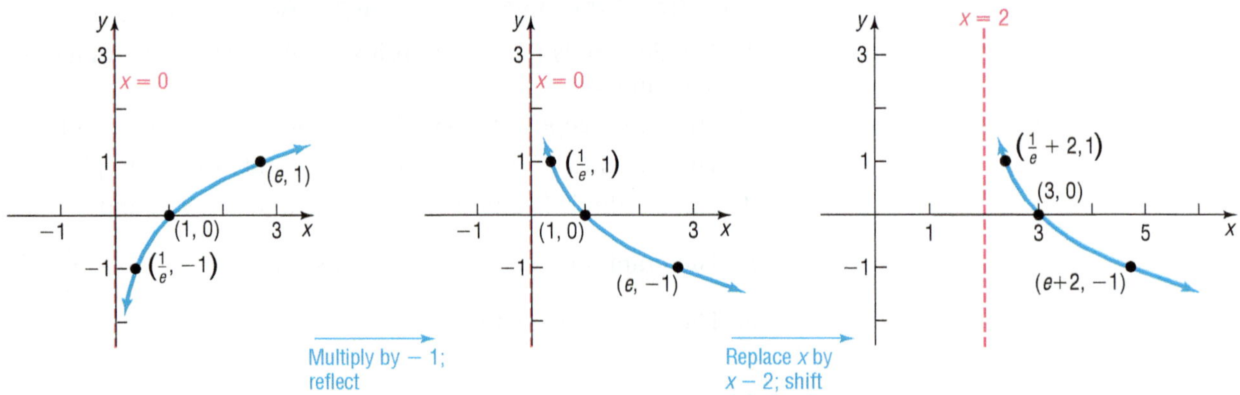

(a) $y = \ln x$ (b) $y = -\ln x$ (c) $y = -\ln(x - 2)$

(c) The range of $f(x) = -\ln(x - 2)$ is the set of all real numbers. The vertical asymptote is $x = 2$. [Do you see why? The original asymptote ($x = 0$) is shifted to the right 2 units.]

(d) We begin with $y = -\ln(x - 2)$. The inverse function is defined (implicitly) by the equation

$$x = -\ln(y - 2)$$

We proceed to solve for y.

$$\begin{aligned} -x &= \ln(y - 2) && \text{Isolate the logarithm.} \\ e^{-x} &= y - 2 && \text{Change to an exponential expression.} \\ y &= e^{-x} + 2 && \text{Solve for } y. \end{aligned}$$

The inverse of f is $f^{-1}(x) = e^{-x} + 2$.

(e) The range of f is the domain of f^{-1}, which is the set of all real numbers, confirming what we found from the graph of f. The range of f^{-1} is the domain of f, which is $(2, \infty)$

(f) To graph f^{-1}, we use the graph of f in Figure 39(c) and reflect it about the line $y = x$. See Figure 40. We could also graph $f^{-1}(x) = e^{-x} + 2$ using transformations.

Figure 40

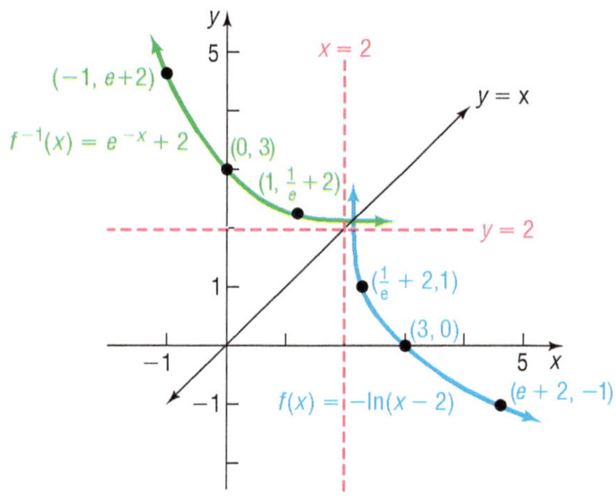

Figure 41

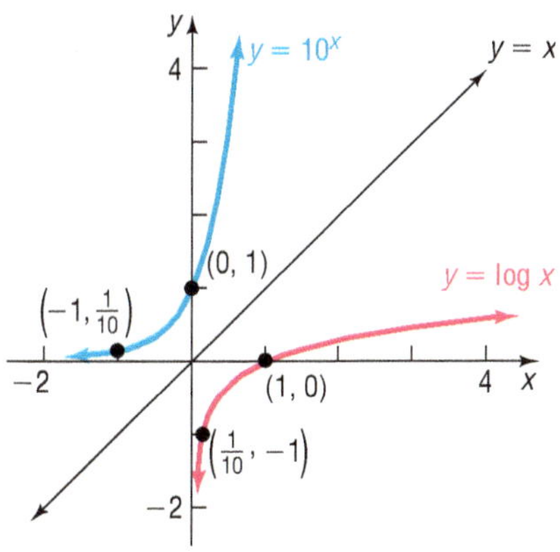

Now Work PROBLEM 71

If the base of a logarithmic function is the number 10, then we have the **common logarithm function.** If the base a of the logarithmic function is not indicated, it is understood to be 10. That is,

$$y = \log x \quad \text{if and only if} \quad x = 10^y$$

Since $y = \log x$ and the exponential function $y = 10^x$ are inverse functions, we can obtain the graph of $y = \log x$ by reflecting the graph of $y = 10^x$ about the line $y = x$. See Figure 41.

EXAMPLE 7 Graphing a Logarithmic Function and Its Inverse

(a) Find the domain of the logarithmic function $f(x) = 3\log(x - 1)$.
(b) Graph f.
(c) From the graph, determine the range and vertical asymptote of f.
(d) Find f^{-1}, the inverse of f.
(e) Use f^{-1} to confirm the range of f found in part (c). From the domain of f, find the range of f^{-1}.
(f) Graph f^{-1}.

Solution

(a) The domain of f consists of all x for which $x - 1 > 0$ or, equivalently, $x > 1$. The domain of f is $\{x \mid x > 1\}$ or $(1, \infty)$.

(b) To obtain the graph of $y = 3\log(x - 1)$, we begin with the graph of $y = \log x$ and use transformations. See Figure 42 on page 446.

Figure 42

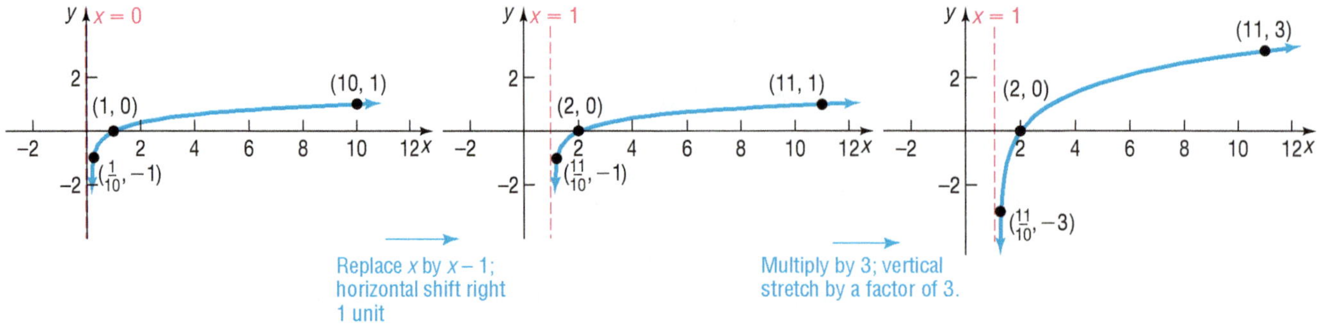

(a) $y = \log x$ **(b)** $y = \log(x - 1)$ **(c)** $y = 3\log(x - 1)$

(c) The range of $f(x) = 3\log(x - 1)$ is the set of all real numbers. The vertical asymptote is $x = 1$.

(d) We begin with $y = 3\log(x - 1)$. The inverse function is defined (implicitly) by the equation

$$x = 3\log(y - 1)$$

We proceed to solve for y.

$$\frac{x}{3} = \log(y - 1) \quad \text{Isolate the logarithm.}$$

$$10^{x/3} = y - 1 \quad \text{Change to an exponential expression.}$$

$$y = 10^{x/3} + 1 \quad \text{Solve for } y.$$

The inverse of f is $f^{-1}(x) = 10^{x/3} + 1$.

(e) The range of f is the domain of f^{-1}, which is the set of all real numbers, confirming what we found from the graph of f. The range of f^{-1} is the domain of f, which is $(1, \infty)$.

(f) To graph f^{-1}, we use the graph of f in Figure 42(c) and reflect it about the line $y = x$. See Figure 43. We could also graph $f^{-1}(x) = 10^{x/3} + 1$ using transformations.

Figure 43

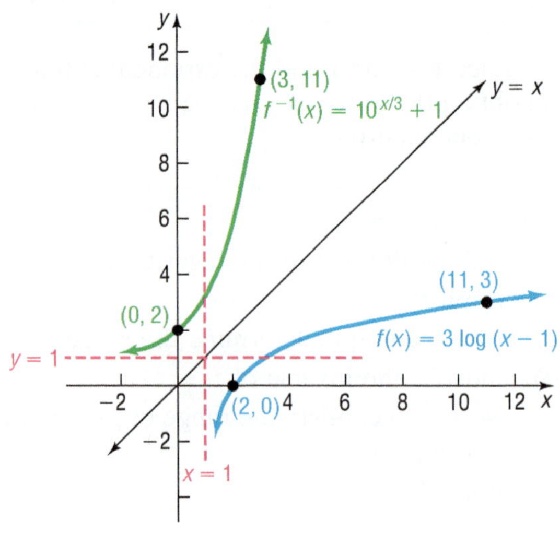

Now Work PROBLEM 79

5 Solve Logarithmic Equations

Equations that contain logarithms are called **logarithmic equations.** Care must be taken when solving logarithmic equations algebraically. In the expression $\log_a M$, remember that a and M are positive and $a \neq 1$. Be sure to check each apparent solution in the original equation and discard any that are extraneous.

Some logarithmic equations can be solved by changing from a logarithmic expression to an exponential expression.

EXAMPLE 8 Solving a Logarithmic Equation

Solve: (a) $\log_3(4x - 7) = 2$ (b) $\log_x 64 = 2$

Solution (a) We can obtain an exact solution by changing the logarithmic equation to exponential form.

$$\log_3(4x - 7) = 2$$
$$4x - 7 = 3^2 \quad \text{Change to an exponential equation.}$$
$$4x - 7 = 9$$
$$4x = 16$$
$$x = 4$$

Check: $\log_3(4x - 7) = \log_3(4 \cdot 4 - 7) = \log_3 9 = 2 \quad 3^2 = 9$

The solution set is $\{4\}$.

(b) We can obtain an exact solution by changing the logarithmic equation to exponential form.

$$\log_x 64 = 2$$
$$x^2 = 64 \quad \text{Change to an exponential equation.}$$
$$x = \pm\sqrt{64} = \pm 8 \quad \text{Square Root Method}$$

The base of a logarithm is always positive. As a result, we discard -8. We check the solution 8.

Check: $\log_8 64 = 2 \quad 8^2 = 64$

The solution set is $\{8\}$.

EXAMPLE 9 Using Logarithms to Solve Exponential Equations

Solve: $e^{2x} = 5$

Solution We can obtain an exact solution by changing the exponential equation to logarithmic form.

$$e^{2x} = 5$$
$$\ln 5 = 2x \quad \text{Change to a logarithmic equation using the fact that if } e^y = x \text{ then } y = \ln x.$$
$$x = \frac{\ln 5}{2} \quad \text{Exact solution}$$
$$\approx 0.805 \quad \text{Approximate solution}$$

The solution set is $\left\{\frac{\ln 5}{2}\right\}$.

Now Work PROBLEMS 87 AND 99

EXAMPLE 10 Alcohol and Driving

The blood alcohol concentration (BAC) is the amount of alcohol in a person's bloodstream. A BAC of 0.04% means that a person has 4 parts alcohol per 10,000 parts blood in the body. Relative risk is defined as the likelihood of one event occurring divided by the likelihood of a second event occurring. For example, if an individual with a BAC of 0.02% is 1.4 times as likely to have a car accident as an individual that has not been drinking, the relative risk of an accident with a BAC of 0.02% is 1.4. Recent medical research suggests that the relative risk R of having an accident while driving a car can be modeled by the equation

$$R = e^{kx}$$

where x is the percent of concentration of alcohol in the bloodstream and k is a constant.

(a) Research indicates that the relative risk of a person having an accident with a BAC of 0.02% is 1.4. Find the constant k in the equation.

(b) Using this value of k, what is the relative risk if the concentration is 0.17%?

(c) Using this same value of k, what BAC corresponds to a relative risk of 100?

(d) If the law asserts that anyone with a relative risk of 5 or more should not have driving privileges, at what concentration of alcohol in the bloodstream should a driver be arrested and charged with a DUI (driving under the influence)?

Solution

(a) For a concentration of alcohol in the blood of 0.02% and a relative risk of 1.4, we let $x = 0.02$ and $R = 1.4$ in the equation and solve for k.

$$R = e^{kx}$$

$$1.4 = e^{k(0.02)} \qquad R = 1.4;\ x = 0.02$$

$$0.02k = \ln 1.4 \qquad \text{Change to a logarithmic expression.}$$

$$k = \frac{\ln 1.4}{0.02} \approx 16.82 \qquad \text{Solve for } k.$$

(b) For a concentration of 0.17%, we have $x = 0.17$. Using $k = 16.82$ in the equation, we find the relative risk R to be

$$R = e^{kx} = e^{(16.82)(0.17)} \approx 17.5$$

For a concentration of alcohol in the blood of 0.17%, the relative risk of an accident is about 17.5. That is, a person with a BAC of 0.17% is 17.5 times as likely to have a car accident as a person with no alcohol in the bloodstream.

(c) For a relative risk of 100, we have $R = 100$. Using $k = 16.82$ in the equation $R = e^{kx}$, we find the concentration x of alcohol in the blood obeys

$$100 = e^{16.82x} \qquad R = e^{kx},\ R = 100;\ k = 16.82$$

$$16.82x = \ln 100 \qquad \text{Change to a logarithmic expression.}$$

$$x = \frac{\ln 100}{16.82} \approx 0.27 \qquad \text{Solve for } x.$$

NOTE A BAC of 0.30% results in a loss of consciousness in most people. ■

For a concentration of alcohol in the blood of 0.27%, the relative risk of an accident is 100.

(d) For a relative risk of 5, we have $R = 5$. Using $k = 16.82$ in the equation $R = e^{kx}$, we find the concentration x of alcohol in the bloodstream obeys

$$5 = e^{16.82x}$$

$$16.82x = \ln 5$$

$$x = \frac{\ln 5}{16.82} \approx 0.096$$

NOTE Most states use 0.08% or 0.10% as the blood alcohol content at which a DUI citation is given. ■

A driver with a BAC of 0.096% or more should be arrested and charged with DUI.

SUMMARY Properties of the Logarithmic Function

$f(x) = \log_a x, \quad a > 1$ $(y = \log_a x$ means $x = a^y)$	Domain: the interval $(0, \infty)$; Range: the interval $(-\infty, \infty)$ x-intercept: 1; y-intercept: none; vertical asymptote: $x = 0$ (y-axis); increasing; one-to-one See Figure 44(a) for a typical graph.
$f(x) = \log_a x, \quad 0 < a < 1$ $(y = \log_a x$ means $x = a^y)$	Domain: the interval $(0, \infty)$; Range: the interval $(-\infty, \infty)$ x-intercept: 1; y-intercept: none; vertical asymptote: $x = 0$ (y-axis); decreasing; one-to-one See Figure 44(b) for a typical graph.

Figure 44

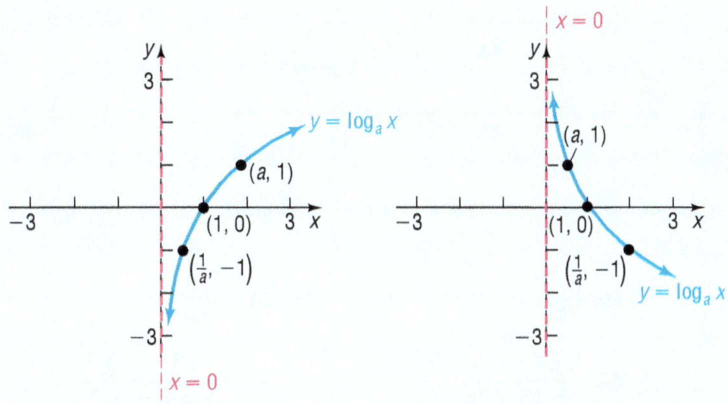

(a) $a > 1$ **(b)** $0 < a < 1$

6.4 Assess Your Understanding

'Are You Prepared?'

Answers are given at the end of these exercises. If you get a wrong answer, read the pages listed in red.

1. Solve the inequality: $3x - 7 \le 8 - 2x$ (pp. 153–155)

2. Solve the inequality: $x^2 - x - 6 > 0$ (pp. 320–322)

3. Solve the inequality: $\dfrac{x-1}{x+4} > 0$ (pp. 371–373)

Concepts and Vocabulary

4. The domain of the logarithmic function $f(x) = \log_a x$ is ______.

5. The graph of every logarithmic function $f(x) = \log_a x$, $a > 0, a \ne 1$, passes through three points: ______, ______, and ______.

6. If the graph of a logarithmic function $f(x) = \log_a x$, $a > 0, a \ne 1$, is increasing, then its base must be larger than ______.

7. ***True or False*** If $y = \log_a x$, then $y = a^x$.

8. ***True or False*** The graph of $f(x) = \log_a x$, $a > 0, a \ne 1$, has an x-intercept equal to 1 and no y-intercept.

Skill Building

In Problems 9–16, change each exponential statement to an equivalent statement involving a logarithm.

9. $9 = 3^2$ **10.** $16 = 4^2$ **11.** $a^2 = 1.6$ **12.** $a^3 = 2.1$

13. $2^x = 7.2$ **14.** $3^x = 4.6$ **15.** $e^x = 8$ **16.** $e^{2.2} = M$

In Problems 17–24, change each logarithmic statement to an equivalent statement involving an exponent.

17. $\log_2 8 = 3$ **18.** $\log_3\left(\dfrac{1}{9}\right) = -2$ **19.** $\log_a 3 = 6$ **20.** $\log_b 4 = 2$

21. $\log_3 2 = x$ **22.** $\log_2 6 = x$ **23.** $\ln 4 = x$ **24.** $\ln x = 4$

In Problems 25–36, find the exact value of each logarithm without using a calculator.

25. $\log_2 1$ **26.** $\log_8 8$ **27.** $\log_5 25$ **28.** $\log_3\left(\frac{1}{9}\right)$

29. $\log_{1/2} 16$ **30.** $\log_{1/3} 9$ **31.** $\log_{10}\sqrt{10}$ **32.** $\log_5 \sqrt[3]{25}$

33. $\log_{\sqrt{2}} 4$ **34.** $\log_{\sqrt{3}} 9$ **35.** $\ln\sqrt{e}$ **36.** $\ln e^3$

In Problems 37–48, find the domain of each function.

37. $f(x) = \ln(x - 3)$ **38.** $g(x) = \ln(x - 1)$ **39.** $F(x) = \log_2 x^2$

40. $H(x) = \log_5 x^3$ **41.** $f(x) = 3 - 2\log_4\left[\frac{x}{2} - 5\right]$ **42.** $g(x) = 8 + 5\ln(2x + 3)$

43. $f(x) = \ln\left(\frac{1}{x + 1}\right)$ **44.** $g(x) = \ln\left(\frac{1}{x - 5}\right)$ **45.** $g(x) = \log_5\left(\frac{x + 1}{x}\right)$

46. $h(x) = \log_3\left(\frac{x}{x - 1}\right)$ **47.** $f(x) = \sqrt{\ln x}$ **48.** $g(x) = \frac{1}{\ln x}$

In Problems 49–56, use a calculator to evaluate each expression. Round your answer to three decimal places.

49. $\ln\frac{5}{3}$ **50.** $\frac{\ln 5}{3}$ **51.** $\frac{\ln\frac{10}{3}}{0.04}$ **52.** $\frac{\ln\frac{2}{3}}{-0.1}$

53. $\frac{\ln 4 + \ln 2}{\log 4 + \log 2}$ **54.** $\frac{\log 15 + \log 20}{\ln 15 + \ln 20}$ **55.** $\frac{2\ln 5 + \log 50}{\log 4 - \ln 2}$ **56.** $\frac{3\log 80 - \ln 5}{\log 5 + \ln 20}$

57. Find a so that the graph of $f(x) = \log_a x$ contains the point $(2, 2)$.

58. Find a so that the graph of $f(x) = \log_a x$ contains the point $\left(\frac{1}{2}, -4\right)$.

In Problems 59–62, graph each function and its inverse on the same Cartesian plane.

59. $f(x) = 3^x;\ f^{-1}(x) = \log_3 x$ **60.** $f(x) = 4^x;\ f^{-1}(x) = \log_4 x$

61. $f(x) = \left(\frac{1}{2}\right)^x;\ f^{-1}(x) = \log_{\frac{1}{2}} x$ **62.** $f(x) = \left(\frac{1}{3}\right)^x;\ f^{-1}(x) = \log_{\frac{1}{3}} x$

In Problems 63–70, the graph of a logarithmic function is given. Match each graph to one of the following functions:

(a) $y = \log_3 x$ *(b)* $y = \log_3(-x)$ *(c)* $y = -\log_3 x$ *(d)* $y = -\log_3(-x)$

(e) $y = \log_3 x - 1$ *(f)* $y = \log_3(x - 1)$ *(g)* $y = \log_3(1 - x)$ *(h)* $y = 1 - \log_3 x$

63.

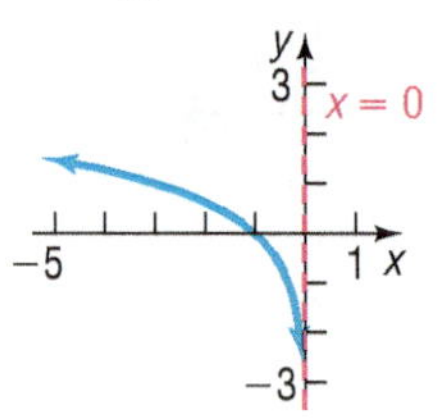

64.

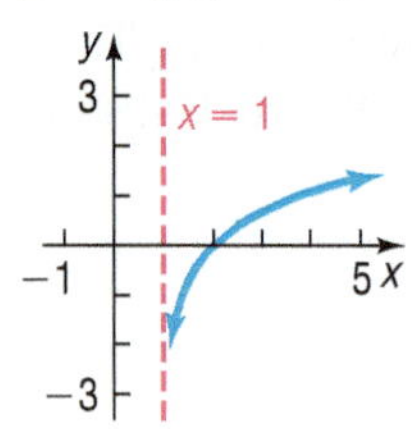

65.

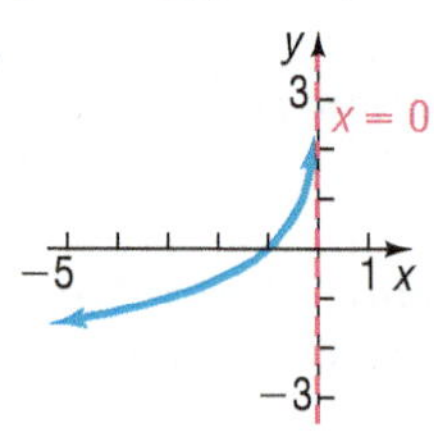

66.

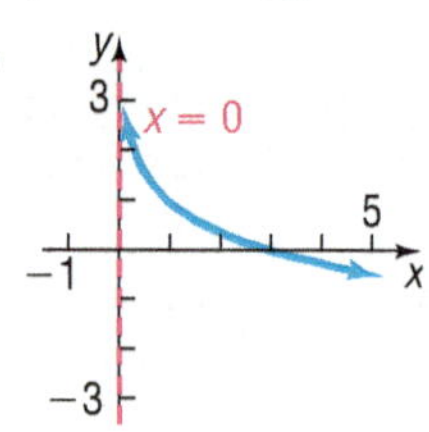

67.

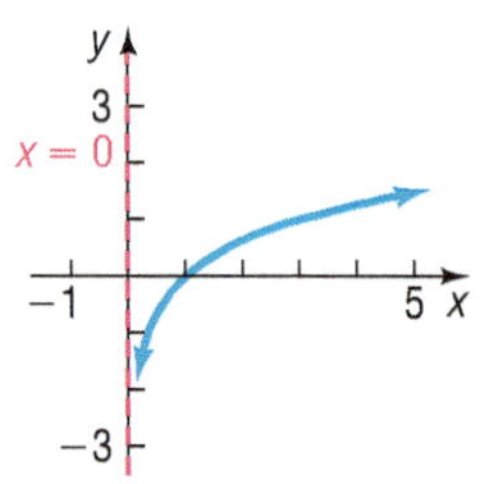

68.

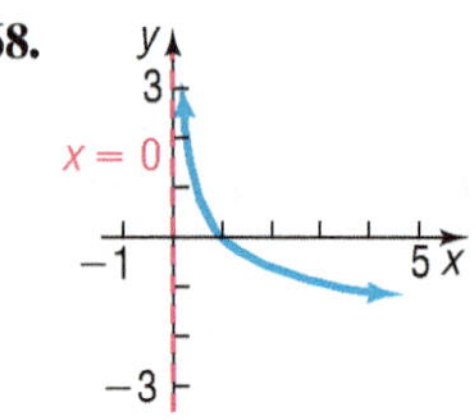

69.

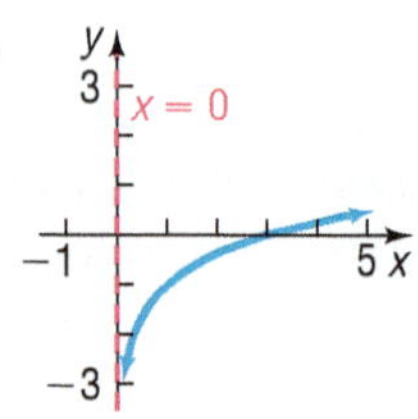

70.

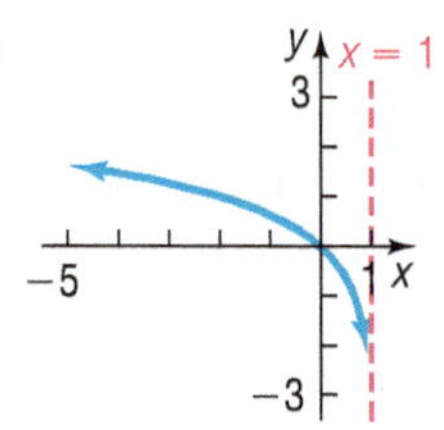

In Problems 71–86, use the given function f to:

(a) Find the domain of f.
(b) Graph f.
(c) From the graph, determine the range and any asymptotes of f.

(d) *Find f^{-1}, the inverse of f.*
(e) *Use f^{-1} to confirm the range of f found in part (c). From the domain of f, find the range of f^{-1}.*
(f) *Graph f^{-1}.*

71. $f(x) = \ln(x + 4)$ **72.** $f(x) = \ln(x - 3)$ **73.** $f(x) = 2 + \ln x$ **74.** $f(x) = -\ln(-x)$

75. $f(x) = \ln(2x) - 3$ **76.** $f(x) = -2\ln(x + 1)$ **77.** $f(x) = \log(x - 4) + 2$ **78.** $f(x) = \frac{1}{2}\log x - 5$

79. $f(x) = \frac{1}{2}\log(2x)$ **80.** $f(x) = \log(-2x)$ **81.** $f(x) = 3 + \log_3(x + 2)$ **82.** $f(x) = 2 - \log_3(x + 1)$

83. $f(x) = e^{x+2} - 3$ **84.** $f(x) = 3e^x + 2$ **85.** $f(x) = 2^{x/3} + 4$ **86.** $f(x) = -3^{x+1}$

In Problems 87–110, solve each equation.

87. $\log_3 x = 2$ **88.** $\log_5 x = 3$ **89.** $\log_2(2x + 1) = 3$ **90.** $\log_3(3x - 2) = 2$

91. $\log_x 4 = 2$ **92.** $\log_x\left(\frac{1}{8}\right) = 3$ **93.** $\ln e^x = 5$ **94.** $\ln e^{-2x} = 8$

95. $\log_4 64 = x$ **96.** $\log_5 625 = x$ **97.** $\log_3 243 = 2x + 1$ **98.** $\log_6 36 = 5x + 3$

99. $e^{3x} = 10$ **100** $e^{-2x} = \frac{1}{3}$ **101.** $e^{2x+5} = 8$ **102.** $e^{-2x+1} = 13$

103. $\log_3(x^2 + 1) = 2$ **104.** $\log_5(x^2 + x + 4) = 2$ **105.** $\log_2 8^x = -3$ **106.** $\log_3 3^x = -1$

107. $5e^{0.2x} = 7$ **108.** $8 \cdot 10^{2x-7} = 3$ **109.** $2 \cdot 10^{2-x} = 5$ **110.** $4e^{x+1} = 5$

Mixed Practice

111. Suppose that $G(x) = \log_3(2x + 1) - 2$.
(a) What is the domain of G?
(b) What is $G(40)$? What point is on the graph of G?
(c) If $G(x) = 3$, what is x? What point is on the graph of G?
(d) What is the zero of G?

112. Suppose that $F(x) = \log_2(x + 1) - 3$.
(a) What is the domain of F?
(b) What is $F(7)$? What point is on the graph of F?
(c) If $F(x) = -1$, what is x? What point is on the graph of F?
(d) What is the zero of F?

In Problems 113–116, graph each function. Based on the graph, state the domain and the range and find any intercepts.

113. $f(x) = \begin{cases} \ln(-x) & \text{if } x < 0 \\ \ln x & \text{if } x > 0 \end{cases}$

114. $f(x) = \begin{cases} \ln(-x) & \text{if } x \le -1 \\ -\ln(-x) & \text{if } -1 < x < 0 \end{cases}$

115. $f(x) = \begin{cases} -\ln x & \text{if } 0 < x < 1 \\ \ln x & \text{if } x \ge 1 \end{cases}$

116. $f(x) = \begin{cases} \ln x & \text{if } 0 < x < 1 \\ -\ln x & \text{if } x \ge 1 \end{cases}$

Applications and Extensions

117. Chemistry The pH of a chemical solution is given by the formula

$$\text{pH} = -\log_{10}[\text{H}^+]$$

where $[\text{H}^+]$ is the concentration of hydrogen ions in moles per liter. Values of pH range from 0 (acidic) to 14 (alkaline).
(a) What is the pH of a solution for which $[\text{H}^+]$ is 0.1?
(b) What is the pH of a solution for which $[\text{H}^+]$ is 0.01?
(c) What is the pH of a solution for which $[\text{H}^+]$ is 0.001?
(d) What happens to pH as the hydrogen ion concentration decreases?
(e) Determine the hydrogen ion concentration of an orange (pH = 3.5).
(f) Determine the hydrogen ion concentration of human blood (pH = 7.4).

118. Diversity Index **Shannon's diversity index** is a measure of the diversity of a population. The diversity index is given by the formula

$$H = -(p_1 \log p_1 + p_2 \log p_2 + \cdots + p_n \log p_n)$$

where p_1 is the proportion of the population that is species 1, p_2 is the proportion of the population that is species 2, and so on.
(a) According to the U.S. Census Bureau, the distribution of race in the United States in 2000 was as follows:

Race	Proportion
American Indian or Native Alaskan	0.014
Asian	0.041
Black or African American	0.128
Hispanic	0.124
Native Hawaiian or Pacific Islander	0.003
White	0.690

Source: U.S. Census Bureau

Compute the diversity index of the United States in 2000.

(b) The largest value of the diversity index is given by $H_{\max} = \log(S)$, where S is the number of categories of race. Compute $H_{\max}$.

(c) The **evenness ratio** is given by $E_H = \dfrac{H}{H_{\max}}$, where $0 \le E_H \le 1$. If $E_H = 1$, there is complete evenness. Compute the evenness ratio for the United States.

(d) Obtain the distribution of race for the United States in 1990 from the Census Bureau. Compute Shannon's diversity index. Is the United States becoming more diverse? Why?

119. Atmospheric Pressure The atmospheric pressure p on a balloon or an aircraft decreases with increasing height. This pressure, measured in millimeters of mercury, is related to the height h (in kilometers) above sea level by the function

$$p(h) = 760e^{-0.145h}$$

(a) Find the height of an aircraft if the atmospheric pressure is 320 millimeters of mercury.

(b) Find the height of a mountain if the atmospheric pressure is 667 millimeters of mercury.

120. Healing of Wounds The normal healing of wounds can be modeled by an exponential function. If A_0 represents the original area of the wound and if A equals the area of the wound, then the function

$$A(n) = A_0e^{-0.35n}$$

describes the area of a wound after n days following an injury when no infection is present to retard the healing. Suppose that a wound initially had an area of 100 square millimeters.

(a) If healing is taking place, after how many days will the wound be one-half its original size?

(b) How long before the wound is 10% of its original size?

121. Exponential Probability Between 12:00 PM and 1:00 PM, cars arrive at Citibank's drive-thru at the rate of 6 cars per hour (0.1 car per minute). The following formula from statistics can be used to determine the probability that a car will arrive within t minutes of 12:00 PM.

$$F(t) = 1 - e^{-0.1t}$$

(a) Determine how many minutes are needed for the probability to reach 50%.

(b) Determine how many minutes are needed for the probability to reach 80%.

(c) Is it possible for the probability to equal 100%? Explain.

122. Exponential Probability Between 5:00 PM and 6:00 PM, cars arrive at Jiffy Lube at the rate of 9 cars per hour (0.15 car per minute). The following formula from statistics can be used to determine the probability that a car will arrive within t minutes of 5:00 PM.

$$F(t) = 1 - e^{-0.15t}$$

(a) Determine how many minutes are needed for the probability to reach 50%.

(b) Determine how many minutes are needed for the probability to reach 80%.

123. Drug Medication The formula

$$D = 5e^{-0.4h}$$

can be used to find the number of milligrams D of a certain drug that is in a patient's bloodstream h hours after the drug was administered. When the number of milligrams reaches 2, the drug is to be administered again. What is the time between injections?

124. Spreading of Rumors A model for the number N of people in a college community who have heard a certain rumor is

$$N = P(1 - e^{-0.15d})$$

where P is the total population of the community and d is the number of days that have elapsed since the rumor began. In a community of 1000 students, how many days will elapse before 450 students have heard the rumor?

125. Current in a *RL* Circuit The equation governing the amount of current I (in amperes) after time t (in seconds) in a simple RL circuit consisting of a resistance R (in ohms), an inductance L (in henrys), and an electromotive force E (in volts) is

$$I = \frac{E}{R}[1 - e^{-(R/L)t}]$$

If $E = 12$ volts, $R = 10$ ohms, and $L = 5$ henrys, how long does it take to obtain a current of 0.5 ampere? Of 1.0 ampere? Graph the equation.

126. Learning Curve Psychologists sometimes use the function

$$L(t) = A(1 - e^{-kt})$$

to measure the amount L learned at time t. The number A represents the amount to be learned, and the number k measures the rate of learning. Suppose that a student has an amount A of 200 vocabulary words to learn. A psychologist determines that the student learned 20 vocabulary words after 5 minutes.

(a) Determine the rate of learning k.

(b) Approximately how many words will the student have learned after 10 minutes?

(c) After 15 minutes?

(d) How long does it take for the student to learn 180 words?

Loudness of Sound *Problems 127–130 use the following discussion: The **loudness** $L(x)$, measured in decibels, of a sound of intensity x, measured in watts per square meter, is defined as $L(x) = 10 \log \dfrac{x}{I_0}$, where $I_0 = 10^{-12}$ watt per square meter is the least intense sound that a human ear can detect. Determine the loudness, in decibels, of each of the following sounds.*

127. Normal conversation: intensity of $x = 10^{-7}$ watt per square meter

128. Amplified rock music: intensity of 10^{-1} watt per square meter

129. Heavy city traffic: intensity of $x = 10^{-3}$ watt per square meter

130. Diesel truck traveling 40 miles per hour 50 feet away: intensity 10 times that of a passenger car traveling 50 miles per hour 50 feet away whose loudness is 70 decibels

The Richter Scale *Problems 131 and 132 use the following discussion: The* ***Richter scale*** *is one way of converting seismographic readings into numbers that provide an easy reference for measuring the magnitude M of an earthquake. All earthquakes are compared to a* ***zero-level earthquake*** *whose seismographic reading measures 0.001 millimeter at a distance of 100 kilometers from the epicenter. An earthquake whose seismographic reading measures x millimeters has* ***magnitude*** *$M(x)$, given by*

$$M(x) = \log\left(\frac{x}{x_0}\right)$$

where $x_0 = 10^{-3}$ is the reading of a zero-level earthquake the same distance from its epicenter. In Problems 131 and 132, determine the magnitude of each earthquake.

131. Magnitude of an Earthquake Mexico City in 1985: seismographic reading of 125,892 millimeters 100 kilometers from the center

132. Magnitude of an Earthquake San Francisco in 1906: seismographic reading of 50,119 millimeters 100 kilometers from the center

133. Alcohol and Driving The concentration of alcohol in a person's bloodstream is measurable. Suppose that the relative risk R of having an accident while driving a car can be modeled by the equation

$$R = e^{kx}$$

where x is the percent of concentration of alcohol in the bloodstream and k is a constant.

(a) Suppose that a concentration of alcohol in the bloodstream of 0.03 percent results in a relative risk of an accident of 1.4. Find the constant k in the equation.

(b) Using this value of k, what is the relative risk if the concentration is 0.17 percent?

(c) Using the same value of k, what concentration of alcohol corresponds to a relative risk of 100?

(d) If the law asserts that anyone with a relative risk of having an accident of 5 or more should not have driving privileges, at what concentration of alcohol in the bloodstream should a driver be arrested and charged with a DUI?

(e) Compare this situation with that of Example 10. If you were a lawmaker, which situation would you support? Give your reasons.

Discussion and Writing

134. Is there any function of the form $y = x^{\alpha}, 0 < \alpha < 1$, that increases more slowly than a logarithmic function whose base is greater than 1? Explain.

135. In the definition of the logarithmic function, the base a is not allowed to equal 1. Why?

136. Critical Thinking In buying a new car, one consideration might be how well the price of the car holds up over time. Different makes of cars have different depreciation rates. One way to compute a depreciation rate for a car is given here. Suppose that the current prices of a certain automobile are as shown in the table.

	Age in Years				
New	**1**	**2**	**3**	**4**	**5**
\$38,000	\$36,600	\$32,400	\$28,750	\$25,400	\$21,200

Use the formula New = Old(e^{Rt}) to find R, the annual depreciation rate, for a specific time t. When might be the best time to trade in the car? Consult the NADA ("blue") book and compare two like models that you are interested in. Which has the better depreciation rate?

'Are You Prepared?' Answers

1. $x \le 3$ **2.** $x < -2$ or $x > 3$ **3.** $x < -4$ or $x > 1$

6.5 Properties of Logarithms

OBJECTIVES

1 Work with the Properties of Logarithms (p. 453)
2 Write a Logarithmic Expression as a Sum or Difference of Logarithms (p. 455)
3 Write a Logarithmic Expression as a Single Logarithm (p. 456)
4 Evaluate Logarithms Whose Base Is Neither 10 Nor e (p. 457)
5 Graph Logarithmic Functions Whose Base Is Neither 10 Nor e (p. 459)

1 Work with the Properties of Logarithms

Logarithms have some very useful properties that can be derived directly from its definition and the laws of exponents.

EXAMPLE 1 **Establishing Properties of Logarithms**

(a) Show that $\log_a 1 = 0$. (b) Show that $\log_a a = 1$.

Solution (a) This fact was established when we graphed $y = \log_a x$ (see Figure 35). To show the result algebraically, let $y = \log_a 1$. Then

$$y = \log_a 1$$
$$a^y = 1 \qquad \text{Change to an exponential expression.}$$
$$a^y = a^0 \qquad a^0 = 1 \text{ since } a > 0, a \neq 1$$
$$y = 0 \qquad \text{Solve for } y.$$
$$\log_a 1 = 0 \qquad y = \log_a 1$$

(b) Let $y = \log_a a$. Then

$$y = \log_a a$$
$$a^y = a \qquad \text{Change to an exponential expression.}$$
$$a^y = a^1 \qquad a = a^1$$
$$y = 1 \qquad \text{Solve for } y.$$
$$\log_a a = 1 \qquad y = \log_a a$$

To summarize:

$$\log_a 1 = 0 \qquad \log_a a = 1$$

THEOREM

Properties of Logarithms

In the properties given next, M and a are positive real numbers, $a \neq 1$, and r is any real number.

The number $\log_a M$ is the exponent to which a must be raised to obtain M. That is,

$$a^{\log_a M} = M \qquad \textbf{(1)}$$

The logarithm to the base a of a raised to a power equals that power. That is,

$$\log_a a^r = r \qquad \textbf{(2)}$$

The proof uses the fact that $y = a^x$ and $y = \log_a x$ are inverses.

Proof of Property (1) For inverse functions,

$$f(f^{-1}(x)) = x \quad \text{for all } x \text{ in the domain of } f^{-1}$$

Using $f(x) = a^x$ and $f^{-1}(x) = \log_a x$, we find

$$f(f^{-1}(x)) = a^{\log_a x} = x \quad \text{for } x > 0$$

Now let $x = M$ to obtain $a^{\log_a M} = M$, where $M > 0$. ■

Proof of Property (2) For inverse functions,

$$f^{-1}(f(x)) = x \quad \text{for all } x \text{ in the domain of } f$$

Using $f(x) = a^x$ and $f^{-1}(x) = \log_a x$, we find

$$f^{-1}(f(x)) = \log_a a^x = x \quad \text{for all real numbers } x$$

Now let $x = r$ to obtain $\log_a a^r = r$, where r is any real number. ■

EXAMPLE 2 Using Properties (1) and (2)

(a) $2^{\log_2 \pi} = \pi$ (b) $\log_{0.2} 0.2^{-\sqrt{2}} = -\sqrt{2}$ (c) $\ln e^{kt} = kt$

Now Work PROBLEM 9

Other useful properties of logarithms are given next.

THEOREM

Properties of Logarithms

In the following properties, M, N, and a are positive real numbers, $a \neq 1$, and r is any real number.

The Log of a Product Equals the Sum of the Logs

$$\log_a(MN) = \log_a M + \log_a N \quad (3)$$

The Log of a Quotient Equals the Difference of the Logs

$$\log_a\left(\frac{M}{N}\right) = \log_a M - \log_a N \quad (4)$$

The Log of a Power Equals the Product of the Power and the Log

$$\log_a M^r = r \log_a M \quad (5)$$

We shall derive properties (3) and (5) and leave the derivation of property (4) as an exercise (see Problem 103).

Proof of Property (3) Let $A = \log_a M$ and let $B = \log_a N$. These expressions are equivalent to the exponential expressions

$$a^A = M \quad \text{and} \quad a^B = N$$

Now

$$\begin{aligned} \log_a(MN) = \log_a(a^A a^B) &= \log_a a^{A+B} && \text{Law of Exponents} \\ &= A + B && \text{Property (2) of logarithms} \\ &= \log_a M + \log_a N \end{aligned}$$

■

Proof of Property (5) Let $A = \log_a M$. This expression is equivalent to

$$a^A = M$$

Now

$$\begin{aligned} \log_a M^r = \log_a(a^A)^r &= \log_a a^{rA} && \text{Law of Exponents} \\ &= rA && \text{Property (2) of logarithms} \\ &= r \log_a M \end{aligned}$$

■

Now Work PROBLEM 13

2 Write a Logarithmic Expression as a Sum or Difference of Logarithms

Logarithms can be used to transform products into sums, quotients into differences, and powers into factors. Such transformations prove useful in certain types of calculus problems.

EXAMPLE 3 **Writing a Logarithmic Expression as a Sum of Logarithms**

Write $\log_a\left(x\sqrt{x^2+1}\right)$, $x > 0$, as a sum of logarithms. Express all powers as factors.

Solution

$$\begin{aligned}\log_a\left(x\sqrt{x^2+1}\right) &= \log_a x + \log_a \sqrt{x^2+1} && \log_a(M \cdot N) = \log_a M + \log_a N\\ &= \log_a x + \log_a(x^2+1)^{1/2}\\ &= \log_a x + \frac{1}{2}\log_a(x^2+1) && \log_a M^r = r\log_a M\end{aligned}$$

EXAMPLE 4 **Writing a Logarithmic Expression as a Difference of Logarithms**

Write

$$\ln\frac{x^2}{(x-1)^3} \qquad x > 1$$

as a difference of logarithms. Express all powers as factors.

Solution

$$\ln\frac{x^2}{(x-1)^3} = \ln x^2 - \ln(x-1)^3 = 2\ln x - 3\ln(x-1)$$

$\log_a\left(\frac{M}{N}\right) = \log_a M - \log_a N$ $\qquad$ $\log_a M^r = r\log_a M$

EXAMPLE 5 **Writing a Logarithmic Expression as a Sum and Difference of Logarithms**

Write

$$\log_a\frac{\sqrt{x^2+1}}{x^3(x+1)^4} \qquad x > 0$$

as a sum and difference of logarithms. Express all powers as factors.

Solution

$$\begin{aligned}\log_a\frac{\sqrt{x^2+1}}{x^3(x+1)^4} &= \log_a\sqrt{x^2+1} - \log_a[x^3(x+1)^4] && \text{Property (4)}\\ &= \log_a\sqrt{x^2+1} - [\log_a x^3 + \log_a(x+1)^4] && \text{Property (3)}\\ &= \log_a(x^2+1)^{1/2} - \log_a x^3 - \log_a(x+1)^4\\ &= \frac{1}{2}\log_a(x^2+1) - 3\log_a x - 4\log_a(x+1) && \text{Property (5)}\end{aligned}$$

WARNING In using properties (3) through (5), be careful about the values that the variable may assume. For example, the domain of the variable for $\log_a x$ is $x > 0$ and for $\log_a(x-1)$ it is $x > 1$. If we add these functions, the domain is $x > 1$. That is, the equality

$$\log_a x + \log_a(x-1) = \log_a[x(x-1)]$$

is true only for $x > 1$.

Now Work PROBLEM 45

3 Write a Logarithmic Expression as a Single Logarithm

Another use of properties (3) through (5) is to write sums and/or differences of logarithms with the same base as a single logarithm. This skill will be needed to solve certain logarithmic equations discussed in the next section.

EXAMPLE 6 **Writing Expressions as a Single Logarithm**

Write each of the following as a single logarithm.

(a) $\log_a 7 + 4\log_a 3$

(b) $\frac{2}{3}\ln 8 - \ln(3^4 - 8)$

(c) $\log_a x + \log_a 9 + \log_a(x^2+1) - \log_a 5$

Solution (a) $\log_a 7 + 4\log_a 3 = \log_a 7 + \log_a 3^4$ — $r\log_a M = \log_a M^r$

$= \log_a 7 + \log_a 81$

$= \log_a(7 \cdot 81)$ — $\log_a M + \log_a N = \log_a(M \cdot N)$

$= \log_a 567$

(b) $\frac{2}{3}\ln 8 - \ln(3^4 - 8) = \ln 8^{2/3} - \ln(81 - 8)$ — $r\log_a M = \log_a M^r$

$= \ln 4 - \ln 73$ — $8^{2/3} = (\sqrt[3]{8})^2 = 2^2 = 4$

$= \ln\left(\frac{4}{73}\right)$ — $\log_a M - \log_a N = \log_a\left(\frac{M}{N}\right)$

(c) $\log_a x + \log_a 9 + \log_a(x^2+1) - \log_a 5 = \log_a(9x) + \log_a(x^2+1) - \log_a 5$

$= \log_a[9x(x^2+1)] - \log_a 5$

$= \log_a\left[\frac{9x(x^2+1)}{5}\right]$

WARNING A common error made by some students is to express the logarithm of a sum as the sum of logarithms.

$$\log_a(M + N) \quad \textit{is not equal to} \quad \log_a M + \log_a N$$

Correct statement $\log_a(MN) = \log_a M + \log_a N$ Property (3)

Another common error is to express the difference of logarithms as the quotient of logarithms.

$$\log_a M - \log_a N \quad \textit{is not equal to} \quad \frac{\log_a M}{\log_a N}$$

Correct statement $\log_a M - \log_a N = \log_a\left(\frac{M}{N}\right)$ Property (4)

A third common error is to express a logarithm raised to a power as the product of the power times the logarithm.

$$(\log_a M)^r \quad \textit{is not equal to} \quad r\log_a M$$

Correct statement $\log_a M^r = r\log_a M$ Property (5)

Now Work PROBLEM 51

Two other properties of logarithms that we need to know are consequences of the fact that the logarithmic function $y = \log_a x$ is a one-to-one function.

THEOREM

Properties of Logarithms

In the following properties, M, N, and a are positive real numbers, $a \neq 1$.

If $M = N$, then $\log_a M = \log_a N$. **(6)**

If $\log_a M = \log_a N$, then $M = N$. **(7)**

When property (6) is used, we start with the equation $M = N$ and say "take the logarithm of both sides" to obtain $\log_a M = \log_a N$.

Properties (6) and (7) are useful for solving *exponential and logarithmic equations,* a topic discussed in the next section.

4 Evaluate Logarithms Whose Base Is Neither 10 Nor *e*

Logarithms to the base 10, common logarithms, were used to facilitate arithmetic computations before the widespread use of calculators. (See the Historical Feature at the end of this section.) Natural logarithms, that is, logarithms whose base is the number e, remain very important because they arise frequently in the study of natural phenomena.

Common logarithms are usually abbreviated by writing **log,** with the base understood to be 10, just as natural logarithms are abbreviated by **ln,** with the base understood to be e.

Most calculators have both [log] and [ln] keys to calculate the common logarithm and natural logarithm of a number. Let's look at an example to see how to approximate logarithms having a base other than 10 or e.

EXAMPLE 7 Approximating a Logarithm Whose Base Is Neither 10 Nor e

Approximate $\log_2 7$. Round the answer to four decimal places.

Solution Let $y = \log_2 7$. Then $2^y = 7$, so

$$\begin{aligned} 2^y &= 7 \\ \ln 2^y &= \ln 7 && \text{Property (6)} \\ y \ln 2 &= \ln 7 && \text{Property (5)} \\ y &= \frac{\ln 7}{\ln 2} && \text{Exact value} \\ y &\approx 2.8074 && \text{Approximate value rounded to four decimal places} \end{aligned}$$

Example 7 shows how to approximate a logarithm whose base is 2 by changing to logarithms involving the base e. In general, we use the **Change-of-Base Formula.**

THEOREM

Change-of-Base Formula

If $a \neq 1$, $b \neq 1$, and M are positive real numbers, then

$$\log_a M = \frac{\log_b M}{\log_b a} \tag{8}$$

Proof We derive this formula as follows: Let $y = \log_a M$. Then

$$\begin{aligned} a^y &= M \\ \log_b a^y &= \log_b M && \text{Property (6)} \\ y \log_b a &= \log_b M && \text{Property (5)} \\ y &= \frac{\log_b M}{\log_b a} && \text{Solve for } y. \\ \log_a M &= \frac{\log_b M}{\log_b a} && y = \log_a M \end{aligned}$$

■

Since calculators have keys only for [log] and [ln], in practice, the Change-of-Base Formula uses either $b = 10$ or $b = e$. That is,

$$\log_a M = \frac{\log M}{\log a} \quad \text{and} \quad \log_a M = \frac{\ln M}{\ln a} \tag{9}$$

EXAMPLE 8 Using the Change-of-Base Formula

Approximate: (a) $\log_5 89$
(b) $\log_{\sqrt{2}} \sqrt{5}$

Round answers to four decimal places.

Solution

(a) $\log_5 89 = \dfrac{\log 89}{\log 5} \approx \dfrac{1.949390007}{0.6989700043} \approx 2.7889$

or

$$\log_5 89 = \frac{\ln 89}{\ln 5} \approx \frac{4.48863637}{1.609437912} \approx 2.7889$$

(b) $\log_{\sqrt{2}} \sqrt{5} = \dfrac{\log \sqrt{5}}{\log \sqrt{2}} = \dfrac{\frac{1}{2}\log 5}{\frac{1}{2}\log 2} = \dfrac{\log 5}{\log 2} \approx 2.3219$

or

$$\log_{\sqrt{2}} \sqrt{5} = \frac{\ln \sqrt{5}}{\ln \sqrt{2}} = \frac{\frac{1}{2}\ln 5}{\frac{1}{2}\ln 2} = \frac{\ln 5}{\ln 2} \approx 2.3219$$

Now Work PROBLEMS 17 AND 65

5 Graph Logarithmic Functions Whose Base Is Neither 10 Nor *e*

We also use the Change-of-Base Formula to graph logarithmic functions whose base is neither 10 nor e.

EXAMPLE 9 **Graphing a Logarithmic Function Whose Base Is Neither 10 Nor e**

Use a graphing utility to graph $y = \log_2 x$.

Solution Since graphing utilities only have logarithms with the base 10 or the base e, we need to use the Change-of-Base Formula to express $y = \log_2 x$ in terms of logarithms with base 10 or base e. We can graph either $y = \dfrac{\ln x}{\ln 2}$ or $y = \dfrac{\log x}{\log 2}$ to obtain the graph of $y = \log_2 x$. See Figure 45.

Figure 45

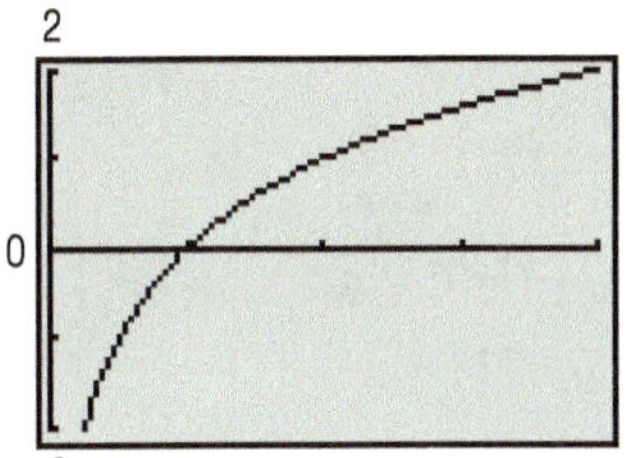

Check: Verify that $y = \dfrac{\ln x}{\ln 2}$ and $y = \dfrac{\log x}{\log 2}$ result in the same graph by graphing each on the same screen.

Now Work PROBLEM 73

SUMMARY Properties of Logarithms

In the list that follows, a, b, M, N, and r are real numbers. Also, $a > 0$, $a \neq 1$, $b > 0$, $b \neq 1$, $M > 0$, and $N > 0$.

Definition $y = \log_a x$ means $x = a^y$

Properties of logarithms

$\log_a 1 = 0;\quad \log_a a = 1$

$a^{\log_a M} = M;\quad \log_a a^r = r$

$\log_a(MN) = \log_a M + \log_a N$

$\log_a\left(\dfrac{M}{N}\right) = \log_a M - \log_a N$

$\log_a M^r = r \log_a M$

If $M = N$, then $\log_a M = \log_a N$.

If $\log_a M = \log_a N$, then $M = N$.

Change-of-Base Formula $\log_a M = \dfrac{\log_b M}{\log_b a}$

Historical Feature

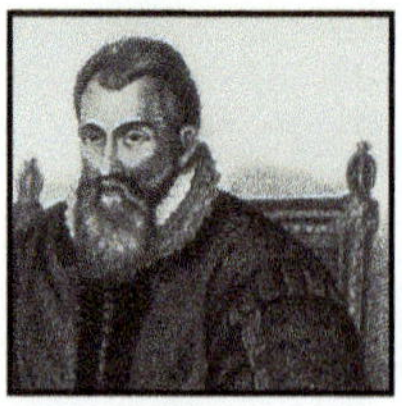

John Napier (1550–1617)

Logarithms were invented about 1590 by John Napier (1550–1617) and Joost Bürgi (1552–1632), working independently. Napier, whose work had the greater influence, was a Scottish lord, a secretive man whose neighbors were inclined to believe him to be in league with the devil. His approach to logarithms was very different from ours; it was based on the relationship between arithmetic and geometric sequences, discussed in a later chapter, and not on the inverse function relationship of logarithms to exponential functions (described in Section 6.4). Napier's tables, published in 1614, listed what would now be called *natural logarithms* of sines and were rather difficult to use. A London professor, Henry Briggs, became interested in the tables and visited Napier. In their conversations, they developed the idea of common logarithms, which were published in 1617. Their importance for calculation was immediately recognized, and by 1650 they were being printed as far away as China. They remained an important calculation tool until the advent of the inexpensive handheld calculator about 1972, which has decreased their calculational, but not their theoretical, importance.

A side effect of the invention of logarithms was the popularization of the decimal system of notation for real numbers.

6.5 Assess Your Understanding

Concepts and Vocabulary

1. The logarithm of a product equals the _____ of the logarithms.

2. If $\log_8 M = \dfrac{\log_5 7}{\log_5 8}$, then $M =$ _____.

3. $\log_a M^r =$ _____.

4. ***True or False*** $\ln(x+3) - \ln(2x) = \dfrac{\ln(x+3)}{\ln(2x)}$

5. ***True or False*** $\log_2(3x^4) = 4\log_2(3x)$

6. ***True or False*** $\log_2 16 = \dfrac{\ln 16}{\ln 2}$

Skill Building

In Problems 7–22, use properties of logarithms to find the exact value of each expression. Do not use a calculator.

7. $\log_3 3^{71}$ **8.** $\log_2 2^{-13}$ **9.** $\ln e^{-4}$ **10.** $\ln e^{\sqrt{2}}$

11. $2^{\log_2 7}$ **12.** $e^{\ln 8}$ **13.** $\log_8 2 + \log_8 4$ **14.** $\log_6 9 + \log_6 4$

15. $\log_6 18 - \log_6 3$ **16.** $\log_8 16 - \log_8 2$ **17.** $\log_2 6 \cdot \log_6 4$ **18.** $\log_3 8 \cdot \log_8 9$

19. $3^{\log_3 5 - \log_3 4}$ **20.** $5^{\log_5 6 + \log_5 7}$ **21.** $e^{\log_{e^2} 16}$ **22.** $e^{\log_{e^2} 9}$

In Problems 23–30, suppose that $\ln 2 = a$ *and* $\ln 3 = b$. *Use properties of logarithms to write each logarithm in terms of a and b.*

23. $\ln 6$ **24.** $\ln \dfrac{2}{3}$ **25.** $\ln 1.5$ **26.** $\ln 0.5$

27. $\ln 8$ **28.** $\ln 27$ **29.** $\ln \sqrt[5]{6}$ **30.** $\ln \sqrt[4]{\dfrac{2}{3}}$

In Problems 31–50, write each expression as a sum and/or difference of logarithms. Express powers as factors.

31. $\log_5(25x)$ **32.** $\log_3 \dfrac{x}{9}$ **33.** $\log_2 z^3$ **34.** $\log_7(x^5)$

35. $\ln(ex)$ **36.** $\ln \dfrac{e}{x}$ **37.** $\ln(xe^x)$ **38.** $\ln \dfrac{x}{e^x}$

39. $\log_a(u^2v^3)$ $u > 0, v > 0$ **40.** $\log_2\left(\dfrac{a}{b^2}\right)$ $a > 0, b > 0$ **41.** $\ln\left(x^2\sqrt{1-x}\right)$ $0 < x < 1$ **42.** $\ln\left(x\sqrt{1+x^2}\right)$ $x > 0$

43. $\log_2\left(\dfrac{x^3}{x-3}\right)$ $x > 3$ **44.** $\log_5\left(\dfrac{\sqrt[3]{x^2+1}}{x^2-1}\right)$ $x > 1$ **45.** $\log\left[\dfrac{x(x+2)}{(x+3)^2}\right]$ $x > 0$ **46.** $\log\left[\dfrac{x^3\sqrt{x+1}}{(x-2)^2}\right]$ $x > 2$

47. $\ln\left[\dfrac{x^2-x-2}{(x+4)^2}\right]^{1/3}$ $x > 2$ **48.** $\ln\left[\dfrac{(x-4)^2}{x^2-1}\right]^{2/3}$ $x > 4$

49. $\ln \dfrac{5x\sqrt{1+3x}}{(x-4)^3}$ $x > 4$ **50.** $\ln\left[\dfrac{5x^2\sqrt[3]{1-x}}{4(x+1)^2}\right]$ $0 < x < 1$

In Problems 51–64, write each expression as a single logarithm.

51. $3\log_5 u + 4\log_5 v$ **52.** $2\log_3 u - \log_3 v$ **53.** $\log_3 \sqrt{x} - \log_3 x^3$

54. $\log_2\left(\dfrac{1}{x}\right) + \log_2\left(\dfrac{1}{x^2}\right)$ **55.** $\log_4(x^2-1) - 5\log_4(x+1)$ **56.** $\log(x^2+3x+2) - 2\log(x+1)$

57. $\ln\left(\frac{x}{x-1}\right) + \ln\left(\frac{x+1}{x}\right) - \ln(x^2-1)$

58. $\log\left(\frac{x^2+2x-3}{x^2-4}\right) - \log\left(\frac{x^2+7x+6}{x+2}\right)$

59. $8\log_2\sqrt{3x-2} - \log_2\left(\frac{4}{x}\right) + \log_2 4$

60. $21\log_3\sqrt[3]{x} + \log_3(9x^2) - \log_3 9$

61. $2\log_a(5x^3) - \frac{1}{2}\log_a(2x+3)$

62. $\frac{1}{3}\log(x^3+1) + \frac{1}{2}\log(x^2+1)$

63. $2\log_2(x+1) - \log_2(x+3) - \log_2(x-1)$

64. $3\log_5(3x+1) - 2\log_5(2x-1) - \log_5 x$

In Problems 65–72, use the Change-of-Base Formula and a calculator to evaluate each logarithm. Round your answer to three decimal places.

65. $\log_3 21$

66. $\log_5 18$

67. $\log_{1/3} 71$

68. $\log_{1/2} 15$

69. $\log_{\sqrt{2}} 7$

70. $\log_{\sqrt{5}} 8$

71. $\log_\pi e$

72. $\log_\pi \sqrt{2}$

In Problems 73–78, graph each function using a graphing utility and the Change-of-Base Formula.

73. $y = \log_4 x$

74. $y = \log_5 x$

75. $y = \log_2(x+2)$

76. $y = \log_4(x-3)$

77. $y = \log_{x-1}(x+1)$

78. $y = \log_{x+2}(x-2)$

Mixed Practice

79. If $f(x) = \ln x$, $g(x) = e^x$, and $h(x) = x^2$, find:
(a) $(f \circ g)(x)$. What is the domain of $f \circ g$?
(b) $(g \circ f)(x)$. What is the domain of $g \circ f$?
(c) $(f \circ g)(5)$
(d) $(f \circ h)(x)$. What is the domain of $f \circ h$?
(e) $(f \circ h)(e)$

80. If $f(x) = \log_2 x$, $g(x) = 2^x$, and $h(x) = 4x$, find:
(a) $(f \circ g)(x)$. What is the domain of $f \circ g$?
(b) $(g \circ f)(x)$. What is the domain of $g \circ f$?
(c) $(f \circ g)(3)$
(d) $(f \circ h)(x)$. What is the domain of $f \circ h$?
(e) $(f \circ h)(8)$

Applications and Extensions

In Problems 81–90, express y as a function of x. The constant C is a positive number.

81. $\ln y = \ln x + \ln C$

82. $\ln y = \ln(x + C)$

83. $\ln y = \ln x + \ln(x+1) + \ln C$

84. $\ln y = 2\ln x - \ln(x+1) + \ln C$

85. $\ln y = 3x + \ln C$

86. $\ln y = -2x + \ln C$

87. $\ln(y-3) = -4x + \ln C$

88. $\ln(y+4) = 5x + \ln C$

89. $3\ln y = \frac{1}{2}\ln(2x+1) - \frac{1}{3}\ln(x+4) + \ln C$

90. $2\ln y = -\frac{1}{2}\ln x + \frac{1}{3}\ln(x^2+1) + \ln C$

91. Find the value of $\log_2 3 \cdot \log_3 4 \cdot \log_4 5 \cdot \log_5 6 \cdot \log_6 7 \cdot \log_7 8$.

92. Find the value of $\log_2 4 \cdot \log_4 6 \cdot \log_6 8$.

93. Find the value of $\log_2 3 \cdot \log_3 4 \cdot \cdots \cdot \log_n(n+1) \cdot \log_{n+1} 2$.

94. Find the value of $\log_2 2 \cdot \log_2 4 \cdot \cdots \cdot \log_2 2^n$.

95. Show that $\log_a\left(x + \sqrt{x^2-1}\right) + \log_a\left(x - \sqrt{x^2-1}\right) = 0$.

96. Show that $\log_a\left(\sqrt{x} + \sqrt{x-1}\right) + \log_a\left(\sqrt{x} - \sqrt{x-1}\right) = 0$.

97. Show that $\ln(1 + e^{2x}) = 2x + \ln(1 + e^{-2x})$.

98. Difference Quotient If $f(x) = \log_a x$, show that $\frac{f(x+h) - f(x)}{h} = \log_a\left(1 + \frac{h}{x}\right)^{1/h}$, $h \neq 0$.

99. If $f(x) = \log_a x$, show that $-f(x) = \log_{1/a} x$.

100. If $f(x) = \log_a x$, show that $f(AB) = f(A) + f(B)$.

101. If $f(x) = \log_a x$, show that $f\left(\frac{1}{x}\right) = -f(x)$.

102. If $f(x) = \log_a x$, show that $f(x^\alpha) = \alpha f(x)$.

103. Show that $\log_a\left(\frac{M}{N}\right) = \log_a M - \log_a N$, where a, M, and N are positive real numbers and $a \neq 1$.

104. Show that $\log_a\left(\frac{1}{N}\right) = -\log_a N$, where a and N are positive real numbers and $a \neq 1$.

Discussion and Writing

105. Graph $Y_1 = \log(x^2)$ and $Y_2 = 2\log(x)$ using a graphing utility. Are they equivalent? What might account for any differences in the two functions?

106. Write an example that illustrates why $(\log_a x)^r \neq r\log_a x$.

107. Write an example that illustrates why $\log_2(x+y) \neq \log_2 x + \log_2 y$.

108. Does $3^{\log_3(-5)} = -5$? Why or why not?

6.6 Logarithmic and Exponential Equations

PREPARING FOR THIS SECTION *Before getting started, review the following:*

- Solving Equations Using a Graphing Utility (Section 1.2, pp. 103–104)
- Solving Quadratic Equations (Section 1.3, pp. 113–120)
- Equations Quadratic in Form (Section 1.5, pp. 135–136)

Now Work the 'Are You Prepared?' problems on page 466.

OBJECTIVES
1. Solve Logarithmic Equations (p. 462)
2. Solve Exponential Equations (p. 464)
3. Solve Logarithmic and Exponential Equations Using a Graphing Utility (p. 466)

1 Solve Logarithmic Equations

In Section 6.4 we solved logarithmic equations by changing a logarithmic equation to an exponential equation. That is, we used the definition of a logarithm:

$$y = \log_a x \quad \text{is equivalent to} \quad x = a^y \qquad a > 0, a \neq 1$$

For example, to solve the equation $\log_2(1 - 2x) = 3$, we use the equivalent exponential equation $1 - 2x = 2^3$ and solve for x.

$$\log_2(1 - 2x) = 3$$

$$1 - 2x = 2^3 \quad \text{Change to an exponential expression.}$$

$$-2x = 7 \quad \text{Simplify.}$$

$$x = -\frac{7}{2} \quad \text{Divide both sides by } -2.$$

You should check this solution for yourself.

For most logarithmic equations, some manipulation of the equation (usually using properties of logarithms) is required to obtain a solution. Also, to avoid extraneous solutions with logarithmic equations, we determine the domain of the variable first.

Our practice will be to solve equations, whenever possible, by finding exact solutions using algebraic methods and exact or approximate solutions using a graphing utility. When algebraic methods cannot be used, approximate solutions will be obtained using a graphing utility. The reader is encouraged to pay particular attention to the form of equations for which exact solutions are possible.

We begin with an example of a logarithmic equation that requires using the fact that a logarithmic function is a one-to-one function.

$$\text{If } \log_a M = \log_a N, \text{ then } M = N \qquad M, N, \text{ and } a \text{ are positive and } a \neq 1$$

EXAMPLE 1 **Solving a Logarithmic Equation**

Solve: $2 \log_5 x = \log_5 9$

Algebraic Solution

We note that the domain of the variable in this equation is $x > 0$. Because each logarithm is to the same base, 5, we can obtain an exact solution as follows:

$$2\log_5 x = \log_5 9$$

$$\log_5 x^2 = \log_5 9 \qquad \log_a M^r = r\log_a M$$

$$x^2 = 9 \qquad \text{If } \log_a M = \log_a N, \text{ then } M = N.$$

$$x = 3 \quad \text{or} \quad \cancel{x = -3} \qquad \text{Recall that the domain of the variable is } x > 0. \text{ Therefore, } -3 \text{ is extraneous and we discard it.}$$

Check:

$$2\log_5 3 \stackrel{?}{=} \log_5 9$$

$$\log_5 3^2 \stackrel{?}{=} \log_5 9 \qquad r\log_a M = \log_a M^r$$

$$\log_5 9 = \log_5 9$$

The solution set is $\{3\}$.

Graphing Solution

To solve the equation using a graphing utility, graph $Y_1 = 2\log_5 x = \dfrac{2\log x}{\log 5}$ and $Y_2 = \log_5 9 = \dfrac{\log 9}{\log 5}$, and determine the point of intersection. See Figure 46. The point of intersection is (3, 1.3652124); so $x = 3$ is the only solution. The solution set is $\{3\}$.

Figure 46

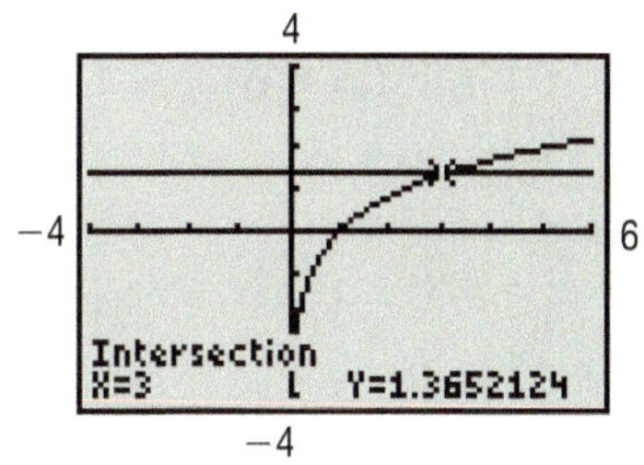

Now Work PROBLEM 13

EXAMPLE 2 Solving a Logarithmic Equation

Solve: $\log_5(x + 6) + \log_5(x + 2) = 1$

Algebraic Solution

The domain of the variable requires that $x + 6 > 0$ and $x + 2 > 0$, so $x > -6$ and $x > -2$. This means any solution must satisfy $x > -2$. To obtain an exact solution, we express the left side as a single logarithm. Then we will change the equation to exponential form.

$$\log_5(x + 6) + \log_5(x + 2) = 1$$

$$\log_5[(x + 6)(x + 2)] = 1 \qquad \log_a M + \log_a N = \log_a(MN)$$

$$(x + 6)(x + 2) = 5^1 = 5 \qquad \text{Change to an exponential equation.}$$

$$x^2 + 8x + 12 = 5 \qquad \text{Simplify.}$$

$$x^2 + 8x + 7 = 0 \qquad \text{Place the quadratic equation in standard form.}$$

$$(x + 7)(x + 1) = 0 \qquad \text{Factor.}$$

$$x = -7 \quad \text{or} \quad x = -1 \qquad \text{Zero-Product Property}$$

Only $x = -1$ satisfies the restriction that $x > -2$, so $x = -7$ is extraneous. The solution set is $\{-1\}$, which you should check.

Graphing Solution

Graph $Y_1 = \log_5(x + 6) + \log_5(x + 2) = \dfrac{\log(x + 6)}{\log 5} + \dfrac{\log(x + 2)}{\log 5}$ and $Y_2 = 1$ and determine the point(s) of intersection. See Figure 47. The point of intersection is $(-1, 1)$, so $x = -1$ is the only solution. The solution set is $\{-1\}$.

Figure 47

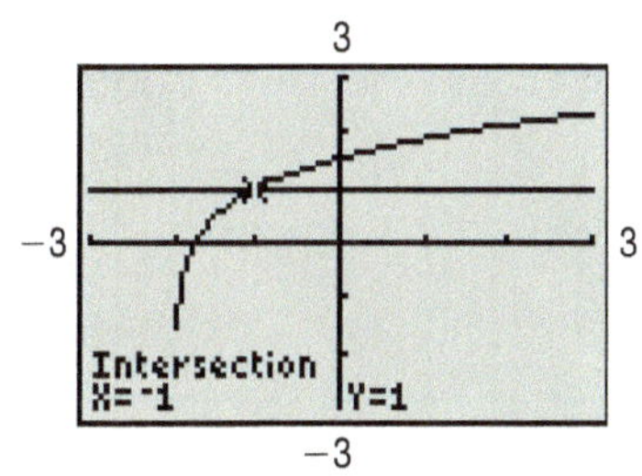

WARNING A negative solution is not automatically extraneous. You must determine whether the potential solution causes the argument of any logarithmic expression in the equation to be negative.

Now Work PROBLEM 21

EXAMPLE 3 Solving a Logarithmic Equation

Solve: $\ln x + \ln(x - 4) = \ln(x + 6)$

Algebraic Solution

The domain of the variable requires that $x > 0$, $x - 4 > 0$, and $x + 6 > 0$. As a result, the domain of the variable is $x > 4$. We begin the solution using the log of a product property.

$$\begin{aligned}
\ln x + \ln(x - 4) &= \ln(x + 6) \\
\ln[x(x - 4)] &= \ln(x + 6) && \ln M + \ln N = \ln(MN) \\
x(x - 4) &= x + 6 && \text{If } \ln M = \ln N \text{, then } M = N. \\
x^2 - 4x &= x + 6 && \text{Simplify.} \\
x^2 - 5x - 6 &= 0 && \text{Place the quadratic equation in standard form.} \\
(x - 6)(x + 1) &= 0 && \text{Factor.} \\
x = 6 \quad \text{or} \quad x &= -1 && \text{Zero-Product Property}
\end{aligned}$$

Since the domain of the variable is $x > 4$, we discard -1 as extraneous. The solution set is $\{6\}$, which you should check.

Graphing Solution

Graph $Y_1 = \ln x + \ln(x - 4)$ and $Y_2 = \ln(x + 6)$ and determine the point(s) of intersection. See Figure 48. The x-coordinate of the point of intersection is 6, so the solution set is $\{6\}$.

Figure 48

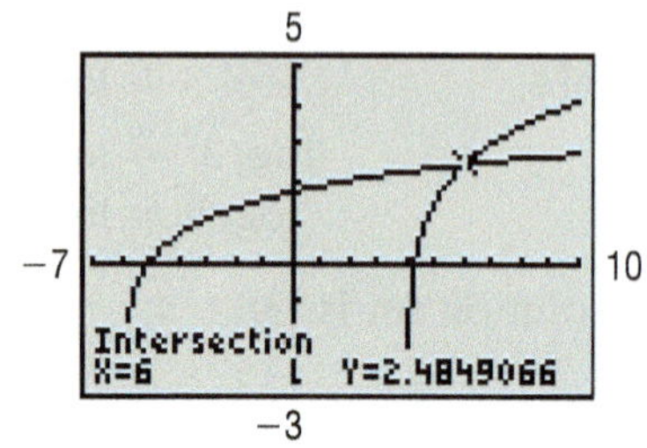

Now Work PROBLEM 31

2 Solve Exponential Equations

In Sections 6.3 and 6.4, we solved exponential equations algebraically by expressing each side of the equation using the same base. That is, we used the one-to-one property of the exponential function:

$$\text{If } a^u = a^v, \quad \text{then} \quad u = v \quad a > 0, a \neq 1$$

For example, to solve the exponential equation $4^{2x+1} = 16$, we notice that $16 = 4^2$ and apply the property above to obtain $2x + 1 = 2$, from which we find $x = \frac{1}{2}$.

For most exponential equations, we cannot express each side of the equation using the same base. In such cases, algebraic techniques can sometimes be used to obtain exact solutions. When algebraic techniques cannot be used, we use a graphing utility to obtain approximate solutions. You should pay particular attention to the form of equations for which exact solutions are obtained.

In the next two examples we solve exponential equations by changing the exponential expression to a logarithmic expression.

EXAMPLE 4 Solving an Exponential Equation

Solve: $2^x = 5$

Algebraic Solution

Since 5 cannot be written as an integer power of 2, we write the exponential equation as the equivalent logarithmic equation.

$$2^x = 5$$

$$x = \log_2 5 = \frac{\ln 5}{\ln 2}$$

↑ Change-of-Base Formula (9), Section 6.5

Graphing Solution

Graph $Y_1 = 2^x$ and $Y_2 = 5$ and determine the x-coordinate of the point of intersection. See Figure 49.

Figure 49

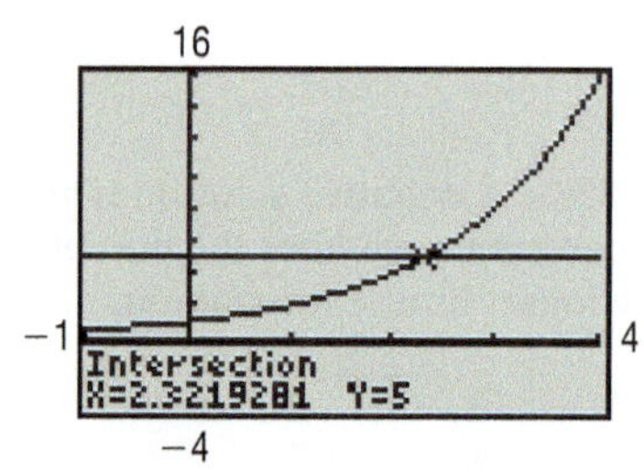

Alternatively, we can solve the equation $2^x = 5$ by taking the natural logarithm (or common logarithm) of each side. Taking the natural logarithm,

$$2^x = 5$$

$$\ln 2^x = \ln 5 \quad \text{If } M = N, \text{ then } \ln M = \ln N.$$

$$x \ln 2 = \ln 5 \quad \ln M^r = r \ln M$$

$$x = \frac{\ln 5}{\ln 2} \quad \text{Exact solution}$$

$$\approx 2.322 \quad \text{Approximate solution}$$

The solution set is $\left\{\frac{\ln 5}{\ln 2}\right\}$.

The approximate solution, rounded to three decimal places, is 2.322.

Now Work PROBLEM 35

EXAMPLE 5 Solving an Exponential Equation

Solve: $8 \cdot 3^x = 5$

Algebraic Solution

We wish to isolate the exponential expression and then rewrite the statement as an equivalent logarithm.

$$8 \cdot 3^x = 5$$

$$3^x = \frac{5}{8} \quad \text{Solve for } 3^x.$$

$$x = \log_3\left(\frac{5}{8}\right) = \frac{\ln\left(\frac{5}{8}\right)}{\ln 3} \quad \text{Exact solution}$$

$$\approx -0.428 \quad \text{Approximate solution}$$

The solution set is $\left\{\log_3\left(\frac{5}{8}\right)\right\}$.

Graphing Solution

Graph $Y_1 = 8 \cdot 3^x$ and $Y_2 = 5$ and determine the x-coordinate of the point of intersection. See Figure 50.

Figure 50

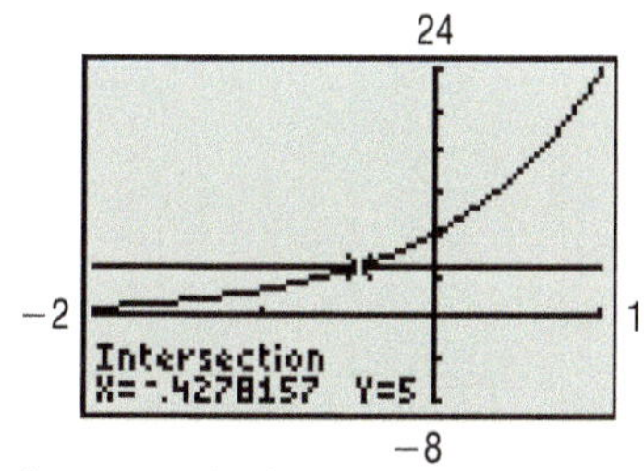

The approximate solution, rounded to three decimal places, is –0.428.

EXAMPLE 6 Solving an Exponential Equation

Solve: $5^{x-2} = 3^{3x+2}$

Algebraic Solution

Because the bases are different, we first apply Property (6), Section 6.5 (take the natural logarithm of each side), and then use appropriate properties of logarithms. The result is a linear equation in x that we can solve.

$$5^{x-2} = 3^{3x+2}$$

$$\ln 5^{x-2} = \ln 3^{3x+2} \quad \text{If } M = N, \ln M = \ln N.$$

$$(x - 2) \ln 5 = (3x + 2) \ln 3 \quad \ln M^r = r \ln M$$

$$(\ln 5)x - 2 \ln 5 = (3 \ln 3)x + 2 \ln 3 \quad \text{Distribute. The equation is now linear in } x.$$

$$(\ln 5)x - (3 \ln 3)x = 2 \ln 3 + 2 \ln 5 \quad \text{Place terms involving } x \text{ on the left.}$$

$$(\ln 5 - 3 \ln 3)x = 2(\ln 3 + \ln 5) \quad \text{Factor.}$$

$$x = \frac{2(\ln 3 + \ln 5)}{\ln 5 - 3 \ln 3} \quad \text{Exact solution}$$

$$\approx -3.212 \quad \text{Approximate solution}$$

Graphing Solution

Graph $Y_1 = 5^{x-2}$ and $Y_2 = 3^{3x+2}$ and determine the x-coordinate of the point of intersection. See Figure 51.

Figure 51

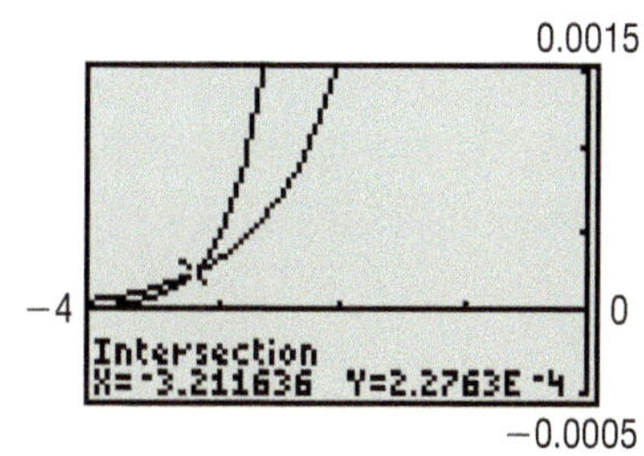

The approximate solution, rounded to three decimal places, is -3.212.

Now Work PROBLEM 41

The next example deals with an exponential equation that is quadratic in form.

EXAMPLE 7 **Solving an Exponential Equation That Is Quadratic in Form**

Solve: $4^x - 2^x - 12 = 0$

Algebraic Solution

We note that $4^x = (2^2)^x = 2^{2x} = (2^x)^2$, so the equation is actually quadratic in form, and we can rewrite it as

$$(2^x)^2 - 2^x - 12 = 0 \quad \text{Let } u = 2^x\text{; then } u^2 - u - 12 = 0.$$

Now we can factor as usual.

$$(2^x - 4)(2^x + 3) = 0 \qquad (u - 4)(u + 3) = 0$$
$$2^x - 4 = 0 \quad \text{or} \quad 2^x + 3 = 0 \qquad u - 4 = 0 \quad \text{or} \quad u + 3 = 0$$
$$2^x = 4 \qquad 2^x = -3 \qquad u = 2^x = 4 \qquad u = 2^x = -3$$

The equation on the left has the solution $x = 2$, since $2^x = 4 = 2^2$; the equation on the right has no solution, since $2^x > 0$ for all x. The only solution is 2. The solution set is $\{2\}$.

Graphing Solution

Graph $Y_1 = 4^x - 2^x - 12$ and determine the x-intercept. See Figure 52. The x-intercept is 2, so the solution set is $\{2\}$.

Figure 52

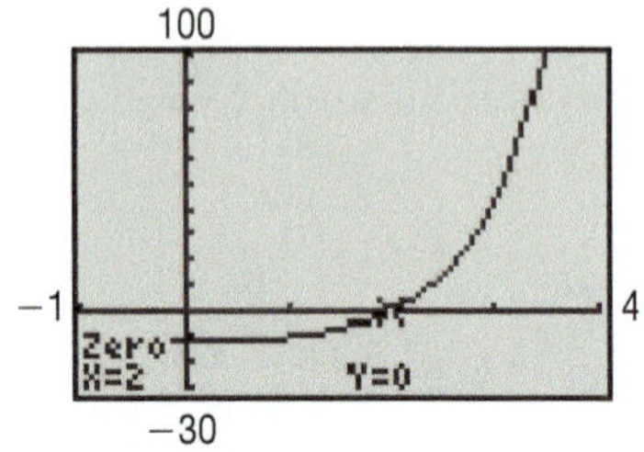

Now Work PROBLEM 53

3 Solve Logarithmic and Exponential Equations Using a Graphing Utility

The algebraic techniques introduced in this section to obtain exact solutions apply only to certain types of logarithmic and exponential equations. Solutions for other types are usually studied in calculus, using numerical methods. For such types, we can use a graphing utility to approximate the solution.

EXAMPLE 8 **Solving Equations Using a Graphing Utility**

Solve: $x + e^x = 2$

Express the solution(s) rounded to two decimal places.

Solution The solution is found by graphing $Y_1 = x + e^x$ and $Y_2 = 2$. Since Y_1 is an increasing function (do you know why?), there is only one point of intersection for Y_1 and Y_2. Figure 53 shows the graphs of Y_1 and Y_2. Using the INTERSECT command, the solution is 0.44 rounded to two decimal places.

Figure 53

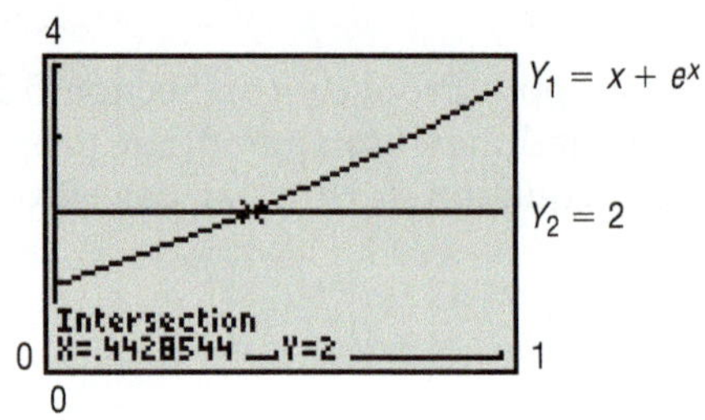

Now Work PROBLEM 63

6.6 Assess Your Understanding

'Are You Prepared?' *Answers are given at the end of these exercises. If you get a wrong answer, read the pages listed in red.*

1. Solve $x^2 - 7x - 30 = 0$. (pp. 113–114)
2. Solve $(x + 3)^2 - 4(x + 3) + 3 = 0$. (pp. 135–136)
3. Approximate the solution(s) to $x^3 = x^2 - 5$ using a graphing utility. (pp. 103–104)
4. Approximate the solution(s) to $x^3 - 2x + 2 = 0$ using a graphing utility. (pp. 103–104)

Skill Building

In Problems 5–32, solve each logarithmic equation. Express irrational solutions in exact form and as a decimal rounded to 3 decimal places. Verify your results using a graphing utility.

5. $\log_4 x = 2$
6. $\log(x + 6) = 1$
7. $\log_2(5x) = 4$
8. $\log_3(3x - 1) = 2$
9. $\log_4(x + 2) = \log_4 8$
10. $\log_5(2x + 3) = \log_5 3$
11. $\frac{1}{2}\log_3 x = 2\log_3 2$
12. $-2\log_4 x = \log_4 9$
13. $3\log_2 x = -\log_2 27$
14. $2\log_5 x = 3\log_5 4$
15. $3\log_2(x - 1) + \log_2 4 = 5$
16. $2\log_3(x + 4) - \log_3 9 = 2$
17. $\log x + \log(x + 15) = 2$
18. $\log x + \log(x - 21) = 2$
19. $\log(2x + 1) = 1 + \log(x - 2)$
20. $\log(2x) - \log(x - 3) = 1$
21. $\log_2(x + 7) + \log_2(x + 8) = 1$
22. $\log_6(x + 4) + \log_6(x + 3) = 1$
23. $\log_8(x + 6) = 1 - \log_8(x + 4)$
24. $\log_5(x + 3) = 1 - \log_5(x - 1)$
25. $\ln x + \ln(x + 2) = 4$
26. $\ln(x + 1) - \ln x = 2$
27. $\log_3(x + 1) + \log_3(x + 4) = 2$
28. $\log_2(x + 1) + \log_2(x + 7) = 3$
29. $\log_{1/3}(x^2 + x) - \log_{1/3}(x^2 - x) = -1$
30. $\log_4(x^2 - 9) - \log_4(x + 3) = 3$
31. $\log_a(x - 1) - \log_a(x + 6) = \log_a(x - 2) - \log_a(x + 3)$
32. $\log_a x + \log_a(x - 2) = \log_a(x + 4)$

In Problems 33–60, solve each exponential equation. Express irrational solutions in exact form and as a decimal rounded to 3 decimal places. Verify your results using a graphing utility.

33. $2^{x-5} = 8$
34. $5^{-x} = 25$
35. $2^x = 10$
36. $3^x = 14$
37. $8^{-x} = 1.2$
38. $2^{-x} = 1.5$
39. $5(2^{3x}) = 8$
40. $0.3(4^{0.2x}) = 0.2$
41. $3^{1-2x} = 4^x$
42. $2^{x+1} = 5^{1-2x}$
43. $\left(\frac{3}{5}\right)^x = 7^{1-x}$
44. $\left(\frac{4}{3}\right)^{1-x} = 5^x$
45. $1.2^x = (0.5)^{-x}$
46. $0.3^{1+x} = 1.7^{2x-1}$
47. $\pi^{1-x} = e^x$
48. $e^{x+3} = \pi^x$
49. $2^{2x} + 2^x - 12 = 0$
50. $3^{2x} + 3^x - 2 = 0$
51. $3^{2x} + 3^{x+1} - 4 = 0$
52. $2^{2x} + 2^{x+2} - 12 = 0$
53. $16^x + 4^{x+1} - 3 = 0$
54. $9^x - 3^{x+1} + 1 = 0$
55. $25^x - 8 \cdot 5^x = -16$
56. $36^x - 6 \cdot 6^x = -9$
57. $3 \cdot 4^x + 4 \cdot 2^x + 8 = 0$
58. $2 \cdot 49^x + 11 \cdot 7^x + 5 = 0$
59. $4^x - 10 \cdot 4^{-x} = 3$
60. $3^x - 14 \cdot 3^{-x} = 5$

In Problems 61–74, use a graphing utility to solve each equation. Express your answer rounded to two decimal places.

61. $\log_5(x + 1) - \log_4(x - 2) = 1$
62. $\log_2(x - 1) - \log_6(x + 2) = 2$
63. $e^x = -x$
64. $e^{2x} = x + 2$
65. $e^x = x^2$
66. $e^x = x^3$
67. $\ln x = -x$
68. $\ln(2x) = -x + 2$
69. $\ln x = x^3 - 1$
70. $\ln x = -x^2$
71. $e^x + \ln x = 4$
72. $e^x - \ln x = 4$
73. $e^{-x} = \ln x$
74. $e^{-x} = -\ln x$

In Problems 75–86, solve each equation. Express irrational solutions in exact form and as a decimal rounded to 3 decimal places. Verify your results using a graphing utility.

75. $\log_2(x + 1) - \log_4 x = 1$
[**Hint:** Change $\log_4 x$ to base 2.]
76. $\log_2(3x + 2) - \log_4 x = 3$
77. $\log_{16} x + \log_4 x + \log_2 x = 7$
78. $\log_9 x + 3\log_3 x = 14$
79. $\left(\sqrt[3]{2}\right)^{2-x} = 2^{x^2}$
80. $\log_2 x^{\log_2 x} = 4$
81. $\dfrac{e^x + e^{-x}}{2} = 1$
[**Hint:** Multiply each side by e^x.]
82. $\dfrac{e^x + e^{-x}}{2} = 3$
83. $\dfrac{e^x - e^{-x}}{2} = 2$
84. $\dfrac{e^x - e^{-x}}{2} = -2$
85. $\log_5 x + \log_3 x = 1$
[**Hint:** Use the Change-of-Base Formula and factor out $\ln x$.]
86. $\log_2 x + \log_6 x = 3$

Mixed Practice

87. $f(x) = \log_2(x + 3)$ and $g(x) = \log_2(3x + 1)$.
(a) Solve $f(x) = 3$. What point is on the graph of f?
(b) Solve $g(x) = 4$. What point is on the graph of g?
(c) Solve $f(x) = g(x)$. Do the graphs of f and g intersect? If so, where?
(d) Solve $(f + g)(x) = 7$.
(e) Solve $(f - g)(x) = 2$.

88. $f(x) = \log_3(x + 5)$ and $g(x) = \log_3(x - 1)$.
(a) Solve $f(x) = 2$. What point is on the graph of f?
(b) Solve $g(x) = 3$. What point is on the graph of g?
(c) Solve $f(x) = g(x)$. Do the graphs of f and g intersect? If so, where?
(d) Solve $(f + g)(x) = 3$.
(e) Solve $(f - g)(x) = 2$.

89. (a) If $f(x) = 3^{x+1}$ and $g(x) = 2^{x+2}$, graph f and g on the same Cartesian plane.
(b) Find the point(s) of intersection of the graphs of f and g by solving $f(x) = g(x)$. Round answers to three decimal places. Label this point on the graph drawn in part (a).
(c) Based on the graph, solve $f(x) > g(x)$.

90. (a) If $f(x) = 5^{x-1}$ and $g(x) = 2^{x+1}$, graph f and g on the same Cartesian plane.
(b) Find the point(s) of intersection of the graphs of f and g by solving $f(x) = g(x)$. Label this point on the graph drawn in part (a).
(c) Based on the graph, solve $f(x) > g(x)$.

91. (a) Graph $f(x) = 3^x$ and $g(x) = 10$ on the same Cartesian plane.
(b) Shade the region bounded by the y-axis, $f(x) = 3^x$, and $g(x) = 10$ on the graph drawn in part (a).
(c) Solve $f(x) = g(x)$ and label the point of intersection on the graph drawn in part (a).

92. (a) Graph $f(x) = 2^x$ and $g(x) = 12$ on the same Cartesian plane.
(b) Shade the region bounded by the y-axis, $f(x) = 2^x$, and $g(x) = 12$ on the graph drawn in part (a).
(c) Solve $f(x) = g(x)$ and label the point of intersection on the graph drawn in part (a).

93. (a) Graph $f(x) = 2^{x+1}$ and $g(x) = 2^{-x+2}$ on the same Cartesian plane.
(b) Shade the region bounded by the y-axis, $f(x) = 2^{x+1}$, and $g(x) = 2^{-x+2}$ on the graph drawn in part (a).
(c) Solve $f(x) = g(x)$ and label the point of intersection on the graph drawn in part (a).

94. (a) Graph $f(x) = 3^{-x+1}$ and $g(x) = 3^{x-2}$ on the same Cartesian plane.
(b) Shade the region bounded by the y-axis, $f(x) = 3^{-x+1}$, and $g(x) = 3^{x-2}$ on the graph drawn in part (a).
(c) Solve $f(x) = g(x)$ and label the point of intersection on the graph drawn in part (a).

95. (a) Graph $f(x) = 2^x - 4$.
(b) Find the zero of f.
(c) Based on the graph, solve $f(x) < 0$.

96. (a) Graph $g(x) = 3^x - 9$.
(b) Find the zero of g.
(c) Based on the graph, solve $g(x) > 0$.

Applications and Extensions

97. A Population Model The resident population of the United States in 2006 was 298 million people and was growing at a rate of 0.9% per year. Assuming that this growth rate continues, the model $P(t) = 298(1.009)^{t-2006}$ represents the population P (in millions of people) in year t.
(a) According to this model, when will the population of the United States be 310 million people?
(b) According to this model, when will the population of the United States be 360 million people?
Source: Statistical Abstract of the United States, 125th ed., 2006.

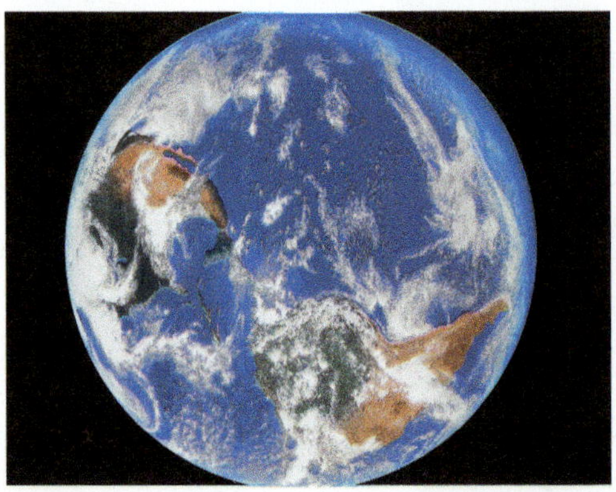

98. A Population Model The population of the world in 2006 was 6.53 billion people and was growing at a rate of 1.14% per year. Assuming that this growth rate continues, the model $P(t) = 6.53(1.0114)^{t-2006}$ represents the population P (in billions of people) in year t.
(a) According to this model, when will the population of the world be 9.25 billion people?
(b) According to this model, when will the population of the world be 11.75 billion people?
Source: U.S. Census Bureau.

99. Depreciation The value V of a Chevy Cobalt that is t years old can be modeled by $V(t) = 14{,}512(0.82)^t$.
(a) According to the model, when will the car be worth \$9000?
(b) According to the model, when will the car be worth \$4000?
(c) According to the model, when will the car be worth \$2000?
Source: Kelley Blue Book

100. Depreciation The value V of a Dodge Stratus that is t years old can be modeled by $V(t) = 19{,}282(0.84)^t$.

(a) According to the model, when will the car be worth \$15,000?

(b) According to the model, when will the car be worth \$8000?

(c) According to the model, when will the car be worth \$2000?

Source: *Kelley Blue Book*

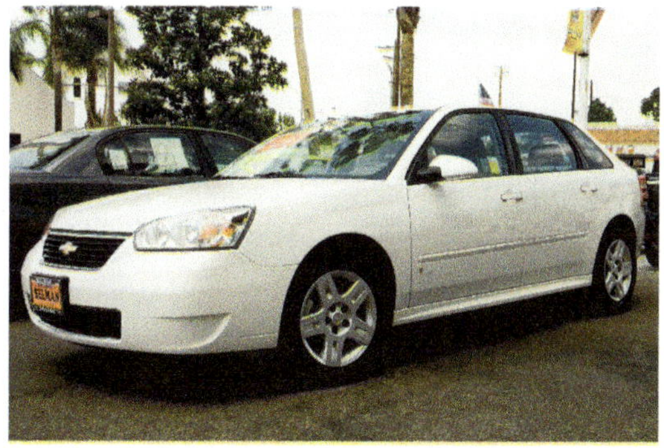

Discussion and Writing

101. Fill in reasons for each step in the following two solutions.

Solve: $\log_3(x - 1)^2 = 2$

Solution A	**Solution B**
$\log_3(x - 1)^2 = 2$	$\log_3(x - 1)^2 = 2$
$(x - 1)^2 = 3^2 = 9$ _____	$2\log_3(x - 1) = 2$ _____
$(x - 1) = \pm 3$ _____	$\log_3(x - 1) = 1$ _____
$x - 1 = -3$ or $x - 1 = 3$ _____	$x - 1 = 3^1 = 3$ _____
$x = -2$ or $x = 4$ _____	$x = 4$ _____

Both solutions given in Solution A check. Explain what caused the solution $x = -2$ to be lost in Solution B.

'Are You Prepared?' Answers

1. $\{-3, 10\}$ **2.** $\{-2, 0\}$ **3.** $\{-1.43\}$ **4.** $\{-1.77\}$

6.7 Financial Models

PREPARING FOR THIS SECTION *Before getting started, review the following:*

- Simple Interest (Section 1.6, p. 142)

Now Work the 'Are You Prepared?' problems on page 476.

OBJECTIVES
1 Determine the Future Value of a Lump Sum of Money (p. 469)
2 Calculate Effective Rates of Return (p. 473)
3 Determine the Present Value of a Lump Sum of Money (p. 474)
4 Determine the Rate of Interest or Time Required to Double a Lump Sum of Money (p. 475)

1 Determine the Future Value of a Lump Sum of Money

Interest is money paid for the use of money. The total amount borrowed (whether by an individual from a bank in the form of a loan or by a bank from an individual in the form of a savings account) is called the **principal.** The **rate of interest,** expressed as a percent, is the amount charged for the use of the principal for a given period of time, usually on a yearly (that is, per annum) basis.

THEOREM

Simple Interest Formula

If a principal of P dollars is borrowed for a period of t years at a per annum interest rate r, expressed as a decimal, the interest I charged is

$$I = Prt \tag{1}$$

Interest charged according to formula (1) is called **simple interest.**

In working with problems involving interest, we define the term **payment period** as follows:

Annually:	Once per year	**Monthly:**	12 times per year
Semiannually:	Twice per year	**Daily:**	365 times per year*
Quarterly:	Four times per year		

When the interest due at the end of a payment period is added to the principal so that the interest computed at the end of the next payment period is based on this new principal amount (old principal + interest), the interest is said to have been **compounded. Compound interest** is interest paid on principal and previously earned interest.

EXAMPLE 1 Computing Compound Interest

A credit union pays interest of 8% per annum compounded quarterly on a certain savings plan. If \$1000 is deposited in such a plan and the interest is left to accumulate, how much is in the account after 1 year?

Solution We use the simple interest formula, $I = Prt$. The principal P is \$1000 and the rate of interest is $8\% = 0.08$. After the first quarter of a year, the time t is $\frac{1}{4}$ year, so the interest earned is

$$I = Prt = (\$1000)(0.08)\left(\frac{1}{4}\right) = \$20$$

The new principal is $P + I = \$1000 + \$20 = \$1020$. At the end of the second quarter, the interest on this principal is

$$I = (\$1020)(0.08)\left(\frac{1}{4}\right) = \$20.40$$

At the end of the third quarter, the interest on the new principal of $\$1020 + \$20.40 = \$1040.40$ is

$$I = (\$1040.40)(0.08)\left(\frac{1}{4}\right) = \$20.81$$

Finally, after the fourth quarter, the interest is

$$I = (\$1061.21)(0.08)\left(\frac{1}{4}\right) = \$21.22$$

After 1 year the account contains $\$1061.21 + \$21.22 = \$1082.43$.

The pattern of the calculations performed in Example 1 leads to a general formula for compound interest. To fix our ideas, let P represent the principal to be invested at a per annum interest rate r that is compounded n times per year, so the time of each compounding period is $\frac{1}{n}$ years. (For computing purposes, r is expressed as a decimal.) The interest earned after each compounding period is given by formula (1).

$$\text{Interest} = \text{principal} \times \text{rate} \times \text{time} = P \cdot r \cdot \frac{1}{n} = P \cdot \left(\frac{r}{n}\right)$$

The amount A after one compounding period is

$$A = P + P \cdot \left(\frac{r}{n}\right) = P \cdot \left(1 + \frac{r}{n}\right)$$

*Most banks use a 360-day "year." Why do you think they do?

After two compounding periods, the amount A, based on the new principal $P\cdot\left(1+\frac{r}{n}\right)$, is

$$A = \underbrace{P\cdot\left(1+\frac{r}{n}\right)}_{\text{New principal}} + \underbrace{P\cdot\left(1+\frac{r}{n}\right)\left(\frac{r}{n}\right)}_{\text{Interest on new principal}} \underset{\uparrow \text{ Factor out } P\cdot(1+\frac{r}{n})}{=} P\cdot\left(1+\frac{r}{n}\right)\left(1+\frac{r}{n}\right) = P\cdot\left(1+\frac{r}{n}\right)^2$$

After three compounding periods, the amount A is

$$A = P\cdot\left(1+\frac{r}{n}\right)^2 + P\cdot\left(1+\frac{r}{n}\right)^2\left(\frac{r}{n}\right) = P\cdot\left(1+\frac{r}{n}\right)^2\cdot\left(1+\frac{r}{n}\right) = P\cdot\left(1+\frac{r}{n}\right)^3$$

Continuing this way, after n compounding periods (1 year), the amount A is

$$A = P\cdot\left(1+\frac{r}{n}\right)^n$$

Because t years will contain $n\cdot t$ compounding periods, after t years we have

$$A = P\cdot\left(1+\frac{r}{n}\right)^{nt}$$

THEOREM

Compound Interest Formula

The amount A after t years due to a principal P invested at an annual interest rate r compounded n times per year is

$$A = P\cdot\left(1+\frac{r}{n}\right)^{nt} \quad \textbf{(2)}$$

Exploration

To see the effects of compounding interest monthly on an initial deposit of \$1, graph $Y_1 = \left(1+\frac{r}{12}\right)^{12x}$ with $r = 0.06$ and $r = 0.12$ for $0 \le x \le 30$. What is the future value of \$1 in 30 years when the interest rate per annum is $r = 0.06$ (6%)? What is the future value of \$1 in 30 years when the interest rate per annum is $r = 0.12$ (12%)? Does doubling the interest rate double the future value?

For example, to rework Example 1, we would use $P = \$1000$, $r = 0.08$, $n = 4$ (quarterly compounding), and $t = 1$ year to obtain

$$A = P\cdot\left(1+\frac{r}{n}\right)^{nt} = 1000\left(1+\frac{0.08}{4}\right)^{4\cdot 1} = \$1082.43$$

In equation (2), the amount A is typically referred to as the **future value** of the account, while P is called the **present value.**

Now Work PROBLEM 3

EXAMPLE 2 Comparing Investments Using Different Compounding Periods

Investing \$1000 at an annual rate of 10% compounded annually, semiannually, quarterly, monthly, and daily will yield the following amounts after 1 year:

Annual compounding ($n = 1$):
$$A = P\cdot(1+r) = (\$1000)(1+0.10) = \$1100.00$$

Semiannual compounding ($n = 2$):
$$A = P\cdot\left(1+\frac{r}{2}\right)^2 = (\$1000)(1+0.05)^2 = \$1102.50$$

Quarterly compounding ($n = 4$):
$$A = P\cdot\left(1+\frac{r}{4}\right)^4 = (\$1000)(1+0.025)^4 = \$1103.81$$

Monthly compounding ($n = 12$):
$$A = P\cdot\left(1+\frac{r}{12}\right)^{12} = (\$1000)\left(1+\frac{0.10}{12}\right)^{12} = \$1104.71$$

Daily compounding ($n = 365$): $$A = P \cdot \left(1 + \frac{r}{365}\right)^{365} = (\$1000)\left(1 + \frac{0.10}{365}\right)^{365} = \$1105.16$$

From Example 2, we can see that the effect of compounding more frequently is that the amount after 1 year is higher: \$1000 compounded 4 times a year at 10% results in \$1103.81, \$1000 compounded 12 times a year at 10% results in \$1104.71, and \$1000 compounded 365 times a year at 10% results in \$1105.16. This leads to the following question: What would happen to the amount after 1 year if the number of times that the interest is compounded were increased without bound?

Let's find the answer. Suppose that P is the principal, r is the per annum interest rate, and n is the number of times that the interest is compounded each year. The amount after 1 year is

$$A = P \cdot \left(1 + \frac{r}{n}\right)^{n}$$

Rewrite this expression as follows:

$$A = P \cdot \left(1 + \frac{r}{n}\right)^{n} = P \cdot \left(1 + \frac{1}{\frac{n}{r}}\right)^{n} = P \cdot \left[\left(1 + \frac{1}{\frac{n}{r}}\right)^{n/r}\right]^{r} \underset{h = \frac{n}{r}}{=} P \cdot \left[\left(1 + \frac{1}{h}\right)^{h}\right]^{r} \quad \textbf{(3)}$$

Now suppose that the number n of times that the interest is compounded per year gets larger and larger; that is, suppose that $n \to \infty$. Then $h = \frac{n}{r} \to \infty$, and the expression in brackets in Equation (3) equals e. That is, $A \to Pe^r$.

Table 9 compares $\left(1 + \frac{r}{n}\right)^{n}$, for large values of n, to e^r for $r = 0.05$, $r = 0.10$, $r = 0.15$, and $r = 1$. The larger that n gets, the closer $\left(1 + \frac{r}{n}\right)^{n}$ gets to e^r. No matter how frequent the compounding, the amount after 1 year has the definite ceiling Pe^r.

Table 9

	$\left(1 + \frac{r}{n}\right)^{n}$			
	$n = 100$	$n = 1000$	$n = 10{,}000$	e^r
$r = 0.05$	1.0512580	1.0512698	1.051271	1.0512711
$r = 0.10$	1.1051157	1.1051654	1.1051704	1.1051709
$r = 0.15$	1.1617037	1.1618212	1.1618329	1.1618342
$r = 1$	2.7048138	2.7169239	2.7181459	2.7182818

When interest is compounded so that the amount after 1 year is Pe^r, we say the interest is **compounded continuously.**

THEOREM

Continuous Compounding

The amount A after t years due to a principal P invested at an annual interest rate r compounded continuously is

$$A = Pe^{rt} \quad \textbf{(4)}$$

EXAMPLE 3

Using Continuous Compounding

The amount A that results from investing a principal P of \$1000 at an annual rate r of 10% compounded continuously for a time t of 1 year is

$$A = \$1000e^{0.10} = (\$1000)(1.10517) = \$1105.17$$

Now Work PROBLEM 11

2 Calculate Effective Rates of Return

Suppose you have \$1000 and a bank offers to pay you 3% annual interest on a savings account compounded monthly. What annual interest rate would you need to earn if the interest was compounded annually (once per year)? To answer this question, we first determine the value of the \$1000 in the account that earns 3% compounded monthly.

$$\begin{aligned} A &= \$1000\left(1 + \frac{0.03}{12}\right)^{12} \qquad \text{Use } A = P\left(1 + \frac{r}{n}\right)^n \text{ with } P = \$1000,\ r = 0.03,\ n = 12. \\ &= \$1030.42 \end{aligned}$$

So the interest earned is \$30.42. Using $I = Prt$ with $t = 1$, $I = \$30.42$, and $P = \$1000$, we find the annual simple interest rate is $0.03042 = 3.042\%$. This interest rate is known as the *effective rate of interest.*

The **effective rate of interest** is the equivalent annual simple interest rate that would yield the same amount as compounding n times per year, or continuously, after 1 year. We can determine the effective rate of interest using the following.

THEOREM

Effective Rate of Interest

The effective rate of interest r_e of an investment earning an annual interest rate r is given by

Compounding n times per year: $r_e = \left(1 + \frac{r}{n}\right)^n - 1$

Continuous compounding: $r_e = e^r - 1$

EXAMPLE 4

Computing the Effective Rate of Interest—Which Is the Best Deal?

Suppose you want to open a money market account. You visit three banks to determine their money market rates. Bank A offers you 6% annual interest compounded daily and Bank B offers you 6.02% compounded quarterly. Bank C offers 5.98% compounded continuously. Determine which bank is offering the best deal.

Solution The bank that offers the best deal is the one with the highest effective interest rate.

Bank A

$$\begin{aligned} r_e &= \left(1 + \frac{0.06}{365}\right)^{365} - 1 \\ &\approx 1.06183 - 1 \\ &= 0.06183 \\ &= 6.183\% \end{aligned}$$

Bank B

$$\begin{aligned} r_e &= \left(1 + \frac{0.0602}{4}\right)^{4} - 1 \\ &\approx 1.06157 - 1 \\ &= 0.06157 \\ &= 6.157\% \end{aligned}$$

Bank C

$$\begin{aligned} r_e &= e^{0.0598} - 1 \\ &\approx 1.06162 - 1 \\ &= 0.06162 \\ &= 6.162\% \end{aligned}$$

Since the effective rate of interest is highest for Bank A, Bank A is offering the best deal.

Now Work PROBLEM 23

3 Determine the Present Value of a Lump Sum of Money

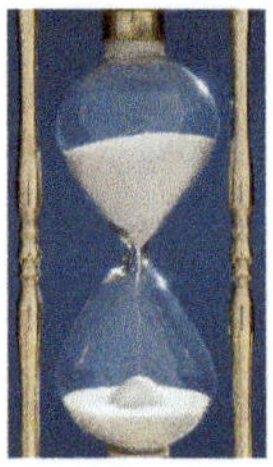

When people engaged in finance speak of the "time value of money," they are usually referring to the *present value* of money. The **present value** of A dollars to be received at a future date is the principal that you would need to invest now so that it will grow to A dollars in the specified time period. The present value of money to be received at a future date is always less than the amount to be received, since the amount to be received will equal the present value (money invested now) *plus* the interest accrued over the time period.

We use the compound interest formula (2) to get a formula for present value. If P is the present value of A dollars to be received after t years at a per annum interest rate r compounded n times per year, then, by formula (2),

$$A = P\cdot\left(1 + \frac{r}{n}\right)^{nt}$$

To solve for P, we divide both sides by $\left(1 + \frac{r}{n}\right)^{nt}$. The result is

$$\frac{A}{\left(1 + \frac{r}{n}\right)^{nt}} = P \quad \text{or} \quad P = A\cdot\left(1 + \frac{r}{n}\right)^{-nt}$$

THEOREM

Present Value Formulas

The present value P of A dollars to be received after t years, assuming a per annum interest rate r compounded n times per year, is

$$P = A\cdot\left(1 + \frac{r}{n}\right)^{-nt} \tag{5}$$

If the interest is compounded continuously,

$$P = Ae^{-rt} \tag{6}$$

To prove (6), solve formula (4) for P.

EXAMPLE 5 Computing the Value of a Zero-coupon Bond

A zero-coupon (noninterest-bearing) bond can be redeemed in 10 years for \$1000. How much should you be willing to pay for it now if you want a return of

(a) 8% compounded monthly?

(b) 7% compounded continuously?

Solution

(a) We are seeking the present value of \$1000. We use formula (5) with $A = \$1000$, $n = 12$, $r = 0.08$, and $t = 10$.

$$P = A\cdot\left(1 + \frac{r}{n}\right)^{-nt} = \$1000\left(1 + \frac{0.08}{12}\right)^{-12(10)} = \$450.52$$

For a return of 8% compounded monthly, you should pay \$450.52 for the bond.

(b) Here we use formula (6) with $A = \$1000$, $r = 0.07$, and $t = 10$.

$$P = Ae^{-rt} = \$1000e^{-(0.07)(10)} = \$496.59$$

For a return of 7% compounded continuously, you should pay \$496.59 for the bond.

Now Work PROBLEM 13

4 Determine the Rate of Interest or Time Required to Double a Lump Sum of Money

EXAMPLE 6 **Rate of Interest Required to Double an Investment**

What annual rate of interest compounded annually should you seek if you want to double your investment in 5 years?

Solution If P is the principal and we want P to double, the amount A will be $2P$. We use the compound interest formula with $n = 1$ and $t = 5$ to find r.

$$A = P \cdot \left(1 + \frac{r}{n}\right)^{nt}$$

$$2P = P \cdot (1 + r)^5 \qquad A = 2P, n = 1, t = 5$$

$$2 = (1 + r)^5 \qquad \text{Divide both sides by } P.$$

$$1 + r = \sqrt[5]{2} \qquad \text{Take the fifth root of each side.}$$

$$r = \sqrt[5]{2} - 1 \approx 1.148698 - 1 = 0.148698$$

The annual rate of interest needed to double the principal in 5 years is 14.87%.

Now Work PROBLEM 31

EXAMPLE 7 **Time Required to Double or Triple an Investment**

(a) How long will it take for an investment to double in value if it earns 5% compounded continuously?
(b) How long will it take to triple at this rate?

Solution (a) If P is the initial investment and we want P to double, the amount A will be $2P$. We use formula (4) for continuously compounded interest with $r = 0.05$. Then

$$A = Pe^{rt}$$

$$2P = Pe^{0.05t} \qquad A = 2P, r = 0.05$$

$$2 = e^{0.05t} \qquad \text{Cancel the } P\text{'s.}$$

$$0.05t = \ln 2 \qquad \text{Rewrite as a logarithm.}$$

$$t = \frac{\ln 2}{0.05} \approx 13.86 \qquad \text{Solve for } t.$$

It will take about 14 years to double the investment.

(b) To triple the investment, we set $A = 3P$ in formula (4).

$$A = Pe^{rt}$$

$$3P = Pe^{0.05t} \qquad A = 3P, r = 0.05$$

$$3 = e^{0.05t} \qquad \text{Cancel the } P\text{'s.}$$

$$0.05t = \ln 3 \qquad \text{Rewrite as a logarithm.}$$

$$t = \frac{\ln 3}{0.05} \approx 21.97 \qquad \text{Solve for } t.$$

It will take about 22 years to triple the investment.

Now Work PROBLEM 35

6.7 Assess Your Understanding

'Are You Prepared?' *Answers are given at the end of these exercises. If you get a wrong answer, read the pages listed in red.*

1. What is the interest due if \$500 is borrowed for 6 months at a simple interest rate of 6% per annum? (p. 142)

2. If you borrow \$5000 and, after 9 months, pay off the loan in the amount of \$5500, what per annum rate of interest was charged? (p. 142)

Skill Building

In Problems 3–12, find the amount that results from each investment.

3. \$100 invested at 4% compounded quarterly after a period of 2 years

4. \$50 invested at 6% compounded monthly after a period of 3 years

5. \$500 invested at 8% compounded quarterly after a period of $2\frac{1}{2}$ years

6. \$300 invested at 12% compounded monthly after a period of $1\frac{1}{2}$ years

7. \$600 invested at 5% compounded daily after a period of 3 years

8. \$700 invested at 6% compounded daily after a period of 2 years

9. \$10 invested at 11% compounded continuously after a period of 2 years

10. \$40 invested at 7% compounded continuously after a period of 3 years

11. \$100 invested at 10% compounded continuously after a period of $2\frac{1}{4}$ years

12. \$100 invested at 12% compounded continuously after a period of $3\frac{3}{4}$ years

In Problems 13–22, find the principal needed now to get each amount; that is, find the present value.

13. To get \$100 after 2 years at 6% compounded monthly

14. To get \$75 after 3 years at 8% compounded quarterly

15. To get \$1000 after $2\frac{1}{2}$ years at 6% compounded daily

16. To get \$800 after $3\frac{1}{2}$ years at 7% compounded monthly

17. To get \$600 after 2 years at 4% compounded quarterly

18. To get \$300 after 4 years at 3% compounded daily

19. To get \$80 after $3\frac{1}{4}$ years at 9% compounded continuously

20. To get \$800 after $2\frac{1}{2}$ years at 8% compounded continuously

21. To get \$400 after 1 year at 10% compounded continuously

22. To get \$1000 after 1 year at 12% compounded continuously

In Problems 23–26, find the effective rate of interest.

23. For 5% compounded quarterly

24. For 6% compounded monthly

25. For 5% compounded continuously

26. For 6% compounded continuously

In Problems 27–30, determine the rate that represents the better deal.

27. 6% compounded quarterly or $6\frac{1}{4}$% compounded annually

28. 9% compounded quarterly of $9\frac{1}{4}$% compounded annually

29. 9% compounded monthly or 8.8% compounded daily

30. 8% compounded semiannually or 7.9% compounded daily

31. What rate of interest compounded annually is required to double an investment in 3 years?

32. What rate of interest compounded annually is required to double an investment in 6 years?

33. What rate of interest compounded annually is required to triple an investment in 5 years?

34. What rate of interest compounded annually is required to triple an investment in 10 years?

35. (a) How long does it take for an investment to double in value if it is invested at 8% compounded monthly?
(b) How long does it take if the interest is compounded continuously?

36. (a) How long does it take for an investment to triple in value if it is invested at 6% compounded monthly?
(b) How long does it take if the interest is compounded continuously?

37. What rate of interest compounded quarterly will yield an effective interest rate of 7%?

38. What rate of interest compounded continuously will yield an effective interest rate of 6%?

Applications and Extensions

39. Time Required to Reach a Goal If Tanisha has \$100 to invest at 8% per annum compounded monthly, how long will it be before she has \$150? If the compounding is continuous, how long will it be?

40. Time Required to Reach a Goal If Angela has \$100 to invest at 10% per annum compounded monthly, how long will it be before she has \$175? If the compounding is continuous, how long will it be?

41. **Time Required to Reach a Goal** How many years will it take for an initial investment of \$10,000 to grow to \$25,000? Assume a rate of interest of 6% compounded continuously.

42. **Time Required to Reach a Goal** How many years will it take for an initial investment of \$25,000 to grow to \$80,000? Assume a rate of interest of 7% compounded continuously.

43. **Price Appreciation of Homes** What will a \$90,000 condominium cost 5 years from now if the price appreciation for condos over that period averages 3% compounded annually?

44. **Credit Card Interest** A department store charges 1.25% per month on the unpaid balance for customers with charge accounts (interest is compounded monthly). A customer charges \$200 and does not pay her bill for 6 months. What is the bill at that time?

45. **Saving for a Car** Jerome will be buying a used car for \$15,000 in 3 years. How much money should he ask his parents for now so that, if he invests it at 5% compounded continuously, he will have enough to buy the car?

46. **Paying off a Loan** John requires \$3000 in 6 months to pay off a loan that has no prepayment privileges. If he has the \$3000 now, how much of it should he save in an account paying 3% compounded monthly so that in 6 months he will have exactly \$3000?

47. **Return on a Stock** George contemplates the purchase of 100 shares of a stock selling for \$15 per share. The stock pays no dividends. The history of the stock indicates that it should grow at an annual rate of 15% per year. How much will the 100 shares of stock be worth in 5 years?

48. **Return on an Investment** A business purchased for \$650,000 in 2005 is sold in 2008 for \$850,000. What is the annual rate of return for this investment?

49. **Comparing Savings Plans** Jim places \$1000 in a bank account that pays 5.6% compounded continuously. After 1 year, will he have enough money to buy a computer system that costs \$1060? If another bank will pay Jim 5.9% compounded monthly, is this a better deal?

50. **Savings Plans** On January 1, Kim places \$1000 in a certificate of deposit that pays 6.8% compounded continuously and matures in 3 months. Then Kim places the \$1000 and the interest in a passbook account that pays 5.25% compounded monthly. How much does Kim have in the passbook account on May 1?

51. **Comparing IRA Investments** Will invests \$2000 in his IRA in a bond trust that pays 9% interest compounded semiannually. His friend Henry invests \$2000 in his IRA in a certificate of deposit that pays $8\frac{1}{2}\%$ compounded continuously. Who has more money after 20 years, Will or Henry?

52. **Comparing Two Alternatives** Suppose that April has access to an investment that will pay 10% interest compounded continuously. Which is better: to be given \$1000 now so that she can take advantage of this investment opportunity or to be given \$1325 after 3 years?

53. **College Costs** The average annual cost of college at 4-year private colleges was \$29,026 in 2005. This was a 5.5% increase from the previous year.

Source: Trends in College Pricing 2005, College Board

(a) If the cost of college increases by 5.5% each year, what will be the average cost of college at a 4-year private college in 2015?

(b) College savings plans, such as a 529 plan, allow individuals to put money aside now to help pay for college later. If one such plan offers a rate of 4% compounded continuously, how much should be put in a college savings plan in 2005 to pay for 1 year of the cost of college at a 4-year private college in 2015?

54. **Analyzing Interest Rates on a Mortgage** Colleen and Bill have just purchased a house for \$650,000, with the seller holding a second mortgage of \$100,000. They promise to pay the seller \$100,000 plus all accrued interest 5 years from now. The seller offers them three interest options on the second mortgage:

(a) Simple interest at 12% per annum

(b) $11\frac{1}{2}\%$ interest compounded monthly

(c) $11\frac{1}{4}\%$ interest compounded continuously

Which option is best; that is, which results in the least interest on the loan?

55. **Federal Deficit** At the end of fiscal year 2005, the federal budget deficit was \$319 billion. At that time, 20-year Series EE bonds had a fixed rate of 3.2% compounded semiannually. If the federal government financed this deficit through EE bonds, how much would it have to pay back in 2025?

Source: U.S. Treasury Department

56. **Federal Deficit** On February 6, 2006, President Bush proposed the fiscal year 2007 federal budget. The proposal projected a fiscal year 2006 deficit of \$423 billion and a fiscal year 2007 deficit of \$354 billion. Assuming the deficit decreases at the same rate each year, when will the deficit be cut to \$100 billion?

Source: Office of Management and Budget

Inflation *Problems 57–62 require the following discussion.* ***Inflation*** *is a term used to describe the erosion of the purchasing power of money. For example, suppose the annual inflation rate is 3%. Then \$1000 worth of purchasing power now will have only \$970 worth of purchasing power in 1 year because 3% of the original \$1000 ($0.03 \times 1000 = 30$) has been eroded due to inflation. In general, if the rate of inflation averages r% over n years, the amount A that \$P will purchase after n years is*

$$A = P \cdot (1 - r)^n$$

where r is expressed as a decimal.

57. **Inflation** If the inflation rate averages 3%, how much will \$1000 purchase in 2 years?

58. **Inflation** If the inflation rate averages 2%, how much will \$1000 purchase in 3 years?

59. **Inflation** If the amount that \$1000 will purchase is only \$950 after 2 years, what was the average inflation rate?

60. **Inflation** If the amount that \$1000 will purchase is only \$930 after 2 years, what was the average inflation rate?

61. **Inflation** If the average inflation rate is 2%, how long is it until purchasing power is cut in half?

62. **Inflation** If the average inflation rate is 4%, how long is it until purchasing power is cut in half?

*Problems 63–66 involve zero-coupon bonds. A **zero-coupon bond** is a bond that is sold now at a discount and will pay its face value at the time when it matures; no interest payments are made.*

63. Zero-Coupon Bonds A zero-coupon bond can be redeemed in 20 years for \$10,000. How much should you be willing to pay for it now if you want a return of:

(a) 10% compounded monthly?

(b) 10% compounded continuously?

64. Zero-Coupon Bonds A child's grandparents are considering buying a \$40,000 face-value, zero-coupon bond at birth so that she will have enough money for her college education 17 years later. If they want a rate of return of 8% compounded annually, what should they pay for the bond?

65. Zero-Coupon Bonds How much should a \$10,000 face-value, zero-coupon bond, maturing in 10 years, be sold for now if its rate of return is to be 8% compounded annually?

66. Zero-Coupon Bonds If Pat pays \$12,485.52 for a \$25,000 face-value, zero-coupon bond that matures in 8 years, what is his annual rate of return?

67. Time to Double or Triple an Investment The formula

$$t = \frac{\ln m}{n \ln\left(1 + \frac{r}{n}\right)}$$

can be used to find the number of years t required to multiply an investment m times when r is the per annum interest rate compounded n times a year.

(a) How many years will it take to double the value of an IRA that compounds annually at the rate of 12%?

(b) How many years will it take to triple the value of a savings account that compounds quarterly at an annual rate of 6%?

(c) Give a derivation of this formula.

68. Time to Reach an Investment Goal The formula

$$t = \frac{\ln A - \ln P}{r}$$

can be used to find the number of years t required for an investment P to grow to a value A when compounded continuously at an annual rate r.

(a) How long will it take to increase an initial investment of \$1000 to \$8000 at an annual rate of 10%?

(b) What annual rate is required to increase the value of a \$2000 IRA to \$30,000 in 35 years?

(c) Give a derivation of this formula.

*Problems 69–72, require the following discussion. The **Consumer Price Index (CPI)** indicates the relative change in price over time for a fixed basket of goods and services. It is a cost of living index that helps measure the effect of inflation on the cost of goods and services. The CPI uses the base period 1982–1984 for comparison (the CPI for this period is 100). The CPI for January 2006 was 198.3. This means that \$100 in the period 1982–1984 had the same purchasing power as \$198.30 in January 2006. In general, if the rate of inflation averages r% over n years, then the CPI index after n years is*

$$\text{CPI} = \text{CPI}_0\left(1 + \frac{r}{100}\right)^n$$

where CPI_0 is the CPI index at the beginning of the n-year period. Source: U.S. Bureau of Labor Statistics

69. Consumer Price Index

(a) The CPI was 152.4 for 1995 and 195.3 for 2005. Assuming that annual inflation remained constant for this time period, determine the average annual inflation rate.

(b) Using the inflation rate from part (a), in what year will the CPI reach 300?

70. Consumer Price Index If the current CPI is 234.2 and the average annual inflation rate is 2.8%, what will be the CPI in 5 years?

71. Consumer Price Index If the average annual inflation rate is 3.1%, how long will it take for the CPI index to double? (A doubling of the CPI index means purchasing power is cut in half).

72. Consumer Price Index The base period for the CPI changed in 1998. Under the previous weight and item structure, the CPI for 1995 was 456.5. If the average annual inflation rate was 5.57%, what year was used as the base period for the CPI?

Discussion and Writing

73. Explain in your own words what the term *compound interest* means. What does *continuous compounding* mean?

74. Explain in your own words the meaning of *present value*.

75. **Critical Thinking** You have just contracted to buy a house and will seek financing in the amount of \$100,000. You go to several banks. Bank 1 will lend you \$100,000 at the rate of 8.75% amortized over 30 years with a loan origination fee of 1.75%. Bank 2 will lend you \$100,000 at the rate of 8.375% amortized over 15 years with a loan origination fee of 1.5%. Bank 3 will lend you \$100,000 at the rate of 9.125% amortized over 30 years with no loan origination fee. Bank 4 will lend you \$100,000 at the rate of 8.625% amortized over 15 years with no loan origination fee. Which loan would you take? Why? Be sure to have sound reasons for your choice. Use the information in the table to assist you. If the amount of the monthly payment does not matter to you, which loan would you take? Again, have sound reasons for your choice. Compare your final decision with others in the class. Discuss.

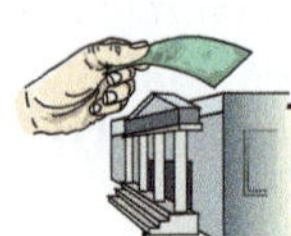

	Monthly Payment	Loan Origination Fee
Bank 1	\$786.70	\$1,750.00
Bank 2	\$977.42	\$1,500.00
Bank 3	\$813.63	\$0.00
Bank 4	\$990.68	\$0.00

'Are You Prepared?' Answers

1. \$15 **2.** $13\frac{1}{3}\%$

6.8 Exponential Growth and Decay Models; Newton's Law; Logistic Growth and Decay Models

OBJECTIVES

1 Find Equations of Populations That Obey the Law of Uninhibited Growth

Many natural phenomena have been found to follow the law that an amount A varies with time t according to the function

$$A(t) = A_0 e^{kt} \qquad \textbf{(1)}$$

Here A_0 is the original amount ($t = 0$) and $k \neq 0$ is a constant.

If $k > 0$, then equation (1) states that the amount A is increasing over time; if $k < 0$, the amount A is decreasing over time. In either case, when an amount A varies over time according to equation (1), it is said to follow the **exponential law** or the **law of uninhibited growth** ($k > 0$) **or decay** ($k < 0$). See Figure 54.

Figure 54

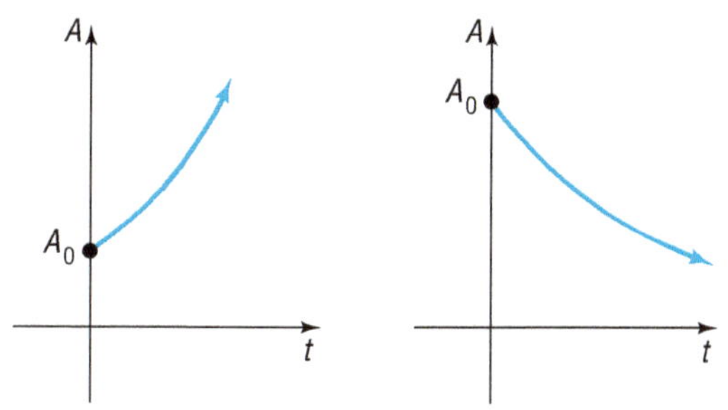

(a) $A(t) = A_0e^{kt}, k > 0$ **(b)** $A(t) = A_0e^{kt}, k < 0$

For example, we saw in Section 6.7 that continuously compounded interest follows the law of uninhibited growth. In this section we shall look at three additional phenomena that follow the exponential law.

Cell division is the growth process of many living organisms, such as amoebas, plants, and human skin cells. Based on an ideal situation in which no cells die and no by-products are produced, the number of cells present at a given time follows the law of uninhibited growth. Actually, however, after enough time has passed, growth at an exponential rate will cease due to the influence of factors such as lack of living space and dwindling food supply. The law of uninhibited growth accurately models only the early stages of the cell division process.

The cell division process begins with a culture containing N_0 cells. Each cell in the culture grows for a certain period of time and then divides into two identical cells. We assume that the time needed for each cell to divide in two is constant and does not change as the number of cells increases. These new cells then grow, and eventually each divides in two, and so on.

Uninhibited Growth of Cells

A model that gives the number N of cells in a culture after a time t has passed (in the early stages of growth) is

$$N(t) = N_0 e^{kt} \qquad k > 0 \qquad \textbf{(2)}$$

where N_0 is the initial number of cells and k is a positive constant that represents the growth rate of the cells.

In using formula (2) to model the growth of cells, we are using a function that yields positive real numbers, even though we are counting the number of cells, which must be an integer. This is a common practice in many applications.

EXAMPLE 1 Bacterial Growth

A colony of bacteria grows according to the law of uninhibited growth according to the function $N(t) = 100e^{0.045t}$, where N is measured in grams and t is measured in days.

(a) Determine the initial amount of bacteria.
(b) What is the growth rate of the bacteria?
(c) Graph the function using a graphing utility.
(d) What is the population after 5 days?
(e) How long will it take for the population to reach 140 grams?
(f) What is the doubling time for the population?

Solution (a) The initial amount of bacteria, N_0, is obtained when $t = 0$, so

$$N_0 = N(0) = 100e^{0.045(0)} = 100 \text{ grams}$$

(b) Compare $N(t) = 100e^{0.045t}$ to $N(t) = 100e^{kt}$. The value of k, 0.045, indicates a growth rate of 4.5%.

(c) Figure 55 shows the graph of $N(t) = 100e^{0.045t}$.

Figure 55

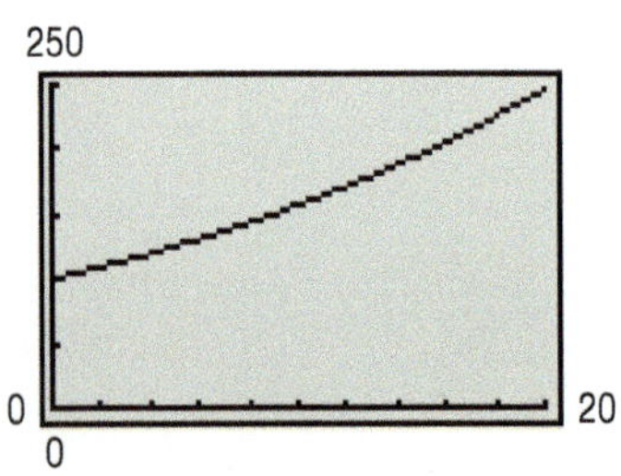

(d) The population after 5 days is $N(5) = 100e^{0.045(5)} \approx 125.2$ grams.

(e) To find how long it takes for the population to reach 140 grams, we solve the equation $N(t) = 140$.

$$\begin{aligned} 100e^{0.045t} &= 140 \\ e^{0.045t} &= 1.4 && \text{Divide both sides of the equation by 100.} \\ 0.045t &= \ln 1.4 && \text{Rewrite as a logarithm.} \\ t &= \frac{\ln 1.4}{0.045} && \text{Divide both sides of the equation by 0.045.} \\ &\approx 7.5 \text{ days} \end{aligned}$$

(f) The population doubles when $N(t) = 200$ grams, so we find the doubling time by solving the equation $200 = 100e^{0.045t}$ for t.

$$\begin{aligned} 200 &= 100e^{0.045t} \\ 2 &= e^{0.045t} && \text{Divide both sides of the equation by 100.} \\ \ln 2 &= 0.045t && \text{Rewrite as a logarithm.} \\ t &= \frac{\ln 2}{0.045} && \text{Divide both sides of the equation by 0.045.} \\ &\approx 15.4 \text{ days} \end{aligned}$$

The population doubles approximately every 15.4 days.

Now Work PROBLEM 1

EXAMPLE 2 **Bacterial Growth**

A colony of bacteria grows according to the law of uninhibited growth.

(a) If N is the number of cells and t is the time in hours, express N as a function of t.
(b) If the number of bacteria doubles in 3 hours, find the function that gives the number of cells in the culture.
(c) How long will it take for the size of the colony to triple?
(d) How long will it take for the population to double a second time (that is, increase four times)?

Solution

(a) Using formula (2), the number N of cells at a time t is

$$N(t) = N_0 e^{kt}$$

where N_0 is the initial number of bacteria present and k is a positive number.

(b) We seek the number k. The number of cells doubles in 3 hours, so we have

$$N(3) = 2N_0$$

But $N(3) = N_0 e^{k(3)}$, so

$$N_0 e^{k(3)} = 2N_0$$
$$e^{3k} = 2 \qquad \text{Divide both sides by } N_0.$$
$$3k = \ln 2 \qquad \text{Write the exponential equation as a logarithm.}$$
$$k = \frac{1}{3}\ln 2$$

The function that models this growth process is therefore

$$N(t) = N_0 e^{\left(\frac{1}{3}\ln 2\right)t}$$

(c) The time t needed for the size of the colony to triple requires that $N = 3N_0$. We substitute $3N_0$ for N to get

$$3N_0 = N_0 e^{\left(\frac{1}{3}\ln 2\right)t}$$
$$3 = e^{\left(\frac{1}{3}\ln 2\right)t} \qquad \text{Divide both sides by } N_0.$$
$$\left(\frac{1}{3}\ln 2\right)t = \ln 3 \qquad \text{Write the exponential equation as a logarithm.}$$
$$t = \frac{3\ln 3}{\ln 2} \approx 4.755 \text{ hours}$$

It will take about 4.755 hours or 4 hours, 45 minutes for the size of the colony to triple.

(d) If a population doubles in 3 hours, it will double a second time in 3 more hours, for a total time of 6 hours.

2 Find Equations of Populations That Obey the Law of Decay

Radioactive materials follow the law of uninhibited decay.

Uninhibited Radioactive Decay

The amount A of a radioactive material present at time t is given by

$$A(t) = A_0 e^{kt} \qquad k < 0 \qquad (3)$$

where A_0 is the original amount of radioactive material and k is a negative number that represents the rate of decay.

All radioactive substances have a specific **half-life,** which is the time required for half of the radioactive substance to decay. In **carbon dating,** we use the fact that all living organisms contain two kinds of carbon, carbon 12 (a stable carbon) and carbon 14 (a radioactive carbon with a half-life of 5600 years). While an organism is living, the ratio of carbon 12 to carbon 14 is constant. But when an organism dies, the original amount of carbon 12 present remains unchanged, whereas the amount of carbon 14 begins to decrease. This change in the amount of carbon 14 present relative to the amount of carbon 12 present makes it possible to calculate when the organism died.

EXAMPLE 3 Estimating the Age of Ancient Tools

Traces of burned wood along with ancient stone tools in an archeological dig in Chile were found to contain approximately 1.67% of the original amount of carbon 14.

(a) If the half-life of carbon 14 is 5600 years, approximately when was the tree cut and burned?

(b) Using a graphing utility, graph the relation between the percentage of carbon 14 remaining and time.

(c) Determine the time that elapses until half of the carbon 14 remains. This answer should equal the half-life of carbon 14.

(d) Use a graphing utility to verify the answer found in part (a).

Solution (a) Using formula (3), the amount A of carbon 14 present at time t is

$$A(t) = A_0 e^{kt}$$

where A_0 is the original amount of carbon 14 present and k is a negative number. We first seek the number k. To find it, we use the fact that after 5600 years half of the original amount of carbon 14 remains, so $A(5600) = \frac{1}{2}A_0$. Then

$$\frac{1}{2}A_0 = A_0 e^{k(5600)}$$

$$\frac{1}{2} = e^{5600k} \quad \text{Divide both sides of the equation by } A_0.$$

$$5600k = \ln\frac{1}{2} \quad \text{Rewrite as a logarithm.}$$

$$k = \frac{1}{5600}\ln\frac{1}{2} \approx -0.000124$$

Formula (3) therefore becomes

$$A(t) = A_0 e^{\frac{\ln\frac{1}{2}}{5600}t}$$

If the amount A of carbon 14 now present is 1.67% of the original amount, it follows that

$$0.0167A_0 = A_0 e^{\frac{\ln\frac{1}{2}}{5600}t}$$

$$0.0167 = e^{\frac{\ln\frac{1}{2}}{5600}t} \quad \text{Divide both sides of the equation by } A_0.$$

$$\frac{\ln\frac{1}{2}}{5600}t = \ln 0.0167 \quad \text{Rewrite as a logarithm.}$$

$$t = \frac{5600}{\ln\frac{1}{2}}\ln 0.0167 \approx 33{,}062 \text{ years}$$

The tree was cut and burned about 33,062 years ago. Some archeologists use this conclusion to argue that humans lived in the Americas 33,000 years ago, much earlier than is generally accepted.

(b) Figure 56 shows the graph of $y = e^{\frac{\ln \frac{1}{2}}{5600}x}$, where y is the fraction of carbon 14 present and x is the time.

Figure 56

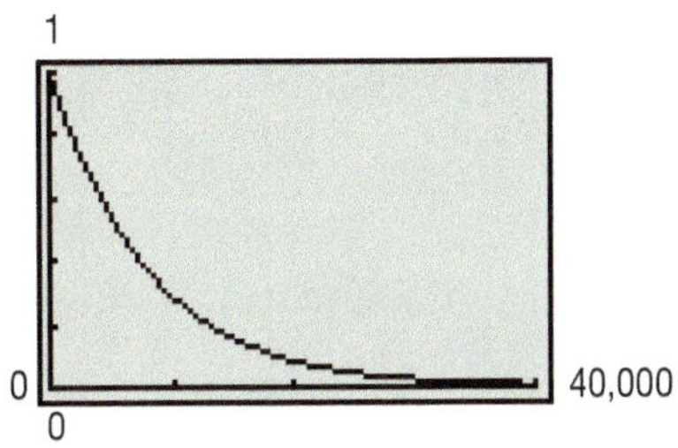

(c) By graphing $Y_1 = 0.5$ and $Y_2 = e^{\frac{\ln \frac{1}{2}}{5600}x}$, where x is time, and using INTERSECT, we find that it takes 5600 years until half the carbon 14 remains. The half-life of carbon 14 is 5600 years.

(d) By graphing $Y_1 = 0.0167$ and $Y_2 = e^{\frac{\ln \frac{1}{2}}{5600}x}$, where x is time, and using INTERSECT, we find that it takes 33,062 years until 1.67% of the carbon 14 remains.

Now Work PROBLEM 3

3 Use Newton's Law of Cooling

Newton's Law of Cooling* states that the temperature of a heated object decreases exponentially over time toward the temperature of the surrounding medium.

Newton's Law of Cooling

The temperature u of a heated object at a given time t can be modeled by the following function:

$$u(t) = T + (u_0 - T)e^{kt} \qquad k < 0 \qquad \textbf{(4)}$$

where T is the constant temperature of the surrounding medium, u_0 is the initial temperature of the heated object, and k is a negative constant.

EXAMPLE 4 Using Newton's Law of Cooling

An object is heated to 100°C (degrees Celsius) and is then allowed to cool in a room whose air temperature is 30°C.

(a) If the temperature of the object is 80°C after 5 minutes, when will its temperature be 50°C?

(b) Using a graphing utility, graph the relation found between the temperature and time.

(c) Using a graphing utility, verify the results from part (a).

(d) Using a graphing utility, determine the elapsed time before the object is 35°C.

(e) What do you notice about the temperature as time passes?

Solution (a) Using formula (4) with $T = 30$ and $u_0 = 100$, the temperature (in degrees Celsius) of the object at time t (in minutes) is

$$u(t) = 30 + (100 - 30)e^{kt} = 30 + 70e^{kt}$$

where k is a negative constant. To find k, we use the fact that $u = 80$ when $t = 5$. Then

$$80 = 30 + 70e^{k(5)}$$
$$50 = 70e^{5k}$$
$$e^{5k} = \frac{50}{70}$$
$$5k = \ln \frac{5}{7}$$
$$k = \frac{1}{5} \ln \frac{5}{7} \approx -0.0673$$

*Named after Sir Isaac Newton (1642–1727), one of the cofounders of calculus.

Formula (4) therefore becomes

$$u(t) = 30 + 70e^{\frac{\ln\frac{5}{7}}{5}t}$$

We want to find t when $u = 50°C$, so

$$50 = 30 + 70e^{\frac{\ln\frac{5}{7}}{5}t}$$

$$20 = 70e^{\frac{\ln\frac{5}{7}}{5}t}$$

$$e^{\frac{\ln\frac{5}{7}}{5}t} = \frac{20}{70}$$

$$\frac{\ln\frac{5}{7}}{5}t = \ln\frac{2}{7}$$

$$t = \frac{5}{\ln\frac{5}{7}}\ln\frac{2}{7} \approx 18.6 \text{ minutes}$$

The temperature of the object will be 50°C after about 18.6 minutes or 18 minutes, 37 seconds.

Figure 57

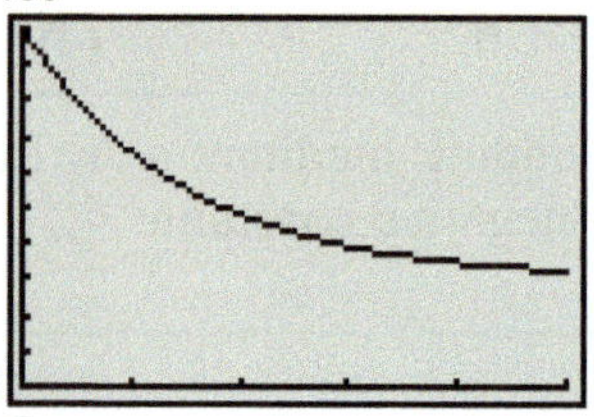

(b) Figure 57 shows the graph of $y = 30 + 70e^{\frac{\ln\frac{5}{7}}{5}x}$, where y is the temperature and x is the time.

(c) By graphing $Y_1 = 50$ and $Y_2 = 30 + 70e^{\frac{\ln\frac{5}{7}}{5}x}$, where x is time, and using INTERSECT, we find that it takes $x = 18.6$ minutes (18 minutes, 37 seconds) for the temperature to cool to 50°C.

(d) By graphing $Y_1 = 35$ and $Y_2 = 30 + 70e^{\frac{\ln\frac{5}{7}}{5}x}$, where x is time, and using INTERSECT, we find that it takes $x = 39.22$ minutes (39 minutes, 13 seconds) for the temperature to cool to 35°C.

(e) As x increases, the value of $e^{\frac{\ln\frac{5}{7}}{5}x}$ approaches zero, so the value of y, the temperature of the object, approaches 30°C, the air temperature of the room.

Now Work PROBLEM 13

4 Use Logistic Models

The exponential growth model $A(t) = A_0e^{kt}, k > 0$, assumes uninhibited growth, meaning that the value of the function grows without limit. Recall that we stated that cell division could be modeled using this function, assuming that no cells die and no by-products are produced. However, cell division eventually is limited by factors such as living space and food supply. The **logistic model,** given next, can describe situations where the growth or decay of the dependent variable is limited.

Logistic Model

In a logistic model, the population P after time t is given by the function

$$P(t) = \frac{c}{1 + ae^{-bt}} \quad (5)$$

where a, b, and c are constants with $c > 0$. The model is a growth model if $b > 0$; the model is a decay model if $b < 0$.

The number c is called the **carrying capacity** (for growth models) because the value $P(t)$ approaches c as t approaches infinity; that is, $\lim_{t \to \infty} P(t) = c$. The number $|b|$ is the growth rate for $b > 0$ and the decay rate for $b < 0$. Figure 58(a) shows the graph of a typical logistic growth function, and Figure 58(b) shows the graph of a typical logistic decay function.

Figure 58

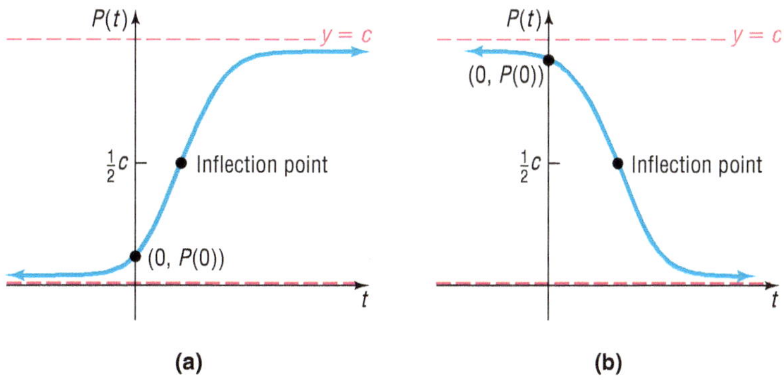

Based on the figures, we have the following properties of logistic growth functions.

Properties of the Logistic Model

1. The domain is the set of all real numbers. The range is the interval $(0, c)$, where c is the carrying capacity.
2. There are no x-intercepts; the y-intercept is $P(0)$.
3. There are two horizontal asymptotes: $y = 0$ and $y = c$.
4. $P(t)$ is an increasing function if $b > 0$ and a decreasing function if $b < 0$.
5. There is an **inflection point** where $P(t)$ equals $\frac{1}{2}$ of the carrying capacity. The inflection point is the point on the graph where the graph changes from being curved upward to curved downward for growth functions and the point where the graph changes from being curved downward to curved upward for decay functions.
6. The graph is smooth and continuous, with no corners or gaps.

EXAMPLE 5 Fruit Fly Population

Fruit flies are placed in a half-pint milk bottle with a banana (for food) and yeast plants (for food and to provide a stimulus to lay eggs). Suppose that the fruit fly population after t days is given by

$$P(t) = \frac{230}{1 + 56.5e^{-0.37t}}$$

(a) State the carrying capacity and the growth rate.
(b) Determine the initial population.
(c) What is the population after 5 days?
(d) How long does it take for the population to reach 180?
(e) Use a graphing utility to determine how long it takes for the population to reach one-half of the carrying capacity.

Solution (a) As $t \to \infty$, $e^{-0.37t} \to 0$ and $P(t) \to \frac{230}{1}$. The carrying capacity of the half-pint bottle is 230 fruit flies. The growth rate is $|b| = |0.37| = 37\%$ per day.

(b) To find the initial number of fruit flies in the half-pint bottle, we evaluate $P(0)$.

$$P(0) = \frac{230}{1 + 56.5e^{-0.37(0)}} = \frac{230}{1 + 56.5} = 4$$

So, initially, there were 4 fruit flies in the half-pint bottle.

(c) To find the number of fruit flies in the half-pint bottle after 5 days, we evaluate $P(5)$.

$$P(5) = \frac{230}{1 + 56.5e^{-0.37(5)}} \approx 23 \text{ fruit flies}$$

After 5 days, there are approximately 23 fruit flies in the bottle.

(d) To determine when the population of fruit flies will be 180, we solve the equation $P(t) = 180$.

$$\frac{230}{1 + 56.5e^{-0.37t}} = 180$$

$$230 = 180(1 + 56.5e^{-0.37t})$$

$$1.2778 = 1 + 56.5e^{-0.37t} \quad \text{Divide both sides by 180.}$$

$$0.2778 = 56.5e^{-0.37t} \quad \text{Subtract 1 from both sides.}$$

$$0.0049 = e^{-0.37t} \quad \text{Divide both sides by 56.5.}$$

$$\ln(0.0049) = -0.37t \quad \text{Rewrite as a logarithmic expression.}$$

$$t \approx 14.4 \text{ days} \quad \text{Divide both sides by } -0.37.$$

It will take approximately 14.4 days (14 days, 10 hours) for the population to reach 180 fruit flies.

(e) One-half of the carrying capacity is 115 fruit flies. We solve $P(t) = 115$ by graphing $Y_1 = \frac{230}{1 + 56.5e^{-0.37x}}$ and $Y_2 = 115$ and using INTERSECT. See Figure 59. The population will reach one-half of the carrying capacity in about 10.9 days (10 days, 22 hours).

Figure 59

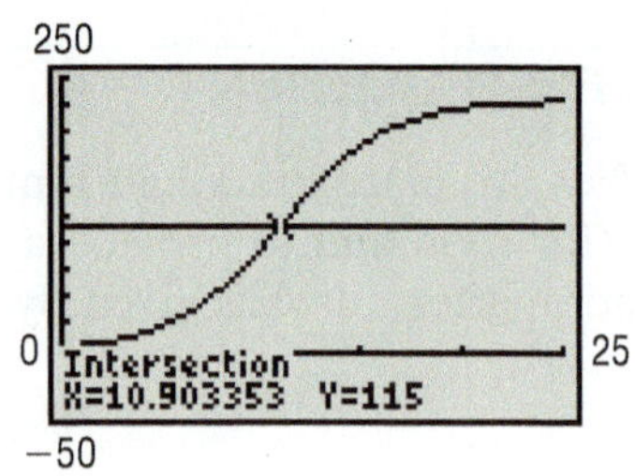

Look back at Figure 59. Notice the point where the graph reaches 115 fruit flies (one-half of the carrying capacity): the graph changes from being curved upward to being curved downward. Using the language of calculus, we say the graph changes from increasing at an increasing rate to increasing at a decreasing rate. For any logistic growth function, when the population reaches one-half the carrying capacity, the population growth starts to slow down.

Exploration

On the same viewing rectangle, graph

$Y_1 = \frac{500}{1 + 24e^{-0.03x}}$ and $Y_2 = \frac{500}{1 + 24e^{-0.08x}}$

What effect does the growth rate $|b|$ have on the logistic growth function?

Now Work PROBLEM 23

EXAMPLE 6 Wood Products

The EFISCEN wood product model classifies wood products according to their life-span. There are four classifications: short (1 year), medium short (4 years), medium long (16 years), and long (50 years). Based on data obtained from the European Forest Institute, the percentage of remaining wood products after t years for wood products with long life-spans (such as those used in the building industry) is given by

$$P(t) = \frac{100.3952}{1 + 0.0316e^{0.0581t}}$$

(a) What is the decay rate?
(b) What is the percentage of remaining wood products after 10 years?
(c) How long does it take for the percentage of remaining wood products to reach 50%?
(d) Explain why the numerator given in the model is reasonable.

Solution

(a) The decay rate is $|b| = |-0.0581| = 5.81\%$.
(b) Evaluate $P(10)$.

$$P(10) = \frac{100.3952}{1 + 0.0316e^{0.0581(10)}} \approx 95.0$$

So 95% of long-life-span wood products remains after 10 years.
(c) Solve the equation $P(t) = 50$.

$$\frac{100.3952}{1 + 0.0316e^{0.0581t}} = 50$$

$$100.3952 = 50(1 + 0.0316e^{0.0581t})$$

$$2.0079 = 1 + 0.0316e^{0.0581t} \quad \text{Divide both sides by 50.}$$

$$1.0079 = 0.0316e^{0.0581t} \quad \text{Subtract 1 from both sides.}$$

$$31.8957 = e^{0.0581t} \quad \text{Divide both sides by 0.0316.}$$

$$\ln(31.8957) = 0.0581t \quad \text{Rewrite as a logarithmic expression.}$$

$$t \approx 59.6 \text{ years} \quad \text{Divide both sides by 0.0581.}$$

It will take approximately 59.6 years for the percentage of long-life-span wood products remaining to reach 50%.
(d) The numerator of 100.3952 is reasonable because the maximum percentage of wood products remaining that is possible is 100%.

Now Work PROBLEM 27

6.8 Assess Your Understanding

Applications and Extensions

1. **Growth of an Insect Population** The size P of a certain insect population at time t (in days) obeys the model $P(t) = 500e^{0.02t}$.
(a) Determine the number of insects at $t = 0$ days.
(b) What is the growth rate of the insect population?
(c) Graph the function using a graphing utility.
(d) What is the population after 10 days?
(e) When will the insect population reach 800?
(f) When will the insect population double?

2. **Growth of Bacteria** The number N of bacteria present in a culture at time t (in hours) obeys the model $N(t) = 1000e^{0.01t}$.
(a) Determine the number of bacteria at $t = 0$ hours.
(b) What is the growth rate of the bacteria?
(c) Graph the function using a graphing utility.
(d) What is the population after 4 hours?
(e) When will the number of bacteria reach 1700?
(f) When will the number of bacteria double?

3. **Radioactive Decay** Strontium 90 is a radioactive material that decays according to the function $A(t) = A_0e^{-0.0244t}$, where A_0 is the initial amount present and A is the amount present at time t (in years). Assume that a scientist has a sample of 500 grams of strontium 90.
(a) What is the decay rate of strontium 90?
(b) Graph the function using a graphing utility.
(c) How much strontium 90 is left after 10 years?
(d) When will 400 grams of strontium 90 be left?
(e) What is the half-life of strontium 90?

4. **Radioactive Decay** Iodine 131 is a radioactive material that decays according to the function $A(t) = A_0 e^{-0.087t}$, where A_0 is the initial amount present and A is the amount present at time t (in days). Assume that a scientist has a sample of 100 grams of iodine 131.
   (a) What is the decay rate of iodine 131?
   (b) Graph the function using a graphing utility.
   (c) How much iodine 131 is left after 9 days?
   (d) When will 70 grams of iodine 131 be left?
   (e) What is the half-life of iodine 131?

5. **Growth of a Colony of Mosquitoes** The population of a colony of mosquitoes obeys the law of uninhibited growth.
   (a) If N is the population of the colony and t is the time in days, express N as a function of t.
   (b) If there are 1000 mosquitoes initially and there are 1800 after 1 day, what is the size of the colony after 3 days?
   (c) How long is it until there are 10,000 mosquitoes?

6. **Bacterial Growth** A culture of bacteria obeys the law of uninhibited growth.
   (a) If N is the number of bacteria in the culture and t is the time in hours, express N as a function of t.
   (b) If 500 bacteria are present initially and there are 800 after 1 hour, how many will be present in the culture after 5 hours?
   (c) How long is it until there are 20,000 bacteria?

7. **Population Growth** The population of a southern city follows the exponential law.
   (a) If N is the population of the city and t is the time in years, express N as a function of t.
   (b) If the population doubled in size over an 18-month period and the current population is 10,000, what will the population be 2 years from now?

8. **Population Decline** The population of a midwestern city follows the exponential law.
   (a) If N is the population of the city and t is the time in years, express N as a function of t.
   (b) If the population decreased from 900,000 to 800,000 from 2005 to 2007, what will the population be in 2009?

9. **Radioactive Decay** The half-life of radium is 1690 years. If 10 grams are present now, how much will be present in 50 years?

10. **Radioactive Decay** The half-life of radioactive potassium is 1.3 billion years. If 10 grams are present now, how much will be present in 100 years? In 1000 years?

11. **Estimating the Age of a Tree** A piece of charcoal is found to contain 30% of the carbon 14 that it originally had.
   (a) When did the tree from which the charcoal came die? Use 5600 years as the half-life of carbon 14.
   (b) Using a graphing utility, graph the relation between the percentage of carbon 14 remaining and time.
   (c) Using INTERSECT, determine the time that elapses until half of the carbon 14 remains.
   (d) Verify the answer found in part (a).

12. **Estimating the Age of a Fossil** A fossilized leaf contains 70% of its normal amount of carbon 14.
   (a) How old is the fossil?
   (b) Using a graphing utility, graph the relation between the percentage of carbon 14 remaining and time.
   (c) Using INTERSECT, determine the time that elapses until half of the carbon 14 remains.
   (d) Verify the answer found in part (a).

13. **Cooling Time of a Pizza** A pizza baked at 450°F is removed from the oven at 5:00 PM into a room that is a constant 70°F. After 5 minutes, the pizza is at 300°F.
   (a) At what time can you begin eating the pizza if you want its temperature to be 135°F?
   (b) Using a graphing utility, graph the relation between temperature and time.
   (c) Using INTERSECT, determine the time that needs to elapse before the pizza is 160°F.
   (d) TRACE the function for large values of time. What do you notice about y, the temperature?

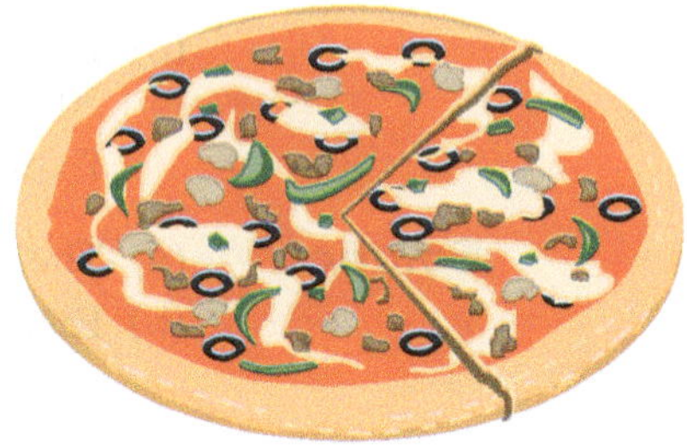

14. **Newton's Law of Cooling** A thermometer reading 72°F is placed in a refrigerator where the temperature is a constant 38°F.
   (a) If the thermometer reads 60°F after 2 minutes, what will it read after 7 minutes?
   (b) How long will it take before the thermometer reads 39°F?
   (c) Using a graphing utility, graph the relation between temperature and time.
   (d) Using INTERSECT, determine the time needed to elapse before the thermometer reads 45°F.
   (e) TRACE the function for large values of time. What do you notice about y, the temperature?

15. **Newton's Law of Heating** A thermometer reading 8°C is brought into a room with a constant temperature of 35°C.
   (a) If the thermometer reads 15°C after 3 minutes, what will it read after being in the room for 5 minutes? For 10 minutes?
   (b) Graph the relation between temperature and time. TRACE to verify that your answers are correct.
   [**Hint:** You need to construct a formula similar to equation (4).]

16. **Warming Time of a Beer Stein** A beer stein has a temperature of 28°F. It is placed in a room with a constant temperature of 70°F. After 10 minutes, the temperature of the stein has risen to 35°F. What will the temperature of the stein be after 30 minutes? How long will it take the stein to reach a temperature of 45°F? (See the hint given for Problem 15.)

17. **Decomposition of Chlorine in a Pool** Under certain water conditions, the free chlorine (hypochlorous acid, HOCl) in a swimming pool decomposes according to the law of uninhibited decay. After shocking his pool, Ben tested the water and found the amount of free chlorine to be 2.5 parts per million (ppm). Twenty-four hours later, Ben tested the water again and found the amount of free chlorine to be 2.2 ppm. What will be the reading after 3 days (that is, 72 hours)? When the chlorine level reaches 1.0 ppm, Ben must shock the pool again. How long can Ben go before he must shock the pool again?

18. Decomposition of Dinitrogen Pentoxide At 45°C, dinitrogen pentoxide (N_2O_5) decomposes into nitrous dioxide (NO_2) and oxygen (O_2) according to the law of uninhibited decay. An initial amount of 0.25 mole of dinitrogen pentoxide decomposes to 0.15 mole in 17 minutes. How much dinitrogen pentoxide will remain after 30 minutes? How long will it take until 0.01 mole of dinitrogen pentoxide remains?

19. Decomposition of Sucrose Reacting with water in an acidic solution at 35°C, sucrose ($C_{12}H_{22}O_{11}$) decomposes into glucose ($C_6H_{12}O_6$) and fructose ($C_6H_{12}O_6$)* according to the law of uninhibited decay. An initial amount of 0.40 mole of sucrose decomposes to 0.36 mole in 30 minutes. How much sucrose will remain after 2 hours? How long will it take until 0.10 mole of sucrose remains?

20. Decomposition of Salt in Water Salt (NaCl) decomposes in water into sodium (Na^+) and chloride (Cl^-) ions according to the law of uninhibited decay. If the initial amount of salt is 25 kilograms and, after 10 hours, 15 kilograms of salt is left, how much salt is left after 1 day? How long does it take until $\frac{1}{2}$ kilogram of salt is left?

21. Radioactivity from Chernobyl After the release of radioactive material into the atmosphere from a nuclear power plant at Chernobyl (Ukraine) in 1986, the hay in Austria was contaminated by iodine 131 (half-life 8 days). If it is safe to feed the hay to cows when 10% of the iodine 131 remains, how long did the farmers need to wait to use this hay?

22. Pig Roasts The hotel Bora-Bora is having a pig roast. At noon, the chef put the pig in a large earthen oven. The pig's original temperature was 75°F. At 2:00 PM the chef checked the pig's temperature and was upset because it had reached only 100°F. If the oven's temperature remains a constant 325°F, at what time may the hotel serve its guests, assuming that pork is done when it reaches 175°F?

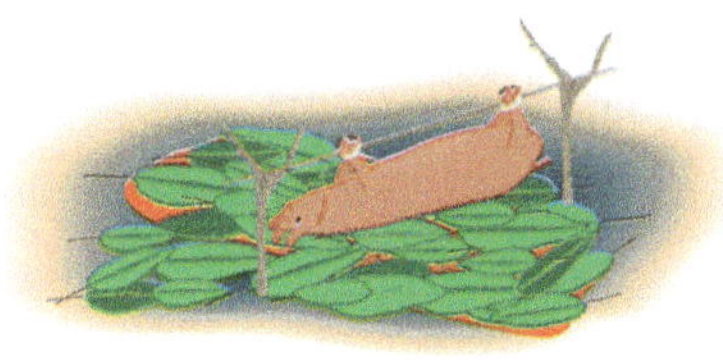

23. Population of a Bacteria Culture The logistic growth model

$$P(t) = \frac{1000}{1 + 32.33e^{-0.439t}}$$

represents the population (in grams) of a bacterium after t hours.

(a) Determine the carrying capacity of the environment.
(b) What is the growth rate of the bacteria?
(c) Determine the initial population size.
(d) Use a graphing utility to graph $P = P(t)$.
(e) What is the population after 9 hours?
(f) When will the population be 700 grams?
(g) How long does it take for the population to reach one-half of the carrying capacity?

24. Population of an Endangered Species Often environmentalists capture an endangered species and transport the species to a controlled environment where the species can produce offspring and regenerate its population. Suppose that six American bald eagles are captured, transported to Montana, and set free. Based on experience, the environmentalists expect the population to grow according to the model

$$P(t) = \frac{500}{1 + 82.33e^{-0.162t}}$$

where t is measured in years.

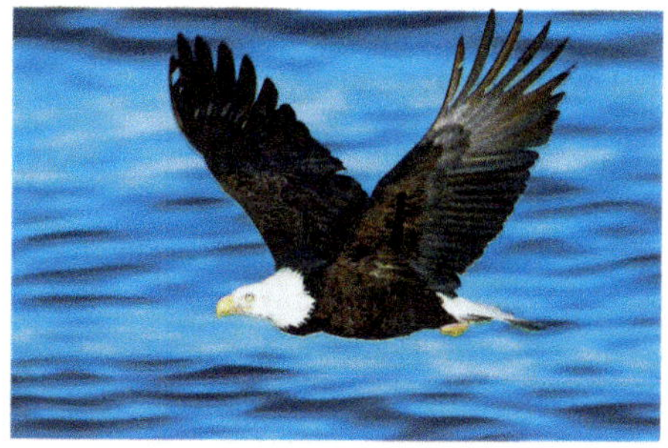

(a) Determine the carrying capacity of the environment.
(b) What is the growth rate of the bald eagle?
(c) Use a graphing utility to graph $P = P(t)$.
(d) What is the population after 3 years?
(e) When will the population be 300 eagles?
(f) How long does it take for the population to reach one-half of the carrying capacity?

25. The *Challenger* Disaster After the *Challenger* disaster in 1986, a study was made of the 23 launches that preceded the fatal flight. A mathematical model was developed involving the relationship between the Fahrenheit temperature x around the O-rings and the number y of eroded or leaky primary O-rings. The model stated that

$$y = \frac{6}{1 + e^{-(5.085 - 0.1156x)}}$$

where the number 6 indicates the 6 primary O-rings on the spacecraft.

(a) What is the predicted number of eroded or leaky primary O-rings at a temperature of 100°F?
(b) What is the predicted number of eroded or leaky primary O-rings at a temperature of 60°F?
(c) What is the predicted number of eroded or leaky primary O-rings at a temperature of 30°F?
(d) Graph the equation. At what temperature is the predicted number of eroded or leaky O-rings 1? 3? 5?

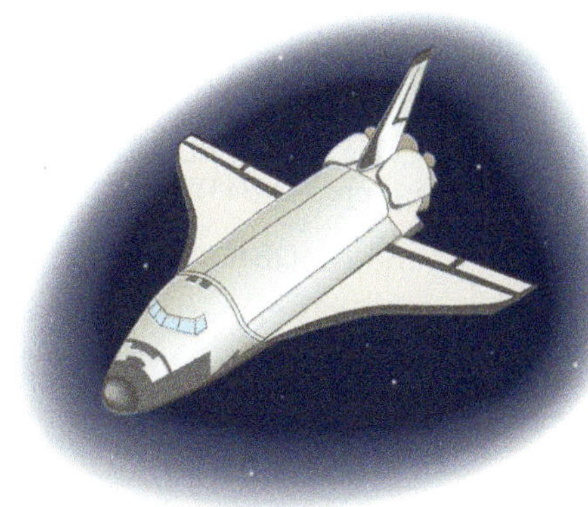

Source: *Linda Tappin, "Analyzing Data Relating to the* Challenger *Disaster,"* Mathematics Teacher, *Vol. 87, No. 6, September 1994, pp. 423–426.*

*Author's Note: Surprisingly, the chemical formulas for glucose and fructose are the same. This is not a typo.

26. Word Users According to a survey by Olsten Staffing Services, the percentage of companies reporting usage of Microsoft Word t years since 1984 is given by

$$P(t) = \frac{99.744}{1 + 3.014e^{-0.799t}}$$

(a) What is the growth rate in the percentage of Microsoft Word users?
(b) Use a graphing utility to graph $P = P(t)$.
(c) What was the percentage of Microsoft Word users in 1990?
(d) During what year did the percentage of Microsoft Word users reach 90%?
(e) Explain why the numerator given in the model is reasonable. What does it imply?

27. Home Computers The logistic model

$$P(t) = \frac{95.4993}{1 + 0.0405e^{0.1968t}}$$

represents the percentage of households that do not own a personal computer t years since 1984.
(a) Evaluate and interpret $P(0)$.
(b) Use a graphing utility to graph $P = P(t)$.
(c) What percentage of households did not own a personal computer in 1995?
(d) In what year will the percentage of households that do not own a personal computer reach 10%?

Source: U.S. Department of Commerce

28. Farmers The logistic model

$$W(t) = \frac{14{,}656.248}{1 + 0.059e^{0.057t}}$$

represents the number of farm workers in the United States t years after 1910.
(a) Evaluate and interpret $W(0)$.
(b) Use a graphing utility to graph $W = W(t)$.
(c) How many farm workers were there in the United States in 1990?
(d) When did the number of farm workers in the United States reach 10,000?
(e) According to this model, what happens to the number of farm workers in the United States as t approaches ∞? Based on this result, do you think that it is reasonable to use this model to predict the number of farm workers in the United States in 2060? Why?

Source: U.S. Department of Agriculture

29. Birthdays The logistic model

$$P(n) = \frac{113.3198}{1 + 0.115e^{0.0912n}}$$

models the probability that, in a room of n people, no two people share the same birthday.
(a) Use a graphing utility to graph $P = P(n)$.
(b) In a room of $n = 15$ people, what is the probability that no two share the same birthday?
(c) How many people must be in a room before the probability that no two people share the same birthday falls below 10%?
(d) What happens to the probability as n increases? Explain what this result means.

30. Bank Failures The logistic model

$$F(t) = \frac{130.118}{1 + 0.00022e^{3.042t}}$$

models the number of bank failures in the United States t years since 1991.
(a) Evaluate and interpret $F(0)$.
(b) Use a graphing utility to graph $F = F(t)$.
(c) How many bank failures were there in 1994 ($t = 3$)?
(d) In what year did the number of bank failures reach 10?
(e) What does the model imply will happen to the number of bank failures as time passes?

Source: Federal Deposit Insurance Corporation

6.9 Building Exponential, Logarithmic, and Logistic Models from Data

PREPARING FOR THIS SECTION *Before getting started, review the following:*

- Building Linear Models from Data (Section 4.2, pp. 292–295)
- Building Quadratic Models from Data (Section 4.4, pp. 314–316)
- Building Cubic Models from Data (Section 5.1, pp. 342–343)

OBJECTIVES
1 Build an Exponential Model from Data (p. 491)
2 Build a Logarithmic Model from Data (p. 492)
3 Build a Logistic Model from Data (p. 493)

In Section 4.2, we discussed how to find the linear function of best fit ($y = ax + b$), in Section 4.4, we discussed how to find the quadratic function of best fit ($y = ax^2 + bx + c$), and in Section 5.1, we discussed how to find the cubic function of best fit ($y = ax^3 + bx^2 + cx + d$).

In this section we will discuss how to use a graphing utility to find equations of best fit that describe the relation between two variables when the relation is thought to be exponential ($y = ab^x$), logarithmic ($y = a + b \ln x$), or logistic $\left(y = \dfrac{c}{1 + ae^{-bx}}\right)$. As before, we draw a scatter diagram of the data to help to determine the appropriate model to use.

Figure 60 shows scatter diagrams that will typically be observed for the three models. Below each scatter diagram are any restrictions on the values of the parameters.

Figure 60

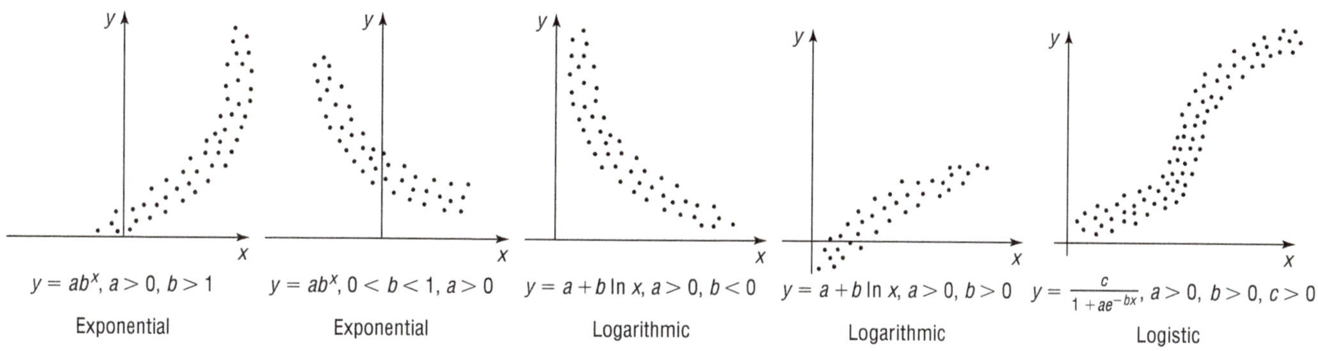

Most graphing utilities have REGression options that fit data to a specific type of curve. Once the data have been entered and a scatter diagram obtained, the type of curve that you want to fit to the data is selected. Then that REGression option is used to obtain the curve of *best fit* of the type selected.

The correlation coefficient r will appear only if the model can be written as a linear expression. As it turns out, r will appear for the linear, power, exponential, and logarithmic models, since these models can be written as a linear expression. Remember, the closer $|r|$ is to 1, the better the fit.

Let's look at some examples.

1 Build an Exponential Model from Data

We saw in Section 6.7 that the future value of money behaves exponentially, and we saw in Section 6.8 that growth and decay models also behave exponentially. The next example shows how data can lead to an exponential model.

EXAMPLE 1 Fitting an Exponential Function to Data

Kathleen is interested in finding a function that explains the growth of cell phone usage in the United States. She gathers data on the number (in millions) of U.S. cell phone subscribers from 1985 through 2005. The data are shown in Table 10.

(a) Using a graphing utility, draw a scatter diagram with year as the independent variable.

(b) Using a graphing utility, build an exponential model from the data.

(c) Express the function found in part (b) in the form $A = A_0e^{kt}$.

(d) Graph the exponential function found in part (b) or (c) on the scatter diagram.

(e) Using the solution to part (b) or (c), predict the number of U.S. cell phone subscribers in 2009.

(f) Interpret the value of k found in part (c).

Solution

(a) Enter the data into the graphing utility, letting 1 represent 1985, 2 represent 1986, and so on. We obtain the scatter diagram shown in Figure 61.

(b) A graphing utility fits the data in Figure 61 to an exponential function of the form $y = ab^x$ using the EXPonential REGression option. From Figure 62 we find that $y = ab^x = 0.72673(1.3470)^x$. Notice that $|r|$ is close to 1, indicating a good fit.

Table 10

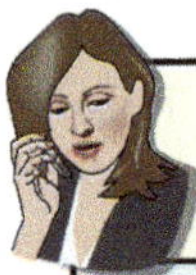

Year, x	Number of Subscribers (in millions), y
1985 ($x = 1$)	0.34
1986 ($x = 2$)	0.68
1987 ($x = 3$)	1.23
1988 ($x = 4$)	2.07
1989 ($x = 5$)	3.51
1990 ($x = 6$)	5.28
1991 ($x = 7$)	7.56
1992 ($x = 8$)	11.03
1993 ($x = 9$)	16.01
1994 ($x = 10$)	24.13
1995 ($x = 11$)	33.76
1996 ($x = 12$)	44.04
1997 ($x = 13$)	55.31
1998 ($x = 14$)	69.21
1999 ($x = 15$)	86.05
2000 ($x = 16$)	109.48
2001 ($x = 17$)	128.37
2002 ($x = 18$)	140.77
2003 ($x = 19$)	158.72
2004 ($x = 20$)	182.14
2005 ($x = 21$)	207.90
2006 ($x = 22$)	233.00

Source: ©2006 CTIA–The Wireless Association®. All Rights Reserved.

Figure 61

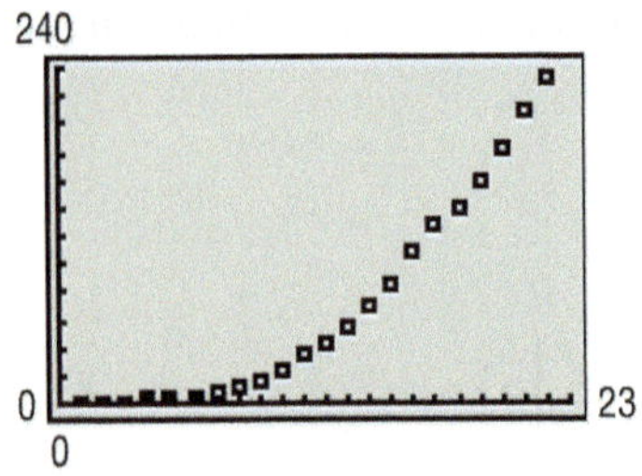

Figure 62

(c) To express $y = ab^x$ in the form $A = A_0e^{kt}$, where $x = t$ and $y = A$, we proceed as follows:

$$ab^x = A_0e^{kt} \quad x = t$$

When $x = t = 0$, we find that $a = A_0$. This leads to

$$a = A_0 \qquad b^x = e^{kt}$$
$$b^x = (e^k)^t$$
$$b = e^k \qquad x = t$$

Since $y = ab^x = 0.72673(1.3470)^x$, we find that $a = 0.72673$ and $b = 1.3470$.

$$a = A_0 = 0.72673 \quad \text{and} \quad b = e^k = 1.3470$$

We want to find k, so we rewrite $e^k = 1.3470$ as a logarithm and obtain

$$k = \ln(1.3470) \approx 0.2979$$

As a result, $A = A_0e^{kt} = 0.72673e^{0.2979t}$.

(d) See Figure 63 for the graph of the exponential function of best fit.

Figure 63

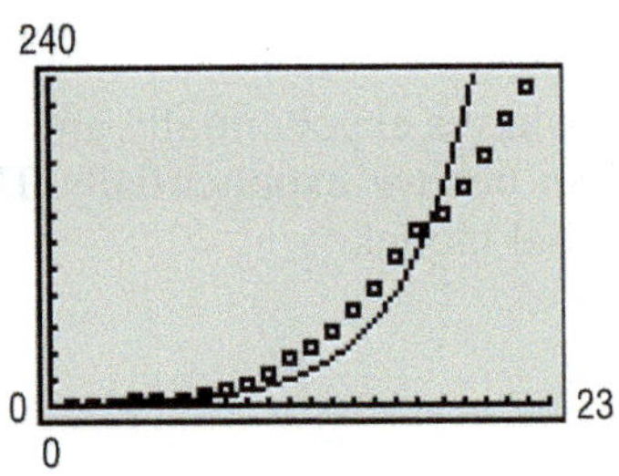

(e) Let $t = 25$ (end of 2009) in the function found in part (c). The predicted number (in millions) of cell phone subscribers in the United States in 2009 is

$$A_0e^{kt} = 0.72673e^{0.2979(25)} \approx 1247$$

This prediction (1247 millon) far exceeds what the U.S. population will be in 2009 (currently the U.S. population is about 300 million). See the answer in part (f).

(f) The value of $k = 0.2979$ represents the growth rate of the number of cell phone subscribers in the United States. Over the period 1985 through 2006, the number of cell phone subscribers has grown at an annual rate of 29.79% compounded continuously. This growth rate is not sustainable as we learned in part (e). In Problem 10 you are asked to build a better model from this data.

Now Work PROBLEM 1

2 Build a Logarithmic Model from Data

Many relations between variables do not follow an exponential model; instead, the independent variable is related to the dependent variable using a logarithmic model.

EXAMPLE 2 Fitting a Logarithmic Function to Data

Jodi, a meteorologist, is interested in finding a function that explains the relation between the height of a weather balloon (in kilometers) and the atmospheric pressure (measured in millimeters of mercury) on the balloon. She collects the data shown in Table 11.

(a) Using a graphing utility, draw a scatter diagram of the data with atmospheric pressure as the independent variable.

(b) It is known that the relation between atmospheric pressure and height follows a logarithmic model. Using a graphing utility, build a logarithmic model from the data.

(c) Draw the logarithmic function found in part (b) on the scatter diagram.

(d) Use the function found in part (b) to predict the height of the weather balloon if the atmospheric pressure is 560 millimeters of mercury.

Table 11

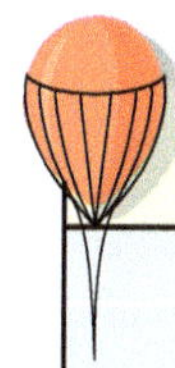

Atmospheric Pressure, p	Height, h
760	0
740	0.184
725	0.328
700	0.565
650	1.079
630	1.291
600	1.634
580	1.862
550	2.235

Solution

(a) After entering the data into the graphing utility, we obtain the scatter diagram shown in Figure 64.

Figure 64

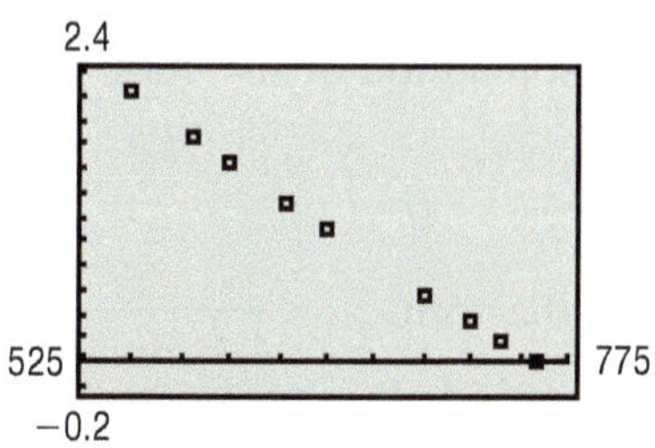

(b) A graphing utility fits the data in Figure 64 to a logarithmic function of the form $y = a + b \ln x$ by using the LOGarithm REGression option. See Figure 65. The logarithmic model from the data is

$$h(p) = 45.7863 - 6.9025 \ln p$$

where h is the height of the weather balloon and p is the atmospheric pressure. Notice that $|r|$ is close to 1, indicating a good fit.

(c) Figure 66 shows the graph of $h(p) = 45.7863 - 6.9025 \ln p$ on the scatter diagram.

Figure 65

Figure 66

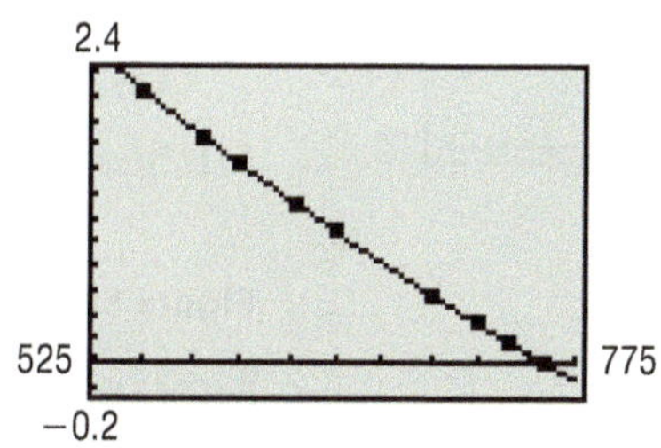

(d) Using the function found in part (b), Jodi predicts the height of the weather balloon when the atmospheric pressure is 560 to be

$$h(560) = 45.7863 - 6.9025 \ln 560$$
$$\approx 2.108 \text{ kilometers}$$

Now Work PROBLEM 5

3 Build a Logistic Model from Data

Logistic growth models can be used to model situations for which the value of the dependent variable is limited. Many real-world situations conform to this scenario. For example, the population of the human race is limited by the availability of natural resources such as food and shelter. When the value of the dependent variable is limited, a logistic growth model is often appropriate.

EXAMPLE 3 Fitting a Logistic Function to Data

The data in Table 12 represent the amount of yeast biomass in a culture after t hours.

Table 12

Time (in hours)	Yeast Biomass	Time (in hours)	Yeast Biomass
0	9.6	10	513.3
1	18.3	11	559.7
2	29.0	12	594.8
3	47.2	13	629.4
4	71.1	14	640.8
5	119.1	15	651.1
6	174.6	16	655.9
7	257.3	17	659.6
8	350.7	18	661.8
9	441.0		

Source: Tor Carlson (Über Geschwindigkeit und Grösse der Hefevermehrung in Würze, Biochemische Zeitschrift, Bd. 57, pp. 313–334, 1913)

(a) Using a graphing utility, draw a scatter diagram of the data with time as the independent variable.

(b) Using a graphing utility, build a logistic model from the data.

(c) Using a graphing utility, graph the function found in part (b) on the scatter diagram.

(d) What is the predicted carrying capacity of the culture?

(e) Use the function found in part (b) to predict the population of the culture at $t = 19$ hours.

Solution

(a) See Figure 67 for a scatter diagram of the data.

Figure 67

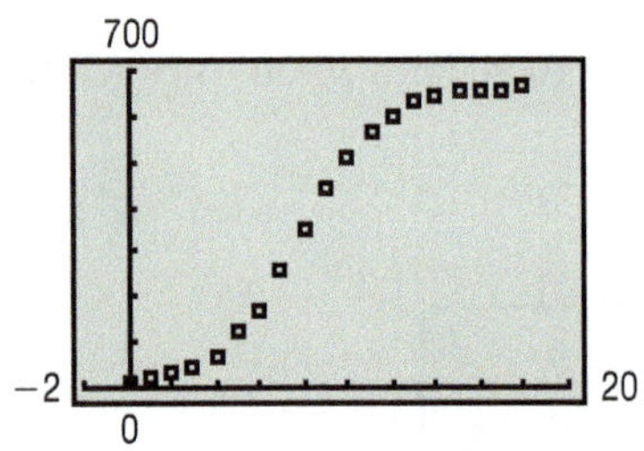

(b) A graphing utility fits a logistic growth model of the form $y = \dfrac{c}{1 + ae^{-bx}}$ by using the LOGISTIC regression option. See Figure 68. The logistic model from the data is

$$y = \frac{663.0}{1 + 71.6e^{-0.5470x}}$$

where y is the amount of yeast biomass in the culture and x is the time.

(c) See Figure 69 for the graph of the logistic model.

Figure 68

Figure 69

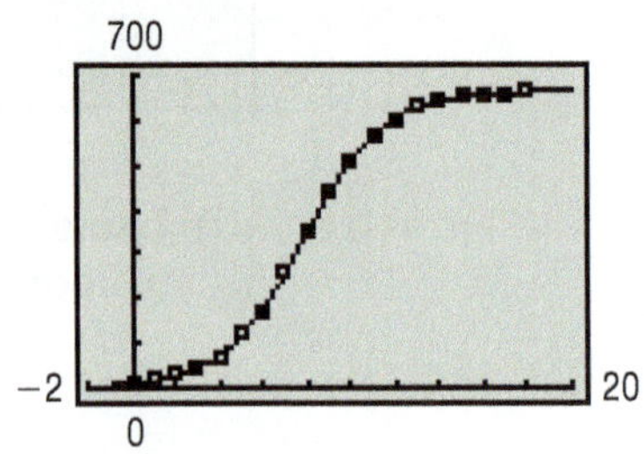

(d) Based on the logistic growth model found in part (b), the carrying capacity of the culture is 663.

(e) Using the logistic growth model found in part (b), the predicted amount of yeast biomass at $t = 19$ hours is

$$y = \frac{663.0}{1 + 71.6e^{-0.5470(19)}} \approx 661.5$$

Now Work PROBLEM 7

6.9 Assess Your Understanding

Applications and Extensions

1. Biology A strain of E-coli Beu 397-recA441 is placed into a nutrient broth at 30° Celsius and allowed to grow. The following data are collected. Theory states that the number of bacteria in the petri dish will initially grow according to the law of uninhibited growth. The population is measured using an optical device in which the amount of light that passes through the petri dish is measured.

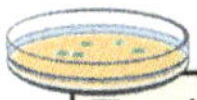

Time (hours), x	Population, y
0	0.09
2.5	0.18
3.5	0.26
4.5	0.35
6	0.50

Source: Dr. Polly Lavery, Joliet Junior College

(a) Draw a scatter diagram treating time as the independent variable.
(b) Using a graphing utility, build an exponential model from the data.
(c) Express the function found in part (b) in the form $N(t) = N_0e^{kt}$.
(d) Graph the exponential function found in part (b) or (c) on the scatter diagram.
(e) Use the exponential function from part (b) or (c) to predict the population at $x = 7$ hours.
(f) Use the exponential function from part (b) or (c) to predict when the population will reach 0.75.

2. Biology A strain of E-coli SC18del-recA718 is placed into a nutrient broth at 30° Celsius and allowed to grow. The following data are collected. Theory states that the number of bacteria in the petri dish will initially grow according to the law of uninhibited growth. The population is measured using an optical device in which the amount of light that passes through the petri dish is measured.

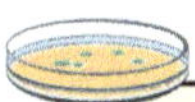

Time (hours), x	Population, y
2.5	0.175
3.5	0.38
4.5	0.63
4.75	0.76
5.25	1.20

Source: Dr. Polly Lavery, Joliet Junior College

(a) Draw a scatter diagram treating time as the independent variable.
(b) Using a graphing utility, build an exponential model from the data.
(c) Express the function found in part (b) in the form $N(t) = N_0e^{kt}$.
(d) Graph the exponential function found in part (b) or (c) on the scatter diagram.
(e) Use the exponential function from part (b) or (c) to predict the population at $x = 6$ hours.
(f) Use the exponential function from part (b) or (c) to predict when the population will reach 2.1.

3. Chemistry A chemist has a 100-gram sample of a radioactive material. He records the amount of radioactive material every week for 6 weeks and obtains the following data:

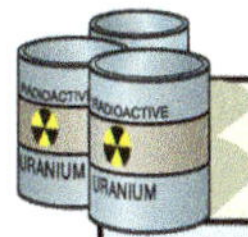

Week	Weight (in Grams)
0	100.0
1	88.3
2	75.9
3	69.4
4	59.1
5	51.8
6	45.5

(a) Using a graphing utility, draw a scatter diagram with week as the independent variable.
(b) Using a graphing utility, build an exponential model from the data.
(c) Express the function found in part (b) in the form $A(t) = A_0e^{kt}$.
(d) Graph the exponential function found in part (b) or (c) on the scatter diagram.
(e) From the result found in part (b), determine the half-life of the radioactive material.
(f) How much radioactive material will be left after 50 weeks?
(g) When will there be 20 grams of radioactive material?

4. Cigarette Exports The following data represent the number of cigarettes (in billions) exported from the United States by year.

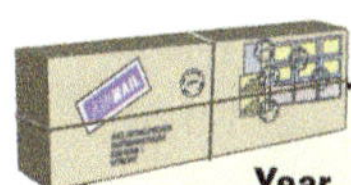

Year	Cigarette Exports (in billions of pieces)
1995	231.1
1998	201.3
1999	151.4
2000	147.9
2001	133.9
2002	127.4
2003	121.5
2004	118.7

Source: Statistical Abstract of the United States, 2006

(a) Let t = the number of years since 1995. Using a graphing utility, draw a scatter diagram of the data using t as the independent variable and number of cigarettes as the dependent variable.

(b) Using a graphing utility, build an exponential model from the data.

(c) Express the function found in part (b) in the form $A(t) = A_0e^{kt}$.

(d) Graph the exponential function found in part (b) or (c) on the scatter diagram.

(e) Use the exponential function from part (b) or (c) to predict the number of cigarettes that will be exported from the United States in 2010.

(f) Use the exponential function from part (b) or (c) to predict when the number of cigarettes exported from the United States will decrease to 50 billion.

5. Economics and Marketing The following data represent the price and quantity demanded in 2007 for Dell personal computers.

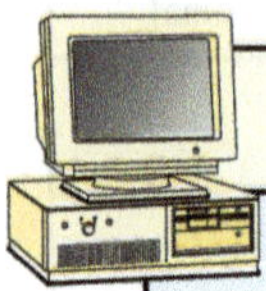

Price ($/Computer)	Quantity Demanded
2300	152
2000	159
1700	164
1500	171
1300	176
1200	180
1000	189

(a) Using a graphing utility, draw a scatter diagram of the data with price as the dependent variable.

(b) Using a graphing utility, build a logarithmic model from the data.

(c) Using a graphing utility, draw the logarithmic function found in part (b) on the scatter diagram.

(d) Use the function found in part (b) to predict the number of Dell personal computers that will be demanded if the price is $1650.

6. Economics and Marketing The following data represent the price and quantity supplied in 2007 for Dell personal computers.

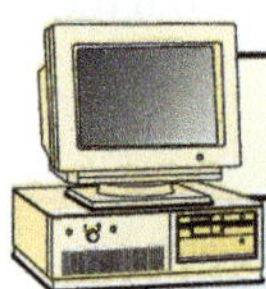

Price ($/Computer)	Quantity Supplied
2300	180
2000	173
1700	160
1500	150
1300	137
1200	130
1000	113

(a) Using a graphing utility, draw a scatter diagram of the data with price as the dependent variable.

(b) Using a graphing utility, build a logarithmic model from the data.

(c) Using a graphing utility, draw the logarithmic function found in part (b) on the scatter diagram.

(d) Use the function found in part (b) to predict the number of Dell personal computers that will be supplied if the price is $1650.

7. Population Model The following data represent the population of the United States. An ecologist is interested in building a model that describes the population of the United States.

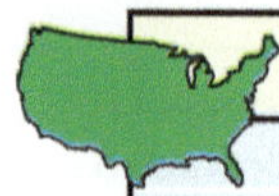

Year	Population
1900	76,212,168
1910	92,228,496
1920	106,021,537
1930	123,202,624
1940	132,164,569
1950	151,325,798
1960	179,323,175
1970	203,302,031
1980	226,542,203
1990	248,709,873
2000	281,421,906

Source: U.S. Census Bureau

(a) Using a graphing utility, draw a scatter diagram of the data using years since 1900 as the independent variable and population as the dependent variable.

(b) Using a graphing utility, build a logistic model from the data.

(c) Using a graphing utility, draw the function found in part (b) on the scatter diagram.

(d) Based on the function found in part (b), what is the carrying capacity of the United States?

(e) Use the function found in part (b) to predict the population of the United States in 2004.

(f) When will the United States population be 300,000,000?

(g) Compare actual U.S. Census figures to the predictions found in parts (e) and (f). Discuss any differences.

8. Population Model The following data represent the world population. An ecologist is interested in building a model that describes the world population.

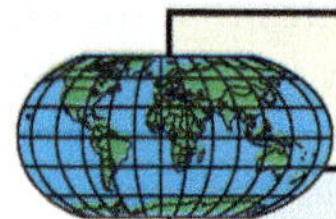

Year	Population (in Billions)
1993	5.531
1994	5.611
1995	5.691
1996	5.769
1997	5.847
1998	5.925
1999	6.003
2000	6.080
2001	6.157

Source: U.S. Census Bureau

(a) Using a graphing utility, draw a scatter diagram of the data using years since 1992 as the independent variable and population as the dependent variable.

(b) Using a graphing utility, build a logistic model from the data.

(c) Using a graphing utility, draw the function found in part (b) on the scatter diagram.

(d) Based on the function found in part (b), what is the carrying capacity of the world?

(e) Use the function found in part (b) to predict the population of the world in 2004.

(f) When will world population be 7 billion?

(g) Compare actual U.S. Census figures to the prediction found in part (e). Discuss any differences.

9. Cable Subscribers The following data represent the number of basic cable TV subscribers in the United States. A market researcher believes that external factors, such as satellite TV, have affected the growth of cable subscribers. She is interested in building a model that can be used to describe the number of cable TV subscribers in the United States.

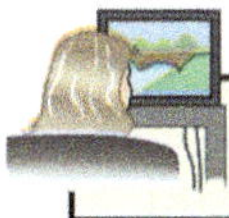

Year	Subscribers (1,000)
1975 ($t = 5$)	9,800
1980 ($t = 10$)	17,500
1985 ($t = 15$)	35,440
1990 ($t = 20$)	50,520
1992 ($t = 22$)	54,300
1994 ($t = 24$)	58,373
1996 ($t = 26$)	62,300
1998 ($t = 28$)	64,650
2000 ($t = 30$)	66,250
2002 ($t = 32$)	66,472
2004 ($t = 34$)	65,853

Source: Statistical Abstract of the United States, 2006

(a) Using a graphing utility, draw a scatter diagram of the data using the number of years after 1970, t, as the independent variable and number of subscribers as the dependent variable.

(b) Using a graphing utility, build a logistic model from the data.

(c) Using a graphing utility, draw the function found in part (b) on the scatter diagram.

(d) Based on the model found in part (b), what is the carrying capacity of the cable TV market in the United States?

(e) Use the model found in part (b) to predict the number of cable TV subscribers in the United States in 2015.

10. Cell Phone Users Refer to the data in Table 10 (p. 492).

(a) Using a graphing utility, build a logistic model from the data.

(b) Graph the logistic function found in part (b) on a scatter diagram of the data.

(c) What is the predicted carrying capacity of U.S. cell phone subscribers?

(d) Use the model found in part (b) to predict the number of U.S. cell phone subscribers at the end of 2009.

(e) Compare the answer to part (d) above with the answer to Example 1, part (e). How do you explain the different predictions?

Mixed Practice

11. Age versus Total Cholesterol The following data represent the age and average total cholesterol for adult males at various ages.

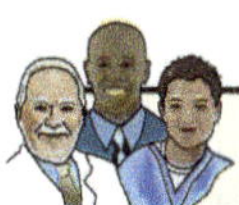

Age	Total Cholesterol
27	189
40	205
50	215
60	210
70	210
80	194

(a) Using a graphing utility, draw a scatter diagram of the data using age, x, as the independent variable and total cholesterol, y, as the dependent variable.

(b) Based on the scatter diagram drawn in part (a), decide on a model (linear, quadratic, cubic, exponential, logarithmic, or logistic) that you think best describes the relation between age and total cholesterol. Be sure to justify your choice of model.

(c) Using a graphing utility, find the model of best fit.

(d) Using a graphing utility, draw the model of best fit on the scatter diagram drawn in part (a).

(e) Use your model to predict the total cholesterol of a 35-year-old male.

12. Income versus Crime Rate The following data represent crime rate against individuals (crimes per 1000 population) and their income in the United States in 2003.

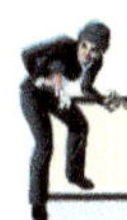

Income	Crime Rate
\$5000	51.1
11,250	31.9
20,000	27.0
30,000	25.8
42,500	22.0
62,500	23.3
85,000	18.5

Source: Statistical Abstract of the United States, 2006

(a) Using a graphing utility, draw a scatter diagram of the data using income, x, as the independent variable and crime rate, y, as the dependent variable.

(b) Based on the scatter diagram drawn in part (a), decide on a model (linear, quadratic, cubic, exponential, logarithmic, or logistic) that you think best describes the relation between income and crime rate. Be sure to justify your choice of model.

(c) Using a graphing utility, find the model of best fit.

(d) Using a graphing utility, draw the model of best fit on the scatter diagram drawn in part (a).

(e) Use your model to predict the crime rate of an individual whose income is \$55,000.

13. Depreciation of a Chevrolet Impala The following data represent the asking price and age of a Chevrolet Impala SS.

Age	Asking Price
1	$27,417
1	26,750
2	22,995
2	23,195
3	17,999
4	16,995
4	16,490

Source: cars.com

(a) Using a graphing utility, draw a scatter diagram of the data using age, x, as the independent variable and asking price, y, as the dependent variable.
(b) Based on the scatter diagram drawn in part (a), decide on a model (linear, quadratic, cubic, exponential, logarithmic, or logistic) that you think best describes the relation between age and asking price. Be sure to justify your choice of model.
(c) Using a graphing utility, find the model of best fit.
(d) Using a graphing utility, draw the model of best fit on the scatter diagram drawn in part (a).
(e) Use your model to predict the asking price of a Chevrolet Impala SS that is 5 years old.

CHAPTER REVIEW

Things to Know

Composite function (p. 404)
$(f \circ g)(x) = f(g(x))$ The domain of $f \circ g$ is the set of all numbers x in the domain of g for which $g(x)$ is in the domain of f.

One-to-one function f (p. 412)
A function for which any two different inputs in the domain correspond to two different outputs in the range.
For any choice of elements x_1, x_2 in the domain of f, if $x_1 \neq x_2$, then $f(x_1) \neq f(x_2)$.

Horizontal-line test (p. 413)
If every horizontal line intersects the graph of a function f in at most one point, f is one-to-one.

Inverse function f^{-1} of f (pp. 414–417)
Domain of f = range of f^{-1}; range of f = domain of f^{-1}
$f^{-1}(f(x)) = x$ for all x in the domain of f
$f(f^{-1}(x)) = x$ for all x in the domain of f^{-1}
The graphs of f and f^{-1} are symmetric with respect to the line $y = x$.

Properties of the exponential function (pp. 429–431)

$f(x) = Ca^x, \quad a > 1$

Domain: the interval $(-\infty, \infty)$
Range: the interval $(0, \infty)$
x-intercepts: none; y-intercept: 1
Horizontal asymptote: x-axis ($y = 0$) as $x \to -\infty$
Increasing; one-to-one; smooth; continuous
See Figure 24 for a typical graph.

$f(x) = Ca^x, \quad 0 < a < 1$

Domain: the interval $(-\infty, \infty)$
Range: the interval $(0, \infty)$
x-intercepts: none; y-intercept: 1
Horizontal asymptote: x-axis ($y = 0$) as $x \to \infty$
Decreasing; one-to-one; smooth; continuous
See Figure 28 for a typical graph.

Number e (p. 432)
Value approached by the expression $\left(1 + \frac{1}{n}\right)^n$ as $n \to \infty$; that is, $\lim_{n\to\infty}\left(1 + \frac{1}{n}\right)^n = e$

Property of exponents (p. 433)
If $a^u = a^v$, then $u = v$.

Properties of the logarithmic function (p. 449)

$f(x) = \log_a x, \quad a > 1$
($y = \log_a x$ means $x = a^y$)

Domain: the interval $(0, \infty)$
Range: the interval $(-\infty, \infty)$
x-intercept: 1; y-intercept; none
Vertical asymptote: $x = 0$ (y-axis)
Increasing; one-to-one; smooth; continuous
See Figure 44(a) for a typical graph.

$f(x) = \log_a x, \quad 0 < a < 1$ $(y = \log_a x \text{ means } x = a^y)$	Domain: the interval $(0, \infty)$ Range: the interval $(-\infty, \infty)$ x-intercept: 1; y-intercept; none Vertical asymptote: $x = 0$ (y-axis) Decreasing; one-to-one; smooth; continuous See Figure 44(b) for a typical graph.

Natural logarithm (p. 443) $y = \ln x$ means $x = e^y$.

Properties of logarithms (pp. 453–455, 457)

$\log_a 1 = 0 \qquad \log_a a = 1 \qquad a^{\log_a M} = M \qquad \log_a a^r = r$

$\log_a(MN) = \log_a M + \log_a N \qquad \log_a\left(\frac{M}{N}\right) = \log_a M - \log_a N$

$\log_a M^r = r \log_a M$

If $M = N$, then $\log_a M = \log_a N$.

If $\log_a M = \log_a N$, then $M = N$.

Formulas

Change-of-Base Formula (p. 458) $\log_a M = \dfrac{\log_b M}{\log_b a}$

Compound Interest Formula (p. 471) $A = P\cdot\left(1 + \dfrac{r}{n}\right)^{nt}$

Continuous compounding (p. 472) $A = Pe^{rt}$

Effective rate of interest (p. 473) Compounding n times per year: $r_e = \left(1 + \dfrac{r}{n}\right)^n - 1$

Continuous Compounding $r_e = e^r - 1$

Present Value Formulas (p. 474) $P = A\cdot\left(1 + \dfrac{r}{n}\right)^{-nt}$ or $P = Ae^{-rt}$

Growth and decay (p. 479) $A(t) = A_0e^{kt}$

Newton's Law of Cooling (p. 483) $u(t) = T + (u_0 - T)e^{kt} \quad k < 0$

Logistic model (p. 484) $P(t) = \dfrac{c}{1 + ae^{-bt}}$

Objectives

Section		You should be able to . . .	Example(s)	Review Exercises
6.1	1	Form a composite function (p. 404)	1, 4, 5	1–12
	2	Find the domain of a composite function (p. 405)	2–4	7–12
6.2	1	Determine whether a function is one-to-one (p. 412)	1, 2	13(a), 14(a), 15, 16
	2	Determine the inverse of a function defined by a map or a set of ordered pairs (p. 414)	3, 4	13(b), 14(b)
	3	Obtain the graph of the inverse function from the graph of the function (p. 416)	7	15, 16
	4	Find the inverse of a function defined by an equation (p. 417)	8, 9, 11	17–22
6.3	1	Evaluate exponential functions (p. 425)	1	23(a), (c), 24(a), (c), 87(a)
	2	Graph exponential functions (p. 429)	3–6	55–60
	3	Define the number e (p. 432)	pg. 432	59, 60
	4	Solve exponential equations (p. 434)	7, 8	63–66, 71, 72, 74–76, 87(b)
6.4	1	Change exponential statements to logarithmic statements and logarithmic statements to exponential statements (p. 441)	2, 3	25–28
	2	Evaluate logarithmic expressions (p. 441)	4	23(b), (d), 24(b), (d), 33, 34, 83(b), 84(b), 85, 86, 88(a), 89
	3	Determine the domain of a logarithmic function (p. 442)	5	29–32, 61(a), 62(a)
	4	Graph logarithmic functions (p. 443)	6, 7	61, 62, 83(a), 84(a)
	5	Solve logarithmic equations (p. 447)	8, 9	67, 68, 73, 83(c), 84(c), 88(b)

6.5	1	Work with the properties of logarithms (p. 453)	1, 2	35–38
	2	Write a logarithmic expression as a sum or difference of logarithms (p. 455)	3–5	39–44
	3	Write a logarithmic expression as a single logarithm (p. 456)	6	45–50
	4	Evaluate logarithms whose base is neither 10 nor *e* (p. 457)	7, 8	51, 52
	5	Graph logarithmic functions whose base is neither 10 nor *e* (p.459)	9	53, 54
6.6	1	Solve logarithmic equations (p. 462)	1–3	67, 77, 78
	2	Solve exponential equations (p. 464)	4–7	63–66, 69, 70, 74–76, 79–82
	3	Solve logarithmic and exponential equations using a graphing utility (p. 466)	8	69–82
6.7	1	Determine the future value of a lump sum of money (p. 469)	1–3	90, 92, 97
	2	Calculate effective rates of return (p. 473)	4	90
	3	Determine the present value of a lump sum of money (p. 474)	5	91
	4	Determine the rate of interest or time required to double a lump sum of money (p. 475)	6, 7	90
6.8	1	Find equations of populations that obey the law of uninhibited growth (p. 479)	1, 2	95
	2	Find equations of populations that obey the law of decay (p. 481)	3	93, 96
	3	Use Newton's Law of Cooling (p. 483)	4	94
	4	Use logistic models (p. 484)	5, 6	98
6.9	1	Build an exponential model from data (p. 491)	1	99
	2	Build a logarithmic model from data (p. 492)	2	100
	3	Build a logistic model from data (p. 493)	3	101

Review Exercises

In Problems 1–6, for the given functions f and g find:

(a) $(f \circ g)(2)$ *(b)* $(g \circ f)(-2)$ *(c)* $(f \circ f)(4)$ *(d)* $(g \circ g)(-1)$

1. $f(x) = 3x - 5;\ \ g(x) = 1 - 2x^2$

2. $f(x) = 4 - x;\ \ g(x) = 1 + x^2$

3. $f(x) = \sqrt{x + 2};\ \ g(x) = 2x^2 + 1$

4. $f(x) = 1 - 3x^2;\ \ g(x) = \sqrt{4 - x}$

5. $f(x) = e^x;\ \ g(x) = 3x - 2$

6. $f(x) = \dfrac{2}{1 + 2x^2};\ \ g(x) = 3x$

In Problems 7–12, find $f \circ g$, $g \circ f$, $f \circ f$, *and* $g \circ g$ *for each pair of functions. State the domain of each composite function.*

7. $f(x) = 2 - x;\ \ g(x) = 3x + 1$

8. $f(x) = 2x - 1;\ \ g(x) = 2x + 1$

9. $f(x) = 3x^2 + x + 1;\ \ g(x) = |3x|$

10. $f(x) = \sqrt{3x};\ \ g(x) = 1 + x + x^2$

11. $f(x) = \dfrac{x + 1}{x - 1};\ \ g(x) = \dfrac{1}{x}$

12. $f(x) = \sqrt{x - 3};\ \ g(x) = \dfrac{3}{x}$

In Problems 13 and 14, (a) verify that the function is one-to-one, and (b) find the inverse of the given function.

13. $\{(1, 2), (3, 5), (5, 8), (6, 10)\}$

14. $\{(-1, 4), (0, 2), (1, 5), (3, 7)\}$

In Problems 15 and 16, state why the graph of the function is one-to-one. Then draw the graph of the inverse function f^{-1}. *For convenience (and as a hint), the graph of* $y = x$ *is also given.*

15.

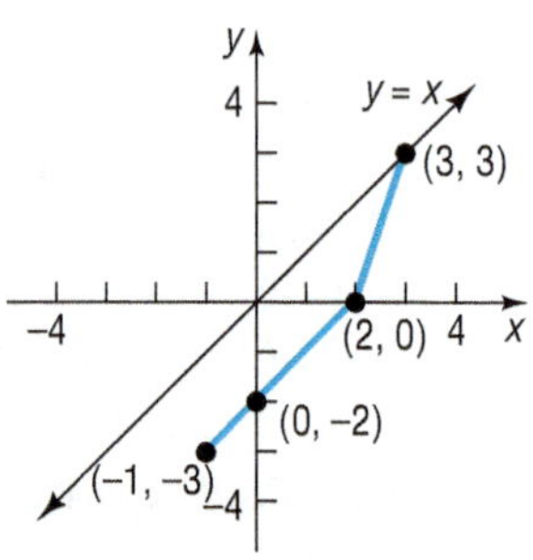

16.

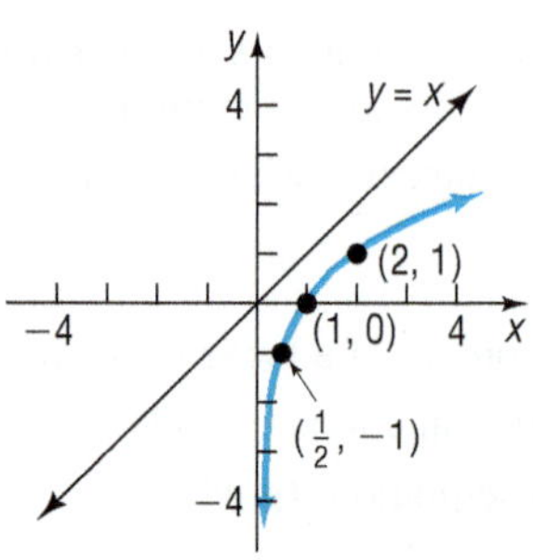

In Problems 17–22, the function f is one-to-one. Find the inverse of each function and check your answer. Find the domain and the range of f and f^{-1}.

17. $f(x) = \dfrac{2x + 3}{5x - 2}$
18. $f(x) = \dfrac{2 - x}{3 + x}$
19. $f(x) = \dfrac{1}{x - 1}$
20. $f(x) = \sqrt{x - 2}$
21. $f(x) = \dfrac{3}{x^{1/3}}$
22. $f(x) = x^{1/3} + 1$

In Problems 23 and 24, $f(x) = 3^x$ and $g(x) = \log_3 x$.

23. Evaluate: (a) $f(4)$ (b) $g(9)$ (c) $f(-2)$ (d) $g\left(\dfrac{1}{27}\right)$

24. Evaluate: (a) $f(1)$ (b) $g(81)$ (c) $f(-4)$ (d) $g\left(\dfrac{1}{243}\right)$

In Problems 25 and 26, convert each exponential statement to an equivalent statement involving a logarithm. In Problems 27 and 28, convert each logarithmic statement to an equivalent statement involving an exponent.

25. $5^2 = z$
26. $a^5 = m$
27. $\log_5 u = 13$
28. $\log_a 4 = 3$

In Problems 29–32, find the domain of each logarithmic function.

29. $f(x) = \log(3x - 2)$
30. $F(x) = \log_5(2x + 1)$
31. $H(x) = \log_2(x^2 - 3x + 2)$
32. $F(x) = \ln(x^2 - 9)$

In Problems 33–38, evaluate each expression. Do not use a calculator.

33. $\log_2\left(\dfrac{1}{8}\right)$
34. $\log_3 81$
35. $\ln e^{\sqrt{2}}$
36. $e^{\ln 0.1}$
37. $2^{\log_2 0.4}$
38. $\log_2 2^{\sqrt{3}}$

In Problems 39–44, write each expression as the sum and/or difference of logarithms. Express powers as factors.

39. $\log_3\left(\dfrac{uv^2}{w}\right)$, $u > 0, v > 0, w > 0$
40. $\log_2\left(a^2\sqrt{b}\right)^4$, $a > 0, b > 0$
41. $\log\left(x^2\sqrt{x^3 + 1}\right)$, $x > 0$
42. $\log_5\left(\dfrac{x^2 + 2x + 1}{x^2}\right)$, $x > 0$
43. $\ln\left(\dfrac{x\sqrt[3]{x^2 + 1}}{x - 3}\right)$, $x > 3$
44. $\ln\left(\dfrac{2x + 3}{x^2 - 3x + 2}\right)^2$, $x > 2$

In Problems 45–50, write each expression as a single logarithm.

45. $3 \log_4 x^2 + \dfrac{1}{2} \log_4 \sqrt{x}$
46. $-2 \log_3\left(\dfrac{1}{x}\right) + \dfrac{1}{3} \log_3 \sqrt{x}$
47. $\ln\left(\dfrac{x - 1}{x}\right) + \ln\left(\dfrac{x}{x + 1}\right) - \ln(x^2 - 1)$
48. $\log(x^2 - 9) - \log(x^2 + 7x + 12)$
49. $2 \log 2 + 3 \log x - \dfrac{1}{2}[\log(x + 3) + \log(x - 2)]$
50. $\dfrac{1}{2} \ln(x^2 + 1) - 4 \ln \dfrac{1}{2} - \dfrac{1}{2}[\ln(x - 4) + \ln x]$

In Problems 51 and 52, use the Change-of-Base Formula and a calculator to evaluate each logarithm. Round your answer to three decimal places.

51. $\log_4 19$
52. $\log_2 21$

In Problems 53 and 54, graph each function using a graphing utility and the Change-of-Base Formula.

53. $y = \log_3 x$
54. $y = \log_7 x$

In Problems 55–62, use the given function f to:
(a) Find the domain of f.
(b) Graph f.
(c) From the graph, determine the range and any asymptotes of f.
(d) Find f^{-1}, the inverse of f.
(e) Use f^{-1} to find the range of f.
(f) Graph f^{-1}.

55. $f(x) = 2^{x-3}$
56. $f(x) = -2^x + 3$
57. $f(x) = \dfrac{1}{2}(3^{-x})$
58. $f(x) = 1 + 3^{-x}$
59. $f(x) = 1 - e^{-x}$
60. $f(x) = 3e^{x-2}$
61. $f(x) = \dfrac{1}{2}\ln(x + 3)$
62. $f(x) = 3 + \ln(2x)$

In Problems 63–82, solve each equation. Express irrational solutions in exact form and as a decimal rounded to 3 decimal places.

63. $4^{1-2x} = 2$ **64.** $8^{6+3x} = 4$ **65.** $3^{x^2+x} = \sqrt{3}$ **66.** $4^{x-x^2} = \frac{1}{2}$

67. $\log_x 64 = -3$ **68.** $\log_{\sqrt{2}} x = -6$ **69.** $5^x = 3^{x+2}$ **70.** $5^{x+2} = 7^{x-2}$

71. $9^{2x} = 27^{3x-4}$ **72.** $25^{2x} = 5^{x^2-12}$ **73.** $\log_3 \sqrt{x-2} = 2$ **74.** $2^{x+1} \cdot 8^{-x} = 4$

75. $8 = 4^{x^2} \cdot 2^{5x}$ **76.** $2^x \cdot 5 = 10^x$ **77.** $\log_6(x+3) + \log_6(x+4) = 1$

78. $\log(7x - 12) = 2\log x$ **79.** $e^{1-x} = 5$ **80.** $e^{1-2x} = 4$

81. $9^x + 4 \cdot 3^x - 3 = 0$ **82.** $4^x - 14 \cdot 4^{-x} = 5$

83. Suppose that $f(x) = \log_2(x-2) + 1$.
(a) Graph f.
(b) What is $f(6)$? What point is on the graph of f?
(c) Solve $f(x) = 4$. What point is on the graph of f?
(d) Based on the graph drawn in part (a), solve $f(x) > 0$.
(e) Find $f^{-1}(x)$. Graph f^{-1} on the same Cartesian plane as f.

84. Suppose that $f(x) = \log_3(x+1) - 4$.
(a) Graph f.
(b) What is $f(8)$? What point is on the graph of f?
(c) Solve $f(x) = -3$. What point is on the graph of f?
(d) Based on the graph drawn in part (a), solve $f(x) < 0$.
(e) Find $f^{-1}(x)$. Graph f^{-1} on the same Cartesian plane as f.

In Problems 85 and 86, use the following result: If x is the atmospheric pressure (measured in millimeters of mercury), then the formula for the altitude h(x) (measured in meters above sea level) is

$$h(x) = (30T + 8000)\log\left(\frac{P_0}{x}\right)$$

where T is the temperature (in degrees Celsius) and P_0 is the atmospheric pressure at sea level, which is approximately 760 millimeters of mercury.

85. Finding the Altitude of an Airplane At what height is a Piper Cub whose instruments record an outside temperature of 0°C and a barometric pressure of 300 millimeters of mercury?

86. Finding the Height of a Mountain How high is a mountain if instruments placed on its peak record a temperature of 5°C and a barometric pressure of 500 millimeters of mercury?

87. Amplifying Sound An amplifier's power output P (in watts) is related to its decibel voltage gain d by the formula

$$P = 25e^{0.1d}$$

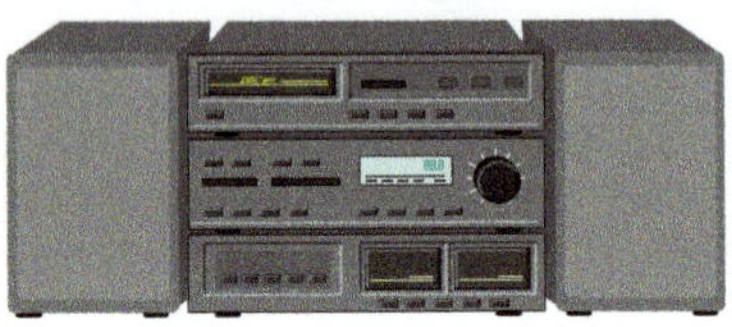

(a) Find the power output for a decibel voltage gain of 4 decibels.
(b) For a power output of 50 watts, what is the decibel voltage gain?

88. Limiting Magnitude of a Telescope A telescope is limited in its usefulness by the brightness of the star that it is aimed at and by the diameter of its lens. One measure of a star's brightness is its *magnitude;* the dimmer the star, the larger its magnitude. A formula for the limiting magnitude L of a telescope, that is, the magnitude of the dimmest star that it can be used to view, is given by

$$L = 9 + 5.1\log d$$

where d is the diameter (in inches) of the lens.
(a) What is the limiting magnitude of a 3.5-inch telescope?
(b) What diameter is required to view a star of magnitude 14?

89. Salvage Value The number of years n for a piece of machinery to depreciate to a known salvage value can be found using the formula

$$n = \frac{\log s - \log i}{\log(1-d)}$$

where s is the salvage value of the machinery, i is its initial value, and d is the annual rate of depreciation.
(a) How many years will it take for a piece of machinery to decline in value from \$90,000 to \$10,000 if the annual rate of depreciation is 0.20 (20%)?
(b) How many years will it take for a piece of machinery to lose half of its value if the annual rate of depreciation is 15%?

90. Funding a College Education A child's grandparents purchase a \$10,000 bond fund that matures in 18 years to be used for her college education. The bond fund pays 4% interest compounded semiannually. How much will the bond fund be worth at maturity? What is the effective rate of interest? How long will it take the bond to double in value under these terms?

91. Funding a College Education A child's grandparents wish to purchase a bond that matures in 18 years to be used for her college education. The bond pays 4% interest compounded semiannually. How much should they pay so that the bond will be worth \$85,000 at maturity?

92. Funding an IRA First Colonial Bankshares Corporation advertised the following IRA investment plans.

Target IRA Plans

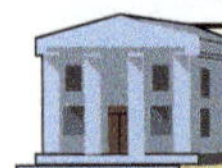

For each \$5000 Maturity Value Desired

Deposit:	At a Term of:
\$620.17	20 Years
\$1045.02	15 Years
\$1760.92	10 Years
\$2967.26	5 Years

(a) Assuming continuous compounding, what annual rate of interest did they offer?
(b) First Colonial Bankshares claims that \$4000 invested today will have a value of over \$32,000 in 20 years. Use the answer found in part (a) to find the actual value of \$4000 in 20 years. Assume continuous compounding.

93. Estimating the Date That a Prehistoric Man Died The bones of a prehistoric man found in the desert of New Mexico contain approximately 5% of the original amount of carbon 14. If the half-life of carbon 14 is 5600 years, approximately how long ago did the man die?

94. Temperature of a Skillet A skillet is removed from an oven whose temperature is 450°F and placed in a room whose temperature is 70°F. After 5 minutes, the temperature of the skillet is 400°F. How long will it be until its temperature is 150°F?

95. World Population The annual growth rate of the world's population in 2005 was $k = 1.15\% = 0.0115$. The population of the world in 2005 was 6,451,058,790. Letting $t = 0$ represent 2005, use the uninhibited growth model to predict the world's population in the year 2015.

Source: *U.S. Census Bureau*

96. Radioactive Decay The half-life of radioactive cobalt is 5.27 years. If 100 grams of radioactive cobalt is present now, how much will be present in 20 years? In 40 years?

97. Federal Deficit In fiscal year 2005, the federal deficit was \$319 billion. At that time, 10-year treasury notes were paying 4.25% interest per annum. If the federal government financed this deficit through 10-year notes, how much would it have to pay back in 2015?

Source: *U.S. Treasury Department*

98. Logistic Growth The logistic growth model

$$P(t) = \frac{0.8}{1 + 1.67e^{-0.16t}}$$

represents the proportion of new cars with a global positioning system (GPS). Let $t = 0$ represent 2006, $t = 1$ represent 2007, and so on.

(a) What proportion of new cars in 2006 had a GPS?
(b) Determine the maximum proportion of new cars that have a GPS.
(c) Using a graphing utility, graph $P = P(t)$.
(d) When will 75% of new cars have a GPS?

99. CBL Experiment The following data were collected by placing a temperature probe in a portable heater, removing the probe, and then recording temperature over time.

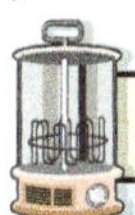

Time (sec.)	Temperature (°F)
0	165.07
1	164.77
2	163.99
3	163.22
4	162.82
5	161.96
6	161.20
7	160.45
8	159.35
9	158.61
10	157.89
11	156.83
12	156.11
13	155.08
14	154.40
15	153.72

According to Newton's Law of Cooling, these data should follow an exponential model.

(a) Using a graphing utility, draw a scatter diagram for the data.
(b) Using a graphing utility, build an exponential model from the data.
(c) Graph the exponential function found in part (b) on the scatter diagram.
(d) Predict how long it will take for the probe to reach a temperature of 110°F.

100. Wind Chill Factor The following data represent the wind speed (mph) and wind chill factor at an air temperature of 15°F.

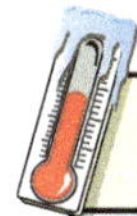

Wind Speed (mph)	Wind Chill Factor (°F)
5	7
10	3
15	0
20	−2
25	−4
30	−5
35	−7

Source: U.S. National Weather Service

(a) Using a graphing utility, draw a scatter diagram with wind speed as the independent variable.
(b) Using a graphing utility, build a logarithmic model from the data.
(c) Using a graphing utility, draw the logarithmic function found in part (b) on the scatter diagram.
(d) Use the function found in part (b) to predict the wind chill factor if the air temperature is 15°F and the wind speed is 23 mph.

101. Spreading of a Disease Jack and Diane live in a small town of 50 people. Unfortunately, both Jack and Diane have a cold. Those who come in contact with someone who has this cold will themselves catch the cold. The following data represent the number of people in the small town who have caught the cold after t days.

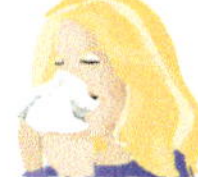

Days, t	Number of People with Cold, C
0	2
1	4
2	8
3	14
4	22
5	30
6	37
7	42
8	44

(a) Using a graphing utility, draw a scatter diagram of the data. Comment on the type of relation that appears to exist between the days and number of people with a cold.
(b) Using a graphing utility, build a logistic model from the data.
(c) Graph the function found in part (b) on the scatter diagram.
(d) According to the function found in part (b), what is the maximum number of people who will catch the cold? In reality, what is the maximum number of people who could catch the cold?
(e) Sometime between the second and third day, 10 people in the town had a cold. According to the model found in part (b), when did 10 people have a cold?
(f) How long will it take for 46 people to catch the cold?

CHAPTER TEST

1. Given $f(x) = \frac{x+2}{x-2}$ and $g(x) = 2x + 5$, find:
   (a) $f \circ g$ and state its domain
   (b) $(g \circ f)(-2)$
   (c) $(f \circ g)(-2)$

2. Determine whether the function is one-to-one.
   (a) $y = 4x^2 + 3$
   (b) $y = \sqrt{x+3} - 5$

3. Find the inverse of $f(x) = \frac{2}{3x-5}$ and check your answer. State the domain and the range of f and f^{-1}.

4. If the point $(3, -5)$ is on the graph of a one-to-one function f, what point must be on the graph of f^{-1}?

In Problems 5–7, solve each equation algebraically.

5. $3^x = 243$ 6. $\log_b 16 = 2$ 7. $\log_5 x = 4$

In Problems 8–11, use a calculator to evaluate each expression. Round your answer to three decimal places.

8. $e^3 + 2$ 9. $\log 20$

10. $\log_3 21$ 11. $\ln 133$

In Problems 12 and 13, use the given function f to:
   (a) Find the domain of f.
   (b) Graph f.
   (c) From the graph, determine the range and any asymptotes of f.
   (d) Find f^{-1}, the inverse of f.
   (e) Use f^{-1} to find the range of f.
   (f) Graph f^{-1}.

12. $f(x) = 4^{x+1} - 2$

13. $f(x) = 1 - \log_5(x - 2)$

In Problems 14–19, solve each equation.

14. $5^{x+2} = 125$ 15. $\log(x + 9) = 2$

16. $8 - 2e^{-x} = 4$ 17. $\log(x^2 + 3) = \log(x + 6)$

18. $7^{x+3} = e^x$ 19. $\log_2(x - 4) + \log_2(x + 4) = 3$

20. Write $\log_2\left(\frac{4x^3}{x^2 - 3x - 18}\right)$ as the sum and/or difference of logarithms. Express powers as factors.

21. A 50-mg sample of a radioactive substance decays to 34 mg after 30 days. How long will it take for there to be 2 mg remaining?

22. (a) If \$1000 is invested at 5% compounded monthly, how much is there after 8 months?
    (b) If you want to have \$1000 in 9 months, how much do you need to place in a savings account now that pays 5% compounded quarterly?
    (c) How long does it take to double your money if you can invest it at 6% compounded annually?

23. The decibel level, D, of sound is given by the equation $D = 10\log\left(\frac{I}{I_0}\right)$, where I is the intensity of the sound and $I_0 = 10^{-12}$ watt per square meter.
    (a) If the shout of a single person measures 80 decibels, how loud will the sound be if two people shout at the same time? That is, how loud would the sound be if the intensity doubled?
    (b) The pain threshold for sound is 125 decibels. If the Athens Olympic Stadium 2004 (Olympiako Stadio Athinas 'Spyros Louis') can seat 74,400 people, how many people in the crowd need to shout at the same time for the resulting sound level to meet or exceed the pain threshold? (Ignore any possible sound dampening.)

CUMULATIVE REVIEW

1. Is the following graph the graph of a function? If it is, is the function one-to-one?

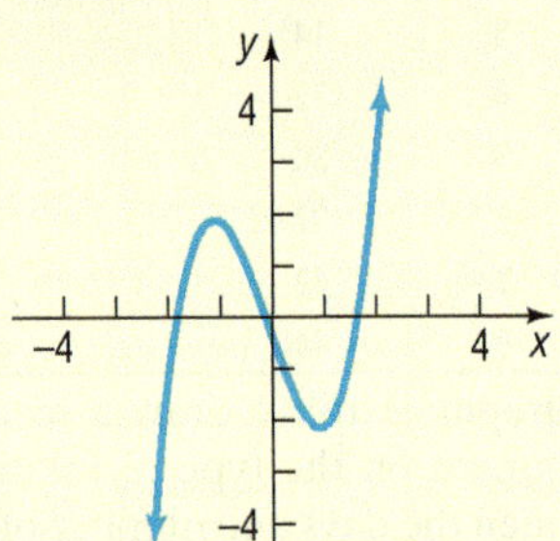

2. For the function $f(x) = 2x^2 - 3x + 1$, find the following:
   (a) $f(3)$ (b) $f(-x)$ (c) $f(x + h)$

3. Determine which of the following points are on the graph of $x^2 + y^2 = 1$.
   (a) $\left(\frac{1}{2}, \frac{1}{2}\right)$ (b) $\left(\frac{1}{2}, \frac{\sqrt{3}}{2}\right)$

4. Solve the equation $3(x - 2) = 4(x + 5)$.

5. Graph the line $2x - 4y = 16$.

6. (a) Graph the quadratic function $f(x) = -x^2 + 2x - 3$ by determining whether its graph opens up or down and by finding its vertex, axis of symmetry, y-intercept, and x-intercept(s), if any.
   (b) Solve $f(x) \leq 0$.

7. Determine the quadratic function whose graph is given in the figure.

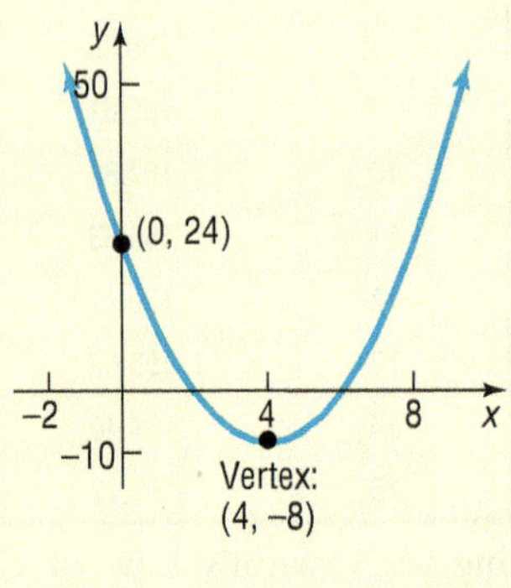

8. Graph $f(x) = 3(x + 1)^3 - 2$ using transformations.

9. Given that $f(x) = x^2 + 2$ and $g(x) = \dfrac{2}{x-3}$, find $f(g(x))$ and state its domain. What is $f(g(5))$?

10. For the polynomial function $f(x) = 4x^3 + 9x^2 - 30x - 8$:
(a) Find the real zeros of f.
(b) Determine the intercepts of the graph of f.
(c) Use a graphing utility to approximate the local maxima and local minima.
(d) Draw a complete graph of f. Be sure to label the intercepts and turning points.

11. For the function $g(x) = 3^x + 2$:
(a) Graph g using transformations. State the domain, range, and horizontal asymptote of g.
(b) Determine the inverse of g. State the domain, range, and vertical asymptote of g^{-1}.
(c) On the same graph as g, graph g^{-1}.

12. Solve the equation $4^{x-3} = 8^{2x}$.

13. Solve the equation: $\log_3(x + 1) + \log_3(2x - 3) = \log_9 9$

14. Suppose that $f(x) = \log_3(x + 2)$. Solve:
(a) $f(x) = 0$ (b) $f(x) > 0$ (c) $f(x) = 3$

15. **Data Analysis** The following data represent the percent of all drivers by age that have been stopped by the police for any reason within the past year. The median age represents the midpoint of the upper and lower limit for the age range.

Age Range	Median Age, x	Percentage Stopped, y
16–19	17.5	18.2
20–29	24.5	16.8
30–39	34.5	11.3
40–49	44.5	9.4
50–59	54.5	7.7
≥60	69.5	3.8

(a) Using your graphing utility, draw a scatter diagram of the data treating median age, x, as the independent variable.
(b) Determine a model that you feel best describes the relation between median age and percentage stopped. You may choose from among linear, quadratic, cubic, exponential, logarithmic, or logistic models.
(c) Provide a justification for the model that you selected in part (b).

CHAPTER PROJECTS

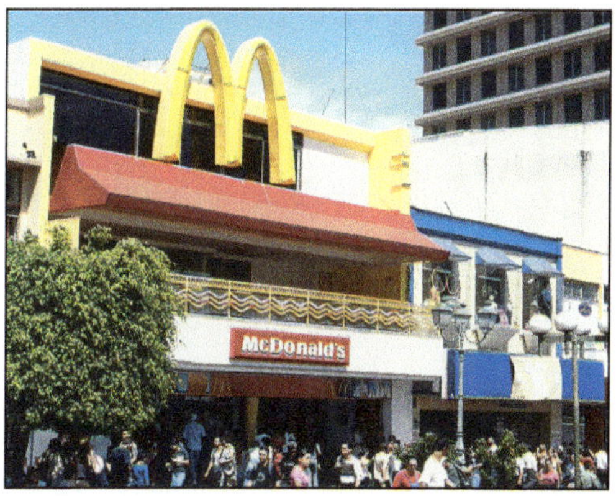

I. **Hot Coffee** A fast-food restaurant wants a special container to hold coffee. The restaurant wishes the container to quickly cool the coffee from 200° to 130°F and keep the liquid between 110° and 130°F as long as possible. The restaurant has three containers to select from.

1. The CentiKeeper Company has a container that reduces the temperature of a liquid from 200° to 100°F in 30 minutes by maintaining a constant temperature of 70°F.
2. The TempControl Company has a container that reduces the temperature of a liquid from 200° to 110°F in 25 minutes by maintaining a constant temperature of 60°F.
3. The Hot'n'Cold Company has a container that reduces the temperature of a liquid from 200° to 120°F in 20 minutes by maintaining a constant temperature of 65°F.

You need to recommend which container the restaurant should purchase.

(a) Use Newton's Law of Cooling to find a function relating the temperature of the liquid over time for each container.
(b) How long does it take each container to lower the coffee temperature from 200° to 130°F?
(c) How long will the coffee temperature remain between 110° and 130°F? This temperature is considered the optimal drinking temperature.
(d) Graph each function using a graphing utility.
(e) Which company would you recommend to the restaurant? Why?
(f) How might the cost of the container affect your decision?

The following projects are available on the Instructor's Resource Center (IRC):

II. **Project at Motorola** ***Thermal Fatigue of Solder Connections*** Product reliability is a major concern of a manufacturer. Here a logarithmic transformation is used to simplify the analysis of a cell phone's ability to withstand temperature change.

III. **Depreciation of a New Car** Resale value is a factor to consider when purchasing a car, and exponential functions provide a way to compare the depreciation rates of different makes and models.

Analytic Geometry

7

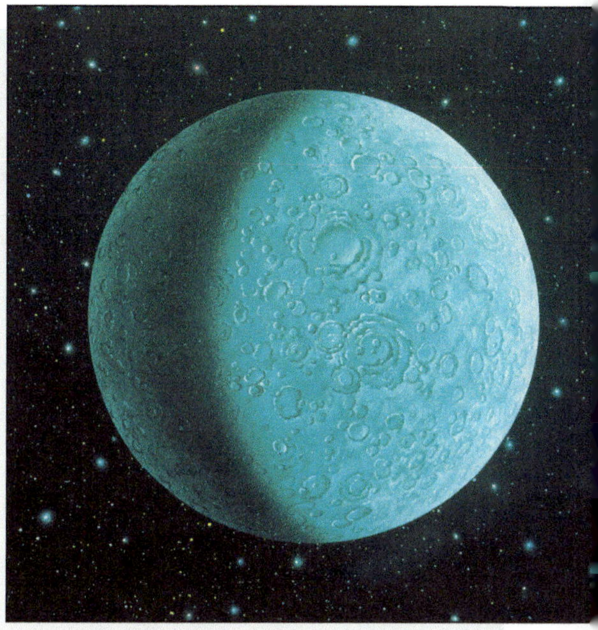

Pluto's Unusual Orbit

Pluto is about 39 times as far from the Sun as Earth is. Its average distance from the Sun is about 3,647,240,000 miles (5,869,660,000 kilometers). Pluto travels around the Sun in an elliptical (oval-shaped) orbit. At some point in its orbit, it comes closer to the Sun than Neptune, the outermost planet. It stays inside Neptune's orbit for about 20 Earth-years. This event occurs every 248 Earth-years, which is about the same number of Earth-years it takes Pluto to travel once around the Sun. Pluto entered Neptune's orbit on January 23, 1979, and remained there until February 11, 1999.

Source: Spinrad, Hyron. "Pluto." World Book Online Reference Center. 2004. World Book, Inc. http://www.worldbookonline.com/wb/Article?id=ar435500. Used with permission of World Book, Inc.

—See Chapter Project I—

A Look Back

In Chapter 1, we introduced rectangular coordinates and showed how geometry problems can be solved algebraically. In Section 2.3, we defined a circle geometrically and then used the distance formula and rectangular coordinates to obtain an equation for a circle.

A Look Ahead

In this chapter we give geometric definitions for the conics and use the distance formula and rectangular coordinates to obtain their equations. We will learn that a circle is a conic.

Historically, Apollonius (200 BC) was among the first to study *conics* and discover some of their interesting properties. Today, conics are still studied because of their many uses. *Paraboloids of revolution* (parabolas rotated about their axes of symmetry) are used as signal collectors (the satellite dishes used with radar and dish TV, for example), as solar energy collectors, and as reflectors (telescopes, light projection, and so on). The planets circle the Sun in approximately *elliptical* orbits. Elliptical surfaces can be used to reflect signals such as light and sound from one place to another. A third conic, the *hyperbola,* can be used to determine the location of lightning strikes.

The Greeks used the methods of Euclidean geometry to study conics. However, we shall use the more powerful methods of analytic geometry, bringing to bear both algebra and geometry, for our study of conics.

Outline

7.1 Conics

OBJECTIVE 1 Know the Names of the Conics (p. 508)

1 Know the Names of the Conics

The word *conic* derives from the word *cone,* which is a geometric figure that can be constructed in the following way: Let a and g be two distinct lines that intersect at a point V. Keep the line a fixed. Now rotate the line g about a while maintaining the same angle between a and g. The collection of points swept out (generated) by the line g is called a **(right circular) cone.** See Figure 1. The fixed line a is called the **axis** of the cone; the point V is its **vertex;** the lines that pass through V and make the same angle with a as g are **generators** of the cone. Each generator is a line that lies entirely on the cone. The cone consists of two parts, called **nappes,** that intersect at the vertex.

Figure 1

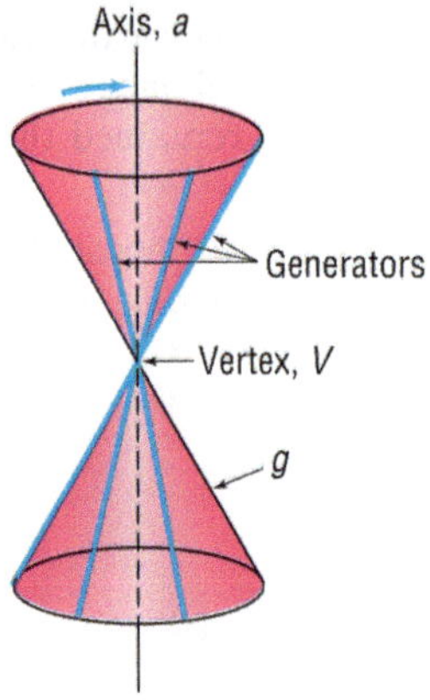

Conics, an abbreviation for **conic sections,** are curves that result from the intersection of a right circular cone and a plane. The conics we shall study arise when the plane does not contain the vertex, as shown in Figure 2. These conics are **circles** when the plane is perpendicular to the axis of the cone and intersects each generator; **ellipses** when the plane is tilted slightly so that it intersects each generator, but intersects only one nappe of the cone; **parabolas** when the plane is tilted farther so that it is parallel to one (and only one) generator and intersects only one nappe of the cone; and **hyperbolas** when the plane intersects both nappes.

If the plane does contain the vertex, the intersection of the plane and the cone is a point, a line, or a pair of intersecting lines. These are usually called **degenerate conics.**

Figure 2

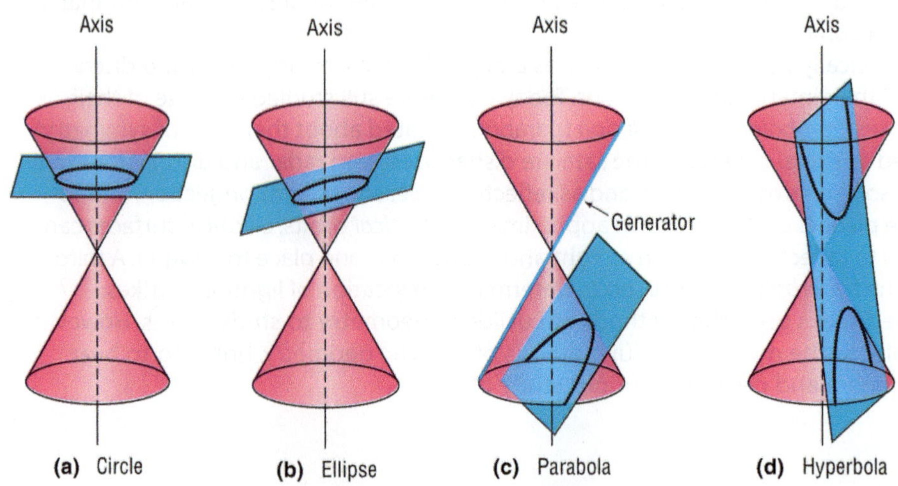

7.2 The Parabola

PREPARING FOR THIS SECTION *Before getting started, review the following:*

- Distance Formula (Section 1.1, p. 89)
- Symmetry (Section 2.1, pp. 171–173)
- Square Root Method (Section 1.3, pp. 114–115)
- Completing the Square (Section 1.3, pp. 115–116)
- Graphing Techniques: Transformations (Section 3.5, pp. 256–264)

Now Work the 'Are You Prepared?' problems on page 516.

OBJECTIVES
1. Analyze Parabolas with Vertex at the Origin (p. 509)
2. Analyze Parabolas with Vertex at (h, k) (p. 513)
3. Solve Applied Problems Involving Parabolas (p. 514)

We stated earlier (Section 4.3) that the graph of a quadratic function is a parabola. In this section, we begin with a geometric definition of a parabola and use it to obtain an equation.

DEFINITION

A **parabola** is the collection of all points P in the plane that are the same distance from a fixed point F as they are from a fixed line D. The point F is called the **focus** of the parabola, and the line D is its **directrix.** As a result, a parabola is the set of points P for which

$$d(F, P) = d(P, D) \quad \textbf{(1)}$$

Figure 3

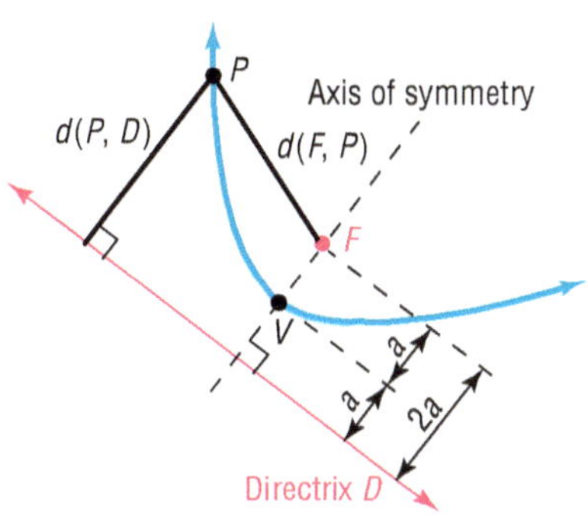

Figure 3 shows a parabola (in blue). The line through the focus F and perpendicular to the directrix D is called the **axis of symmetry** of the parabola. The point of intersection of the parabola with its axis of symmetry is called the **vertex** V.

Because the vertex V lies on the parabola, it must satisfy equation (1): $d(F, V) = d(V, D)$. The vertex is midway between the focus and the directrix. We shall let a equal the distance $d(F, V)$ from F to V. Now we are ready to derive an equation for a parabola. To do this, we use a rectangular system of coordinates positioned so that the vertex V, focus F, and directrix D of the parabola are conveniently located.

1 Analyze Parabolas with Vertex at the Origin

If we choose to locate the vertex V at the origin $(0, 0)$, we can conveniently position the focus F on either the x-axis or the y-axis. First, we consider the case where the focus F is on the positive x-axis, as shown in Figure 4. Because the distance from F to V is a, the coordinates of F will be $(a, 0)$ with $a > 0$. Similarly, because the distance from V to the directrix D is also a and, because D must be perpendicular to the x-axis (since the x-axis is the axis of symmetry), the equation of the directrix D must be $x = -a$.

Figure 4

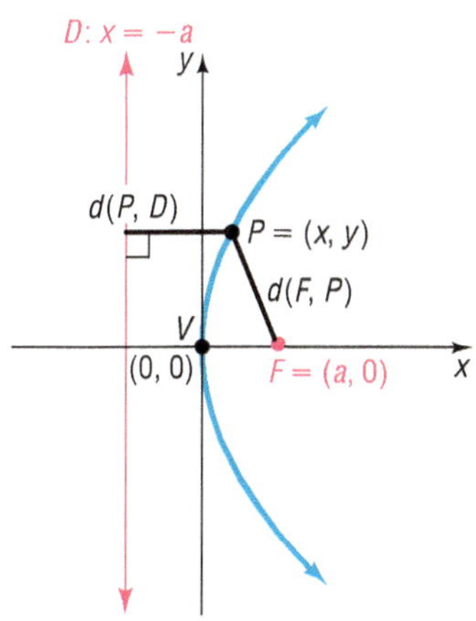

Now, if $P = (x, y)$ is any point on the parabola, P must obey equation (1):

$$d(F, P) = d(P, D)$$

So we have

$$\sqrt{(x - a)^2 + (y - 0)^2} = |x + a| \quad \text{Use the Distance Formula.}$$

$$(x - a)^2 + y^2 = (x + a)^2 \quad \text{Square both sides.}$$

$$x^2 - 2ax + a^2 + y^2 = x^2 + 2ax + a^2 \quad \text{Remove parentheses.}$$

$$y^2 = 4ax \quad \text{Simplify.}$$

THEOREM

Equation of a Parabola: Vertex at (0, 0), Focus at (a, 0), $a > 0$

The equation of a parabola with vertex at $(0, 0)$, focus at $(a, 0)$, and directrix $x = -a$, $a > 0$, is

$$y^2 = 4ax \quad (2)$$

EXAMPLE 1 Finding the Equation of a Parabola and Graphing It

Find an equation of the parabola with vertex at $(0, 0)$ and focus at $(3, 0)$. Graph the equation.

Solution The distance from the vertex $(0, 0)$ to the focus $(3, 0)$ is $a = 3$. Based on equation (2), the equation of this parabola is

$$y^2 = 4ax$$
$$y^2 = 12x \quad a = 3$$

Figure 5

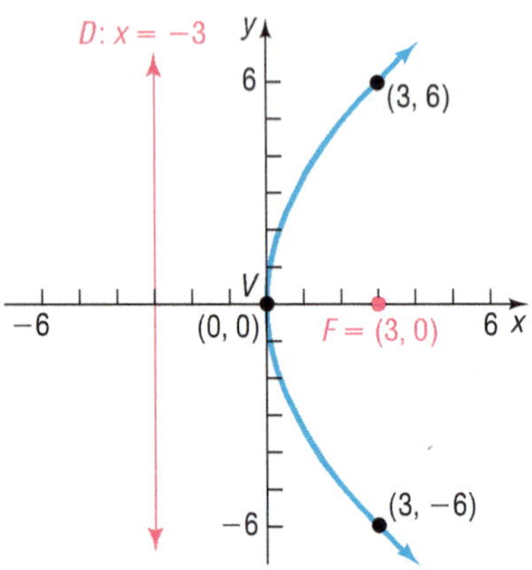

To graph this parabola, it is helpful to plot the points on the graph directly above or below the focus. To locate these two points, we let $x = 3$. Then

$$y^2 = 12x = 12(3) = 36$$
$$y = \pm 6 \quad \text{Solve for } y.$$

The points on the parabola directly above or below the focus are $(3, 6)$ and $(3, -6)$. These points help in graphing the parabola because they determine the "opening." See Figure 5.

In general, the points on a parabola $y^2 = 4ax$ that lie above and below the focus $(a, 0)$ are each at a distance $2a$ from the focus. This follows from the fact that if $x = a$ then $y^2 = 4ax = 4a^2$, so $y = \pm 2a$. The line segment joining these two points is called the **latus rectum;** its length is $4a$.

Now Work PROBLEM 19

EXAMPLE 2 Graphing a Parabola Using a Graphing Utility

Graph the parabola $y^2 = 12x$.

Solution To graph the parabola $y^2 = 12x$, we need to graph the two functions $Y_1 = \sqrt{12x}$ and $Y_2 = -\sqrt{12x}$ on a square screen. Figure 6 shows the graph of $y^2 = 12x$. Notice that the graph fails the vertical line test, so $y^2 = 12x$ is not a function.

Figure 6

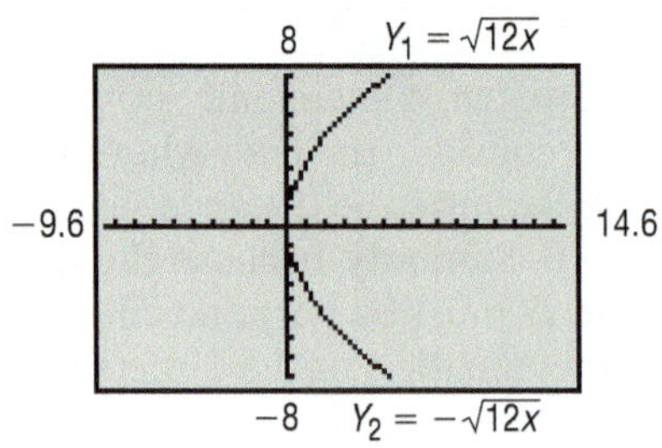

By reversing the steps we used to obtain equation (2), it follows that the graph of an equation of the form of equation (2), $y^2 = 4ax$, is a parabola; its vertex is at $(0, 0)$, its focus is at $(a, 0)$, its directrix is the line $x = -a$, and its axis of symmetry is the x-axis.

For the remainder of this section, the direction **"Analyze the equation"** will mean to find the vertex, focus, and directrix of the parabola and graph it.

EXAMPLE 3 Analyzing the Equation of a Parabola

Analyze the equation: $y^2 = 8x$

Solution The equation $y^2 = 8x$ is of the form $y^2 = 4ax$, where $4a = 8$, so $a = 2$. Consequently, the graph of the equation is a parabola with vertex at $(0, 0)$ and focus on the positive x-axis at $(2, 0)$. The directrix is the vertical line $x = -2$. The two points defining the latus rectum are obtained by letting $x = 2$. Then $y^2 = 16$, so $y = \pm 4$. See Figure 7(a) for the graph drawn by hand. Figure 7(b) shows the graph obtained using a graphing utility.

Figure 7

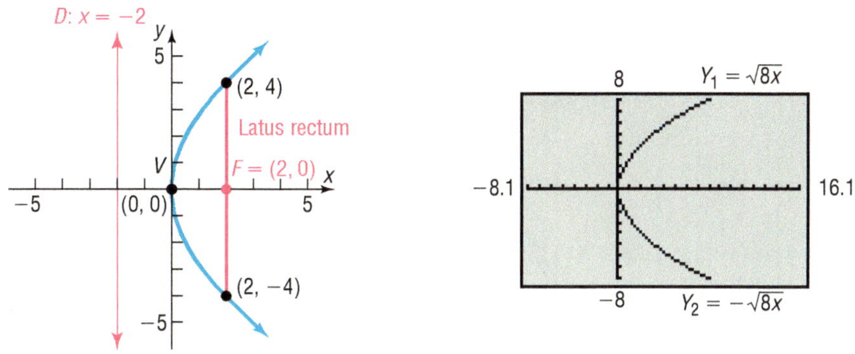

(a) (b)

Recall that we arrived at equation (2) after placing the focus on the positive x-axis. If the focus is placed on the negative x-axis, positive y-axis, or negative y-axis, a different form of the equation for the parabola results. The four forms of the equation of a parabola with vertex at $(0, 0)$ and focus on a coordinate axis a distance a from $(0, 0)$ are given in Table 1, and their graphs are given in Figure 8. Notice that each graph is symmetric with respect to its axis of symmetry.

Table 1

EQUATIONS OF A PARABOLA VERTEX AT (0,0); FOCUS ON AN AXIS; $a > 0$				
Vertex	**Focus**	**Directrix**	**Equation**	**Description**
$(0, 0)$	$(a, 0)$	$x = -a$	$y^2 = 4ax$	Parabola, axis of symmetry is the x-axis, opens right
$(0, 0)$	$(-a, 0)$	$x = a$	$y^2 = -4ax$	Parabola, axis of symmetry is the x-axis, opens left
$(0, 0)$	$(0, a)$	$y = -a$	$x^2 = 4ay$	Parabola, axis of symmetry is the y-axis, opens up
$(0, 0)$	$(0, -a)$	$y = a$	$x^2 = -4ay$	Parabola, axis of symmetry is the y-axis, opens down

Figure 8

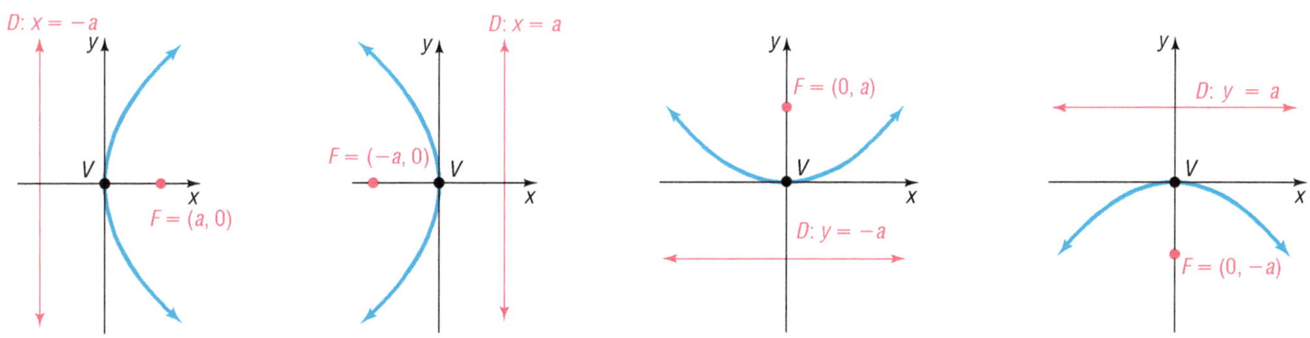

(a) $y^2 = 4ax$ (b) $y^2 = -4ax$ (c) $x^2 = 4ay$ (d) $x^2 = -4ay$

EXAMPLE 4 Analyzing the Equation of a Parabola

Analyze the equation: $x^2 = -12y$

Solution The equation $x^2 = -12y$ is of the form $x^2 = -4ay$, with $a = 3$. Consequently, the graph of the equation is a parabola with vertex at $(0, 0)$, focus at $(0, -3)$ and directrix the line $y = 3$. The parabola opens down, and its axis of symmetry is the y-axis. To obtain the points defining the latus rectum, let $y = -3$. Then $x^2 = 36$, so $x = \pm 6$. See Figure 9(a) for the graph drawn by hand. Figure 9(b) shows the graph using a graphing utility.

Figure 9

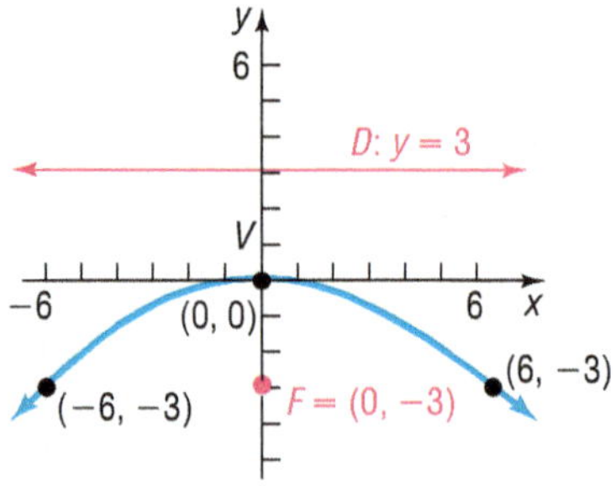

(a)

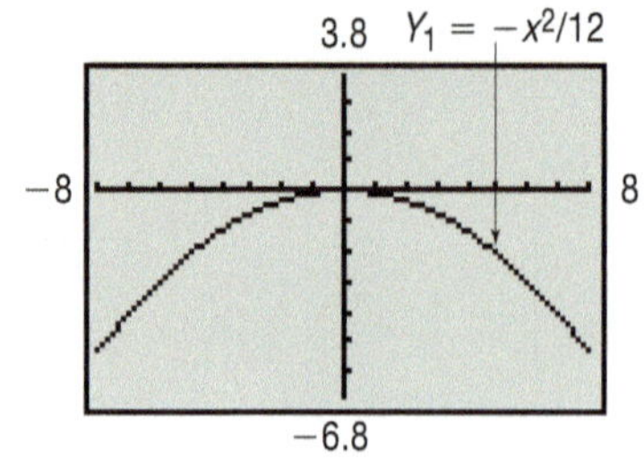

(b)

Now Work PROBLEM 39

EXAMPLE 5 **Finding the Equation of a Parabola**

Find the equation of the parabola with focus at (0, 4) and directrix the line $y = -4$. Graph the equation.

Solution A parabola whose focus is at (0, 4) and whose directrix is the horizontal line $y = -4$ will have its vertex at (0, 0). (Do you see why? The vertex is midway between the focus and the directrix.) Since the focus is on the positive y-axis at (0, 4), the equation of this parabola is of the form $x^2 = 4ay$, with $a = 4$; that is,

$$x^2 = 4ay = 4(4)y = 16y$$

$a = 4$

Letting $y = 4$, we find $x^2 = 64$, so $x = \pm 8$. The points (8, 4) and (–8, 4) determine the latus rectum. Figure 10 shows the graph of $x^2 = 16y$.

Figure 10

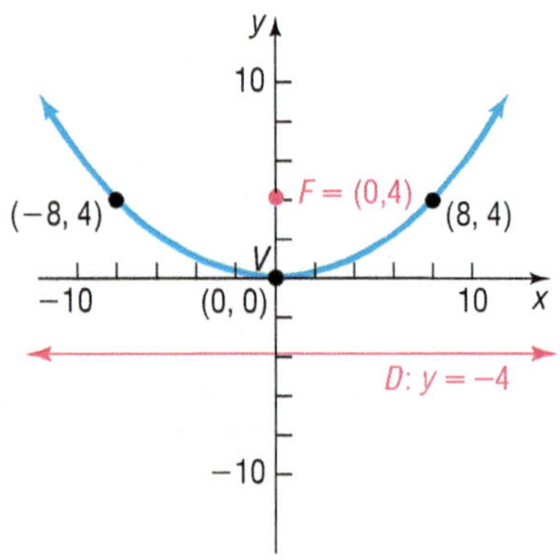

Check: Verify the graph drawn in Figure 10 by graphing $Y_1 = \frac{x^2}{16}$ using a graphing utility.

EXAMPLE 6 **Finding the Equation of a Parabola**

Find the equation of a parabola with vertex at (0, 0) if its axis of symmetry is the x-axis and its graph contains the point $\left(-\frac{1}{2}, 2\right)$. Find its focus and directrix, and graph the equation.

Solution The vertex is at the origin, the axis of symmetry is the x-axis, and the graph contains a point in the second quadrant, so the parabola opens to the left. We see from Table 1 that the form of the equation is

$$y^2 = -4ax$$

Because the point $\left(-\frac{1}{2}, 2\right)$ is on the parabola, the coordinates $x = -\frac{1}{2}$, $y = 2$ must satisfy the equation $y^2 = -4ax$. Substituting $x = -\frac{1}{2}$ and $y = 2$ into the equation, we find that

$$2^2 = -4a\left(-\frac{1}{2}\right) \qquad y^2 = -4ax;\ x = -\frac{1}{2},\ y = 2$$

$$a = 2$$

The equation of the parabola is

$$y^2 = -4(2)x = -8x$$

Figure 11

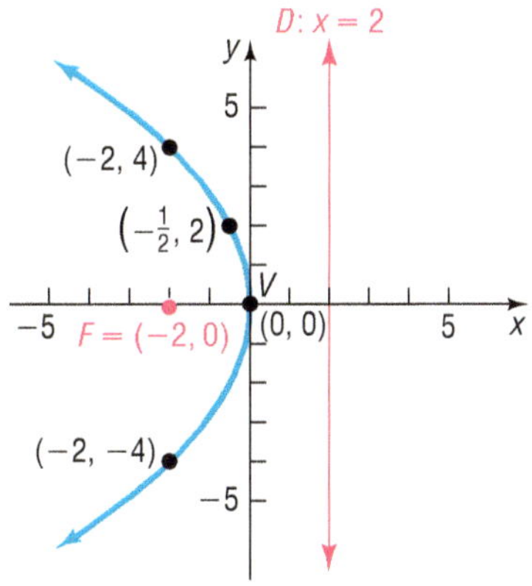

The focus is at $(-2, 0)$ and the directrix is the line $x = 2$. Letting $x = -2$, we find $y^2 = 16$, so $y = \pm 4$. The points $(-2, 4)$ and $(-2, -4)$ define the latus rectum. See Figure 11.

Now Work PROBLEM 27

2 Analyze Parabolas with Vertex at (h, k)

If a parabola with vertex at the origin and axis of symmetry along a coordinate axis is shifted horizontally h units and then vertically k units, the result is a parabola with vertex at (h, k) and axis of symmetry parallel to a coordinate axis. The equations of such parabolas have the same forms as those in Table 1, but with x replaced by $x - h$ (the horizontal shift) and y replaced by $y - k$ (the vertical shift). Table 2 gives the forms of the equations of such parabolas. Figures 12(a)–(d) illustrate the graphs for $h > 0, k > 0$.

Table 2

PARABOLAS WITH VERTEX AT (h, k); AXIS OF SYMMETRY PARALLEL TO A COORDINATE AXIS; $a > 0$				
Vertex	**Focus**	**Directrix**	**Equation**	**Description**
(h, k)	$(h + a, k)$	$x = h - a$	$(y - k)^2 = 4a(x - h)$	Parabola, axis of symmetry parallel to x-axis, opens right
(h, k)	$(h - a, k)$	$x = h + a$	$(y - k)^2 = -4a(x - h)$	Parabola, axis of symmetry parallel to x-axis, opens left
(h, k)	$(h, k + a)$	$y = k - a$	$(x - h)^2 = 4a(y - k)$	Parabola, axis of symmetry parallel to y-axis, opens up
(h, k)	$(h, k - a)$	$y = k + a$	$(x - h)^2 = -4a(y - k)$	Parabola, axis of symmetry parallel to y-axis, opens down

Figure 12

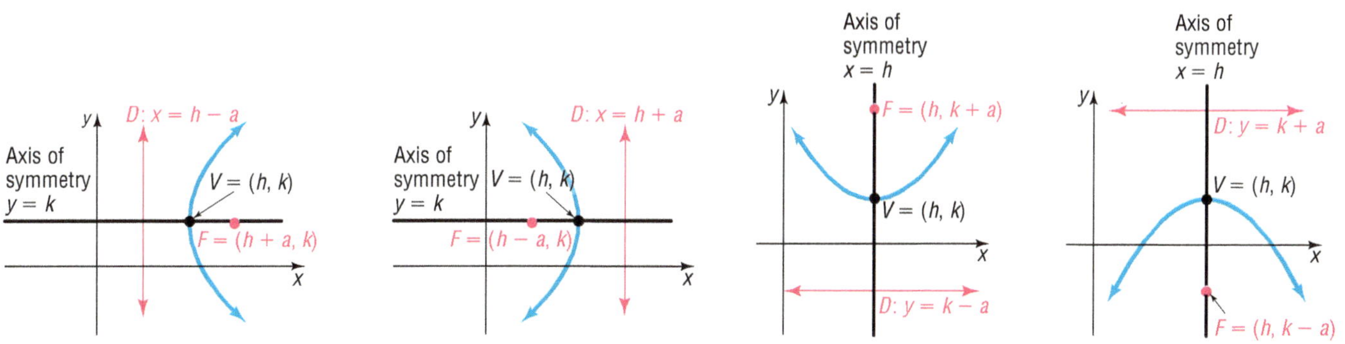

(a) $(y - k)^2 = 4a(x - h)$ (b) $(y - k)^2 = -4a(x - h)$ (c) $(x - h)^2 = 4a(y - k)$ (d) $(x - h)^2 = -4a(y - k)$

EXAMPLE 7 Finding the Equation of a Parabola, Vertex Not at Origin

Find an equation of the parabola with vertex at $(-2, 3)$ and focus at $(0, 3)$. Graph the equation.

Solution The vertex $(-2, 3)$ and focus $(0, 3)$ both lie on the horizontal line $y = 3$ (the axis of symmetry). The distance a from the vertex $(-2, 3)$ to the focus $(0, 3)$ is $a = 2$. Also, because the focus lies to the right of the vertex, we know that the parabola opens to the right. Consequently, the form of the equation is

$$(y - k)^2 = 4a(x - h)$$

Figure 13

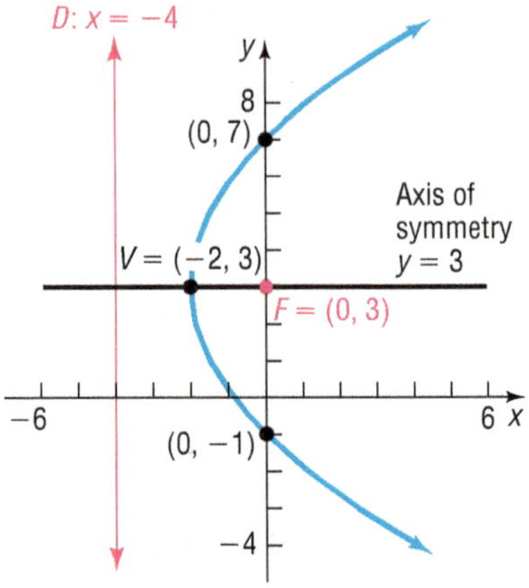

where $(h, k) = (-2, 3)$ and $a = 2$. Therefore, the equation is

$$(y - 3)^2 = 4 \cdot 2[x - (-2)]$$

$$(y - 3)^2 = 8(x + 2)$$

If $x = 0$, then $(y - 3)^2 = 16$. Then $y - 3 = \pm 4$, so $y = -1$ or $y = 7$. The points $(0, -1)$ and $(0, 7)$ define the latus rectum; the line $x = -4$ is the directrix. See Figure 13.

Now Work PROBLEM 29

EXAMPLE 8 **Using a Graphing Utility to Graph a Parabola, Vertex Not at Origin**

Using a graphing utility, graph the equation $(y - 3)^2 = 8(x + 2)$.

Solution First, we must solve the equation for y.

$$(y - 3)^2 = 8(x + 2)$$

$$y - 3 = \pm\sqrt{8(x + 2)} \quad \text{Use the Square Root Method.}$$

$$y = 3 \pm \sqrt{8(x + 2)} \quad \text{Add 3 to both sides.}$$

Figure 14

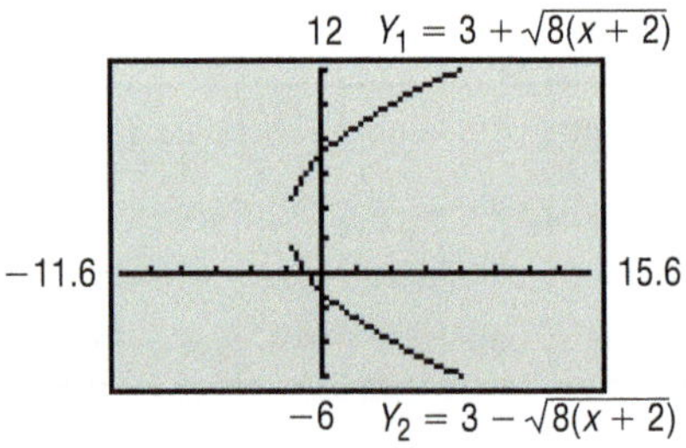

Figure 14 shows the graphs of the equations $Y_1 = 3 + \sqrt{8(x + 2)}$ and $Y_2 = 3 - \sqrt{8(x + 2)}$.

Polynomial equations define parabolas whenever they involve two variables that are quadratic in one variable and linear in the other. To analyze this type of equation, we first complete the square of the variable that is quadratic.

EXAMPLE 9 **Analyzing the Equation of a Parabola**

Analyze the equation: $x^2 + 4x - 4y = 0$

Solution To analyze the equation $x^2 + 4x - 4y = 0$, we complete the square involving the variable x.

$$x^2 + 4x - 4y = 0$$

$$x^2 + 4x = 4y \quad \text{Isolate the terms involving } x \text{ on the left side.}$$

$$x^2 + 4x + 4 = 4y + 4 \quad \text{Complete the square on the left side.}$$

$$(x + 2)^2 = 4(y + 1) \quad \text{Factor.}$$

Figure 15

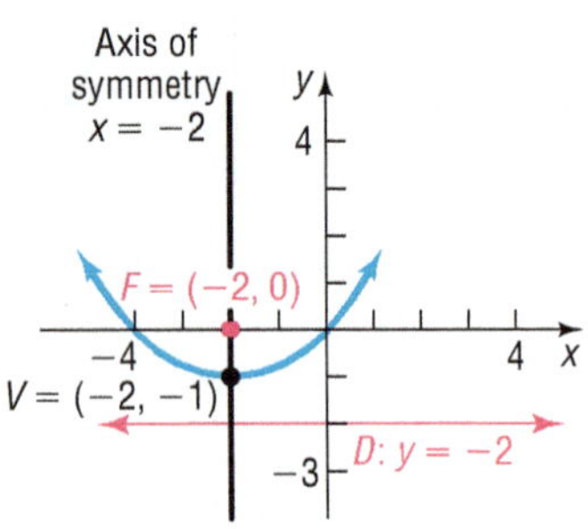

This equation is of the form $(x - h)^2 = 4a(y - k)$, with $h = -2$, $k = -1$, and $a = 1$. The graph is a parabola with vertex at $(h, k) = (-2, -1)$ that opens up. The focus is $a = 1$ unit above the vertex at $(-2, 0)$, and the directrix is the line $y = -2$. See Figure 15.

Now Work PROBLEM 47

3 Solve Applied Problems Involving Parabolas

Parabolas find their way into many applications. For example, as we discussed in Section 4.4, suspension bridges have cables in the shape of a parabola. Another property of parabolas that is used in applications is their reflecting property.

Suppose that a mirror is shaped like a **paraboloid of revolution,** a surface formed by rotating a parabola about its axis of symmetry. If a light (or any other emitting source) is placed at the focus of the parabola, all the rays emanating from the light will reflect off the mirror in lines parallel to the axis of symmetry. This principle is used in the design of searchlights, flashlights, certain automobile headlights, and other such devices. See Figure 16.

Conversely, suppose that rays of light (or other signals) emanate from a distant source so that they are essentially parallel. When these rays strike the surface of a parabolic mirror whose axis of symmetry is parallel to these rays, they are reflected to a single point at the focus. This principle is used in the design of some solar energy devices, satellite dishes, and the mirrors used in some types of telescopes. See Figure 17.

Figure 16
Searchlight

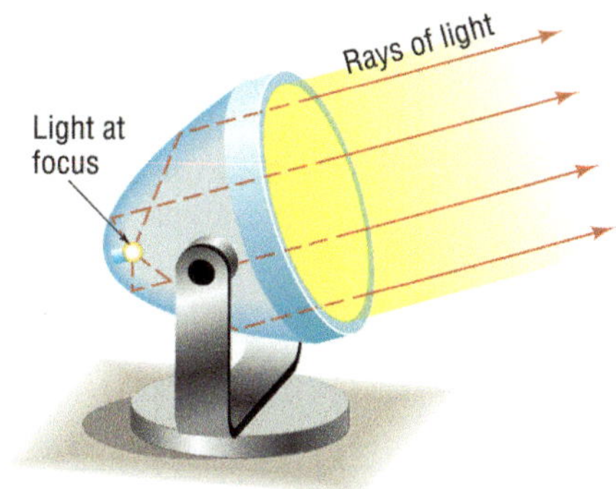

Figure 17
Telescope

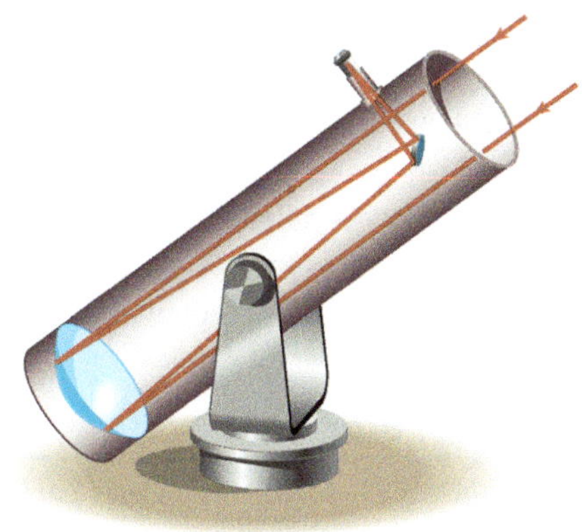

EXAMPLE 10 Satellite Dish

A satellite dish is shaped like a paraboloid of revolution. The signals that emanate from a satellite strike the surface of the dish and are reflected to a single point, where the receiver is located. If the dish is 8 feet across at its opening and 3 feet deep at its center, at what position should the receiver be placed? That is, where is the focus?

Solution Figure 18(a) shows the satellite dish. We draw the parabola used to form the dish on a rectangular coordinate system so that the vertex of the parabola is at the origin and its focus is on the positive y-axis. See Figure 18(b).

Figure 18

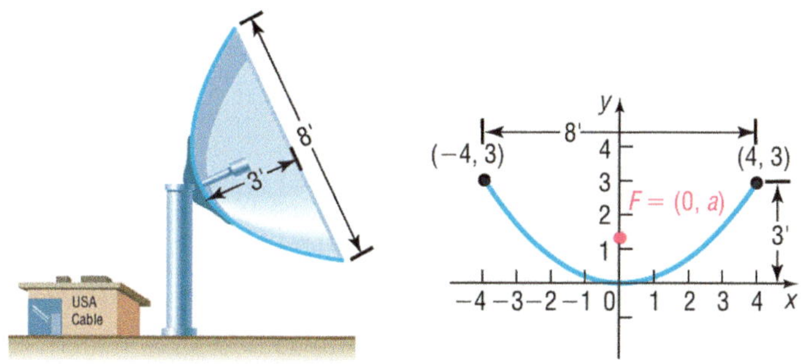

(a) (b)

The form of the equation of the parabola is

$$x^2 = 4ay$$

and its focus is at $(0, a)$. Since $(4, 3)$ is a point on the graph, we have

$$4^2 = 4a(3) \qquad x = 4, y = 3$$

$$a = \frac{4}{3} \qquad \text{Solve for } a.$$

The receiver should be located $1\frac{1}{3}$ feet from the base of the dish, along its axis of symmetry.

Now Work PROBLEM 63

7.2 Assess Your Understanding

'Are You Prepared?' *Answers are given at the end of these exercises. If you get a wrong answer, read the pages listed in red.*

1. The formula for the distance d from $P_1 = (x_1, y_1)$ to $P_2 = (x_2, y_2)$ is $d =$ _____. (p. 89)
2. To complete the square of $x^2 - 4x$, add _____. (pp. 115–116)
3. Use the Square Root Method to find the real solutions of $(x + 4)^2 = 9$. (pp. 114–115)
4. The point that is symmetric with respect to the x-axis to the point $(-2, 5)$ is _____. (pp. 171–172)
5. To graph $y = (x - 3)^2 + 1$, shift the graph of $y = x^2$ to the right _____ units and then _____ 1 unit. (pp. 256–258)

Concepts and Vocabulary

6. A(n) _____ is the collection of all points in the plane such that the distance from each point to a fixed point equals its distance to a fixed line.
7. The surface formed by rotating a parabola about its axis of symmetry is called a _____ _____ _____.
8. ***True or False*** The vertex of a parabola is a point on the parabola that also is on its axis of symmetry.
9. ***True or False*** If a light is placed at the focus of a parabola, all the rays reflected off the parabola will be parallel to the axis of symmetry.
10. ***True or False*** The graph of a quadratic function is a parabola.

Skill Building

In Problems 11–18, the graph of a parabola is given. Match each graph to its equation.

(A) $y^2 = 4x$
(B) $x^2 = 4y$
(C) $y^2 = -4x$
(D) $x^2 = -4y$
(E) $(y - 1)^2 = 4(x - 1)$
(F) $(x + 1)^2 = 4(y + 1)$
(G) $(y - 1)^2 = -4(x - 1)$
(H) $(x + 1)^2 = -4(y + 1)$

11.
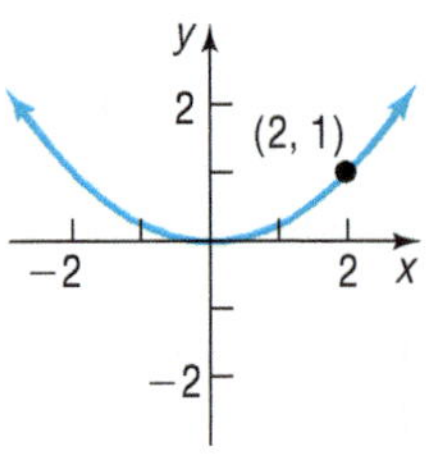

12.
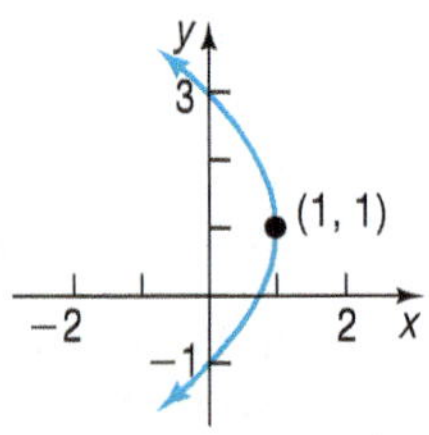

13.
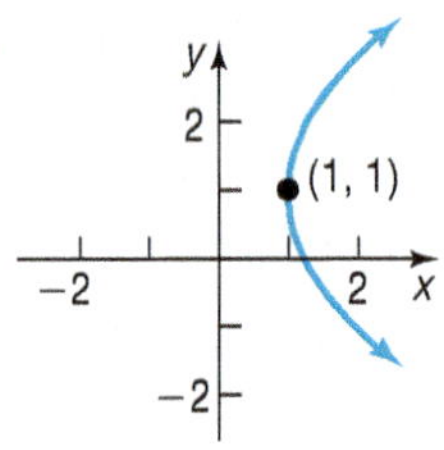

14.
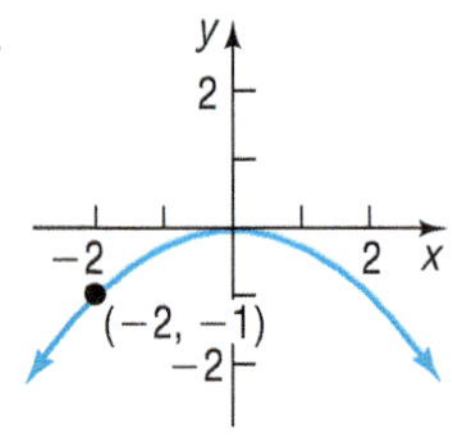

15.
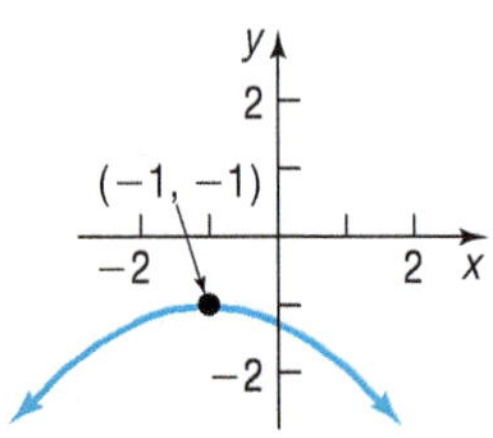

16.
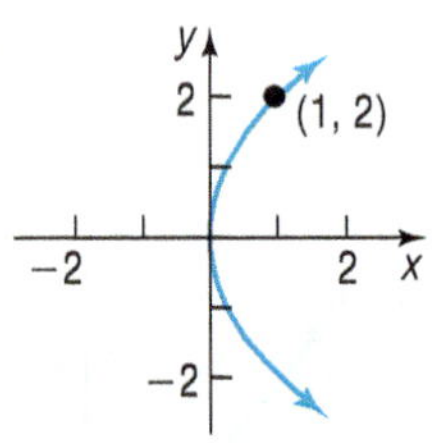

17.
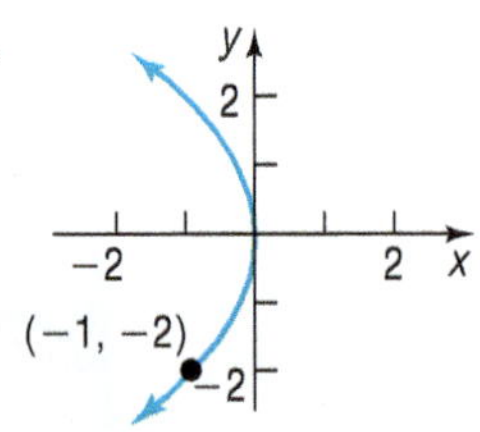

18.
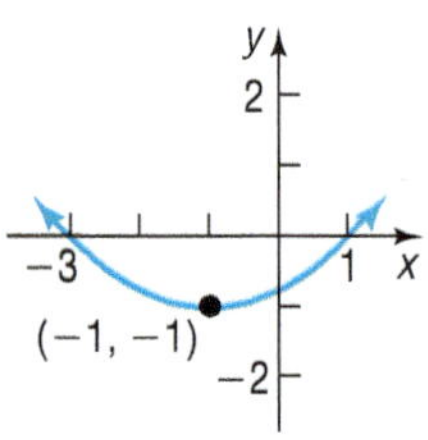

In Problems 19–36, find the equation of the parabola described. Find the two points that define the latus rectum, and graph the equation by hand.

19. Focus at $(4, 0)$; vertex at $(0, 0)$
20. Focus at $(0, 2)$; vertex at $(0, 0)$
21. Focus at $(0, -3)$; vertex at $(0, 0)$
22. Focus at $(-4, 0)$; vertex at $(0, 0)$
23. Focus at $(-2, 0)$; directrix the line $x = 2$
24. Focus at $(0, -1)$; directrix the line $y = 1$
25. Directrix the line $y = -\frac{1}{2}$; vertex at $(0, 0)$
26. Directrix the line $x = -\frac{1}{2}$; vertex at $(0, 0)$
27. Vertex at $(0, 0)$; axis of symmetry the y-axis; containing the point $(2, 3)$
28. Vertex at $(0, 0)$; axis of symmetry the x-axis; containing the point $(2, 3)$
29. Vertex at $(2, -3)$; focus at $(2, -5)$
30. Vertex at $(4, -2)$; focus at $(6, -2)$
31. Vertex at $(-1, -2)$; focus at $(0, -2)$
32. Vertex at $(3, 0)$; focus at $(3, -2)$
33. Focus at $(-3, 4)$; directrix the line $y = 2$
34. Focus at $(2, 4)$; directrix the line $x = -4$
35. Focus at $(-3, -2)$; directrix the line $x = 1$
36. Focus at $(-4, 4)$; directrix the line $y = -2$

In Problems 37–54, find the vertex, focus, and directrix of each parabola. Graph the equation by hand. Verify your graph using a graphing utility.

37. $x^2 = 4y$ **38.** $y^2 = 8x$ **39.** $y^2 = -16x$ **40.** $x^2 = -4y$

41. $(y - 2)^2 = 8(x + 1)$ **42.** $(x + 4)^2 = 16(y + 2)$ **43.** $(x - 3)^2 = -(y + 1)$ **44.** $(y + 1)^2 = -4(x - 2)$

45. $(y + 3)^2 = 8(x - 2)$ **46.** $(x - 2)^2 = 4(y - 3)$ **47.** $y^2 - 4y + 4x + 4 = 0$ **48.** $x^2 + 6x - 4y + 1 = 0$

49. $x^2 + 8x = 4y - 8$ **50.** $y^2 - 2y = 8x - 1$ **51.** $y^2 + 2y - x = 0$ **52.** $x^2 - 4x = 2y$

53. $x^2 - 4x = y + 4$ **54.** $y^2 + 12y = -x + 1$

In Problems 55–62, write an equation for each parabola.

55.

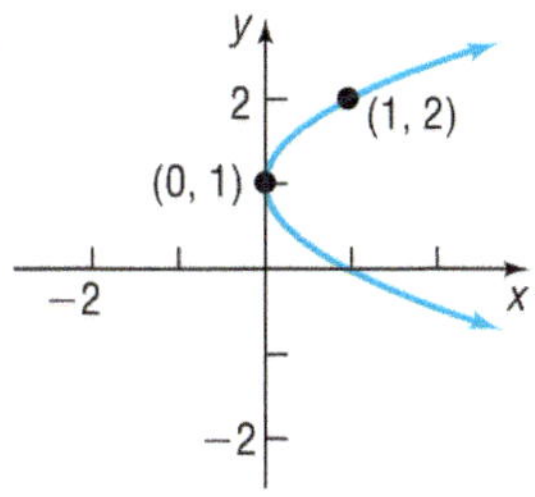

56.

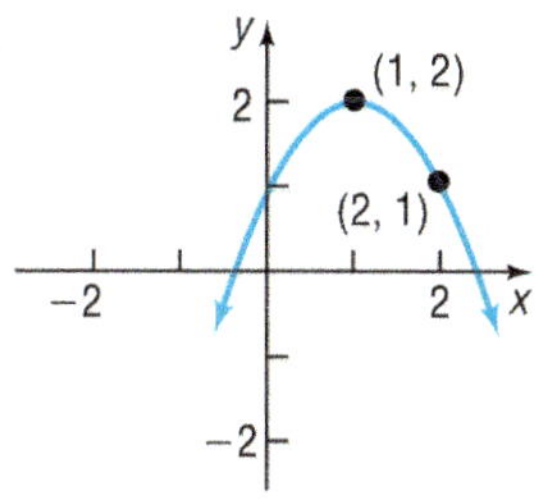

57.

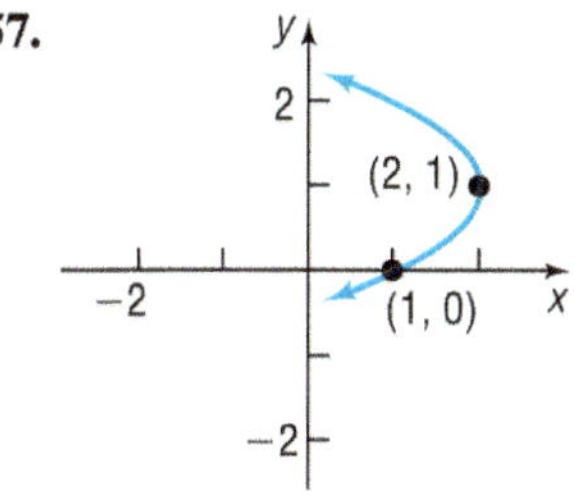

58.

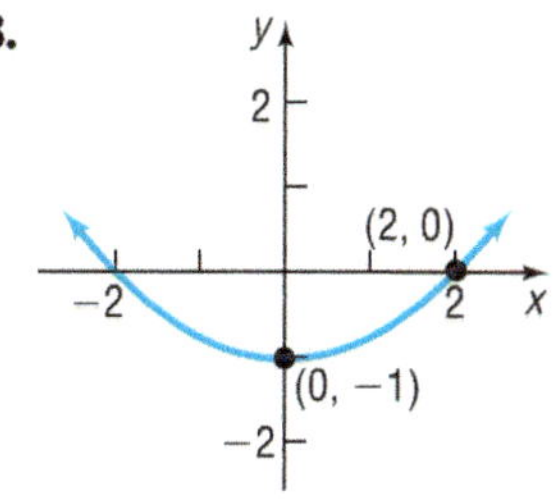

59.

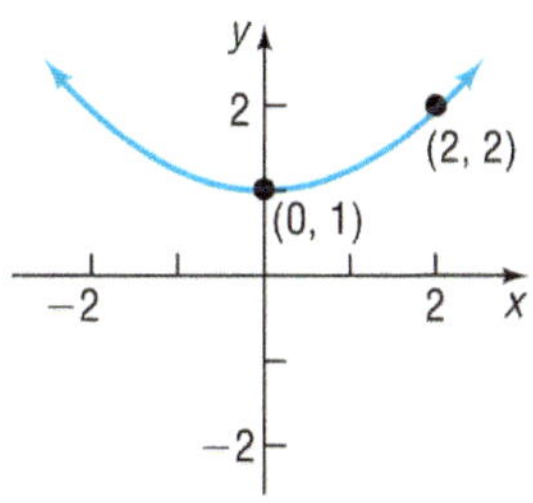

60.

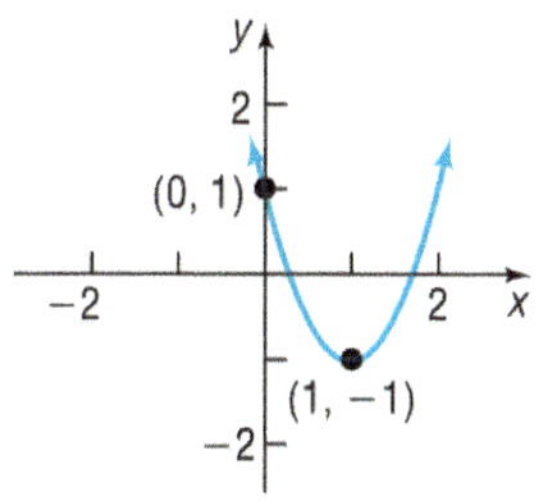

61.

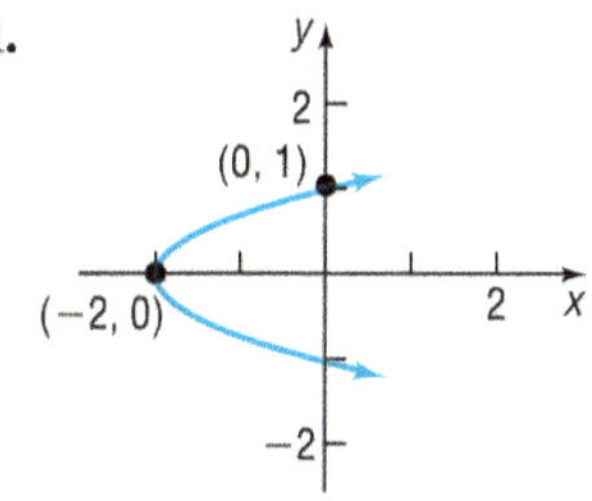

62.

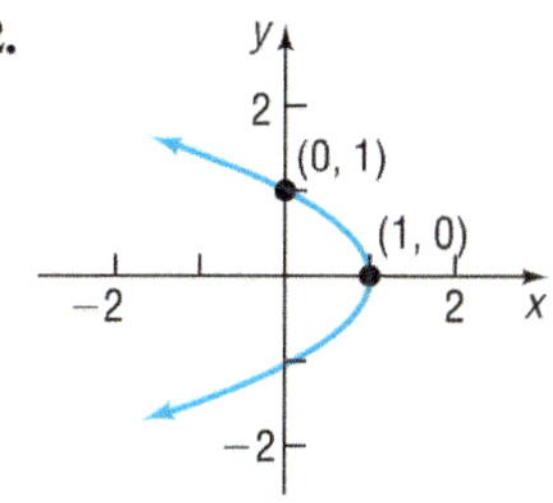

Applications and Extensions

63. Satellite Dish A satellite dish is shaped like a paraboloid of revolution. The signals that emanate from a satellite strike the surface of the dish and are reflected to a single point, where the receiver is located. If the dish is 10 feet across at its opening and 4 feet deep at its center, at what position should the receiver be placed?

64. Constructing a TV Dish A cable TV receiving dish is in the shape of a paraboloid of revolution. Find the location of the receiver, which is placed at the focus, if the dish is 6 feet across at its opening and 2 feet deep.

65. Constructing a Flashlight The reflector of a flashlight is in the shape of a paraboloid of revolution. Its diameter is 4 inches and its depth is 1 inch. How far from the vertex should the light bulb be placed so that the rays will be reflected parallel to the axis?

66. Constructing a Headlight A sealed-beam headlight is in the shape of a paraboloid of revolution. The bulb, which is placed at the focus, is 1 inch from the vertex. If the depth is to be 2 inches, what is the diameter of the headlight at its opening?

67. Suspension Bridge The cables of a suspension bridge are in the shape of a parabola, as shown in the figure. The towers supporting the cable are 600 feet apart and 80 feet high. If the cables touch the road surface midway between the towers, what is the height of the cable from the road at a point 150 feet from the center of the bridge?

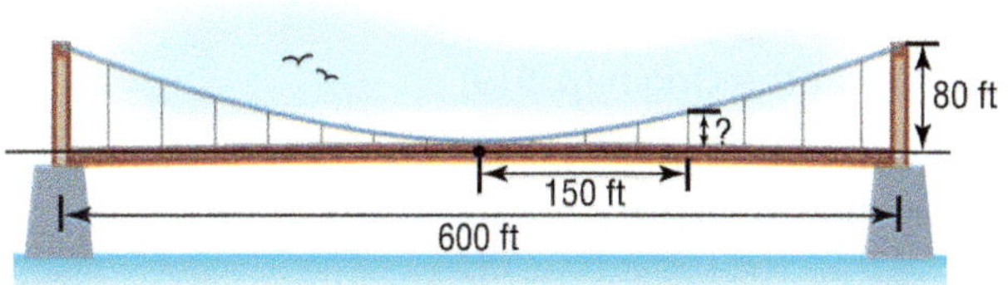

68. Suspension Bridge The cables of a suspension bridge are in the shape of a parabola. The towers supporting the cable are 400 feet apart and 100 feet high. If the cables are at a height of 10 feet midway between the towers, what is the height of the cable at a point 50 feet from the center of the bridge?

69. Searchlight A searchlight is shaped like a paraboloid of revolution. If the light source is located 2 feet from the base along the axis of symmetry and the opening is 5 feet across, how deep should the searchlight be?

70. Searchlight A searchlight is shaped like a paraboloid of revolution. If the light source is located 2 feet from the base along the axis of symmetry and the depth of the searchlight is 4 feet, what should the width of the opening be?

71. Solar Heat A mirror is shaped like a paraboloid of revolution and will be used to concentrate the rays of the sun at its focus, creating a heat source. See the figure. If the mirror is 20 feet across at its opening and is 6 feet deep, where will the heat source be concentrated?

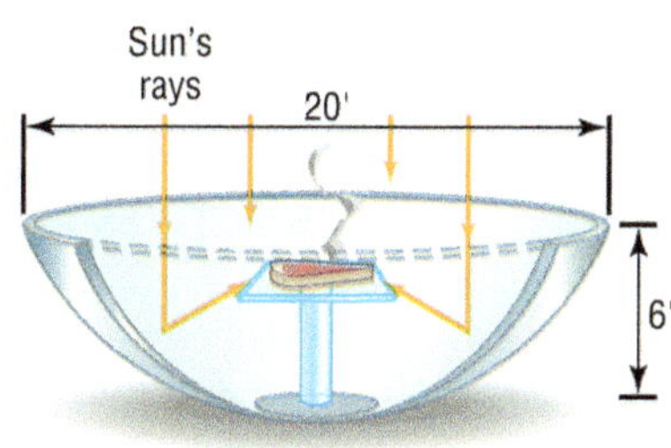

72. Reflecting Telescope A reflecting telescope contains a mirror shaped like a paraboloid of revolution. If the mirror is 4 inches across at its opening and is 3 inches deep, where will the collected light be concentrated?

73. Parabolic Arch Bridge A bridge is built in the shape of a parabolic arch. The bridge has a span of 120 feet and a maximum height of 25 feet. See the illustration. Choose a suitable rectangular coordinate system and find the height of the arch at distances of 10, 30, and 50 feet from the center.

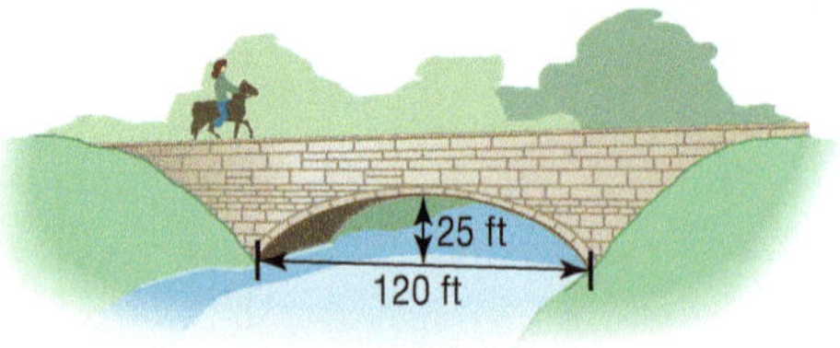

74. Parabolic Arch Bridge A bridge is to be built in the shape of a parabolic arch and is to have a span of 100 feet. The height of the arch a distance of 40 feet from the center is to be 10 feet. Find the height of the arch at its center.

75. Gateway Arch The Gateway Arch in St. Louis is often mistaken to be parabolic is shape. In fact, it is a *catenary,* which has a more complicated formula than a parabola. The Arch is 625 feet high and 598 feet wide at its base.

(a) Find the equation of a parabola with the same dimensions. Let x equal the horizontal distance from the center of the arc.

(b) The table gives the height of the Arch at various widths; find the corresponding heights for the parabola found in part (a).

Width (ft)	Height (ft)
567	100
478	312.5
308	525

(c) Do the data support the notion that the Arch is in the shape of a parabola?

Source: Wikipedia, the free encyclopedia

76. Show that an equation of the form

$$Ax^2 + Ey = 0 \qquad A \neq 0, E \neq 0$$

is the equation of a parabola with vertex at (0, 0) and axis of symmetry the y-axis. Find its focus and directrix.

77. Show that an equation of the form

$$Cy^2 + Dx = 0 \qquad C \neq 0, D \neq 0$$

is the equation of a parabola with vertex at (0, 0) and axis of symmetry the x-axis. Find its focus and directrix.

78. Show that the graph of an equation of the form

$$Ax^2 + Dx + Ey + F = 0 \qquad A \neq 0$$

(a) Is a parabola if $E \neq 0$.
(b) Is a vertical line if $E = 0$ and $D^2 - 4AF = 0$.
(c) Is two vertical lines if $E = 0$ and $D^2 - 4AF > 0$.
(d) Contains no points if $E = 0$ and $D^2 - 4AF < 0$.

79. Show that the graph of an equation of the form

$$Cy^2 + Dx + Ey + F = 0 \qquad C \neq 0$$

(a) Is a parabola if $D \neq 0$.
(b) Is a horizontal line if $D = 0$ and $E^2 - 4CF = 0$.
(c) Is two horizontal lines if $D = 0$ and $E^2 - 4CF > 0$.
(d) Contains no points if $D = 0$ and $E^2 - 4CF < 0$.

'Are You Prepared?' Answers

1. $d = \sqrt{(x_2 - x_1)^2 + (y_2 - y_1)^2}$ **2.** 4 **3.** $x + 4 = \pm 3; \{-7, -1\}$ **4.** $(-2, -5)$ **5.** 3; up

7.3 The Ellipse

PREPARING FOR THIS SECTION *Before getting started, review the following:*

- Distance Formula (Section 1.1, p. 89)
- Completing the Square (Section 1.3, pp. 115–116)
- Intercepts (Section 2.1, pp. 170–171)
- Symmetry (Section 2.1, pp. 171–173)
- Circles (Section 2.3, pp. 193–197)
- Graphing Techniques: Transformations (Section 3.5, pp. 256–264)

Now Work the 'Are You Prepared?' problems on page 527.

OBJECTIVES
1 Analyze Ellipses with Center at the Origin (p. 519)
2 Analyze Ellipses with Center at (h, k) (p. 523)
3 Solve Applied Problems Involving Ellipses (p. 526)

DEFINITION

An **ellipse** is the set of all points P in the plane the sum of whose distances from two fixed points, called the **foci,** is a constant.

The definition contains within it a physical means for drawing an ellipse. Find a piece of string (the length of this string is the constant referred to in the definition). Then take two thumbtacks (the foci) and stick them into a piece of cardboard so that the distance between them is less than the length of the string. Now attach the ends of the string to the thumbtacks and, using the point of a pencil, pull the string taut. See Figure 19. Keeping the string taut, rotate the pencil around the two thumbtacks. The pencil traces out an ellipse, as shown in Figure 19.

Figure 19

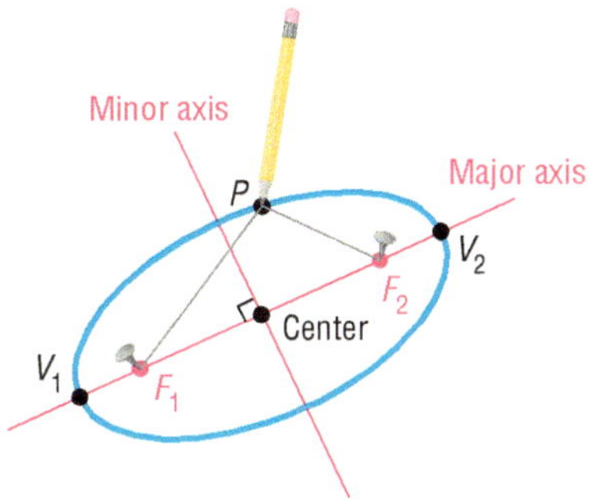

In Figure 19, the foci are labeled F_1 and F_2. The line containing the foci is called the **major axis.** The midpoint of the line segment joining the foci is the **center** of the ellipse. The line through the center and perpendicular to the major axis is the **minor axis.**

The two points of intersection of the ellipse and the major axis are the **vertices,** V_1 and V_2, of the ellipse. The distance from one vertex to the other is the **length of the major axis.** The ellipse is symmetric with respect to its major axis, with respect to its minor axis, and with respect to its center.

Figure 20

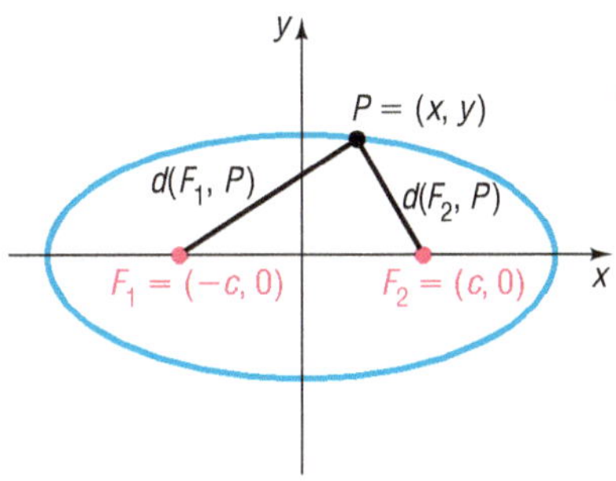

1 Analyze Ellipses with Center at the Origin

With these ideas in mind, we are now ready to find the equation of an ellipse in a rectangular coordinate system. First, we place the center of the ellipse at the origin. Second, we position the ellipse so that its major axis coincides with a coordinate axis. Suppose that the major axis coincides with the x-axis, as shown in Figure 20. If c is the distance from the center to a focus, one focus will be at $F_1 = (-c, 0)$ and the other at $F_2 = (c, 0)$. As we shall see, it is convenient to let $2a$ denote the constant distance referred to in the definition. Then, if $P = (x, y)$ is any point on the ellipse, we have

$$d(F_1, P) + d(F_2, P) = 2a$$ Sum of the distances from P to the foci equals a constant, $2a$

$$\sqrt{(x + c)^2 + (y - 0)^2} + \sqrt{(x - c)^2 + (y - 0)^2} = 2a$$ Use the Distance Formula.

$$\sqrt{(x + c)^2 + y^2} = 2a - \sqrt{(x - c)^2 + y^2}$$ Isolate one radical.

$$(x + c)^2 + y^2 = 4a^2 - 4a\sqrt{(x - c)^2 + y^2} + (x - c)^2 + y^2$$ Square both sides.

$$x^2 + 2cx + c^2 + y^2 = 4a^2 - 4a\sqrt{(x - c)^2 + y^2} + x^2 - 2cx + c^2 + y^2$$ Remove parentheses.

$$4cx - 4a^2 = -4a\sqrt{(x - c)^2 + y^2}$$ Simplify; isolate the radical.

$$cx - a^2 = -a\sqrt{(x - c)^2 + y^2}$$ Divide each side by 4.

$$(cx - a^2)^2 = a^2[(x - c)^2 + y^2]$$ Square both sides again.

$$c^2x^2 - 2a^2cx + a^4 = a^2(x^2 - 2cx + c^2 + y^2)$$ Remove parentheses.

$$(c^2 - a^2)x^2 - a^2y^2 = a^2c^2 - a^4$$ Rearrange the terms.

$$(a^2 - c^2)x^2 + a^2y^2 = a^2(a^2 - c^2) \qquad \textbf{(1)}$$ Multiply each side by -1; factor a^2 on the right side.

To obtain points on the ellipse off the x-axis, it must be that $a > c$. To see why, look again at Figure 20. Then

$$d(F_1, P) + d(F_2, P) > d(F_1, F_2)$$ The sum of the lengths of two sides of a triangle is greater than the length of the third side.

$$2a > 2c$$ $d(F_1, P) + d(F_2, P) = 2a, \quad d(F_1, F_2) = 2c$

$$a > c$$

Since $a > c$, we also have $a^2 > c^2$, so $a^2 - c^2 > 0$. Let $b^2 = a^2 - c^2, b > 0$. Then $a > b$ and equation (1) can be written as

$$b^2x^2 + a^2y^2 = a^2b^2$$

$$\frac{x^2}{a^2} + \frac{y^2}{b^2} = 1 \qquad \text{Divide each side by } a^2b^2.$$

As you can verify, this equation is symmetric with respect to the x-axis, y-axis, and origin.

Because the major axis is the x-axis, we find the vertices of this ellipse by letting $y = 0$. The vertices satisfy the equation $\frac{x^2}{a^2} = 1$, the solutions of which are $x = \pm a$. Consequently, the vertices of this ellipse are $V_1 = (-a, 0)$ and $V_2 = (a, 0)$. The y-intercepts of the ellipse, found by letting $x = 0$, have coordinates $(0, -b)$ and $(0, b)$. These four intercepts, $(a, 0)$, $(-a, 0)$, $(0, b)$, and $(0, -b)$, are used to graph the ellipse.

Figure 21

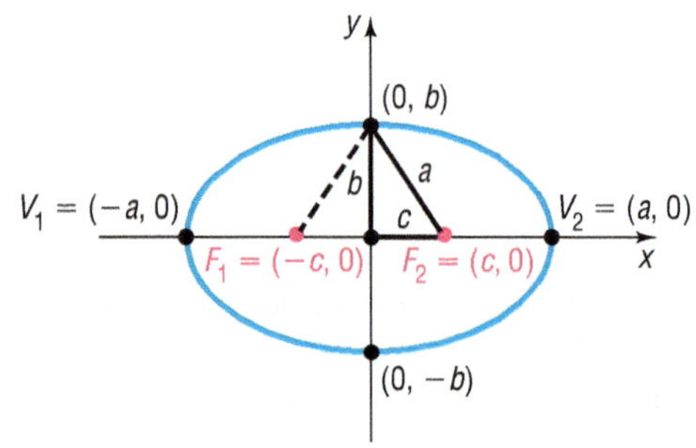

THEOREM

Equation of an Ellipse: Center at (0, 0); Major Axis along the x-Axis

An equation of the ellipse with center at $(0, 0)$, foci at $(-c, 0)$ and $(c, 0)$, and vertices at $(-a, 0)$ and $(a, 0)$ is

$$\frac{x^2}{a^2} + \frac{y^2}{b^2} = 1 \qquad \text{where } a > b > 0 \text{ and } b^2 = a^2 - c^2 \qquad \textbf{(2)}$$

The major axis is the x-axis. See Figure 21.

Notice in Figure 21 the right triangle formed with the points $(0, 0)$, $(c, 0)$, and $(0, b)$. Because $b^2 = a^2 - c^2$ (or $b^2 + c^2 = a^2$), the distance from the focus at $(c, 0)$ to the point $(0, b)$ is a.

This can be seen another way. Look at the two right triangles in Figure 21. They are congruent. Do you see why? Because the sum of the distances from the foci to a point on the ellipse is $2a$, it follows that the distance from $(c, 0)$ to $(0, b)$ is a.

EXAMPLE 1 Finding an Equation of an Ellipse

Find an equation of the ellipse with center at the origin, one focus at $(3, 0)$, and a vertex at $(-4, 0)$. Graph the equation.

Figure 22 $\frac{x^2}{16} + \frac{y^2}{7} = 1$

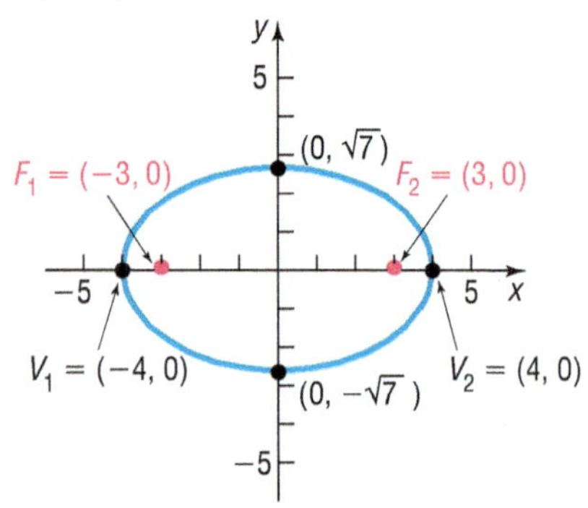

Solution The ellipse has its center at the origin and, since the given focus and vertex lie on the x-axis, the major axis is the x-axis. The distance from the center, $(0, 0)$, to one of the foci, $(3, 0)$, is $c = 3$. The distance from the center, $(0, 0)$, to one of the vertices, $(-4, 0)$, is $a = 4$. Because the center is the origin, the other focus is $(-3, 0)$ and the other vertex is $(4, 0)$. With $a = 4$ and $c = 3$, from equation (2), it follows that

$$b^2 = a^2 - c^2 = 16 - 9 = 7$$

so an equation of the ellipse is

$$\frac{x^2}{16} + \frac{y^2}{7} = 1$$

The y-intercepts are $(0, \pm b) = (0, \pm\sqrt{7})$. Figure 22 shows the graph.

Notice in Figure 22 how we used the intercepts of the equation to graph the ellipse. Following this practice will make it easier for you to obtain an accurate graph of an ellipse when graphing by hand. The intercepts also tell you how to set the initial viewing window when using a graphing utility.

Now Work PROBLEM 27

EXAMPLE 2 Graphing an Ellipse Using a Graphing Utility

Use a graphing utility to graph the ellipse $\frac{x^2}{16} + \frac{y^2}{7} = 1$.

Solution First, we must solve $\frac{x^2}{16} + \frac{y^2}{7} = 1$ for y.

$$\frac{y^2}{7} = 1 - \frac{x^2}{16} \qquad \text{Subtract } \frac{x^2}{16} \text{ from each side.}$$

$$y^2 = 7\left(1 - \frac{x^2}{16}\right) \qquad \text{Multiply both sides by 7.}$$

$$y = \pm\sqrt{7\left(1 - \frac{x^2}{16}\right)} \qquad \text{Apply the Square Root Method.}$$

Figure 23

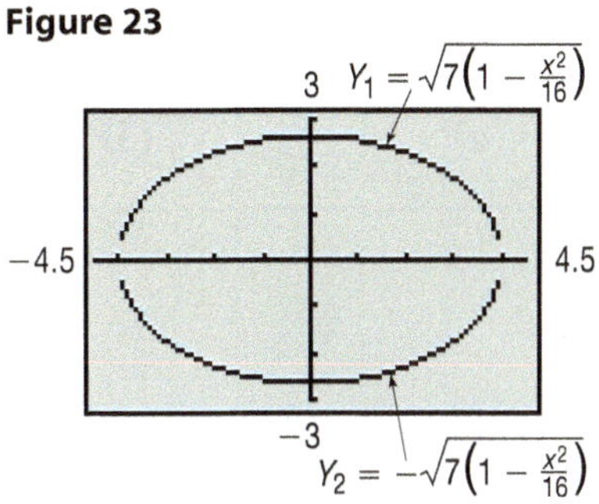

Figure 23* shows the graphs of $Y_1 = \sqrt{7\left(1 - \frac{x^2}{16}\right)}$ and $Y_2 = -\sqrt{7\left(1 - \frac{x^2}{16}\right)}$.

Notice in Figure 23 that we used a square screen. As with circles and parabolas, this is done to avoid a distorted view of the graph.

An equation of the form of equation (2), with $a > b$, is the equation of an ellipse with center at the origin, foci on the x-axis at $(-c, 0)$ and $(c, 0)$, where $c^2 = a^2 - b^2$, and major axis along the x-axis.

For the remainder of this section, the direction **"Analyze the equation"** will mean to find the center, major axis, foci, and vertices of the ellipse and graph it.

EXAMPLE 3 Analyzing the Equation of an Ellipse

Analyze the equation: $\frac{x^2}{25} + \frac{y^2}{9} = 1$

Solution The given equation is of the form of equation (2), with $a^2 = 25$ and $b^2 = 9$. The equation is that of an ellipse with center $(0, 0)$ and major axis along the x-axis. The vertices are at $(\pm a, 0) = (\pm 5, 0)$. Because $b^2 = a^2 - c^2$, we find that

$$c^2 = a^2 - b^2 = 25 - 9 = 16$$

The foci are at $(\pm c, 0) = (\pm 4, 0)$. The y-intercepts are $(0, \pm b) = (0, \pm 3)$. Figure 24(a) shows the graph drawn by hand. Figure 24(b) shows the graph obtained using a graphing utility.

Figure 24

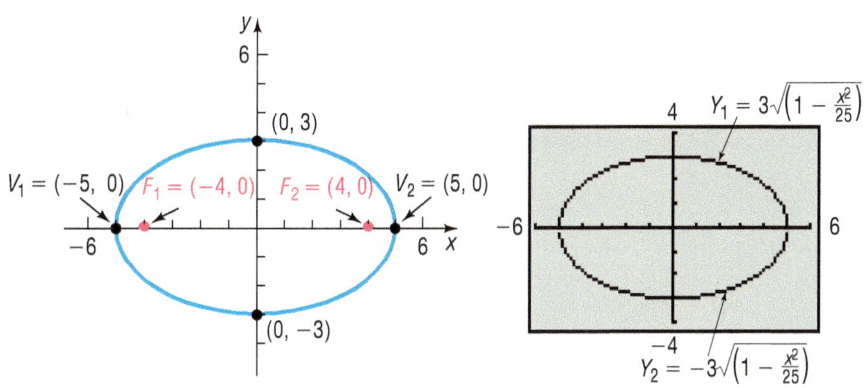

(a) (b)

Now Work PROBLEM 17

*The initial viewing window selected was $X\min = -4$, $X\max = 4$, $Y\min = -3$, $Y\max = 3$. Then we used the ZOOM-SQUARE option to obtain the window shown.

If the major axis of an ellipse with center at (0, 0) lies on the y-axis, then the foci are at $(0, -c)$ and $(0, c)$. Using the same steps as before, the definition of an ellipse leads to the following result:

THEOREM

Equation of an Ellipse; Center at (0, 0); Major Axis along the *y*-Axis

An equation of the ellipse with center at (0, 0), foci at $(0, -c)$ and $(0, c)$, and vertices at $(0, -a)$ and $(0, a)$ is

$$\frac{x^2}{b^2} + \frac{y^2}{a^2} = 1 \quad \text{where } a > b > 0 \text{ and } b^2 = a^2 - c^2 \qquad \textbf{(3)}$$

The major axis is the y-axis.

Figure 25

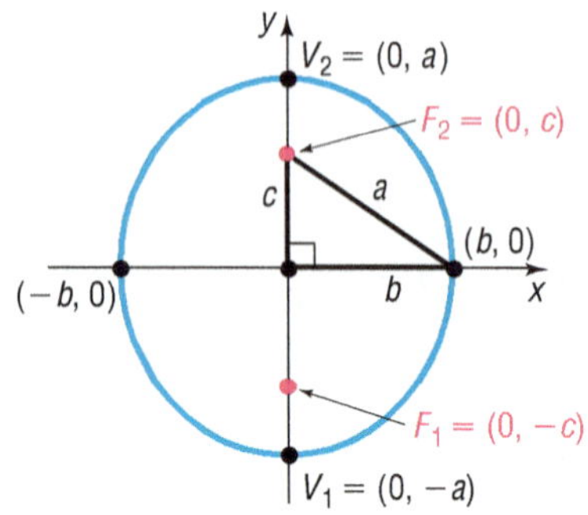

Figure 25 illustrates the graph of such an ellipse. Again, notice the right triangle with the points at (0, 0), $(b, 0)$, and $(0, c)$.

Look closely at equations (2) and (3). Although they may look alike, there is a difference! In equation (2), the larger number, a^2, is in the denominator of the x^2-term, so the major axis of the ellipse is along the x-axis. In equation (3), the larger number, a^2, is in the denominator of the y^2-term, so the major axis is along the y-axis.

EXAMPLE 4 Analyzing the Equation of an Ellipse

Analyze the equation: $9x^2 + y^2 = 9$

Solution To put the equation in proper form, we divide each side by 9.

$$x^2 + \frac{y^2}{9} = 1$$

The larger number, 9, is in the denominator of the y^2 term so, based on equation (3), this is the equation of an ellipse with center at the origin and major axis along the y-axis. Also, we conclude that $a^2 = 9$, $b^2 = 1$, and $c^2 = a^2 - b^2 = 9 - 1 = 8$. The vertices are at $(0, \pm a) = (0, \pm 3)$, and the foci are at $(0, \pm c) = (0, \pm 2\sqrt{2})$. The x-intercepts are at $(\pm b, 0) = (\pm 1, 0)$. Figure 26(a) shows the graph drawn by hand. Figure 26(b) shows the graph obtained using a graphing utility.

Figure 26

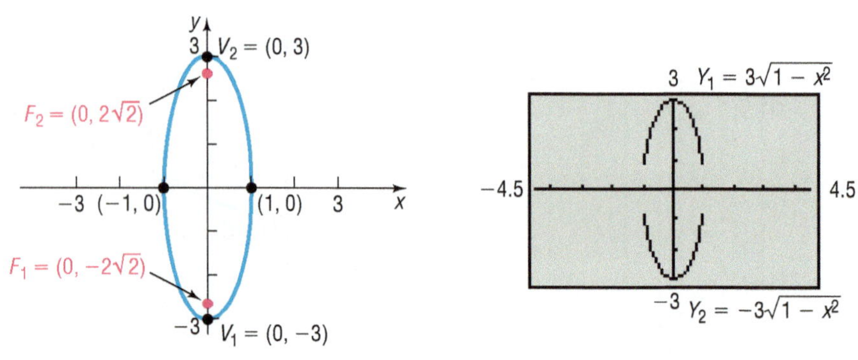

(a) (b)

Now Work PROBLEM 21

EXAMPLE 5 **Finding an Equation of an Ellipse**

Find an equation of the ellipse having one focus at $(0, 2)$ and vertices at $(0, -3)$ and $(0, 3)$. Graph the equation by hand.

Solution Because the vertices are at $(0, -3)$ and $(0, 3)$, the center of this ellipse is at their midpoint, the origin. The vertices and foci lie on the y-axis, so the major axis lies on the y-axis. The distance from the center, $(0, 0)$, to one of the foci, $(0, 2)$, is $c = 2$. The distance from the center, $(0, 0)$, to one of the vertices, $(0, 3)$, is $a = 3$. So $b^2 = a^2 - c^2 = 9 - 4 = 5$. The form of the equation of this ellipse is given by equation (3).

Figure 27

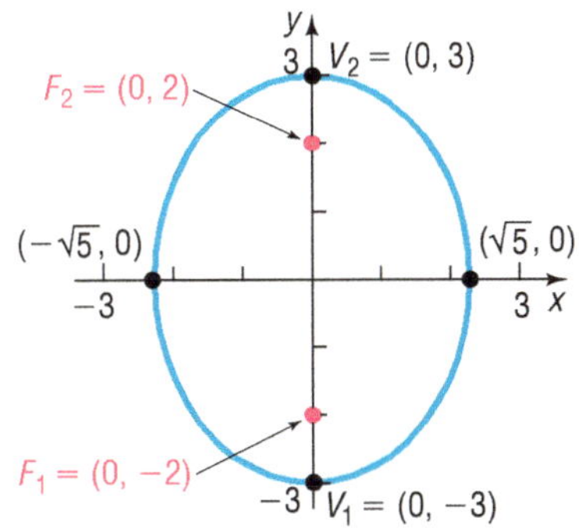

$$\frac{x^2}{b^2} + \frac{y^2}{a^2} = 1$$

$$\frac{x^2}{5} + \frac{y^2}{9} = 1$$

Figure 27 shows the graph.

Now Work PROBLEM 29

The circle may be considered a special kind of ellipse. To see why, let $a = b$ in equation (2) or (3). Then

$$\frac{x^2}{a^2} + \frac{y^2}{a^2} = 1$$

$$x^2 + y^2 = a^2$$

This is the equation of a circle with center at the origin and radius a. The value of c is

$$c^2 = a^2 - b^2 = 0$$

We conclude that the closer the two foci of an ellipse are to the center the more the ellipse will look like a circle.

2 Analyze Ellipses with Center at (h, k)

If an ellipse with center at the origin and major axis coinciding with a coordinate axis is shifted horizontally h units and then vertically k units, the result is an ellipse with center at (h, k) and major axis parallel to a coordinate axis. The equations of such ellipses have the same forms as those given in equations (2) and (3), except that x is replaced by $x - h$ (the horizontal shift) and y is replaced by $y - k$ (the vertical shift). Table 3 gives the forms of the equations of such ellipses and Figure 28 on the next page shows their graphs.

Table 3

ELLIPSES WITH CENTER AT (h, k) AND MAJOR AXIS PARALLEL TO A COORDINATE AXIS

Center	Major Axis	Foci	Vertices	Equation
(h, k)	Parallel to x-axis	$(h + c, k)$	$(h + a, k)$	$\frac{(x - h)^2}{a^2} + \frac{(y - k)^2}{b^2} = 1,$
		$(h - c, k)$	$(h - a, k)$	$a > b$ and $b^2 = a^2 - c^2$
(h, k)	Parallel to y-axis	$(h, k + c)$	$(h, k + a)$	$\frac{(x - h)^2}{b^2} + \frac{(y - k)^2}{a^2} = 1,$
		$(h, k - c)$	$(h, k - a)$	$a > b$ and $b^2 = a^2 - c^2$

Figure 28

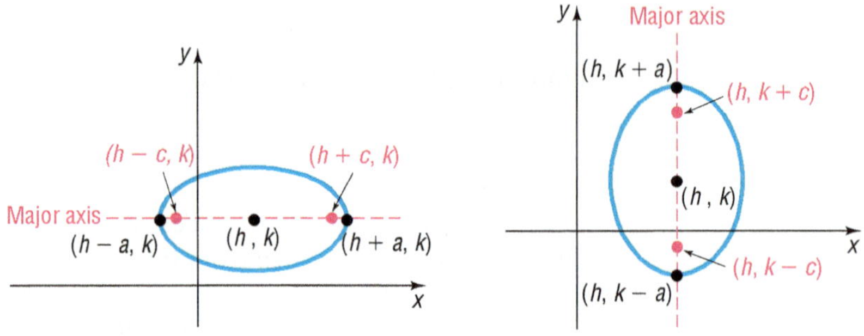

(a) $\dfrac{(x-h)^2}{a^2} + \dfrac{(y-k)^2}{b^2} = 1$ (b) $\dfrac{(x-h)^2}{b^2} + \dfrac{(y-k)^2}{a^2} = 1$

EXAMPLE 6 Finding an Equation of an Ellipse, Center Not at the Origin

Find an equation for the ellipse with center at $(2, -3)$, one focus at $(3, -3)$, and one vertex at $(5, -3)$. Graph the equation by hand.

Solution The center is at $(h, k) = (2, -3)$, so $h = 2$ and $k = -3$. Since the center, focus, and vertex all lie on the line $y = -3$, the major axis is parallel to the x-axis. The distance from the center $(2, -3)$ to a focus $(3, -3)$ is $c = 1$; the distance from the center $(2, -3)$ to a vertex $(5, -3)$ is $a = 3$. Then $b^2 = a^2 - c^2 = 9 - 1 = 8$. The form of the equation is

$$\frac{(x-h)^2}{a^2} + \frac{(y-k)^2}{b^2} = 1 \qquad \text{where } h = 2, k = -3, a = 3, b = 2\sqrt{2}$$

$$\frac{(x-2)^2}{9} + \frac{(y+3)^2}{8} = 1$$

To graph the equation, we use the center $(h, k) = (2, -3)$ to locate the vertices. The major axis is parallel to the x-axis, so the vertices are $a = 3$ units left and right of the center $(2, -3)$. Therefore, the vertices are

$$V_1 = (2-3, -3) = (-1, -3) \quad \text{and} \quad V_2 = (2+3, -3) = (5, -3)$$

Since $c = 1$ and the major axis is parallel to the x-axis, the foci are 1 unit left and right of the center. Therefore, the foci are

$$F_1 = (2-1, -3) = (1, -3) \quad \text{and} \quad F_2 = (2+1, -3) = (3, -3)$$

Finally, we use the value of $b = 2\sqrt{2}$ to find the two points above and below the center.

$$\left(2, -3 - 2\sqrt{2}\right) \quad \text{and} \quad \left(2, -3 + 2\sqrt{2}\right)$$

Figure 29 shows the graph.

Figure 29

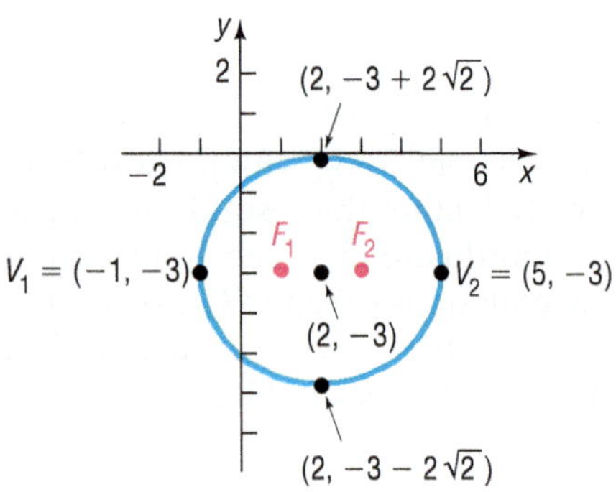

Now Work PROBLEM 55

EXAMPLE 7 Using a Graphing Utility to Graph an Ellipse, Center Not at the Origin

Using a graphing utility, graph the ellipse: $\dfrac{(x-2)^2}{9} + \dfrac{(y+3)^2}{8} = 1$

Solution First, we must solve the equation $\frac{(x-2)^2}{9} + \frac{(y+3)^2}{8} = 1$ for y.

$$\frac{(y+3)^2}{8} = 1 - \frac{(x-2)^2}{9} \qquad \text{Subtract } \frac{(x-2)^2}{9} \text{ from each side.}$$

$$(y+3)^2 = 8\left[1 - \frac{(x-2)^2}{9}\right] \qquad \text{Multiply each side by 8.}$$

$$y + 3 = \pm\sqrt{8\left[1 - \frac{(x-2)^2}{9}\right]} \qquad \text{Apply the Square Root Method.}$$

$$y = -3 \pm \sqrt{8\left[1 - \frac{(x-2)^2}{9}\right]} \qquad \text{Subtract 3 from each side.}$$

Figure 30 shows the graphs of $Y_1 = -3 + \sqrt{8\left[1 - \frac{(x-2)^2}{9}\right]}$ and $Y_2 = -3 - \sqrt{8\left[1 - \frac{(x-2)^2}{9}\right]}$.

Figure 30

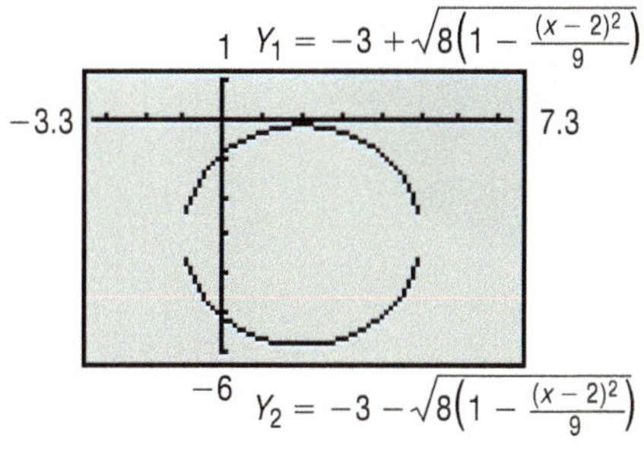

EXAMPLE 8 Analyzing the Equation of an Ellipse

Analyze the equation: $4x^2 + y^2 - 8x + 4y + 4 = 0$

Solution We proceed to complete the squares in x and in y.

$$4x^2 + y^2 - 8x + 4y + 4 = 0$$

$$4x^2 - 8x + y^2 + 4y = -4 \qquad \text{Group like variables; place the constant on the right side.}$$

$$4(x^2 - 2x) + (y^2 + 4y) = -4 \qquad \text{Factor out 4 from the first two terms.}$$

$$4(x^2 - 2x + 1) + (y^2 + 4y + 4) = -4 + 4 + 4 \qquad \text{Complete each square.}$$

$$4(x-1)^2 + (y+2)^2 = 4 \qquad \text{Factor.}$$

$$(x-1)^2 + \frac{(y+2)^2}{4} = 1 \qquad \text{Divide each side by 4.}$$

This is the equation of an ellipse with center at $(1, -2)$ and major axis parallel to the y-axis. Since $a^2 = 4$ and $b^2 = 1$, we have $c^2 = a^2 - b^2 = 4 - 1 = 3$. The vertices are at $(h, k \pm a) = (1, -2 \pm 2)$ or $(1, 0)$ and $(1, -4)$. The foci are at $(h, k \pm c) = (1, -2 \pm \sqrt{3})$ or $(1, -2 - \sqrt{3})$ and $(1, -2 + \sqrt{3})$. Figure 31(a) shows the graph drawn by hand. Figure 31(b) shows the graph obtained using a graphing utility.

Figure 31

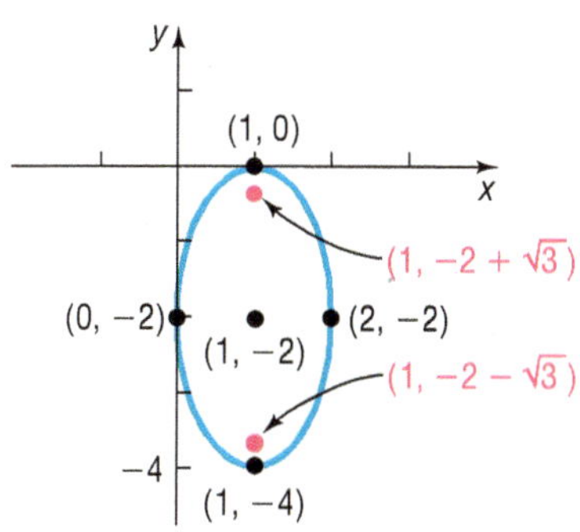

(a)

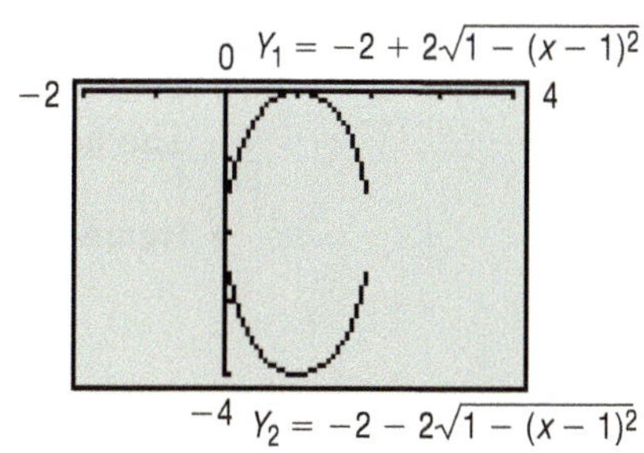

(b)

Now Work PROBLEM 47

3 Solve Applied Problems Involving Ellipses

Ellipses are found in many applications in science and engineering. For example, the orbits of the planets around the Sun are elliptical, with the Sun's position at a focus. See Figure 32.

Figure 32

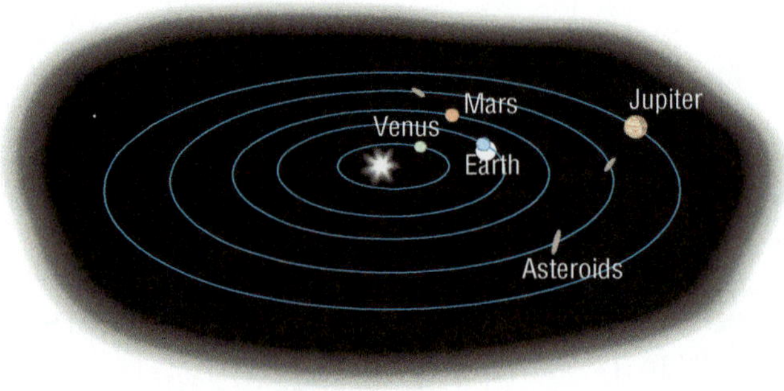

Stone and concrete bridges are often shaped as semielliptical arches. Elliptical gears are used in machinery when a variable rate of motion is required.

Ellipses also have an interesting reflection property. If a source of light (or sound) is placed at one focus, the waves transmitted by the source will reflect off the ellipse and concentrate at the other focus. This is the principle behind *whispering galleries,* which are rooms designed with elliptical ceilings. A person standing at one focus of the ellipse can whisper and be heard by a person standing at the other focus, because all the sound waves that reach the ceiling are reflected to the other person.

EXAMPLE 9 A Whispering Gallery

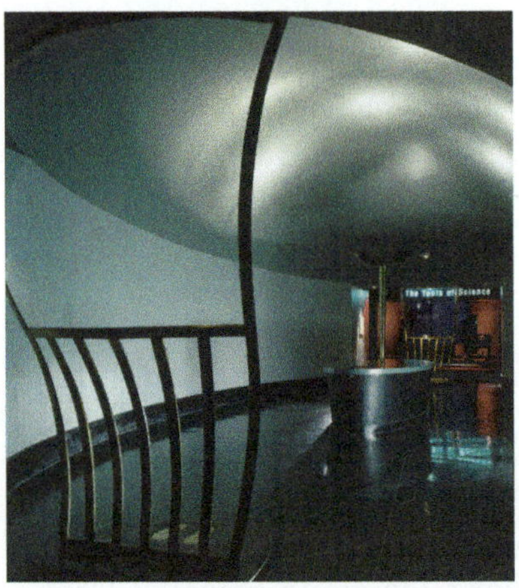

The whispering gallery in the Museum of Science and Industry in Chicago is 47.3 feet long. The distance from the center of the room to the foci is 20.3 feet. Find an equation that describes the shape of the room. How high is the room at its center?

Source: Chicago Museum of Science and Industry Web site; www.msichicago.org

Solution We set up a rectangular coordinate system so that the center of the ellipse is at the origin and the major axis is along the x-axis. The equation of the ellipse is

$$\frac{x^2}{a^2} + \frac{y^2}{b^2} = 1$$

Since the length of the room is 47.3 feet, the distance from the center of the room to each vertex (the end of the room) will be $\frac{47.3}{2} = 23.65$ feet; so $a = 23.65$ feet. The distance from the center of the room to each focus is $c = 20.3$ feet. See Figure 33.

Since $b^2 = a^2 - c^2$, we find $b^2 = 23.65^2 - 20.3^2 = 147.2325$. An equation that describes the shape of the room is given by

$$\frac{x^2}{23.65^2} + \frac{y^2}{147.2325} = 1$$

The height of the room at its center is $b = \sqrt{147.2325} \approx 12.1$ feet.

Figure 33

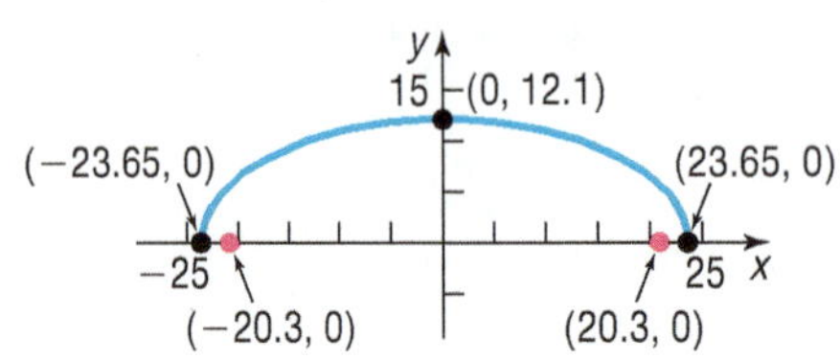

Now Work PROBLEM 71

7.3 Assess Your Understanding

'Are You Prepared?'

Answers are given at the end of these exercises. If you get a wrong answer, read the pages listed in red.

1. The distance d from $P_1 = (2, -5)$ to $P_2 = (4, -2)$ is $d =$ _____. (pp. 88–90)

2. To complete the square of $x^2 - 3x$, add _____. (pp. 115–116)

3. Find the intercepts of the equation $y^2 = 16 - 4x^2$. (pp. 170–171)

4. The point that is symmetric with respect to the y-axis to the point $(-2, 5)$ is _____. (pp. 171–172)

5. To graph $y = (x + 1)^2 - 4$, shift the graph of $y = x^2$ to the (left/right) _____ unit(s) and then (up/down) _____ unit(s). (pp. 256–258)

6. The standard equation of a circle with center at $(2, -3)$ and radius 1 is _____. (pp. 193–194)

Concepts and Vocabulary

7. A(n) _____ is the collection of all points in the plane the sum of whose distances from two fixed points is a constant.

8. For an ellipse, the foci lie on a line called the _____ axis.

9. For the ellipse $\frac{x^2}{4} + \frac{y^2}{25} = 1$, the vertices are the points _____ and _____.

10. ***True or False*** The foci, vertices, and center of an ellipse lie on a line called the axis of symmetry.

11. ***True or False*** If the center of an ellipse is at the origin and the foci lie on the y-axis, the ellipse is symmetric with respect to the x-axis, the y-axis, and the origin.

12. ***True or False*** A circle is a certain type of ellipse.

Skill Building

In Problems 13–16, the graph of an ellipse is given. Match each graph to its equation.

(A) $\frac{x^2}{4} + y^2 = 1$ (B) $x^2 + \frac{y^2}{4} = 1$ (C) $\frac{x^2}{16} + \frac{y^2}{4} = 1$ (D) $\frac{x^2}{4} + \frac{y^2}{16} = 1$

13.

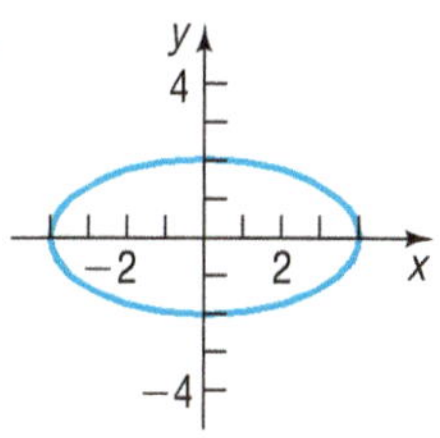

14.

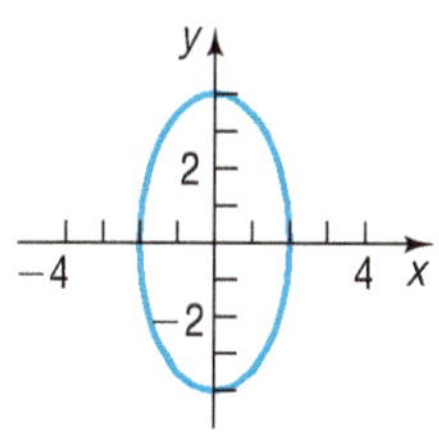

15.

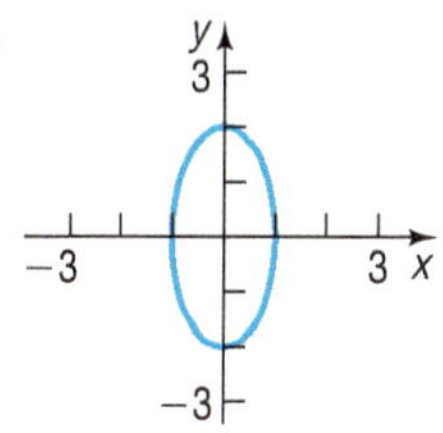

16.

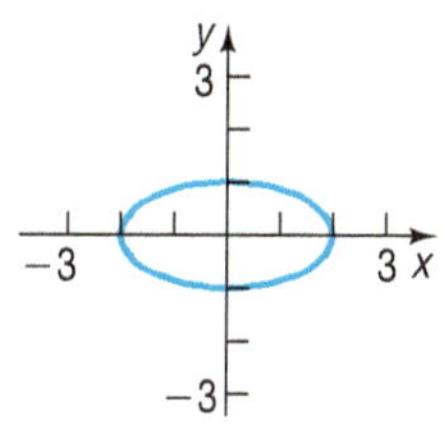

In Problems 17–26, find the vertices and foci of each ellipse. Graph each equation by hand. Verify your graph using a graphing utility.

17. $\frac{x^2}{25} + \frac{y^2}{4} = 1$

18. $\frac{x^2}{9} + \frac{y^2}{4} = 1$

19. $\frac{x^2}{9} + \frac{y^2}{25} = 1$

20. $x^2 + \frac{y^2}{16} = 1$

21. $4x^2 + y^2 = 16$

22. $x^2 + 9y^2 = 18$

23. $4y^2 + x^2 = 8$

24. $4y^2 + 9x^2 = 36$

25. $x^2 + y^2 = 16$

26. $x^2 + y^2 = 4$

In Problems 27–38, find an equation for each ellipse. Graph the equation by hand.

27. Center at $(0, 0)$; focus at $(3, 0)$; vertex at $(5, 0)$

28. Center at $(0, 0)$; focus at $(-1, 0)$; vertex at $(3, 0)$

29. Center at $(0, 0)$; focus at $(0, -4)$; vertex at $(0, 5)$

30. Center at $(0, 0)$; focus at $(0, 1)$; vertex at $(0, -2)$

31. Foci at $(\pm 2, 0)$; length of the major axis is 6

32. Foci at $(0, \pm 2)$; length of the major axis is 8

33. Focus at $(-4, 0)$; vertices at $(\pm 5, 0)$

34. Focus at $(0, -4)$; vertices at $(0, \pm 8)$

35. Foci at $(0, \pm 3)$; x-intercepts are ± 2

36. Vertices at $(\pm 4, 0)$; y-intercepts are ± 1

37. Center at $(0, 0)$; vertex at $(0, 4)$; $b = 1$

38. Vertices at $(\pm 5, 0)$; $c = 2$

In Problems 39–42, write an equation for each ellipse.

39.

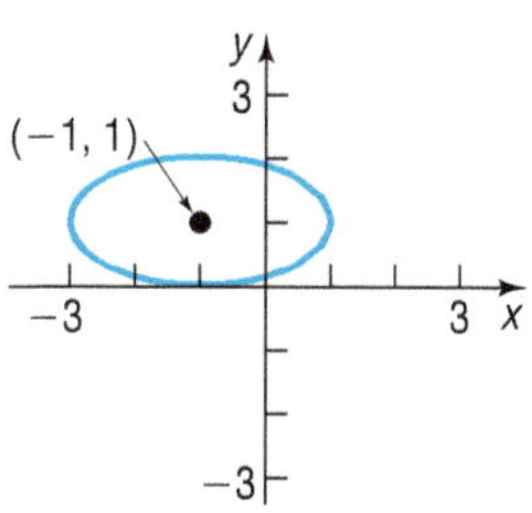

40.

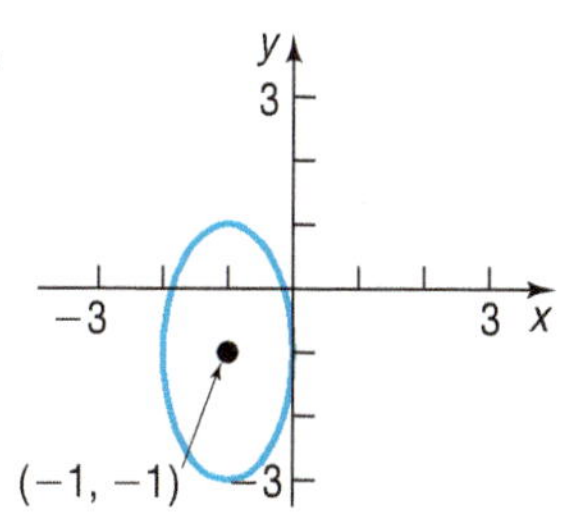

41.

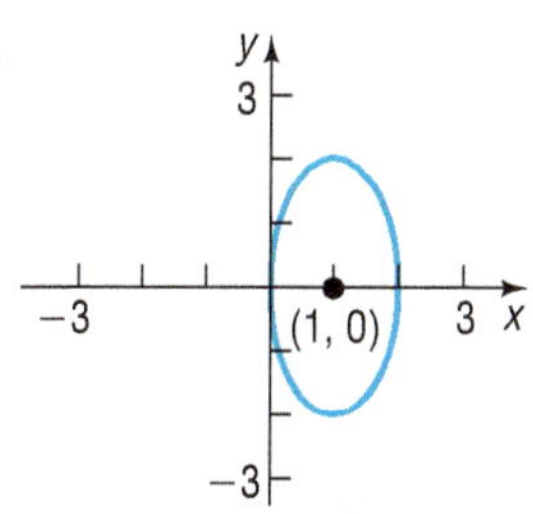

42.

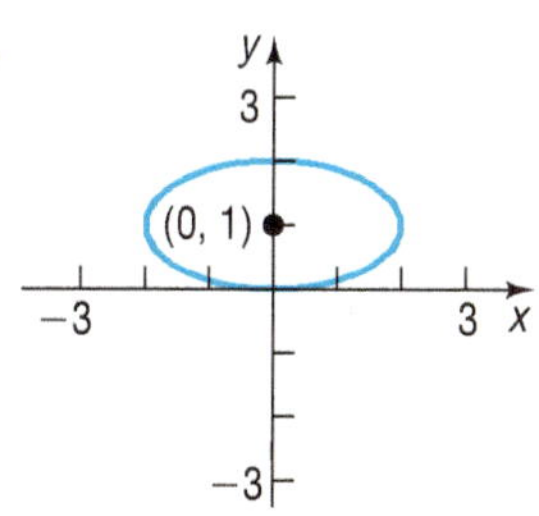

In Problems 43–54, analyze each equation; that is, find the center, foci, and vertices of each ellipse. Graph each equation by hand. Verify your graph using a graphing utility.

43. $\frac{(x-3)^2}{4} + \frac{(y+1)^2}{9} = 1$

44. $\frac{(x+4)^2}{9} + \frac{(y+2)^2}{4} = 1$

45. $(x+5)^2 + 4(y-4)^2 = 16$

46. $9(x-3)^2 + (y+2)^2 = 18$

47. $x^2 + 4x + 4y^2 - 8y + 4 = 0$

48. $x^2 + 3y^2 - 12y + 9 = 0$

49. $2x^2 + 3y^2 - 8x + 6y + 5 = 0$

50. $4x^2 + 3y^2 + 8x - 6y = 5$

51. $9x^2 + 4y^2 - 18x + 16y - 11 = 0$

52. $x^2 + 9y^2 + 6x - 18y + 9 = 0$

53. $4x^2 + y^2 + 4y = 0$

54. $9x^2 + y^2 - 18x = 0$

In Problems 55–64, find an equation for each ellipse. Graph the equation by hand.

55. Center at $(2, -2)$; vertex at $(7, -2)$; focus at $(4, -2)$

56. Center at $(-3, 1)$; vertex at $(-3, 3)$; focus at $(-3, 0)$

57. Vertices at $(4, 3)$ and $(4, 9)$; focus at $(4, 8)$

58. Foci at $(1, 2)$ and $(-3, 2)$; vertex at $(-4, 2)$

59. Foci at $(5, 1)$ and $(-1, 1)$; length of the major axis is 8

60. Vertices at $(2, 5)$ and $(2, -1)$; $c = 2$

61. Center at $(1, 2)$; focus at $(4, 2)$; contains the point $(1, 3)$

62. Center at $(1, 2)$; focus at $(1, 4)$; contains the point $(2, 2)$

63. Center at $(1, 2)$; vertex at $(4, 2)$; contains the point $(1, 3)$

64. Center at $(1, 2)$; vertex at $(1, 4)$; contains the point $(2, 2)$

In Problems 65–68, graph each function. Be sure to label all the intercepts.
[**Hint:** *Notice that each function is half an ellipse.*]

65. $f(x) = \sqrt{16 - 4x^2}$

66. $f(x) = \sqrt{9 - 9x^2}$

67. $f(x) = -\sqrt{64 - 16x^2}$

68. $f(x) = -\sqrt{4 - 4x^2}$

Applications and Extensions

69. Semielliptical Arch Bridge An arch in the shape of the upper half of an ellipse is used to support a bridge that is to span a river 20 meters wide. The center of the arch is 6 meters above the center of the river. See the figure. Write an equation for the ellipse in which the x-axis coincides with the water level and the y-axis passes through the center of the arch.

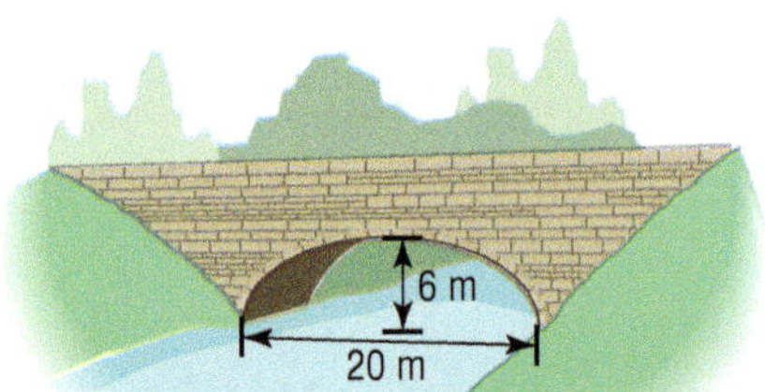

70. Semielliptical Arch Bridge The arch of a bridge is a semiellipse with a horizontal major axis. The span is 30 feet, and the top of the arch is 10 feet above the major axis. The roadway is horizontal and is 2 feet above the top of the arch. Find the vertical distance from the roadway to the arch at 5-foot intervals along the roadway.

71. Whispering Gallery A hall 100 feet in length is to be designed as a whispering gallery. If the foci are located 25 feet from the center, how high will the ceiling be at the center?

72. Whispering Gallery Jim, standing at one focus of a whispering gallery, is 6 feet from the nearest wall. His friend is standing at the other focus, 100 feet away. What is the length of this whispering gallery? How high is its elliptical ceiling at the center?

73. Semielliptical Arch Bridge A bridge is built in the shape of a semielliptical arch. The bridge has a span of 120 feet and a maximum height of 25 feet. Choose a suitable rectangular coordinate system and find the height of the arch at distances of 10, 30, and 50 feet from the center.

74. Semielliptical Arch Bridge A bridge is to be built in the shape of a semielliptical arch and is to have a span of 100 feet. The height of the arch, at a distance of 40 feet from the center, is to be 10 feet. Find the height of the arch at its center.

75. Racetrack Design Consult the figure. A racetrack is in the shape of an ellipse, 100 feet long and 50 feet wide. What is the width 10 feet from a vertex?

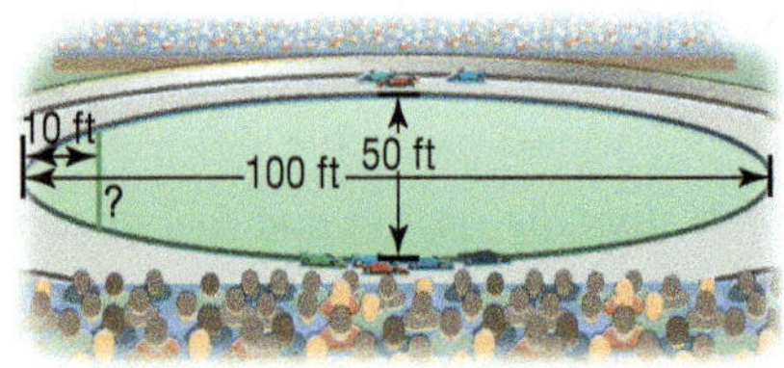

76. Semielliptical Arch Bridge An arch for a bridge over a highway is in the form of half an ellipse. The top of the arch is 20 feet above the ground level (the major axis). The highway has four lanes, each 12 feet wide; a center safety strip 8 feet wide; and two side strips, each 4 feet wide. What should the span of the bridge be (the length of its major axis) if the height 28 feet from the center is to be 13 feet?

77. Installing a Vent Pipe A homeowner is putting in a fireplace that has a 4-inch-radius vent pipe. He needs to cut an elliptical hole in his roof to accommodate the pipe. If the pitch of his roof is $\frac{5}{4}$ (a rise of 5, run of 4), what are the dimensions of the hole?

Source: www.pen.k12.va.us

78. Volume of a Football A football is in the shape of a **prolate spheroid,** which is simply a solid obtained by rotating an ellipse $\left(\frac{x^2}{a^2} + \frac{y^2}{b^2} = 1\right)$ about its major axis. An inflated NFL football averages 11.125 inches in length and 28.25 inches in center circumference. If the volume of a prolate spheroid is $\frac{4}{3}\pi ab^2$, how much air does the football contain? (Neglect material thickness.)

Source: www.answerbag.com

In Problems 79–82, use the fact that the orbit of a planet about the Sun is an ellipse, with the Sun at one focus. The ***aphelion*** *of a planet is its greatest distance from the Sun, and the* ***perihelion*** *is its shortest distance. The* ***mean distance*** *of a planet from the Sun is the length of the semimajor axis of the elliptical orbit. See the illustration.*

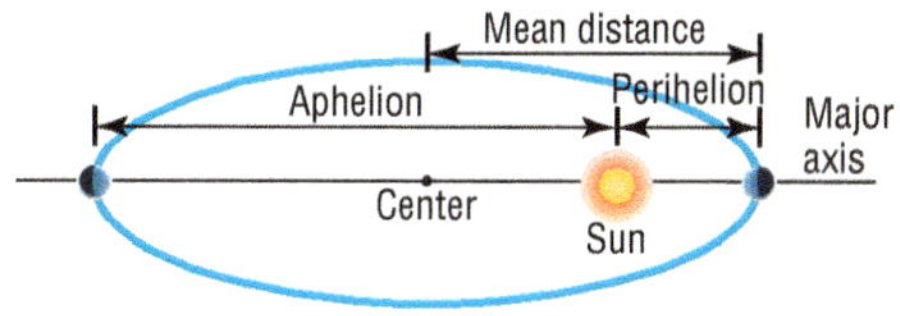

79. Earth The mean distance of Earth from the Sun is 93 million miles. If the aphelion of Earth is 94.5 million miles, what is the perihelion? Write an equation for the orbit of Earth around the Sun.

80. Mars The mean distance of Mars from the Sun is 142 million miles. If the perihelion of Mars is 128.5 million miles, what is the aphelion? Write an equation for the orbit of Mars about the Sun.

81. Jupiter The aphelion of Jupiter is 507 million miles. If the distance from the center of its elliptical orbit to the Sun is 23.2 million miles, what is the perihelion? What is the mean distance? Write an equation for the orbit of Jupiter around the Sun.

82. Pluto The perihelion of Pluto is 4551 million miles, and the distance from the center of its elliptical orbit to the Sun is 897.5 million miles. Find the aphelion of Pluto. What is the mean distance of Pluto from the Sun? Write an equation for the orbit of Pluto about the Sun.

83. Show that an equation of the form

$$Ax^2 + Cy^2 + F = 0 \qquad A \neq 0, C \neq 0, F \neq 0$$

where A and C are of the same sign and F is of opposite sign,

(a) Is the equation of an ellipse with center at $(0, 0)$ if $A \neq C$.

(b) Is the equation of a circle with center $(0, 0)$ if $A = C$.

84. Show that the graph of an equation of the form

$$Ax^2 + Cy^2 + Dx + Ey + F = 0 \qquad A \neq 0, C \neq 0$$

where A and C are of the same sign,

(a) Is an ellipse if $\dfrac{D^2}{4A} + \dfrac{E^2}{4C} - F$ is the same sign as A.

(b) Is a point if $\dfrac{D^2}{4A} + \dfrac{E^2}{4C} - F = 0$.

(c) Contains no points if $\dfrac{D^2}{4A} + \dfrac{E^2}{4C} - F$ is of opposite sign to A.

Discussion and Writing

85. The **eccentricity** e of an ellipse is defined as the number $\dfrac{c}{a}$, where a and c are the numbers given in equation (2). Because $a > c$, it follows that $e < 1$. Write a brief paragraph about the general shape of each of the following ellipses. Be sure to justify your conclusions.

(a) Eccentricity close to 0 (b) Eccentricity $= 0.5$ (c) Eccentricity close to 1

'Are You Prepared?' Answers

1. $\sqrt{13}$ **2.** $\dfrac{9}{4}$ **3.** $(-2, 0), (2, 0), (0, -4), (0, 4)$ **4.** $(2, 5)$ **5.** left; 1; down: 4 **6.** $(x - 2)^2 + (y + 3)^2 = 1$

7.4 The Hyperbola

PREPARING FOR THIS SECTION *Before getting started, review the following:*

- Distance Formula (Section 1.1, p. 89)
- Completing the Square (Section 1.3, pp. 115–116)
- Intercepts (Section 2.1, pp. 170–171)
- Symmetry (Section 2.1, pp. 171–173)
- Asymptotes (Section 5.2, pp. 350–355)
- Graphing Techniques: Transformations (Section 3.5, pp. 256–264)
- Square Root Method (Section 1.3, pp. 114–115)

Now Work the 'Are You Prepared?' problems on page 540.

OBJECTIVES
1. Analyze Hyperbolas with Center at the Origin (p. 530)
2. Find the Asymptotes of a Hyperbola (p. 535)
3. Analyze Hyperbolas with Center at (h, k) (p. 536)
4. Solve Applied Problems Involving Hyperbolas (p. 538)

DEFINITION A **hyperbola** is the set of all points P in the plane, the difference of whose distances from two fixed points, called the **foci,** is a constant.

Figure 34

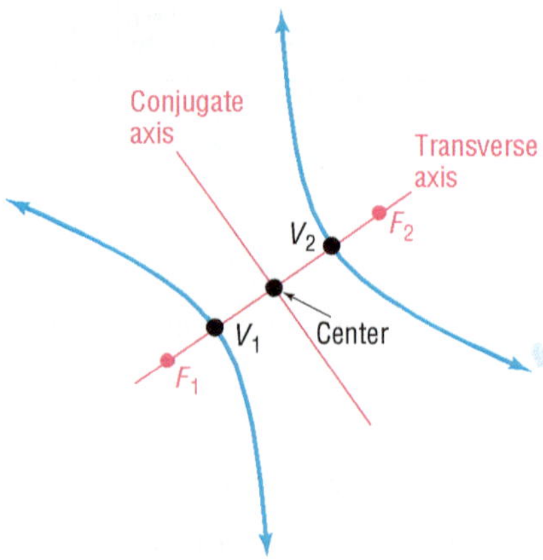

Figure 34 illustrates a hyperbola with foci F_1 and F_2. The line containing the foci is called the **transverse axis.** The midpoint of the line segment joining the foci is the **center** of the hyperbola. The line through the center and perpendicular to the transverse axis is the **conjugate axis.** The hyperbola consists of two separate curves, called **branches,** that are symmetric with respect to the transverse axis, conjugate axis, and center. The two points of intersection of the hyperbola and the transverse axis are the **vertices,** V_1 and V_2, of the hyperbola.

1 Analyze Hyperbolas with Center at the Origin

With these ideas in mind, we are now ready to find the equation of a hyperbola in the rectangular coordinate system. First, we place the center at the origin. Next, we position the hyperbola so that its transverse axis coincides with a coordinate axis. Suppose that the transverse axis coincides with the x-axis, as shown in Figure 35.

If c is the distance from the center to a focus, one focus will be at $F_1 = (-c, 0)$ and the other at $F_2 = (c, 0)$. Now we let the constant difference of the distances from any point $P = (x, y)$ on the hyperbola to the foci F_1 and F_2 be denoted by $\pm 2a$. (If P is on the right branch, the + sign is used; if P is on the left branch, the − sign is used.) The coordinates of P must satisfy the equation

Figure 35 $d(F_1, P) - d(F_2, P) = \pm 2a$

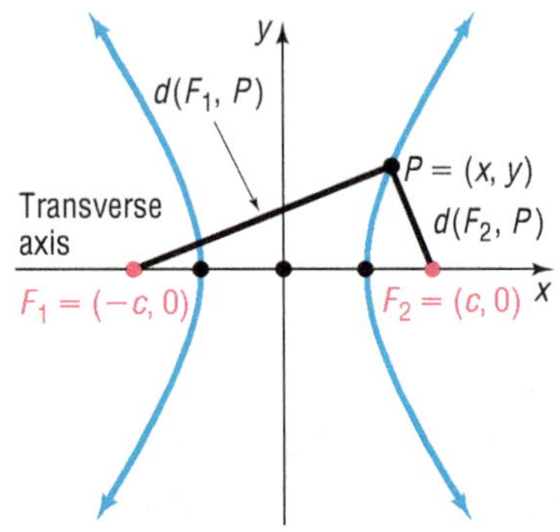

$$d(F_1, P) - d(F_2, P) = \pm 2a$$ Difference of the distances from P to the foci equals $\pm 2a$.

$$\sqrt{(x + c)^2 + y^2} - \sqrt{(x - c)^2 + y^2} = \pm 2a$$ Use the Distance Formula.

$$\sqrt{(x + c)^2 + y^2} = \pm 2a + \sqrt{(x - c)^2 + y^2}$$ Isolate one radical.

$$(x + c)^2 + y^2 = 4a^2 \pm 4a\sqrt{(x - c)^2 + y^2} + (x - c)^2 + y^2$$ Square both sides.

Next we remove the parentheses by expanding the binomials.

$$x^2 + 2cx + c^2 + y^2 = 4a^2 \pm 4a\sqrt{(x - c)^2 + y^2} + x^2 - 2cx + c^2 + y^2$$

$$4cx - 4a^2 = \pm 4a\sqrt{(x - c)^2 + y^2}$$ Simplify; isolate the radical.

$$cx - a^2 = \pm a\sqrt{(x - c)^2 + y^2}$$ Divide each side by 4.

$$(cx - a^2)^2 = a^2[(x - c)^2 + y^2]$$ Square both sides.

$$c^2x^2 - 2ca^2x + a^4 = a^2(x^2 - 2cx + c^2 + y^2)$$ Simplify.

$$c^2x^2 + a^4 = a^2x^2 + a^2c^2 + a^2y^2$$ Distribute and simplify.

$$(c^2 - a^2)x^2 - a^2y^2 = a^2c^2 - a^4$$ Rearrange terms.

$$(c^2 - a^2)x^2 - a^2y^2 = a^2(c^2 - a^2) \quad (1)$$ Factor a^2 on the right side.

To obtain points on the hyperbola off the x-axis, it must be that $a < c$. To see why, look again at Figure 35.

$$d(F_1, P) < d(F_2, P) + d(F_1, F_2)$$ Use triangle F_1PF_2.

$$d(F_1, P) - d(F_2, P) < d(F_1, F_2)$$

$$2a < 2c$$ P is on the right branch, so $d(F_1, P) - d(F_2, P) = 2a$; $d(F_1, F_2) = 2c$.

$$a < c$$

Since $a < c$, we also have $a^2 < c^2$, so $c^2 - a^2 > 0$. Let $b^2 = c^2 - a^2, b > 0$. Then equation (1) can be written as

$$b^2x^2 - a^2y^2 = a^2b^2$$

$$\frac{x^2}{a^2} - \frac{y^2}{b^2} = 1 \quad \text{Divide each side by } a^2b^2.$$

To find the vertices of the hyperbola defined by this equation, let $y = 0$. The vertices satisfy the equation $\frac{x^2}{a^2} = 1$, the solutions of which are $x = \pm a$. Consequently, the vertices of the hyperbola are $V_1 = (-a, 0)$ and $V_2 = (a, 0)$. Notice that the distance from the center $(0, 0)$ to either vertex is a.

THEOREM

Equation of a Hyperbola: Center at (0, 0); Transverse Axis along the x-Axis

An equation of the hyperbola with center at $(0, 0)$, foci at $(-c, 0)$ and $(c, 0)$, and vertices at $(-a, 0)$ and $(a, 0)$ is

$$\frac{x^2}{a^2} - \frac{y^2}{b^2} = 1 \quad \text{where } b^2 = c^2 - a^2 \qquad \textbf{(2)}$$

The transverse axis is the x-axis.

Figure 36

$\frac{x^2}{a^2} - \frac{y^2}{b^2} = 1, \quad b^2 = c^2 - a^2$

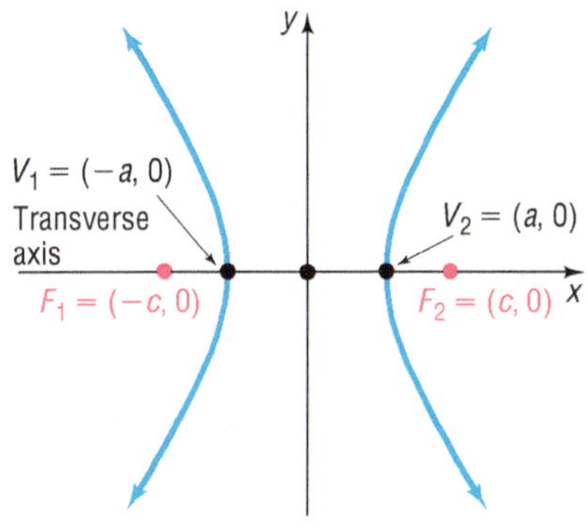

See Figure 36. As you can verify, the hyperbola defined by equation (2) is symmetric with respect to the x-axis, y-axis, and origin. To find the y-intercepts, if any, let $x = 0$ in equation (2). This results in the equation $\frac{y^2}{b^2} = -1$, which has no real solution. We conclude that the hyperbola defined by equation (2) has no y-intercepts. In fact, since $\frac{x^2}{a^2} - 1 = \frac{y^2}{b^2} \geq 0$, it follows that $\frac{x^2}{a^2} \geq 1$. There are no points on the graph for $-a < x < a$.

EXAMPLE 1 Finding and Graphing an Equation of a Hyperbola

Find an equation of the hyperbola with center at the origin, one focus at $(3, 0)$, and one vertex at $(-2, 0)$. Graph the equation.

Solution The hyperbola has its center at the origin. Because the focus and vertex lie on the x-axis, the transverse axis coincides with the x-axis. One focus is at $(c, 0) = (3, 0)$, so $c = 3$. One vertex is at $(-a, 0) = (-2, 0)$, so $a = 2$. From equation (2), it follows that $b^2 = c^2 - a^2 = 9 - 4 = 5$, so an equation of the hyperbola is

$$\frac{x^2}{4} - \frac{y^2}{5} = 1$$

To graph a hyperbola, it is helpful to locate and plot other points on the graph. For example, to find the points above and below the foci, we let $x = \pm 3$. Then

$$\frac{x^2}{4} - \frac{y^2}{5} = 1$$

$$\frac{(\pm 3)^2}{4} - \frac{y^2}{5} = 1 \quad x = \pm 3$$

$$\frac{9}{4} - \frac{y^2}{5} = 1$$

Figure 37

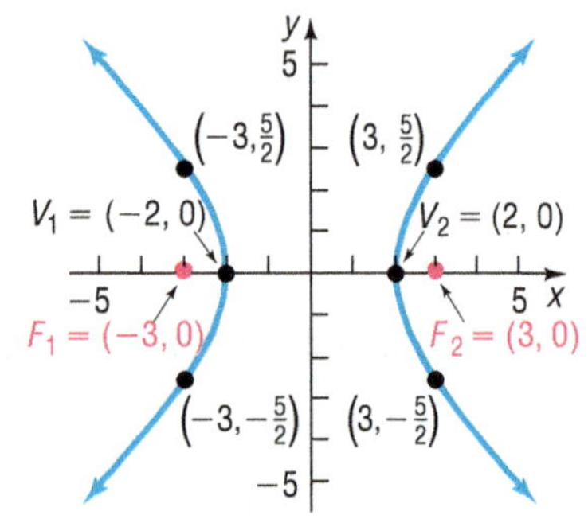

$$\frac{y^2}{5} = \frac{5}{4}$$
$$y^2 = \frac{25}{4}$$
$$y = \pm\frac{5}{2}$$

The points above and below the foci are $\left(\pm 3, \frac{5}{2}\right)$ and $\left(\pm 3, -\frac{5}{2}\right)$. These points determine the "opening" of the hyperbola. See Figure 37.

Now Work PROBLEM 17

EXAMPLE 2 Using a Graphing Utility to Graph a Hyperbola

Using a graphing utility, graph the hyperbola: $\frac{x^2}{4} - \frac{y^2}{5} = 1$

Solution To graph the hyperbola $\frac{x^2}{4} - \frac{y^2}{5} = 1$, we need to graph the two functions $Y_1 = \sqrt{5}\sqrt{\frac{x^2}{4} - 1}$ and $Y_2 = -\sqrt{5}\sqrt{\frac{x^2}{4} - 1}$. As with graphing circles, parabolas, and ellipses on a graphing utility, we use a square screen setting so that the graph is not distorted. Figure 38 shows the graph of the hyperbola.

Figure 38

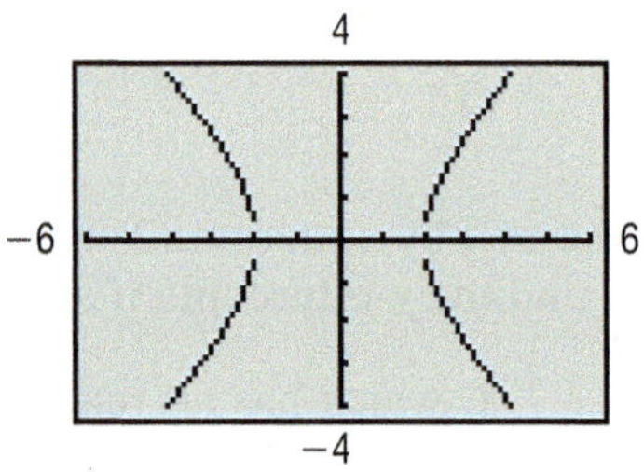

An equation of the form of equation (2) is the equation of a hyperbola with center at the origin, foci on the x-axis at $(-c, 0)$ and $(c, 0)$, where $c^2 = a^2 + b^2$, and transverse axis along the x-axis.

For the next two examples of this section, the direction **"Analyze the equation"** will mean to find the center, transverse axis, vertices, and foci of the hyperbola and graph it.

EXAMPLE 3 Analyzing the Equation of a Hyperbola

Analyze the equation: $\frac{x^2}{16} - \frac{y^2}{4} = 1$

Solution The given equation is of the form of equation (2), with $a^2 = 16$ and $b^2 = 4$. The graph of the equation is a hyperbola with center at $(0, 0)$ and transverse axis along the x-axis. Also, we know that $c^2 = a^2 + b^2 = 16 + 4 = 20$. The vertices are at $(\pm a, 0) = (\pm 4, 0)$, and the foci are at $(\pm c, 0) = (\pm 2\sqrt{5}, 0)$.

To locate the points on the graph above and below the foci, we let $x = \pm 2\sqrt{5}$. Then

$$\frac{x^2}{16} - \frac{y^2}{4} = 1$$
$$\frac{(\pm 2\sqrt{5})^2}{16} - \frac{y^2}{4} = 1 \qquad x = \pm 2\sqrt{5}$$
$$\frac{20}{16} - \frac{y^2}{4} = 1$$
$$\frac{5}{4} - \frac{y^2}{4} = 1$$
$$\frac{y^2}{4} = \frac{1}{4}$$
$$y = \pm 1$$

The points above and below the foci are $(\pm 2\sqrt{5}, 1)$ and $(\pm 2\sqrt{5}, -1)$. See Figure 39(a) for the graph drawn by hand. Figure 39(b) shows the graph obtained using a graphing utility, where $Y_1 = \sqrt{4\left(\frac{x^2}{16} - 1\right)}$ and $Y_2 = -\sqrt{4\left(\frac{x^2}{16} - 1\right)}$.

Figure 39

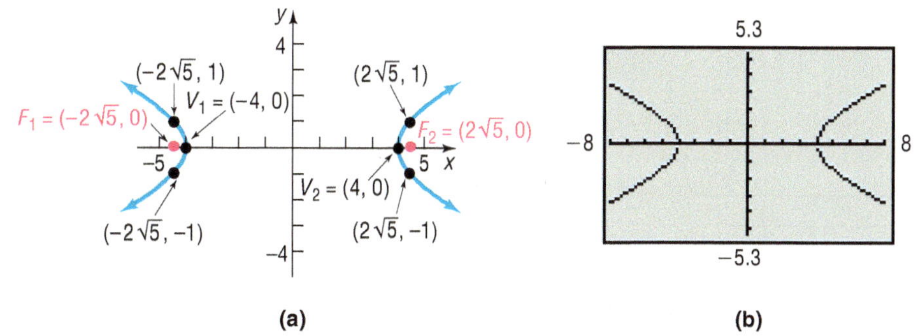

The next result gives the form of the equation of a hyperbola with center at the origin and transverse axis along the y-axis.

THEOREM

Equation of a Hyperbola; Center at (0, 0); Transverse Axis along the *y*-Axis

An equation of the hyperbola with center at $(0, 0)$, foci at $(0, -c)$ and $(0, c)$, and vertices at $(0, -a)$ and $(0, a)$ is

$$\frac{y^2}{a^2} - \frac{x^2}{b^2} = 1 \quad \text{where } b^2 = c^2 - a^2 \qquad (3)$$

The transverse axis is the y-axis.

Figure 40

$\frac{y^2}{a^2} - \frac{x^2}{b^2} = 1, b^2 = c^2 - a^2$

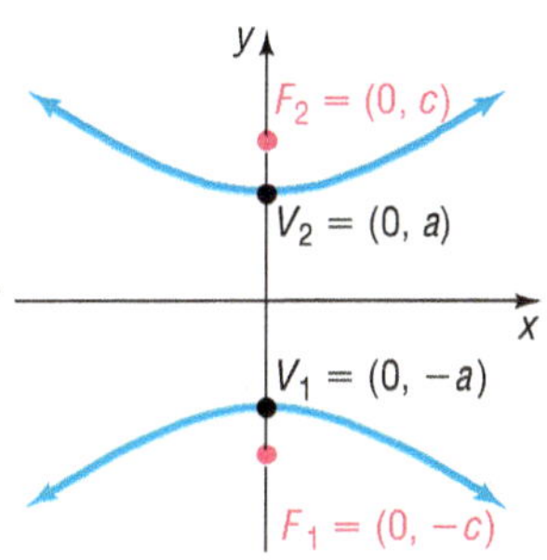

Figure 40 shows the graph of a typical hyperbola defined by equation (3).

An equation of the form of equation (2), $\frac{x^2}{a^2} - \frac{y^2}{b^2} = 1$, is the equation of a hyperbola with center at the origin, foci on the x-axis at $(-c, 0)$ and $(c, 0)$, where $c^2 = a^2 + b^2$, and transverse axis along the x-axis.

An equation of the form of equation (3), $\frac{y^2}{a^2} - \frac{x^2}{b^2} = 1$, is the equation of a hyperbola with center at the origin, foci on the y-axis at $(0, -c)$ and $(0, c)$, where $c^2 = a^2 + b^2$, and transverse axis along the y-axis.

Notice the difference in the forms of equations (2) and (3). When the y^2-term is subtracted from the x^2-term, the transverse axis is along the x-axis. When the x^2-term is subtracted from the y^2-term, the transverse axis is along the y-axis.

EXAMPLE 4 Analyzing the Equation of a Hyperbola

Analyze the equation: $y^2 - 4x^2 = 4$

Solution To put the equation in proper form, we divide each side by 4:

$$\frac{y^2}{4} - x^2 = 1$$

Since the x^2-term is subtracted from the y^2-term, the equation is that of a hyperbola with center at the origin and transverse axis along the y-axis. Also, comparing the above equation to equation (3), we find $a^2 = 4$, $b^2 = 1$, and $c^2 = a^2 + b^2 = 5$. The vertices are at $(0, \pm a) = (0, \pm 2)$, and the foci are at $(0, \pm c) = (0, \pm\sqrt{5})$.

To locate other points on the graph, we let $x = \pm 2$. Then

$$y^2 - 4x^2 = 4$$
$$y^2 - 4(\pm 2)^2 = 4 \qquad x = \pm 2$$
$$y^2 - 16 = 4$$
$$y^2 = 20$$
$$y = \pm 2\sqrt{5}$$

Four other points on the graph are $(\pm 2, 2\sqrt{5})$ and $(\pm 2, -2\sqrt{5})$. See Figure 41(a) for the graph drawn by hand. Figure 41(b) shows the graph obtained using a graphing utility.

Figure 41

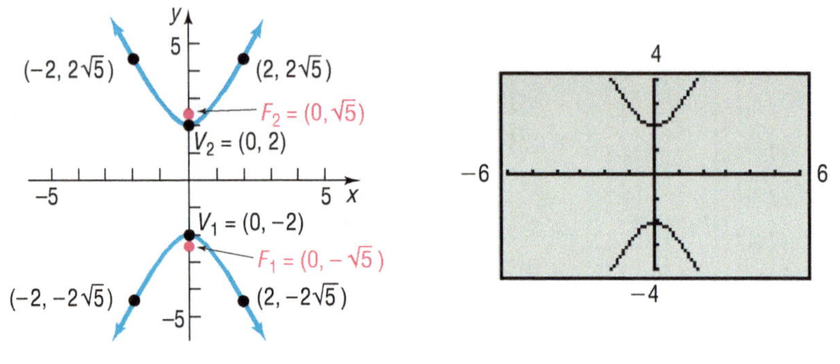

(a) (b)

EXAMPLE 5 Finding an Equation of a Hyperbola

Find an equation of the hyperbola having one vertex at $(0, 2)$ and foci at $(0, -3)$ and $(0, 3)$. Graph the equation.

Solution Since the foci are at $(0, -3)$ and $(0, 3)$, the center of the hyperbola, which is at their midpoint, is the origin. Because the foci lie on the y-axis, the transverse axis is along the y-axis. The given information also reveals that $c = 3$, $a = 2$, and $b^2 = c^2 - a^2 = 9 - 4 = 5$. The form of the equation of the hyperbola is given by equation (3):

$$\frac{y^2}{a^2} - \frac{x^2}{b^2} = 1$$
$$\frac{y^2}{4} - \frac{x^2}{5} = 1$$

Let $y = \pm 3$ to obtain points on the graph on either side of the foci. See Figure 42.

Figure 42

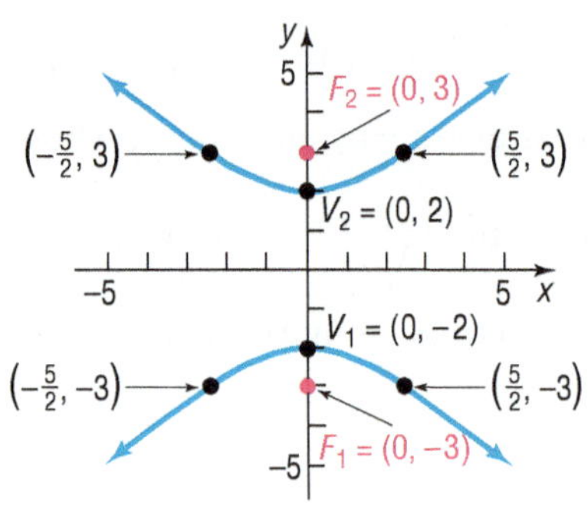

Now Work PROBLEM 19

Look at the equations of the hyperbolas in Examples 3 and 5. For the hyperbola in Example 3, $a^2 = 16$ and $b^2 = 4$, so $a > b$; for the hyperbola in Example 5, $a^2 = 4$ and $b^2 = 5$, so $a < b$. We conclude that, for hyperbolas, there are no requirements involving the relative sizes of a and b. Contrast this situation to the case of an ellipse, in which the relative sizes of a and b dictate which axis is the major axis. Hyperbolas have another feature to distinguish them from ellipses and parabolas: Hyperbolas have asymptotes.

2 Find the Asymptotes of a Hyperbola

Recall from Section 5.2 that a horizontal or oblique asymptote of a graph is a line with the property that the distance from the line to points on the graph approaches 0 as $x \to -\infty$ or as $x \to \infty$. The asymptotes provide information about the end behavior of the graph of a hyperbola.

THEOREM

Asymptotes of a Hyperbola

The hyperbola $\frac{x^2}{a^2} - \frac{y^2}{b^2} = 1$ has the two oblique asymptotes

$$y = \frac{b}{a}x \quad \text{and} \quad y = -\frac{b}{a}x \tag{4}$$

Proof We begin by solving for y in the equation of the hyperbola.

$$\frac{x^2}{a^2} - \frac{y^2}{b^2} = 1$$

$$\frac{y^2}{b^2} = \frac{x^2}{a^2} - 1$$

$$y^2 = b^2\left(\frac{x^2}{a^2} - 1\right)$$

Since $x \neq 0$, we can rearrange the right side in the form

$$y^2 = \frac{b^2x^2}{a^2}\left(1 - \frac{a^2}{x^2}\right)$$

$$y = \pm\frac{bx}{a}\sqrt{1 - \frac{a^2}{x^2}}$$

Now, as $x \to -\infty$ or as $x \to \infty$, the term $\frac{a^2}{x^2}$ approaches 0, so the expression under the radical approaches 1. So, as $x \to -\infty$ or as $x \to \infty$, the value of y approaches $\pm\frac{bx}{a}$; that is, the graph of the hyperbola approaches the lines

$$y = -\frac{b}{a}x \quad \text{and} \quad y = \frac{b}{a}x$$

These lines are oblique asymptotes of the hyperbola. ■

The asymptotes of a hyperbola are not part of the hyperbola, but they do serve as a guide for graphing a hyperbola. For example, suppose that we want to graph the equation

$$\frac{x^2}{a^2} - \frac{y^2}{b^2} = 1$$

We begin by plotting the vertices $(-a, 0)$ and $(a, 0)$. Then we plot the points $(0, -b)$ and $(0, b)$ and use these four points to construct a rectangle, as shown in Figure 43. The diagonals of this rectangle have slopes $\frac{b}{a}$ and $-\frac{b}{a}$, and their extensions are the asymptotes $y = \frac{b}{a}x$ and $y = -\frac{b}{a}x$ of the hyperbola. If we graph the asymptotes, we can use them to establish the "opening" of the hyperbola and avoid plotting other points.

Figure 43

$\frac{x^2}{a^2} - \frac{y^2}{b^2} = 1$

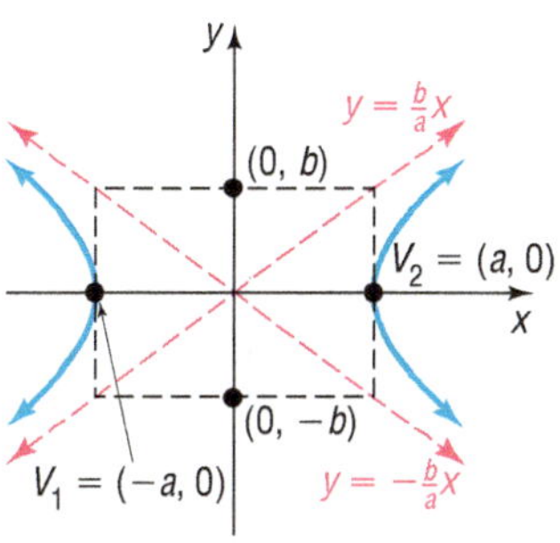

THEOREM

Asymptotes of a Hyperbola

The hyperbola $\frac{y^2}{a^2} - \frac{x^2}{b^2} = 1$ has the two oblique asymptotes:

$$y = \frac{a}{b}x \quad \text{and} \quad y = -\frac{a}{b}x \tag{5}$$

You are asked to prove this result in Problem 82.

For the remainder of this section, the direction **"Analyze the equation"** will mean to find the center, transverse axis, vertices, foci, and asymptotes of the hyperbola and graph it.

EXAMPLE 6 Analyzing the Equation of a Hyperbola

Analyze the equation: $9x^2 - 4y^2 = 36$

Solution Divide each side of the equation by 36 to put the equation in proper form.

$$\frac{x^2}{4} - \frac{y^2}{9} = 1$$

We now proceed to analyze the equation. The center of the hyperbola is the origin. Since the x^2-term is first in the equation, we know that the transverse axis is along the x-axis and the vertices and foci will lie on the x-axis. Using equation (2), we find $a^2 = 4$, $b^2 = 9$, and $c^2 = a^2 + b^2 = 13$. The vertices are $a = 2$ units left and right of the center at $(\pm a, 0) = (\pm 2, 0)$, the foci are $c = \sqrt{13}$ units left and right of the center at $(\pm c, 0) = (\pm\sqrt{13}, 0)$, and the asymptotes have the equations

$$y = \frac{b}{a}x = \frac{3}{2}x \quad \text{and} \quad y = -\frac{b}{a}x = -\frac{3}{2}x$$

To graph the hyperbola by hand, form the rectangle containing the points $(\pm a, 0)$ and $(0, \pm b)$, that is, $(-2, 0)$, $(2, 0)$, $(0, -3)$, and $(0, 3)$. The extensions of the diagonals of this rectangle are the asymptotes. See Figure 44(a) for the graph drawn by hand. Figure 44(b) shows the graph obtained using a graphing utility.

Figure 44

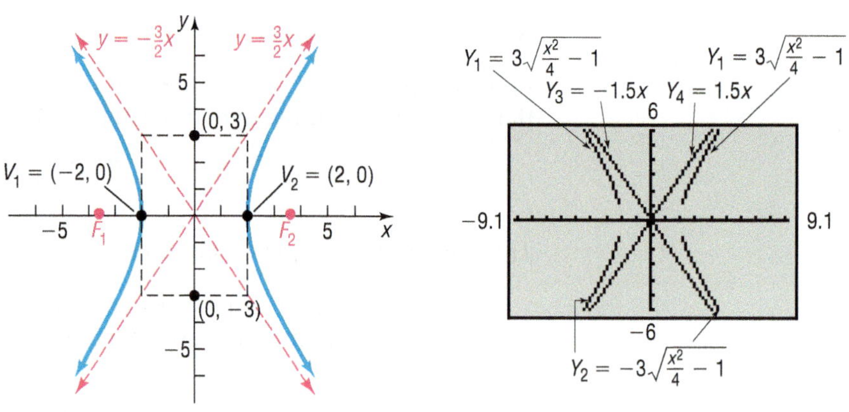

(a) (b)

Seeing the Concept

Refer to Figure 44(b). Create a TABLE using Y_1 and Y_4 with $x = 10, 100, 1000$, and $10{,}000$. Compare the values of Y_1 and Y_4. Repeat for Y_1 and Y_3, Y_2 and Y_3, and Y_2 and Y_4.

Now Work PROBLEM 29

3 Analyze Hyperbolas with Center at (h, k)

If a hyperbola with center at the origin and transverse axis coinciding with a coordinate axis is shifted horizontally h units and then vertically k units, the result is a hyperbola with center at (h, k) and transverse axis parallel to a coordinate axis. The equations of such hyperbolas have the same forms as those given in equations (2) and (3), except that x is replaced by $x - h$ (the horizontal shift) and y is replaced by $y - k$ (the vertical shift). Table 4 gives the forms of the equations of such hyperbolas. See Figure 45 for typical graphs.

Table 4

HYPERBOLAS WITH CENTER AT (h, k) AND TRANSVERSE AXIS PARALLEL TO A COORDINATE AXIS

Center	Transverse Axis	Foci	Vertices	Equation	Asymptotes
(h, k)	Parallel to the x-axis	$(h \pm c, k)$	$(h \pm a, k)$	$\dfrac{(x-h)^2}{a^2} - \dfrac{(y-k)^2}{b^2} = 1, \quad b^2 = c^2 - a^2$	$y - k = \pm\dfrac{b}{a}(x - h)$
(h, k)	Parallel to the y-axis	$(h, k \pm c)$	$(h, k \pm a)$	$\dfrac{(y-k)^2}{a^2} - \dfrac{(x-h)^2}{b^2} = 1, \quad b^2 = c^2 - a^2$	$y - k = \pm\dfrac{a}{b}(x - h)$

Figure 45

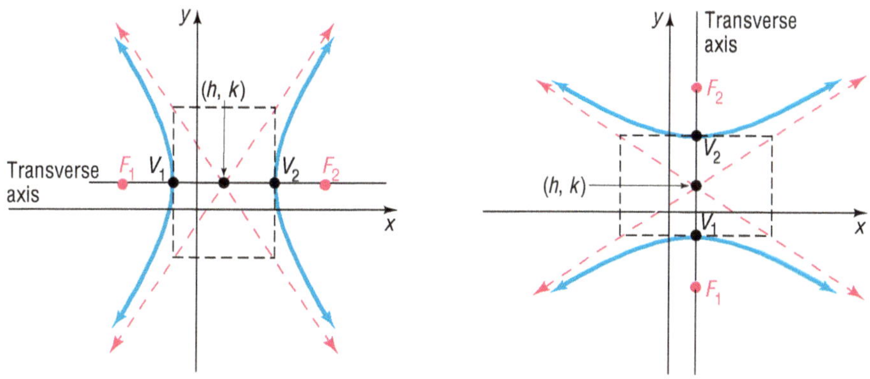

(a) $\dfrac{(x-h)^2}{a^2} - \dfrac{(y-k)^2}{b^2} = 1$ **(b)** $\dfrac{(y-k)^2}{a^2} - \dfrac{(x-h)^2}{b^2} = 1$

EXAMPLE 7 Finding an Equation of a Hyperbola, Center Not at the Origin

Find an equation for the hyperbola with center at $(1, -2)$, one focus at $(4, -2)$, and one vertex at $(3, -2)$. Graph the equation by hand.

Figure 46

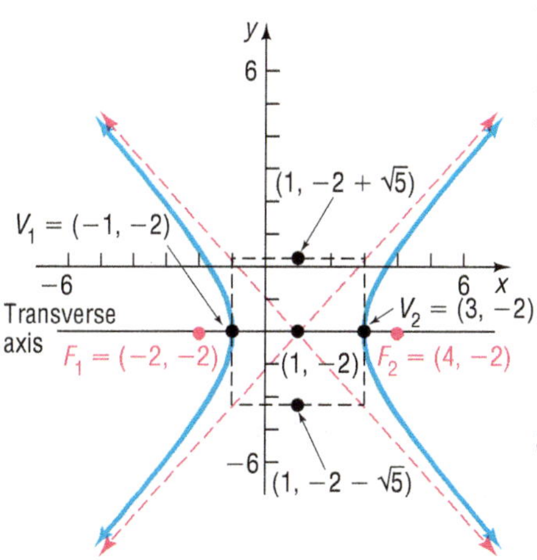

Solution The center is at $(h, k) = (1, -2)$, so $h = 1$ and $k = -2$. Since the center, focus, and vertex all lie on the line $y = -2$, the transverse axis is parallel to the x-axis. The distance from the center $(1, -2)$ to the focus $(4, -2)$ is $c = 3$; the distance from the center $(1, -2)$ to the vertex $(3, -2)$ is $a = 2$. Thus, $b^2 = c^2 - a^2 = 9 - 4 = 5$. The equation is

$$\frac{(x-h)^2}{a^2} - \frac{(y-k)^2}{b^2} = 1$$

$$\frac{(x-1)^2}{4} - \frac{(y+2)^2}{5} = 1$$

See Figure 46.

Now Work PROBLEM 39

EXAMPLE 8 **Analyzing the Equation of a Hyperbola**

Analyze the equation: $-x^2 + 4y^2 - 2x - 16y + 11 = 0$

Solution We complete the squares in x and in y.

$$-x^2 + 4y^2 - 2x - 16y + 11 = 0$$

$$-(x^2 + 2x) + 4(y^2 - 4y) = -11 \quad \text{Group terms.}$$

$$-(x^2 + 2x + 1) + 4(y^2 - 4y + 4) = -11 - 1 + 16 \quad \text{Complete each square.}$$

$$-(x + 1)^2 + 4(y - 2)^2 = 4$$

$$(y - 2)^2 - \frac{(x + 1)^2}{4} = 1 \quad \text{Divide each side by 4.}$$

This is the equation of a hyperbola with center at $(-1, 2)$ and transverse axis parallel to the y-axis. Also, $a^2 = 1$ and $b^2 = 4$, so $c^2 = a^2 + b^2 = 5$. Since the transverse axis is parallel to the y-axis, the vertices and foci are located a and c units above and below the center, respectively. The vertices are at $(h, k \pm a) = (-1, 2 \pm 1)$, or $(-1, 1)$ and $(-1, 3)$. The foci are at $(h, k \pm c) = \left(-1, 2 \pm \sqrt{5}\right)$. The asymptotes are $y - 2 = \frac{1}{2}(x + 1)$ and $y - 2 = -\frac{1}{2}(x + 1)$. Figure 47(a) shows the graph drawn by hand. Figure 47(b) shows the graph obtained using a graphing utility.

Figure 47

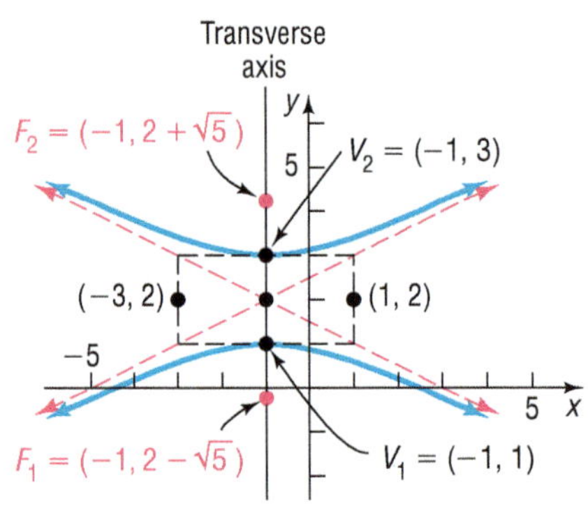

(a)

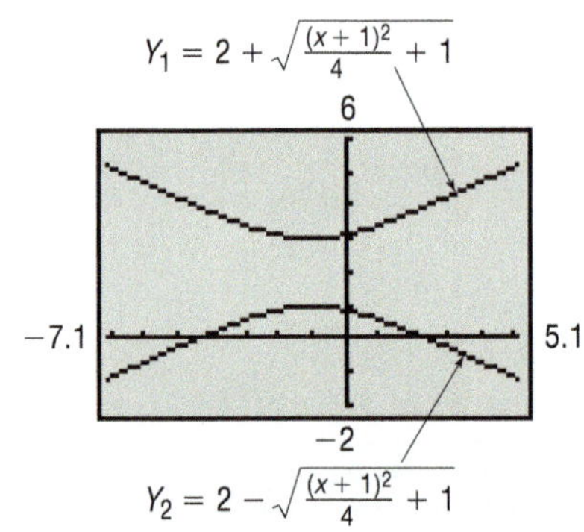

(b)

Now Work PROBLEM 53

4 Solve Applied Problems Involving Hyperbolas

Look at Figure 48. Suppose that three microphones are located at points O_1, O_2, and O_3 (the foci of the two hyperbolas). In addition, suppose that a gun is fired at S and the microphone at O_1 records the gun shot 1 second after the microphone at O_2. Because sound travels at about 1100 feet per second, we conclude that the microphone at O_1 is 1100 feet farther from the gunshot than O_2. We can model this situation by saying that S lies on a branch of a hyperbola with foci at O_1 and O_2. (Do you see why? The difference of the distances from S to O_1 and from S to O_2 is the constant 1100.) If the third microphone at O_3 records the gunshot 2 seconds after O_1, then S will lie on a branch of a second hyperbola with foci at O_1 and O_3. In this case, the constant difference will be 2200. The intersection of the two hyperbolas will identify the location of S.

Figure 48

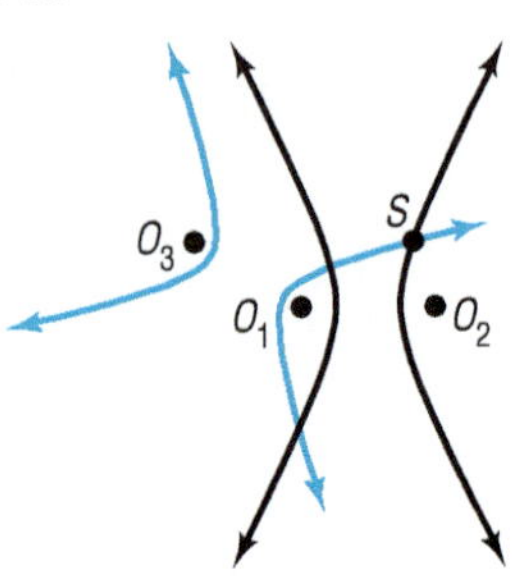

EXAMPLE 9 Lightning Strikes

Suppose that two people standing 1 mile apart both see a flash of lightning. After a period of time, the person standing at point A hears the thunder. One second later, the person standing at point B hears the thunder. If the person at B is due west of the person at A and the lightning strike is known to occur due north of the person standing at point A, where did the lightning strike?

Solution See Figure 49 in which the ordered pair (x, y) represents the location of the lightning strike. We know that sound travels at 1100 feet per second, so the person at point A is 1100 feet closer to the lightning strike than the person at point B. Since the difference of the distance from (x, y) to A and the distance from (x, y) to B is the constant 1100, the point (x, y) lies on a hyperbola whose foci are at A and B.

Figure 49

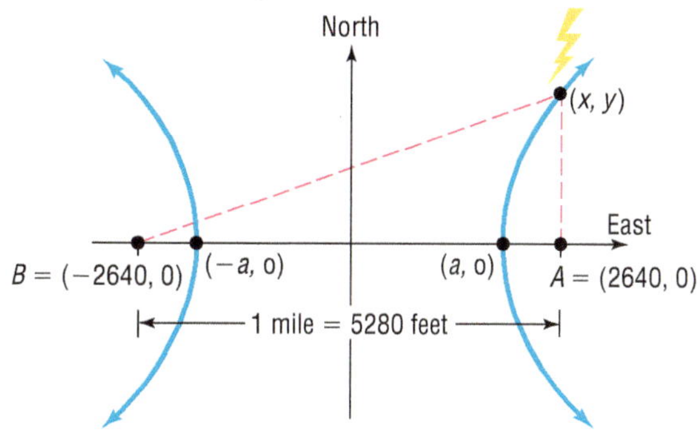

An equation of the hyperbola is

$$\frac{x^2}{a^2} - \frac{y^2}{b^2} = 1$$

where $2a = 1100$, so $a = 550$.

Because the distance between the two people is 1 mile (5280 feet) and each person is at a focus of the hyperbola, we have

$$2c = 5280$$
$$c = \frac{5280}{2} = 2640$$

Since $b^2 = c^2 - a^2 = 2640^2 - 550^2 = 6{,}667{,}100$, the equation of the hyperbola that describes the location of the lightning strike is

$$\frac{x^2}{550^2} - \frac{y^2}{6{,}667{,}100} = 1$$

Refer again to Figure 49. Since the lightning strikes due north of the individual at the point $A = (2640, 0)$, we let $x = 2640$ and solve the resulting equation.

$$\frac{2640^2}{550^2} - \frac{y^2}{6{,}667{,}100} = 1$$

$$-\frac{y^2}{6{,}667{,}100} = -22.04 \qquad \text{Subtract } \frac{2640^2}{550^2} \text{ from both sides.}$$

$$y^2 = 146{,}942{,}884 \qquad \text{Multiply both sides by } -6{,}667{,}100.$$

$$y = 12{,}122 \qquad y > 0 \text{ since the lightning strike occured in quadrant I.}$$

Check: The difference between the distance from (2640, 12,122) to the person at the point $B = (-2640, 0)$ and the distance from (2640, 12,122) to the person at the point $A = (2640, 0)$ should be 1100. Using the distance formula, we find the difference in the distances is

$$\sqrt{[2640 - (-2640)]^2 + (12{,}122 - 0)^2} - \sqrt{(2640 - 2640)^2 + (12{,}122 - 0)^2} = 1100$$

as required.

The lightning strike is 12,122 feet north of the person standing at point A.

Now Work PROBLEM 73

7.4 Assess Your Understanding

'Are You Prepared?' *Answers are given at the end of these exercises. If you get a wrong answer, read the pages listed in red.*

1. The distance d from $P_1 = (3, -4)$ to $P_2 = (-2, 1)$ is $d =$ ______. (pp. 88–90)
2. To complete the square of $x^2 + 5x$, add ______. (pp. 115–116)
3. Find the intercepts of the equation $y^2 = 9 + 4x^2$. (pp. 170–171)
4. ***True or False*** The equation $y^2 = 9 + x^2$ is symmetric with respect to the x-axis, the y-axis, and the origin. (pp. 171–173)
5. To graph $y = (x - 5)^3 - 4$, shift the graph of $y = x^3$ to the (left/right) ______ unit(s) and then (up/down) ______ unit(s). (pp. 256–258)
6. Find the vertical asymptotes, if any, and the horizontal or oblique asymptotes, if any, of $y = \dfrac{x^2 - 9}{x^2 - 4}$. (pp. 350–355)

Concepts and Vocabulary

7. A(n) ______ is the collection of points in the plane the difference of whose distances from two fixed points is a constant.
8. For a hyperbola, the foci lie on a line called the ______ ______.
9. The asymptotes of the hyperbola $\dfrac{x^2}{4} - \dfrac{y^2}{9} = 1$ are ______ and ______.
10. ***True or False*** The foci of a hyperbola lie on a line called the axis of symmetry.
11. ***True or False*** Hyperbolas always have asymptotes.
12. ***True or False*** A hyperbola will never intersect its transverse axis.

Skill Building

In Problems 13–16, the graph of a hyperbola is given. Match each graph to its equation.

(A) $\dfrac{x^2}{4} - y^2 = 1$ (B) $x^2 - \dfrac{y^2}{4} = 1$ (C) $\dfrac{y^2}{4} - x^2 = 1$ (D) $y^2 - \dfrac{x^2}{4} = 1$

13.
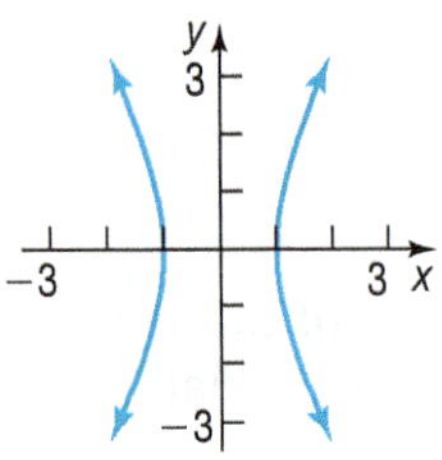

14.
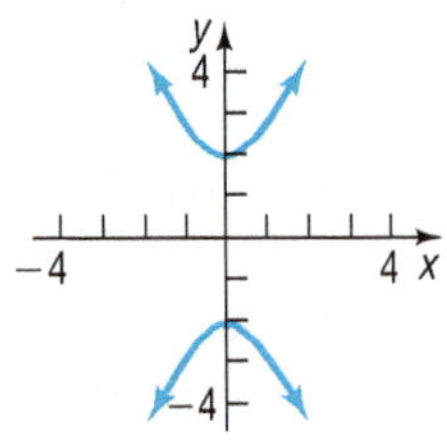

15.
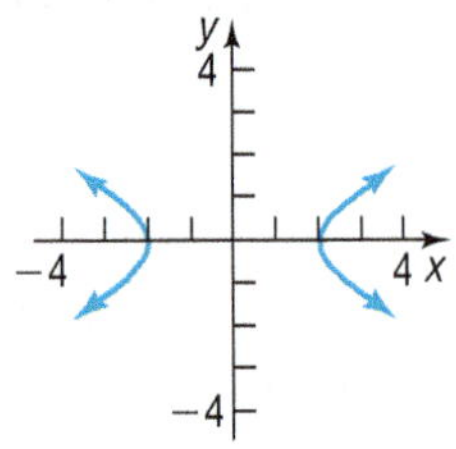

16.
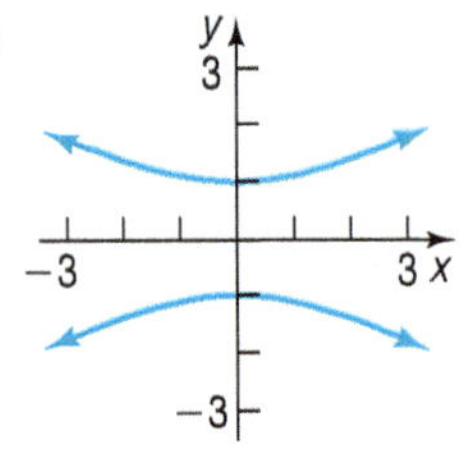

In Problems 17–26, find an equation for the hyperbola described. Graph the equation by hand.

17. Center at (0, 0); focus at (3, 0); vertex at (1, 0)
18. Center at (0, 0); focus at (0, 5); vertex at (0, 3)
19. Center at (0, 0); focus at (0, −6); vertex at (0, 4)
20. Center at (0, 0); focus at (−3, 0); vertex at (2, 0)
21. Foci at (−5, 0) and (5, 0); vertex at (3, 0)
22. Focus at (0, 6); vertices at (0, −2) and (0, 2)
23. Vertices at (0, −6) and (0, 6); asymptote the line $y = 2x$
24. Vertices at (−4, 0) and (4, 0); asymptote the line $y = 2x$
25. Foci at (−4, 0) and (4, 0); asymptote the line $y = -x$
26. Foci at (0, −2) and (0, 2); asymptote the line $y = -x$

In Problems 27–34, find the center, transverse axis, vertices, foci, and asymptotes. Graph each equation by hand. Verify your graph using a graphing utility.

27. $\dfrac{x^2}{25} - \dfrac{y^2}{9} = 1$ **28.** $\dfrac{y^2}{16} - \dfrac{x^2}{4} = 1$ **29.** $4x^2 - y^2 = 16$ **30.** $4y^2 - x^2 = 16$

31. $y^2 - 9x^2 = 9$ **32.** $x^2 - y^2 = 4$ **33.** $y^2 - x^2 = 25$ **34.** $2x^2 - y^2 = 4$

In Problems 35–38, write an equation for each hyperbola.

35.

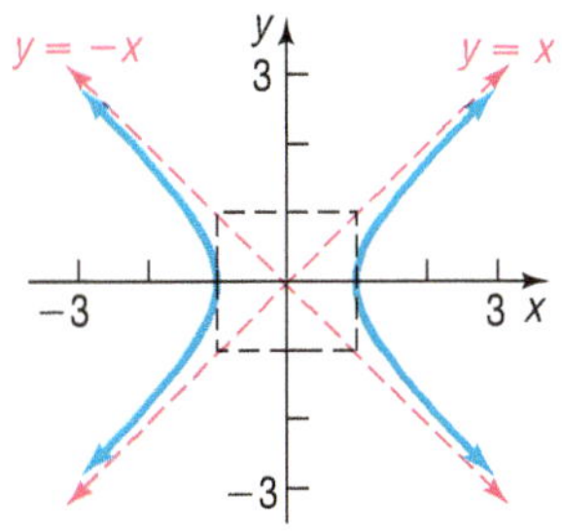

36.

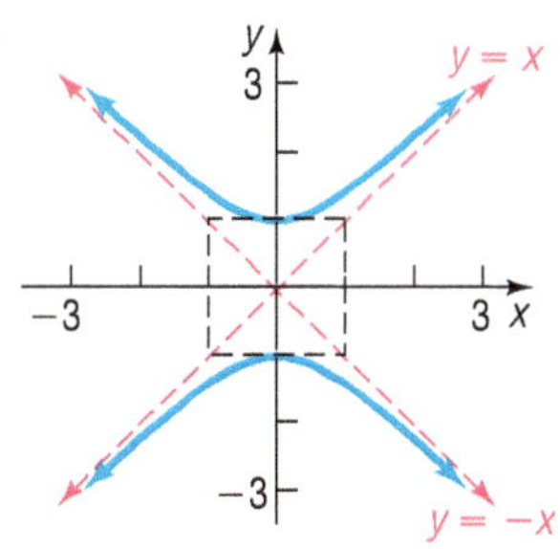

37.

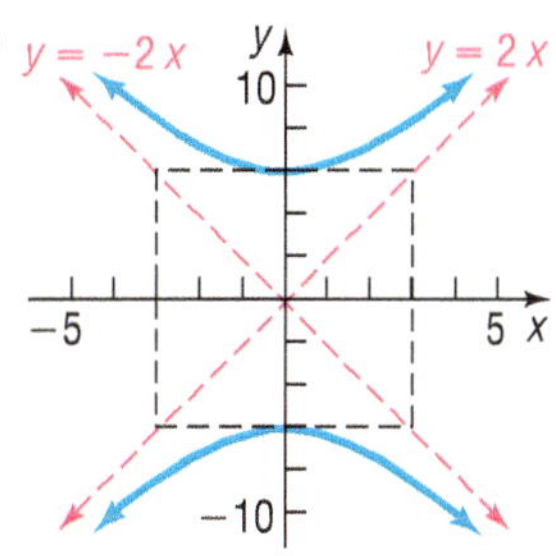

38.

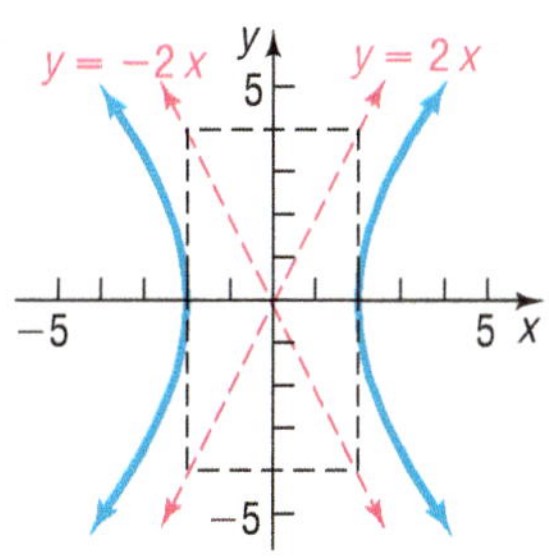

In Problems 39–46, find an equation for the hyperbola described. Graph each equation by hand.

39. Center at $(4, -1)$; focus at $(7, -1)$; vertex at $(6, -1)$

40. Center at $(-3, 1)$; focus at $(-3, 6)$; vertex at $(-3, 4)$

41. Center at $(-3, -4)$; focus at $(-3, -8)$; vertex at $(-3, -2)$

42. Center at $(1, 4)$; focus at $(-2, 4)$; vertex at $(0, 4)$

43. Foci at $(3, 7)$ and $(7, 7)$; vertex at $(6, 7)$

44. Focus at $(-4, 0)$ vertices at $(-4, 4)$ and $(-4, 2)$

45. Vertices at $(-1, -1)$ and $(3, -1)$; asymptote the line $y + 1 = \dfrac{3}{2}(x - 1)$

46. Vertices at $(1, -3)$ and $(1, 1)$; asymptote the line $y + 1 = \dfrac{3}{2}(x - 1)$

In Problems 47–60, find the center, transverse axis, vertices, foci, and asymptotes. Graph each equation by hand. Verify your graph using a graphing utility.

47. $\dfrac{(x-2)^2}{4} - \dfrac{(y+3)^2}{9} = 1$ **48.** $\dfrac{(y+3)^2}{4} - \dfrac{(x-2)^2}{9} = 1$ **49.** $(y-2)^2 - 4(x+2)^2 = 4$

50. $(x+4)^2 - 9(y-3)^2 = 9$ **51.** $(x+1)^2 - (y+2)^2 = 4$ **52.** $(y-3)^2 - (x+2)^2 = 4$

53. $x^2 - y^2 - 2x - 2y - 1 = 0$ **54.** $y^2 - x^2 - 4y + 4x - 1 = 0$ **55.** $y^2 - 4x^2 - 4y - 8x - 4 = 0$

56. $2x^2 - y^2 + 4x + 4y - 4 = 0$ **57.** $4x^2 - y^2 - 24x - 4y + 16 = 0$ **58.** $2y^2 - x^2 + 2x + 8y + 3 = 0$

59. $y^2 - 4x^2 - 16x - 2y - 19 = 0$ **60.** $x^2 - 3y^2 + 8x - 6y + 4 = 0$

In Problems 61–64, graph each function. Be sure to label any intercepts.
[**Hint:** *Notice that each function is half a hyperbola.*]

61. $f(x) = \sqrt{16 + 4x^2}$ **62.** $f(x) = -\sqrt{9 + 9x^2}$ **63.** $f(x) = -\sqrt{-25 + x^2}$ **64.** $f(x) = \sqrt{-1 + x^2}$

Mixed Practice

In Problems 65–72, analyze each conic section.

65. $\dfrac{(x-3)^2}{4} - \dfrac{y^2}{25} = 1$ **66.** $\dfrac{(y+2)^2}{16} - \dfrac{(x-2)^2}{4} = 1$ **67.** $x^2 = 16(y - 3)$

68. $y^2 = -12(x + 1)$ **69.** $25x^2 + 9y^2 - 250x + 400 = 0$ **70.** $x^2 + 36y^2 - 2x + 288y + 541 = 0$

71. $x^2 - 6x - 8y - 31 = 0$ **72.** $9x^2 - y^2 - 18x - 8y - 88 = 0$

Applications and Extensions

73. Fireworks Display Suppose that two people standing 2 miles apart both see the burst from a fireworks display. After a period of time, the first person standing at point A hears the burst. One second later, the second person standing at point B hears the burst. If the person at point B is due west of the person at point A and if the display is known to occur due north of the person at point A, where did the fireworks display occur?

74. Lightning Strikes Suppose that two people standing 1 mile apart both see a flash of lightning. After a period of time, the first person standing at point A hears the thunder. Two seconds later, the second person standing at point B hears the thunder. If the person at point B is due west of the person at point A and if the lightning strike is known to occur due north of the person standing at point A, where did the lightning strike?

75. Nuclear Power Plant Some nuclear power plants utilize "natural draft" cooling towers in the shape of a **hyperboloid,** a solid obtained by rotating a hyperbola about its conjugate axis. Suppose such a cooling tower has a base diameter of 400 feet and the diameter at its narrowest point, 360 feet above the ground, is 200 feet. If the diameter at the top of the tower is 300 feet, how tall is the tower?

Source: *Bay Area Air Quality Management District*

76. An Explosion Two recording devices are set 2400 feet apart, with the device at point A to the west of the device at point B. At a point between the devices, 300 feet from point B, a small amount of explosive is detonated. The recording devices record the time until the sound reaches each. How far directly north of point B should a second explosion be done so that the measured time difference recorded by the devices is the same as that for the first detonation?

77. Rutherford's Experiment In May 1911, Ernest Rutherford published a paper in *Philosophical Magazine*. In this article, he described the motion of alpha particles as they are shot at a piece of gold foil 0.00004 cm thick. Before conducting this experiment, Rutherford expected that the alpha particles would shoot through the foil just as a bullet would shoot through snow. Instead, a small fraction of the alpha particles bounced off the foil. This led to the conclusion that the nucleus of an atom is dense, while the remainder of the atom is sparse. Only the density of the nucleus could cause the alpha particles to deviate from their path. The figure shows a diagram from Rutherford's paper that indicates that the deflected alpha particles follow the path of one branch of a hyperbola.

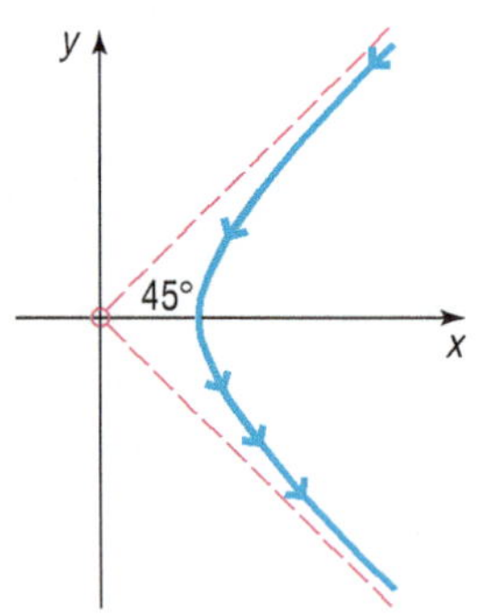

(a) Find an equation of the asymptotes under this scenario.
(b) If the vertex of the path of the alpha particles is 10 cm from the center of the hyperbola, find a model that describes the path of the particle.

78. Hyperbolic Mirrors Hyperbolas have interesting reflective properties that make them useful for lenses and mirrors. For example, if a ray of light strikes a convex hyperbolic mirror on a line that would (theoretically) pass through its rear focus, it is reflected through the front focus. This property, and that of the parabola, were used to develop the *Cassegrain* telescope in 1672. The focus of the parabolic mirror and the rear focus of the hyperbolic mirror are the same point. The rays are collected by the parabolic mirror, reflected toward the (common) focus, and thus are reflected by the hyperbolic mirror through the opening to its front focus, where the eyepiece is located. If the equation of the hyperbola is $\frac{y^2}{9} - \frac{x^2}{16} = 1$ and the focal length (distance from the vertex to the focus) of the parabola is 6, find the equation of the parabola.

Source: *www.enchantedlearning.com*

79. The **eccentricity** e of a hyperbola is defined as the number $\frac{c}{a}$, where a and c are the numbers given in equation (2). Because $c > a$, it follows that $e > 1$. Describe the general shape of a hyperbola whose eccentricity is close to 1. What is the shape if e is very large?

80. A hyperbola for which $a = b$ is called an **equilateral hyperbola.** Find the eccentricity e of an equilateral hyperbola.

[**Note:** The eccentricity of a hyperbola is defined in Problem 79.]

81. Two hyperbolas that have the same set of asymptotes are called **conjugate.** Show that the hyperbolas

$$\frac{x^2}{4} - y^2 = 1 \quad \text{and} \quad y^2 - \frac{x^2}{4} = 1$$

are conjugate. Graph each hyperbola on the same set of coordinate axes.

82. Prove that the hyperbola

$$\frac{y^2}{a^2} - \frac{x^2}{b^2} = 1$$

has the two oblique asymptotes

$$y = \frac{a}{b}x \quad \text{and} \quad y = -\frac{a}{b}x$$

83. Show that the graph of an equation of the form

$$Ax^2 + Cy^2 + F = 0 \qquad A \neq 0, C \neq 0, F \neq 0$$

where A and C are of opposite sign, is a hyperbola with center at $(0, 0)$.

84. Show that the graph of an equation of the form

$$Ax^2 + Cy^2 + Dx + Ey + F = 0 \qquad A \neq 0, C \neq 0$$

where A and C are of opposite sign,

(a) is a hyperbola if $\frac{D^2}{4A} + \frac{E^2}{4C} - F \neq 0$

(b) is two intersecting lines if $\frac{D^2}{4A} + \frac{E^2}{4C} - F = 0$

'Are You Prepared?' Answers

1. $5\sqrt{2}$ **2.** $\frac{25}{4}$ **3.** $(0, -3), (0, 3)$ **4.** True **5.** right; 5; down; 4 **6.** Vertical: $x = -2$, $x = 2$; horizontal: $y = 1$

CHAPTER REVIEW

Things to Know

Equations

Parabola (pp. 509–514)	See Tables 1 and 2 (pp. 511 and 513).
Ellipse (pp. 518–525)	See Table 3 (p. 523).
Hyperbola (pp. 529–538)	See Table 4 (p. 537).

Definitions

Parabola (p. 509)	Set of points P in the plane for which $d(F, P) = d(P, D)$, where F is the focus and D is the directrix
Ellipse (p. 519)	Set of points P in the plane, the sum of whose distances from two fixed points (the foci) is a constant
Hyperbola (p. 529)	Set of points P in the plane, the difference of whose distances from two fixed points (the foci) is a constant

Objectives

Section		You should be able to . . .	Example(s)	Review Exercises
7.1	1	Know the names of the conics (p. 508)		1–32
7.2	1	Analyze parabolas with vertex at the origin (p. 509)	1–6	1, 2, 21, 24
	2	Analyze parabolas with vertex at (h, k) (p. 513)	7–9	7, 11, 12, 17, 18, 27, 30
	3	Solve applied problems involving parabolas (p. 514)	10	41, 42
7.3	1	Analyze ellipses with center at the origin (p. 519)	1–5	5, 6, 10, 22, 25
	2	Analyze ellipses with center at (h, k) (p. 523)	6–8	14–16, 19, 28, 31
	3	Solve applied problems involving ellipses (p. 526)	9	43, 44
7.4	1	Analyze hyperbolas with center at the origin (p. 530)	1–5	3, 4, 8, 9, 23, 26
	2	Find the asymptotes of a hyperbola (p. 535)	6	3, 4, 8, 9
	3	Analyze hyperbolas with center at (h, k) (p. 536)	7, 8	13, 20, 29, 32–36
	4	Solve applied problems involving hyperbolas (p. 538)	9	45

Review Exercises

In Problems 1–20, identify each equation. If it is a parabola, give its vertex, focus, and directrix; if it is an ellipse, give its center, vertices, and foci; if it is a hyperbola, give its center, vertices, foci, and asymptotes.

1. $y^2 = -16x$

2. $16x^2 = y$

3. $\frac{x^2}{25} - y^2 = 1$

4. $\frac{y^2}{25} - x^2 = 1$

5. $\frac{y^2}{25} + \frac{x^2}{16} = 1$

6. $\frac{x^2}{9} + \frac{y^2}{16} = 1$

7. $x^2 + 4y = 4$

8. $3y^2 - x^2 = 9$

9. $4x^2 - y^2 = 8$

10. $9x^2 + 4y^2 = 36$

11. $x^2 - 4x = 2y$

12. $2y^2 - 4y = x - 2$

13. $y^2 - 4y - 4x^2 + 8x = 4$

14. $4x^2 + y^2 + 8x - 4y + 4 = 0$

15. $4x^2 + 9y^2 - 16x - 18y = 11$

16. $4x^2 + 9y^2 - 16x + 18y = 11$

17. $4x^2 - 16x + 16y + 32 = 0$

18. $4y^2 + 3x - 16y + 19 = 0$

19. $9x^2 + 4y^2 - 18x + 8y = 23$

20. $x^2 - y^2 - 2x - 2y = 1$

In Problems 21–36, find an equation of the conic described. Graph the equation by hand.

21. Parabola; focus at $(-2, 0)$; directrix the line $x = 2$

22. Ellipse; center at $(0, 0)$; focus at $(0, 3)$; vertex at $(0, 5)$

23. Hyperbola; center at (0,0); focus at (0,4); vertex at $(0, -2)$

24. Parabola; vertex at $(0, 0)$; directrix the line $y = -3$

25. Ellipse; foci at $(-3, 0)$ and $(3, 0)$; vertex at $(4, 0)$

26. Hyperbola; vertices at $(-2, 0)$ and $(2, 0)$; focus at $(4, 0)$

27. Parabola; vertex at $(2, -3)$; focus at $(2, -4)$

28. Ellipse; center at $(-1, 2)$; focus at $(0, 2)$; vertex at $(2, 2)$

29. Hyperbola; center at $(-2, -3)$; focus at $(-4, -3)$; vertex at $(-3, -3)$

30. Parabola; focus at $(3, 6)$; directrix the line $y = 8$

31. Ellipse; foci at $(-4, 2)$ and $(-4, 8)$; vertex at $(-4, 10)$

32. Hyperbola; vertices at $(-3, 3)$ and $(5, 3)$; focus at $(7, 3)$

33. Center at $(-1, 2)$; $a = 3$; $c = 4$; transverse axis parallel to the x-axis

34. Center at $(4, -2)$; $a = 1$; $c = 4$; transverse axis parallel to the y-axis

35. Vertices at $(0, 1)$ and $(6, 1)$; asymptote the line $3y + 2x = 9$

36. Vertices at $(4, 0)$ and $(4, 4)$; asymptote the line $y + 2x = 10$

37. Find an equation of the hyperbola whose foci are the vertices of the ellipse $4x^2 + 9y^2 = 36$ and whose vertices are the foci of this ellipse.

38. Find an equation of the ellipse whose foci are the vertices of the hyperbola $x^2 - 4y^2 = 16$ and whose vertices are the foci of this hyperbola.

39. Describe the collection of points in a plane so that the distance from each point to the point $(3, 0)$ is three-fourths of its distance from the line $x = \frac{16}{3}$.

40. Describe the collection of points in a plane so that the distance from each point to the point $(5, 0)$ is five-fourths of its distance from the line $x = \frac{16}{5}$.

41. Searchlight A searchlight is shaped like a paraboloid of revolution. If a light source is located 1 foot from the vertex along the axis of symmetry and the opening is 2 feet across, how deep should the mirror be in order to reflect the light rays parallel to the axis of symmetry?

42. Parabolic Arch Bridge A bridge is built in the shape of a parabolic arch. The bridge has a span of 60 feet and a maximum height of 20 feet. Find the height of the arch at distances of 5, 10, and 20 feet from the center.

43. Semielliptical Arch Bridge A bridge is built in the shape of a semielliptical arch. The bridge has a span of 60 feet and a maximum height of 20 feet. Find the height of the arch at distances of 5, 10, and 20 feet from the center.

44. Whispering Gallery The figure shows the specifications for an elliptical ceiling in a hall designed to be a whispering gallery. Where are the foci located in the hall?

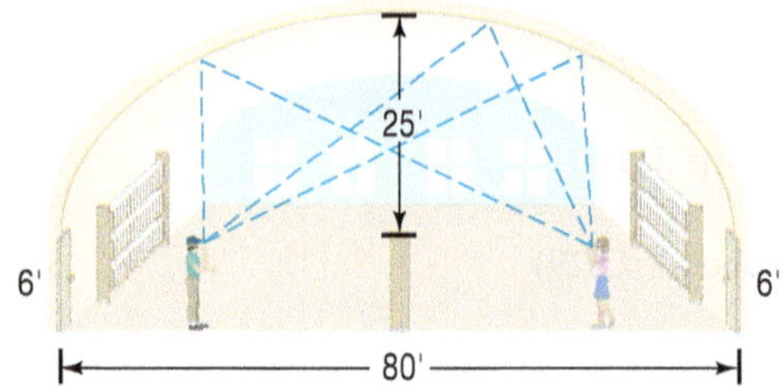

45. Calibrating Instruments In a test of their recording devices, a team of seismologists positioned two of the devices 2000 feet apart, with the device at point A to the west of the device at point B. At a point between the devices and 200 feet from point B, a small amount of explosive was detonated and a note made of the time at which the sound reached each device. A second explosion is to be carried out at a point directly north of point B. How far north should the site of the second explosion be chosen so that the measured time difference recorded by the devices for the second detonation is the same as that recorded for the first detonation?

CHAPTER TEST

In Problems 1–3, identify each equation. If it is a parabola, give its vertex, focus, and directrix; if an ellipse, give its center, vertices, and foci; if a hyperbola, give its center, vertices, foci, and asymptotes.

1. $\frac{(x+1)^2}{4} - \frac{y^2}{9} = 1$

2. $8y = (x-1)^2 - 4$

3. $2x^2 + 3y^2 + 4x - 6y = 13$

In Problems 4–6, find an equation of the conic described; graph the equation by hand.

4. Parabola: focus $(-1, 4.5)$, vertex $(-1, 3)$

5. Ellipse: center $(0, 0)$, vertex $(0, -4)$, focus $(0, 3)$

6. Hyperbola: center $(2, 2)$, vertex $(2, 4)$, contains the point $\left(2 + \sqrt{10}, 5\right)$

7. A parabolic reflector (paraboloid of revolution) is used by TV crews at football games to pick up the referee's announcements, quarterback signals, and so on. A microphone is placed at the focus of the parabola. If a certain reflector is 4 feet wide and 1.5 feet deep, where should the microphone be placed?

CUMULATIVE REVIEW

1. For $f(x) = -3x^2 + 5x - 2$, find

$$\frac{f(x+h) - f(x)}{h} \qquad h \neq 0$$

2. In the complex number system, solve the equation

$$9x^4 + 33x^3 - 71x^2 - 57x - 10 = 0$$

3. For what numbers x is $6 - x \geq x^2$?

4. (a) Find the domain and range of $y = 3^x + 2$.
(b) Find the inverse of $y = 3^x + 2$ and state its domain and range.

5. $f(x) = \log_4(x - 2)$
(a) Solve $f(x) = 2$.
(b) Solve $f(x) \leq 2$.

6. Find an equation for each of the following graphs:

(a) Line:

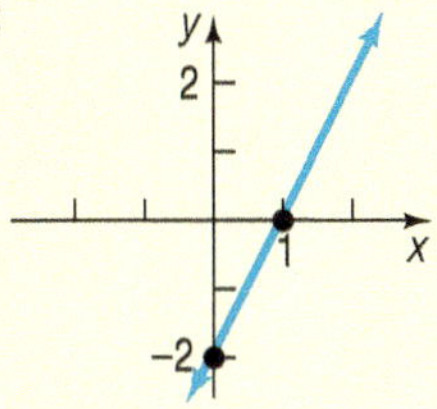

(b) Circle:

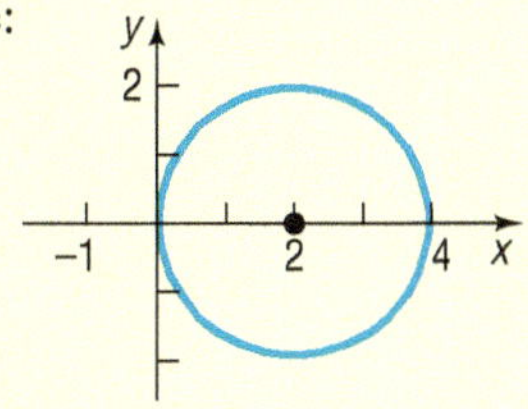

(c) Ellipse:

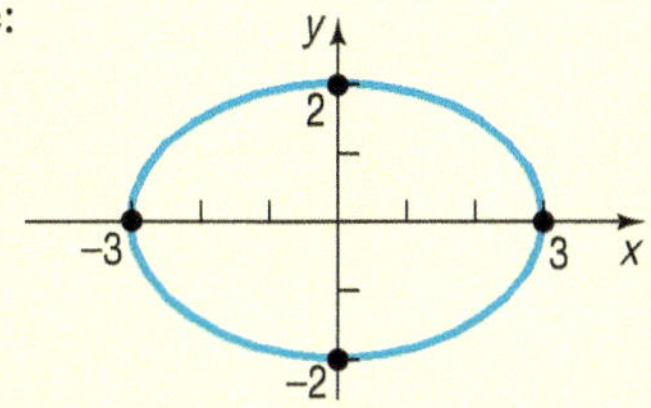

(d) Parabola:

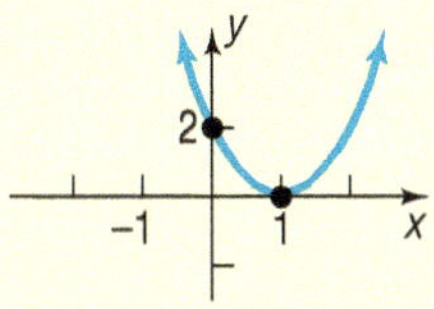

(e) Hyperbola:

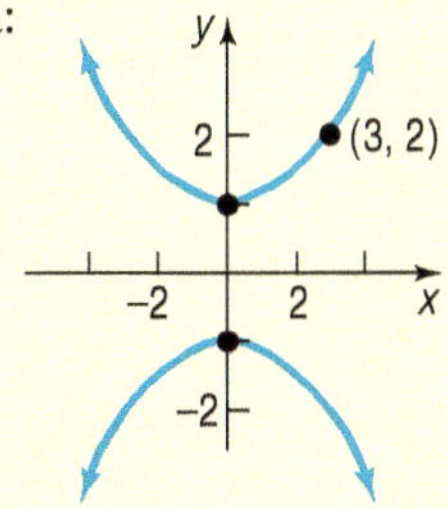

(f) Exponential:

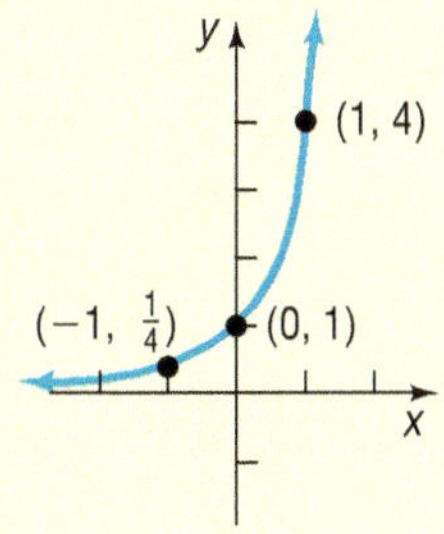

CHAPTER PROJECTS

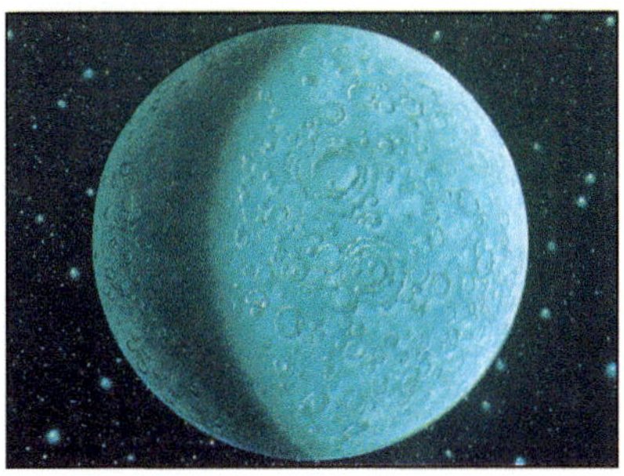

I. The Orbits of Neptune and Pluto The orbit of a planet about the Sun is an ellipse, with the Sun at one focus. The **aphelion** of a planet is its greatest distance from the Sun and the **perihelion** is its shortest distance. The **mean distance** of a planet from the Sun is the length of the semimajor axis of the elliptical orbit. See the illustration.

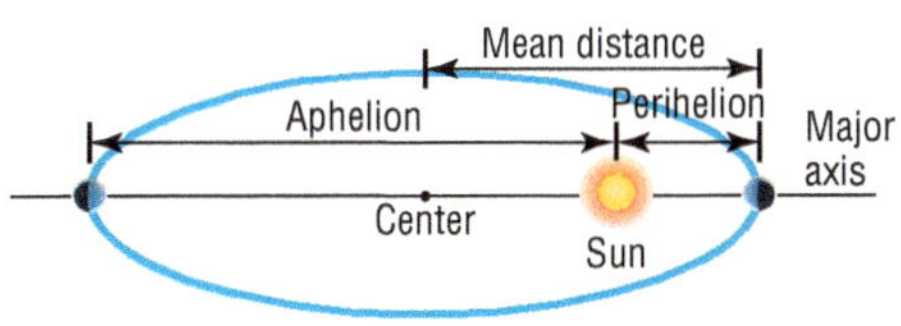

1. The aphelion of Neptune is 4532.2×10^6 kilometers (km) and its perihelion is 4458.0×10^6 km. Find a model for the orbit of Neptune around the Sun.
2. The aphelion of Pluto* is 7381.2×10^6 km and its perihelion is 4445.8×10^6 km. Find a model for the orbit of Pluto around the Sun.
3. Graph the orbits of Pluto and Neptune on a graphing utility. Knowing that the orbits of the planets intersect, what is wrong with the graphs you obtained?
4. The graphs of the orbits drawn in part 3 have the same center, but their foci lie in different locations. To see an accurate representation, the location of the Sun (a focus) needs to be the same for both graphs. This can be accomplished by shifting Pluto's orbit to the left. The shift amount is equal to Pluto's distance from the center (in the graph in part 3) to the Sun minus Neptune's distance from the center to the Sun. Find the new model representing the orbit of Pluto.
5. Graph the equation for the orbit of Pluto found in part 4 along with the equation of the orbit of Neptune. Do you see that Pluto's orbit is sometimes inside Neptune's?
6. Find the point(s) of intersection of the two orbits.
7. Do you think two planets will ever collide?

The following projects can be found on the Instructor's Resource Center (IRC):

II. Project at Motorola ***Distorted Deployable Space Reflector Antennas*** An engineer designs an antenna that will deploy in space to collect sunlight.

III. Constructing a Bridge over the East River The size of ships using a river and fluctuations in water height due to tides or flooding must be considered when designing a bridge that will cross a major waterway.

*Pluto's status was reduced to a dwarf planet in September 2006.

Systems of Equations and Inequalities

8

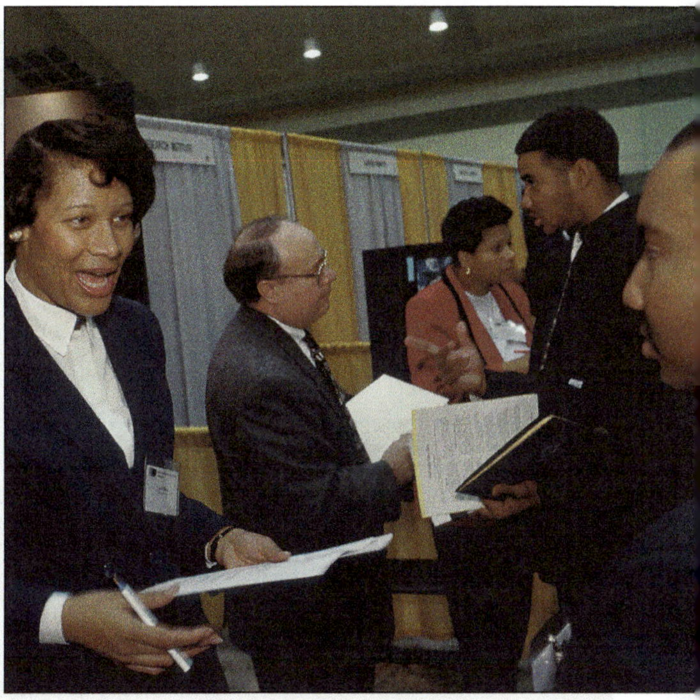

Economic Outcomes

Annual Earnings of Young Adults

For both males and females, earnings increase with education: full-time workers with at least a bachelor's degree have higher median earnings than those with less education. For example, in 2003, male college graduates earned 93 percent more than male high school completers. Females with a bachelor's or higher degree earned 91 percent more than female high school completers. Males and females who dropped out of high school earned 37 and 39 percent less, respectively, than male and female high school completers.

The median earnings of young adults who have at least a bachelor's degree declined in the 1970s relative to their counterparts who were high school completers, before increasing between 1980 and 2003. Males with a bachelor's degree or higher had earnings 19 percent higher than male high school completers in 1980 and had earnings 93 percent higher in 2003. Among females, those with at least a bachelor's degree had earnings 34 percent higher than female high school completers in 1980, compared with earnings 91 percent higher in 2003.

—See Chapter Project I—

A Look Back

In Chapters 1–6, we solved various kinds of equations and inequalities involving a single variable.

A Look Ahead

In this chapter we take up the problem of solving equations and inequalities containing two or more variables. There are various ways to solve such problems.

The *method of substitution* for solving equations in several unknowns goes back to ancient times.

The *method of elimination*, although it had existed for centuries, was put into systematic order by Karl Friedrich Gauss (1777–1855) and by Camille Jordan (1838–1922).

The theory of *matrices* was developed in 1857 by Arthur Cayley (1821–1895), although only later were matrices used as we use them in this chapter. Matrices have become a very flexible instrument, useful in almost all areas of mathematics.

The method of *determinants* was invented by Takakazu Seki Kôwa (1642–1708) in 1683 in Japan and by Gottfried Wilhelm von Leibniz (1646–1716) in 1693 in Germany. *Cramer's Rule* is named after Gabriel Cramer (1704–1752) of Switzerland, who popularized the use of determinants for solving linear systems.

Section 8.5, on *partial fraction decomposition*, provides an application of systems of equations. This particular application is one that is used in integral calculus.

Section 8.8 introduces *linear programming*, a modern application of linear inequalities. This topic is particularly useful for students interested in operations research.

Outline

8.1 Systems of Linear Equations: Substitution and Elimination

PREPARING FOR THIS SECTION *Before getting started, review the following:*

- Linear Equations (Section 1.2, pp. 105–106)
- Lines (Section 2.2, pp. 178–187)

Now Work the 'Are You Prepared?' problems on page 558.

OBJECTIVES
1. Solve Systems of Equations by Substitution (p. 550)
2. Solve Systems of Equations by Elimination (p. 552)
3. Identify Inconsistent Systems of Equations Containing Two Variables (p. 553)
4. Express the Solution of a System of Dependent Equations Containing Two Variables (p. 554)
5. Solve Systems of Three Equations Containing Three Variables (p. 554)
6. Identify Inconsistent Systems of Equations Containing Three Variables (p. 556)
7. Express the Solution of a System of Dependent Equations Containing Three Variables (p. 557)

We begin with an example.

EXAMPLE 1 Movie Theater Ticket Sales

A movie theater sells tickets for \$8.00 each, with seniors receiving a discount of \$2.00. One evening the theater took in \$3580 in revenue. If x represents the number of tickets sold at \$8.00 and y the number of tickets sold at the discounted price of \$6.00, write an equation that relates these variables.

Solution Each nondiscounted ticket brings in \$8.00, so x tickets will bring in $8x$ dollars. Similarly, y discounted tickets bring in $6y$ dollars. Since the total brought in is \$3580, we must have

$$8x + 6y = 3580$$

In Example 1, suppose that we also know that 525 tickets were sold that evening. Then we have another equation relating the variables x and y:

$$x + y = 525$$

The two equations

$$\begin{aligned} 8x + 6y &= 3580 \\ x + y &= 525 \end{aligned}$$

form a *system* of equations.

In general, a **system of equations** is a collection of two or more equations, each containing one or more variables. Example 2 gives some illustrations of systems of equations.

EXAMPLE 2 Examples of Systems of Equations

(a) $\begin{cases} 2x + y = 5 & (1) \\ -4x + 6y = -2 & (2) \end{cases}$ Two equations containing two variables, x and y

(b) $\begin{cases} x + y^2 = 5 & (1) \\ 2x + y = 4 & (2) \end{cases}$ Two equations containing two variables, x and y

(c) $\begin{cases} x + y + z = 6 & (1) \\ 3x - 2y + 4z = 9 & (2) \\ x - y - z = 0 & (3) \end{cases}$ Three equations containing three variables, x, y, and z

(d) $\begin{cases} x + y + z = 5 & (1) \\ x - y = 2 & (2) \end{cases}$ Two equations containing three variables, x, y, and z

(e) $\begin{cases} x + y + z = 6 & (1) \\ 2x + 2z = 4 & (2) \\ y + z = 2 & (3) \\ x = 4 & (4) \end{cases}$ Four equations containing three variables, x, y, and z

We use a brace, as shown, to remind us that we are dealing with a system of equations. We also will find it convenient to number each equation in the system.

A **solution** of a system of equations consists of values for the variables that are solutions of each equation of the system. To **solve** a system of equations means to find all solutions of the system.

For example, $x = 2$, $y = 1$ is a solution of the system in Example 2(a), because

$$\begin{cases} 2x + y = 5 & (1) \\ -4x + 6y = -2 & (2) \end{cases} \qquad \begin{cases} 2(2) + 1 = 4 + 1 = 5 \\ -4(2) + 6(1) = -8 + 6 = -2 \end{cases}$$

We may also write this solution as the ordered pair (2, 1).

A solution of the system in Example 2(b) is $x = 1$, $y = 2$, because

$$\begin{cases} x + y^2 = 5 & (1) \\ 2x + y = 4 & (2) \end{cases} \qquad \begin{cases} 1 + 2^2 = 1 + 4 = 5 \\ 2(1) + 2 = 2 + 2 = 4 \end{cases}$$

Another solution of the system in Example 2(b) is $x = \frac{11}{4}$, $y = -\frac{3}{2}$, which you can check for yourself.

A solution of the system in Example 2(c) is $x = 3$, $y = 2$, $z = 1$, because

$$\begin{cases} x + y + z = 6 & (1) \\ 3x - 2y + 4z = 9 & (2) \\ x - y - z = 0 & (3) \end{cases} \qquad \begin{cases} 3 + 2 + 1 = 6 & (1) \\ 3(3) - 2(2) + 4(1) = 9 - 4 + 4 = 9 & (2) \\ 3 - 2 - 1 = 0 & (3) \end{cases}$$

We may also write this solution as the ordered triplet (3, 2, 1).

Note that $x = 3$, $y = 3$, $z = 0$ is not a solution of the system in Example 2(c).

$$\begin{cases} x + y + z = 6 & (1) \\ 3x - 2y + 4z = 9 & (2) \\ x - y - z = 0 & (3) \end{cases} \qquad \begin{cases} 3 + 3 + 0 = 6 & (1) \\ 3(3) - 2(3) + 4(0) = 3 \neq 9 & (2) \\ 3 - 3 - 0 = 0 & (3) \end{cases}$$

Although $x = 3$, $y = 3$, and $z = 0$ satisfy equations (1) and (3), they do not satisfy equation (2). Any solution of the system must satisfy *each* equation of the system.

Now Work PROBLEM 9

When a system of equations has at least one solution, it is said to be **consistent;** otherwise, it is called **inconsistent.**

An equation in n variables is said to be **linear** if it is equivalent to an equation of the form

$$a_1x_1 + a_2x_2 + \cdots + a_nx_n = b$$

where $x_1, x_2, \ldots, x_n$ are n distinct variables, $a_1, a_2, \ldots, a_n, b$ are constants, and at least one of the a's is not 0.

Some examples of linear equations are

$$2x + 3y = 2 \qquad 5x - 2y + 3z = 10 \qquad 8x + 8y - 2z + 5w = 0$$

If each equation in a system of equations is linear, we have a **system of linear equations.** The systems in Examples 2(a), (c), (d), and (e) are linear, whereas the system in Example 2(b) is nonlinear. In this chapter we shall solve linear systems in Sections 8.1 to 8.3. We discuss nonlinear systems in Section 8.6.

We begin by discussing a system of two linear equations containing two variables. We can view the problem of solving such a system as a geometry problem. The graph of each equation in such a system is a line. So a system of two linear equations containing two variables represents a pair of lines. The lines either (1) intersect or (2) are parallel or (3) are **coincident** (that is, identical).

1. If the lines intersect, the system of equations has one solution, given by the point of intersection. The system is **consistent** and the equations are **independent.** See Figure 1(a).
2. If the lines are parallel, the system of equations has no solution, because the lines never intersect. The system is **inconsistent.** See Figure 1(b).
3. If the lines are coincident, the system of equations has infinitely many solutions, represented by the totality of points on the line. The system is **consistent** and the equations are **dependent.** See Figure 1(c).

Figure 1

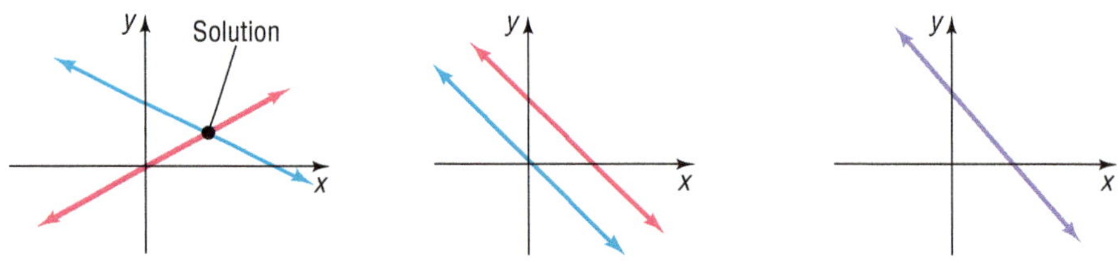

(a) Intersecting lines; system has one solution **(b)** Parallel lines; system has no solution **(c)** Coincident lines; system has infinitely many solutions

EXAMPLE 3 Solving a System of Linear Equations Using a Graphing Utility

$$\text{Solve: } \begin{cases} 2x + y = 5 & (1) \\ -4x + 6y = 12 & (2) \end{cases}$$

Solution First, we solve each equation for y. This is equivalent to writing each equation in slope–intercept form. Equation (1) in slope–intercept form is $Y_1 = -2x + 5$. Equation (2) in slope–intercept form is $Y_2 = \frac{2}{3}x + 2$. Figure 2 shows the graphs using a graphing utility. From the graph in Figure 2, we see that the lines intersect, so the system is consistent and the equations are independent. Using INTERSECT, we obtain the solution (1.125, 2.75).

Figure 2

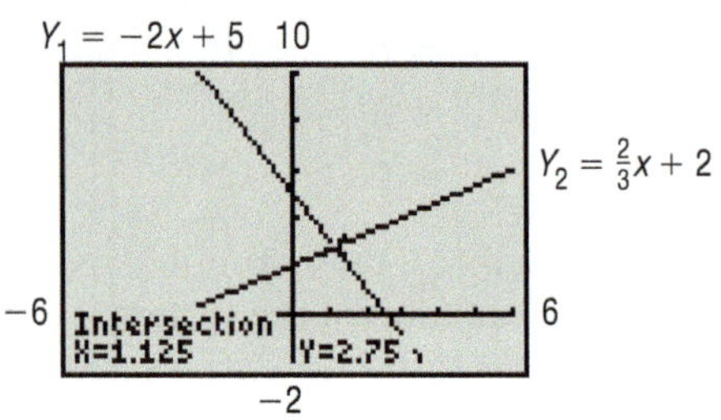

1 Solve Systems of Equations by Substitution

Sometimes to obtain exact solutions, we must use algebraic methods. A number of methods are available to us for solving systems of linear equations algebraically. In this section, we introduce two methods: *substitution* and *elimination*. We illustrate the **method of substitution** by solving the system given in Example 3.

EXAMPLE 4 How to Solve a System of Linear Equations by Substitution

$$\text{Solve: } \begin{cases} 2x + y = 5 & (1) \\ -4x + 6y = 12 & (2) \end{cases}$$

Step-by-Step Solution

STEP 1 Pick one of the equations and solve for one of the variables in terms of the remaining variable(s).

We solve equation (1) for y.

$$2x + y = 5 \quad \text{Equation (1)}$$
$$y = -2x + 5 \quad \text{Subtract } 2x \text{ from each side of (1).}$$

STEP 2 Substitute the result into the remaining equation(s).

We substitute $-2x + 5$ for y in equation (2). The result is an equation containing just the variable x, which we can solve.

$$-4x + 6y = 12 \quad \text{Equation (2)}$$
$$-4x + 6(-2x + 5) = 12 \quad \text{Substitute } -2x + 5 \text{ for } y \text{ in (2).}$$

STEP 3 If one equation in one variable results, solve this equation. Otherwise, repeat Steps 1 and 2 until a single equation with one variable remains.

$$-4x - 12x + 30 = 12 \quad \text{Distribute.}$$
$$-16x + 30 = 12 \quad \text{Combine like terms.}$$
$$-16x = -18 \quad \text{Subtract 30 from both sides.}$$
$$x = \frac{-18}{-16} \quad \text{Divide both sides by } -16.$$
$$= \frac{9}{8} \quad \text{Simplify.}$$

STEP 4 Find the values of the remaining variables by back-substitution.

Because we know that $x = \frac{9}{8}$, we can find the value of y by **back-substitution,** that is, by substituting $\frac{9}{8}$ for x in one of the original equations. Equation (1) seems easier to work with, so we will back-substitute into equation (1).

$$2x + y = 5 \quad \text{Equation (1)}$$
$$2\left(\frac{9}{8}\right) + y = 5 \quad \text{Substitute } x = \frac{9}{8} \text{ into equation (1).}$$
$$\frac{9}{4} + y = 5 \quad \text{Simplify.}$$
$$y = 5 - \frac{9}{4} \quad \text{Subtract } \frac{9}{4} \text{ from both sides.}$$
$$= \frac{11}{4} \quad 5 - \frac{9}{4} = \frac{20}{4} - \frac{9}{4} = \frac{11}{4}$$

STEP 5 Check the solution found.

We have $x = \frac{9}{8}$ and $y = \frac{11}{4}$. We verify that both equations are satisfied (true) for these values.

$$\begin{cases} 2x + y = 5: & 2\left(\frac{9}{8}\right) + \frac{11}{4} = \frac{9}{4} + \frac{11}{4} = \frac{20}{4} = 5 \\ -4x + 6y = 12: & -4\left(\frac{9}{8}\right) + 6\left(\frac{11}{4}\right) = -\frac{9}{2} + \frac{33}{2} = \frac{24}{2} = 12 \end{cases}$$

The solution of the system is $x = \frac{9}{8} = 1.125$ and $y = \frac{11}{4} = 2.75$. We can also write the solution as the ordered pair $\left(\frac{9}{8}, \frac{11}{4}\right)$.

Now Use Substitution to Work PROBLEM 19

2 Solve Systems of Equations by Elimination

A second method for solving a system of linear equations is the *method of elimination*. This method is usually preferred over substitution if substitution leads to fractions or if the system contains more than two variables. Elimination also provides the necessary motivation for solving systems using matrices (the subject of Section 8.2).

In Words
When using elimination, we want to get the coefficients of one of the variables to be negatives of one another.

The idea behind the **method of elimination** is to replace the original system of equations by an equivalent system so that adding two of the equations eliminates a variable. The rules for obtaining equivalent equations are the same as those studied earlier. However, we may also interchange any two equations of the system and/or replace any equation in the system by the sum (or difference) of that equation and a nonzero multiple of any other equation in the system.

Rules for Obtaining an Equivalent System of Equations

1. Interchange any two equations of the system.
2. Multiply (or divide) each side of an equation by the same nonzero constant.
3. Replace any equation in the system by the sum (or difference) of that equation and a nonzero multiple of any other equation in the system.

An example will give you the idea. As you work through the example, pay particular attention to the pattern being followed.

EXAMPLE 5 How to Solve a System of Linear Equations by Elimination

Solve: $\begin{cases} 2x + 3y = 1 & (1) \\ -x + y = -3 & (2) \end{cases}$

Step-by-Step Solution

Step 1 Multiply both sides of one or both equations by a nonzero constant so that the coefficients of one of the variables are additive inverses.

We will multiply both sides of equation (2) by 2 so that the coefficients of x in the two equations are additive inverses.

$$\begin{cases} 2x + 3y = 1 & (1) \\ -x + y = -3 & (2) \end{cases}$$

$$\begin{cases} 2x + 3y = 1 & (1) \\ 2(-x + y) = 2(-3) & (2) \end{cases}$$

$$\begin{cases} 2x + 3y = 1 & (1) \\ -2x + 2y = -6 & (2) \end{cases}$$

Step 2 Add the equations to eliminate the variable. Solve the resulting equation for the remaining unknown.

$$\begin{cases} 2x + 3y = 1 & (1) \\ -2x + 2y = -6 & (2) \end{cases}$$

$$5y = -5 \quad \text{Add equations (1) and (2).}$$

$$y = -1 \quad \text{Divide both sides by 5.}$$

Step 3 Back-substitute the value of the variable found in Step 2 into one of the *original* equations to find the value of the remaining variable.

We back-substitute $y = -1$ into equation (1) and solve for x.

$$2x + 3y = 1 \quad \text{Equation (1)}$$

$$2x + 3(-1) = 1 \quad \text{Substitute } y = -1 \text{ into equation (1).}$$

$$2x - 3 = 1 \quad \text{Simplify.}$$

$$2x = 4 \quad \text{Add 3 to both sides.}$$

$$x = 2 \quad \text{Divide both sides by 2.}$$

Step 4 Check the solution found.

We leave the check to you.

The solution of the system is $x = 2$ and $y = -1$. We can also write the solution as the ordered pair $(2, -1)$.

Now Use Elimination to Work PROBLEM 19

EXAMPLE 6 Movie Theater Ticket Sales

A movie theater sells tickets for \$8.00 each, with seniors receiving a discount of \$2.00. One evening the theater sold 525 tickets and took in \$3580 in revenue. How many of each type of ticket were sold?

Solution If x represents the number of tickets sold at \$8.00 and y the number of tickets sold at the discounted price of \$6.00, then the given information results in the system of equations

$$\begin{cases} 8x + 6y = 3580 & (1) \\ x + y = 525 & (2) \end{cases}$$

We use the method of elimination. First, multiply the second equation by -6, and then add the equations.

$$\begin{cases} 8x + 6y = 3580 \\ -6x - 6y = -3150 \end{cases}$$
$$2x = 430 \quad \text{Add the equations.}$$
$$x = 215$$

Since $x + y = 525$, then $y = 525 - x = 525 - 215 = 310$. So 215 nondiscounted tickets and 310 senior discount tickets were sold.

3 Identify Inconsistent Systems of Equations Containing Two Variables

The previous examples dealt with consistent systems of equations that had a single solution. The next two examples deal with two other possibilities that may occur, the first being a system that has no solution.

EXAMPLE 7 An Inconsistent System of Linear Equations

Solve: $\begin{cases} 2x + y = 5 & (1) \\ 4x + 2y = 8 & (2) \end{cases}$

Solution We choose to use the method of substitution and solve equation (1) for y.

$$2x + y = 5 \quad (1)$$
$$y = -2x + 5 \quad \text{Subtract } 2x \text{ from each side.}$$

Now substitute $-2x + 5$ for y in equation (2) and solve for x.

$$4x + 2y = 8 \quad (2)$$
$$4x + 2(-2x + 5) = 8 \quad \text{Substitute } y = -2x + 5 \text{ in (2).}$$
$$4x - 4x + 10 = 8 \quad \text{Remove parentheses.}$$
$$0 = -2 \quad \text{Subtract 10 from both sides.}$$

This statement is false. We conclude that the system has no solution and is therefore inconsistent.

Figure 3

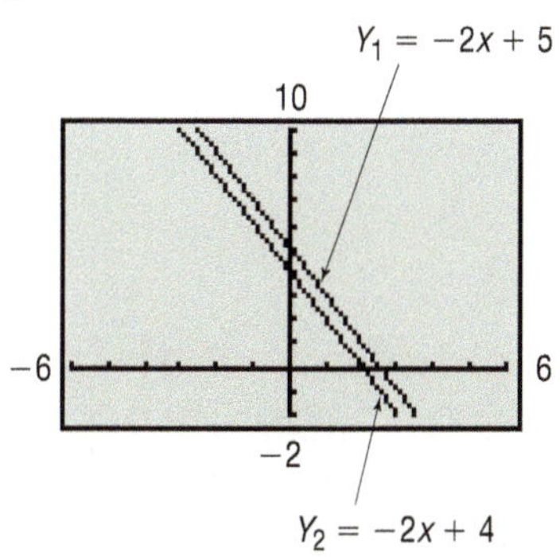

Figure 3 illustrates the pair of lines whose equations form the system in Example 7. Notice that the graphs of the two equations are lines, each with slope -2;

one has a y-intercept of 5, the other a y-intercept of 4. The lines are parallel and have no point of intersection. This geometric statement is equivalent to the algebraic statement that the system has no solution.

4 Express the Solution of a System of Dependent Equations Containing Two Variables

EXAMPLE 8 **Solving a System of Dependent Equations**

Solve: $\begin{cases} 2x + y = 4 & (1) \\ -6x - 3y = -12 & (2) \end{cases}$

Solution We choose to use the method of elimination.

$$\begin{cases} 2x + y = 4 & (1) \\ -6x - 3y = -12 & (2) \end{cases}$$

$$\begin{cases} 6x + 3y = 12 & (1) \quad \text{Multiply each side of equation (1) by 3.} \\ -6x - 3y = -12 & (2) \end{cases}$$

$$\begin{cases} 6x + 3y = 12 & (1) \\ 0 = 0 & (2) \quad \text{Replace equation (2) by the sum of equations (1) and (2).} \end{cases}$$

The original system is equivalent to a system containing one equation, so the equations are dependent. This means that any values of x and y which satisfy $6x + 3y = 12$ or, equivalently, $2x + y = 4$ are solutions. For example, $x = 2, y = 0$; $x = 0, y = 4$; $x = -2, y = 8$; $x = 4, y = -4$; and so on, are solutions. There are, in fact, infinitely many values of x and y for which $2x + y = 4$, so the original system has infinitely many solutions. We will write the solution of the original system either as

$$y = -2x + 4 \quad \text{where } x \text{ can be any real number}$$

or as

$$x = -\frac{1}{2}y + 2 \quad \text{where } y \text{ can be any real number.}$$

We can also express the solution as $\{(x, y) | y = -2x + 4, x \text{ is any real number}\}$ or as $\{(x, y) | x = -\frac{1}{2}y + 2, y \text{ is any real number}\}$.

Figure 4

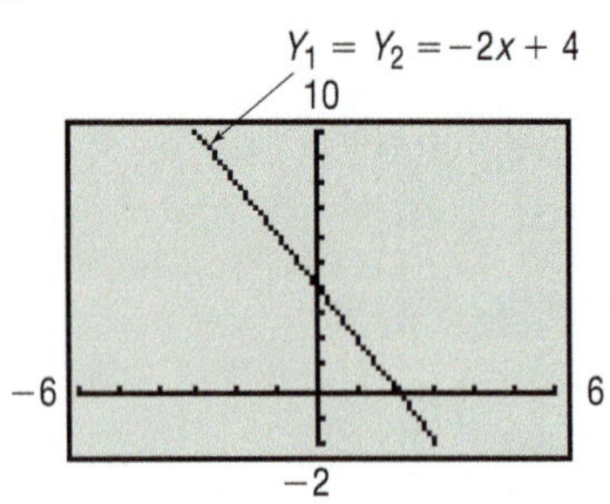

Figure 4 illustrates the situation presented in Example 8. Notice that the graphs of the two equations are lines, each with slope -2 and each with y-intercept 4. The lines are coincident. Notice also that equation (2) in the original system is -3 times equation (1), indicating that the two equations are dependent.

For the system in Example 8, we can write down some of the infinite number of solutions by assigning values to x and then finding $y = -2x + 4$.

If $x = -2$, then $y = -2(-2) + 4 = 8$.
If $x = 0$, then $y = 4$.
If $x = 2$, then $y = 0$.

The ordered pairs $(-2, 8)$, $(0, 4)$, and $(2, 0)$ are three of the points on the line in Figure 4.

Now Work PROBLEMS 25 AND 29

5 Solve Systems of Three Equations Containing Three Variables

Just as with a system of two linear equations containing two variables, a system of three linear equations containing three variables has either (1) exactly one solution (a consistent system with independent equations), or (2) no solution (an inconsistent system), or (3) infinitely many solutions (a consistent system with dependent equations).

We can view the problem of solving a system of three linear equations containing three variables as a geometry problem. The graph of each equation in such a system is a plane in space. A system of three linear equations containing three variables represents three planes in space. Figure 5 illustrates some of the possibilities.

Figure 5

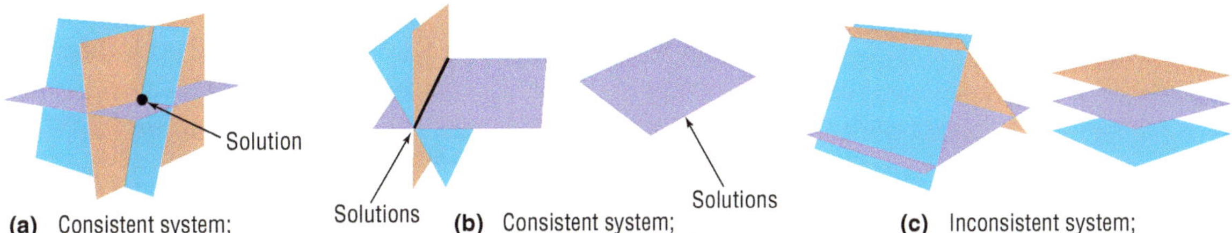

(a) Consistent system; one solution

(b) Consistent system; infinite number of solutions

(c) Inconsistent system; no solution

Recall that a **solution** to a system of equations consists of values for the variables that are solutions of each equation of the system. For example, $x = 3$, $y = -1$, $z = -5$ or, using ordered triplets, $(3, -1, -5)$ is a solution to the system of equations

$$\begin{cases} x + y + z = -3 & (1) \\ 2x - 3y + 6z = -21 & (2) \\ -3x + 5y = -14 & (3) \end{cases} \qquad \begin{aligned} &3 + (-1) + (-5) = -3 \\ &2(3) - 3(-1) + 6(-5) = 6 + 3 - 30 = -21 \\ &-3(3) + 5(-1) = -9 - 5 = -14 \end{aligned}$$

because these values of the variables are solutions of each equation.

Typically, when solving a system of three linear equations containing three variables, we use the method of elimination. Recall that the idea behind the method of elimination is to form equivalent equations so that adding two of the equations eliminates a variable.

Let's see how elimination works on a system of three equations containing three variables.

EXAMPLE 9 Solving a System of Three Linear Equations with Three Variables

Use the method of elimination to solve the system of equations.

$$\begin{cases} x + y - z = -1 & (1) \\ 4x - 3y + 2z = 16 & (2) \\ 2x - 2y - 3z = 5 & (3) \end{cases}$$

Solution For a system of three equations, we attempt to eliminate one variable at a time, using pairs of equations until an equation with a single variable remains. Our strategy for solving this system will be to use equation (1) to eliminate the variable x from equations (2) and (3). We can then treat the new equations (2) and (3) as a system with two unknowns. Alternatively, we could use equation (1) to eliminate either y or z from equations (2) and (3). Try one of these approaches for yourself.

We begin by multiplying each side of equation (1) by -4 and adding the result to equation (2). (Do you see why? The coefficients of x are now negatives of one another.) We also multiply equation (1) by -2 and add the result to equation (3). Notice that these two procedures result in the elimination of the variable x from equations (2) and (3).

$$\begin{aligned} x + y - z &= -1 \quad (1) \quad \text{Multiply by } -4 \\ 4x - 3y + 2z &= 16 \quad (2) \end{aligned} \qquad \begin{aligned} -4x - 4y + 4z &= 4 \quad (1) \\ 4x - 3y + 2z &= 16 \quad (2) \\ \hline -7y + 6z &= 20 \quad \text{Add.} \end{aligned}$$

$$\begin{aligned} x + y - z &= -1 \quad (1) \quad \text{Multiply by } -2 \\ 2x - 2y - 3z &= 5 \quad (3) \end{aligned} \qquad \begin{aligned} -2x - 2y + 2z &= 2 \quad (1) \\ 2x - 2y - 3z &= 5 \quad (3) \\ \hline -4y - z &= 7 \quad \text{Add.} \end{aligned}$$

$$\begin{cases} x + y - z = -1 & (1) \\ -7y + 6z = 20 & (2) \\ -4y - z = 7 & (3) \end{cases}$$

We now concentrate on the new equations (2) and (3), treating them as a system of two equations containing two variables. It is easiest to eliminate z. We multiply each side of equation (3) by 6 and add equations (2) and (3). The result is the new equation (3).

$$\begin{array}{ll} -7y + 6z = 20 & (2) \\ -4y - z = 7 & (3) \end{array} \quad \text{Multiply by 6} \quad \begin{array}{ll} -7y + 6z = 20 & (2) \\ -24y - 6z = 42 & (3) \\ \hline -31y \qquad = 62 & \text{Add.} \end{array} \longrightarrow \begin{cases} x + y - z = -1 & (1) \\ -7y + 6z = 20 & (2) \\ -31y \qquad = 62 & (3) \end{cases}$$

We now solve equation (3) for y by dividing both sides of the equation by -31.

$$\begin{cases} x + y - z = -1 & (1) \\ -7y + 6z = 20 & (2) \\ y \qquad = -2 & (3) \end{cases}$$

Back-substitute $y = -2$ in equation (2) and solve for z.

$$\begin{aligned} -7y + 6z &= 20 && (2) \\ -7(-2) + 6z &= 20 && \text{Substitute } y = -2 \text{ in (2).} \\ 6z &= 6 && \text{Subtract 14 from both sides of the equation.} \\ z &= 1 && \text{Divide both sides of the equation by 6.} \end{aligned}$$

Finally, we back-substitute $y = -2$ and $z = 1$ in equation (1) and solve for x.

$$\begin{aligned} x + y - z &= -1 && (1) \\ x + (-2) - 1 &= -1 && \text{Substitute } y = -2 \text{ and } z = 1 \text{ in (1).} \\ x - 3 &= -1 && \text{Simplify.} \\ x &= 2 && \text{Add 3 to both sides.} \end{aligned}$$

The solution of the original system is $x = 2$, $y = -2$, $z = 1$ or, using ordered triplets, $(2, -2, 1)$. You should check this solution.

Look back over the solution given in Example 9. Note the pattern of removing one of the variables from two of the equations, followed by solving this system of two equations and two unknowns. Although which variables to remove is your choice, the methodology remains the same for all systems.

Now Work PROBLEM 43

6 Identify Inconsistent Systems of Equations Containing Three Variables

EXAMPLE 10 An Inconsistent System of Linear Equations

Solve: $\begin{cases} 2x + y - z = -2 & (1) \\ x + 2y - z = -9 & (2) \\ x - 4y + z = 1 & (3) \end{cases}$

Solution Our strategy is the same as in Example 9. However, in this system, it seems easiest to eliminate the variable z first. Do you see why?

Multiply each side of equation (1) by -1 and add the result to equation (2). Also, add equations (2) and (3).

$$\begin{array}{ll} -2x - y + z = 2 & (1) \text{ Multiply by } -1. \\ x + 2y - z = -9 & (2) \\ \hline -x + y \qquad = -7 & \text{Add.} \end{array}$$

$$\begin{array}{ll} x + 2y - z = -9 & (2) \\ x - 4y + z = 1 & (3) \\ \hline 2x - 2y \qquad = -8 & \text{Add.} \end{array}$$

$$\longrightarrow \begin{cases} 2x + y - z = -2 & (1) \\ -x + y \qquad = -7 & (2) \\ 2x - 2y \qquad = -8 & (3) \end{cases}$$

We now concentrate on the new equations (2) and (3), treating them as a system of two equations containing two variables. Multiply each side of equation (2) by 2 and add the result to equation (3).

$$\begin{array}{ll} -x + y = -7 & (2) \text{ Multiply by 2.} \\ 2x - 2y = -8 & (3) \end{array} \quad \begin{array}{rl} -2x + 2y = -14 & (2) \\ 2x - 2y = -8 & (3) \\ \hline 0 = -22 & \text{Add.} \end{array} \longrightarrow \begin{cases} 2x + y - z = -2 & (1) \\ -x + y = -7 & (2) \\ 0 = -22 & (3) \end{cases}$$

Equation (3) has no solution so the system is inconsistent.

7 Express the Solution of a System of Dependent Equations Containing Three Variables

EXAMPLE 11 Solving a System of Dependent Equations

$$\text{Solve: } \begin{cases} x - 2y - z = 8 & (1) \\ 2x - 3y + z = 23 & (2) \\ 4x - 5y + 5z = 53 & (3) \end{cases}$$

Solution Our plan is to eliminate x from equations (2) and (3). Multiply each side of equation (1) by -2 and add the result to equation (2). Also, multiply each side of equation (1) by -4 and add the result to equation (3).

$$\begin{array}{ll} x - 2y - z = 8 & (1) \text{ Multiply by } -2. \\ 2x - 3y + z = 23 & (2) \end{array} \quad \begin{array}{rl} -2x + 4y + 2z = -16 & (1) \\ 2x - 3y + z = 23 & (2) \\ \hline y + 3z = 7 & \text{Add.} \end{array}$$

$$\begin{array}{ll} x - 2y - z = 8 & (1) \text{ Multiply by } -4. \\ 4x - 5y + 5z = 53 & (3) \end{array} \quad \begin{array}{rl} -4x + 8y + 4z = -32 & (1) \\ 4x - 5y + 5z = 53 & (2) \\ \hline 3y + 9z = 21 & \text{Add.} \end{array}$$

$$\begin{cases} x - 2y - z = 8 & (1) \\ y + 3z = 7 & (2) \\ 3y + 9z = 21 & (3) \end{cases}$$

Treat equations (2) and (3) as a system of two equations containing two variables, and eliminate the variable y by multiplying both sides of equation (2) by -3 and adding the result to equation (3).

$$\begin{array}{ll} y + 3z = 7 & \text{Multiply by } -3. \\ 3y + 9z = 21 & \end{array} \quad \begin{array}{rl} -3y - 9z = -21 & \\ 3y + 9z = 21 & \text{Add.} \\ \hline 0 = 0 & \end{array} \longrightarrow \begin{cases} x - 2y - z = 8 & (1) \\ y + 3z = 7 & (2) \\ 0 = 0 & (3) \end{cases}$$

The original system is equivalent to a system containing two equations, so the equations are dependent and the system has infinitely many solutions. If we solve equation (2) for y, we can express y in terms of z as $y = -3z + 7$. Substitute this expression into equation (1) to determine x in terms of z.

$$\begin{array}{rl} x - 2y - z = 8 & (1) \\ x - 2(-3z + 7) - z = 8 & \text{Substitute } y = -3z + 7 \text{ in (1).} \\ x + 6z - 14 - z = 8 & \text{Remove parentheses.} \\ x + 5z = 22 & \text{Combine like terms.} \\ x = -5z + 22 & \text{Solve for } x. \end{array}$$

We will write the solution to the system as

$$\begin{cases} x = -5z + 22 \\ y = -3z + 7 \end{cases} \quad \text{where } z \text{ can be any real number.}$$

This way of writing the solution makes it easier to find specific solutions of the system. To find specific solutions, choose any value of z and use the equations

$x = -5z + 22$ and $y = -3z + 7$ to determine x and y. For example, if $z = 0$, then $x = 22$ and $y = 7$, and if $z = 1$, then $x = 17$ and $y = 4$.

Using ordered triplets, the solution is

$$\{(x, y, z) \mid x = -5z + 22,\ y = -3z + 7,\ z \text{ is any real number}\}$$

Now Work PROBLEM 45

Two distinct points in the Cartesian plane determine a unique line. Given three noncollinear points, we can find the (unique) quadratic function whose graph contains these three points.

EXAMPLE 12 Curve Fitting

Find real numbers a, b, and c so that the graph of the quadratic function $y = ax^2 + bx + c$ contains the points $(-1, -4)$, $(1, 6)$, and $(3, 0)$.

Solution We require that the three points satisfy the equation $y = ax^2 + bx + c$.

For the point $(-1, -4)$ we have: $-4 = a(-1)^2 + b(-1) + c \quad -4 = a - b + c$

For the point $(1, 6)$ we have: $6 = a(1)^2 + b(1) + c \quad 6 = a + b + c$

For the point $(3, 0)$ we have: $0 = a(3)^2 + b(3) + c \quad 0 = 9a + 3b + c$

We wish to determine a, b, and c so that each equation is satisfied. That is, we want to solve the following system of three equations containing three variables:

$$\begin{cases} a - b + c = -4 & (1) \\ a + b + c = 6 & (2) \\ 9a + 3b + c = 0 & (3) \end{cases}$$

Figure 6

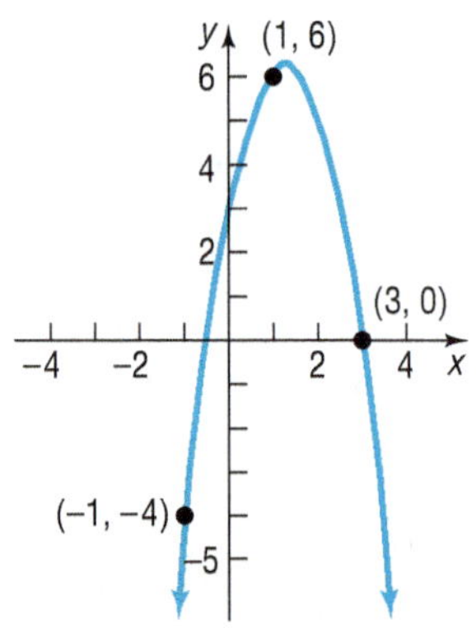

Solving this system of equations, we obtain $a = -2$, $b = 5$, and $c = 3$. So the quadratic function whose graph contains the points $(-1, -4)$, $(1, 6)$, and $(3, 0)$ is

$$y = -2x^2 + 5x + 3 \quad y = ax^2 + bx + c,\ a = -2, b = 5, c = 3$$

Figure 6 shows the graph of the function along with the three points.

Now Work PROBLEM 69

8.1 Assess Your Understanding

'Are You Prepared?' *Answers are given at the end of these exercises. If you get a wrong answer, read the pages listed in red.*

1. Solve the equation: $3x + 4 = 8 - x$. (pp. 105–106)

2. (a) Graph the line: $3x + 4y = 12$.
(b) What is the slope of a line parallel to this line? (pp. 178–187)

Concepts and Vocabulary

3. If a system of equations has no solution, it is said to be ______.

4. If a system of equations has one or more solutions, the system is said to be ______.

5. ***True or False*** A system of two linear equations containing two variables always has at least one solution.

6. ***True or False*** A solution of a system of equations consists of values for the variables that are solutions of each equation of the system.

Skill Building

In Problems 7–16, verify that the values of the variables listed are solutions of the system of equations.

7. $\begin{cases} 2x - y = 5 \\ 5x + 2y = 8 \end{cases}$
$x = 2, y = -1; (2, -1)$

8. $\begin{cases} 3x + 2y = 2 \\ x - 7y = -30 \end{cases}$
$x = -2, y = 4; (-2, 4)$

9. $\begin{cases} 3x - 4y = 4 \\ \frac{1}{2}x - 3y = -\frac{1}{2} \end{cases}$
$x = 2, y = \frac{1}{2}; \left(2, \frac{1}{2}\right)$

10. $\begin{cases} 2x + \frac{1}{2}y = 0 \\ 3x - 4y = -\frac{19}{2} \end{cases}$
$x = -\frac{1}{2}, y = 2; \left(-\frac{1}{2}, 2\right)$

11. $\begin{cases} x - y = 3 \\ \frac{1}{2}x + y = 3 \end{cases}$
$x = 4, y = 1; (4, 1)$

12. $\begin{cases} x - y = 3 \\ -3x + y = 1 \end{cases}$
$x = -2, y = -5; (-2, -5)$

13. $\begin{cases} 3x + 3y + 2z = 4 \\ x - y - z = 0 \\ 2y - 3z = -8 \end{cases}$
$x = 1, y = -1, z = 2;$ $(1, -1, 2)$

14. $\begin{cases} 4x - z = 7 \\ 8x + 5y - z = 0 \\ -x - y + 5z = 6 \end{cases}$
$x = 2, y = -3, z = 1;$ $(2, -3, 1)$

15. $\begin{cases} 3x + 3y + 2z = 4 \\ x - 3y + z = 10 \\ 5x - 2y - 3z = 8 \end{cases}$
$x = 2, y = -2, z = 2; (2, -2, 2)$

16. $\begin{cases} 4x - 5z = 6 \\ 5y - z = -17 \\ -x - 6y + 5z = 24 \end{cases}$
$x = 4, y = -3, z = 2; (4, -3, 2)$

In Problems 17–54, solve each system of equations. If the system has no solution, say that it is inconsistent.

17. $\begin{cases} x + y = 8 \\ x - y = 4 \end{cases}$

18. $\begin{cases} x + 2y = -7 \\ x + y = -3 \end{cases}$

19. $\begin{cases} 5x - y = 21 \\ 2x + 3y = -12 \end{cases}$

20. $\begin{cases} x + 3y = 5 \\ 2x - 3y = -8 \end{cases}$

21. $\begin{cases} 3x = 24 \\ x + 2y = 0 \end{cases}$

22. $\begin{cases} 4x + 5y = -3 \\ -2y = -8 \end{cases}$

23. $\begin{cases} 3x - 6y = 2 \\ 5x + 4y = 1 \end{cases}$

24. $\begin{cases} 2x + 4y = \frac{2}{3} \\ 3x - 5y = -10 \end{cases}$

25. $\begin{cases} 2x + y = 1 \\ 4x + 2y = 3 \end{cases}$

26. $\begin{cases} x - y = 5 \\ -3x + 3y = 2 \end{cases}$

27. $\begin{cases} 2x - y = 0 \\ 4x + 2y = 12 \end{cases}$

28. $\begin{cases} 3x + 3y = -1 \\ 4x + y = \frac{8}{3} \end{cases}$

29. $\begin{cases} x + 2y = 4 \\ 2x + 4y = 8 \end{cases}$

30. $\begin{cases} 3x - y = 7 \\ 9x - 3y = 21 \end{cases}$

31. $\begin{cases} 2x - 3y = -1 \\ 10x + y = 11 \end{cases}$

32. $\begin{cases} 3x - 2y = 0 \\ 5x + 10y = 4 \end{cases}$

33. $\begin{cases} 2x + 3y = 6 \\ x - y = \frac{1}{2} \end{cases}$

34. $\begin{cases} \frac{1}{2}x + y = -2 \\ x - 2y = 8 \end{cases}$

35. $\begin{cases} \frac{1}{2}x + \frac{1}{3}y = 3 \\ \frac{1}{4}x - \frac{2}{3}y = -1 \end{cases}$

36. $\begin{cases} \frac{1}{3}x - \frac{3}{2}y = -5 \\ \frac{3}{4}x + \frac{1}{3}y = 11 \end{cases}$

37. $\begin{cases} 3x - 5y = 3 \\ 15x + 5y = 21 \end{cases}$

38. $\begin{cases} 2x - y = -1 \\ x + \frac{1}{2}y = \frac{3}{2} \end{cases}$

39. $\begin{cases} \frac{1}{x} + \frac{1}{y} = 8 \\ \frac{3}{x} - \frac{5}{y} = 0 \end{cases}$

40. $\begin{cases} \frac{4}{x} - \frac{3}{y} = 0 \\ \frac{6}{x} + \frac{3}{2y} = 2 \end{cases}$

[**Hint:** Let $u = \frac{1}{x}$ and $v = \frac{1}{y}$, and solve for u and v. Then $x = \frac{1}{u}$ and $y = \frac{1}{v}$.]

41. $\begin{cases} x - y = 6 \\ 2x - 3z = 16 \\ 2y + z = 4 \end{cases}$

42. $\begin{cases} 2x + y = -4 \\ -2y + 4z = 0 \\ 3x - 2z = -11 \end{cases}$

43. $\begin{cases} x - 2y + 3z = 7 \\ 2x + y + z = 4 \\ -3x + 2y - 2z = -10 \end{cases}$

44. $\begin{cases} 2x + y - 3z = 0 \\ -2x + 2y + z = -7 \\ 3x - 4y - 3z = 7 \end{cases}$

45. $\begin{cases} x - y - z = 1 \\ 2x + 3y + z = 2 \\ 3x + 2y = 0 \end{cases}$

46. $\begin{cases} 2x - 3y - z = 0 \\ -x + 2y + z = 5 \\ 3x - 4y - z = 1 \end{cases}$

47. $\begin{cases} x - y - z = 1 \\ -x + 2y - 3z = -4 \\ 3x - 2y - 7z = 0 \end{cases}$

48. $\begin{cases} 2x - 3y - z = 0 \\ 3x + 2y + 2z = 2 \\ x + 5y + 3z = 2 \end{cases}$

49. $\begin{cases} 2x - 2y + 3z = 6 \\ 4x - 3y + 2z = 0 \\ -2x + 3y - 7z = 1 \end{cases}$

50. $\begin{cases} 3x - 2y + 2z = 6 \\ 7x - 3y + 2z = -1 \\ 2x - 3y + 4z = 0 \end{cases}$

51. $\begin{cases} x + y - z = 6 \\ 3x - 2y + z = -5 \\ x + 3y - 2z = 14 \end{cases}$

52. $\begin{cases} x - y + z = -4 \\ 2x - 3y + 4z = -15 \\ 5x + y - 2z = 12 \end{cases}$

53. $\begin{cases} x + 2y - z = -3 \\ 2x - 4y + z = -7 \\ -2x + 2y - 3z = 4 \end{cases}$

54. $\begin{cases} x + 4y - 3z = -8 \\ 3x - y + 3z = 12 \\ x + y + 6z = 1 \end{cases}$

Applications and Extensions

55. The perimeter of a rectangular floor is 90 feet. Find the dimensions of the floor if the length is twice the width.

56. The length of fence required to enclose a rectangular field is 3000 meters. What are the dimensions of the field if it is known that the difference between its length and width is 50 meters?

57. **Orbital Launches** In 2005 there was a total of 55 commercial and noncommercial orbital launches worldwide. In addition, the number of noncommercial orbital launches was one more than twice the number of commercial orbital launches. Determine the number of commercial and noncommercial orbital launches in 2005.

Source: Federal Aviation Administration

58. **Movie Theater Tickets** A movie theater charges \$9.00 for adults and \$7.00 for senior citizens. On a day when 325 people paid an admission, the total receipts were \$2495. How many who paid were adults? How many were seniors?

59. **Mixing Nuts** A store sells cashews for \$5.00 per pound and peanuts for \$1.50 per pound. The manager decides to mix 30 pounds of peanuts with some cashews and sell the mixture for \$3.00 per pound. How many pounds of cashews should be mixed with the peanuts so that the mixture will produce the same revenue as would selling the nuts separately?

60. **Financial Planning** A recently retired couple needs \$12,000 per year to supplement their Social Security. They have \$150,000 to invest to obtain this income. They have decided on two investment options: AA bonds yielding 10% per annum and a Bank Certificate yielding 5%.

(a) How much should be invested in each to realize exactly \$12,000?

(b) If, after 2 years, the couple requires \$14,000 per year in income, how should they reallocate their investment to achieve the new amount?

61. **Computing Wind Speed** With a tail wind, a small Piper aircraft can fly 600 miles in 3 hours. Against this same wind, the Piper can fly the same distance in 4 hours. Find the average wind speed and the average airspeed of the Piper.

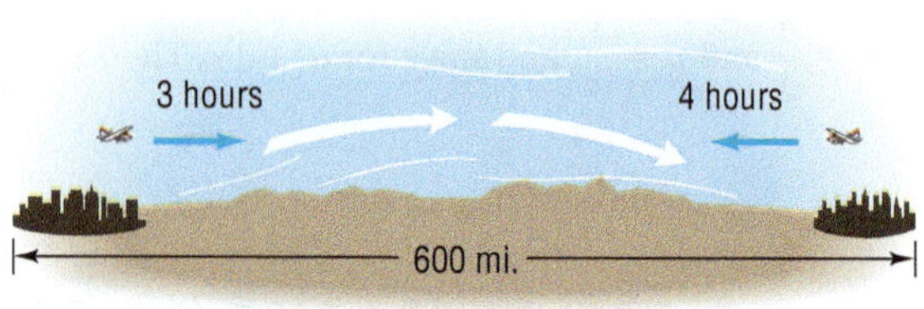

62. **Computing Wind Speed** The average airspeed of a single-engine aircraft is 150 miles per hour. If the aircraft flew the same distance in 2 hours with the wind as it flew in 3 hours against the wind, what was the wind speed?

63. **Restaurant Management** A restaurant manager wants to purchase 200 sets of dishes. One design costs \$25 per set, while another costs \$45 per set. If she only has \$7400 to spend, how many of each design should be ordered?

64. **Cost of Fast Food** One group of people purchased 10 hot dogs and 5 soft drinks at a cost of \$35.00. A second bought 7 hot dogs and 4 soft drinks at a cost of \$25.25. What is the cost of a single hot dog? A single soft drink?

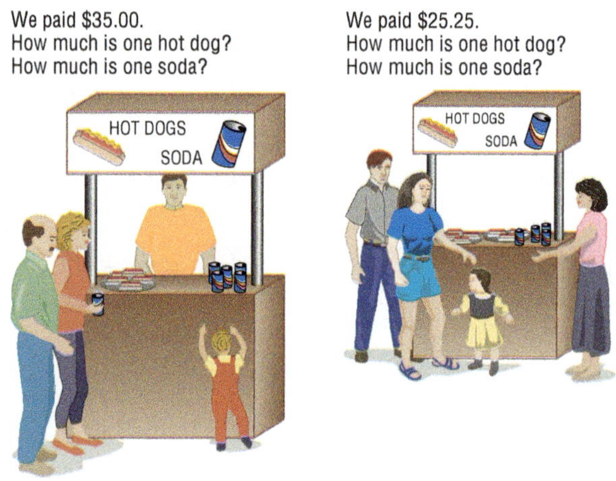

65. **Computing a Refund** The grocery store we use does not mark prices on its goods. My wife went to this store, bought three 1-pound packages of bacon and two cartons of eggs, and paid a total of \$13.45. Not knowing that she went to the store, I also went to the same store, purchased two 1-pound packages of bacon and three cartons of eggs, and paid a total of \$11.45. Now we want to return two 1-pound packages of bacon and two cartons of eggs. How much will be refunded?

66. **Finding the Current of a Stream** Pamela requires 3 hours to swim 15 miles downstream on the Illinois River. The return trip upstream takes 5 hours. Find Pamela's average speed in still water. How fast is the current? (Assume that Pamela's speed is the same in each direction.)

67. **Pharmacy** A doctor's prescription calls for a daily intake containing 40 milligrams (mg) of vitamin C and 30 mg of vitamin D. Your pharmacy stocks two liquids that can be used: one contains 20% vitamin C and 30% vitamin D, the other 40% vitamin C and 20% vitamin D. How many milligrams of each compound should be mixed to fill the prescription?

68. **Pharmacy** A doctor's prescription calls for the creation of pills that contain 12 units of vitamin B_{12} and 12 units of vitamin E. Your pharmacy stocks two powders that can be used to make these pills: one contains 20% vitamin B_{12} and 30% vitamin E, the other 40% vitamin B_{12} and 20% vitamin E. How many units of each powder should be mixed in each pill?

69. **Curve Fitting** Find real numbers $a, b,$ and c so that the graph of the function $y = ax^2 + bx + c$ contains the points $(-1, 4), (2, 3),$ and $(0, 1)$.

70. **Curve Fitting** Find real numbers $a, b,$ and c so that the graph of the function $y = ax^2 + bx + c$ contains the points $(-1, -2), (1, -4),$ and $(2, 4)$.

71. **IS–LM Model in Economics** In economics, the IS curve is a linear equation that represents all combinations of income Y and interest rates r that maintain an equilibrium in the market for goods in the economy. The LM curve is a linear equation that represents all combinations of income Y and interest rates r that maintain an equilibrium in the market for money in the economy. In an economy, suppose the equilibrium level of income (in millions of dollars) and interest rates satisfy the system of equations

$$\begin{cases} 0.06Y - 5000r = 240 \\ 0.06Y + 6000r = 900 \end{cases}$$

Find the equilibrium level of income and interest rates.

72. **IS–LM Model in Economics** In economics, the IS curve is a linear equation that represents all combinations of income Y and interest rates r that maintain an equilibrium in the market for goods in the economy. The LM curve is a linear equation that represents all combinations of income Y and interest rates r that maintain an equilibrium in the market for money in the economy. In an economy, suppose the equilibrium level of income (in millions of dollars) and interest rates satisfy the system of equations

$$\begin{cases} 0.05Y - 1000r = 10 \\ 0.05Y + 800r = 100 \end{cases}$$

Find the equilibrium level of income and interest rates.

73. **Electricity: Kirchhoff's Rules** An application of Kirchhoff's Rules to the circuit shown results in the following system of equations:

$$\begin{cases} I_2 = I_1 + I_3 \\ 5 - 3I_1 - 5I_2 = 0 \\ 10 - 5I_2 - 7I_3 = 0 \end{cases}$$

Find the currents I_1, I_2, and I_3.

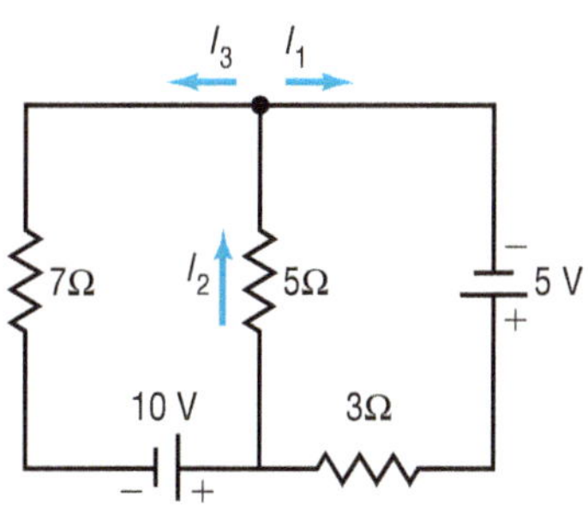

Source: Physics for Scientists & Engineers, 3rd ed., by Serway. © 1990 Brooks/Cole, a division of Thomson Learning.

74. **Electricity: Kirchhoff's Rules** An application of Kirchhoff's Rules to the circuit shown results in the following system of equations:

$$\begin{cases} I_3 = I_1 + I_2 \\ 8 = 4I_3 + 6I_2 \\ 8I_1 = 4 + 6I_2 \end{cases}$$

Find the currents I_1, I_2, and I_3.

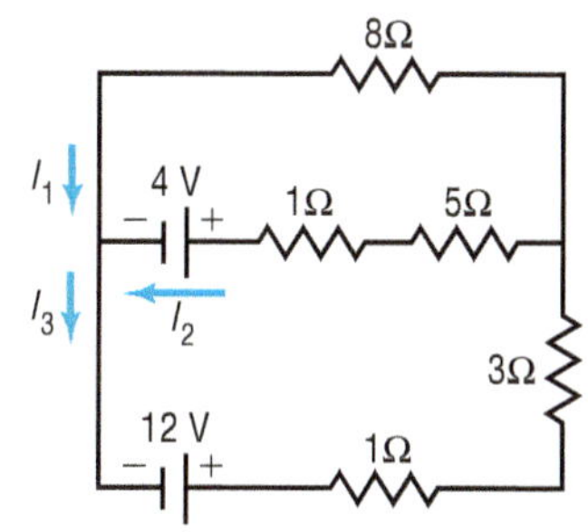

Source: Physics for Scientists & Engineers, 3rd ed., by Serway. © 1990 Brooks/Cole, a division of Thomson Learning.

75. **Theater Revenues** A Broadway theater has 500 seats, divided into orchestra, main, and balcony seating. Orchestra seats sell for \$50, main seats for \$35, and balcony seats for \$25. If all the seats are sold, the gross revenue to the theater is \$17,100. If all the main and balcony seats are sold, but only half the orchestra seats are sold, the gross revenue is \$14,600. How many are there of each kind of seat?

76. **Theater Revenues** A movie theater charges \$8.00 for adults, \$4.50 for children, and \$6.00 for senior citizens. One day the theater sold 405 tickets and collected \$2320 in receipts. Twice as many children's tickets were sold as adult tickets. How many adults, children, and senior citizens went to the theater that day?

77. **Nutrition** A dietitian wishes a patient to have a meal that has 66 grams (g) of protein, 94.5 g of carbohydrates, and 910 milligrams (mg) of calcium. The hospital food service tells the dietitian that the dinner for today is chicken, corn, and 2% milk. Each serving of chicken has 30 g of protein, 35 g of carbohydrates, and 200 mg of calcium. Each serving of corn has 3 g of protein, 16 g of carbohydrates, and 10 mg of calcium. Each glass of 2% milk has 9 g of protein, 13 g of carbohydrates, and 300 mg of calcium. How many servings of each food should the dietitian provide for the patient?

78. **Investments** Kelly has \$20,000 to invest. As her financial planner, you recommend that she diversify into three investments: Treasury bills that yield 5% simple interest, Treasury bonds that yield 7% simple interest, and corporate bonds that yield 10% simple interest. Kelly wishes to earn \$1390 per year in income. Also, Kelly wants her investment in Treasury bills to be \$3000 more than her investment in corporate bonds. How much money should Kelly place in each investment?

79. **Prices of Fast Food** One group of customers bought 8 deluxe hamburgers, 6 orders of large fries, and 6 large colas for \$26.10. A second group ordered 10 deluxe hamburgers, 6 large fries, and 8 large colas and paid \$31.60. Is there sufficient information to determine the price of each food item? If not, construct a table showing the various possibilities. Assume that the hamburgers cost between \$1.75 and \$2.25, the fries between \$0.75 and \$1.00, and the colas between \$0.60 and \$0.90.

80. **Prices of Fast Food** Use the information given in Problem 79. Suppose that a third group purchased 3 deluxe hamburgers, 2 large fries, and 4 large colas for \$10.95. Now is there sufficient information to determine the price of each food item? If so, determine each price.

81. **Painting a House** Three painters, Beth, Bill, and Edie, working together, can paint the exterior of a home in 10 hours (hr). Bill and Edie together have painted a similar house in 15 hr. One day, all three worked on this same kind of house for 4 hr, after which Edie left. Beth and Bill required 8 more hr to finish. Assuming no gain or loss in efficiency, how long should it take each person to complete such a job alone?

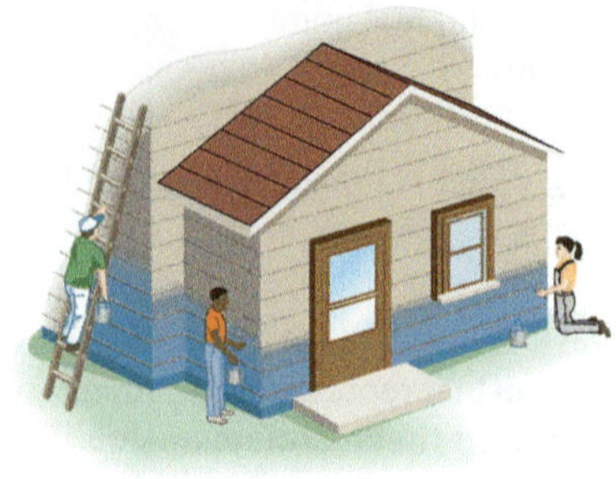

Discussion and Writing

82. Make up a system of three linear equations containing three variables that has:
(a) No solution
(b) Exactly one solution
(c) Infinitely many solutions
Give the three systems to a friend to solve and critique.

83. Write a brief paragraph outlining your strategy for solving a system of two linear equations containing two variables.

84. Do you prefer the method of substitution or the method of elimination for solving a system of two linear equations containing two variables? Give reasons.

'Are You Prepared?' Answers

1. $\{1\}$

2. (a)

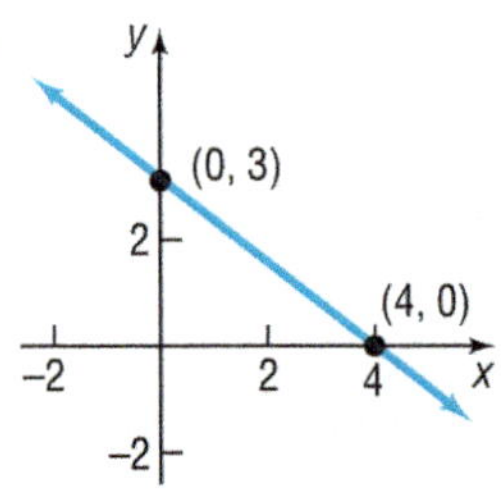

(b) $-\frac{3}{4}$

8.3 Systems of Linear Equations: Determinants

OBJECTIVES

1. Evaluate 2 by 2 Determinants (p. 578)
2. Use Cramer's Rule to Solve a System of Two Equations Containing Two Variables (p. 578)
3. Evaluate 3 by 3 Determinants (p. 581)
4. Use Cramer's Rule to Solve a System of Three Equations Containing Three Variables (p. 583)
5. Know Properties of Determinants (p. 584)

In the preceding section, we described a method of using matrices to solve a system of linear equations. This section deals with yet another method for solving systems of linear equations; however, it can be used only when the number of equations equals the number of variables. Although the method will work for any system (provided that the number of equations equals the number of variables), it is most often used for systems of two equations containing two variables or three equations containing three variables. This method, called *Cramer's Rule,* is based on the concept of a *determinant.*

1 Evaluate 2 by 2 Determinants

DEFINITION

If a, b, c, and d are four real numbers, the symbol

$$D = \begin{vmatrix} a & b \\ c & d \end{vmatrix}$$

is called a **2 by 2 determinant.** Its value is the number $ad - bc$; that is,

$$D = \begin{vmatrix} a & b \\ c & d \end{vmatrix} = ad - bc \qquad \textbf{(1)}$$

The following device may be helpful for remembering the value of a 2 by 2 determinant:

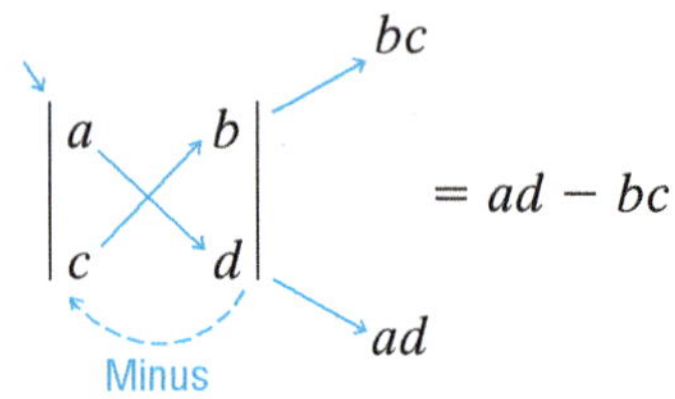

EXAMPLE 1 **Evaluating a 2 × 2 Determinant**

Evaluate: $\begin{vmatrix} 3 & -2 \\ 6 & 1 \end{vmatrix}$

Algebraic Solution

$$\begin{vmatrix} 3 & -2 \\ 6 & 1 \end{vmatrix} = (3)(1) - (6)(-2)$$
$$= 3 - (-12)$$
$$= 15$$

Graphing Solution

First, we enter the matrix whose entries are those of the determinant into the graphing utility and name it A. Using the determinant command, we obtain the result shown in Figure 9.

Figure 9

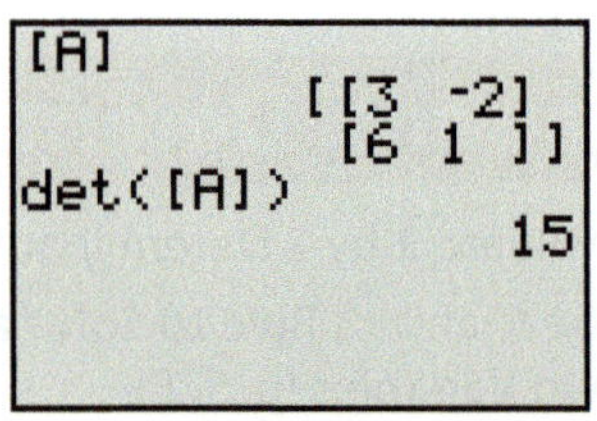

Now Work PROBLEM 7

2 Use Cramer's Rule to Solve a System of Two Equations Containing Two Variables

Let's now see the role that a 2 by 2 determinant plays in the solution of a system of two equations containing two variables. Consider the system

$$\begin{cases} ax + by = s & (1) \\ cx + dy = t & (2) \end{cases} \qquad \textbf{(2)}$$

We shall use the method of elimination to solve this system.

Provided $d \neq 0$ and $b \neq 0$, this system is equivalent to the system

$$\begin{cases} adx + bdy = sd & (1) \text{ Multiply by } d. \\ bcx + bdy = tb & (2) \text{ Multiply by } b. \end{cases}$$

Subtract the second equation from the first equation and obtain

$$\begin{cases} (ad - bc)x + 0 \cdot y = sd - tb & (1) \\ bcx + bdy = tb & (2) \end{cases}$$

Now the first equation can be rewritten using determinant notation.

$$\begin{vmatrix} a & b \\ c & d \end{vmatrix} x = \begin{vmatrix} s & b \\ t & d \end{vmatrix}$$

If $D = \begin{vmatrix} a & b \\ c & d \end{vmatrix} = ad - bc \neq 0$, we can solve for x to get

$$x = \frac{\begin{vmatrix} s & b \\ t & d \end{vmatrix}}{\begin{vmatrix} a & b \\ c & d \end{vmatrix}} = \frac{\begin{vmatrix} s & b \\ t & d \end{vmatrix}}{D} \qquad (3)$$

Return now to the original system (2). Provided that $a \neq 0$ and $c \neq 0$, the system is equivalent to

$$\begin{cases} acx + bcy = cs & (1) \text{ Multiply by } c. \\ acx + ady = at & (2) \text{ Multiply by } a. \end{cases}$$

Subtract the first equation from the second equation and obtain

$$\begin{cases} acx + \quad bcy = cs & (1) \\ 0 \cdot x + (ad - bc)y = at - cs & (2) \end{cases}$$

The second equation can now be rewritten using determinant notation.

$$\begin{vmatrix} a & b \\ c & d \end{vmatrix} y = \begin{vmatrix} a & s \\ c & t \end{vmatrix}$$

If $D = \begin{vmatrix} a & b \\ c & d \end{vmatrix} = ad - bc \neq 0$, we can solve for y to get

$$y = \frac{\begin{vmatrix} a & s \\ c & t \end{vmatrix}}{\begin{vmatrix} a & b \\ c & d \end{vmatrix}} = \frac{\begin{vmatrix} a & s \\ c & t \end{vmatrix}}{D} \qquad (4)$$

Equations (3) and (4) lead us to the following result, called **Cramer's Rule.**

THEOREM

Cramer's Rule for Two Equations Containing Two Variables

The solution to the system of equations

$$\begin{cases} ax + by = s & (1) \\ cx + dy = t & (2) \end{cases} \qquad (5)$$

is given by

$$x = \frac{\begin{vmatrix} s & b \\ t & d \end{vmatrix}}{\begin{vmatrix} a & b \\ c & d \end{vmatrix}} \qquad y = \frac{\begin{vmatrix} a & s \\ c & t \end{vmatrix}}{\begin{vmatrix} a & b \\ c & d \end{vmatrix}} \qquad (6)$$

provided that

$$D = \begin{vmatrix} a & b \\ c & d \end{vmatrix} = ad - bc \neq 0$$

In the derivation given for Cramer's Rule above, we assumed that none of the numbers a, b, c, and d was 0. In Problem 61 you will be asked to complete the proof under the less stringent condition that $D = ad - bc \neq 0$.

Now look carefully at the pattern in Cramer's Rule. The denominator in the solution (6) is the determinant of the coefficients of the variables.

$$\begin{cases} ax + by = s \\ cx + dy = t \end{cases} \qquad D = \begin{vmatrix} a & b \\ c & d \end{vmatrix}$$

In the solution for x, the numerator is the determinant, denoted by D_x, formed by replacing the entries in the first column (the coefficients of x) of D by the constants on the right side of the equal sign.

$$D_x = \begin{vmatrix} s & b \\ t & d \end{vmatrix}$$

In the solution for y, the numerator is the determinant, denoted by D_y, formed by replacing the entries in the second column (the coefficients in y) of D by the constants on the right side of the equal sign.

$$D_y = \begin{vmatrix} a & s \\ c & t \end{vmatrix}$$

Cramer's Rule then states that, if $D \neq 0$,

$$x = \frac{D_x}{D} \qquad y = \frac{D_y}{D} \qquad \textbf{(7)}$$

EXAMPLE 2 Solving a System of Linear Equations Using Determinants

Use Cramer's Rule, if applicable, to solve the system

$$\begin{cases} 3x - 2y = 4 & (1) \\ 6x + y = 13 & (2) \end{cases}$$

Algebraic Solution

The determinant D of the coefficients of the variables is

$$D = \begin{vmatrix} 3 & -2 \\ 6 & 1 \end{vmatrix} = (3)(1) - (6)(-2) = 15$$

Because $D \neq 0$, Cramer's Rule (7) can be used.

$$x = \frac{D_x}{D} = \frac{\begin{vmatrix} 4 & -2 \\ 13 & 1 \end{vmatrix}}{15} = \frac{(4)(1) - (13)(-2)}{15} = \frac{30}{15} = 2$$

$$y = \frac{D_y}{D} = \frac{\begin{vmatrix} 3 & 4 \\ 6 & 13 \end{vmatrix}}{15} = \frac{(3)(13) - (6)(4)}{15} = \frac{15}{15} = 1$$

Graphing Solution

We enter the coefficient matrix into our graphing utility. Call it A and evaluate $\det[A]$. Since $\det[A] \neq 0$, we can use Cramer's Rule. We enter the matrices D_x and D_y into our graphing utility and call them B and C, respectively. Finally, we find x by calculating $\frac{\det[B]}{\det[A]}$ and y by calculating $\frac{\det[C]}{\det[A]}$. The results are shown in Figure 10.

Figure 10

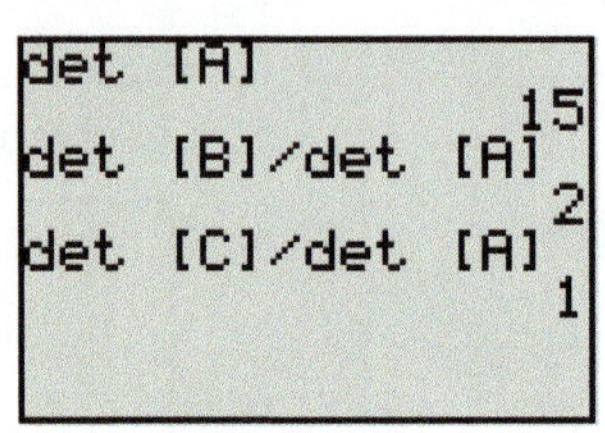

The solution is $x = 2$, $y = 1$ or, using ordered pairs, $(2, 1)$.

In attempting to use Cramer's Rule, if the determinant D of the coefficients of the variables is found to equal 0 (so that Cramer's Rule is not applicable), then the system either is inconsistent or has infinitely many solutions.

Now Work PROBLEM 15

3 Evaluate 3 by 3 Determinants

To use Cramer's Rule to solve a system of three equations containing three variables, we need to define a 3 by 3 determinant.

A **3 by 3 determinant** is symbolized by

$$\begin{vmatrix} a_{11} & a_{12} & a_{13} \\ a_{21} & a_{22} & a_{23} \\ a_{31} & a_{32} & a_{33} \end{vmatrix} \qquad \textbf{(8)}$$

in which $a_{11}, a_{12}, \ldots,$ are real numbers.

As with matrices, we use a double subscript to identify an entry by indicating its row and column numbers. For example, the entry a_{23} is in row 2, column 3.

The value of a 3 by 3 determinant may be defined in terms of 2 by 2 determinants by the following formula:

$$\begin{vmatrix} a_{11} & a_{12} & a_{13} \\ a_{21} & a_{22} & a_{23} \\ a_{31} & a_{32} & a_{33} \end{vmatrix} = a_{11}\begin{vmatrix} a_{22} & a_{23} \\ a_{32} & a_{33} \end{vmatrix} - a_{12}\begin{vmatrix} a_{21} & a_{23} \\ a_{31} & a_{33} \end{vmatrix} + a_{13}\begin{vmatrix} a_{21} & a_{22} \\ a_{31} & a_{32} \end{vmatrix} \qquad \textbf{(9)}$$

Minus (the sign before a_{12}); Plus (the sign before a_{13})

2 by 2 determinant left after removing the row and column containing a_{11}; 2 by 2 determinant left after removing the row and column containing a_{12}; 2 by 2 determinant left after removing the row and column containing a_{13}

The 2 by 2 determinants shown in formula (9) are called **minors** of the 3 by 3 determinant. For an n by n determinant, the **minor** M_{ij} of entry a_{ij} is the determinant resulting from removing the ith row and jth column.

EXAMPLE 3 Finding Minors of a 3 by 3 Determinant

For the determinant $A = \begin{vmatrix} 2 & -1 & 3 \\ -2 & 5 & 1 \\ 0 & 6 & -9 \end{vmatrix}$, find: (a) M_{12} (b) M_{23}

Solution (a) M_{12} is the determinant that results from removing the first row and second column from A.

$$A = \begin{vmatrix} 2 & -1 & 3 \\ -2 & 5 & 1 \\ 0 & 6 & -9 \end{vmatrix} \qquad M_{12} = \begin{vmatrix} -2 & 1 \\ 0 & -9 \end{vmatrix} = (-2)(-9) - (0)(1) = 18$$

(b) M_{23} is the determinant that results from removing the second row and third column from A.

$$A = \begin{vmatrix} 2 & -1 & 3 \\ -2 & 5 & 1 \\ 0 & 6 & -9 \end{vmatrix} \qquad M_{23} = \begin{vmatrix} 2 & -1 \\ 0 & 6 \end{vmatrix} = (2)(6) - (0)(-1) = 12$$

Referring back to formula (9), we see that each element a_{ij} is multiplied by its minor, but sometimes this term is added and other times, subtracted. To determine whether to add or subtract a term, we must consider the *cofactor*.

DEFINITION

For an n by n determinant A, the **cofactor** of entry a_{ij}, denoted by A_{ij}, is given by

$$A_{ij} = (-1)^{i+j}M_{ij}$$

where M_{ij} is the minor of entry a_{ij}.

The exponent of $(-1)^{i+j}$ is the sum of the row and column of the entry a_{ij}, so if $i + j$ is even, $(-1)^{i+j}$ will equal 1, and if $i + j$ is odd, $(-1)^{i+j}$ will equal -1.

To find the value of a determinant, multiply each entry in any row or column by its cofactor and sum the results. This process is referred to as **expanding across a row or column.** For example, the value of the 3 by 3 determinant in formula (9) was found by expanding across row 1.

If we choose to expand down column 2, we obtain

$$\begin{vmatrix} a_{11} & a_{12} & a_{13} \\ a_{21} & a_{22} & a_{23} \\ a_{31} & a_{32} & a_{33} \end{vmatrix} = (-1)^{1+2}a_{12}\begin{vmatrix} a_{21} & a_{23} \\ a_{31} & a_{33} \end{vmatrix} + (-1)^{2+2}a_{22}\begin{vmatrix} a_{11} & a_{13} \\ a_{31} & a_{33} \end{vmatrix} + (-1)^{3+2}a_{32}\begin{vmatrix} a_{11} & a_{13} \\ a_{21} & a_{23} \end{vmatrix}$$

Expand down column 2.

If we choose to expand across row 3, we obtain

$$\begin{vmatrix} a_{11} & a_{12} & a_{13} \\ a_{21} & a_{22} & a_{23} \\ a_{31} & a_{32} & a_{33} \end{vmatrix} = (-1)^{3+1}a_{31}\begin{vmatrix} a_{12} & a_{13} \\ a_{22} & a_{23} \end{vmatrix} + (-1)^{3+2}a_{32}\begin{vmatrix} a_{11} & a_{13} \\ a_{21} & a_{23} \end{vmatrix} + (-1)^{3+3}a_{33}\begin{vmatrix} a_{11} & a_{12} \\ a_{21} & a_{22} \end{vmatrix}$$

Expand across row 3.

It can be shown that the value of a determinant does not depend on the choice of the row or column used in the expansion. However, expanding across a row or column that has an element equal to 0 reduces the amount of work needed to compute the value of the determinant.

EXAMPLE 4 Evaluating a 3 × 3 Determinant

Find the value of the 3 by 3 determinant: $\begin{vmatrix} 3 & 0 & -1 \\ 4 & 6 & 2 \\ 8 & -2 & 3 \end{vmatrix}$

Solution Because of the 0, it is easiest to expand across row 1 or down row 2. We choose to expand across row 1.

$$\begin{vmatrix} 3 & 0 & -1 \\ 4 & 6 & 2 \\ 8 & -2 & 3 \end{vmatrix} = (-1)^{1+1} \cdot 3 \cdot \begin{vmatrix} 6 & 2 \\ -2 & 3 \end{vmatrix} + (-1)^{1+2} \cdot 0 \cdot \begin{vmatrix} 4 & 2 \\ 8 & 3 \end{vmatrix} + (-1)^{1+3} \cdot (-1) \cdot \begin{vmatrix} 4 & 6 \\ 8 & -2 \end{vmatrix}$$

$$= 3(18 - (-4)) - 0 + (-1)(-8 - 48)$$

$$= 3(22) + (-1)(-56)$$

$$= 66 + 56 = 122$$

Check: Use a graphing utility to verify the result.

Now Work PROBLEM 11

4 Use Cramer's Rule to Solve a System of Three Equations Containing Three Variables

Consider the following system of three equations containing three variables.

$$\begin{cases} a_{11}x + a_{12}y + a_{13}z = c_1 \\ a_{21}x + a_{22}y + a_{23}z = c_2 \\ a_{31}x + a_{32}y + a_{33}z = c_3 \end{cases} \qquad \textbf{(10)}$$

It can be shown that if the determinant D of the coefficients of the variables is not 0, that is, if

$$D = \begin{vmatrix} a_{11} & a_{12} & a_{13} \\ a_{21} & a_{22} & a_{23} \\ a_{31} & a_{32} & a_{33} \end{vmatrix} \neq 0$$

the unique solution of system (10) is given by

THEOREM

Cramer's Rule for Three Equations Containing Three Variables

$$x = \frac{D_x}{D} \qquad y = \frac{D_y}{D} \qquad z = \frac{D_z}{D}$$

where

$$D_x = \begin{vmatrix} c_1 & a_{12} & a_{13} \\ c_2 & a_{22} & a_{23} \\ c_3 & a_{32} & a_{33} \end{vmatrix} \qquad D_y = \begin{vmatrix} a_{11} & c_1 & a_{13} \\ a_{21} & c_2 & a_{23} \\ a_{31} & c_3 & a_{33} \end{vmatrix} \qquad D_z = \begin{vmatrix} a_{11} & a_{12} & c_1 \\ a_{21} & a_{22} & c_2 \\ a_{31} & a_{32} & c_3 \end{vmatrix}$$

Do you see the similarity of this pattern and the pattern observed earlier for a system of two equations containing two variables?

EXAMPLE 5 Using Cramer's Rule

Use Cramer's Rule, if applicable, to solve the following system:

$$\begin{cases} 2x + y - z = 3 & (1) \\ -x + 2y + 4z = -3 & (2) \\ x - 2y - 3z = 4 & (3) \end{cases}$$

Solution The value of the determinant D of the coefficients of the variables is

$$D = \begin{vmatrix} 2 & 1 & -1 \\ -1 & 2 & 4 \\ 1 & -2 & -3 \end{vmatrix} = (-1)^{1+1} \cdot 2 \cdot \begin{vmatrix} 2 & 4 \\ -2 & -3 \end{vmatrix} + (-1)^{1+2} \cdot 1 \cdot \begin{vmatrix} -1 & 4 \\ 1 & -3 \end{vmatrix} + (-1)^{1+3}(-1)\begin{vmatrix} -1 & 2 \\ 1 & -2 \end{vmatrix}$$

$$= 2(2) - 1(-1) + (-1)(0)$$
$$= 4 + 1 = 5$$

Because $D \neq 0$, we proceed to find the values of D_x, D_y, and D_z.

$$D_x = \begin{vmatrix} 3 & 1 & -1 \\ -3 & 2 & 4 \\ 4 & -2 & -3 \end{vmatrix} = (-1)^{1+1} \cdot 3 \cdot \begin{vmatrix} 2 & 4 \\ -2 & -3 \end{vmatrix} + (-1)^{1+2} \cdot 1 \cdot \begin{vmatrix} -3 & 4 \\ 4 & -3 \end{vmatrix} + (-1)^{1+3}(-1)\begin{vmatrix} -3 & 2 \\ 4 & -2 \end{vmatrix}$$

$$= 3(2) - 1(-7) + (-1)(-2) = 15$$

$$D_y = \begin{vmatrix} 2 & 3 & -1 \\ -1 & -3 & 4 \\ 1 & 4 & -3 \end{vmatrix} = (-1)^{1+1} \cdot 2 \cdot \begin{vmatrix} -3 & 4 \\ 4 & -3 \end{vmatrix} + (-1)^{1+2} \cdot 3 \cdot \begin{vmatrix} -1 & 4 \\ 1 & -3 \end{vmatrix} + (-1)^{1+3}(-1) \begin{vmatrix} -1 & -3 \\ 1 & 4 \end{vmatrix}$$

$$= 2(-7) - 3(-1) + (-1)(-1) = -10$$

$$D_z = \begin{vmatrix} 2 & 1 & 3 \\ -1 & 2 & -3 \\ 1 & -2 & 4 \end{vmatrix} = (-1)^{1+1} \cdot 2 \cdot \begin{vmatrix} 2 & -3 \\ -2 & 4 \end{vmatrix} + (-1)^{1+2} \cdot 1 \cdot \begin{vmatrix} -1 & -3 \\ 1 & 4 \end{vmatrix} + (-1)^{1+3} \cdot 3 \cdot \begin{vmatrix} -1 & 2 \\ 1 & -2 \end{vmatrix}$$

$$= 2(2) - 1(-1) + 3(0) = 5$$

As a result,

$$x = \frac{D_x}{D} = \frac{15}{5} = 3 \qquad y = \frac{D_y}{D} = \frac{-10}{5} = -2 \qquad z = \frac{D_z}{D} = \frac{5}{5} = 1$$

The solution is $x = 3$, $y = -2$, $z = 1$, or, using ordered triplets, $(3, -2, 1)$.

We already know that Cramer's Rule does not apply when the determinant of the coefficients on the variables, D, is 0. But can we learn anything about the system other than it is not a consistent and independent system if $D = 0$? The answer is yes!

Cramer's Rule with Inconsistent or Dependent Systems

- If $D = 0$ and at least one of the determinants D_x, D_y, or D_z is different from 0, then the system is inconsistent and the solution set is $\varnothing$ or $\{\ \}$.
- If $D = 0$ and all the determinants D_x, D_y, and D_z equal 0, then the system is consistent and dependent so that there are infinitely many solutions. The system must be solved using row reduction techniques.

Now Work PROBLEM 33

5 Know Properties of Determinants

Determinants have several properties that are sometimes helpful for obtaining their value. We list some of them here.

THEOREM The value of a determinant changes sign if any two rows (or any two columns) are interchanged. **(11)**

Proof for 2 by 2 Determinants

$$\begin{vmatrix} a & b \\ c & d \end{vmatrix} = ad - bc \quad \text{and} \quad \begin{vmatrix} c & d \\ a & b \end{vmatrix} = bc - ad = -(ad - bc)$$ ■

EXAMPLE 6 Demonstrating Theorem (11)

$$\begin{vmatrix} 3 & 4 \\ 1 & 2 \end{vmatrix} = 6 - 4 = 2 \qquad \begin{vmatrix} 1 & 2 \\ 3 & 4 \end{vmatrix} = 4 - 6 = -2$$

THEOREM If all the entries in any row (or any column) equal 0, the value of the determinant is 0. **(12)**

Proof Expand across the row (or down the column) containing the 0's. ■

THEOREM If any two rows (or any two columns) of a determinant have corresponding entries that are equal, the value of the determinant is 0. **(13)**

You are asked to prove this result for a 3 by 3 determinant in which the entries in column 1 equal the entries in column 3 in Problem 64.

EXAMPLE 7 **Demonstrating Theorem (13)**

$$\begin{vmatrix} 1 & 2 & 3 \\ 1 & 2 & 3 \\ 4 & 5 & 6 \end{vmatrix} = (-1)^{1+1} \cdot 1 \cdot \begin{vmatrix} 2 & 3 \\ 5 & 6 \end{vmatrix} + (-1)^{1+2} \cdot 2 \cdot \begin{vmatrix} 1 & 3 \\ 4 & 6 \end{vmatrix} + (-1)^{1+3} \cdot 3 \cdot \begin{vmatrix} 1 & 2 \\ 4 & 5 \end{vmatrix}$$

$$= 1(-3) - 2(-6) + 3(-3) = -3 + 12 - 9 = 0$$

THEOREM If any row (or any column) of a determinant is multiplied by a nonzero number k, the value of the determinant is also changed by a factor of k. **(14)**

You are asked to prove this result for a 3 by 3 determinant using row 2 in Problem 63.

EXAMPLE 8 **Demonstrating Theorem (14)**

$$\begin{vmatrix} 1 & 2 \\ 4 & 6 \end{vmatrix} = 6 - 8 = -2$$

$$\begin{vmatrix} k & 2k \\ 4 & 6 \end{vmatrix} = 6k - 8k = -2k = k(-2) = k\begin{vmatrix} 1 & 2 \\ 4 & 6 \end{vmatrix}$$

THEOREM If the entries of any row (or any column) of a determinant are multiplied by a nonzero number k and the result is added to the corresponding entries of another row (or column), the value of the determinant remains unchanged. **(15)**

In Problem 65, you are asked to prove this result for a 3 by 3 determinant using rows 1 and 2.

EXAMPLE 9 **Demonstrating Theorem (15)**

$$\begin{vmatrix} 3 & 4 \\ 5 & 2 \end{vmatrix} = -14 \qquad \begin{vmatrix} 3 & 4 \\ 5 & 2 \end{vmatrix} \rightarrow \begin{vmatrix} -7 & 0 \\ 5 & 2 \end{vmatrix} = -14$$

↑ Multiply row 2 by −2 and add to row 1.

8.3 Assess Your Understanding

Concepts and Vocabulary

1. Cramer's Rule uses _____ to solve a system of linear equations.

2. $D = \begin{vmatrix} a & b \\ c & d \end{vmatrix} =$ _____

3. ***True or False*** A 3 by 3 determinant can never equal 0.

4. ***True or False*** The value of a determinant remains unchanged if any two rows or any two columns are interchanged.

Skill Building

In Problems 5–14, find the value of each determinant by hand. Verify your results using a graphing utility.

5. $\begin{vmatrix} 3 & 1 \\ 4 & 2 \end{vmatrix}$ **6.** $\begin{vmatrix} 6 & 1 \\ 5 & 2 \end{vmatrix}$ **7.** $\begin{vmatrix} 6 & 4 \\ -1 & 3 \end{vmatrix}$ **8.** $\begin{vmatrix} 8 & -3 \\ 4 & 2 \end{vmatrix}$ **9.** $\begin{vmatrix} -3 & -1 \\ 4 & 2 \end{vmatrix}$

10. $\begin{vmatrix} -4 & 2 \\ -5 & 3 \end{vmatrix}$ **11.** $\begin{vmatrix} 3 & 4 & 2 \\ 1 & -1 & 5 \\ 1 & 2 & -2 \end{vmatrix}$ **12.** $\begin{vmatrix} 1 & 3 & -2 \\ 6 & 1 & -5 \\ 8 & 2 & 3 \end{vmatrix}$ **13.** $\begin{vmatrix} 4 & -1 & 2 \\ 6 & -1 & 0 \\ 1 & -3 & 4 \end{vmatrix}$ **14.** $\begin{vmatrix} 3 & -9 & 4 \\ 1 & 4 & 0 \\ 8 & -3 & 1 \end{vmatrix}$

In Problems 15–42, solve each system of equations using Cramer's Rule if it is applicable. If Cramer's Rule is not applicable, say so.

15. $\begin{cases} x + y = 8 \\ x - y = 4 \end{cases}$ **16.** $\begin{cases} x + 2y = 5 \\ x - y = 3 \end{cases}$ **17.** $\begin{cases} 5x - y = 13 \\ 2x + 3y = 12 \end{cases}$ **18.** $\begin{cases} x + 3y = 5 \\ 2x - 3y = -8 \end{cases}$

19. $\begin{cases} 3x = 24 \\ x + 2y = 0 \end{cases}$ **20.** $\begin{cases} 4x + 5y = -3 \\ -2y = -4 \end{cases}$ **21.** $\begin{cases} 3x - 6y = 24 \\ 5x + 4y = 12 \end{cases}$ **22.** $\begin{cases} 2x + 4y = 16 \\ 3x - 5y = -9 \end{cases}$

23. $\begin{cases} 3x - 2y = 4 \\ 6x - 4y = 0 \end{cases}$ **24.** $\begin{cases} -x + 2y = 5 \\ 4x - 8y = 6 \end{cases}$ **25.** $\begin{cases} 2x - 4y = -2 \\ 3x + 2y = 3 \end{cases}$ **26.** $\begin{cases} 3x + 3y = 3 \\ 4x + 2y = \frac{8}{3} \end{cases}$

27. $\begin{cases} 2x - 3y = -1 \\ 10x + 10y = 5 \end{cases}$ **28.** $\begin{cases} 3x - 2y = 0 \\ 5x + 10y = 4 \end{cases}$ **29.** $\begin{cases} 2x + 3y = 6 \\ x - y = \frac{1}{2} \end{cases}$ **30.** $\begin{cases} \frac{1}{2}x + y = -2 \\ x - 2y = 8 \end{cases}$

31. $\begin{cases} 3x - 5y = 3 \\ 15x + 5y = 21 \end{cases}$ **32.** $\begin{cases} 2x - y = -1 \\ x + \frac{1}{2}y = \frac{3}{2} \end{cases}$ **33.** $\begin{cases} x + y - z = 6 \\ 3x - 2y + z = -5 \\ x + 3y - 2z = 14 \end{cases}$ **34.** $\begin{cases} x - y + z = -4 \\ 2x - 3y + 4z = -15 \\ 5x + y - 2z = 12 \end{cases}$

35. $\begin{cases} x + 2y - z = -3 \\ 2x - 4y + z = -7 \\ -2x + 2y - 3z = 4 \end{cases}$ **36.** $\begin{cases} x + 4y - 3z = -8 \\ 3x - y + 3z = 12 \\ x + y + 6z = 1 \end{cases}$ **37.** $\begin{cases} x - 2y + 3z = 1 \\ 3x + y - 2z = 0 \\ 2x - 4y + 6z = 2 \end{cases}$ **38.** $\begin{cases} x - y + 2z = 5 \\ 3x + 2y = 4 \\ -2x + 2y - 4z = -10 \end{cases}$

39. $\begin{cases} x + 2y - z = 0 \\ 2x - 4y + z = 0 \\ -2x + 2y - 3z = 0 \end{cases}$ **40.** $\begin{cases} x + 4y - 3z = 0 \\ 3x - y + 3z = 0 \\ x + y + 6z = 0 \end{cases}$ **41.** $\begin{cases} x - 2y + 3z = 0 \\ 3x + y - 2z = 0 \\ 2x - 4y + 6z = 0 \end{cases}$ **42.** $\begin{cases} x - y + 2z = 0 \\ 3x + 2y = 0 \\ -2x + 2y - 4z = 0 \end{cases}$

In Problems 43–50, use properties of determinants to find the value of each determinant if it is known that

$$\begin{vmatrix} x & y & z \\ u & v & w \\ 1 & 2 & 3 \end{vmatrix} = 4$$

43. $\begin{vmatrix} 1 & 2 & 3 \\ u & v & w \\ x & y & z \end{vmatrix}$ **44.** $\begin{vmatrix} x & y & z \\ u & v & w \\ 2 & 4 & 6 \end{vmatrix}$ **45.** $\begin{vmatrix} x & y & z \\ -3 & -6 & -9 \\ u & v & w \end{vmatrix}$ **46.** $\begin{vmatrix} 1 & 2 & 3 \\ x - u & y - v & z - w \\ u & v & w \end{vmatrix}$

47. $\begin{vmatrix} 1 & 2 & 3 \\ x - 3 & y - 6 & z - 9 \\ 2u & 2v & 2w \end{vmatrix}$ **48.** $\begin{vmatrix} x & y & z - x \\ u & v & w - u \\ 1 & 2 & 2 \end{vmatrix}$ **49.** $\begin{vmatrix} 1 & 2 & 3 \\ 2x & 2y & 2z \\ u - 1 & v - 2 & w - 3 \end{vmatrix}$ **50.** $\begin{vmatrix} x + 3 & y + 6 & z + 9 \\ 3u - 1 & 3v - 2 & 3w - 3 \\ 1 & 2 & 3 \end{vmatrix}$

Mixed Practice

In Problems 51–56, solve for x.

51. $\begin{vmatrix} x & x \\ 4 & 3 \end{vmatrix} = 5$ **52.** $\begin{vmatrix} x & 1 \\ 3 & x \end{vmatrix} = -2$ **53.** $\begin{vmatrix} x & 1 & 1 \\ 4 & 3 & 2 \\ -1 & 2 & 5 \end{vmatrix} = 2$

54. $\begin{vmatrix} 3 & 2 & 4 \\ 1 & x & 5 \\ 0 & 1 & -2 \end{vmatrix} = 0$ **55.** $\begin{vmatrix} x & 2 & 3 \\ 1 & x & 0 \\ 6 & 1 & -2 \end{vmatrix} = 7$ **56.** $\begin{vmatrix} x & 1 & 2 \\ 1 & x & 3 \\ 0 & 1 & 2 \end{vmatrix} = -4x$

Applications and Extensions

57. Geometry: Equation of a Line An equation of the line containing the two points (x_1, y_1) and (x_2, y_2) may be expressed as the determinant

$$\begin{vmatrix} x & y & 1 \\ x_1 & y_1 & 1 \\ x_2 & y_2 & 1 \end{vmatrix} = 0$$

Prove this result by expanding the determinant and comparing the result to the two-point form of the equation of a line.

58. Geometry: Collinear Points Using the result obtained in Problem 57, show that three distinct points (x_1, y_1), (x_2, y_2), and (x_3, y_3) are collinear (lie on the same line) if and only if

$$\begin{vmatrix} x_1 & y_1 & 1 \\ x_2 & y_2 & 1 \\ x_3 & y_3 & 1 \end{vmatrix} = 0$$

59. Geometry: Area of a Triangle A triangle has vertices (x_1, y_1),(x_2, y_2), and (x_3, y_3). Show that the area of the triangle is given by the absolute value of D, where $D = \frac{1}{2}\begin{vmatrix} x_1 & x_2 & x_3 \\ y_1 & y_2 & y_3 \\ 1 & 1 & 1 \end{vmatrix}$. Use this formula to find the area of a triangle with vertices (2, 3), (5, 2), and (6, 5).

60. Show that $\begin{vmatrix} x^2 & x & 1 \\ y^2 & y & 1 \\ z^2 & z & 1 \end{vmatrix} = (y - z)(x - y)(x - z)$.

61. Complete the proof of Cramer's Rule for two equations containing two variables.

[**Hint:** In system (5), page 579, if $a = 0$, then $b \neq 0$ and $c \neq 0$, since $D = -bc \neq 0$. Now show that equation (6) provides a solution of the system when $a = 0$. Then three cases remain: $b = 0$, $c = 0$, and $d = 0$.]

62. Interchange columns 1 and 3 of a 3 by 3 determinant. Show that the value of the new determinant is -1 times the value of the original determinant.

63. Multiply each entry in row 2 of a 3 by 3 determinant by the number k, $k \neq 0$. Show that the value of the new determinant is k times the value of the original determinant.

64. Prove that a 3 by 3 determinant in which the entries in column 1 equal those in column 3 has the value 0.

65. Prove that, if row 2 of a 3 by 3 determinant is multiplied by k, $k \neq 0$, and the result is added to the entries in row 1, there is no change in the value of the determinant.

Answers

CHAPTER R Review

Historical Problems (page 15)

1. (a) 1,20 **(b)** 2,50 **2. (a)** $\frac{7}{3} = 2.333\ldots$ **(b)** $\frac{39}{8} = 4.875$ **(c)** $\frac{84{,}823}{27{,}000} = 3.141592592\ldots$

R.1 Assess Your Understanding (page 15)

1. rational **2.** 31 **3.** Distributive **4.** $5(x + 3) = 6$ **5.** T **6.** F **7.** F **8.** T **9.** $\{1, 2, 3, 4, 5, 6, 7, 8, 9\}$ **11.** $\{4\}$ **13.** $\{1, 3, 4, 6\}$ **15.** $\{0, 2, 6, 7, 8\}$ **17.** $\{0, 1, 2, 3, 5, 6, 7, 8, 9\}$ **19.** $\{0, 1, 2, 3, 5, 6, 7, 8, 9\}$ **21. (a)** $\{2, 5\}$ **(b)** $\{-6, 2, 5\}$ **(c)** $\left\{-6, \frac{1}{2}, -1.\overline{3}, 2, 5\right\}$ **(d)** $\{\pi\}$ **(e)** $\left\{-6, \frac{1}{2}, -1.\overline{3}, \pi, 2, 5\right\}$ **23. (a)** $\{1\}$ **(b)** $\{0, 1\}$ **(c)** $\left\{0, 1, \frac{1}{2}, \frac{1}{3}, \frac{1}{4}\right\}$ **(d)** None **(e)** $\left\{0, 1, \frac{1}{2}, \frac{1}{3}, \frac{1}{4}\right\}$ **25. (a)** None **(b)** None **(c)** None **(d)** $\left\{\sqrt{2}, \pi, \sqrt{2} + 1, \pi + \frac{1}{2}\right\}$ **(e)** $\left\{\sqrt{2}, \pi, \sqrt{2} + 1, \pi + \frac{1}{2}\right\}$ **27. (a)** 18.953 **(b)** 18.952 **29. (a)** 28.653 **(b)** 28.653 **31. (a)** 0.063 **(b)** 0.062 **33. (a)** 9.999 **(b)** 9.998 **35. (a)** 0.429 **(b)** 0.428 **37. (a)** 34.733 **(b)** 34.733 **39.** $3 + 2 = 5$ **41.** $x + 2 = 3 \cdot 4$ **43.** $3y = 1 + 2$ **45.** $x - 2 = 6$ **47.** $\frac{x}{2} = 6$ **49.** 7 **51.** 6 **53.** 1 **55.** $\frac{13}{3}$ **57.** -11 **59.** 11 **61.** -4 **63.** 1 **65.** 6 **67.** $\frac{2}{7}$ **69.** $\frac{4}{45}$ **71.** $\frac{23}{20}$ **73.** $\frac{79}{30}$ **75.** $\frac{13}{36}$ **77.** $-\frac{16}{45}$ **79.** $\frac{1}{60}$ **81.** $\frac{15}{22}$ **83.** 1 **85.** $\frac{15}{8}$ **87.** $6x + 24$ **89.** $x^2 - 4x$ **91.** $\frac{3}{2}x - 1$ **93.** $x^2 + 6x + 8$ **95.** $x^2 - x - 2$ **97.** $x^2 - 10x + 16$ **99.** $2x + 3x = (2 + 3)x = 5x$ **101.** $2(3 \cdot 4) = 2 \cdot 12 = 24; (2 \cdot 3) \cdot (2 \cdot 4) = 6 \cdot 8 = 48$ **103.** No; $2 - 3 \neq 3 - 2$ **105.** No; $\frac{2}{3} \neq \frac{3}{2}$ **107.** Symmetric Property **109.** No; no

R.2 Assess Your Understanding (page 26)

1. variable **2.** origin **3.** strict **4.** base; exponent or power **5.** 1.2345678×10^3 **6.** T **7.** T **8.** F **9.** F **10.** F

11. 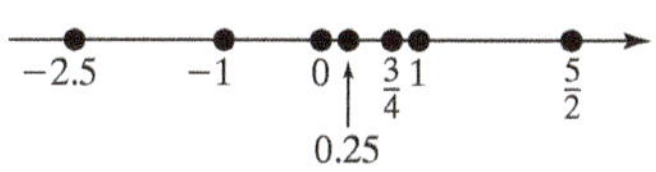

13. $>$ **15.** $>$ **17.** $>$ **19.** $=$ **21.** $<$ **23.** $x > 0$ **25.** $x < 2$ **27.** $x \leq 1$

29. 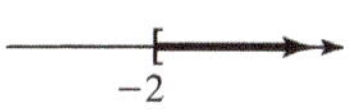

31. 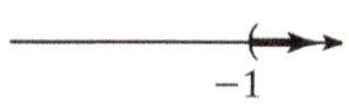

33. 1 **35.** 2 **37.** 6 **39.** 4 **41.** -28 **43.** $\frac{4}{5}$ **45.** 0 **47.** 1 **49.** 5 **51.** 1 **53.** 22 **55.** 2 **57.** $x = 0$ **59.** $x = 3$ **61.** None **63.** $x = 0, x = 1, x = -1$ **65.** $\{x \mid x \neq 5\}$ **67.** $\{x \mid x \neq -4\}$ **69.** 0°C **71.** 25°C **73.** 16 **75.** $\frac{1}{16}$ **77.** $\frac{1}{9}$ **79.** 9 **81.** 5 **83.** 4 **85.** $64x^6$ **87.** $\frac{x^4}{y^2}$ **89.** $\frac{x}{y}$ **91.** $-\frac{8x^3z}{9y}$ **93.** $\frac{16x^2}{9y^2}$ **95.** -4 **97.** 5 **99.** 4 **101.** 2 **103.** $\sqrt{5}$ **105.** $\frac{1}{2}$ **107.** 10; 0 **109.** 81 **111.** 304,006.671 **113.** 0.004 **115.** 481.890 **117.** 0.000 **119.** 4.542×10^2 **121.** 1.3×10^{-2} **123.** 3.2155×10^4 **125.** 4.23×10^{-4} **127.** 61,500 **129.** 0.001214 **131.** 110,000,000 **133.** 0.081 **135.** $A = lw$ **137.** $C = \pi d$ **139.** $A = \frac{\sqrt{3}}{4}x^2$ **141.** $V = \frac{4}{3}\pi r^3$ **143.** $V = x^3$ **145. (a)** \$6000 **(b)** \$8000 **147.** $|x - 4| \geq 6$ **149. (a)** $2 \leq 5$ **(b)** $6 > 5$ **151. (a)** Yes **(b)** No **153.** 400,000,000 m **155.** 0.0000005 m **157.** 5×10^{-4} in. **159.** 5.865696×10^{12} mi **161.** No; $\frac{1}{3}$ is larger; 0.000333 ... **163.** No

R.3 Assess Your Understanding (page 36)

1. right; hypotenuse **2.** $A = \frac{1}{2}bh$ **3.** $C = 2\pi r$ **4.** similar **5.** T **6.** T **7.** F **8.** T **9.** T **10.** F **11.** 13 **13.** 26 **15.** 25 **17.** Right triangle; 5 **19.** Not a right triangle **21.** Right triangle; 25 **23.** Not a right triangle **25.** 8 in.^{2} **27.** 4 in.^{2} **29.** $A = 25\pi$ m^2; $C = 10\pi$ m **31.** $V = 224$ ft^3; $S = 232$ ft^2 **33.** $V = \frac{256}{3}\pi$ cm^3; $S = 64\pi$ cm^2 **35.** $V = 648\pi$ in.^{3}; $S = 306\pi$ in.^{2} **37.** π square units **39.** 2π square units **41.** $x = 4$ units; $A = 90°$; $B = 60°$; $C = 30°$ **43.** $x = 67.5$ units; $A = 60°$; $B = 95°$; $C = 25°$ **45.** About 16.8 ft **47.** 64 ft^2 **49.** $24 + 2\pi \approx 30.28$ ft^2; $16 + 2\pi \approx 22.28$ ft **51.** 160 paces **53.** About 5.477 mi **55.** From 100 ft: 12.2 mi; From 150 ft: 15.0 mi

R.4 Assess Your Understanding (page 47)

1. 4; 3 **2.** $x^4 - 16$ **3.** $x^3 - 8$ **4.** F **5.** T **6.** F **7.** Monomial; variable: x; coefficient: 2; degree: 3 **9.** Not a monomial; the exponent of the variable is not a nonnegative integer **11.** Monomial; variables: x, y; coefficient: -2; degree: 3 **13.** Not a monomial; the exponent of one of the variables is not a nonnegative integer **15.** Not a monomial; it has more than one term **17.** Yes; 2 **19.** Yes; 0
21. No; the variable of one of the terms is not a nonnegative integer **23.** Yes; 3 **25.** No; the polynomial of the denominator has a degree greater than 0 **27.** $x^2 + 7x + 2$ **29.** $x^3 - 4x^2 + 9x + 7$ **31.** $6x^5 + 5x^4 + 3x^2 + x$ **33.** $7x^2 - x - 7$ **35.** $-2x^3 + 18x^2 - 18$ **37.** $2x^2 - 4x + 6$ **39.** $15y^2 - 27y + 30$ **41.** $x^3 + x^2 - 4x$ **43.** $-8x^5 - 10x^2$ **45.** $x^3 + 3x^2 - 2x - 4$ **47.** $x^2 + 6x + 8$ **49.** $2x^2 + 9x + 10$ **51.** $x^2 - 2x - 8$

53. $x^2 - 5x + 6$ **55.** $2x^2 - x - 6$ **57.** $-2x^2 + 11x - 12$ **59.** $2x^2 + 8x + 8$ **61.** $x^2 - xy - 2y^2$ **63.** $-6x^2 - 13xy - 6y^2$ **65.** $x^2 - 49$ **67.** $4x^2 - 9$ **69.** $x^2 + 8x + 16$ **71.** $x^2 - 8x + 16$ **73.** $9x^2 - 16$ **75.** $4x^2 - 12x + 9$ **77.** $x^2 - y^2$ **79.** $9x^2 - y^2$ **81.** $x^2 + 2xy + y^2$ **83.** $x^2 - 4xy + 4y^2$ **85.** $x^3 - 6x^2 + 12x - 8$ **87.** $8x^3 + 12x^2 + 6x + 1$ **89.** $4x^2 - 11x + 23$; remainder -45 **91.** $4x - 3$; remainder $x + 1$ **93.** $5x^2 - 13$; remainder $x + 27$ **95.** $2x^2$; remainder $-x^2 + x + 1$ **97.** $x^2 - 2x + \frac{1}{2}$; remainder $\frac{5}{2}x + \frac{1}{2}$ **99.** $-4x^2 - 3x - 3$; remainder -7 **101.** $x^2 - x - 1$; remainder $2x + 2$ **103.** $x^2 + ax + a^2$; remainder 0

R.5 Assess Your Understanding *(page 56)*

1. $3x(x - 2)(x + 2)$ **2.** prime **3.** T **4.** F **5.** $3(x + 2)$ **7.** $a(x^2 + 1)$ **9.** $x(x^2 + x + 1)$ **11.** $2x(x - 1)$ **13.** $3xy(x - 2y + 4)$ **15.** $(x + 1)(x - 1)$ **17.** $(2x + 1)(2x - 1)$ **19.** $(x + 4)(x - 4)$ **21.** $(5x + 2)(5x - 2)$ **23.** $(x + 1)^2$ **25.** $(x + 2)^2$ **27.** $(x - 5)^2$ **29.** $(2x + 1)^2$ **31.** $(4x + 1)^2$ **33.** $(x - 3)(x^2 + 3x + 9)$ **35.** $(x + 3)(x^2 - 3x + 9)$ **37.** $(2x + 3)(4x^2 - 6x + 9)$ **39.** $(x + 2)(x + 3)$ **41.** $(x + 6)(x + 1)$ **43.** $(x + 5)(x + 2)$ **45.** $(x - 8)(x - 2)$ **47.** $(x - 8)(x + 1)$ **49.** $(x + 8)(x - 1)$ **51.** $(x + 2)(2x + 3)$ **53.** $(x - 2)(2x + 1)$ **55.** $(2x + 3)(3x + 2)$ **57.** $(3x + 1)(x + 1)$ **59.** $(z + 1)(2z + 3)$ **61.** $(x + 2)(3x - 4)$ **63.** $(x - 2)(3x + 4)$ **65.** $(x + 4)(3x + 2)$ **67.** $(x + 4)(3x - 2)$ **69.** $(x + 6)(x - 6)$ **71.** $2(1 + 2x)(1 - 2x)$ **73.** $(x + 1)(x + 10)$ **75.** $(x - 7)(x - 3)$ **77.** $4(x^2 - 2x + 8)$ **79.** Prime **81.** $-(x - 5)(x + 3)$ **83.** $3(x + 2)(x - 6)$ **85.** $y^2(y + 5)(y + 6)$ **87.** $(2x + 3)^2$ **89.** $2(3x + 1)(x + 1)$ **91.** $(x - 3)(x + 3)(x^2 + 9)$ **93.** $(x - 1)^2(x^2 + x + 1)^2$ **95.** $x^5(x - 1)(x + 1)$ **97.** $(4x + 3)^2$ **99.** $-(4x - 5)(4x + 1)$ **101.** $(2y - 5)(2y - 3)$ **103.** $-(3x - 1)(3x + 1)(x^2 + 1)$ **105.** $(x + 3)(x - 6)$ **107.** $(x + 2)(x - 3)$ **109.** $(3x - 5)(9x^2 - 3x + 7)$ **111.** $(x + 5)(3x + 11)$ **113.** $(x - 1)(x + 1)(x + 2)$ **115.** $(x - 1)(x + 1)(x^2 - x + 1)$ **117.** $2(3x + 4)(9x + 13)$ **119.** $2x(3x + 5)$ **121.** $5(x + 3)(x - 2)^2(x + 1)$ **123.** $3(4x - 3)(4x - 1)$ **125.** $6(3x - 5)(2x + 1)^2(5x - 4)$ **127.** The possibilities are $(x \pm 1)(x \pm 4) = x^2 \pm 5x + 4$ or $(x \pm 2)(x \pm 2) = x^2 \pm 4x + 4$, none of which equals $x^2 + 4$.

R.6 Assess Your Understanding *(page 60)*

1. quotient; divisor; remainder **2.** $-3\overline{)2 \ \ 0 \ -5 \ \ 1}$ **3.** T **4.** T **5.** $x^2 + x + 4$; remainder 12 **7.** $3x^2 + 11x + 32$; remainder 99 **9.** $x^4 - 3x^3 + 5x^2 - 15x + 46$; remainder -138 **11.** $4x^5 + 4x^4 + x^3 + x^2 + 2x + 2$; remainder 7 **13.** $0.1x^2 - 0.11x + 0.321$; remainder -0.3531 **15.** $x^4 + x^3 + x^2 + x + 1$; remainder 0 **17.** No **19.** Yes **21.** Yes **23.** No **25.** Yes **27.** -9

R.7 Assess Your Understanding *(page 69)*

1. lowest terms **2.** least common multiple **3.** T **4.** F **5.** $\frac{3}{x - 3}$ **7.** $\frac{x}{3}$ **9.** $\frac{4x}{2x - 1}$ **11.** $\frac{y + 5}{2(y + 1)}$ **13.** $\frac{x + 5}{x - 1}$ **15.** $-(x + 7)$ **17.** $\frac{3}{5x(x - 2)}$ **19.** $\frac{2x(x^2 + 4x + 16)}{x + 4}$ **21.** $\frac{8}{3x}$ **23.** $\frac{x - 3}{x + 7}$ **25.** $\frac{4x}{(x - 2)(x - 3)}$ **27.** $\frac{4}{5(x - 1)}$ **29.** $-\frac{(x - 4)^2}{4x}$ **31.** $\frac{(x + 3)^2}{(x - 3)^2}$ **33.** $\frac{(x - 4)(x + 3)}{(x - 1)(2x + 1)}$ **35.** $\frac{x + 5}{2}$ **37.** $\frac{(x - 2)(x + 2)}{2x - 3}$ **39.** $\frac{3x - 2}{x - 3}$ **41.** $\frac{x + 9}{2x - 1}$ **43.** $\frac{4 - x}{x - 2}$ **45.** $\frac{2(x + 5)}{(x - 1)(x + 2)}$ **47.** $\frac{3x^2 - 2x - 3}{(x + 1)(x - 1)}$ **49.** $\frac{-(11x + 2)}{(x + 2)(x - 2)}$ **51.** $\frac{2(x^2 - 2)}{x(x - 2)(x + 2)}$ **53.** $(x - 2)(x + 2)(x + 1)$ **55.** $x(x - 1)(x + 1)$ **57.** $x^3(2x - 1)^2$ **59.** $x(x - 1)^2(x + 1)(x^2 + x + 1)$ **61.** $\frac{5x}{(x - 6)(x - 1)(x + 4)}$ **63.** $\frac{2(2x^2 + 5x - 2)}{(x - 2)(x + 2)(x + 3)}$ **65.** $\frac{5x + 1}{(x - 1)^2(x + 1)^2}$ **67.** $\frac{-x^2 + 3x + 13}{(x - 2)(x + 1)(x + 4)}$ **69.** $\frac{x^3 - 2x^2 + 4x + 3}{x^2(x + 1)(x - 1)}$ **71.** $\frac{-1}{x(x + h)}$ **73.** $\frac{x + 1}{x - 1}$ **75.** $\frac{(x - 1)(x + 1)}{2x(2x + 1)}$ **77.** $\frac{2(5x - 1)}{(x - 2)(x + 1)^2}$ **79.** $\frac{-2x(x^2 - 2)}{(x + 2)(x^2 - x - 3)}$ **81.** $\frac{-1}{x - 1}$ **83.** $\frac{3x - 1}{2x + 1}$ **85.** $\frac{19}{(3x - 5)^2}$ **87.** $\frac{(x + 1)(x - 1)}{(x^2 + 1)^2}$ **89.** $\frac{x(3x + 2)}{(3x + 1)^2}$ **91.** $-\frac{(x + 3)(3x - 1)}{(x^2 + 1)^2}$ **93.** $f = \frac{R_1 \cdot R_2}{(n - 1)(R_1 + R_2)}; \frac{2}{15}$ m

R.8 Assess Your Understanding *(page 77)*

3. index **4.** T **5.** cube root **6.** F **7.** 3 **9.** -2 **11.** $2\sqrt{2}$ **13.** $-2x\sqrt[3]{x}$ **15.** x^3y^2 **17.** x^2y **19.** $6\sqrt{x}$ **21.** $6x\sqrt{x}$ **23.** $15\sqrt[3]{3}$ **25.** $12\sqrt{3}$ **27.** $7\sqrt{2}$ **29.** $\sqrt{2}$ **31.** $2\sqrt{3}$ **33.** $-\sqrt[3]{2}$ **35.** $x - 2\sqrt{x} + 1$ **37.** $(2x - 1)\sqrt[3]{2x}$ **39.** $(2x - 15)\sqrt{2x}$ **41.** $-(x + 5y)\sqrt[3]{2xy}$ **43.** $\frac{\sqrt{2}}{2}$ **45.** $-\frac{\sqrt{15}}{5}$ **47.** $\frac{(5 + \sqrt{2})\sqrt{3}}{23}$ **49.** $\frac{8\sqrt{5} - 19}{41}$ **51.** $\frac{5\sqrt[3]{4}}{2}$ **53.** $\frac{2x + h - 2\sqrt{x^2 + xh}}{h}$ **55.** 4 **57.** -3 **59.** 64 **61.** $\frac{1}{27}$ **63.** $\frac{27\sqrt{2}}{32}$ **65.** $\frac{27\sqrt{2}}{32}$ **67.** $x^{7/12}$ **69.** xy^2 **71.** $x^{2/3}y$ **73.** $\frac{8x^{5/4}}{y^{3/4}}$ **75.** $\frac{3x + 2}{(1 + x)^{1/2}}$ **77.** $\frac{x(3x^2 + 2)}{(x^2 + 1)^{1/2}}$ **79.** $\frac{22x + 5}{10\sqrt{(x - 5)(4x + 3)}}$ **81.** $\frac{2 + x}{2(1 + x)^{3/2}}$

83. $\frac{4-x}{(x+4)^{3/2}}$ **85.** $\frac{1}{x^2(x^2-1)^{1/2}}$ **87.** $\frac{1-3x^2}{2\sqrt{x}(1+x^2)^2}$ **89.** $\frac{1}{2}(5x+2)(x+1)^{1/2}$ **91.** $2x^{1/2}(3x-4)(x+1)$ **93.** $(x^2+4)^{1/3}(11x^2+12)$ **95.** $(3x+5)^{1/3}(2x+3)^{1/2}(17x+27)$ **97.** $\frac{3(x+2)}{2x^{1/2}}$ **99.** 1.41 **101.** 1.59 **103.** 4.89 **105.** 2.15 **107. (a)** 15,660.4 gal **(b)** 390.7 gal **109.** $2\pi\sqrt{2} \approx 8.89$ s **111.** $\frac{\pi\sqrt{3}}{6} \approx 0.91$ s

Review Exercises *(page 81)*

1. $\{1, 3, 5, 6, 7, 8\}$ **3.** $\{3, 7\}$ **5.** $\{2, 4, 6, 8, 9\}$ **7.** $\{1, 2, 4, 5, 6, 9\}$ **9. (a)** None **(b)** $\{-10\}$ **(c)** $\left\{-10, 0.65, 1.\overline{34}, \frac{1}{9}\right\}$ **(d)** $\{\sqrt{7}\}$ **(e)** $\left\{-10, 0.65, 1.\overline{34}, \sqrt{7}, \frac{1}{9}\right\}$ **11.** 14 **13.** $\frac{3}{4}$ **15.** 3 **17.** $4x-12$ **19.** 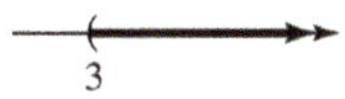

21. 5 **23.** -10 **25.** -49 **27.** 5 **29.** $\{x \mid x \neq 6\}$ **31.** $x = 3\sqrt{2}$ units; $A = 90°$; $B = 45°$; $C = 45°$ **33.** 5.0625 **35.** Coefficients: 3, 4, −2, 0, 5, −12; degree: 5 **37.** $2x^4 - 2x^3 + x^2 + 5x + 3$ **39.** $6x^2 - 7xy - 5y^2$ **41.** $16x^2 - 1$ **43.** $x^3 - 7x - 6$ **45.** $3x^2 + 8x + 25$; remainder 79 **47.** $-3x^2 + 4$; remainder -2 **49.** $x^4 - x^3 + x^2 - x + 1$; remainder 0 **51.** $(x+7)(x-2)$ **53.** $(3x+2)(2x-3)$ **55.** $3(x+2)(x-7)$ **57.** $(2x+1)(4x^2-2x+1)$ **59.** $(2x+3)(x-1)(x+1)$ **61.** $(5x-2)(5x+2)$ **63.** Prime **65.** $(x+4)^2$ **67.** $\frac{2x+7}{x-2}$ **69.** $\frac{3(3x-1)}{(x+3)(3x+1)}$ **71.** $\frac{4x}{(x+1)(x-1)}$ **73.** $\frac{x^2+17x+2}{(x-2)(x+2)^2}$ **75.** $\frac{x-1}{x-3}$ **77.** $4\sqrt{2}$ **79.** $-2\sqrt[3]{2}$ **81.** $2\sqrt{2}$ **83.** $\frac{1}{x^5y}$ **85.** $\frac{125}{x^2y}$ **87.** $\frac{3x}{5y^2}$ **89.** $3xy^4\sqrt[3]{x}$ **91.** $\frac{4\sqrt{5}}{5}$ **93.** $-2(1+\sqrt{2})$ **95.** $-\frac{3+\sqrt{5}}{2}$ **97.** $\frac{2(1+x^2)}{(2+x^2)^{1/2}}$ **99.** $\frac{x(3x+16)}{2(x+4)^{3/2}}$ **101.** $\frac{1}{x^2\sqrt{x^2-1}}$ **103.** $(x^2+4)^{1/3}(11x^2+12)$ **105.** 2.81421906×10^8 **107.** Yes **109.** \$0.35 per share **111.** $16\pi \approx 50.27$ ft²; $10\pi \approx 31.42$ ft

Chapter Test *(page 84)*

1. (a) $\{7\}$ **(b)** $\{0, 7\}$ **(c)** $\left\{0, 1.2, 7, \frac{1}{2}\right\}$ **(d)** $\left\{\sqrt{2}, \pi\right\}$ **(e)** $\left\{0, 1.2, \sqrt{2}, 7, \frac{1}{2}, \pi\right\}$ **2. (a)** -16 **(b)** 18 **(c)** 5 **(d)** -174 **3.** $x = \frac{3}{2}$; $A = 40°$; $B = 95°$; $C = 45°$ **4. (a)** $-8x^3 + 11x^2 - 14x + 11$ **(b)** $-10x^2 + 19x - 6$ **5. (a)** $(x-4)(x-2)$ **(b)** $(2x-5)(2x+5)$ **(c)** $(3x+1)(2x-7)$ **6.** $x^2 - x + 6$; remainder 2 **7. (a)** $3\sqrt{3}$ **(b)** -2 **(c)** $\frac{x}{y^4}$ **(d)** $\frac{64x^2}{y}$ **8.** $\frac{15-3\sqrt{3}}{22}$ **9.** 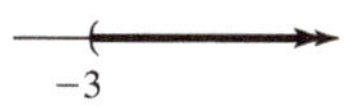

10. $a = 0$; $b = 0$ **11.** 216 ft² 84 feet of fence

CHAPTER 1 Graphs, Equations, and Inequalities

1.1 Assess Your Understanding *(page 98)*

5. abscissa; ordinate **6.** quadrants **7.** midpoint **8.** F **9.** F **10.** T **11. (a)** Quadrant II **(b)** Positive x-axis **(c)** Quadrant III **(d)** Quadrant I **(e)** Negative y-axis **(f)** Quadrant IV

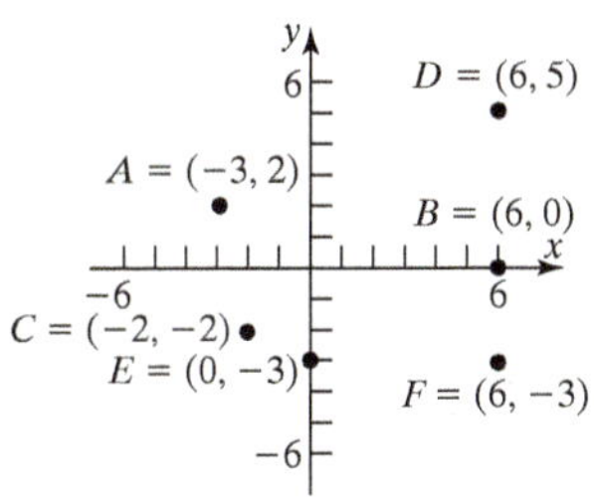

13. The points will be on a vertical line that is 2 units to the right of the y-axis.

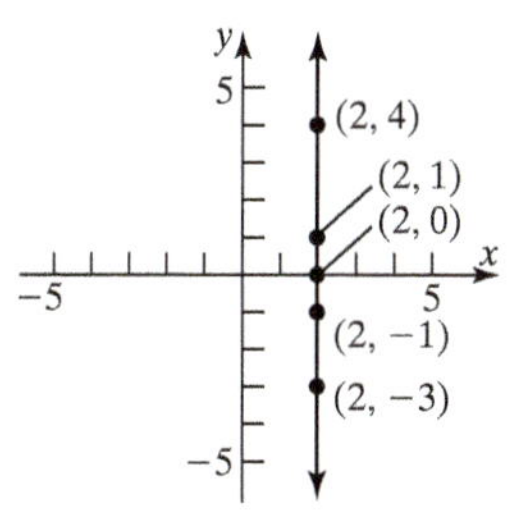

15. $(-1, 4)$; Quadrant II **17.** $(3, 1)$; Quadrant I **19.** $X\text{min} = -11$, $X\text{max} = 5$, $X\text{scl} = 1$, $Y\text{min} = -3$, $Y\text{max} = 6$, $Y\text{scl} = 1$ **21.** $X\text{min} = -30$, $X\text{max} = 50$, $X\text{scl} = 10$, $Y\text{min} = -90$, $Y\text{max} = 50$, $Y\text{scl} = 10$ **23.** $X\text{min} = -10$, $X\text{max} = 110$, $X\text{scl} = 10$, $Y\text{min} = -10$, $Y\text{max} = 160$, $Y\text{scl} = 10$ **25.** $X\text{min} = -6$, $X\text{max} = 6$, $X\text{scl} = 2$, $Y\text{min} = -4$, $Y\text{max} = 4$, $Y\text{scl} = 2$ **27.** $X\text{min} = -6$, $X\text{max} = 6$, $X\text{scl} = 2$, $Y\text{min} = -1$, $Y\text{max} = 3$, $Y\text{scl} = 1$

29. $X\text{min} = 3$, $X\text{max} = 9$, $X\text{scl} = 1$, $Y\text{min} = 2$, $Y\text{max} = 10$, $Y\text{scl} = 2$ **31.** $\sqrt{5}$ **33.** $\sqrt{10}$ **35.** $2\sqrt{17}$ **37.** 20 **39.** $\sqrt{53}$ **41.** $\sqrt{5.49} \approx 2.34$ **43.** $\sqrt{a^2 + b^2}$ **45.** $4\sqrt{10}$ **47.** $2\sqrt{65}$

49. $d(A, B) = \sqrt{13}$
$d(B, C) = \sqrt{13}$
$d(A, C) = \sqrt{26}$
$(\sqrt{13})^2 + (\sqrt{13})^2 = (\sqrt{26})^2$
Area $= \frac{13}{2}$ square units

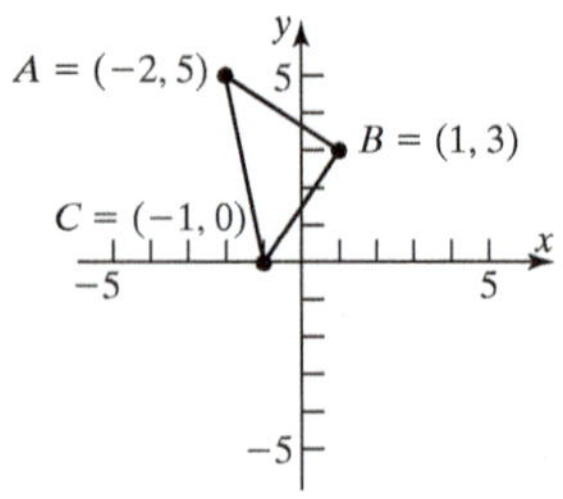

51. $d(A, B) = \sqrt{130}$
$d(B, C) = \sqrt{26}$
$d(A, C) = 2\sqrt{26}$
$(\sqrt{26})^2 + (2\sqrt{26})^2 = (\sqrt{130})^2$
Area = 26 square units

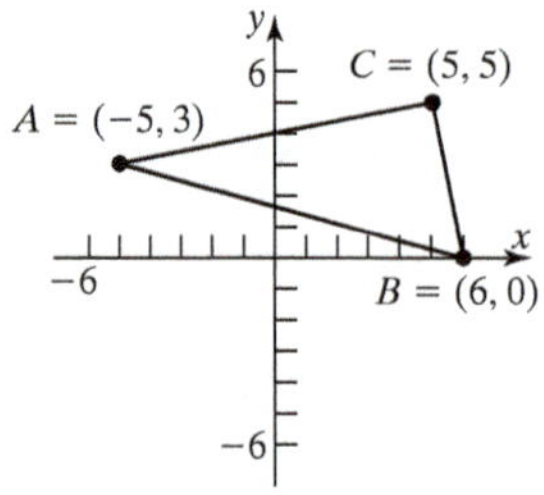

53. $d(A, B) = 4$
$d(A, C) = 5$
$d(B, C) = \sqrt{41}$
$4^2 + 5^2 = 16 + 25 = (\sqrt{41})^2$
Area = 10 square units

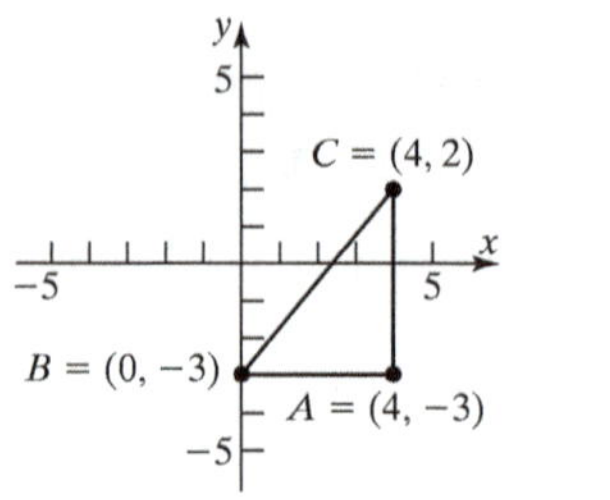

55. (4, 0) **57.** (3, 3) **59.** (5, −1) **61.** (0.55, 1.2) **63.** $\left(\frac{a}{2}, \frac{b}{2}\right)$ **65.** (0, 0) is on the graph. **67.** (0, 3) is on the graph.

69. (0, 2) and $(\sqrt{2}, \sqrt{2})$ are on the graph. **71.** (−1, 0), (1, 0) **73.** $\left(-\frac{\pi}{2}, 0\right), (0, 1), \left(\frac{\pi}{2}, 0\right)$ **75.** (0, 2), (1, 0), (0, −2)

77. (−4, 0), (−1, 0), (0, −3), (4, 0) **79.**

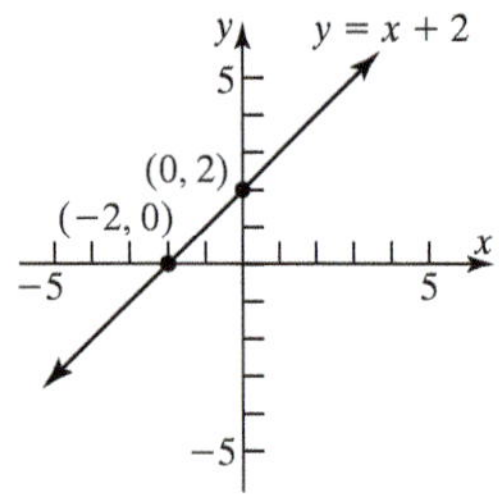

81.

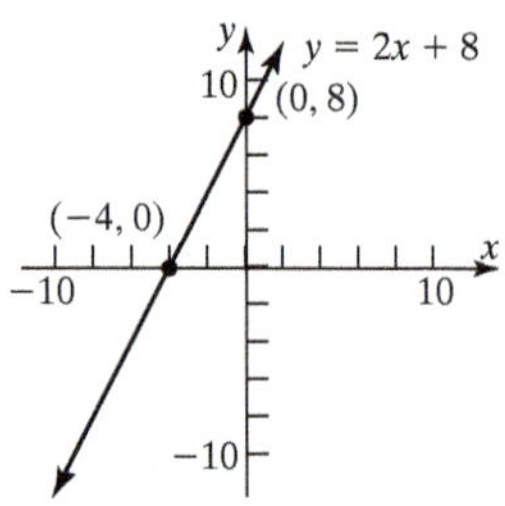

83.

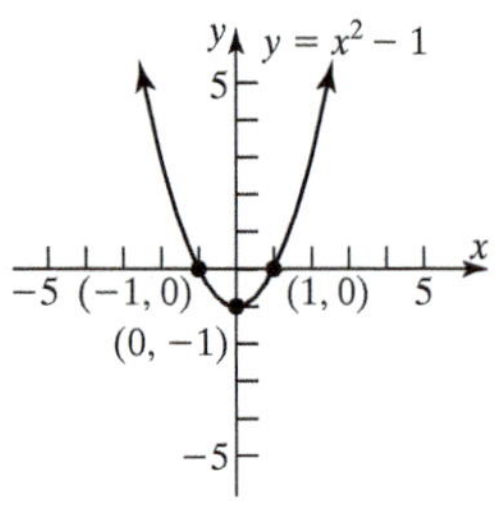

85.

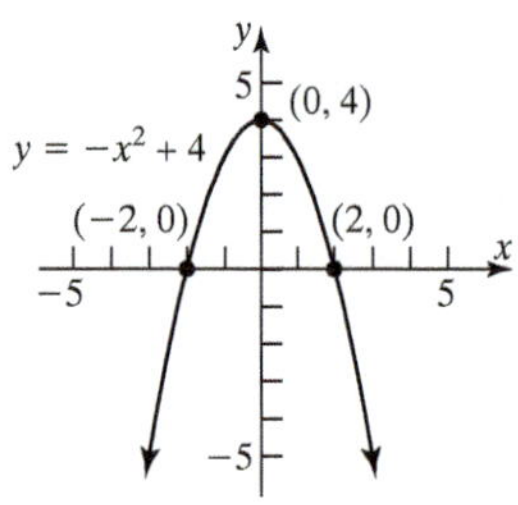

87.

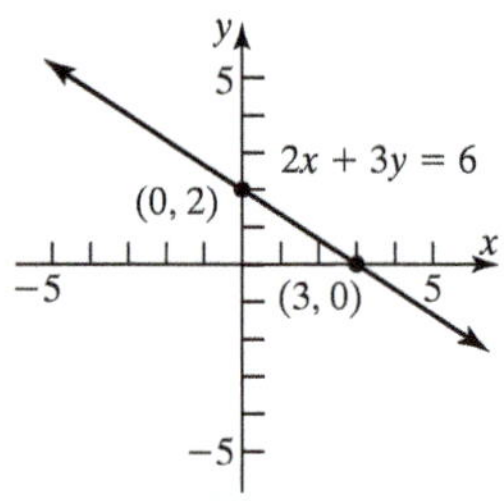

89.

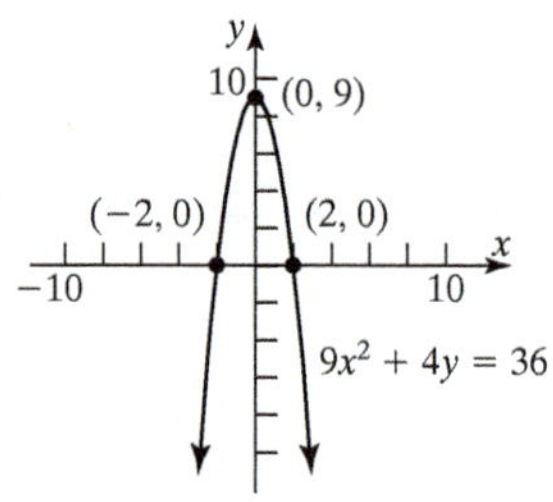

91. x-intercept: 6.5
y-intercept: −13

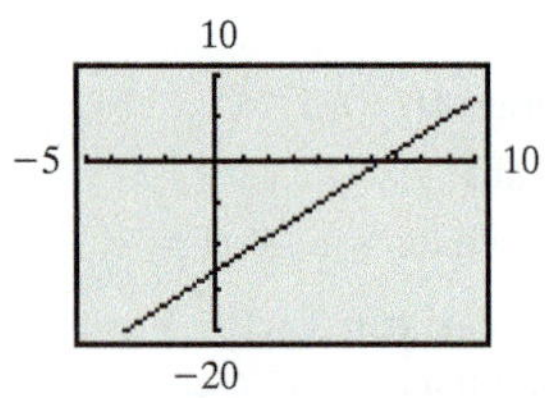

93. x-intercepts: −2.74, 2.74
y-intercept: −15

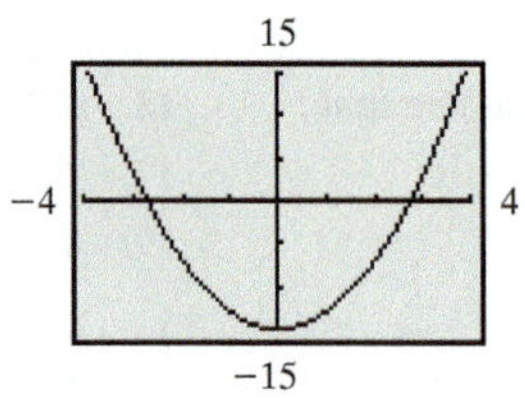

95. x-intercept: 14.33
y-intercept: −21.5

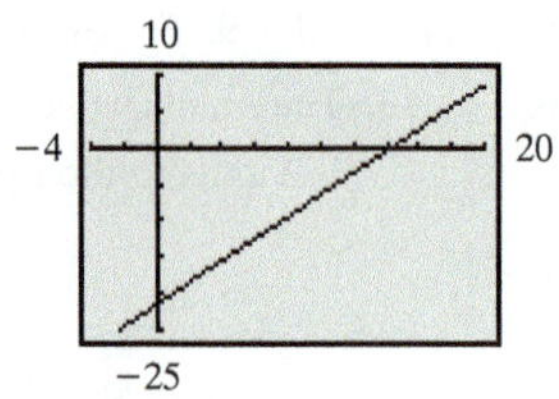

97. x-intercepts: −2.72, 2.72
y-intercept: 16.4

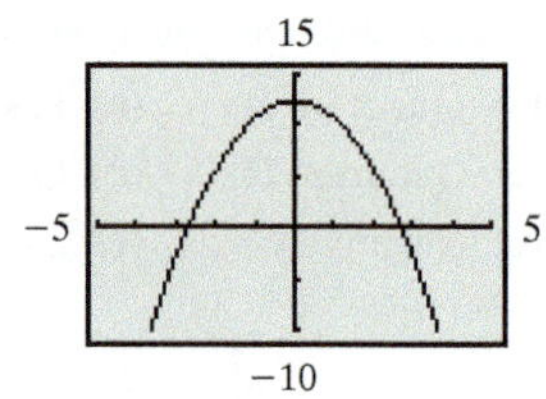

99. $\sqrt{17}; 2\sqrt{5}; \sqrt{29}$ **101.** $d(P_1, P_2) = 6; d(P_2, P_3) = 4; d(P_1, P_3) = 2\sqrt{13}$; right triangle **103.** $d(P_1, P_2) = 2\sqrt{17}$; $d(P_2, P_3) = \sqrt{34}$; $d(P_1, P_3) = \sqrt{34}$; isosceles right triangle **105.** $90\sqrt{2} \approx 127.28$ ft **107.** **(a)** (90, 0), (90, 90), (0, 90) **(b)** $5\sqrt{2161} \approx 232.43$ ft **(c)** $30\sqrt{149} \approx 366.20$ ft **109.** $d = 50t$ **111.** **(a)** (2.65, 1.6) **(b)** ≈1.285 units **113.** \$17,630.50 **115.** $B = (5, -2)$

1.2 Assess Your Understanding *(page 110)*

3. F **4.** identity **5.** linear; first-degree **6.** F **7.** T **8.** F **9.** {3} **11.** {−5} **13.** $\left\{\frac{3}{2}\right\}$ **15.** $\left\{\frac{5}{4}\right\}$ **17.** {−2.21, 0.54, 1.68} **19.** {−1.55, 1.15} **21.** {−1.12, 0.36} **23.** {−2.69, −0.49, 1.51} **25.** {−2.86, −1.34, 0.20, 1.00} **27.** No real solutions **29.** {−2} **31.** {3} **33.** {−1} **35.** {−18} **37.** {−4} **39.** $\left\{-\frac{5}{4}\right\}$ **41.** $\left\{-\frac{3}{4}\right\}$ **43.** {−20} **45.** {−2} **47.** {0.5} **49.** $\left\{\frac{29}{10}\right\}$ **51.** {2} **53.** {8} **55.** {2} **57.** {−1} **59.** {3} **61.** No solution **63.** No solution **65.** {−6} **67.** {34} **69.** $\left\{-\frac{20}{39}\right\}$ **71.** {−1} **73.** $\left\{-\frac{11}{6}\right\}$ **75.** {−6} **77.** $-\frac{2}{5}$ **79.** $2a + 3b = 6$

81. $x = \frac{b+c}{a}$ **83.** $x = \frac{abc}{a+b}$ **85.** $x = a^2$ **87.** $a = 3$ **89.** $R = \frac{R_1 R_2}{R_1 + R_2}$ **91.** $R = \frac{mv^2}{F}$ **93.** $r = \frac{S-a}{S}$ **95.** \$11,500 will be invested in bonds and \$8500 in CDs. **97.** Scott will receive \$400,000, Alice \$300,000, and Tricia \$200,000. **99.** The regular hourly rate is \$11.50. **101.** Brooke needs a score of 85. **103.** The original price was \$200,000; purchasing the model saves \$30,000. **105.** The bookstore paid \$80 for the book. **107.** There were 400 adults. **109.** The length is 19 ft; the width is 11 ft. **111.** Approximately 38.559 billion hits

Historical Problems *(page 121)*

1. The area of each shaded square is 9, so the larger square will have area $85 + 4(9) = 121$. The area of the larger square is also given by the expression $(x+6)^2$, so $(x+6)^2 = 121$. Taking the positive square root of each side, $x + 6 = 11$ or $x = 5$.

2. Let $z = -6$, so $z^2 + 12z - 85 = -121$. We get the equation $u^2 - 121 = 0$ or $u^2 = 121$. Thus $u = \pm 11$, so $x = \pm 11 - 6$. $x = -17$ or $x = 5$.

3.

$$\left(x + \frac{b}{2a}\right)^2 = \left(\frac{\sqrt{b^2-4ac}}{2a}\right)^2$$

$$\left(x + \frac{b}{2a}\right)^2 - \left(\frac{\sqrt{b^2-4ac}}{2a}\right)^2 = 0$$

$$\left(x + \frac{b}{2a} - \frac{\sqrt{b^2-4ac}}{2a}\right)\left(x + \frac{b}{2a} + \frac{\sqrt{b^2-4ac}}{2a}\right) = 0$$

$$\left(x + \frac{b - \sqrt{b^2-4ac}}{2a}\right)\left(x + \frac{b + \sqrt{b^2-4ac}}{2a}\right) = 0$$

$$x = \frac{-b + \sqrt{b^2-4ac}}{2a} \text{ or } x = \frac{-b - \sqrt{b^2-4ac}}{2a}$$

1.3 Assess Your Understanding *(page 121)*

5. repeated; multiplicity two **6.** discriminant; negative **7.** second **8.** T **9.** F **10.** T **11.** $\{0, 9\}$ **13.** $\{-5, 5\}$ **15.** $\{-3, 2\}$ **17.** $\left\{-\frac{1}{2}, 3\right\}$ **19.** $\{-4, 4\}$ **21.** $\{-6, -2\}$ **23.** $\left\{\frac{3}{2}\right\}$ **25.** $\left\{-\frac{2}{3}, \frac{3}{2}\right\}$ **27.** $\left\{-\frac{5}{2}, \frac{4}{3}\right\}$ **29.** $\left\{-\frac{3}{4}, 2\right\}$ **31.** $\{-5\sqrt{2}, 5\sqrt{2}\}$ **33.** $\{-1, 3\}$ **35.** $\{-3, 0\}$ **37.** 16 **39.** $\frac{1}{16}$ **41.** $\{-2 - 2\sqrt{3}, -2 + 2\sqrt{3}\}$ **43.** $\left\{-\frac{1}{4}, \frac{3}{4}\right\}$ **45.** $\left\{\frac{-1-\sqrt{7}}{6}, \frac{-1+\sqrt{7}}{6}\right\}$ **47.** $\{2 - \sqrt{2}, 2 + \sqrt{2}\}$ **49.** $\{2 - \sqrt{5}, 2 + \sqrt{5}\}$ **51.** $\left\{1, \frac{3}{2}\right\}$ **53.** No real solution **55.** $\left\{\frac{-1-\sqrt{5}}{4}, \frac{-1+\sqrt{5}}{4}\right\}$ **57.** $\left\{\frac{-2-\sqrt{10}}{2}, \frac{-2+\sqrt{10}}{2}\right\}$ **59.** $\left\{\frac{1}{3}\right\}$ **61.** $\left\{-\frac{2}{3}, 1\right\}$ **63.** $\left\{\frac{1-\sqrt{33}}{8}, \frac{1+\sqrt{33}}{8}\right\}$ **65.** $\left\{\frac{9-\sqrt{73}}{2}, \frac{9+\sqrt{73}}{2}\right\}$ **67.** No real solution **69.** Repeated real solution **71.** Two unequal real solutions **73.** $\{-2\sqrt{2}, 2\sqrt{2}\}$ **75.** $\left\{\frac{1}{4}\right\}$ **77.** $\left\{-\frac{3}{5}, \frac{5}{2}\right\}$ **79.** $\left\{-\frac{1}{2}, \frac{2}{3}\right\}$ **81.** $\left\{\frac{7}{3}\right\}$ **83.** $\left\{\frac{-1-\sqrt{17}}{2}, \frac{-1+\sqrt{17}}{2}\right\}$ **85.** $\left\{-\frac{1}{4}, \frac{2}{3}\right\}$ **87.** $\left\{\frac{5}{3}\right\}$ **89.** The dimensions are 11 ft by 13 ft. **91.** The dimensions are 5 m by 8 m. **93.** 2; 5 meters, 12 meters, 13 meters; 20 meters, 21 meters, 29 meters **95.** The screen of a 37-inch TV in 4:3 format has an area of 657.12 square inches; the screen of a 37-inch TV in 16:9 format has an area of 584.97 square inches. The traditional TV has a larger screen. **97.** The dimensions should be 4 ft by 4 ft. **99. (a)** The ball strikes the ground after 6 s. **(b)** The ball passes the top of the building on its way down after 5 s. **101.** The dimensions should be 11.55 cm by 6.55 cm by 3 cm. **103.** The border will be 2.71 ft wide. **105.** The border will be 2.56 ft wide. **107.** 36 consecutive integers must be added.

109. $\frac{-b+\sqrt{b^2-4ac}}{2a} + \frac{-b-\sqrt{b^2-4ac}}{2a} = \frac{-2b}{2a} = -\frac{b}{a}$ **111.** $k = \frac{1}{2}$ or $k = -\frac{1}{2}$

113. $ax^2 + bx + c = 0, x = \frac{-b \pm \sqrt{b^2-4ac}}{2a}; ax^2 - bx + c = 0, x = \frac{b \pm \sqrt{(-b)^2 - 4ac}}{2a} = \frac{b \pm \sqrt{b^2-4ac}}{2a} = -\frac{-b \pm \sqrt{b^2-4ac}}{2a}$

115. (b)

1.4 Assess Your Understanding *(page 131)*

4. real; imaginary; imaginary unit **5.** $-2i, 2i$ **6.** F **7.** T **8.** F **9.** $8 + 5i$ **11.** $-7 + 6i$ **13.** $-6 - 11i$ **15.** $6 - 18i$ **17.** $6 + 4i$ **19.** $10 - 5i$ **21.** 37 **23.** $\frac{6}{5} + \frac{8}{5}i$ **25.** $1 - 2i$ **27.** $\frac{5}{2} - \frac{7}{2}i$ **29.** $-\frac{1}{2} + \frac{\sqrt{3}}{2}i$ **31.** $2i$ **33.** $-i$ **35.** i **37.** -6 **39.** $-10i$ **41.** $-2 + 2i$ **43.** 0 **45.** 0 **47.** $2i$ **49.** $5i$ **51.** $5i$ **53.** $\{-2i, 2i\}$ **55.** $\{-4, 4\}$ **57.** $\{3 - 2i, 3 + 2i\}$ **59.** $\{3 - i, 3 + i\}$ **61.** $\left\{\frac{1}{4} - \frac{1}{4}i, \frac{1}{4} + \frac{1}{4}i\right\}$ **63.** $\left\{-\frac{1}{5} - \frac{2}{5}i, -\frac{1}{5} + \frac{2}{5}i\right\}$ **65.** $\left\{-\frac{1}{2} - \frac{\sqrt{3}}{2}i, -\frac{1}{2} + \frac{\sqrt{3}}{2}i\right\}$ **67.** $\{2, -1 - \sqrt{3}i, -1 + \sqrt{3}i\}$ **69.** $\{-2, 2, -2i, 2i\}$ **71.** $\{-3i, -2i, 2i, 3i\}$ **73.** Two complex solutions **75.** Two unequal real solutions **77.** A repeated real solution **79.** $2 - 3i$ **81.** 6 **83.** 25 **85.** $2 + 3i$

87. $z + \bar{z} = (a + bi) + (a - bi) = 2a$; $z - \bar{z} = (a + bi) - (a - bi) = 2bi$
89. $\overline{z + w} = \overline{(a + bi) + (c + di)} = \overline{(a + c) + (b + d)i} = (a + c) - (b + d)i = (a - bi) + (c - di) = \bar{z} + \bar{w}$
95. $\sqrt{a} \cdot \sqrt{b} = \sqrt{ab}$ only when $\sqrt{a}, \sqrt{b}$ are real numbers.

1.5 Assess Your Understanding *(page 138)*

6. extraneous **7.** quadratic in form **8.** T **9.** $\{22\}$ **11.** $\{1\}$ **13.** No real solution **15.** $\{-13\}$ **17.** $\{4\}$ **19.** $\{-1\}$ **21.** $\{0, 64\}$ **23.** $\{3\}$ **25.** $\left\{-\frac{8}{5}\right\}$ **27.** $\{8\}$ **29.** $\{-1, 3\}$ **31.** $\{1, 5\}$ **33.** $\{1\}$ **35.** $\{5\}$ **37.** $\{2\}$ **39.** $\{-4, 4\}$ **41.** $\{-2, 2\}$ **43.** $\{-2, -1, 1, 2\}$ **45.** $\{-1, 1\}$ **47.** $\{-2, 1\}$ **49.** $\{-6, -5\}$ **51.** $\left\{-\frac{3}{2}, 2\right\}$ **53.** $\{0, 16\}$ **55.** $\{16\}$ **57.** $\{1\}$ **59.** $\left\{\left(\frac{9 - \sqrt{17}}{8}\right)^4, \left(\frac{9 + \sqrt{17}}{8}\right)^4\right\}$ **61.** $\left\{-2, -\frac{1}{2}\right\}$ **63.** $\left\{-\frac{3}{2}, \frac{1}{3}\right\}$ **65.** $\left\{-\frac{1}{8}, 27\right\}$ **67.** $\{-4, 1\}$ **69.** $\left\{-1, \frac{3}{2}\right\}$ **71.** $\{-4, 4\}$ **73.** $\left\{-\frac{1}{2}, \frac{1}{2}\right\}$ **75.** $\left\{-\frac{27}{2}, \frac{27}{2}\right\}$ **77.** $\left\{-\frac{36}{5}, \frac{24}{5}\right\}$ **79.** No real solution **81.** $\{-3, 3\}$ **83.** $\{-1, 3\}$ **85.** $\{-3, 0, 3\}$ **87.** $\{-5, 0, 4\}$ **89.** $\{-1, 1\}$ **91.** $\{-2, 2, 3\}$ **93.** $\left\{-2, \frac{1}{2}, 2\right\}$ **95.** $\{0.34, 11.66\}$ **97.** $\{-1.03, 1.03\}$ **99.** $\left\{-4, \frac{5}{3}\right\}$ **101.** $\left\{-3, -\frac{2}{5}, 3\right\}$ **103.** $\left\{-\frac{1}{5}, 1\right\}$ **105.** $\{5\}$ **107.** $\left\{-2, -\frac{4}{5}\right\}$ **109.** $\{-2, 6\}$ **111.** $\{2\}$ **113.** $\left\{\frac{-3 - \sqrt{6}}{3}, \frac{-3 + \sqrt{6}}{3}\right\}$ **115.** $\{-2, -1, 0, 1\}$ **117.** $\{\sqrt{2}, \sqrt{3}\}$ **119.** $\{-2, 2, -2i, 2i\}$ **121.** $\left\{1, 2, -1 - \sqrt{3}i, -1 + \sqrt{3}i, -\frac{1}{2} - \frac{\sqrt{3}}{2}i, -\frac{1}{2} + \frac{\sqrt{3}}{2}i\right\}$ **123.** $\left\{\frac{3}{2}, 5\right\}$ **125.** $(2, 2)$; $(2, -4)$ **127.** $(0, 0)$; $(8, 0)$ **129.** The distance is approximately 229.94 ft. **131.** approx. 221 feet

1.6 Assess Your Understanding *(page 146)*

1. mathematical modeling **2.** interest **3.** uniform motion **4.** F **5.** T **6.** $100 - x$ **7.** $A = \pi r^2$; r = radius, A = area **9.** $A = s^2$; A = area, s = length of a side **11.** $F = ma$; F = force, m = mass, a = acceleration **13.** $W = Fd$; W = work, F = force, d = distance **15.** $C = 150x$; C = total variable cost, x = number of dishwashers **17.** Invest \$48,000 in bonds and \$36,000 in CDs. **19.** \$85,000 was loaned out at 8%. **21.** Mix 75 lb of Earl Grey tea with 25 lb of Orange Pekoe tea. **23.** Mix 160 lb of cashews with the almonds. **25.** The speed of the current is 2.286 mi/hr. **27.** The speed of the current is 5 mi/hr. **29.** Working together, it takes 12 min. **31. (a)** The dimensions are 10 ft by 5 ft. **(b)** The area is 50 sq ft. **(c)** The dimensions would be 7.5 ft by 7.5 ft. **(d)** The area would be 56.25 sq ft. **33.** The defensive back catches up to the tight end at the tight end's 45-yd line. **35.** Add $\frac{2}{3}$ gal of water. **37.** Approximately 4.05 ft s. **39.** hp 1300; 30 min; 20 pages/min; hp 2420; 20 min; 30 pages/min **41.** Evaporate 10.67 oz of water. **43.** 40 g of 12-karat gold should be mixed with 20 g of pure gold. **45.** Mike passes Dan $\frac{1}{3}$ mi from the start, 2 min from the time Mike started to race. **47.** Start the auxiliary pump at 9:45 AM. **49.** The tub will fill in 1 hr. **51.** The distance of the run is 12 mi. The distance of the bicycle race is 75 mi. **53.** Lewis would beat Burke by 16.75 m. **57.** The average speed is 49.5 mi/hr.

1.7 Assess Your Understanding *(page 158)*

4. negative **5.** closed interval **6.** $-a \le u \le a$ **7.** T **8.** T **9.** T **10.** F **11.** $[0, 2]$; $0 \le x \le 2$ **13.** $[2, \infty)$; $x \ge 2$ **15.** $[0, 3)$; $0 \le x < 3$
17. (a) $6 < 8$ **(b)** $-2 < 0$ **(c)** $9 < 15$ **(d)** $-6 > -10$ **19. (a)** $2x + 4 < 5$ **(b)** $2x - 4 < -3$ **(c)** $6x + 3 < 6$ **(d)** $-4x - 2 > -4$

21. $[0, 4]$

23. $[4, 6)$
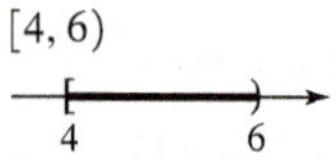

25. $[4, \infty)$
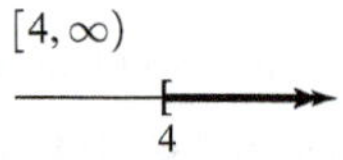

27. $(-\infty, -4)$
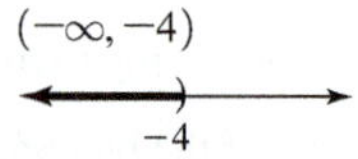

29. $2 \le x \le 5$
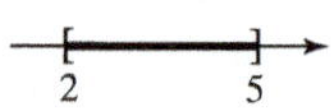

31. $-3 < x < -2$
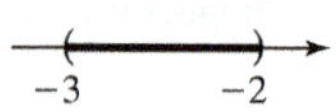

33. $x \ge 4$
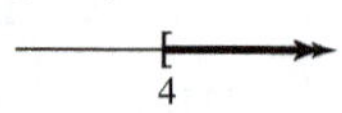

35. $x < -3$
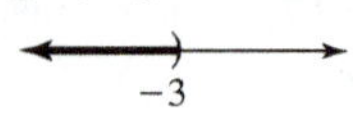

37. $<$ **39.** $>$ **41.** $\ge$ **43.** $<$ **45.** $<$ **47.** $\ge$

49. $\{x \mid x > 3\}$; $(3, \infty)$
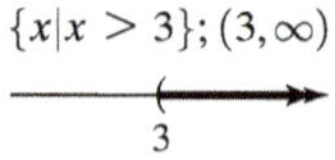

51. $\{x \mid x \ge -1\}$; $[-1, \infty)$
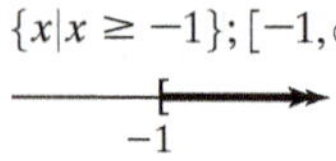

53. $\{x \mid x \ge 2\}$; $[2, \infty)$
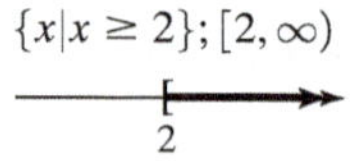

55. $\{x \mid x > -7\}$; $(-7, \infty)$
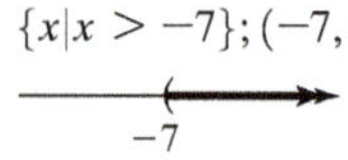

57. $\left\{x \middle| x \le \frac{2}{3}\right\}$; $\left(-\infty, \frac{2}{3}\right]$
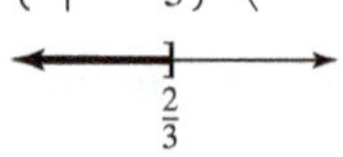

59. $\{x \mid x < -20\}$; $(-\infty, -20)$
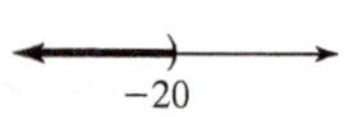

61. $\{x \mid 3 \le x \le 5\}$; $[3, 5]$
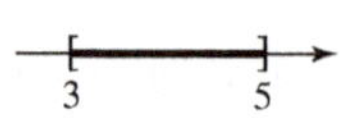

63. $\left\{x \middle| \frac{2}{3} \le x \le 3\right\}$; $\left[\frac{2}{3}, 3\right]$
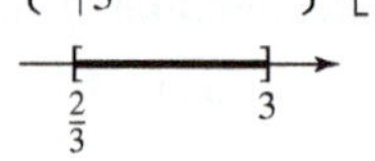

65. $\left\{x \middle| -\frac{11}{2} < x < \frac{1}{2}\right\}$; $\left(-\frac{11}{2}, \frac{1}{2}\right)$
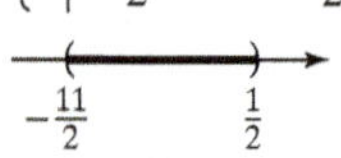

67. $\{x \mid -6 < x < 0\}$; $(-6, 0)$
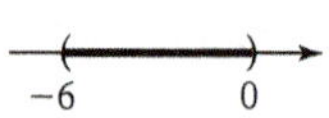

69. $\{x \mid x < -5\}$; $(-\infty, -5)$
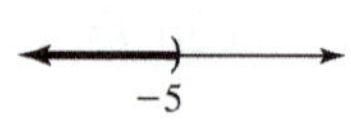

71. $\{x \mid -4 < x < 4\}; (-4, 4)$

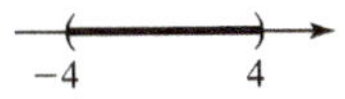

73. $\{x \mid x < -4 \text{ or } x > 4\}; (-\infty, -4) \cup (4, \infty)$

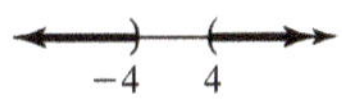

75. $\left\{t \,\middle|\, -\frac{2}{3} \le t \le 2\right\}; \left[-\frac{2}{3}, 2\right]$

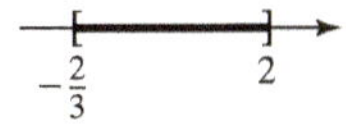

77. $\{x \mid 1 < x < 3\}; (1, 3)$

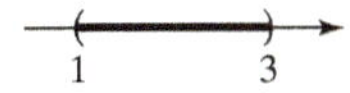

79. $\{x \mid x \le 1 \text{ or } x \ge 5\}; (-\infty, 1] \cup [5, \infty)$

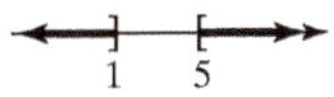

81. $\{x \mid x < -1 \text{ or } x > 2\}; (-\infty, -1) \cup (2, \infty)$

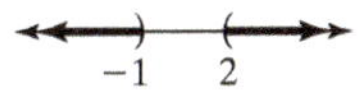

83. $\left\{x \,\middle|\, -1 < x < \frac{3}{2}\right\}; \left(-1, \frac{3}{2}\right)$

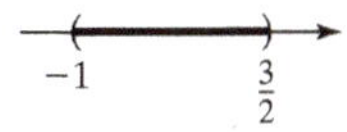

85. No real solution; $\varnothing$

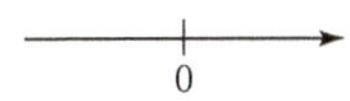

87. $\{x \mid x > -2\}; (-2, \infty)$

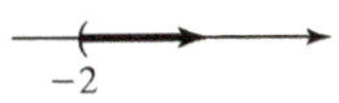

89. $\left\{x \,\middle|\, x \le -\frac{5}{2} \text{ or } x \ge \frac{3}{2}\right\}; \left(-\infty, -\frac{5}{2}\right] \cup \left[\frac{3}{2}, \infty\right)$

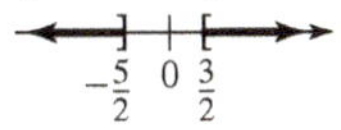

91. $\left\{x \,\middle|\, x \ge \frac{4}{3}\right\}; \left[\frac{4}{3}, \infty\right)$

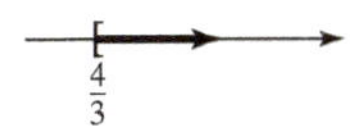

93. $\{x \mid -3 \le x < 7\}; [-3, 7)$

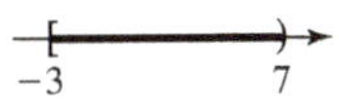

95. $\{x \mid x \ge -1\}; [-1, \infty)$

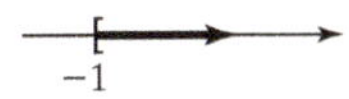

97. $\left\{x \,\middle|\, \frac{17}{6} < x < \frac{19}{6}\right\}; \left(\frac{17}{6}, \frac{19}{6}\right)$

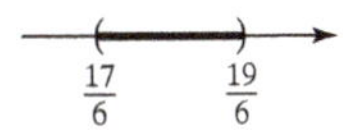

99. $\{x \mid -3 \le x < 4\}; [-3, 4)$

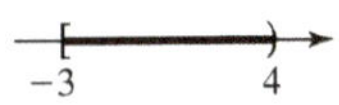

101. $\{x \mid -2 < x < 4\}; (-2, 4)$

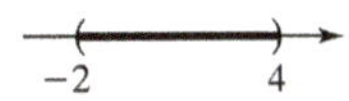

103. $\{x \mid x > 5\}; (5, \infty)$

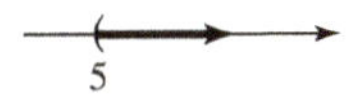

105. $\{x \mid x > 3\}; (3, \infty)$

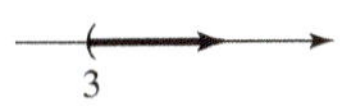

107. $\left|x - 2\right| < \frac{1}{2}; \left\{x \,\middle|\, \frac{3}{2} < x < \frac{5}{2}\right\}$

109. $|x + 3| > 2; \{x \mid x < -5 \text{ or } x > -1\}$

111. $21 < \text{age} < 30$ **113.** $|x - 98.6| \ge 1.5; \{x \mid x \le 97.1 \text{ or } x \ge 100.1\}$ **115. (a)** Male ≥ 79.66 **(b)** Female ≥ 83.58 **(c)** A female can expect to live at least 3.92 years longer. **117.** The agent's commission ranges from \$45,000 to \$95,000, inclusive. As a percent of selling price, the commission ranges from 5% to approximately 8.6%, inclusive. **119.** The amount withheld varies from \$98.30 to \$148.30, inclusive. **121.** The usage varies from approximately 710.94 to 2536.44 kilowatt-hours, inclusive. **123.** The dealer's cost varies from \$7457.63 to \$7857.14, inclusive. **125. (a)** You need at least a 74 on the last test. **(b)** You need at least a 77 on the last test. **127.** $|x - 13.4| < 1.35$; between 12.05 and 14.75 books per year are read, on average.

129. $\frac{a+b}{2} - a = \frac{a+b-2a}{2} = \frac{b-a}{2} > 0$; therefore, $a < \frac{a+b}{2}$.

$b - \frac{a+b}{2} = \frac{2b-a-b}{2} = \frac{b-a}{2} > 0$; therefore, $b > \frac{a+b}{2}$.

131. $(\sqrt{ab})^2 - a^2 = ab - a^2 = a(b-a) > 0$; thus, $(\sqrt{ab})^2 > a^2$ and $\sqrt{ab} > a$.

$b^2 - (\sqrt{ab})^2 = b^2 - ab = b(b-a) > 0$; thus, $b^2 > (\sqrt{ab})^2$ and $b > \sqrt{ab}$.

133. $\frac{1}{h} = \frac{1}{2}\left(\frac{1}{a} + \frac{1}{b}\right) = \frac{1}{2}\left(\frac{b+a}{ab}\right); h = \frac{2ab}{b+a}$

$h - a = \frac{2ab}{b+a} - a = \frac{2ab - a(b+a)}{b+a} = \frac{2ab - ab - a^2}{b+a} = \frac{ab - a^2}{b+a} = \frac{a(b-a)}{b+a} > 0$, since $0 < a < b$.

$b - h = b - \frac{2ab}{b+a} = \frac{b(b+a) - 2ab}{b+a} = \frac{b^2 + ab - 2ab}{b+a} = \frac{b^2 - ab}{b+a} = \frac{b(b-a)}{b+a} > 0$, since $0 < a < b$.

Review Exercises *(page 162)*

1. $\{-18\}$ **3.** $\{6\}$ **5.** $\left\{\frac{1}{5}\right\}$ **7.** $\{6\}$ **9.** No real solution **11.** $\left\{\frac{11}{8}\right\}$ **13.** $\left\{-2, \frac{3}{2}\right\}$ **15.** $\left\{\frac{1-\sqrt{13}}{4}, \frac{1+\sqrt{13}}{4}\right\}$ **17.** $\{-3, 3\}$

19. No real solution **21.** $\{-2, -1, 1, 2\}$ **23.** $\{2\}$ **25.** $\left\{\frac{13}{2}\right\}$ **27.** $\left\{\frac{\sqrt{5}}{2}\right\}$ **29.** $\left\{\frac{9}{4}\right\}$ **31.** $\left\{-1, \frac{1}{2}\right\}$ **33.** $\left\{\frac{m}{1-n}, \frac{m}{1+n}\right\}$ **35.** $\left\{-\frac{9b}{5a}, \frac{2b}{a}\right\}$

37. $\left\{-\frac{9}{5}\right\}$ **39.** $\{-5, 2\}$ **41.** $\left\{-\frac{5}{3}, 3\right\}$ **43.** $\left\{0, \frac{3}{2}\right\}$ **45.** $\left\{-\frac{5}{2}, -2, 2\right\}$ **47.** $\{3\}$ **49.** $\{-1, 5\}$ **51.** $\{-2.49, 0.66, 1.83\}$ **53.** $\{-1.14, 1.64\}$

55. $\{x \mid x \ge 14\}; [14, \infty)$

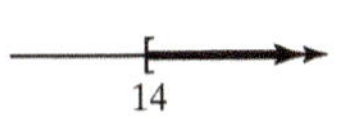

57. $\left\{x \,\middle|\, -\frac{31}{2} \le x \le \frac{33}{2}\right\}; \left[-\frac{31}{2}, \frac{33}{2}\right]$

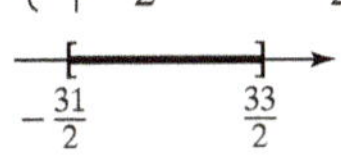

59. $\{x \mid -23 < x < -7\}; (-23, -7)$

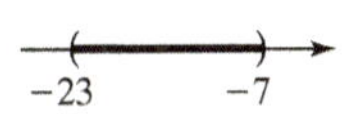

61. $\left\{x \middle| -\frac{3}{2} < x < -\frac{7}{6}\right\}; \left(-\frac{3}{2}, -\frac{7}{6}\right)$

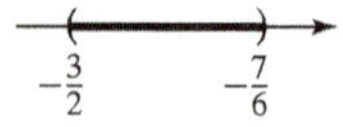

63. $\{x \mid x \le -2 \text{ or } x \ge 7\}; (-\infty, -2] \cup [7, \infty)$

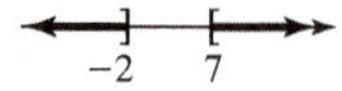

65. $\left\{x \middle| 0 \le x \le \frac{4}{3}\right\}; \left[0, \frac{4}{3}\right]$

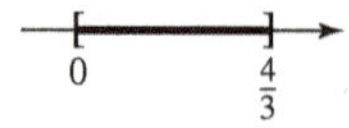

67. $\left\{x \middle| x < -1 \text{ or } x > \frac{7}{3}\right\}; (-\infty, -1) \cup \left(\frac{7}{3}, \infty\right)$

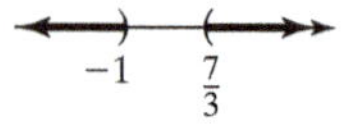

69. 9 **71.** $\frac{4}{9}$ **73.** $4 + 7i$ **75.** $-3 + 2i$ **77.** $\frac{9}{10} - \frac{3}{10}i$ **79.** -1 **81.** $-46 + 9i$ **83.** $\left\{-\frac{1}{2} - \frac{\sqrt{3}}{2}i, -\frac{1}{2} + \frac{\sqrt{3}}{2}i\right\}$ **85.** $\left\{\frac{-1 - \sqrt{17}}{4}, \frac{-1 + \sqrt{17}}{4}\right\}$

87. $\left\{\frac{1}{2} - \frac{\sqrt{11}}{2}i, \frac{1}{2} + \frac{\sqrt{11}}{2}i\right\}$ **89.** $\left\{\frac{1}{2} - \frac{\sqrt{23}}{2}i, \frac{1}{2} + \frac{\sqrt{23}}{2}i\right\}$ **91. (a)** $2\sqrt{5}$ **(b)** $(2, 1)$ **93. (a)** 5 **(b)** $\left(-\frac{1}{2}, 1\right)$ **95. (a)** 12 **(b)** $(4, 2)$

97.

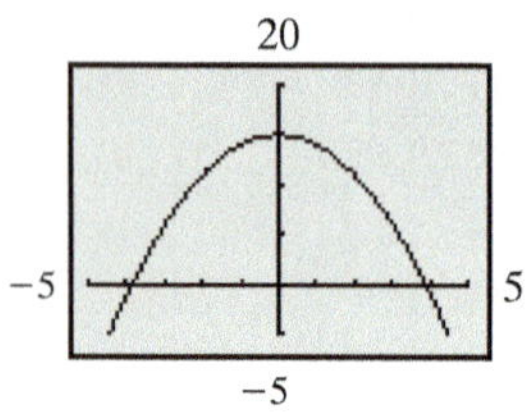

99.

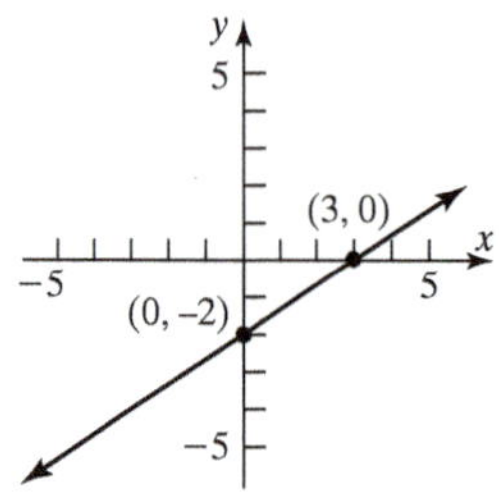

101.

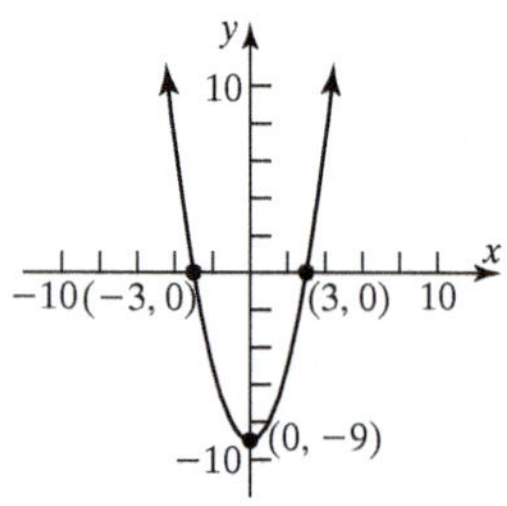

103.

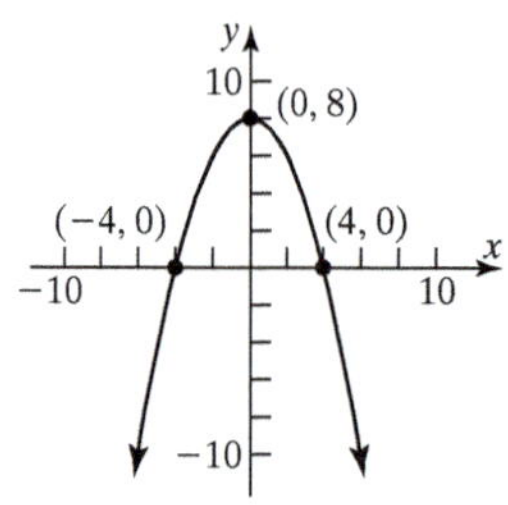

105. $d(A, B) = \sqrt{13}; d(B, C) = \sqrt{13}$ **107.** −4 and 8 **109.** $C = 50{,}000 + 95x$ **111.** He should invest $50,000 in A-rated bonds and $20,000 in the CD.

113. The range of distances is from 0.5 to 0.75 m, $0.5 \le x \le 0.75$. **115.** The plane can extend its search 246.4 mi. **117.** The bees meet for the first time in 18.75 sec. The bees meet for the second time 37.5 sec. later. **119. (a)** The object will strike the ground in 8 s. **(b)** The height is 896 ft.

121. It will take about 12 hr 23 min. **123.** Add $6\frac{2}{3}$ lb of $8/lb coffee to get $26\frac{2}{3}$ lb of $5/lb coffee. **125.** Evaporate 51.2 oz of water. **127.** 6 in. by 8 in.

129. (a) 6.5 in. by 6.5 in.; 12.5 in. by 12.5 in. by $4\frac{1}{3}$ in.; $14\frac{2}{3}$ in. by $10\frac{1}{3}$ in. **131.** The length should be approximately 6.47 ft.

133. It would take the older copier 180 min or 3 hr.

Chapter Test *(page 166)*

1. (a) $d = 10$ **(b)** $M = (1, 1)$ **2.** $\left\{-\frac{3}{2}, -1\right\}$ **3.** $\{2\}$ **4.** $\left\{-\frac{3}{2}, 2\right\}$ **5.** $\{2\}$ **6.** $\{0, 3\}$ **7.** $\{-2, 2\}$ **8.** $\{2 - \sqrt{2}, 2 + \sqrt{2}\}$

9. $\{3 - 2\sqrt{3}, 3 + 2\sqrt{3}\}$ **10.**

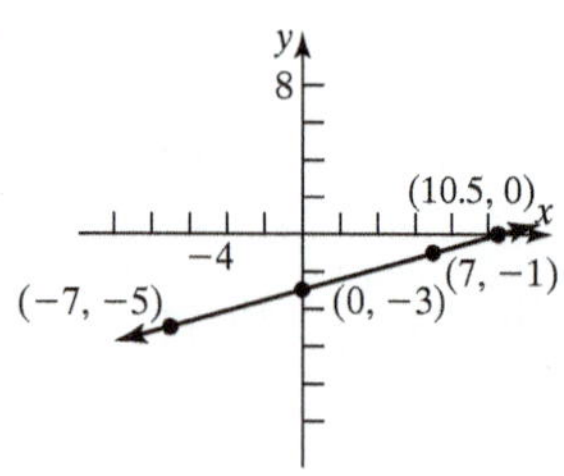

11.

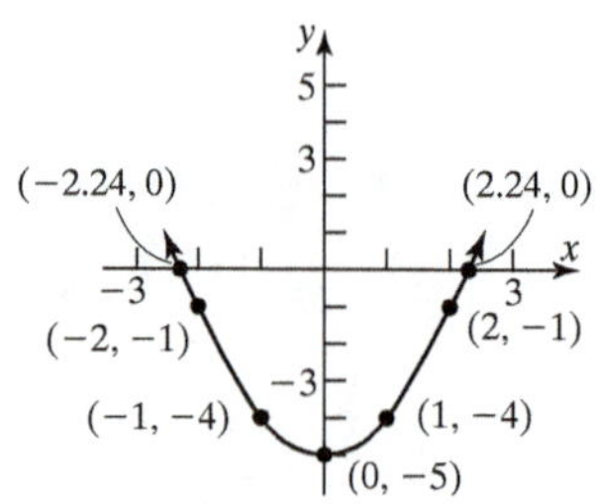

12. −1, 0.5, and 1 **13.** −2.50 and 2.50 **14.** −2.46, −0.24, and 1.70

15. $\left\{x \middle| x < -\frac{11}{2}\right\}; \left(-\infty, -\frac{11}{2}\right)$

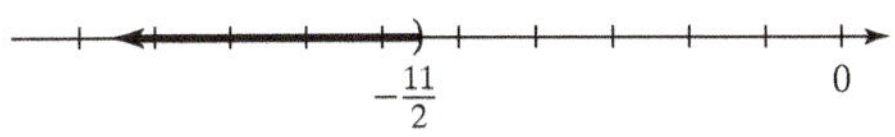

16. $\{x|x \le -5 \text{ or } x \ge 2\}; (-\infty, -5] \cup [2, \infty)$

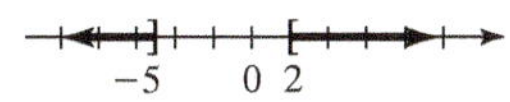

17. $\{x|-1 \le x < 2\}; [-1, 2)$

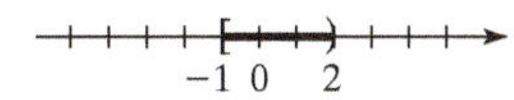

18. $\left\{x \middle| -4 < x < \frac{4}{3}\right\}; \left(-4, \frac{4}{3}\right)$ **19.** $2 - 25i$ **20.** $14 + 83i$ **21.** $\frac{7}{34} + \frac{11}{34}i$ **22.** $\left\{\frac{1}{2} - i, \frac{1}{2} + i\right\}$ **23.** About 204.63 min (3.41 hr)

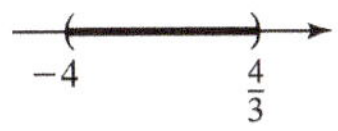

24. 23.75 lb of banana chips **25.** The sale price is \$159.50. **26.** Glenn will make \$100 in interest after 3 months.

CHAPTER 2 Graphs

2.1 Assess Your Understanding *(page 175)*

3. $y = 0$ **4.** y-axis **5.** 4 **6.** $(-3, 4)$ **7.** T **8.** F

9. $(-2, 0), (0, 2)$

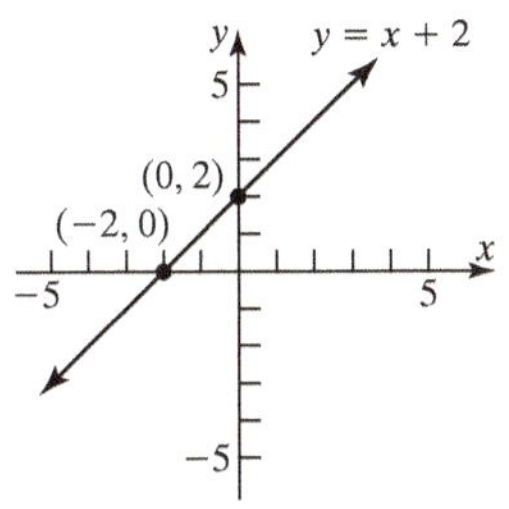

11. $(-4, 0), (0, 8)$

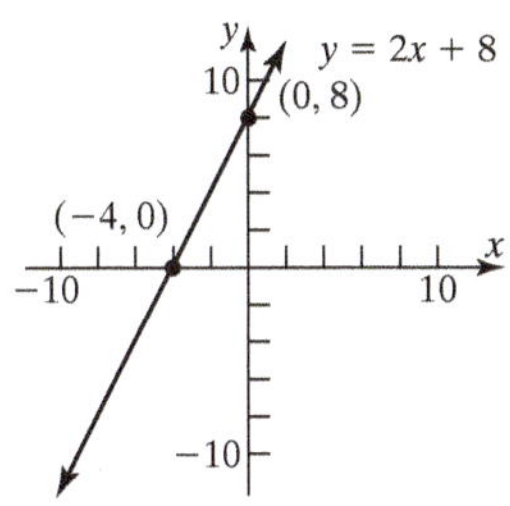

13. $(-1, 0), (1, 0), (0, -1)$

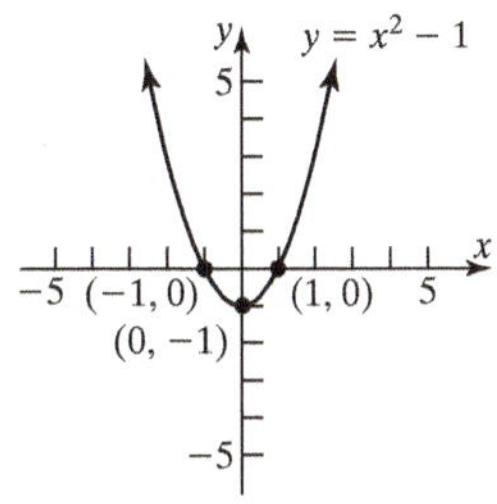

15. $(-2, 0), (2, 0), (0, 4)$

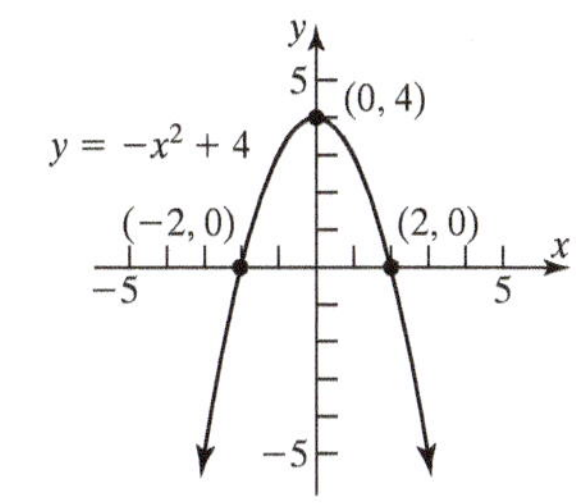

17. $(3, 0), (0, 2)$

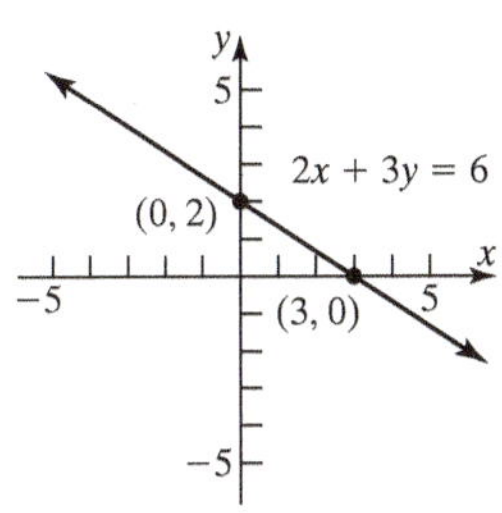

19. $(-2, 0), (2, 0), (0, 9)$

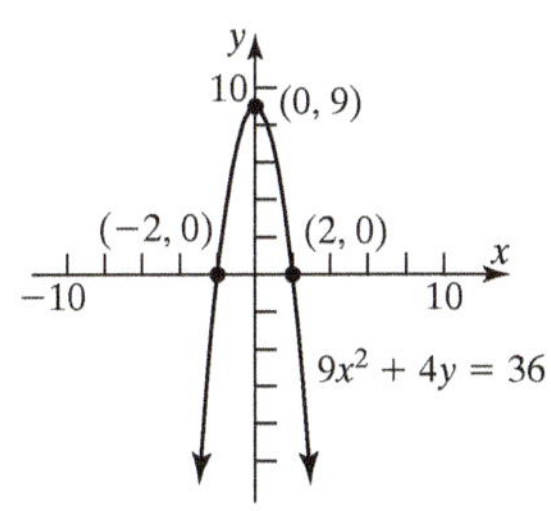

21.

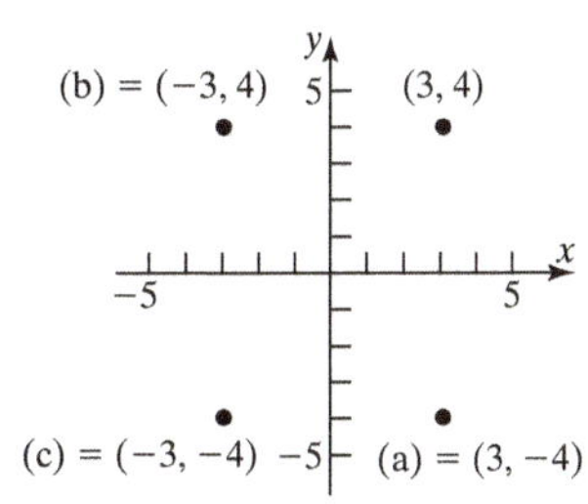

23.

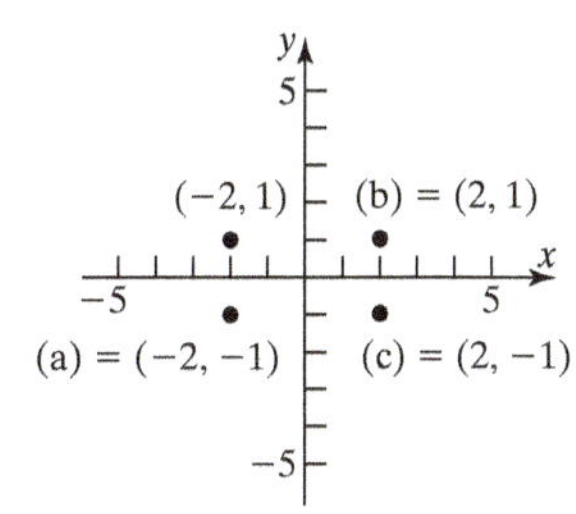

25.

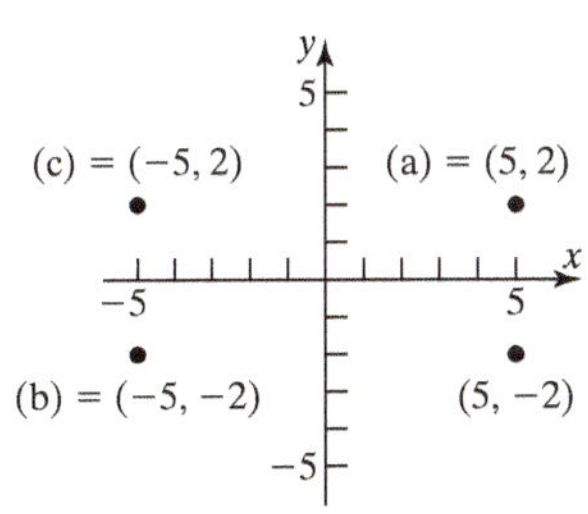

27.

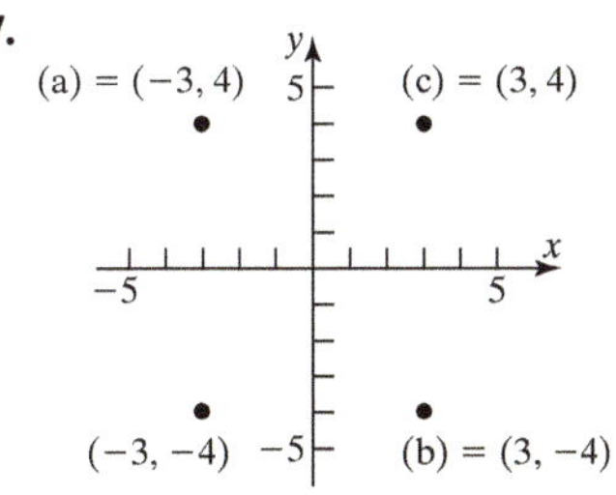

29.

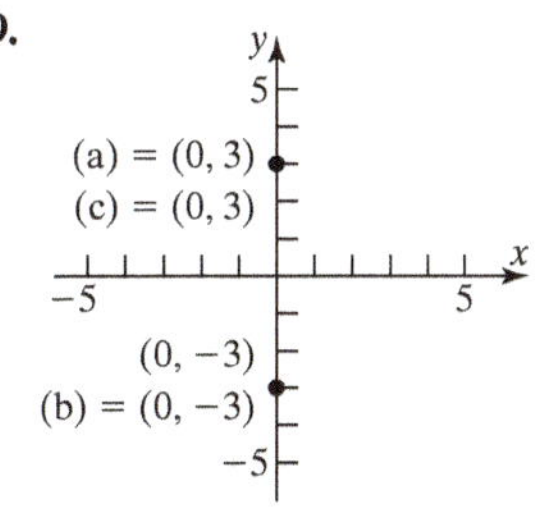

31. Symmetric with respect to the x-axis, the y-axis, and the origin **33.** Symmetric with respect to the y-axis
35. Symmetric with respect to the x-axis **37.** Symmetric with respect to the origin

39.

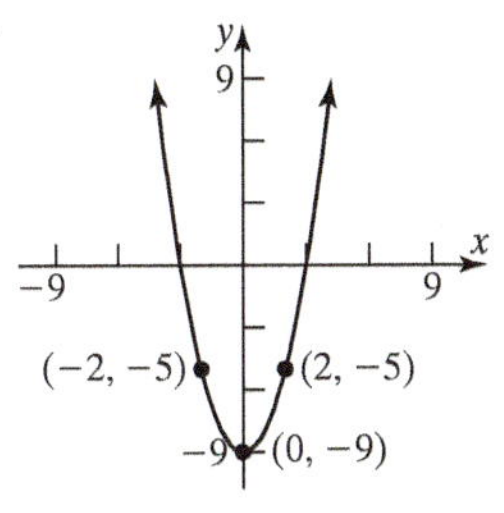

41.

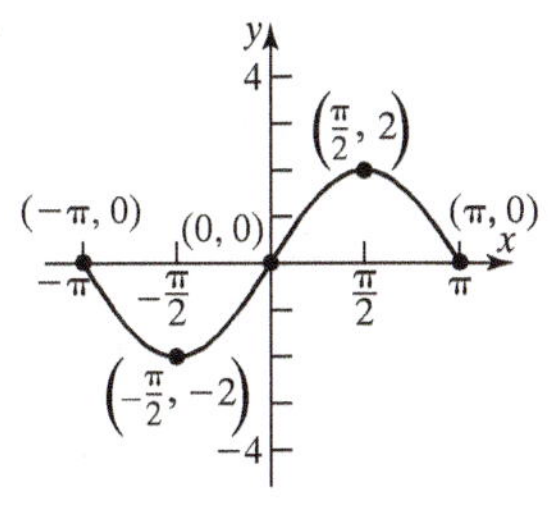

43.

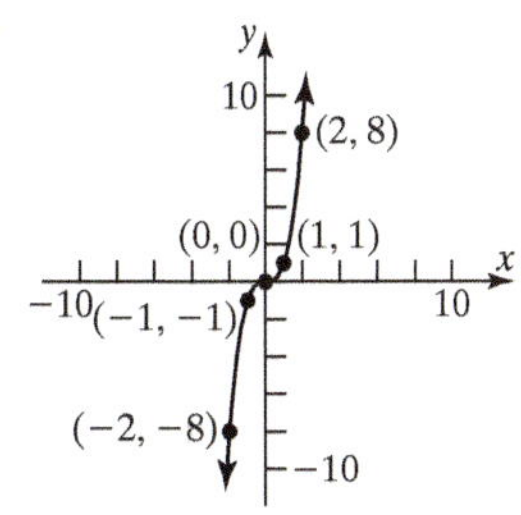

45.

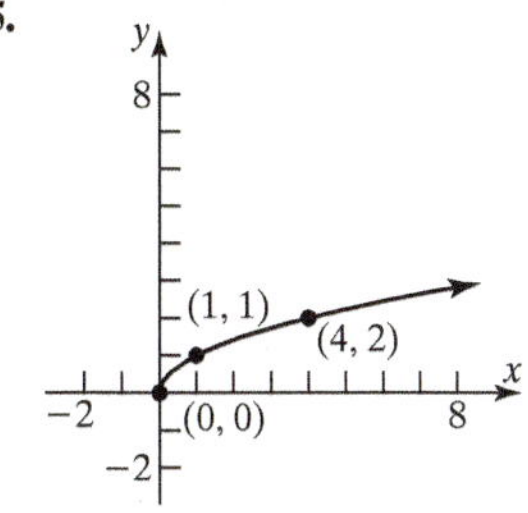

47. $(-4, 0), (0, -2), (0, 2)$; symmetric with respect to the x-axis **49.** $(0, 0)$; symmetric with respect to the origin **51.** $(0, -9), (3, 0), (-3, 0)$; symmetric with respect to the y-axis **53.** $(-2, 0), (2, 0), (0, -3), (0, 3)$; symmetric with respect to the x-axis, y-axis, and origin **55.** $(0, -9), (3, 0), (-3, 0), (-1, 0)$; no symmetry **57.** $(0, -4), (4, 0), (-4, 0)$; symmetric with respect to the y-axis **59.** $(0, 0)$; symmetric with respect to the origin **61.** $(0, 0)$; symmetric with respect to the origin

63. (a) $(0, -5), (-\sqrt{5}, 0), (\sqrt{5}, 0)$

(b) symmetric with respect to the y-axis

(c)

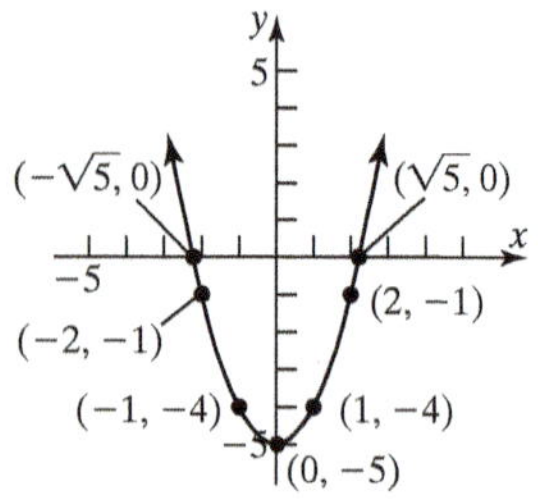

65. (a) $(-9, 0), (0, -3), (0, 3)$

(b) symmetric with respect to the x-axis

(c)

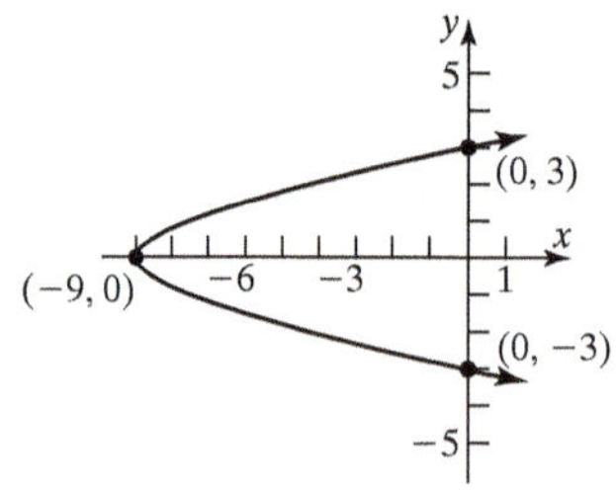

67. (a) $(0, 3), (0, -3), (-3, 0), (3, 0)$

(b) symmetric with respect to the x-axis, y-axis, and origin

(c)

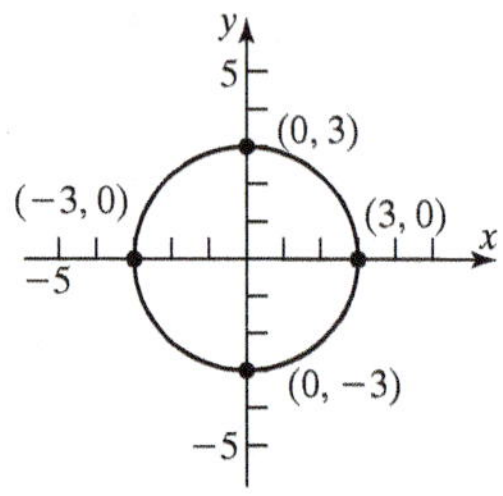

69. (a) $(0, 0), (-2, 0), (2, 0)$

(b) symmetric with respect to the origin

(c)

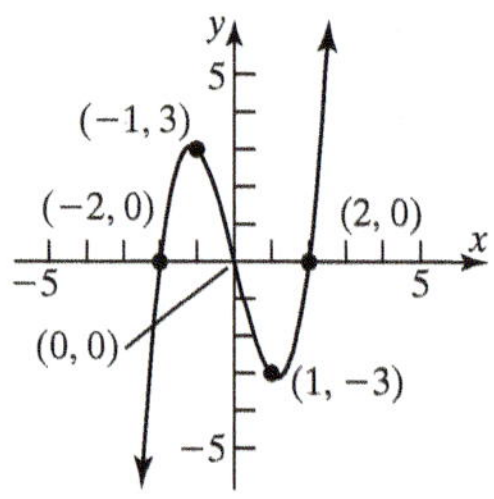

71. $(-1, -2)$ **73.** 4 **75. (a)** $(0, 0), (2, 0), (0, 1), (0, -1)$ **(b)** x-axis symmetry

2.2 Assess Your Understanding *(page 189)*

1. undefined; 0 **2.** 3; 2 **3.** $y = b$; y-intercept **4.** T **5.** F **6.** T **7.** $m_1 = m_2$; y-intercepts; $m_1 m_2 = -1$ **8.** 2 **9.** $-\frac{1}{2}$ **10.** F

11. (a) Slope $= \frac{1}{2}$

(b) If x increases by 2 units, y will increase by 1 unit.

13. (a) Slope $= -\frac{1}{3}$

(b) If x increases by 3 units, y will decrease by 1 unit

15. Slope $= -\frac{3}{2}$

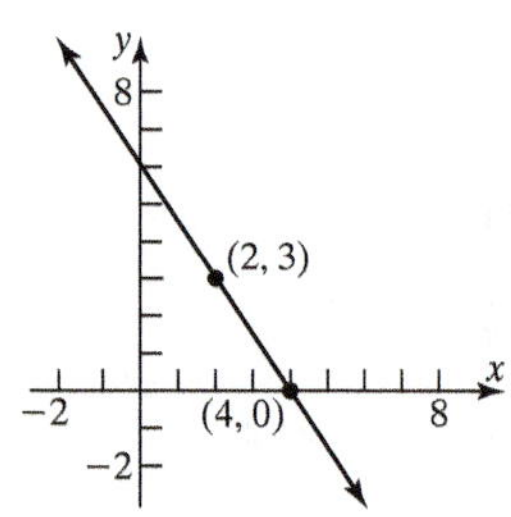

17. Slope $= -\frac{1}{2}$

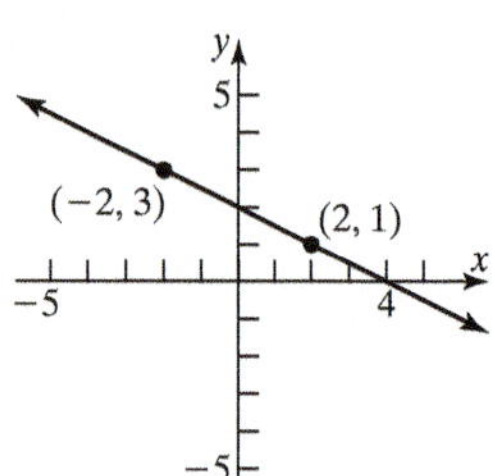

19. Slope $= 0$

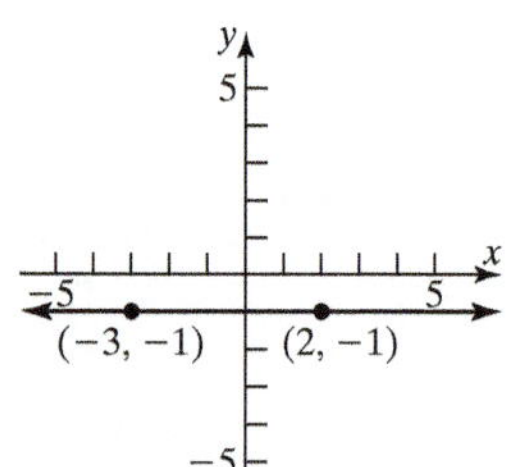

21. Slope undefined

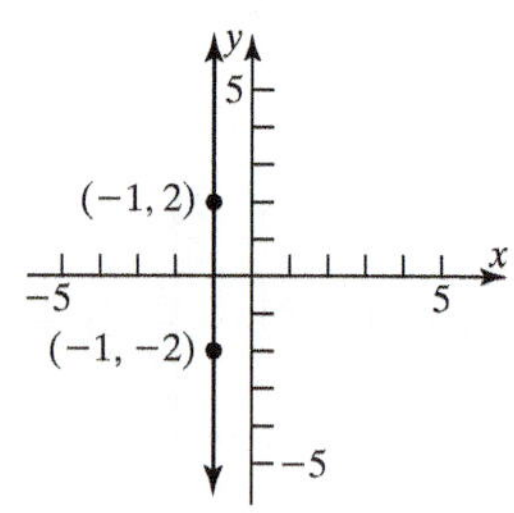

23.

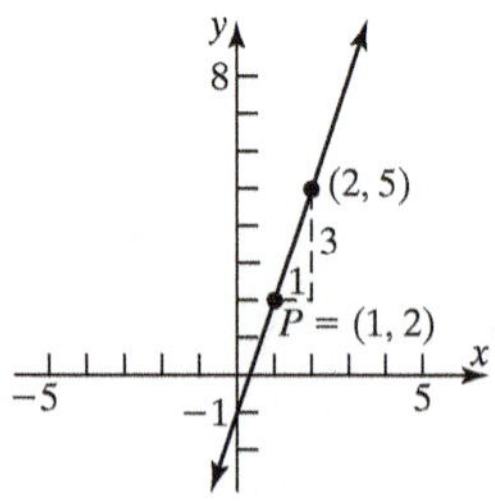

25.

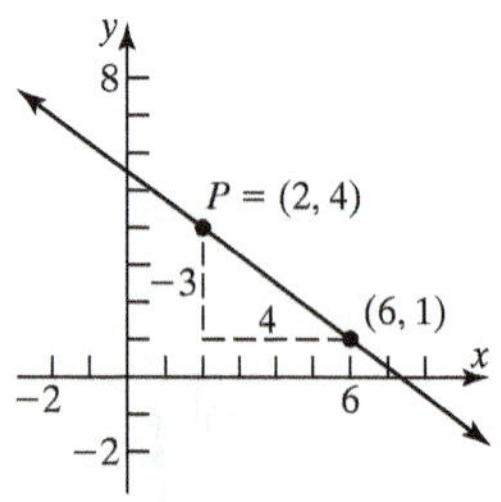

27.

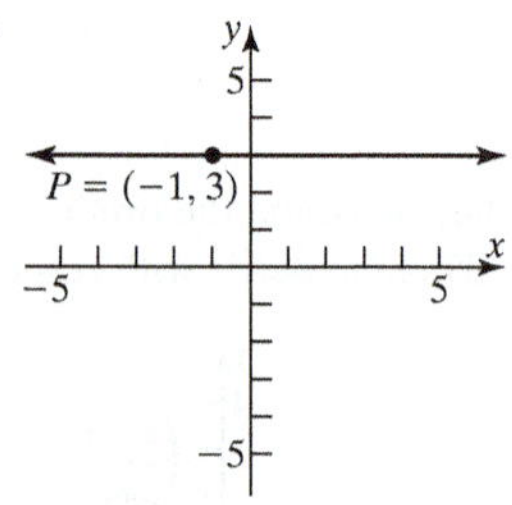

29.

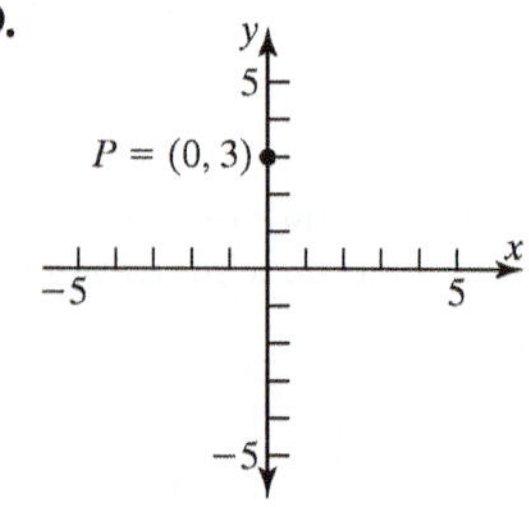

31. $(2, 6); (3, 10); (4, 14)$ **33.** $(4, -7); (6, -10); (8, -13)$ **35.** $(-1, -5); (0, -7); (1, -9)$ **37.** $x - 2y = 0$ or $y = \frac{1}{2}x$ **39.** $x + y = 2$ or $y = -x + 2$ **41.** $2x - y = 3$ or $y = 2x - 3$ **43.** $x + 2y = 5$ or $y = -\frac{1}{2}x + \frac{5}{2}$ **45.** $3x - y = -9$ or $y = 3x + 9$ **47.** $2x + 3y = -1$ or $y = -\frac{2}{3}x - \frac{1}{3}$ **49.** $x - 2y = -5$ or $y = \frac{1}{2}x + \frac{5}{2}$ **51.** $3x + y = 3$ or $y = -3x + 3$ **53.** $x - 2y = 2$ or $y = \frac{1}{2}x - 1$

55. $x = 2$; no slope–intercept form **57.** $y = 2$ **59.** $2x - y = -4$ or $y = 2x + 4$ **61.** $2x - y = 0$ or $y = 2x$ **63.** $x = 4$; no slope–intercept form **65.** $2x + y = 0$ or $y = -2x$ **67.** $x - 2y = -3$ or $y = \frac{1}{2}x + \frac{3}{2}$ **69.** $y = 4$

71. Slope = 2; y-intercept = 3

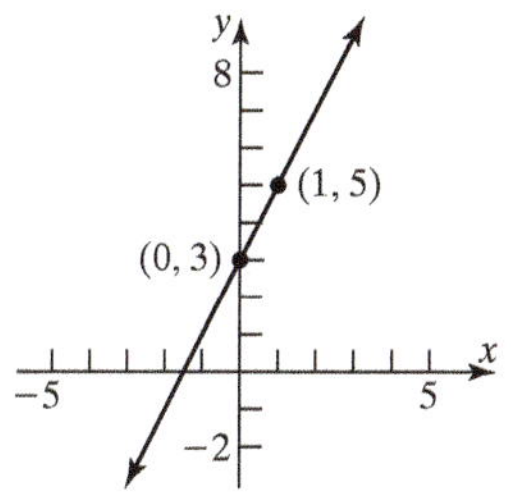

73. Slope = 2; y-intercept = −2

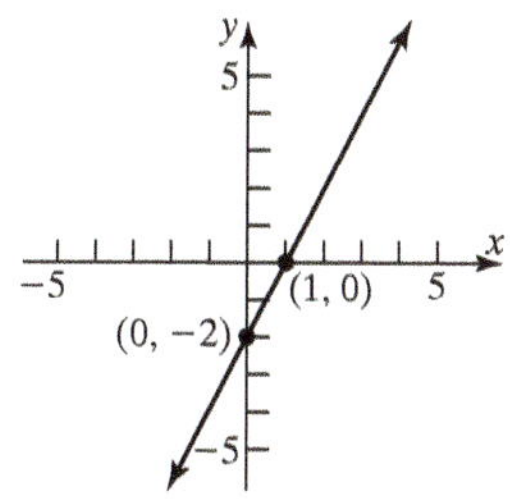

75. Slope = $\frac{1}{2}$; y-intercept = 2

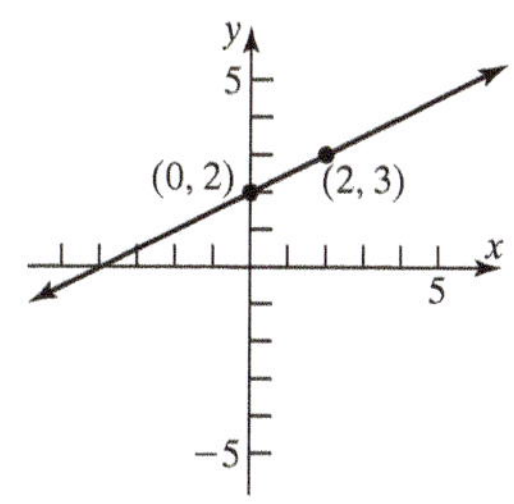

77. Slope = $-\frac{1}{2}$; y-intercept = 2

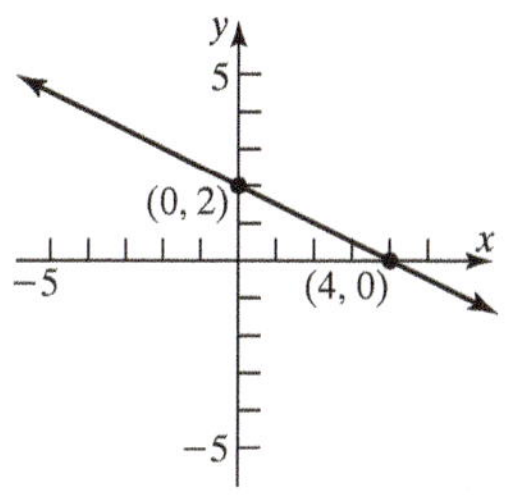

79. Slope = $\frac{2}{3}$; y-intercept = −2

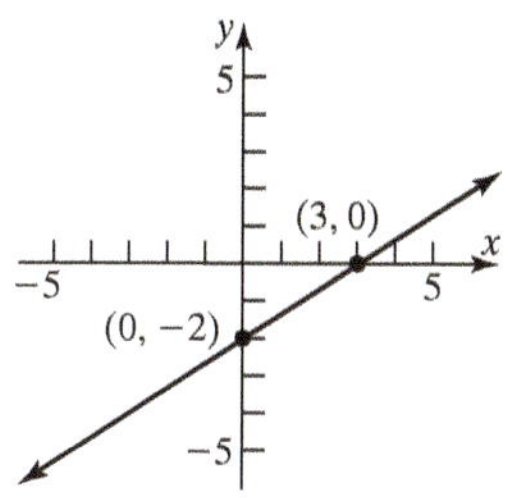

81. Slope = −1; y-intercept = 1

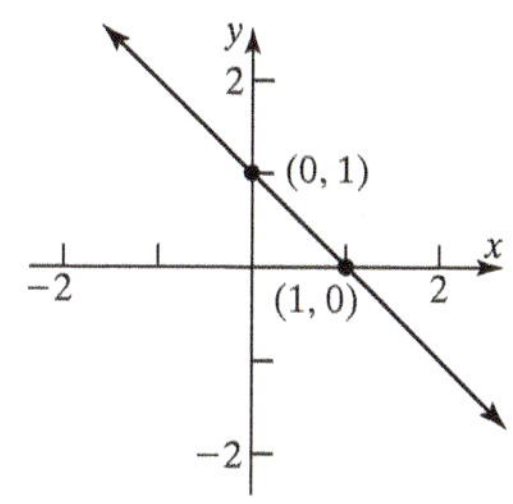

83. Slope undefined; no y-intercept

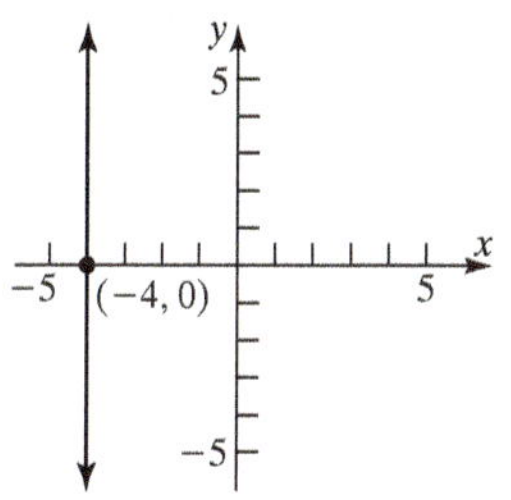

85. Slope = 0; y-intercept = 5

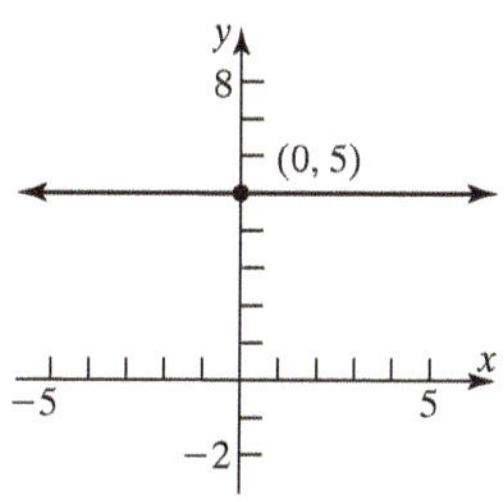

87. Slope = 1; y-intercept = 0

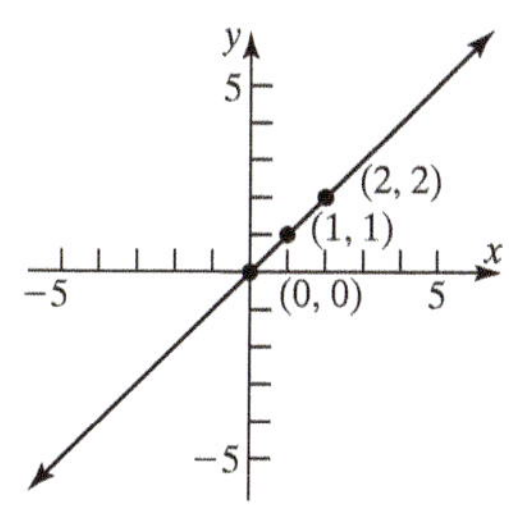

89. Slope = $\frac{3}{2}$; y-intercept = 0

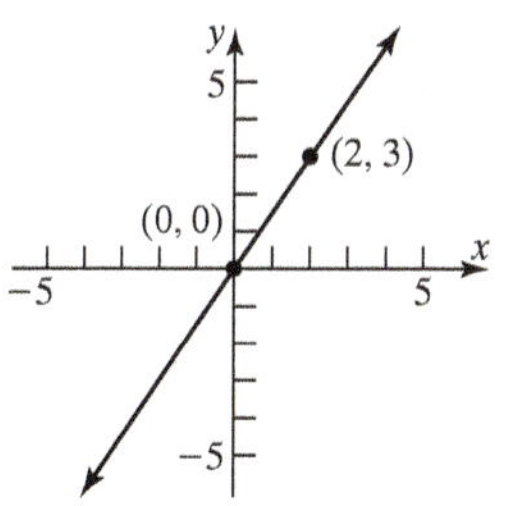

91. (a) x-intercept: 3; y-intercept: 2
(b)

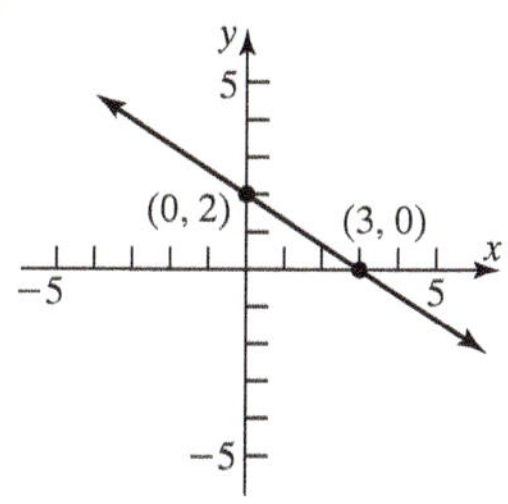

93. (a) x-intercept: −10; y-intercept: 8
(b)

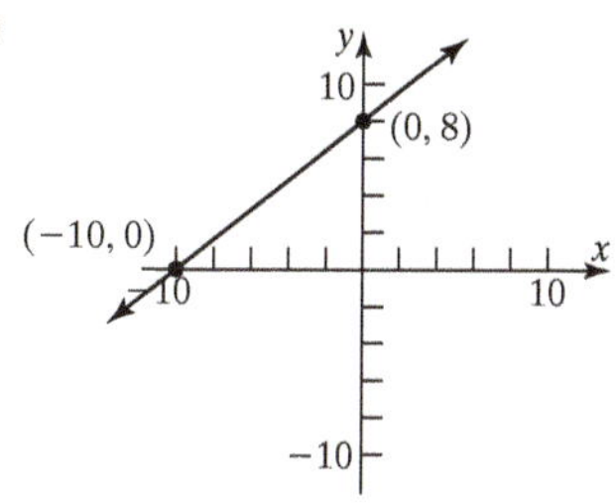

95. (a) x-intercept: 3; y-intercept: $\frac{21}{2}$
(b)

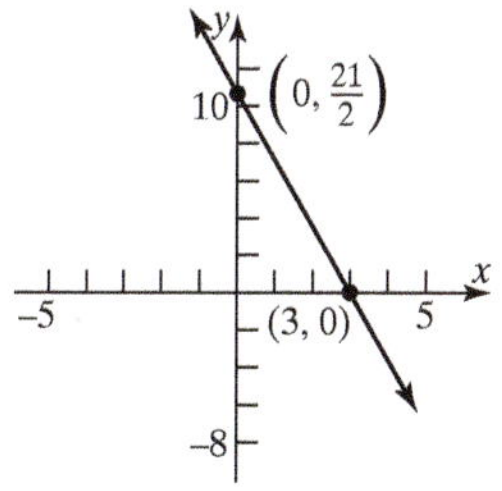

97. (a) x-intercept: 2; y-intercept: 3
(b)

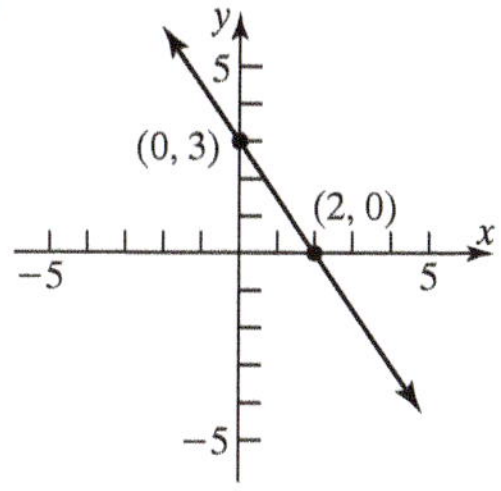

99. (a) x-intercept: 5; y-intercept: −2
(b)

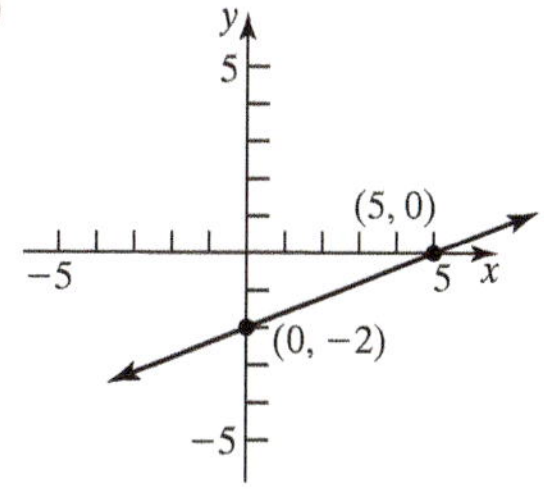

101. $y = 0$ **103.** Parallel **105.** Neither **107.** (b) **109.** (d)

111. $P_1 = (-2, 5)$, $P_2 = (1, 3)$, $m_1 = -\frac{2}{3}$; $P_2 = (1, 3)$, $P_3 = (-1, 0)$, $m_2 = \frac{3}{2}$; because $m_1 m_2 = -1$, the lines are perpendicular and the points $(-2, 5)$, $(1, 3)$, and $(-1, 0)$ are the vertices of a right triangle; thus, the points P_1, P_2, and P_3 are the vertices of a right triangle.

113. $P_1 = (-1, 0)$, $P_2 = (2, 3)$, $m = 1$; $P_3 = (1, -2)$, $P_4 = (4, 1)$, $m = 1$; $P_1 = (-1, 0)$, $P_3 = (1, -2)$, $m = -1$; $P_2 = (2, 3)$, $P_4 = (4, 1)$, $m = -1$; opposite sides are parallel, and adjacent sides are perpendicular; the points are the vertices of a rectangle.

115. $C = 0.20x + 29$; \$51.00; \$75.00 **117.** $C = 0.53x + 1{,}070{,}000$

119. **(a)** $C = 0.08275x + 7.58,\ 0 \le x \le 400$
(b)

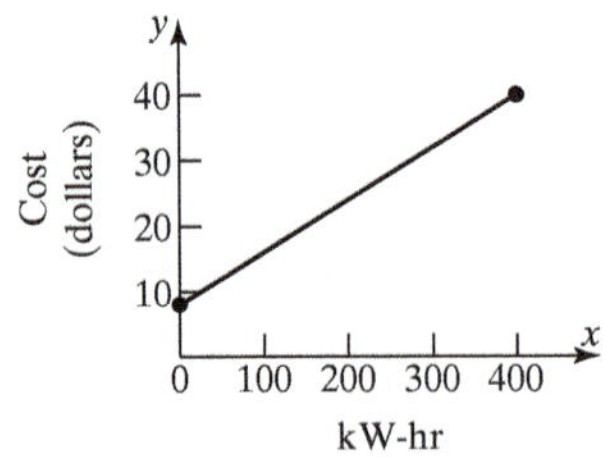

(c) \$15.86
(d) \$32.41
(e) Each additional kW-hr used adds \$0.08275 to the bill.

121. $°C = \frac{5}{9}(°F - 32)$; approximately 21.1°C

123. **(a)** $y = -\frac{2}{25}x + 30$ **(b)** x-intercept: 375; the ramp meets the floor 375 in. (31.25 ft) from the base of the platform.

(c) The ramp does not meet design requirements. It has a run of 31.25 ft long.

(d) The only slope possible for the ramp to comply with the requirement is for it to drop 1 in. for every 12-in. run.

125. **(a)** $A = \frac{1}{5}x + 20{,}000$ **(b)** \$80,000 **(c)** Each additional box sold requires an additional \$0.20 in advertising.

127. All have the same slope, 2; the lines are parallel.

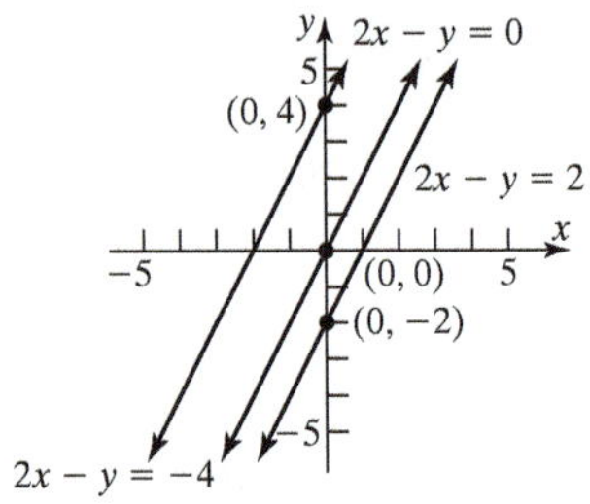

129. (b), (c), (e), (g) **131.** (c) **137.** No; no **139.** They are the same line. **141.** Yes, if the y-intercept is 0.

2.3 Assess Your Understanding *(page 198)*

3. F **4.** radius **5.** T **6.** F **7.** Center (2, 1); radius = 2; $(x - 2)^2 + (y - 1)^2 = 4$ **9.** Center $\left(\frac{5}{2}, 2\right)$; radius $= \frac{3}{2}$; $\left(x - \frac{5}{2}\right)^2 + (y - 2)^2 = \frac{9}{4}$

11. $x^2 + y^2 = 4$;
$x^2 + y^2 - 4 = 0$

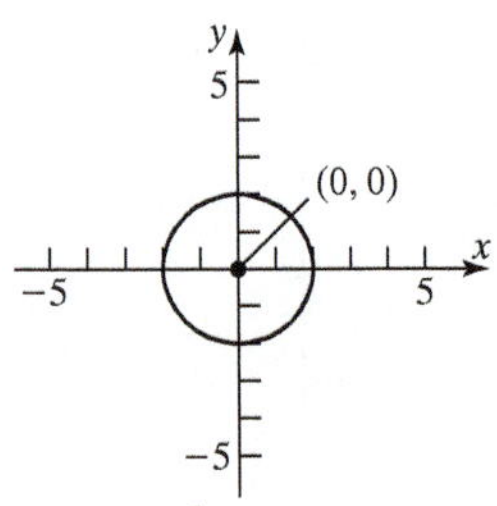

13. $x^2 + (y - 2)^2 = 4$;
$x^2 + y^2 - 4y = 0$

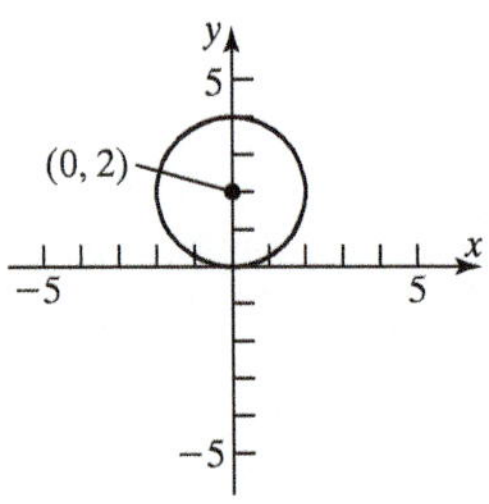

15. $(x - 4)^2 + (y + 3)^2 = 25$;
$x^2 + y^2 - 8x + 6y = 0$

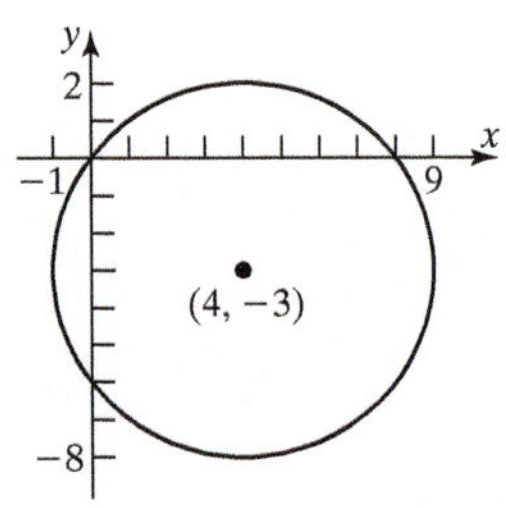

17. $(x + 2)^2 + (y - 1)^2 = 16$;
$x^2 + y^2 + 4x - 2y - 11 = 0$

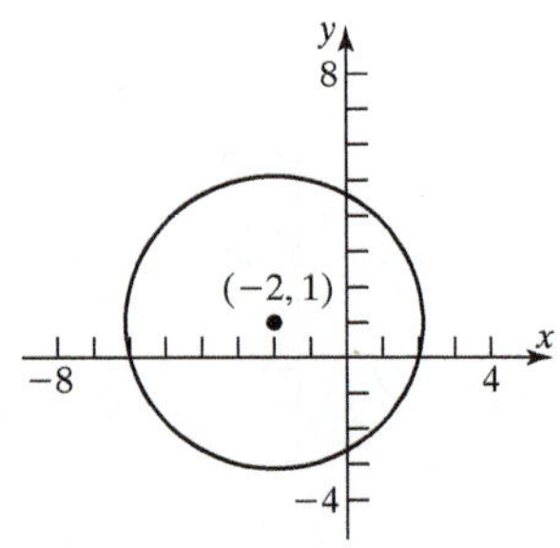

19. $\left(x - \frac{1}{2}\right)^2 + y^2 = \frac{1}{4}$;
$x^2 + y^2 - x = 0$

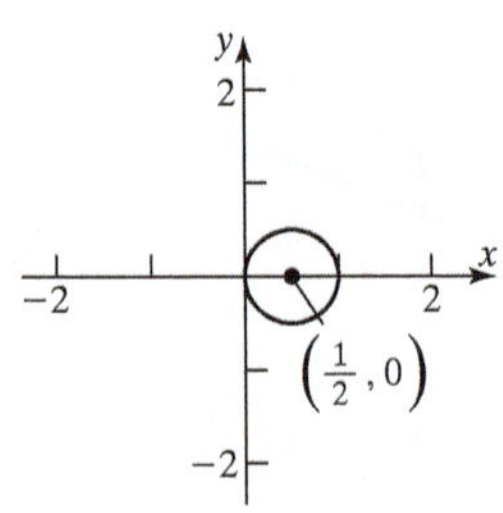

21. **(a)** $(h, k) = (0, 0)$; $r = 2$
(b)

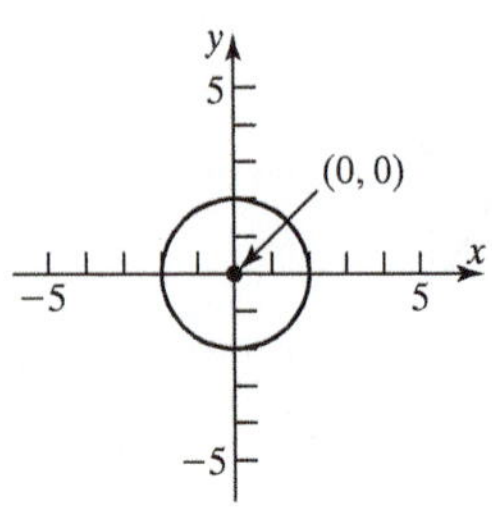

(c) $(\pm 2, 0)$; $(0, \pm 2)$

23. **(a)** $(h, k) = (3, 0)$; $r = 2$
(b)

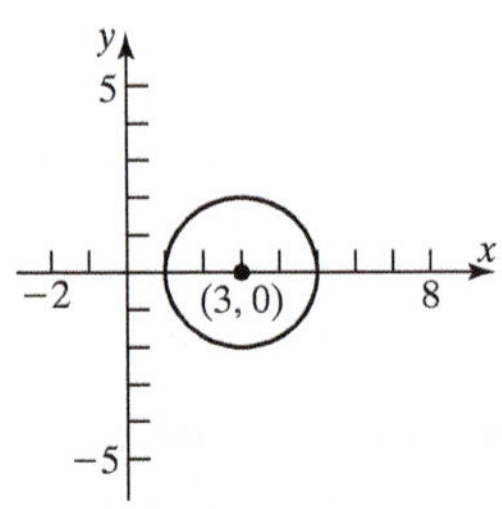

(c) (1, 0); (5, 0)

25. **(a)** $(h, k) = (1, 2)$; $r = 3$
(b)

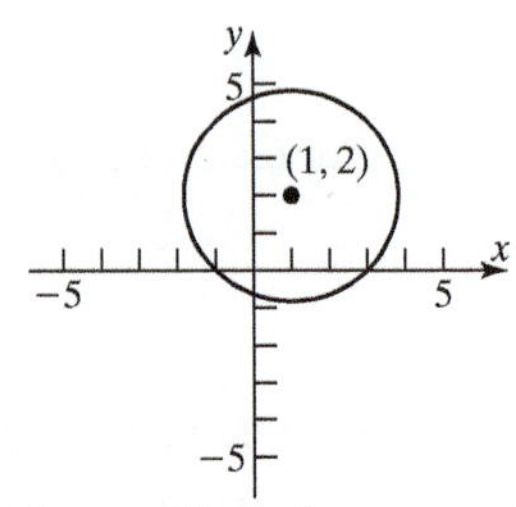

(c) $(1 \pm \sqrt{5}, 0)$; $(0, 2 \pm 2\sqrt{2})$

27. (a) $(h, k) = (-2, 2); r = 3$

(b)

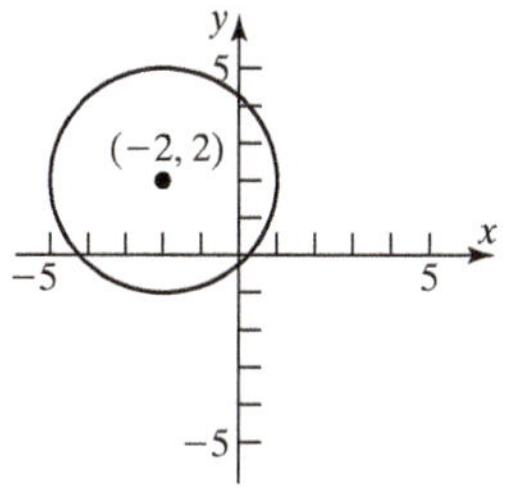

(c) $\left(-2 \pm \sqrt{5}, 0\right); \left(0, 2 \pm \sqrt{5}\right)$

29. (a) $(h, k) = \left(\frac{1}{2}, -1\right); r = \frac{1}{2}$

(b)

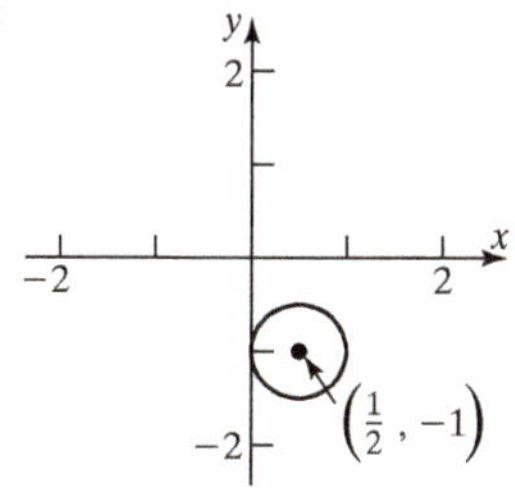

(c) $(0, -1)$

31. (a) $(h, k) = (3, -2); r = 5$

(b)

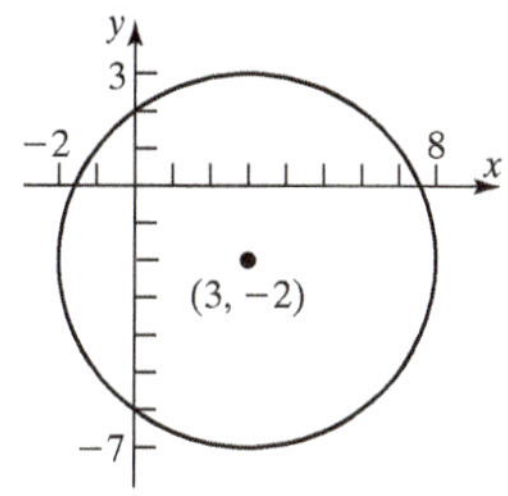

(c) $\left(3 \pm \sqrt{21}, 0\right); (0, -6), (0, 2)$

33. (a) $(h, k) = (-2, 0); r = 2$

(b)

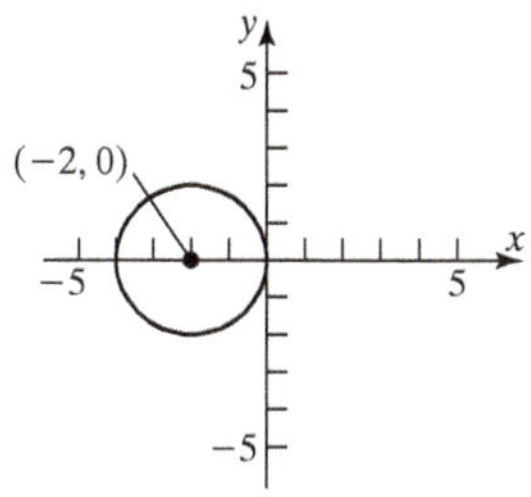

(c) $(0, 0), (-4, 0)$

35. $x^2 + y^2 = 13$ **37.** $(x - 2)^2 + (y - 3)^2 = 9$ **39.** $(x + 1)^2 + (y - 3)^2 = 5$ **41.** $(x + 1)^2 + (y - 3)^2 = 1$ **43.** (c) **45.** (b) **47.** 18 units^2 **49.** $x^2 + (y - 139)^2 = 15{,}625$ **51.** $x^2 + y^2 + 2x + 4y - 4168.16 = 0$ **53.** $\sqrt{2}x + 4y = 9\sqrt{2}$ **55.** $(1, 0)$ **57.** $y = 2$ **59.** (b), (c), (e), (g)

2.4 Assess Your Understanding *(page 203)*

1. $y = kx$ **2.** F **3.** $y = \frac{1}{5}x$ **5.** $A = \pi x^2$ **7.** $F = \frac{250}{d^2}$ **9.** $z = \frac{1}{5}(x^2 + y^2)$ **11.** $M = \frac{9d^2}{2\sqrt{x}}$ **13.** $T^2 = \frac{8a^3}{d^2}$ **15.** $V = \frac{4\pi}{3}r^3$ **17.** $A = \frac{1}{2}bh$ **19.** $F = 6.67 \times 10^{-11}\left(\frac{mM}{d^2}\right)$ **21.** $p = 0.00649B$; \$941.05 **23.** 144 ft; 2 s **25.** 2.25 **27.** $R = 3.95g$; \$41.48 **29. (a)** $D = \frac{429}{p}$ **(b)** 143 bags **31.** 450 cm^3 **33.** 124.76 lb **35.** $V = \pi r^2 h$ **37.** 54.86 lb **39.** $\sqrt[3]{6} \approx 1.82$ in. **41.** 2812.5 joules **43.** 384 psi

Review Exercises *(page 206)*

1. (a) $\frac{1}{2}$ **(b)** For every 2-unit change in x, y will change by 1 unit. **3. (a)** $-\frac{4}{3}$ **(b)** For every 3-unit change in x, y will change by -4 units.

5. (a) undefined **(b)** For any change in y, x will not change. **7.** $(0, 0)$; symmetric with respect to the x-axis **9.** $(\pm 4, 0), (0, \pm 2)$; symmetric with respect to the x-axis, y-axis, and origin **11.** $(0, 1)$; symmetric with respect to the y-axis **13.** $(0, 0), (-1, 0), (0, -2)$; no symmetry

15. $(x + 2)^2 + (y - 3)^2 = 16$ **17.** $(x + 1)^2 + (y + 2)^2 = 1$

19. Center $(0, 1)$; radius $= 2$

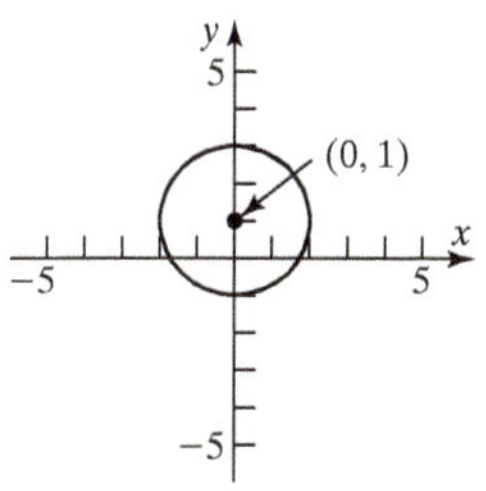

Intercepts: $(-\sqrt{3}, 0), (\sqrt{3}, 0), (0, -1), (0, 3)$

21. Center $(1, -2)$; radius $= 3$

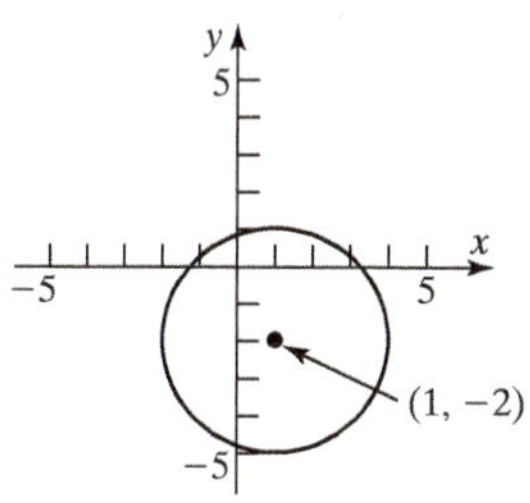

Intercepts: $(1 - \sqrt{5}, 0), (1 + \sqrt{5}, 0), (0, -2 - 2\sqrt{2}), (0, -2 + 2\sqrt{2})$

23. Center $(1, -2)$; radius $= \sqrt{5}$

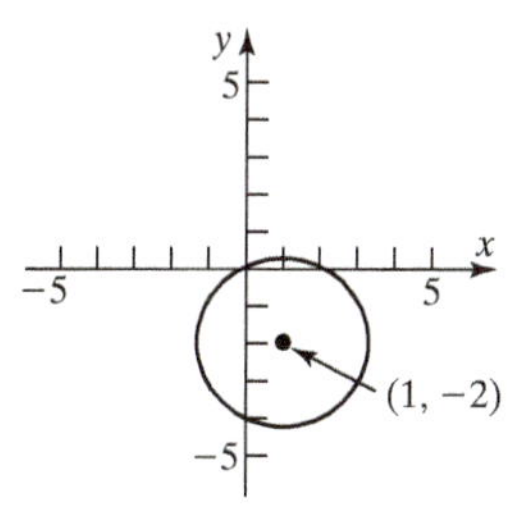

Intercepts: $(0, 0), (2, 0), (0, -4)$

25. $2x + y = 5$ or $y = -2x + 5$ **27.** $x = -3$; no slope–intercept form **29.** $x + 5y = -10$ or $y = -\frac{1}{5}x - 2$

31. $2x - 3y = -19$ or $y = \frac{2}{3}x + \frac{19}{3}$ **33.** $x - y = 7$ or $y = x - 7$

35. Slope $= \frac{4}{5}$; y-intercept $= 4$

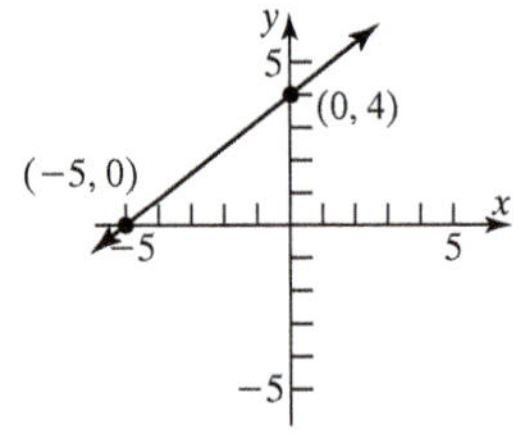

37. Slope $= \frac{3}{2}$; y-intercept $= \frac{1}{2}$

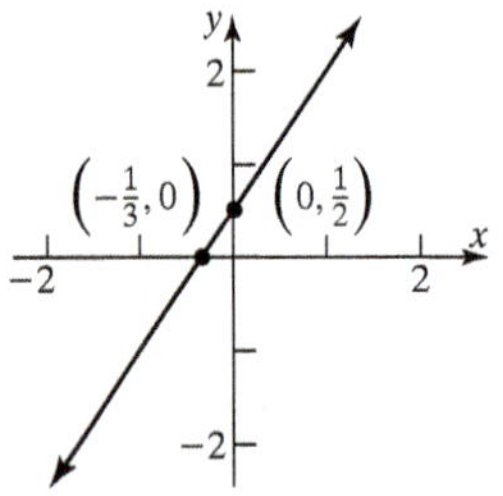

39. Intercepts: $(6, 0)$, $(0, -4)$

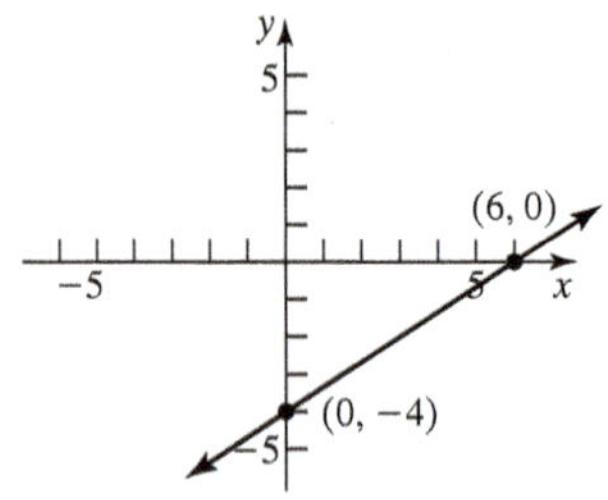

41. Intercepts: $(4, 0)$, $(0, 6)$

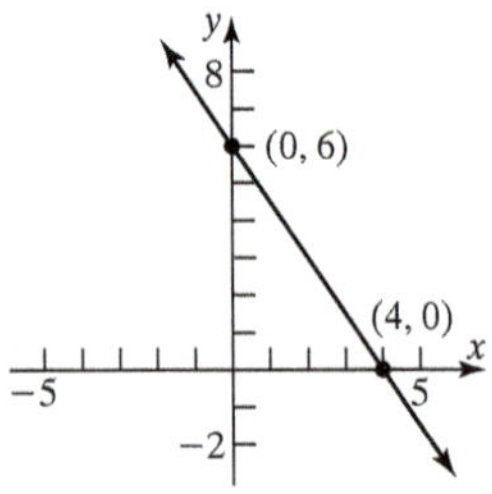

43.

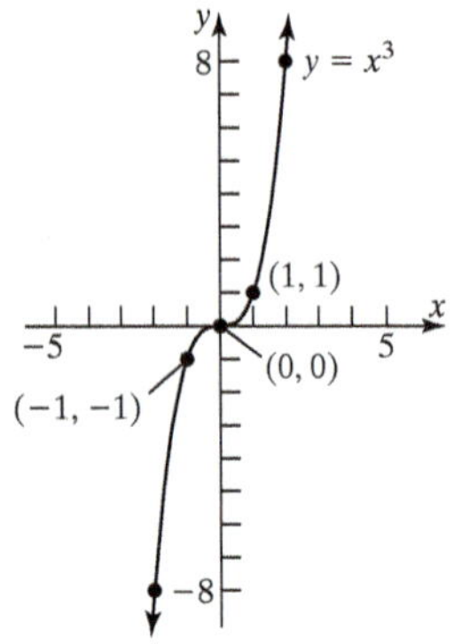

45.

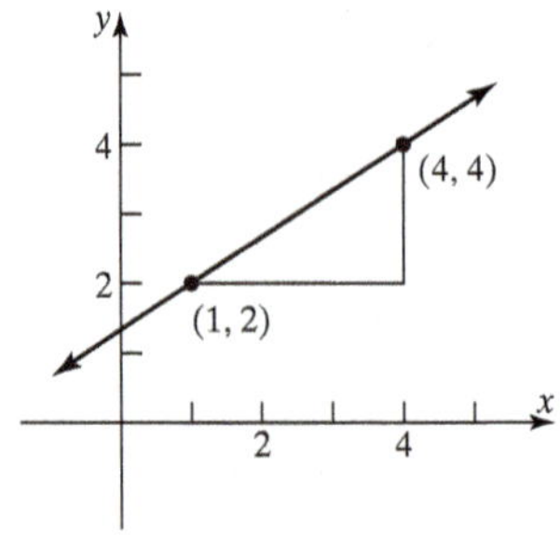

47. Slope from A to B is -1; slope from A to C is -1. **49.** $R = 3.59g$; \$40.21 **51.** Approximately 36 million miles

Chapter Test *(page 208)*

1. (a) $m = -\frac{2}{3}$

(b) For every 3-unit change in x, y will change by -2 units.

2.

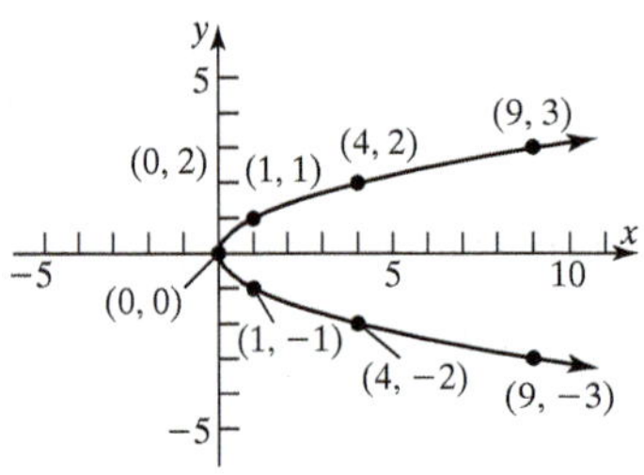

3. Intercepts: $(-3, 0)$, $(3, 0)$, $(0, 9)$; symmetric with respect to the y-axis

4. $y = -2x + 2$

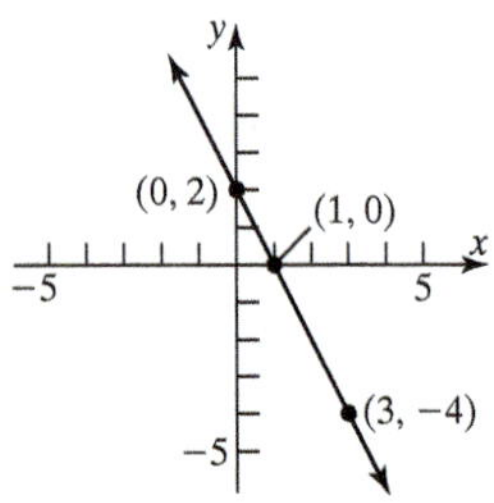

5. $x^2 + y^2 - 8x + 6y = 0$

6. Center: $(-2, 1)$; radius: 3

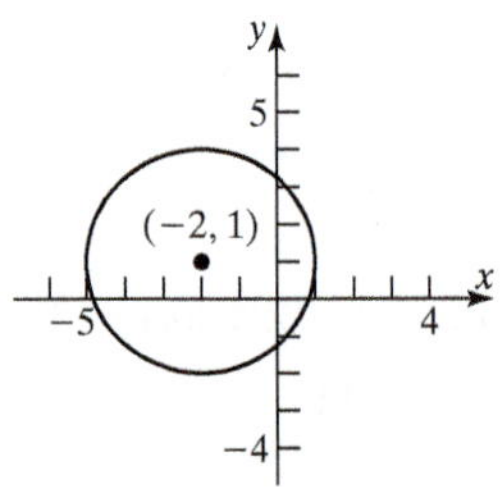

7. Parallel line: $y = -\frac{2}{3}x - \frac{1}{3}$; perpendicular line: $y = \frac{3}{2}x + 3$

8. 14.69 ohms

Cumulative Review *(page 208)*

1. $\left\{\frac{5}{3}\right\}$ **2.** $\{-3, 4\}$ **3.** $\left\{-\frac{1}{2}, 3\right\}$ **4.** $\{1 - \sqrt{3}, 1 + \sqrt{3}\}$ **5.** No real solution **6.** $\{4\}$ **7.** $\{1, 3\}$ **8.** $\{-2 - 2\sqrt{2}, -2 + 2\sqrt{2}\}$ **9.** $\{-3i, 3i\}$

10. $\{1 - 2i, 1 + 2i\}$ **11.** $\{x|x \le 5\}$, or $(-\infty, 5]$;

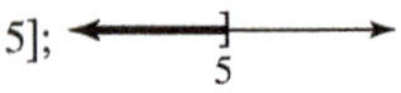

12. $\{x|-5 < x < 1\}$ or $(-5, 1)$;

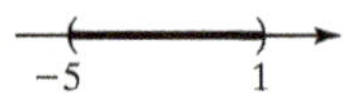

13. $\{x|1 \le x \le 3\}$ or $[1, 3]$;

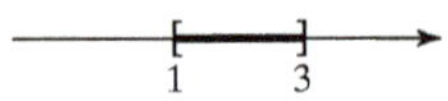

14. $\{x|x < -5 \text{ or } x > 1\}$ or $(-\infty, -5) \cup (1, \infty)$;

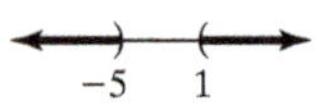

15. $5\sqrt{2}$; $\left(\frac{3}{2}, \frac{1}{2}\right)$ **16.** (a), (b)

17.

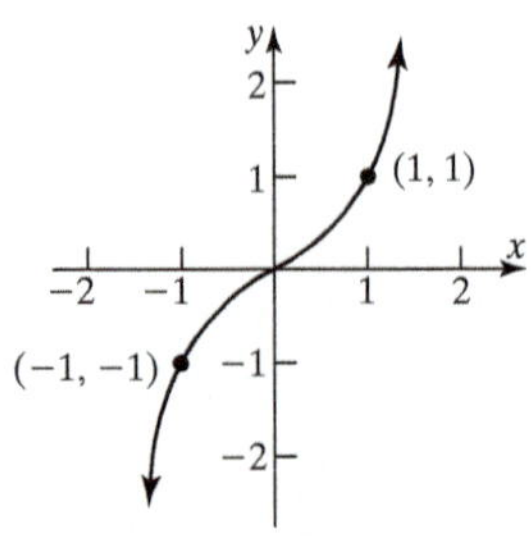

18. $y = -2x + 2$

19. $y = -\frac{1}{2}x + \frac{13}{2}$

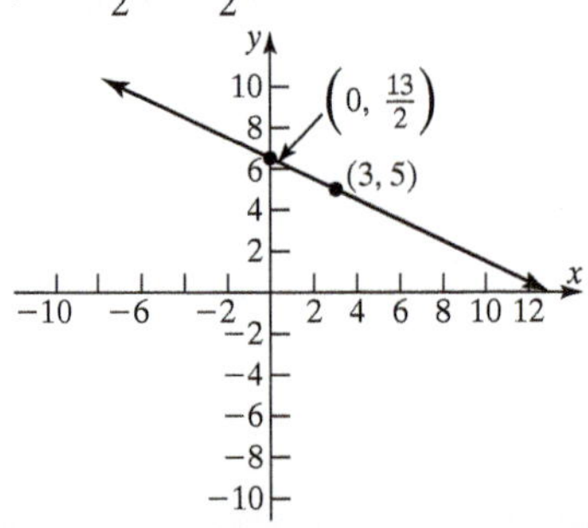

20.

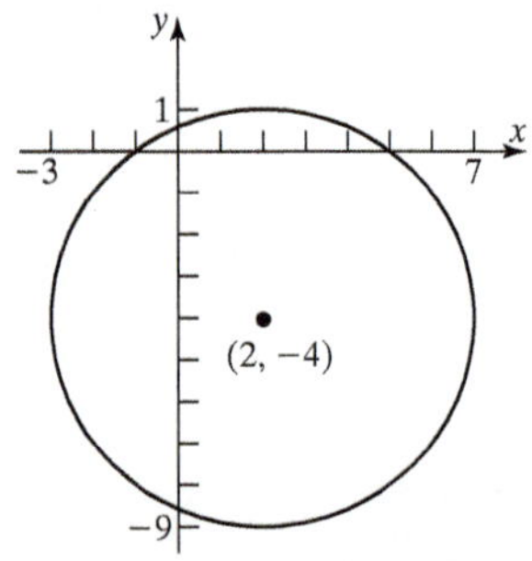

CHAPTER 3 Functions and Their Graphs

3.1 Assess Your Understanding *(page 223)*

5. independent; dependent **6.** range **7.** $[0, 5]$ **8.** image **9.** $(g - f)(x)$ **10.** F **11.** T **12.** T **13.** F **14.** F
15. Function; Domain: {Elvis, Colleen, Kaleigh, Marissa}, Range: {January 8, March 15, September 17} **17.** Not a function **19.** Not a function
21. Function; Domain: $\{1, 2, 3, 4\}$; Range: $\{3\}$ **23.** Not a function **25.** Function; Domain: $\{-2, -1, 0, 1\}$, Range: $\{0, 1, 4\}$ **27.** Function
29. Function **31.** Not a function **33.** Not a function **35.** Function **37.** Not a function **39.** **(a)** -4 **(b)** 1 **(c)** -3 **(d)** $3x^2 - 2x - 4$
(e) $-3x^2 - 2x + 4$ **(f)** $3x^2 + 8x + 1$ **(g)** $12x^2 + 4x - 4$ **(h)** $3x^2 + 6xh + 3h^2 + 2x + 2h - 4$ **41.** **(a)** 0 **(b)** $\frac{1}{2}$ **(c)** $-\frac{1}{2}$ **(d)** $\frac{-x}{x^2 + 1}$
(e) $\frac{-x}{x^2 + 1}$ **(f)** $\frac{x + 1}{x^2 + 2x + 2}$ **(g)** $\frac{2x}{4x^2 + 1}$ **(h)** $\frac{x + h}{x^2 + 2xh + h^2 + 1}$ **43.** **(a)** 4 **(b)** 5 **(c)** 5 **(d)** $|x| + 4$ **(e)** $-|x| - 4$ **(f)** $|x + 1| + 4$
(g) $2|x| + 4$ **(h)** $|x + h| + 4$ **45.** **(a)** $-\frac{1}{5}$ **(b)** $-\frac{3}{2}$ **(c)** $\frac{1}{8}$ **(d)** $\frac{2x - 1}{3x + 5}$ **(e)** $\frac{-2x - 1}{3x - 5}$ **(f)** $\frac{2x + 3}{3x - 2}$ **(g)** $\frac{4x + 1}{6x - 5}$ **(h)** $\frac{2x + 2h + 1}{3x + 3h - 5}$
47. All real numbers **49.** All real numbers **51.** $\{x | x \neq -4, x \neq 4\}$ **53.** $\{x | x \neq 0\}$ **55.** $\{x | x \geq 4\}$ **57.** $\{x | x > 9\}$ **59.** $\{x | x > 1\}$
61. **(a)** $(f + g)(x) = 5x + 1$; All real numbers **(b)** $(f - g)(x) = x + 7$; All real numbers **(c)** $(f \cdot g)(x) = 6x^2 - x - 12$; All real numbers
(d) $\left(\frac{f}{g}\right)(x) = \frac{3x + 4}{2x - 3}; \left\{x \middle| x \neq \frac{3}{2}\right\}$ **(e)** 16 **(f)** 11 **(g)** 10 **(h)** -7 **63.** **(a)** $(f + g)(x) = 2x^2 + x - 1$; All real numbers
(b) $(f - g)(x) = -2x^2 + x - 1$; All real numbers **(c)** $(f \cdot g)(x) = 2x^3 - 2x^2$; All real numbers **(d)** $\left(\frac{f}{g}\right)(x) = \frac{x - 1}{2x^2}; \{x | x \neq 0\}$ **(e)** 20
(f) -29 **(g)** 8 **(h)** 0 **65.** **(a)** $(f + g)(x) = \sqrt{x} + 3x - 5; \{x | x \geq 0\}$ **(b)** $(f - g)(x) = \sqrt{x} - 3x + 5; \{x | x \geq 0\}$
(c) $(f \cdot g)(x) = 3x\sqrt{x} - 5\sqrt{x}; \{x | x \geq 0\}$ **(d)** $\left(\frac{f}{g}\right)(x) = \frac{\sqrt{x}}{3x - 5} \left\{x \middle| x \geq 0, x \neq \frac{5}{3}\right\}$ **(e)** $\sqrt{3} + 4$ **(f)** -5 **(g)** $\sqrt{2}$ **(h)** $-\frac{1}{2}$
67. **(a)** $(f + g)(x) = 1 + \frac{2}{x}; \{x | x \neq 0\}$ **(b)** $(f - g)(x) = 1; \{x | x \neq 0\}$ **(c)** $(f \cdot g)(x) = \frac{1}{x} + \frac{1}{x^2}; \{x | x \neq 0\}$
(d) $\left(\frac{f}{g}\right)(x) = x + 1; \{x | x \neq 0\}$ **(e)** $\frac{5}{3}$ **(f)** 1 **(g)** $\frac{3}{4}$ **(h)** 2 **69.** **(a)** $(f + g)(x) = \frac{6x + 3}{3x - 2}; \left\{x \middle| x \neq \frac{2}{3}\right\}$
(b) $(f - g)(x) = \frac{-2x + 3}{3x - 2}; \left\{x \middle| x \neq \frac{2}{3}\right\}$ **(c)** $(f \cdot g)(x) = \frac{8x^2 + 12x}{(3x - 2)^2}; \left\{x \middle| x \neq \frac{2}{3}\right\}$ **(d)** $\left(\frac{f}{g}\right)(x) = \frac{2x + 3}{4x}; \left\{x \middle| x \neq 0; x \neq \frac{2}{3}\right\}$ **(e)** 3
(f) $-\frac{1}{2}$ **(g)** $\frac{7}{2}$ **(h)** $\frac{5}{4}$ **71.** $g(x) = 5 - \frac{7}{2}x$ **73.** 4 **75.** $2x + h - 1$ **77.** $6x + 3h - 2$ **79.** $3x^2 + 3xh + h^2$ **81.** $A = -\frac{7}{2}$ **83.** $A = -4$
85. $A = 8$; undefined at $x = 3$ **87.** $A(x) = \frac{1}{2}x^2$ **89.** $G(x) = 10x$ **91.** **(a)** P is the dependent variable; a is the independent variable.
(b) $P(20) = 197.34$; In 2005, there are 197.34 million people 20 years of age or older. **(c)** $P(0) = 290.580$; In 2005, there are 290.580 million people.
93. **(a)** 15.1 m, 14.071 m, 12.944 m, 11.719 m **(b)** 1.01 s, 1.43 s, 1.75 s **(c)** 2.02 s **95.** **(a)** \$222 **(b)** \$225 **(c)** \$220 **(d)** \$230
97. $R(x) = \frac{L(x)}{P(x)}$ **99.** $H(x) = P(x) \cdot I(x)$ **101.** **(a)** $P(x) = -0.05x^3 + 0.8x^2 + 155x - 500$ **(b)** $P(15) = \$1836.25$
(c) When 15 hundred cell phones are sold, the profit is \$1836.25. **103.** Only $h(x) = 2x$ **107.** Square the real number t and then subtract 2.

3.2 Assess Your Understanding *(page 230)*

3. vertical **4.** 5; -3 **5.** $(-2, 7)$ **6.** F **7.** F **8.** T **9.** **(a)** $f(0) = 3; f(-6) = -3$ **(b)** $f(6) = 0; f(11) = 1$ **(c)** Positive **(d)** Negative
(e) -3, 6, and 10 **(f)** $-3 < x < 6; 10 < x \leq 11$ **(g)** $\{x | -6 \leq x \leq 11\}$ **(h)** $\{y | -3 \leq y \leq 4\}$ **(i)** $-3, 6, 10$ **(j)** 3 **(k)** 3 times **(l)** Once
(m) 0, 4 **(n)** $-5, 8$ **11.** Not a function **13.** Function **(a)** Domain: $\{x | -\pi \leq x \leq \pi\}$; Range: $\{y | -1 \leq y \leq 1\}$ **(b)** $\left(-\frac{\pi}{2}, 0\right), \left(\frac{\pi}{2}, 0\right), (0, 1)$
(c) y-axis **15.** Not a function **17.** Function **(a)** Domain: $\{x | x > 0\}$; Range: all real numbers **(b)** $(1, 0)$ **(c)** None
19. Function **(a)** Domain: all real numbers; Range: $\{y | y \leq 2\}$ **(b)** $(-3, 0), (3, 0), (0, 2)$ **(c)** y-axis **21.** Function **(a)** Domain: all real numbers; Range: $\{y | y \geq -3\}$ **(b)** $(1, 0), (3, 0), (0, 9)$ **(c)** None **23.** **(a)** Yes **(b)** $f(-2) = 9; (-2, 9)$ **(c)** $0, \frac{1}{2}; (0, -1), \left(\frac{1}{2}, -1\right)$ **(d)** All real numbers
(e) $-\frac{1}{2}, 1$ **(f)** -1 **25.** **(a)** No **(b)** $f(4) = -3; (4, -3)$ **(c)** 14; $(14, 2)$ **(d)** $\{x | x \neq 6\}$ **(e)** -2 **(f)** $-\frac{1}{3}$ **27.** **(a)** Yes **(b)** $f(2) = \frac{8}{17}; \left(2, \frac{8}{17}\right)$
(c) $-1, 1; (-1, 1), (1, 1)$ **(d)** All real numbers **(e)** 0 **(f)** 0

29. **(a)** Approximately 10.4 ft high **(b)** Approximately 9.9 ft high

(c)

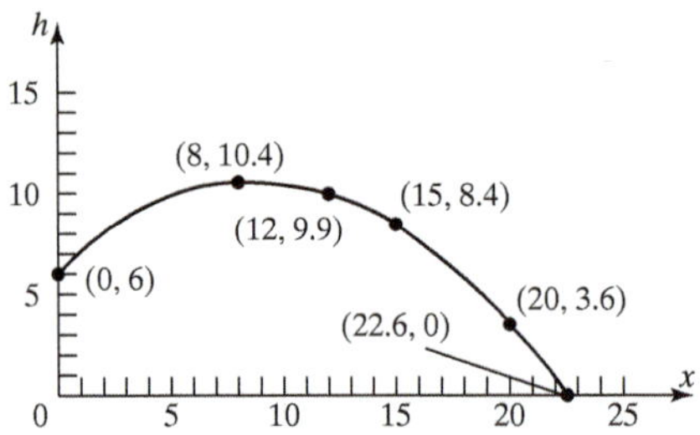

(d) The ball will not go through the hoop; $h(15) \approx 8.4$ ft.
If $v = 30$ ft/s, $h(15) = 10$ ft.

31. **(a)**

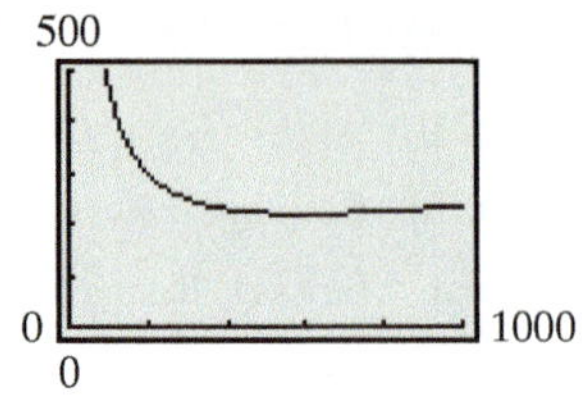

33. **(a)** 3 **(b)** -2 **(c)** -1 **(d)** 1 **(e)** 2 **(f)** $-\frac{1}{3}$ **35.** **(a)** \$80; it costs \$80 if you use 0 miniutes. **(b)** \$80; it costs \$80 if you use 1000 minutes. **(c)** \$210; it costs \$210 if you use 2000 minutes. **(d)** $\{m \mid 0 \le m \le 14{,}400\}$. There are at most 14,400 anytime minutes in a month. **37.** The x-intercepts can number anywhere from 0 to infinitely many. There is at most one y-intercept. **39.** **(a)** III **(b)** IV **(c)** I **(d)** V **(e)** II

41.

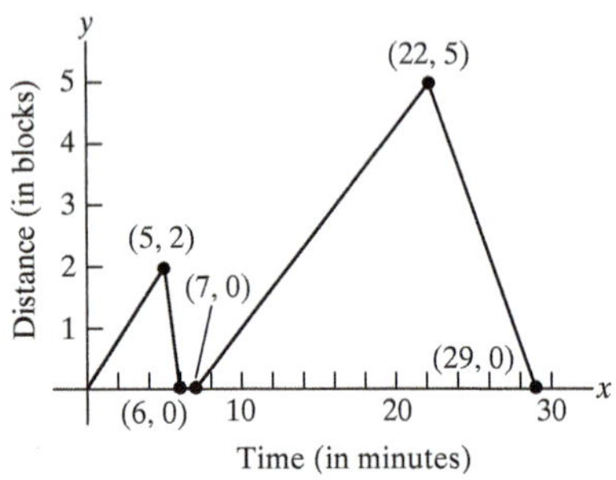

43. **(a)** 2 hr elapsed during which Kevin was between 0 and 3 mi from home. **(b)** 0.5 hr elapsed during which Kevin was 3 mi from home. **(c)** 0.3 hr elapsed during which Kevin was between 0 and 3 mi from home. **(d)** 0.2 hr elapsed during which Kevin was 0 mi from home. **(e)** 0.9 hr elapsed during which Kevin was between 0 and 2.8 mi from home. **(f)** 0.3 hr elapsed during which Kevin was 2.8 mi from home. **(g)** 1.1 hr elapsed during which Kevin was between 0 and 2.8 mi from home. **(h)** 3 mi **(i)** 2 times

45. No points whose x-coordinate is 5 or whose y-coordinate is 0 can be on the graph.

3.3 Assess Your Understanding *(page 242)*

6. increasing **7.** even; odd **8.** T **9.** T **10.** F **11.** Yes **13.** No **15.** $(-8, -2)$; $(0, 2)$; $(5, \infty)$ **17.** Yes; 10 **19.** $-2, 2$; $6, 10$

21. **(a)** $(-2, 0)$, $(0, 3)$, $(2, 0)$ **(b)** Domain: $\{x \mid -4 \le x \le 4\}$ or $[-4, 4]$; Range: $\{y \mid 0 \le y \le 3\}$ or $[0, 3]$ **(c)** Increasing on $(-2, 0)$ and $(2, 4)$; Decreasing on $(-4, -2)$ and $(0, 2)$ **(d)** Even **23.** **(a)** $(0, 1)$ **(b)** Domain: all real numbers; Range: $\{y \mid y > 0\}$ or $(0, \infty)$ **(c)** Increasing on $(-\infty, \infty)$ **(d)** Neither **25.** **(a)** $(-\pi, 0)$, $(0, 0)$, $(\pi, 0)$ **(b)** Domain: $\{x \mid -\pi \le x \le \pi\}$ or $[-\pi, \pi]$; Range: $\{y \mid -1 \le y \le 1\}$ or $[-1, 1]$ **(c)** Increasing on $\left(-\frac{\pi}{2}, \frac{\pi}{2}\right)$; Decreasing on $\left(-\pi, -\frac{\pi}{2}\right)$ and $\left(\frac{\pi}{2}, \pi\right)$ **(d)** Odd

27. **(a)** $\left(0, \frac{1}{2}\right)$, $\left(\frac{1}{3}, 0\right)$, $\left(\frac{5}{2}, 0\right)$ **(b)** Domain: $\{x \mid -3 \le x \le 3\}$ or $[-3, 3]$; Range: $\{y \mid -1 \le y \le 2\}$ or $[-1, 2]$ **(c)** Increasing on $(2, 3)$; Decreasing on $(-1, 1)$; Constant on $(-3, -1)$ and $(1, 2)$ **(d)** Neither

29. **(a)** 0; 3 **(b)** $-2, 2$; $0, 0$

31. **(a)** $\frac{\pi}{2}$; 1 **(b)** $-\frac{\pi}{2}$; -1 **33.** Odd **35.** Even **37.** Odd **39.** Neither **41.** Even **43.** Odd

45.

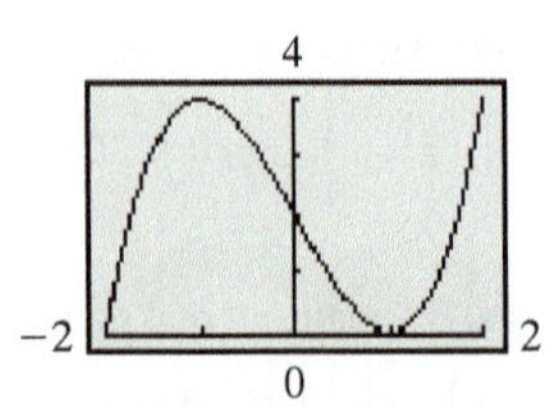

Increasing: $(-2, -1)$, $(1, 2)$
Decreasing: $(-1, 1)$
Local maximum: 4 at $x = -1$
Local minimum: 0 at $x = 1$

47.

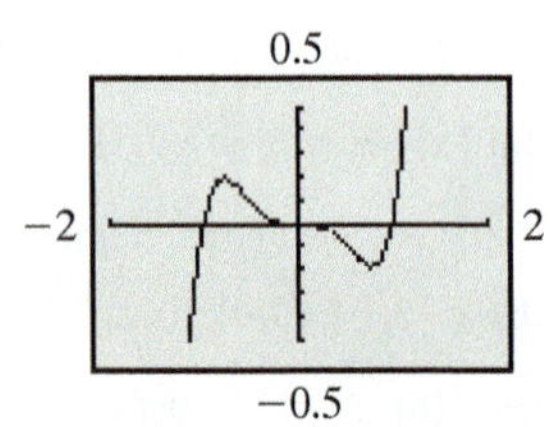

Increasing: $(-2, -0.77)$, $(0.77, 2)$
Decreasing: $(-0.77, 0.77)$
Local maximum: 0.19 at $x = -0.77$
Local minimum: -0.19 at $x = 0.77$

49.

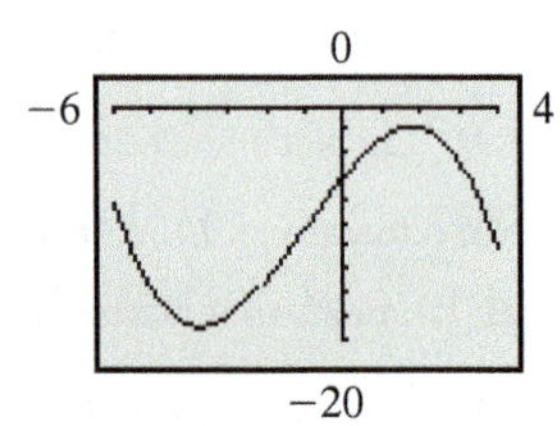

Increasing: $(-3.77, 1.77)$
Decreasing: $(-6, -3.77)$, $(1.77, 4)$
Local maximum: -1.91 at $x = 1.77$
Local minimum: -18.89 at $x = -3.77$

51.

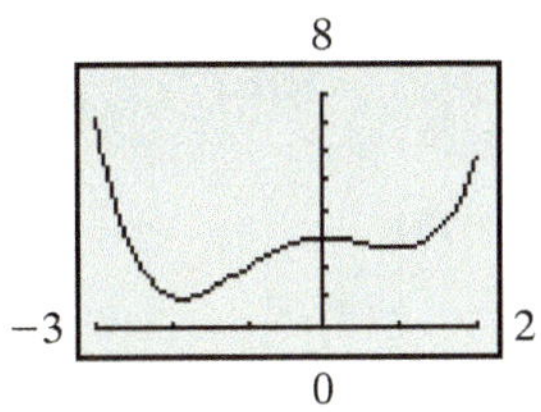

Increasing: $(-1.87, 0)$, $(0.97, 2)$
Decreasing: $(-3, -1.87)$, $(0, 0.97)$
Local maximum: 3 at $x = 0$
Local minimum: 0.95 at $x = -1.87$
Local minimum: 2.65 at $x = 0.97$

53. **(a)** -4 **(b)** -8 **(c)** -10 **55.** **(a)** 17 **(b)** -1 **(c)** 11
57. **(a)** 5 **(b)** $y = 5x - 2$ **59.** **(a)** -1; **(b)** $y = -x$
61. **(a)** 4 **(b)** $y = 4x - 8$

63. **(a)** Odd

(b)

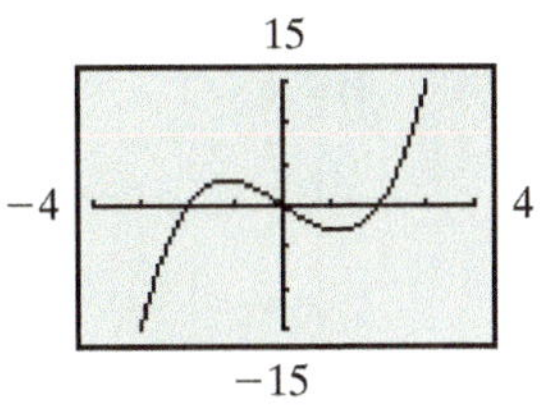

(c) Local minimum: -3.08 at $x = 1.15$
(d) Local maximum: 3.08 at $x = -1.15$

65. **(a)** Even

(b)

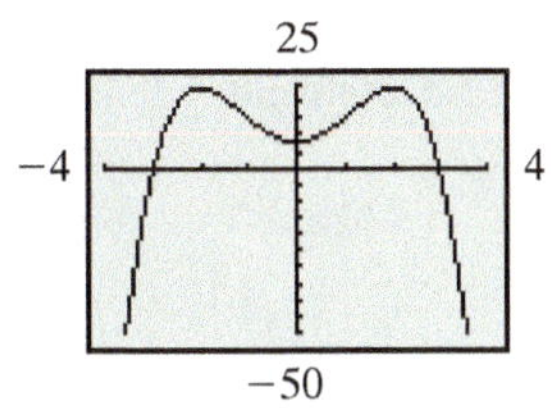

(c) Local maximum: 24 at $x = 2$
(d) Local maximum: 24 at $x = -2$
(e) 47.4 sq. units

67. **(a)**

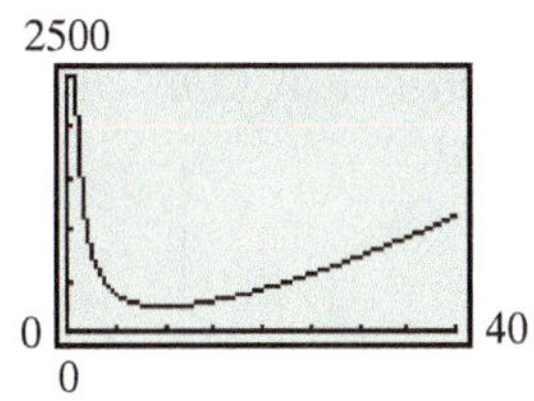

(b) 10 riding lawn mowers/hr
(c) \$239/mower

69. **(a)** On average, the population is increasing at a rate of 0.036 g/hr from 0 to 2.5 hr.
(b) On average, from 4.5 to 6 hr, the population is increasing at a rate of 0.1 g/hr. **(c)** The average rate of change is increasing over time.

71. **(a)** 1 **(b)** 0.5 **(c)** 0.1 **(d)** 0.01 **(e)** 0.001
(f)

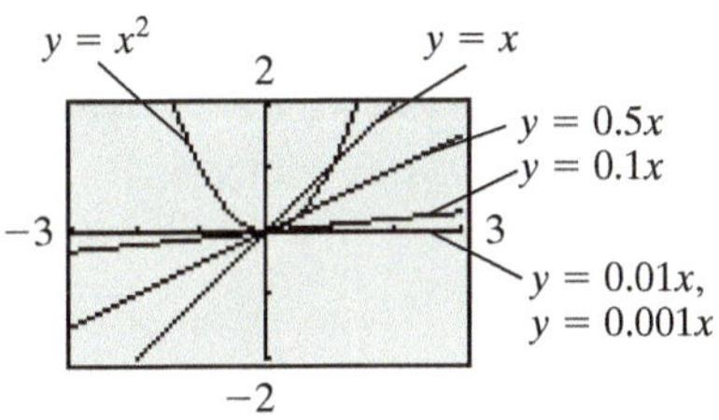

(g) They are getting closer to the tangent line at $(0, 0)$.
(h) They are getting closer to 0.

73. **(a)** 2 **(b)** 2; 2; 2; 2
(c) $y = 2x + 5$
(d)

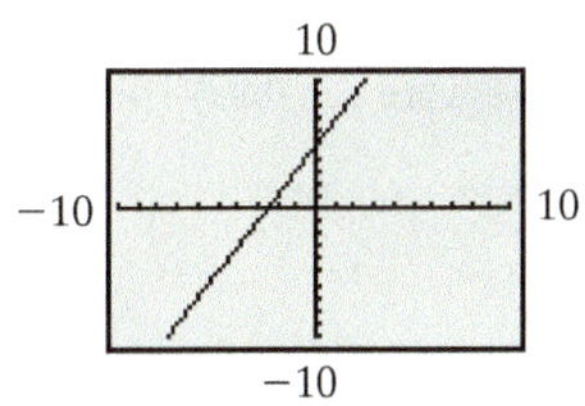

75. **(a)** $2x + h + 2$ **(b)** 4.5; 4.1; 4.01; 4
(c) $y = 4.01x - 1.01$
(d)

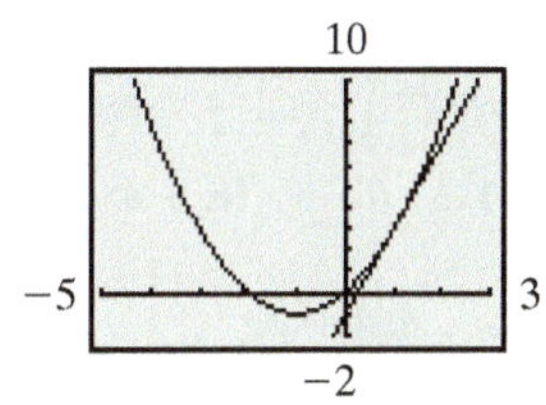

77. **(a)** $4x + 2h - 3$ **(b)** 2; 1.2; 1.02; 1
(c) $y = 1.02x - 1.02$
(d)

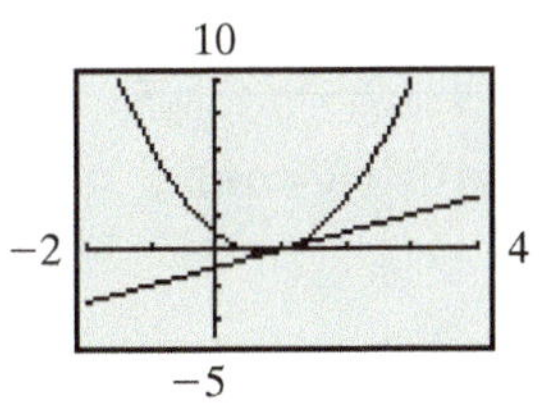

79. **(a)** $-\dfrac{1}{(x + h)x}$ **(b)** $-\dfrac{2}{3}; -\dfrac{10}{11}; -\dfrac{100}{101}; -1$ **(c)** $y = -\dfrac{100}{101}x + \dfrac{201}{101}$

(d)

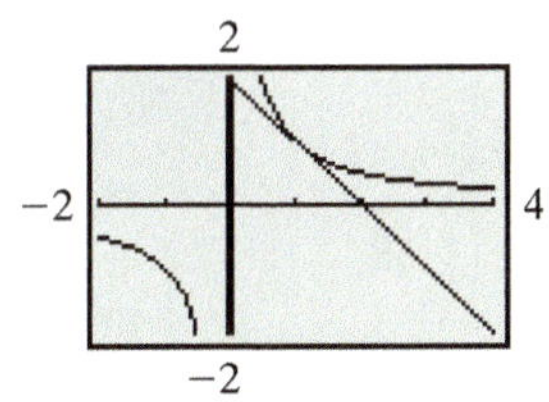

83. At most one **85.** Yes; the function $f(x) = 0$ is both even and odd.

3.4 Assess Your Understanding *(page 252)*

4. $(-\infty, 0)$ **5.** piecewise-defined **6.** T **7.** F **8.** F **9.** C **11.** E **13.** B **15.** F

17.

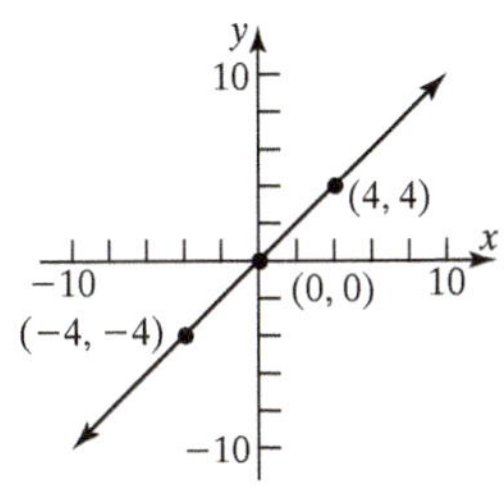

19.

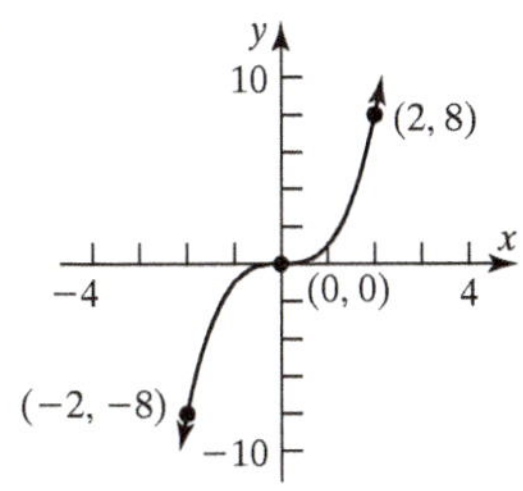

21.

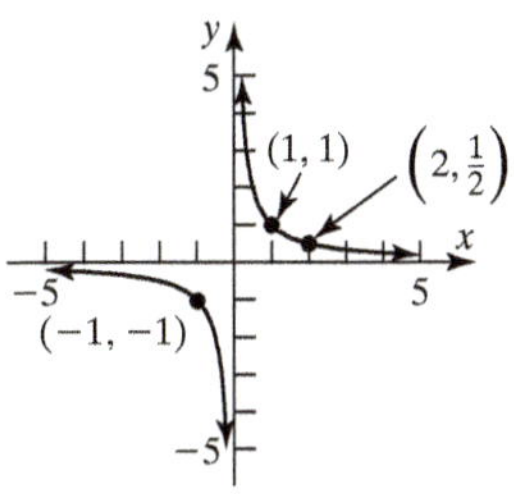

23.

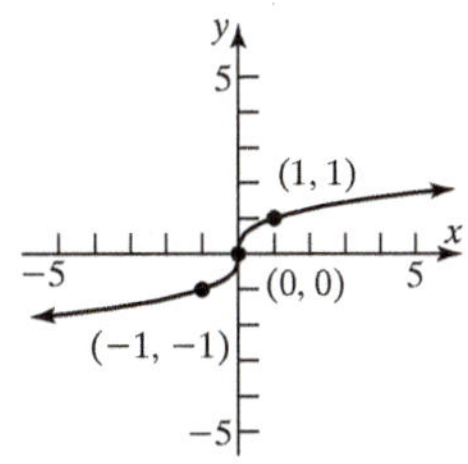

25. **(a)** 4 **(b)** 2 **(c)** 5 **27.** **(a)** −4 **(b)** −2 **(c)** 0 **(d)** 25

29. **(a)** All real numbers
(b) (0, 1)
(c)

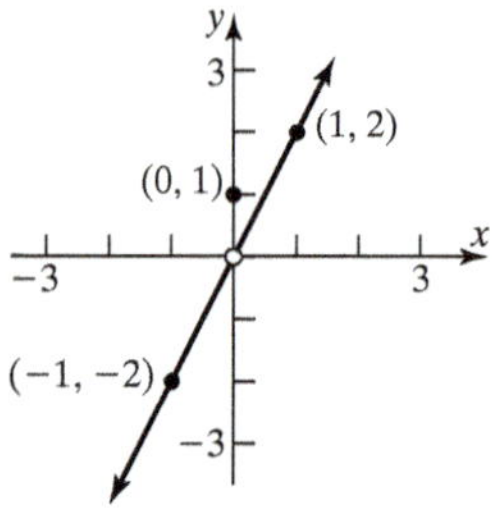

(d) $\{y | y \neq 0\}; (-\infty, 0) \cup (0, \infty)$
(e) Discontinuous at $x = 0$

31. **(a)** All real numbers
(b) (0, 3)
(c)

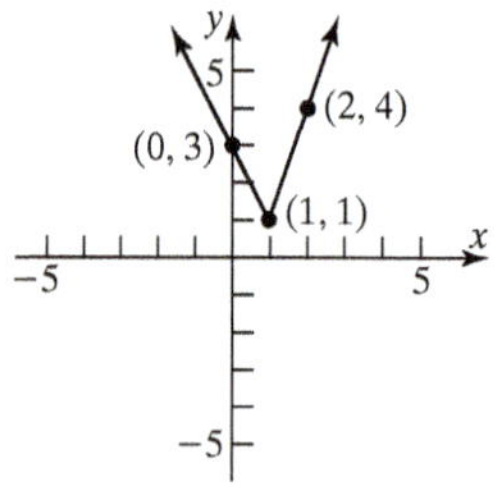

(d) $\{y | y \geq 1\}; [1, \infty)$
(e) Continuous

33. **(a)** $\{x | x \geq -2\}; [-2, \infty)$
(b) (0, 3), (2, 0)
(c)

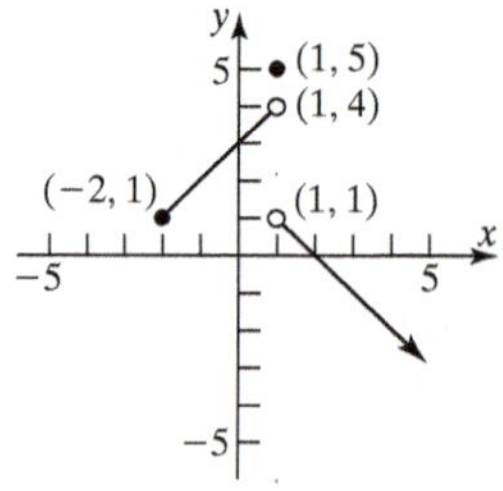

(d) $\{y | y < 4, y = 5\}; (-\infty, 4) \cup \{5\}$
(e) Discontinuous at $x = 1$

35. **(a)** All real numbers
(b) (−1, 0), (0, 0)
(c)

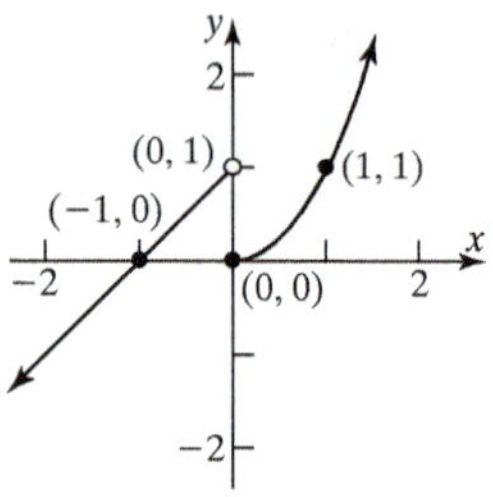

(d) All real numbers
(e) Discontinuous at $x = 0$

37. **(a)** $\{x | x \geq -2, x \neq 0\}; [-2, 0) \cup (0, \infty)$
(b) No intercepts
(c)

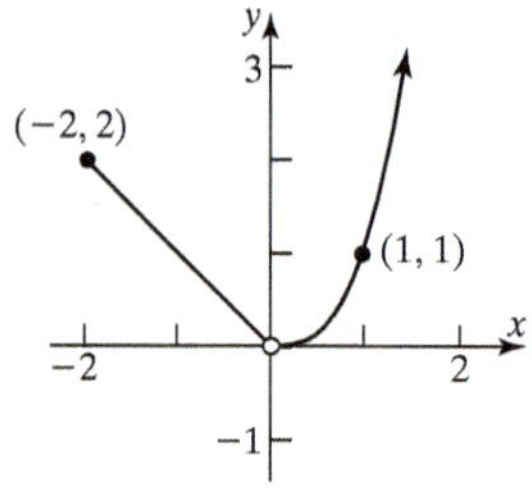

(d) $\{y | y > 0\}; (0, \infty)$
(e) Discontinuous at $x = 0$

39. **(a)** All real numbers
(b) $(x, 0)$ for $0 \leq x < 1$
(c)

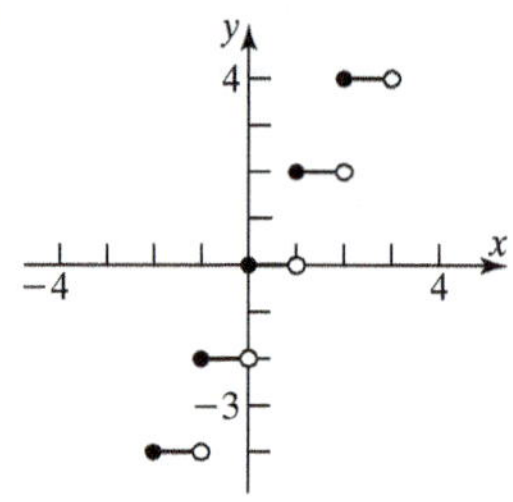

(d) Set of even integers
(e) Discontinuous at $\{x | x \text{ is an integer}\}$

41. $f(x) = \begin{cases} -x & \text{if } -1 \leq x \leq 0 \\ \frac{1}{2}x & \text{if } 0 < x \leq 2 \end{cases}$ (Other answers are possible.)

43. $f(x) = \begin{cases} -x & \text{if } x \leq 0 \\ -x + 2 & \text{if } 0 < x \leq 2 \end{cases}$ (Other answers are possible.)

45. **(a)** 2 **(b)** 3 **(c)** −4 **47.** **(a)** \$39.99 **(b)** \$46.74 **(c)** \$40.44

49. **(a)** \$71.33 **(b)** \$516.04

(c) $C(x) = \begin{cases} 1.23755x + 9.45 & \text{if } 0 \leq x \leq 50 \\ 0.98825x + 21.915 & \text{if } x > 50 \end{cases}$

(d)

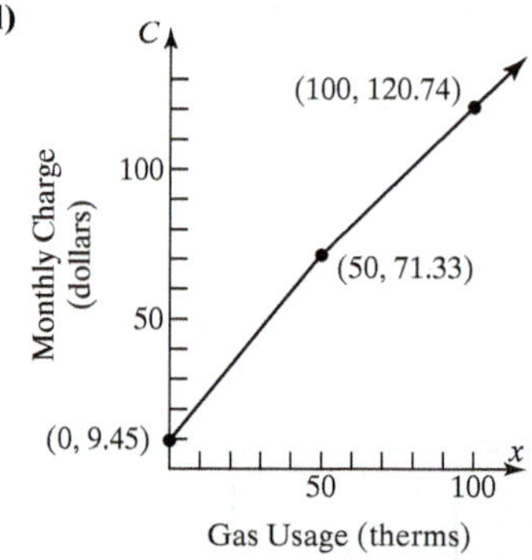

51. For schedule X:

$$f(x) = \begin{cases} 0.10x & \text{if } 0 < x \leq 7825 \\ 782.50 + 0.15(x - 7825) & \text{if } 7825 < x \leq 31{,}850 \\ 4386.25 + 0.25(x - 31{,}850) & \text{if } 31{,}850 < x \leq 77{,}100 \\ 15{,}698.75 + 0.28(x - 77{,}100) & \text{if } 77{,}100 < x \leq 160{,}850 \\ 39{,}148.75 + 0.33(x - 160{,}850) & \text{if } 160{,}850 < x \leq 349{,}700 \\ 101{,}469.25 + 0.35(x - 349{,}700) & \text{if } x > 349{,}700 \end{cases}$$

53. **(a)**

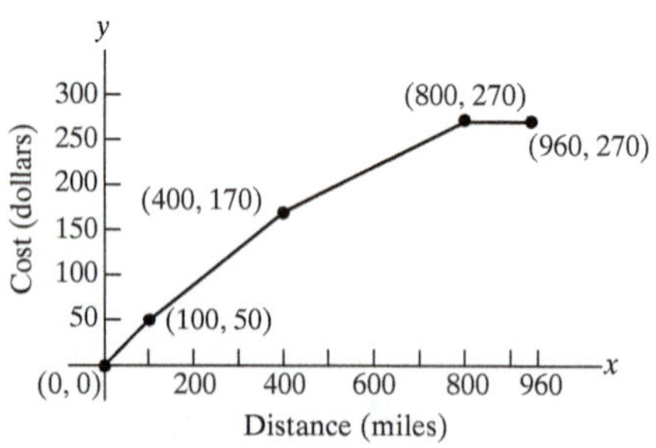

(b) $C(x) = 10 + 0.4x$

(c) $C(x) = 70 + 0.25x$

55. $f(x) = \begin{cases} x & \text{if } 0 \leq x < 10 \\ 10 & \text{if } 10 \leq x < 500 \\ 30 & \text{if } 500 \leq x < 1000 \\ 50 & \text{if } 1000 \leq x < 1500 \\ 70 & \text{if } 1500 \leq x \end{cases}$

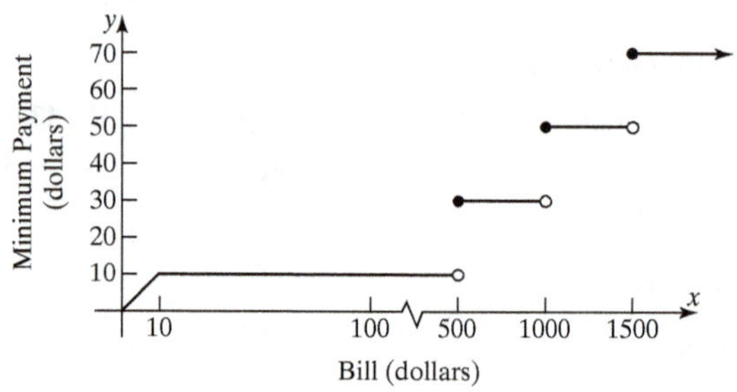

57. **(a)** 10°C

(b) 4°C

(c) −3°C

(d) −4°C

(e) The wind chill is equal to the air temperature.

(f) At wind speed greater than 20 m/s, the wind chill factor depends only on the air temperature.

59.
$$C(x) = \begin{cases} 0.80 & 0 < x \le 1 \\ 0.97 & 1 < x \le 2 \\ 1.14 & 2 < x \le 3 \\ 1.31 & 3 < x \le 4 \\ 1.48 & 4 < x \le 5 \\ 1.65 & 5 < x \le 6 \\ 1.82 & 6 < x \le 7 \\ 1.99 & 7 < x \le 8 \\ 2.16 & 8 < x \le 9 \\ 2.33 & 9 < x \le 10 \\ 2.50 & 10 < x \le 11 \\ 2.67 & 11 < x \le 12 \\ 2.84 & 12 < x \le 13 \end{cases}$$

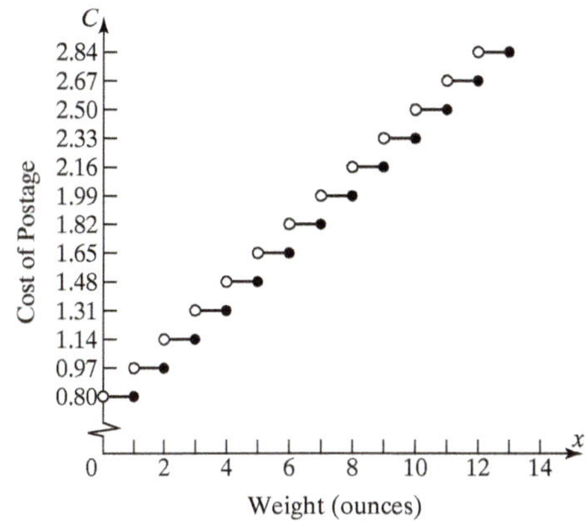

61. Each graph is that of $y = x^2$, but shifted horizontally. If $y = (x - k)^2$, $k > 0$, the shift is right k units; if $y = (x + k)^2$, $k > 0$, the shift is left k units. **63.** The graph of $y = -f(x)$ is the reflection about the x-axis of the graph of $y = f(x)$. **65.** Yes. The graph of $y = (x - 1)^3 + 2$ is the graph of $y = x^3$ shifted right 1 unit and up 2 units. **67.** They all have the same general shape. All three go through the points $(-1, -1)$, $(0, 0)$, and $(1, 1)$. As the exponent increases, the steepness of the curve increases (except near $x = 0$).

3.5 Assess Your Understanding *(page 264)*

1. horizontal; right **2.** y **3.** −5; −2; 2 **4.** T **5.** F **6.** T **7.** B **9.** H **11.** I **13.** L **15.** F **17.** G **19.** $y = (x - 4)^3$ **21.** $y = x^3 + 4$ **23.** $y = -x^3$ **25.** $y = 4x^3$ **27.** (1) $y = \sqrt{x} + 2$; (2) $y = -(\sqrt{x} + 2)$; (3) $y = -(\sqrt{-x} + 2)$ **29.** (1) $y = -\sqrt{x}$; (2) $y = -\sqrt{x} + 2$; (3) $y = -\sqrt{x + 3} + 2$ **31.** (c) **33.** (c) **35.** **(a)** −7 and 1 **(b)** −3 and 5 **(c)** −5 and 3 **(d)** −3 and 5 **37** **(a)** $(-3, 3)$ **(b)** $(4, 10)$ **(c)** Decreasing on $(-1, 5)$ **(d)** Decreasing on $(-5, 1)$

39.

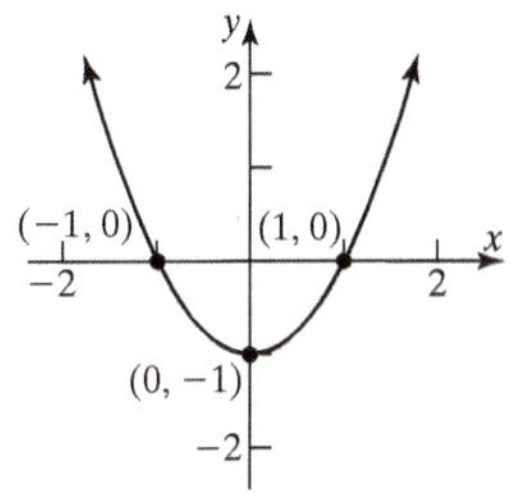

Domain: $(-\infty, \infty)$; Range: $[-1, \infty)$

41.

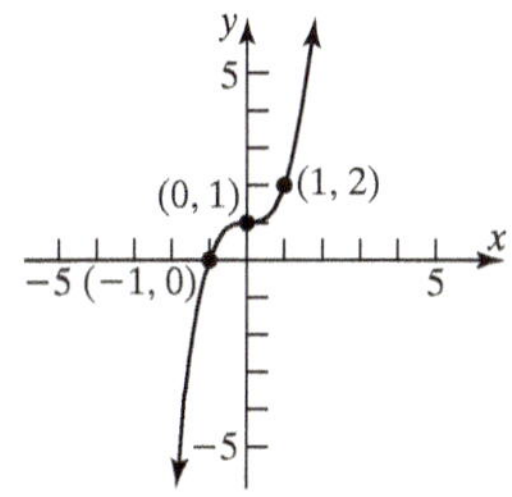

Domain: $(-\infty, \infty)$; Range: $(-\infty, \infty)$

43.

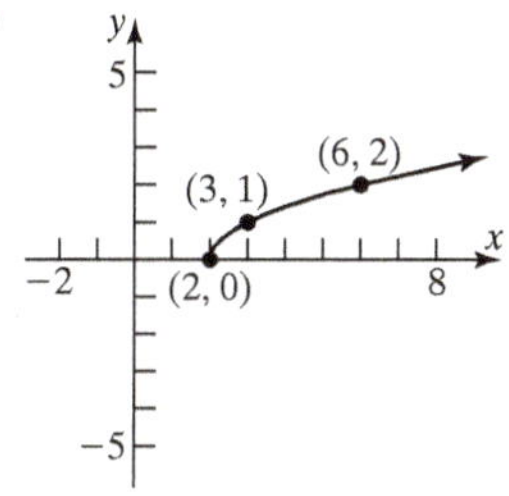

Domain: $[2, \infty)$; Range: $[0, \infty)$

45.

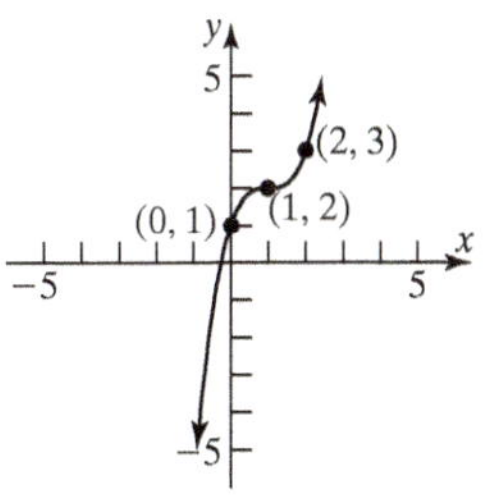

Domain: $(-\infty, \infty)$; Range: $(-\infty, \infty)$

47.

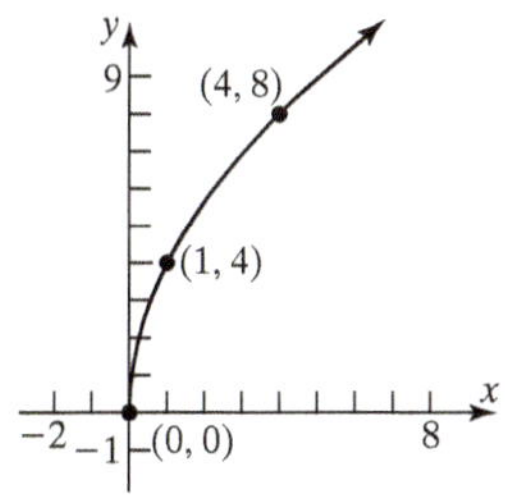

Domain: $[0, \infty)$; Range: $[0, \infty)$

49.

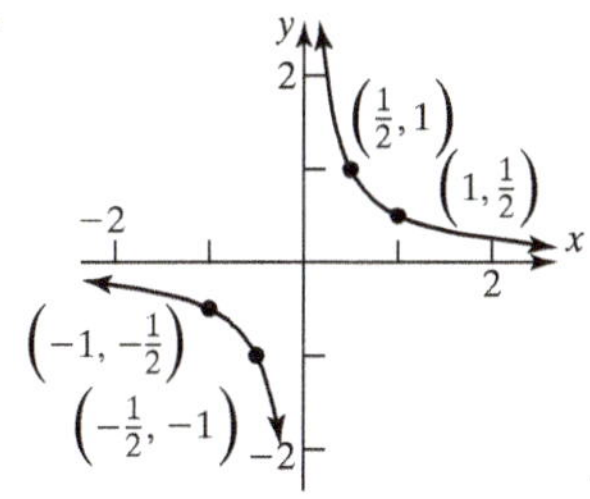

Domain: $(-\infty, 0) \cup (0, \infty)$

Range: $(-\infty, 0) \cup (0, \infty)$

51.

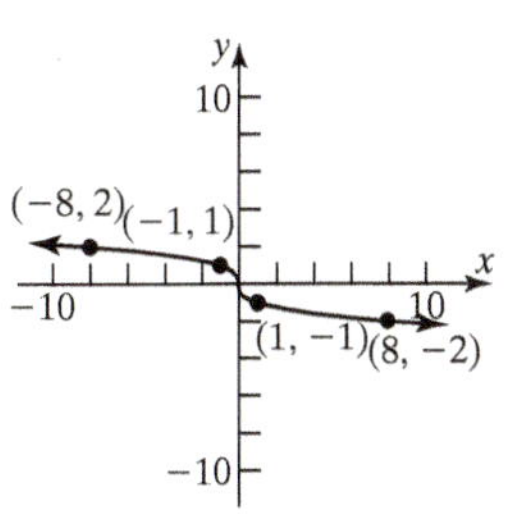

Domain: $(-\infty, \infty)$; Range: $(-\infty, \infty)$

53.

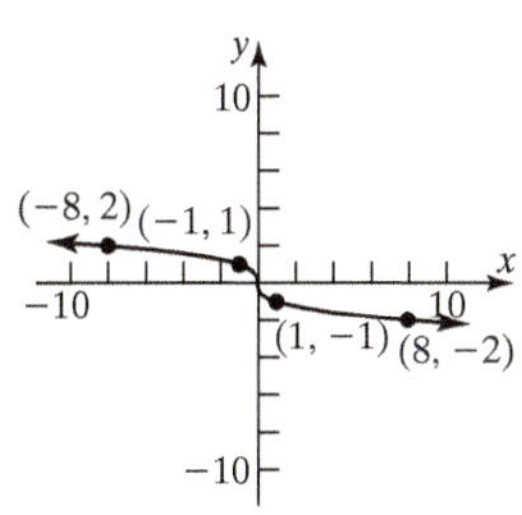

Domain: $(-\infty, \infty)$; Range: $(-\infty, \infty)$

55.

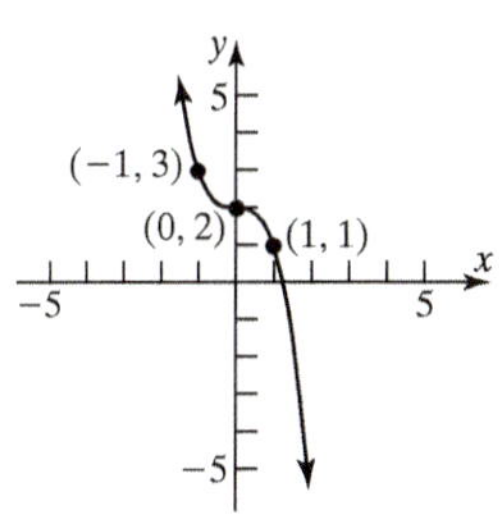

Domain: $(-\infty, \infty)$; Range: $(-\infty, \infty)$

57.

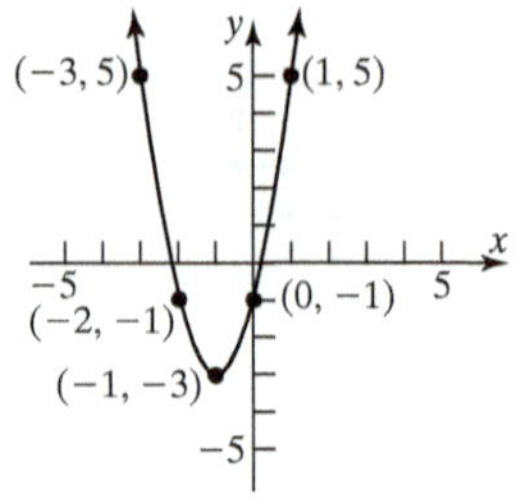

Domain: $(-\infty, \infty)$; Range: $[-3, \infty)$

59.

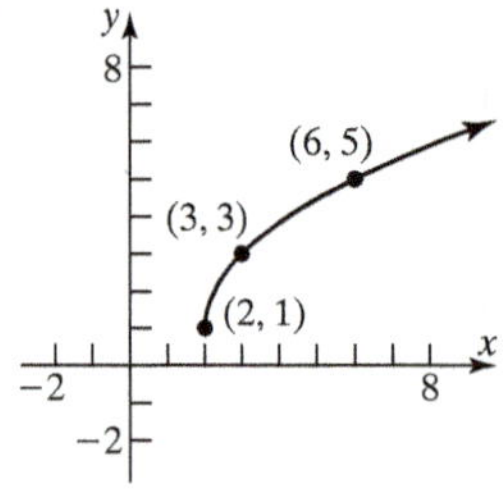

Domain: $[2, \infty)$; Range: $[1, \infty)$

61.

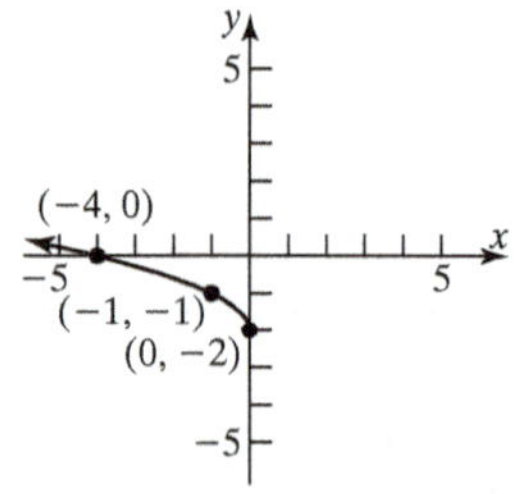

Domain: $(-\infty, 0]$; Range: $[-2, \infty)$

63.

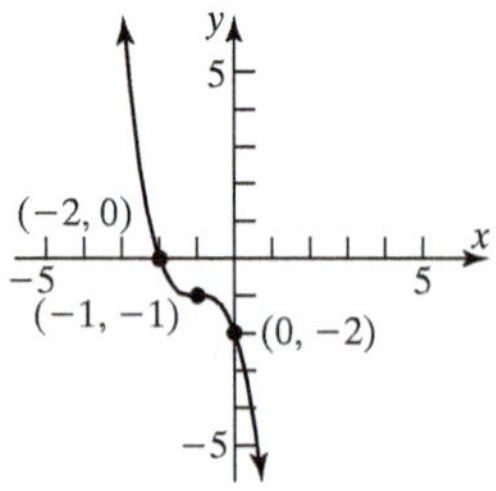

Domain: $(-\infty, \infty)$; Range: $(-\infty, \infty)$

65.

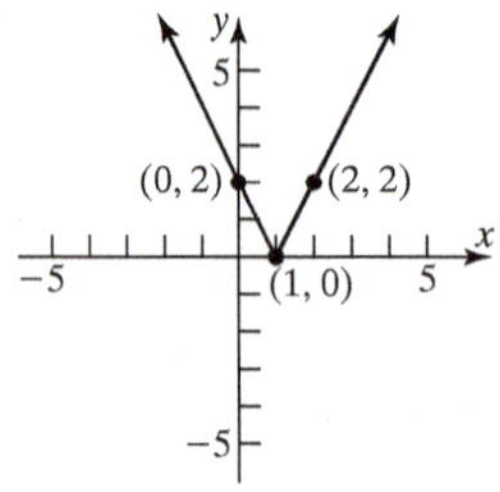

Domain: $(-\infty, \infty)$; Range: $[0, \infty)$

67.

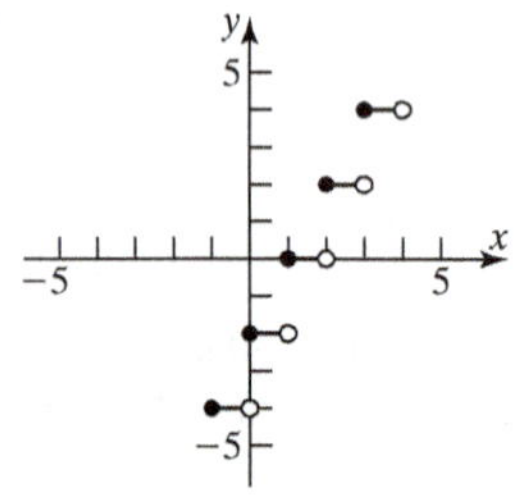

Domain: $(-\infty, \infty)$;

Range: $\{y \mid y \text{ is an even integer}\}$

69. **(a)** $F(x) = f(x) + 3$

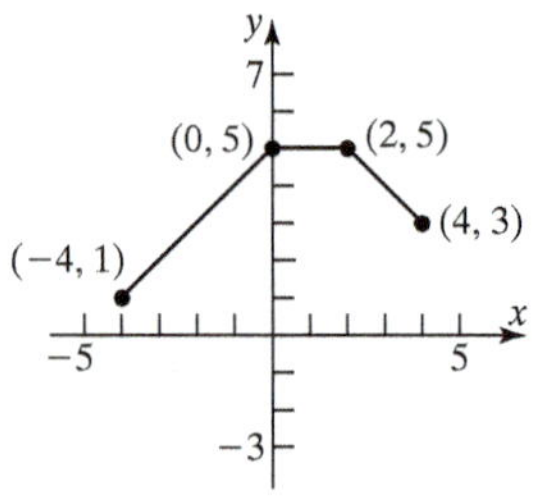

(b) $G(x) = f(x + 2)$

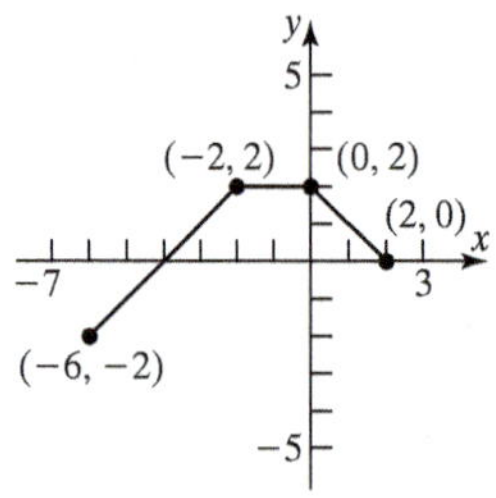

(c) $P(x) = -f(x)$

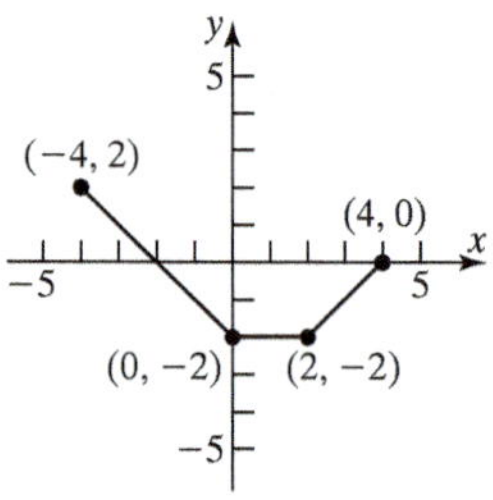

(d) $H(x) = f(x + 1) - 2$

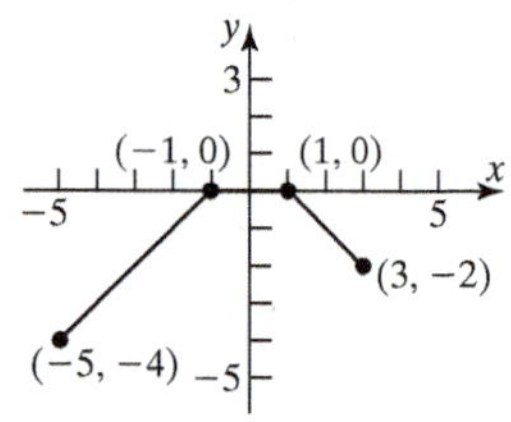

(e) $Q(x) = \frac{1}{2}f(x)$

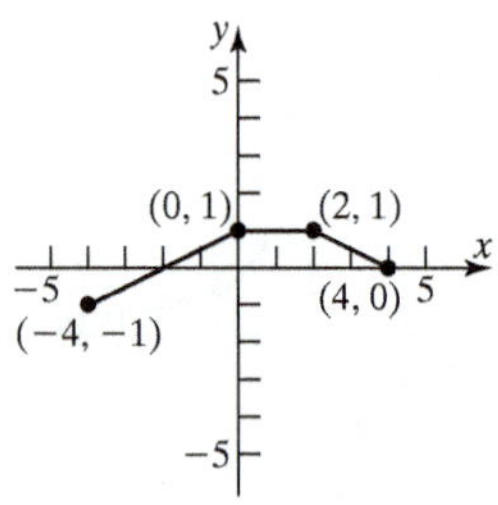

(f) $g(x) = f(-x)$

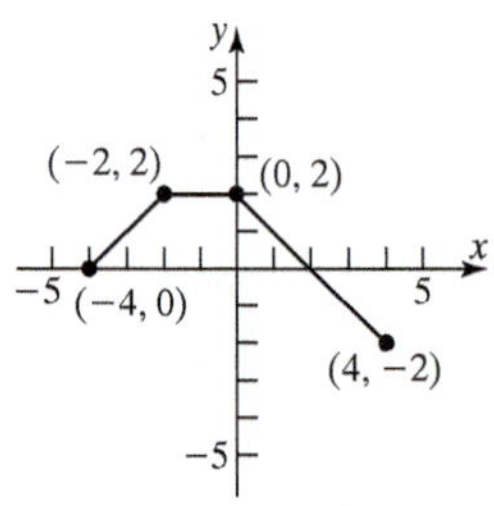

(g) $h(x) = f(2x)$

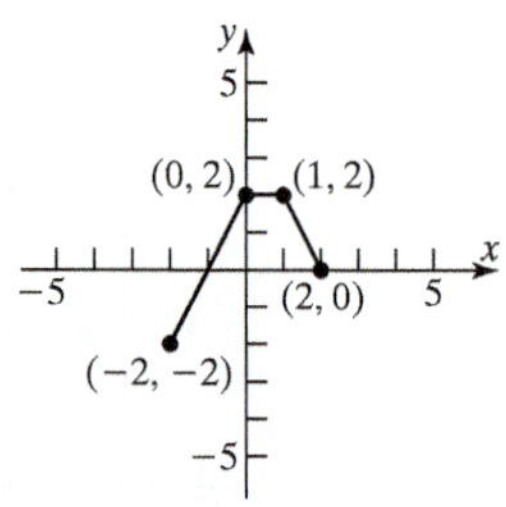

71. **(a)** $F(x) = f(x) + 3$

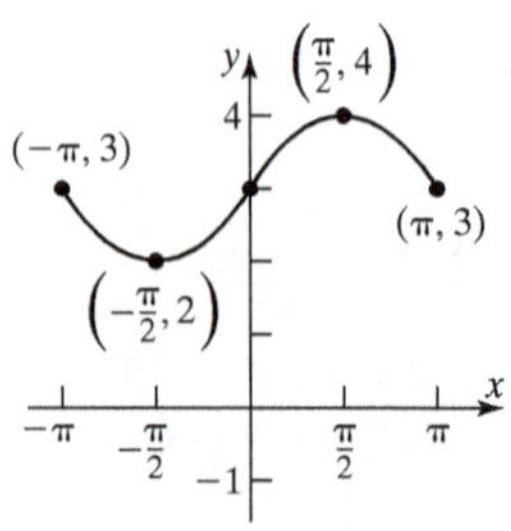

(b) $G(x) = f(x + 2)$

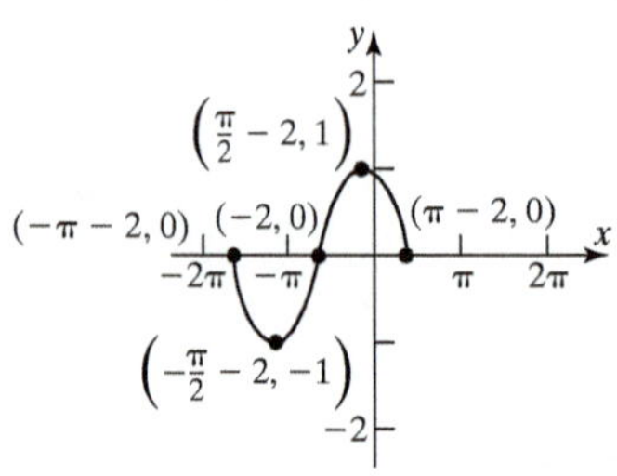

(c) $P(x) = -f(x)$

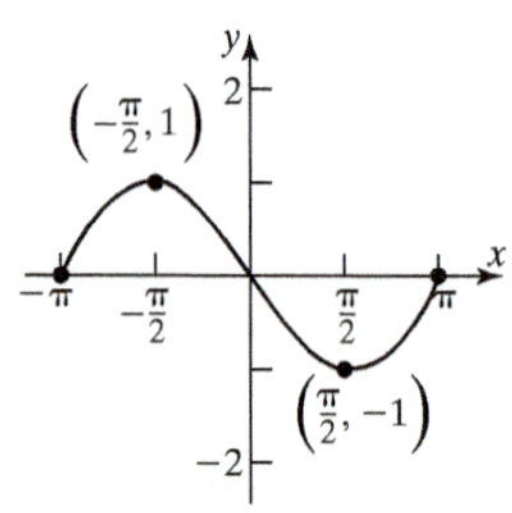

(d) $H(x) = f(x + 1) - 2$

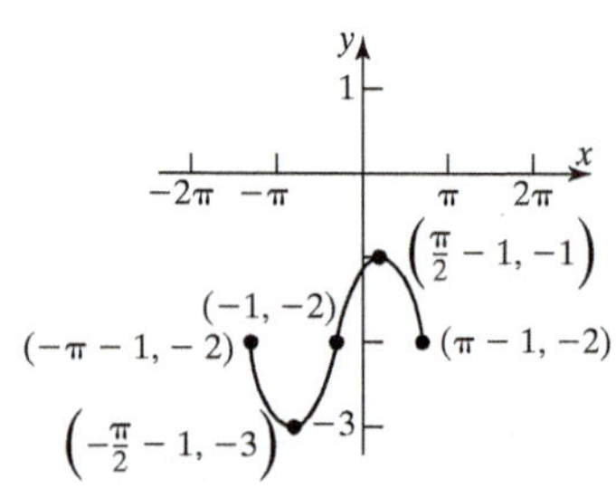

(e) $Q(x) = \frac{1}{2}f(x)$

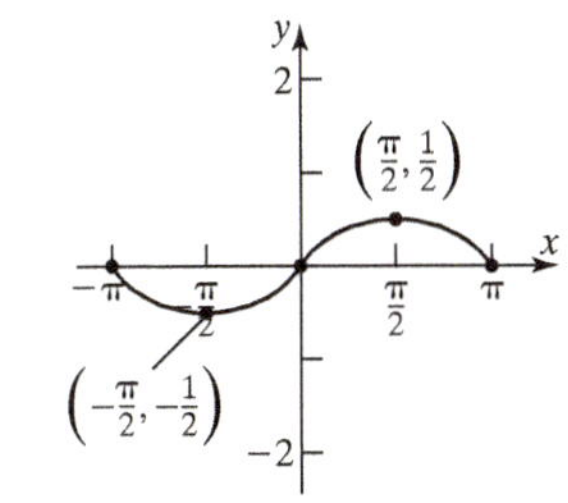

(f) $g(x) = f(-x)$

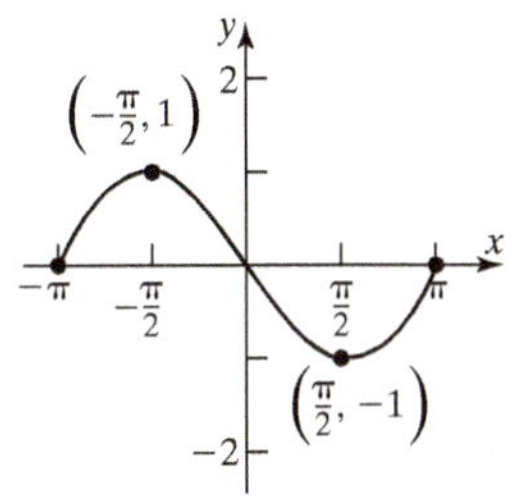

(g) $h(x) = f(2x)$

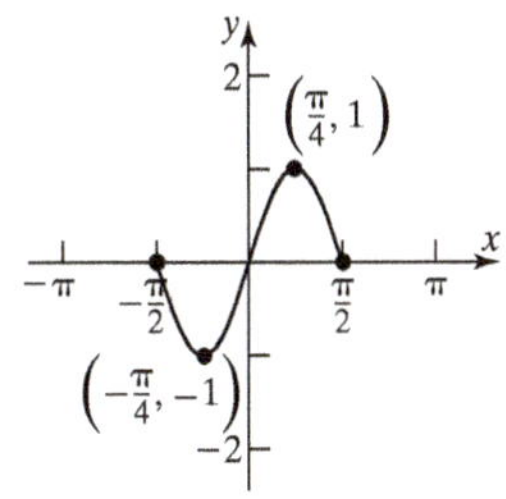

73. (a)

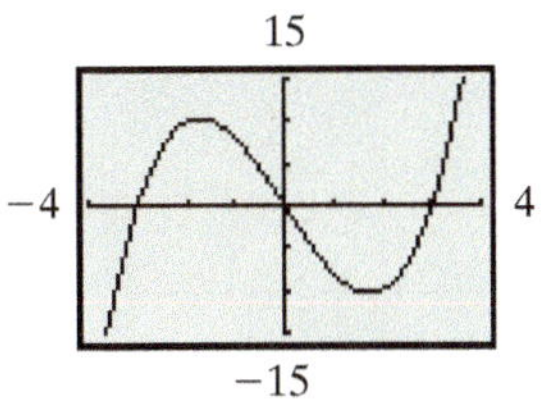

(b) $-3, 0, 3$

(c) Local maximum: 10.39 at $x = -1.73$
Local minimum: -10.39 at $x = 1.73$

(d) Increasing: $(-4, -1.73)$, $(1.73, 4)$
Decreasing: $(-1.73, 1.73)$

(e) Intercepts: $-5, -2, 1$;
Local maximum: 10.39 at $x = -3.73$;
Local minimum: -10.39 at $x = -0.27$;
Increasing: $(-6, -3.73)$, $(-0.27, 2)$;
Decreasing: $(-3.73, -0.27)$

(f) Intercepts: $-3, 0, 3$;
Local maximum: 20.78 at $x = -1.73$;
Local minimum: -20.78 at $x = 1.73$;
Increasing: $(-4, -1.73)$, $(1.73, 4)$;
Decreasing: $(-1.73, 1.73)$

(g) Intercepts: $-3, 0, 3$;
Local maximum: 10.39 at $x = 1.73$;
Local minimum: -10.39 at $x = -1.73$;
Decreasing: $(-4, -1.73)$, $(1.73, 4)$;
Increasing: $(-1.73, 1.73)$

75. $f(x) = (x + 1)^2 - 1$

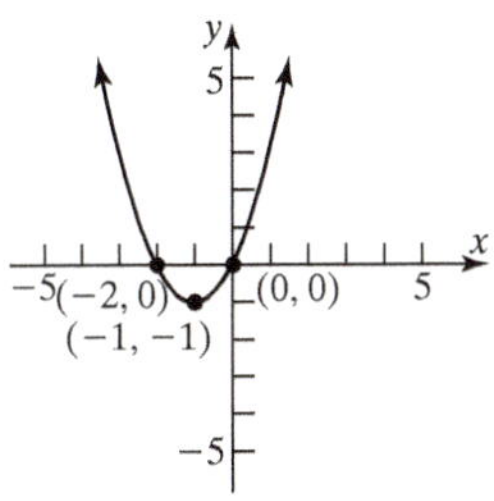

77. $f(x) = (x - 4)^2 - 15$

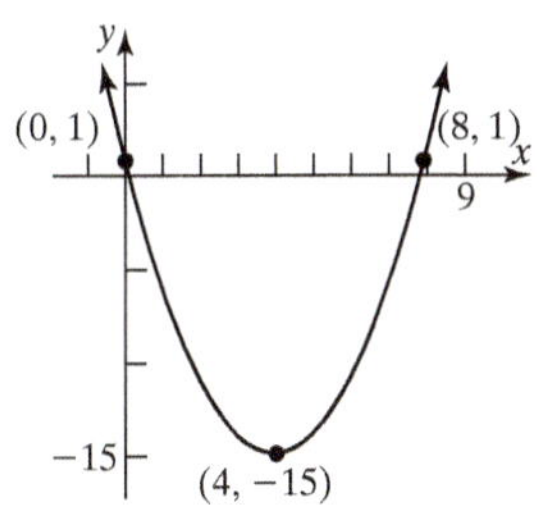

79. $f(x) = \left(x + \frac{1}{2}\right)^2 + \frac{3}{4}$

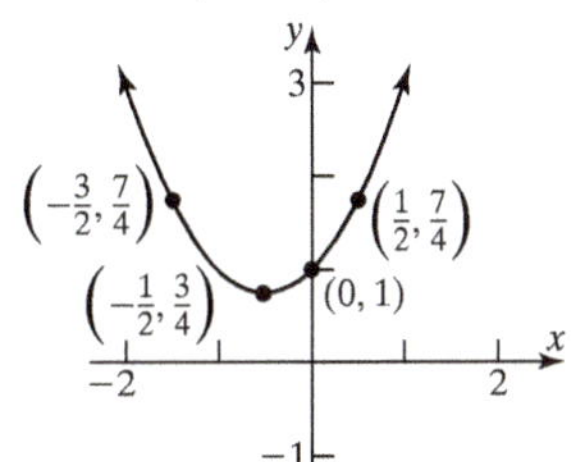

81. $f(x) = 2(x - 3)^2 + 1$

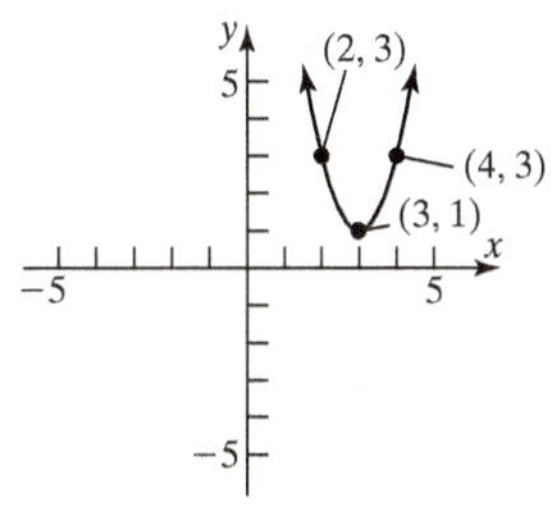

83. $f(x) = -3(x + 2)^2 - 5$

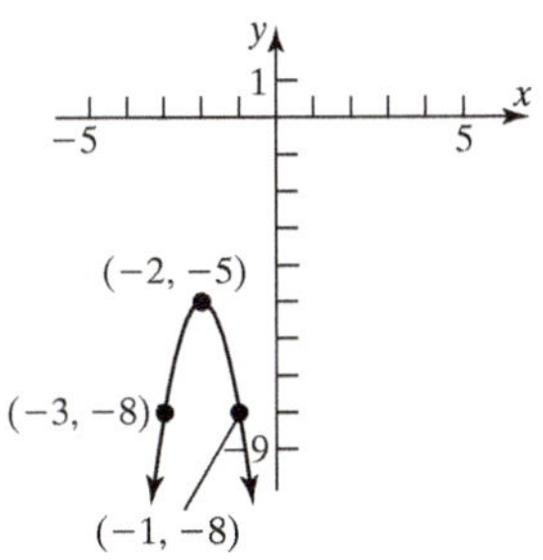

85.

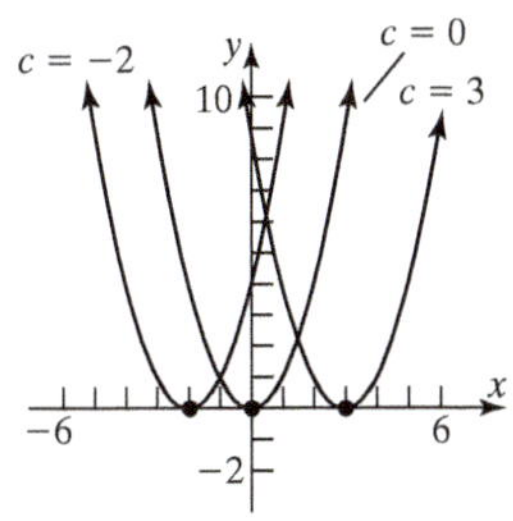

87. (a) 72°F; 65°F

(b) The temperature decreases by 2° to 70°F during the day and 63°F overnight

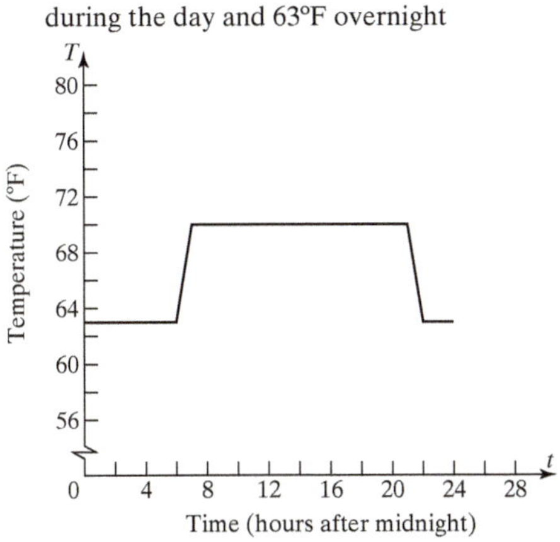

(c) The time at which the temperature adjusts between the daytime and overnight settings is moved to 1 hr sooner. It begins warming up at 5:00 AM instead of 6:00 AM, and it begins cooling down at 8:00 PM in stead of 9:00 PM.

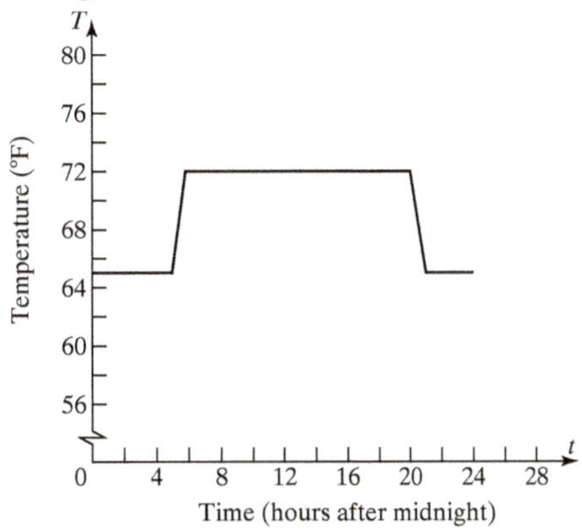

89.

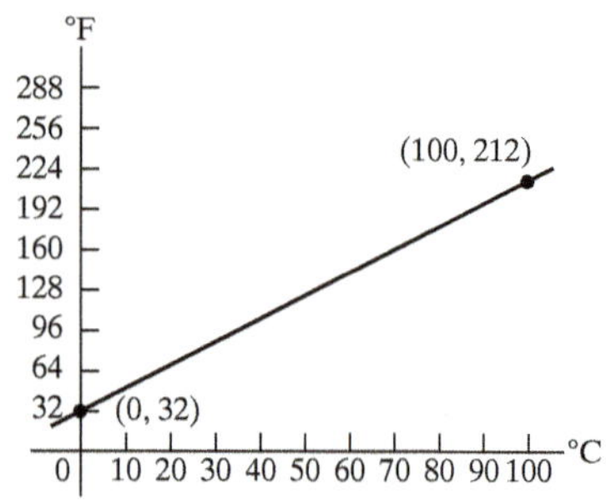

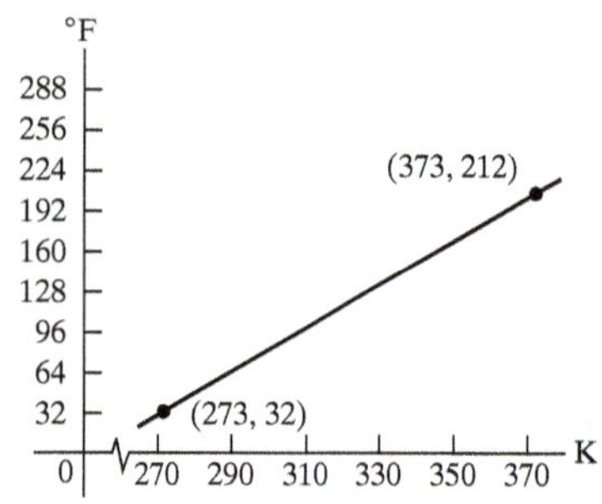

91. **(a)**

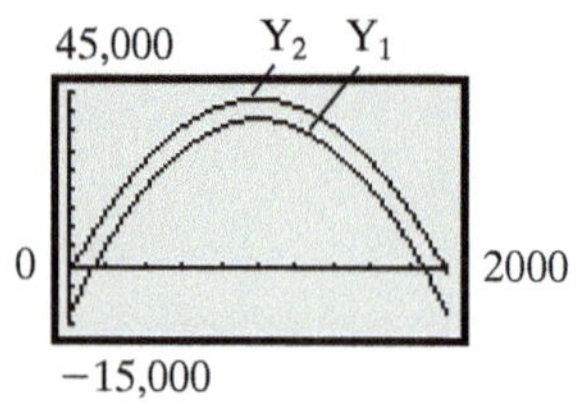

(b) 10% tax

(c) Y_1 is the graph of $p(x)$ shifted down vertically 10,000 units.
Y_2 is the graph of $p(x)$ vertically compressed by a factor of 0.9.

(d) 10% tax

93. **(a)**

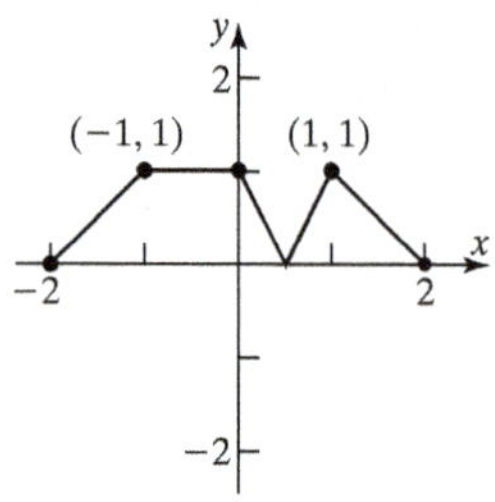

(b)

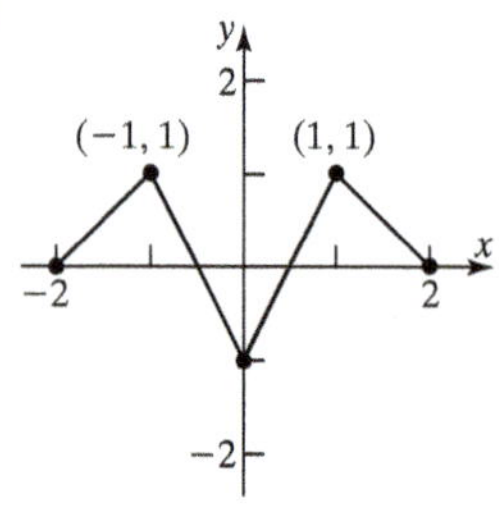

95. **(a)** $(-4, 2)$ **(b)** $(1, -12)$ **(c)** $(-4, 5)$ **97.** $\frac{32}{3}$ square units

3.6 Assess Your Understanding *(page 271)*

1. **(a)** $d(x) = \sqrt{x^4 - 15x^2 + 64}$

(b) $d(0) = 8$

(c) $d(1) = \sqrt{50} \approx 7.07$

(d)

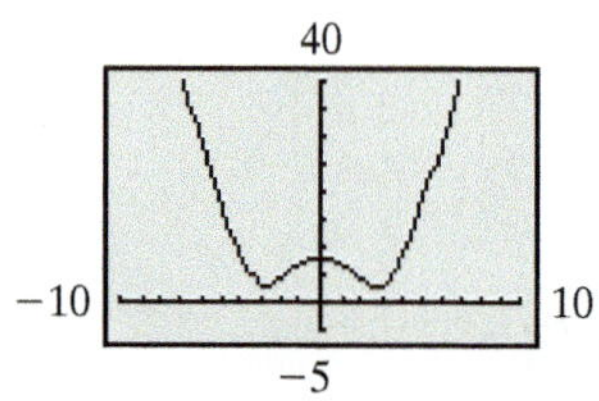

(e) d is smallest when $x \approx -2.74$ or $x \approx 2.74$.

3. **(a)** $d(x) = \sqrt{x^2 - x + 1}$

(b)

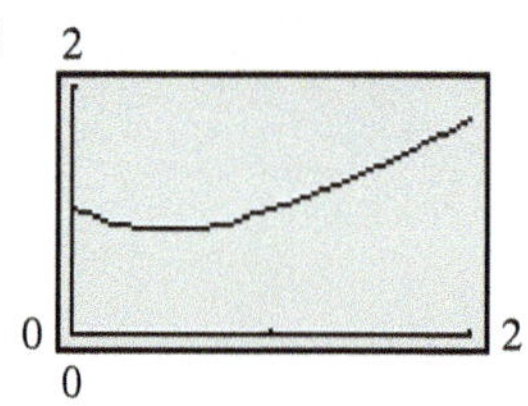

(c) d is smallest when $x = 0.5$.

5. $A(x) = \frac{1}{2}x^4$

7. **(a)** $A(x) = x(16 - x^2)$

(b) Domain: $\{x \mid 0 < x < 4\}$

(c) The area is largest when $x \approx 2.31$.

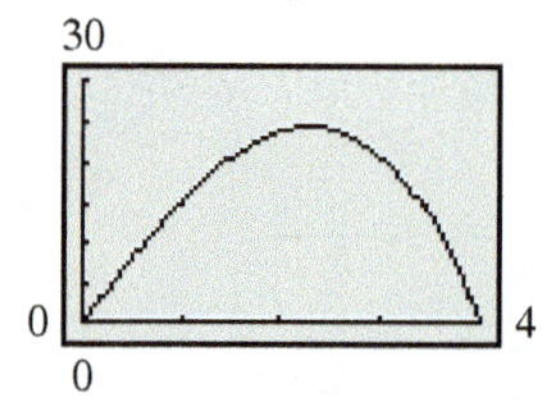

9. **(a)** $A(x) = 4x\sqrt{4 - x^2}$

(b) $p(x) = 4x + 4\sqrt{4 - x^2}$

(c) A is largest when $x \approx 1.41$.

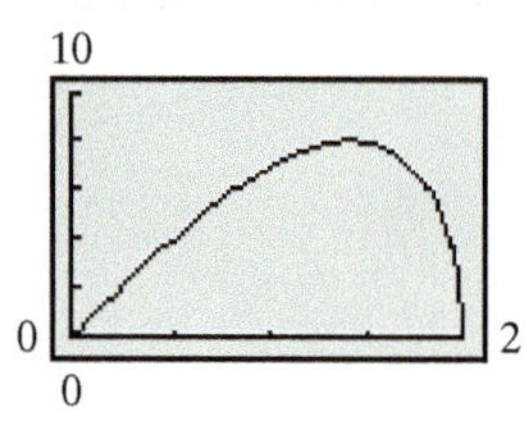

(d) p is largest when $x \approx 1.41$.

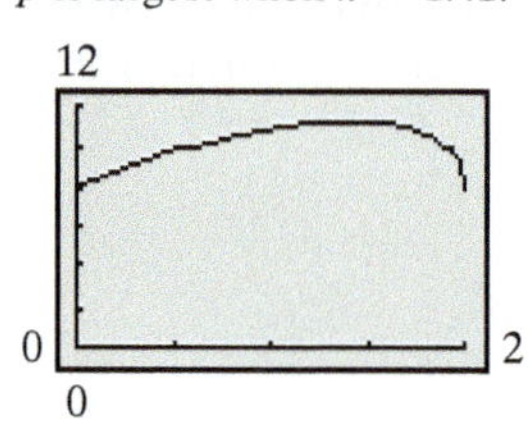

11. **(a)** $A(x) = x^2 + \frac{25 - 20x + 4x^2}{\pi}$

(b) Domain: $\{x \mid 0 < x < 2.5\}$

(c) A is smallest when $x \approx 1.40$ m.

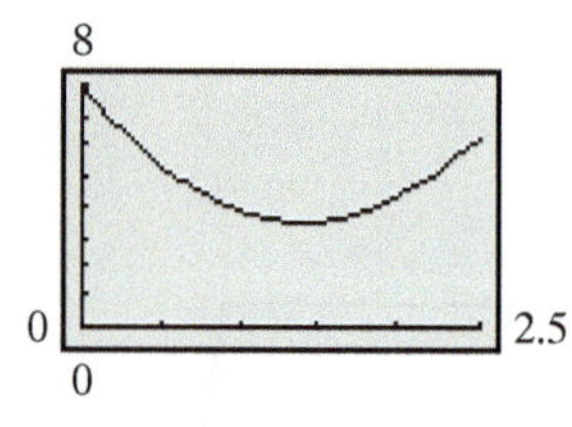

13. **(a)** $C(x) = x$ **(b)** $A(x) = \frac{x^2}{4\pi}$ **15.** **(a)** $A(r) = 2r^2$ **(b)** $p(r) = 6r$ **17.** $A(x) = \left(\frac{\pi}{3} - \frac{\sqrt{3}}{4}\right)x^2$

19. (a) $d(t) = \sqrt{2500t^2 - 360t + 13}$ **(b)** d is smallest when $t \approx 0.07$ hr.

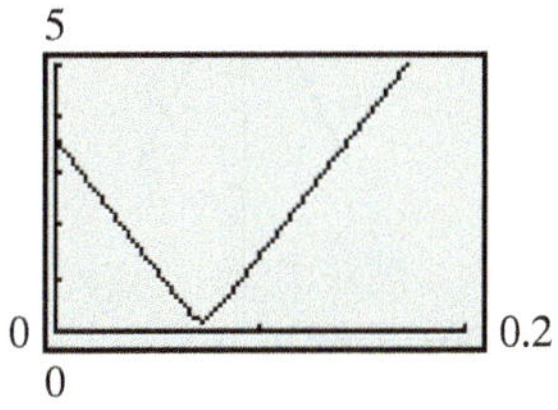

21. $V(r) = \dfrac{\pi H(R - r)r^2}{R}$

23. (a) $T(x) = \dfrac{12 - x}{5} + \dfrac{\sqrt{x^2 + 4}}{3}$ **(b)** $\{x|0 \le x \le 12\}$ **(c)** 3.09 hr **(d)** 3.55 hr

25. (a) $V(x) = x(24 - 2x)^2$ **(b)** 972 in.^{3} **(c)** 160 in.^{3} **(d)** V is largest when $x = 4$.

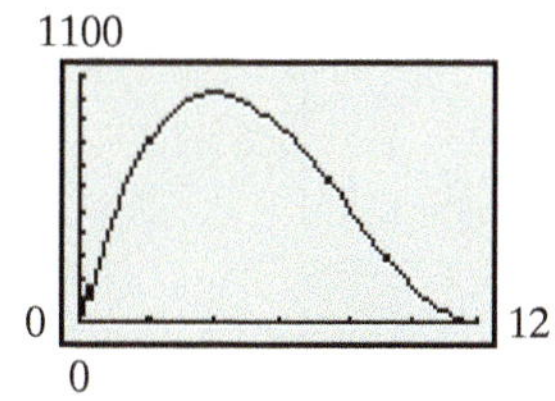

Review Exercises *(page 275)*

1. Function; domain $\{-1, 2, 4\}$, range $\{0, 3\}$ **3. (a)** 2 **(b)** -2 **(c)** $-\dfrac{3x}{x^2 - 1}$ **(d)** $-\dfrac{3x}{x^2 - 1}$ **(e)** $\dfrac{3(x - 2)}{x^2 - 4x + 3}$ **(f)** $\dfrac{6x}{4x^2 - 1}$

5. (a) 0 **(b)** 0 **(c)** $\sqrt{x^2 - 4}$ **(d)** $-\sqrt{x^2 - 4}$ **(e)** $\sqrt{x^2 - 4x}$ **(f)** $2\sqrt{x^2 - 1}$

7. (a) 0 **(b)** 0 **(c)** $\dfrac{x^2 - 4}{x^2}$ **(d)** $-\dfrac{x^2 - 4}{x^2}$ **(e)** $\dfrac{x(x - 4)}{(x - 2)^2}$ **(f)** $\dfrac{x^2 - 1}{x^2}$

9. $\{x|x \ne -3, x \ne 3\}$ **11.** $\{x|x \le 2\}$ **13.** $\{x|x > 0\}$ **15.** $\{x|x \ne -3, x \ne 1\}$

17. $(f + g)(x) = 2x + 3$; Domain: all real numbers
$(f - g)(x) = -4x + 1$; Domain: all real numbers
$(f \cdot g)(x) = -3x^2 + 5x + 2$; Domain: all real numbers
$\left(\dfrac{f}{g}\right)(x) = \dfrac{2 - x}{3x + 1}$; Domain: $\left\{x \middle| x \ne -\dfrac{1}{3}\right\}$

19. $(f + g)(x) = 3x^2 + 4x + 1$; Domain: all real numbers
$(f - g)(x) = 3x^2 - 2x + 1$; Domain: all real numbers
$(f \cdot g)(x) = 9x^3 + 3x^2 + 3x$; Domain: all real numbers
$\left(\dfrac{f}{g}\right)(x) = \dfrac{3x^2 + x + 1}{3x}$; Domain: $\{x|x \ne 0\}$

21. $(f + g)(x) = \dfrac{x^2 + 2x - 1}{x(x - 1)}$; Domain: $\{x|x \ne 0, x \ne 1\}$
$(f - g)(x) = \dfrac{x^2 + 1}{x(x - 1)}$; Domain: $\{x|x \ne 0, x \ne 1\}$
$(f \cdot g)(x) = \dfrac{x + 1}{x(x - 1)}$; Domain: $\{x|x \ne 0, x \ne 1\}$
$\left(\dfrac{f}{g}\right)(x) = \dfrac{x(x + 1)}{x - 1}$; Domain: $\{x|x \ne 0, x \ne 1\}$

23. $-4x + 1 - 2h$

25. (a) Domain: $\{x|-4 \le x \le 3\}$; Range: $\{y|-3 \le y \le 3\}$ **(b)** $(0, 0)$ **(c)** -1 **(d)** -4 **(e)** $\{x|0 < x \le 3\}$

(f)

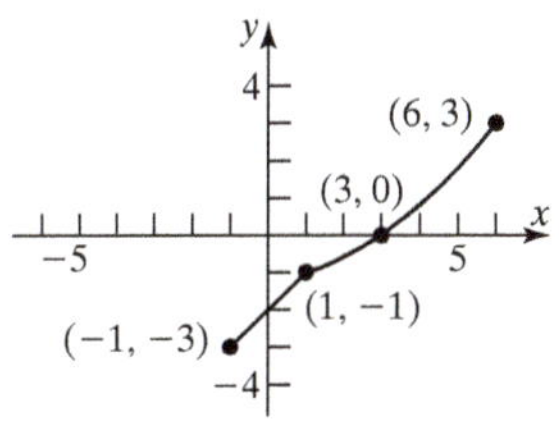

(g)

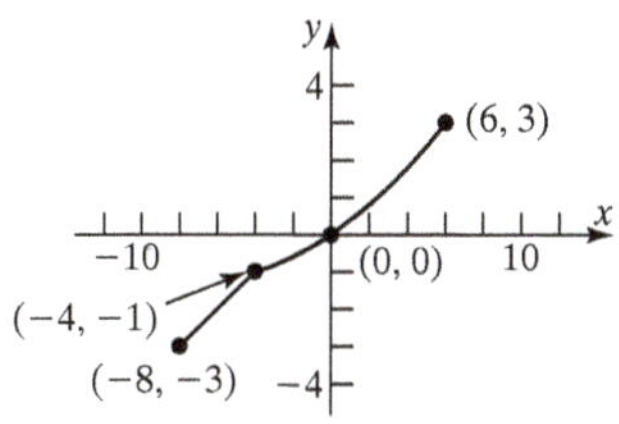

(h)

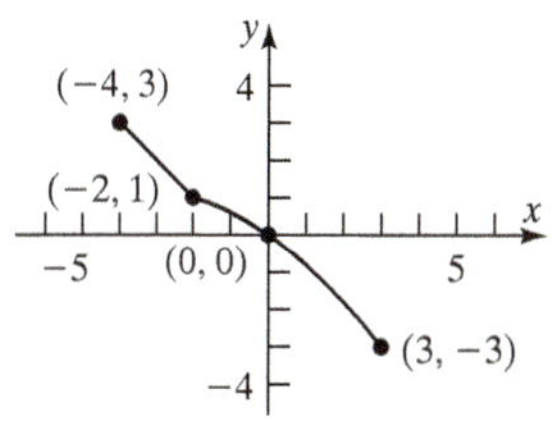

27. (a) Domain: $\{x|-4 \le x \le 4\}$ or $[-4, 4]$
Range: $\{y|-3 \le y \le 1\}$ or $[-3, 1]$
(b) Increasing on $(-4, -1)$ and $(3, 4)$
Decreasing on $(-1, 3)$
(c) Local maximum is 1 and occurs at $x = -1$.
Local minimum is -3 and occurs at $x = 3$.
(d) No symmetry **(e)** Neither **(f)** x-intercepts: $-2, 0, 4$; y-intercept: 0

29. Odd **31.** Even **33.** Neither **35.** Odd

37.

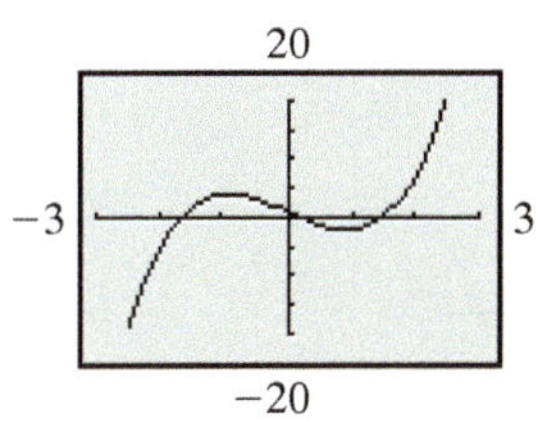

Local maximum: 4.04 at $x = -0.91$
Local minimum: -2.04 at $x = 0.91$

Increasing: $(-3, -0.91)$; $(0.91, 3)$
Decreasing: $(-0.91, 0.91)$

39.

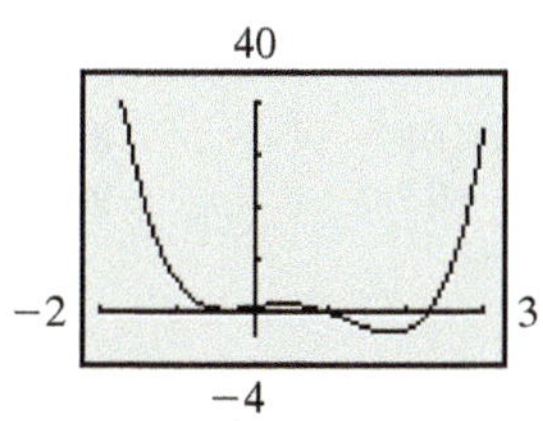

Local maximum: 1.53 at $x = 0.41$
Local minima: 0.54 at $x = -0.34$
and -3.56 at $x = 1.80$

Increasing: $(-0.34, 0.41)$; $(1.80, 3)$
Decreasing: $(-2, -0.34)$; $(0.41, 1.80$

41. (a) 23 **(b)** 7 **(c)** 47

43. -5

45. -17

47. No

49. Yes

51.

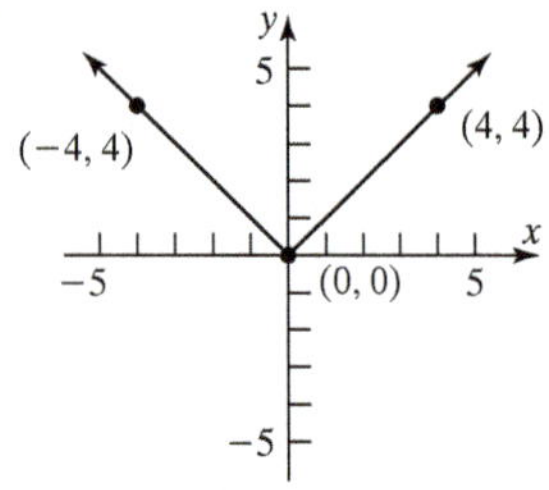

53.

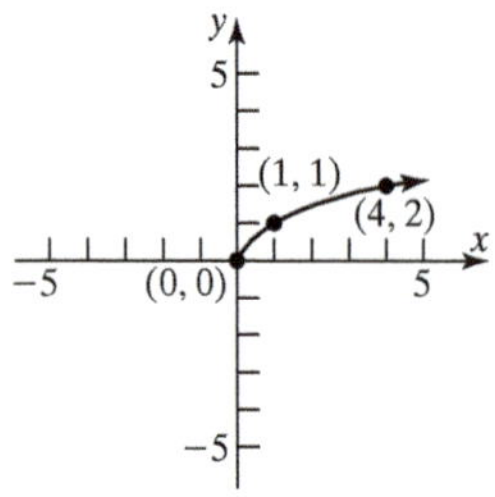

55.

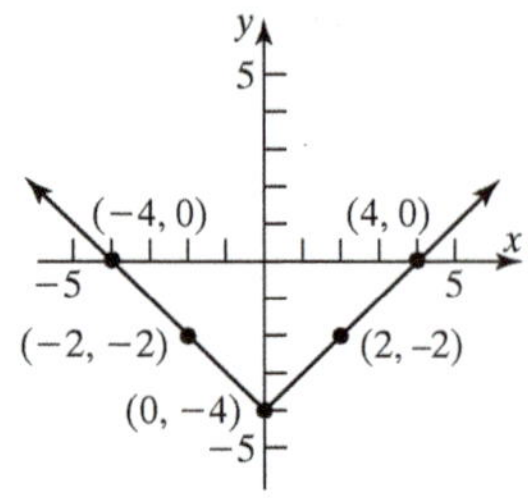

Intercepts: $(-4, 0), (4, 0), (0, -4)$
Domain: all real numbers
Range: $\{y|y \geq -4\}$ or $[-4, \infty)$

57.

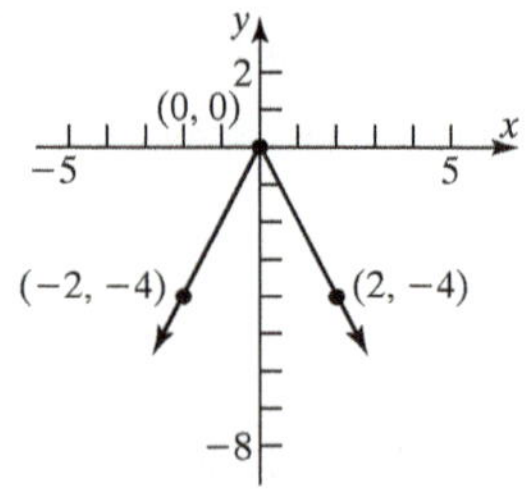

Intercept: $(0, 0)$
Domain: all real numbers
Range: $\{y|y \leq 0\}$ or $(-\infty, 0]$

59.

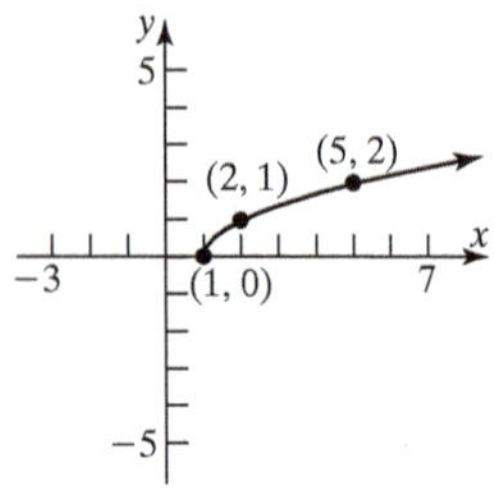

Intercept: $(1, 0)$

Domain: $\{x|x \geq 1\}$ or $[1, \infty)$
Range: $\{y|y \geq 0\}$ or $[0, \infty)$

61.

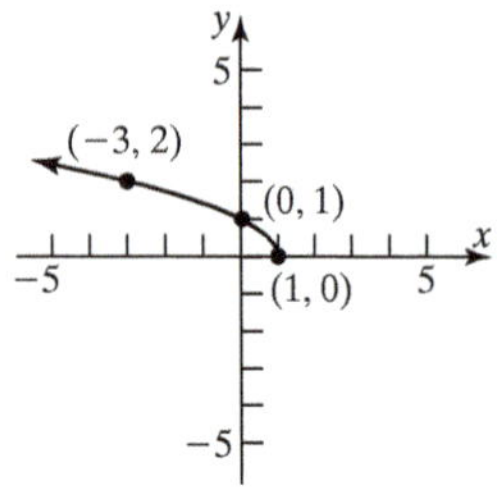

Intercepts: $(0, 1), (1, 0)$

Domain: $\{x|x \leq 1\}$ or $(-\infty, 1]$
Range: $\{y|y \geq 0\}$ or $[0, \infty)$

63.

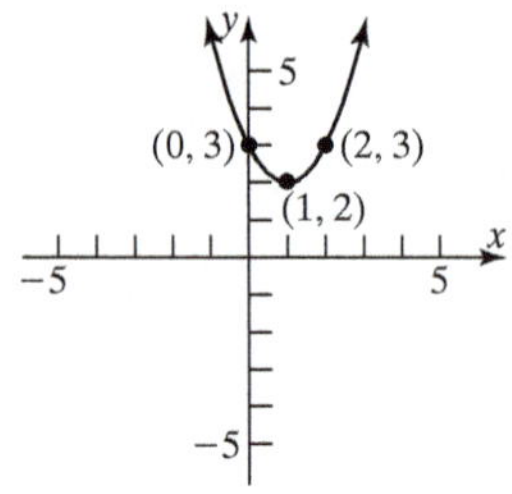

Intercept: $(0, 3)$

Domain: all real numbers
Range: $\{y|y \geq 2\}$ or $[2, \infty)$

65.

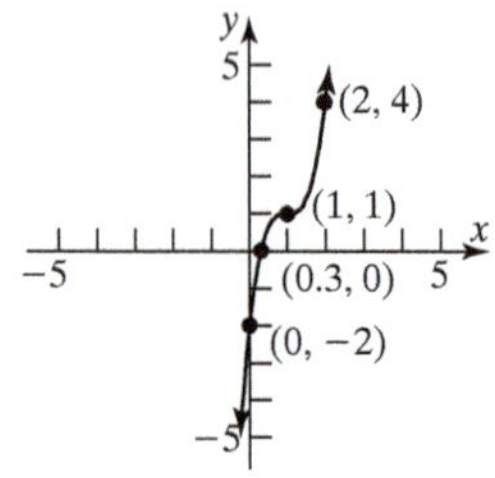

Intercepts: $(0, -2)$,
$\left(1 - \frac{\sqrt[3]{9}}{3}, 0\right)$ or about $(0.3, 0)$
Domain: all real numbers
Range: all real numbers

67. (a) $\{x|x > -2\}$ or $(-2, \infty)$
(b) $(0, 0)$
(c)

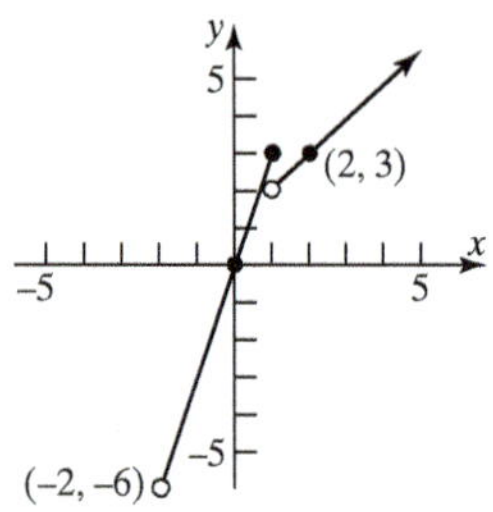

(d) $\{y|y > -6\}$ or $(-6, \infty)$
(e) Discontinuous at $x = 1$

69. (a) $\{x|x \geq -4\}$ or $[-4, \infty)$
(b) $(0, 1)$
(c)

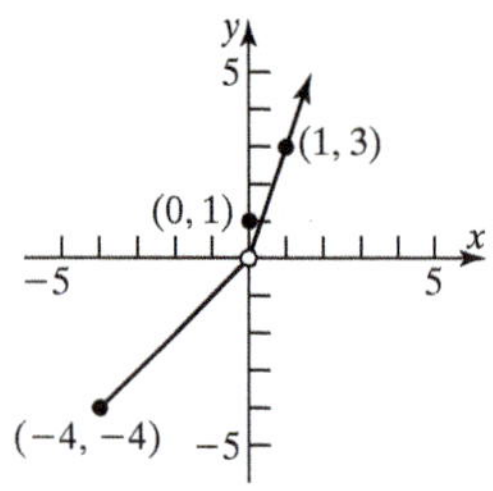

(d) $\{y| -4 \leq y < 0 \text{ or } y > 0\}$ or $[-4, 0) \cup (0, \infty)$
(e) Discontinuous at $x = 0$

71. $A = 11$

73. (a) $A(x) = (8.5 - 2x)(11 - 2x)$
(b) $0 \leq x < 4.25, 0 < A \leq 93.5$
(c) $A(1) = 58.5 \text{ in.}^2$, $A(1.2) = 52.46 \text{ in.}^2$, $A(1.5) = 44 \text{ in.}^2$
(d)

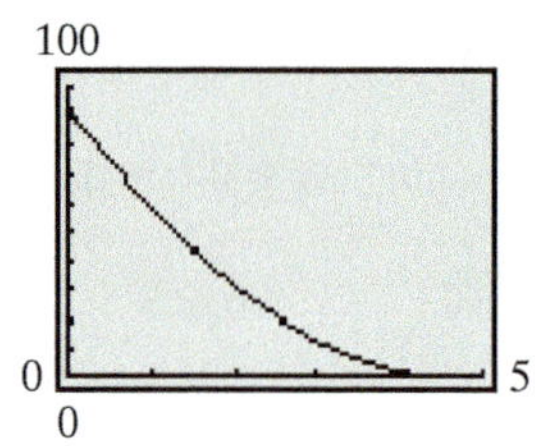

(e) $A(x) \approx 70$ when $x \approx 0.64$ in.; $A(x) \approx 50$ when $x \approx 1.28$ in.

75. (a) $A(x) = 10x - x^3$ **(b)** The largest area that can be enclosed by the rectangle is approximately 12.17 square units.

Chapter Test *(page 278)*

1. (a) Function; domain: $\{2, 4, 6, 8\}$; range: $\{5, 6, 7, 8\}$ **(b)** Not a function **(c)** Not a function **(d)** Function; domain: all real numbers; range: $\{y|y \geq 2\}$ **2.** Domain: $\left\{x \middle| x \leq \frac{4}{5}\right\}$; $f(-1) = 3$ **3.** Domain: $\{x|x \neq -2\}$; $g(-1) = 1$ **4.** Domain: $\{x|x \neq -9, x \neq 4\}$; $h(-1) = \frac{1}{8}$

5. (a) Domain: $\{x| -5 \leq x \leq 5\}$; range: $\{y| -3 \leq y \leq 3\}$ **(b)** $(0, 2), (-2, 0)$, and $(2, 0)$ **(c)** $f(1) = 3$ **(d)** $x = -5$ and $x = 3$
(e) $\{x| -5 \leq x < -2 \text{ or } 2 < x \leq 5\}$ or $[-5, -2) \cup (2, 5]$ **6.** Local maxima: $f(-0.85) \approx -0.86$; $f(2.35) \approx 15.55$; local minima: $f(0) = -2$; thefunction is increasing on the intervals $(-5, -0.85)$ and $(0, 2.35)$ and decreasing on the intervals $(-0.85, 0)$ and $(2.35, 5)$.

7. (a)

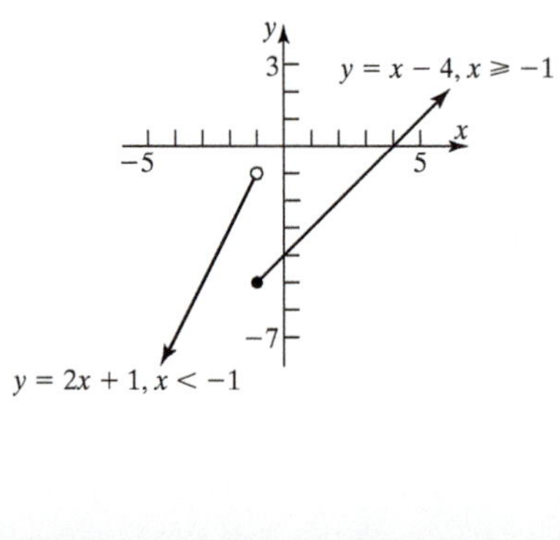

(b) $(0, -4), (4, 0)$
(c) $g(-5) = -9$
(d) $g(2) = -2$

8. 19

9. (a) $(f - g) = 2x^2 - 3x + 3$
(b) $(f \cdot g) = 6x^3 - 4x^2 + 3x - 2$
(c) $f(x + h) - f(x) = 4xh + 2h^2$

10. (a)

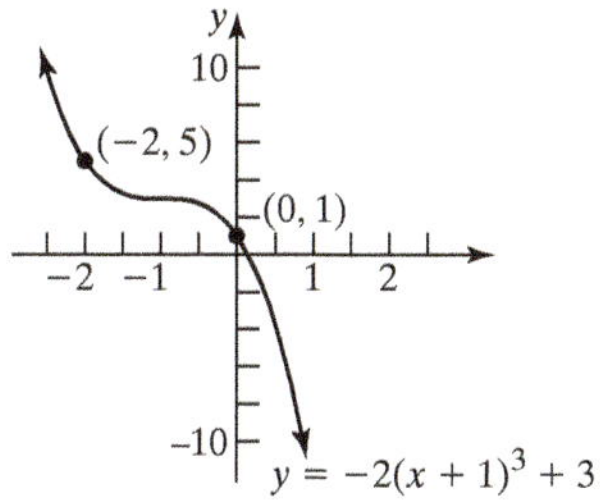

(b)

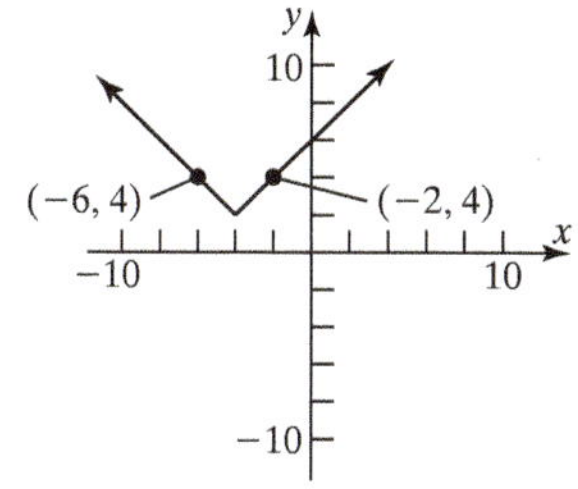

11. (a) 8.67% occurring in 1997 ($x \approx 5$) **(b)** The model predicts that the interest rate will be −10.343%. This is not reasonable.

12. (a) $V(x) = \frac{x^2}{8} - \frac{5x}{4} + \frac{\pi x^2}{64}$ **(b)** 1297.61 ft^3

Cumulative Review *(page 279)*

1. $\left\{\frac{4}{5}\right\}$ **2.** $\{3, 4\}$ **3.** $\left\{-\frac{1}{3}, 2\right\}$ **4.** $\left\{-\frac{1}{2}\right\}$ **5.** No real solution **6.** $\{-7\}$ **7.** $\{-31\}$ **8.** $\left\{\frac{1}{3}, 1\right\}$ **9.** $\left\{\frac{1 - \sqrt{15}i}{4}, \frac{1 + \sqrt{15}i}{4}\right\}$

10. $\{x|1 < x < 4\}$

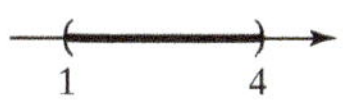

11.

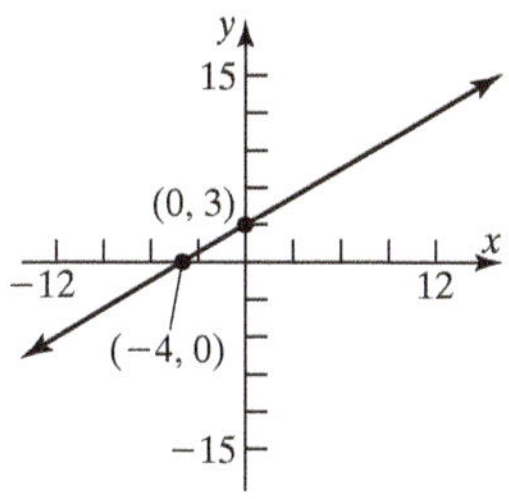

12.

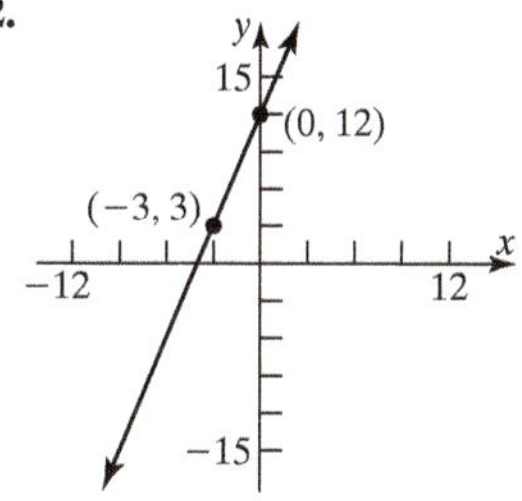

13.

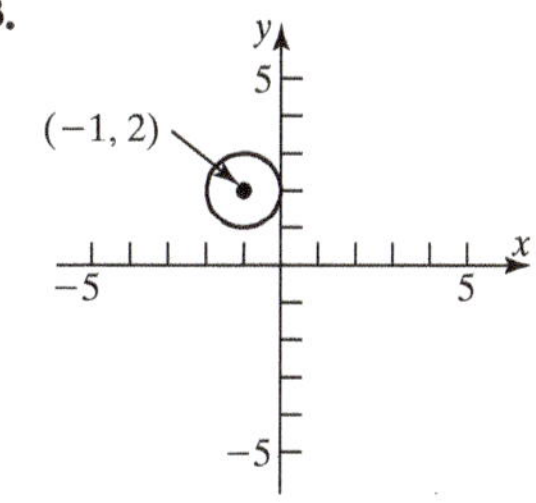

14.

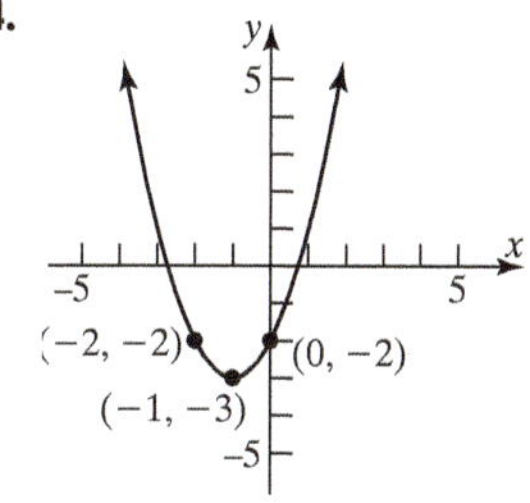

15. (a) Domain: $\{x|-4 \le x \le 4\}$; range: $\{y|-1 \le y \le 3\}$ **(b)** $(-1, 0), (0, -1), (1, 0)$ **(c)** y-axis **(d)** 1 **(e)** −4 and 4 **(f)** $\{x|-1 < x < 1\}$

(g)

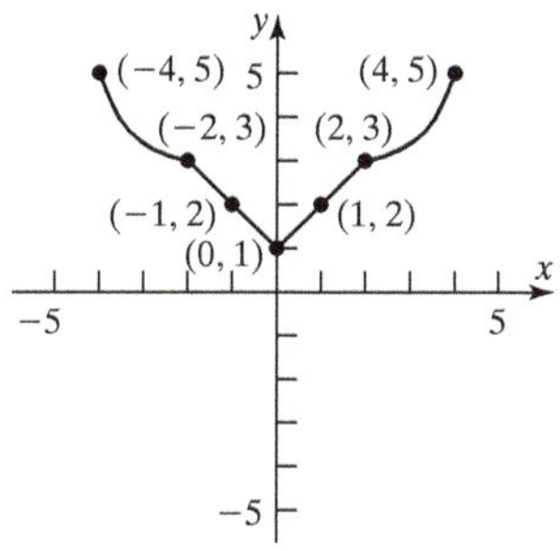

(h)

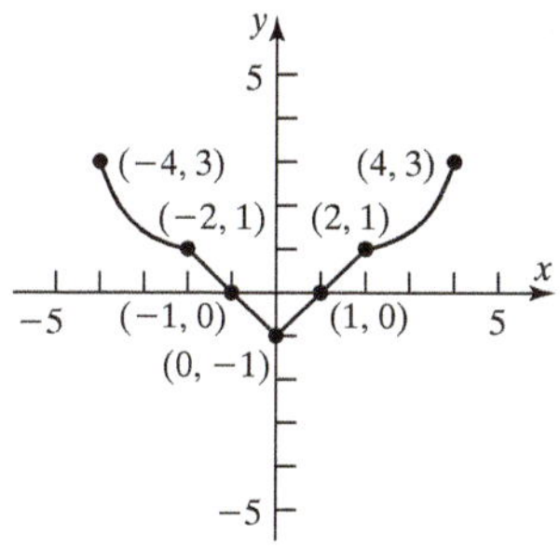

(i)

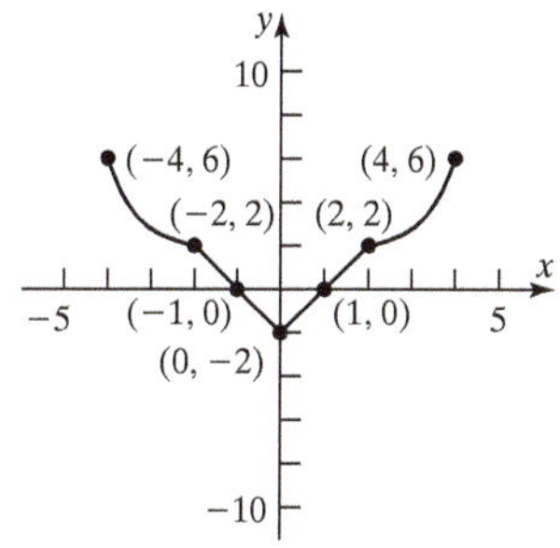

(j) even **(k)** $(0, 4)$ **(l)** $(-4, 0)$ **(m)** f has a local minimum of −1 at $x = 0$ **(n)** 1

16. $5\sqrt{2}$ **17.** $(-2, -1)$ and $(2, 3)$ are on the graph. **18.** x-intercepts: $-5, \frac{1}{3}$; y-intercept: −5 **19.** $\{-1.10, 0.26, 1.48, 2.36\}$

20. $y = -\frac{1}{2}x + \frac{13}{2}$

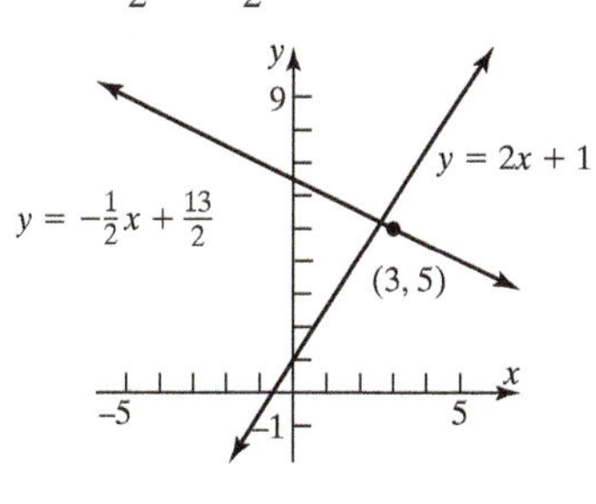

21. Yes, a function **22. (a)** −3 **(b)** $x^2 - 4x - 2$ **(c)** $x^2 + 4x + 1$ **(d)** $-x^2 + 4x - 1$ **(e)** $x^2 - 3$ **(f)** $2x + h - 4$ **23.** $\{z|z \ne -1, z \ne 7\}$ **24.** Yes, a function **25. (a)** No **(b)** −1; $(-2, -1)$ is on the graph. **(c)** −8; $(-8, 2)$ is on the graph.

26. $2(x^2 + 1)^{1/2}(4x^2 + 1)$ **27.** $\frac{(3x + 10)(x + 5)^{1/2}}{x + 5}$

CHAPTER 4 Linear and Quadratic Functions

4.1 Assess Your Understanding *(page 288)*

7. slope; y-intercept **8.** $-4; 3$ **9.** positive **10.** T **11.** F **12.** F

13. (a) $m = 2; b = 3$

(b)

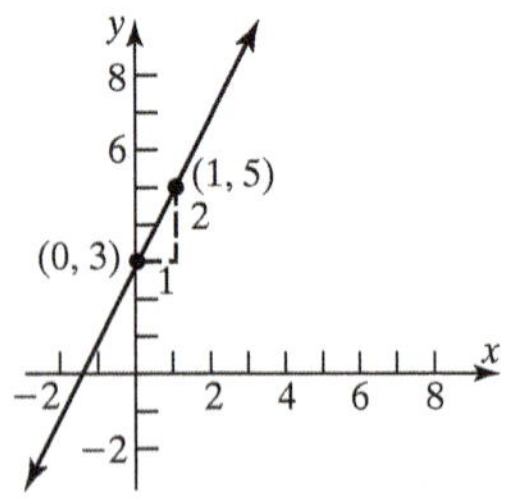

(c) Domain and Range: $(-\infty, \infty)$

(d) 2 **(e)** Increasing

15. (a) $m = -3; b = 4$

(b)

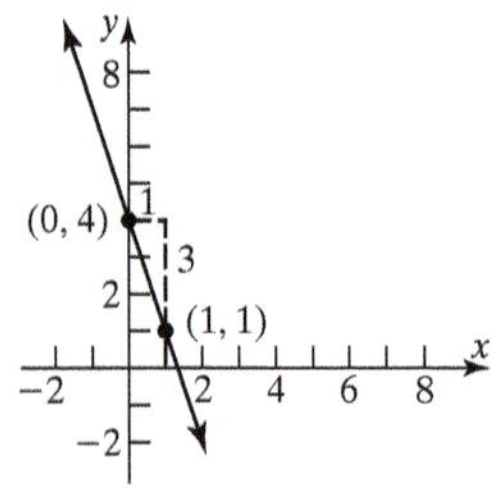

(c) Domain and Range: $(-\infty, \infty)$

(d) -3 **(e)** Decreasing

17. (a) $m = \frac{1}{4}; b = -3$

(b)

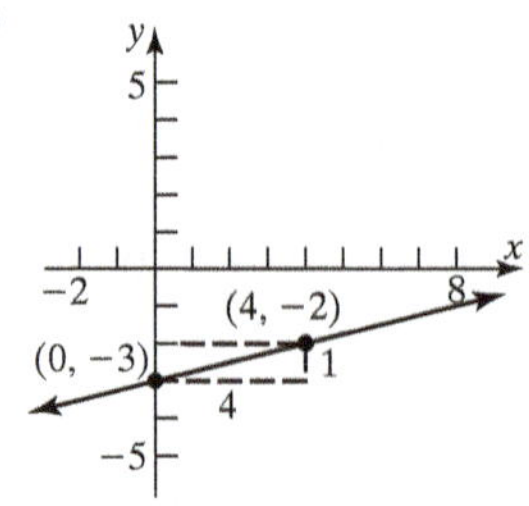

(c) Domain and Range: $(-\infty, \infty)$

(d) $\frac{1}{4}$ **(e)** Increasing

19. (a) $m = 0; b = 4$

(b)

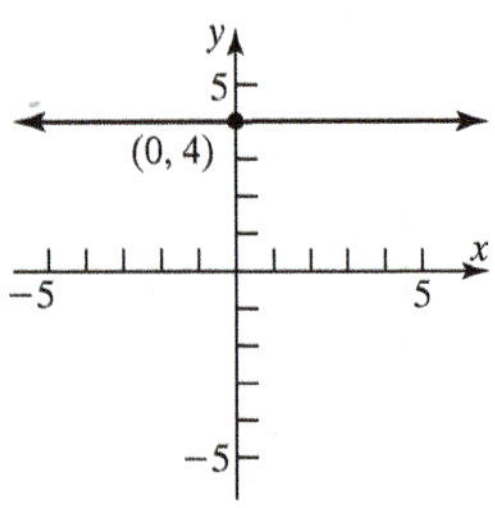

(c) Domain: $(-\infty, \infty)$

Range: $\{y \mid y = 4\}$

(d) 0 **(e)** Constant

21. Linear; $y = -3x - 2$

23. Nonlinear

25. Nonlinear

27. Linear; $y = 8$

29. (a) $\frac{1}{4}$

(b) $\left\{x \,\middle|\, x > \frac{1}{4}\right\}$ or $\left(\frac{1}{4}, \infty\right)$

(c) 1

(d) $\{x \mid x \le 1\}$ or $(-\infty, 1]$

(e)

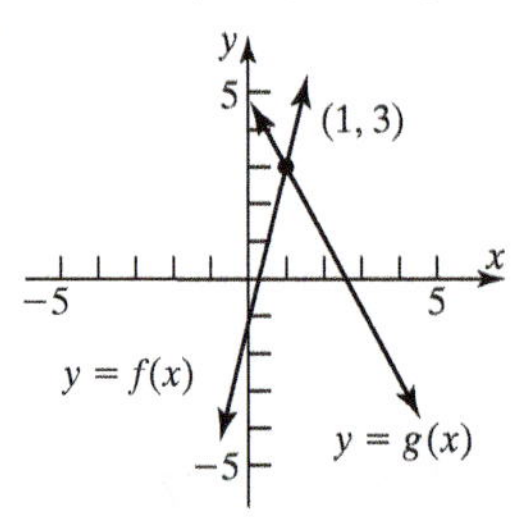

31. (a) 40 **(b)** 88 **(c)** -40 **(d)** $\{x \mid x > 40\}$ or $(40, \infty)$ **(e)** $\{x \mid x \le 88\}$ or $(-\infty, 88]$ **(f)** $\{x \mid -40 < x < 88\}$ or $(-40, 88)$

33. (a) -4 **(b)** $\{x \mid x < -4\}$ or $(-\infty, -4)$ **35. (a)** -6 **(b)** $\{x \mid -6 \le x < 5\}$ or $[-6, 5)$

37. (a) \$45 **(b)** 180 mi **(c)** 260 mi **(d)** $\{x \mid x \ge 0\}$ or $[0, \infty)$ **39. (a)** \$16; 600 T-shirts **(b)** $\$0 \le p < \$16$ **(c)** The price will increase.

41. (a) $\{x \mid 7825 \le x \le 31{,}850\}$ or $[7825, 31{,}850]$ **(b)** \$2308.75

(c) The independent variable is adjusted gross income, x. The dependent variable is the tax bill, T.

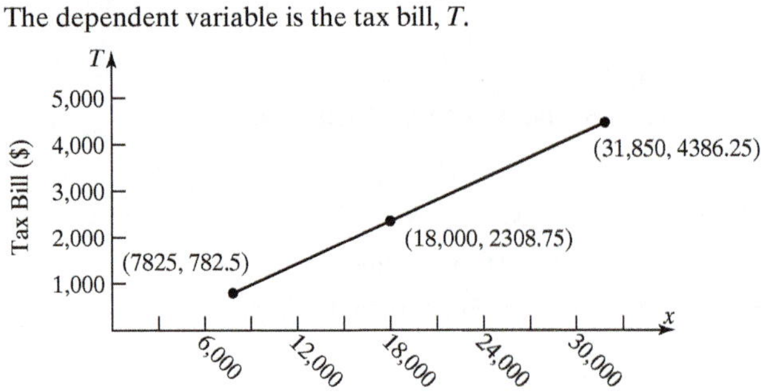

(e) \$25,000

43. (a) $x = 5000$

(b) $x > 5000$

45. (a) $V(x) = -1000x + 3000$ **(b)** $\{x \mid 0 \le x \le 3\}$ or $[0, 3]$

(c)

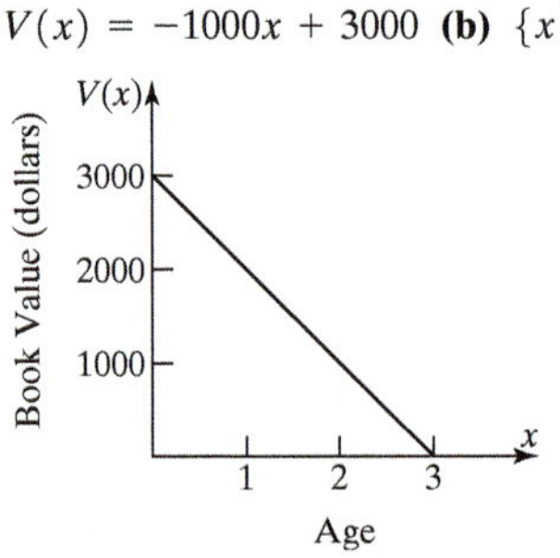

(d) \$1000 **(e)** After 1 year

47. (a) $C(x) = 90x + 1800$

(b)

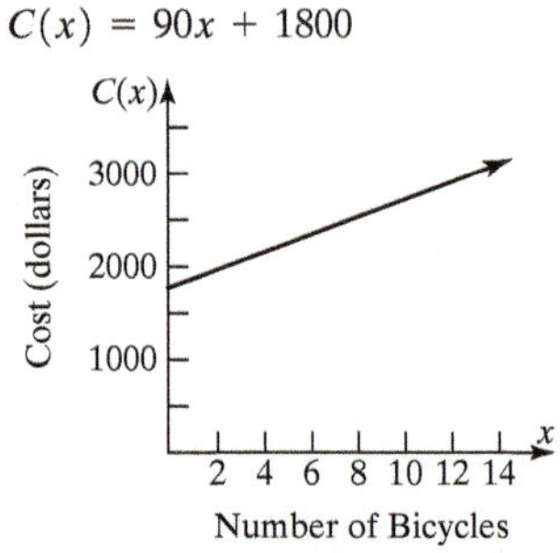

(c) \$3060 **(d)** 22 bicycles

49. (a) $C(x) = 0.07x + 29$ **(b)** \$36.70; \$45.10

51. (a)

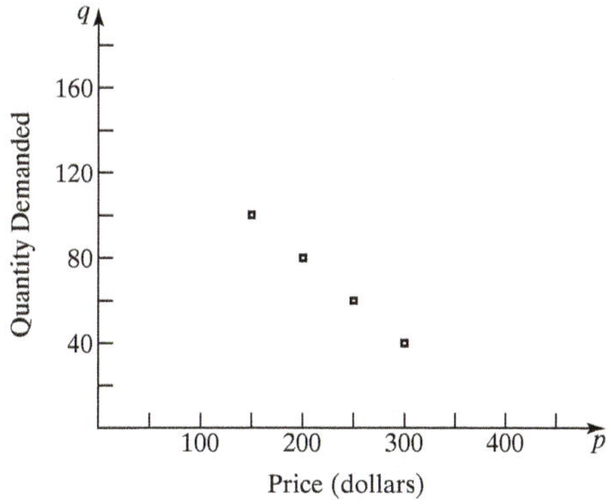

(b) Since each input (price) corresponds to a single output (quantity demanded), we know that quantity demanded is a function of price. Also, because the average rate of change is a constant −\$0.4 per MP3 player, the function is linear.

(c) $q(p) = -0.4p + 160$

(d) $\{p|0 \le p \le 400\}$ or $[0, 400]$

(e)

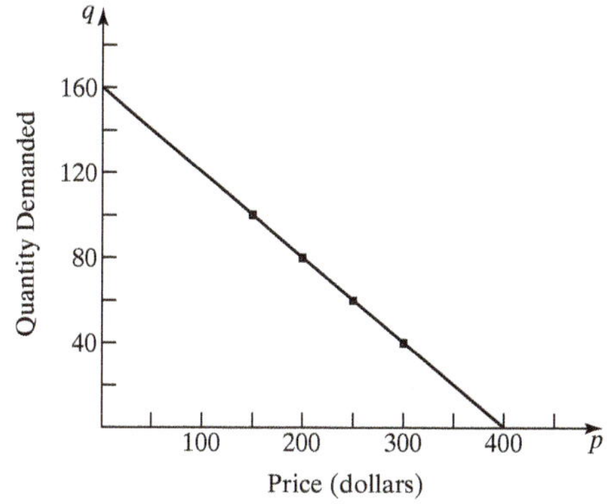

(f) If price increases by \$1, quantity demanded of MP3 players decreases by 0.4 player.

(g) q-intercept: When the price is \$0, 160 MP3 players will be demanded. p-intercept: There will be 0 MP3 players demanded when the price is \$400.

53. (d), (e) **55.** $b = 0$; yes

4.2 Assess Your Understanding *(page 295)*

3. scatter diagram **4.** T **5.** Linear relation, $m > 0$ **7.** Linear relation, $m < 0$ **9.** Nonlinear relation

11. (a)

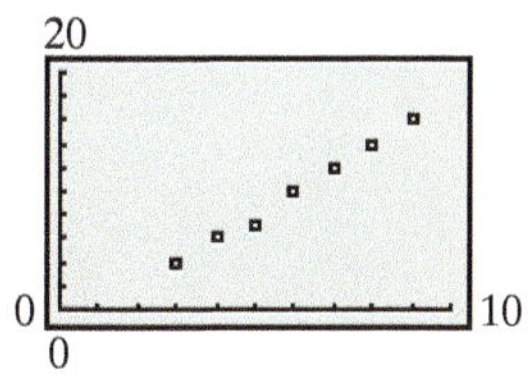

(b) Answers vary. Using (4, 6) and (8, 14), $y = 2x - 2$.

(c)

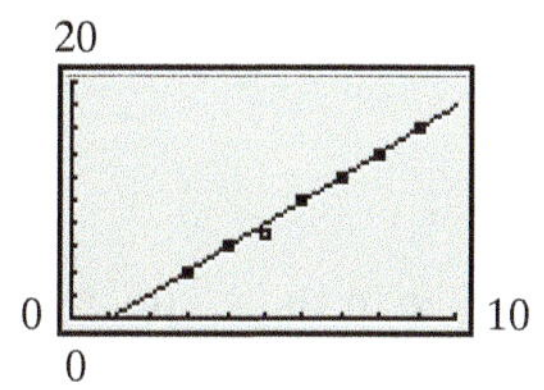

(d) $y = 2.0357x - 2.3571$

(e)

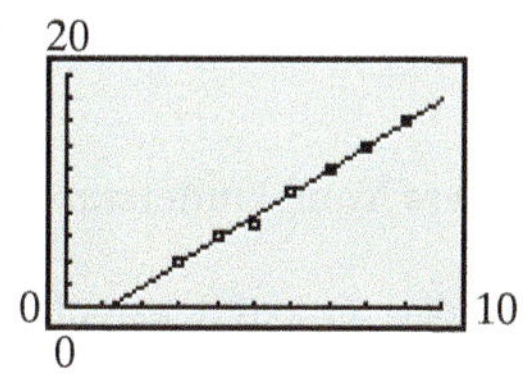

13. (a)

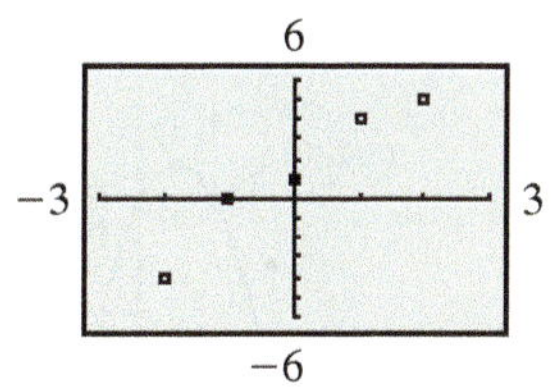

(b) Answers will vary. Using $(-2, -4)$ and $(2, 5)$, $y = \frac{9}{4}x + \frac{1}{2}$.

(c)

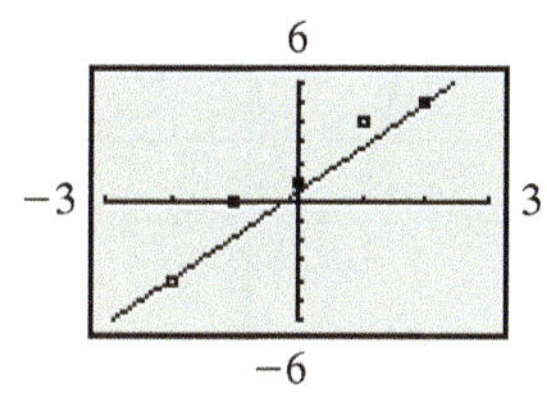

(d) $y = 2.2x + 1.2$

(e)

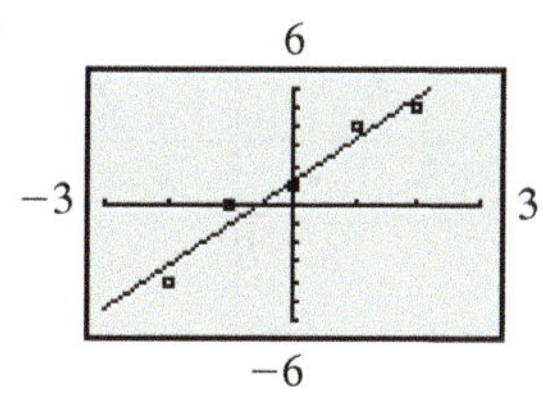

15. (a)

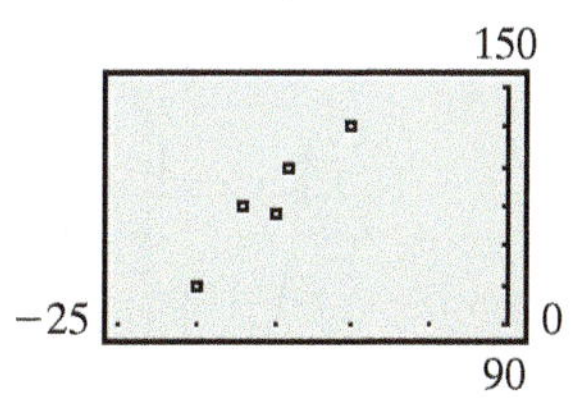

(b) Answers will vary. Using $(-20, 100)$ and $(-10, 140)$, $y = 4x + 180$

(c)

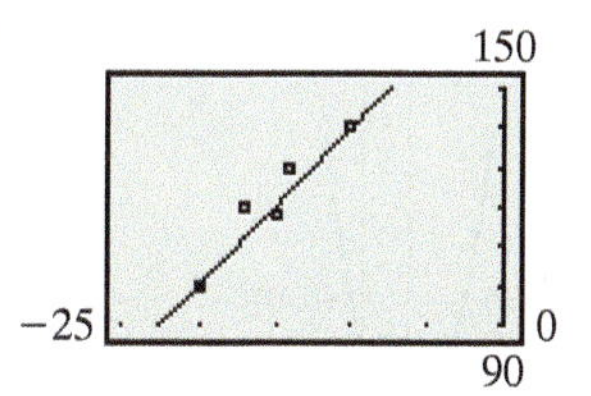

(d) $y = 3.8613x + 180.2920$

(e)

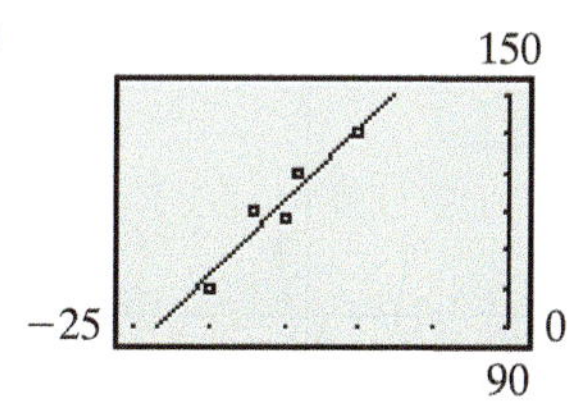

17. (a)

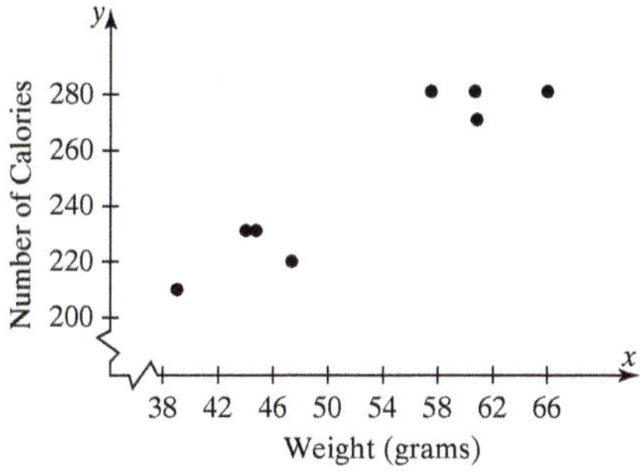

(b) Linear

(c) Answers will vary. Using the points (39.52, 210) and (66.45, 280), $y = 2.599x + 107.288$.

(d)

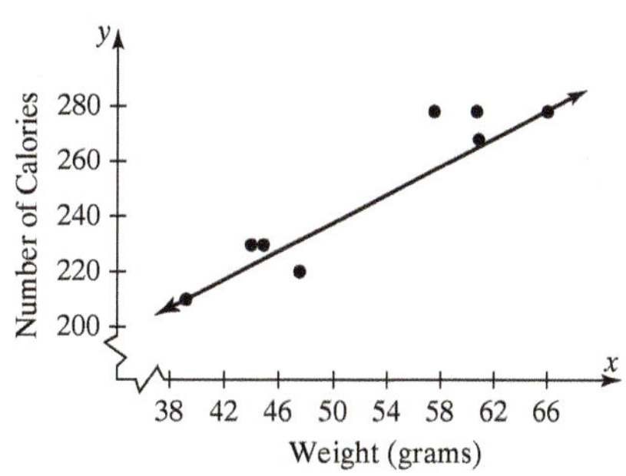

(e) 269 calories

(f) If the weight of a candy bar is increased by 1 gram, the number of calories will increase by 2.599.

19. (a)

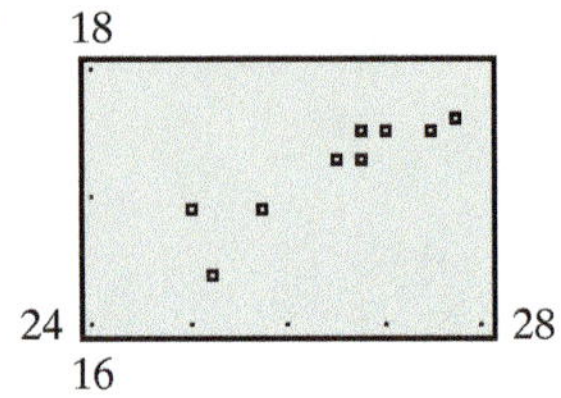

(b) $C(H) = 0.3734H + 7.3268$
(c) If height increases by 1 inch, head circumference increases by about 0.3734 inch.
(d) About 17.0 inches **(e)** About 26.98 inches

21. (a) No
(b)

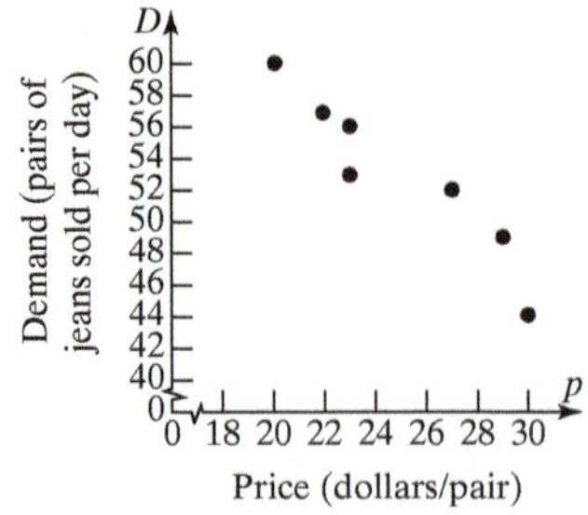

(c) $D = -1.3355p + 86.1974$
(d) If the price increases $1, the quantity sold per day decreases by about 1.34 pairs of jeans.
(e) $D(p) = -1.3355p + 86.1974$ **(f)** $\{p \mid 0 < p \le 64\}$
(g) About 49 pairs

23.

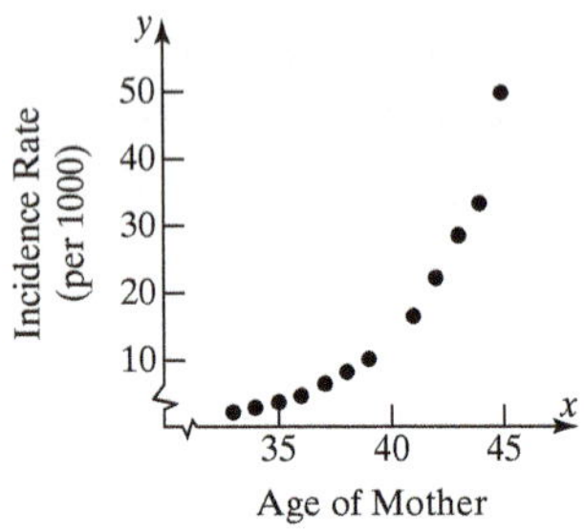

No, the data do not follow a linear pattern.

25. No linear relation

4.3 Assess Your Understanding *(page 307)*

5. parabola **6.** axis or axis of symmetry **7.** $-\dfrac{b}{2a}$ **8.** T **9.** T **10.** T **11.** C **13.** F **15.** G **17.** H

19.

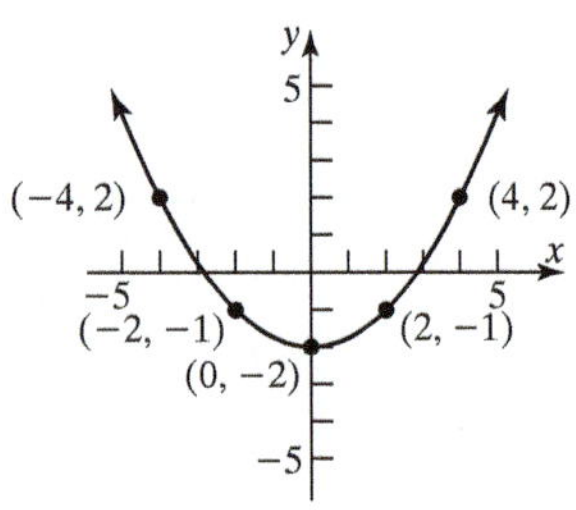

21.

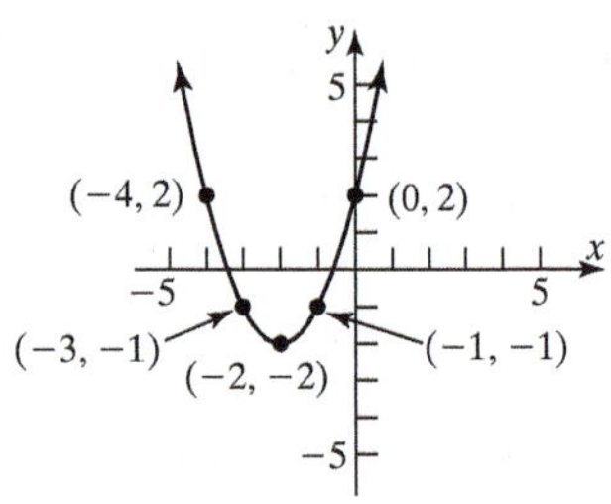

23.

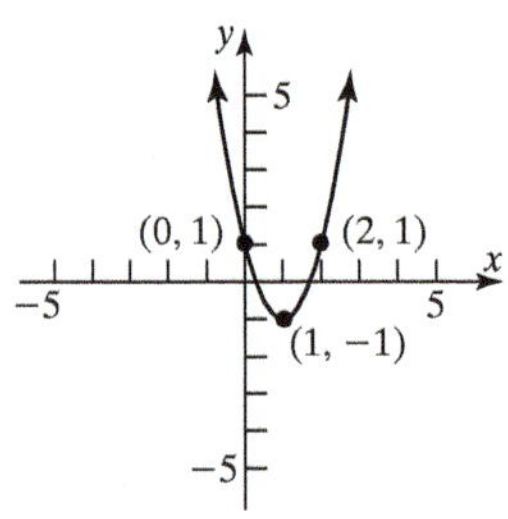

25. $f(x) = -(x + 1)^2 + 1$

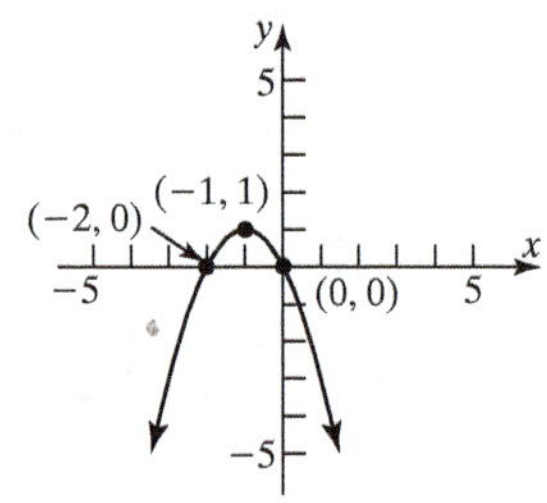

27. $f(x) = 2(x + 1)^2 - 4$

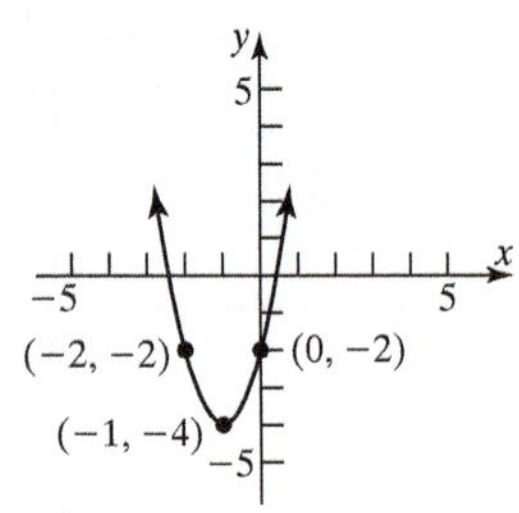

29. (a)

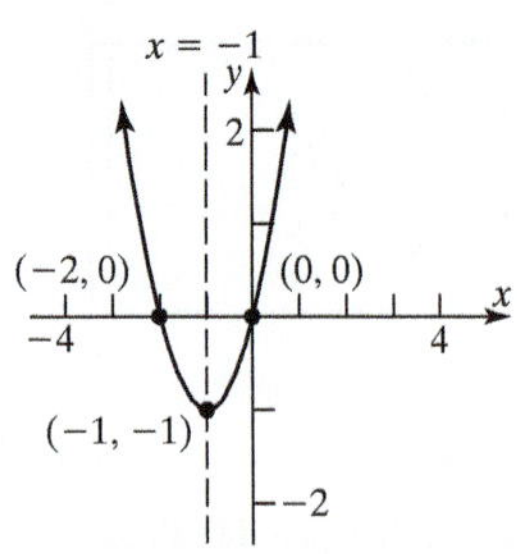

(b) Domain: $(-\infty, \infty)$
Range: $[-1, \infty)$
(c) Decreasing: $(-\infty, -1)$
Increasing: $(-1, \infty)$

31. (a)

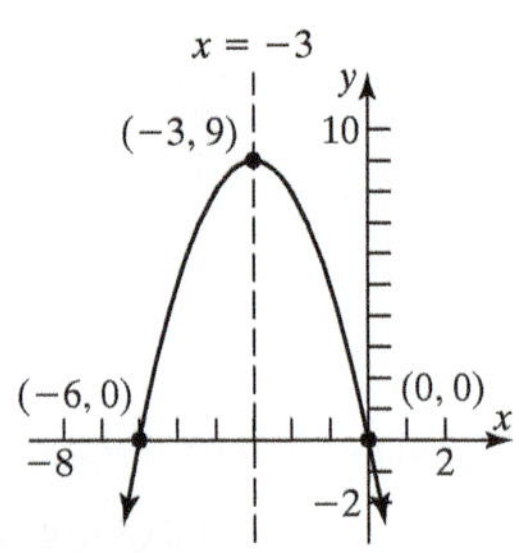

(b) Domain: $(-\infty, \infty)$
Range: $(-\infty, 9]$
(c) Increasing: $(-\infty, -3)$
Decreasing: $(-3, \infty)$

33. (a)

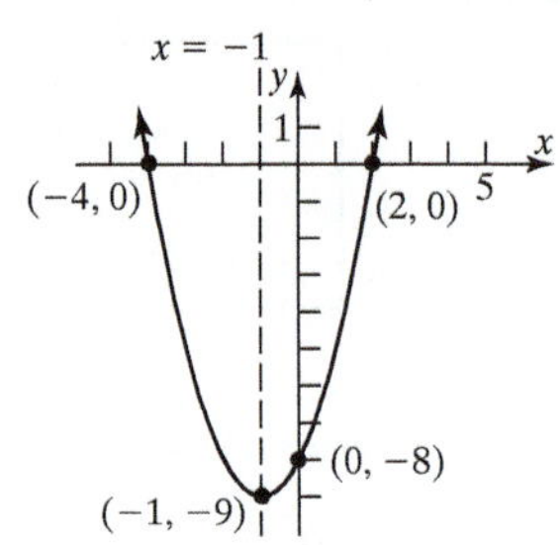

(b) Domain: $(-\infty, \infty)$
Range: $[-9, \infty)$
(c) Decreasing: $(-\infty, -1)$
Increasing: $(-1, \infty)$

35. (a)
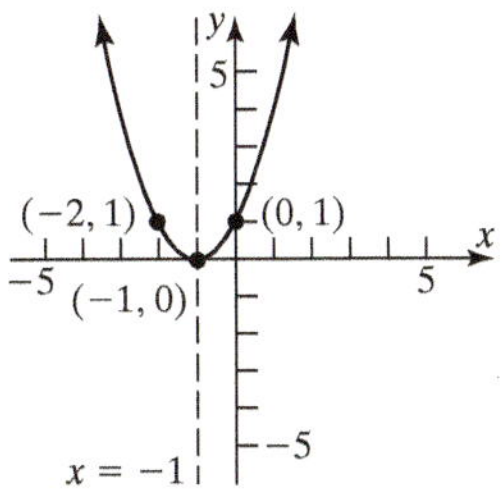

(b) Domain: $(-\infty, \infty)$
Range: $[0, \infty)$

(c) Decreasing: $(-\infty, -1)$
Increasing: $(-1, \infty)$

37. (a)
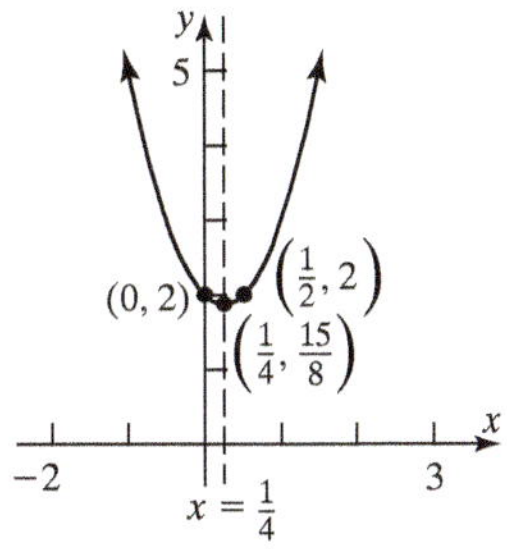

(b) Domain: $(-\infty, \infty)$
Range: $\left[\frac{15}{8}, \infty\right)$

(c) Decreasing: $\left(-\infty, \frac{1}{4}\right)$
Increasing: $\left(\frac{1}{4}, \infty\right)$

39. (a)
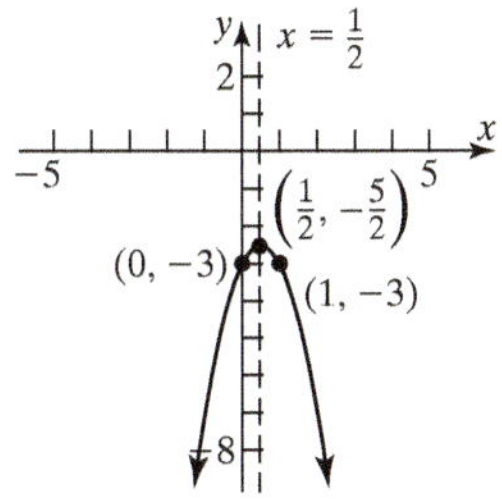

(b) Domain: $(-\infty, \infty)$
Range: $\left(-\infty, -\frac{5}{2}\right]$

(c) Increasing: $\left(-\infty, \frac{1}{2}\right)$
Decreasing: $\left(\frac{1}{2}, \infty\right)$

41. (a)
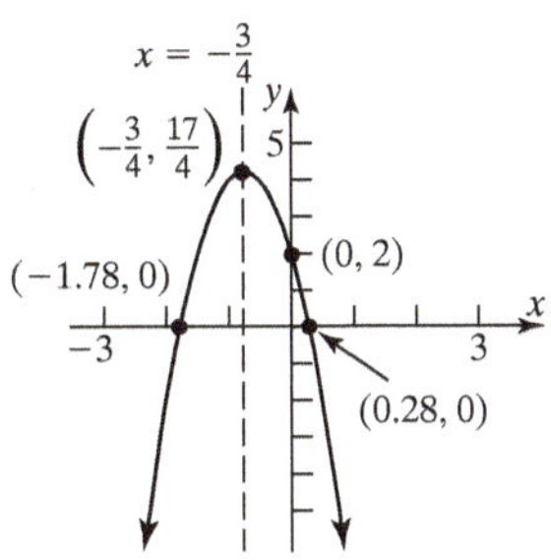

(b) Domain: $(-\infty, \infty)$
Range: $\left(-\infty, \frac{17}{4}\right]$

(c) Increasing: $\left(-\infty, -\frac{3}{4}\right)$
Decreasing: $\left(-\frac{3}{4}, \infty\right)$

43. $f(x) = (x + 1)^2 - 2 = x^2 + 2x - 1$

45. $f(x) = -(x + 3)^2 + 5 = -x^2 - 6x - 4$

47. $f(x) = 2(x - 1)^2 - 3 = 2x^2 - 4x - 1$

49. Minimum value; −18

51. Minimum value; −21

53. Maximum value; 21

55. Maximum value; 13

57. (a)
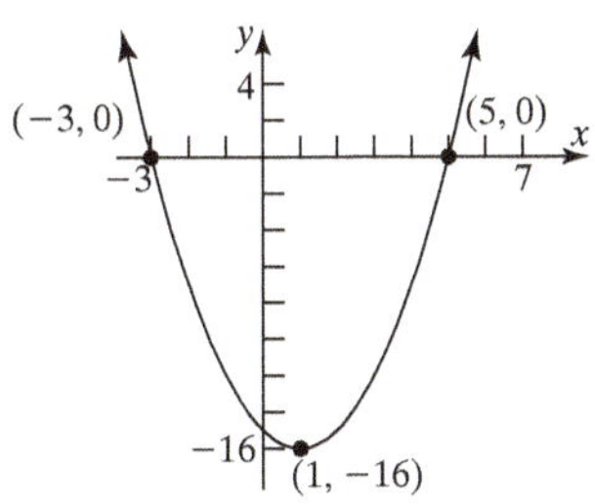

(b) Domain: $(-\infty, \infty)$
Range: $[-16, \infty)$
(c) Decreasing: $(-\infty, 1)$
Increasing: $(1, \infty)$

59. (a)
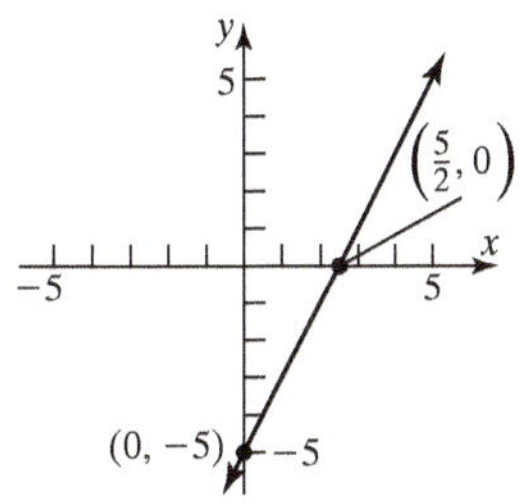

(b) Domain: $(-\infty, \infty)$
Range: $(-\infty, \infty)$
(c) Increasing: $(-\infty, \infty)$

61. (a)
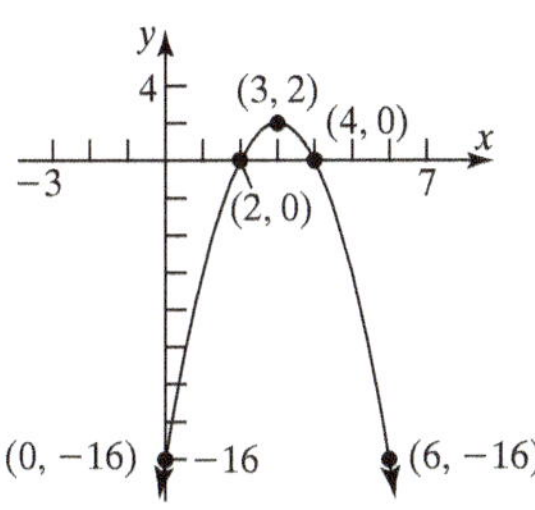

(b) Domain: $(-\infty, \infty)$
Range: $(-\infty, 2]$
(c) Increasing: $(-\infty, 3)$
Decreasing: $(3, \infty)$

63. (a)
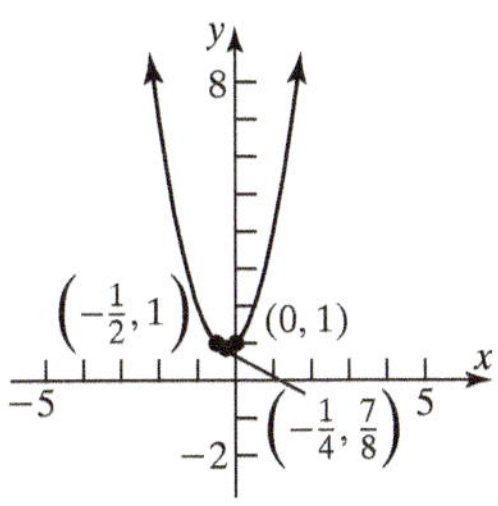

(b) Domain: $(-\infty, \infty)$
Range: $\left[\frac{7}{8}, \infty\right)$

(c) Decreasing: $\left(-\infty, -\frac{1}{4}\right)$
Increasing: $\left(-\frac{1}{4}, \infty\right)$

65. (a)
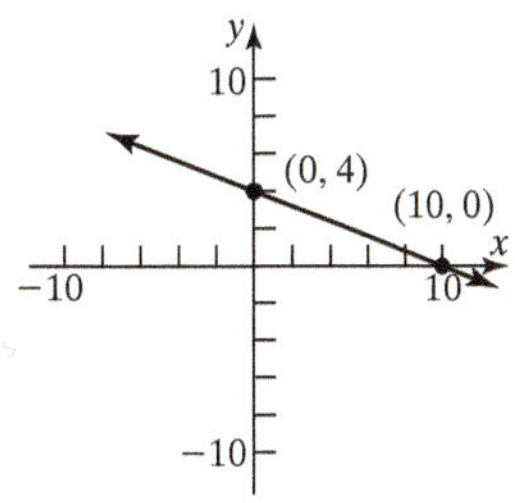

(b) Domain: $(-\infty, \infty)$
Range: $(-\infty, \infty)$

(c) Decreasing: $(-\infty, \infty)$

67. (a)
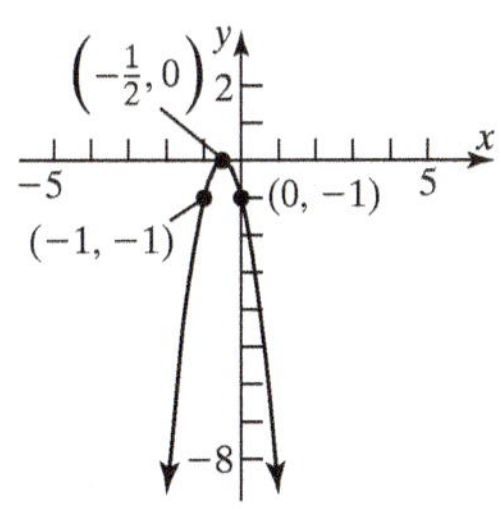

(b) Domain: $(-\infty, \infty)$
Range: $(-\infty, 0]$

(c) Increasing: $\left(-\infty, -\frac{1}{2}\right)$
Decreasing: $\left(-\frac{1}{2}, \infty\right)$

69. $a = 6, b = 0, c = 2$

71. **(a)**, **(c)**, **(d)**

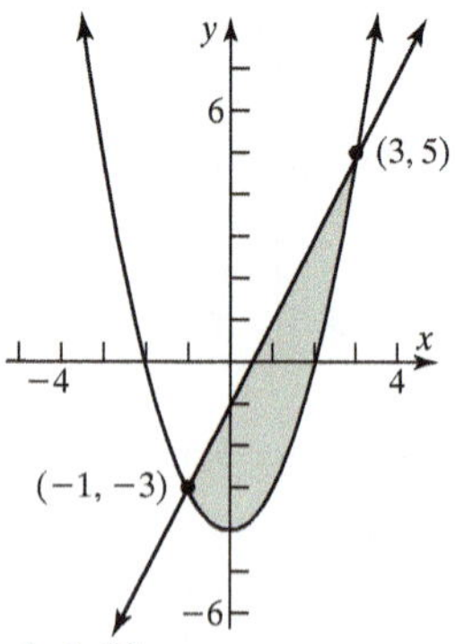

(b) $\{-1, 3\}$

73. **(a)**, **(c)**, **(d)**

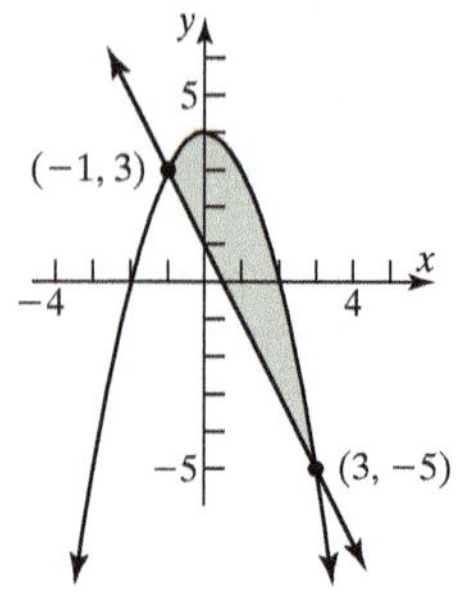

(b) $\{-1, 3\}$

75. **(a)**, **(c)**, **(d)**

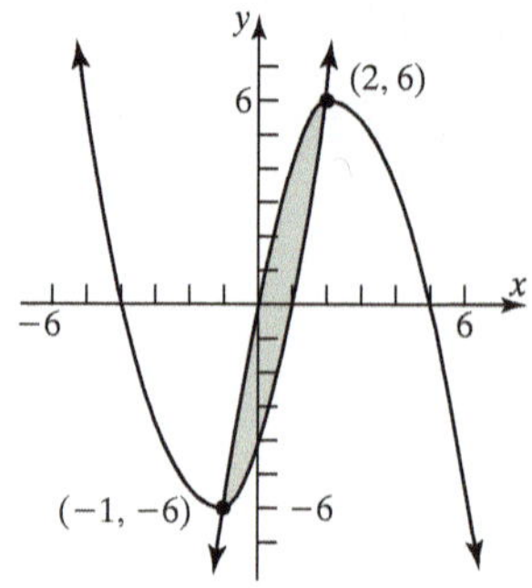

(b) $\{-1, 2\}$

77. **(a)** $a = 1: f(x) = (x + 3)(x - 1) = x^2 + 2x - 3$

$a = 2: f(x) = 2(x + 3)(x - 1) = 2x^2 + 4x - 6$

$a = -2: f(x) = -2(x + 3)(x - 1) = -2x^2 - 4x + 6$

$a = 5: f(x) = 5(x + 3)(x - 1) = 5x^2 + 10x - 15$

(b) The value of a does not affect the x-intercepts, but it changes the y-intercept by a factor of a.

(c) The value of a does not affect the axis of symmetry. It is $x = -1$ for all values of a.

(d) The value of a does not affect the x-coordinate of the vertex. However, the y-coordinate of the vertex is multiplied by a.

(e) The mean of the x-intercepts is the x-coordinate of the vertex.

79. **(a)** $(-2, -25)$ **(b)** $-7, 3$

(c) $-4, 0; (-4, -21), (0, -21)$

(d)

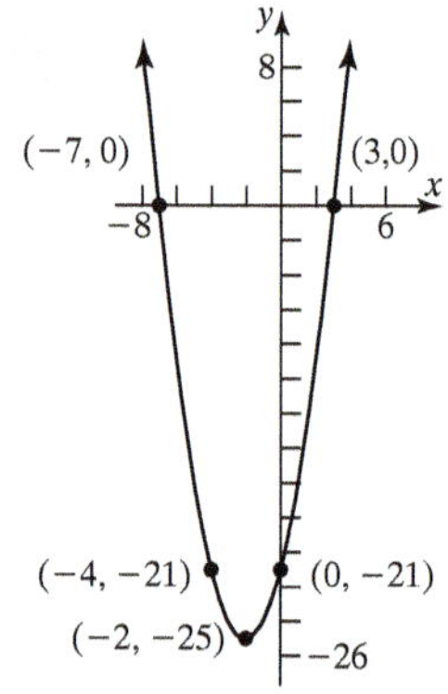

81. $(2, 2)$ **83.** \$500; \$1,000,000 **85.** **(a)** 70,000 mp3 players **(b)** \$2500 **87.** **(a)** 37.4 years old; 3784 **(b)** Decreasing

89. **(a)** 187 or 188 watches; \$7031.20 **(b)** $P(x) = -0.2x^2 + 43x - 1750$ **(c)** 107 or 108 watches; \$561.20

91. If x is even, then ax^2 and bx are even and $ax^2 + bx$ is even, which means that $ax^2 + bx + c$ is odd. If x is odd, then ax^2 and bx are odd and $ax^2 + bx$ is even, which means that $ax^2 + bx + c$ is odd. In either case, $f(x)$ is odd.

93.

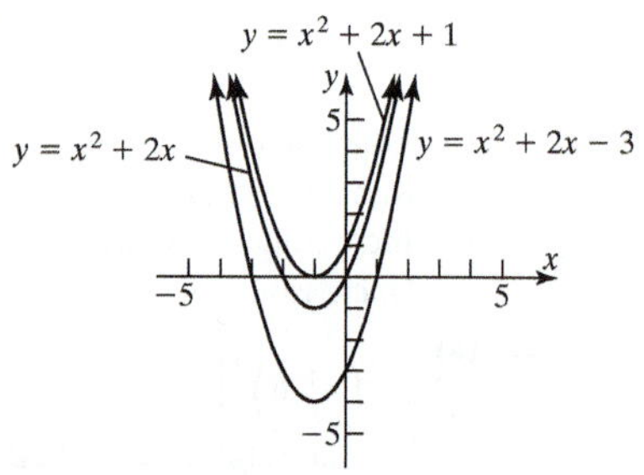

95. $b^2 - 4ac < 0$

97. No

4.4 Assess Your Understanding *(page 316)*

3. **(a)** $R(x) = -\frac{1}{6}x^2 + 100x$ **(b)** $\{x|0 \le x < 600\}$ **(c)** \$13,333.33 **(d)** 300; \$15,000 **(e)** \$50 **5.** **(a)** $R(x) = -\frac{1}{5}x^2 + 20x$ **(b)** \$255 **(c)** 50; \$500 **(d)** \$10 **7.** **(a)** $A(w) = -w^2 + 200w$ **(b)** A is largest when $w = 100$ yd. **(c)** 10,000 yd^2 **9.** 2,000,000 m^2

11. **(a)** $\frac{625}{16} \approx 39$ ft **(b)** $\frac{7025}{32} \approx 219.5$ ft

(c) About 170 ft

(d)

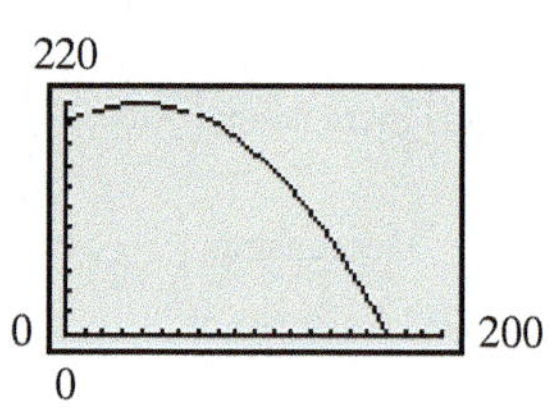

(f) When the height is 100 ft, the projectile is about 135.7 ft from the cliff.

13. 18.75 m

15. 3 in.

17. $\frac{750}{\pi} \approx 238.73$ m by 375 m

19. **(a)** 171 ft **(b)** 49 mph

21. $x = \frac{a}{2}$

23. $\frac{38}{3}$

25. $\frac{248}{3}$

27. **(a)**

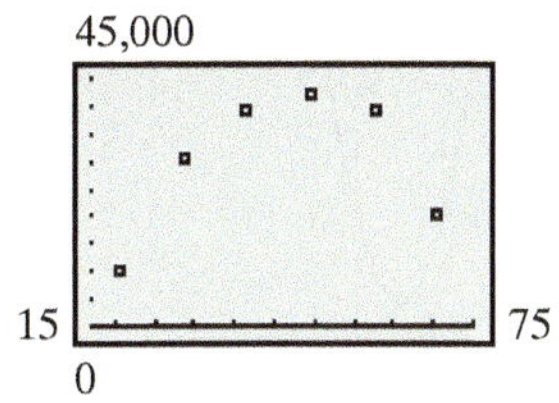

The data appear to follow a quadratic relation with $a < 0$.

(b) $I(x) = -43.384x^2 + 4089.654x - 53{,}244.570$ **(c)** About 47 years of age **(d)** Approximately \$43,134 **(e)**

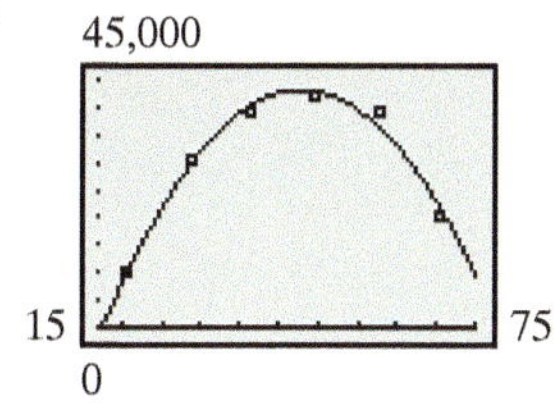

29. **(a)**

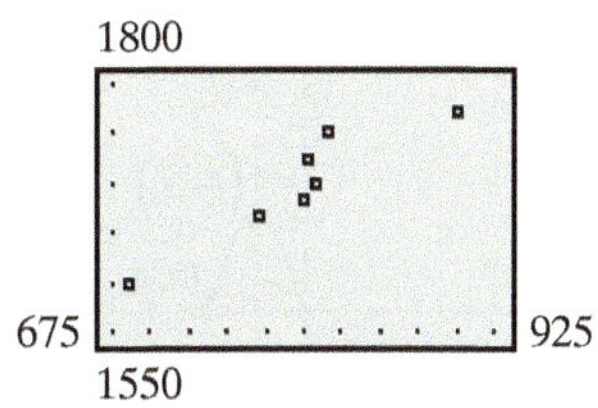

The data appear to be linearly related with positive slope.

(b) $R(x) = 0.836x + 1032.273$ **(c)** \$1743

31. **(a)**

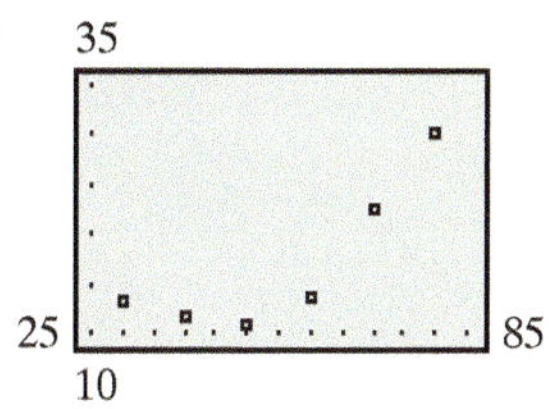

The data appear to follow a quadratic relation with $a > 0$.

(b) $P(a) = 0.015a^2 - 1.332a + 39.823$ **(c)** 11.6%

4.5 Assess Your Understanding *(page 322)*

3. **(a)** $\{x|x < -2 \text{ or } x > 2\}; (-\infty, -2) \cup (2, \infty)$ **(b)** $\{x|-2 \le x \le 2\}; [-2, 2]$

5. **(a)** $\{x|-2 \le x \le 1\}; [-2, 1]$ **(b)** $\{x|x < -2 \text{ or } x > 1\}; (-\infty, -2) \cup (1, \infty)$ **7.** $\{x|-2 \le x \le 5\}; [-2, 5]$

9. $\{x|x < 0 \text{ or } x > 4\}; (-\infty, 0) \cup (4, \infty)$ **11.** $\{x|-3 < x < 3\}; (-3, 3)$ **13.** $\{x|x < -4 \text{ or } x > 3\}; (-\infty, -4) \cup (3, \infty)$

15. $\left\{x\left|-\frac{1}{2} \le x \le 3\right.\right\}; \left[-\frac{1}{2}, 3\right]$ **17.** $\{x|x < -1 \text{ or } x > 8\}; (-\infty, -1) \cup (8, \infty)$ **19.** No real solution **21.** All real numbers; $(-\infty, \infty)$

23. $\{x|x \le -4 \text{ or } x \ge 4\}; (-\infty, -4] \cup [4, \infty)$ **25.** **(a)** $\{-1, 1\}$ **(b)** $\{-1\}$ **(c)** $\{-1, 4\}$ **(d)** $\{x|x < -1 \text{ or } x > 1\}; (-\infty, -1) \cup (1, \infty)$ **(e)** $\{x|x \le -1\}; (-\infty, -1]$ **(f)** $\{x|x < -1 \text{ or } x > 4\}; (-\infty, -1) \cup (4, \infty)$ **(g)** $\{x|x \le -\sqrt{2} \text{ or } x \ge \sqrt{2}\}; (-\infty, -\sqrt{2}] \cup [\sqrt{2}, \infty)$

27. **(a)** $\{-1, 1\}$ **(b)** $\left\{-\frac{1}{4}\right\}$ **(c)** $\{-4, 0\}$ **(d)** $\{x|-1 < x < 1\}; (-1, 1)$ **(e)** $\left\{x\left|x \le -\frac{1}{4}\right.\right\}$ or $\left(-\infty, -\frac{1}{4}\right]$ **(f)** $\{x|-4 < x < 0\}; (-4, 0)$ **(g)** $\{0\}$ **29.** **(a)** $\{-2, 2\}$ **(b)** $\{-2, 2\}$ **(c)** $\{-2, 2\}$ **(d)** $\{x|x < -2 \text{ or } x > 2\}; (-\infty, -2) \cup (2, \infty)$ **(e)** $\{x|x \le -2 \text{ or } x \ge 2\}; (-\infty, -2] \cup [2, \infty)$ **(f)** $\{x|x < -2 \text{ or } x > 2\}; (-\infty, -2) \cup (2, \infty)$ **(g)** $\{x|x \le -\sqrt{5} \text{ or } x \ge \sqrt{5}\}; (-\infty, -\sqrt{5}] \cup [\sqrt{5}, \infty)$ **31.** **(a)** $\{-1, 2\}$ **(b)** $\{-2, 1\}$ **(c)** $\{0\}$ **(d)** $\{x|x < -1 \text{ or } x > 2\}; (-\infty, -1) \cup (2, \infty)$ **(e)** $\{x|-2 \le x \le 1\}; [-2, 1]$ **(f)** $\{x|x < 0\}; (-\infty, 0)$ **(g)** $\left\{x\left|x \le \frac{1-\sqrt{13}}{2} \text{ or } x \ge \frac{1+\sqrt{13}}{2}\right.\right\}; \left(-\infty, \frac{1-\sqrt{13}}{2}\right] \cup \left[\frac{1+\sqrt{13}}{2}, \infty\right)$

33. **(a)** 5 sec **(b)** The ball is more than 96 ft above the ground from time t between 2 and 3 sec, $2 < t < 3$.

35. **(a)** \$0, \$1000 **(b)** The revenue is more than \$800,000 for prices between \$276.39 and \$723.61, $\$276.39 < p < \$723.61$.

37. **(a)** $\{c \mid 0.112 < c < 81.907\}; (0.112, 81.907)$ **(b)** It is possible to hit a target 75 km away if $c = 0.651$ or $c = 1.536$.

Review Exercises *(page 324)*

1. **(a)** $m = 2; b = -5$

(b)

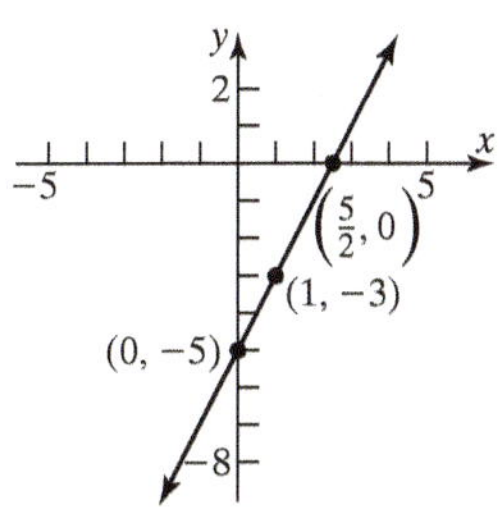

(c) Domain and Range: $(-\infty, \infty)$

(d) 2 **(e)** Increasing

3. **(a)** $m = \frac{4}{5}; b = -6$

(b)

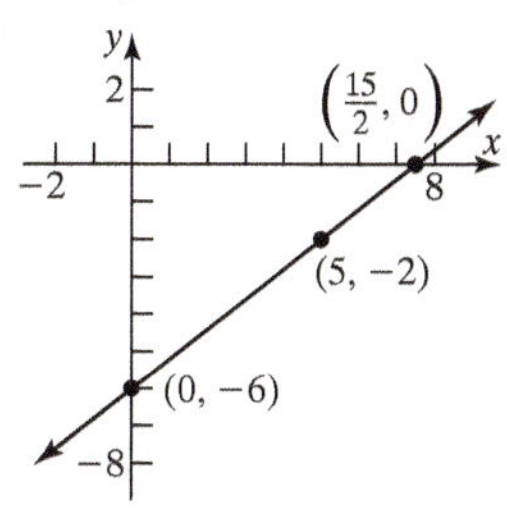

(c) Domain and Range: $(-\infty, \infty)$

(d) $\frac{4}{5}$ **(e)** Increasing

5. **(a)** $m = 0; b = 4$

(b)

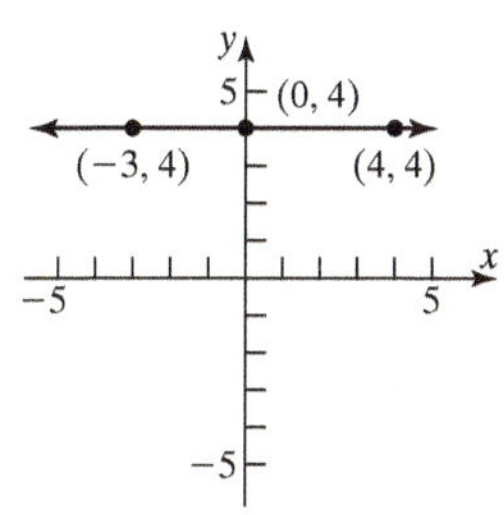

(c) Domain: $(-\infty, \infty)$

Range: $\{y \mid y = 4\}$

(d) 0 **(e)** Constant

7. Linear; $y = 5x + 3$

9.

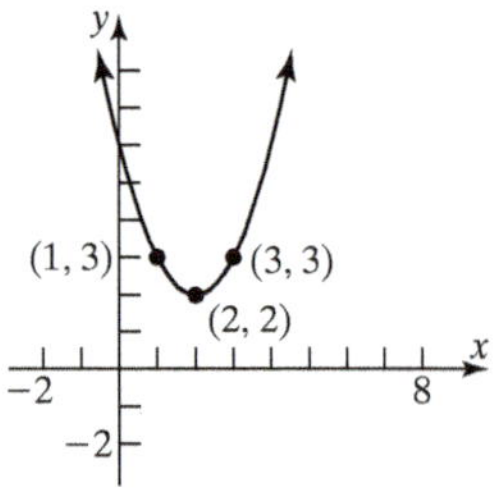

11.

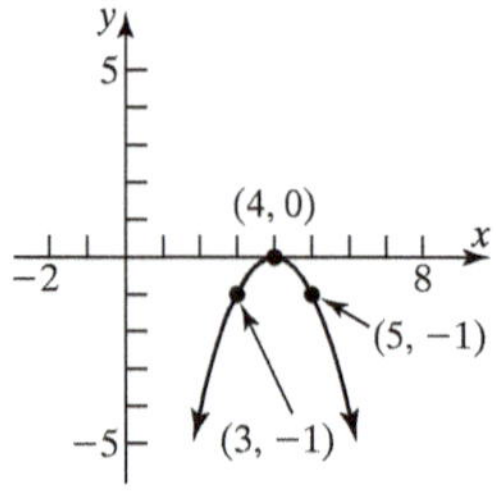

13.

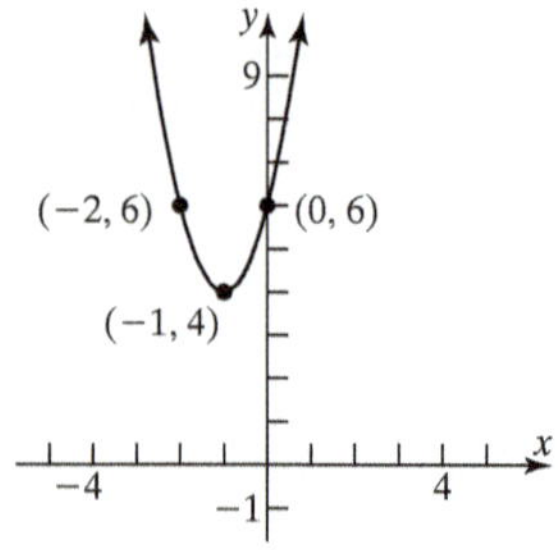

15. (a)

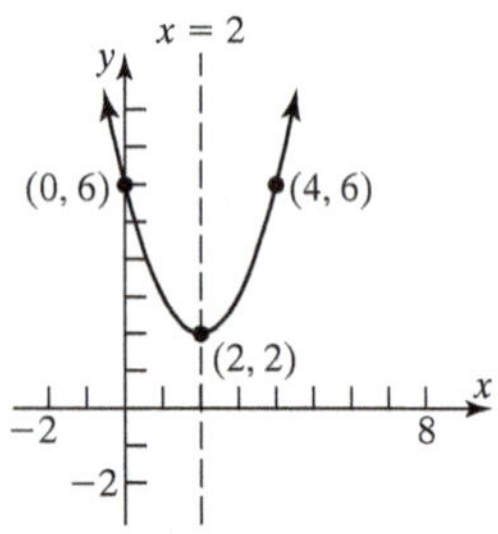

(b) Domain: $(-\infty, \infty)$

Range: $[2, \infty)$

(c) Decreasing: $(-\infty, 2)$

Increasing: $(2, \infty)$

17. (a)

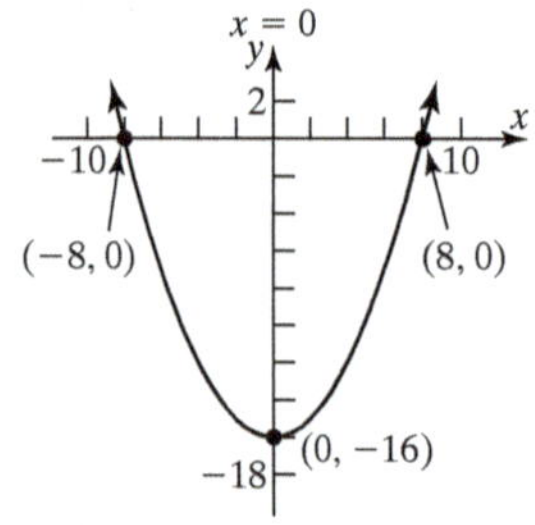

(b) Domain: $(-\infty, \infty)$

Range: $[-16, \infty)$

(c) Decreasing: $(-\infty, 0)$

Increasing: $(0, \infty)$

19. (a)

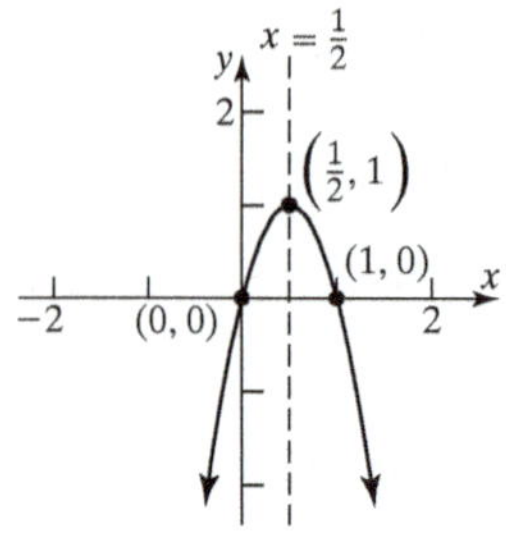

(b) Domain: $(-\infty, \infty)$

Range: $(-\infty, 1]$

(c) Increasing: $\left(-\infty, \frac{1}{2}\right)$

Decreasing: $\left(\frac{1}{2}, \infty\right)$

21. (a)

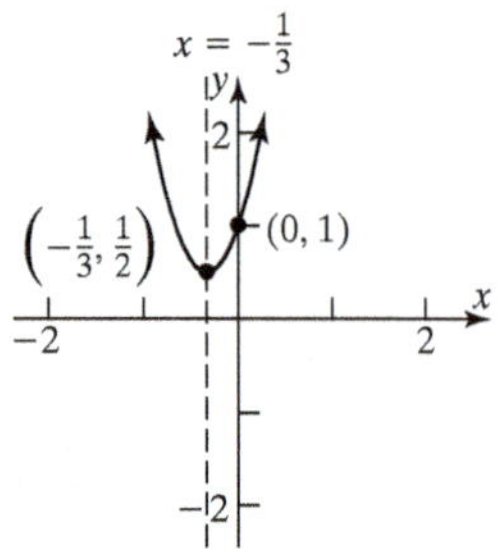

(b) Domain: $(-\infty, \infty)$

Range: $\left[\frac{1}{2}, \infty\right)$

(c) Decreasing: $\left(-\infty, -\frac{1}{3}\right)$

Increasing: $\left(-\frac{1}{3}, \infty\right)$

23. (a)

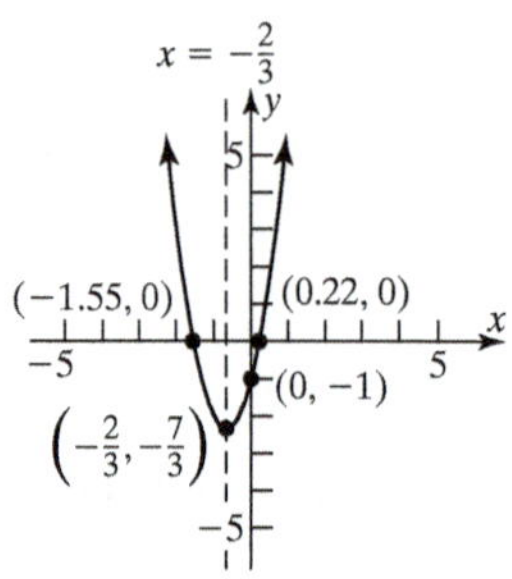

(b) Domain: $(-\infty, \infty)$

Range: $\left[-\frac{7}{3}, \infty\right)$

(c) Decreasing: $\left(-\infty, -\frac{2}{3}\right)$

Increasing: $\left(-\frac{2}{3}, \infty\right)$

25. Minimum value; 1 **27.** Maximum value; 12 **29.** Maximum value; 16 **31.** $\{x \mid -8 < x < 2\}$; $(-8, 2)$

33. $\left\{x \mid x \leq -\frac{1}{3} \text{ or } x \geq 5\right\}$; $\left(-\infty, -\frac{1}{3}\right] \cup [5, \infty)$ **35. (a)** Company A: $C(x) = 0.06x + 7$; Company B: $C(x) = 0.08x$ **(b)** 350 min **(c)** $0 \leq x < 350$ **37. (a)** $R(x) = -\frac{1}{10}x^2 + 150x$ **(b)** \$14,000 **(c)** 750; \$56,250 **(d)** \$75 **39.** 4,166,666.7 m^2

41. (a) 63 clubs **(b)** \$151.90 **43.** $A(x) = -x^2 + 10x, 0 < x < 10$; 25 units^2

45. (a)

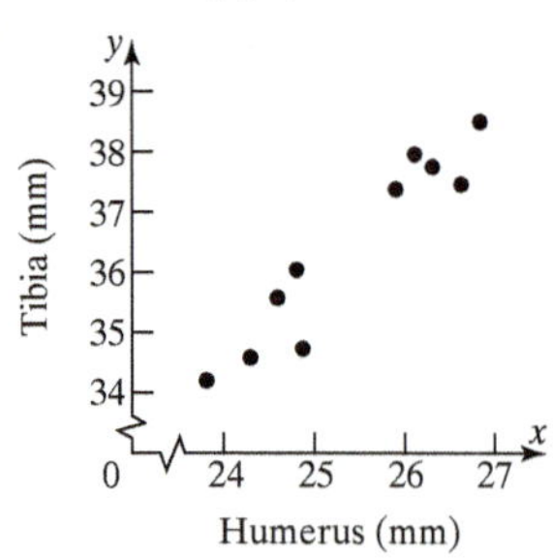

(b) Yes; the two variables appear to be linearly related with positive slope.

(c) $y = 1.3902x + 1.1140$

(d) 37.95 mm

Chapter Test *(page 327)*

1. (a) Slope: -4; y-intercept: 3 **(b)** Decreasing **(c)**

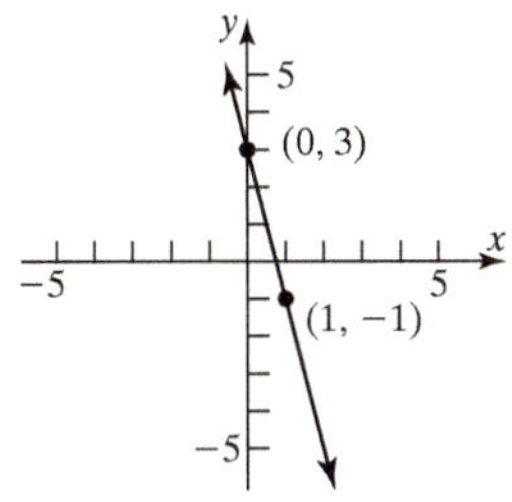

2. Linear; $y = -5x + 2$ **3.**

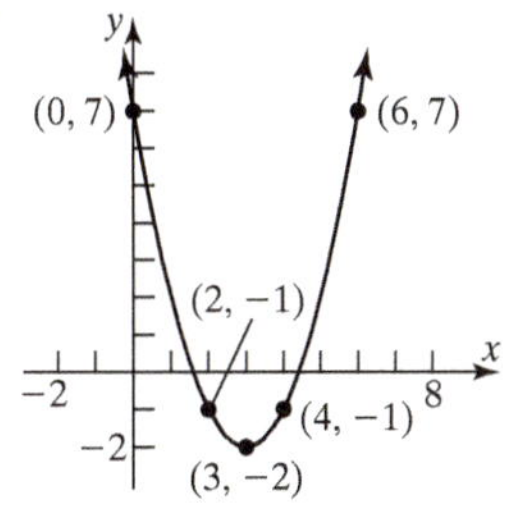

4. (a) Opens up **(b)** $(2, -8)$ **(c)** $x = 2$ **(d)** x-intercepts: $\frac{6 - 2\sqrt{6}}{3}, \frac{6 + 2\sqrt{6}}{3}$; y-intercept: 4 **(e)**

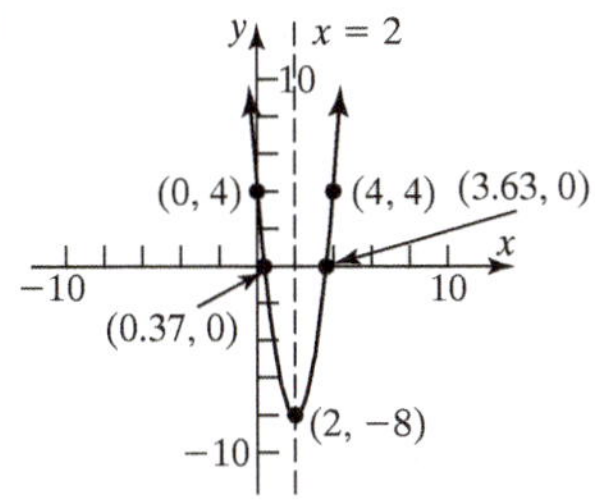

(f) Domain: All real numbers; $(-\infty, \infty)$ Range: $\{y|y \geq -8\}$; $[-8, \infty)$ **(g)** Decreasing: $(-\infty, 2)$; Increasing: $(2, \infty)$

5. (a) Opens down **(b)** $(1, -3)$ **(c)** $x = 1$ **(d)** No x-intercepts: y-intercept: -5 **(e)**

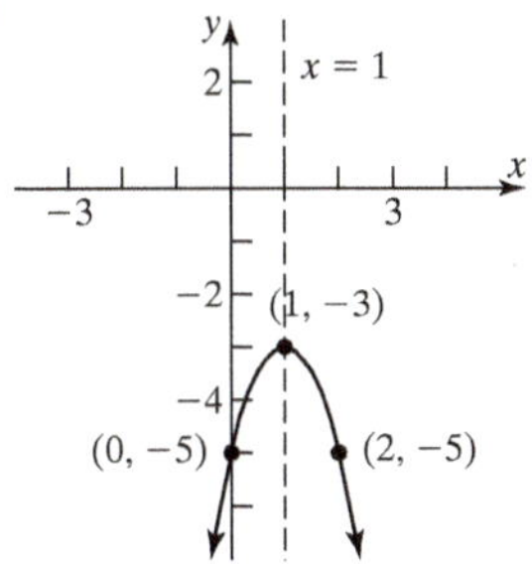

(f) Domain: All real numbers; $(-\infty, \infty)$ Range: $\{y \mid y \leq -3\}$; $(-\infty, -3]$ **(g)** Increasing: $(-\infty, 1)$; Decreasing: $(1, \infty)$

6. $f(x) = 2x^2 - 4x - 30$ **7.** Maximum value; 21 **8.** $\{x|x \leq 4 \text{ or } x \geq 6\}$; $(-\infty, 4] \cup [6, \infty)$

9. (a) $C(m) = 0.15m + 129.50$ **(b)** \$258.50 **(c)** 562 miles **10. (a)** $R(x) = -\frac{1}{10}x^2 + 1000x$ **(b)** \$384,000 **(c)** 5000 units; \$2,500,000 **(d)** \$500

11. (a)

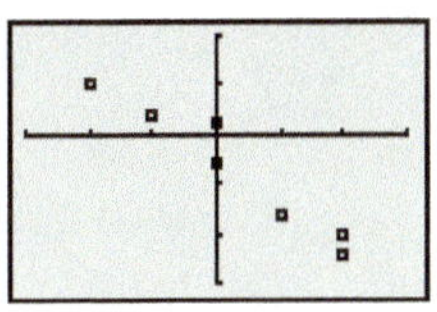

Linear with negative slope

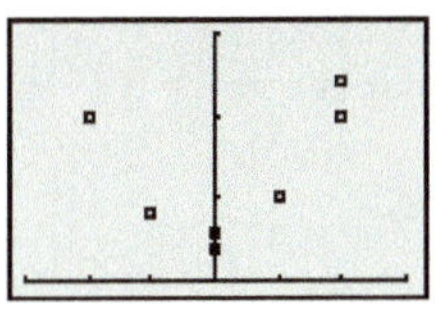

Quadratic that opens up

(b) $y = -4.234x - 2.362$ **(c)** $y = 1.993x^2 + 0.289x + 2.503$

Cumulative Review *(page 327)*

1. (a) -3 **(b)** $x^2 - 4x - 2$ **(c)** $x^2 + 4x + 1$ **(d)** $-x^2 + 4x - 1$ **(e)** $x^2 - 3$ **(f)** $2x + h - 4$ **2.** $\left\{z \middle| z \neq \frac{7}{6}\right\}$ **3.** Yes **4. (a)** No **(b)** -1; $(-2, -1)$ is on the graph. **(c)** -8; $(-8, 2)$ is on the graph. **5.** Neither **6.** Local maximum is 5.30 and occurs at $x = -1.29$. Local minimum is -3.30 and occurs at $x = 1.29$. Increasing: $(-4, -1.29)$ and $(1.29, 4)$; Decreasing: $(-1.29, 1.29)$ **7. (a)** $\{-4\}$ **(b)** $\{x|x > -4\}$ or $(-4, \infty)$ **8. (a)** $\{-1, 3\}$ **(b)** $\{x|x < -1 \text{ or } x > 3\}$ or $(-\infty, -1) \cup (3, \infty)$

9. (a) Domain: $\{x|-4 \leq x \leq 4\}$ or $[-4, 4]$; Range: $\{y|-1 \leq y \leq 3\}$ or $[-1, 3]$ **(b)** $(-1, 0), (0, -1), (1, 0)$ **(c)** y-axis **(d)** 1 **(e)** -4 and 4 **(f)** $\{x|-1 < x < 1\}$

(g)

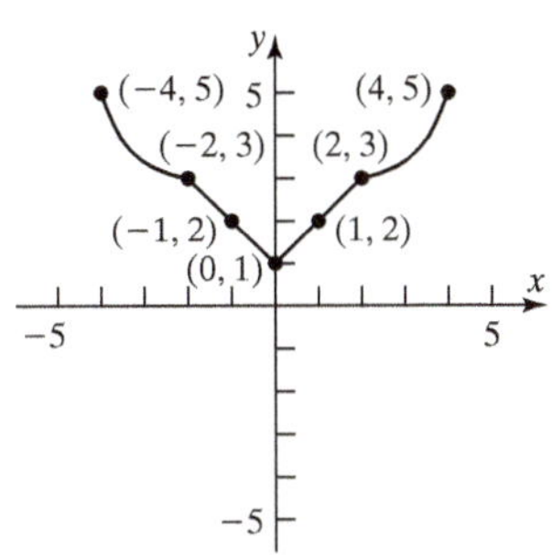

(h)

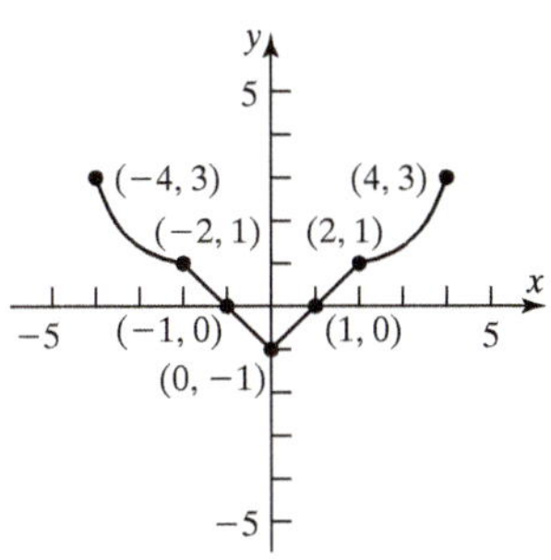

(i)

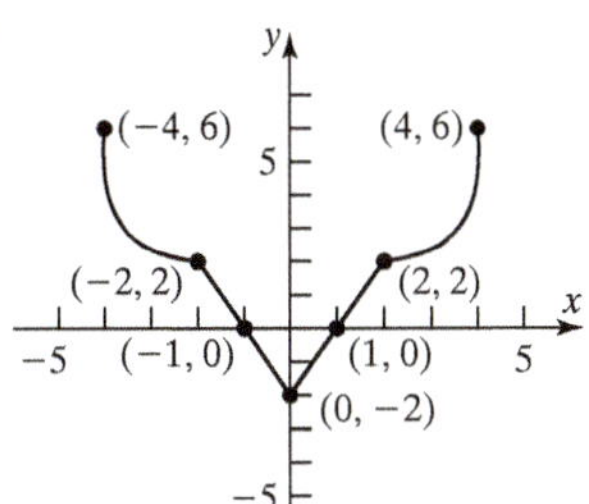

(j) Even **(k)** $(0, 4)$

CHAPTER 5 Polynomial and Rational Functions

5.1 Assess Your Understanding *(page 343)*

5. smooth; continuous **6.** zero or root **7.** touches **8.** T **9.** F **10.** F **11.** Yes; degree 3 **13.** Yes; degree 2 **15.** No; x is raised to the -1 power. **17.** No; x is raised to the $\frac{3}{2}$ power. **19.** Yes; degree 4 **21.** Yes; degree 4

23.
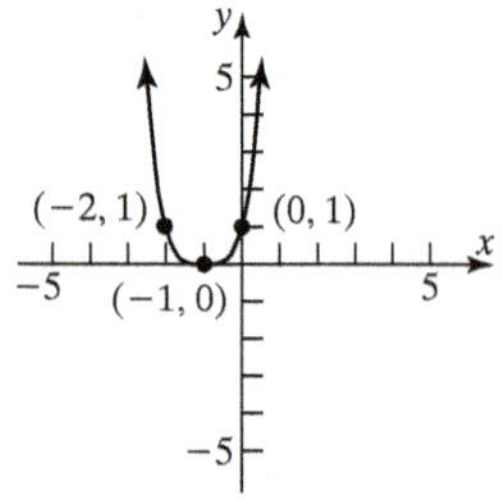

25.
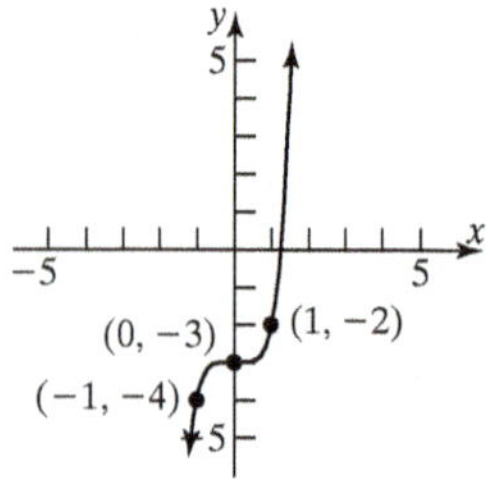

27.
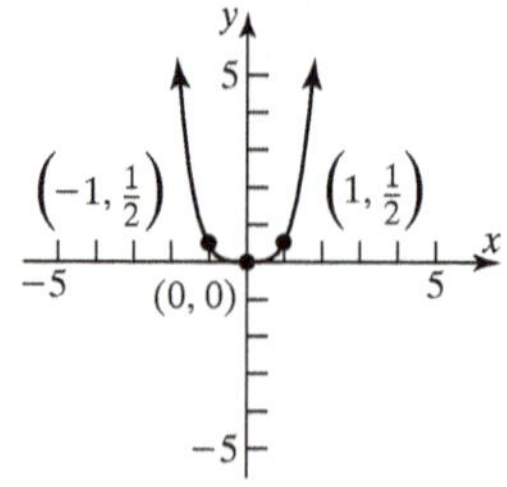

29.
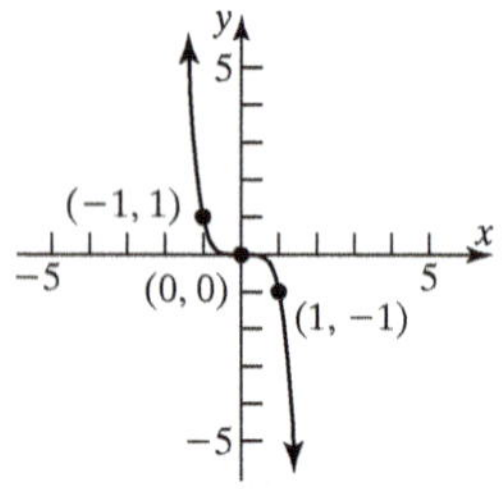

31.
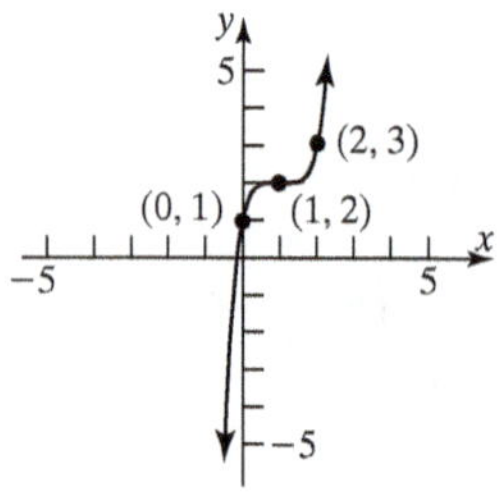

33.
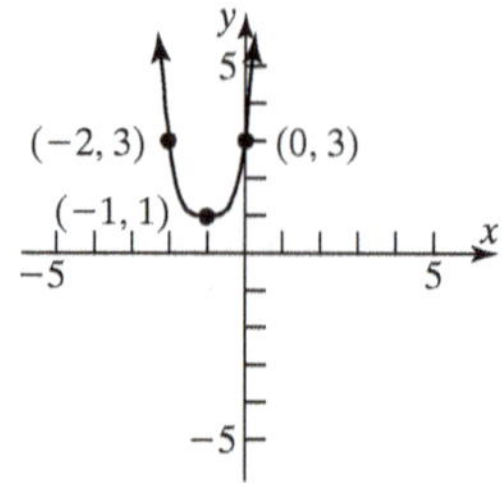

35.
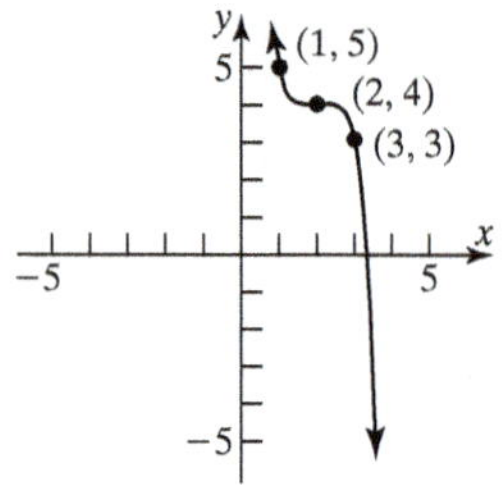

37. $f(x) = x^3 - 3x^2 - x + 3$ for $a = 1$
39. $f(x) = x^3 - x^2 - 12x$ for $a = 1$
41. $f(x) = x^4 - 15x^2 + 10x + 24$ for $a = 1$
43. $f(x) = x^3 - 5x^2 + 3x + 9$ for $a = 1$
45. **(a)** 7, multiplicity 1; −3, multiplicity 2
(b) Graph touches the x-axis at −3 and crosses it at 7.
(c) 2 **(d)** $y = 3x^3$

47. **(a)** 2, multiplicity 3 **(b)** Graph crosses the x-axis at 2 **(c)** 4 **(d)** $y = 4x^5$ **49.** **(a)** $-\frac{1}{2}$, multiplicity 2; −4, multiplicity 3 **(b)** Graph touches the x-axis at $-\frac{1}{2}$ and crosses it at −4. **(c)** 4 **(d)** $y = -2x^5$ **51.** **(a)** 5, multiplicity 3; −4, multiplicity 2 **(b)** Graph touches the x-axis at −4 and crosses it at 5. **(c)** 4 **(d)** $y = x^5$ **53.** **(a)** No real zeros **(b)** Graph neither crosses nor touches the x-axis. **(c)** 5 **(d)** $y = 3x^6$
55. **(a)** 0, multiplicity 2; $-\sqrt{2}, \sqrt{2}$, multiplicity 1 **(b)** Graph touches the x-axis at 0 and crosses it at $-\sqrt{2}$ and $\sqrt{2}$. **(c)** 3 **(d)** $y = -2x^4$
57. Could be; zeros: −1, 1, 2; least degree is 3. **59.** Cannot be the graph of a polynomial; gap at $x = -1$ **61.** $f(x) = x(x - 1)(x - 2)$
63. $f(x) = -\frac{1}{2}(x + 1)(x - 1)(x - 2)$

65. Step 1: $y = x^3$
Step 2: x-intercepts: 0, 3; y-intercept: 0
Step 3: 0: multiplicity 2; touches;
3: multiplicity 1; crosses
Step 4:
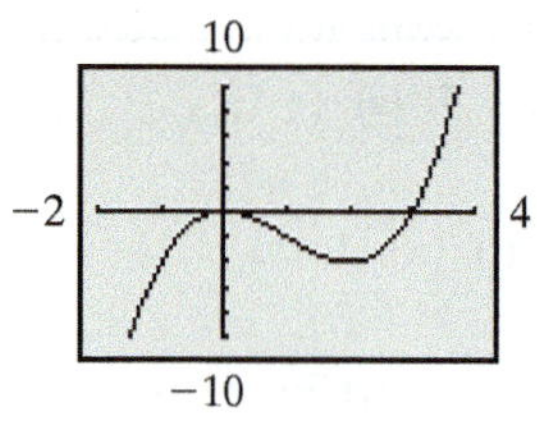

Step 5: (2, −4); (0, 0)
Step 6:
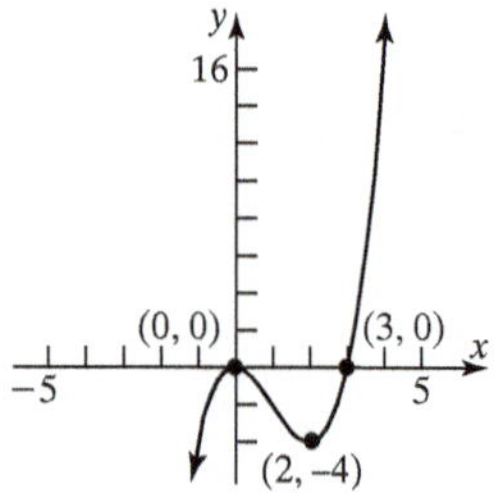

Step 7: Domain: $(-\infty, \infty)$;
Range: $(-\infty, \infty)$
Step 8: Increasing on $(-\infty, 0)$ and $(2, \infty)$
Decreasing on (0, 2)

67. Step 1: $y = x^3$
Step 2: x-intercepts: −4, 2
y-intercept: 16
Step 3: −4: multiplicity 1, crosses; 2: multiplicity 2, touches
Step 4:
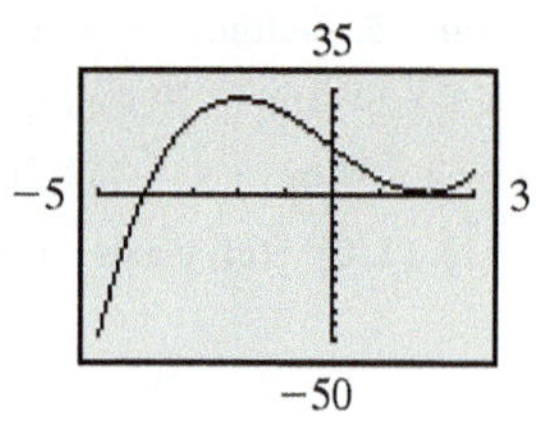

Step 5: (−2, 32); (2, 0)
Step 6:
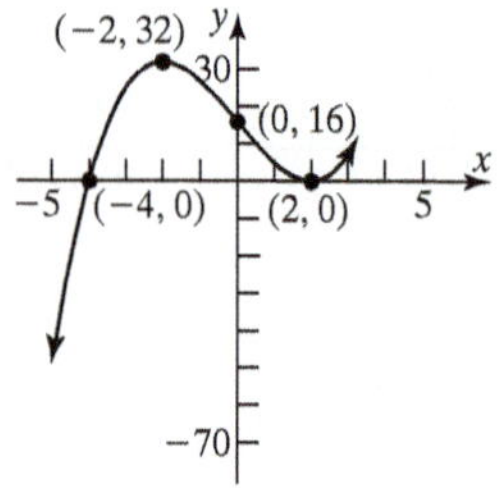

Step 7: Domain: $(-\infty, \infty)$; Range: $(-\infty, \infty)$
Step 8: Increasing on $(-\infty, -2)$ and $(2, \infty)$
Decreasing on (−2, 2)

69. Step 1: $y = -2x^4$
Step 2: x-intercepts: −2, 2
y-intercept: 32
Step 3: −2: multiplicity 1, crosses;
2: multiplicity 3, crosses
Step 4:
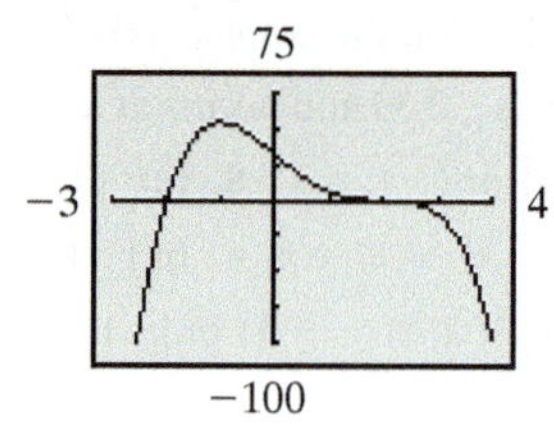

Step 5: (−1, 54)
Step 6:
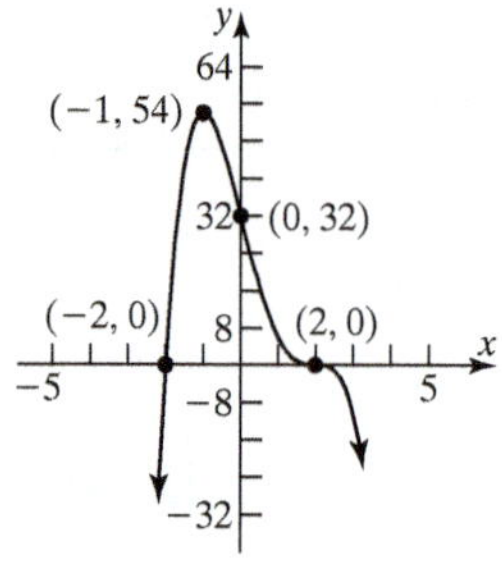

Step 7: Domain: $(-\infty, \infty)$; Range: $(-\infty, 54]$
Step 8: Increasing on $(-\infty, -1)$
Decreasing on $(-1, \infty)$

71. Step 1: $y = x^3$

Step 2: x-intercepts: $-4, -1, 2$ y-intercept: -8

Step 3: $-4, -1, 2$: multiplicity 1, crosses

Step 4:

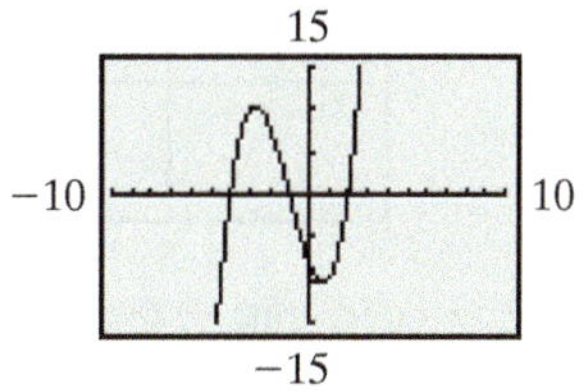

Step 5: $(-2.73, 10.39)$; $(0.73, -10.39)$

Step 6:

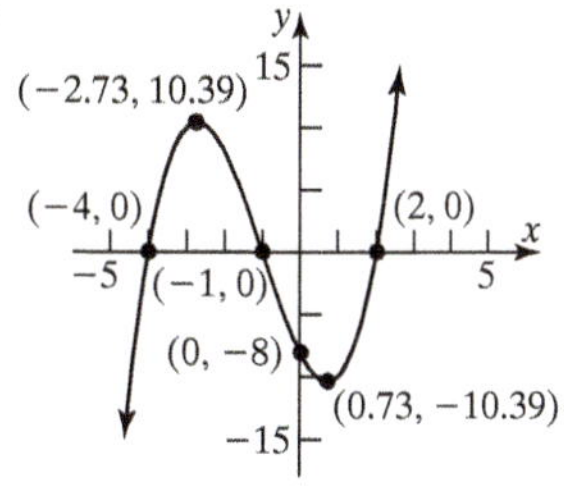

Step 7: Domain: $(-\infty, \infty)$; Range: $(-\infty, \infty)$

Step 8: Increasing on $(-\infty, -2.73)$ and $(0.73, \infty)$

Decreasing on $(-2.73, 0.73)$

73. Step 1: $y = x^4$

Step 2: x-intercepts: $-2, 0, 2$ y-intercept: 0

Step 3: $-2, 2$: multiplicity 1, crosses; 0: multiplicity 2, touches

Step 4:

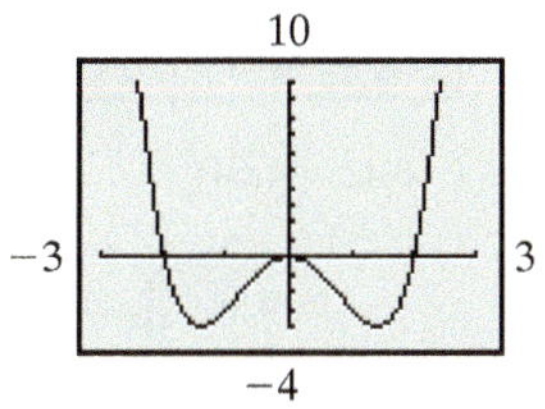

Step 5: $(-1.41, -4)$; $(1.41, -4)$; $(0, 0)$

Step 6:

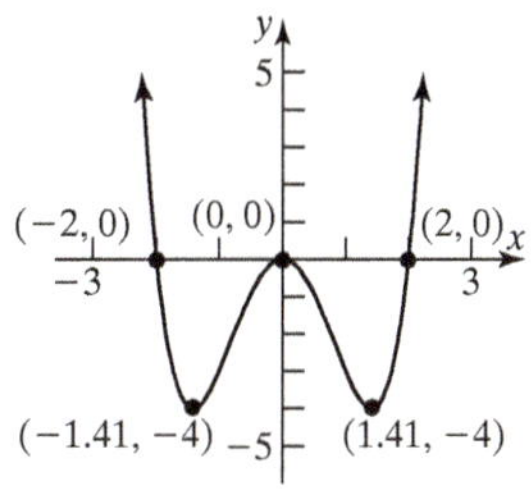

Step 7: Domain: $(-\infty, \infty)$; Range: $[-4, \infty)$

Step 8: Increasing on $(-1.41, 0)$ and $(1.41, \infty)$

Decreasing on $(-\infty, -1.41)$ and $(0, 1.41)$

75. Step 1: $y = x^4$

Step 2: x-intercepts: $-1, 2$ y-intercept: 4

Step 3: $-1, 2$: multiplicity 2, touches

Step 4:

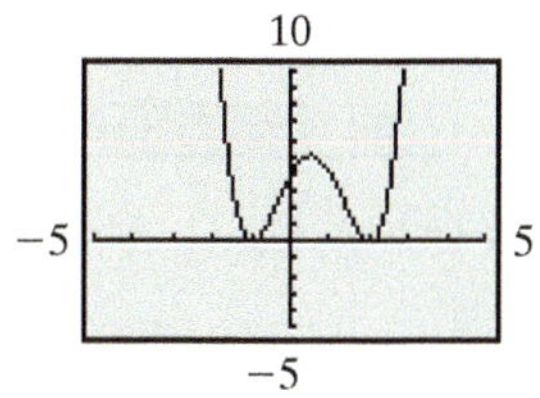

Step 5: $(-1, 0)$; $(2, 0)$; $(0.5, 5.06)$

Step 6:

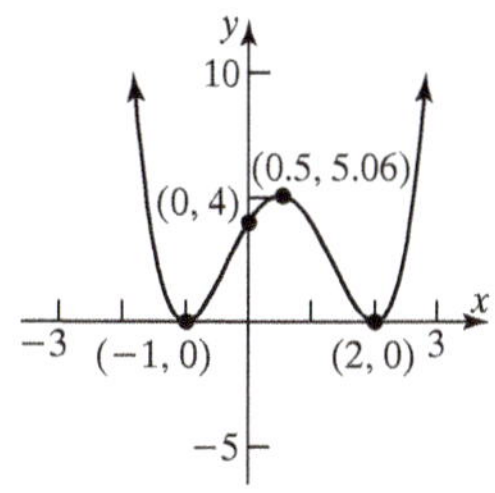

Step 7: Domain: $(-\infty, \infty)$; Range: $[0, \infty)$

Step 8: Increasing on $(-1, 0.5)$ and $(2, \infty)$

Decreasing on $(-\infty, -1)$ and $(0.5, 2)$

77. Step 1: $y = x^4$

Step 2: x-intercepts: $-1, 0, 3$; y-intercept: 0

Step 3: $-1, 3$: multiplicity 1, crosses; 0: multiplicity 2, touches

Step 4:

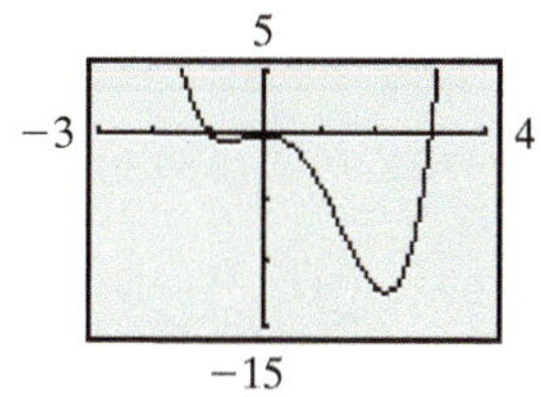

Step 5: $(-0.69, -0.54)$; $(2.19, -12.39)$; $(0, 0)$

Step 6:

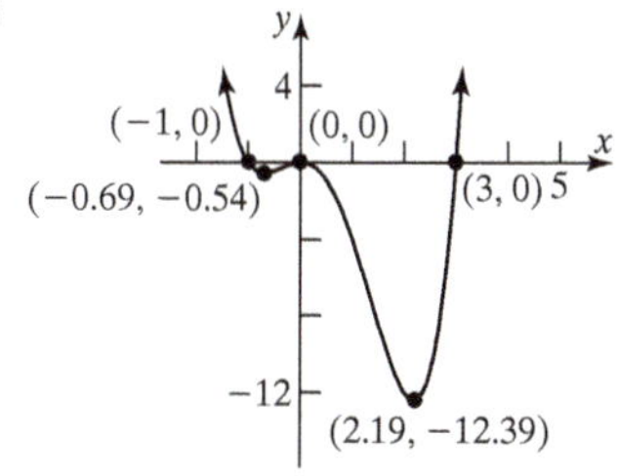

Step 7: Domain: $(-\infty, \infty)$; Range: $[-12.39, \infty)$

Step 8: Increasing on $(-0.69, 0)$ and $(2.19, \infty)$

Decreasing on $(-\infty, -0.69)$ and $(0, 2.19)$

79. Step 1: $y = x^4$

Step 2: x-intercepts: $-2, 4$; y-intercept: 64

Step 3: $-2, 4$: multiplicity 2, touches

Step 4:

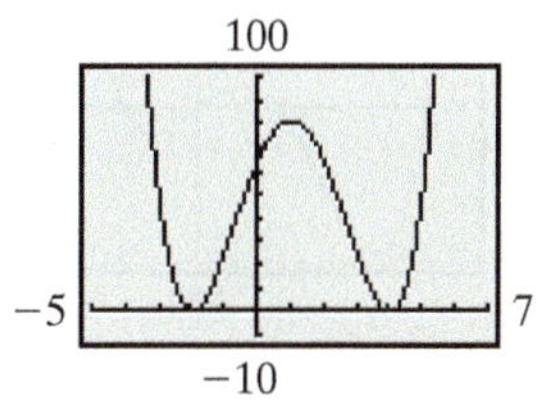

Step 5: $(-2, 0)$; $(4, 0)$; $(1, 81)$

Step 6:

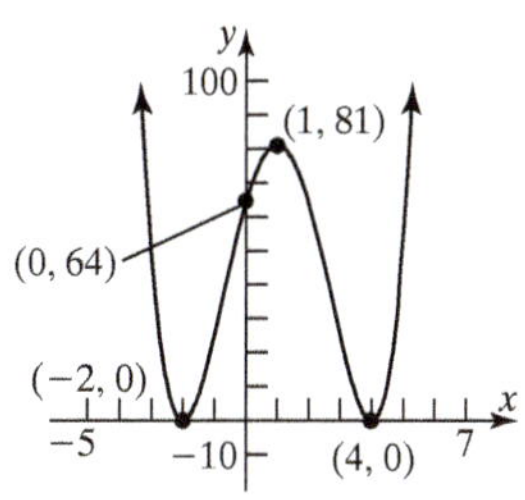

Step 7: Domain: $(-\infty, \infty)$; Range: $[0, \infty)$

Step 8: Increasing on $(-2, 1)$ and $(4, \infty)$

Decreasing on $(-\infty, -2)$ and $(1, 4)$

81. Step 1: $y = x^5$

Step 2: x-intercepts: $0, 2$; y-intercept: 0

Step 3: 0: multiplicity 2, touches; 2: multiplicity 1, crosses

Step 4:

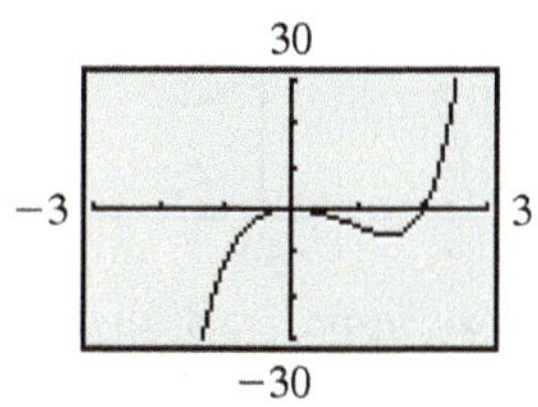

Step 5: $(0, 0)$; $(1.48, -5.91)$

Step 6:

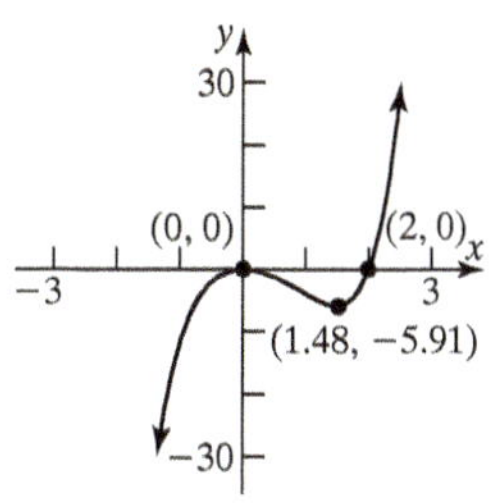

Step 7: Domain: $(-\infty, \infty)$; Range: $(-\infty, \infty)$

Step 8: Increasing on $(-\infty, 0)$ and $(1.48, \infty)$

Decreasing on $(0, 1.48)$

83. Step 1: $y = x^3$

Step 2:

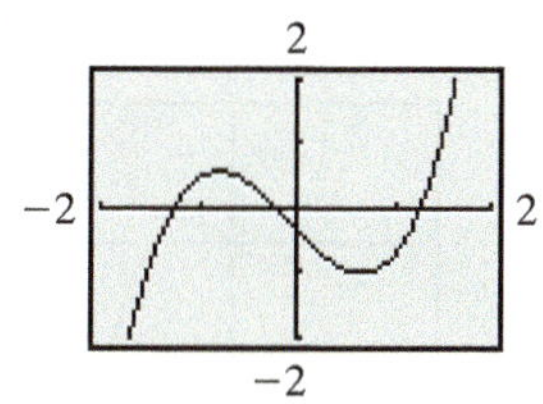

Step 3: x-intercepts: $-1.26, -0.20, 1.26$

y-intercept: -0.31752

Step 4:

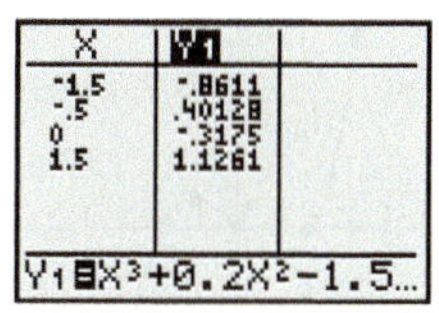

Step 5: $(-0.80, 0.57)$; $(0.66, -0.99)$

Step 6:

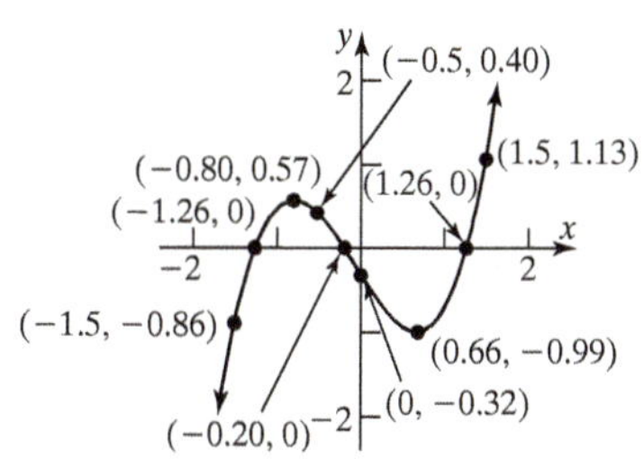

Step 7: Domain: $(-\infty, \infty)$; Range: $(-\infty, \infty)$

Step 8: Increasing on $(-\infty, -0.80)$ and $(0.66, \infty)$

Decreasing on $(-0.80, 0.66)$

85. Step 1: $y = x^3$

Step 2:

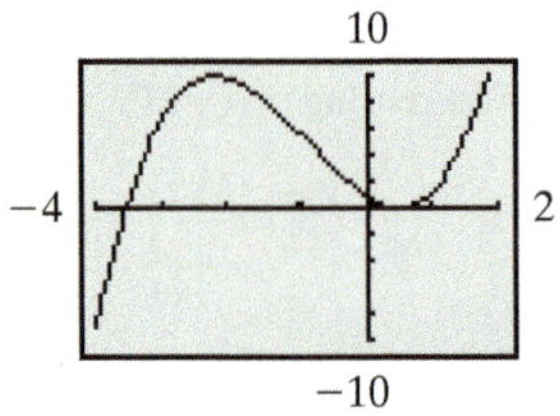

Step 3: x-intercepts: −3.56, 0.50
y-intercept: 0.89

Step 4:

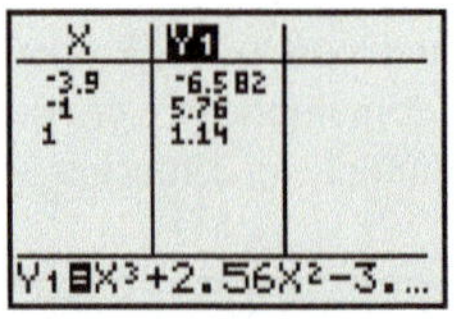

Step 5: (−2.21, 9.91); (0.50, 0)

Step 6:

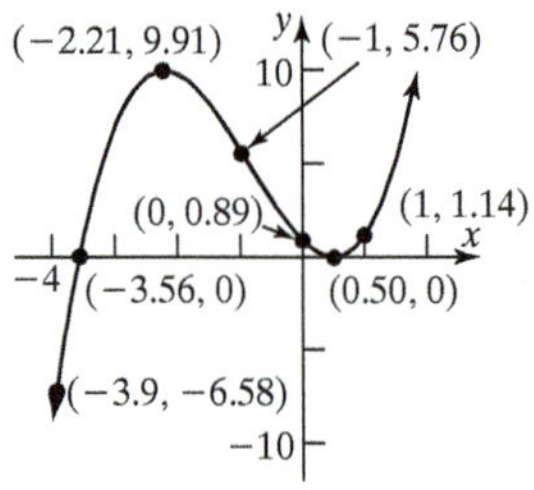

Step 7: Domain: $(-\infty, \infty)$; Range: $(-\infty, \infty)$

Step 8: Increasing on $(-\infty, -2.21)$ and $(0.50, \infty)$
Decreasing on $(-2.21, 0.50)$

87. Step 1: $y = x^3$

Step 2:

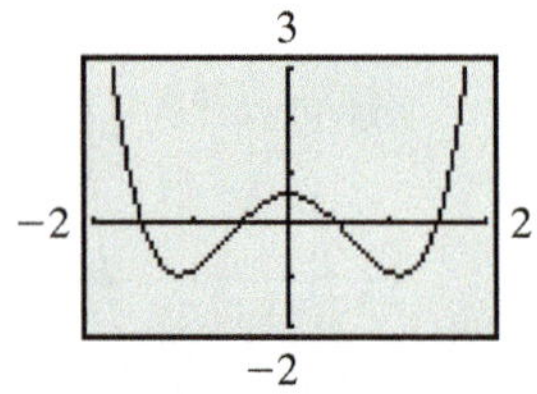

Step 3: x-intercepts: −1.5, −0.5, 0.5, 1.5
y-intercept: 0.5625

Step 4:

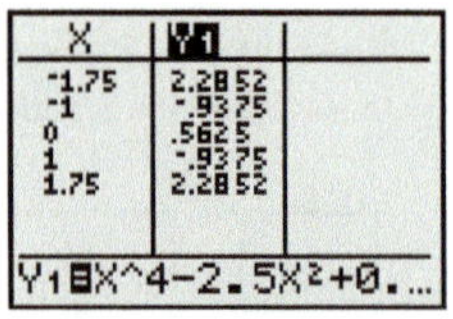

Step 5: (−1.12, −1); (1.12, −1); (0, 0.56)

Step 6:

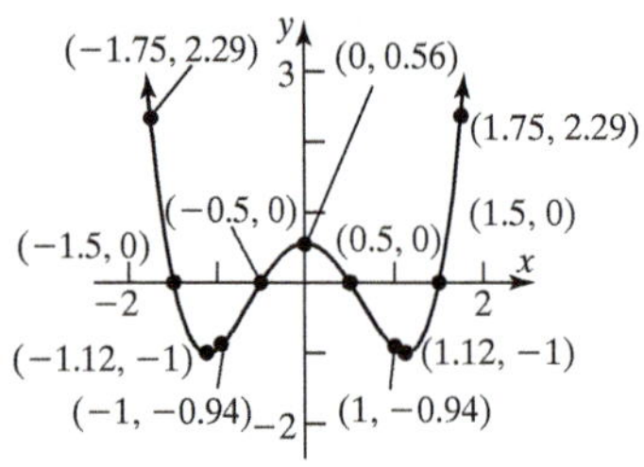

Step 7: Domain: $(-\infty, \infty)$; Range: $[-1, \infty)$

Step 8: Increasing on $(-1.12, 0)$ and $(1.12, \infty)$
Decreasing on $(-2.21, 0.50)$

89. Step 1: $y = 2x^4$

Step 2:

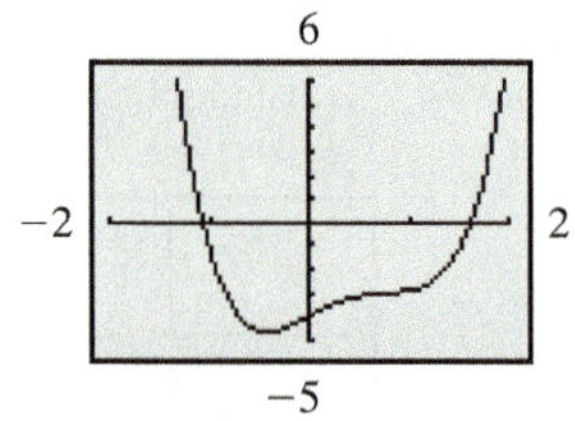

Step 3: x-intercepts: −1.07, 1.62
y-intercept: −4

Step 4:

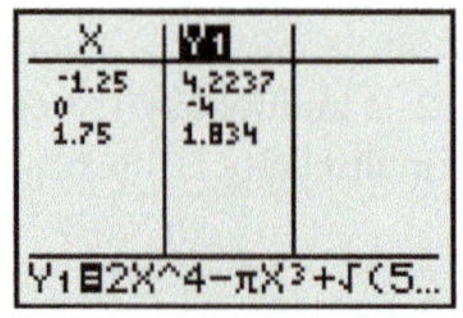

Step 5: (−0.42, −4.64)

Step 6:

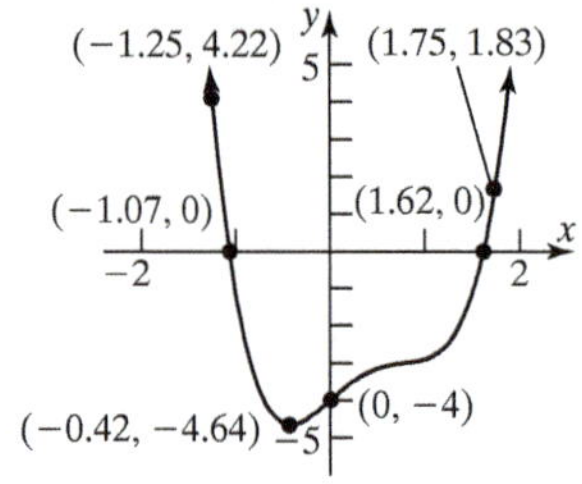

Step 7: Domain: $(-\infty, \infty)$; Range: $[-4.64, \infty)$

Step 8: Increasing on $(-0.42, \infty)$
Decreasing on $(-\infty, -0.42)$

91. $f(x) = -x(x + 2)(x - 2)$
Step 1: $y = -x^3$
Step 2: x-intercepts: −2, 0, 2; y-intercept: 0
Step 3: −2, 0, 2: multiplicity 1, crosses
Step 4:

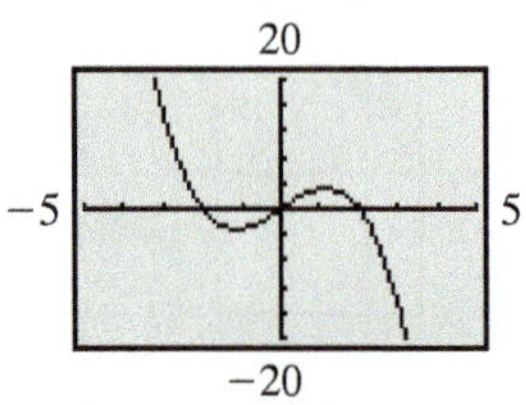

Step 5: (−1.15, −3.08); (1.15, 3.08)
Step 6:

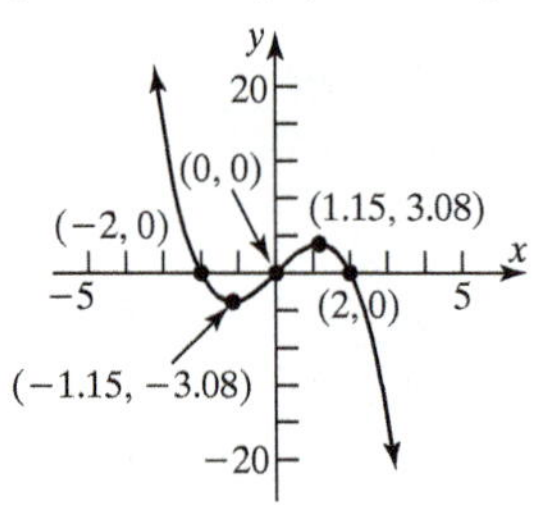

Step 7: Domain: $(-\infty, \infty)$;
Range: $(-\infty, \infty)$
Step 8: Increasing on $(-1.15, 1.15)$
Decreasing on $(-\infty, -1.15)$ and $(1.15, \infty)$

93. $f(x) = x(x + 4)(x - 3)$
Step 1: $y = x^3$
Step 2: x-intercepts: −4, 0, 3; y-intercept: 0
Step 3: −4, 0, 3: multiplicity 1, crosses
Step 4:

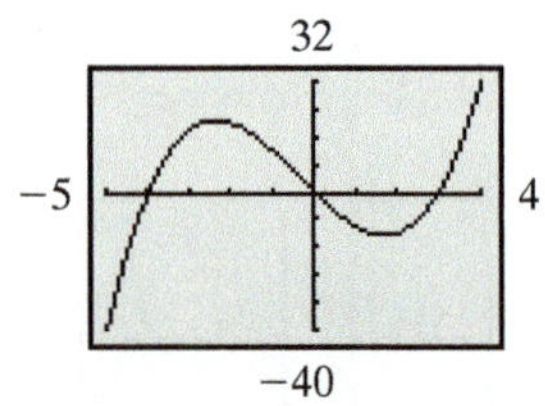

Step 5: (−2.36, 20.75); (1.69, −12.60)
Step 6:

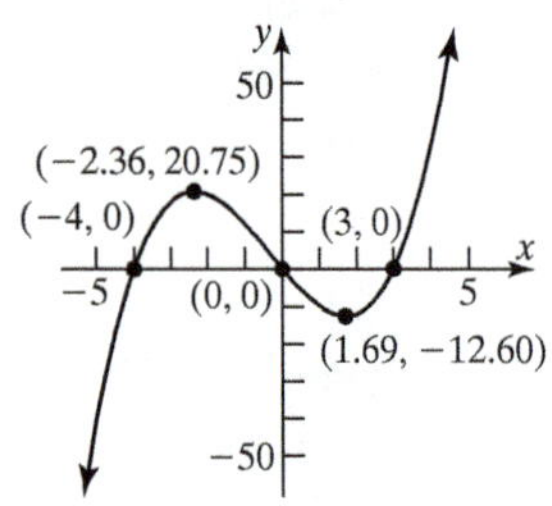

Step 7: Domain: $(-\infty, \infty)$;
Range: $(-\infty, \infty)$
Step 8: Increasing on $(-\infty, -2.36)$ and $(1.69, \infty)$
Decreasing on $(-2.36, 1.69)$

95. $f(x) = 2x(x + 6)(x - 2)(x + 2)$
Step 1: $y = 2x^4$
Step 2: x-intercepts: −6, −2, 0, 2; y-intercept: 0
Step 3: −6, −2, 0, 2: multiplicity 1, crosses
Step 4:

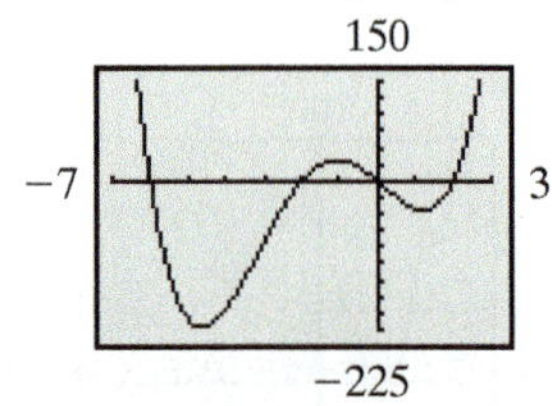

Step 5: (−4.65, −221.25); (−1.06, 30.12);
(1.21, −44.25)
Step 6:

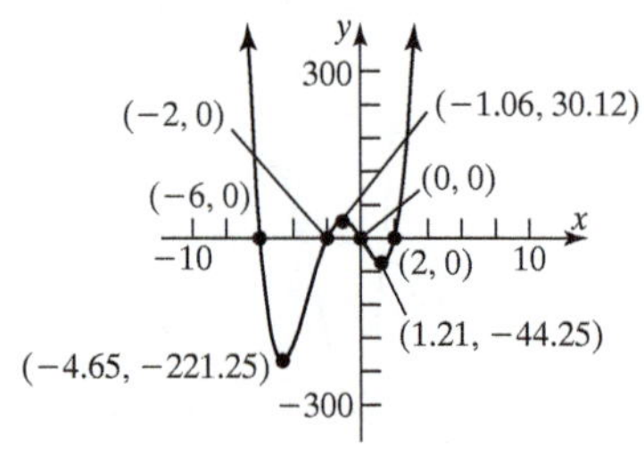

Step 7: Domain: $(-\infty, \infty)$;
Range: $[-221.25, \infty)$
Step 8: Decreasing on $(-\infty, -4.65)$ and $(-1.06, 1.21)$
and $(-1.06, 1.21)$
Increasing on $(-4.65, -1.06)$ and $(1.21, \infty)$

97. $f(x) = -x^2(x+1)^2(x-1)$

Step 1: $y = -x^5$

Step 2: x-intercepts: $-1, 0, 1$; y-intercept: 0

Step 3: 1: multiplicity 1, crosses; $-1, 0$: multiplicity 2, touches

Step 4:

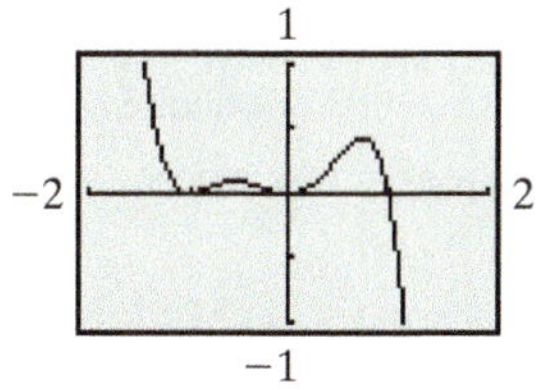

Step 5: $(-1, 0)$; $(-0.54, 0.10)$; $(0, 0)$; $(0.74, 0.43)$

Step 6:

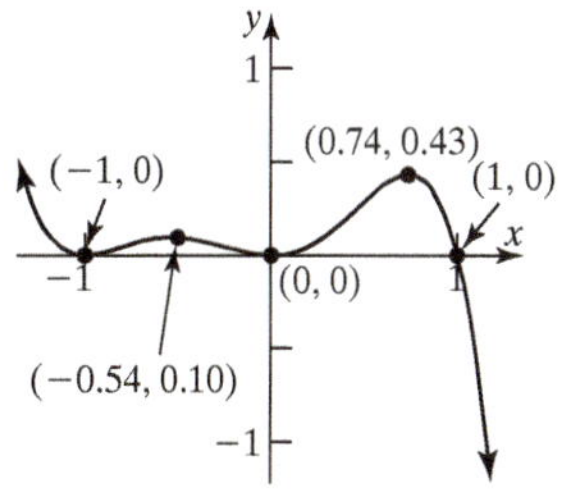

Step 7: Domain: $(-\infty, \infty)$; Range: $(-\infty, \infty)$

Step 8: Increasing on $(-1, -0.54)$ and $(0, 0.74)$
Decreasing on $(-\infty, -1)$, $(-0.54, 0)$ and $(0.74, \infty)$

99. **(a)** $T(r) = \underbrace{500(1+r)(1+r)}_{\text{Account value of deposit 1}} + \underbrace{500(1+r)}_{\text{Account value of deposit 2}} + \underbrace{500}_{\text{Deposit 3}}$

$= 500(1 + 2r + r^2) + 500 + 500r + 500 = 500 + 1000r + 500r^2 + 500 + 500r + 500 = 500r^2 + 1500r + 1500$

(b) \$2155.06

101. **(a)**

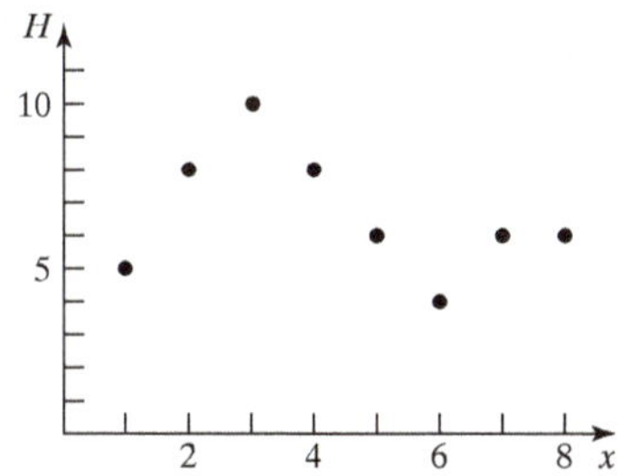

The relation appears to be cubic.

(b) $H(x) = 0.1591x^3 - 2.3203x^2 + 9.3301x - 2.2143$

(c) About 6 major hurricanes

(d)

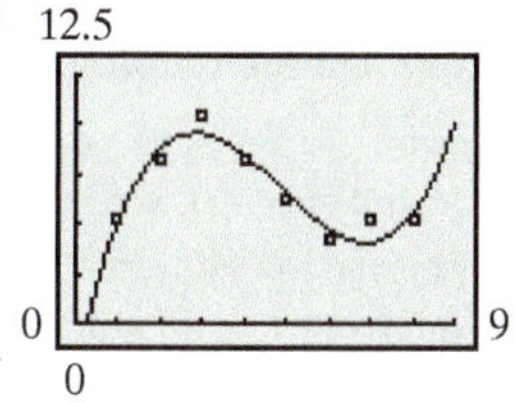

(e) Approximately 10 major hurricanes

103. **(a)**

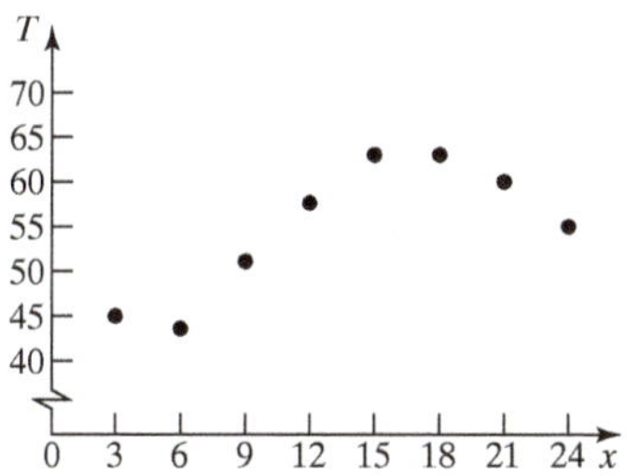

The relation appears to be cubic.

(b) Average rate of change: 2.27° /h

(c) Average rate of change: 0° /h

(d) $T(x) = -0.0103x^3 + 0.3174x^2 - 1.3742x + 45.3929$; The predicted temperature at 5 PM is 63.2°F.

(e)

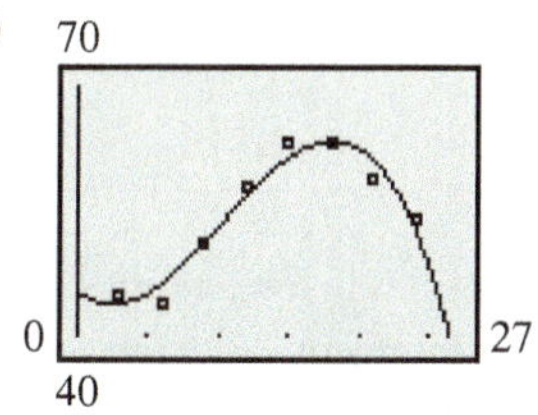

(f) The y-intercept, 45.4°, is the predicted temperature at midnight.

109. (a)–(d)

5.2 Assess Your Understanding *(page 356)*

5. $y = 1$ **6.** $x = -1$ **7.** Proper **8.** F **9.** T **10.** T **11.** All real numbers except 3; $\{x \mid x \neq 3\}$

13. All real numbers except 2 and -4; $\{x \mid x \neq 2, x \neq -4\}$ **15.** All real numbers except $-\frac{1}{2}$ and 3; $\left\{x \,\middle|\, x \neq -\frac{1}{2}, x \neq 3\right\}$

17. All real numbers except 2; $\{x \mid x \neq 2\}$ **19.** All real numbers **21.** All real numbers except -3 and 3; $\{x \mid x \neq -3, x \neq 3\}$

23. **(a)** Domain: $\{x \mid x \neq 2\}$; Range: $\{y \mid y \neq 1\}$ **(b)** $(0, 0)$ **(c)** $y = 1$ **(d)** $x = 2$ **(e)** None

25. **(a)** Domain: $\{x \mid x \neq 0\}$; Range: all real numbers **(b)** $(-1, 0), (1, 0)$ **(c)** None **(d)** $x = 0$ **(e)** $y = 2x$

27. **(a)** Domain: $\{x \mid x \neq -2, x \neq 2\}$; Range: $\{y \mid y \leq 0, y > 1\}$ **(b)** $(0, 0)$ **(c)** $y = 1$ **(d)** $x = -2, x = 2$ **(e)** None

29.

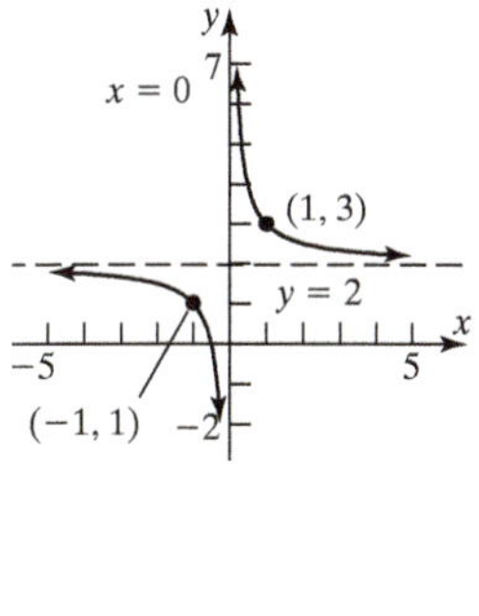

31.

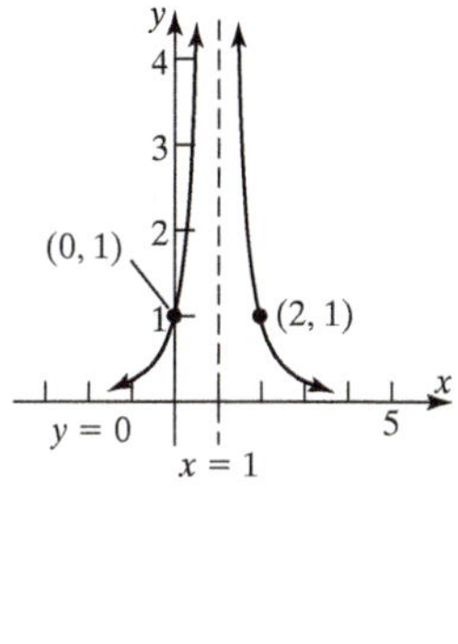

33.

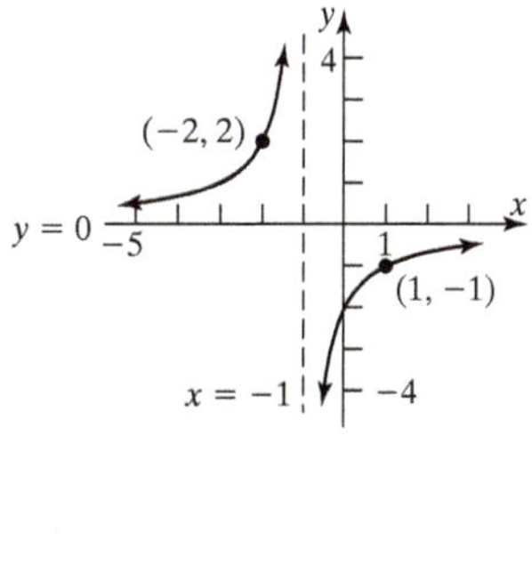

35.

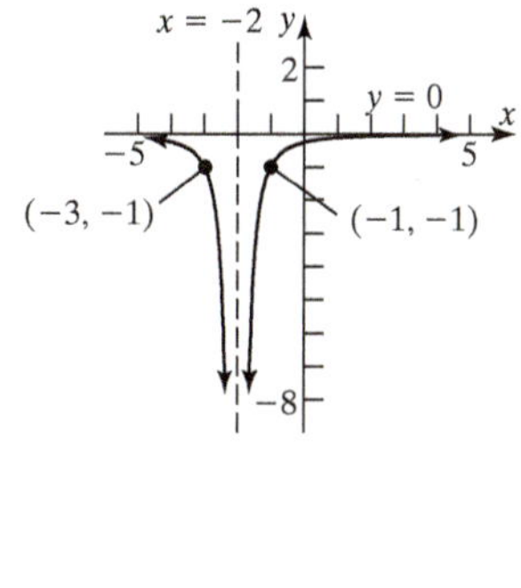

37.

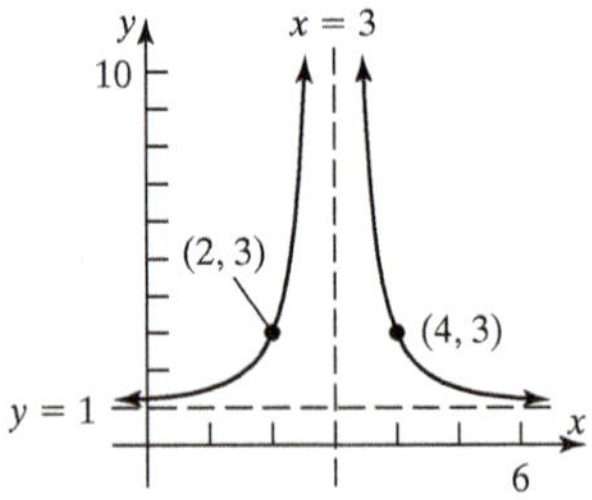

39.

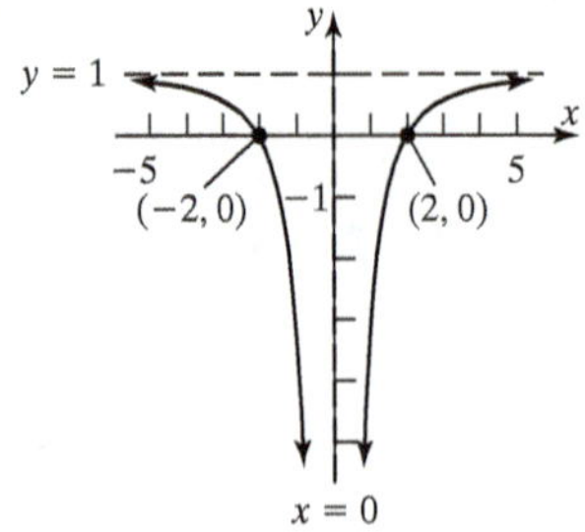

41. Vertical asymptote: $x = -4$; horizontal asymptote: $y = 3$
43. Vertical asymptote: $x = 3$; oblique asymptote: $y = x + 5$
45. Vertical asymptotes: $x = 1, x = -1$; horizontal asymptote: $y = 0$
47. Vertical asymptote: $x = -\frac{1}{3}$; horizontal asymptote: $y = \frac{2}{3}$
49. Vertical asymptote: none; oblique asymptote: $y = 2x - 1$
51. Vertical asymptote: $x = 0$; no horizontal or oblique asymptotes
53. **(a)** 9.8208 m/sec^2 **(b)** 9.8195 m/sec^2 **(c)** 9.7936 m/sec^2
(d) h-axis **(e)** $\varnothing$

55. (a)

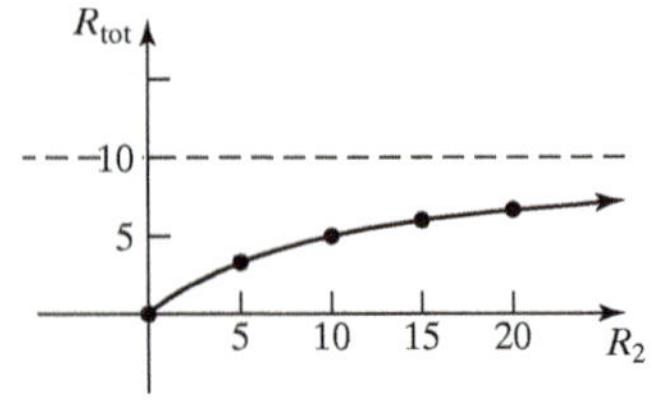

(b) Horizontal: $R_{tot} = 10$; as the resistance of R_2 increases without bound, the total resistance approaches 10 ohms, the resistance R_1.
(c) $R_1 \approx 103.5$ ohms

5.3 Assess Your Understanding *(page 366)*

3. in lowest terms **4.** F **5.** F **6.** T

7. Step 1: Domain: $\{x | x \neq 0, x \neq -4\}$
Step 2: $R(x)$ is in lowest terms.
Step 3: x-intercept: -1; no y-intercept
Step 4: No symmetry
Step 5: Vertical asymptotes: $x = 0, x = -4$
Step 6: Horizontal asymptote: $y = 0$, intersected at $(-1, 0)$

Step 7:

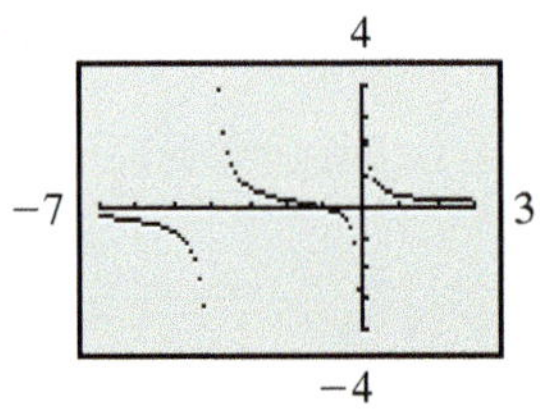

Step 8:

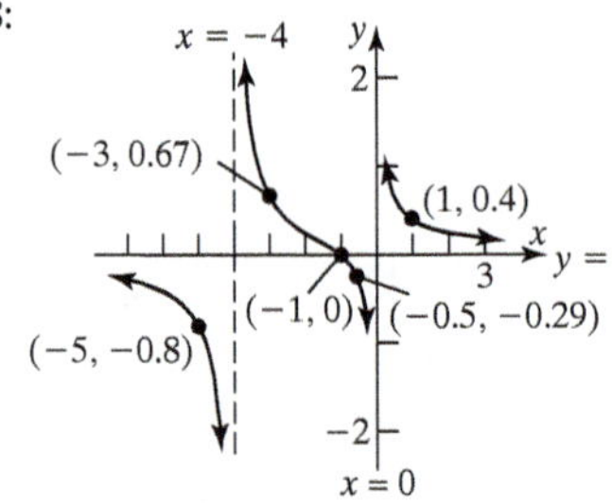

9. Step 1: $R(x) = \frac{3(x+1)}{2(x+2)}$; Domain: $\{x | x \neq -2\}$
Step 2: $R(x)$ is in lowest terms.
Step 3: x-intercept: -1; y-intercept: $\frac{3}{4}$
Step 4: No symmetry
Step 5: Vertical asymptote: $x = -2$
Step 6: Horizontal asymptote: $y = \frac{3}{2}$, not intersected

Step 7:

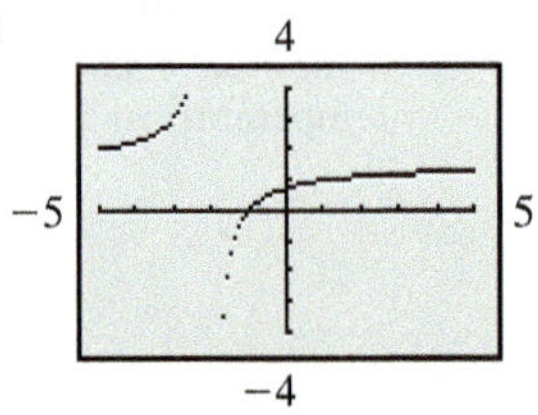

Step 8:

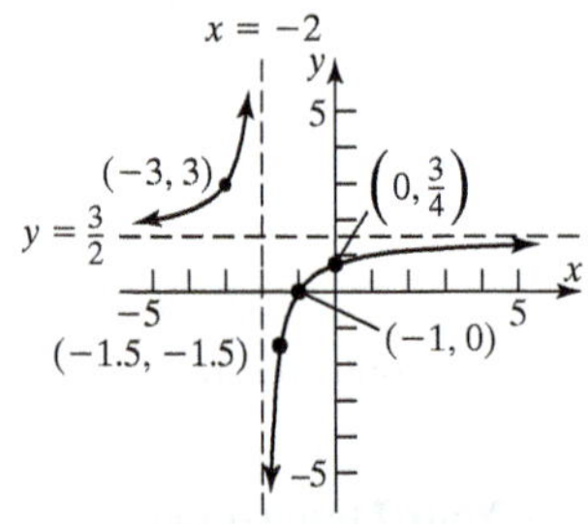

11. Step 1: $R(x) = \frac{3}{(x+2)(x-2)}$;
Domain: $\{x | x \neq -2, x \neq 2\}$
Step 2: $R(x)$ is in lowest terms.
Step 3: No x-intercept; y-intercept: $-\frac{3}{4}$
Step 4: Symmetric with respect to the y-axis
Step 5: Vertical asymptotes: $x = 2, x = -2$
Step 6: Horizontal asymptote: $y = 0$, not intersected

Step 7:

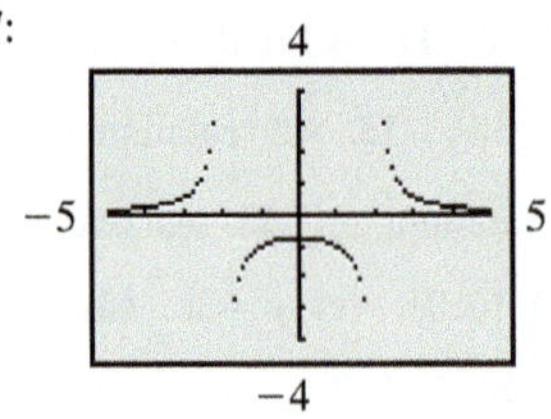

Step 8:

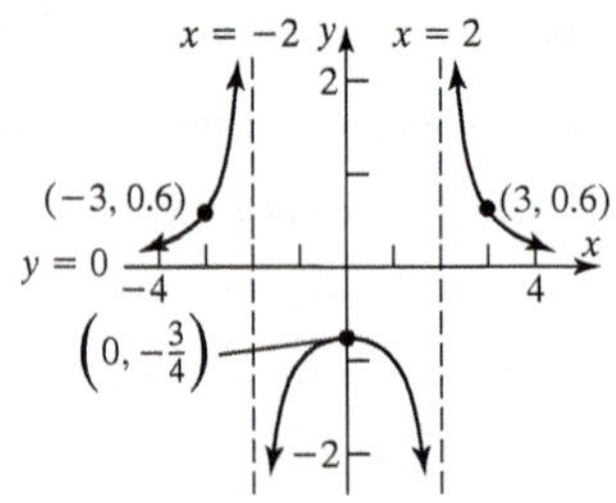

13. Step 1: $P(x) = \frac{x^4 + x^2 + 1}{(x+1)(x-1)}$; Domain: $\{x | x \neq -1, x \neq 1\}$
Step 2: $P(x)$ is in lowest terms.
Step 3: No x-intercept; y-intercept: -1
Step 4: Symmetric with respect to the y-axis
Step 5: Vertical asymptotes: $x = -1, x = 1$
Step 6: No horizontal or oblique asymptotes

Step 7:

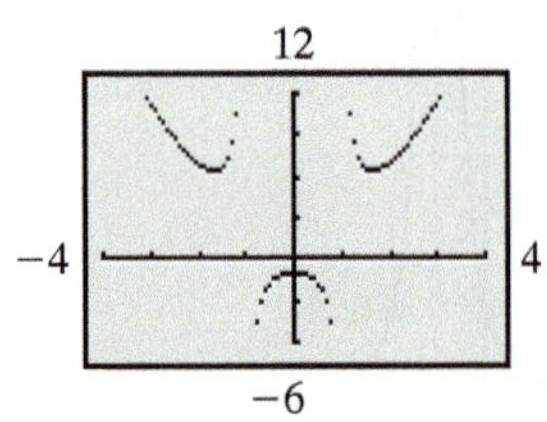

Step 8:

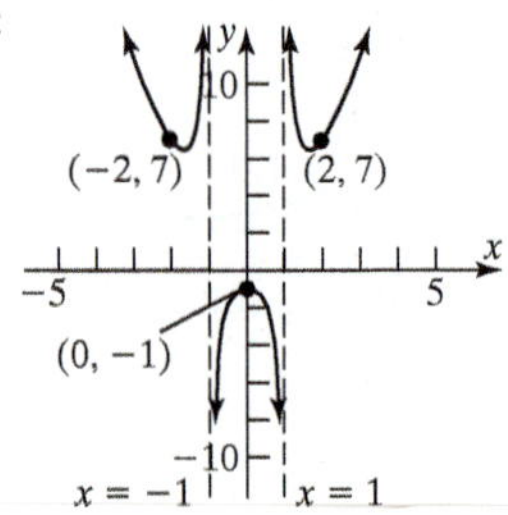

15. Step 1: $H(x) = \dfrac{(x-1)(x^2+x+1)}{(x+3)(x-3)}$; Domain: $\{x \mid x \neq -3, x \neq 3\}$

Step 2: $H(x)$ is in lowest terms.

Step 3: x-intercept: 1; y-intercept: $\dfrac{1}{9}$

Step 4: No symmetry

Step 5: Vertical asymptotes: $x = 3$, $x = -3$

Step 6: Oblique asymptote: $y = x$, intersected at $\left(\dfrac{1}{9}, \dfrac{1}{9}\right)$

Step 7:

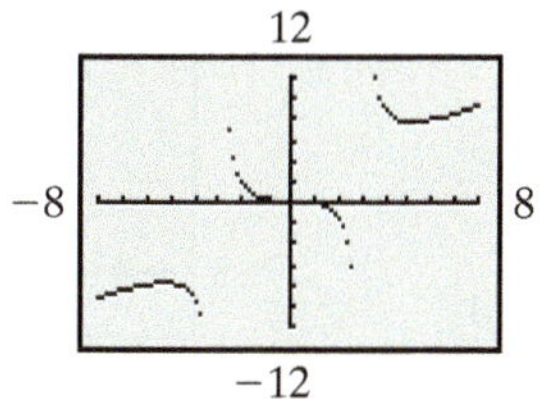

Step 8:

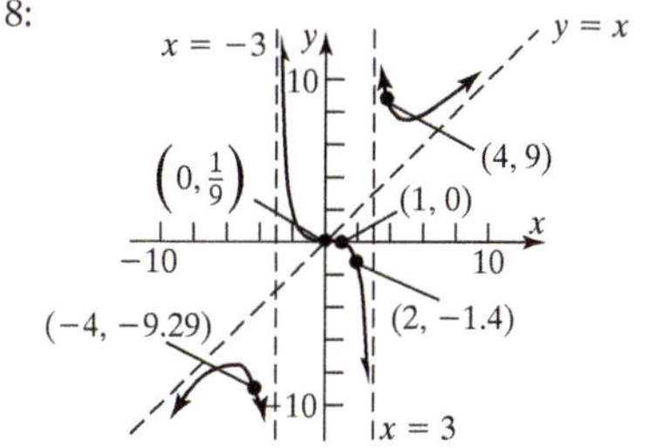

17. Step 1: $R(x) = \dfrac{x^2}{(x+3)(x-2)}$; Domain: $\{x \neq -3, x \neq 2\}$

Step 2: $R(x)$ is in lowest terms.

Step 3: Intercept: $(0, 0)$

Step 4: No symmetry

Step 5: Vertical asymptotes: $x = 2$, $x = -3$

Step 6: Horizontal asymptote: $y = 1$, intersected at $(6, 1)$

Step 7:

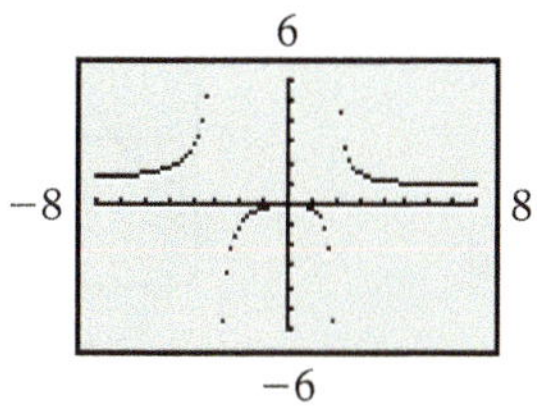

Step 8:

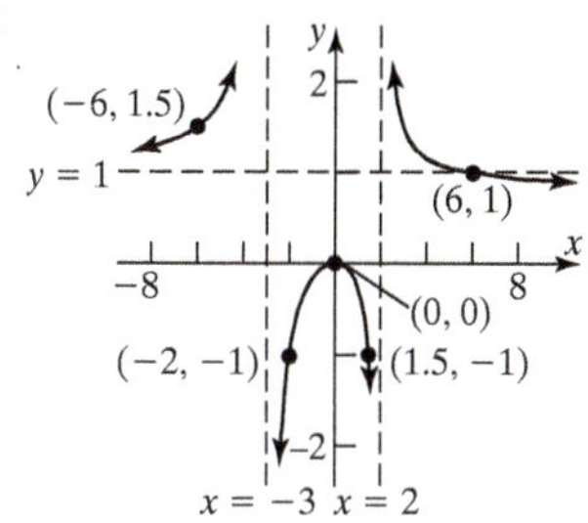

19. Step 1: $G(x) = \dfrac{x}{(x+2)(x-2)}$; Domain: $\{x \mid x \neq -2, x \neq 2\}$

Step 2: $G(x)$ is in lowest terms.

Step 3: Intercept: $(0, 0)$

Step 4: Symmetry with respect to the origin

Step 5: Vertical asymptotes: $x = -2$, $x = 2$

Step 6: Horizontal asymptote: $y = 0$, intersected at $(0, 0)$

Step 7:

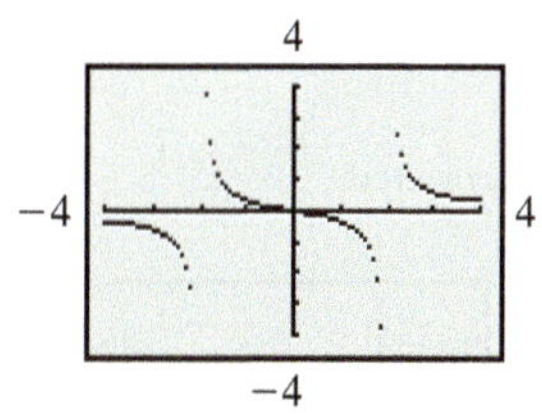

Step 8:

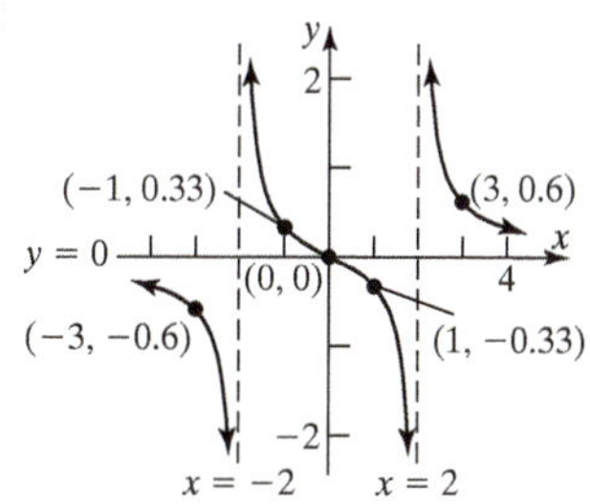

21. Step 1: $R(x) = \dfrac{3}{(x-1)(x-2)(x+2)}$; Domain: $\{x \mid x \neq 1, x \neq -2, x \neq 2\}$

Step 2: $R(x)$ is in lowest terms.

Step 3: No x-intercept; y-intercept: $\dfrac{3}{4}$

Step 4: No symmetry

Step 5: Vertical asymptotes: $x = -2$, $x = 1$, $x = 2$

Step 6: Horizontal asymptote: $y = 0$, not intersected

Step 7:

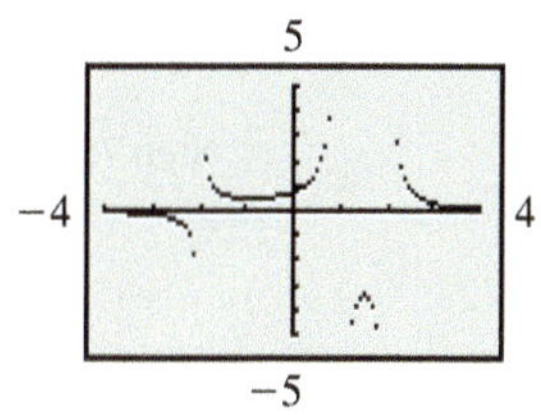

Step 8:

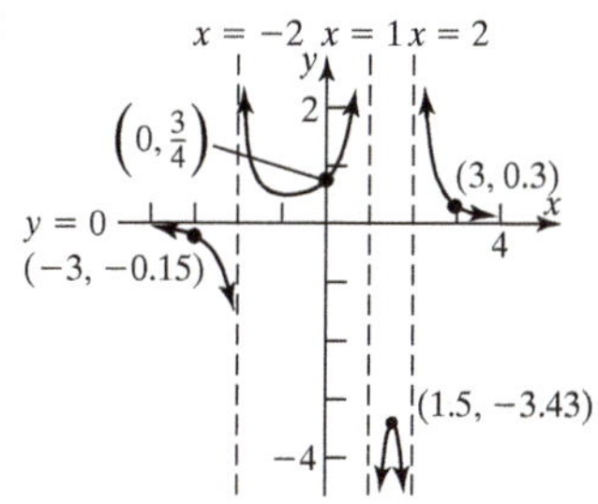

23. Step 1: $H(x) = \dfrac{4(x+1)(x-1)}{(x^2+4)(x+2)(x-2)}$; Domain: $\{x \mid x \neq -2, x \neq 2\}$

Step 2: $H(x)$ is in lowest terms.

Step 3: x-intercepts: $-1, 1$; y-intercept: $\dfrac{1}{4}$

Step 4: Symmetry with respect to the y-axis

Step 5: Vertical asymptotes: $x = -2$, $x = 2$

Step 6: Horizontal asymptote: $y = 0$, intersected at $(-1, 0)$ and $(1, 0)$

Step 7:

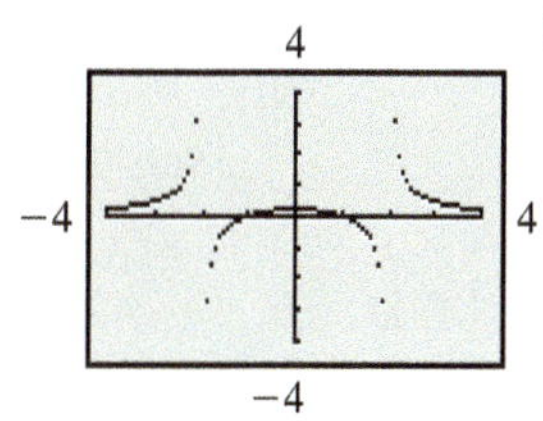

Step 8:

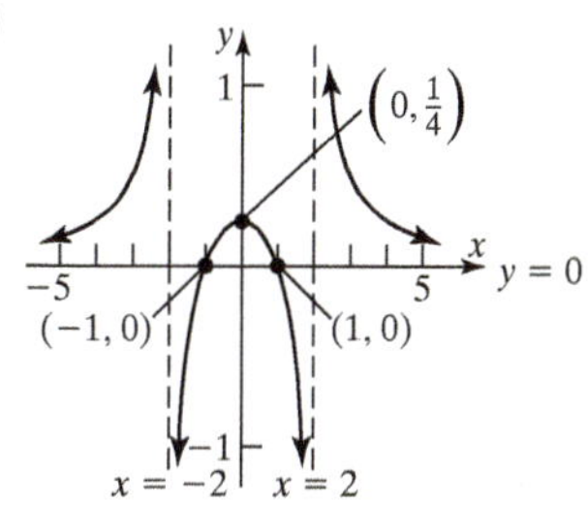

25. Step 1: $F(x) = \dfrac{(x-4)(x+1)}{x+2}$; Domain: $\{x \mid x \neq -2\}$

Step 2: $F(x)$ is in lowest terms.

Step 3: x-intercepts: $-1, 4$; y-intercept: -2

Step 4: No symmetry

Step 5: Vertical asymptote: $x = -2$

Step 6: Oblique asymptote: $y = x - 5$, not intersected

Step 7:

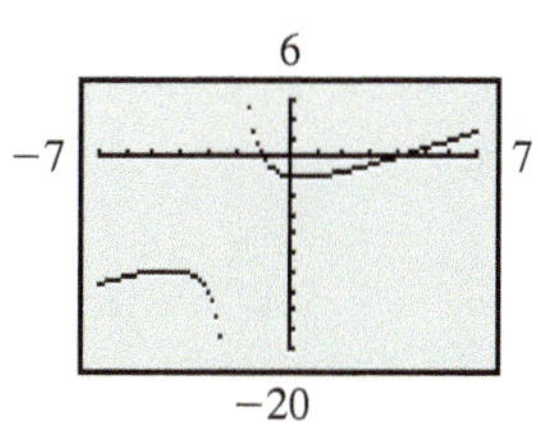

Step 8:

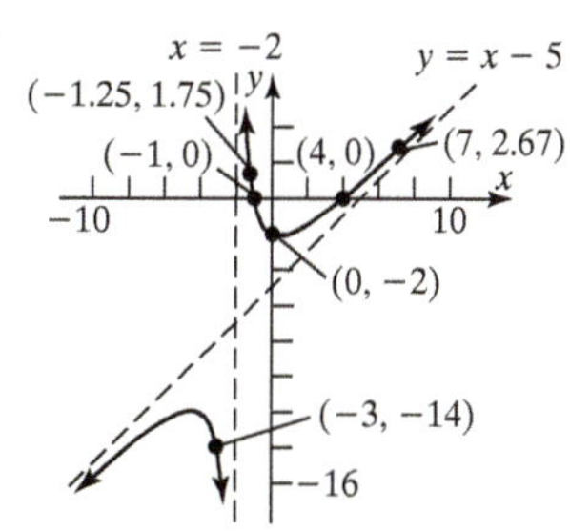

27. Step 1: $R(x) = \dfrac{(x+4)(x-3)}{x-4}$; Domain: $\{x|x \neq 4\}$
Step 2: $R(x)$ is in lowest terms.
Step 3: x-intercepts: $-4, 3$; y-intercept: 3
Step 4: No symmetry
Step 5: Vertical asymptote: $x = 4$
Step 6: Oblique asymptote: $y = x + 5$, not intersected

Step 7:

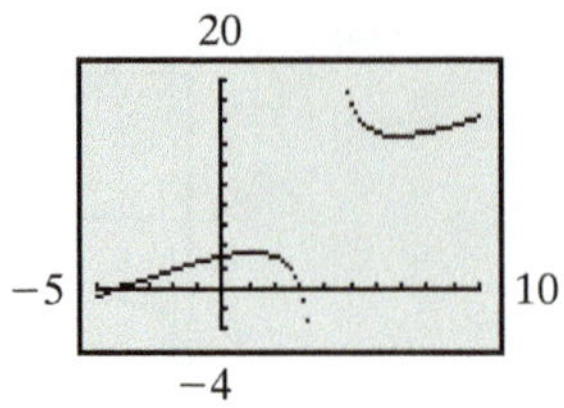

Step 8.

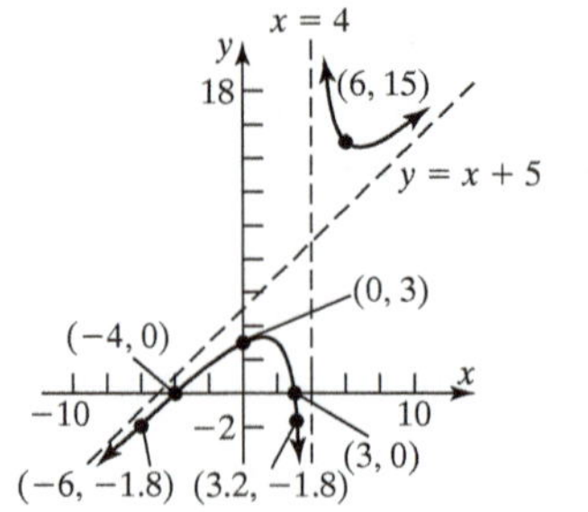

29. Step 1: $F(x) = \dfrac{(x+4)(x-3)}{x+2}$; Domain: $\{x|x \neq -2\}$
Step 2: $F(x)$ is in lowest terms.
Step 3: x-intercepts: $-4, 3$; y-intercept: -6
Step 4: No symmetry
Step 5: Vertical asymptote: $x = -2$
Step 6: Oblique asymptote: $y = x - 1$, not intersected

Step 7:

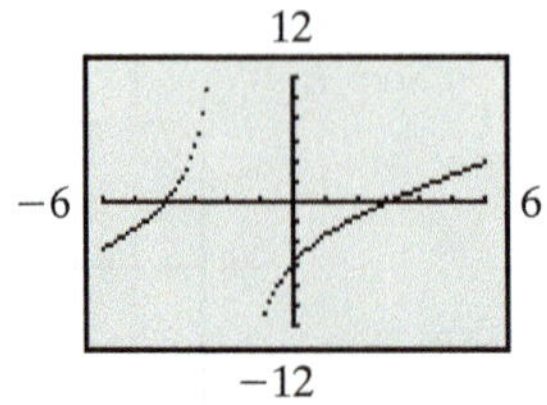

Step 8:

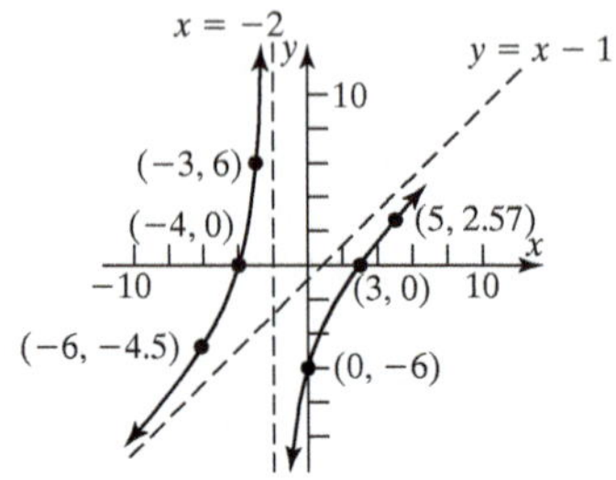

31. Step 1: Domain: $\{x|x \neq -3\}$
Step 2: $R(x)$ is in lowest terms.
Step 3: x-intercepts: 0, 1; y-intercept: 0
Step 4: No symmetry
Step 5: Vertical asymptote: $x = -3$
Step 6: Horizontal asymptote: $y = 1$, not intersected
Step 7:

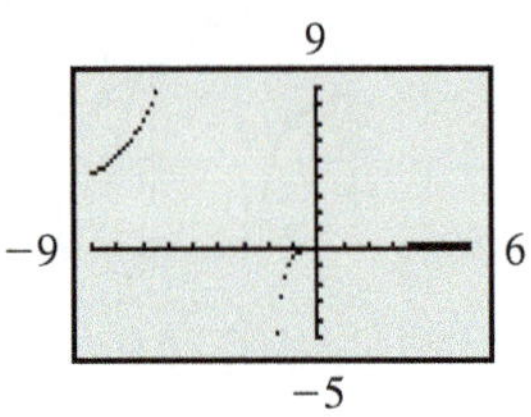

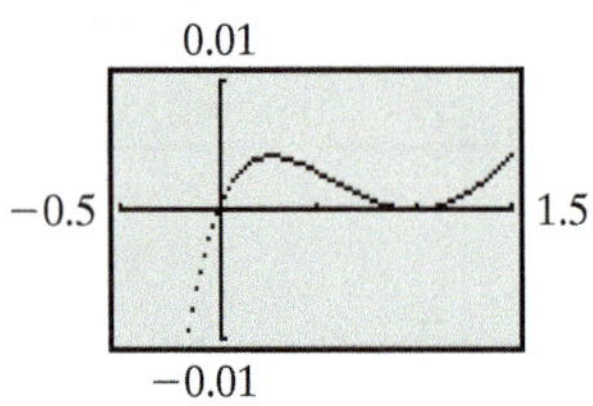

Step 8:

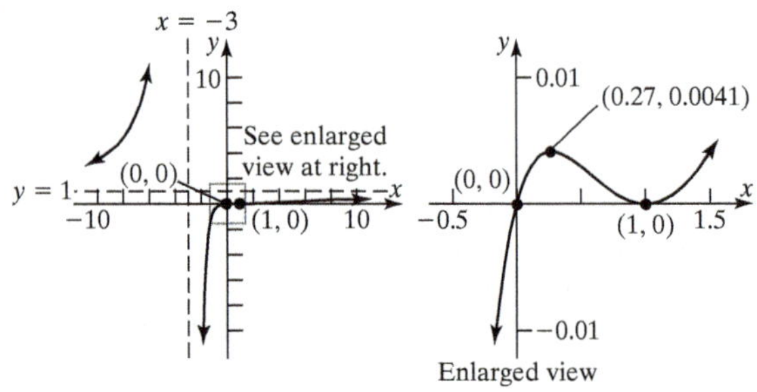

33. Step 1: $R(x) = \dfrac{(x+4)(x-3)}{(x+2)(x-3)}$;
Domain: $\{x|x \neq -2, x \neq 3\}$
Step 2: $R(x) = \dfrac{x+4}{x+2}$
Step 3: x-intercept: -4; y-intercept: 2
Step 4: No symmetry
Step 5: Vertical asymptote: $x = -2$; hole at $\left(3, \dfrac{7}{5}\right)$
Step 6: Horizontal asymptote: $y = 1$, not intersected

Step 7:

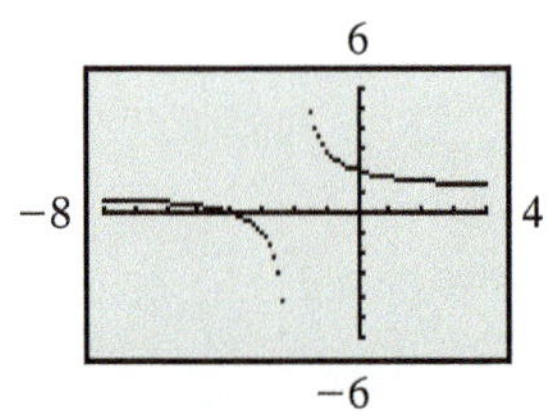

Step 8:

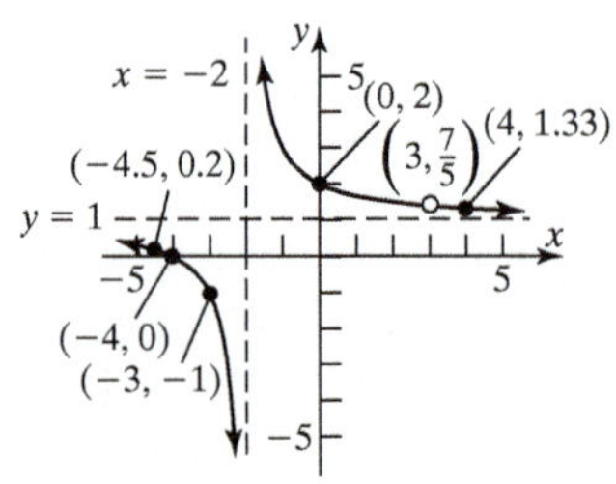

35. Step 1: $R(x) = \dfrac{(3x+1)(2x-3)}{(x-2)(2x-3)}$;
Domain: $\left\{x \middle| x \neq \dfrac{3}{2}, x \neq 2\right\}$
Step 2: $R(x) = \dfrac{3x+1}{x-2}$
Step 3: x-intercept: $-\dfrac{1}{3}$; y-intercept: $-\dfrac{1}{2}$
Step 4: No symmetry
Step 5: Vertical asymptote: $x = 2$; hole at $\left(\dfrac{3}{2}, -11\right)$
Step 6: Horizontal asymptote: $y = 3$, not intersected

Step 7:

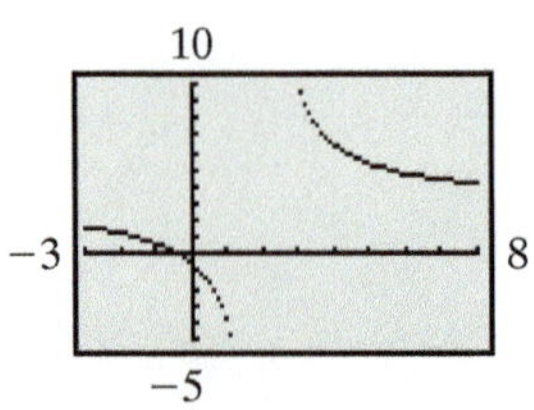

Step 8:

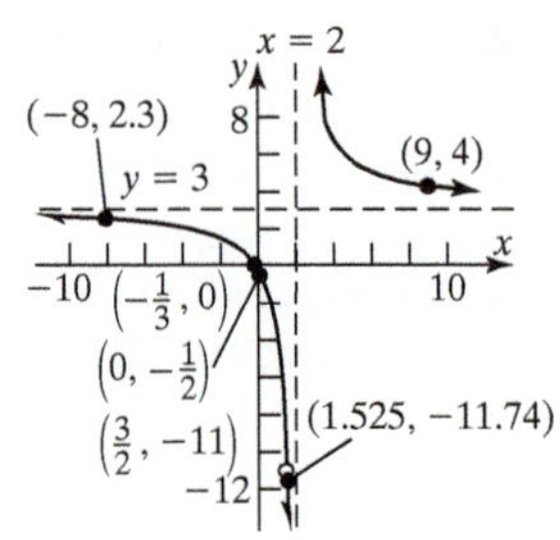

37. Step 1: $R(x) = \dfrac{(x+2)(x+3)}{(x+3)}$; Domain: $\{x \mid x \neq -3\}$

Step 2: $R(x) = x + 2$

Step 3: x-intercept: -2; y-intercept: 2

Step 4: No symmetry

Step 5: Vertical asymptote: none; hole at $(-3, -1)$

Step 6: Oblique asymptote: $y = x + 2$, intersected over entire domain

Step 7:

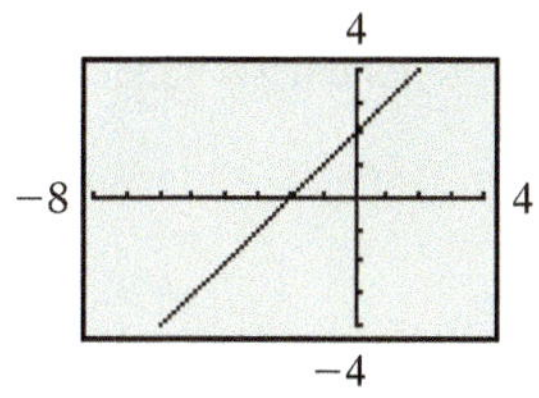

Step 8:

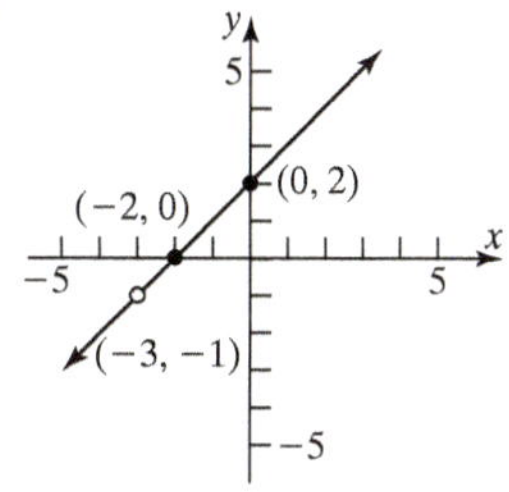

39. Step 1: $f(x) = \dfrac{x^2+1}{x}$; Domain: $\{x \mid x \neq 0\}$

Step 2: $f(x)$ is in lowest terms.

Step 3: No x-intercepts; no y-intercept

Step 4: Symmetric with respect to the origin

Step 5: Vertical asymptote: $x = 0$

Step 6: Oblique asymptote: $y = x$, not intersected

Step 7:

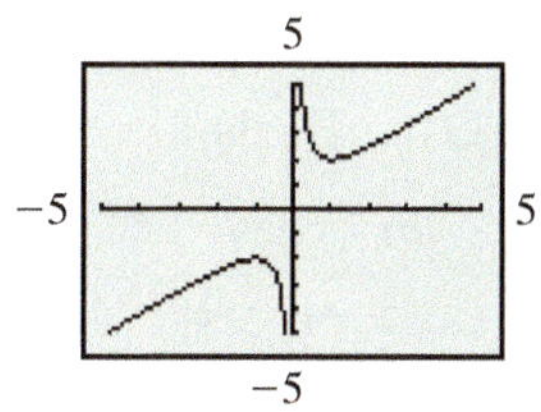

Step 8:

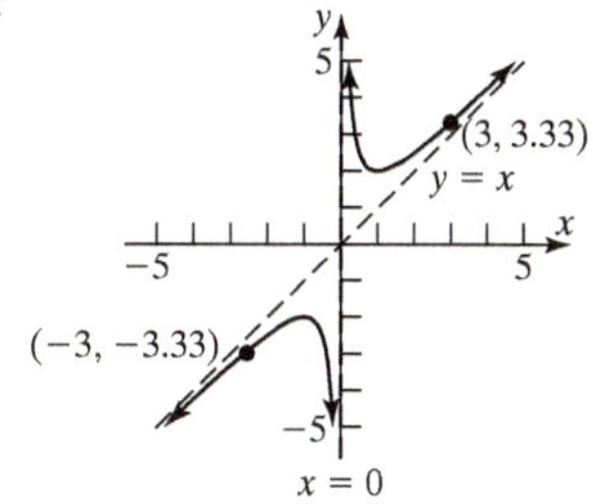

41. Step 1: $f(x) = \dfrac{x^3+1}{x} = \dfrac{(x+1)(x^2-x+1)}{x}$;

Domain: $\{x \mid x \neq 0\}$

Step 2: $f(x)$ is in lowest terms.

Step 3: x-intercept: -1; no y-intercept

Step 4: No symmetry

Step 5: Vertical asymptote: $x = 0$

Step 6: No horizontal or oblique asymptotes

Step 7:

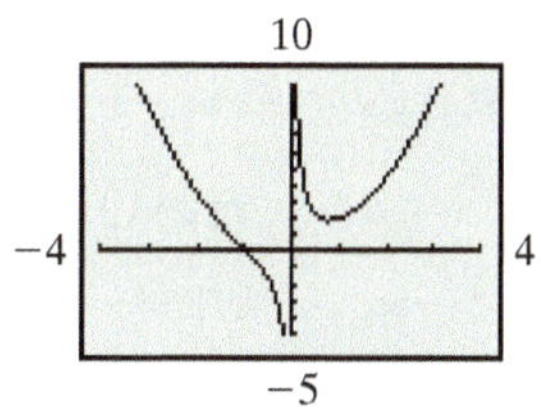

Step 8:

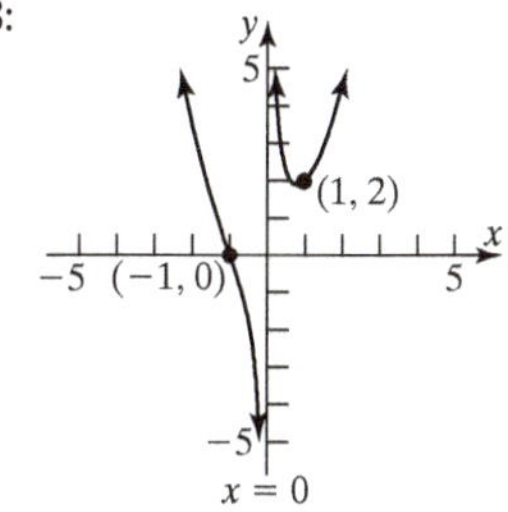

43. Step 1: $f(x) = \dfrac{x^4+1}{x^3}$; Domain: $\{x \mid x \neq 0\}$

Step 2: $f(x)$ is in lowest terms.

Step 3: No x-intercepts; no y-intercept

Step 4: Symmetric with respect to the origin

Step 5: Vertical asymptote: $x = 0$

Step 6: Oblique asymptote: $y = x$, not intersected.

Step 7:

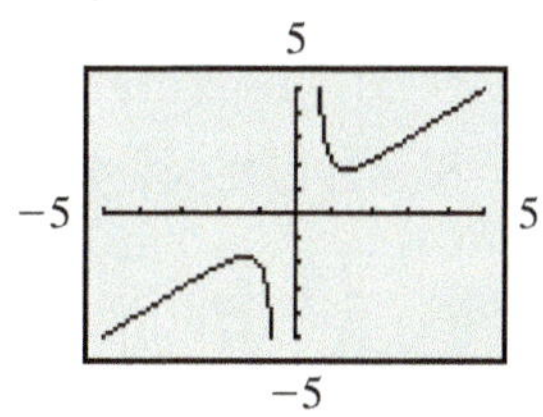

Step 8:

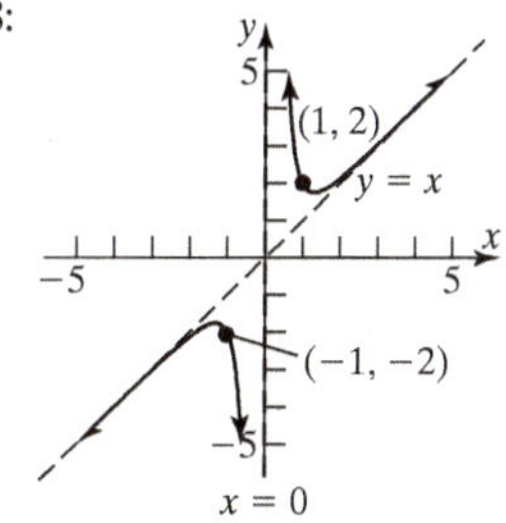

45. Minimum value: 2.00 at $x = 1.00$

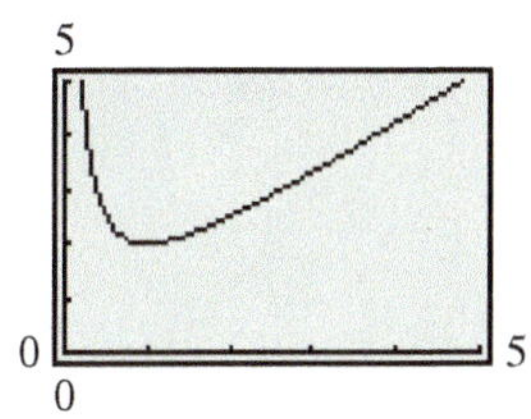

47. Minimum value: 1.89 at $x = 0.79$

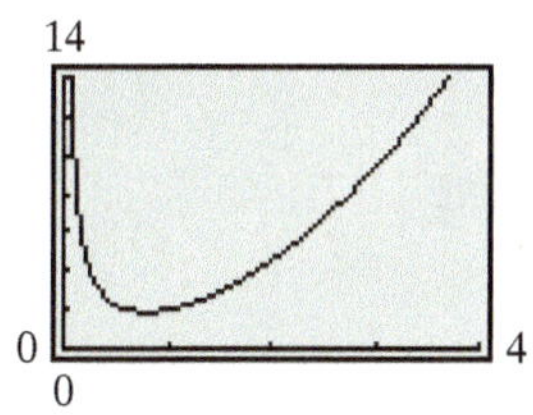

49. Minimum value: 1.75 at $x = 1.32$

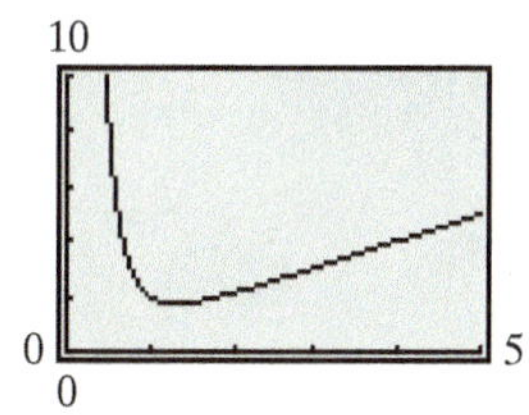

51. One possibility: $R(x) = \dfrac{x^2}{x^2 - 4}$

53. One possibility: $R(x) = \dfrac{(x-1)(x-3)\left(x^2 + \frac{4}{3}\right)}{(x+1)^2(x-2)^2}$

55. **(a)** t-axis; $C(t) \to 0$

(b)

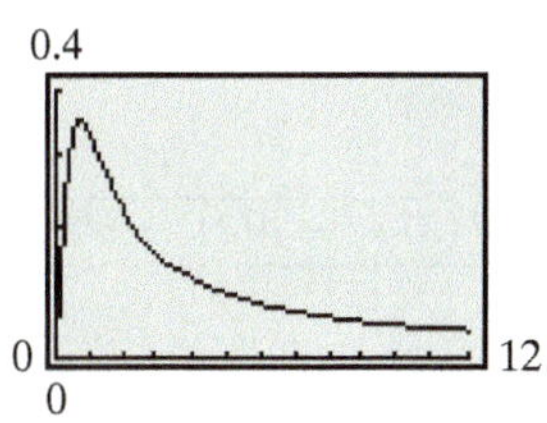

(c) 0.71 h after injection

57. **(a)** $C(x) = 16x + \dfrac{5000}{x} + 100$

(b) $x > 0$

(c)

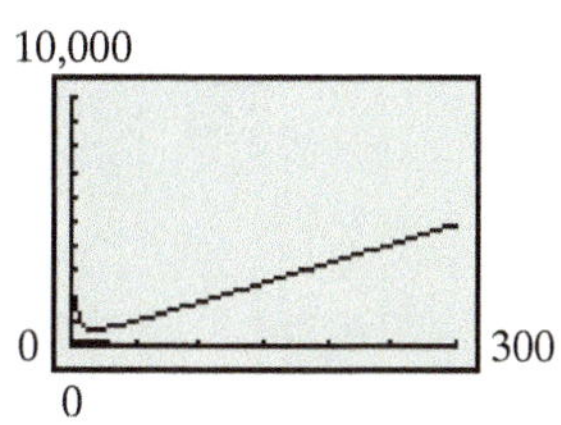

(d) Approximately 17.7 ft by 56.6 ft (longer side parallel to river)

59. **(a)** $S(x) = 2x^2 + \dfrac{40{,}000}{x}$

(b)

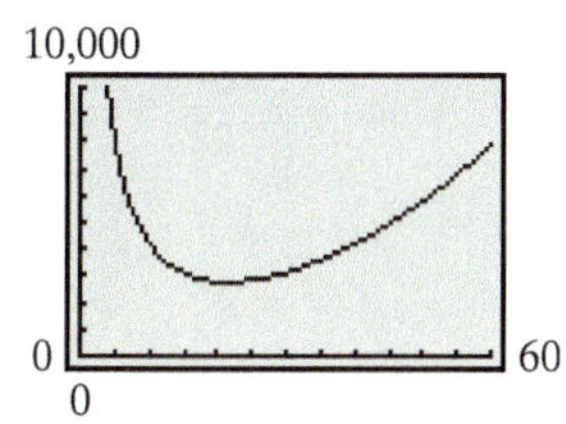

(c) 2784.95 in.^{2}

(d) 21.54 in. × 21.54 in. × 21.54 in.

(e) To minimize the cost of materials needed for construction

61. (a) $C(r) = 12\pi r^2 + \dfrac{4000}{r}$

(b)

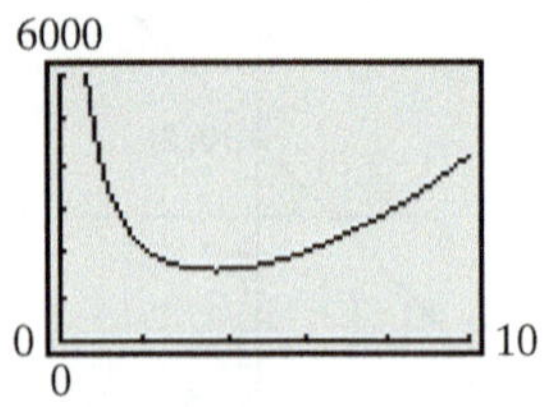

The cost is smallest when $r = 3.76$ cm.

63. No. Each function is a quotient of polynomials, but it is not written in lowest terms Each function is undefined for $x = 1$; each graph has a hole at $x = 1$.

5.4 Assess Your Understanding *(page 373)*

3. T **4.** F **5. (a)** $\{x|0 < x < 1 \text{ or } x > 2\}; (0, 1) \cup (2, \infty)$ **(b)** $\{x|x \le 0 \text{ or } 1 \le x \le 2\}; (-\infty, 0] \cup [1, 2]$

7. (a) $\{x|-1 < x < 0 \text{ or } x > 1\}; (-1, 0) \cup (1, \infty)$ **(b)** $\{x|x < -1 \text{ or } 0 \le x < 1\}; (-\infty, -1) \cup [0, 1)$

9. $\{x|x < 0 \text{ or } 0 < x < 3\}; (-\infty, 0) \cup (0, 3)$ **11.** $\{x|x \ge -4\}; [-4, \infty)$ **13.** $\{x|x \le -2 \text{ or } x \ge 2\}; (-\infty, -2] \cup [2, \infty)$

15. $\{x|-4 < x < -1 \text{ or } x > 0\}; (-4, -1) \cup (0, \infty)$ **17.** $\{x|-2 < x \le -1\}; (-2, -1]$ **19.** $\{x|x < -2\}; (-\infty, -2)$ **21.** $\{x|x > 4\}; (4, \infty)$

23. $\{x|-4 < x < 0 \text{ or } x > 0\}; (-4, 0) \cup (0, \infty)$ **25.** $\{x|x \le 1 \text{ or } 2 \le x \le 3\}; (-\infty, 1] \cup [2, 3]$ **27.** $\{x|-1 < x < 0 \text{ or } x > 3\}; (-1, 0) \cup (3, \infty)$

29. $\{x|x < -1 \text{ or } x > 1\}; (-\infty, -1) \cup (1, \infty)$ **31.** $\{x|x < -1 \text{ or } x > 1\}; (-\infty, -1) \cup (1, \infty)$ **33.** $\{x|x < -1 \text{ or } x > 1\}; (-\infty, -1) \cup (1, \infty)$

35. $\{x|x \le -1 \text{ or } 0 < x \le 1\}; (-\infty, -1] \cup (0, 1]$ **37.** $\{x|x < -1 \text{ or } x > 1\}; (-\infty, -1) \cup (1, \infty)$ **39.** $\{x|x < 2\}; (-\infty, 2)$

41. $\{x|-2 < x \le 9\}; (-2, 9]$ **43.** $\{x|x < 2 \text{ or } 3 < x < 5\}; (-\infty, 2) \cup (3, 5)$

45. $\{x|x < -5 \text{ or } -4 \le x \le -3 \text{ or } x = 0 \text{ or } x > 1\}; (-\infty, -5) \cup [-4, -3] \cup \{0\} \cup (1, \infty)$ **47.** $\left\{x\middle|-\frac{1}{2} < x < 1 \text{ or } x > 3\right\}; \left(-\frac{1}{2}, 1\right) \cup (3, \infty)$

49. $\{x|-1 < x < 3 \text{ or } x > 5\}; (-1, 3) \cup (5, \infty)$ **51.** $\left\{x\middle|x \le -4 \text{ or } x \ge \frac{1}{2}\right\}; (-\infty, -4] \cup \left[\frac{1}{2}, \infty\right)$ **53.** $\{x|x < 3 \text{ or } x \ge 7\}; (-\infty, 3) \cup [7, \infty)$

55. $\{x|x < 2\}; (-\infty, 2)$ **57.** $\left\{x\middle|x < -\frac{2}{3} \text{ or } 0 < x < \frac{3}{2}\right\}; \left(-\infty, -\frac{2}{3}\right) \cup \left(0, \frac{3}{2}\right)$ **59.** $\{x|x \le -3 \text{ or } 0 \le x \le 3\}; (-\infty, -3] \cup [0, 3]$

61. $\{x|x > 4\}; (4, \infty)$ **63.** $\{x|x \le -2 \text{ or } x \ge 2\}; (-\infty, -2] \cup [2, \infty)$ **65.** $\{x|x < -4 \text{ or } x \ge 2\}; (-\infty, -4) \cup [2, \infty)$

67.

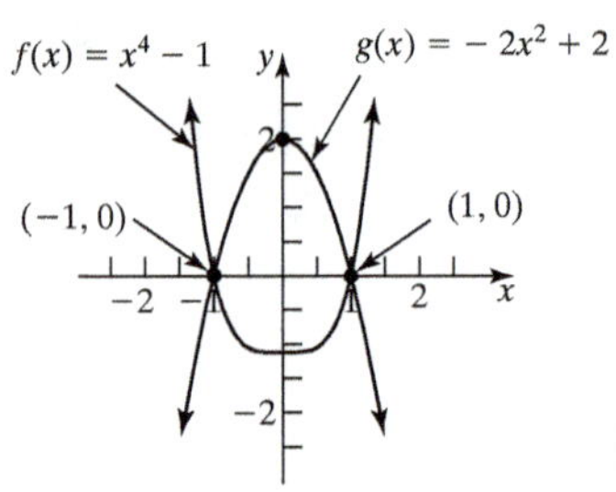

$f(x) \le g(x)$ if $-1 \le x \le 1$

69.

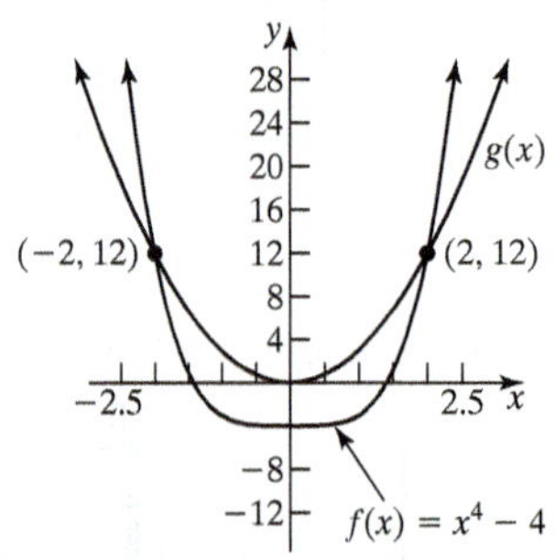

$f(x) \le g(x)$ if $-2 \le x \le 2$

71. Produce at least 250 bicycles

73. (a) The stretch is less than 39 ft.

(b) The ledge should be at least 84 ft above the ground for a 150-lb jumper.

75. At least 50 students must attend.

Historical Problems *(page 387)*

1.

$$\left(x - \frac{b}{3}\right)^3 + b\left(x - \frac{b}{3}\right)^2 + c\left(x - \frac{b}{3}\right) + d = 0$$

$$x^3 - bx^2 + \frac{b^2x}{3} - \frac{b^3}{27} + bx^2 - \frac{2b^2x}{3} + \frac{b^3}{9} + cx - \frac{bc}{3} + d = 0$$

$$x^3 + \left(c - \frac{b^2}{3}\right)x + \left(\frac{2b^3}{27} - \frac{bc}{3} + d\right) = 0$$

Let $p = c - \dfrac{b^2}{3}$ and $q = \dfrac{2b^3}{27} - \dfrac{bc}{3} + d$. Then $x^3 + px + q = 0$.

2.

$$(H + K)^3 + p(H + K) + q = 0$$

$$H^3 + 3H^2K + 3HK^2 + K^3 + pH + pK + q = 0$$

Let $3HK = -p$.

$$H^3 - pH - pK + K^3 + pH + pK + q = 0,$$

$$H^3 + K^3 = -q$$

3.

$$3HK = -p$$

$$K = -\frac{p}{3H}$$

$$H^3 + \left(-\frac{p}{3H}\right)^3 = -q$$

$$H^3 - \frac{p^3}{27H^3} = -q$$

$$27H^6 - p^3 = -27qH^3$$

$$27H^6 + 27qH^3 - p^3 = 0$$

$$H^3 = \frac{-27q \pm \sqrt{(27q)^2 - 4(27)(-p^3)}}{2 \cdot 27}$$

$$H^3 = \frac{-q}{2} \pm \sqrt{\frac{27^2q^2}{2^2(27^2)} + \frac{4(27)p^3}{2^2(27^2)}}$$

$$H^3 = \frac{-q}{2} \pm \sqrt{\frac{q^2}{4} + \frac{p^3}{27}}$$

$$H = \sqrt[3]{\frac{-q}{2} + \sqrt{\frac{q^2}{4} + \frac{p^3}{27}}}$$

Choose the positive root for now.

4. $H^3 + K^3 = -q$

$$K^3 = -q - H^3$$

$$K^3 = -q - \left[\frac{-q}{2} + \sqrt{\frac{q^2}{4} + \frac{p^3}{27}}\right]$$

$$K^3 = \frac{-q}{2} - \sqrt{\frac{q^2}{4} + \frac{p^3}{27}}$$

$$K = \sqrt[3]{\frac{-q}{2} - \sqrt{\frac{q^2}{4} + \frac{p^3}{27}}}$$

5. $x = H + K$

$$x = \sqrt[3]{\frac{-q}{2} + \sqrt{\frac{q^2}{4} + \frac{p^3}{27}}} + \sqrt[3]{\frac{-q}{2} - \sqrt{\frac{q^2}{4} + \frac{p^3}{27}}}$$

(Note that had we used the negative root in 3 the result would be the same.)

6. $x = 3$ **7.** $x = 2$ **8.** $x = 2$

5.5 Assess Your Understanding *(page 387)*

5. Remainder; Dividend **6.** $f(c)$ **7.** -4 **8.** F **9.** F **10.** T **11.** $R = f(2) = 8$; no **13.** $R = f(2) = 0$; yes; $f(x) = (x - 2)(3x^3 - 5)$ **15.** $R = f(-3) = 0$; yes; $f(x) = (x + 3)(3x^5 - 9x^4 + 27x^3 + x^2 - 3x + 9)$ **17.** $R = f(-4) = 1$; no **19.** $R = f\left(\frac{1}{2}\right) = 0$; yes; $f(x) = \left(x - \frac{1}{2}\right)(2x^3 + 2)$ **21.** 4; $\pm 1, \pm\frac{1}{3}$ **23.** 5; $\pm 1, \pm 3$ **25.** 3; $\pm 1, \pm 2, \pm\frac{1}{4}, \pm\frac{1}{2}$ **27.** 4; $\pm 1, \pm 3, \pm 9, \pm\frac{1}{2}, \pm\frac{1}{3}, \pm\frac{1}{6}, \pm\frac{3}{2}, \pm\frac{9}{2}$ **29.** 5; $\pm 1, \pm 2, \pm 3, \pm 4, \pm 6, \pm 12, \pm\frac{1}{2}, \pm\frac{3}{2}$ **31.** 4; $\pm 1, \pm 2, \pm 4, \pm 5, \pm 10, \pm 20, \pm\frac{1}{2}, \pm\frac{5}{2}, \pm\frac{1}{3}, \pm\frac{2}{3}, \pm\frac{4}{3}, \pm\frac{5}{3}, \pm\frac{10}{3}, \pm\frac{20}{3}, \pm\frac{1}{6}, \pm\frac{5}{6}$

33. -1 and 1

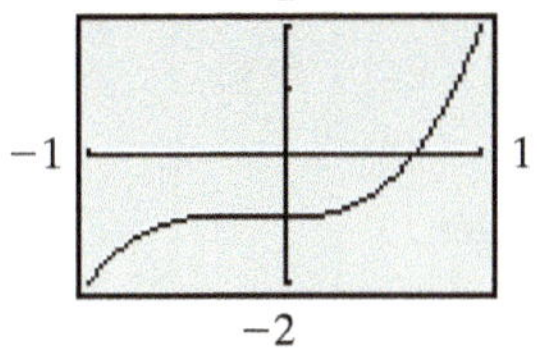

35. -12 and 12

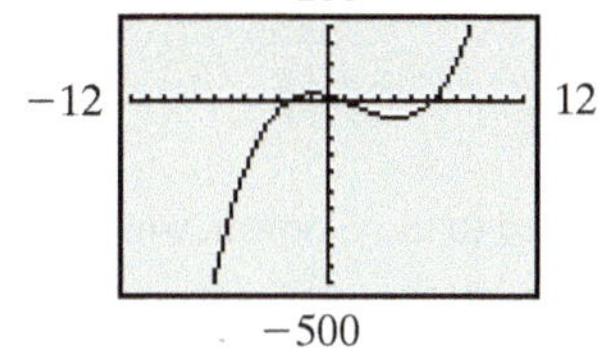

37. -10 and 10

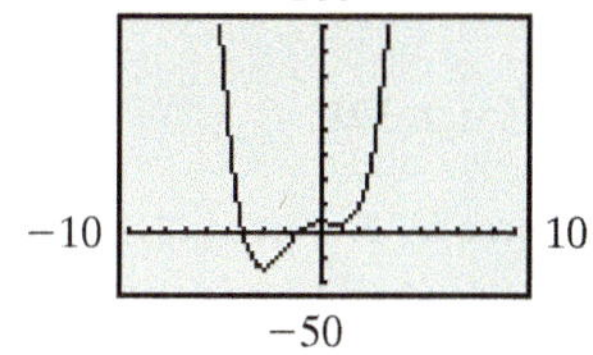

39. $-3, -1, 2$; $f(x) = (x + 3)(x + 1)(x - 2)$ **41.** $\frac{1}{2}, 3, 3$; $f(x) = (2x - 1)(x - 3)^2$ **43.** $-\frac{1}{3}$; $f(x) = (3x + 1)(x^2 + x + 1)$ **45.** $3, \frac{5 + \sqrt{17}}{2}, \frac{5 - \sqrt{17}}{2}$; $f(x) = (x - 3)\left(x - \left(\frac{5 + \sqrt{17}}{2}\right)\right)\left(x - \left(\frac{5 - \sqrt{17}}{2}\right)\right)$ **47.** $-2, -1, 1, 1$; $f(x) = (x + 2)(x + 1)(x - 1)^2$ **49.** $-5, -3, -\frac{3}{2}, 1$; $f(x) = (x + 5)(x + 3)(2x + 3)(x - 1)$ **51.** $-2, -\frac{3}{2}, 1, 4$; $f(x) = (x + 2)(2x + 3)(x - 1)(x - 4)$ **53.** $-\frac{1}{2}, \frac{1}{2}$; $f(x) = (2x + 1)(2x - 1)(x^2 + 2)$ **55.** $\frac{\sqrt{2}}{2}, -\frac{\sqrt{2}}{2}, 2$; $f(x) = (x - 2)(\sqrt{2}x - 1)(\sqrt{2}x + 1)(2x^2 + 1)$ **57.** $-5.9, -0.3, 3$ **59.** $-3.8, 4.5$ **61.** $-43.5, 1, 23$ **63.** $\{-1, 2\}$ **65.** $\left\{\frac{2}{3}, -1 + \sqrt{2}, -1 - \sqrt{2}\right\}$ **67.** $\left\{\frac{1}{3}, \sqrt{5}, -\sqrt{5}\right\}$ **69.** $\{-3, -2\}$ **71.** $\left\{-\frac{1}{3}\right\}$ **73.** $f(0) = -1$; $f(1) = 10$; Zero: 0.22 **75.** $f(-5) = -58$; $f(-4) = 2$; Zero: -4.05 **77.** $f(1.4) = -0.17536$; $f(1.5) = 1.40625$; Zero: 1.41

79. Step 1: $y = x^3$
Step 2: x-intercepts: $-3, -1, 2$; y-intercept: -6
Step 3: $-3, -1, 2$: multiplicity 1, crosses
1: multiplicity 2, touches
Step 4:

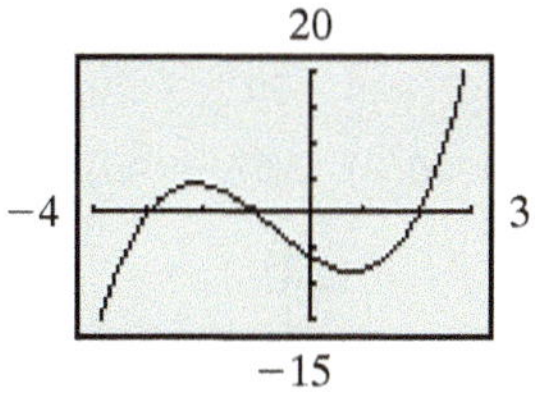

Step 5: $(-2.12, 4.06)$; $(0.79, -8.21)$
Step 6:

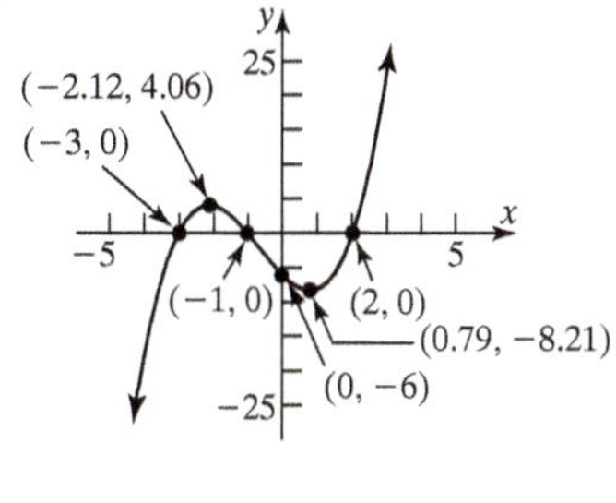

Step 7: Domain and Range: $(-\infty, \infty)$
Step 8: Increasing: $(-\infty, -2.12)$, $(0.79, \infty)$; Decreasing: $(-2.12, 0.79)$

81. Step 1: $y = x^4$
Step 2: x-intercepts: $-2, -1, 1$; y-intercept: 2
Step 3: $-2, -1$: multiplicity 1, crosses;
1: multiplicity 2, touches
Step 4:

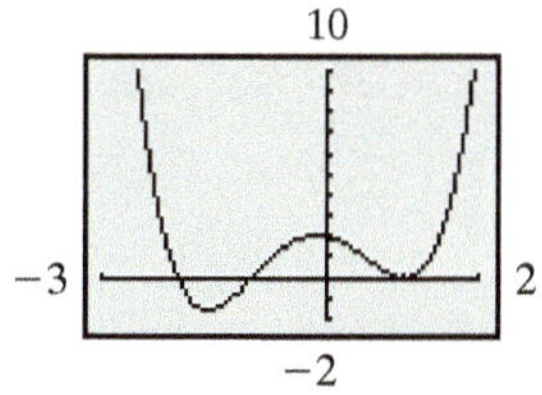

Step 5: $(-1.59, -1.62)$, $(-0.16, 2.08)$, $(1, 0)$
Step 6:

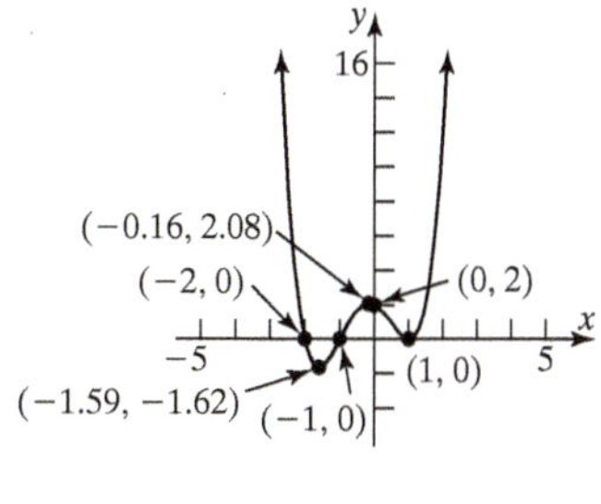

Step 7: Domain: $[-\infty, \infty)$; Range: $[-1.62, \infty)$
Step 8: Increasing: $(-1.59, -0.16)$, $(1, \infty)$; Decreasing: $(-\infty, -1.59)$, $(-0.16, 1)$

83. Step 1: $y = 4x^5$
Step 2: x-intercepts: $-\frac{\sqrt{2}}{2}, \frac{\sqrt{2}}{2}, 2$; y-intercept: 2
Step 3: $-\frac{\sqrt{2}}{2}, \frac{\sqrt{2}}{2}, 2$: multiplicity 1, crosses
Step 4:

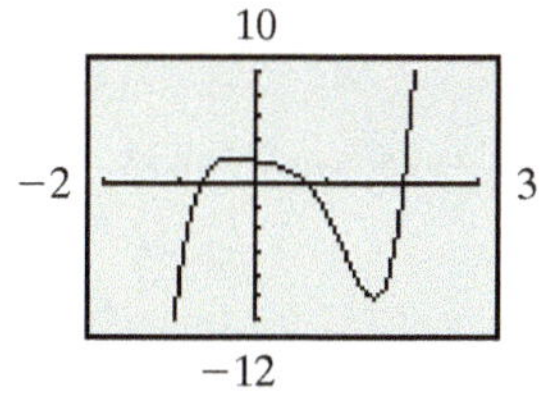

Step 5: $(-0.30, 2.23)$; $(1.61, -10.09)$
Step 6:

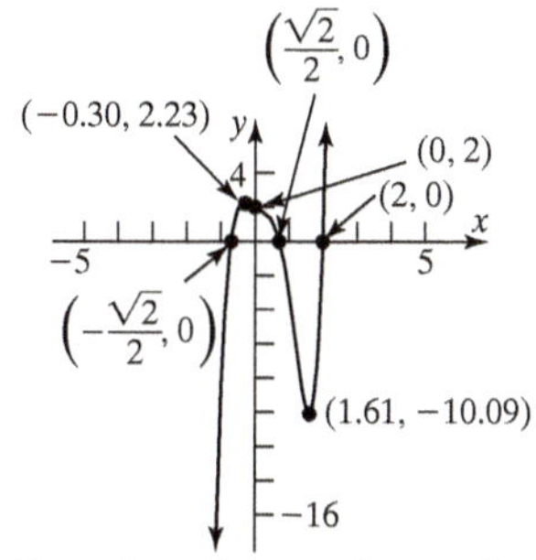

Step 7: Domain and range: $(-\infty, \infty)$
Step 8: Increasing: $(-\infty, -0.30)$, $(1.61, \infty)$; Decreasing: $(-0.30, 1.61)$

85. $k = 5$ **87.** -7 **89.** If $f(x) = x^n - c^n$, then $f(c) = c^n - c^n = 0$; so $x - c$ is a factor of f.
91. 5 **93.** 7 in. **95.** All the potential rational zeros are integers, so r is either an integer or is not a rational zero (and is therefore irrational).
97. 0.215 **99.** No; by the Rational Zeros Theorem, $\frac{1}{3}$ is not a potential rational zero. **101.** No; by the Rational Zeros Theorem, $\frac{2}{3}$ is not a potential rational zero.

5.6 Assess Your Understanding *(page 394)*

3. one **4.** $3 - 4i$ **5.** T **6.** F **7.** $4 + i$ **9.** $-i, 1 - i$ **11.** $-i, -2i$ **13.** $-i$ **15.** $2 - i, -3 + i$
17. $f(x) = x^4 - 14x^3 + 77x^2 - 200x + 208; a = 1$ **19.** $f(x) = x^5 - 4x^4 + 7x^3 - 8x^2 + 6x - 4; a = 1$
21. $f(x) = x^4 - 6x^3 + 10x^2 - 6x + 9; a = 1$ **23.** $-2i, 4$ **25.** $2i, -3, \frac{1}{2}$ **27.** $3 + 2i, -2, 5$ **29.** $4i, -\sqrt{11}, \sqrt{11}, -\frac{2}{3}$
31. $1, -\frac{1}{2} - \frac{\sqrt{3}}{2}i, -\frac{1}{2} + \frac{\sqrt{3}}{2}i; f(x) = (x - 1)\left(x + \frac{1}{2} + \frac{\sqrt{3}}{2}i\right)\left(x + \frac{1}{2} - \frac{\sqrt{3}}{2}i\right)$
33. $2, 3 - 2i, 3 + 2i; f(x) = (x - 2)(x - 3 + 2i)(x - 3 - 2i)$ **35.** $-i, i, -2i, 2i; f(x) = (x + i)(x - i)(x + 2i)(x - 2i)$
37. $-5i, 5i, -3, 1; f(x) = (x + 5i)(x - 5i)(x + 3)(x - 1)$ **39.** $-4, \frac{1}{3}, 2 - 3i, 2 + 3i; f(x) = 3(x + 4)\left(x - \frac{1}{3}\right)(x - 2 + 3i)(x - 2 - 3i)$
41. Zeros that are complex numbers must occur in conjugate pairs; or a polynomial with real coefficients of odd degree must have at least one real zero.
43. If the remaining zero were a complex number, its conjugate would also be a zero, creating a polynomial of degree 5.

Review Exercises *(page 397)*

1. Polynomial of degree 5 **3.** Not a polynomial; The variable x is raised to the $\frac{1}{2}$ power, which is not a nonnegative integer.

5.

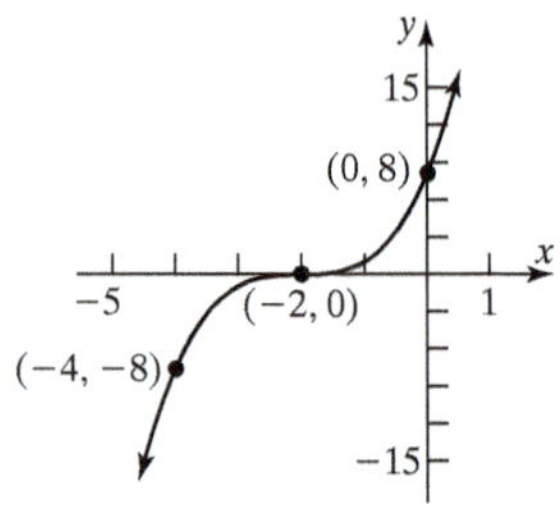

7.

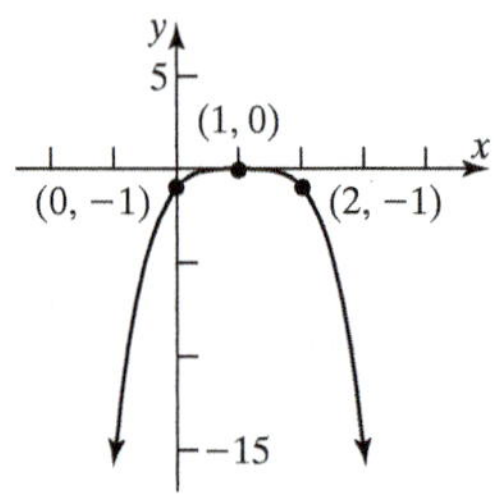

9.

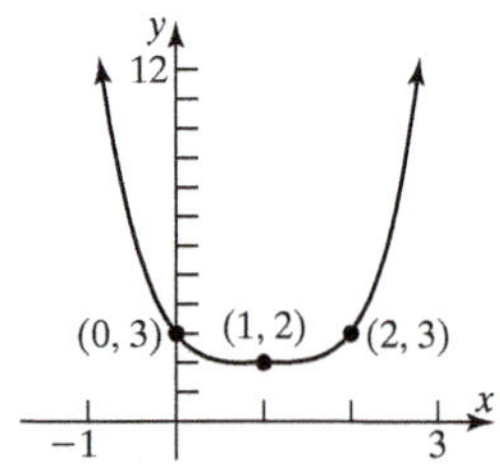

11. Step 1: $y = x^3$
Step 2: x-intercepts: $-4, -2, 0$; y-intercept: 0
Step 3: $-4, -2, 0$: multiplicity 1; crosses
Step 4:

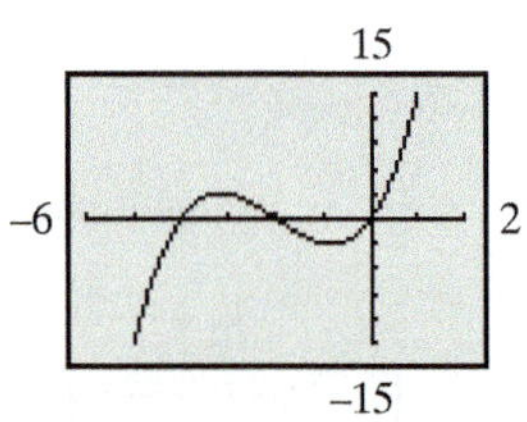

Step 5: $(-3.15, 3.08), (-0.85, -3.08)$
Step 6:

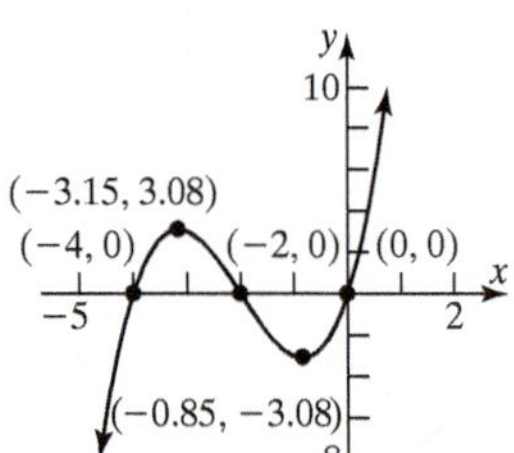

Step 7: Domain: $(-\infty, \infty)$; Range: $(-\infty, \infty)$
Step 8: Increasing on $(-\infty, -3.15)$ and $(-0.85, \infty)$
Decreasing on $(-3.15, -0.85)$

13. Step 1: $y = x^3$
Step 2: x-intercepts: $-4, 2$; y-intercept: 16
Step 3: -4: multiplicity 1; crosses;
2: multiplicity 2; touches
Step 4:

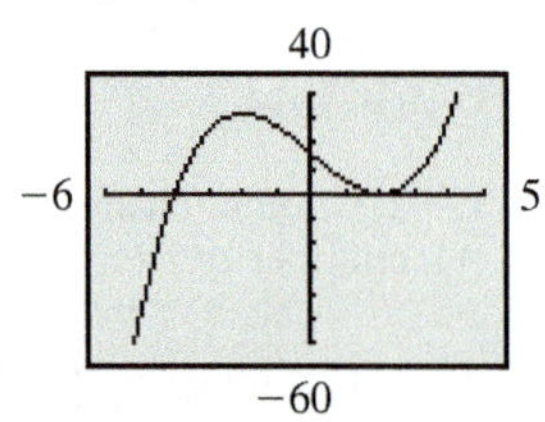

Step 5: $(-2, 32), (2, 0)$
Step 6:

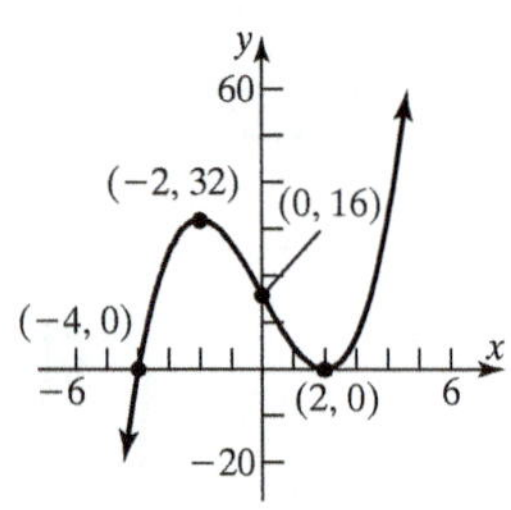

Step 7: Domain: $(-\infty, \infty)$; Range: $(-\infty, \infty)$
Step 8: Increasing on $(-\infty, 0.85)$ and $(3.15, \infty)$
Decreasing on $(0.85, 3.15)$

15. Step 1: $y = x^3$
Step 2: x-intercepts: 0, 4; y-intercept: 0
Step 3: 0: multiplicity 2; touches; 4: multiplicity 1; crosses
Step 4:

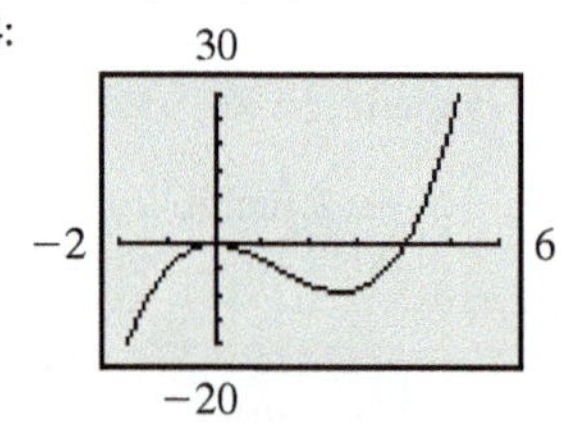

Step 5: $(0, 0), (2.67, -9.48)$
Step 6:

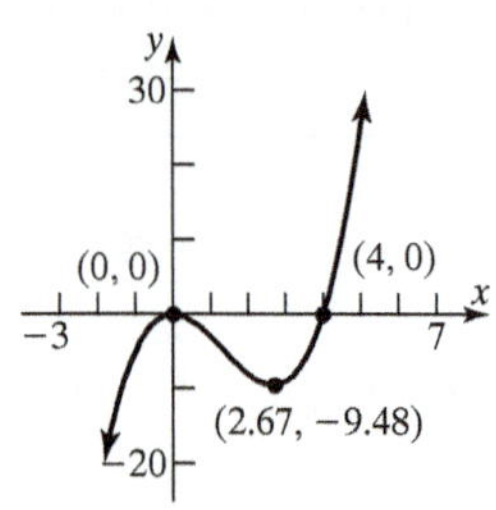

Step 7: Domain: $(-\infty, \infty)$; Range: $(-\infty, \infty)$
Step 8: Increasing on $(-\infty, 0)$ and $(2.67, \infty)$
Decreasing on $(0, 2.67)$

17. Step 1: $y = x^4$
Step 2: x-intercepts: $-3, -1, 1$; y-intercept: 3
Step 3: $-3, -1$: multiplicity 1; crosses; 1: multiplicity 2; touches
Step 4:

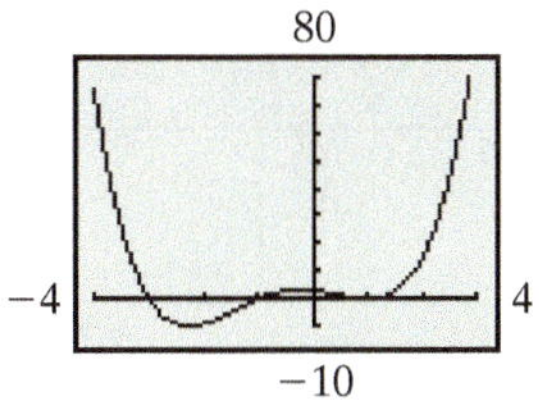

Step 5: $(-2.28, -9.91), (-0.22, 3.23), (1, 0)$
Step 6:

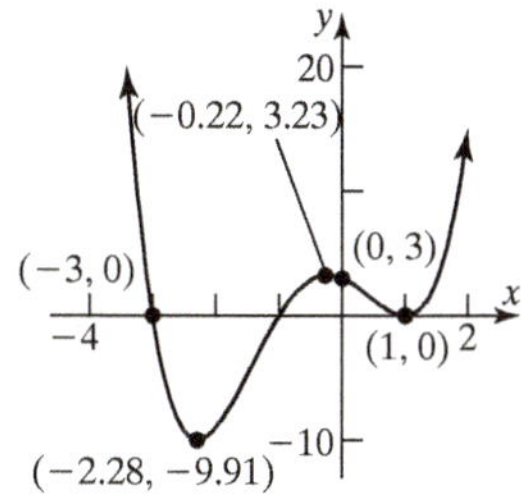

Step 7: Domain: $(-\infty, \infty)$; Range: $(-\infty, \infty)$
Step 8: Increasing on $(-2.28, -0.22)$ and $(1, \infty)$
Decreasing on $(-\infty, -2.28)$ and $(-0.22, 1)$

19. Domain: $\{x \mid x \neq -3, x \neq 3\}$; Horizontal asymptote: $y = 0$; Vertical asymptotes: $x = -3, x = 3$

21. Domain: $\{x \mid x \neq -2\}$; Horizontal asymptote: $y = 2$; Vertical asymptote: $x = -2$

23. Step 1: $R(x) = \dfrac{2(x-3)}{x}$; Domain: $\{x \mid x \neq 0\}$
Step 2: $R(x)$ is in lowest terms.
Step 3: x-intercept: 3; no y-intercept
Step 4: No symmetry
Step 5: Vertical asymptote: $x = 0$
Step 6: Horizontal asymptote: $y = 2$, not intersected
Step 7:

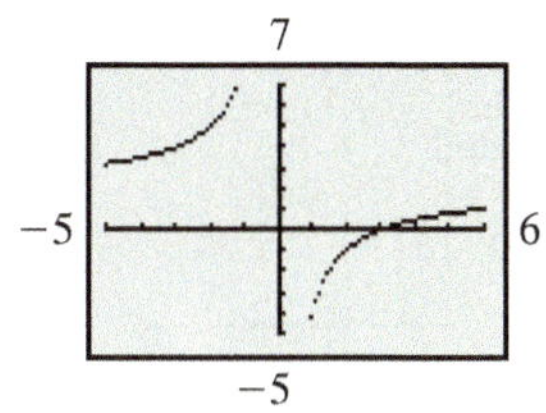

Step 8:

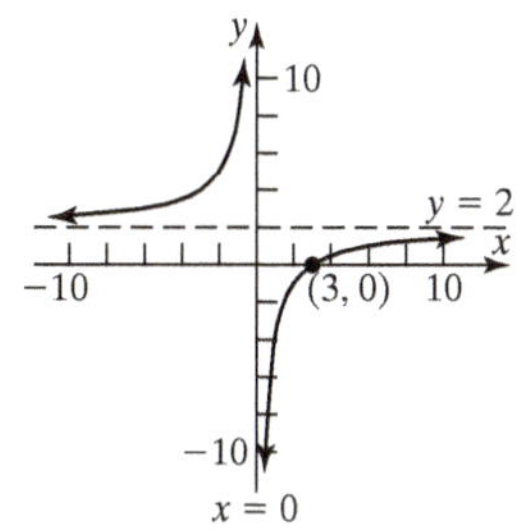

25. Step 1: Domain: $\{x \mid x \neq 0, x \neq 2\}$
Step 2: $H(x)$ is in lowest terms.
Step 3: x-intercept: -2; no y-intercept
Step 4: No symmetry
Step 5: Vertical asymptotes: $x = 0, x = 2$
Step 6: Horizontal asymptote: $y = 0$, intersected at $(-2, 0)$
Step 7:

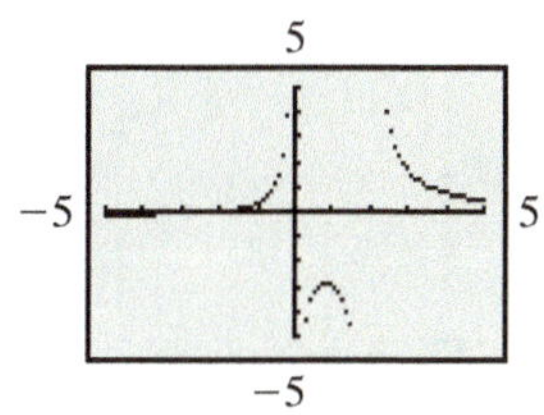

Step 8:

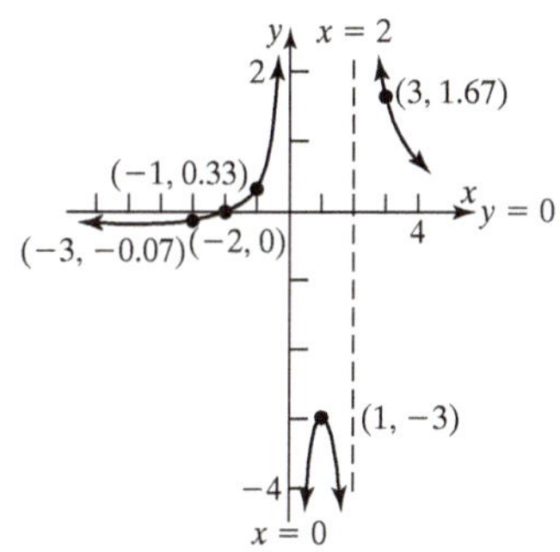

27. Step 1: $R(x) = \dfrac{(x+3)(x-2)}{(x-3)(x+2)}$;
Domain: $\{x \mid x \neq -2, x \neq 3\}$
Step 2: $R(x)$ is in lowest terms.
Step 3: x-intercepts: $-3, 2$; y-intercept: 1
Step 4: No symmetry
Step 5: Vertical asymptotes: $x = -2, x = 3$
Step 6: Horizontal asymptote: $y = 1$, intersected at $(0, 1)$
Step 7:

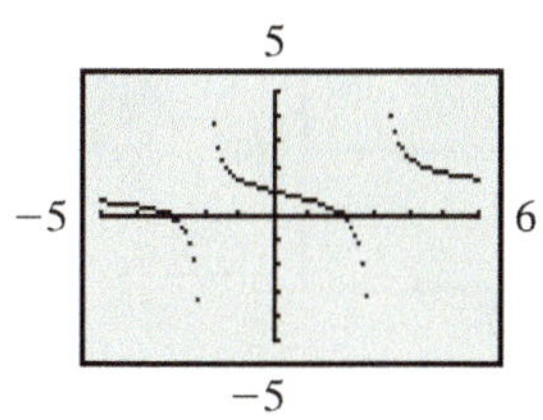

Step 8:

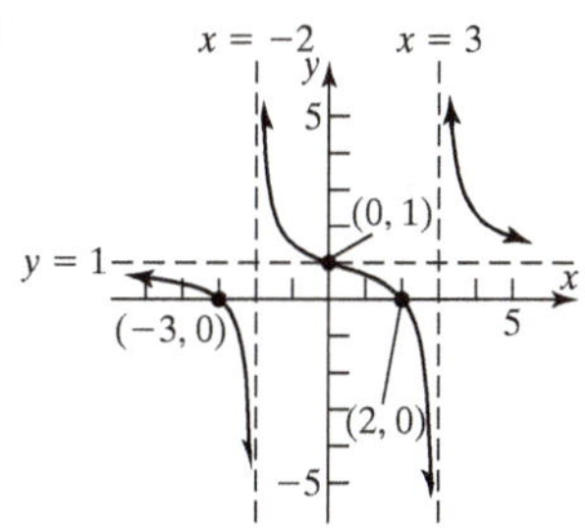

29. Step 1: $F(x) = \dfrac{x^3}{(x+2)(x-2)}$;
Domain: $\{x \mid x \neq -2, x \neq 2\}$
Step 2: $F(x)$ is in lowest terms.
Step 3: Intercept: $(0, 0)$
Step 4: Symmetric with respect to the origin
Step 5: Vertical asymptotes: $x = -2, x = 2$
Step 6: Oblique asymptote: $y = x$, intersected at $(0, 0)$
Step 7:

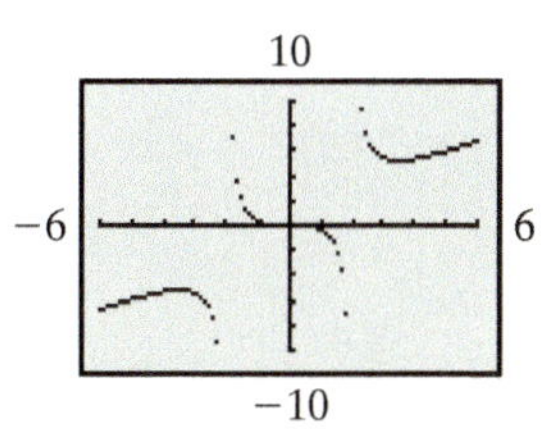

Step 8:

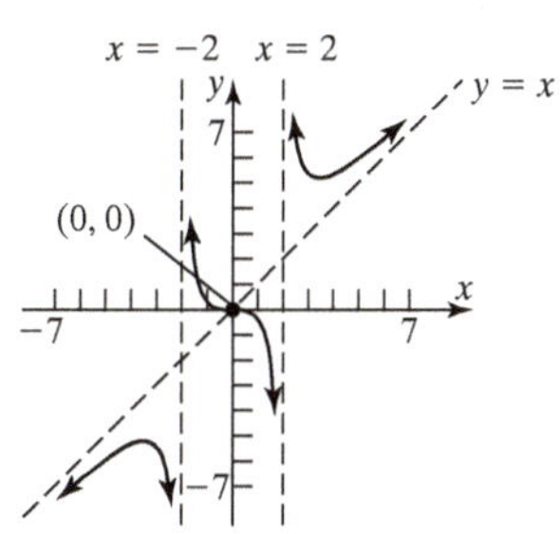

31. Step 1: Domain: $\{x|x \neq 1\}$
Step 2: $R(x)$ is in lowest terms.
Step 3: Intercept: $(0, 0)$
Step 4: No symmetry
Step 5: Vertical asymptote: $x = 1$
Step 6: No oblique or horizontal asymptote

Step 7:

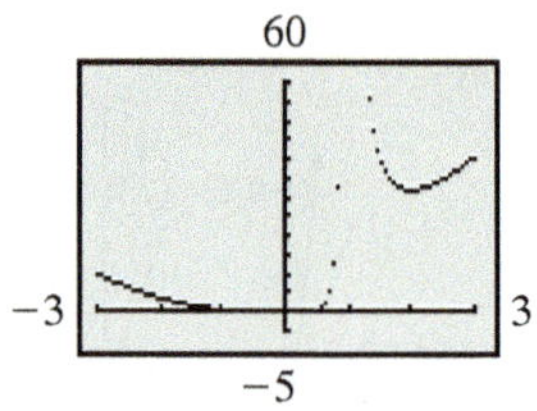

Step 8:

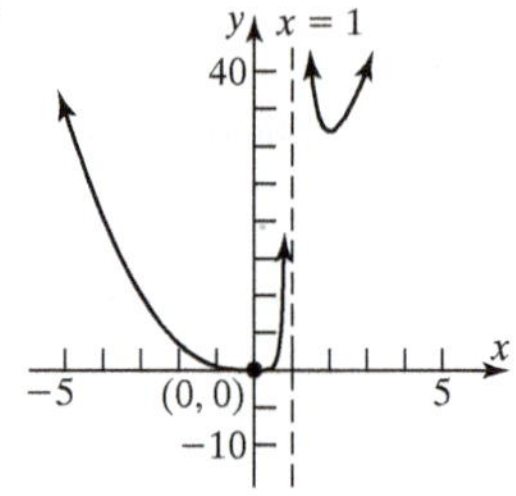

33. Step 1: $G(x) = \dfrac{(x + 2)(x - 2)}{(x + 1)(x - 2)}$;
Domain: $\{x|x \neq -1, x \neq 2\}$
Step 2: $G(x) = \dfrac{x + 2}{x + 1}$
Step 3: x-intercept: -2; y-intercept: 2
Step 4: No symmetry
Step 5: Vertical asymptote: $x = -1$; hole at $\left(2, \frac{4}{3}\right)$
Step 6: Horizontal asymptote: $y = 1$, not intersected

Step 7:

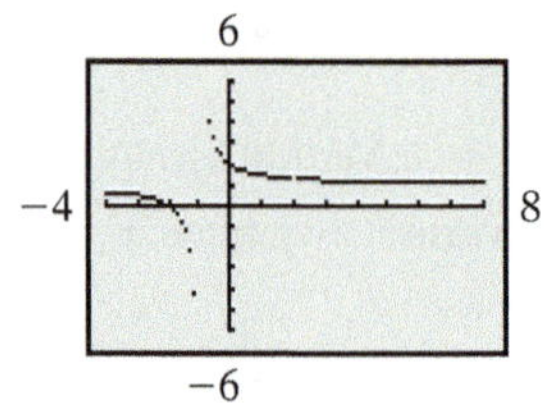

Step 8:

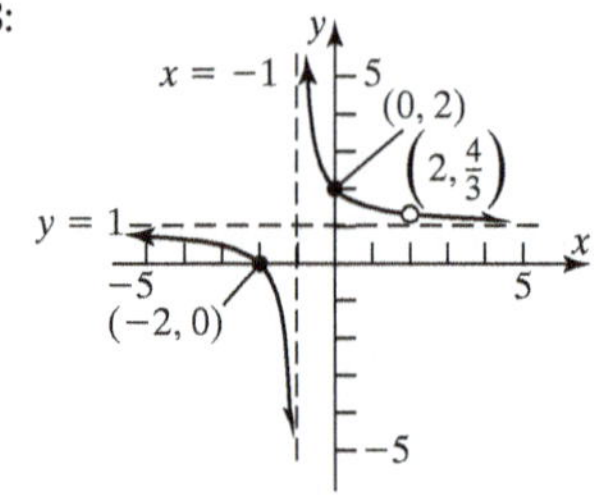

35. $\{x|x < -2 \text{ or } -1 < x < 2\}$;
$(-\infty, -2) \cup (-1, 2)$

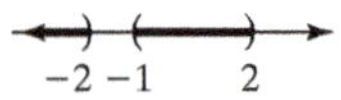

37. $\{x|-3 < x \leq 3\}$; $(-3, 3]$

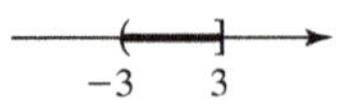

39. $\{x|x < 1 \text{ or } x > 2\}$; $(-\infty, 1) \cup (2, \infty)$

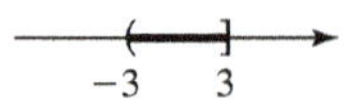

41. $\{x|1 \leq x \leq 2 \text{ or } x > 3\}$; $[1, 2] \cup (3, \infty)$

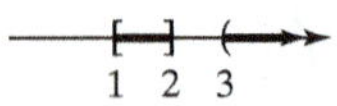

43. $\{x|x < -4 \text{ or } 2 < x < 4 \text{ or } x > 6\}$;
$(-\infty, -4) \cup (2, 4) \cup (6, \infty)$

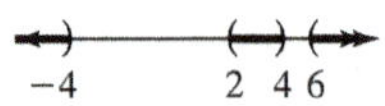

45. $R = 10$; g is not a factor of f.
47. $R = 0$; g is a factor of f.

49. $f(4) = 47{,}105$ **51.** 8; $\pm 1, \pm 3, \pm\frac{1}{2}, \pm\frac{3}{2}, \pm\frac{1}{3}, \pm\frac{1}{4}, \pm\frac{3}{4}, \pm\frac{1}{6}, \pm\frac{1}{12}$ **53.** $-2, 1, 4$; $f(x) = (x + 2)(x - 1)(x - 4)$

55. $\frac{1}{2}$, multiplicity 2; -2; $f(x) = 4\left(x - \frac{1}{2}\right)^2(x + 2)$ **57.** 2, multiplicity 2; $f(x) = (x - 2)^2(x^2 + 5)$ **59.** $-2.5, 3.1, 5.32$ **61.** $-11.3, -0.6, 4, 9.33$

63. $\{-3, 2\}$ **65.** $\left\{-3, -1, -\frac{1}{2}, 1\right\}$

67. -5 and 5

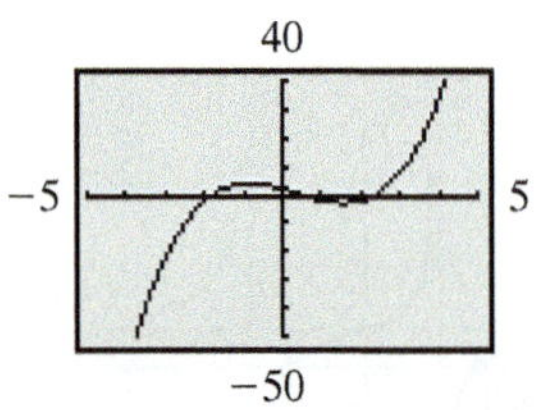

69. $-\frac{37}{2}$ and $\frac{37}{2}$

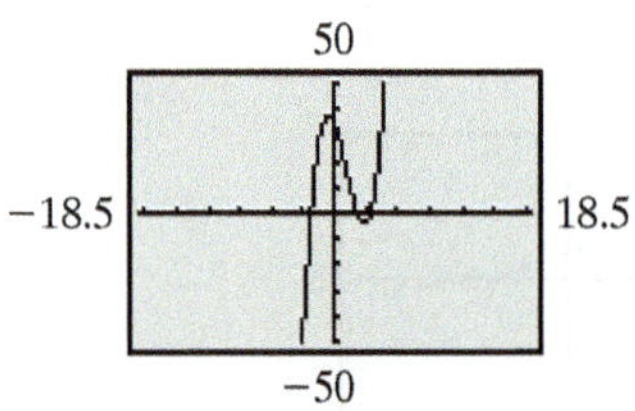

71. $f(0) = -1$; $f(1) = 1$ **73.** $f(0) = -1$; $f(1) = 1$
75. $4 - i$; $f(x) = x^3 - 14x^2 + 65x - 102$
77. $-i, 1 - i$; $f(x) = x^4 - 2x^3 + 3x^2 - 2x + 2$
79. $-2, 1, 4$; $f(x) = (x + 2)(x - 1)(x - 4)$
81. $-2, \frac{1}{2}$ (multiplicity 2); $f(x) = 4(x + 2)\left(x - \frac{1}{2}\right)^2$

83. 2 (multiplicity 2), $-\sqrt{5}i, \sqrt{5}i$; $f(x) = (x + \sqrt{5}i)(x - \sqrt{5}i)(x - 2)^2$

85. $-3, 2, -\frac{\sqrt{2}}{2}i, \frac{\sqrt{2}}{2}i$; $f(x) = 2(x + 3)(x - 2)\left(x + \frac{\sqrt{2}}{2}i\right)\left(x - \frac{\sqrt{2}}{2}i\right)$

87. **(a)** $A(r) = 2\pi r^2 + \dfrac{500}{r}$
(b) 223.22 cm^2
(c) 257.08 cm^2
(d)

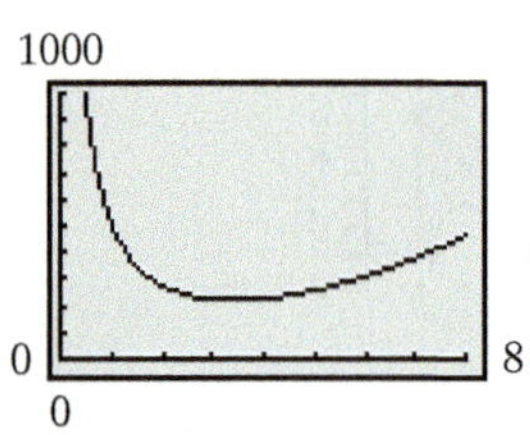

A is smallest when $r \approx 3.41$ cm.

91. **(a)** Even **(b)** Positive **(c)** Even **(d)** The graph touches the x-axis at $x = 0$. **(e)** 8

Chapter Test *(page 399)*

1.

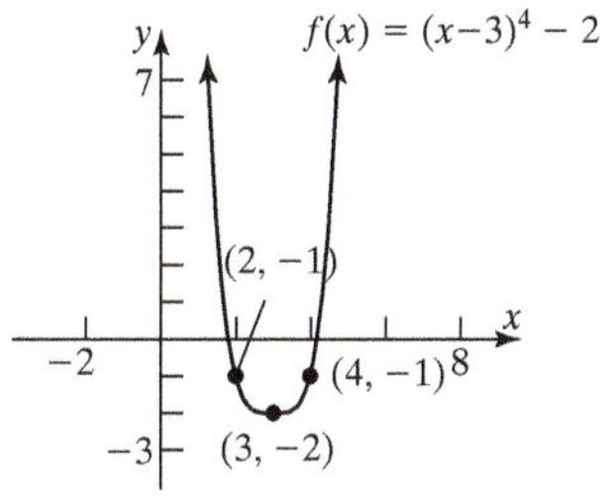

2. (a) 3 **(b)** Every zero of g lies between −15 and 15.
(c) $\pm\frac{1}{2}, \pm 1, \pm\frac{3}{2}, \pm\frac{5}{2}, \pm 3, \pm 5, \pm\frac{15}{2}, \pm 15$
(d) $-5, -\frac{1}{2}, 3; g(x) = (x + 5)(2x + 1)(x - 3)$
(e) y-intercept: −15; x-intercepts: $-5, -\frac{1}{2}, 3$
(f) Crosses at $-5, -\frac{1}{2}, 3$ **(g)** $y = 2x^3$
(h) (−3.15, 60.30), (1.48, −39.00)
(i)

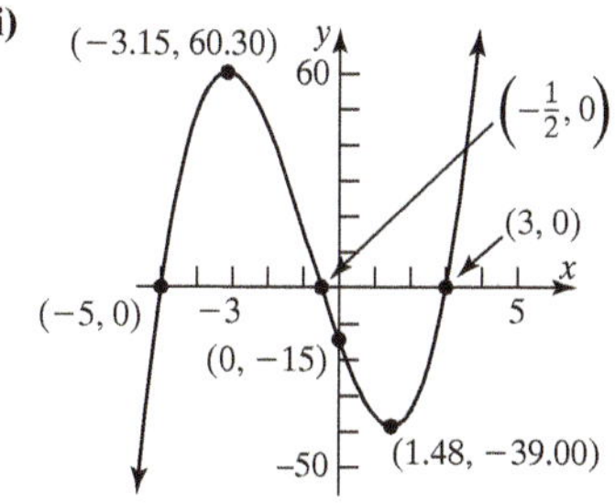

3. $4, -5i, 5i$ **4.** $\left\{1, \frac{5 - \sqrt{61}}{6}, \frac{5 + \sqrt{61}}{6}\right\}$ **5.** Domain: $\{x \mid x \neq -10, x \neq 4\}$; asymptotes: $x = -10, y = 2$

6. Domain: $\{x \mid x \neq -1\}$; asymptotes: $x = -1, y = x + 1$

7.

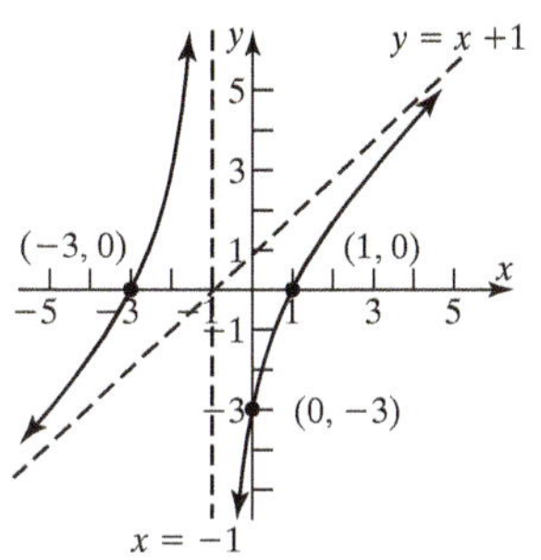

8. Answers may vary. One possibility is $f(x) = x^4 - 4x^3 - 2x^2 + 20x$.

9. Answers may vary. One possibility is $r(x) = \frac{2(x - 9)(x - 1)}{(x - 4)(x - 9)}$.

10. $f(0) = 8; f(4) = -36$
Since $f(0) = 8 > 0$ and $f(4) = -36 < 0$, the Intermediate Value Theorem guarantees that there is at least one real zero between 0 and 4.

11. $\{x \mid x < 3 \text{ or } x > 8\}; (-\infty, 3) \cup (8, \infty)$

Cumulative Review *(page 399)*

1. $\sqrt{26}$

2. $\{x \mid x \leq 0 \text{ or } x \geq 1\}$; $(-\infty, 0]$ or $[1, \infty)$

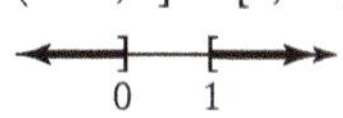

3. $\{x \mid -1 < x < 4\}$; $(-1, 4)$

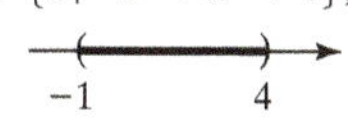

4. $f(x) = -3x + 1$

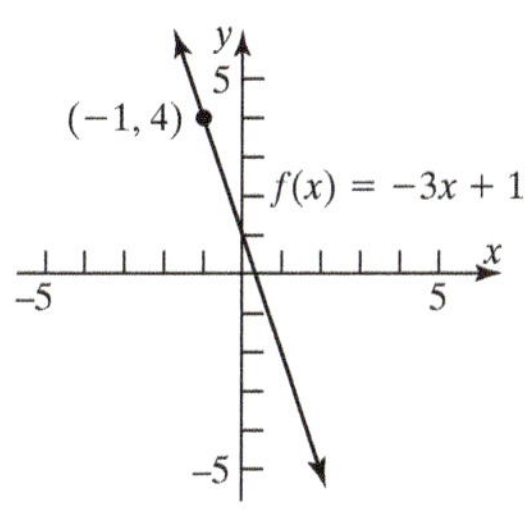

5. $y = 2x - 1$

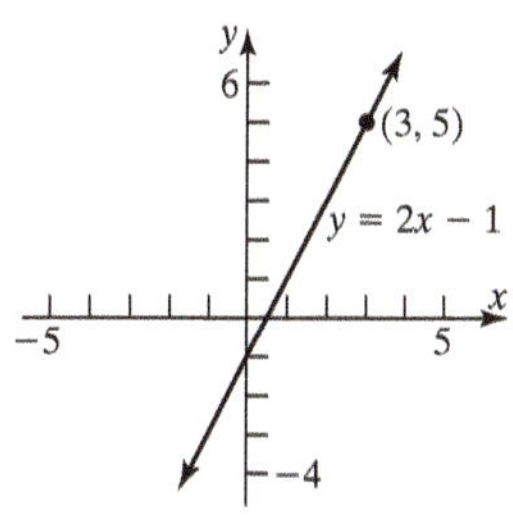

6.

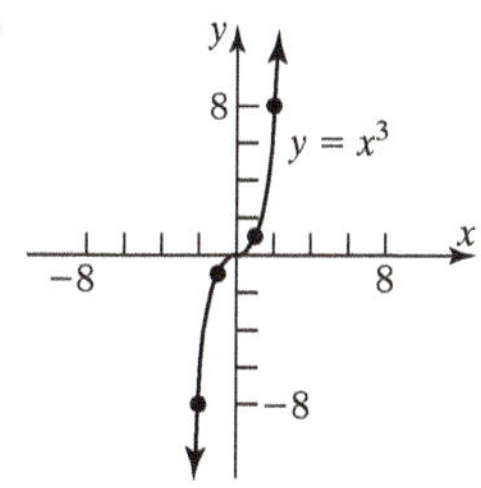

7. Not a function; 3 has two images. **8.** $\{0, 2, 4\}$ **9.** $\left\{x \,\middle|\, x \geq \frac{3}{2}\right\}; \left[\frac{3}{2}, \infty\right)$

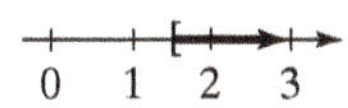

10. Center: (−2, 1); radius: 3

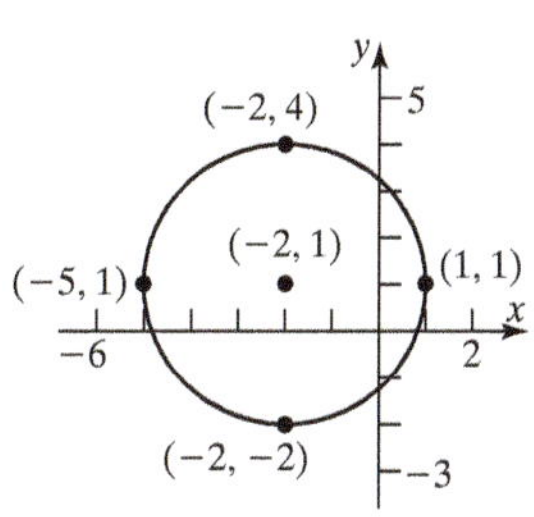

11. x-intercepts: −3, 0, 3; y-intercept: 0; symmetric with respect to the origin.

12. $y = -\frac{2}{3}x + \frac{17}{3}$

13. Not a function; it fails the vertical-line test.

14. (a) 22 **(b)** $x^2 - 5x - 2$ **(c)** $-x^2 - 5x + 2$ **(d)** $9x^2 + 15x - 2$ **(e)** $2x + h + 5$

15. (a) $\{x \mid x \neq 1\}$ **(b)** No; (2, 7) is on the graph. **(c)** 4; (3, 4) is on the graph.
(d) $\frac{7}{4}; \left(\frac{7}{4}, 9\right)$ is on the graph.

16.

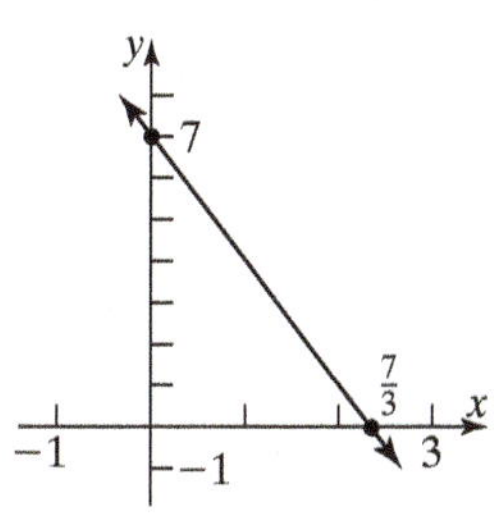

17.

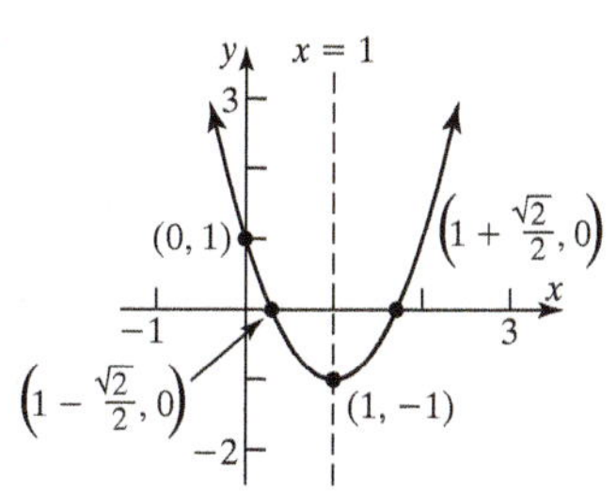

18. 6; $y = 6x - 1$

19. (a) x-intercepts: −5, −1, 5; y-intercept: −3
(b) No symmetry
(c) Neither
(d) Increasing: $(-\infty, -3)$ and $(2, \infty)$;
Decreasing: $(-3, 2)$
(e) Local maximum is 5 and occurs at $x = -3$.
(f) Local minimum is −6 and occurs at $x = 2$.

20. Odd

21. (a) Domain: $\{x|x > -3\}$ or $(-3, \infty)$ **(b)** x-intercept: $-\frac{1}{2}$; y-intercept: 1

(c)

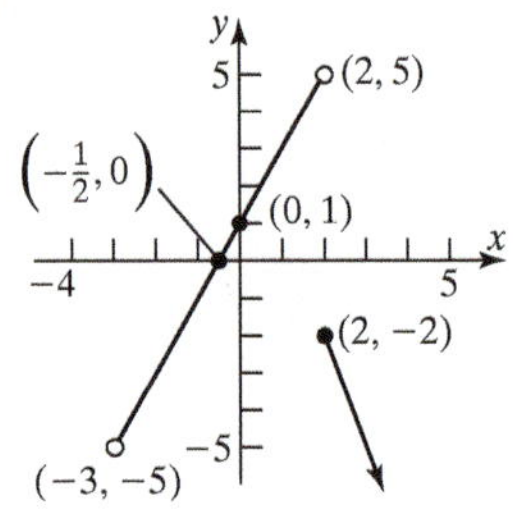

(d) Range: $\{y|y < 5\}$ or $(-\infty, 5)$

22.

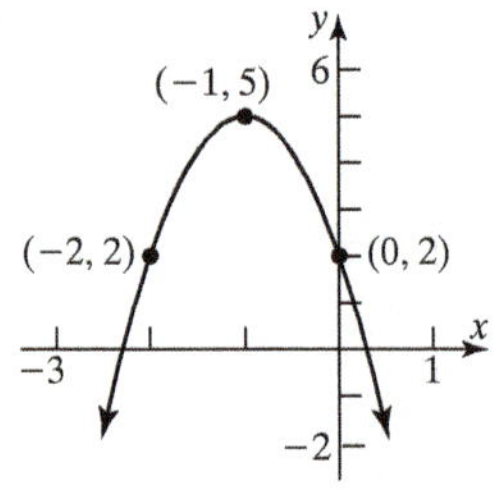

23. (a) $(f + g)(x) = x^2 - 9x - 6$; domain: all real numbers **(b)** $\left(\frac{f}{g}\right)(x) = \frac{x^2 - 5x + 1}{-4x - 7}$; domain: $\left\{x \middle| x \neq -\frac{7}{4}\right\}$

24. (a) $R(x) = -\frac{1}{10}x^2 + 150x$ **(b)** \$14,000 **(c)** 750; \$56,250 **(d)** \$75

CHAPTER 6 Exponential and Logarithmic Functions

6.1 Assess Your Understanding *(page 409)*

4. $(g \circ f)(x)$ **5.** F **6.** F **7. (a)** -1 **(b)** -1 **(c)** 8 **(d)** 0 **(e)** 8 **(f)** -7 **9. (a)** 4 **(b)** 5 **(c)** -1 **(d)** -2 **11. (a)** 98 **(b)** 49 **(c)** 4 **(d)** 4 **13. (a)** 97 **(b)** $-\frac{163}{2}$ **(c)** 1 **(d)** $-\frac{3}{2}$ **15. (a)** $2\sqrt{2}$ **(b)** $2\sqrt{2}$ **(c)** 1 **(d)** 0 **17. (a)** $\frac{1}{17}$ **(b)** $\frac{1}{5}$ **(c)** 1 **(d)** $\frac{1}{2}$ **19. (a)** $\frac{3}{\sqrt[3]{4} + 1}$ **(b)** 1 **(c)** $\frac{6}{5}$ **(d)** 0 **21.** $\{x|x \neq 0, x \neq 2\}$ **23.** $\{x|x \neq -4, x \neq 0\}$ **25.** $\left\{x \middle| x \geq -\frac{3}{2}\right\}$ **27.** $\{x|x \geq 1\}$ **29. (a)** $(f \circ g)(x) = 6x + 3$; all real numbers **(b)** $(g \circ f)(x) = 6x + 9$; all real numbers **(c)** $(f \circ f)(x) = 4x + 9$; all real numbers **(d)** $(g \circ g)(x) = 9x$; all real numbers

31. (a) $(f \circ g)(x) = 3x^2 + 1$; all real numbers **(b)** $(g \circ f)(x) = 9x^2 + 6x + 1$; all real numbers **(c)** $(f \circ f)(x) = 9x + 4$; all real numbers **(d)** $(g \circ g)(x) = x^4$; all real numbers **33. (a)** $(f \circ g)(x) = x^4 + 8x^2 + 16$; all real numbers **(b)** $(g \circ f)(x) = x^4 + 4$; all real numbers **(c)** $(f \circ f)(x) = x^4$; all real numbers **(d)** $(g \circ g)(x) = x^4 + 8x^2 + 20$; all real numbers **35. (a)** $(f \circ g)(x) = \frac{3x}{2 - x}$; $\{x|x \neq 0, x \neq 2\}$ **(b)** $(g \circ f)(x) = \frac{2(x - 1)}{3}$; $\{x|x \neq 1\}$ **(c)** $(f \circ f)(x) = \frac{3(x - 1)}{4 - x}$; $\{x|x \neq 1, x \neq 4\}$ **(d)** $(g \circ g)(x) = x$; $\{x|x \neq 0\}$

37. (a) $(f \circ g)(x) = \frac{4}{4 + x}$; $\{x|x \neq -4, x \neq 0\}$ **(b)** $(g \circ f)(x) = \frac{-4(x - 1)}{x}$; $\{x|x \neq 0, x \neq 1\}$ **(c)** $(f \circ f)(x) = x$; $\{x|x \neq 1\}$ **(d)** $(g \circ g)(x) = x$; $\{x|x \neq 0\}$ **39. (a)** $(f \circ g)(x) = \sqrt{2x + 3}$; $\left\{x \middle| x \geq -\frac{3}{2}\right\}$ **(b)** $(g \circ f)(x) = 2\sqrt{x} + 3$; $\{x|x \geq 0\}$ **(c)** $(f \circ f)(x) = \sqrt[4]{x}$; $\{x|x \geq 0\}$ **(d)** $(g \circ g)(x) = 4x + 9$; all real numbers **41. (a)** $(f \circ g)(x) = x$; $\{x|x \geq 1\}$ **(b)** $(g \circ f)(x) = |x|$; all real numbers **(c)** $(f \circ f)(x) = x^4 + 2x^2 + 2$; all real numbers **(d)** $(g \circ g)(x) = \sqrt{\sqrt{x - 1} - 1}$; $\{x|x \geq 2\}$

43. (a) $(f \circ g)(x) = -\frac{4x - 17}{2x - 1}$; $\left\{x \middle| x \neq 3;\ x \neq \frac{1}{2}\right\}$ **(b)** $(g \circ f)(x) = -\frac{3x - 3}{2x + 8}$; $\{x|x \neq -4;\ x \neq -1\}$ **(c)** $(f \circ f)(x) = -\frac{2x + 5}{x - 2}$; $\{x|x \neq -1;\ x \neq 2\}$ **(d)** $(g \circ g)(x) = -\frac{3x - 4}{2x - 11}$; $\left\{x \middle| x \neq \frac{11}{2};\ x \neq 3\right\}$

45. $(f \circ g)(x) = f(g(x)) = f\left(\frac{1}{2}x\right) = 2\left(\frac{1}{2}x\right) = x$; $(g \circ f)(x) = g(f(x)) = g(2x) = \frac{1}{2}(2x) = x$

47. $(f \circ g)(x) = f(g(x)) = f(\sqrt[3]{x}) = (\sqrt[3]{x})^3 = x$; $(g \circ f)(x) = g(f(x)) = g(x^3) = \sqrt[3]{x^3} = x$

49. $(f \circ g)(x) = f(g(x)) = f\left(\frac{1}{2}(x + 6)\right) = 2\left[\frac{1}{2}(x + 6)\right] - 6 = x + 6 - 6 = x$;
$(g \circ f)(x) = g(f(x)) = g(2x - 6) = \frac{1}{2}(2x - 6 + 6) = \frac{1}{2}(2x) = x$

51. $(f \circ g)(x) = f(g(x)) = f\left(\frac{1}{a}(x - b)\right) = a\left[\frac{1}{a}(x - b)\right] + b = x$; $(g \circ f)(x) = g(f(x)) = g(ax + b) = \frac{1}{a}(ax + b - b) = x$

53. $f(x) = x^4$; $g(x) = 2x + 3$ (Other answers are possible.) **55.** $f(x) = \sqrt{x}$; $g(x) = x^2 + 1$ (Other answers are possible.) **57.** $f(x) = |x|$; $g(x) = 2x + 1$ (Other answers are possible.) **59.** $(f \circ g)(x) = 11$; $(g \circ f)(x) = 2$ **61.** $-3, 3$ **63. (a)** $(f \circ g)(x) = acx + ad + b$ **(b)** $(g \circ f)(x) = acx + bc + d$ **(c)** The domains of both $f \circ g$ and $g \circ f$ are all real numbers. **(d)** $f \circ g = g \circ f$ when $ad + b = bc + d$

65. $S(t) = \frac{16}{9}\pi t^6$ **67.** $C(t) = 15{,}000 + 800{,}000t - 40{,}000t^2$ **69.** $C(p) = \frac{2\sqrt{100 - p}}{25} + 600, 0 \leq p \leq 100$ **71.** $V(r) = 2\pi r^3$

73. (a) $f(x) = 0.7480x$ **(b)** $g(x) = 158.7160x$ **(c)** $g(f(x)) = g(0.7480x) = 118.719568x$ **(d)** 118,719.568 yen **75. (a)** $f(p) = p - 200$ **(b)** $g(p) = 0.8p$ **(c)** $(f \circ g)(p) = 0.8p - 200$; $(g \circ f)(p) = 0.8p - 160$; The 20% discount followed by the \$200 rebate is the better deal.

77. f is an odd function, so $f(-x) = -f(x)$. g is an even function, so $g(-x) = g(x)$. Then $(f \circ g)(-x) = f(g(-x)) = f(g(x)) = (f \circ g)(x)$. So $f \circ g$ is even. Also, $(g \circ f)(-x) = g(f(-x)) = g(-f(x)) = g(f(x)) = (g \circ f)(x)$, so $g \circ f$ is even.

6.2 Assess Your Understanding *(page 421)*

4. one-to-one **5.** $y = x$ **6.** $[4, \infty)$ **7.** F **8.** T **9.** one-to-one **11.** not one-to-one **13.** not one-to-one **15.** one-to-one **17.** one-to-one **19.** not one-to-one **21.** one-to-one

23.

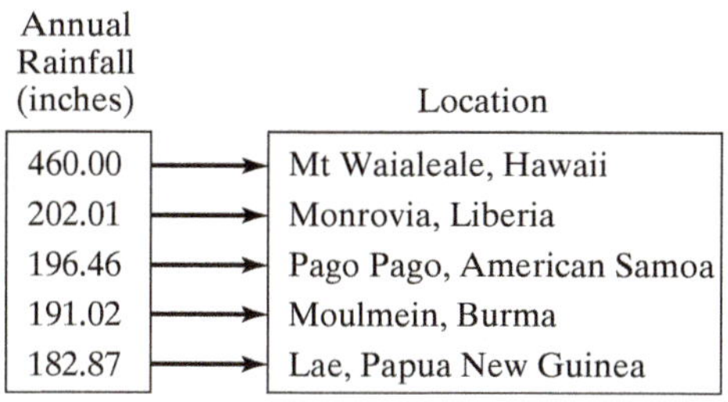

Domain: {460.00, 202.01, 196.46, 191.02, 182.87}
Range: {Mt Waialeale, Monrovia, Pago Pago, Moulmein, Lae}

25.

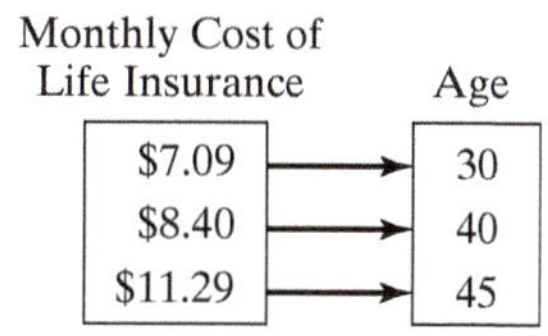

Domain: {$7.09, $8.40, $11.29}
Range: {30, 40, 45}

27. $\{(5, -3), (9, -2), (2, -1), (11, 0), (-5, 1)\}$
Domain: $\{5, 9, 2, 11, -5\}$
Range: $\{-3, -2, -1, 0, 1\}$

29. $\{(1, -2), (2, -3), (0, -10), (9, 1), (4, 2)\}$
Domain: $\{1, 2, 0, 9, 4\}$
Range: $\{-2, -3, -10, 1, 2\}$

31. $f(g(x)) = f\left(\frac{1}{3}(x-4)\right) = 3\left[\frac{1}{3}(x-4)\right] + 4 = (x-4) + 4 = x$

$g(f(x)) = g(3x+4) = \frac{1}{3}[(3x+4) - 4] = \frac{1}{3}(3x) = x$

33. $f(g(x)) = f\left(\frac{x}{4} + 2\right) = 4\left[\frac{x}{4} + 2\right] - 8 = (x+8) - 8 = x$

$g(f(x)) = g(4x-8) = \frac{4x-8}{4} + 2 = (x-2) + 2 = x$

35. $f(g(x)) = f\left(\sqrt[3]{x+8}\right) = \left(\sqrt[3]{x+8}\right)^3 - 8 = (x+8) - 8 = x$

$g(f(x)) = g(x^3 - 8) = \sqrt[3]{(x^3-8)+8} = \sqrt[3]{x^3} = x$

37. $f(g(x)) = f\left(\frac{1}{x}\right) = \frac{1}{\left(\frac{1}{x}\right)} = x$; $g(f(x)) = g\left(\frac{1}{x}\right) = \frac{1}{\left(\frac{1}{x}\right)} = x$

39. $f(g(x)) = f\left(\frac{4x-3}{2-x}\right) = \frac{2\left(\frac{4x-3}{2-x}\right) + 3}{\frac{4x-3}{2-x} + 4}$

$= \frac{2(4x-3) + 3(2-x)}{4x - 3 + 4(2-x)} = \frac{5x}{5} = x, x \neq 2$

$g(f(x)) = g\left(\frac{2x+3}{x+4}\right) = \frac{4\left(\frac{2x+3}{x+4}\right) - 3}{2 - \frac{2x+3}{x+4}}$

$= \frac{4(2x+3) - 3(x+4)}{2(x+4) - (2x+3)} = \frac{5x}{5} = x, x \neq -4$

41.

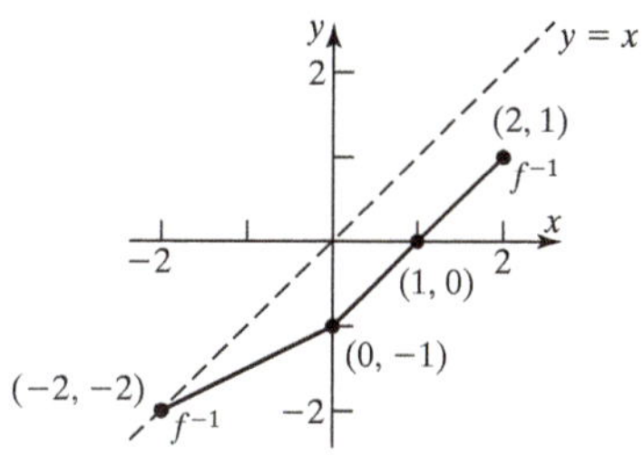

43.

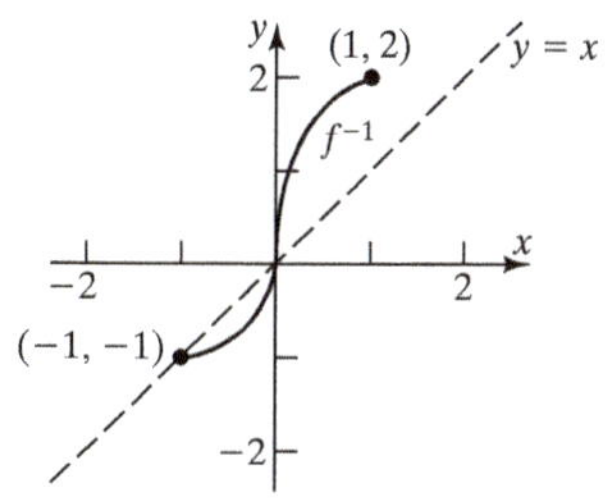

45.

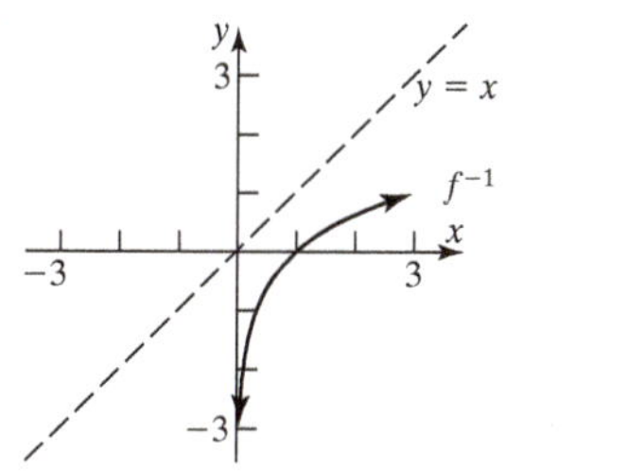

47. $f^{-1}(x) = \frac{1}{3}x$

$f(f^{-1}(x)) = f\left(\frac{1}{3}x\right) = 3\left(\frac{1}{3}x\right) = x$

$f^{-1}(f(x)) = f^{-1}(3x) = \frac{1}{3}(3x) = x$

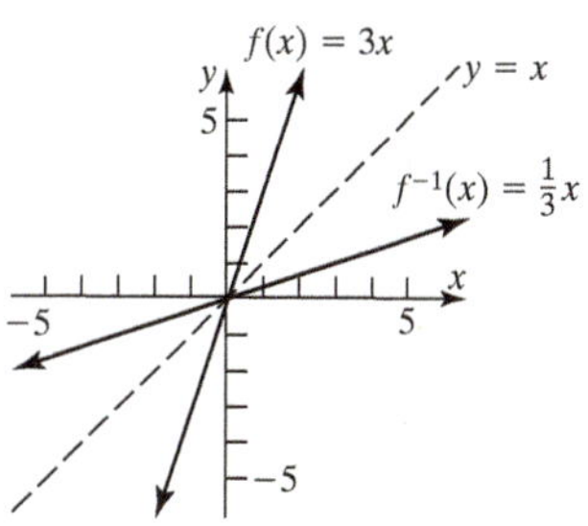

49. $f^{-1}(x) = \frac{x}{4} - \frac{1}{2}$

$f(f^{-1}(x)) = f\left(\frac{x}{4} - \frac{1}{2}\right) = 4\left(\frac{x}{4} - \frac{1}{2}\right) + 2$

$= (x-2) + 2 = x$

$f^{-1}(f(x)) = f^{-1}(4x+2) = \frac{4x+2}{4} - \frac{1}{2}$

$= \left(x + \frac{1}{2}\right) - \frac{1}{2} = x$

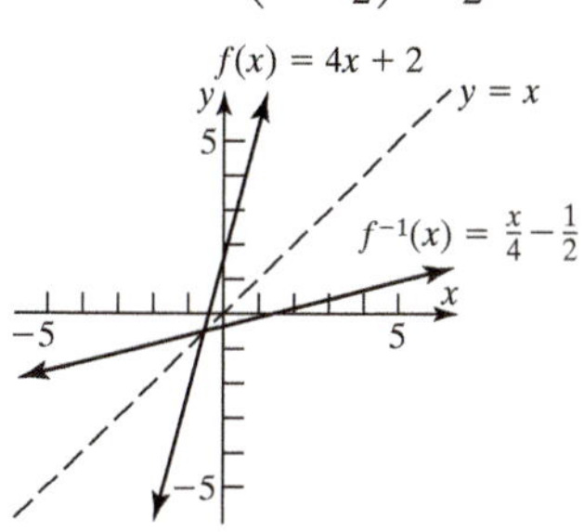

51. $f^{-1}(x) = \sqrt[3]{x+1}$

$f(f^{-1}(x)) = f\left(\sqrt[3]{x+1}\right)$

$= \left(\sqrt[3]{x+1}\right)^3 - 1 = x$

$f^{-1}(f(x)) = f^{-1}(x^3 - 1)$

$= \sqrt[3]{(x^3-1)+1} = x$

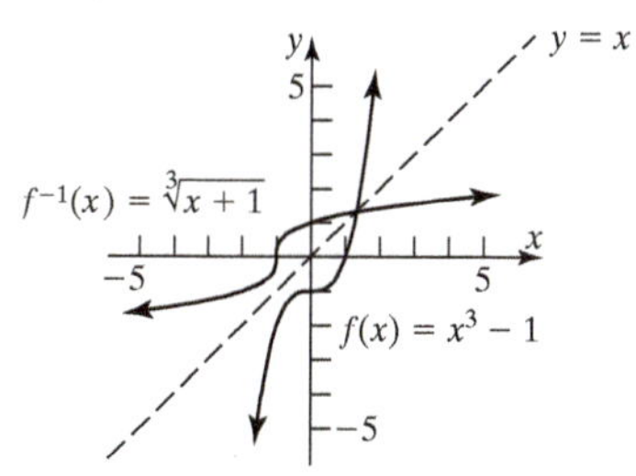

53. $f^{-1}(x) = \sqrt{x-4},\ x \geq 4$

$$f(f^{-1}(x)) = f(\sqrt{x-4}) = (\sqrt{x-4})^2 + 4 = x$$

$$f^{-1}(f(x)) = f^{-1}(x^2+4) = \sqrt{(x^2+4)-4} = \sqrt{x^2} = x,\ x \geq 0$$

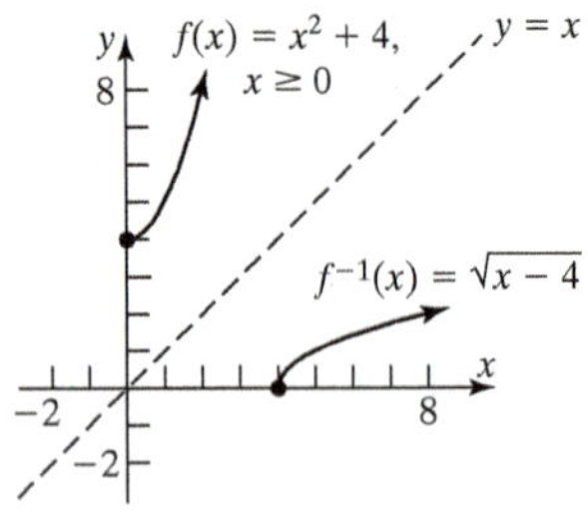

55. $f^{-1}(x) = \dfrac{4}{x}$

$$f(f^{-1}(x)) = f\left(\frac{4}{x}\right) = \frac{4}{\left(\frac{4}{x}\right)} = x$$

$$f^{-1}(f(x)) = f^{-1}\left(\frac{4}{x}\right) = \frac{4}{\left(\frac{4}{x}\right)} = x$$

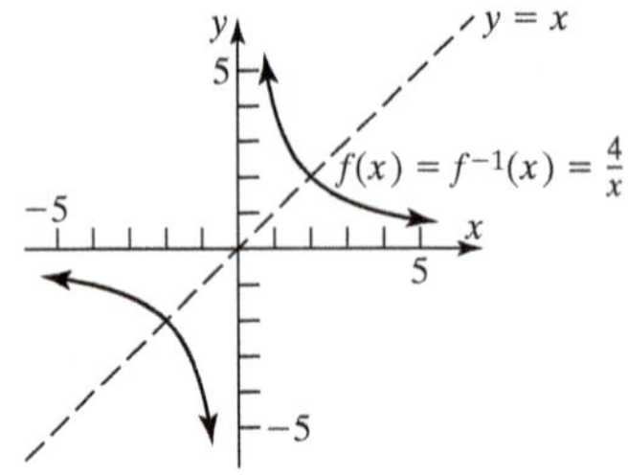

57. $f^{-1}(x) = \dfrac{2x+1}{x}$

$$f(f^{-1}(x)) = f\left(\frac{2x+1}{x}\right) = \frac{1}{\frac{2x+1}{x} - 2} = \frac{x}{(2x+1) - 2x} = x$$

$$f^{-1}(f(x)) = f^{-1}\left(\frac{1}{x-2}\right) = \frac{2\left(\frac{1}{x-2}\right) + 1}{\frac{1}{x-2}} = \frac{2 + (x-2)}{1} = x$$

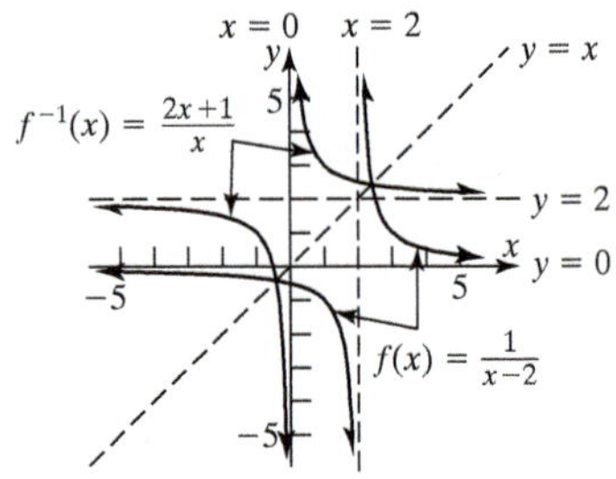

59. $f^{-1}(x) = \dfrac{2-3x}{x}$

$$f(f^{-1}(x)) = f\left(\frac{2-3x}{x}\right) = \frac{2}{3 + \frac{2-3x}{x}} = \frac{2x}{3x + 2 - 3x} = \frac{2x}{2} = x$$

$$f^{-1}(f(x)) = f^{-1}\left(\frac{2}{3+x}\right) = \frac{2 - 3\left(\frac{2}{3+x}\right)}{\frac{2}{3+x}} = \frac{2(3+x) - 3\cdot 2}{2} = \frac{2x}{2} = x$$

Domain f = range of f^{-1} = all real numbers except -3

Range f = domain f^{-1} = all real numbers except 0

61. $f^{-1}(x) = \dfrac{-2x}{x-3}$

$$f(f^{-1}(x)) = f\left(\frac{-2x}{x-3}\right) = \frac{3\left(\frac{-2x}{x-3}\right)}{\frac{-2x}{x-3} + 2} = \frac{3(-2x)}{-2x + 2(x-3)} = \frac{-6x}{-6} = x$$

$$f^{-1}(f(x)) = f^{-1}\left(\frac{3x}{x+2}\right) = \frac{-2\left(\frac{3x}{x+2}\right)}{\frac{3x}{x+2} - 3} = \frac{-2(3x)}{3x - 3(x+2)} = \frac{-6x}{-6} = x$$

Domain f = range of f^{-1} = all real numbers except -2

Range f = domain f^{-1} = all real numbers except 3

63. $f^{-1}(x) = \dfrac{x}{3x-2}$

$$f(f^{-1}(x)) = f\left(\frac{x}{3x-2}\right) = \frac{2\left(\frac{x}{3x-2}\right)}{3\left(\frac{x}{3x-2}\right) - 1} = \frac{2x}{3x - (3x-2)} = \frac{2x}{2} = x$$

$$f^{-1}(f(x)) = f^{-1}\left(\frac{2x}{3x-1}\right) = \frac{\frac{2x}{3x-1}}{3\left(\frac{2x}{3x-1}\right) - 2} = \frac{2x}{6x - 2(3x-1)} = \frac{2x}{2} = x$$

Domain f = range of f^{-1} = all real numbers except $\dfrac{1}{3}$

Range f = domain f^{-1} = all real numbers except $\dfrac{2}{3}$

65. $f^{-1}(x) = \dfrac{3x+4}{2x-3}$

$$f(f^{-1}(x)) = f\left(\frac{3x+4}{2x-3}\right) = \frac{3\left(\frac{3x+4}{2x-3}\right)+4}{2\left(\frac{3x+4}{2x-3}\right)-3}$$
$$= \frac{3(3x+4)+4(2x-3)}{2(3x+4)-3(2x-3)}$$
$$= \frac{17x}{17} = x$$
$$f^{-1}(f(x)) = f^{-1}\left(\frac{3x+4}{2x-3}\right) = \frac{3\left(\frac{3x+4}{2x-3}\right)+4}{2\left(\frac{3x+4}{2x-3}\right)-3}$$
$$= \frac{3(3x+4)+4(2x-3)}{2(3x+4)-3(2x-3)} = \frac{17x}{17} = x$$

Domain f = range of f^{-1} = all real numbers except $\frac{3}{2}$

Range f = domain f^{-1} = all real numbers except $\frac{3}{2}$

67. $f^{-1}(x) = \dfrac{-2x+3}{x-2}$

$$f(f^{-1}(x)) = f\left(\frac{-2x+3}{x-2}\right) = \frac{2\left(\frac{-2x+3}{x-2}\right)+3}{\frac{-2x+3}{x-2}+2}$$
$$= \frac{2(-2x+3)+3(x-2)}{-2x+3+2(x-2)} = \frac{-x}{-1} = x$$
$$f^{-1}(f(x)) = f^{-1}\left(\frac{2x+3}{x+2}\right) = \frac{-2\left(\frac{2x+3}{x+2}\right)+3}{\frac{2x+3}{x+2}-2}$$
$$= \frac{-2(2x+3)+3(x+2)}{2x+3-2(x+2)} = \frac{-x}{-1} = x$$

Domain f = range of f^{-1} = all real numbers except -2

Range f = domain f^{-1} = all real numbers except 2

69. $f^{-1}(x) = \dfrac{2}{\sqrt{1-2x}}$

$$f(f^{-1}(x)) = f\left(\frac{2}{\sqrt{1-2x}}\right) = \frac{\frac{4}{1-2x}-4}{2\cdot\frac{4}{1-2x}} = \frac{4-4(1-2x)}{2\cdot 4}$$
$$= \frac{8x}{8} = x$$
$$f^{-1}(f(x)) = f^{-1}\left(\frac{x^2-4}{2x^2}\right) = \frac{2}{\sqrt{1-2\left(\frac{x^2-4}{2x^2}\right)}} = \frac{2}{\sqrt{\frac{4}{x^2}}} = \sqrt{x^2}$$
$$= x, \text{ since } x > 0$$

Domain f = range of f^{-1} = $(0, \infty)$

Range f = domain f^{-1} = $\left(-\infty, \frac{1}{2}\right)$

71. **(a)** 0 **(b)** 2 **(c)** 0 **(d)** 1

73. 7 **75.** Domain of f^{-1}: $[-2, \infty)$; range of f^{-1}: $[5, \infty)$

77. Domain of g^{-1}: $[0, \infty)$; range of g^{-1}: all real numbers

79. Increasing on the interval $(f(0), f(5))$

81. $f^{-1}(x) = \frac{1}{m}(x-b), m \neq 0$

83. Quadrant I

85. Possible answer: $f(x) = |x|, x \geq 0$, is one-to-one; $f^{-1}(x) = x, x \geq 0$

87. **(a)** $r(d) = \dfrac{d+90.39}{6.97}$

(b) $r(d(r)) = \dfrac{6.97r - 90.39 + 90.39}{6.97} = \dfrac{6.97r}{6.97} = r$

$d(r(d)) = 6.97\left(\dfrac{d+90.39}{6.97}\right) - 90.39 = d + 90.39 - 90.39 = d$

(c) 56 miles per hour

89. **(a)** 77.6 kg **(b)** $h(W) = \dfrac{W-50}{2.3} + 60 = \dfrac{W+88}{2.3}$ **(c)** $h(W(h)) = \dfrac{50+2.3(h-60)+88}{2.3} = \dfrac{2.3h}{2.3} = h$ **(d)** 73 inches

$W(h(W)) = 50 + 2.3\left(\dfrac{W+88}{2.3} - 60\right)$

$= 50 + W + 88 - 138 = W$

91. **(a)** $\{g \mid 31{,}850 \leq g \leq 77{,}100\}$

(b) $\{T \mid 4386.25 \leq T \leq 15{,}698.75\}$

(c) $g(T) = \dfrac{T - 4386.25}{0.25} + 31{,}850$

Domain: $\{T \mid 4386.25 \leq T \leq 15{,}698.75\}$

Range: $\{g \mid 31{,}850 \leq g \leq 77{,}100\}$

93. **(a)** t represents time, so $t \geq 0$.

(b) $t(H) = \sqrt{\dfrac{H-100}{-4.9}} = \sqrt{\dfrac{100-H}{4.9}}$

(c) 2.02 seconds

95. $f^{-1}(x) = \dfrac{-dx+b}{cx-a}$; $f = f^{-1}$ if $a = -d$ **99.** No

6.3 Assess Your Understanding *(page 435)*

6. $\left(-1, \frac{1}{a}\right), (0, 1), (1, a)$ **7.** 1 **8.** 4 **9.** F **10.** F **11.** **(a)** 11.212 **(b)** 11.587 **(c)** 11.664 **(d)** 11.665 **13.** **(a)** 8.815 **(b)** 8.821 **(c)** 8.824 **(d)** 8.825 **15.** **(a)** 21.217 **(b)** 22.217 **(c)** 22.440 **(d)** 22.459 **17.** 3.320 **19.** 0.427 **21.** Neither **23.** Exponential; $H(x) = 4^x$ **25.** Exponential; $f(x) = 3(2^x)$ **27.** Linear; $H(x) = 2x + 4$ **29.** B **31.** D **33.** F **35.** E

37.

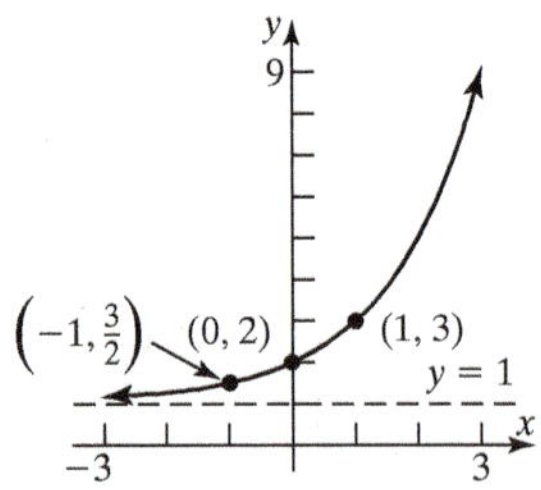

Domain: All real numbers
Range: $\{y|y > 1\}$ or $(1, \infty)$
Horizontal asymptote: $y = 1$

39.

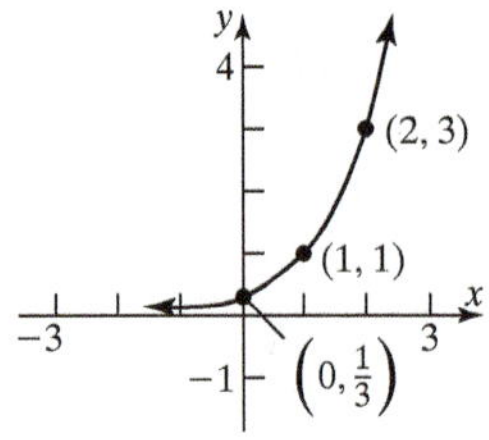

Domain: All real numbers
Range: $\{y|y > 0\}$ or $(0, \infty)$
Horizontal asymptote: $y = 0$

41.

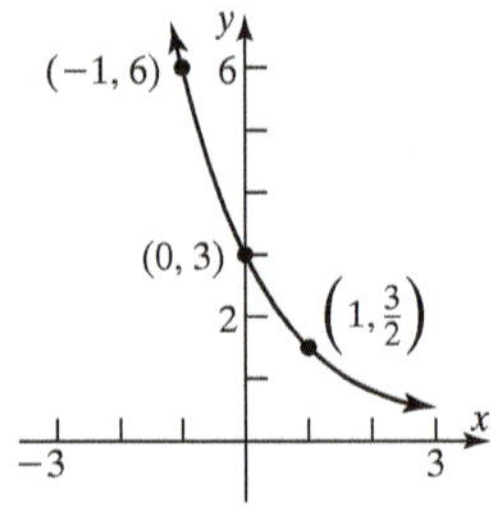

Domain: All real numbers
Range: $\{y|y > 0\}$ or $(0, \infty)$
Horizontal asymptote: $y = 0$

43.

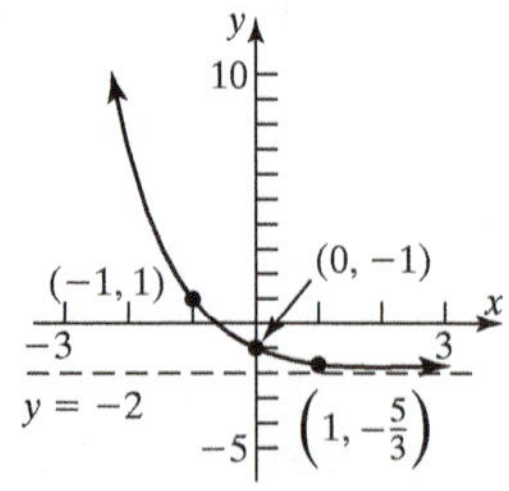

Domain: All real numbers
Range: $\{y|y > -2\}$ or $(-2, \infty)$
Horizontal asymptote: $y = -2$

45.

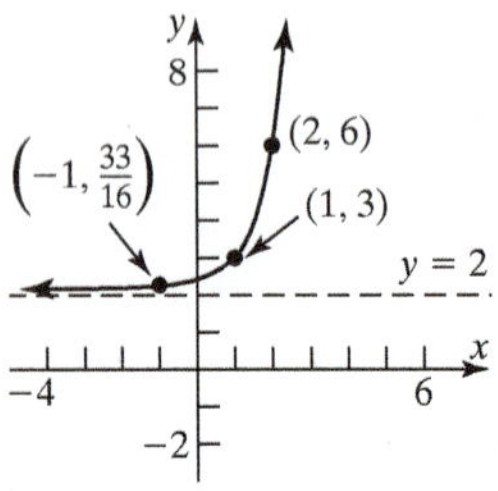

Domain: All real numbers
Range: $\{y|y > 2\}$ or $(2, \infty)$
Horizontal asymptote: $y = 2$

47.

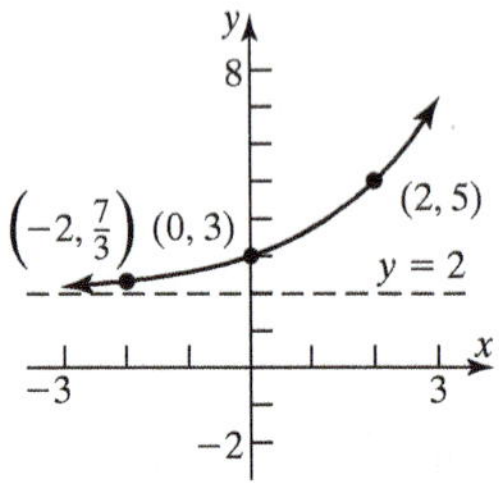

Domain: All real numbers
Range: $\{y|y > 2\}$ or $(2, \infty)$
Horizontal asymptote: $y = 2$

49.

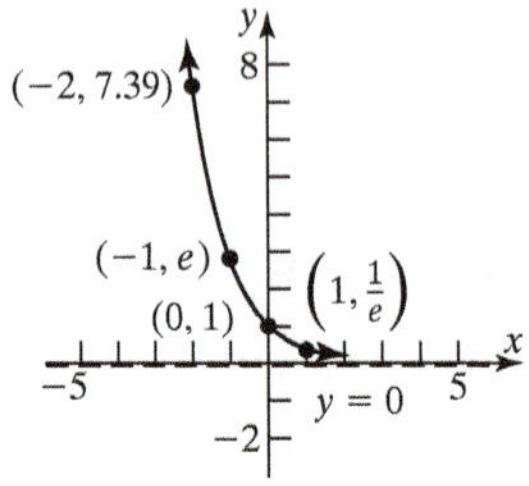

Domain: All real numbers
Range: $\{y|y > 0\}$ or $(0, \infty)$
Horizontal asymptote: $y = 0$

51.

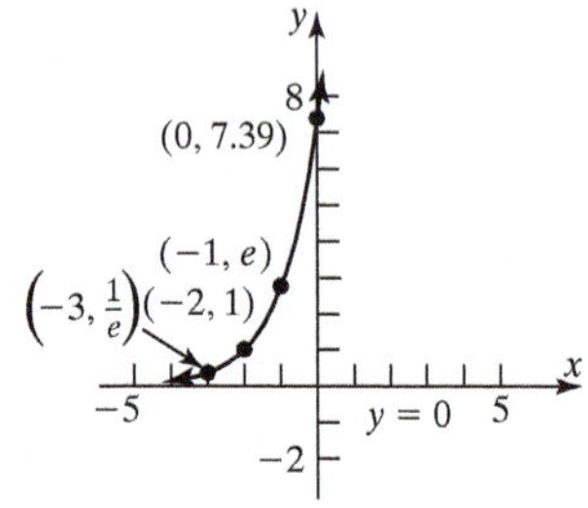

Domain: All real numbers
Range: $\{y|y > 0\}$ or $(0, \infty)$
Horizontal asymptote: $y = 0$

53.

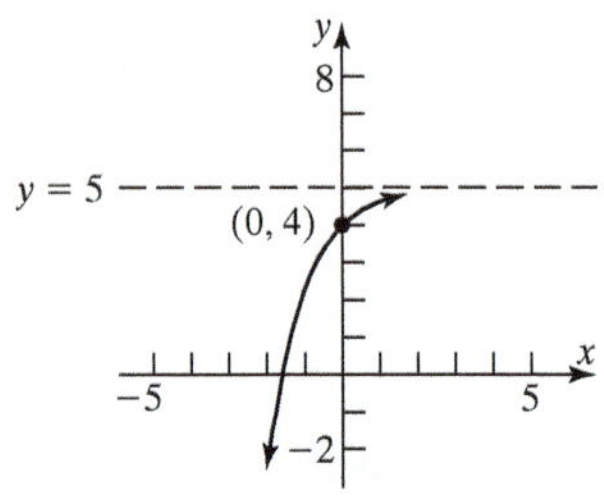

Domain: All real numbers
Range: $\{y|y < 5\}$ or $(-\infty, 5)$
Horizontal asymptote: $y = 5$

55.

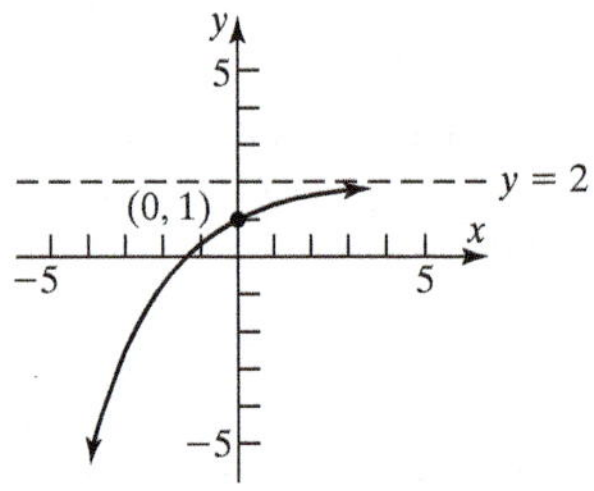

Domain: All real numbers
Range: $\{y|y < 2\}$ or $(-\infty, 2)$
Horizontal asymptote: $y = 2$

57. $\{3\}$ **59.** $\{-4\}$ **61.** $\{2\}$ **63.** $\left\{\frac{3}{2}\right\}$
65. $\left\{-\sqrt{2}, 0, \sqrt{2}\right\}$ **67.** $\{6\}$ **69.** $\{-1, 7\}$ **71.** $\{-4, 2\}$
73. $\{-4\}$ **75.** $\{1, 2\}$ **77.** $\frac{1}{49}$ **79.** $\frac{1}{4}$ **81.** $f(x) = 3^x$
83. $f(x) = -6^x$ **85.** **(a)** 16; (4, 16) **(b)** -4; $\left(-4, \frac{1}{16}\right)$
87. **(a)** $\frac{9}{4}$; $\left(-1, \frac{9}{4}\right)$ **(b)** 3; (3, 66)
89. **(a)** 60; (−6, 60) **(b)** −4: (−4, 12) **(c)** −2

91.

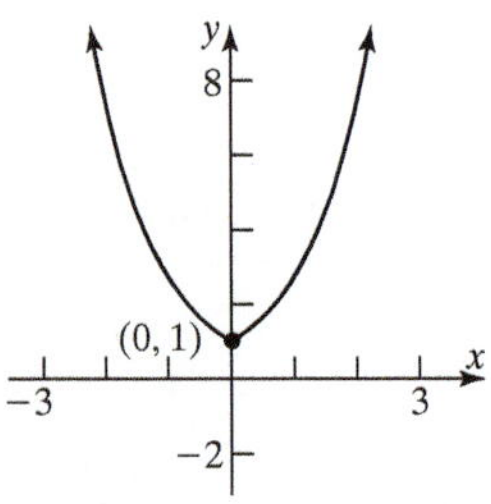

Domain: $(-\infty, \infty)$
Range: $[1, \infty)$
Intercept: (0, 1)

93.

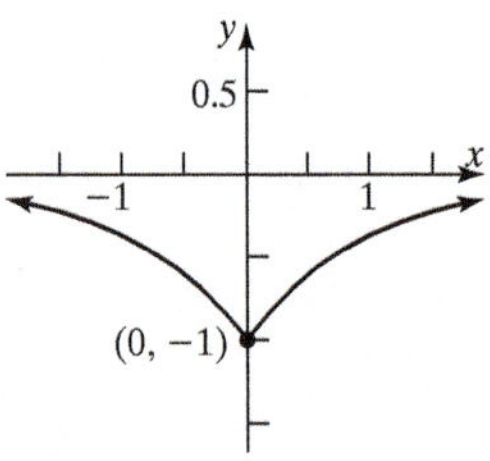

Domain: $(-\infty, \infty)$
Range: $[-1, 0)$
Intercept: (0, −1)

95. **(a)** 74% **(b)** 47%
97. **(a)** \$12,123 **(b)** \$6443
99. 3.35 mg; 0.45 mg

101. **(a)** 0.632 **(b)** 0.982 **(c)** 1
(d)

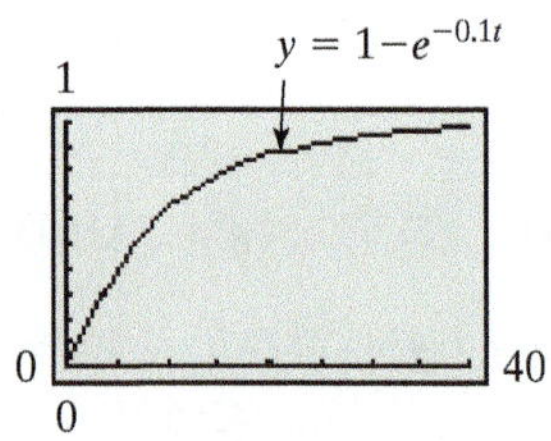

(e) About 7 min

103. **(a)** 0.0516 **(b)** 0.0888 **105.** **(a)** 70.95% **(b)** 72.62% **(c)** 100%
107. **(a)** 5.41 amp, 7.59 amp, 10.38 amp **(b)** 12 amp **(d)** 3.34 amp, 5.31 amp, 9.44 amp **(e)** 24 amp
(c), (f)

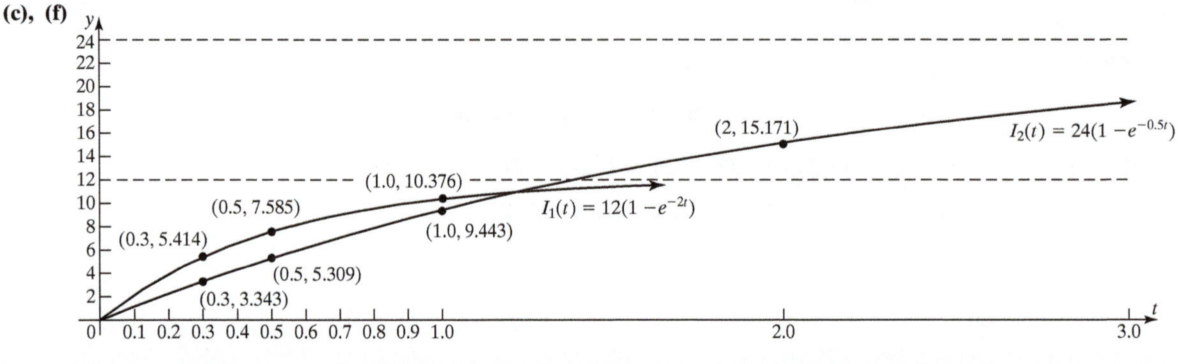

109. $n = 4$: 2.7083; $n = 6$: 2.7181; $n = 8$: 2.7182788; $n = 10$: 2.7182818

111. $\dfrac{f(x+h) - f(x)}{h} = \dfrac{a^{x+h} - a^x}{h} = \dfrac{a^x a^h - a^x}{h} = \dfrac{a^x(a^h - 1)}{h}$

113. $f(-x) = a^{-x} = \dfrac{1}{a^x} = \dfrac{1}{f(x)}$

115. **(a)** $f(-x) = \dfrac{1}{2}(e^{-x} - e^{-(-x)})$
$= \dfrac{1}{2}(e^{-x} - e^{x})$
$= -\dfrac{1}{2}(e^{x} - e^{-x})$
$= -f(x)$

(b)

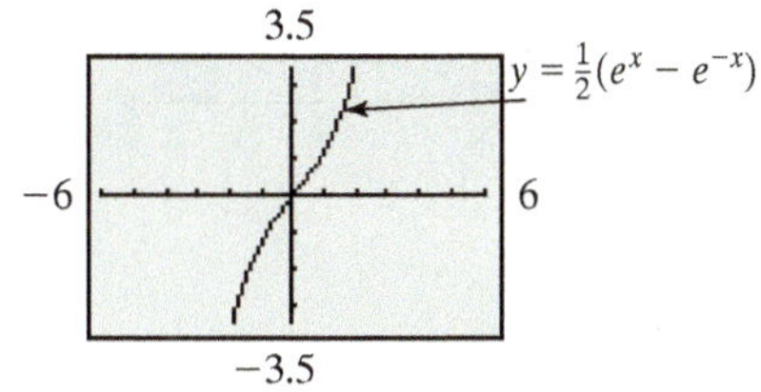

117. $f(1) = 5, f(2) = 17, f(3) = 257, f(4) = 65{,}537; f(5) = 4{,}294{,}967{,}297 = 641 \times 6{,}700{,}417$

6.4 Assess Your Understanding *(page 449)*

4. $\{x|x > 0\}$ or $(0, \infty)$ **5.** $\left(\dfrac{1}{a}, -1\right), (1, 0), (a, 1)$ **6.** 1 **7.** F **8.** T **9.** $2 = \log_3 9$ **11.** $2 = \log_a 1.6$ **13.** $x = \log_2 7.2$ **15.** $x = \ln 8$ **17.** $2^3 = 8$ **19.** $a^6 = 3$ **21.** $3^x = 2$ **23.** $e^x = 4$ **25.** 0 **27.** 2 **29.** −4 **31.** $\dfrac{1}{2}$ **33.** 4 **35.** $\dfrac{1}{2}$ **37.** $\{x|x > 3\}; (3, \infty)$ **39.** All real numbers except 0; $\{x|x \neq 0\}$ **41.** $\{x|x > 10\}; (10, \infty)$ **43.** $\{x|x > -1\}; (-1, \infty)$ **45.** $\{x|x < -1 \text{ or } x > 0\}; (-\infty, -1) \cup (0, \infty)$ **47.** $\{x|x \geq 1\}; [1, \infty)$ **49.** 0.511 **51.** 30.099 **53.** 2.303 **55.** −53.991 **57.** $\sqrt{2}$

59.

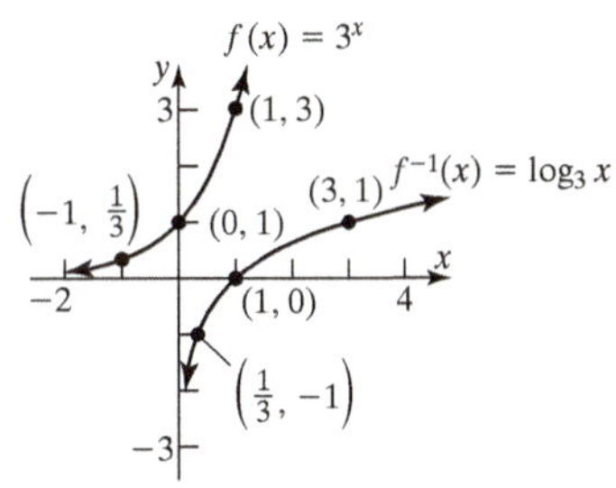

61.

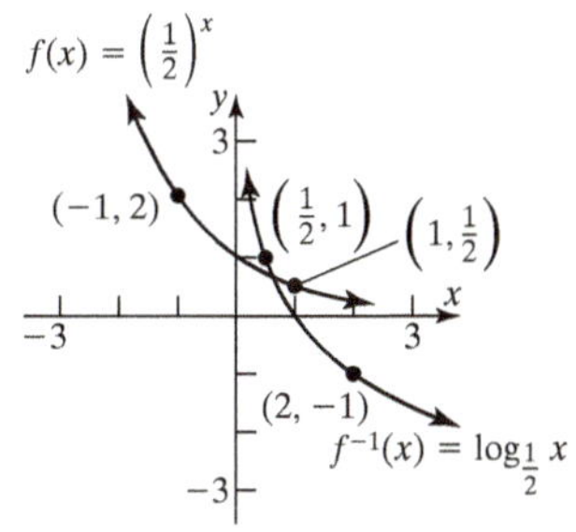

63. B **65.** D **67.** A **69.** E

71. **(a)** Domain: $(-4, \infty)$
(b)

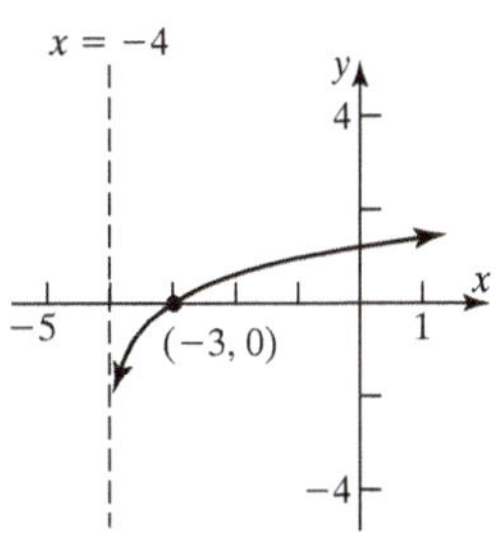

(c) Range: $(-\infty, \infty)$
Vertical asymptote: $x = -4$
(d) $f^{-1}(x) = e^x - 4$
(e) Range of f: $(-\infty, \infty)$
Range of f^{-1}: $(-4, \infty)$
(f)

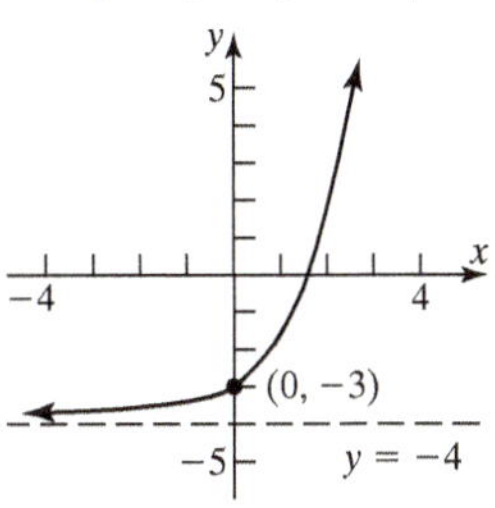

73. **(a)** Domain: $(0, \infty)$
(b)

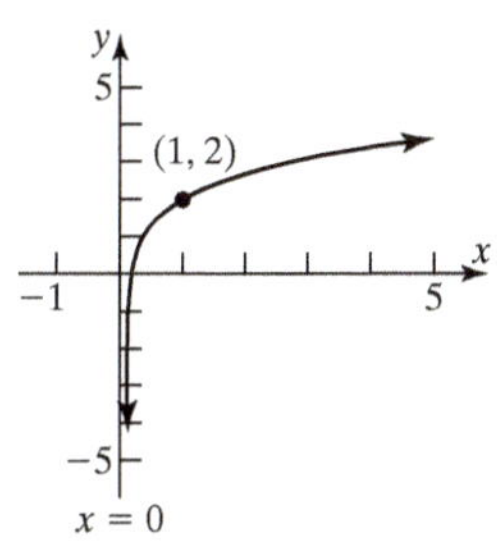

(c) Range: $(-\infty, \infty)$
Vertical asymptote: $x = 0$
(d) $f^{-1}(x) = e^{x-2}$
(e) Range of f: $(-\infty, \infty)$
Range of f^{-1}: $(0, \infty)$
(f)

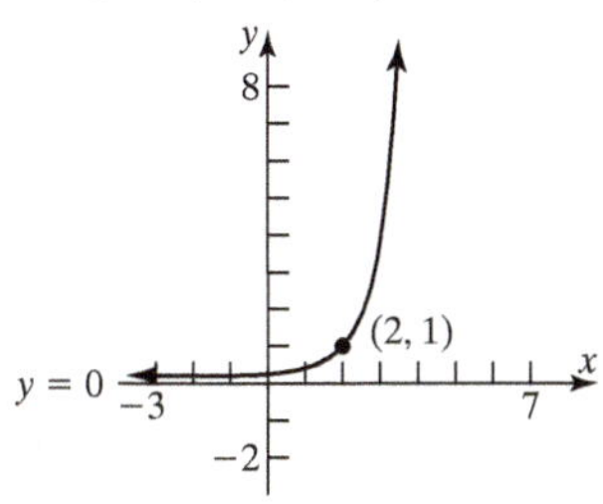

75. **(a)** Domain: $(0, \infty)$
(b)

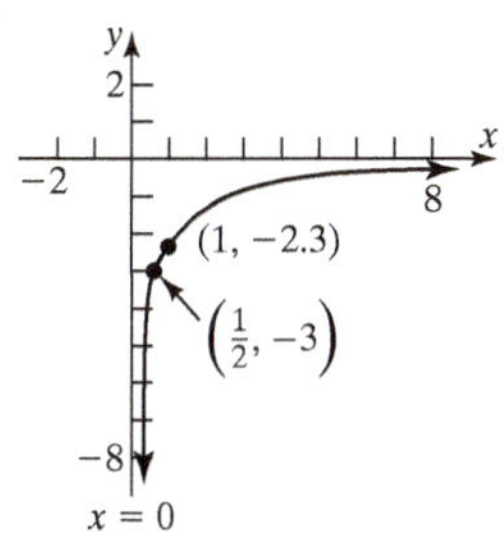

(c) Range: $(-\infty, \infty)$
Vertical asymptote: $x = 0$
(d) $f^{-1}(x) = \dfrac{1}{2}e^{x+3}$
(e) Range of f: $(-\infty, \infty)$
Range of f^{-1}: $(0, \infty)$
(f)

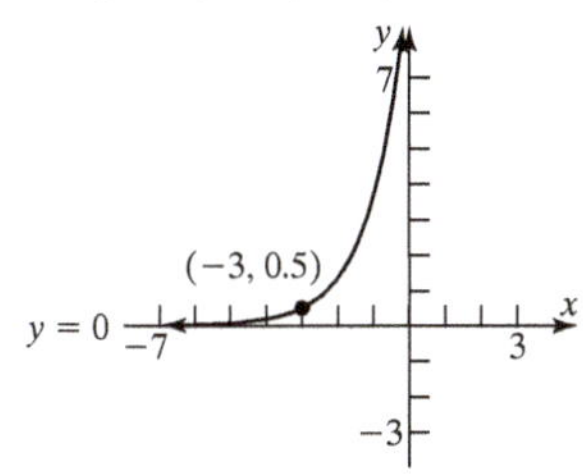

77. **(a)** Domain: $(4, \infty)$

(b)

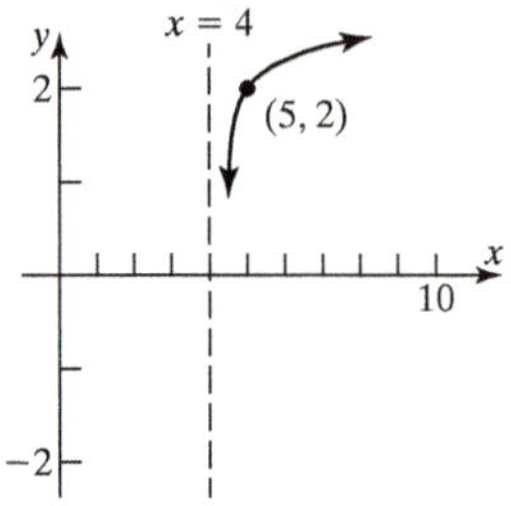

(c) Range: $(-\infty, \infty)$
Vertical asymptote: $x = 4$

(d) $f^{-1}(x) = 10^{x-2} + 4$

(e) Range of f: $(-\infty, \infty)$
Range of f^{-1}: $(4, \infty)$

(f)

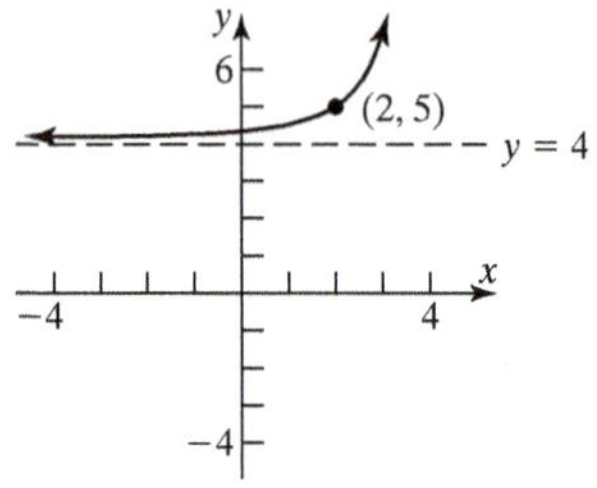

79. **(a)** Domain: $(0, \infty)$

(b)

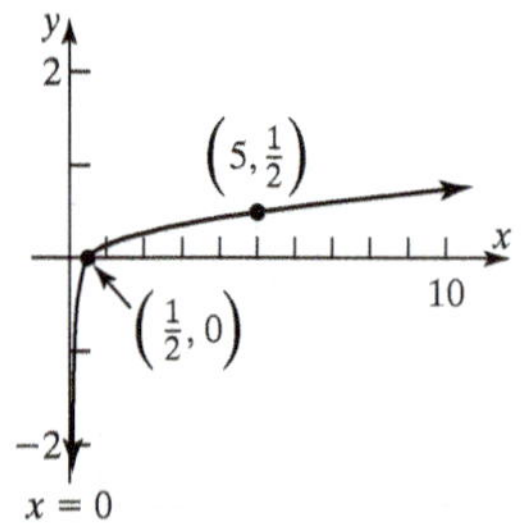

(c) Range: $(-\infty, \infty)$
Vertical asymptote: $x = 0$

(d) $f^{-1}(x) = \frac{1}{2} \cdot 10^{2x}$

(e) Range of f: $(-\infty, \infty)$
Range of f^{-1}: $(0, \infty)$

(f)

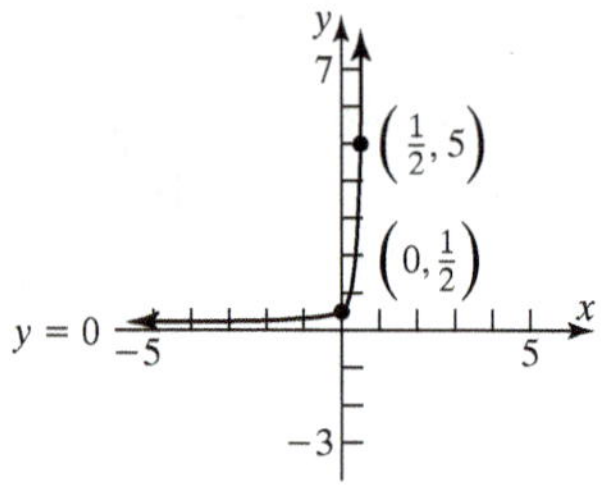

81. **(a)** Domain: $(-2, \infty)$

(b)

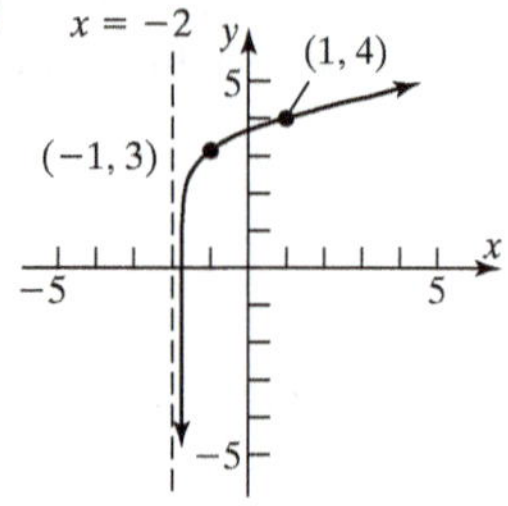

(c) Range: $(-\infty, \infty)$
Vertical asymptote: $x = -2$

(d) $f^{-1}(x) = 3^{x-3} - 2$

(e) Range of f: $(-\infty, \infty)$
Range of f^{-1}: $(-2, \infty)$

(f)

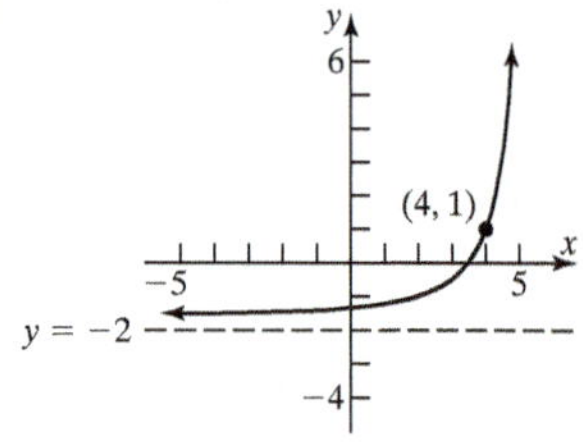

83. **(a)** Domain: $(-\infty, \infty)$

(b)

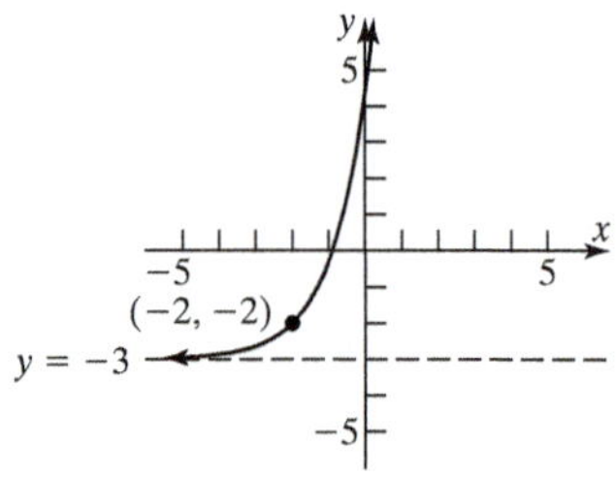

(c) Range: $(-3, \infty)$
Horizontal asymptote: $y = -3$

(d) $f^{-1}(x) = \ln(x + 3) - 2$

(e) Range of f: $(-3, \infty)$
Range of f^{-1}: $(-\infty, \infty)$

(f)

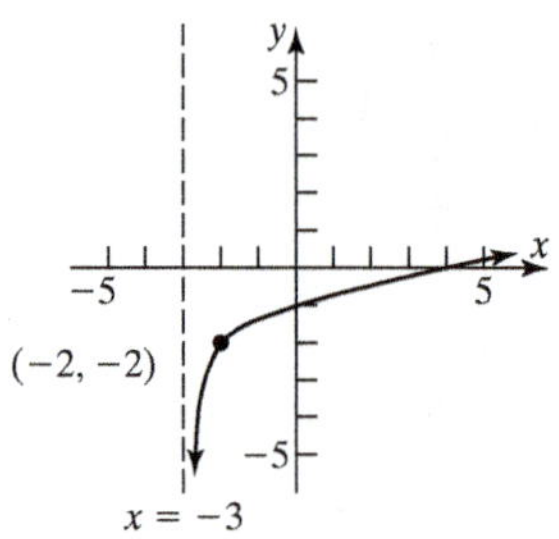

85. **(a)** Domain: $(-\infty, \infty)$

(b)

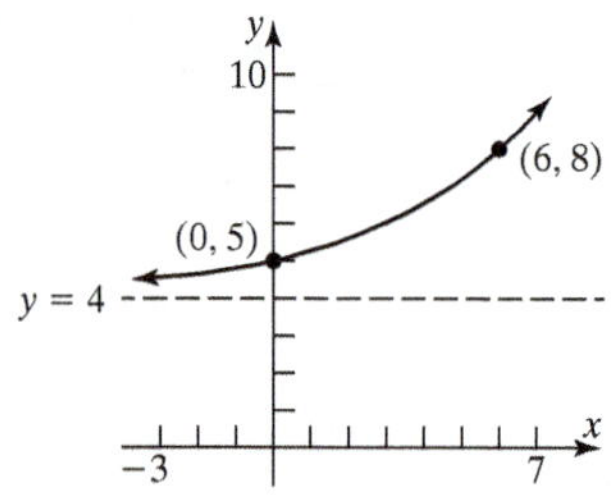

(c) Range: $(4, \infty)$
Horizontal asymptote: $y = 4$

(d) $f^{-1}(x) = 3\log_2(x - 4)$

(e) Range of f: $(4, \infty)$
Range of f^{-1}: $(-\infty, \infty)$

(f)

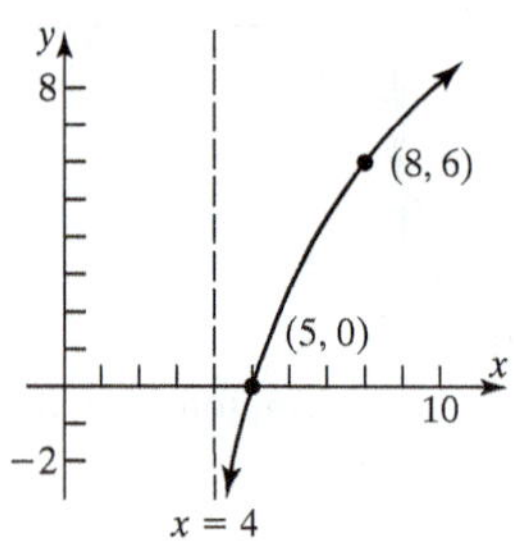

87. $\{9\}$ **89.** $\left\{\frac{7}{2}\right\}$ **91.** $\{2\}$ **93.** $\{5\}$

95. $\{3\}$ **97.** $\{2\}$ **99.** $\left\{\frac{\ln 10}{3}\right\}$

101. $\left\{\frac{\ln 8 - 5}{2}\right\}$

103. $\{-2\sqrt{2}, 2\sqrt{2}\}$ **105.** $\{-1\}$

107. $\left\{5 \ln \frac{7}{5}\right\}$ **109.** $\left\{2 - \log \frac{5}{2}\right\}$

111. **(a)** $\left\{x \middle| x > -\frac{1}{2}\right\}; \left(-\frac{1}{2}, \infty\right)$

(b) 2; (40, 2) **(c)** 121; (121, 3) **(d)** 4

113.

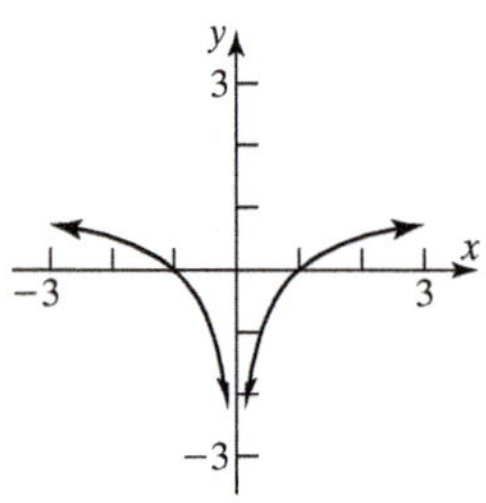

Domain: $\{x | x \neq 0\}$
Range: $(-\infty, \infty)$
Intercepts: $(-1, 0), (1, 0)$

115.

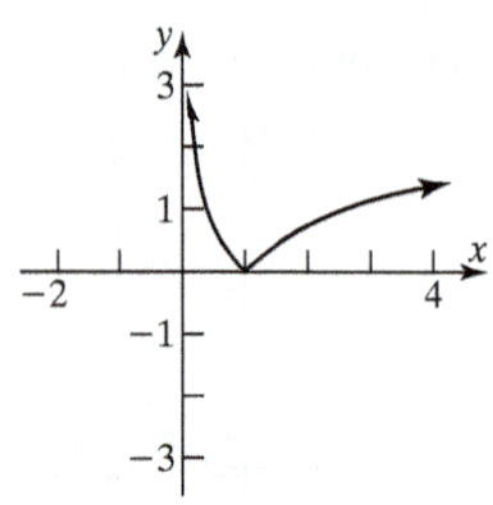

Domain: $\{x | x > 0\}$
Range: $\{y | y \geq 0\}$
Intercept: $(1, 0)$

117. **(a)** 1 **(b)** 2 **(c)** 3 **(d)** It increases.
(e) 0.000316 **(f)** 3.981×10^{-8}

119. **(a)** 5.97 km **(b)** 0.90 km

121. **(a)** 6.93 min **(b)** 16.09 min

123. $h \approx 2.29$, so the time between injections is about 2 h, 17 min.

125. 0.2695 s
0.8959 s

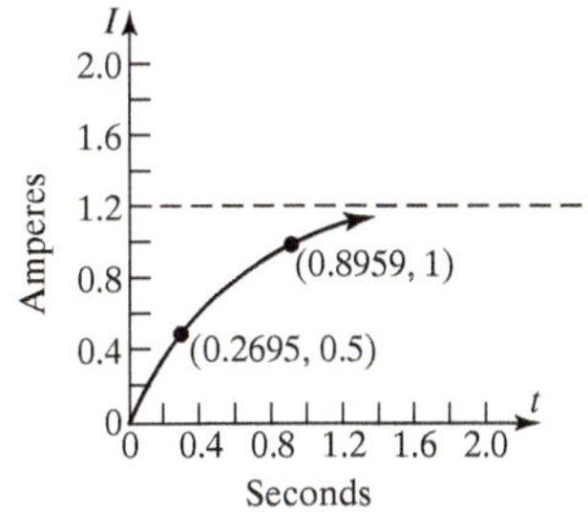

127. 50 decibels (dB) **129.** 90 dB **131.** 8.1
133. (a) $k \approx 11.216$ **(b)** 6.73 **(c)** 0.41% **(d)** 0.14%
135. Because $y = \log_1 x$ means $1^y = 1 = x$, which cannot be true for $x \neq 1$

6.5 Assess Your Understanding *(page 460)*

1. sum **2.** 7 **3.** $r \log_a M$ **4.** F **5.** F **6.** T **7.** 71 **9.** -4 **11.** 7 **13.** 1 **15.** 1 **17.** 2 **19.** $\frac{5}{4}$ **21.** 4 **23.** $a + b$ **25.** $b - a$
27. $3a$ **29.** $\frac{1}{5}(a + b)$ **31.** $2 + \log_5 x$ **33.** $3 \log_2 z$ **35.** $1 + \ln x$ **37.** $\ln x + x$ **39.** $2 \log_a u + 3 \log_a v$ **41.** $2 \ln x + \frac{1}{2}\ln(1 - x)$
43. $3 \log_2 x - \log_2(x - 3)$ **45.** $\log x + \log(x + 2) - 2 \log(x + 3)$ **47.** $\frac{1}{3}\ln(x - 2) + \frac{1}{3}\ln(x + 1) - \frac{2}{3}\ln(x + 4)$
49. $\ln 5 + \ln x + \frac{1}{2}\ln(1 + 3x) - 3 \ln(x - 4)$ **51.** $\log_5 u^3v^4$ **53.** $\log_3\left(\frac{1}{x^{5/2}}\right)$ **55.** $\log_4\left[\frac{x - 1}{(x + 1)^4}\right]$ **57.** $-2 \ln(x - 1)$ **59.** $\log_2[x(3x - 2)^4]$
61. $\log_a\left(\frac{25x^6}{\sqrt{2x + 3}}\right)$ **63.** $\log_2\left[\frac{(x + 1)^2}{(x + 3)(x - 1)}\right]$ **65.** 2.771 **67.** -3.880 **69.** 5.615 **71.** 0.874

73. $y = \frac{\log x}{\log 4}$

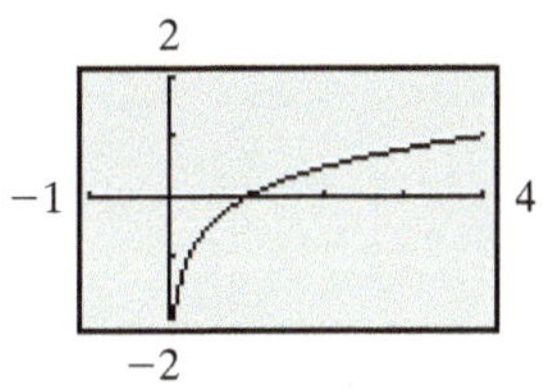

75. $y = \frac{\log(x + 2)}{\log 2}$

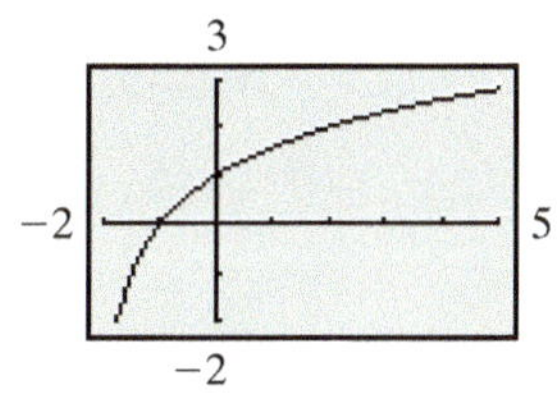

77. $y = \frac{\log(x + 1)}{\log(x - 1)}$

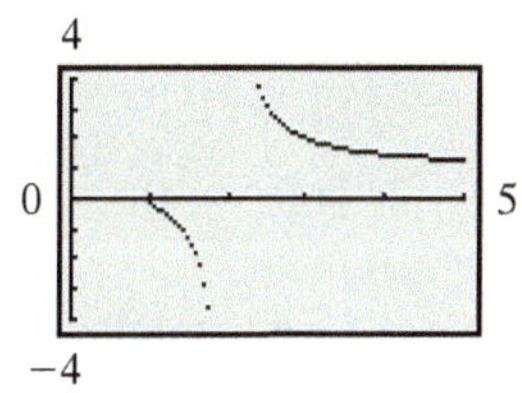

79. (a) $(f \circ g)(x) = x$; $\{x | x \text{ is any real number}\}$ or $(-\infty, \infty)$
(b) $(g \circ f)(x) = x$; $\{x | x > 0\}$ or $(0, \infty)$ **(c)** 5
(d) $(f \circ h)(x) = \ln x^2$; $\{x | x \neq 0\}$ or $(-\infty, 0) \cup (0, \infty)$ **(e)** 2

81. $y = Cx$ **83.** $y = Cx(x + 1)$ **85.** $y = Ce^{3x}$ **87.** $y = Ce^{-4x} + 3$

89. $y = \frac{\sqrt[3]{C}(2x + 1)^{1/6}}{(x + 4)^{1/9}}$ **91.** 3 **93.** 1

95. $\log_a\left(x + \sqrt{x^2 - 1}\right) + \log_a\left(x - \sqrt{x^2 - 1}\right) = \log_a\left[\left(x + \sqrt{x^2 - 1}\right)\left(x - \sqrt{x^2 - 1}\right)\right] = \log_a[x^2 - (x^2 - 1)] = \log_a 1 = 0$

97. $\ln(1 + e^{2x}) = \ln[e^{2x}(e^{-2x} + 1)] = \ln e^{2x} + \ln(e^{-2x} + 1) = 2x + \ln(1 + e^{-2x})$

99. $y = f(x) = \log_a x$; $a^y = x$ implies $a^y = \left(\frac{1}{a}\right)^{-y} = x$, so $-y = \log_{1/a} x = -f(x)$.

101. $f(x) = \log_a x$; $f\left(\frac{1}{x}\right) = \log_a \frac{1}{x} = \log_a 1 - \log_a x = -f(x)$

103. $\log_a \frac{M}{N} = \log_a(M \cdot N^{-1}) = \log_a M + \log_a N^{-1} = \log_a M - \log_a N$, since $a^{\log_a N^{-1}} = N^{-1}$ implies $a^{-\log_a N^{-1}} = N$; i.e., $\log_a N = -\log_a N^{-1}$.

6.6 Assess Your Understanding *(page 466)*

5. $\{16\}$ **7.** $\left\{\frac{16}{5}\right\}$ **9.** $\{6\}$ **11.** $\{16\}$ **13.** $\left\{\frac{1}{3}\right\}$ **15.** $\{3\}$ **17.** $\{5\}$ **19.** $\left\{\frac{21}{8}\right\}$ **21.** $\{-6\}$ **23.** $\{-2\}$ **25.** $\left\{-1 + \sqrt{1 + e^4}\right\} \approx \{6.456\}$

27. $\left\{\frac{-5 + 3\sqrt{5}}{2}\right\} \approx \{0.854\}$ **29.** $\{2\}$ **31.** $\left\{\frac{9}{2}\right\}$ **33.** $\{8\}$ **35.** $\{\log_2 10\} = \left\{\frac{\ln 10}{\ln 2}\right\} \approx \{3.322\}$ **37.** $\{-\log_8 1.2\} = \left\{-\frac{\ln 1.2}{\ln 8}\right\} \approx \{-0.088\}$

39. $\left\{\frac{1}{3}\log_2 \frac{8}{5}\right\} = \left\{\frac{\ln \frac{8}{5}}{3 \ln 2}\right\} \approx \{0.226\}$ **41.** $\left\{\frac{\ln 3}{2 \ln 3 + \ln 4}\right\} \approx \{0.307\}$ **43.** $\left\{\frac{\ln 7}{\ln 0.6 + \ln 7}\right\} \approx \{1.356\}$

45. $\{0\}$ **47.** $\left\{\frac{\ln \pi}{1 + \ln \pi}\right\} \approx \{0.534\}$ **49.** $\left\{\frac{\ln 3}{\ln 2}\right\} \approx \{1.585\}$ **51.** $\{0\}$ **53.** $\left\{\log_4\left(-2 + \sqrt{7}\right)\right\} \approx \{-0.315\}$

55. $\{\log_5 4\} \approx \{0.861\}$ **57.** No real solution **59.** $\{\log_4 5\} \approx \{1.161\}$ **61.** {2.79} **63.** {–0.57} **65.** {–0.70} **67.** {0.57} **69.** {0.39, 1.00}

71. {1.32} **73.** {1.31} **75.** {1} **77.** {16} **79.** $\left\{-1, \frac{2}{3}\right\}$ **81.** {0} **83.** $\{\ln(2+\sqrt{5})\} \approx \{1.444\}$ **85.** $\left\{e^{\frac{\ln 5 \cdot \ln 3}{\ln 15}}\right\} \approx \{1.921\}$

87. **(a)** $\{5\}; (5, 3)$ **(b)** $\{5\}; (5, 4)$ **(c)** $\{1\}$; yes, at $(1, 2)$ **(d)** $\{5\}$ **(e)** $\left\{-\frac{1}{11}\right\}$

89. **(a)**

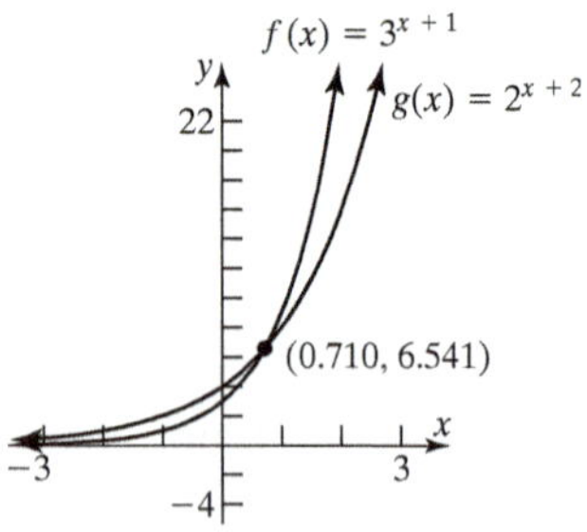

(b) $(0.710, 6.541)$
(c) $\{x \mid x > 0.710\}$ or $(0.710, \infty)$

91. **(a), (b)**

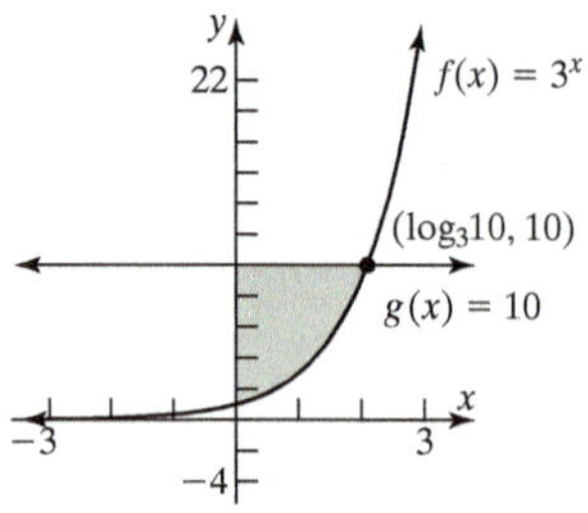

(c) $(\log_3 10, 10)$

93. **(a), (b)**

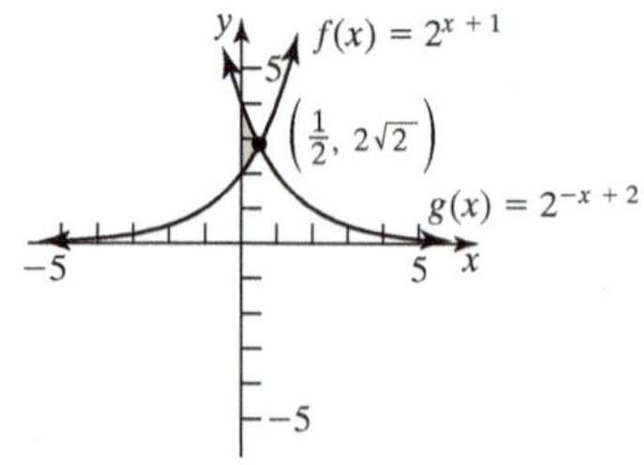

(c) $\left(\frac{1}{2}, 2\sqrt{2}\right)$

95. **(a)**

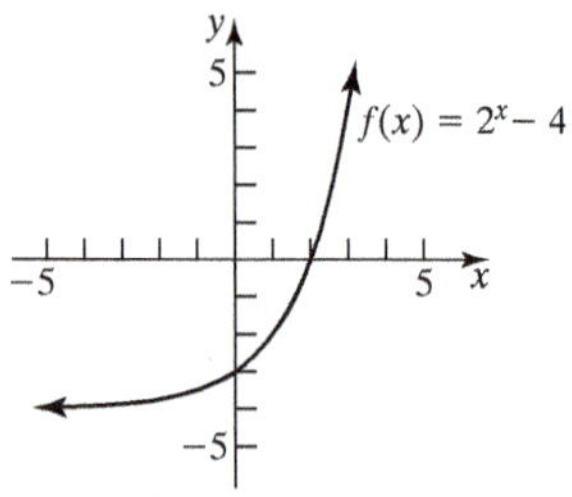

(b) 2
(c) $\{x \mid x < 2\}$ or $(-\infty, 2)$

97. **(a)** 2010 **(b)** 2027
99. **(a)** After 2.4 yr
(b) After 6.5 yr
(c) After 10 yr

6.7 Assess Your Understanding *(page 476)*

3. \$108.29 **5.** \$609.50 **7.** \$697.09 **9.** \$12.46 **11.** \$125.23 **13.** \$88.72 **15.** \$860.72 **17.** \$554.09 **19.** \$59.71 **21.** \$361.93 **23.** 5.095% **25.** 5.127% **27.** $6\frac{1}{4}$% compounded annually **29.** 9% compounded monthly **31.** 25.992% **33.** 24.573% **35.** **(a)** About 8.69 yr **(b)** About 8.66 yr **37.** 6.823% **39.** 5.09 yr; 5.07 yr **41.** 15.27 yr or 15 yr, 3 mo **43.** \$104,335 **45.** \$12,910.62 **47.** About \$30.17 per share or \$3017
49. Not quite. Jim will have \$1057.60. The second bank gives a better deal, since Jim will have \$1060.62 after 1 yr.
51. Will has \$11,632.73; Henry has \$10,947.89. **53.** **(a)** \$49,581 **(b)** \$33,235 **55.** Approximately \$602 billion
57. \$940.90 **59.** 2.53% **61.** 34.31 yr **63.** **(a)** \$1364.62 **(b)** \$1353.35 **65.** \$4631.93

67. **(a)** 6.12 yr
(b) 18.45 yr
(c)
$$mP = P\left(1 + \frac{r}{n}\right)^{nt}$$
$$m = \left(1 + \frac{r}{n}\right)^{nt}$$
$$\ln m = \ln\left(1 + \frac{r}{n}\right)^{nt} = nt \ln\left(1 + \frac{r}{n}\right)$$
$$t = \frac{\ln m}{n \ln\left(1 + \frac{r}{n}\right)}$$

69. **(a)** 2.51% **(b)** In 2022 or after 27 yr **71.** 22.7 yr

6.8 Assess Your Understanding *(page 487)*

1. **(a)** 500 insects
(b) $0.02 = 2\%$
(c)

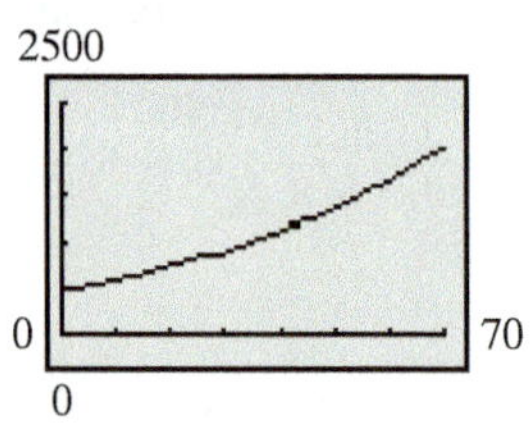

(d) About 611 insects
(e) After about 23.5 days
(f) After about 34.7 days

3. **(a)** $-0.0244 = -2.44\%$
(b)

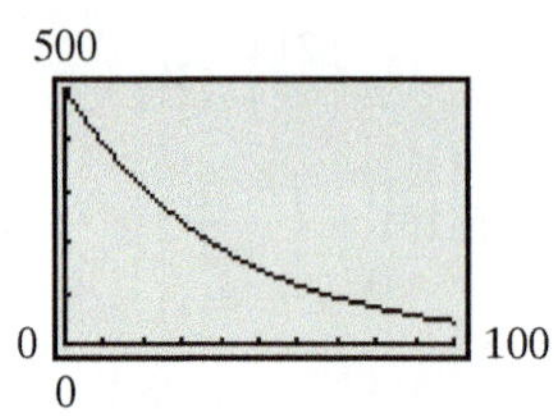

(c) About 391.7 g
(d) After about 9.1 yr
(e) 28.4 yr

5. **(a)** $N(t) = N_0 e^{kt}$
(b) 5832
(c) 3.9 days
7. **(a)** $N(t) = N_0 e^{kt}$
(b) 25,198
9. 9.797 g

11. (a) 9727 yr ago **(b)**

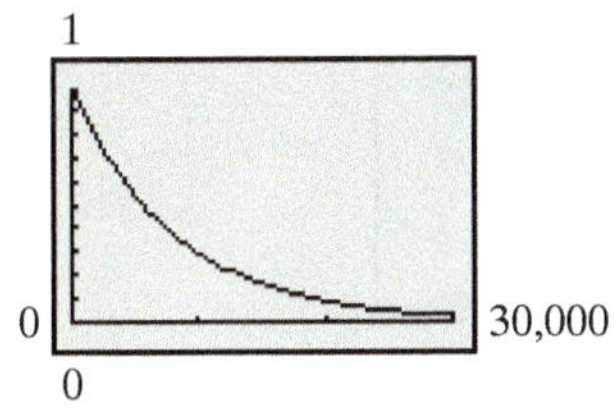

(c) 5600 yr

13. (a) About 5:18 PM **(b)**

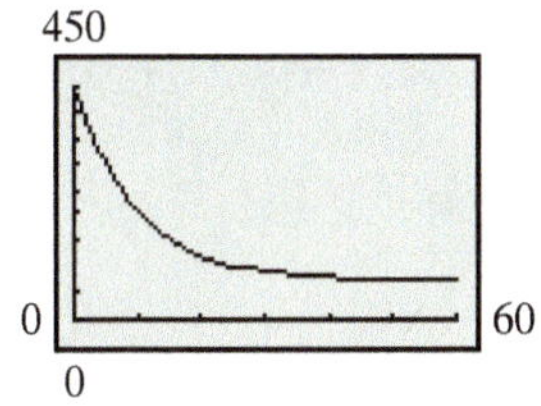

(c) About 14.3 min
(d) The temperature of the pizza approaches 70°F.

15. (a) 18.63°C; 25.07°C **(b)**

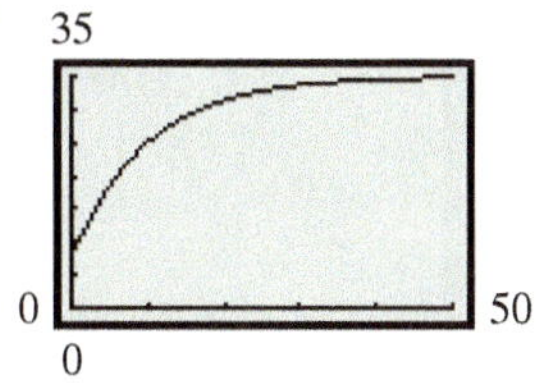

17. 1.7 ppm; 7.17 days or 172 h **19.** 0.26 mol; 6.58 h or 395 min **21.** 26.6 days

23. (a) 1000 g **(b)** 43.9% **(c)** 30 g **(d)**

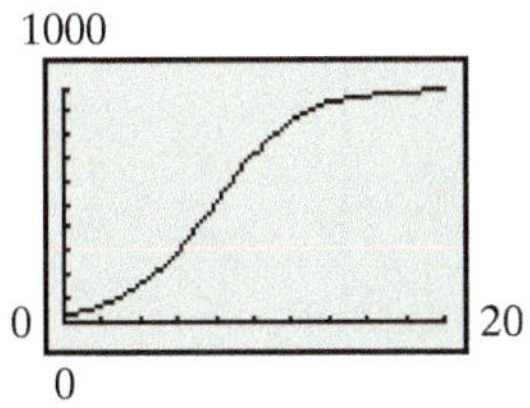

(e) 616.6 g **(f)** After 9.85 h
(g) About 7.9 h

25. (a) 9.23×10^{-3}, or about 0
(b) 0.81, or about 1
(c) 5.01, or about 5
(d) 57.91°, 43.99°, 30.07°

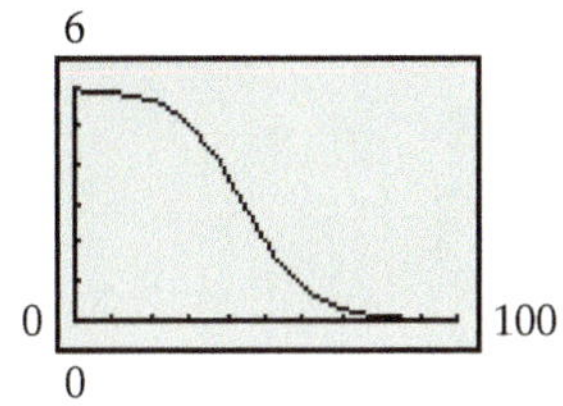

27. (a) In 1984, 91.8% of households did not own a personal computer.
(b)

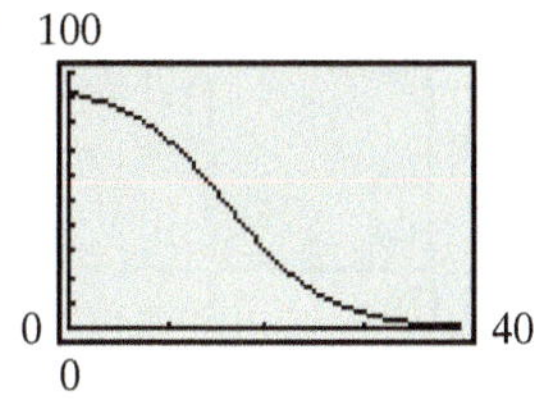

(c) 70.6% **(d)** During 2011

29. (a)

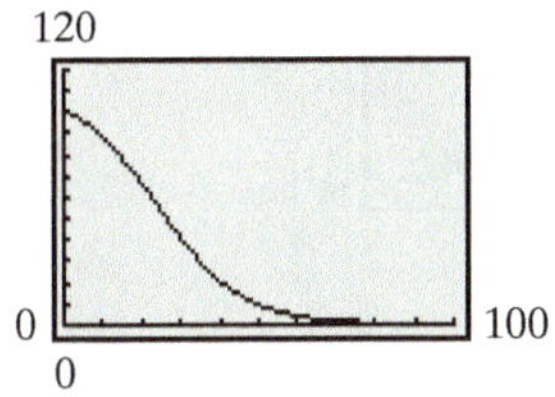

(b) 0.78 or 78% **(c)** 50 people
(d) As n increases, the probability decreases.

6.9 Assess Your Understanding *(page 495)*

1. (a)

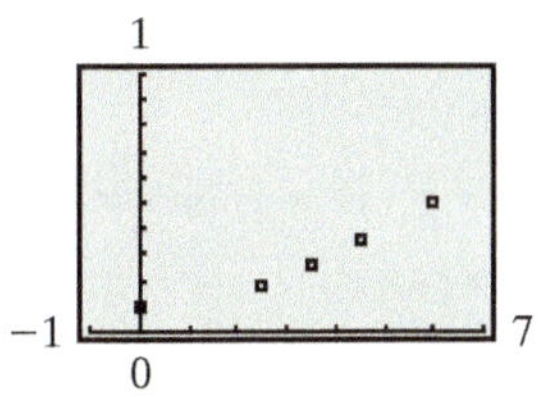

(b) $y = 0.0903(1.3384)^x$
(c) $N(t) = 0.0903e^{0.2915t}$

(d)

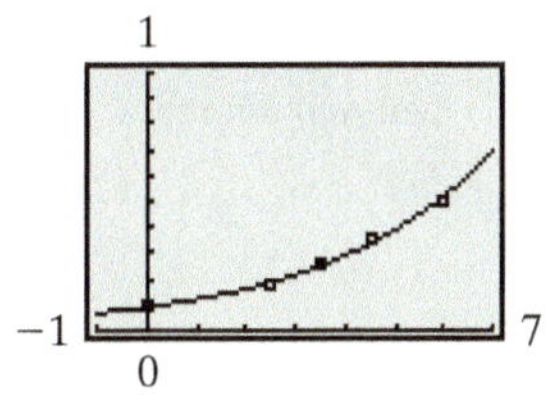

(e) 0.69
(f) After about 7.26 hr

3. (a)

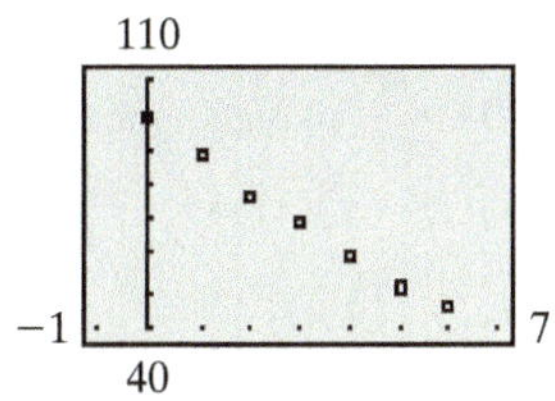

(b) $y = 100.326(0.8769)^x$
(c) $A(t) = 100.326e^{-0.1314t}$

(d)

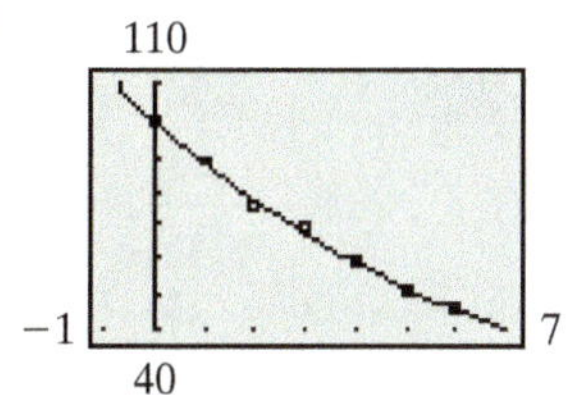

(e) 5.3 weeks **(f)** 0.14 g
(g) After about 12.3 weeks

5. (a)

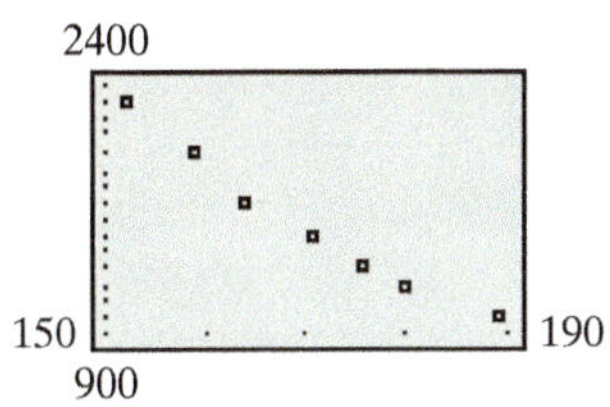

(b) $y = 32{,}741.02 - 6070.96 \ln x$

(c)

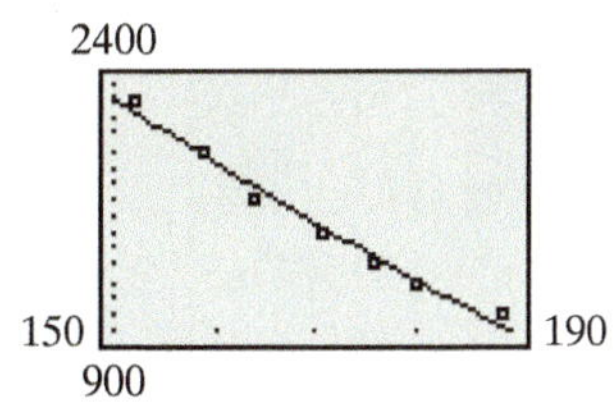

(d) Approximately 168 computers

7. (a)

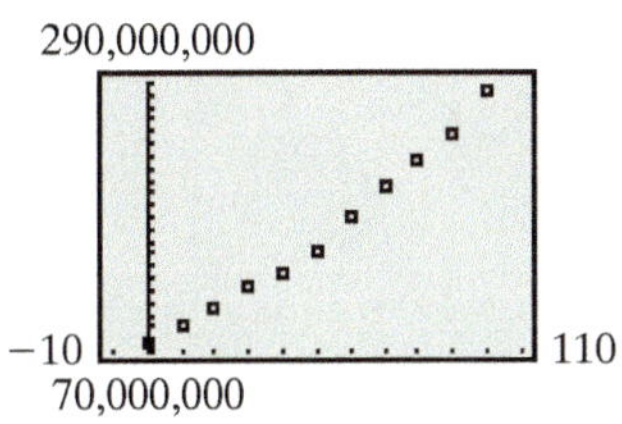

(b) $y = \dfrac{799{,}475{,}916.5}{1 + 9.1968e^{-0.0160x}}$

(c)

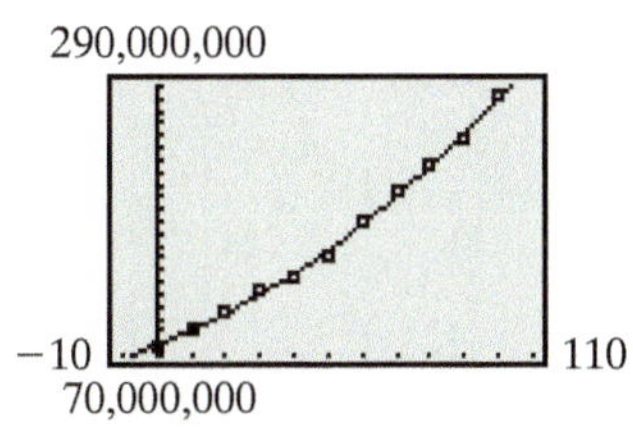

(d) 799,475,917
(e) Approximately 291,599,733
(f) 2006

9. (a)

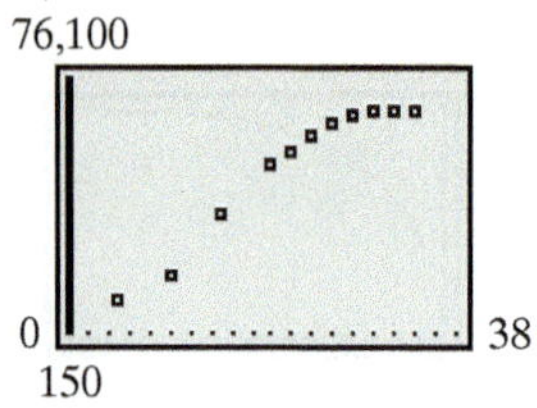

(b) $y = \dfrac{68{,}684.8}{1 + 18.9416e^{-0.1974x}}$

(c)

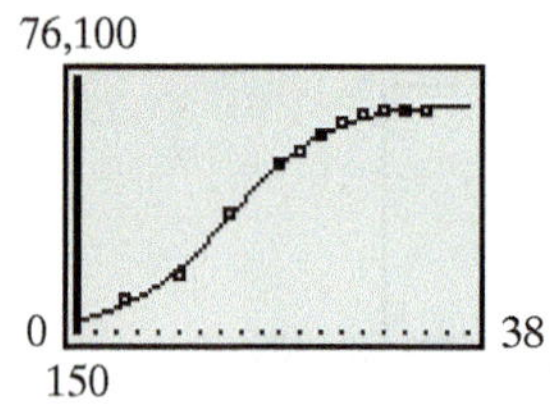

(d) About 68,685,000 subscribers
(e) About 68,505,000 subscribers

11. (a)

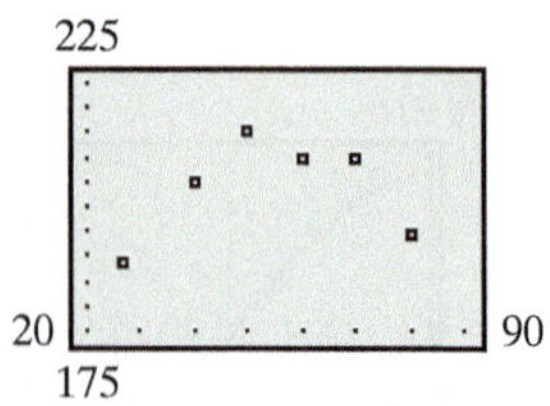

(b) Quadratic with $a < 0$ because of the "upside down U-shape" of the data
(c) $y = -0.0311x^2 + 3.4444x + 118.2493$

(d)

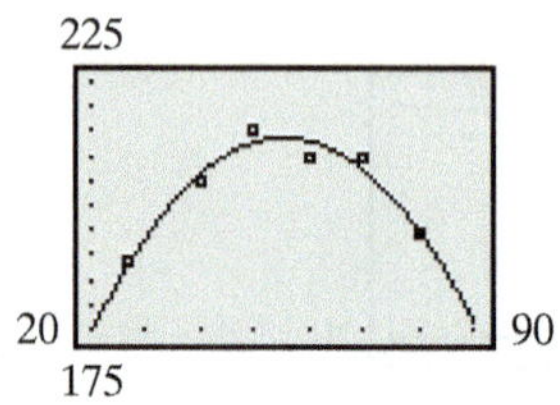

(e) 201

13. (a)

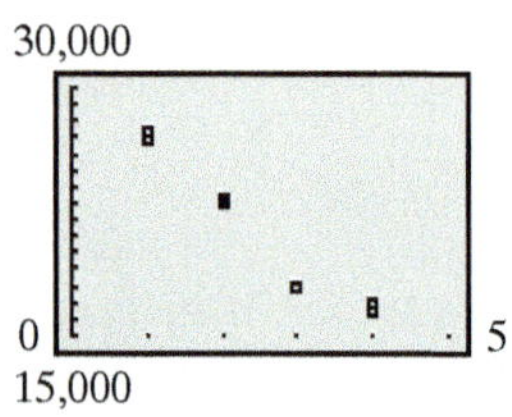

(b) Exponential because depreciation of a car is described by exponential models in the theory of finance
(c) $y = 31{,}808.51\,(0.8474)^x$

(d)

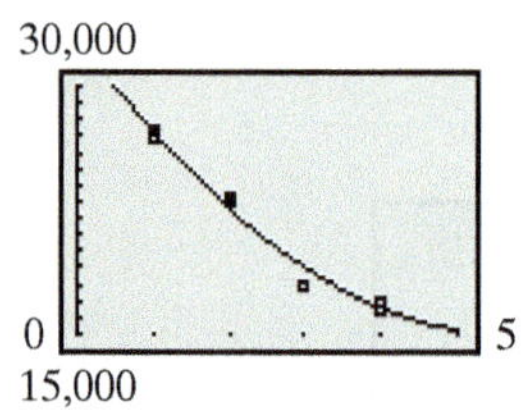

(e) \$13,899

Review Exercises *(page 500)*

1. (a) -26 **(b)** -241 **(c)** 16 **(d)** -1 **3. (a)** $\sqrt{11}$ **(b)** 1 **(c)** $\sqrt{\sqrt{6} + 2}$ **(d)** 19 **5. (a)** e^4 **(b)** $3e^{-2} - 2$ **(c)** e^{e^4} **(d)** -17

7. $(f \circ g)(x) = 1 - 3x$, all real numbers; $(g \circ f)(x) = 7 - 3x$, all real numbers; $(f \circ f)(x) = x$, all real numbers; $(g \circ g)(x) = 9x + 4$, all real numbers

9. $(f \circ g)(x) = 27x^2 + 3|x| + 1$, all real numbers; $(g \circ f)(x) = 3|3x^2 + x + 1|$, all real numbers; $(f \circ f)(x) = 27x^4 + 18x^3 + 24x^2 + 7x + 5$, all real numbers; $(g \circ g)(x) = 9|x|$, all real numbers

11. $(f \circ g)(x) = \dfrac{1 + x}{1 - x}$, $\{x|x \neq 0, x \neq 1\}$; $(g \circ f)(x) = \dfrac{x - 1}{x + 1}$, $\{x|x \neq -1, x \neq 1\}$; $(f \circ f)(x) = x$, $\{x|x \neq 1\}$; $(g \circ g)(x) = x$, $\{x|x \neq 0\}$

13. (a) one-to-one **(b)** $\{(2, 1), (5, 3), (8, 5), (10, 6)\}$

15.

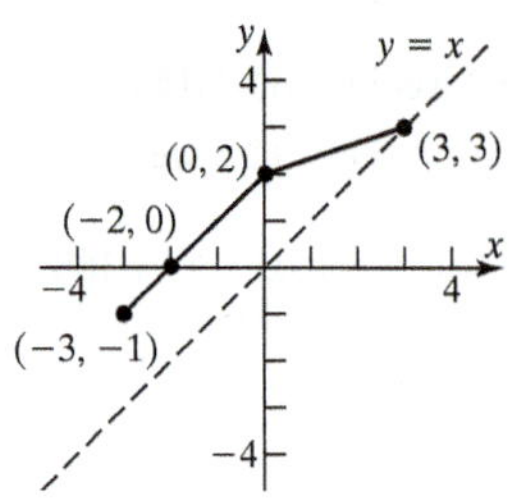

17. $f^{-1}(x) = \dfrac{2x + 3}{5x - 2}$

$$f(f^{-1}(x)) = \frac{2\left(\dfrac{2x + 3}{5x - 2}\right) + 3}{5\left(\dfrac{2x + 3}{5x - 2}\right) - 2} = x$$

$$f^{-1}(f(x)) = \frac{2\left(\dfrac{2x + 3}{5x - 2}\right) + 3}{5\left(\dfrac{2x + 3}{5x - 2}\right) - 2} = x$$

Domain of f = range of f^{-1} = all real numbers except $\frac{2}{5}$

Range of f = domain of f^{-1} = all real numbers except $\frac{2}{5}$

19. $f^{-1}(x) = \dfrac{x + 1}{x}$

$$f(f^{-1}(x)) = \frac{1}{\dfrac{x + 1}{x} - 1} = x$$

$$f^{-1}(f(x)) = \frac{\dfrac{1}{x - 1} + 1}{\dfrac{1}{x - 1}} = x$$

Domain of f = range of f^{-1} = all real numbers except 1

Range of f = domain of f^{-1} = all real numbers except 0

21. $f^{-1}(x) = \dfrac{27}{x^3}$

$f(f^{-1}(x)) = \dfrac{3}{\left(\dfrac{27}{x^3}\right)^{1/3}} = x$

$f^{-1}(f(x)) = \dfrac{27}{\left(\dfrac{3}{x^{1/3}}\right)^3} = x$

Domain of f = range of f^{-1} = all real numbers except 0
Range of f = domain of f^{-1} = all real numbers except 0

23. **(a)** 81 **(b)** 2 **(c)** $\dfrac{1}{9}$ **(d)** -3 **25.** $\log_5 z = 2$ **27.** $5^{13} = u$

29. $\left\{x \middle| x > \dfrac{2}{3}\right\}; \left(\dfrac{2}{3}, \infty\right)$ **31.** $\{x|x < 1 \text{ or } x > 2\}; (-\infty, 1) \cup (2, \infty)$

33. -3 **35.** $\sqrt{2}$ **37.** 0.4 **39.** $\log_3 u + 2\log_3 v - \log_3 w$

41. $2\log x + \dfrac{1}{2}\log(x^3 + 1)$ **43.** $\ln x + \dfrac{1}{3}\ln(x^2 + 1) - \ln(x - 3)$

45. $\dfrac{25}{4}\log_4 x$ **47.** $-2\ln(x + 1)$ **49.** $\log\left(\dfrac{4x^3}{[(x + 3)(x - 2)]^{1/2}}\right)$

51. 2.124

53.

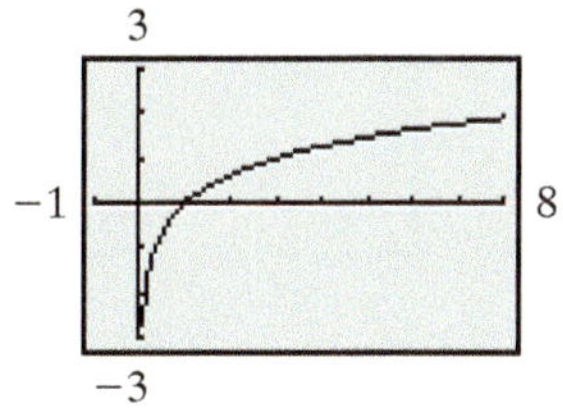

55. **(a)** Domain of f: $(-\infty, \infty)$

(b)

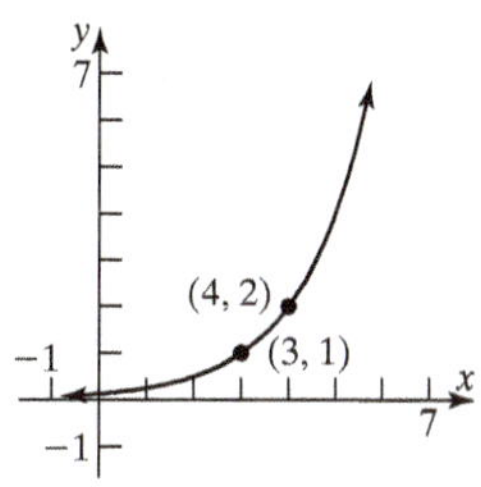

(c) Range of f: $(0, \infty)$
Horizontal asymptote: $y = 0$
(d) $f^{-1}(x) = 3 + \log_2 x$
(e) Range of f: $(0, \infty)$
(f)

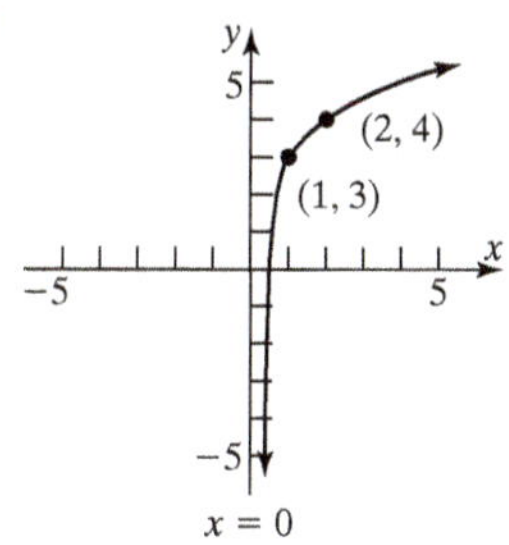

57. **(a)** Domain of f: $(-\infty, \infty)$

(b)

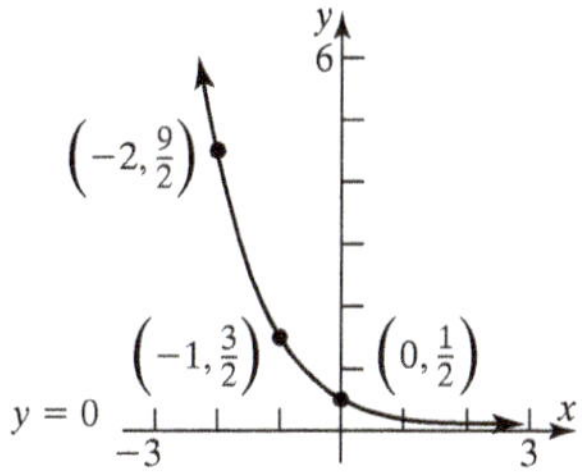

(c) Range of f: $(0, \infty)$
Horizontal asymptote: $y = 0$
(d) $f^{-1}(x) = -\log_3(2x)$

(e) Range of f: $(0, \infty)$
(f)

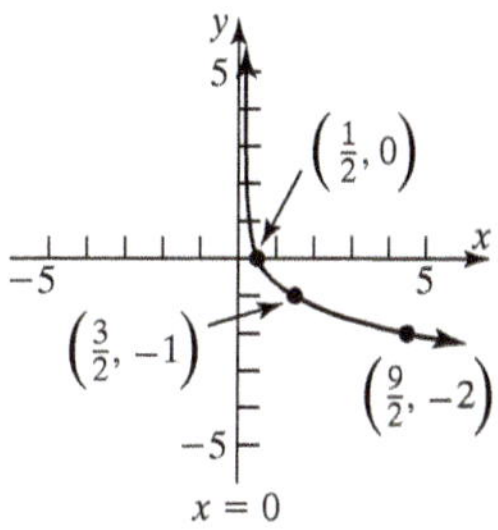

59. **(a)** Domain of f: $(-\infty, \infty)$

(b)

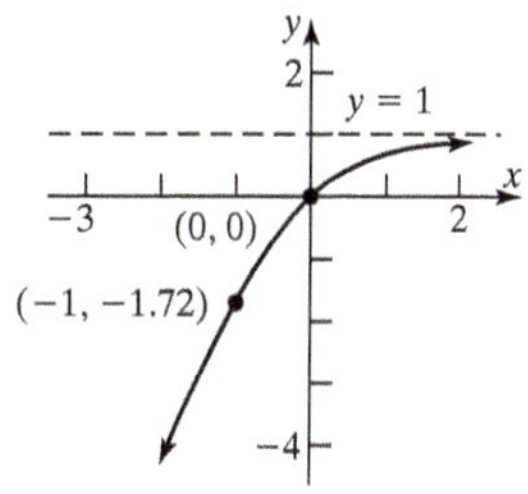

(c) Range of f: $(-\infty, 1)$
Horizontal asymptote: $y = 1$
(d) $f^{-1}(x) = -\ln(1 - x)$

(e) Range of f: $(-\infty, 1)$
(f)

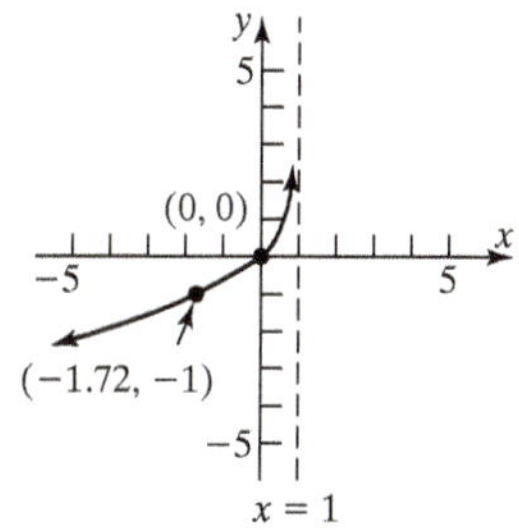

61. **(a)** Domain of f: $(-3, \infty)$

(b)

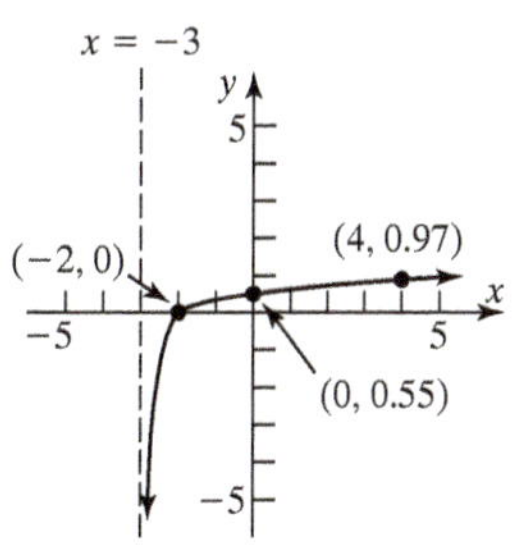

(c) Range of f: $(-\infty, \infty)$
Vertical asymptote: $x = -3$
(d) $f^{-1}(x) = e^{2x} - 3$

(e) Range of f: $(-\infty, \infty)$
(f)

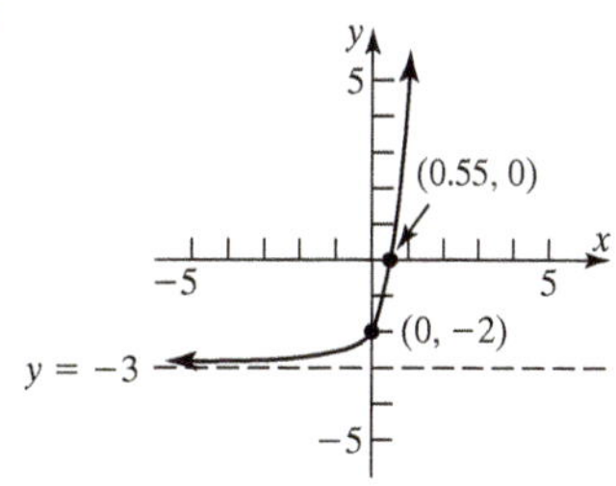

63. $\left\{\dfrac{1}{4}\right\}$ **65.** $\left\{\dfrac{-1 - \sqrt{3}}{2}, \dfrac{-1 + \sqrt{3}}{2}\right\} \approx \{-1.366, 0.366\}$ **67.** $\left\{\dfrac{1}{4}\right\}$ **69.** $\left\{\dfrac{2\ln 3}{\ln 5 - \ln 3}\right\} \approx \{4.301\}$ **71.** $\left\{\dfrac{12}{5}\right\}$ **73.** $\{83\}$ **75.** $\left\{\dfrac{1}{2}, -3\right\}$

77. $\{-1\}$ **79.** $\{1 - \ln 5\} \approx \{-0.609\}$ **81.** $\left\{\log_3\left(-2 + \sqrt{7}\right)\right\} = \left\{\dfrac{\ln\left(-2 + \sqrt{7}\right)}{\ln 3}\right\} \approx \{-0.398\}$

83. (a), (e)

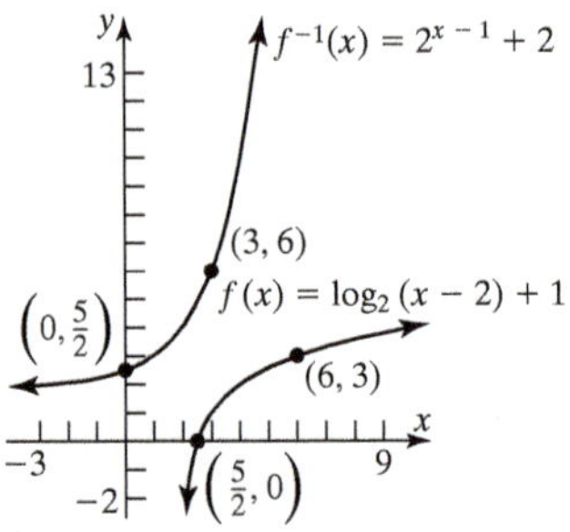

(b) 3; (6, 3) **(c)** 10; (10, 4)
(d) $\left\{x \middle| x > \frac{5}{2}\right\}$ or $\left(\frac{5}{2}, \infty\right)$
(e) $f^{-1}(x) = 2^{x-1} + 2$

85. 3229.5 m **87. (a)** 37.3 W **(b)** 6.9 dB **89. (a)** 9.85 yr **(b)** 4.27 yr **91.** $41,668.97 **93.** 24,203 yr ago **95.** 7,237,271,501 **97.** $483.67 billion

99. (a)

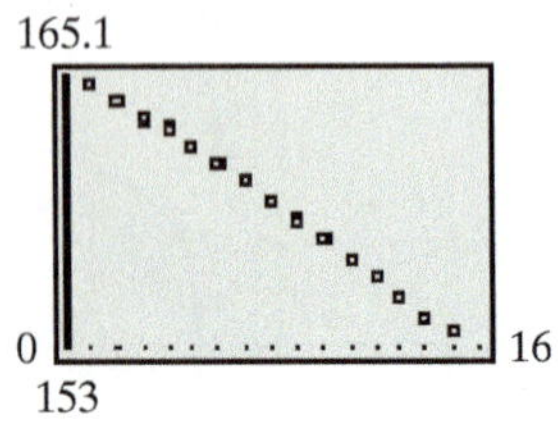

(b) $y = 165.73(0.9951)^x$

(c)

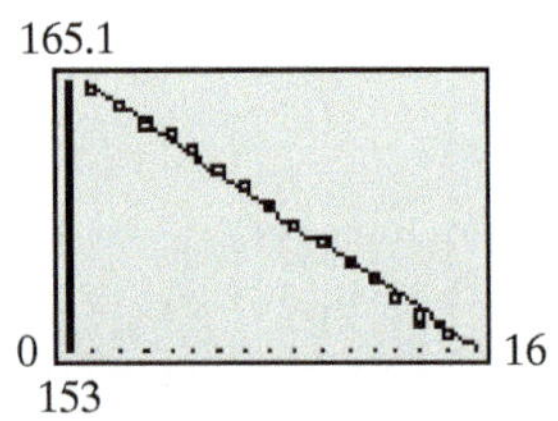

(d) Approximately 83 s

101. (a)

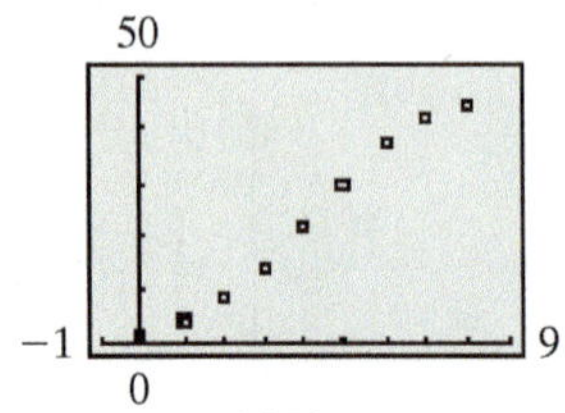

(b) $C = \dfrac{46.93}{1 + 21.273e^{-0.7306t}}$

(c)

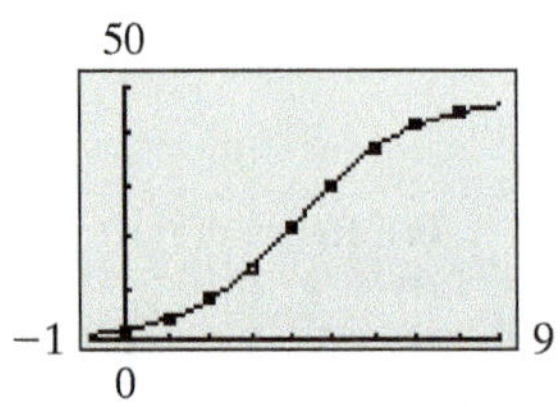

(d) About 47 people; 50 people

(e) 2.4 days; during the tenth hour of day 3

(f) 9.5 days

Chapter Test *(page 504)*

1. (a) $f \circ g = \dfrac{2x + 7}{2x + 3}$; domain: $\left\{x \middle| x \neq -\frac{3}{2}\right\}$ **(b)** $(g \circ f)(-2) = 5$ **(c)** $(f \circ g)(-2) = -3$

2. (a) The function is not one-to-one. **(b)** The function is one-to-one.

3. $f^{-1}(x) = \dfrac{2 + 5x}{3x}$; domain of $f = \left\{x \middle| x \neq \frac{5}{3}\right\}$, range of $f = \{y | y \neq 0\}$; domain of $f^{-1} = \{x | x \neq 0\}$; range of $f^{-1} = \left\{y \middle| y \neq \frac{5}{3}\right\}$

4. The point $(-5, 3)$ must be on the graph of f^{-1}. **5.** $x = 5$ **6.** $b = 4$ **7.** $x = 625$ **8.** $e^3 + 2 \approx 22.086$ **9.** $\log 20 \approx 1.301$

10. $\log_3 21 = \dfrac{\ln 21}{\ln 3} \approx 2.771$ **11.** $\ln 133 \approx 4.890$

12. (a) Domain of f: $\{x | -\infty < x < \infty\}$ or $(-\infty, \infty)$ **(e)** Range of f: $\{y | y > -2\}$ or $(-2, \infty)$

(b)

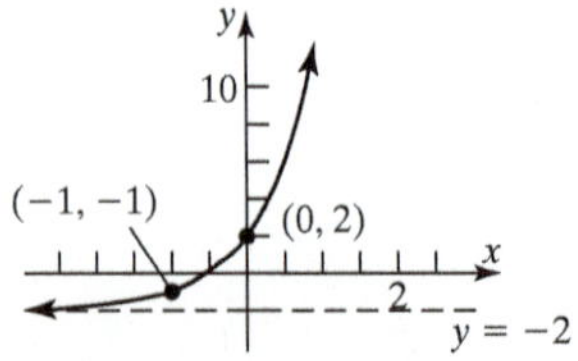

(f)

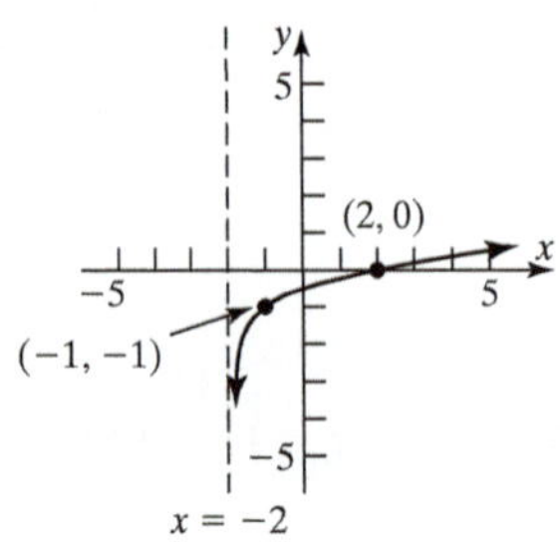

(c) Range of f: $\{y | y > -2\}$ or $(-2, \infty)$; Horizontal asymptote: $y = -2$

(d) $f^{-1}(x) = \log_4(x + 2) - 1$

13. (a) Domain of f: $\{x | x > 2\}$ or $(2, \infty)$ **(e)** Range of f: $\{y | -\infty < y < \infty\}$ or $(-\infty, \infty)$

(b)

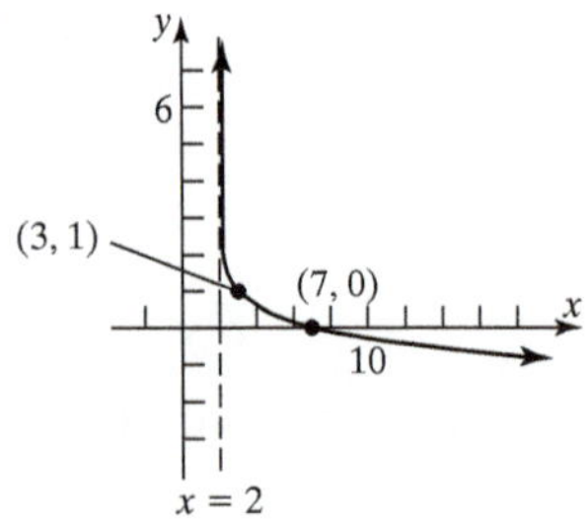

(f)

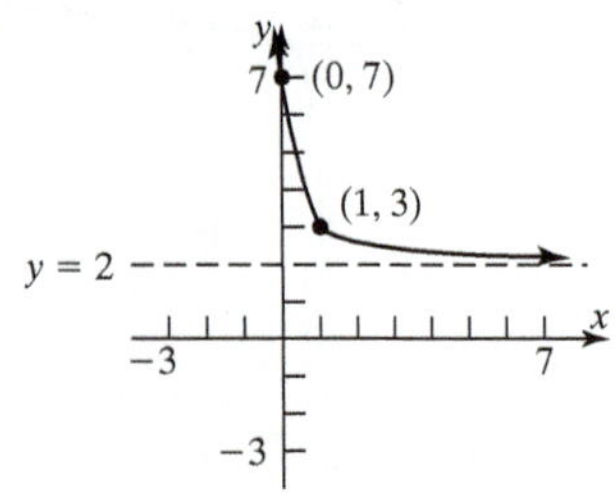

(c) Range of f: $\{y | -\infty < y < \infty\}$ or $(-\infty, \infty)$; vertical asymptote: $x = 2$

(d) $f^{-1}(x) = 5^{1-x} + 2$

14. $\{1\}$ **15.** $\{91\}$ **16.** $\{-\ln 2\} \approx \{-0.693\}$ **17.** $\left\{\dfrac{1 - \sqrt{13}}{2}, \dfrac{1 + \sqrt{13}}{2}\right\} \approx \{-1.303, 2.303\}$ **18.** $\left\{\dfrac{3 \ln 7}{1 - \ln 7}\right\} \approx \{-6.172\}$ **19.** $\{2\sqrt{6}\} \approx \{4.899\}$

20. $2 + 3\log_2 x - \log_2(x - 6) - \log_2(x + 3)$ **21.** About 250.39 days **22. (a)** \$1033.82 **(b)** \$963.42 **(c)** 11.9 yr
23. (a) About 83 dB **(b)** The pain threshold will be exceeded if 31,623 people shouted at the same time.

Cumulative Review *(page 504)*

1. Yes; no **2. (a)** 10 **(b)** $2x^2 + 3x + 1$ **(c)** $2x^2 + 4xh + 2h^2 - 3x - 3h + 1$ **3.** $\left(\frac{1}{2}, \frac{\sqrt{3}}{2}\right)$ is on the graph. **4.** $\{-26\}$

5.

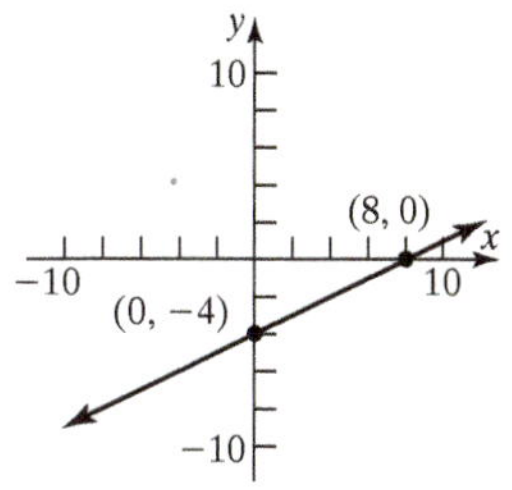

6. (a)

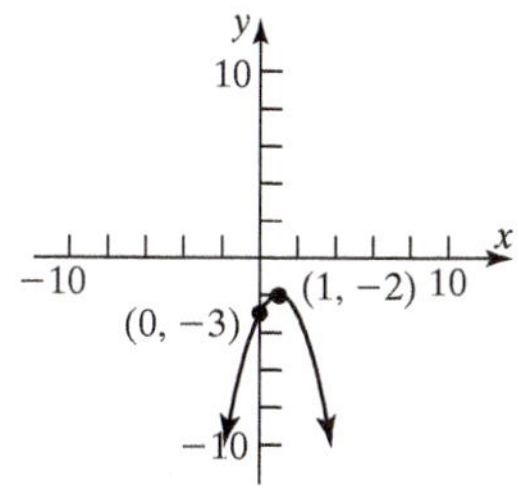

(b) $\{x \mid -\infty < x < \infty\}$

7. $f(x) = 2(x - 4)^2 - 8 = 2x^2 - 16x + 24$

8.

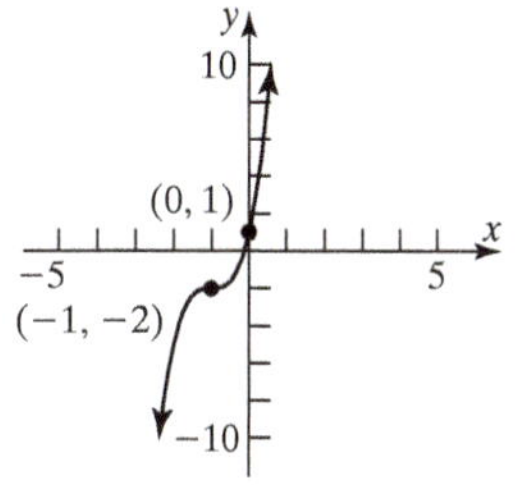

9. $f(g(x)) = \dfrac{4}{(x - 3)^2} + 2$; domain: $\{x \mid x \neq 3\}$; 3

10. (a) Zeros: $-4, -\frac{1}{4}, 2$ **(b)** x-intercepts: $-4, -\frac{1}{4}, 2$; y-intercept: -8 **(c)** Local maximum of 60.75 occurs at $x = -2.5$. Local minimum of -25 occurs at $x = 1$. **(d)**

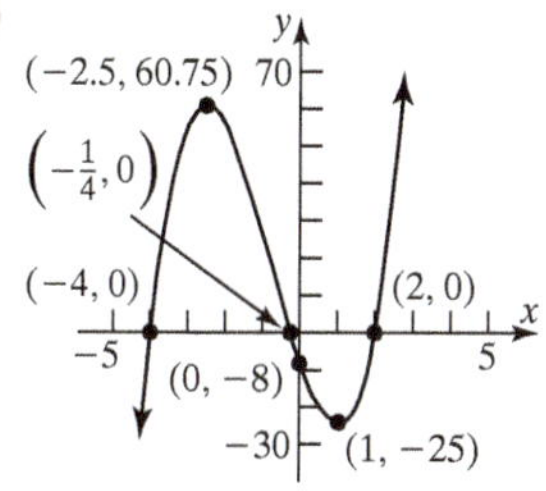

11. (a), (c)

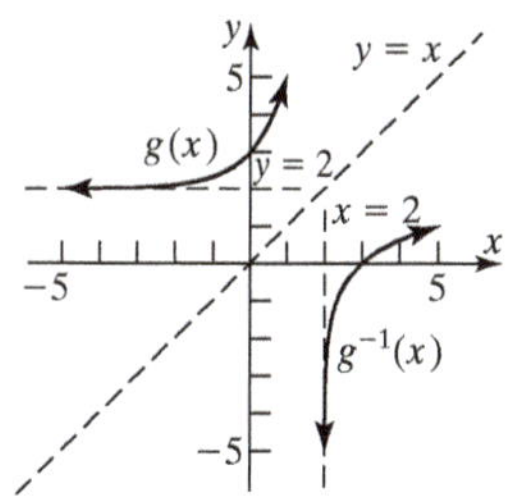

Domain g = range $g^{-1} = (-\infty, \infty)$
Range g = domain $g^{-1} = (2, \infty)$

(b) $g^{-1}(x) = \log_3(x - 2)$

12. $\left\{-\frac{3}{2}\right\}$ **13.** $\{2\}$ **14. (a)** $\{-1\}$ **(b)** $\{x \mid x > -1\}$ or $(-1, \infty)$ **(c)** $\{25\}$

15. (a)

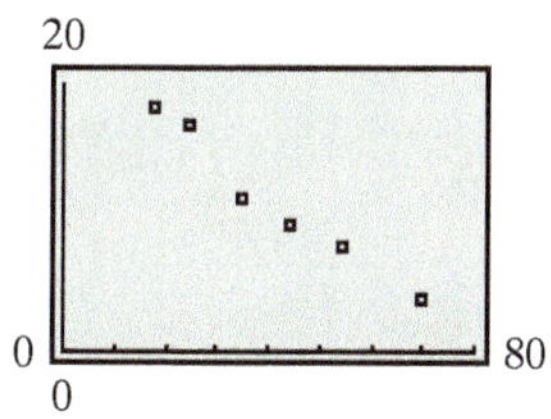

CHAPTER 7 Analytic Geometry

7.2 Assess Your Understanding *(page 516)*

6. parabola **7.** paraboloid of revolution **8.** T **9.** T **10.** T **11.** *B* **13.** *E* **15.** *H* **17.** *C*

19. $y^2 = 16x$

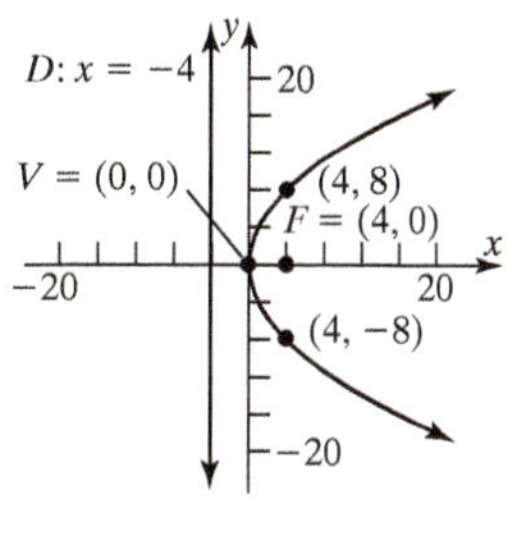

21. $x^2 = -12y$

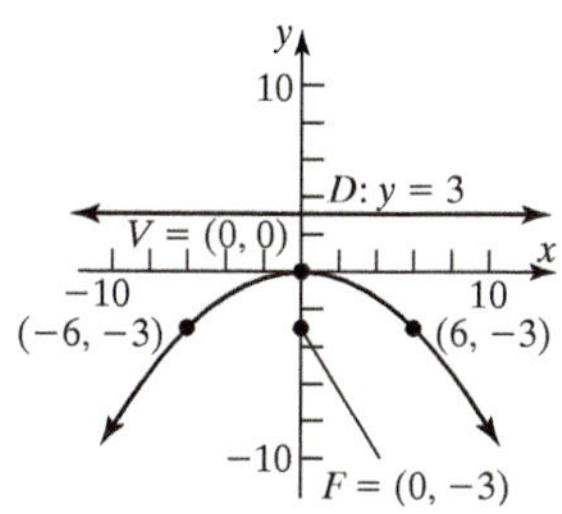

23. $y^2 = -8x$

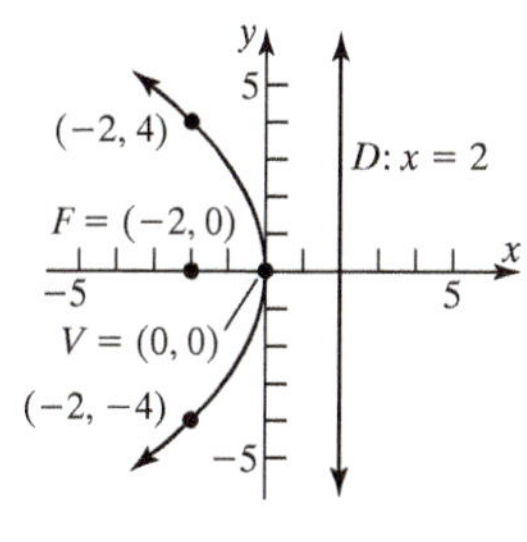

25. $x^2 = 2y$

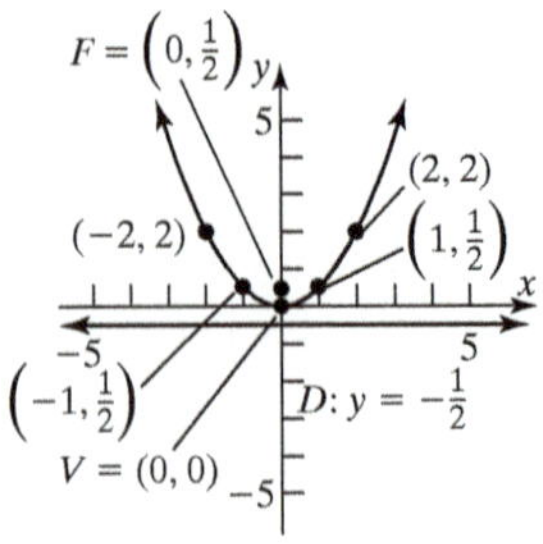

27. $x^2 = \frac{4}{3}y$

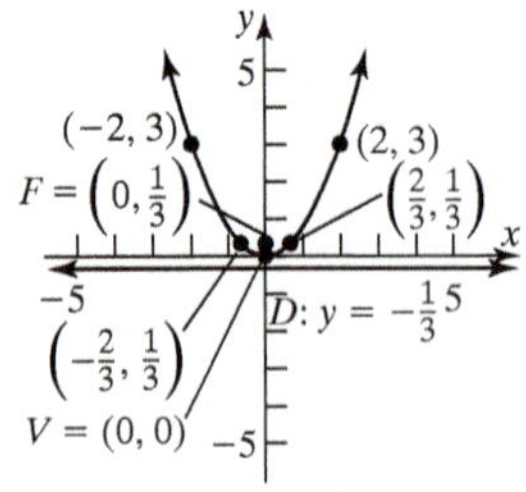

29. $(x - 2)^2 = -8(y + 3)$

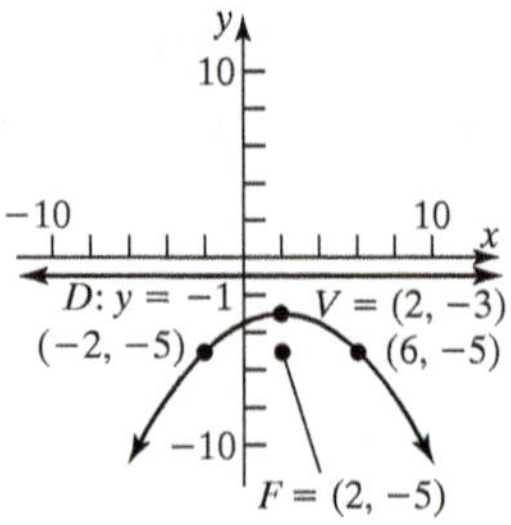

31. $(y + 2)^2 = 4(x + 1)$

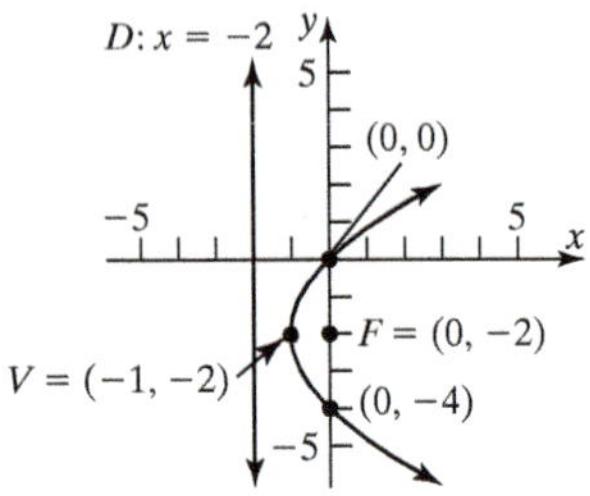

33. $(x + 3)^2 = 4(y - 3)$

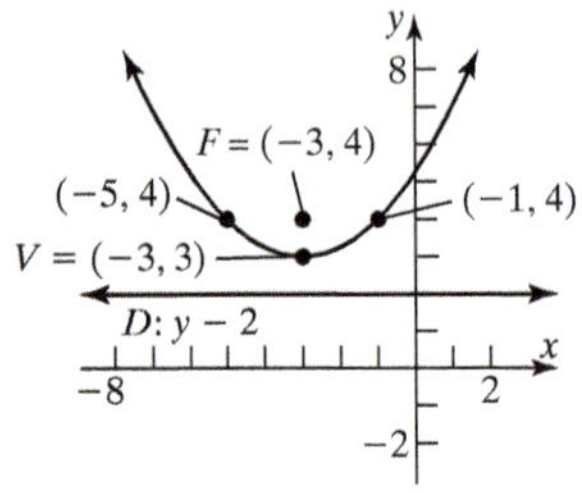

35. $(y + 2)^2 = -8(x + 1)$

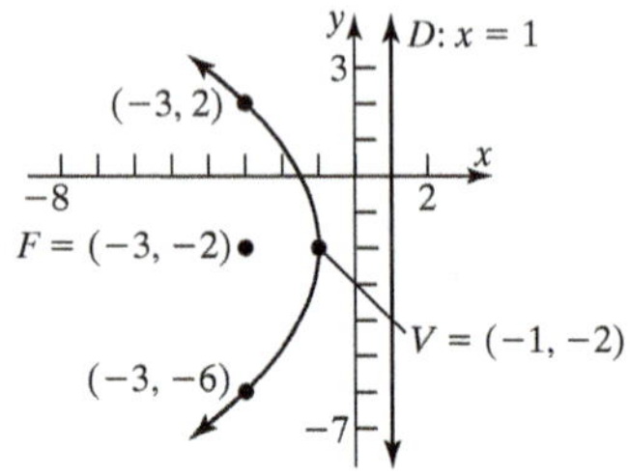

37. Vertex: (0, 0); focus: (0, 1); directrix: $y = -1$

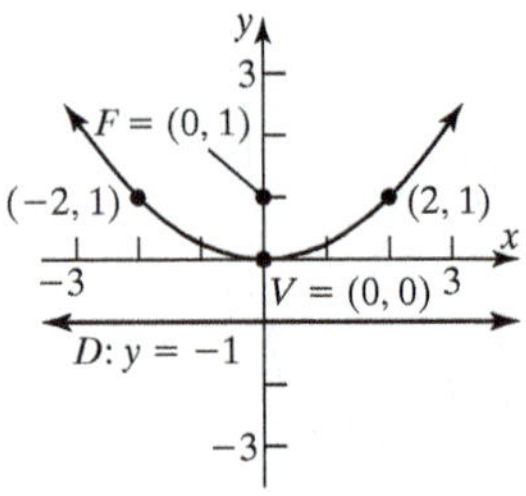

39. Vertex: (0, 0); focus: (−4, 0); directrix: $x = 4$

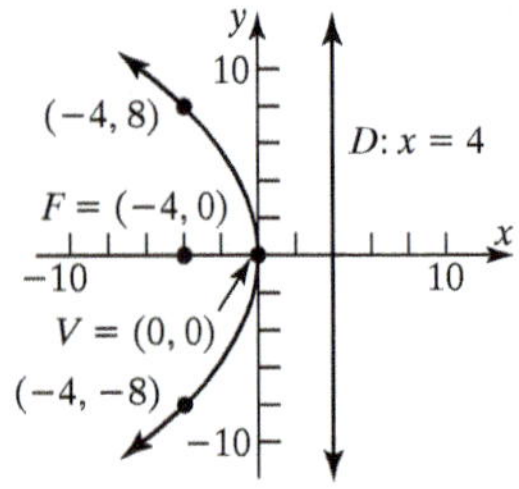

41. Vertex: (−1, 2); focus: (1, 2); directrix: $x = -3$

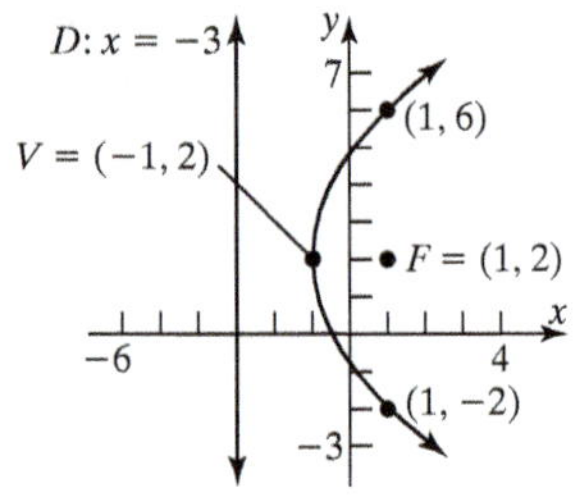

43. Vertex: (3, −1); focus: $\left(3, -\frac{5}{4}\right)$; directrix: $y = -\frac{3}{4}$

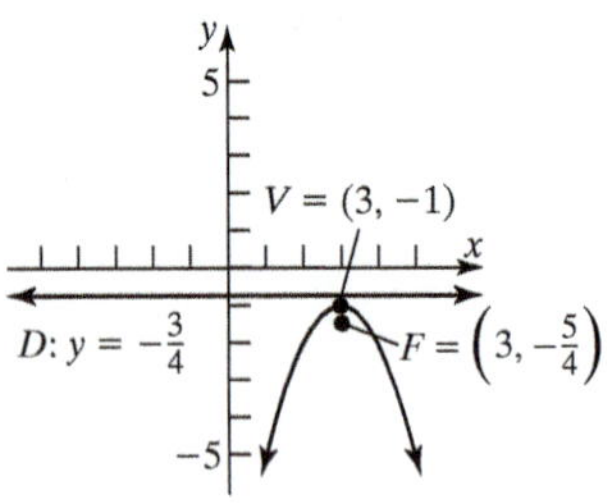

45. Vertex: (2, −3); focus: (4, −3); directrix: $x = 0$

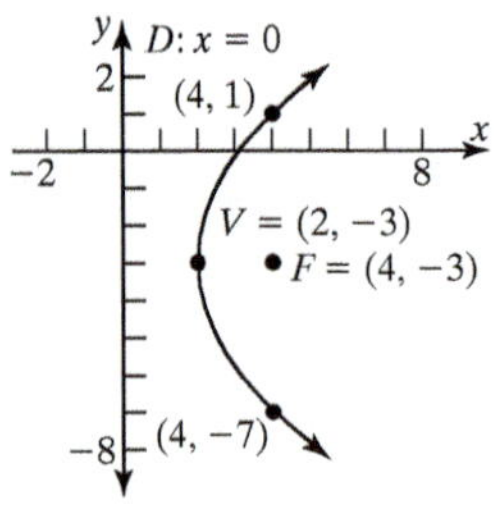

47. Vertex: (0, 2); focus: (−1, 2); directrix: $x = 1$

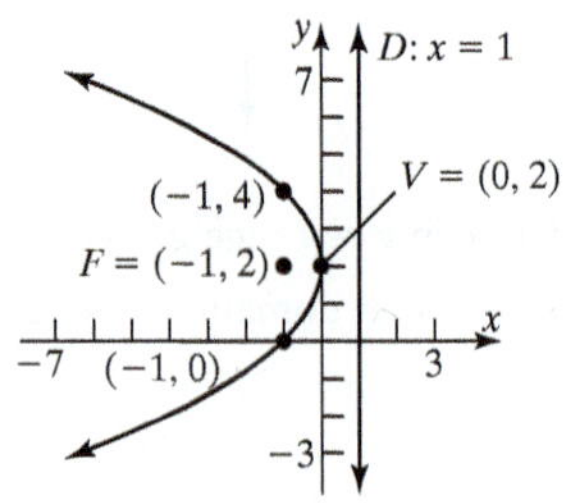

49. Vertex: (−4, −2); focus: (−4, −1); directrix: $y = -3$

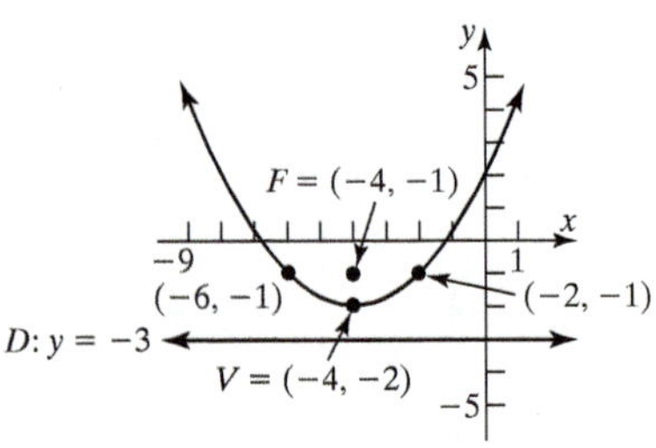

51. Vertex: (−1, −1); focus: $\left(-\frac{3}{4}, -1\right)$; directrix: $x = -\frac{5}{4}$

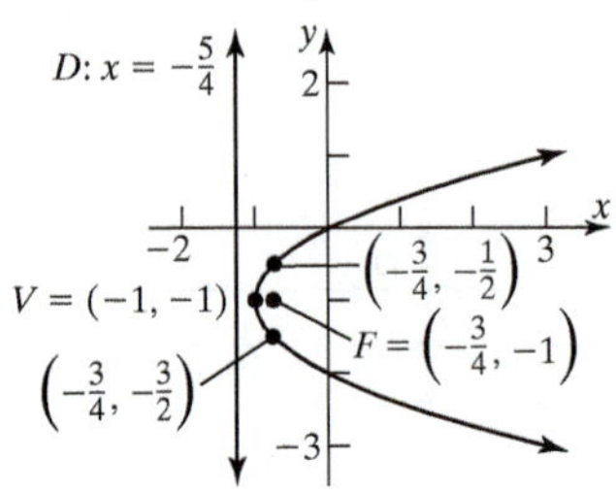

53. Vertex: (2, −8); focus: $\left(2, -\frac{31}{4}\right)$; directrix: $y = -\frac{33}{4}$

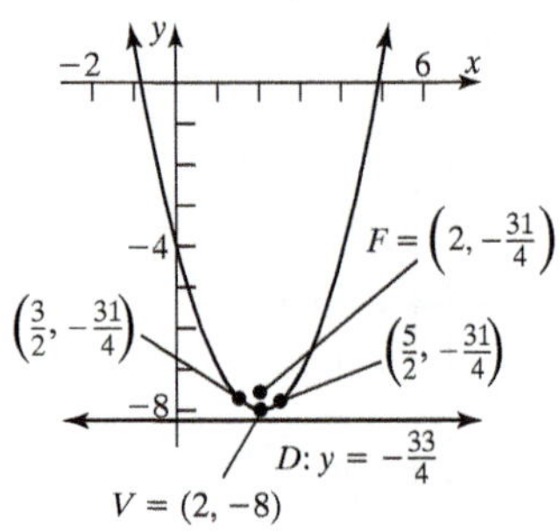

55. $(y-1)^2 = x$ **57.** $(y-1)^2 = -(x-2)$ **59.** $x^2 = 4(y-1)$ **61.** $y^2 = \frac{1}{2}(x+2)$

63. 1.5625 ft from the base of the dish, along the axis of symmetry **65.** 1 in. from the vertex, along the axis of symmetry **67.** 20 ft

69. 0.78125 ft **71.** 4.17 ft from the base, along the axis of symmetry **73.** 24.31 ft, 18.75 ft, 7.64 ft

75. **(a)** $y = -\frac{625}{(299)^2}x^2 + 625$ **(b)** 567 ft: 63.12 ft; 478 ft: 225.67 ft; 308 ft: 459.2 ft **(c)** No

77. $Cy^2 + Dx = 0, C \neq 0, D \neq 0$

$Cy^2 = -Dx$

$y^2 = -\frac{D}{C}x$

This is the equation of a parabola with vertex at $(0, 0)$ and axis of symmetry the x-axis. The focus is $\left(-\frac{D}{4C}, 0\right)$; the directrix is the line $x = \frac{D}{4C}$. The parabola opens to the right if $-\frac{D}{C} > 0$ and to the left if $-\frac{D}{C} < 0$.

79. $Cy^2 + Dx + Ey + F = 0, C \neq 0$

$Cy^2 + Ey = -Dx - F$

$y^2 + \frac{E}{C}y = -\frac{D}{C}x - \frac{F}{C}$

$\left(y + \frac{E}{2C}\right)^2 = -\frac{D}{C}x - \frac{F}{C} + \frac{E^2}{4C^2}$

$\left(y + \frac{E}{2C}\right)^2 = -\frac{D}{C}x + \frac{E^2 - 4CF}{4C^2}$

(a) If $D \neq 0$, then the equation may be written as $\left(y + \frac{E}{2C}\right)^2 = -\frac{D}{C}\left(x - \frac{E^2 - 4CF}{4CD}\right)$. This is the equation of a parabola with vertex at $\left(\frac{E^2 - 4CF}{4CD}, -\frac{E}{2C}\right)$ and axis of symmetry parallel to the x-axis.

(b)–(d) If $D = 0$, the graph of the equation contains no points if $E^2 - 4CF < 0$, is a single horizontal line if $E^2 - 4CF = 0$, and is two horizontal lines if $E^2 - 4CF > 0$.

7.3 Assess Your Understanding *(page 527)*

7. ellipse **8.** major **9.** $(0, -5)$; $(0, 5)$ **10.** F **11.** T **12.** T **13.** *C* **15.** *B*

17. Vertices: $(-5, 0)$, $(5, 0)$

Foci: $(-\sqrt{21}, 0)$, $(\sqrt{21}, 0)$

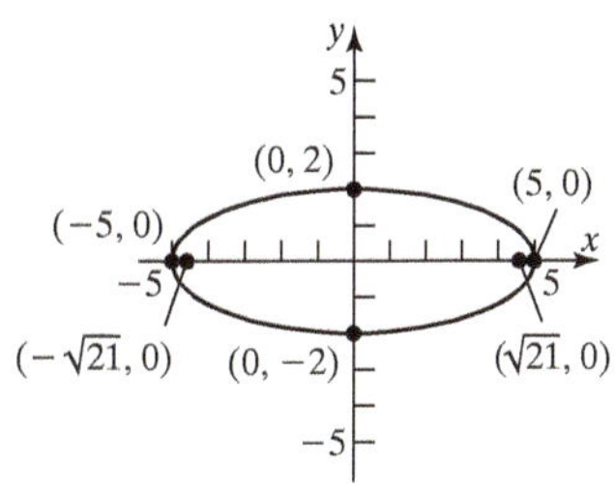

19. Vertices: $(0, -5)$, $(0, 5)$

Foci: $(0, -4)$, $(0, 4)$

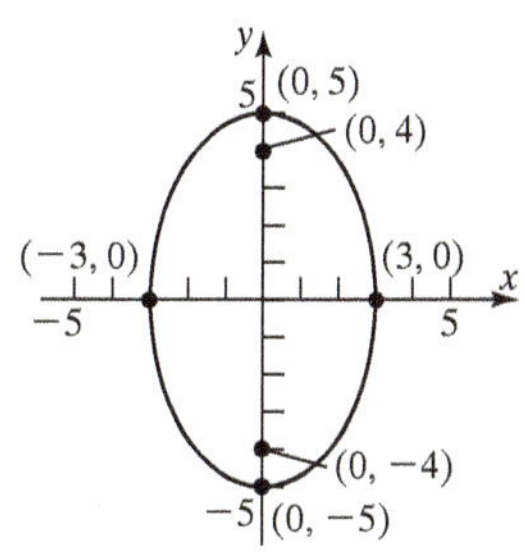

21. $\frac{x^2}{4} + \frac{y^2}{16} = 1$

Vertices: $(0, -4)$, $(0, 4)$

Foci: $(0, -2\sqrt{3})$, $(0, 2\sqrt{3})$

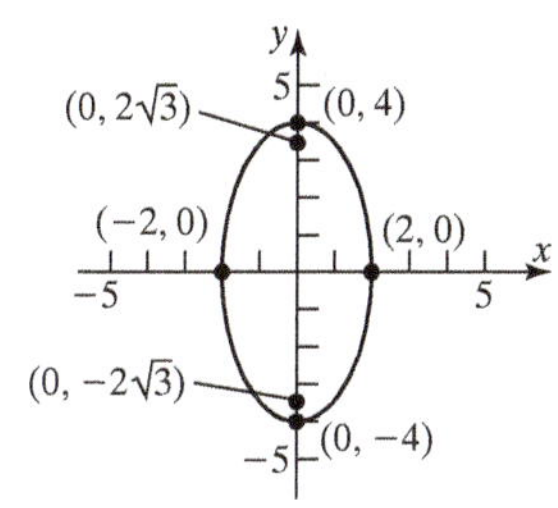

23. $\frac{x^2}{8} + \frac{y^2}{2} = 1$

Vertices: $(-2\sqrt{2}, 0)$, $(2\sqrt{2}, 0)$

Foci: $(-\sqrt{6}, 0)$, $(\sqrt{6}, 0)$

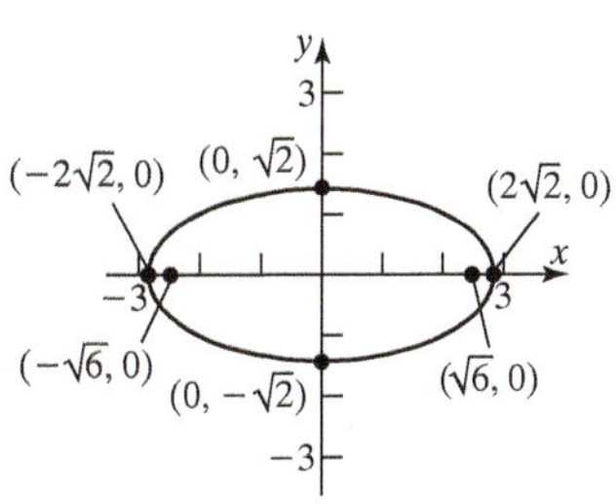

25. $\frac{x^2}{16} + \frac{y^2}{16} = 1$

Vertices: $(-4, 0)$, $(4, 0)$, $(0, -4)$, $(0, 4)$

Focus: $(0, 0)$

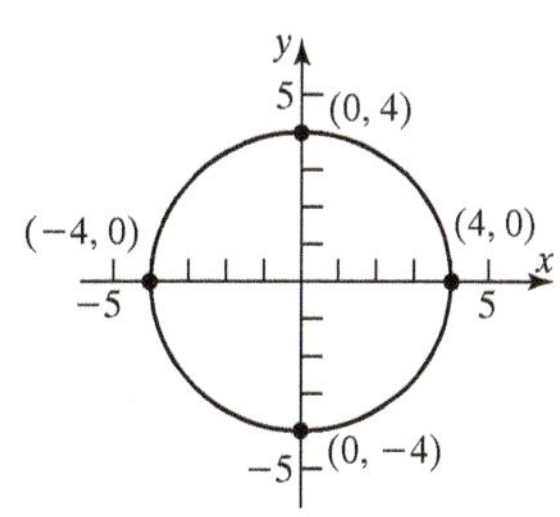

27. $\frac{x^2}{25} + \frac{y^2}{16} = 1$

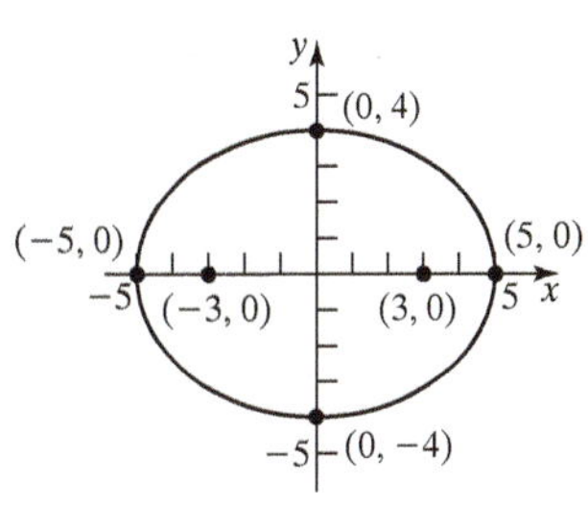

29. $\frac{x^2}{9} + \frac{y^2}{25} = 1$

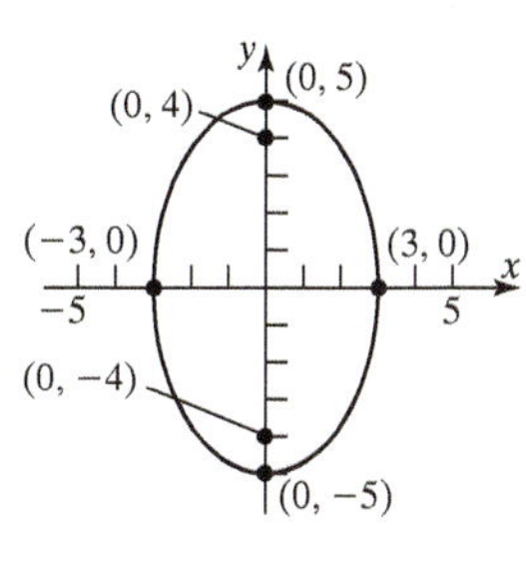

31. $\frac{x^2}{9} + \frac{y^2}{5} = 1$

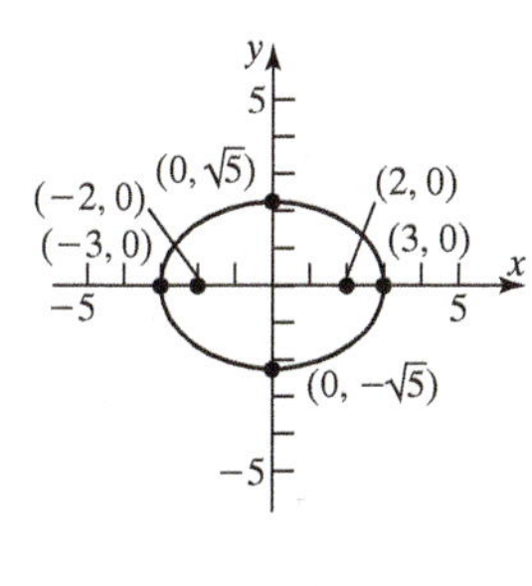

33. $\frac{x^2}{25} + \frac{y^2}{9} = 1$

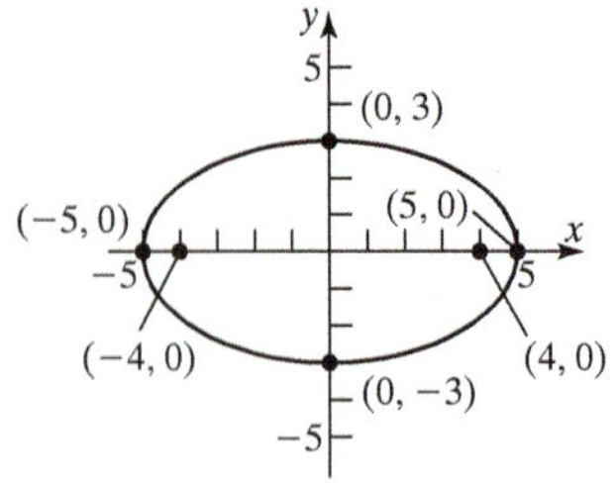

35. $\frac{x^2}{4} + \frac{y^2}{13} = 1$

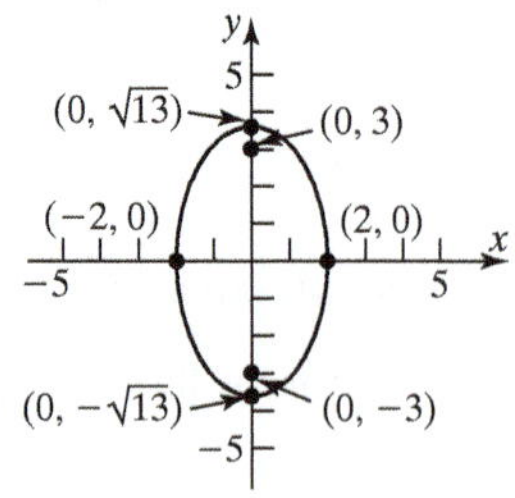

37. $x^2 + \frac{y^2}{16} = 1$

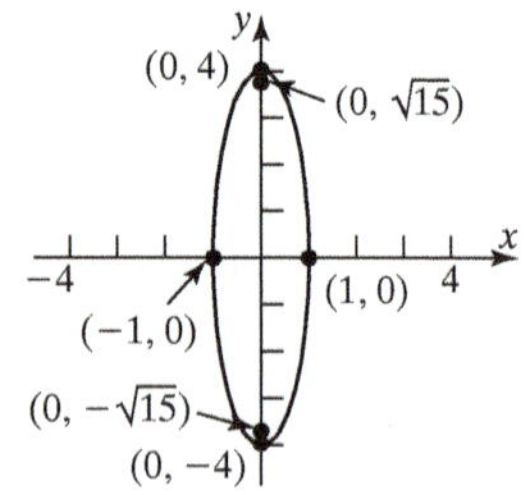

39. $\frac{(x+1)^2}{4} + (y-1)^2 = 1$ **41.** $(x-1)^2 + \frac{y^2}{4} = 1$

43. Center: $(3, -1)$; vertices: $(3, -4)$, $(3, 2)$; foci: $\left(3, -1 - \sqrt{5}\right)$, $\left(3, -1 + \sqrt{5}\right)$

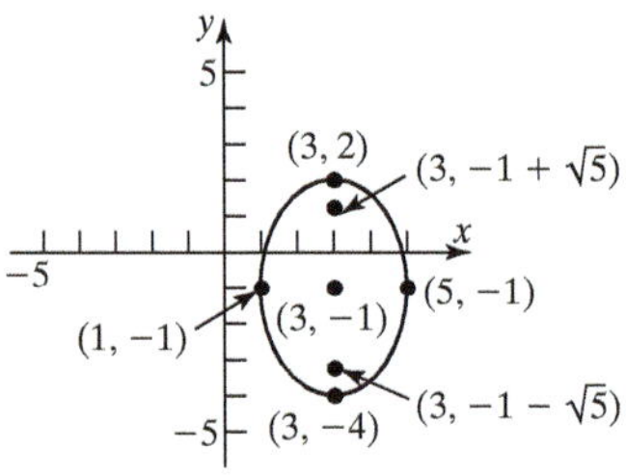

45. $\frac{(x+5)^2}{16} + \frac{(y-4)^2}{4} = 1$

Center: $(-5, 4)$; vertices: $(-9, 4)$, $(-1, 4)$; foci: $\left(-5 - 2\sqrt{3}, 4\right)$, $\left(-5 + 2\sqrt{3}, 4\right)$

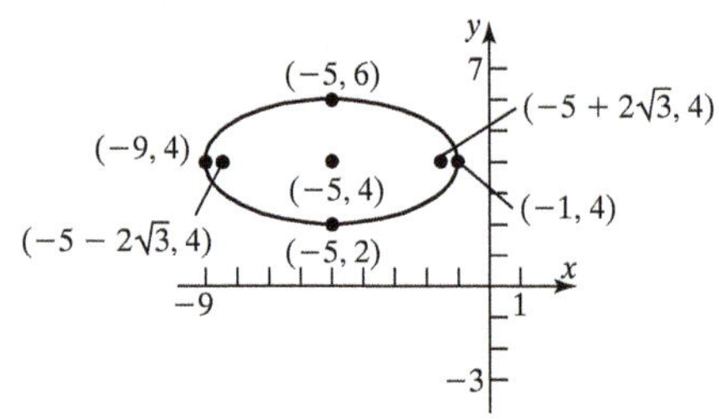

47. $\frac{(x+2)^2}{4} + (y-1)^2 = 1$

Center: $(-2, 1)$; vertices: $(-4, 1)$, $(0, 1)$; foci: $\left(-2 - \sqrt{3}, 1\right)$, $\left(-2 + \sqrt{3}, 1\right)$

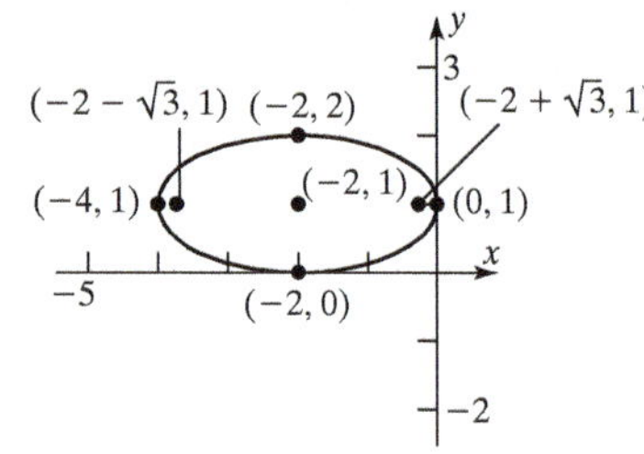

49. $\frac{(x-2)^2}{3} + \frac{(y+1)^2}{2} = 1$

Center: $(2, -1)$; vertices: $\left(2 - \sqrt{3}, -1\right)$; $\left(2 + \sqrt{3}, -1\right)$; foci: $(1, -1)$, $(3, -1)$

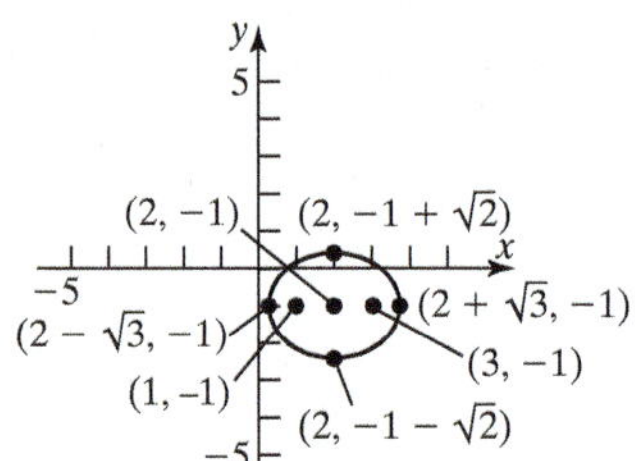

51. $\frac{(x-1)^2}{4} + \frac{(y+2)^2}{9} = 1$

Center: $(1, -2)$; vertices: $(1, -5)$, $(1, 1)$; foci: $\left(1, -2 - \sqrt{5}\right)$, $\left(1, -2 + \sqrt{5}\right)$

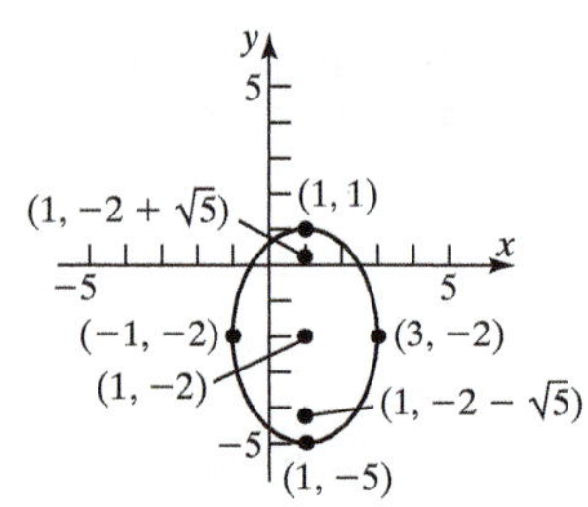

53. $x^2 + \frac{(y+2)^2}{4} = 1$

Center: $(0, -2)$; vertices: $(0, -4)$, $(0, 0)$; foci: $\left(0, -2 - \sqrt{3}\right)$, $\left(0, -2 + \sqrt{3}\right)$

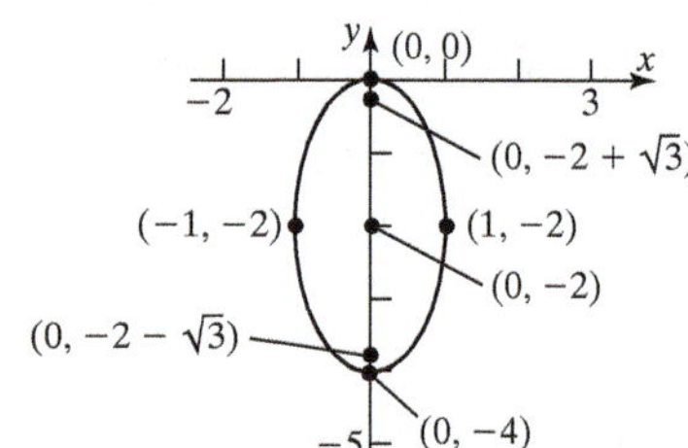

55. $\frac{(x-2)^2}{25} + \frac{(y+2)^2}{21} = 1$

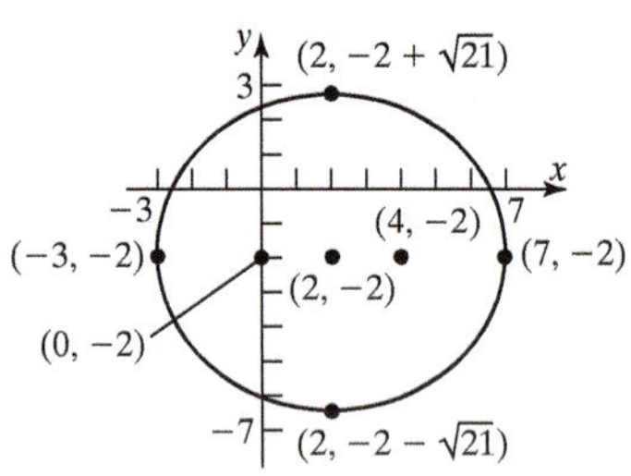

57. $\frac{(x-4)^2}{5} + \frac{(y-6)^2}{9} = 1$

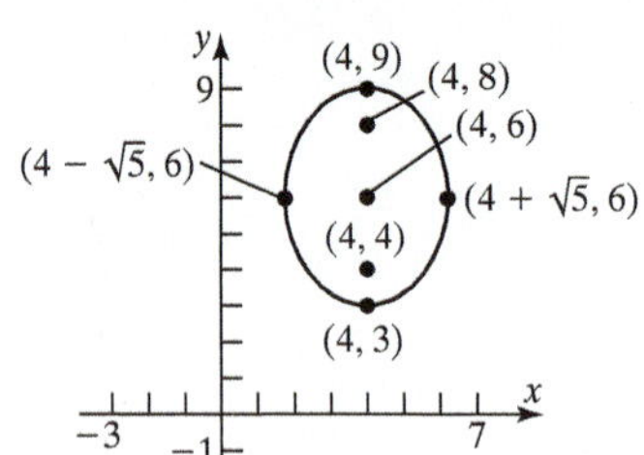

59. $\frac{(x-2)^2}{16} + \frac{(y-1)^2}{7} = 1$

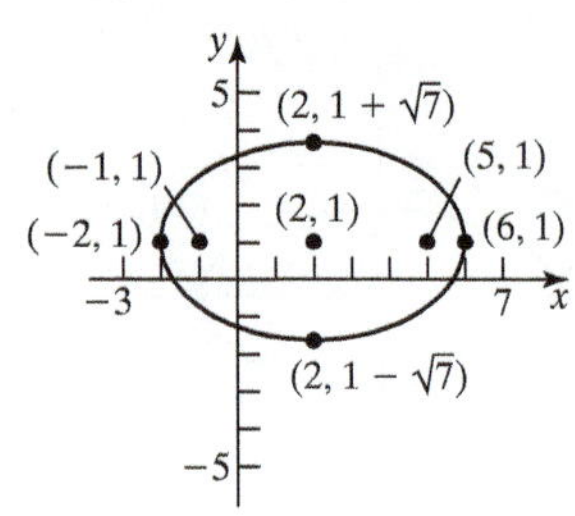

61. $\frac{(x-1)^2}{10} + (y-2)^2 = 1$

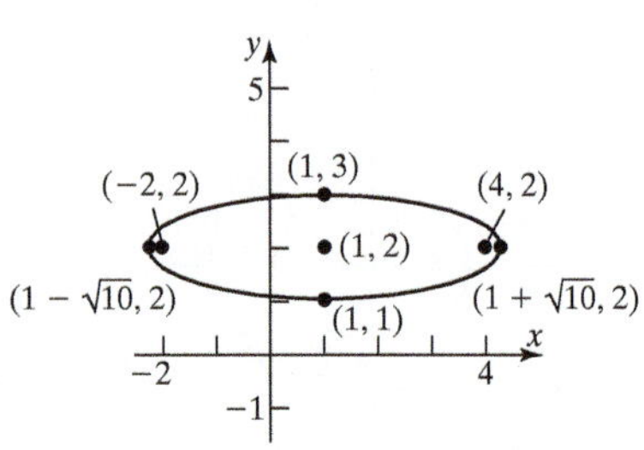

63. $\frac{(x-1)^2}{9} + (y-2)^2 = 1$

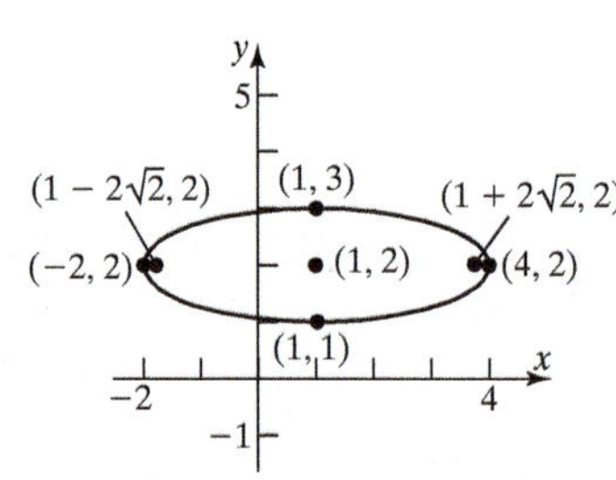

65.

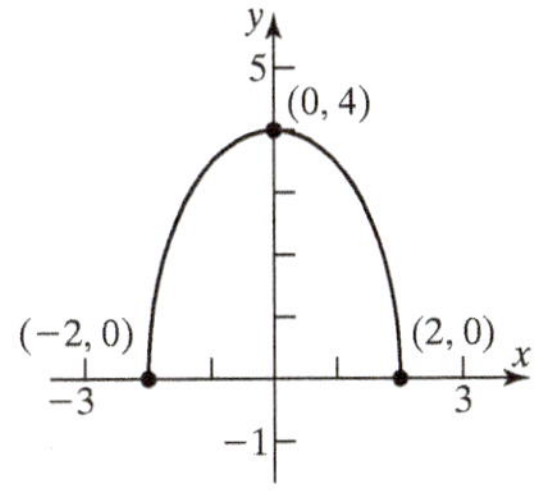

67.

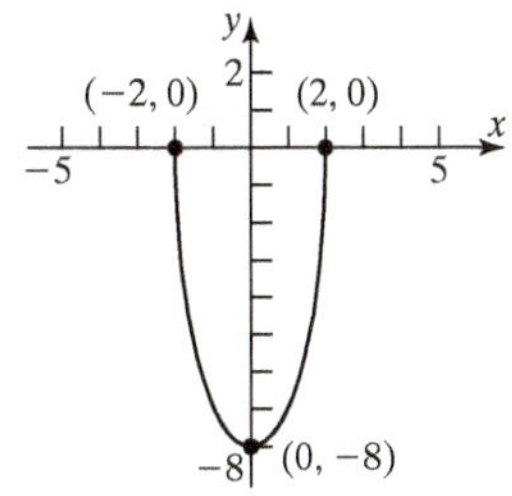

69. $\frac{x^2}{100} + \frac{y^2}{36} = 1$ **71.** 43.3 ft **73.** 24.65 ft, 21.65 ft, 13.82 ft **75.** 30 ft

77. The elliptical hole will have a major axis of length $2\sqrt{41}$ in. and a minor axis of length 8 in. **79.** 91.5 million mi; $\frac{x^2}{8649} + \frac{y^2}{8646.75} = 1$

81. Perihelion: 460.6 million mi; mean distance: 483.8 million mi; $\frac{x^2}{233{,}482} + \frac{y^2}{233{,}524.2} = 1$

83. (a) $Ax^2 + Cy^2 + F = 0$

$Ax^2 + Cy^2 = -F$

If A and C are of the same sign and F is of opposite sign, then the equation takes the form $\frac{x^2}{\left(-\frac{F}{A}\right)} + \frac{y^2}{\left(-\frac{F}{C}\right)} = 1$, where $-\frac{F}{A}$ and $-\frac{F}{C}$ are positive. This is the equation of an ellipse with center at (0, 0).

(b) If $A = C$, the equation may be written as $x^2 + y^2 = \frac{-F}{A}$. This is the equation of a circle with center at (0, 0) and radius equal to $\sqrt{-\frac{F}{A}}$.

7.4 Assess Your Understanding *(page 540)*

7. hyperbola **8.** transverse axis **9.** $y = \frac{3}{2}x$; $y = -\frac{3}{2}x$ **10.** F **11.** T **12.** F **13.** *B* **15.** *A*

17. $x^2 - \frac{y^2}{8} = 1$

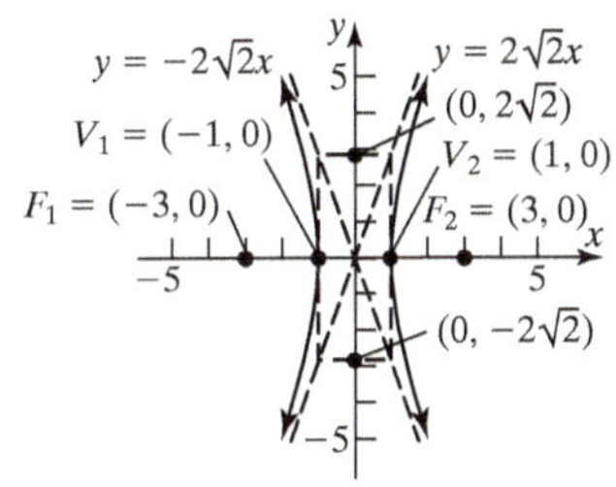

19. $\frac{y^2}{16} - \frac{x^2}{20} = 1$

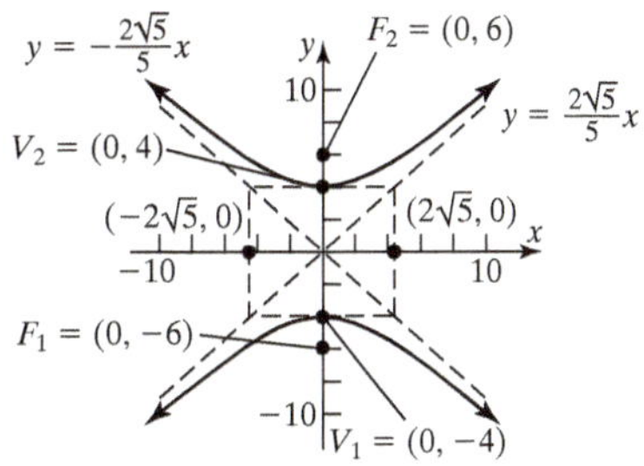

21. $\frac{x^2}{9} - \frac{y^2}{16} = 1$

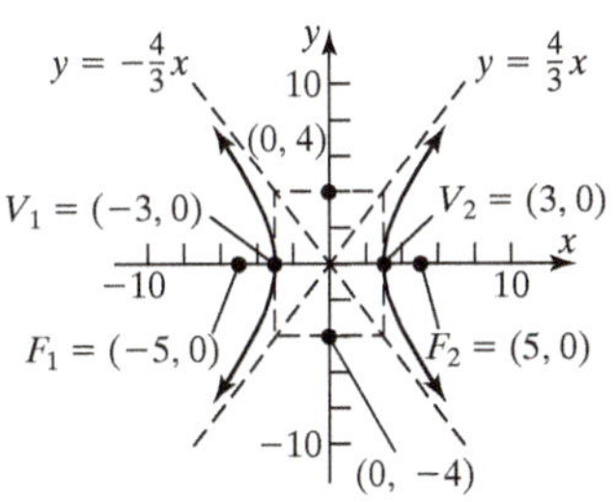

23. $\frac{y^2}{36} - \frac{x^2}{9} = 1$

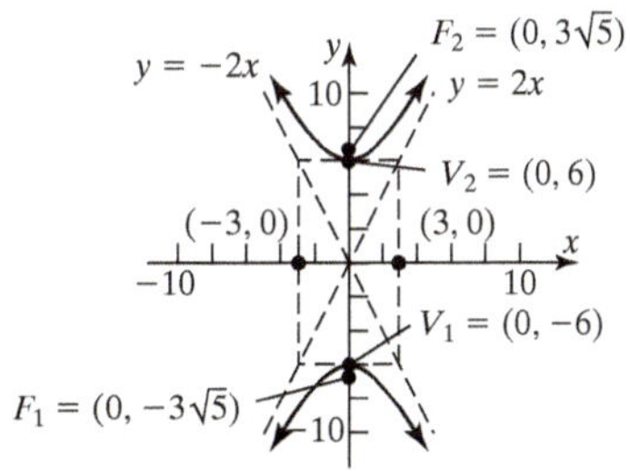

25. $\frac{x^2}{8} - \frac{y^2}{8} = 1$

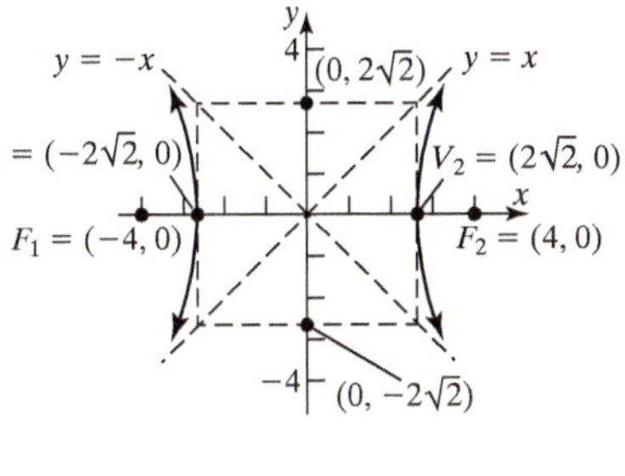

27. $\dfrac{x^2}{25} - \dfrac{y^2}{9} = 1$

Center: $(0, 0)$
Transverse axis: x-axis
Vertices: $(-5, 0)$, $(5, 0)$
Foci: $\left(-\sqrt{34}, 0\right)$, $\left(\sqrt{34}, 0\right)$
Asymptotes: $y = \pm\dfrac{3}{5}x$

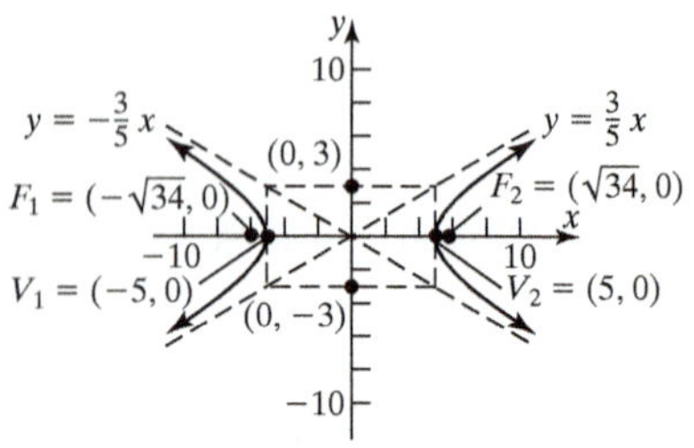

29. $\dfrac{x^2}{4} - \dfrac{y^2}{16} = 1$

Center: $(0, 0)$
Transverse axis: x-axis
Vertices: $(-2, 0)$, $(2, 0)$
Foci: $\left(-2\sqrt{5}, 0\right)$, $\left(2\sqrt{5}, 0\right)$
Asymptotes: $y = \pm 2x$

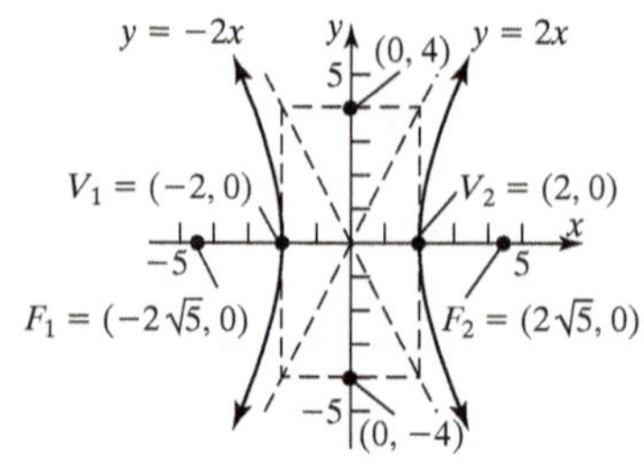

31. $\dfrac{y^2}{9} - x^2 = 1$

Center: $(0, 0)$
Transverse axis: y-axis
Vertices: $(0, -3)$, $(0, 3)$
Foci: $\left(0, -\sqrt{10}\right)$, $\left(0, \sqrt{10}\right)$
Asymptotes: $y = \pm 3x$

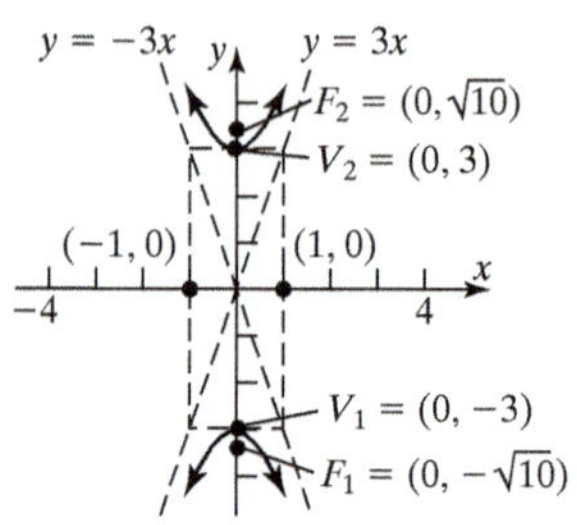

33. $\dfrac{y^2}{25} - \dfrac{x^2}{25} = 1$

Center: $(0, 0)$
Transverse axis: y-axis
Vertices: $(0, -5)$, $(0, 5)$
Foci: $\left(0, -5\sqrt{2}\right)$, $\left(0, 5\sqrt{2}\right)$
Asymptotes: $y = \pm x$

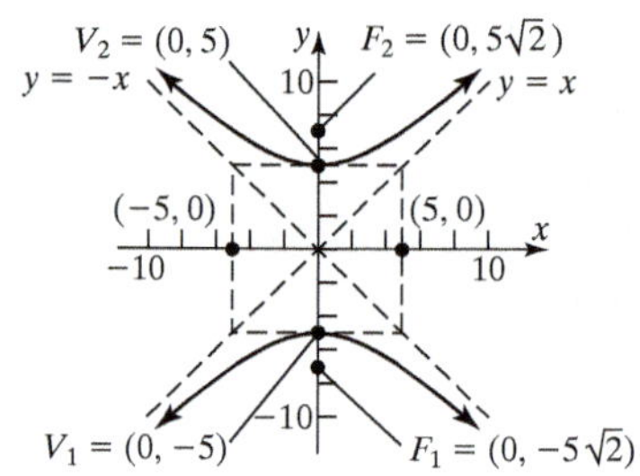

35. $x^2 - y^2 = 1$ **37.** $\dfrac{y^2}{36} - \dfrac{x^2}{9} = 1$

39. $\dfrac{(x-4)^2}{4} - \dfrac{(y+1)^2}{5} = 1$

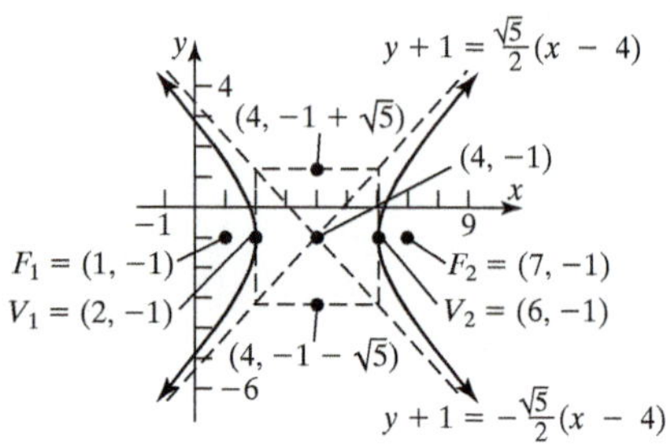

41. $\dfrac{(y+4)^2}{4} - \dfrac{(x+3)^2}{12} = 1$

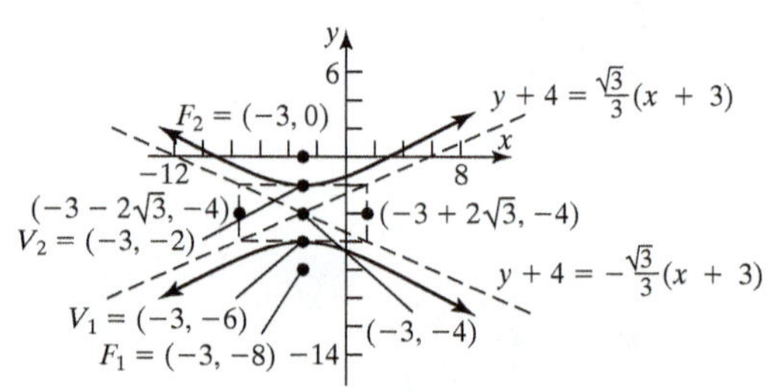

43. $(x-5)^2 - \dfrac{(y-7)^2}{3} = 1$

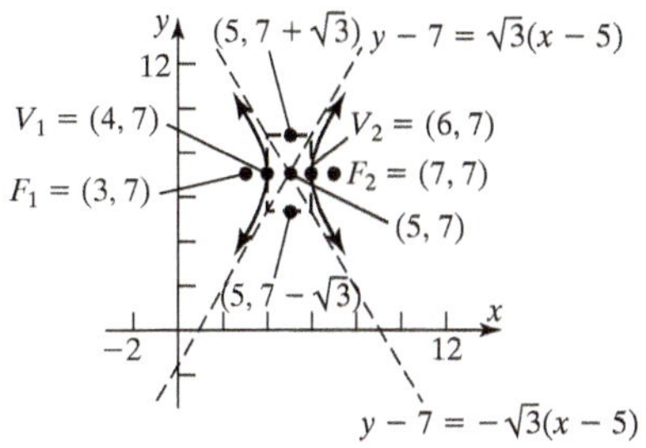

45. $\dfrac{(x-1)^2}{4} - \dfrac{(y+1)^2}{9} = 1$

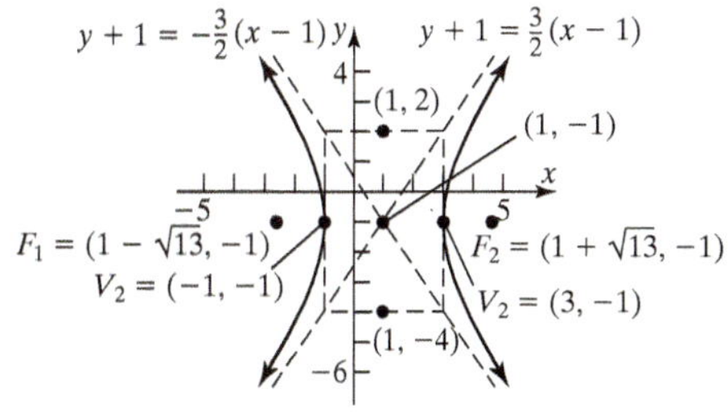

47. $\dfrac{(x-2)^2}{4} - \dfrac{(y+3)^2}{9} = 1$

Center: $(2, -3)$

Transverse axis: parallel to x-axis

Vertices: $(0, -3), (4, -3)$

Foci: $\left(2 - \sqrt{13}, -3\right), \left(2 + \sqrt{13}, -3\right)$

Asymptotes: $y + 3 = \pm\dfrac{3}{2}(x - 2)$

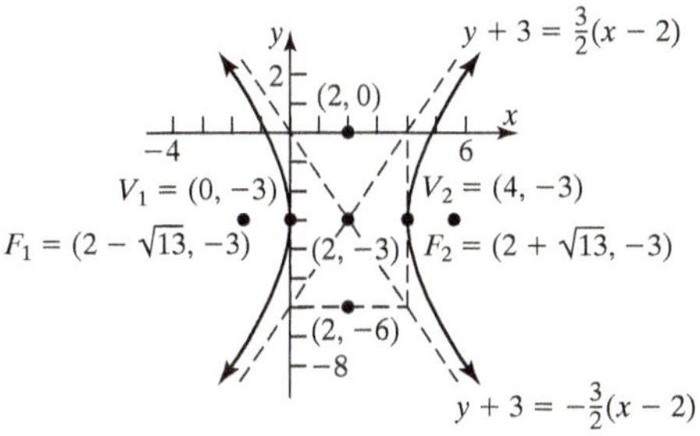

49. $\dfrac{(y-2)^2}{4} - (x+2)^2 = 1$

Center: $(-2, 2)$

Transverse axis: parallel to y-axis

Vertices: $(-2, 0), (-2, 4)$

Foci: $\left(-2, 2 - \sqrt{5}\right), \left(-2, 2 + \sqrt{5}\right)$

Asymptotes: $y - 2 = \pm 2(x + 2)$

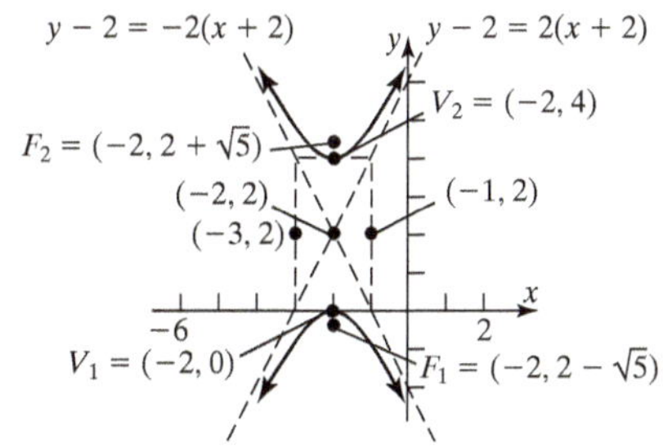

51. $\dfrac{(x+1)^2}{4} - \dfrac{(y+2)^2}{4} = 1$

Center: $(-1, -2)$

Transverse axis: parallel to x-axis

Vertices: $(-3, -2), (1, -2)$

Foci: $\left(-1 - 2\sqrt{2}, -2\right), \left(-1 + 2\sqrt{2}, -2\right)$

Asymptotes: $y + 2 = \pm(x + 1)$

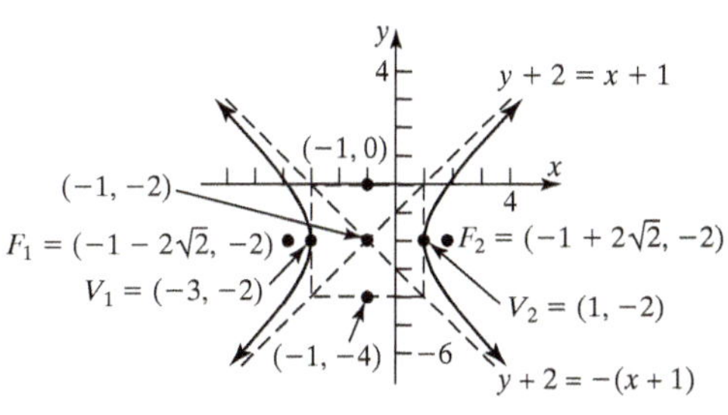

53. $(x-1)^2 - (y+1)^2 = 1$

Center: $(1, -1)$

Transverse axis: parallel to x-axis

Vertices: $(0, -1), (2, -1)$

Foci: $\left(1 - \sqrt{2}, -1\right), \left(1 + \sqrt{2}, -1\right)$

Asymptotes: $y + 1 = \pm(x - 1)$

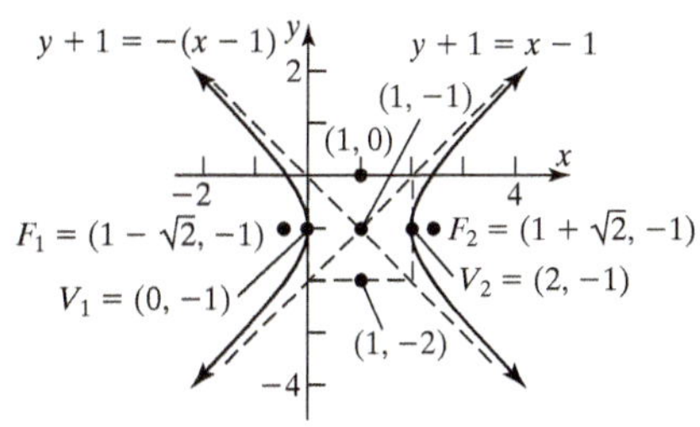

55. $\dfrac{(y-2)^2}{4} - (x+1)^2 = 1$

Center: $(-1, 2)$
Transverse axis: parallel to y-axis
Vertices: $(-1, 0), (-1, 4)$
Foci: $\left(-1, 2 - \sqrt{5}\right), \left(-1, 2 + \sqrt{5}\right)$
Asymptotes: $y - 2 = \pm 2(x + 1)$

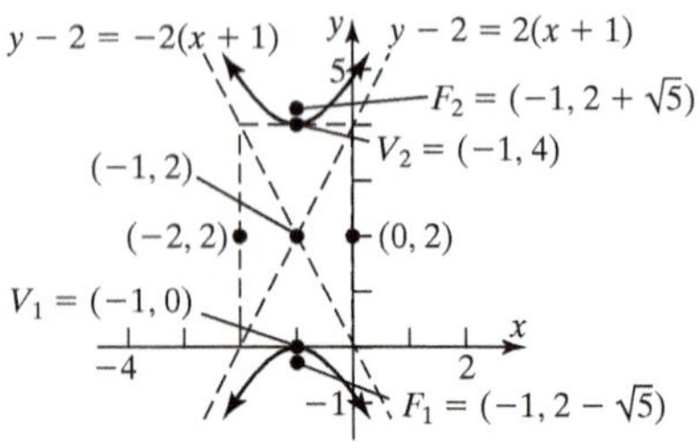

57. $\dfrac{(x-3)^2}{4} - \dfrac{(y+2)^2}{16} = 1$

Center: $(3, -2)$
Transverse axis: parallel to x-axis
Vertices: $(1, -2), (5, -2)$
Foci: $\left(3 - 2\sqrt{5}, -2\right), \left(3 + 2\sqrt{5}, -2\right)$
Asymptotes: $y + 2 = \pm 2(x - 3)$

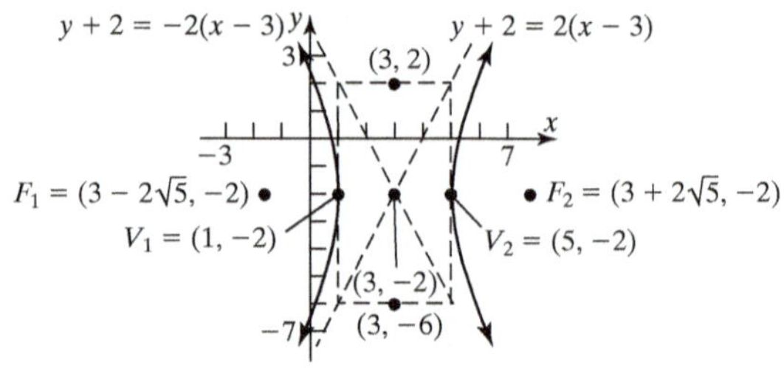

59. $\dfrac{(y-1)^2}{4} - (x+2)^2 = 1$

Center: $(-2, 1)$
Transverse axis: parallel to y-axis
Vertices: $(-2, -1), (-2, 3)$
Foci: $\left(-2, 1 - \sqrt{5}\right), \left(-2, 1 + \sqrt{5}\right)$
Asymptotes: $y - 1 = \pm 2(x + 2)$

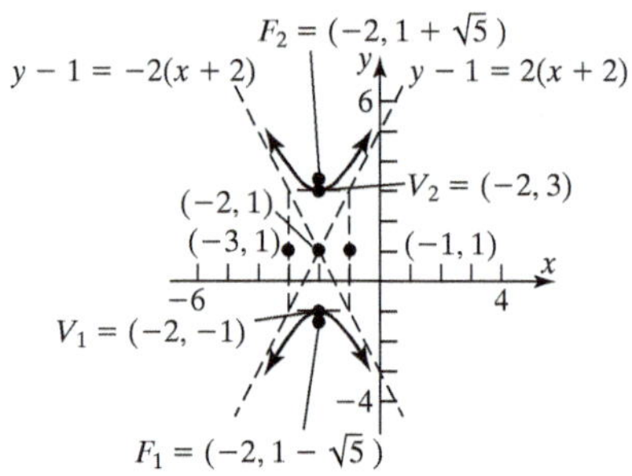

61.

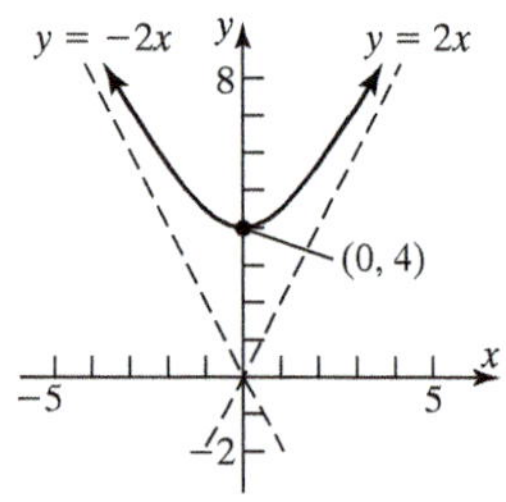

63.

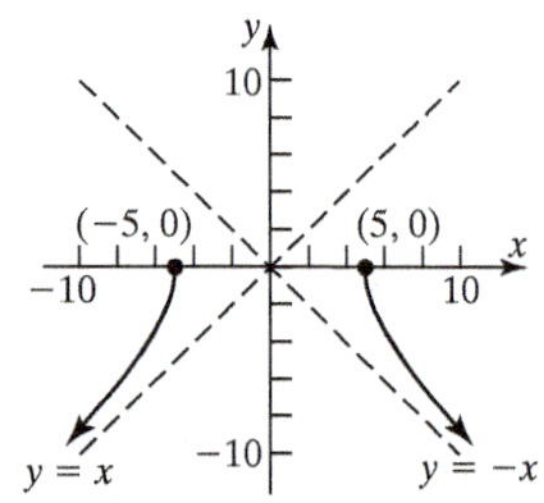

65. Center: $(3, 0)$
Transverse axis: parallel to x-axis
Vertices: $(1, 0), (5, 0)$
Foci: $\left(3 - \sqrt{29}, 0\right), \left(3 + \sqrt{29}, 0\right)$
Asymptotes: $y = \pm\dfrac{5}{2}(x - 3)$

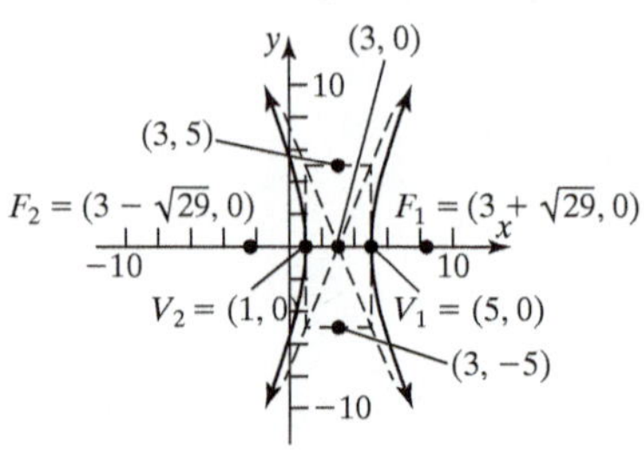

67. Vertex: $(0, 3)$; focus: $(0, 7)$; directrix: $y = -1$

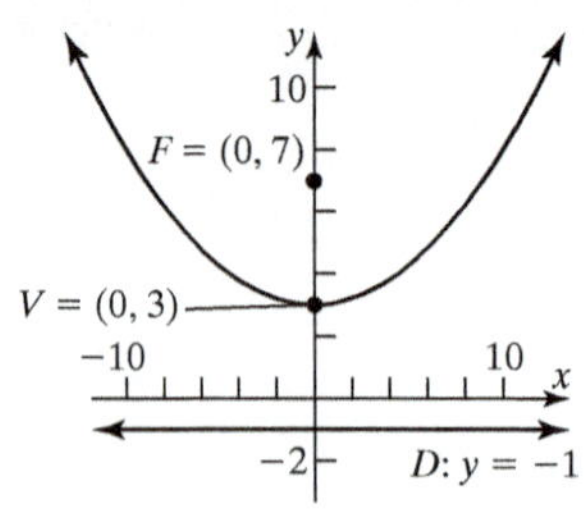

69. $\dfrac{(x-5)^2}{9} + \dfrac{y^2}{25} = 1$

Center: $(5, 0)$; vertices: $(5, 5), (5, -5)$; foci: $(5, -4), (5, 4)$

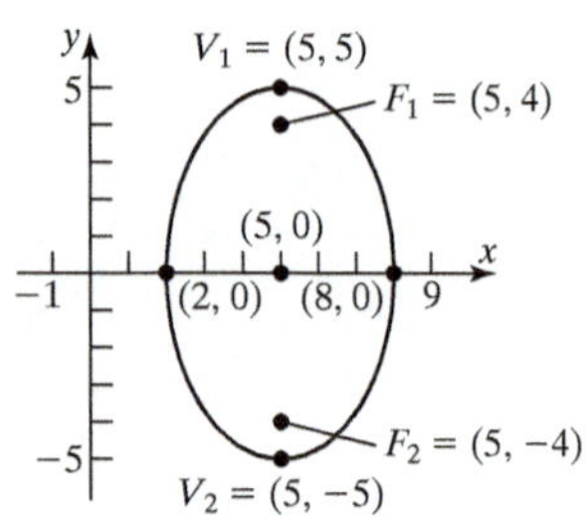

71. $(x-3)^2 = 8(y+5)$
Vertex: $(3, -5)$; foci: $(3, -3)$;
directrix: $y = -7$

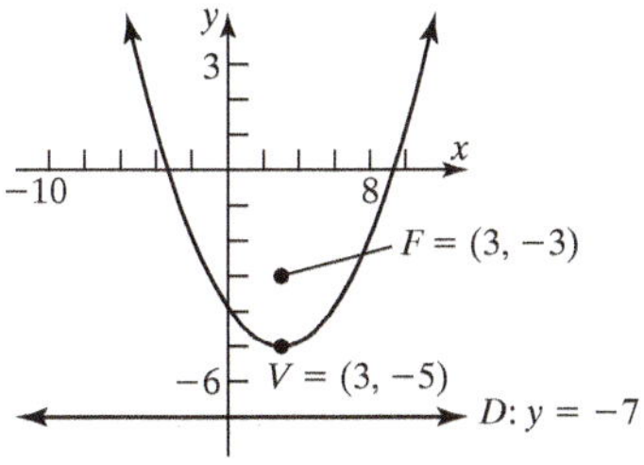

73. The fireworks display is 50,138 ft north of the person at point A.

75. The tower is 592.4 ft tall.

77. **(a)** $y = \pm x$

(b) $\dfrac{x^2}{100} - \dfrac{y^2}{100} = 1,\ x \geq 0$

81. $\dfrac{x^2}{4} - y^2 = 1$; asymptotes $y = \pm\dfrac{1}{2}x$

$y^2 - \dfrac{x^2}{4} = 1$; asymptotes $y = \pm\dfrac{1}{2}x$

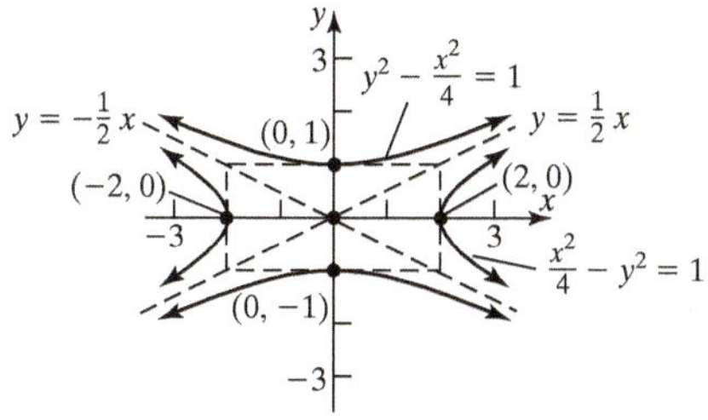

83. $Ax^2 + Cy^2 + F = 0$

$Ax^2 + Cy^2 = -F$

If A and C are of opposite sign and $F \neq 0$, this equation may be written as $\dfrac{x^2}{\left(-\dfrac{F}{A}\right)} + \dfrac{y^2}{\left(-\dfrac{F}{C}\right)} = 1$, where $-\dfrac{F}{A}$ and $-\dfrac{F}{C}$ are opposite in sign. This is the equation of a hyperbola with center $(0, 0)$. The transverse axis is the x-axis if $-\dfrac{F}{A} > 0$; the transverse axis is the y-axis if $-\dfrac{F}{A} < 0$.

Review Exercises *(page 543)*

1. Parabola; vertex $(0, 0)$, focus $(-4, 0)$, directrix $x = 4$ **3.** Hyperbola; center $(0, 0)$, vertices $(5, 0)$ and $(-5, 0)$, foci $(\sqrt{26}, 0)$ and $(-\sqrt{26}, 0)$, asymptotes $y = \dfrac{1}{5}x$ and $y = -\dfrac{1}{5}x$ **5.** Ellipse; center $(0, 0)$, vertices $(0, 5)$ and $(0, -5)$, foci $(0, 3)$ and $(0, -3)$ **7.** $x^2 = -4(y-1)$: Parabola; vertex $(0, 1)$, focus $(0, 0)$, directrix $y = 2$ **9.** $\dfrac{x^2}{2} - \dfrac{y^2}{8} = 1$: Hyperbola; center $(0, 0)$, vertices $(\sqrt{2}, 0)$ and $(-\sqrt{2}, 0)$, foci $(\sqrt{10}, 0)$ and $(-\sqrt{10}, 0)$, asymptotes $y = 2x$ and $y = -2x$ **11.** $(x-2)^2 = 2(y+2)$: Parabola; vertex $(2, -2)$, focus $\left(2, -\dfrac{3}{2}\right)$, directrix $y = -\dfrac{5}{2}$

13. $\dfrac{(y-2)^2}{4} - (x-1)^2 = 1$: Hyperbola; center $(1, 2)$, vertices $(1, 4)$ and $(1, 0)$, foci $(1, 2+\sqrt{5})$ and $(1, 2-\sqrt{5})$, asymptotes $y - 2 = \pm 2(x-1)$

15. $\dfrac{(x-2)^2}{9} + \dfrac{(y-1)^2}{4} = 1$: Ellipse; center $(2, 1)$, vertices $(5, 1)$ and $(-1, 1)$, foci $(2+\sqrt{5}, 1)$ and $(2-\sqrt{5}, 1)$

17. $(x-2)^2 = -4(y+1)$: Parabola; vertex $(2, -1)$, focus $(2, -2)$, directrix $y = 0$

19. $\dfrac{(x-1)^2}{4} + \dfrac{(y+1)^2}{9} = 1$: Ellipse; center $(1, -1)$, vertices $(1, 2)$ and $(1, -4)$, foci $(1, -1+\sqrt{5})$ and $(1, -1-\sqrt{5})$

21. $y^2 = -8x$

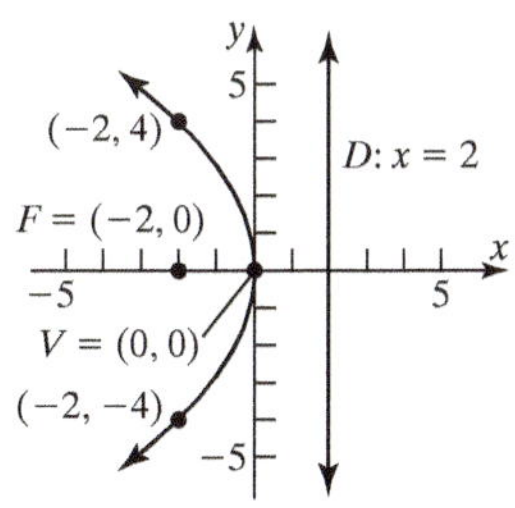

23. $\dfrac{y^2}{4} - \dfrac{x^2}{12} = 1$

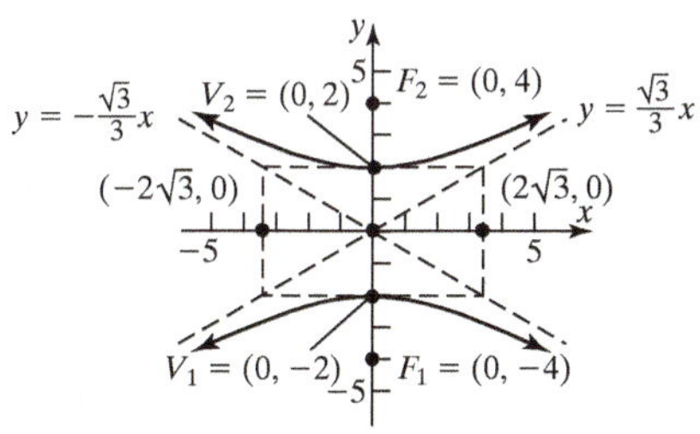

25. $\dfrac{x^2}{16} + \dfrac{y^2}{7} = 1$

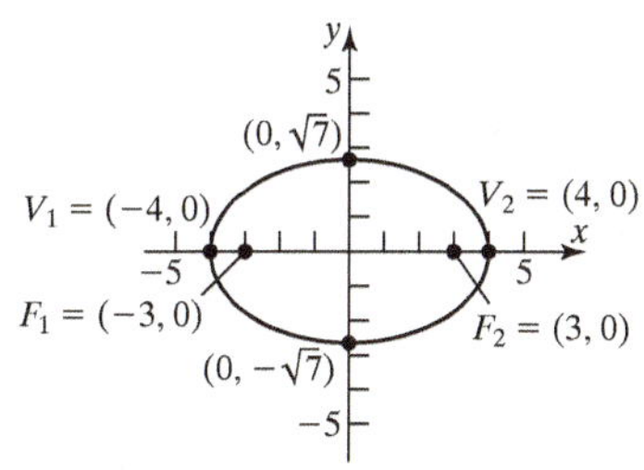

27. $(x-2)^2 = -4(y+3)$

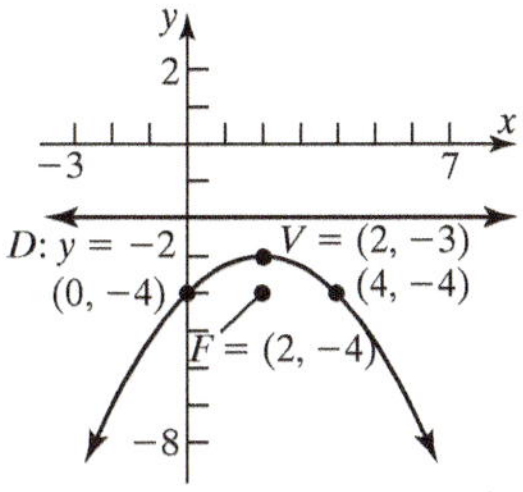

29. $(x+2)^2 - \dfrac{(y+3)^2}{3} = 1$

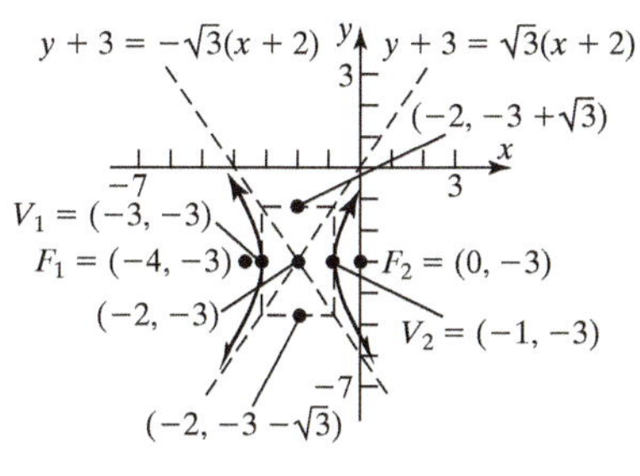

31. $\dfrac{(x+4)^2}{16} + \dfrac{(y-5)^2}{25} = 1$

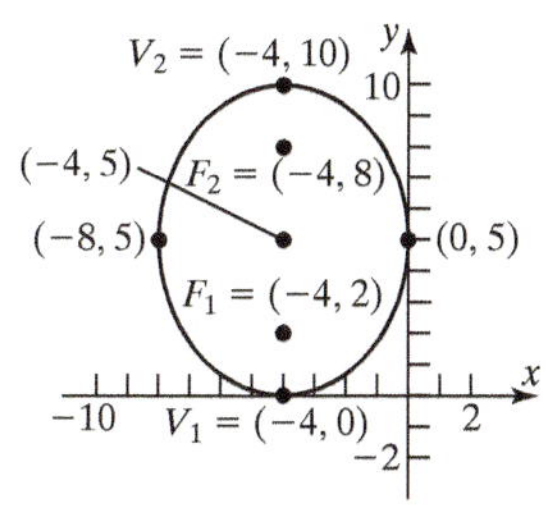

33. $\dfrac{(x+1)^2}{9} - \dfrac{(y-2)^2}{7} = 1$

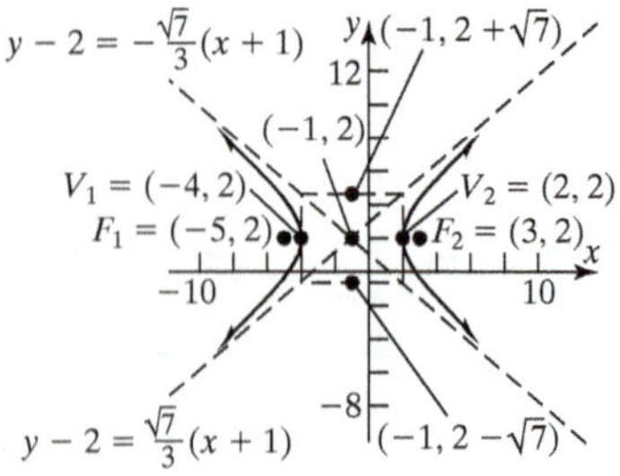

35. $\dfrac{(x-3)^2}{9} - \dfrac{(y-1)^2}{4} = 1$

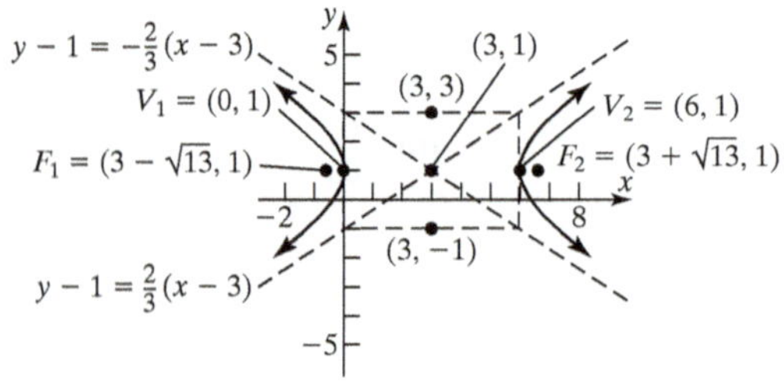

37. $\dfrac{x^2}{5} - \dfrac{y^2}{4} = 1$ **39.** The ellipse $\dfrac{x^2}{16} + \dfrac{y^2}{7} = 1$ **41.** $\dfrac{1}{4}$ ft or 3 in. **43.** 19.72 ft, 18.86 ft, 14.91 ft **45.** 450 ft

Chapter Test *(page 544)*

1. Hyperbola; center: $(-1, 0)$; vertices: $(-3, 0)$ and $(1, 0)$; foci: $\left(-1 - \sqrt{13}, 0\right)$ and $\left(-1 + \sqrt{13}, 0\right)$; asymptotes: $y = -\dfrac{3}{2}(x+1)$ and $y = \dfrac{3}{2}(x+1)$
2. Parabola; vertex: $\left(1, -\dfrac{1}{2}\right)$; focus: $\left(1, \dfrac{3}{2}\right)$; directrix: $y = -\dfrac{5}{2}$
3. Ellipse; center: $(-1, 1)$; foci: $\left(-1 - \sqrt{3}, 1\right)$ and $\left(-1 + \sqrt{3}, 1\right)$; vertices: $(-4, 1)$ and $(2, 1)$

4. $(x+1)^2 = 6(y-3)$

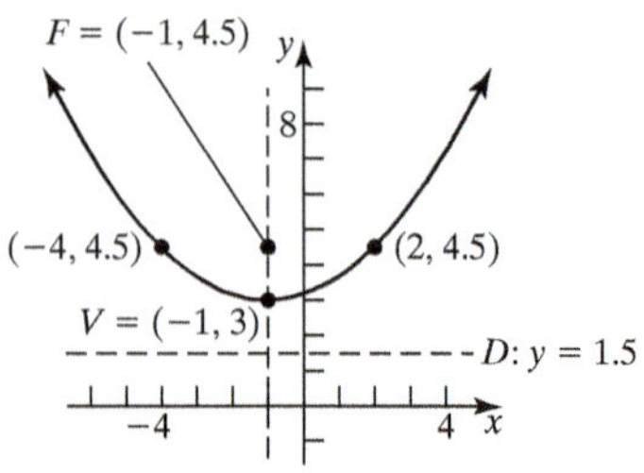

5. $\dfrac{x^2}{7} + \dfrac{y^2}{16} = 1$

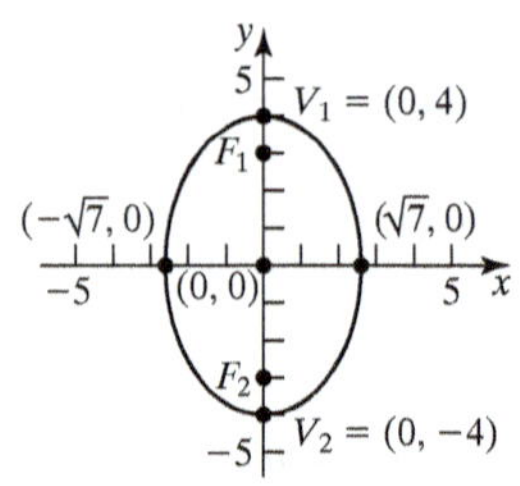

6. $\dfrac{(y-2)^2}{4} - \dfrac{(x-2)^2}{8} = 1$

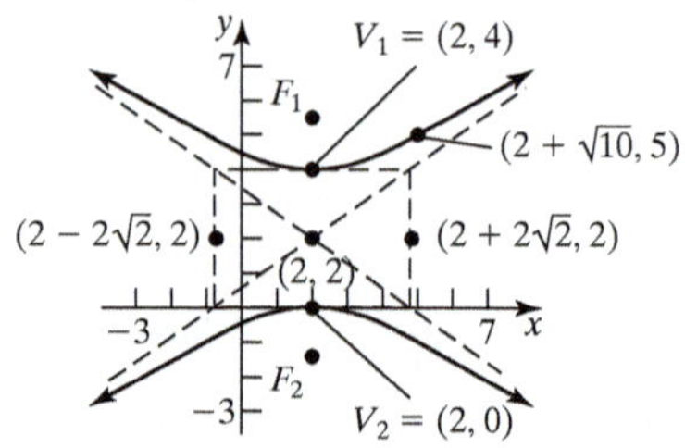

7. The microphone should be located $\dfrac{2}{3}$ ft from the base of the reflector, along its axis of symmetry.

Cumulative Review *(page 544)*

1. $-6x + 5 - 3h$ **2.** $\left\{-5, -\dfrac{1}{3}, 2\right\}$ **3.** $\{x \mid -3 \le x \le 2\}$ or $[-3, 2]$
4. (a) Domain: $(-\infty, \infty)$; range: $(2, \infty)$ **(b)** $y = \log_3(x-2)$; domain: $(2, \infty)$; range: $(-\infty, \infty)$ **5. (a)** $\{18\}$ **(b)** $(2, 18]$
6. (a) $y = 2x - 2$ **(b)** $(x-2)^2 + y^2 = 4$ **(c)** $\dfrac{x^2}{9} + \dfrac{y^2}{4} = 1$ **(d)** $y = 2(x-1)^2$ **(e)** $y^2 - \dfrac{x^2}{3} = 1$ **(f)** $y = 4^x$

CHAPTER 8 Systems of Equations and Inequalities

8.1 Assess Your Understanding *(page 558)*

3. inconsistent **4.** consistent **5.** F **6.** T **7.** $\begin{cases} 2(2) - (-1) = 5 \\ 5(2) + 2(-1) = 8 \end{cases}$ **9.** $\begin{cases} 3(2) - 4\left(\frac{1}{2}\right) = 4 \\ \frac{1}{2}(2) - 3\left(\frac{1}{2}\right) = -\frac{1}{2} \end{cases}$ **11.** $\begin{cases} 4 - 1 = 3 \\ \frac{1}{2}(4) + 1 = 3 \end{cases}$

13. $\begin{cases} 3(1) + 3(-1) + 2(2) = 4 \\ 1 - (-1) - 2 = 0 \\ 2(-1) - 3(2) = -8 \end{cases}$ **15.** $\begin{cases} 3(2) + 3(-2) + 2(2) = 4 \\ 2 - 3(-2) + 2 = 10 \\ 5(2) - 2(-2) - 3(2) = 8 \end{cases}$ **17.** $x = 6, y = 2; (6, 2)$ **19.** $x = 3, y = -6; (3, -6)$

21. $x = 8, y = -4; (8, -4)$ **23.** $x = \dfrac{1}{3}, y = -\dfrac{1}{6}; \left(\dfrac{1}{3}, -\dfrac{1}{6}\right)$ **25.** Inconsistent **27.** $x = \dfrac{3}{2}, y = 3; \left(\dfrac{3}{2}, 3\right)$

29. $\{(x, y) \mid x = 4 - 2y, y \text{ is any real number}\}$ or $\left\{(x, y) \mid y = \dfrac{4-x}{2}, x \text{ is any real number}\right\}$ **31.** $x = 1, y = 1; (1, 1)$ **33.** $x = \dfrac{3}{2}, y = 1; \left(\dfrac{3}{2}, 1\right)$

35. $x = 4, y = 3; (4, 3)$ **37.** $x = \dfrac{4}{3}, y = \dfrac{1}{5}; \left(\dfrac{4}{3}, \dfrac{1}{5}\right)$ **39.** $x = \dfrac{1}{5}, y = \dfrac{1}{3}; \left(\dfrac{1}{5}, \dfrac{1}{3}\right)$ **41.** $x = 8, y = 2, z = 0; (8, 2, 0)$

43. $x = 2, y = -1, z = 1; (2, -1, 1)$ **45.** Inconsistent **47.** $\{(x, y, z) \mid x = 5z - 2, y = 4z - 3; z \text{ is any real number}\}$

49. Inconsistent **51.** $x = 1, y = 3, z = -2; (1, 3, -2)$ **53.** $x = -3, y = \frac{1}{2}, z = 1; \left(-3, \frac{1}{2}, 1\right)$ **55.** Length 30 ft; width 15 ft

57. There were 18 commercial launches and 37 noncommercial launches in 2005. **59.** 22.5 lb

61. Average wind speed 25 mph; average airspeed 175 mph **63.** 80 \$25 sets and 120 \$45 sets **65.** \$9.96

67. Mix 50 mg of first compound with 75 mg of second. **69.** $a = \frac{4}{3}, b = -\frac{5}{3}, c = 1$ **71.** $Y = 9000, r = 0.06$ **73.** $I_1 = \frac{10}{71}, I_2 = \frac{65}{71}, I_3 = \frac{55}{71}$

75. 100 orchestra, 210 main, and 190 balcony seats **77.** 1.5 chicken, 1 corn, 2 milk

79. If x = price of hamburgers, y = price of fries, and z = price of colas, then $x = 2.75 - z, y = \frac{41}{60} + \frac{1}{3}z, \$0.60 \le z \le \$0.90$. There is not sufficient information:

x	\$2.13	\$2.01	\$1.86
y	\$0.89	\$0.93	\$0.98
z	\$0.62	\$0.74	\$0.89

81. It will take Beth 30 hr, Bill 24 hr, and Edie 40 hr.

8.2 Assess Your Understanding *(page 574)*

1. matrix **2.** augmented **3.** T **4.** T **5.** $\left[\begin{array}{cc|c} 1 & -5 & 5 \\ 4 & 3 & 6 \end{array}\right]$ **7.** $\left[\begin{array}{cc|c} 2 & 3 & 6 \\ 4 & -6 & -2 \end{array}\right]$ **9.** $\left[\begin{array}{cc|c} 0.01 & -0.03 & 0.06 \\ 0.13 & 0.10 & 0.20 \end{array}\right]$

11. $\left[\begin{array}{ccc|c} 1 & -1 & 1 & 10 \\ 3 & 3 & 0 & 5 \\ 1 & 1 & 2 & 2 \end{array}\right]$ **13.** $\left[\begin{array}{cccc|c} 1 & 1 & -1 & 2 \\ 3 & -2 & 0 & 2 \\ 5 & 3 & -1 & 1 \end{array}\right]$ **15.** $\left[\begin{array}{ccc|c} 1 & -1 & -1 & 10 \\ 2 & 1 & 2 & -1 \\ -3 & 4 & 0 & 5 \\ 4 & -5 & 1 & 0 \end{array}\right]$ **17.** $\begin{cases} x - 3y = -2 & (1) \\ 2x - 5y = 5 & (2) \end{cases}$ $\left[\begin{array}{cc|c} 1 & -3 & -2 \\ 0 & 1 & 9 \end{array}\right]$

19. $\begin{cases} x - 3y + 4z = 3 & (1) \\ 3x - 5y + 6z = 6 & (2) \\ -5x + 3y + 4z = 6 & (3) \end{cases}$ **(a)** $\left[\begin{array}{ccc|c} 1 & -3 & 4 & 3 \\ 0 & 4 & -6 & -3 \\ -5 & 3 & 4 & 6 \end{array}\right]$ **(b)** $\left[\begin{array}{ccc|c} 1 & -3 & 4 & 3 \\ 3 & -5 & 6 & 6 \\ 0 & -12 & 24 & 21 \end{array}\right]$

21. $\begin{cases} x - 3y + 2z = -6 & (1) \\ 2x - 5y + 3z = -4 & (2) \\ -3x - 6y + 4z = 6 & (3) \end{cases}$ **(a)** $\left[\begin{array}{ccc|c} 1 & -3 & 2 & -6 \\ 0 & 1 & -1 & 8 \\ -3 & -6 & 4 & 6 \end{array}\right]$ **(b)** $\left[\begin{array}{ccc|c} 1 & -3 & 2 & -6 \\ 2 & -5 & 3 & -4 \\ 0 & -15 & 10 & -12 \end{array}\right]$

23. $\begin{cases} 5x - 3y + z = -2 & (1) \\ 2x - 5y + 6z = -2 & (2) \\ -4x + y + 4z = 6 & (3) \end{cases}$ **(a)** $\left[\begin{array}{ccc|c} 1 & 7 & -11 & 2 \\ 2 & -5 & 6 & -2 \\ -4 & 1 & 4 & 6 \end{array}\right]$ **(b)** $\left[\begin{array}{ccc|c} 1 & -2 & 5 & 4 \\ 2 & -5 & 6 & -2 \\ -4 & 1 & 4 & 6 \end{array}\right]$

25. $\begin{cases} x = 5 \\ y = -1 \end{cases}$ Consistent; $x = 5, y = -1$ or $(5, -1)$

27. $\begin{cases} x = 1 \\ y = 2 \\ 0 = 3 \end{cases}$ Inconsistent

29. $\begin{cases} x + 2z = -1 \\ y - 4z = -2 \\ 0 = 0 \end{cases}$ Consistent: $\begin{cases} x = -1 - 2z \\ y = -2 + 4z \\ z \text{ is any real number or } \{(x, y, z) \mid x = -1 - 2z, y = -2 + 4z, z \text{ is any real number}\} \end{cases}$

31. $\begin{cases} x_1 = 1 \\ x_2 + x_4 = 2 \\ x_3 + 2x_4 = 3 \end{cases}$ Consistent: $\begin{cases} x_1 = 1, x_2 = 2 - x_4 \\ x_3 = 3 - 2x_4 \\ x_4 \text{ is any real number or } \{(x_1, x_2, x_3, x_4) \mid x_1 = 1, x_2 = 2 - x_4, x_3 = 3 - 2x_4, x_4 \text{ is any real number}\} \end{cases}$

33. $\begin{cases} x_1 + 4x_4 = 2 \\ x_2 + x_3 + 3x_4 = 3 \\ 0 = 0 \end{cases}$ Consistent: $\begin{cases} x_1 = 2 - 4x_4 \\ x_2 = 3 - x_3 - 3x_4 \\ x_3, x_4 \text{ are any real numbers or } \{(x_1, x_2, x_3, x_4) \mid x_1 = 2 - 4x_4, x_2 = 3 - x_3 - 3x_4, x_3, x_4 \text{ are any real numbers}\} \end{cases}$

35. $\begin{cases} x_1 + x_4 = -2 \\ x_2 + 2x_4 = 2 \\ x_3 - x_4 = 0 \\ 0 = 0 \end{cases}$ Consistent: $\begin{cases} x_1 = -2 - x_4 \\ x_2 = 2 - 2x_4 \\ x_3 = x_4 \\ x_4 \text{ is any real number or } \{(x_1, x_2, x_3, x_4) \mid x_1 = -2 - x_4, x_2 = 2 - 2x_4, x_3 = x_4, x_4 \text{ is any real number}\} \end{cases}$

37. $x = 6, y = 2; (6, 2)$ **39.** $x = \frac{1}{2}, y = \frac{3}{4}; \left(\frac{1}{2}, \frac{3}{4}\right)$ **41.** $x = 4 - 2y$, y is any real number; $\{(x, y) \mid x = 4 - 2y, y \text{ is any real number}\}$

43. $x = \frac{3}{2}, y = 1; \left(\frac{3}{2}, 1\right)$ **45.** $x = \frac{4}{3}, y = \frac{1}{5}; \left(\frac{4}{3}, \frac{1}{5}\right)$ **47.** $x = 8, y = 2, z = 0; (8, 2, 0)$ **49.** $x = 2, y = -1, z = 1; (2, -1, 1)$ **51.** Inconsistent

53. $x = 5z - 2, y = 4z - 3$, where z is any real number; $\{(x, y, z) \mid x = 5z - 2, y = 4x - 3, z \text{ is any real number}\}$

55. Inconsistent **57.** $x = 1, y = 3, z = -2; (1, 3, -2)$ **59.** $x = -3, y = \frac{1}{2}, z = 1; \left(-3, \frac{1}{2}, 1\right)$ **61.** $x = \frac{1}{3}, y = \frac{2}{3}, z = 1; \left(\frac{1}{3}, \frac{2}{3}, 1\right)$

63. $x = 1, y = 2, z = 0, w = 1; (1, 2, 0, 1)$ **65.** $y = 0, z = 1 - x$, x is any real number; $\{(x, y, z) \mid y = 0, z = 1 - x, x \text{ is any real number}\}$

67. $x = 2, y = z - 3$, z is any real number; $\{(x, y, z) \mid x = 2, y = z - 3, z \text{ is any real number}\}$ **69.** $x = \frac{13}{9}, y = \frac{7}{18}, z = \frac{19}{18}; \left(\frac{13}{9}, \frac{7}{18}, \frac{19}{18}\right)$

71. $x = \frac{7}{5} - \frac{3}{5}z - \frac{2}{5}w, y = -\frac{8}{5} + \frac{7}{5}z + \frac{13}{5}w$, where z and w are any real numbers;
$\left\{(x, y, z, w) \mid x = \frac{7}{5} - \frac{3}{5}z - \frac{2}{5}w, y = -\frac{8}{5} + \frac{7}{5}z + \frac{13}{5}w, \text{ z and w are any real numbers}\right\}$

73. $y = -2x^2 + x + 3$ **75.** $f(x) = 3x^3 - 4x^2 + 5$ **77.** 1.5 salmon steak, 2 baked eggs, 1 acorn squash

79. \$4000 in Treasury bills, \$4000 in Treasury bonds, \$2000 in corporate bonds **81.** 8 Deltas, 5 Betas, 10 Sigmas **83.** $I_1 = \frac{44}{23}, I_2 = 2, I_3 = \frac{16}{23}, I_4 = \frac{28}{23}$

85. (a)

Amount Invested At		
7%	**9%**	**11%**
0	10,000	10,000
1000	8000	11,000
2000	6000	12,000
3000	4000	13,000
4000	2000	14,000
5000	0	15,000

(b)

Amount Invested At		
7%	**9%**	**11%**
12,500	12,500	0
14,500	8500	2000
16,500	4500	4000
18,750	0	6250

(c) All the money invested at 7% provides \$2100, more than what is required.

87.

First Supplement	Second Supplement	Third Supplement
50 mg	75 mg	0 mg
36 mg	76 mg	8 mg
22 mg	77 mg	16 mg
8 mg	78 mg	24 mg

8.3 Assess Your Understanding *(page 585)*

1. determinants **2.** $ad - bc$ **3.** F **4.** F **5.** 2 **7.** 22 **9.** -2 **11.** 10 **13.** -26 **15.** $x = 6, y = 2; (6, 2)$ **17.** $x = 3, y = 2; (3, 2)$
19. $x = 8, y = -4; (8, -4)$ **21.** $x = 4, y = -2; (4, -2)$ **23.** Not applicable **25.** $x = \frac{1}{2}, y = \frac{3}{4}; \left(\frac{1}{2}, \frac{3}{4}\right)$ **27.** $x = \frac{1}{10}, y = \frac{2}{5}; \left(\frac{1}{10}, \frac{2}{5}\right)$
29. $x = \frac{3}{2}, y = 1; \left(\frac{3}{2}, 1\right)$ **31.** $x = \frac{4}{3}, y = \frac{1}{5}; \left(\frac{4}{3}, \frac{1}{5}\right)$ **33.** $x = 1, y = 3, z = -2; (1, 3, -2)$ **35.** $x = -3, y = \frac{1}{2}, z = 1; \left(-3, \frac{1}{2}, 1\right)$
37. Not applicable **39.** $x = 0, y = 0, z = 0; (0,0,0)$ **41.** Not applicable **43.** -4 **45.** 12 **47.** 8 **49.** 8 **51.** -5 **53.** $\frac{13}{11}$ **55.** 0 or -9

57.
$$(y_1 - y_2)x - (x_1 - x_2)y + (x_1y_2 - x_2y_1) = 0$$
$$(y_1 - y_2)x + (x_2 - x_1)y = x_2y_1 - x_1y_2$$
$$(x_2 - x_1)y - (x_2 - x_1)y_1 = (y_2 - y_1)x + x_2y_1 - x_1y_2 - (x_2 - x_1)y_1$$
$$(x_2 - x_1)(y - y_1) = (y_2 - y_1)x - (y_2 - y_1)x_1$$
$$y - y_1 = \frac{y_2 - y_1}{x_2 - x_1}(x - x_1)$$

59. The triangle has an area of 5 square units.

61. If $a = 0$, we have
$$by = s$$
$$cx + dy = t$$
Thus, $y = \frac{s}{b}$ and
$$x = \frac{t - dy}{c} = \frac{tb - ds}{bc}$$
Using Cramer's Rule, we get
$$x = \frac{sd - tb}{-bc} = \frac{tb - sd}{bc}$$
$$y = \frac{-sc}{-bc} = \frac{s}{b}$$

If $b = 0$, we have
$$ax = s$$
$$cx + dy = t$$
Since $D = ad \neq 0$, then $a \neq 0$ and $d \neq 0$.
Thus, $x = \frac{s}{a}$ and
$$y = \frac{t - cx}{d} = \frac{ta - cs}{ad}$$
Using Cramer's Rule, we get
$$x = \frac{sd}{ad} = \frac{s}{a}$$
$$y = \frac{ta - cs}{ad}$$

If $c = 0$, we have
$$ax + by = s$$
$$dy = t$$
Since $D = ad \neq 0$, then $a \neq 0$ and $d \neq 0$.
Thus, $y = \frac{t}{d}$ and
$$x = \frac{s - by}{a} = \frac{sd - bt}{ad}$$
Using Cramer's Rule, we get
$$x = \frac{sd - bt}{ad}$$
$$y = \frac{at}{ad} = \frac{t}{d}$$

If $d = 0$, we have
$$ax + by = s$$
$$cx = t$$
Since $D = -bc \neq 0$, then $b \neq 0$ and $c \neq 0$.
Thus, $x = \frac{t}{c}$ and
$$y = \frac{s - ax}{b} = \frac{sc - at}{bc}$$
Using Cramer's Rule, we get
$$x = \frac{-tb}{-bc} = \frac{t}{c}$$
$$y = \frac{at - sc}{-bc} = \frac{sc - at}{bc}$$

63.
$$\begin{vmatrix} a_{11} & a_{12} & a_{13} \\ ka_{21} & ka_{22} & ka_{23} \\ a_{31} & a_{32} & a_{33} \end{vmatrix} = -ka_{21}(a_{12}a_{33} - a_{32}a_{13}) + ka_{22}(a_{11}a_{33} - a_{31}a_{13}) - ka_{23}(a_{11}a_{32} - a_{31}a_{12})$$
$$= k[-a_{21}(a_{12}a_{33} - a_{32}a_{13}) + a_{22}(a_{11}a_{33} - a_{31}a_{13}) - a_{23}(a_{11}a_{32} - a_{31}a_{12})] = k\begin{vmatrix} a_{11} & a_{12} & a_{13} \\ a_{21} & a_{22} & a_{23} \\ a_{31} & a_{32} & a_{33} \end{vmatrix}$$

65. $\begin{vmatrix} a_{11}+ka_{21} & a_{12}+ka_{22} & a_{13}+ka_{23} \\ a_{21} & a_{22} & a_{23} \\ a_{31} & a_{32} & a_{33} \end{vmatrix} = (a_{11}+ka_{21})(a_{22}a_{33}-a_{32}a_{23}) - (a_{12}+ka_{22})(a_{21}a_{33}-a_{31}a_{23}) + (a_{13}+ka_{23})(a_{21}a_{32}-a_{31}a_{22})$

$= a_{11}a_{22}a_{33} - a_{11}a_{32}a_{23} + \cancel{ka_{21}a_{22}a_{33}} - \cancel{ka_{21}a_{32}a_{23}} - a_{12}a_{21}a_{33} + a_{12}a_{31}a_{23}$
$- \cancel{ka_{22}a_{21}a_{33}} + \cancel{ka_{22}a_{31}a_{23}} + a_{13}a_{21}a_{32} - a_{13}a_{31}a_{22} + \cancel{ka_{23}a_{21}a_{32}} - \cancel{ka_{23}a_{31}a_{22}}$

$= a_{11}a_{22}a_{33} - a_{11}a_{32}a_{23} - a_{12}a_{21}a_{33} + a_{12}a_{31}a_{23} + a_{13}a_{21}a_{32} - a_{13}a_{31}a_{22}$

$= a_{11}(a_{22}a_{33} - a_{32}a_{23}) - a_{12}(a_{21}a_{33} - a_{31}a_{23}) + a_{13}(a_{21}a_{32} - a_{31}a_{22})$

$= \begin{vmatrix} a_{11} & a_{12} & a_{13} \\ a_{21} & a_{22} & a_{23} \\ a_{31} & a_{32} & a_{33} \end{vmatrix}$

Historical Problems *(page 601)*

1. (a) $2 - 5i \longleftrightarrow \begin{bmatrix} 2 & -5 \\ 5 & 2 \end{bmatrix}, 1 + 3i \longleftrightarrow \begin{bmatrix} 1 & 3 \\ -3 & 1 \end{bmatrix}$ **(b)** $\begin{bmatrix} 2 & -5 \\ 5 & 2 \end{bmatrix}\begin{bmatrix} 1 & 3 \\ -3 & 1 \end{bmatrix} = \begin{bmatrix} 17 & 1 \\ -1 & 17 \end{bmatrix}$ **(c)** $17 + i$ **(d)** $17 + i$

2. $\begin{bmatrix} a & b \\ -b & a \end{bmatrix}\begin{bmatrix} a & -b \\ b & a \end{bmatrix} = \begin{bmatrix} a^2+b^2 & 0 \\ 0 & b^2+a^2 \end{bmatrix}$; the product is a real number.

3. (a) $x = k(ar + bs) + l(cr + ds) = r(ka + lc) + s(kb + ld)$
$y = m(ar + bs) + n(cr + ds) = r(ma + nc) + s(mb + nd)$
(b) $A = \begin{bmatrix} ka + lc & kb + ld \\ ma + nc & mb + nd \end{bmatrix}$

8.4 Assess Your Understanding *(page 601)*

1. inverse **2.** square **3.** identity **4.** F **5.** F **6.** F

7. $\begin{bmatrix} 4 & 4 & -5 \\ -1 & 5 & 4 \end{bmatrix}$ **9.** $\begin{bmatrix} 0 & 12 & -20 \\ 4 & 8 & 24 \end{bmatrix}$ **11.** $\begin{bmatrix} -8 & 7 & -15 \\ 7 & 0 & 22 \end{bmatrix}$ **13.** $\begin{bmatrix} 28 & -9 \\ 4 & 23 \end{bmatrix}$ **15.** $\begin{bmatrix} 1 & 14 & -14 \\ 2 & 22 & -18 \\ 3 & 0 & 28 \end{bmatrix}$ **17.** $\begin{bmatrix} 15 & 21 & -16 \\ 22 & 34 & -22 \\ -11 & 7 & 22 \end{bmatrix}$ **19.** $\begin{bmatrix} 25 & -9 \\ 4 & 20 \end{bmatrix}$

21. $\begin{bmatrix} -13 & 7 & -12 \\ -18 & 10 & -14 \\ 17 & -7 & 34 \end{bmatrix}$ **23.** $\begin{bmatrix} -2 & 4 & 2 & 8 \\ 2 & 1 & 4 & 6 \end{bmatrix}$ **25.** $\begin{bmatrix} 5 & 14 \\ 9 & 16 \end{bmatrix}$ **27.** $\begin{bmatrix} 9 & 2 \\ 34 & 13 \\ 47 & 20 \end{bmatrix}$ **29.** $\begin{bmatrix} 1 & -1 \\ -1 & 2 \end{bmatrix}$ **31.** $\begin{bmatrix} 1 & -\frac{5}{2} \\ -1 & 3 \end{bmatrix}$ **33.** $\begin{bmatrix} 1 & -\frac{1}{a} \\ -1 & \frac{2}{a} \end{bmatrix}$

35. $\begin{bmatrix} 3 & -3 & 1 \\ -2 & 2 & -1 \\ -4 & 5 & -2 \end{bmatrix}$ **37.** $\begin{bmatrix} -\frac{5}{7} & \frac{1}{7} & \frac{3}{7} \\ \frac{9}{7} & \frac{1}{7} & -\frac{4}{7} \\ \frac{3}{7} & -\frac{2}{7} & \frac{1}{7} \end{bmatrix}$ **39.** $x = 3, y = 2; (3, 2)$ **41.** $x = -5, y = 10; (-5, 10)$ **43.** $x = 2, y = -1; (2, -1)$

45. $x = \frac{1}{2}, y = 2; \left(\frac{1}{2}, 2\right)$ **47.** $x = -2, y = 1; (-2, 1)$ **49.** $x = \frac{2}{a}, y = \frac{3}{a}; \left(\frac{2}{a}, \frac{3}{a}\right)$ **51.** $x = -2, y = 3, z = 5; (-2, 3, 5)$

53. $x = \frac{1}{2}, y = -\frac{1}{2}, z = 1; \left(\frac{1}{2}, -\frac{1}{2}, 1\right)$ **55.** $x = -\frac{34}{7}, y = \frac{85}{7}, z = \frac{12}{7}; \left(-\frac{34}{7}, \frac{85}{7}, \frac{12}{7}\right)$ **57.** $x = \frac{1}{3}, y = 1, z = \frac{2}{3}; \left(\frac{1}{3}, 1, \frac{2}{3}\right)$

59. $\left[\begin{array}{cc|cc} 4 & 2 & 1 & 0 \\ 2 & 1 & 0 & 1 \end{array}\right] \rightarrow \left[\begin{array}{cc|cc} 1 & \frac{1}{2} & \frac{1}{4} & 0 \\ 2 & 1 & 0 & 1 \end{array}\right] \rightarrow \left[\begin{array}{cc|cc} 1 & \frac{1}{2} & \frac{1}{4} & 0 \\ 0 & 0 & -\frac{1}{2} & 1 \end{array}\right]$ **61.** $\left[\begin{array}{cc|cc} 15 & 3 & 1 & 0 \\ 10 & 2 & 0 & 1 \end{array}\right] \rightarrow \left[\begin{array}{cc|cc} 1 & \frac{1}{5} & \frac{1}{15} & 0 \\ 10 & 2 & 0 & 1 \end{array}\right] \rightarrow \left[\begin{array}{cc|cc} 1 & \frac{1}{5} & \frac{1}{15} & 0 \\ 0 & 0 & -\frac{2}{3} & 1 \end{array}\right]$

63. $\left[\begin{array}{ccc|ccc} -3 & 1 & -1 & 1 & 0 & 0 \\ 1 & -4 & -7 & 0 & 1 & 0 \\ 1 & 2 & 5 & 0 & 0 & 1 \end{array}\right] \rightarrow \left[\begin{array}{ccc|ccc} 1 & 2 & 5 & 0 & 0 & 1 \\ 1 & -4 & -7 & 0 & 1 & 0 \\ -3 & 1 & -1 & 1 & 0 & 0 \end{array}\right] \rightarrow \left[\begin{array}{ccc|ccc} 1 & 2 & 5 & 0 & 0 & 1 \\ 0 & -6 & -12 & 0 & 1 & -1 \\ 0 & 7 & 14 & 1 & 0 & 3 \end{array}\right]$

$\rightarrow \left[\begin{array}{ccc|ccc} 1 & 2 & 5 & 0 & 0 & 1 \\ 0 & 1 & 2 & 0 & -\frac{1}{6} & \frac{1}{6} \\ 0 & 1 & 2 & \frac{1}{7} & 0 & \frac{3}{7} \end{array}\right] \rightarrow \left[\begin{array}{ccc|ccc} 1 & 2 & 5 & 0 & 0 & 1 \\ 0 & 1 & 2 & 0 & -\frac{1}{6} & \frac{1}{6} \\ 0 & 0 & 0 & \frac{1}{7} & \frac{1}{6} & \frac{11}{42} \end{array}\right]$

65. $\begin{bmatrix} 0.01 & 0.05 & -0.01 \\ 0.01 & -0.02 & 0.01 \\ -0.02 & 0.01 & 0.03 \end{bmatrix}$ **67.** $\begin{bmatrix} 0.02 & -0.04 & -0.01 & 0.01 \\ -0.02 & 0.05 & 0.03 & -0.03 \\ 0.02 & 0.01 & -0.04 & 0.00 \\ -0.02 & 0.06 & 0.07 & 0.06 \end{bmatrix}$ **69.** $x = 4.57, y = -6.44, z = -24.07$ or $(4.57, -6.44, -24.07)$

CONICS

Parabola

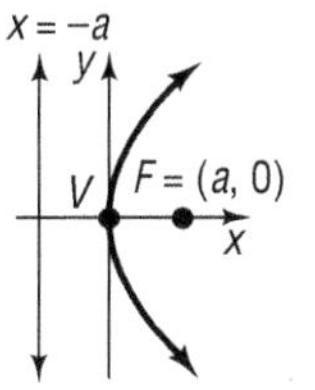

$$y^2 = 4ax$$

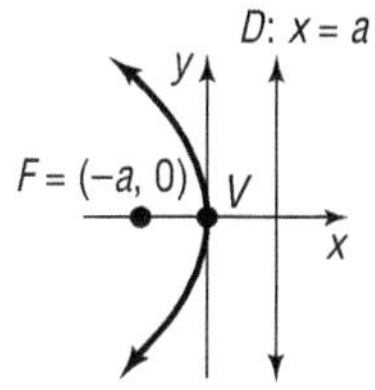

$$y^2 = -4ax$$

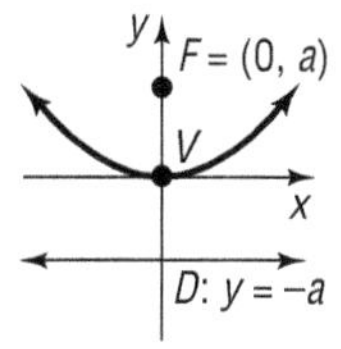

$$x^2 = 4ay$$

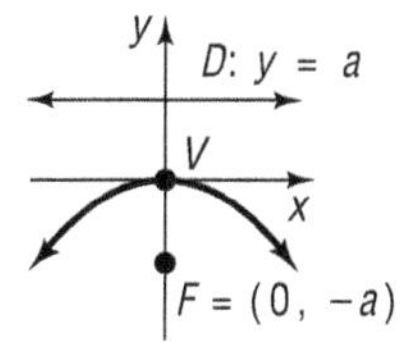

$$x^2 = -4ay$$

Ellipse

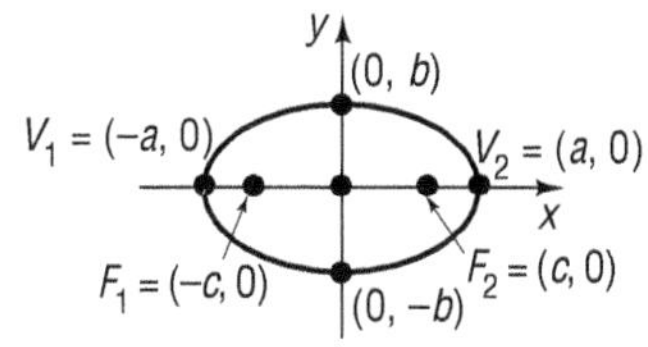

$$\frac{x^2}{a^2} + \frac{y^2}{b^2} = 1, \quad a > b, \quad c^2 = a^2 - b^2$$

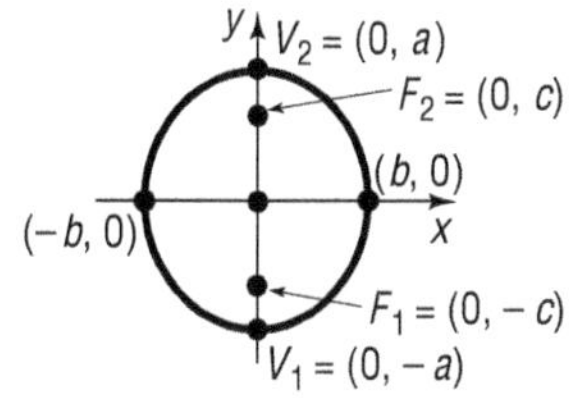

$$\frac{x^2}{b^2} + \frac{y^2}{a^2} = 1, \quad a > b, \quad c^2 = a^2 - b^2$$

Hyperbola

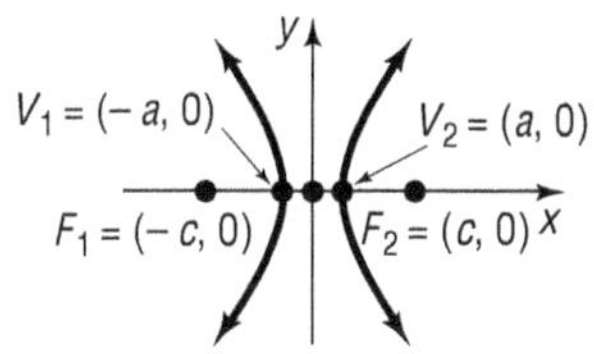

$$\frac{x^2}{a^2} - \frac{y^2}{b^2} = 1, \quad c^2 = a^2 + b^2$$

Asymptotes: $y = \frac{b}{a}x, \quad y = -\frac{b}{a}x$

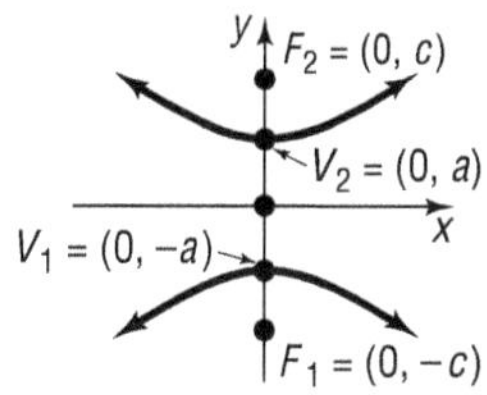

$$\frac{y^2}{a^2} - \frac{x^2}{b^2} = 1, \quad c^2 = a^2 + b^2$$

Asymptotes: $y = \frac{a}{b}x, \quad y = -\frac{a}{b}x$

PROPERTIES OF LOGARITHMS

$$\log_a(MN) = \log_a M + \log_a N$$

$$\log_a\left(\frac{M}{N}\right) = \log_a M - \log_a N$$

$$\log_a M^r = r \log_a M$$

$$\log_a M = \frac{\log M}{\log a} = \frac{\ln M}{\ln a}$$

PERMUTATIONS/COMBINATIONS

$0! = 1 \qquad 1! = 1$

$n! = n(n-1) \cdot \ldots \cdot (3)(2)(1)$

$$P(n, r) = \frac{n!}{(n-r)!}$$

$$C(n, r) = \binom{n}{r} = \frac{n!}{(n-r)!\,r!}$$

BINOMIAL THEOREM

$$(a+b)^n = a^n + \binom{n}{1}ba^{n-1} + \binom{n}{2}b^2a^{n-2} + \cdots + \binom{n}{n-1}b^{n-1}a + b^n$$

ARITHMETIC SEQUENCE

$$a_1 + (a_1 + d) + (a_1 + 2d) + \cdots + [a_1 + (n-1)d] = \frac{n}{2}[2a_1 + (n-1)d] = \frac{n}{2}[a_1 + a_n]$$

GEOMETRIC SEQUENCE

$$a_1 + a_1r + a_1r^2 + \cdots + a_1r^{n-1} = a_1 \cdot \frac{1 - r^n}{1 - r}$$

GEOMETRIC SERIES

If $|r| < 1, a_1 + a_1r + a_1r^2 + \cdots = \sum_{k=1}^{\infty} a_1r^{k-1} = \frac{a_1}{1 - r}$

FORMULAS/EQUATIONS

Distance Formula

If $P_1 = (x_1, y_1)$ and $P_2 = (x_2, y_2)$, the distance from P_1 to P_2 is

$$d(P_1, P_2) = \sqrt{(x_2 - x_1)^2 + (y_2 - y_1)^2}$$

Standard Equation of a Circle

The standard equation of a circle of radius r with center at (h, k) is

$$(x - h)^2 + (y - k)^2 = r^2$$

Slope Formula

The slope m of the line containing the points $P_1 = (x_1, y_1)$ and $P_2 = (x_2, y_2)$ is

$$m = \frac{y_2 - y_1}{x_2 - x_1} \qquad \text{if } x_1 \neq x_2$$

$$m \text{ is undefined} \qquad \text{if } x_1 = x_2$$

Point–Slope Equation of a Line

The equation of a line with slope m containing the point (x_1, y_1) is

$$y - y_1 = m(x - x_1)$$

Slope–Intercept Equation of a Line

The equation of a line with slope m and y-intercept b is

$$y = mx + b$$

Quadratic Formula

The solutions of the equation $ax^2 + bx + c = 0$, $a \neq 0$, are

$$x = \frac{-b \pm \sqrt{b^2 - 4ac}}{2a}$$

If $b^2 - 4ac > 0$, there are two unequal real solutions.
If $b^2 - 4ac = 0$, there is a repeated real solution.
If $b^2 - 4ac < 0$, there are two complex solutions that are not real.

GEOMETRY FORMULAS

Circle

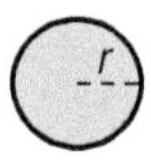

r = Radius, A = Area, C = Circumference

$$A = \pi r^2 \qquad C = 2\pi r$$

Triangle

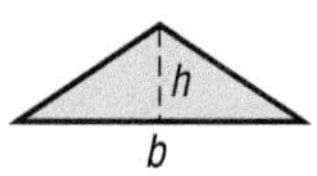

b = Base, h = Altitude (Height), A = area

$$A = \tfrac{1}{2}bh$$

Rectangle

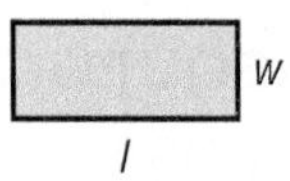

l = Length, w = Width, A = area, P = perimeter

$$A = lw \qquad P = 2l + 2w$$

Rectangular Box

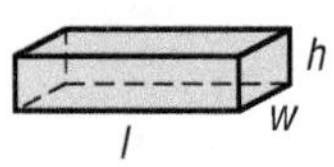

l = Length, w = Width, h = Height, V = Volume, S = Surface area

$$V = lwh \qquad S = 2lw + 2lh + 2wh$$

Sphere

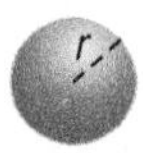

r = Radius, V = Volume, S = Surface area

$$V = \tfrac{4}{3}\pi r^3 \qquad S = 4\pi r^2$$

Right Circular Cylinder

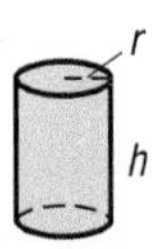

r = Radius, h = Height, V = Volume, S = Surface area

$$V = \pi r^2 h \qquad S = 2\pi r^2 + 2\pi rh$$